# CONTENTS—1986 REFRIGERATION VOLUME



# CONTENTS—1985 FUNDAMENTALS VOLUME

# 1988 ASHRAE HANDBOOK

# EQUIPMENT

**American Society of Heating, Refrigerating and Air-Conditioning Engineers, Inc.**
1791 Tullie Circle, N.E., Atlanta, GA 30329

**DEDICATED**

**TO THE ADVANCEMENT OF**

**THE PROFESSION**

**AND ITS ALLIED INDUSTRIES**



Although great care has been taken in the compilation and publication of this volume, no warranties, express or implied, are given in connection herewith and no responsibility can be taken for any claims arising herewith.

Comments, criticisms, and suggestions regarding the subject matter are invited. Any errors or omissions in the data should be brought to the attention of the Editor. If required, an errata sheet will be issued at approximately the same time as the next Handbook. Notice of any significant errors found after that time will be published in the ASHRAE JOURNAL.

ISBN: 0-910110-52-2

# CONTENTS

# PREFACE

The EQUIPMENT Volume of the ASHRAE Handbook series describes the components and assemblies that perform various heating, ventilating, air-conditioning, or refrigeration functions. The information is particularly for the system designer rather than the designer of specific components of equipment. These chapters generally survey available equipment and cover (1) principles of operation, (2) types of construction, (3) performance characteristics, (4) methods for testing and rating, (5) pertinent standards, and (6) factors for selection.

The Society works through its Handbook Committee and Technical Committees to develop handbooks that are reliable sources of authoritative technical information. To achieve this goal, the technical committees reviewed and revised all chapters in this volume, as well as included major additions or changes to about half the chapters. The table of contents and a footnote on the first page of each chapter list the technical committee that revised or prepared the chapter.

The 45 chapters are arranged in six major sections. The first covers the equipment that moves, cleans, or modifies the temperature or humidity of process or ventilation air. The second section describes the components and controls for refrigeration equipment. The performance and features of furnaces, boilers, chimneys and other heating devices are covered in the third section. The fourth section describes the pumps, motors, engines, and piping involved in HVAC installations. The fifth section describes unitary refrigeration equipment, air-conditioners, and heat pumps. The last section includes a new chapter, which describes solar energy equipment and how it is tested.

This volume includes both inch-pound (I-P) and the International System (SI) units of measurement. Future volumes of the Handbook should be available in separate I-P and SI editions, which will lessen the inconvenience of two set of units in one volume.

Errata for the 1985, 1986, and 1987 volumes precede the index. Errata for this volume will be included in the 1989 FUNDAMENTALS Volume.

During the past two years, both the presidents of ASHRAE and the Handbook Committee have encouraged more members to review chapters in the Handbooks to insure they are correct and reflect current practices. The Handbook Committee would like your help as well. If you have suggestions and comments on improving a chapter or would like more information on how you can help revise a chapter, please call or write the Handbook Editor, ASHRAE, 1791 Tullie Circle, Atlanta, GA 30329.

Robert A. Parsons
Handbook Editor

# ASHRAE TECHNICAL COMMITTEES AND TASK GROUPS

**SECTION 1.0—FUNDAMENTALS AND GENERAL**

1.1 Thermodynamics and Psychrometrics
1.2 Instruments and Measurements
1.3 Heat Transfer and Fluid Flow
1.4 Control Theory and Application
1.5 Computer Applications
1.6 Terminology
1.7 Operation and Maintenance
1.8 Owning and Operating Costs
1.9 Electrical Systems

**SECTION 2.0—ENVIRONMENTAL QUALITY**

2.1 Physiology and Human Environment
2.2 Plant and Animal Environment
2.3 Gaseous Air Contaminants and Gas Contaminant Removal Equipment
2.4 Particulate Air Contaminants and Particulate Contaminant Removal Equipment
2.5 Air Flow Around Buildings
2.6 Sound and Vibration Control
TG Safety
TG Halocarbon Emission
TG Seismic Restraint Design

**SECTION 3.0—MATERIALS AND PROCESSES**

3.1 Refrigerant and Brines
3.2 Refrigerant System Chemistry
3.3 Contaminant Control in Refrigerating Systems
3.4 Lubrication
3.5 Sorption
3.6 Corrosion and Water Treatment
3.7 Fuels and Combustion

**SECTION 4.0—LOAD CALCULATIONS AND ENERGY REQUIREMENTS**

4.1 Load Calculation Data and Procedures
4.2 Weather Data
4.3 Ventilation Requirements and Infiltration
4.4 Thermal Insulation and Moisture Retarders
4.5 Fenestration
4.6 Building Operation Dynamics
4.7 Energy Calculations
4.8 Energy Resources
4.9 Building Envelope Systems
TG Indoor Environmental Calculations
TG Cold Climate Design

**SECTION 5.0—VENTILATION AND AIR DISTRIBUTION**

5.1 Fans
5.2 Duct Design
5.3 Room Air Distribution
5.4 Industrial Process Air Cleaning (Air Pollution Control)
5.5 Air-to-Air Energy Recovery
5.6 Control of Fire and Smoke
5.7 Evaporative Cooling
5.8 Industrial Ventilation
5.9 Enclosed Vehicular Facilities

**SECTION 6.0—HEATING EQUIPMENT, HEATING AND COOLING SYSTEMS AND APPLICATIONS**

6.1 Hot Water and Steam Heating Equipment and Systems
6.2 District Heating and Cooling
6.3 Central Forced Air Heating and Cooling Systems
6.4 In-Space Convection Heating
6.5 Radiant Space Heating and Cooling
6.6 Service Water Heating
6.7 Solar Energy Utilization
6.8 Geothermal Energy Utilization
6.9 Thermal Storage

**SECTION 7.0—PACKAGED AIR-CONDITIONING AND REFRIGERATION EQUIPMENT**

7.1 Residential Refrigerators, Food Freezers and Drinking Water Coolers
7.2 Beverage Coolers
7.5 Room Air Conditioners and Dehumidifiers
7.6 Unitary Air Conditioners and Heat Pumps

**SECTION 8.0—AIR-CONDITIONING AND REFRIGERATION SYSTEM COMPONENTS**

8.1 Positive Displacement Compressors
8.2 Centrifugal Machines
8.3 Absorption and Heat Operated Machines
8.4 Air-to-Refrigerant Heat Transfer Equipment
8.5 Liquid-to-Refrigerant Heat Exchangers
8.6 Cooling Towers and Evaporative Condensers
8.7 Humidifying Equipment
8.8 Refrigerant System Controls and Accessories
8.10 Pumps and Hydronic Piping
8.11 Electric Motors—Open and Hermetic
TG Unitary Combustion-Engine-Driven Heat Pumps

**SECTION 9.0—AIR-CONDITIONING SYSTEMS AND APPLICATIONS**

9.1 Large Building Air-Conditioning Systems
9.2 Industrial Air Conditioning
9.3 Transportation Air Conditioning
9.4 Applied Heat Pump/Heat Recovery Systems
9.5 Cogeneration Systems
9.6 Systems Energy Utilization
9.7 Testing and Balancing
9.8 Large Building Air-Conditioning Applications

**SECTION 10.0—REFRIGERATION SYSTEMS AND APPLICATIONS**

10.1 Custom Engineered Refrigeration Systems
10.2 Automatic Ice-Making Plants and Skating Rinks
10.3 Refrigerant Piping
10.4 Ultra-Low Temperature Systems and Cryogenics
10.5 Refrigerated Distribution and Storage Facilities
10.6 Transport Refrigeration
10.7 Commercial Food Display and Storage Equipment
10.8 Refrigeration Load Calculations

**SECTION 11.0—REFRIGERATED FOOD TECHNOLOGY AND PROCESSING**

11.1 Meat, Fish and Poultry Products
11.3 Dairy Products
11.5 Fruits, Vegetables and Other Products
11.6 Prepared Food Products
11.9 Thermal Properties of Foods

# CONTRIBUTORS

In addition to the Technical Committees, the following individuals contributed significantly to this volume. The appropriate chapter numbers follow each contributor's name.

---

**Herman Behls** (1)
Sargent and Lundy

**John Stratton** (1)
Sheet Metal and Air Conditioning Contractors' National Association

**Phillip Berger** (2)
Krueger Manufacturing Company

**J. Barrie Graham** (3)
Graham Consultants

**G.P. Jolette** (3)
Air Movement and Control Association

**Calvin H. McClellan** (4)
cba Company

**John L. Peterson** (4)
University of Texas at Austin

**Bernard Morton** (5)
Dri-Steem Humidifier Company

**Hong Chun Kim** (6, 15-II)
Bohn Heat Transfer

**Joseph K. Thornton** (7)

**Roland Ares** (8, 9)
Hussmann Refrigeration Corporation

**J.E. Sjordal** (9)
McQuay Air Conditioning

**Carl J. Bauder** (10)
Cambridge Filter Corporation

**Robert L. Greenwell** (10)
Owens Corning

**Eugene L. Valerio** (10)
Air Filter/Control

**Leonard J. O'Dell** (11)
Consultant

**Stanton V. Sheppard** (11)
Ceilcote Company, Inc.

**Gunther T. Jensen** (12, 17, 21)
The Trane Company

**Joseph W. Pillis** (12)
Frick Company

**Kenneth N. Puetzer** (12)
Sullair Refrigeration, Inc.

**Richard S. Sweetser** (12)
Single Screw Compressor, Inc.

**Fred W. Bawel** (13)
Preway Industries, Inc.

**Reinhard Radermacher** (13)
University of Maryland

**Gary Vliet** (13)
University of Texas at Austin

**James F. Gausling** (14)
U.S. Air Force/Aeronautical Systems Division

**George C. Letton, Jr.** (14)
U.S. Air Force/Aeronautical Systems Division

**Jon M. Edmonds** (15-III)
St. Onge Ruff and Associates

**Arthur G. Fovargue** (15-I)
Dunham-Bush, Inc.

**Donald E. Enslen** (16)
Dunham-Bush, Inc.

**Jerry Errath** (17)
Vilter Manufacturing Corporation

**Ralph Wright** (17)
Vilter Manufacturing Corporation

**Donald K. Miller** (18)
York International

**Richard J. Buck** (19)
Sporlan Valve Company

**David P. Hargraves** (19)
Emerson Electric

**Richard Krause** (19)
Henry Valve Company

**Lee A. White** (19)
Ranco Controls, Inc.

**James I. Lanoue** (20)
Marley Cooling Tower Company

**William Axtman** (22)
American Boiler Manufacturers Association

**Earl M. Clark** (22)
DuPont Company

**William F. Sell** (22, 26)
The Brooklyn Union Gas Company

**Hall Virgil** (22, 26)
Carrier Corporation

**J. Burt Rishel** (23, 27)
Systecon, Inc.

**John I. Woodworth** (23)
The Hydronics Institute

**Lawrence R. Brand** (24)
Gas Research Institute

**Lorne W. Nelson** (24)
Honeywell, Inc.

**R.J. Kolodgy** (25)
American Gas Association Laboratories

**Esher R. Kweller** (25)
U.S. Department of Energy

**Clifton F. Briner** (26)
Purdue University at Fort Wayne

**Douglas W. DeWerth** (26)
American Gas Association Laboratories

**Richard L. Stone** (26)

**Donald J. Clinton** (28)
Kagan & Clinton

**Norman A. Buckley** (29)
Buckley Associates

**Joseph P. Neises** (29)
Perfection Products

**John M. Noble** (29)
Aztech International, Inc.

**Gene M. Meyer** (30)
Kansas State University

**Walter C. Stethem** (30)
Chapple Engineering, Ltd.

**K.W. Lichius** (31)
White Consolidated Industries, Inc.

**Thomas I. Wetherington** (32)
Florida Power Corporation

**Richard E. Batherman** (33)
Vico, Inc.

**Paul A. Bourquin** (33)
Wolff and Munier, Inc.

**Joseph B. O'Brien** (34)
Somerset Technologies, Inc.

**Fayez F. Ibrahim** (35, 36)
Bangor Cooler Company

**Peter W. Likes** (35, 36)
Hussman Refrigeration, Inc.

**Robert L. Cushman** (37)
Amana Refrigeration, Inc.

**Ronald I. Greenwald** (38)
EBCO Manufacturing Company

**John R. McMillin** (39)
The Cornelius Company

**Jeff J. Anselmino** (40)
Whirlpool Corporation

**Ron R. Huffman** (41)
Whirlpool Corporation

**Duane L. Lom** (42, 43)
The Trane Company

**Nance C. Lovvorn** (42)
Alabama Power Company

**Joseph A. Pietsch** (42, 43)
Arco Comfort Products Company

**B.H. Prasad** (42)
Oklahoma Gas & Electric Company

**Mark S. Menzer** (43)
American Gas Association

**Gerald R. Guinn** (44)
University of Alabama

ASHRAE HANDBOOK COMMITTEE

**Donald E. Ross,** Chairman

1988 EQUIPMENT VOLUME SUBCOMMITTEE

**Kennard L. Bowlen,** Chairman
**Darwin R. Grahl**
**Otto M. Kershock**
**Richard F. Sharp**
**Carl F. Speich**

ASHRAE HANDBOOK STAFF

Director of Communications
and Publications
**W. Stephen Comstock**

Handbook Editor
**Robert A. Parsons**

Assistant Editor
**Kelley D. Alexander**

Editorial Assistant
**Carolyn H. Baird**

Production Manager
**Stanley S. Beitler**

Typography
**Paul Bratton**
**Brenda C. Magbee**

Graphics
**Roxanne Starr**

Consultant-Handbook
and Technical Publications
**Carl W. MacPhee**

## CHAPTER 1

# DUCT CONSTRUCTION



THIS chapter covers the construction of heating, ventilating, air-conditioning, and exhaust duct systems for residential, commercial and industrial applications. Chapter 33, "Duct Design," of the 1985 FUNDAMENTALS Volume and Chapters 10, "Air Distribution Design for Small Heating and Cooling Systems," and 57, "Testing, Adjusting, and Balancing, of the 1987 HVAC Volume cover duct design methods and testing and balancing of systems. Technological advances in duct construction should be judged relative to the construction requirements herein and to appropriate codes and standards. While the construction details shown in this chapter may coincide, in part, with industry standards, they do not constitute an ASHRAE standard.

## BUILDING CODE REQUIREMENTS

In the private sector, each new construction or renovation project is normally governed by state laws or local ordinances that require compliance with specific health, safety, property protection, and energy conservation regulations. Figure 1 illustrates relationships between laws, ordinances, codes, and standards that can affect the design and construction of HVAC duct systems; however, Figure 1 may not list all applicable regulations and standards for a specific locality. Specifications for federal government construction are promulgated by the Federal Construction Council, the General Services Administration, the Department of the Navy, the Veterans Administration, and other agencies.

Model code changes require long cycles for approval by the consensus process. Since the development of safety codes, energy codes and standards proceed independently; the most recent edition of a code or standard may not have been adopted by a local jurisdiction. HVAC designers must know which code compliance obligations affect their designs. If a provision is in conflict with the design intent, the designer should resolve the issue with local building officials. New or different construction methods can be accommodated by the provisions for equivalency that are incorporated into codes. Staff engineers from the model code agencies are available to assist in the resolution of conflicts, ambiguities, and equivalencies.

Fire and smoke control is covered in Chapter 58 of the 1987 HVAC Volume. The designer should consider flame spread, smoke development, and toxic gas production from duct and duct insulation materials. Code documents for ducts in certain locations within buildings rely on a criterion of *limited combustibility* (see definitions, NFPA 90A) that is independent of the generally accepted criteria of 25 flame spread and 50 smoke development; however, certain duct construction protected by extinguishing systems may be accepted with higher levels of combustibility by code officials.

Combustibilty and tocity ratings are normally based on tests of new materials; little research is reported on ratings of duct materials that are aged or of systems that are poorly maintained for cleanliness. Fibrous and other porous materials exposed to airflow in ducts may accumulate more dirt than nonporous materials.

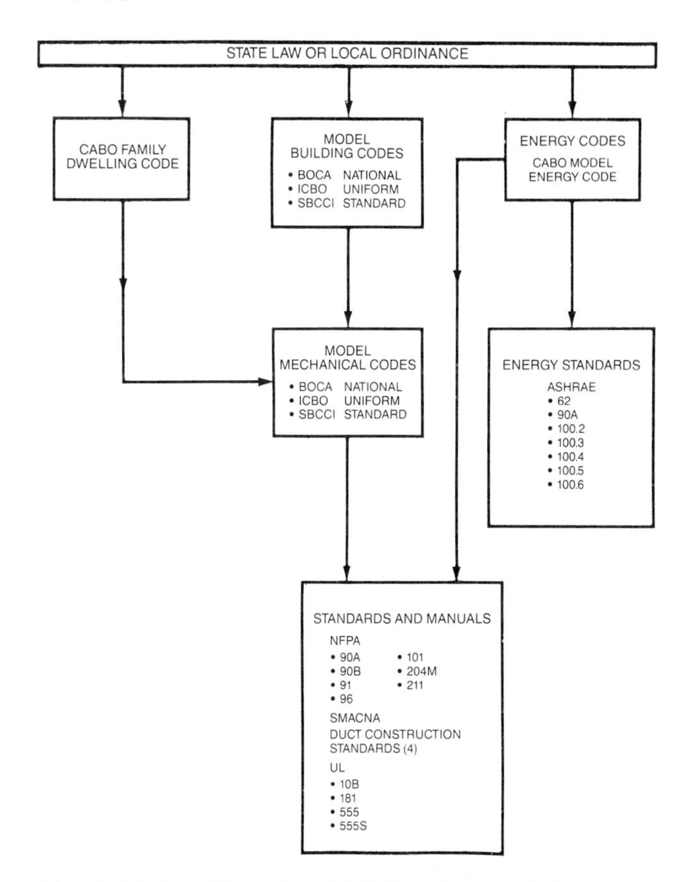


**Fig. 1 Typical Hierarchy of Building Codes and Standards**

---

The preparation of this chapter is assigned to TC 5.2, Duct Design.
Tables in the text are also included in the *Appendix* in SI units.

## CLASSIFICATION OF DUCTS

Duct construction is classified below in terms of application and pressure. HVAC systems in public assembly, business, educational, general factory, and mercantile buildings are usually designed as *commercial systems.* Air pollution control systems, industrial exhaust systems, and systems outside the above pressure range are classified as *industrial systems.*

| | |
|---|---|
| *Residences:* | ±0.5 in. of water (±125 Pa) |
| | ±1 in. of water (±250 Pa) |
| *Commercial Systems:* | ±0.5 in. of water (±125 Pa) |
| | ±1 in. of water (±250 Pa) |
| | ±2 in. of water (±500 Pa) |
| | ±3 in. of water (±750 Pa) |
| | +4 in. of water (+1000 Pa) |
| | +6 in. of water (+1500 Pa) |
| | +10 in. of water (+2500 Pa) |
| *Industrial Systems:* | any pressure |

Air conveyed by a duct imposes both air pressure and velocity pressure loads on the duct's structure. The load resulting from mean static-pressure differential across the duct wall normally dominates, and is generally used for duct classification. Turbulent airflow introduces relatively low but rapidly pulsating loading on the duct wall.

Static pressure at specific points in an air-distribution system is not necessarily the static pressure rating of the fan; the actual static pressure in each duct section must be obtained by computation. Therefore, the designer should specify the pressure classification of the various duct sections in the system. All modes of operation must be taken into account, especially systems used for smoke control.

## DUCT SYSTEM LEAKAGE

Project specifications should define allowable duct leakage, the need for leakage testing, and require the ductwork installer to perform a leakage test after installing an initial portion of the duct system. Procedures in the *HVAC Air Duct Leakage Test Manual* (SMACNA 1985) should be followed for leakage testing. If a test indicates excess leakage, corrective measures should be taken to assure quality control.

Responsibility for proper assembly and sealing belongs to the installing contractor. The most cost-effective way to control leakage is to follow proper installation procedures. Because access for repairs is usually limited, poorly installed duct systems that must later be resealed can cost more than a proper installation.

## RESIDENTIAL DUCT CONSTRUCTION

NFPA *Standard* 90*B*, CABO's *One- and Two- Family Dwelling Code*, or a local code is used for duct systems in single-family dwellings. Generally, local authorities use NFPA *Standard* 90*A* for multi-family homes.

**Table 1 Residential Metal Duct Construction**

| Shape of Duct and Exposure | Galvanized Steel Minimum Thickness, in. | Aluminum (3003) Nominal Thickness, in. |
|---|---|---|
| Enclosed Rectangular Ducts[a] | | |
| 14 in. or less | 0.0127 | 0.016 |
| Over 14 in. | 0.0157 | 0.020 |
| Rectangular and Round Ducts | Consult SMACNA HVAC *Duct Construction Standards—Metal and Flexible* | |

[a]Data based on nominal thickness, NFPA 90B (1984).

Supply ducts may be steel, aluminum, or a material with a UL *Standard* 181 rating. Sheet-metal ducts should be constructed of minimum thickness as shown in Table 1, and installed in accordance with *HVAC Duct Construction Standards—Metal and Flexible* (SMACNA 1985). Fibrous glass ducts should be installed in accordance with the *Fibrous Glass Duct Construction Standards* (SMACNA 1979). For return duct systems, the use of alternate materials, and other exceptions, NFPA *Standard* 90*B* should be consulted.

## COMMERCIAL DUCT CONSTRUCTION

### Materials

NFPA *Standard* 90*A* is frequently used as a guide standard by many building code agencies. NFPA *Standard* 90*A* invokes UL *Standard* 181, which classifies ducts as follows:

Class 0—zero flame spread, zero smoke developed
Class 1—25 flame spread, 50 smoke developed
Class 2—50 flame spread, 100 smoke developed

**Table 2a Galvanized Sheet Thickness**

| Gauge | Thickness, in. Nominal | Minimum | Nominal Weight, lb/ft$^2$ |
|---|---|---|---|
| 30 | 0.0157 | 0.0127 | 0.656 |
| 28 | 0.0187 | 0.0157 | 0.781 |
| 26 | 0.0217 | 0.0187 | 0.906 |
| 24 | 0.0276 | 0.0236 | 1.156 |
| 22 | 0.0336 | 0.0296 | 1.406 |
| 20 | 0.0396 | 0.0356 | 1.656 |
| 18 | 0.0516 | 0.0466 | 2.156 |
| 16 | 0.0635 | 0.0575 | 2.656 |
| 14 | 0.0785 | 0.0705 | 3.281 |
| 13 | 0.0934 | 0.0854 | 3.906 |
| 12 | 0.1084 | 0.0994 | 4.531 |
| 11 | 0.1233 | 0.1143 | 5.156 |
| 10 | 0.1382 | 0.1292 | 5.781 |

*Notes:*

1. Minimum thickness is based on thickness tolerances of hot-dip galvanized sheets in cut lengths and coils (per ASTM *Standard* A525). Tolerance is valid for 48-in. and 60-in. wide sheets.
2. Galvanized sheet gauge.

**Table 2b Uncoated Steel Sheet Thickness**

| Gauge | Thickness, in. Nominal | Minimum Hot-Rolled | Minimum Cold-Rolled | Nominal Weight, lb/ft$^2$ |
|---|---|---|---|---|
| 28 | 0.0149 | | 0.0129 | 0.625 |
| 26 | 0.0179 | | 0.0159 | 0.750 |
| 24 | 0.0239 | | 0.0209 | 1.000 |
| 22 | 0.0299 | | 0.0269 | 1.250 |
| 20 | 0.0359 | | 0.0329 | 1.500 |
| 18 | 0.0478 | 0.0428 | 0.0438 | 2.000 |
| 16 | 0.0598 | 0.0538 | 0.0548 | 2.500 |
| 14 | 0.0747 | 0.0677 | 0.0697 | 3.125 |
| 13 | 0.0897 | 0.0827 | 0.0847 | 3.750 |
| 12 | 0.1046 | 0.0966 | 0.0986 | 4.375 |
| 11 | 0.1196 | 0.1116 | 0.1136 | 5.000 |
| 10 | 0.1345 | 0.1265 | 0.1285 | 5.625 |

*Notes:*

1. Minimum thickness is based on thickness tolerances of hot-rolled and cold-rolled sheets in cut lengths and coils (per ASTM *Standards* A366, A568, and A569).
2. Table is based on 48-in. width coil and sheet stock. Sixty-in. coil has same tolerance, except that 16-gauge is ±0.007 in. in hot-rolled coils and sheets.
3. Manufacturer's standard gauge.

**Table 2c Stainless Steel Sheet Thickness**

| Gauge | Thickness, in. | | Nominal Weight, lb/ft² Stainless Steel | |
|---|---|---|---|---|
| | Nominal | Minimum | 300 Series | 400 Series |
| 28 | 0.0151 | 0.0131 | 0.634 | 0.622 |
| 26 | 0.0178 | 0.0148 | 0.748 | 0.733 |
| 24 | 0.0235 | 0.0205 | 0.987 | 0.968 |
| 22 | 0.0293 | 0.0253 | 1.231 | 1.207 |
| 20 | 0.0355 | 0.0315 | 1.491 | 1.463 |
| 18 | 0.0480 | 0.0430 | 2.016 | 1.978 |
| 16 | 0.0595 | 0.0535 | 2.499 | 2.451 |
| 14 | 0.0751 | 0.0681 | 3.154 | 3.094 |
| 13 | 0.0900 | 0.0820 | 3.780 | 3.708 |
| 12 | 0.1054 | 0.0964 | 4.427 | 4.342 |
| 11 | 0.1200 | 0.1100 | 5.040 | 4.944 |
| 10 | 0.1350 | 0.1230 | 5.670 | 5.562 |

*Notes:*

1. Minimum thickness is based on thickness tolerances for hot-rolled sheets in cut lengths and cold-rolled sheets in cut lengths and coils (per ASTM *Standard* A480).
2. Stainless sheet gauge has no standard nominal thickness among manufacturers.

**Table 3 Steel Angle Weights (Approximate)**

| Angle Size, in. | Weight, lb/ft |
|---|---|
| 3/4 × 3/4 × 1/8 | 0.59 |
| 1 × 1 × 0.0466 (min) | 0.36 |
| 1 × 1 × 0.0575 (min) | 0.44 |
| 1 × 1 × 1/8 | 0.80 |
| 1 1/4 × 1 1/4 × 0.0466 (min) | 0.45 |
| 1 1/4 × 1 1/4 × 0.0575 (min) | 0.55 |
| 1 1/4 × 1 1/4 × 0.0854 (min) | 0.65 |
| 1 1/4 × 1 1/4 × 1/8 | 1.01 |
| 1 1/2 × 1 1/2 × 0.0575 (min) | 0.66 |
| 1 1/2 × 1 1/2 × 1/8 | 1.23 |
| 1 1/2 × 1 1/2 × 3/16 | 1.80 |
| 1 1/2 × 1 1/2 × 1/4 | 2.34 |
| 2 × 2 × 0.0575 (min) | 0.89 |
| 2 × 2 × 1/8 | 1.65 |
| 2 × 2 × 3/16 | 2.44 |
| 2 × 2 × 1/4 | 3.19 |
| 2 1/2 × 2 1/2 × 3/16 | 3.07 |
| 2 1/2 × 2 1/2 × 1/4 | 4.10 |

NFPA *Standard* 90*A* states, in summary, that ducts must be iron, steel, aluminum, concrete, masonry, or clay tile. However, ducts may be UL *Standard* 181 Class 1 materials when they are not used as vertical risers serving more than two stories or in systems with air temperature higher than 250°F (121°C). Many manufactured flexible and fibrous glass ducts are UL approved and listed as Class 1. For galvanized ducts, a G90 (Z275) coating is recommended (see ASTM *A525* [A525M]). The minimum thickness and weight of sheet-metal sheets are given in Table 2.

Duct-reinforcing members are either formed from sheet metal or made from hot-rolled or extruded structural shapes. The size and weights of commonly used members are given in Table 3.

## Rectangular and Round Ducts

**Rectangular Metal Ducts.** Table 4 gives construction requirements for rectangular steel ducts and includes combinations of duct thicknesses, reinforcement, and maximum distance between reinforcements. *HVAC Duct Construction Standards—Metal and Flexible* (SMACNA 1985) gives the functional criteria on which Table 4 is based. Transverse joints (*e.g.*, standing drive slips, pocket locks, and companion angles) and, when necessary, intermediate construction joints are designed to reinforce the duct system. Ducts larger than 96 in. (2400 mm) require internal tie rods to maintain their structural integrity. Table 4 also

**Table 4a Rectangular Ferrous Metal Duct Construction for Commercial Systems*[a,b]**

| Duct Dimensions, in. | Minimum Galvanized Steel Thickness[c], Inches | | | | | | | | | | | | | | | | | | |
|---|---|---|---|---|---|---|---|---|---|---|---|---|---|---|---|---|---|---|---|
| | 0.0575 | | | | | | | 0.0466 | | | | | | | 0.0356 | | | | |
| | Pressure, in. of Water | | | | | | | | | | | | | | | | | | |
| | ±0.5 | ±1 | ±2 | ±3 | +4 | +6 | +10 | ±0.5 | ±1 | ±2 | ±3 | +4 | +6 | +10 | ±0.5 | ±1 | ±2 | ±3 | +4 |
| Up through 10 | Reinforcement not required[g] | | | | | | | Reinforcement not required[g] | | | | | | | Reinforcement not required[g] | | | | |
| 12 | | | | | | | | | | | | | | A-8 | | | | | |
| 14 | | | | | | | B-8 | | | | | | | B-8 | | | | | A-10 |
| 16 | | | | | | | B-8 | | | | | | B-10 | B-5 | | | | A-8 | A-10 |
| 18 | | | | | | C-10 | C-8 | | | | | B-10 | C-10 | C-5 | | | | A-8 | B-8 |
| 20 | | | | | C-10 | C-10 | D-8 | | | | B-10 | C-10 | C-8 | C-5 | | | B-10 | B-8 | C-8 |
| 22 | | | | | C-10 | D-10 | C-5 | | | B-10 | C-10 | C-10 | C-8 | C-5 | | A-10 | B-10 | B-8 | B-5 |
| 24 | | | | | D-10 | D-8 | D-5 | | | C-10 | C-10 | D-10 | D-8 | D-5 | | B-10 | C-10 | B-5 | C-5 |
| 26 | | | C-10 | D-10 | D-10 | D-8 | D-5 | | | C-10 | D-10 | D-10 | D-5 | D-5 | | B-10 | C-10 | C-5 | C-5 |
| 28 | | | C-10 | D-10 | E-10 | E-8 | E-5 | | C-10 | C-10 | D-10 | E-8 | D-5 | E-5 | B-10 | C-10 | C-8 | C-5 | D-5 |
| 30 | | | D-10 | D-10 | E-10 | D-5 | E-5 | | C-10 | D-10 | D-8 | E-8 | D-5 | E-4 | B-10 | C-10 | D-8 | C-5 | D-5 |
| 36 | | D-10 | E-10 | E-8 | E-5 | F-5 | F-5 | C-10 | D-10 | E-8 | E-5 | E-5 | F-5 | F-4 | C-10 | D-10 | D-5 | E-5 | E-5 |
| 42 | D-10 | E-10 | E-8 | E-5 | F-5 | G-5 | H-4 | D-10 | E-10 | E-5 | E-5 | F-5 | G-4 | G-3 | D-10 | D-8 | E-5 | E-5 | F-4 |
| 48 | E-10 | F-10 | G-8 | G-5 | G-5 | H-4 | H-3 | E-10 | E-8 | F-5 | G-5 | G-5 | H-4 | H-3 | E-10 | E-5 | F-5 | F-4 | F-3 |
| 54 | E-10 | G-10[d] | G-5[d] | H-5[d] | H-5[d] | H-4[d] | I-3[d] | E-10 | F-8 | G-5[d] | H-5[d] | H-4[d] | H-3[d] | H-2.5[d] | E-8 | E-5 | F-4 | G-3[d] | G-3[d] |
| 60 | F-10 | G-8[d] | H-5[d] | H-5[d] | I-5[d] | H-3[d] | J-3[d] | F-10 | G-8[d] | H-5[d] | H-4[d] | H-3[d] | H-3[d] | I-2.5[d] | F-8 | F-5 | G-4[d] | G-3[d] | H-3[d] |
| 72 | H-10[d] | H-5[d] | I-5[d] | I-4[d] | I-3[d] | J-3[d] | K-2.5[d] | G-8[d] | H-5[d] | H-4[d] | H-3[d] | I-3[d] | J-2.5[d] | K-2[e] | F-5 | G-4[d] | H-3[d] | H-3[d] | I-2.5[d] |
| 84 | H-8[d] | I-5[d] | J-4[d] | J-3[d] | K-3[d] | L-2.5[d] | H-2 plus rods | H-5[d] | I-5[d] | J-4[d] | J-3[d] | J-2.5[d] | K-2[e] | — | H-5[d] | H-4[d] | I-3[d] | J-2.5[d] | J-2[d] |
| 96 | I-8[d] | J-5[d] | K-4[d] | L-3[d] | L-2.5[d] | L-2[d] | H-2 plus rods | H-5 | I-4 | K-3[d] | K-2.5[d] | K-2[d] | L-2[d] | — | H-5[d] | I-3[d] | J-2.5[d] | J-2[d] | K-2[d] |
| Over 96 | H-5 | H-2.5 plus rods[f] | H-2.5 plus rods[f] | H-2.5[e] plus rods[f] | H-2.5[e] plus rods[f] | H-2[e] plus rods[f] | H-2 plus rods[f] | H-5 | H-2.5 plus rods[f] | H-2.5 plus rods[f] | H-2.5 plus rods[f] | H-2.5 plus rods[f] | H-2 plus rods[f] | — | — | — | — | — | — |

*See Table 4b for notes.

**Table 4b Rectangular Ferrous Metal Duct Construction for Commercial Systems[a,b]**

| Duct Dimensions, in. | Minimum Galvanized Steel Thickness[c], Inches | | | | | | | | | | | | | | | | | | |
|---|---|---|---|---|---|---|---|---|---|---|---|---|---|---|---|---|---|---|---|
| | 0.0356 | | 0.0296 | | | | | | | 0.0236 | | | | | | | 0.0187 | | |
| | Pressure, in. of Water | | | | | | | | | | | | | | | | | | |
| | +6 | +10 | ±0.5 | ±1 | ±2 | ±3 | +4 | +6 | +10 | ±0.5 | ±1 | ±2 | ±3 | +4 | +6 | +10 | ±0.5 | ±1 | ±2 |
| Up through 8 | | | | | Reinforcement not required[g] | | | | | | Reinforcement not required[g] | | | | | A-5 | Reinforcement not required[g] | | |
| 10 | | A-5 | | | | | | A-5 | A-5 | | | | | A-5 | A-5 | A-4 | | | |
| 12 | A-10 | A-5 | | | | | A-10 | A-5 | A-5 | | | | A-8 | A-5 | A-5 | A-4 | | | A-8 |
| 14 | A-10 | A-5 | | | | A-8 | A-8 | A-5 | A-4 | | | A-8 | A-5 | A-5 | A-4 | A-3 | | A-10 | A-5 |
| 16 | A-5 | B-5 | | | A-10 | A-8 | A-5 | A-5 | B-4 | | A-10 | A-8 | A-5 | A-5 | A-4 | B-3 | | A-8 | A-5 |
| 18 | B-5 | C-5 | | | A-10 | A-8 | A-5 | B-5 | B-4 | | A-10 | A-8 | A-5 | A-5 | B-4 | B-3 | | A-8 | A-5 |
| 20 | B-5 | C-4 | | A-10 | B-8 | A-5 | B-5 | B-5 | B-3 | | A-10 | A-5 | A-5 | B-5 | B-4 | B-3 | A-10 | A-8 | A-5 |
| 22 | C-5 | C-4 | | A-10 | B-8 | B-5 | B-5 | C-5 | C-3 | A-10 | A-10 | A-5 | B-5 | B-4 | C-4 | C-3 | A-10 | A-5 | A-5 |
| 24 | C-5 | D-4 | | B-10 | C-8 | B-5 | C-5 | C-5 | C-3 | A-10 | B-10 | B-5 | B-5 | C-4 | C-3 | C-3 | A-10 | A-5 | B-5 |
| 26 | D-5 | D-4 | A-10 | B-10 | C-8 | C-5 | C-5 | C-4 | D-3 | A-10 | B-8 | B-5 | C-5 | C-4 | C-3 | C-2.5 | A-10 | A-5 | B-5 |
| 28 | D-5 | D-4 | B-10 | C-10 | C-5 | C-5 | D-5 | D-4 | D-3 | B-10 | C-8 | C-5 | C-4 | D-4 | C-3 | D-2.5 | B-8 | B-5 | B-4 |
| 30 | D-4 | D-3 | B-10 | C-10 | C-5 | C-5 | D-5 | D-4 | D-3 | B-10 | C-8 | C-5 | C-4 | D-4 | D-3 | D-2.5 | B-8 | B-5 | C-4 |
| 36 | E-4 | F-3 | C-10 | D-8 | D-5 | D-4 | E-4 | E-3 | E-2.5 | C-8 | C-5 | D-4 | D-4 | D-3 | E-2.5 | E-2 | C-5 | C-5 | — |
| 42 | F-3 | G-2.5 | D-8 | D-5 | E-5 | E-4 | E-3 | E-2.5 | F-2 | D-8 | D-5 | E-4 | E-3 | E-2.5 | — | — | D-5 | D-4 | — |
| 48 | G-2.5 | G-2 | D-8 | E-5 | E-4 | E-3 | F-3 | G-2.5 | G-2 | D-5 | E-5 | E-3 | E-2.5 | E-2 | — | — | D-5 | D-4 | — |
| 54 | H-2.5[d] | H-2[d] | D-5 | E-5 | F-3 | G-3[d] | G-2.5[d] | G-2[d] | — | D-5 | E-4 | F-3 | E-2.5 | F-2 | — | — | D-5 | — | — |
| 60 | H-2.5[d] | I-2[d] | E-5 | F-5 | G-3[d] | G-2.5[d] | H-2.5[d] | H-2[d] | — | E-5 | F-4 | G-2.5[d] | G-2.5[d] | G-2[d] | — | — | E-4 | — | — |
| 72 | I-2[d] | — | F-5 | G-4[d] | H-3[d] | H-2.5[d] | H-2[d] | — | — | F-4 | H-2[d] | H-2[d] | H-2[d] | — | — | — | — | — | — |
| 84 | — | — | H-5[d] | — | — | — | — | — | — | G-4[d] | — | — | — | — | — | — | — | — | — |
| 96 | — | — | H-4[d] | — | — | — | — | — | — | — | — | — | dashes indicate not allowed | | | — | — | — | — |
| Over 96 | — | — | — | — | — | — | — | — | — | — | — | — | — | — | — | — | — | — | — |

[a]Table 4 is based on Tables 1-3 through 1-9 in SMACNA's publication, HVAC *Duct Construction Standards—Metal and Flexible*. For tie rod details, refer to this standard.

[b]For a given duct thickness, numbers indicate maximum spacing (feet) between duct reinforcement; letters indicate type (rigidity class) of duct reinforcement (see Tables 5 and 6).

Transverse joint spacing is unrestricted on unreinforced ducts. To qualify joints on reinforced ducts, select transverse joints from Table 5. Select intermediate reinforcement from Table 6. Tables are based on steel construction. Designers should specify galvanized, uncoated, or painted steel joint and intermediate reinforcement.

Use the same metal duct thickness on all duct sides. Evaluate duct reinforcement on each duct side separately. When required on four sides for +4, +6, and +10 in. of water pressure systems, corners must be tied. When required on two sides, corners must be tied with rods or angles at the ends for +4, +6, and +10 in. of water pressure systems.

Duct sides over 18 in. width with less than 0.0356 in. thickness, which have more than 10 $ft^2$ of unbraced panel area, must be cross-broken or beaded, unless they are lined or insulated externally. Lined or externally insulated ducts are not required to have cross-breaking or beading.

[c]The reinforcement tables are based on galvanized steel of the indicated thickness. They apply to galvanized, painted, uncoated, and stainless steel whenever the base metal thickness is not less than 0.0015 inches below that indicated for galvanized steel.

[d]See SMACNA's publication for alternative reinforcements using tie rods or tie straps for positive pressure.

[e]Sheet metal 0.0466 in. thick is acceptable.

[f]Tie rods with a minimum diameter of 0.375 in. (or 0.25 in. if the maximum length is 36 in.) must be used on these constructions. The rods for positive pressure ducts are spaced a maximum of 60 in. apart along joints and reinforcements.

[g]Blank spaces indicate that no reinforcement is required.

shows alternative tie rod construction for positive pressure ducts over 48 in. (1200 mm). Tie rods allow the use of smaller reinforcements than would otherwise be required. *Rectangular Industrial Duct Construction Standards* (SMACNA 1980) gives construction details for ducts up to 168 in. (4270 mm) in width.

Acceptable transverse joints and intermediate reinforcement members are given in Tables 5 and 6 for various duct pressures. Other joint systems and intermediate reinforcement must meet the rigidity requirement given in Table 6 for each class of reinforcement.

Fittings must be reinforced similarly to sections of straight duct. On size change fittings, the greater fitting dimension determines material thickness. Where fitting curvature or internal member attachments provide equivalent rigidity, such features may be credited as reinforcement.

**Round Metal Ducts.** Round ducts are inherently strong, rigid, and generally the most efficient and economical for air systems. The dominant factor in round duct construction is the material's ability to withstand the physical abuse of installation and negative pressure requirements. Table 7 gives construction requirements as a function of static pressure, type of seam (spiral or longitudinal), and diameter.

**Nonferrous Ducts.** *HVAC Duct Construction Standards—Metal and Flexible* (SMACNA 1985) gives construction requirements for rectangular (±3 in. of water [±750 Pa]) and round (±2 in. of water [±500 Pa]) aluminum ducts. *Round Industrial Duct Construction Standards* (SMACNA 1977) gives construction requirements for round aluminum duct systems with pressures greater than ±2 in. of water (±500 Pa).

**Construction Details.** *HVAC Duct Construction Standards—Metal and Flexible* (SMACNA 1985) gives construction details for rectangular and round ducts operating with static pressures from −3 to +10 in. of water (−750 to +2500 Pa).

## Flat-Oval Ducts

Table 8 gives flat-oval duct construction requirements. Seams and transverse joints are as permitted for round ducts. Reinforcement is determined from Tables 4 and 6 using dimension "F" defined in Table 8. Reinforcement ends on opposite sides are tied at pressures of 4 to 10 in. of water (1000 to 2500 Pa).

*HVAC Duct Construction Standards—Metal and Flexible* (SMACNA 1985) has additional construction details. Hanger designs and installation details for rectangular ducts generally apply to flat-oval ducts.

## Fibrous Glass Ducts

Fibrous glass ducts are a composite of rigid fiberglass and a factory-applied facing (typically aluminum or reinforced aluminum), which serves as a finish and vapor barrier. This material is available in molded round sections or in board form for fabrication into rectangular or polygon shapes. Duct systems of round and rectangular fibrous glass are generally limited to 2400 fpm (12 m/s) and ±2 in. of water (±500 Pa). Moulded roun ducts are available in higher pressure ratings. *Fibrous Glass Duct Construction Standards* (SMACNA 1979) and manufacturers' installation instructions give details on fibrous glass duct construction. The SMACNA standard also covers duct and fit-

**Table 5a Transverse Joint Reinforcement*[a]**

| Minimum Rigidity Class[f] | 4 in. of Water Maximum[b] | | | | | |
|---|---|---|---|---|---|---|
| | STANDING DRIVE SLIP | STANDING S | | STANDING S | STANDING S | STANDING S (BAR REINFORCED) / STANDING S (ANGLE REINFORCED) |
| | $H_s$ × T (min), in. | W, in. | $H_s$ × T (min), in. | $H_s$ × T (min), | $H_s$ × T (min), | $H_s$ × T (min) plus Reinforcement (H × T), in.[c] |
| A | Use Class B | Use Class C | | 1/2 × 0.0187 | Use Class D | Use Class F |
| B | 1 1/8 × 0.0187 | | | 1/2 × 0.0296 | | |
| C | 1 1/8 × 0.0296 | — | 1 × 0.0187 | 1 × 0.0187 | | |
| D | — | — | 1 × 0.0236 | 1 × 0.0236 | 1 1/8 × 0.0187 | |
| E | — | 3/16 | 1 1/8 × 0.0356 | — | 1 1/8 × 0.0466 | |
| F | — | 3/16 | 1 5/8 × 0.0296 | — | 1 1/2 × 0.0236 | 1 1/2 × 0.0236 plus 1 1/2 × 1/8 bar |
| G | — | 3/16 | 1 5/8 × 0.0466 | — | 1 1/2 × 0.0466 | 1 1/2 × 0.0296 plus 1 1/2 × 1/8 bar |
| H | — | — | — | — | — | 1 1/2 × 0.0356 plus 1 1/2 × 1 1/2 × 3/16 angle |
| I | — | — | — | — | — | 2 × 0.0356 plus 2 × 2 × 1/8 angle |
| J | — | — | — | — | — | 2 × 0.0356 plus 2 × 2 × 3/16 angle |

*See Table 5c for notes.

ting fabrication, closure, and installation, including installation of duct-mounted HVAC appurtenances (*e.g.*, volume dampers, turning vanes, register and grille connections, diffuser connections, access doors, fire damper connections, and electric heaters).

## Flexible Ducts

Flexible ducts connect mixing boxes, light troffers, diffusers, and other terminals to the air distribution system. *HVAC Duct Construction Standards—Metal and Flexible* (SMACNA 1985) gives installation details. Because unnecessary length, offsetting and compression of these ducts significantly increases airflow resistance (see Chapter 33 of the 1985 FUNDAMENTALS Volume), they should be kept as short as possible and fully extended.

UL *Standard* 181 covers testing of materials used to fabricate flexible ducts categorized separately as *air ducts* and *connectors*. NFPA *Standard* 90*A* defines the acceptable use of these products. The flexible duct connector has less resistance to flame penetration, lower puncture and impact resistance, and is subject to many restrictions listed in NFPA *Standard* 90*A*. *Air duct* rated flexible ducts should be specified. Tested products are listed in UL's *Gas and Oil Directory*.

## Plenums and Apparatus Casings

*HVAC Duct Construction Standards—Metal and Flexible* (SMACNA 1985) shows details on field-fabricated plenum and apparatus casings. Sheet metal thickness and reinforcement for plenum and casing pressures outside the range of −3 to +10 in. of water (−750 to +2500 Pa) can be based on *Rectangular Industrial Duct Construction Standards* (SMACNA 1980).

Plenums and apparatus casings on the discharge side of a fan should be analyzed carefully for maximum operating pressure in relation to the construction detail being specified. On the suction side of a fan, plenums and apparatus casings are normally constructed to withstand negative air pressures at least equal to the total upstream static pressure losses. The accidental stoppage of intake airflow can apply a negative pressure as great as the fan shutoff pressure. Conditions such as clogged louvers, filters, or coils or malfunctioning dampers can collapse a normally adequate casing. To protect large walls or roofs from damage caused by a collapsed casing, it is more economical to provide fan safety interlocks, such as damper end switches or pressure limit switches, instead of heavier sheet-metal construction.

**Table 5b Transverse Joint Reinforcement*[a]**

| Minimum Rigidity Class[f] | Standing Seam | Reinforced Welded or Standing Seam: 0.0187 through 0.0296 Duct, in. | | Reinforced Welded or Standing Seam: 0.0356 through 0.0575 Duct, in. | | Welded Flange | 3 in. of Water Maximum (Pocket Lock Type A; Bar Reinforced Pocket Lock Type B; Angle Reinforced Pocket Lock Type C) | | Thickness, in.[e] | |
|---|---|---|---|---|---|---|---|---|---|---|
| | $H_s$ × T (min), in.[d] | $H_s$, in. | H × T, in.[c] | $H_s$, in. | H × T, in.[c] | $H_s$ × T(min), in.[d] | Lock Type | $H_s$, in. | Lock | Reinforcement (H × T) |
| A | 1/2 × 0.0236 | | | | | 1/2 × 0.0296 | | | | |
| B | 3/4 × 0.0236 | Use Class D | | | | 1/2 × 0.0575<br>3/4 × 0.0296 | Use Class D | | | |
| C | 1 × 0.0236 | | | | | 3/4 × 0.0466<br>1 × 0.0296 | | | | |
| D | 3/4 × 0.0575<br>1 × 0.0356 | 1 | 1 × 0.0575 | Use Class E | | 1 × 0.0466<br>1 1/4 × 0.0296 | A | 1 | 0.0296 | None |
| E | 1 × 0.0575<br>1 1/2 × 0.0236 | 1 | 1 × 1/8 | 1 | 1 × 0.0575 | 1 1/4 × 0.0466<br>1 1/2 × 0.0296 | B | 1 | 0.0296 | 1 × 1/8<br>Bar |
| F | 1 1/2 × 0.0356 | 1 1/2 | 1 1/2 × 0.0575 | 1 1/4 | 1 1/4 × 0.0575 | 1 1/4 × 0.0575<br>1 1/2 × 0.0356 | A | 1 1/2 | 0.0296 | None |
| G | 1 1/2 × 0.0466 | 1 1/2<br>1 1/2 | 1 1/2 × 1/8<br>2 × 0.0575 | 1 1/2 | 1 1/2 × 1/8 | 1 1/2 × 0.0575 | B | 1 1/2 | 0.0296 | 1 1/2 × 1/8<br>Bar |
| H | — | 1 1/2 | 2 × 1/8 | 1 1/2<br>1 1/2 | 1 1/2 × 3/16<br>2 × 0.0575 | — | C | 1 1/2 | 0.0356 | 1 1/2 × 3/16<br>Angle |
| I | — | Use Class J | | 1 1/2 | 2 × 1/8 | — | C | 1 1/2 | 0.0356 | 2 × 1/8<br>Angle |
| J | — | 1 1/2 | 2 × 3/16 | 1 1/2 | 2 × 3/16 | — | C | 1 1/2 | 0.0356 | 2 × 3/16<br>Angle |
| K | — | 1 1/2 | 2 1/2 × 3/16 | Use Class L | | — | C | 1 1/2 | 0.0356 | 2 1/2 × 3/16<br>Angle |
| L | — | 1 1/2 | 2 1/2 × 1/4 | 1 1/2 | 2 1/2 × 3/16 | — | — | — | — | — |

*See Table 5c for notes.

Apparatus casings can perform two acoustical functions. If the fan is completely enclosed within the casing, the transmission of fan noise through the fan room to adjacent areas is reduced substantially. An acoustically lined casing also reduces airborne noise levels in connecting ductwork. Acoustical treatment may consist of a single metal wall with a field-applied acoustical liner or thermal insulation, or a double-wall panel with an acoustical liner and a perforated metal inner liner. Double-wall casings are marketed by many manufacturers who publish data on structural, acoustical, and thermal performance and also prepare designs on a custom basis.

## Acoustical Treatment

Metal ducts are frequently lined with acoustically absorbent materials to reduce aerodynamic noise. Although many materials are acoustically absorbent, duct liners must also be resistant to erosion and fire and have properties compatible with the ductwork fabrication and erection process. For high-velocity duct systems, double-wall construction, using a perforated metal inner liner, is frequently specified. Chapter 52 of the 1987 HVAC Volume addresses design considerations. ASTM *Standard* C423 (1981) covers laboratory testing of duct liner materials to determine their sound absorption coefficients. Designers should review all of the tests incorporated in ASTM *Standard* C1071 (1986). A wide range of performance attributes, including erosion resistance, vapor adsorption, temperature resistance, and fungi resistance, are covered in the standard. Health and safety precautions are addressed, and manufacturer's certifications of compliance are also covered.

Rectangular duct liners should be secured by mechanical fasteners and installed in accordance with *HVAC Duct Construction Standards—Metal and Flexible* (SMACNA 1985). Adhesives should be Type I, in conformance to ASTM *Standard* C916 (1985) and should be applied to the duct, with 100% coverage of mating surfaces. Quality workmanship prevents delamination of the liner and possible blockage of coils, dampers, flow sensors, or terminal devices. Uneven edge alignment at butted joints should be avoided to minimize unnecessary resistance to airflow (Swim 1978).

Rectangular metal ducts are susceptible to rumble from flexure in the duct walls during startup and shutdown. If a designer

**Table 5c Transverse Joint Reinforcement[a]**

| | CAPPED FLANGE | | COMPANION ANGLES | FLANGED | FLANGED / SLIP-ON FLANGE |
|---|---|---|---|---|---|
| **Minimum Rigidity Class[f]** | **$H_s$ × T (min), in.** | **Min Cap Thickness, in.** | **H × T (nominal), in.** | **$H_s$ × T (min), in.** | |
| A | Use Class B | | Use Class E | Use Class C | Consult SMACNA or manufacturers to establish ratings. |
| B | 3/4 × 0.0187 | 0.0236 | | | |
| C | 1 × 0.0236 | 0.0236 | | 1 × 0.0236 | |
| D | 1 × 0.0296 | 0.0296 | | 1 × 0.0296 | |
| E | 1 1/2 × 0.0236 | 0.0296 | 1 × 1/8 | 1 × 0.0575<br>1 1/2 × 0.0236 | |
| F | 1 1/2 × 0.0356 | 0.0356 | Use Class G | 1 1/2 × 0.0296 | |
| G | Use Class H | | 1 1/4 × 1/8 | 1 1/2 × 0.0466 | |
| H | 2 × 0.0575 | 0.0356 | 1 1/2 × 1/8 | 2 × 0.0466 | |
| I | — | — | 1 1/2 × 3/16 | 2 × 0.0575 | |
| J | — | — | 1 1/2 × 1/4 | — | |
| K | — | — | 2 × 3/16 | — | |

[a]Table 5 is based on Tables 1-11 through 1-13 of SMACNA's publication, HVAC *Duct Construction Standards—Metal and Flexible.* For assembly details and other construction alternatives using tie rods, refer to this standard.

[b]Standing S slip length limits: 30 in. at 4 in. of water, 36 in. at 3 in. of water, none at 2 in. of water or less.

[c]T-dimension for formed angles is minimum thickness; nominal for structural angles.

[d]Duct reinforcement (rigidity class) may have to be increased from that obtained from Table 4 to match duct thickness.

[e]T-dimension for locks and ducts is minimum thickness; nominal for bars and angles.

[f]EI index for rigidity class is in Table 6.

wants systems to go on and off frequently (as an energy conservation measure) during the times in which buildings are occupied, he or she should specify duct construction that reduces objectionable noise.

### Hangers

Tables 9, 10, and 11 and *HVAC Duct Construction—Metal and Flexible* (SMACNA 1985) describe commercial HVAC system hangers for rectangular, round, and flat-oval ducts. When special analysis is required for larger ducts or loads than are given in the tables, or for other hanger configurations, the AISC (1980) and AISI (1986) design manuals should be consulted. To hang or support fibrous glass ducts, the methods detailed in *Fibrous Glass Duct Construction Standards* (SMACNA 1979) are recommended. UL *Standard* 181 involves maximum support intervals for UL *listed* ducts.

## INDUSTRIAL DUCT CONSTRUCTION

NFPA *Standard* 91 is widely used for duct systems conveying particulates, removing flammable vapors (including paint-spraying residue) and corrosive fumes. Particulate-conveying duct systems are classified as follows:

**Class 1:** Nonparticulate applications, including makeup air, general ventilation, and gaseous emission control.

**Class 2:** Moderately abrasive particulate in light concentration, such as that produced by buffing and polishing, woodworking, and grain-handling operations.

**Class 3:** Highly abrasive material in low concentration, such as that produced from abrasive cleaning operations, driers and kilns, boiler breaching, and sand handling.

**Class 4:** Highly abrasive particulate in high concentrations, including materials conveying high concentrations of particulates listed under **Class 3.**

For contaminant abrasiveness ratings, see *Round Industrial Duct Construction Standards* (SMACNA 1977).

### Materials

Galvanized steel, uncoated carbon steel, or aluminum are most frequently used for industrial air-handling systems. Aluminum ductwork is not used for systems conveying abrasive materials; when temperatures exceed 400°F (200°C), galvanized steel is not recommended. Ductwork materials for systems handling corrosive gases, vapors, or mists must be selected carefully. For the application of metals and use of protective coatings in corrosive environments consult *Accepted Industry Practice for Industrial Duct Construction* (SMACNA 1975), the *Pollution Engineering Practice Handbook* (Cheremisinoff and Young, 1975), and the publications of the *National Association of Corrosive Engineers* (NACE).

### Round Ducts

*Round Industrial Duct Construction Standards* (SMACNA 1977) is recommended for the selection of material thickness and reinforcement members for non-spiral industrial systems. The tables in this manual are presented as follows:

**Table 6 Intermediate Reinforcement[a]**

| Minimum Rigidity | | ANGLE | ZEE OR CHANNEL | HAT SECTION | CHANNEL |
|---|---|---|---|---|---|
| **Class** | **EI × 10⁵, lb-in.²** | **H × T, in.[b]** | **H × B × T, in.[b]** | **H × B × D × T, in.[b]** | **H × B × T, in.[b]** |
| A | 0.5 | Use Class C | Use Class B | Use Class F | Use Class C |
| B | 1.0 | | 3/4 × 1/2 × 0.0356 | | |
| C | 2.5 | 1 × 0.0466<br>3/4 × 1/8 | 3/4 × 1/2 × 0.0466 | | 3/4 × 3 × 0.0466 |
| D | 5 | 1 1/4 × 0.0466<br>1 × 1/8 | 1 × 3/4 × 0.0356 | | 1 1/8 × 3 1/4 × 0.0466 |
| E | 10 | 1 1/4 × 0.0854<br>1 1/2 × 0.0575 | 1 × 3/4 × 0.0854<br>1 1/2 × 3/4 × 0.0356 | | 1 × 2 × 1/8 |
| F | 15 | 1 1/2 × 1/8 | 1 × 3/4 × 1/8<br>1 1/2 × 3/4 × 0.0466 | 1 1/2 × 3/4 × 5/8 × 0.0356<br>1 1/2 × 1 1/2 × 3/4 × 0.0356 | 1 1/2 × 3 × 0.0575 |
| G | 25 | 1 1/2 × 3/16 | 1 1/2 × 3/4 × 1/8<br>2 × 1 1/8 × 0.0356 | 1 1/2 × 3/4 × 5/8 × 0.0575<br>1 1/2 × 1 1/2 × 3/4 × 0.0466 | 1 1/4 × 3 × 1/8 |
| H | 50 | 2 × 1/8 | 2 × 1 1/8 × 0.0575 | 1 1/2 × 1 1/2 × 3/4 × 0.0854<br>2 × 1 × 3/4 × 0.0466 | 1.4 × 3 × 0.22 |
| I | 75 | 2 × 3/16 | 2 × 1 1/8 × 0.0854 | 2 × 1 × 3/4 × 0.0854<br>2 1/2 × 2 × 3/4 × 0.0575 | 2 × 2 × 1/8 |
| J | 100 | 2 × 1/4<br>2 1/2 × 1/8 | 2 × 1 1/8 × 1/8<br>3 × 1 1/8 × 0.0575 | 2 × 1 × 3/4 × 1/8<br>2 1/2 × 2 × 3/4 × 0.0854 | 1.6 × 4 × 0.28 |
| K | 150 | 2 1/2 × 3/16 | 3 × 1 1/8 × 0.0854 | 2 1/2 × 2 × 3/4 × 1/8<br>3 × 1 1/2 × 3/4 × 0.0575 | |
| L | 200 | 2 1/2 × 1/4 | 3 × 1 1/8 × 1/8 | 3 × 1 1/2 × 3/4 × 0.0854 | |

[a]Table 6 is based on Table 1-10 of SMACNA's publication, HVAC *Duct Construction Standards—Metal and Flexible.*

[b]T-dimension for formed shapes is minimum thickness; nominal for structural shapes.

**Class.** Steel—Classes 1, 2, 3, and 4; Aluminum—Class 1 only.

**Pressure Classes for Steel and Aluminum.** ±2 to ±30 in. of water (500 Pa to 7.5 kPa), in increments of 2 in. of water (500 Pa).

**Duct Diameters for Steel and Aluminum.** 4 to 60 in. (100 to 1500 mm), in increments of 2 in. (50 mm). Equations are available for calculating the construction requirements for diameters greater than 60 in. (1500 mm).

For spiral duct applications, consult manufacturer's construction schedules, such as those listed in the *Industrial Duct Engineering Data and Recommended Design Standards* (United McGill Corp. 1985).

## Rectangular Ducts

*Rectangular Industrial Duct Construction Standards* (SMACNA 1980) is available for selecting material thickness and reinforcement members for industrial systems. The data in this manual give the duct construction for any system pressure class and panel width. Each side of a rectangular duct is considered a panel. Usually, the four sides of a rectangular duct are built of material with the same thickness. Ducts are sometimes built with the bottom plate thicker than the other three sides (usually in ducts with heavy particulate accumulation) to save material.

The designer selects a combination of panel thickness, reinforcement, and reinforcement member spacing to limit the deflection of the duct panel to a design maximum. Any shape transverse joint or intermediate reinforcement member that meets the minimum requirement of both section modulus and moment of inertia may be selected. The SMACNA data, which may also be used for designing apparatus casings, limit the combined stress in either the panel or structural member to 24 kpsi (165 MPa) and the maximum allowable deflection of the reinforcement members to 1/360 of the duct width.

## Construction Details

Recommended manuals for other construction details are *Industrial Ventilation* (ACGIH 1986), NFPA 91 (1983), and *Accepted Industry Practice for Industrial Ventilation* (SMACNA 1975). The transverse reinforcing of ducts subject to negative pressures below −3 in. of water (−750 Pa) should be welded to the duct wall rather than relying on mechanical fasteners to transfer the static loading to the reinforcing.

## Hangers

The *Manual of Steel Construction* (AISC 1980) and the *Cold-Formed Steel Design Manual* (AISI 1986) give design information for industrial duct hangers and supports. The SMACNA standards for rectangular and round industrial ducts (SMACNA 1980, 1977) and manufacturers' schedules include duct design information for supporting ducts at intervals up to 35 ft (10 m).

**Table 7a Round Ferrous Metal Duct Construction for Commercial Systems*[a]**

| Duct Diameter, In. | Minimum Galvanized Steel Thickness[b], in. −2 in. of Water: Spiral Seam Duct | Longitudinal Seam Duct[c] | Fittings | +2 in. of Water: Spiral Seam Duct | Longitudinal Seam Duct[c] | Fittings | Type of Joint[d] |
|---|---|---|---|---|---|---|---|
| Up through 8 | 0.0157 | 0.0236 | 0.0236 | 0.0157 | 0.0157 | 0.0187 | Beaded slip |
| 14 | 0.0187 | 0.0236 | 0.0236 | 0.0157 | 0.0187 | 0.0187 | Beaded slip |
| 26 | 0.0236 | 0.0296 | 0.0296 | 0.0187 | 0.0236 | 0.0236 | Beaded slip |
| 36 | 0.0296 | 0.0356 | 0.0356 | 0.0236 | 0.0296 | 0.0296 | Beaded slip |
| 50 | 0.0356 | 0.0466 | 0.0466 | 0.0296 | 0.0356 | 0.0356 | Flange |
| 60 | 0.0466 | 0.0575 | 0.0575 | 0.0356 | 0.0466 | 0.0466 | Flange |
| 84 | 0.0575 | 0.0705 | 0.0705 | 0.0466 | 0.0575 | 0.0575 | Flange |

*See Table 7b for notes.

**Table 7b Round Ferrous Metal Duct Construction for Commercial Systems[a]**

| Duct Diameter, In. | Minimum Galvanized Steel Thickness[b], in. −3 in. of Water Maximum: Spiral Seam Duct | Longitudinal Seam Duct[e] | Fittings | +10 in. of Water Maximum: Spiral Seam Duct | Longitudinal Seam Duct[e] | Fittings | Type of Joint[d] |
|---|---|---|---|---|---|---|---|
| Up through 8 | 0.0187 | 0.0236 | 0.0236 | 0.0187 | 0.0236 | 0.0236 | Beaded slip |
| 14 | 0.0187 | 0.0236 | 0.0236 | 0.0187 | 0.0236 | 0.0236 | Beaded slip |
| 26 | 0.0236 | 0.0296 | 0.0296 | 0.0236 | 0.0296 | 0.0296 | Beaded slip |
| 36 | 0.0296 | 0.0356 | 0.0356 | 0.0296 | 0.0356 | 0.0356 | Beaded slip |
| 50 | 0.0356 | 0.0356 | 0.0356 | 0.0356 | 0.0356 | 0.0356 | Flange |
| 60 | — | 0.0466 | 0.0466 | 0.0466 | 0.0466 | 0.0466 | Flange |
| 84 | — | 0.0575 | 0.0575 | 0.0466 | 0.0575 | 0.0575 | Flange |

[a]Table 7 is based on SMACNA's publication, HVAC *Duct Construction Standards—Metal and Flexible*, Table 3-2.

[b]Table 7 may be used for galvanized, painted, uncoated, and stainless steel whenever the base metal thickness is not less than 0.0015 in. below those in the table.

[c]For seam details and limitations, refer to SMACNA's publication, HVAC *Duct Construction Standards—Metal and Flexible*, Figure 3-1.

[d]Recommended joints are listed. For alternate joints, details, and limitations, refer to SMACNA's HVAC *Duct Construction Standards—Metal and Flexible*, Figure 3-2.

[e]Only butt weld or flat lock (grooved seam, pipe lock) longitudinal seams are permitted.

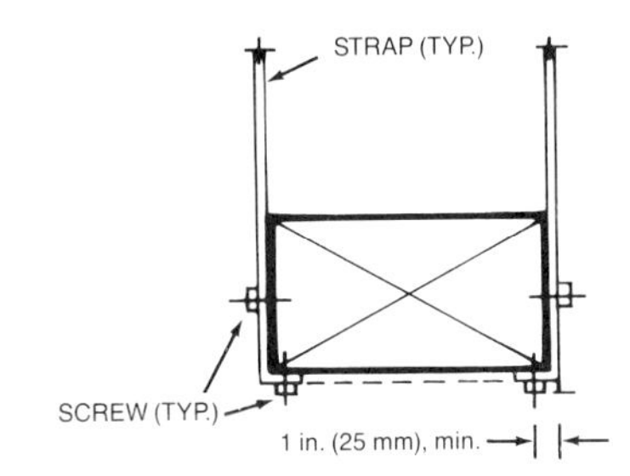


**Fig. 2 Strap Hanger—Rectangular Ducts**

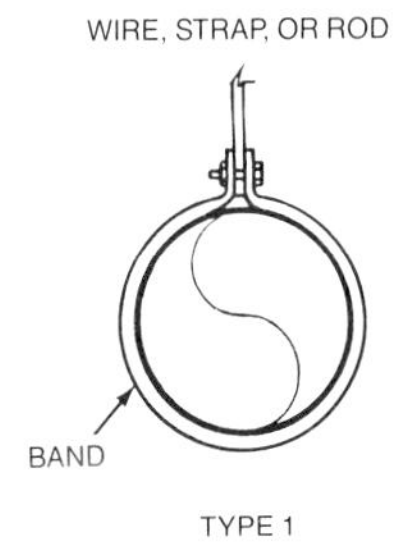


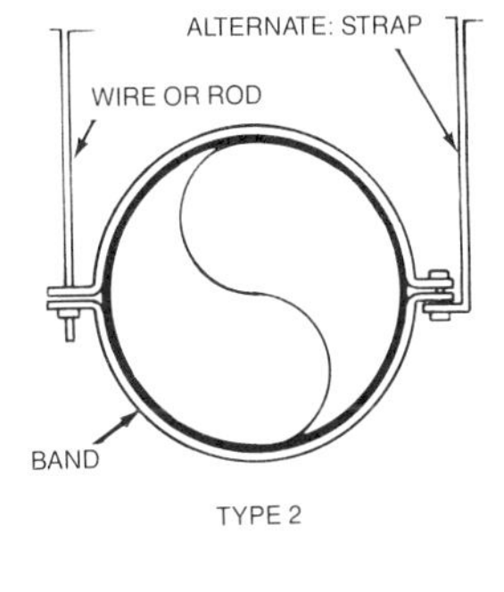


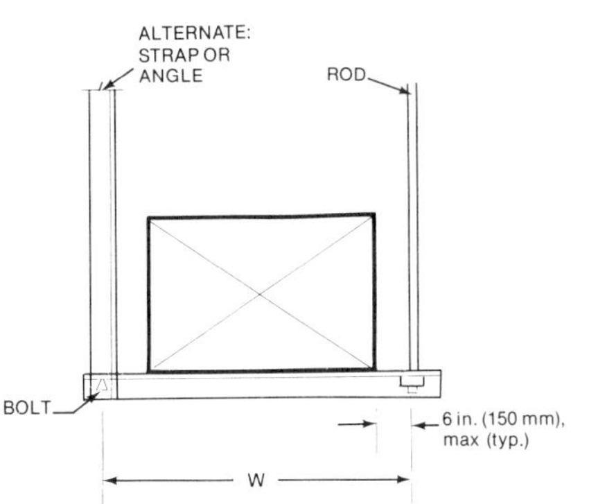


**Fig. 3 Trapeze Hanger—Rectangular Ducts**

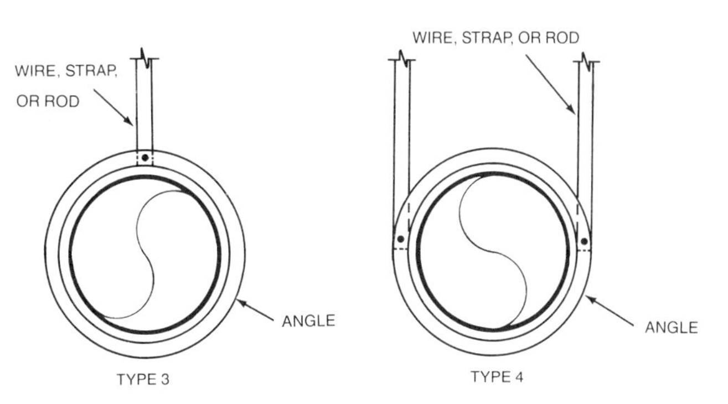


**Fig. 4 Round Duct Hangers**

## DUCT CONSTRUCTION FOR GREASE- AND MOISTURE-LADEN VAPORS

The installation and construction of ducts used for the removal of smoke or grease-laden vapors from cooking equipment should be in accordance with NFPA *Standard* 96 and SMACNA's rectangular and round industrial duct construction standards (SMACNA 1980, 1977). Kitchen exhaust ducts that conform to NFPA 96 must (1) be constructed from carbon steel with a minimum thickness of 0.054 in. (1.37 mm) or stainless-steel sheet with a minimum thickness of 0.043 in. (1.09 mm); (2) have all longitudinal seams and transverse joints continuously welded; and (3) be installed without dips or traps that may collect residues, except where traps with continuous or automatic removal of residues is provided. Since fires may occur in these systems (producing temperatures in excess of 2000 °F (1100 °C), provisions are necessary for expansion in accordance with the following table.

| Kitchen Exhaust Duct Material | Duct Expansion @ 2000 °F, in./ft | Duct Expansion @ 1100 °C, mm/m |
|---|---|---|
| Carbon steel | 0.19 | 15.8 |
| Type 304 stainless steel | 0.23 | 19.2 |
| Type 430 stainless steel | 0.13 | 10.8 |

Ducts that convey moisture-laden air must have construction specifications that properly account for corrosion resistance, drainage, and waterproofing of joints and seams. No nationally recognized standards exist for applications in areas such as kitchens, swimming pools, shower rooms, and steam cleaning or washdown chambers. Galvanized steel, stainless steel, aluminum, and plastic materials have been used. Wet and dry cycles increase corrosion of metals. Chemical concentrations affect corrosion rate significantly. Chapter 53 of the 1987 HVAC Volume addresses material selection for corrosive environments. Conventional duct construction standards are frequently modified to require welded or soldered joints, which are generally more reliable and durable than sealant-filled, mechanically locked joints. The number of transverse joints should be minimized, and longitudinal seams should not be located on the bottom of the duct. Risers should drain and horizontal ducts should pitch in the direction most favorable for moisture control. *Industrial Ventilation* (ACGIH 1986) covers hood design.

**Table 8 Flat-Oval Duct Construction for Positive Pressure Commercial Systems[a]**

| Major Axis (W), in. | Spiral Seam Duct | Longitudinal Seam Duct | Fittings | Type of Joint |
|---|---|---|---|---|
| | Minimum Galvanized Steel Thickness[b], in. | | | |
| Up through 24 | 0.0236 | 0.0356 | 0.0356 | Beaded slip |
| 36 | 0.0296 | 0.0356 | 0.0356 | Beaded slip |
| 48 | 0.0296 | 0.0466 | 0.0466 | Flange |
| 60 | 0.0356 | 0.0466 | 0.0466 | Flange |
| 70 | 0.0356 | 0.0575 | 0.0575 | Flange |
| Over 70 | 0.0466 | 0.0575 | 0.0575 | Flange |

[a]Table 8 is based on SMACNA's publication, HVAC *Duct Construction Standards—Metal and Flexible*, Table 3-4.

[b]Use Table 8 for galvanized, painted, uncoated, and stainless steel whenever the base metal thickness is not less than 0.0015 in. below those in the table.

W = Major Dimension F = W − D = Flat Width

## PLASTIC DUCTS

The *Thermoplastic Duct Construction Manual* (SMACNA 1974) covers thermoplastic (polyvinyl chloride, polyethylene, polypropylene, acrylonitrile butadiene styrene) and thermosetting (glass-fiber-reinforced polyester) plastic ducts used in commercial and industrial installations. SMACNA's manual provides comprehensive construction detail for positive or negative 2, 6, and 10 in. of water (500, 1500, and 2500 Pa) polyvinyl chloride duct systems. NFPA *Standard* 91 provides construction details and application limitations for plastic ducts, and references the Department of Commerce *Standard* PS 15-69 (NBS 1969) for reinforced thermosetting plastic duct fabrication details. For a listing of ASTM and commercial product standards concerning thermoplastic materials, see NFPA *Standard* 91.

**Table 9 Rectangular Duct Hangers—Commercial Systems[a,b]**

| Half of Duct Perimeter (P/2)[d], in. | Type of Hanger[c]: Galvanized Straps (see Figs. 3 and 4) Width × Thickness (min), in. | | | | Rods (see Fig. 4) Diameter, in. | | | |
|---|---|---|---|---|---|---|---|---|
| | Maximum Hanger Spacing, ft. | | | | | | | |
| | 10 | 8 | 5 | 4 | 10 | 8 | 5 | 4 |
| Up through 30 | 1 × 0.0296 | 1 × 0.0296 | 1 × 0.0296 | 1 × 0.0296 | 0.135 | 0.135 | 0.106 | 0.106 |
| 72 | 1 × 0.0466 | 1 × 0.0356 | 1 × 0.0296 | 1 × 0.0296 | 3/8 | 1/4 | 1/4 | 1/4 |
| 96 | 1 × 0.0575 | 1 × 0.0466 | 1 × 0.0356 | 1 × 0.0296 | 3/8 | 3/8 | 3/8 | 1/4 |
| 120 | 1 1/2 × 0.0575 | 1 × 0.0575 | 1 × 0.0466 | 1 × 0.0356 | 1/2 | 3/8 | 3/8 | 1/4 |
| 168 | 1 1/2 × 0.0575 | 1 × 0.0575 | 1 × 0.0575 | 1 × 0.0466 | 1/2 | 1/2 | 3/8 | 3/8 |
| 192 | — | 1 1/2 × 0.0575 | 1 × 0.0575 | 1 × 0.0575 | 1/2 | 1/2 | 3/8 | 3/8 |
| Over 192 | — | — | — | — | ◄— Special Analysis Required[e] —► | | | |

[a]Table 9 is based on SMACNA's publication, HVAC *Duct Construction Standards—Metal and Flexible*, Table 4-1.

[b]Table based on (1) 0.0575-in. thick ducts; (2) reinforcements and trapeze weights; (3) 1 lb/ft$^2$ for insulation; and (4) no external loads.

[c]For trapeze angle size, see Table 10. For hanger attachments to structures, typical upper attachments, lower hanger attachments, and strap splicing details, consult the HVAC *Duct Construction Standards—Metal and Flexible*, Table 4-1 and Figures 4-1 through 4-4.

[d]When dimension (W) exceeds 60 in., P/2 maximum is 1.25 W.

[e]Consult AISC Manual (1980) for hanger design criteria.

Maximum allowable strap, wire, and rod loads are:

| Strap (W × T[min]), in. | Load, lb | Rod or Wire (dia), in. | Load, lb |
|---|---|---|---|
| 1 × 0.0296 | 260 | 0.106 | 80 |
| 1 × 0.0356 | 320 | 0.135 | 120 |
| 1 × 0.0466 | 420 | 0.162 | 160 |
| 1 × 0.0575 | 700 | 1/4 | 270 |
| 1 1/2 × 0.0575 | 1100 | 3/8 | 680 |
| | | 1/2 | 1250 |
| | | 5/8 | 2000 |
| | | 3/4 | 3000 |

**Table 10 Allowable Trapeze Angle Loads[a]**

| Width of Hanger (W)[b], in. | Maximum Hanger Load[c], lb — Angle[e], in. | | | | | | | | | |
|---|---|---|---|---|---|---|---|---|---|---|
| | 1 × 1 × 0.0575[d] | 1 × 1 × 1/8 | 1 1/2 × 1 1/2 × 0.0575[d] | 1 1/2 × 1 1/2 × 1/8 | 1 1/2 × 1 1/2 × 3/16 | 1 1/2 × 1 1/2 × 1/4 or 2 × 2 × 1/8 | 2 × 2 × 3/16 | 2 × 2 × 1/4 | 2 1/2 × 2 1/2 × 3/16 | 2 1/2 × 2 1/2 × 1/4 |
| Up through 18 | 80 | 150 | 180 | 350 | 510 | 650 | 940 | 1230 | 1500 | 1960 |
| 24 | 75 | 150 | 180 | 350 | 510 | 650 | 940 | 1230 | 1500 | 1960 |
| 30 | 70 | 150 | 180 | 350 | 510 | 650 | 940 | 1230 | 1500 | 1960 |
| 36 | 60 | 130 | 160 | 340 | 500 | 620 | 920 | 1200 | 1480 | 1940 |
| 42 | 40 | 110 | 140 | 320 | 480 | 610 | 900 | 1190 | 1470 | 1930 |
| 48 | — | 80 | 110 | 290 | 450 | 580 | 870 | 1160 | 1440 | 1900 |
| 54 | — | 40 | 70 | 250 | 400 | 540 | 840 | 1120 | 1400 | 1860 |
| 60 | — | — | — | 190 | 350 | 490 | 780 | 1060 | 1340 | 1800 |
| 66 | — | — | — | 100 | 270 | 400 | 700 | 980 | 1260 | 1720 |
| 72 | — | — | — | — | 190 | 320 | 620 | 900 | 1180 | 1640 |
| 78 | — | — | — | — | 80 | 210 | 500 | 790 | 1070 | 1530 |
| 84 | — | — | — | — | — | 80 | 380 | 660 | 940 | 1400 |
| 96 | — | — | — | — | — | — | — | 320 | 600 | 1060 |
| 108 | — | — | — | — | — | — | — | — | 150 | 610 |

[a]Table 10 is based on SMACNA's publication, HVAC *Duct Construction Standards—Metal and Flexible*, Table 4-3.

[b]For W-dimension, see Figure 4.

[c]Loads in this table assume rods or other structural members are 6-in. maximum from duct sides (see Figure 4).

[d]Galvanized formed angles; thickness is minimum.

[e]Steel yield stress is based on 25,000 psi minimum.

**Table 11 Round Duct Hangers—Commercial Systems[a,b]**

| Duct Diameter, in. | Max. Hanger Spacing[c], ft. | Hanger Configuration[d] (see Figure 5) | Galvanized Ring Size (min), in. | Minimum Hanger Size, in. — Wire | Strap | Rod |
|---|---|---|---|---|---|---|
| Up through 10 | 12 | Type 1 or 3 | 1 × 0.0296 | 0.106 | 1 × 0.0296 | 1/4 |
| 18 | 12 | Type 1 or 3 | 1 × 0.0296 | 0.162 | 1 × 0.0296 | 1/4 |
| 24 | 12 | Type 1 or 3 | 1 × 0.0296 | — | 1 × 0.0296 | 1/4 |
| | | Type 2 or 4 | 1 × 0.0296 | 0.135 | 1 × 0.0296 | 1/4 |
| 36 | 12 | Type 1 or 3 | 1 × 0.0356 | — | 1 × 0.0356 | 3/8 |
| | | Type 2 or 4 | 1 × 0.0356 | 0.162 | 1 × 0.0296 | 1/4 |
| 50 | 12 | Type 4 | 1 × 0.0356 | — | 1 × 0.0356 | 3/8 |
| 60 | 12 | Type 4 | 1 × 0.0466 | — | 1 × 0.0466 | 3/8 |
| 84 | 12 | Type 4 | 1 × 0.0575 | — | 1 × 0.0575 | 3/8 |

[a]Table 11 is based on SMACNA's publication HVAC *Duct Construction Standards—Metal and Flexible*, Table 4-2.

[b]Table 11 is for duct construction per Table 7 plus one lb/ft$^2$ insulation. For heavier ductwork, size hanger members so allowable load (see Table 9) is not exceeded.

[c]The 12 ft hanger spacing may be increased for risers and other conditions with substantial horizontal support.

[d]Refer to Table 7 for angle flange sizes. For hanger attachments to structures, typical upper attachments, lower hanger attachments, and strap splicing details, consult the HVAC *Duct Construction Standards—Metal and Flexible*, Table 4-1 and Figures 4-1 through 4-4.

## UNDERGROUND DUCTS

No comprehensive standards exist for the construction of underground air ducts. Coated steel, asbestos cement, plastic, tile, concrete, reinforced fiber glass, and other materials have been used. Underground duct and fittings should always be round and have a minimum thickness as listed in Table 7. Thickness above the minimum may be needed for individual applications. Specifications for the construction and installation of underground ducts should account for the following: water tables, ground surface flooding, the need for drainage piping beneath ductwork, temporary or permanent anchorage to resist flotation, frost heave, backfill loading, vehicular traffic load, corrosion, cathodic protection, heat loss or gain, building entry, bacterial organisms, degree of water and air tightness, inspection or testing prior to backfill, and code compliance. Corrosion considerations for soil environments are addressed in Chapter 53 of the 1987 HVAC Volume. *Criteria and Test Pro-*

*cedures for Combustible Materials Used for Warm Air Ducts Encased in Concrete Slab Floors* (NRC) provides criteria and test procedures for fire resistance, crushing strength, bending strength, deterioration, and odor. *Installation Techniques for Perimeter Heating and Cooling* (ACCA 1964) covers residential systems and gives five classifications of duct material related to particular performance characteristics. Residential installations may also be subject to the requirements in NFPA *Standard 90B*. Commercial systems also normally require compliance with NFPA *Standard 90A*.

## DUCTS OUTSIDE OF BUILDINGS

The location and construction of ducts exposed to outdoor atmospheric conditions are generally regulated by building codes. Exposed ducts and their sealant/joining systems must be evaluated for the following: (1) waterproofing; (2) resistance to external loads (wind, snow, and ice); (3) degradation from corrosion, ultraviolet radiation, or thermal cycles; (4) heat transfer; (5) susceptability to physical damage; (6) hazards at air inlets and discharges; and (7) maintenance needs. In addition, support systems must be custom designed for roof-top, wall-mounted, and bridge or ground-based applications. Specific requirements must also be met for insulated and uninsulated ducts.

## SEISMIC QUALIFICATION

Seismic analysis of duct systems may be required by building codes or federal regulations. Provisions for seismic analysis are provided by the Federal Emergency Management Agency (FEMA 1985), the National Bureau of Standards (NBS 1981), and the Department of Defense (NAVFAC 1982). Ducts, duct hangers, fans, fan supports, and other duct-mounted equipment are generally evaluated independently. The Sheet Metal Industry Fund of Los Angeles has published guidelines for seismic restraints of mechanical systems (SMIFLA 1982).

## SHEET-METAL WELDING

*Specification for Welding of Sheet-Metal* (ASW 1984) covers sheet-metal arc welding and braze welding procedures. This specification also addresses the qualification of welders and welding operators, workmanship and the inspection of production welds.

## THERMAL INSULATION

Insulation materials for ducts, plenums, and apparatus casings are covered in Chapter 20 of the 1985 FUNDAMENTALS Volume. Codes generally limit factory insulated ducts to UL 181, Class 0 or Class 1. *Commercial and Industrial Insulation Standards* (MICA 1983) gives insulation details.

## MASTER SPECIFICATIONS

Master specifications for duct construction and most other elements in building construction are produced and regularly updated by several organizations. Two documents are MASTERSPEC by The American Institute of Architects (AIA) and SPECTEXT by the Construction Specifications Institute (CSI). These documents are model project specifications that require little editing to customize each application for a project.

Several potential benefits of using nationally recognized model specifications are (1) they focus industry practice on a uniform set of requirements in a widely known format; (2) they reduce the need to prepare new specifications for each project; (3) they remain current and automatically incorporate new and revised editions of construction, test, and performance standards published by other organizations; (4) they tend to reduce professional liability by reducing errors and omissions and by reflecting normal practice; (5) they have inherent flexibility adaptable to small or large projects; (6) they are performance or prescription oriented as the designer desires; (7) they give lists of products and equipment by name and number or descriptions that are deemed equal; (8) each subsection is coordinated with other sections of related work; (9) they are increasingly being used by government agencies to replace separate and often different agency specifications.

## REFERENCES

ACCA. 1964 *Installation Techniques for Perimeter Heating and Cooling*. Manual No. 4. Air Conditioning Contractors of America, Washington, DC.

ACGIH. 1986. *Industrial Ventilation—A Manual of Recommended Practice*. American Conference of Governmental Industrial Hygienists, Lansing, MI.

AISC. 1980. *Manual of Steel Construction*. American Institute of Steel Construction, Chicago, IL.

AISI. 1986. *Cold-Formed Steel Design Manual*. American Iron and Steel Institute, Washington, DC.

ANSI *Standard* B32.3M-1984. American National Standards Institute, New York, NY.

ANSI. 1984. Metric Sizes for Flat Metal Products, Preferred.

ASHRAE. 1980. Energy Conservation in New Building Design. ANSI/ASHRAE *Standard* 90A-1980. (Cosponsored by IES.)

ASHRAE. 1981. Energy Conservation in Existing Buildings—High Rise Residential. ANSI/ASHRAE *Standard* 100.2-1981. (Cosponsored by IES.)

ASHRAE. 1981. Energy Conservation in Existing Buildings—Institutional. ANSI/ASHRAE *Standard* 100.5-1981. (Cosponsored by IES.)

ASHRAE. 1981. Energy Conservation in Existing Buildings—Public Assembly. ANSI/ASHRAE *Standard* 100.6-1981. (Cosponsored by IES.)

ASHRAE. 1981. Ventilation for Acceptable Indoor Air Quality. ASHRAE *Standard* 62-1981.

ASTM. 1984. Test Method for Sound Absorption and Sound Absorption Coefficients by the Reverberation Room Method. ASTM *Standard* C423-84. American Society for Testing and Materials, Philadelphia, PA.

ASTM. 1985. Specification for Steel, Sheet, Carbon, Cold-Rolled Commercial Quality. ASTM *Standard* A366/A366M-85.

ASTM. 1985. Specification for Steel, Carbon (0.15 Maximum, Percent), Hot-Rolled Sheet and Strip, Commercial Quality. ASTM *Standard* A569/A569M-85.

ASTM. 1985. Specifications for Adhesives for Duct Thermal Insulation. ASTM *Standard* C916-85.

ASTM. 1986. Standard Specification for Thermal and Acoustical Insulation (Mineral Fiber, Duct Lining Material). ASTM *Standard* C1071-86.

ASTM. 1987 General Requirements for Flat-Rolled Stainless and Heat-Resisting Steel Plate, Sheet, and Strip. ASTM *Standard* A480/A480-87.

ASTM. 1987. General Requirements for Steel Sheet, Zinc-Coated (Galvanized ) by the Hot-Dip Process. ASTM *Standard* A525-87.

ASTM. 1987. General Requirements for Steel Sheet, Zinc-Coated (Galvanized) by the Hot-Dip Process (Metric). ASTM *Standard* A525M-87.

ASTM. 1987. Standard Specification for General Requirement for Steel, Sheet, Carbon, and High-Strength, Low-Alloy, Hot-Rolled and Cold-Rolled. ASTM *Standard* A568-87.

ASTM. 1987. Standard Specification for General Requirement for Steel, Carbon and High-Strength, Low-Alloy, Hot-Rolled Sheet and Cold-Rolled Sheet (Metric). ASTM *Standard* A568M-87.

AWS. 1984. Specification for Welding of Sheet Metal. AWS D9-84. American Welding Society, Miami, FL.

BOCA. 1987. *National Building Code.* Building Officials and Code Administrators International, Country Club Hills, IL.
BOCA. 1987. *National Mechanical Code.*
CABO. 1981. *Model Energy Code.* Council of American Building Officials, Falls Church, VA.
CABO. 1986. *One- and Two-Family Dwelling Code.*
Cheremisinoff, P.N., and R.A. Young. 1975. *Pollution Engineering Practice Handbook.* Ann Arbor Science Publishers, Inc.
FEMA. 1985. NEHRP (National Earthquake Hazards Reduction Program) Recommended Provisions for the Development of Seismic Regulations for New Buildings. Building Seismic Safety Council for the Federal Emergency Management Agency, Washington, DC.
ICBO. 1988. *Uniform Building Code.* International Conference of Building Officials, Whittier, CA.
ICBO. 1988. *Uniform Mechanical Code.* International Conference of Building Officials, Whittier, CA.
MICA. 1983. *Commerical and Industrial Insulation Standards.* Midwest Insulation Contractors Association, Omaha, NE.
NAVFAC. 1982. Technical Manual-Seismic Design for Buildings. Army TM 5-809-10, Navy NAVFAC P-355, Air Force ARM 88-3 (Chapter 3). U.S. Government Printing Office, Washington, DC.
NBS. 1969. Voluntary Product Standard for Custom Contact-Molded Reinforced-Polyester Chemical-Resistant Process Equipment. NBS *Standard PS* 15-69. National Bureau of Standards, Department of Commerce, Washington, DC. (Available from the Society of the Plastics Industry, Washington, DC.)
NBS. 1981. Draft Seismic Standard for Federal Buildings. *Publication* NBSIR 81-2195.
NFPA. 1985. Installation of Air Conditioning and Ventilating Systems. NFPA *Standard* 90*A*. National Fire Protection Association, Quincy, MA.
NFPA. 1984. Warm Air Heating and Air Conditioning Systems. ANSI/NFPA *Standard* 90B.
NFPA. 1983. Blower and Exhaust Systems. ANSI/NFPA *Standard* 91.
NFPA. 1984. Vapor Removal from Cooking Equipment. ANSI/NFPA *Standard* 96.
NFPA. 1985. *Life Safety Code.* ANSI/NFPA *Standard* 101.
NFPA. 1985. Smoke and Heat Venting. ANSI/NFPA *Standard* 204M.
NFPA. 1984. Chimneys, Fireplaces, Vents and Solid Fuel Burning Appliances. ANSI/NFPA *Standard* 211.
NRC. *Criteria and Test Procedures for Combustible Materials Used for Warm Air Ducts Encased in Concrete Slab Floors.* National Research Council, National Academy of Sciences, Washington, DC.
SBCCI. 1988. *Standard Building Code.* Southern Building Code Congress International, Inc., Birmingham, AL.
SBCCI. 1988. *Standard Mechanical Code.*
SMACNA. 1974. *Thermoplastic Duct (PVC) Construction Manual.* Sheet Metal and Air Conditioning Contractors' National Association, Inc., Vienna, VA.
SMACNA. 1975. *Accepted Industry Practice for Industrial Duct Construction.*
SMACNA. 1977. *Round Industrial Duct Construction Standards.*
SMACNA. 1979. *Fibrous Glass Duct Construction Standards.*
SMACNA. 1980. *Rectangular Industrial Duct Construction Standards.*
SMACNA. 1985. *HVAC Duct Construction Standards—Metal and Flexible.*
SMACNA. 1985. *HVAC Air Duct Leakage Test Manual.*
SMIFLA. 1982. *Guidelines for Seismic Restraints of Mechanical Systems and Plumbing Piping Systems,* 1st ed. Sheet Metal Industry Fund of Los Angeles. The Plumbing and Piping Industry Council, Los Angeles, CA.
Swim, W.B. 1978. Flow Losses in Rectangular Ducts Lined with Fiberglass. ASHRAE *Transactions,* Vol. 84, Part 2.
UL. 1979. Fire Dampers and Ceiling Dampers. UL *Standard* 555-79. Underwriters Laboratories, Inc. Northbrook, IL.
UL. 1983. Leakage Rated Dampers for Use in Smoke Control Systems. UL *Standard* 555S-83.
UL. 1984. Factory-Made Air Ducts and Connectors. UL *Standard* 181-81.
UL. 1986. Fire Tests of Door Assemblies. ANSI/UL *Standard* 10B-86.
United McGill Corporation. 1985. *Industrial Duct Engineering Data and Recommended Design Standards,* Form No. SMP-IDP, Feb., pp. 24-28. Westerville, OH.

## APPENDIX OF SI TABLES

**Table A-1 Residential Metal Duct Construction**

| Shape of Duct and Exposure | Galvanized Steel Minimum Thickness, mm | Aluminum (3003) Nominal Thickness, mm |
|---|---|---|
| Enclosed Rectangular Ducts[a] | | |
| 360 mm or less | 0.0319 | 0.406 |
| Over 360 mm | 0.0395 | 0.508 |
| Rectangular and Round Ducts | Consult SMACNA HVAC *Duct Construction Standards—Metal and Flexible* | |

[a]Data based on nominal thickness, NFPA 90B (1984).

**Table A-2a Galvanized Sheet Thickness**

| Gauge | Thickness, mm Nominal | Thickness, mm Minimum | Nominal Mass, kg/m² |
|---|---|---|---|
| 30 | 0.399 | 0.319 | 3.20 |
| 28 | 0.475 | 0.395 | 3.81 |
| 26 | 0.551 | 0.471 | 4.42 |
| 24 | 0.701 | 0.601 | 5.64 |
| 22 | 0.853 | 0.753 | 6.86 |
| 20 | 1.006 | 0.906 | 8.08 |
| 18 | 1.311 | 1.181 | 10.52 |
| 16 | 1.613 | 1.463 | 12.96 |
| 14 | 1.994 | 1.784 | 16.01 |
| 13 | 2.372 | 2.162 | 19.07 |
| 12 | 2.753 | 2.523 | 22.12 |
| 11 | 3.132 | 2.902 | 25.16 |
| 10 | 3.510 | 3.280 | 28.21 |

*Notes:*

1. Minimum thickness is based on thickness tolerances of hot-dip galvanized sheets in cut lengths and coils (per ASTM *Standard* A525-80). Tolerance is valid for 1220- and 1520-mm wide sheets.
2. Galvanized sheet gauge is used for convenience. A metric series would conform to ANSI *Standard* B32.3, and the tolerances in ASTM *Standard* A525M apply.

**Table A-2b Uncoated Steel Sheet Thickness**

| Gauge | Thickness, mm Nominal | Thickness, mm Minimum Hot-Rolled | Thickness, mm Minimum Cold-Rolled | Nominal Mass, kg/m² |
|---|---|---|---|---|
| 28 | 0.378 | | 0.328 | 3.05 |
| 26 | 0.455 | | 0.405 | 3.66 |
| 24 | 0.667 | | 0.587 | 4.88 |
| 22 | 0.759 | | 0.679 | 6.10 |
| 20 | 0.912 | | 0.832 | 7.32 |
| 18 | 1.214 | 1.084 | 1.114 | 9.76 |
| 16 | 1.519 | 1.369 | 1.389 | 12.20 |
| 14 | 1.897 | 1.717 | 1.767 | 15.25 |
| 13 | 2.278 | 2.101 | 2.456 | 18.31 |
| 12 | 2.656 | 2.456 | 2.506 | 21.35 |
| 11 | 3.038 | 2.838 | 2.888 | 24.40 |
| 10 | 3.416 | 3.216 | 3.266 | 27.45 |

*Notes:*

1. Minimum thickness is based on thickness tolerances of hot-rolled and coldrolled sheets in cut lengths and coils (per ASTM *Standards* A366, A568, and A569).
2. Table is based on 1220-mm width coil and sheet stock. 1520-mm coil has same tolerance, except that 16-gauge is ±0.18-mm in hot-rolled coils and sheets.
3. Manufacturer's standard gauge is used for convenience. A metric series would conform to ANSI *Standard* B32.3, and the tolerances in ASTM *Standard* A568M apply.

### Table A-2c Stainless Steel Sheet Thickness

| Gauge | Thickness, mm Nominal | Thickness, mm Minimum | Nominal Mass, kg/m$^2$ Stainless Steel 300 Series | Nominal Mass, kg/m$^2$ Stainless Steel 400 Series |
|---|---|---|---|---|
| 28 | 0.384 | 0.334 | 3.09 | 3.04 |
| 26 | 0.452 | 0.372 | 3.65 | 3.58 |
| 24 | 0.597 | 0.517 | 4.82 | 4.72 |
| 22 | 0.744 | 0.644 | 6.01 | 5.89 |
| 20 | 0.902 | 0.802 | 7.28 | 7.14 |
| 18 | 1.219 | 1.089 | 9.84 | 9.65 |
| 16 | 1.511 | 1.361 | 12.20 | 11.96 |
| 14 | 1.908 | 1.728 | 15.39 | 15.10 |
| 13 | 2.286 | 2.056 | 18.45 | 18.10 |
| 12 | 2.677 | 2.447 | 21.60 | 21.19 |
| 11 | 3.048 | 2.798 | 24.60 | 24.13 |
| 10 | 3.429 | 3.129 | 27.67 | 27.14 |

*Notes:*

1. Minimum thickness is based on thickness tolerances for hot-rolled sheets in cut lengths and cold-rolled sheets in cut lengths and coils (per ASTM *Standard* A480).
2. Stainless sheet gauge is not standardized. A metric series would conform to ANSI *Standard* B32.3, and the tolerances in ASTM *Standard* A480M apply.

### Table A-3 Steel Angle Mass per Unit Length (Approximate)

| Angle Size, mm | Mass, kg/m |
|---|---|
| 25 × 25 × 1.181 (min) | 0.53 |
| 25 × 25 × 1.463 (min) | 0.65 |
| 25 × 25 × 3 | 1.11 |
| 30 × 30 × 1.181 (min) | 0.63 |
| 30 × 30 × 1.463 (min) | 0.78 |
| 30 × 30 × 2.162 (min) | 1.14 |
| 30 × 30 × 3 | 1.36 |
| 40 × 40 × 1.463 (min) | 1.04 |
| 40 × 40 × 3.280 (min) | 2.26 |
| 40 × 40 × 4 | 2.42 |
| 40 × 40 × 5 | 2.97 |
| 40 × 40 × 6 | 3.52 |
| 50 × 50 × 1.463 (min) | 1.30 |
| 50 × 50 × 5 | 3.77 |
| 50 × 50 × 6 | 4.47 |
| 60 × 60 × 5 | 4.57 |
| 60 × 60 × 6 | 5.42 |
| 60 × 60 × 8 | 7.09 |
| 70 × 70 × 6 | 6.38 |

### Table A-4a Rectangular Ferrous Metal Duct Construction for Commercial Systems[a,b]

| Duct Dimensions, mm | Minimum Galvanized Steel Thickness[c], mm | | | | | | | | | | | | | | | | | | |
|---|---|---|---|---|---|---|---|---|---|---|---|---|---|---|---|---|---|---|---|
| | 1.463 | | | | | | | 1.181 | | | | | | | 9.906 | | | | |
| | Pressure, Pa | | | | | | | | | | | | | | | | | | |
| | ±125 | ±250 | ±500 | ±750 | +1000 | +1500 | +2500 | ±125 | ±250 | ±500 | ±750 | +1000 | +1500 | +2500 | ±125 | ±250 | ±500 | ±750 | +1000 |
| Up through 250 | | | | Reinforcement not required[g] | | | | | | | Reinforcement not required[g] | | | | | | Reinforcement not required[g] | | |
| 300 | | | | | | | | | | | | | | A-2.4 | | | | | |
| 350 | | | | | | | B-2.4 | | | | | | | B-2.4 | | | | | A-3 |
| 400 | | | | | | | B-2.4 | | | | | | B-3 | B-1.5 | | | | A-2.4 | A-3 |
| 450 | | | | | | C-3 | C-2.4 | | | | | B-3 | C-3 | C-1.5 | | | | A-2.4 | B-2.4 |
| 500 | | | | | C-3 | C-3 | D-2.4 | | | | B-3 | C-3 | C-2.4 | C-1.5 | | | B-3 | B-2.4 | C-2.4 |
| 550 | | | | | C-3 | D-3 | C-1.5 | | | B-3 | C-3 | C-3 | C-2.4 | C-1.5 | | A-3 | B-3 | B-2.4 | B-1.5 |
| 600 | | | | | D-3 | D-2.4 | D-1.5 | | | C-3 | C-3 | D-3 | D-2.4 | D-1.5 | | B-3 | C-3 | B-1.5 | C-1.5 |
| 650 | | | C-3 | D-3 | D-3 | D-2.4 | D-1.5 | | | C-3 | D-3 | D-3 | D-1.5 | D-1.5 | | B-3 | C-3 | C-1.5 | C-1.5 |
| 700 | | | C-3 | D-3 | E-3 | E-2.4 | E-1.5 | | C-3 | C-3 | D-3 | E-2.4 | D-1.5 | E-1.5 | B-3 | C-3 | C-2.4 | C-1.5 | D-1.5 |
| 750 | | | D-3 | D-3 | E-3 | D-1.5 | E-1.5 | | C-3 | D-3 | D-2.4 | E-2.4 | D-1.5 | E-1.2 | B-3 | C-3 | D-2.4 | C-1.5 | D-1.5 |
| 900 | | D-3 | E-3 | E-2.4 | E-1.5 | F-1.5 | F-1.5 | C-3 | D-3 | E-2.4 | E-1.5 | E-1.5 | F-1.5 | F-1.2 | C-3 | D-3 | D-1.5 | E-1.5 | E-1.5 |
| 1050 | D-3 | E-3 | E-2.4 | E-1.5 | F-1.5 | G-1.5 | H-1.2 | D-3 | E-3 | E-1.5 | E-1.5 | F-1.5 | G-1.2 | G-.9 | D-3 | D-2.4 | E-1.5 | E-1.5 | F-1.2 |
| 1200 | E-3 | F-3 | G-2.4 | G-1.5 | G-1.5 | H-1.2 | H-.9 | E-3 | E-2.4 | F-1.5 | G-1.5 | G-1.5 | H-1.2 | H-.9 | E-3 | E-1.5 | F-1.5 | F-1.2 | F-.9 |
| 1350 | E-3 | G-3[d] | G-1.5[d] | H-1.5[d] | H-1.5[d] | H-1.2[d] | I-.9[d] | E-3 | F-2.4 | G-1.5[d] | H-1.5[d] | H-1.2[d] | H-.9[d] | H-.75[d] | E-2.4 | E-1.5 | F-1.2 | G-.9[d] | G-.9[d] |
| 1500 | F-3 | G-2.4[d] | H-1.5[d] | H-1.5[d] | I-1.5[d] | H-.9[d] | J-.9[d] | F-3 | G-2.4[d] | H-1.5[d] | H-1.2[d] | H-.9[d] | H-.9[d] | I-.75[d] | F-2.4 | F-1.5 | G-1.2[d] | G-.9[d] | H-.9[d] |
| 1800 | H-3[d] | H-1.5[d] | I-1.5[d] | I-1.2[d] | I-.9[d] | J-.9[d] | K-.75[d] | G-2.4[d] | H-1.5[d] | H-1.2[d] | H-.9[d] | I-.9[d] | J-.75[d] | K-.6[d] | F-1.5 | G-1.2[d] | H-.9[d] | H-.9[d] | I-.75[d] |
| 2100 | H-2.4[d] | I-1.5[d] | J-1.2[d] | J-.9[d] | K-.9[d] | L-.75[d] | H-.6 plus rods | H-1.5[d] | I-1.5[d] | J-1.2[d] | J-.9[d] | J-.75[d] | K-.6[d] | — | H-1.5[d] | H-1.2[d] | I-.9[d] | J-.75[d] | J-.6[d] |
| 2400 | I-2.4[d] | J-1.5[d] | K-1.2[d] | L-.9[d] | L-.75[d] | L-.6[d] | H-.6 plus rods | H-1.5 | I-1.2 | K-.9[d] | K-.75[d] | K-.6[d] | L-.6[d] | — | H-1.5[d] | I-.9[d] | J-.75[d] | J-.6[d] | K-.6[d] |
| Over 2400 | H-1.5 | H-.75 plus rods[f] | H-.75 plus rods[f] | H-.75[e] plus rods[f] | H-.75[e] plus rods[f] | H-.6[e] plus rods[f] | H-.6 plus rods[f] | H-1.5 | H-.75 plus rods[f] | H-.75 plus rods[f] | H-.75 plus rods[f] | H-.75 plus rods[f] | H-.6 plus rods[f] | — | — | — | — | — | — |

[a]Table A-4 is based on SMACNA's publication, HVAC *Duct Construction Standards—Metal and Flexible*, Tables 1-3 through 1-9. For tie rod details, refer to this standard.

[b]For a given duct thickness, numbers indicate maximum spacing (m) between duct reinforcement; letters indicate type (rigidity class) of duct reinforcement (see Tables A-5 and A-6).

Transverse joint spacing is unrestricted on unreinforced ducts. To qualify joints on reinforced ducts, select transverse joints from Table A-5. Select intermediate reinforcement from Table A-6. Tables are based on steel construction. Designers should specify galvanized, uncoated, or painted steel joint and intermediate reinforcement.

Use the same metal duct thickness on all duct sides. Evaluate duct reinforcement on each duct side separately. When required on four sides for 1000, 1500, and 2500 Pa positive pressure systems, corners must be tied. When required on two sides, corners must be tied with rods or angles at the ends of 1000, 1500, and 2500 Pa positive pressure systems.

Duct sides over 450 mm width with less than 0.906 mm thickness, which have more than 0.93 m$^2$ of unbraced panel area, must be cross-broken or beaded, unless they are lined or insulated externally. Lined or externally insulated ducts are not required to have cross-breaking or beading.

[c]The reinforcement tables are based on galvanized steel of the indicated thickness. They apply to galvanized, painted, uncoated, and stainless steel whenever the base metal thickness is not less than 0.038 mm below that indicated for galvanized steel.

[d]See SMACNA's publication for alternative reinforcements using tie rods or tie straps for positive pressure.

[e]Sheet metal 1.181 mm thick is acceptable.

[f]Tie rods with a minimum diameter of 10 mm must be used on these constructions. The rods for positive pressure ducts only are spaced a maximum of 1500 mm apart along joints and reinforcements.

[g]Blank spaces indicate that no reinforcement is required.

### Table A-4b Rectangular Ferrous Metal Duct Construction for Commercial Systems*[a,b]

| Duct Dimensions, mm | Minimum Galvanized Steel Thickness,[c] mm: 0.906 | | 0.753 | | | | | | | 0.601 | | | | | | | 0.471 | | |
|---|---|---|---|---|---|---|---|---|---|---|---|---|---|---|---|---|---|---|---|
| | Pressure, Pa | | | | | | | | | | | | | | | | | | |
| | +1500 | +2500 | ±125 | ±250 | ±500 | ±750 | +1000 | +1500 | +2500 | ±125 | ±250 | ±500 | ±750 | +1000 | +1500 | +2500 | ±125 | ±250 | ±500 |
| Up through 200 | | | | | Reinforcement not required[g] | | | | | | Reinforcement not required[g] | | | | | A-1.5 | Reinforcement not required[g] | | |
| 250 | | A-1.5 | | | | | | A-1.5 | A-1.5 | | | | | A-1.5 | A-1.5 | A-1.2 | | | |
| 300 | A-3 | A-1.5 | | | | | A-3 | A-1.5 | A-1.5 | | | | A-2.4 | A-1.5 | A-1.5 | A-1.2 | | | A-2.4 |
| 350 | A-3 | A-1.5 | | | | A-2.4 | A-2.4 | A-1.5 | A-1.2 | | | A-2.4 | A-1.5 | A-1.5 | A-1.2 | A-.9 | | A-3 | A-1.5 |
| 400 | A-1.5 | B-1.5 | | | A-3 | A-2.4 | A-1.5 | A-1.5 | B-1.2 | | A-3 | A-2.4 | A-1.5 | A-1.5 | A-1.2 | B-.9 | | A-2.4 | A-1.5 |
| 450 | B-1.5 | C-1.5 | | | A-3 | A-2.4 | A-1.5 | B-1.5 | B-1.2 | | A-3 | A-2.4 | A-1.5 | A-1.5 | B-1.2 | B-.9 | | A-2.4 | A-1.5 |
| 500 | B-1.5 | C-1.2 | | A-3 | B-2.4 | A-1.5 | B-1.5 | B-1.5 | B-.9 | | A-3 | A-1.5 | A-1.5 | B-1.5 | B-1.2 | B-.9 | A-3 | A-2.4 | A-1.5 |
| 550 | C-1.5 | C-1.2 | | A-3 | B-2.4 | B-1.5 | B-1.5 | C-1.5 | C-.9 | A-3 | A-3 | A-1.5 | B-1.5 | B-1.2 | C-1.2 | C-.9 | A-3 | A-1.5 | A-1.5 |
| 600 | C-1.5 | D-1.2 | | B-3 | C-2.4 | B-1.5 | C-1.5 | C-1.5 | C-.9 | A-3 | B-3 | B-1.5 | B-1.5 | C-1.2 | C-.9 | C-.9 | A-3 | A-1.5 | B-1.5 |
| 650 | D-1.5 | D-1.2 | A-3 | B-3 | C-2.4 | C-1.5 | C-1.5 | C-1.2 | D-.9 | A-3 | B-2.4 | B-1.5 | C-1.5 | C-1.2 | C-.9 | C-.75 | A-3 | A-1.5 | B-1.5 |
| 700 | D-1.5 | D-1.2 | B-3 | C-3 | C-1.5 | C-1.5 | D-1.5 | D-1.2 | D-.9 | B-3 | C-2.4 | C-1.5 | C-1.2 | D-1.2 | C-.9 | D-.75 | B-2.4 | B-1.5 | B-1.2 |
| 750 | D-1.2 | D-.9 | B-3 | C-3 | C-1.5 | C-1.5 | D-1.5 | D-1.2 | D-.9 | B-3 | C-2.4 | C-1.5 | C-1.2 | D-1.2 | D-.9 | D-.75 | B-2.4 | B-1.5 | C-1.2 |
| 900 | E-1.2 | F-.9 | C-3 | D-2.4 | D-1.5 | D-1.2 | E-1.2 | E-.9 | E-.75 | C-2.4 | C-1.5 | D-1.2 | D-1.2 | D-.9 | E-.75 | E-.6 | C-1.5 | C-1.5 | — |
| 1050 | F-.9 | G-.75 | D-2.4 | D-1.5 | E-1.5 | E-1.2 | E-.9 | E.75 | F-.6 | D-2.4 | D-1.5 | E-1.2 | E-.9 | E-.75 | — | — | D-1.5 | D-1.2 | — |
| 1200 | G-.75 | G-.6 | D-2.4 | E-1.5 | E-1.2 | E-.9 | F-.9 | G-.75 | G-.6 | D-1.5 | E-1.5 | E-.9 | E-.75 | E-.6 | — | — | D-1.5 | D-1.2 | — |
| 1350 | H-.75[d] | H-.6[d] | D-1.5 | E-1.5 | F-.9 | G-.9[d] | G-.75[d] | G-.6[d] | — | D-1.5 | E-1.2 | F-.9 | E-.75 | F-.6 | — | — | D-1.5 | — | — |
| 1500 | H-.75[d] | I-.6[d] | E-1.5 | F-1.5 | G-.9[d] | G-.75[d] | H-.75[d] | H-.6[d] | — | E-1.5 | F-1.2 | G-.75[d] | G-.75[d] | G-.6[d] | — | — | E-1.2 | — | — |
| 1800 | I-.6[d] | — | F-1.5 | G-1.2[d] | H-.9[d] | H-.75[d] | H-.6[d] | — | — | F-1.2 | H-.6[d] | H-.6[d] | H-.6[d] | — | — | — | — | — | — |
| 2100 | — | — | H-1.5[d] | — | — | — | — | — | — | G-1.2[d] | — | — | — | — | — | — | — | — | — |
| 2400 | — | — | H-1.2[d] | — | — | — | — | — | — | — | — | — | dashes indicate not allowed | | | — | — | — | — |
| Over 2400 | — | — | — | — | — | — | — | — | — | — | — | — | — | — | — | — | — | — | — |

*See Table A-4a for notes.

### Table A-5a Transverse Joint Reinforcement*[a]

| Minimum Rigidity Class[f] | 1000 Pa Maximum[b]: STANDING DRIVE SLIP | STANDING S | | STANDING S | STANDING S | STANDING S (BAR REINFORCED) / STANDING S (ANGLE REINFORCED) |
|---|---|---|---|---|---|---|
| | $H_s \times T$ (min), mm | W, mm | $H_s \times T$ (min), mm | $H_s \times T$ (min), | $H_s \times T$ (min), | $H_s \times T$ (min) plus Reinforcement (H × T), mm[c] |
| A | Use Class B | Use Class C | | 13 × 0.471 | Use Class D | Use Class F |
| B | 30 × 0.471 | | | 13 × 0.753 | | |
| C | 30 × 753 | — | 25 × 0.471 | 25 × 0.471 | | |
| D | — | — | 25 × 0.601 | 25 × 0.601 | 30 × 0.471 | |
| E | — | 5 | 30 × 0.906 | — | 30 × 1.181 | |
| F | — | 5 | 45 × 0.753 | — | 40 × 0.601 | 40 × 0.061 plus<br>40 × 3 bar |
| G | — | 5 | 45 × 1.181 | — | 40 × 1.181 | 40 × 0.753 plus<br>40 × 3 bar |
| H | — | — | — | — | — | 40 × 0.906 plus<br>40 × 40 × 5 angle |
| I | — | — | — | — | — | 50 × 0.906 plus<br>50 × 50 × 5 angle |
| J | — | — | — | — | — | 50 × 0.906 plus<br>50 × 50 × 5 angle |

*See Table A-5c for notes.

**Table A-5b Transverse Joint Reinforcement*[a]**

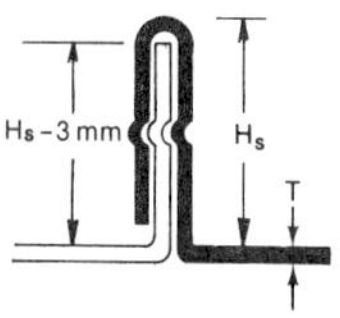

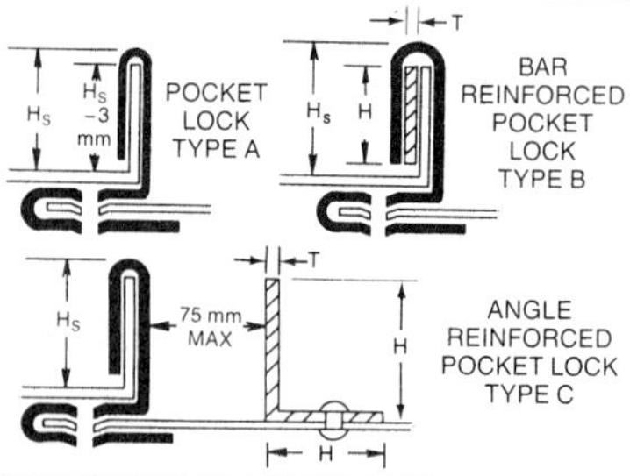

| Minimum Rigidity Class[f] | Standing Seam | Reinforced Welded or Standing Seam: 0.471 through 0.753 Duct, mm | | Reinforced Welded or Standing Seam: 0.906 through 1.463 Duct, mm | | Welded Flange | 750 Pa Maximum (Pocket Lock Type A; Bar Reinforced Pocket Lock Type B; Angle Reinforced Pocket Lock Type C) | | | |
|---|---|---|---|---|---|---|---|---|---|---|
| | $H_s \times T$ (min), mm[d] | $H_s$, mm | $H \times T$, mm[c] | $H_s$, mm | $H \times T$, mm[c] | $H_s \times T$ (min), mm[d] | Lock Type | $H_s$, mm | Thickness, mm[e]: Lock | Thickness, mm[e]: Reinforcement ($H \times T$) |
| A | 13 × 0.601 | Use Class D | | Use Class E | | 13 × 0.753 | Use Class D | | | |
| B | 20 × 0.601 | | | | | 13 × 1.463<br>20 × 0.753 | | | | |
| C | 25 × 0.601 | | | | | 20 × 1.181<br>25 × 0.753 | | | | |
| D | 20 × 1.463<br>25 × 0.906 | 25 | 25 × 1.463 | | | 25 × 1.181<br>30 × 0.753 | A | 25 | 0.753 | None |
| E | 25 × 1.463<br>40 × 0.601 | 25 | 25 × 3 | 25 | 25 × 1.463 | 30 × 1.181<br>40 × 0.753 | B | 25 | 0.753 | 25 × 3<br>Bar |
| F | 40 × 0.906 | 40 | 40 × 1.463 | 30 | 30 × 1.463 | 30 × 1.463<br>40 × 0.906 | A | 40 | 0.753 | None |
| G | 40 × 1.181 | 40<br>40 | 40 × 4<br>50 × 1.463 | 40 | 40 × 4 | 40 × 1.463 | B | 40 | 0.753 | 40 × 3<br>Bar |
| H | — | Use Class J | | 40<br>40 | 40 × 5<br>50 × 1.463 | — | C | 40 | 0.906 | 40 × 5<br>Angle |
| I | — | | | Use Class J | | — | Use Class J | | | |
| J | — | 40 | 50 × 5 | 40 | 50 × 5 | — | C | 40 | 0.906 | 50 × 5<br>Angle |
| K | — | Use Class L | | Use Class L | | — | C | 40 | 0.906 | 70 × 6<br>Angle |
| L | — | 40 | 70 × 6 | 40 | 70 × 6 | — | — | — | — | — |

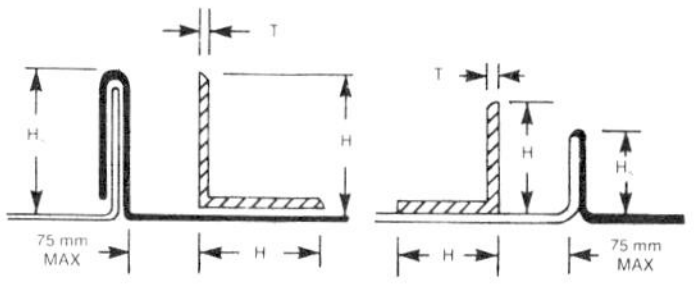

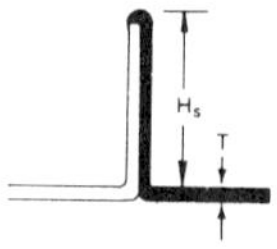

*See Table A-5c for notes.

## Table A-5c Transverse Joint Reinforcement[a]

<table>
<tr><th rowspan="2">Minimum Rigidity Class[f]</th><th colspan="2">CAPPED FLANGE</th><th>COMPANION ANGLES</th><th>FLANGED</th><th>FLANGED</th><th>SLIP-ON FLANGE</th></tr>
<tr><th>$H_s$ × T (min), mm</th><th>Min Cap Thickness, mm</th><th>H × T (nominal), mm</th><th>$H_s$ × T (min), mm</th><th colspan="2"></th></tr>
<tr><td>A</td><td colspan="2">Use Class B</td><td rowspan="4">Use Class E</td><td rowspan="2">Use Class C</td><td colspan="2" rowspan="11">Consult SMACNA or manufacturers to establish ratings.</td></tr>
<tr><td>B</td><td>20 × 0.471</td><td>0.601</td></tr>
<tr><td>C</td><td>25 × 0.601</td><td>0.601</td><td>25 × 0.601</td></tr>
<tr><td>D</td><td>25 × 0.753</td><td>0.753</td><td>25 × 0.753</td></tr>
<tr><td>E</td><td>40 × 0.601</td><td>0.753</td><td>25 × 3</td><td>25 × 1.463<br>40 × 0.601</td></tr>
<tr><td>F</td><td>40 × 0.906</td><td>0.906</td><td>Use Class G</td><td>40 × 0.753</td></tr>
<tr><td>G</td><td colspan="2">Use Class H</td><td>30 × 3</td><td>40 × 1.181</td></tr>
<tr><td>H</td><td>50 × 1.463</td><td>0.906</td><td>40 × 4</td><td>50 × 1.181</td></tr>
<tr><td>I</td><td>—</td><td>—</td><td>40 × 5</td><td>50 × 1.463</td></tr>
<tr><td>J</td><td>—</td><td>—</td><td>40 × 6</td><td>—</td></tr>
<tr><td>K</td><td>—</td><td>—</td><td>50 × 5</td><td>—</td></tr>
</table>

[a]Table A-5 is based on SMACNA's publication, HVAC *Duct Construction Standards—Metal and Flexible*, Tables 1-11 through 1-13. For assembly details and other construction alternatives using tie rods, refer to this standard.

[b]Standing S slip length limits: 750 mm at 1000 Pa, 900 mm at 750 Pa, none at 500 Pa or less.

[c]T-dimension for formed angles is minimum thickness; nominal for structural angles.

[d]Duct reinforcement (rigidity class) may have to be increased from that obtained from Table A-4 to match duct thickness.

[e]T-dimension for locks and ducts is minimum thickness; nominal for bars and angles.

[f]EI index for rigidity class is in Table A-6.

**Table A-6 Intermediate Reinforcement**[a]

| Minimum Rigidity | | ANGLE | ZEE OR CHANNEL | HAT SECTION | CHANNEL |
|---|---|---|---|---|---|
| **Class** | **EI, GN • mm²** | **H × T, mm**[b] | **H × B × T, mm**[b] | **H × B × D × T, mm**[b] | **H × B × T, mm**[b] |
| A | 0.15 | Use Class C | Use Class B | Use Class F | Use Class C |
| B | 0.30 | | 20 × 13 × 0.906 | | |
| C | 0.75 | 25 × 1.181<br>20 × 3 | 20 × 13 × 1.181 | | 20 × 80 × 1.181 |
| D | 1.5 | 30 × 1.181<br>25 × 3 | 25 × 20 × 0.906 | | 30 × 90 × 1.181 |
| E | 3.0 | 30 × 2.162<br>40 × 1.463 | 25 × 20 × 2.162<br>40 × 20 × 0.906 | | 25 × 50 × 3 |
| F | 4.5 | 40 × 4 | 25 × 20 × 3.280<br>40 × 20 × 1.181 | 40 × 20 × 16 × 0.906<br>40 × 40 × 20 × 0.906 | 40 × 80 × 1.463 |
| G | 7.0 | 40 × 5 | 40 × 20 × 3.280<br>50 × 30 × 0.906 | 40 × 20 × 16 × 1.463<br>40 × 40 × 20 × 1.181 | 30 × 80 × 3 |
| H | 15 | 50 × 4 | 50 × 30 × 1.463 | 40 × 40 × 20 × 2.162<br>50 × 25 × 20 × 1.181 | 35 × 80 × 5.6 |
| I | 22 | 50 × 5 | 50 × 30 × 2.162 | 50 × 25 × 20 × 2.162<br>70 × 50 × 20 × 1.463 | 50 × 50 × 3 |
| J | 30 | 50 × 6<br>70 × 3.280 | 50 × 30 × 3.280<br>80 × 30 × 1.463 | 50 × 25 × 20 × 3.280<br>70 × 50 × 20 × 2.162 | 40 × 100 × 7 |
| K | 45 | 70 × 6 | 80 × 30 × 2.162 | 70 × 50 × 20 × 3.280<br>80 × 40 × 20 × 1.463 | |
| L | 60 | 70 × 6 | 80 × 30 × 3.280 | 80 × 40 × 20 × 2.162 | |

[a]Table A-6 is based on SMACNA's publication, HVAC *Duct Construction Standards—Metal and Flexible*, Table 1-10.

[b]T-dimension for formed shapes is minimum thickness; nominal for structural shapes.

**Table A-7a Round Ferrous Metal Duct Construction for Commercial Systems***[a]

| Duct Diameter, mm | Minimum Galvanized Steel Thickness[b], mm | | | | | | |
|---|---|---|---|---|---|---|---|
| | −500 Pa Maximum | | | +500 Pa Maximum | | | |
| | Spiral Seam Duct | Longitudinal Seam Duct[c] | Fittings | Spiral Seam Duct | Longitudinal Seam Duct[c] | Fittings | Type of Joint[d] |
| Up through 200 | 0.395 | 0.601 | 0.601 | 0.395 | 0.395 | 0.471 | Beaded slip |
| 350 | 0.471 | 0.601 | 0.601 | 0.395 | 0.471 | 0.471 | Beaded slip |
| 650 | 0.601 | 0.753 | 0.753 | 0.471 | 0.601 | 0.601 | Beaded slip |
| 900 | 0.753 | 0.906 | 0.906 | 0.601 | 0.753 | 0.753 | Beaded slip |
| 1270 | 0.906 | 1.181 | 1.181 | 0.753 | 0.906 | 0.906 | Flange |
| 1500 | 1.181 | 1.463 | 1.463 | 0.906 | 1.181 | 1.181 | Flange |
| 2100 | 1.463 | 1.784 | 1.784 | 1.181 | 1.463 | 1.463 | Flange |

*See Table A-7b for notes.

[a]Table A-7 is based on SMACNA's publication, HVAC *Duct Construction Standards—Metal and Flexible*, Table 3-2.

[b]Table A-7 may be used for galvanized, painted, uncoated, and stainless steel whenever the base metal thickness is not less than 0.038 mm below those in the table.

[c]For seam details and limitations, refer to SMACNA's publication, HVAC *Duct Construction Standards—Metal and Flexible*, Figure 3-1.

[d]Recommended joints are listed. For alternate joints, details, and limitations, refer to SMACNA's HVAC *Duct Construction Standards—Metal and Flexible*, Figure 3-2.

[e]Only butt weld or pipe lock (grooved seam, pipe lock) longitudinal seams are permitted.

### Table A-7b Round Ferrous Metal Duct Construction for Commercial Systems[a]

| Duct Diameter, mm | Minimum Galvanized Steel Thickness[b], mm | | | | | | |
|---|---|---|---|---|---|---|---|
| | −750 Pa Maximum | | | +2500 Pa Maximum | | | |
| | Spiral Seam Duct | Longitudinal Seam Duct[e] | Fittings | Spiral Seam Duct | Longitudinal Seam Duct[e] | Fittings | Type of Joint[d] |
| Up through 200 | 0.471 | 0.601 | 0.601 | 0.471 | 0.601 | 0.601 | Beaded slip |
| 350 | 0.471 | 0.601 | 0.601 | 0.471 | 0.601 | 0.601 | Beaded slip |
| 650 | 0.601 | 0.753 | 0.753 | 0.601 | 0.753 | 0.753 | Beaded slip |
| 900 | 0.753 | 0.906 | 0.906 | 0.753 | 0.906 | 0.906 | Beaded slip |
| 1270 | 0.906 | 0.906 | 0.906 | 0.906 | 0.906 | 0.906 | Flange |
| 1500 | — | 1.181 | 1.181 | 1.181 | 1.181 | 1.181 | Flange |
| 2100 | — | 1.463 | 1.463 | 1.181 | 1.463 | 1.463 | Flange |

### Table A-8 Flat-Oval Duct Construction for Positive Pressure Commercial Systems[a]

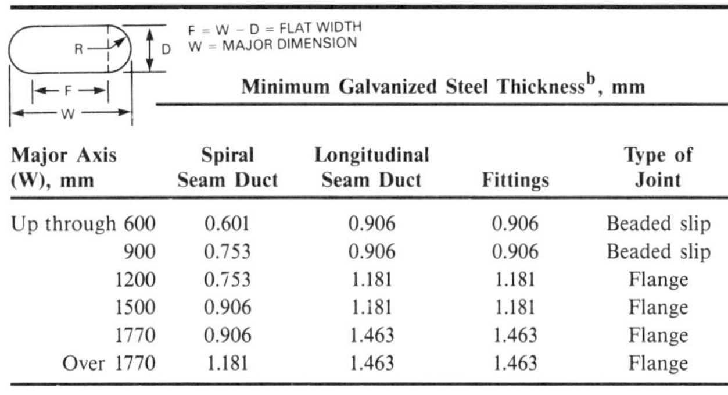


| Major Axis (W), mm | Minimum Galvanized Steel Thickness[b], mm | | | |
|---|---|---|---|---|
| | Spiral Seam Duct | Longitudinal Seam Duct | Fittings | Type of Joint |
| Up through 600 | 0.601 | 0.906 | 0.906 | Beaded slip |
| 900 | 0.753 | 0.906 | 0.906 | Beaded slip |
| 1200 | 0.753 | 1.181 | 1.181 | Flange |
| 1500 | 0.906 | 1.181 | 1.181 | Flange |
| 1770 | 0.906 | 1.463 | 1.463 | Flange |
| Over 1770 | 1.181 | 1.463 | 1.463 | Flange |

[a]Table A-8 is based on SMACNA's publication, HVAC *Duct Construction Standards—Metal and Flexible*, Table 3-4.

[b]Use Table A-8 for galvanized, painted, uncoated, and stainless steel whenever the base metal thickness is not less than 0.038 mm below those in the table.

### Table A-9 Rectangular Duct Hangers—Commercial Systems[a,b]

| Half of Duct Perimeter (P/2), mm[d] | Type of Hanger[c] | | | | | | | |
|---|---|---|---|---|---|---|---|---|
| | Galvanized Straps (see Figs. 3 and 4) Width × Thickness (min), mm | | | | Rods (see Fig. 4) Diameter, mm | | | |
| | Maximum Hanger Spacing, mm | | | | | | | |
| | 3000 | 2400 | 1500 | 1200 | 3000 | 2400 | 1500 | 1200 |
| Up through 750 | 25 × 0.753 | 25 × 0.753 | 25 × 0.753 | 25 × 0.753 | 4.11 | 4.11 | 4.11 | 4.11 |
| 1800 | 25 × 1.181 | 25 × 0.906 | 25 × 0.753 | 25 × 0.753 | 10 | 10 | 10 | 10 |
| 2400 | 25 × 1.463 | 25 × 1.181 | 25 × 0.906 | 25 × 0.753 | 10 | 10 | 10 | 10 |
| 3000 | 40 × 1.463 | 25 × 1.463 | 25 × 1.181 | 25 × 0.906 | 13 | 10 | 10 | 10 |
| 4200 | 40 × 1.463 | 40 × 1.463 | 25 × 1.463 | 25 × 1.181 | 13 | 13 | 10 | 10 |
| 4800 | — | 40 × 1.463 | 25 × 1.463 | 25 × 1.463 | 13 | 13 | 10 | 10 |
| Over 4800 | — | — | — | — | ◄— Special Analysis Required[e] —► | | | |

[a]This table is based on SMACNA's publication, HVAC *Duct Construction Standards—Metal and Flexible*, Table 4-1.

[b]Table based on (1) 1.463 mm thick ducts; (2) reinforcements and trapeze mass; (3) 4.9 kg/m$^2$ for insulation; and (4) no external loads.

[c]For trapeze angle size, See Table 10. For hanger attachments to structures, typical upper attachments, lower hanger attachments, and strap splicing details, consult the HVAC *Duct Construction Standards—Metal and Flexible*, Table 4-1 and Figures 4-1 through 4-4.

[d]When dimension (W) exceeds 1500 mm, P/2 maximum is 1.25 W.

[e]Consult AISC Manual (1980) for hanger design criteria.

Maximum allowable strap, wire, and rod loads are:

| Strap (W × T[min]), mm | Load, kg |
|---|---|
| 25 × 0.753 | 120 |
| 25 × 0.906 | 145 |
| 25 × 1.181 | 190 |
| 25 × 1.463 | 320 |
| 40 × 1.463 | 500 |

| Rod or Wire (dia), mm | Load, kg |
|---|---|
| 2.67 | 35 |
| 3.43 | 55 |
| 4.11 | 70 |
| 10 | 310 |
| 13 | 570 |
| 16 | 910 |
| 20 | 1360 |

Table A-10 Allowable Trapeze Angle Loads[a]

| Width of Hanger (W)[b], mm | Maximum Hanger Load[c], kg — Angle[e], mm | | | | | | | | |
|---|---|---|---|---|---|---|---|---|---|
| | 25 × 25 × 1.463[d] | 25 × 25 × 3 | 40 × 40 × 1.463[d] | 40 × 40 × 4 | 40 × 40 × 5 | 40 × 40 × 6 | 50 × 50 × 5 | 50 × 50 × 6 | 70 × 70 × 6 |
| Up through 450 | 35 | 70 | 80 | 160 | 230 | 295 | 425 | 560 | 890 |
| 600 | 35 | 70 | 80 | 160 | 230 | 295 | 425 | 560 | 890 |
| 750 | 30 | 70 | 80 | 160 | 230 | 295 | 425 | 560 | 890 |
| 900 | 25 | 60 | 70 | 155 | 225 | 280 | 420 | 545 | 880 |
| 1050 | 15 | 50 | 65 | 145 | 220 | 275 | 410 | 540 | 875 |
| 1200 | — | 35 | 50 | 130 | 205 | 260 | 395 | 525 | 860 |
| 1350 | — | 15 | 30 | 115 | 180 | 245 | 380 | 510 | 845 |
| 1500 | — | — | — | 85 | 160 | 220 | 355 | 480 | 815 |
| 1650 | — | — | — | 45 | 120 | 180 | 320 | 445 | 780 |
| 1800 | — | — | — | — | 85 | 145 | 280 | 410 | 745 |
| 2000 | — | — | — | — | 35 | 95 | 230 | 360 | 695 |
| 2100 | — | — | — | — | — | 35 | 170 | 300 | 635 |
| 2400 | — | — | — | — | — | — | — | 145 | 480 |
| 2750 | — | — | — | — | — | — | — | — | 275 |

[a]Information in this table is based on SMACNA's publication, HVAC *Duct Construction Standards—Metal and Flexible*, Table 4-3.

[b]For W-dimension, see Figure 4.

[c]Loads in this table assume rods or other structural members are 150-mm maximum from duct sides (see Figure 4).

[d]Galvanized formed angles; thickness is minimum.

[e]Steel yield stress is based on 172 MPa minimum.

Table A-11 Round Duct Hangers—Commercial Systems[a,b]

| Duct Diameter, mm | Maximum Hanger Spacing[c], mm | Hanger Configuration[d] (see Figure 5) | Galvanized Ring Size (min), mm | Minimum Hanger Size, mm | | |
|---|---|---|---|---|---|---|
| | | | | Wire | Strap | Rod |
| Up through 250 | 3600 | Type 1 or 3 | 25 × 0.753 | 2.67 | 25 × 0.753 | 10 |
| 450 | 3600 | Type 1 or 3 | 25 × 0.753 | 4.11 | 25 × 0.753 | 10 |
| 600 | 3600 | Type 1 or 3 | 25 × 0.753 | — | 25 × 0.753 | 10 |
| | | Type 2 or 4 | 25 × 0.753 | 3.43 | 25 × 0.753 | 10 |
| 900 | 3600 | Type 1 or 3 | 25 × 0.906 | — | 25 × 0.906 | 10 |
| | | Type 2 or 4 | 25 × 0.906 | 4.11 | 25 × 0.753 | 10 |
| 1250 | 3600 | Type 4 | 25 × 0.906 | — | 25 × 0.906 | 10 |
| 1500 | 3600 | Type 4 | 25 × 1.181 | — | 25 × 1.181 | 10 |
| 2100 | 3600 | Type 4 | 25 × 1.463 | — | 25 × 1.463 | 10 |

[a]Table A-11 is based on SMACNA's publication HVAC *Duct Construction Standards—Metal and Flexible*, Table 4-2.

[b]Table A-11 is for duct construction per Table A-7 plus 4.9 kg/m$^2$ insulation. For heavier ductwork, size hanger members so allowable load (see Table A-9) is not exceeded.

[c]The 3600 mm hanger spacing may be increased for risers and other conditions with substantial horizontal support.

[d]Refer to Table A-7 for angle flange sizes. For hanger attachments to structures, typical upper attachments, lower hanger attachments, and strap splicing details, consult the HVAC *Duct Construction Standards—Metal and Flexible*, Table 4-1 and Figures 4-1 through 4-4.

# CHAPTER 2

# AIR-DIFFUSING EQUIPMENT



SUPPLY air outlets and diffusing equipment introduce air into a conditioned space to obtain a desired indoor atmospheric environment. Return and exhaust air is removed from a space through return and exhaust inlets. Various types of diffusing equipment are available as standard manufactured products. This chapter describes this equipment and details its proper use. Chapter 32 in the 1985 FUNDAMENTALS Volume and Chapter 2 in the 1987 HVAC Volume also cover the subject.

## SUPPLY-AIR OUTLETS

The correct types of supply outlets, properly sized and located, control the air pattern within a given space to obtain proper air motion and temperature equalization. The following basic supply outlets are commonly available: (1) grille outlets, (2) slot diffuser outlets, (3) ceiling diffuser outlets, and (4) perforated ceiling panels. These types differ in their construction features, physical configurations, and in the manner in which they diffuse or disperse supply air and induce or entrain room air into a primary airstream.

Accessories used with these outlets regulate the volume of supply air and control its flow pattern. For example, the outlet cannot discharge air properly and uniformly unless the air is conveyed to it in a uniform flow. Accessories may also be necessary for proper air distribution in a space, so they must be selected and used in accordance with manufacturers' recommendations. Outlets should be sized to project air so that its velocity and temperature reach acceptable levels before entering the occupied zone.

Outlets with high induction rates move (throw) air short distances but have rapid temperature equalization. Ceiling diffusers with radial patterns have shorter throws and obtain more rapid temperature equalization than slot diffusers. Grilles, which have long throws, have the lowest diffusion and induction rates. Therefore, in those cases, round or square ceiling diffusers deliver more air to a given space than grilles and slot diffuser outlets that require room velocities of 25 to 35 fpm (0.13 to 0.18 m/s). In some spaces, higher room velocities can be tolerated, or the ceilings may be high enough to permit a throw long enough to result in the recommended room velocities.

Outlets with high induction characteristics can also be used advantageously in air-conditioning systems with low supply-air temperatures and consequent high-temperature differentials between room air temperature and supply-air temperatures. Therefore, ceiling diffusers may be used in systems with cooling temperature differentials up to 30 to 35 °F (17 to 19 °C) and still provide satisfactory temperature equalizations within the spaces. Slot diffusers may be used in systems with cooling temperature differentials as high as 25 °F (14 °C). Grilles may generally be used in well-designed systems with cooling temperature differentials up to 20 °F (11 °C).

### Surface Effect

The induction or entrainment characteristics of a moving airstream cause a *surface effect*. An airstream moving adjacent to, or in contact with, a wall or ceiling surface creates a low-pressure area immediately adjacent to that surface, causing the air to remain in contact with the surface substantially throughout

The preparation of this chapter is assigned to TC 5.3, Room Air Distribution.

the length of throw. The surface effect counteracts the drop of horizontally projected cool airstreams.

Ceiling diffusers exhibit surface effect to a high degree because a circular air pattern blankets the entire ceiling area surrounding each outlet. Slot diffusers, which discharge the airstream across the ceiling, exhibit surface effect only if they are long enough to blanket the ceiling area. Grilles exhibit varying degrees of surface effect, depending on the spread of the particular air pattern.

In many installations, the outlets must be mounted on an exposed duct and discharge the airstream into free space. In this type of installation, the airstream entrains air on both its upper and lower surfaces; as a result, a higher rate of entrainment is obtained and the throw is shortened by about 33%. Airflow per unit area for these types of outlets can, therefore, be increased. Because there is no surface effect from ceiling diffusers installed on exposed ducts, the air drops rapidly to the floor. Therefore, temperature differentials in air-conditioning systems must be restricted to a range of 15 to 20°F (8 to 11°C). Airstreams from slot diffusers and grilles show a marked tendency to drop because of the lack of surface effect.

### Smudging

Smudging is a problem with ceiling and slot diffusers. Dirt particles held in suspension in the room air are subjected to turbulence at the outlet face. This turbulence is primarily responsible for smudging. Smudging can be expected in areas of high pedestrian traffic (lobbies, stores, etc.). When ceiling diffusers are installed on smooth ceilings (such as plaster, mineral tile, and metal pan), smudging is usually in the form of a narrow band of discoloration around the diffuser. Anti-smudge rings reduce this type of smudging effectively. On highly textured ceiling surfaces (such as rough plaster and sprayed-on composition), smudging occurs over a more extensive area.

## PROCEDURE FOR OUTLET SELECTION

The following procedure is generally used in selecting outlet locations and types:

1. Determine the amount of air to be supplied to each room. (Refer to Chapters 25 and 26 in the 1985 FUNDAMENTALS Volume to determine air quantities for heating and cooling.)
2. Select the type and quantity of outlets for each room, considering such factors as air quantity required, distance available for throw or radius of diffusion, structural characteristics, and architectural concepts. Table 1 is based on experience and typical ratings of various outlets. It may be used as a guide to the outlets applicable for use with various room air loadings. Special conditions, such as ceiling heights greater than the normal 8 to 12 ft (2.4 to 3.6 m) and exposed duct mounting, as well as product modifications and unusual conditions of room occupancy, can modify this table. Manufacturers' rating data should be consulted for final determination of the suitability of the outlets used.
3. Locate outlets in the room to distribute the air as uniformly as possible. Outlets may be sized and located to distribute air in proportion to the heat gain or loss in various portions of the room.
4. Select proper outlet size from manufacturers' ratings according to air quantities, discharge velocities, distribution patterns, and sound levels. Note manufacturers' recommendations with regard to use. In an open space configuration, the interaction of airstreams from multiple diffuser sources may alter single diffuser throw data or single diffuser air temperature/air velocity data, and it may not be sufficient to predict particular levels of air motion in a space. Also, obstructions to the primary air distribution pattern require special consideration.
5. Chapter 32 of the 1985 FUNDAMENTALS Volume gives the complete procedure for selecting the size and locations of the diffusers.

**Table 1 Guide to Use of Various Outlets**

| Type of Outlet | Air Loading, cfm/ft$^2$ (L/s per m$^2$) of Floor Space | Approx. Max. Air Changes per Hour for 10 ft (3 m) Ceiling |
|---|---|---|
| Grille | 0.6 to 1.2 (3 to 6) | 7 |
| Slot | 0.8 to 2.0 (4 to 10) | 12 |
| Perforated Panel | 0.9 to 3.0 (5 to 15) | 18 |
| Ceiling Diffuser | 0.9 to 5.0 (5 to 25) | 30 |
| Perforated Ceiling | 1.0 to 10.0 (5 to 50) | 60 |

### Sound Level

The sound level of an outlet is a function of the discharge velocity and the transmission of systemic noise, which is a function of the size of the outlet. Higher frequency sounds can be the result of excessive outlet velocity but may also be generated in the duct by the moving airstream. Lower-pitched sounds are generally the result of mechanical equipment noise transmitted through the duct system and outlet. The cause of higher-frequency sounds can be pinpointed as outlet or systemic sounds by removing the outlet during operation. A reduction in sound level indicates that the outlet is causing noise. If the sound level remains essentially unchanged, the system is at fault. Chapter 52 in the 1987 HVAC Volume has more information on design criteria, acoustic treatment, and selection procedures.

## GRILLE OUTLETS

**Adjustable Bar Grille** is the most common type of grille used as a supply outlet. The single deflection grille consists of a frame enclosing a set of vertical or horizontal vanes. Vertical vanes deflect the airstream in the horizontal plane; horizontal vanes deflect the airstream in the vertical plane. The double deflection grille has a second set of vanes installed behind and at right angles to the face vanes. This grille controls the airstream in both the horizontal and vertical planes.

**Fixed Bar Grille** is similar to the single deflection grille, except that the vanes are not adjustable; the vanes may be straight or set at an angle. The angle at which the air is discharged from this grille depends on the type of deflection vanes.

**Stamped Grille** is stamped from a single sheet of metal to form a fretwork through which air can pass. Various designs are used, varying from square or rectangular holes to intricate ornamental designs.

**Variable Area Grille** is similar to the adjustable double deflection grille but can vary the discharge area to achieve an air volume change (variable volume outlet) at constant pressure, so that the variation in throw is minimized for a given change in supply-air volume.

## Applications

Properly selected grilles operate satisfactorily from high side and perimeter locations in the sill, curb, or floor. Ceiling-mounted grilles, which discharge the airstream down, are generally not acceptable in comfort air-conditioning installations in interior zones and may cause draft in perimeter applications.

**High Side Wall.** The use of a double-deflection grille usually provides the most satisfactory solution. The vertical face louvers of a well-designed grille deflects the air approximately 50 deg to either side and amply covers the conditioned space. The rear louvers deflect the air at least 15 deg in the vertical plane, which is ample to control the elevation of the discharge pattern.

**Perimeter Installations.** The grille selected must fit the specific job. When small grilles are used, adjustable vane grilles improve the coverage of perimeter surfaces. Where the perimeter surface can be covered with long grilles, the fixed vane grille is satisfactory. Where grilles are located more than 8 in. (200 mm) from the perimeter surface, it is usually desirable to deflect the airstream toward the perimeter wall. This can be done with adjustable or fixed deflecting vane grilles.

**Ceiling Installation** is generally limited to grilles having curved vanes, which, because of their design, provide a horizontal pattern. Curved vane grilles may also be used satisfactorily in high side wall or perimeter installations.

## Accessories

Various accessories, designed to improve the performance of grille outlets, are available as standard equipment.

**Opposed-Blade dampers** can be attached to the back of grilles (the combination of a grille and a damper is called a *register*) or installed as separate units in the duct (Figure 1A). Adjacent blades of this damper rotate in opposite directions and may receive air from any direction, discharging it in a series of jets without adversely deflecting the airstream to one side of the duct.

**Multishutter dampers** have a series of gang-operated blades that rotate in the same direction (Figure 1B). This uniform rotation deflects the airstream when the damper is partially open. Most dampers are operated by removable keys or fixed or removable levers.

**Gang-Operated Turning Vanes** are installed in collar connections to branch ducts. The device shown in Figure 1C has vanes that pivot and remain parallel to the duct airflow, regardless of the setting. The second device (Figure 1D) has a set of fixed vanes. Both devices restrict the area of the duct in which they are installed. They should be used only when the duct is wide enough to allow the device to open to its maximum position without causing undue restriction of airflow in the duct, which might limit downstream airflow.

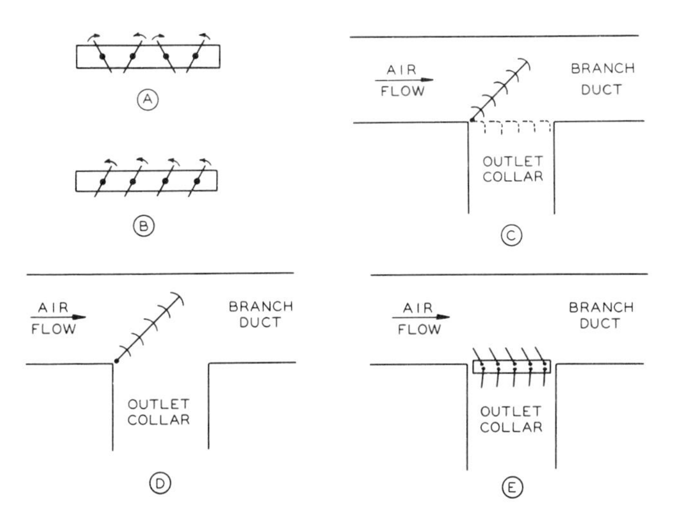


**Fig. 1 Accessory Controls for Grille and Register Outlets**

**Individually Adjusted Turning Vanes** are used in the device shown in Figure 1E. Two sets of vanes are used. The downstream set equalizes flow across the collar, while the upstream one turns the air. They can also be adjusted, at various angles, to act as a damper, but it is not practicable to use this device as a damper because its balancing requires removal of the grille to gain access and reinstallation to measure airflow.

**Other Miscellaneous Accessories** available as standard equipment include remote-control devices to operate the dampers from locations other than the grille face and maximum-minimum stops to limit damper travel.

# SLOT DIFFUSER OUTLETS

A slot diffuser is an elongated outlet consisting of a single or multiple number of slots. It is usually installed in long continuous lengths. Outlets with dimensional aspect ratios of 25 to 1 or greater and a maximum height of approximately 3 in. (80 mm), generally meet the performance criteria for slot diffusers.

## Applications

**High Side Wall Installation.** The perpendicular-flow slot diffuser is best suited to high side wall installations and perimeter installations in sills, curbs, and floors. The air discharged from a perpendicular slot diffuser will not drop if the diffuser is located within 6 to 12 in. (150 to 300 mm) from the ceiling and is long enough to establish surface effect. Under these conditions, air travels along the ceiling to the end of the throw. If the slot diffuser is mounted 1 to 2 ft (300 to 600 mm) below the ceiling, an outlet that deflects the air up to the ceiling must be used to achieve the same result. If the slot is located more than 2 ft (600 mm) below the ceiling, premature drop of cold air into the occupied zone will probably result.

**Ceiling Installation.** The parallel-flow slot diffuser is ideal for ceiling installation because it discharges across the ceiling. The perpendicular-flow slot diffuser may be mounted in the ceiling; however, the downward discharge pattern may cause localized areas of high air motion. This device performs satisfactorily when installed adjacent to a wall or over an unoccupied or transiently occupied area. Care should be exercised in using a perpendicular-flow slot diffuser in a downward discharge pattern because variations of supply air temperature cause large variations in throw.

**Sill Installation.** The perpendicular-flow slot diffuser is well suited to sill installation, but it may also be installed in the curb and floor. When the diffuser is located within 8 in. (200 mm) of the perimeter wall, the discharged air may be either directed straight toward the ceiling or deflected slightly toward the wall. When the diffuser distance from the wall is greater than 8 in. (200 mm), the air should generally be deflected toward the wall at an angle of approximately 15 deg; deflections as great as 30 deg may be desirable in some cases. The air should not be deflected away from the wall into the occupied zone.

To perform satisfactorily, outlets of this type must be used only in installations with carefully designed duct and plenum systems. (Refer to the section, "Discharge from a Long Slot" in Chapter 32 of the 1985 FUNDAMENTALS Volume.)

Slot diffusers are generally equipped with accessory devices for uniform supply air discharge along the entire length of the slot. While accessory devices help correct the airflow pattern, proper approach conditions for the airstream are also important for satisfactory performance. When the plenum supplying a slot diffuser is being designed, the transverse velocity in the plenum should be less than the discharge velocity of the jet, as recommended by the manufacturer and also as shown by experience.

If *Tapered Ducts* are used for introducing supply air into the diffuser, they should be sized to maintain a velocity of approximately 500 fpm (2.5 m/s) and tapered to maintain constant static pressure.

**Air-Light Fixtures.** Slot diffusers, having a single-slot discharge and nominal 2, 3, and 4 ft (600, 900, and 1200 mm) lengths, are available for use in conjunction with recessed fluorescent light troffers. A diffuser mates with a light fixture and is entirely concealed from the room. It discharges air through suitable openings in the fixture and is available with fixed or adjustable air discharge patterns, air distribution plenum, inlet dampers for balancing, and inlet collars suitable for flexible duct connections. Light fixtures adapted for slot diffusers are available in styles to fit common ceiling constructions. Various slot diffuser and light fixture manufacturers can furnish products compatible with one another's equipment.

### Accessories

Dampers are normally available as integral equipment with slot diffusers and are used for minor flow adjustments. Installations of balancing dampers in the distribution plenum is not recommended. For accessibility, balancing dampers should be located as far from the outlet as is practical.

**Flow Equalizing Vanes** each consist of one set of individually adjustable blades installed behind the slot openings at right angles to the slot. If designed and installed correctly, the blades improve the discharge pattern of the diffuser. (See the section, "Discharge from a Long Slot," in Chapter 32 of the 1985 FUNDAMENTALS Volume.) These blades may also be used to increase spread of the airstream for greater coverage. Dampers and flow-equalizing vanes can usually be adjusted without removing the diffuser.

## CEILING DIFFUSER OUTLETS

**Multipassage Ceiling Diffusers** consist of a series of flaring rings or louvers, which form a series of concentric air passages. They may be round, square, or rectangular. For easy installation, these diffusers are usually made in two parts: an outer shell with duct collar and an easily removable inner assembly.

**Flush** and **Stepped-Down Diffusers** are also available. In the flush unit, as the name implies, all rings or louvers project to a plane surface, whereas in the stepped-down unit, they project beyond the surface of the outer shell.

Common variations of this diffuser type are the *adjustable-pattern* diffuser and the *multipattern* diffuser. In the adjustable-pattern diffuser, the air-discharge pattern may be changed from a horizontal to a vertical or downblow pattern. Special construction of the diffuser or separate deflection devices allow adjustment. Multipattern diffusers are square or rectangular and have special louvers to discharge the air in one or more directions.

Other outlets available as standard equipment are half-round diffusers, supply and return diffusers, and light fixture-air diffuser combinations.

**Perforated-Face Ceiling Diffusers** meet architectural demands for air outlets that blend into perforated ceilings. Each has a perforated metal face with an open area of 10 to 50% which determines its capacity. Units are usually equipped with deflection devices to obtain multi-pattern horizontal air discharge.

**Variable Area Ceiling Diffusers** may be round, square, or linear and have parallel or concentric passages or a perforated face. In addition, they feature a means of effectively varying the discharge area to achieve an air volume change (variable volume outlet) at a constant pressure so that the variation in throw is minimized for a given change in supply-air volume.

### Applications

**Ceiling Installations.** Ceiling diffusers should be mounted in the center of the space that they serve when they discharge the supply air in all directions. Multi-pattern diffusers can be used in the center of the space or adjacent to partitions, depending on the discharge pattern. By using different inner assemblies, their air pattern can be changed to suit particular requirements.

**Side Wall Installations.** Half-round diffusers and multi-pattern diffusers, when installed high in side walls, should generally discharge the air toward the ceiling.

**Exposed Duct Installation.** Some ceiling diffuser types, particularly *stepped-down* units, perform satisfactorily on exposed ducts. Consult manufacturers' catalogs for specific types.

**Adjustable-Pattern Diffusers.** Surface effect is important in the performance of adjustable-pattern diffusers. In fact, this effect is so pronounced that usually only two discharge patterns are possible with adjustable-pattern diffusers mounted directly on the ceiling. When a diffuser is changed from a horizontal pattern position toward the downblow pattern position, the surface effect maintains the horizontal discharge pattern until the discharge airstream is effectively deflected at the diffuser face, resulting in a vertical pattern. However, when adjustable-pattern ceiling diffusers are mounted on exposed duct, and there is no surface effect, the air may assume any pattern between horizontal and vertical discharge. Directional or segmented horizontal air patterns can usually be obtained by adjusting internal baffles or deflectors.

### Accessories

Dampering and air-straightening accessories of various types are available as standard equipment.

**Multilouver Dampers** consist of a series of parallel blades mounted inside a round or square frame and are installed in the diffuser collar or the takeoff. The blades are usually arranged in two groups rotating in opposite directions (Figure 2A) and are key-operated from the face of the diffuser. This arrangement equalizes airflow in the diffuser collar.

**Opposed-Blade Dampers** for round ceiling diffusers usually consist of a series of pie-shaped vanes mounted inside a round frame installed in the diffuser collar or the takeoff duct. The vanes pivot about a horizontal axis and are arranged in two groups, with adjacent vanes rotating counter to each other (Figure 2B). The vanes are key-operated from the diffuser face.

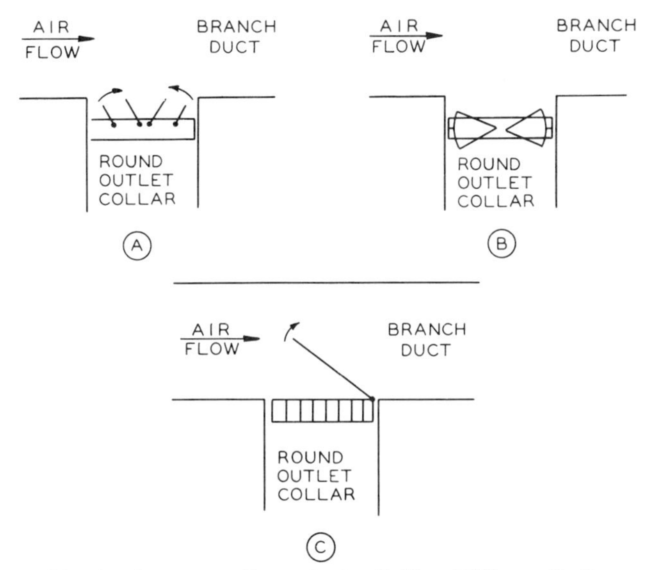


**Fig. 2 Accessory Controls for Ceiling Diffuser Outlets**

Another opposed-blade design is similar in construction to the damper shown in Figure 1A and has either a round or square frame.

**Splitter Damper.** A splitter damper is a single-blade sheet-metal plate hinged at one edge and usually located at the branch connection of a duct or outlet (Figure 2C). It is easy to operate, but often causes irregular airflow in the duct. When used in conjunction with and near an outlet, additional directional control is required.

**Equalizing Devices** provide uniform airflow in the diffuser collar and consist of individually adjustable parallel blades mounted within a frame. They are installed in the diffuser collar or the takeoff duct. If used in pairs at right angles to each other, they serve as both equalizing and directional control devices.

**Blankoff Baffles** are used for minor adjustments of the airflow from a diffuser. These baffles blank off a sector of the diffuser and prevent the supply air from striking an obstruction such as a column, partition, or the wall of the conditioned space by reducing flow in a given direction. Blankoff baffles generally reduce the area and increase supply air velocity, which must be considered in selecting diffuser size. Pattern control in diffusers having removable directional cores may be accomplished by rearranging the cores, generally without a change in area or increase in velocity.

**Anti-smudge Rings** are round or square metal frames attached to and extending 4 to 12 in. (100 to 300 mm) beyond the outer edge of the diffuser. Their purpose is to minimize ceiling smudging.

## AIR-DISTRIBUTING CEILINGS

The air-distributing ceiling uses the confined space above the ceiling as a supply plenum that receives air from stub ducts. The plenum should be designed to achieve uniform plenum pressure, resulting in uniform delivery of air to the conditioned space below.

Air is delivered through round holes or slots in the ceiling material or suspension system. These holes and slots vary in shape and size among manufacturers. Various manufacturers have developed a number of products based on the principle of a supply plenum, with sizes ranging mainly from 1 by 1 ft (300 by 300 mm) tile for a concealed grid to 2 by 4 ft (600 by 1200 mm) lay-in panels for an exposed grid. Sometimes, the slots are equipped with adjustable dampers to facilitate changing the open area after installation.

The upper limit of plenum pressure must be that recommended by the ceiling manufacturer. It generally ranges from 0.10 to 0.15 in. of water (25 to 35 Pa) , dictated by resistance to sag, to a lower pressure limit of about 0.01 in. of water (2.5 Pa), where uniformity of plenum pressure becomes more important. The range of air rates extends from about 15 cfm (7.0 L/s) down to about 1 cfm per ft$^2$ (5 L/s per m$^2$) of floor area. High flow rates are recommended only for low-temperature differentials (Nevins and Ward 1968; Miller and Nevins 1969).

Active portions of the air-distributing ceilings should be located with respect to room load distribution, with higher airflow rates at the exterior exposures. This method of air distribution is particularly suited to large zones of uniform room temperature. Where different room temperatures are desired, a separate ceiling plenum is required for each zone. Construction of the ceiling plenum requires care with regard to air tightness, obstructions causing unequal plenum pressure and temperature, heat storage effect of the structure, and the influence of a roof or the areas surrounding the plenum (Hemphill 1966).

## OUTLETS IN VARIABLE-AIR-VOLUME SYSTEMS

The performance of a particular outlet or diffuser is generally independent of the terminal box that is upstream. For a given supply-air volume and temperature differential (to meet a particular load), a standard outlet does not recognize whether the terminal box is of a constant volume, variable volume, or induction type. However, any diffuser, or system of diffusers, gives optimum air diffusion at some particular load condition and air volume. For a VAV system, as the load changes, so does the level of air movement in the conditioned space. At minimum load condition, when the volume of air is lowest, the level of air movement in the conditioned space may be low.

## RETURN- AND EXHAUST-AIR INLETS

Return-air inlets may either be connected to a duct or be simple vents that transfer air from one area to another. Exhaust-air inlets remove air directly from a building and, therefore, are always connected to a duct. Whatever the arrangement, inlet size and configuration determine velocity and pressure requirements for the required airflow. (See the section, "Return and Exhaust Inlets," in Chapter 32 of the 1985 FUNDAMENTALS Volume for further discussion of these factors and the effect of inlet location on the system.)

In general, the same type of equipment, grilles, slot diffusers, and ceiling diffusers used for supplying air may also be used for air return and exhaust. Inlets do not require the deflection, flow equalizing, and turning devices necessary for supply outlets. Dampers, however, are necessary when it is desirable to balance the airflow in the return-duct system.

### Types of Inlets

**Adjustable Bar Grilles.** The same grilles used for air supply are used to match the deflection setting of the bars with that of the supply outlets.

**Fixed Bar Grilles.** The same grilles described in the section "Grille Outlets" are used. This grille is the most common return-air inlet. Vanes are straight or set at a certain angle, the latter being preferred when appearance is important.

The **V-Bar Grille**, with bars in the shape of inverted *V's* stacked within the grille frame, has the advantage of being *sightproof;* it can be viewed from any angle without detracting from appearance. Door grilles are usually V-Bar grilles. The capacity of the grille decreases with increased *sight tightness.*

A **Lightproof Grille** is used to transfer air to or from darkrooms. The bars of this type of grille form a labyrinth and are painted black. The bars may take the form of several sets of V-bars or be of some special interlocking louver design to provide the required labyrinth.

**Stamped Grilles** are also frequently used as return and exhaust inlets, particularly in rest rooms and utility areas.

**Ceiling and Slot Diffusers** may also be used as return and exhaust inlets.

### Applications

Return and exhaust inlets may be mounted in practically any location: in ceilings, high or low in side walls, in floors, and in doors. (See the "Return and Exhaust Inlets" section in Chapter 32 of the 1985 FUNDAMENTALS Volume for their selection and location.)

The dampers shown in Figure 1A are used in conjunction with grille return and exhaust inlets. The type of damper does not affect the performance of the inlet. Usually, no other accessory devices are required.

## EQUIPMENT FOR AIR-DISTRIBUTION SYSTEMS

In high-velocity pressure and low-velocity pressure air distribution systems, the duct velocities and static pressures are such that special control and acoustical equipment may be required for proper introduction of conditioned air into the space to be served. This section deals with control equipment for single-duct and dual-duct air-conditioning systems. Chapter 33 of the 1985 FUNDAMENTALS Volume has information on the design of high-velocity ducts, and Chapter 1 of this volume has duct construction details. Chapter 2 of the 1987 HVAC Volume has further information on control and functions of terminal boxes. Chapter 52 of the 1987 HVAC Volume includes information on sound control in air-conditioning systems and sound-rating methods for air outlets.

Control equipment for high-velocity pressure and low-velocity pressure systems may be classified as *pressure reducing valves* or as *terminal boxes.* Terminal boxes may be further classified as single- or dual-duct boxes, reheat boxes, variable volume boxes, or ceiling induction boxes. They provide all or some of the following functions: pressure reduction, volume control, temperature modulation, air mixing, and sound attenuation.

### Pressure-Reducing Valves

Pressure-reducing or air valves each consist of a series of gang-operated vane sections mounted within a rigid casing and gasketed to reduce as much air leakage as possible between valve and duct. They are usually installed in rectangular ducts between a high-pressure trunk duct and a low-pressure branch duct for manual, remote manual, pneumatic, or electrical control.

Pressure is reduced by partially closing the valve and results in high-pressure drop through the valve. This action generates noise, which must be attenuated in the low pressure discharge duct. The length and type of duct lining depend on the amount and frequency of noise to be attenuated.

Volume control is obtained by adjustment of the valve manually, mechanically, or automatically. Automatic adjustment is achieved by a pneumatic or electric control motor actuated by a pressure regulator or a thermostat.

Pressure-reducing valves are generally equal in size to the low-pressure branch duct connected to the valve discharge. This arrangement provides minimum pressure drop with valves opened fully.

### Terminal Boxes

A terminal box is a factory-made assembly for air-distribution. This terminal box, without altering the composition of the treated air from the distribution system, manually or automatically fulfills one or more of the following functions: (1) controls the velocity, pressure, or temperature of the air; (2) controls the rate of airflow; (3) mixes airstreams of different temperatures or humidities; and (4) mixes, within the assembly, air at high velocity and/or high pressure with air from the treated space. To achieve these functions, the terminal box assemblies are made from an appropriate selection of the following component parts: casing, mixing section, manual damper, heat exchanger, induction section, and flow rate controller.

A terminal box commonly integrates a sound chamber to reduce noise generated within it by the manual damper or flow rate controller reducing the high-velocity, high-pressure inlet air to low-velocity, low-pressure air. The sound attenuation chamber is typically lined with thermal and sound-insulating material and is equipped with baffles. Special sound attenuation in the air-discharge ducts is not usually required in smaller boxes. In larger capacity boxes, additional sound absorption materials may be required in the low-pressure distribution ducts connected to the discharge of the box. Manufacturers' catalogs should be consulted for specific performance information.

Terminal boxes are typically categorized according to the function of the their flow rate controllers, which are generally categorized as *constant flow rate* or *variable flow rate* devices. They are further categorized as being *pressure dependent,* where the airflow rate through the assembly varies in response to changes in system pressure, or as *pressure independent (or pressure compensating),* where the airflow rate through the device does not vary in response to changes in system pressure. Constant flow rate controllers may be of the *mechanical volume control* type, which are actuated by means of the static pressure in the primary duct system and require no outside source of power. Constant flow rate controllers may also be of the *pneumatic or electric volume regulator* type, which are actuated by an outside motive power and typically require internal differential pressure sensing, selector devices, and pneumatic or electric motors for operation.

Variable flow rate controllers may be of the *mechanical pneumatic or electric volume control* type, which incorporates a means to reset the constant volume regulation automatically to a different control point within the range of the control device in response to an outside signal, such as from a thermostat. Boxes with this feature are pressure independent and may be used with reheat components. Variable flow rate may also be obtained by using a modulating damper ahead of a constant volume regulator. This arrangement typically allows for variations in flow between high and low limits or between a high limit and shutoff. These boxes are pressure dependent and volume limiting in function. Pneumatic variable volume may be either pressure independent, volume limiting, or pressure dependent, according to the equipment selected.

Terminal boxes can be further categorized as being *system powered,* wherein the assembly derives all of the energy necessary for its operation from the supply air within the distribution system, or as *externally powered,* wherein the assembly derives part or all of the energy necessary for its operation from a pneumatic or electric outside source. In addition, assemblies are *self-contained* (when they are furnished with all necessary controls for their operation, including actuators, regulators, motors, and thermostats), as opposed to *non-self-contained* assemblies (where part or all of the necessary controls for operation are furnished by someone other than the assembly manufacturer). In this latter case, the controls may be mounted on the assembly by the assembly manufacturer or may be mounted by others after delivery of the equipment.

The manual damper or flow rate controller within the box can be adjusted manually, automatically, or by a pneumatic or electric motor actuated by a thermostat or pressure regulator, all depending on the desired function of the box.

Air from the box may be discharged through a single rectangular opening suitable for low-pressure rigid branch duct connection or a supply outlet connected directly to the discharge end or bottom of the box. In addition, air may be discharged through several round outlets suitable for connection to flexible ducts.

### Reheat Boxes

Boxes of this type add sensible heat to the supply air. Water or steam coils or electric resistance heaters are located within or attached directly to the air discharge end of the box. These boxes typically are single duct; operation can be either constant or variable volume. However, if they are variable volume, they must maintain some minimum airflow to accomplish the reheat function. This type of equipment can accomplish localized individual reheat without a central equipment station or zone change.

### Dual-Duct Boxes

Dual-duct boxes are typically under control of a room thermostat. They receive warm and cold air from separate air-supply ducts in accordance with room requirements to obtain room control without zoning. Pneumatic and electric volume regulated boxes often have individual modulating dampers and operators to regulate the amount of warm and cool air. When a single modulating damper operator regulates the amount of both warm and cold air, a separate pressure-reducing damper or volume controller (either pneumatic or mechanical) is needed in the box to reduce pressure and limit airflow. Specially designed baffles may be required within the unit or at the box discharge to mix varying amounts of warm and cold air and/or to provide uniform flow downstream. Dual-duct boxes can be equipped with constant flow rate or variable flow rate devices to be either pressure independent or pressure dependent to provide a number of volume and temperature control functions.

### Ceiling Induction Boxes: Air to Air

The air to air ceiling induction box provides either primary air or a mixture of primary air and relatively warm air to the conditioned space. It accomplishes this function by permitting the primary air to induce air from the ceiling plenum or via ducted return air from individual rooms. A single duct supplies primary air at a temperature cool enough to satisfy all zone cooling loads. The ceiling plenum air induced into the primary air is at a higher temperature than the room because heat from recessed lighting fixtures enters the plenum directly.

The induction box contains damper assemblies controlled by an actuator in response to a thermostat to control the amount of cool primary air and warm induced air. As reduction in cooling is required, the primary air flow rate is gradually reduced and the induced air rate is generally increased.

Reheat coils can be used in the primary air supply and/or in the induction opening to meet occasional interior and perimeter load requirements. Either hot water or electric reheat coils may be used.

### Ceiling Induction Boxes: Fan Assisted

Fan assisted ceiling induction boxes differ from air-to-air induction boxes in that they are equipped with a blower. This blower, generally driven by a fractional horsepower motor, draws air from the space or ceiling plenum (secondary air) to be mixed with the cool primary air. The advantage of fan induction boxes over straight VAV boxes is that for a small energy expenditure to the terminal fan, constant air circulation can be maintained in the space. Fan induction boxes operate at lower primary air static pressure than air-to-air induction boxes, and perimeter zones can be heated without operating the primary fan during unoccupied periods. The warm air in the plenum can be used for low- to medium-heating loads (depending on construction of the building envelope). As the load increases, heating coils in the perimeter boxes can be activated to heat the recirculated plenum air to the necessary level. Fan assisted induction boxes can be divided into two categories: constant volume and bypass-type units.

Constant volume, fan assisted induction boxes are used when constant air circulation is desired in the space. The unit has two inlets—one for cool primary air from the central fan system and one for the secondary or plenum air. All air delivered to the space passes through the blower. The blower operates continuously whenever the primary air fan is on and can be cycled to deliver heat, as required, when the primary fan is off.

As the cooling load decreases, a damper throttles the amount of primary air delivered to the blower. The blower makes up for this reduction of primary air by drawing air in from the space or ceiling plenum through the return or secondary air opening.

Bypass-type, fan assisted induction boxes are called bypass boxes because the cool primary air bypasses the blower portion of the unit and is delivered directly to the space. The blower section draws in plenum air only and is mounted in parallel with the primary air damper. A *back draft* damper prevents primary air from flowing into the blower section when the blower is not energized. The blower in these units is generally energized after the damper in the primary air is partially or completely throttled. Some electronically controlled units gradually increase the fan speed as the primary air damper is throttled to maintain constant airflow, while permitting the fan to shut off when it is in the full cooling mode.

### System Static-Pressure Control

To prevent static-pressure imbalance in single-duct and dual-duct high-pressure systems, some control of static pressure is necessary in systems operating with variable air flow.

In single-duct systems, variations in static pressure can be limited by the following: (1) static-pressure controllers, operating dampers in the air-distribution system; (2) static-pressure controllers, operating inlet vane dampers on the fan; (3) zoning and changing air-supply temperature in response to static-pressure changes; (4) constant volume regulators within the terminal device; and (5) mechanical and electrical variable speed fan controls.

In dual-duct systems, the methods of controlling static pressure are as follows: (1) dampers operated by static-pressure regulators located at critical points in the air-distribution system; (2) static-pressure controllers, regulating warm and cold air temperatures to limit the variations in the airflow in individual ducts; and (3) constant-volume regulators in each individual air-mixing valve or acoustical terminal device.

## AIR CURTAINS

In its simplest application, an air curtain is a continuous broad stream of air circulated across a doorway of a conditioned space. It inhibits the penetration of unconditioned air into a conditioned space by forcing an air layer of predetermined thickness and velocity over the entire entrance. The layer moves at such a velocity and angle that the air that tries to penetrate the curtain is entrained or opposed. The air layer or jet can be redirected to compensate for pressure changes across the opening. If air is forced inward because of a difference in pressure, the jet can be redirected outward to equalize the pressure differential.

Both vertical flow (usually downward) and horizontal flow air curtains are available. The vertical flow type may have either a ducted or non-ducted return.

The energy requirements of an air curtain should be considered carefully. The effectiveness of an air curtain in preventing infiltration through an entrance generally ranges from 60 to 80%. The *effectiveness* is the comparison of infiltration rate or heat flux through an opening when using an air curtain as opposed to the transmission that would take place through a simple opening with no restriction.

Two important factors influence the pressure differential against which an air curtain must work the height of the structure and the structure's orientation. In high-rise structures, the practicality of using an air curtain depends mainly on the magnitude of the pressure differential caused by height or the stack effect. The orientation of the particular building and the location of adjacent buildings should also be studied and considered.

## REFERENCES

Hemphill, J.M. 1966. Ventilating Ceiling Application. *Heating, Piping and Air Conditioning.* May, p. 139.

Miller, P.L. and R.G. Nevins. 1969. Room Air Distribution with an Air Distributing Ceiling—Part II. ASHRAE *Transactions,* 75(I):118.

Nevins, R.G. and E.D. Ward. 1968. Room Air Distribution with an Air Distributing Ceiling. ASHRAE *Transactions,* 74(I):VI.2.1-VI.2.14.

# CHAPTER 3

# FANS



THE fan is an air pump that creates a pressure difference and causes airflow. The impeller does work on the air, imparting to it both static and kinetic energy, varying in proportion depending on the fan type.

Fan efficiency ratings are based on ideal conditions; some fans are now rated at more then 90% total efficiency. However, actual connections often make it impossible to achieve ideal efficiencies in the field.

## SYMBOLS AND DEFINITIONS

$A$ = fan outlet area, ft$^2$ (m$^2$)
$D$ = fan size or impeller diameter
$K$ = value for calculating system effect factors
$N$ = rotational speed, revolutions per minute (revolutions per second)
$Q$ = volume flow rate moved by the fan at fan inlet conditions, cfm (L/s)
$P_{tf}$ = fan total pressure rise: the fan total pressure at outlet minus the fan total pressure at inlet, in. of water (Pa)
$P_{vf}$ = fan velocity pressure: the pressure corresponding to the average velocity determined from the volume flow rate and fan outlet area, in. of water (Pa)
$P_{sf}$ = fan static pressure rise: the fan total pressure rise diminished by the fan velocity pressure, in. of water (Pa). The fan inlet velocity head is assumed equal to zero for fan rating purposes
$P_{sx}$ = static pressure at a given point, in. of water (Pa)
$P_{vx}$ = velocity pressure at a given point, in. of water (Pa)
$P_{tx}$ = total pressure at given point, in. of water (Pa)
$SEF$ = system effect factor, in. of water (Pa)
$V$ = fan inlet or outlet velocity, ft/min (m/s)
$W_o$ = power output of a fan: based on the fan volume flow rate and the fan total pressure, horsepower (W)
$W_i$ = power input to a fan: measured by power delivered to the fan shaft, horsepower (W)
$\eta_t$ = mechanical efficiency of a fan (or fan total efficiency): the ratio of power output to power input ($\eta_t = W_o/W_i$)
$\eta_s$ = static efficiency of a fan: the mechanical efficiency multiplied by the ratio of static pressure to fan total pressure, $\eta_s = (P_s/P_t)\eta_t$.
$\varrho$ = gas density, lb/ft$^3$ (kg/m$^3$)

The preparation of this chapter is assigned to TC 5.1, Fans.

## TYPES OF FANS

Fans are generally classified as centrifugal fans or axial flow fans according to the direction of airflow through the impeller. Figure 1 shows the general configuration of a centrifugal fan. A similar description of an axial flow fan is shown in Figure 2. There are a number of subdivisions of each of the general types. Table 1 compares typical characteristics of some of the most common fan types.

## PRINCIPLES OF OPERATION

All fans produce pressure by altering the velocity vector of the flow. The fans produce pressure and/or flow because the rotating blades of the impeller impart kinetic energy to the air by changing its velocity. Velocity change is the result of tangential and radial velocity components in the case of centrifugal fans, and of axial and tangential velocity components in the case of axial flow fans.

Centrifugal fan impellers produce pressure from (1) the centrifugal force created by rotating the air column enclosed be-

## Table 1 Types of Fans

| | TYPE | IMPELLER DESIGN | HOUSING DESIGN |
|---|---|---|---|
| CENTRIFUGAL FANS | AIRFOIL | 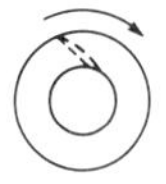 HIGHEST EFFICIENCY OF ALL CENTRIFUGAL FAN DESIGNS. 10 TO 16 BLADES OF AIRFOIL CONTOUR CURVED AWAY FROM THE DIRECTION OF ROTATION. AIR LEAVES THE IMPELLER AT A VELOCITY LESS THAN ITS TIP SPEED AND RELATIVELY DEEP BLADES PROVIDE FOR EFFICIENT EXPANSION WITHIN THE BLADE PASSAGES. FOR GIVEN DUTY THIS WILL BE THE HIGHEST SPEED OF THE CENTRIFUGAL-FAN DESIGNS. | 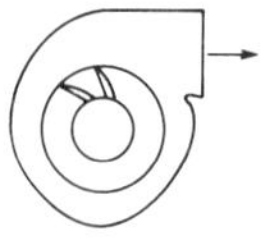 SCROLL-TYPE, USUALLY DESIGNED TO PERMIT EFFICIENT CONVERSION OF VELOCITY PRESSURE TO STATIC PRESSURE, THUS PERMITTING A HIGH STATIC EFFICIENCY; ESSENTIAL THAT CLEARANCE AND ALIGNMENT BETWEEN WHEEL AND INLET BELL BE VERY CLOSE IN ORDER TO REACH THE MAXIMUM EFFICIENCY CAPABILITY.<br>CONCENTRIC HOUSINGS CAN ALSO BE USED AS IN POWER ROOF VENTILATORS, SINCE THERE IS EFFICIENT PRESSURE CONVERSION IN THE WHEEL. |
| | BACKWARD-INCLINED BACKWARD-CURVED | 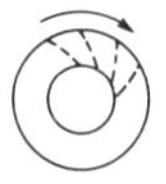 EFFICIENCY IS ONLY SLIGHTLY LESS THAN THAT OF AIRFOIL FANS. BACKWARD-INCLINED OR BACKWARD-CURVED BLADES ARE SINGLE THICKNESS. 10 TO 16 BLADES CURVED OR INCLINED AWAY FROM THE DIRECTION OF ROTATION. EFFICIENT FOR THE SAME REASONS GIVEN FOR THE AIRFOIL FAN ABOVE. | 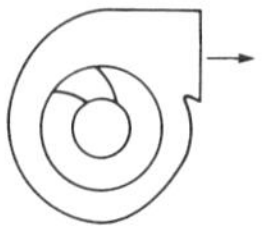 UTILIZES THE SAME HOUSING CONFIGURATION AS THE AIRFOIL DESIGN. |
| | RADIAL | 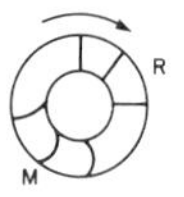 SIMPLEST OF ALL CENTRIFUGAL FANS AND LEAST EFFICIENT. HAS HIGH MECHANICAL STRENGTH AND THE WHEEL IS EASILY REPAIRED. FOR A GIVEN POINT OF RATING THIS FAN REQUIRES MEDIUM SPEED. THIS CLASSIFICATION INCLUDES RADIAL BLADES (R) AND MODIFIED RADIAL BLADES (M). USUALLY 6 TO 10 IN NUMBER. | 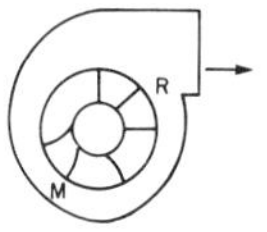 SCROLL-TYPE, USUALLY THE NARROWEST DESIGN OF ALL CENTRIFUGAL-FAN DESIGNS DESCRIBED HERE. BECAUSE OF LESS EFFICIENT WHEEL CAPABILITIES, DIMENSIONAL REQUIREMENTS OF THIS HOUSING ARE NOT AS CRITICAL AS FOR AIRFOIL AND BACKWARD-INCLINED FANS. |
| | FORWARD-CURVED | 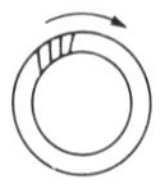 EFFICIENCY IS SOMEWHAT LESS THAN AIRFOIL AND BACKWARD-CURVED BLADED FANS. USUALLY FABRICATED OF LIGHTWEIGHT AND LOW COST CONSTRUCTION. HAS 24 TO 64 SHALLOW BLADES WITH BOTH THE HEEL AND TIP CURVED FORWARD. AIR LEAVES WHEEL AT VELOCITY GREATER THAN WHEEL TIP SPEED AND PRIMARY ENERGY TRANSFERRED TO THE AIR IS BY USE OF HIGH VELOCITY IN THE WHEEL. FOR GIVEN DUTY, WHEEL IS THE SMALLEST OF ALL CENTRIFUGAL TYPES AND OPERATES AT LOWEST SPEED. | 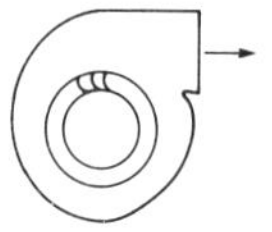 SCROLL IS SIMILAR TO AND OFTEN IDENTICAL TO OTHER CENTRIFUGAL-FAN DESIGNS. THE FIT BETWEEN THE WHEEL AND INLET IS NOT AS CRITICAL AS ON AIRFOIL AND BACKWARD-INCLINED BLADED FANS. |
| AXIAL FANS | PROPELLER | 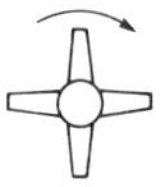 EFFICIENCY IS LOW. IMPELLERS ARE USUALLY OF INEXPENSIVE CONSTRUCTION AND LIMITED TO LOW-PRESSURE APPLICATIONS. IMPELLER IS OF 2 OR MORE BLADES, USUALLY OF SINGLE THICKNESS ATTACHED TO RELATIVELY SMALL HUB. ENERGY TRANSFER IS PRIMARILY IN FORM OF VELOCITY PRESSURE. | 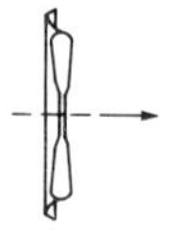 SIMPLE CIRCULAR RING, ORIFICE PLATE, OR VENTURI DESIGN. DESIGN CAN SUBSTANTIALLY INFLUENCE PERFORMANCE AND OPTIMUM DESIGN IS REASONABLY CLOSE TO THE BLADE TIPS AND FORMS A SMOOTH INLET FLOW CONTOUR TO THE WHEEL. |
| | TUBEAXIAL | 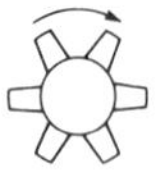 SOMEWHAT MORE EFFICIENT THAN PROPELLER-FAN DESIGN AND IS CAPABLE OF DEVELOPING A MORE USEFUL STATIC PRESSURE RANGE. NUMBER OF BLADES USUALLY FROM 4 TO 8 AND HUB IS USUALLY LESS THAN 50% OF FAN TIP DIAMETER, BLADES CAN BE OF AIRFOIL OR SINGLE THICKNESS CROSS SECTION. | 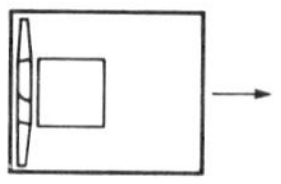 CYLINDRICAL TUBE FORMED SO THAT THE RUNNING CLEARANCE BETWEEN THE WHEEL TIP AND TUBE IS CLOSE. THIS RESULTS IN SIGNIFICANT IMPROVEMENT OVER PROPELLER FANS. |
| | VANEAXIAL | 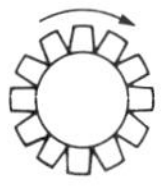 GOOD DESIGN OF BLADES PERMITS MEDIUM- TO HIGH-PRESSURE CAPABILITY AT GOOD EFFICIENCY. THE MOST EFFICIENT FANS OF THIS TYPE HAVE AIRFOIL BLADES. BLADES ARE FIXED, ADJUSTABLE, OR CONTROLLABLE PITCH TYPES AND HUB IS USUALLY GREATER THAN 50% OF FAN TIP DIAMETER. | 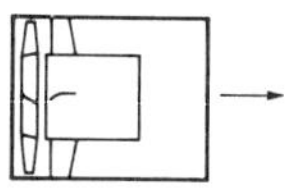 CYLINDRICAL TUBE CLOSELY FITTED TO THE OUTER DIAMETER OF BLADE TIPS AND FITTED WITH A SET OF GUIDE VANES, UPSTREAM OR DOWNSTREAM FROM THE IMPELLER. GUIDE VANES CONVERT THE ROTARY ENERGY IMPARTED TO THE AIR AND INCREASE PRESSURE AND EFFICIENCY OF FAN. |
| SPECIAL DESIGNS | TUBULAR CENTRIFUGAL | 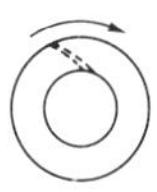 THIS FAN USUALLY HAS A WHEEL SIMILAR TO THE AIRFOIL. BACKWARD-INCLINED OR BACKWARD-CURVED BLADE AS DESCRIBED ABOVE. (HOWEVER THIS FAN WHEEL TYPE IS OF LOWER EFFICIENCY WHEN USED IN FAN OF THIS TYPE.) MIXED FLOW IMPELLERS ARE SOMETIMES USED. | 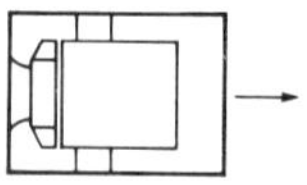 CYLINDRICAL SHELL SIMILAR TO A VANEAXIAL FAN HOUSING, EXCEPT THE OUTER DIAMETER OF THE WHEEL DOES NOT RUN CLOSE TO THE HOUSING. AIR IS DISCHARGED RADIALLY FROM THE WHEEL AND MUST CHANGE DIRECTION BY 90 DEG. TO FLOW THROUGH THE GUIDE VANE SECTION. |
| | POWER ROOF VENTILATORS CENTRIFUGAL | 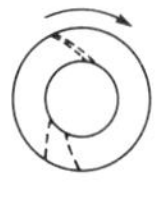 MANY MODELS USE AIRFOIL (A) OR BACKWARD-INCLINED (B) IMPELLER DESIGNS. THESE HAVE BEEN MODIFIED FROM THOSE MENTIONED ABOVE TO PRODUCE A LOW-PRESSURE, HIGH-VOLUME FLOW RATE CHARACTERISTIC. IN ADDITION, MANY SPECIAL CENTRIFUGAL IMPELLER DESIGNS ARE USED, INCLUDING MIXED-FLOW DESIGN. | 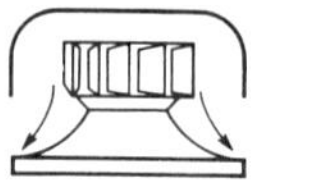 DOES NOT UTILIZE A HOUSING IN A NORMAL SENSE SINCE THE AIR IS SIMPLY DISCHARGED FROM THE IMPELLER IN A 360 DEG. PATTERN AND USUALLY DOES NOT INCLUDE A CONFIGURATION TO RECOVER THE VELOCITY PRESSURE COMPONENT. |
| | POWER ROOF VENTILATORS AXIAL |  A GREAT VARIETY OF PROPELLER DESIGNS ARE EMPLOYED WITH THE OBJECTIVE OF HIGH-VOLUME FLOW RATE AT LOW PRESSURE. | 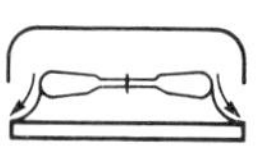 ESSENTIALLY A PROPELLER FAN MOUNTED IN A SUPPORTING STRUCTURE WITH A COVER FOR WEATHER PROTECTION AND SAFETY CONSIDERATIONS. THE AIR IS DISCHARGED THROUGH THE ANNULAR SPACE AROUND THE BOTTOM OF THE WEATHER HOOD. |

## Table 1 Types of Fans *(continued)*

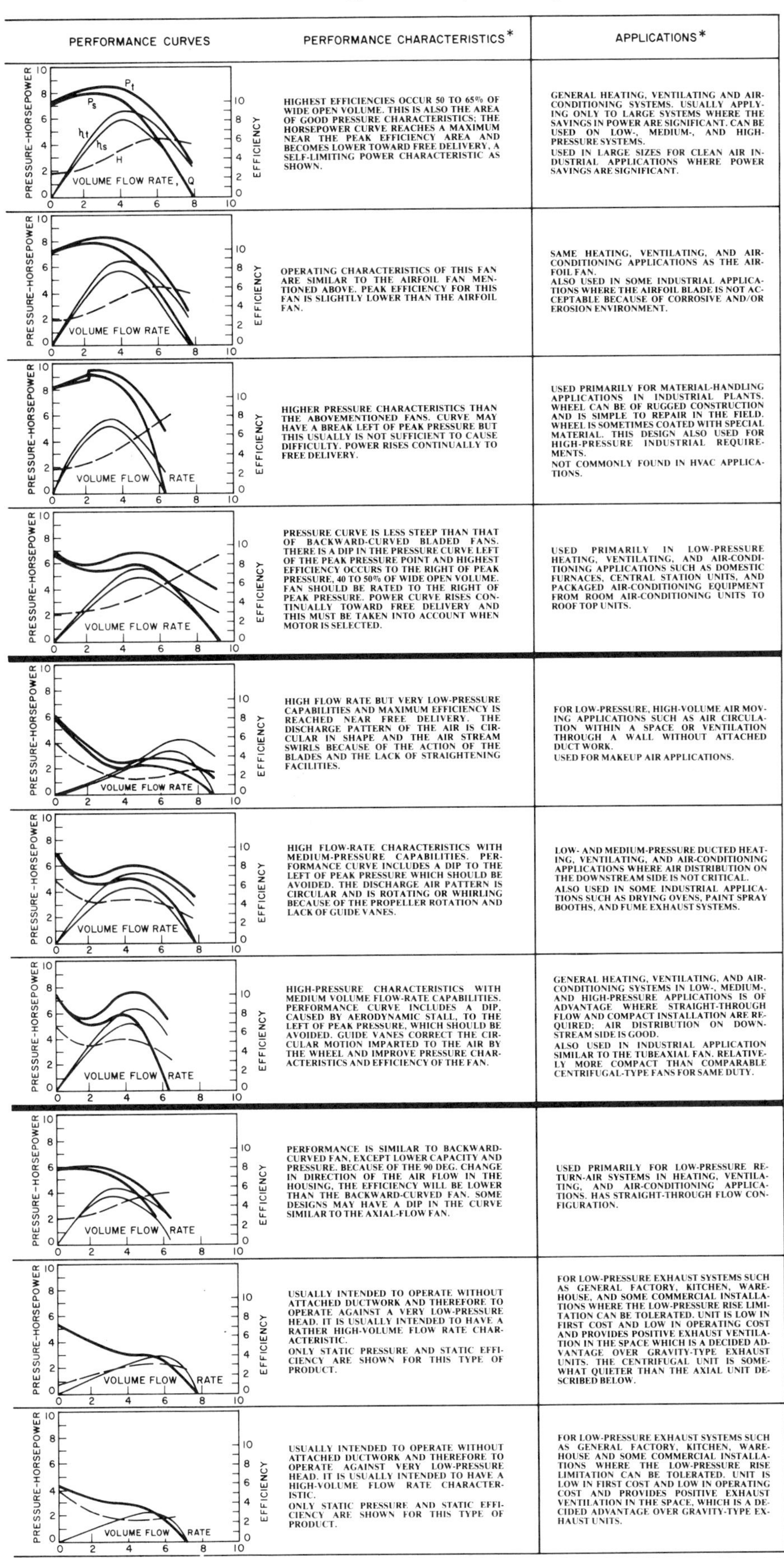

| PERFORMANCE CURVES | PERFORMANCE CHARACTERISTICS* | APPLICATIONS* |
|---|---|---|
| | HIGHEST EFFICIENCIES OCCUR 50 TO 65% OF WIDE OPEN VOLUME. THIS IS ALSO THE AREA OF GOOD PRESSURE CHARACTERISTICS; THE HORSEPOWER CURVE REACHES A MAXIMUM NEAR THE PEAK EFFICIENCY AREA AND BECOMES LOWER TOWARD FREE DELIVERY, A SELF-LIMITING POWER CHARACTERISTIC AS SHOWN. | GENERAL HEATING, VENTILATING AND AIR-CONDITIONING SYSTEMS. USUALLY APPLYING ONLY TO LARGE SYSTEMS WHERE THE SAVINGS IN POWER ARE SIGNIFICANT. CAN BE USED ON LOW-, MEDIUM-, AND HIGH-PRESSURE SYSTEMS.<br>USED IN LARGE SIZES FOR CLEAN AIR INDUSTRIAL APPLICATIONS WHERE POWER SAVINGS ARE SIGNIFICANT. |
| | OPERATING CHARACTERISTICS OF THIS FAN ARE SIMILAR TO THE AIRFOIL FAN MENTIONED ABOVE. PEAK EFFICIENCY FOR THIS FAN IS SLIGHTLY LOWER THAN THE AIRFOIL FAN. | SAME HEATING, VENTILATING, AND AIR-CONDITIONING APPLICATIONS AS THE AIRFOIL FAN.<br>ALSO USED IN SOME INDUSTRIAL APPLICATIONS WHERE THE AIRFOIL BLADE IS NOT ACCEPTABLE BECAUSE OF CORROSIVE AND/OR EROSION ENVIRONMENT. |
| | HIGHER PRESSURE CHARACTERISTICS THAN THE ABOVEMENTIONED FANS. CURVE MAY HAVE A BREAK LEFT OF PEAK PRESSURE BUT THIS USUALLY IS NOT SUFFICIENT TO CAUSE DIFFICULTY. POWER RISES CONTINUALLY TO FREE DELIVERY. | USED PRIMARILY FOR MATERIAL-HANDLING APPLICATIONS IN INDUSTRIAL PLANTS. WHEEL CAN BE OF RUGGED CONSTRUCTION AND IS SIMPLE TO REPAIR IN THE FIELD. WHEEL IS SOMETIMES COATED WITH SPECIAL MATERIAL. THIS DESIGN ALSO USED FOR HIGH-PRESSURE INDUSTRIAL REQUIREMENTS.<br>NOT COMMONLY FOUND IN HVAC APPLICATIONS. |
| | PRESSURE CURVE IS LESS STEEP THAN THAT OF BACKWARD-CURVED BLADED FANS. THERE IS A DIP IN THE PRESSURE CURVE LEFT OF THE PEAK PRESSURE POINT AND HIGHEST EFFICIENCY OCCURS TO THE RIGHT OF PEAK PRESSURE, 40 TO 50% OF WIDE OPEN VOLUME. FAN SHOULD BE RATED TO THE RIGHT OF PEAK PRESSURE. POWER CURVE RISES CONTINUALLY TOWARD FREE DELIVERY AND THIS MUST BE TAKEN INTO ACCOUNT WHEN MOTOR IS SELECTED. | USED PRIMARILY IN LOW-PRESSURE HEATING, VENTILATING, AND AIR-CONDITIONING APPLICATIONS SUCH AS DOMESTIC FURNACES, CENTRAL STATION UNITS, AND PACKAGED AIR-CONDITIONING EQUIPMENT FROM ROOM AIR-CONDITIONING UNITS TO ROOF TOP UNITS. |
| | HIGH FLOW RATE BUT VERY LOW-PRESSURE CAPABILITIES AND MAXIMUM EFFICIENCY IS REACHED NEAR FREE DELIVERY. THE DISCHARGE PATTERN OF THE AIR IS CIRCULAR IN SHAPE AND THE AIR STREAM SWIRLS BECAUSE OF THE ACTION OF THE BLADES AND THE LACK OF STRAIGHTENING FACILITIES. | FOR LOW-PRESSURE, HIGH-VOLUME AIR MOVING APPLICATIONS SUCH AS AIR CIRCULATION WITHIN A SPACE OR VENTILATION THROUGH A WALL WITHOUT ATTACHED DUCT WORK.<br>USED FOR MAKEUP AIR APPLICATIONS. |
| | HIGH FLOW-RATE CHARACTERISTICS WITH MEDIUM-PRESSURE CAPABILITIES. PERFORMANCE CURVE INCLUDES A DIP TO THE LEFT OF PEAK PRESSURE WHICH SHOULD BE AVOIDED. THE DISCHARGE AIR PATTERN IS CIRCULAR AND IS ROTATING OR WHIRLING BECAUSE OF THE PROPELLER ROTATION AND LACK OF GUIDE VANES. | LOW- AND MEDIUM-PRESSURE DUCTED HEATING, VENTILATING, AND AIR-CONDITIONING APPLICATIONS WHERE AIR DISTRIBUTION ON THE DOWNSTREAM SIDE IS NOT CRITICAL.<br>ALSO USED IN SOME INDUSTRIAL APPLICATIONS SUCH AS DRYING OVENS, PAINT SPRAY BOOTHS, AND FUME EXHAUST SYSTEMS. |
| | HIGH-PRESSURE CHARACTERISTICS WITH MEDIUM VOLUME FLOW-RATE CAPABILITIES. PERFORMANCE CURVE INCLUDES A DIP, CAUSED BY AERODYNAMIC STALL, TO THE LEFT OF PEAK PRESSURE, WHICH SHOULD BE AVOIDED. GUIDE VANES CORRECT THE CIRCULAR MOTION IMPARTED TO THE AIR BY THE WHEEL AND IMPROVE PRESSURE CHARACTERISTICS AND EFFICIENCY OF THE FAN. | GENERAL HEATING, VENTILATING, AND AIR-CONDITIONING SYSTEMS IN LOW-, MEDIUM-, AND HIGH-PRESSURE APPLICATIONS IS OF ADVANTAGE WHERE STRAIGHT-THROUGH FLOW AND COMPACT INSTALLATION ARE REQUIRED; AIR DISTRIBUTION ON DOWNSTREAM SIDE IS GOOD.<br>ALSO USED IN INDUSTRIAL APPLICATION SIMILAR TO THE TUBEAXIAL FAN. RELATIVELY MORE COMPACT THAN COMPARABLE CENTRIFUGAL-TYPE FANS FOR SAME DUTY. |
| | PERFORMANCE IS SIMILAR TO BACKWARD-CURVED FAN, EXCEPT LOWER CAPACITY AND PRESSURE. BECAUSE OF THE 90 DEG. CHANGE IN DIRECTION OF THE AIR FLOW IN THE HOUSING, THE EFFICIENCY WILL BE LOWER THAN THE BACKWARD-CURVED FAN. SOME DESIGNS MAY HAVE A DIP IN THE CURVE SIMILAR TO THE AXIAL-FLOW FAN. | USED PRIMARILY FOR LOW-PRESSURE RETURN-AIR SYSTEMS IN HEATING, VENTILATING, AND AIR-CONDITIONING APPLICATIONS. HAS STRAIGHT-THROUGH FLOW CONFIGURATION. |
| | USUALLY INTENDED TO OPERATE WITHOUT ATTACHED DUCTWORK AND THEREFORE TO OPERATE AGAINST A VERY LOW-PRESSURE HEAD. IT IS USUALLY INTENDED TO HAVE A RATHER HIGH-VOLUME FLOW RATE CHARACTERISTIC.<br>ONLY STATIC PRESSURE AND STATIC EFFICIENCY ARE SHOWN FOR THIS TYPE OF PRODUCT. | FOR LOW-PRESSURE EXHAUST SYSTEMS SUCH AS GENERAL FACTORY, KITCHEN, WAREHOUSE, AND SOME COMMERCIAL INSTALLATIONS WHERE THE LOW-PRESSURE RISE LIMITATION CAN BE TOLERATED. UNIT IS LOW IN FIRST COST AND LOW IN OPERATING COST AND PROVIDES POSITIVE EXHAUST VENTILATION IN THE SPACE WHICH IS A DECIDED ADVANTAGE OVER GRAVITY-TYPE EXHAUST UNITS. THE CENTRIFUGAL UNIT IS SOMEWHAT QUIETER THAN THE AXIAL UNIT DESCRIBED BELOW. |
| | USUALLY INTENDED TO OPERATE WITHOUT ATTACHED DUCTWORK AND THEREFORE TO OPERATE AGAINST VERY LOW-PRESSURE HEAD. IT IS USUALLY INTENDED TO HAVE A HIGH-VOLUME FLOW RATE CHARACTERISTIC.<br>ONLY STATIC PRESSURE AND STATIC EFFICIENCY ARE SHOWN FOR THIS TYPE OF PRODUCT. | FOR LOW-PRESSURE EXHAUST SYSTEMS SUCH AS GENERAL FACTORY, KITCHEN, WAREHOUSE AND SOME COMMERCIAL INSTALLATIONS WHERE THE LOW-PRESSURE RISE LIMITATION CAN BE TOLERATED. UNIT IS LOW IN FIRST COST AND LOW IN OPERATING COST AND PROVIDES POSITIVE EXHAUST VENTILATION IN THE SPACE, WHICH IS A DECIDED ADVANTAGE OVER GRAVITY-TYPE EXHAUST UNITS. |

*These performance curves reflect the general characteristics of various fans as commonly employed. They are not intended to provide complete selection criteria for application purposes, since other parameters, such as diameter and speed, are not defined.

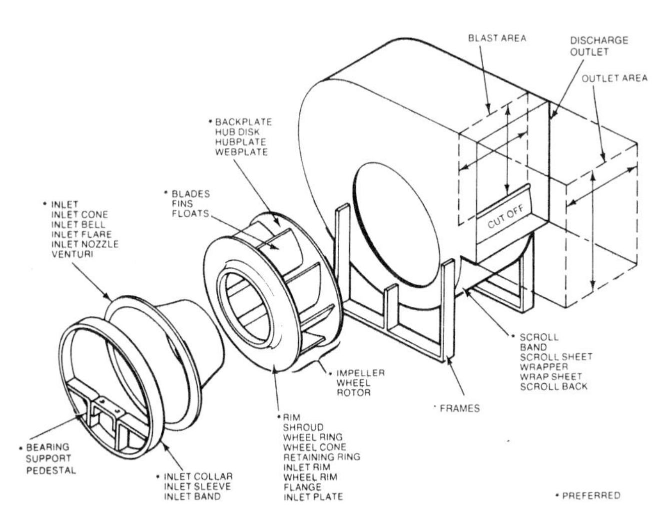


**Fig. 1 Centrifugal Fan Components**

tween the blades and (2) the kinetic energy imparted to the air by virtue of its velocity leaving the impeller. This velocity in turn is a combination of rotative velocity of the impeller and air speed relative to the impeller. When the blades are inclined forward, these two velocities are cumulative; when backward, oppositional. Backward-curved blade fans are generally more efficient than forward-curved blade fans.

Axial flow fans produce pressure from the change in velocity passing through the impeller, with none being produced by centrifugal force. These fans are divided into three types: propeller, tubeaxial, and vaneaxial. Propeller fans, customarily used at or near free air delivery, usually have a small hub-to-tip ratio impeller mounted in an orifice plate or inlet ring. Tubeaxial fans

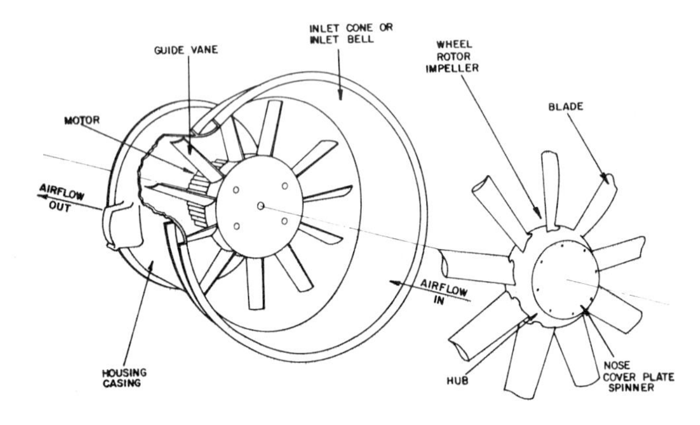


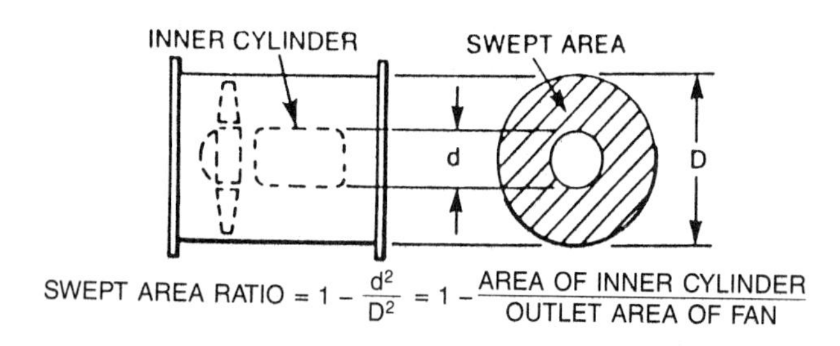


$$\text{SWEPT AREA RATIO} = 1 - \frac{d^2}{D^2} = 1 - \frac{\text{AREA OF INNER CYLINDER}}{\text{OUTLET AREA OF FAN}}$$

Note: The swept area ratio in axial fans is equivalent to the blast area ratio in centrifugal fans.

**Fig. 2 Axial Fan Components**

usually have reduced tip clearance and operate at higher tip speeds, which give them a higher total pressure capability than the propeller fan. Vaneaxial fans are essentially tubeaxial fans with guide vanes and reduced running blade tip clearance, which give improved pressure, efficiency, and noise characteristics.

Table 1 includes typical performance curves for various types of fans. These performance curves show the general characteristics of the various fans as they are normally used; they do not reflect the characteristics of these fans reduced to such common denominators as constant speed or constant propeller diameter, since fans are not selected on the basis of these constants. The efficiencies and power characteristics shown are general indications for each type of fan. A specific fan (size, speed) must be selected by evaluating actual characteristics.

## FAN TESTING AND RATING

Fans are tested in accordance with the strict requirements of ANSI/ASHRAE *Standard* 51-1985 and ANSI/AMCA *Standard* 210-85. The ASHRAE *Standard* specifies the procedures and test setups to be used in testing the various types of fans and other air-moving devices.

Figure 3 diagrams one of the most common procedures for developing the characteristics of a fan in which it is tested from *shutoff* conditions to nearly *free delivery* conditions. At shutoff, the duct is completely blanked off; at free delivery, the outlet resistance is reduced to zero. Between these two conditions, various flow restrictions are placed on the end of the duct to simulate various conditions on the fan. Sufficient points are obtained to define the curve between shutoff point and free delivery conditions. Pitot tube traverses of the test duct are performed with the fan operating at constant speed. The point of rating may be any point on the fan performance curve. For each case, the specific point on the curve must be defined by referring to the flow rate and the corresponding total pressure.

Other test setups, also described in AMCA *Standard* 210-85 and ASHRAE *Standard* 51-1985, should produce the same performance curve.

Fans designed for use with duct systems are tested with a length of duct between the fan and the measuring station. This length of duct smooths the flow of the fan and provides stable, uniform flow conditions at the plane of measurement. The measured pressures are corrected back to fan outlet conditions. Fans designed for use without ducts, including almost all propeller fans and power roof ventilators, are tested without ductwork.

Not all sizes are tested for rating. Test information may be used to calculate the performance of larger fans that are geometrically similar, but such information should not be extrapolated to smaller fans. For the performance of one fan to be determined from the known performance of another, the two fans must by dynamically similar. Strict dynamic similarity requires that the important nondimensional parameters vary in only insignificant ways. These nondimensional parameters include those that affect aerodynamic characteristics such as Mach number, Reynolds number, surface roughness, and gap size. (For more specific information, the manufacturer's application manual or engineering data should be consulted.)

## FAN LAWS

The *Fan Laws* (see Table 2) relate the performance variables for any dynamically similar series of fans. The variables are fan size, $D$; rotational speed, $N$; gas density, $\varrho$; volume flow rate, $Q$; pressure, $P_t$ or $P_s$; power, $W$: either air, $W_e$, or shaft, $W_i$; and mechanical efficiency, $\eta_t$. *Fan Law No. 1* shows the effect of changing size, speed, or density on volume flow, pressure,

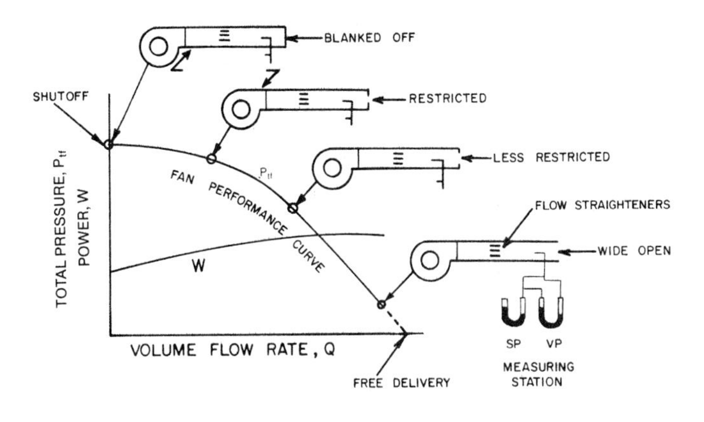


**Fig. 3 Method of Obtaining Fan Performance Curves**

and power level. *Fan Law No. 2* shows the effect of changing size, pressure, or density on volume flow rate, speed, and power. *Fan Law No. 3* shows the effect of changing size, volume flow, or density on speed, pressure, and power.

The Fan Laws apply only to a series of dynamically similar fans at the same point of rating on the performance curve. They can be used to predict the performance of any fan when test

**Table 2 Fan Laws[a,b,c]**

**For All Fan Laws: $\eta_{t1} = \eta_{t2}$ and (Pt. of Rtg.)$_1$ = (Pt. of Rtg.)$_2$**

| No. | Dependent Variables | Independent Variables |
|---|---|---|
| 1a | $Q_1 = Q_2$ | $\times \left(\frac{D_1}{D_2}\right)^3 \left(\frac{N_1}{N_2}\right)$ |
| 1b | $P_1 = P_2$ | $\times \left(\frac{D_1}{D_2}\right)^2 \left(\frac{N_1}{N_2}\right)^2 \frac{\varrho_1}{\varrho_2}$ |
| 1c | $W_1 = W_2$ | $\times \left(\frac{D_1}{D_2}\right)^5 \left(\frac{N_1}{N_2}\right)^3 \frac{\varrho_1}{\varrho_2}$ |
| 2a | $Q_1 = Q_2$ | $\times \left(\frac{D_1}{D_2}\right)^2 \left(\frac{P_1}{P_2}\right)^{1/2} \left(\frac{\varrho_2}{\varrho_1}\right)^{1/2}$ |
| 2b | $N_1 = N_2$ | $\times \left(\frac{D_2}{D_1}\right) \left(\frac{P_1}{P_2}\right)^{1/2} \left(\frac{\varrho_2}{\varrho_1}\right)^{1/2}$ |
| 2c | $W_1 = W_2$ | $\times \left(\frac{D_1}{D_2}\right)^2 \left(\frac{P_1}{P_2}\right)^{3/2} \left(\frac{\varrho_2}{\varrho_1}\right)^{1/2}$ |
| 3a | $N_1 = N_2$ | $\times \left(\frac{D_2}{D_1}\right)^3 \left(\frac{Q_1}{Q_2}\right)$ |
| 3b | $P_1 = P_2$ | $\times \left(\frac{D_2}{D_1}\right)^4 \left(\frac{Q_1}{Q_2}\right)^2 \frac{\varrho_1}{\varrho_2}$ |
| 3c | $W_1 = W_2$ | $\times \left(\frac{D_2}{D_1}\right)^4 \left(\frac{Q_1}{Q_2}\right)^3 \frac{\varrho_1}{\varrho_2}$ |

[a] The subscript 1 denotes the variable for the fan under consideration.
[b] The subscript 2 denotes the variable for the tested fan.
[c] $P$ = either $P_{tf}$ or $P_{sf}$

data are available for any fan of the same series. Fan Laws may also be used with a particular fan to determine the effect of speed change. However, caution should be exercised in these cases, since they apply only when *all* flow conditions are similar. Changing the speed of a given fan changes parameters that may invalidate the Fan Laws.

Unless otherwise identified, fan performance data are based on dry air at standard conditions—14.696 psi and 70°F (0.075 lb/ft$^3$) (101.325 kPa and 20°C [1.204 kg/m$^3$]). In actual applications, the fan may be required to handle air or gas at some other density. The change in density may be because of temperature, composition of the gas, or altitude. As indicated by the Fan Laws, the fan performance is affected by gas density. With constant size and speed, the horsepower and pressure vary in accordance with the ratio of gas density to the standard air density.

Figure 4 illustrates the application of the Fan Laws for a change in fan speed, $N$, for a specific size fan. The computed $P_t$ curve is derived from the base curve. For example, point E($N_1$ = 650) is computed from point D($N_2$ = 600) as follows:

At D,

$$Q_2 = 6000 \text{ and } P_{tf_2} = 1.13$$

Using Fan Law 1a at point E

$$Q_1 = 6000 \times 650/600 = 6500$$

Using Fan Law 1b

$$P_{tf_1} = 1.13 \times (650/600)^2 = 1.33$$

The completed total pressure curve, $P_{tf_1}$ at $N$ = 650 curve thus may be generated by computing additional points from data on the base curve, such as point G from point F.

If equivalent points of rating are joined, as shown by the dotted lines in Figure 4, these points will form parabolas which are defined by the relationship expressed in Eq. (1.)

Each point on the base $P_{tf}$ curve determines only one point on the computed curve. For example, point H cannot be calculated from either point D or point F. Point H is, however, related to some point between these two points on the base curve, and only that point can be used to locate point H. Furthermore, point D cannot be used to calculate point F on the base curve. The entire base curve must be defined by test.

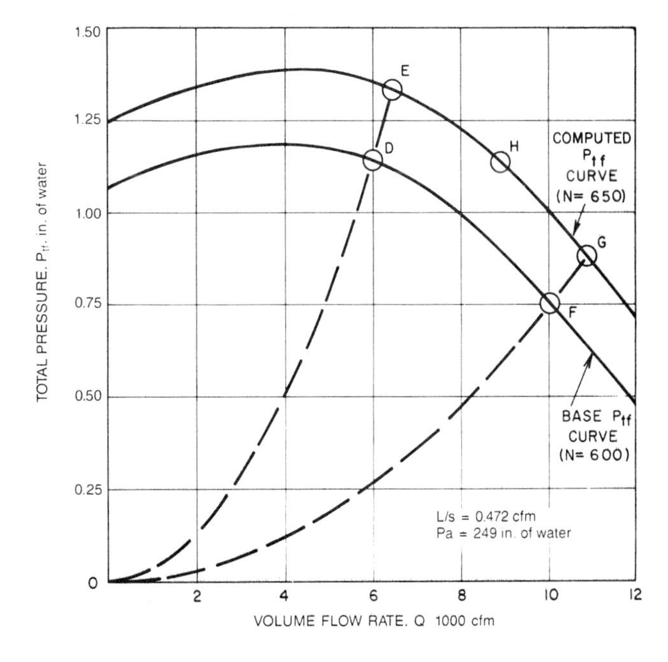


**Fig. 4 Example Application of the Fan Laws**

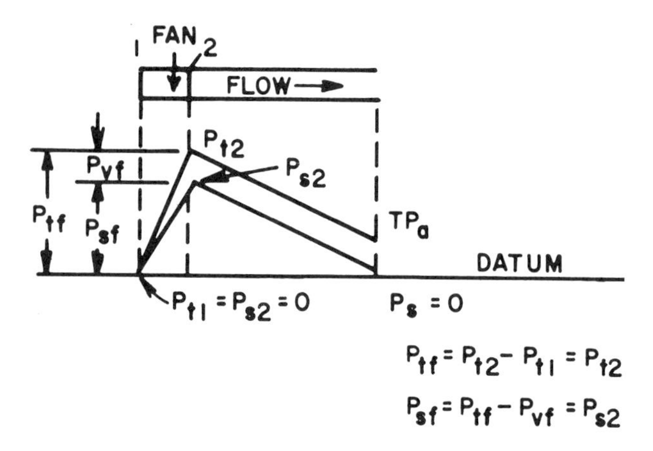


Fig. 5 Pressure Relationships of Fan with an Outlet System Only

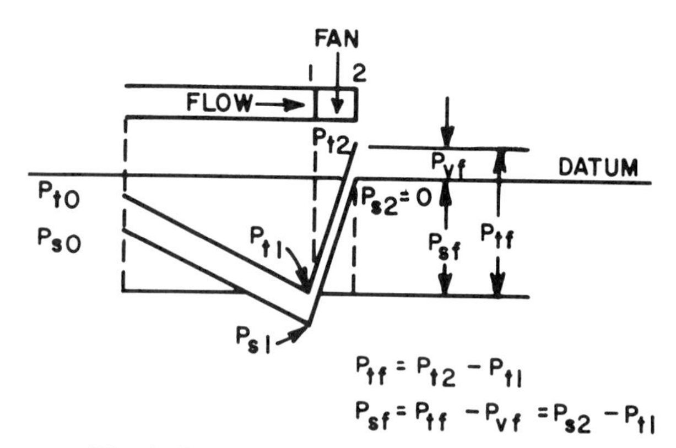


Fig. 6 Pressure Relationships of Fan with an Inlet System Only

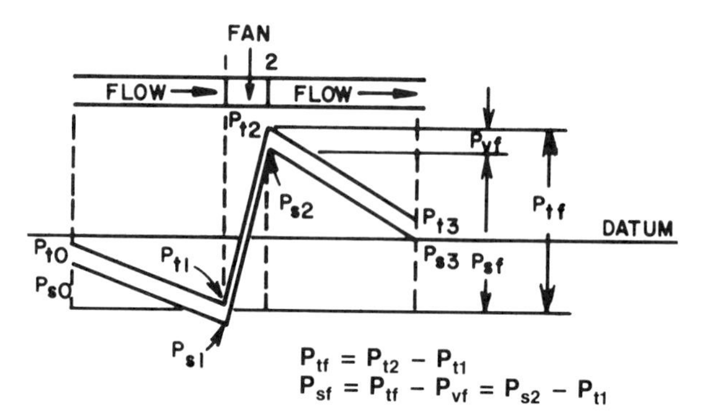


Fig. 7 Pressure Relationships of Fan with Equal-Sized Inlet and Outlet Systems

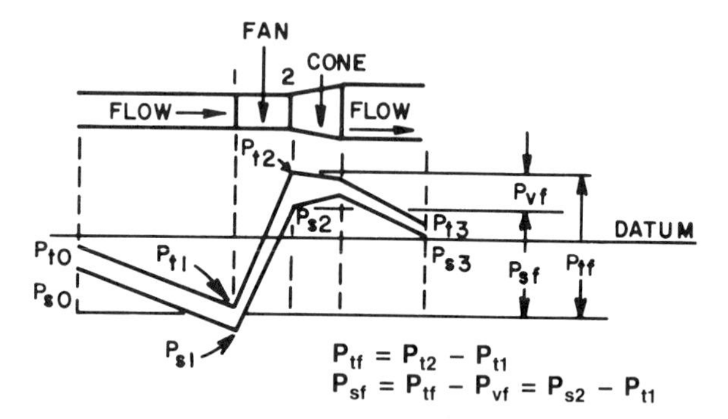


Fig. 8 Pressure Relationships of Fan with Diverging Cone Outlet

## FAN AND SYSTEM PRESSURE RELATIONSHIPS

As stated before, a fan impeller imparts static and kinetic energy to the air. This energy is represented in the increase in total pressure and can be converted to static or velocity pressure. These two quantities are interdependent; fan performance cannot be evaluated by considering one or the other alone. The conversion of energy, indicated by changes in velocity pressure to static pressure and vice versa, depends on the efficiency of conversion. Energy conversion occurs in the discharge duct connected to a fan being tested in accordance with AMCA *Standard* 210-85, and ASHRAE *Standard* 51-1985, and the efficiency is reflected in the rating.

Fan total pressure, $P_{tf}$, is a true indication of the energy imparted to the airstream by the fan. System pressure loss ($\Delta P$) is the sum of all individual total pressure losses imposed by the air distribution system duct elements on both the inlet and outlet sides of the fan. An energy loss in a duct system can be defined only as a total pressure loss. The measured static pressure loss in a duct element equals the total pressure loss only in the special case where air velocities are the same at both the entrance and exit of the duct element. By using total pressure for both fan selection and air distribution system design, the design engineer is assured of proper design. These fundamental principles apply to both high- and low-velocity systems. (Chapter 33 of the 1985 FUNDAMENTALS Volume has further information.)

Fan static pressure rise, $P_{sf}$, is often used in low-velocity ventilating systems where the fan outlet area essentially equals the fan outlet duct area, and little energy conversion occurs. When fan performance data are given in terms of $P_{sf}$, $P_{tf}$ may be calculated from the catalog data.

To specify the pressure performance of a fan, the relationship of $P_t$, $P_s$, and $P_v$ must be understood, expecially when negative pressures are involved. Most importantly, $P_s$ is a defined term in AMCA *Standard* 210 and ASHRAE *Standard* 51 as $P_s = P_t - P_v$. Except in special cases, $P_s$ is not necessarily the measured difference between static pressure on the inlet side and static pressure on the outlet side.

Figures 5 through 8 illustrate the relationships among these various pressures. Note that, as defined, $P_{tf} = P_{t2} - P_{t1}$. Figure 5 illustrates a fan with an outlet system but no connected inlet system. In this particular case, the fan static pressure ($P_{sf}$) equals the static pressure rise across the fan. Figure 6 shows a fan with an inlet system but no outlet system. Figure 7 shows a fan with both an inlet system and an outlet system. In both cases, the measured difference in static pressure across the fan ($P_{s2} - P_{s1}$) is not equal to the fan static pressure.

All of the systems illustrated in Figures 5 to 7 have inlet or outlet ducts that match the fan connections in size. Usually the duct size is not identical to the fan outlet or the fan inlet, so a further complication is introduced. To illustrate the pressure relationships in this case, Figure 8 shows a diverging outlet cone, which is a commonly used type of fan connection. In this case, the pressure relationships at the fan outlet do not match the pressure relationships in the flow section. Furthermore, the static pressure in the cone actually increases in the direction of flow. The static pressure changes throughout the system, depending on velocity. The total pressure, which, as noted in the sketch, decreases in the direction of flow, more truly represents the loss introduced by the cone or by flow in the duct. Only the fan changes this trend (*i.e.*, the decrease of total pressure in the direction of flow). Total pressure, therefore, is a better indication of fan and duct system performance. In this rather normal fan situation, the static pressure across the fan ($P_{s2} - P_{s1}$) does not equal the fan static pressure ($P_{sf}$).

## DUCT SYSTEM CHARACTERISTICS

Figure 9 shows a simplified duct system with three 90-deg elbows. These elbows represent the resistance offered by the ductwork, heat exchangers, cabinets, dampers, grilles, and other system components. A given rate of airflow through a system requires a definite total pressure in the system. If the rate of flow is changed, the resulting total pressure required will vary, as shown in Eq. (1), which is true for turbulent airflow systems. Heating, ventilating, and air-conditioning systems generally follow this law very closely.

$$(\Delta P_2/\Delta P_1) = (Q_2/Q_1)^2 \quad (1)$$

This chapter only covers turbulent flow—the flow regime in which most fans operate. In some systems, particularly constant or variable-volume air conditioning, the air-handling devices and associated controls may produce effective system resistance curves that deviate widely from Eq. (1), even though each element of the system may be described by this equation.

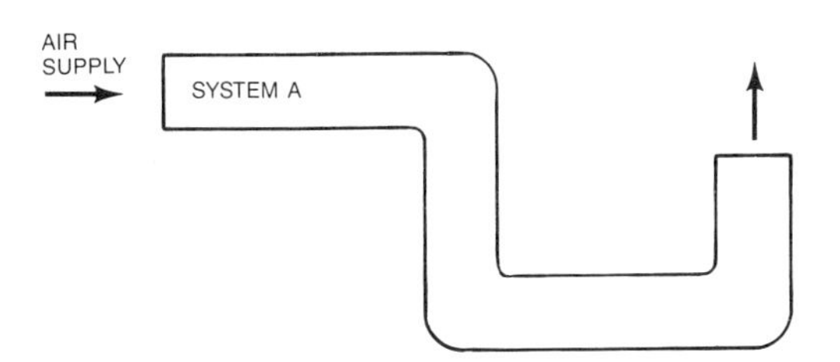


**Fig. 9 Simple Duct System with Resistance to Flow Represented by Three 90-deg Elbows**

Equation (1) permits plotting a turbulent flow system's pressure loss ($\Delta P$) curve from one known operating condition (see Figure 4). The fixed system must operate at some point on this system curve as the volume flow rate changes. As an example, at point A, curve A, Figure 10, when the flow rate through a duct system such as shown in Figure 9 is 10,000 cfm, the total pressure drop is 3 in. of water. If these values are substituted in Eq. (1) for $\Delta P_1$ and $Q_1$, other points of the system's $\Delta P$ curve (see Figure 10) can be determined.

For 6000 cfm (Point D on Figure 10):

$$\Delta P_2 = 3 \text{ in. of water} \left(\frac{6000}{10{,}000}\right)^2 = 1.08 \text{ in. of water}$$

If a change is made within the system so that the total pressure at design flow rate is increased, the system will no longer operate on the previous $\Delta P$ curve, and a new curve will be defined.

For example, in Figure 11, an additional elbow has been added to the duct system shown in Figure 9, which increases the total pressure of the system. If the total pressure at 10,000 cfm is increased by 1.00 in. of water, the system total pressure drop at this point will now be 4.00 in. of water, as shown by point B in Figure 10.

If the system in Figure 9 is changed by removing one of the schematic elbows (see Figure 12), the resulting system total pressure will drop below the total pressure resistance, and the new $\Delta P$ curve will be shown in Curve C of Figure 10. For curve C, a total pressure reduction of 1.00 in. of water has been assumed when 10,000 cfm flows through the system; thus, the point

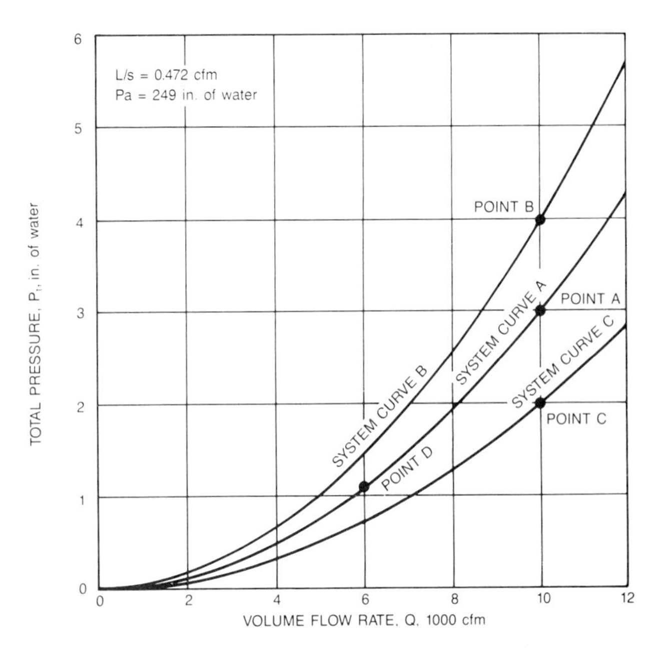


**Fig. 10 Example System Total Pressure Loss ($\Delta P$) Curves**

of operation will be at 2.00 in. of water, as shown by point C.

These three $\Delta P$ curves all follow the relationship expressed in Eq. (1). These curves result from changes within the system itself and do not change the fan performance. During the design phase, such system total pressure changes may occur because of studies of alternate duct routing, studies of differences in duct sizes, allowance for future duct extensions, or the effect of the design safety factor being applied to the system.

In an actual operating system, these three $\Delta P$ curves can represent three system characteristic lines caused by three different positions of a throttling control damper. Curve C is the most open position, and curve B is the most closed position of the three positions illustrated. A control damper forms a continuous series of these $\Delta P$ curves as it moves from a wide open position to a completely closed position and covers a much wider range of operation than is illustrated here. Such curves can also represent the clogging of turbulent flow filters in a system.

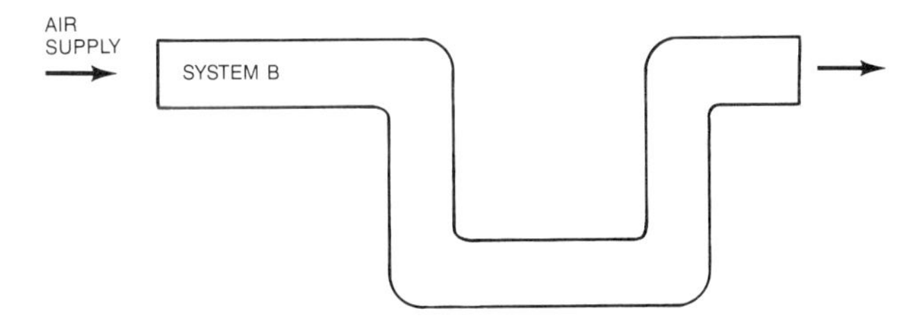


**Fig. 11 Resistance Added to Duct System of Figure 9**

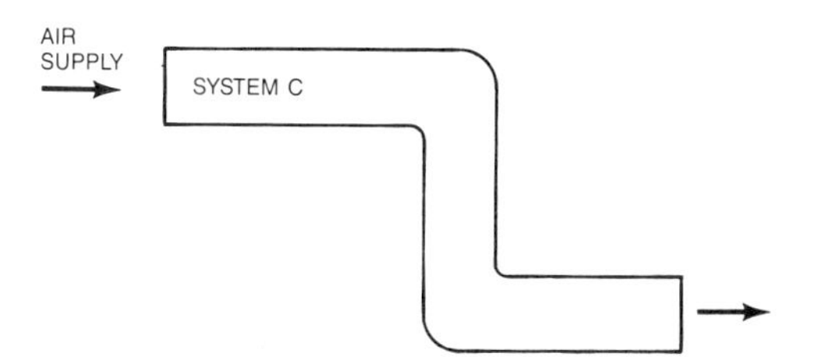


**Fig. 12 Resistance Subtracted from Duct System of Figure 9**

## SYSTEM EFFECTS

The following material on the effect of the system on fan performance is adapted from AMCA Publication 201, *Fans and Systems.* Clarke et al. (1978) also provide further information.

Normally, a fan is tested with open inlets and a section of straight duct is attached to the outlet. This setup results in uniform flow into the fan and efficient static pressure recovery on the fan outlet. If good inlet and outlet conditions are not provided in the actual installation, the performance of the fan will suffer. To select and apply the fan properly, these effects must be considered and the pressure requirements of the fan, as calculated by standard duct design procedures, must be increased. The following procedures may be used to identify these *system effect factors.*

These calculated system effect factors are only an approximation, however. Fans of different types and even fans of the same type, but supplied by different manufacturers, do not necessarily react to a system in the same way. Therefore, judgement based on experience must be applied to any design.

Figure 13 illustrates deficient fan/system performance. The system pressure losses have been determined accurately and a suitable fan selected for operation at Point 1. However, no allowance has been made for the effect of the system connections on the fan performance. To compensate, a system effect factor must be added to the calculated system pressure losses to determine the actual system curve.

The point of intersection between the fan performance curve and the actual system curve is Point 4. The actual flow volume is, therefore, deficient by the difference from 1 to 4. To achieve design flow volume, a system effect factor equal to the pressure difference between Point 1 and 2 should be added to the calculated system pressure losses and the fan selected to operate at Point 2.

The system effect factor is calculated by:

$$SEF = K\left(\frac{V}{4005}\right)^2 \qquad (2\ \text{I-P})$$

$$SEF = K\left(\frac{V}{1.29}\right)^2 \qquad (2\ \text{SI})$$

*where*

$SEF$ = system effect factor, in. of water (Pa)
$K$ = value from Figures 14 through 20

ACTUAL DUCT SYSTEM CURVE
CALCULATED DUCT SYSTEM CURVE WITH NO ALLOWANCE FOR SYSTEM EFFECT
SYSTEM EFFECT LOSS AT DESIGN VOLUME
SYSTEM EFFECT AT ACTUAL FLOW VOLUME
FAN CATALOG PRESSURE-VOLUME CURVE
DEFICIENT PERFORMANCE
DESIGN PRESSURE
DESIGN VOLUME
1 2 3 4

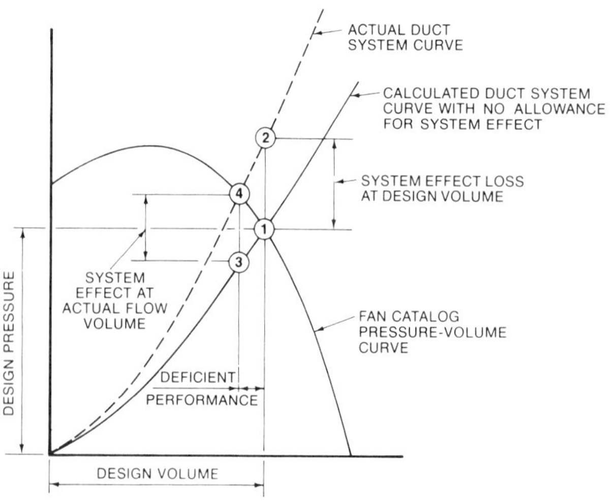

**Fig. 13 Deficient Duct System Performance System Effect Ignored**

(Reprinted by permission from AMCA Publication 201)

$V$ for Centrifugal Fans
$V$ = fan inlet velocity based on area at the inlet collar (see Figure 1), fpm (m/s)
$V$ = fan outlet velocity based on outlet area (see Figure 1), fpm (m/s)

$V$ for Axial Fans
$V$ = fan inlet or velocity based on area calculated from fan diameter, D (see Figure 2), fpm (m/s)

If more than one configuration is included in a system, the system effect factor for each must be determined separately, and the total of these system effects must be added to the total system pressure losses.

### Fan Outlet Conditions

Fans intended primarily for use with duct systems are usually tested with an outlet duct in place. In most cases, the fan manufacturer does not supply this duct as part of the fan, but rated performance will not be achieved unless a comparable duct is included in the system design.

Figure 14 shows the changes in velocity profiles at various distances from the fan outlet. For 100% recovery, the duct, including the transition, must meet the requirements for 100% effective duct length as specified in Figure 14.

If a full length outlet duct cannot be used, a system effect factor calculated with the K values listed in Table 3 must be added to the system resistance losses.

Often, an elbow must be installed near the fan discharge. To obtain the rated performance from the fan, the first elbow should be at 100% effective duct length from the fan outlet. If this length cannot be provided, a system effect factor must be added to the fan total pressure requirements. The system effect factor may be determined by using the K values in Table 4 and the elbow orientations shown in Figure 15.

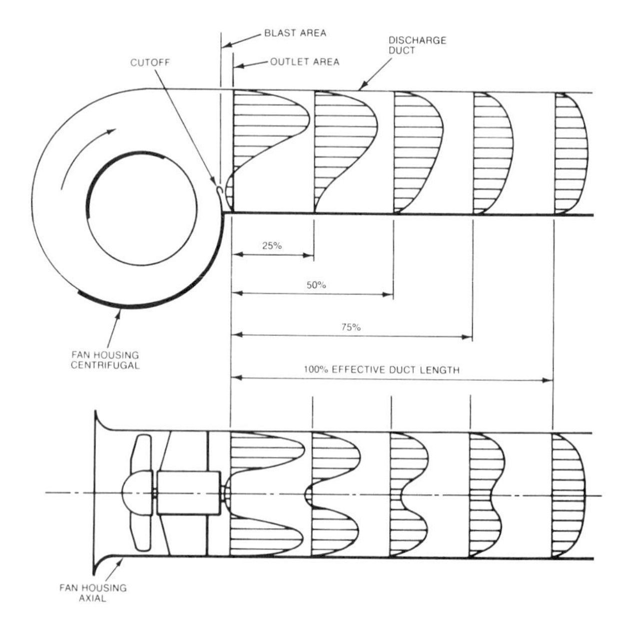


Table 3 shows K Values.
Assume Effective Duct Length is a minimum of 2.5 duct diameters for each 2500 fpm (13 m/s) or less. Add 1 duct diameter for each additional 1000 fpm (5 m/s).

*For example,* 5000 fpm = 5 Equivalent Duct Diameters.

The Equivalent Duct Diameter of a rectangular duct with side dimensions $a$ and $b$ is $\sqrt{4ab/\pi}$

**Fig. 14 Controlled Diffusion and Establishment of a Uniform Velocity Profile in a Straight Outlet Duct**

(Adapted by permission from AMCA Publication 201)

**Table 3 K Values for Outlet Ducts**

| | No Duct | 12% Effective Duct | 25% Effective Duct | 50% Effective Duct | 100% Effective Duct |
|---|---|---|---|---|---|
| Pressure Recovery | 0% | 50% | 80% | 90% | 100% |
| Blast Area / Outlet Area | K Values | | | | |
| 0.4 | 2.0 | 1.00 | 0.40 | 0.18 | No System Effect Factor |
| 0.5 | 2.0 | 1.00 | 0.40 | 0.18 | |
| 0.6 | 1.00 | 0.66 | 0.33 | 0.14 | |
| 0.7 | 0.80 | 0.40 | 0.14 | — | |
| 0.8 | 0.47 | 0.22 | 0.10 | — | |
| 0.9 | 0.22 | 0.14 | — | — | |
| 1.0 | — | — | — | — | |

Fan outlets are sometimes connected directly to a larger duct or a large plenum, without an intermediate transition. This connection causes a pressure loss of up to one velocity head using the highest velocity in the fan outlet. This highest velocity is in the *blast area* (which is the area between the cutoff and the scroll of a centrifugal fan) and in the swept area of a vaneaxial fan. If the average outlet velocity is used, based on the outlet flange area, the pressure loss may exceed one velocity head. A duct of the same size as the fan outlet between the fan outlet and plenum, and one duct diameter or more in length, will reduce this loss materially.

A diffuser with the same inlet size as the fan outlet may be used to connect the fan to the plenum. In this case, the losses must be calculated according to the fan manufacturer's recommendations, since they depend substantially on fan design. An arrangement in which a fan discharges through a coil having a face area considerably greater than the fan outlet area presents an especially difficult problem. To improve the velocity distribution across the coil face, a perforated baffle plate is sometimes located between the fan outlet and coil. This type of baffle causes an increase in system resistance that is almost impossible to predict.

**Table 4 K Values for Single Width Single Inlet (SWSI)[a]**

| Blast Area / Outlet Area | Outlet Elbow Position | No Outlet Duct | 12% Effective Duct | 25% Effective Duct | 50% Effective Duct | 100% Effective Duct |
|---|---|---|---|---|---|---|
| 0.4 | A | 3.20 | 2.68 | 1.76 | 0.84 | No System Effect Factor |
| | B | 4.00 | 3.32 | 2.24 | 1.04 | |
| | C | 5.80 | 4.84 | 3.20 | 1.52 | |
| | D | 5.80 | 4.84 | 3.20 | 1.52 | |
| 0.5 | A | 2.28 | 1.88 | 1.28 | 0.60 | |
| | B | 2.84 | 2.36 | 1.56 | 0.72 | |
| | C | 4.00 | 3.32 | 2.24 | 1.04 | |
| | D | 4.00 | 3.32 | 2.24 | 1.04 | |
| 0.6 | A | 1.60 | 1.32 | 0.88 | 0.40 | |
| | B | 2.00 | 1.68 | 1.12 | 0.52 | |
| | C | 2.92 | 2.44 | 1.64 | 0.76 | |
| | D | 2.92 | 2.44 | 1.64 | 0.76 | |
| 0.7 | A | 1.08 | 0.88 | 0.60 | 0.28 | |
| | B | 1.32 | 1.08 | 0.72 | 0.36 | |
| | C | 1.96 | 1.64 | 1.08 | 0.52 | |
| | D | 1.96 | 1.64 | 1.08 | 0.52 | |
| 0.8 | A | 0.76 | 0.64 | 0.44 | 0.20 | |
| | B | 0.96 | 0.80 | 0.52 | 0.24 | |
| | C | 1.40 | 1.16 | 0.76 | 0.36 | |
| | D | 1.40 | 1.16 | 0.76 | 0.36 | |
| 0.9 | A | 0.60 | 0.48 | 0.32 | 0.16 | |
| | B | 0.76 | 0.64 | 0.44 | 0.20 | |
| | C | 1.12 | 0.92 | 0.64 | 0.28 | |
| | D | 1.12 | 0.92 | 0.64 | 0.28 | |
| 1.0 | A | 0.56 | 0.48 | 0.32 | 0.16 | |
| | B | 0.68 | 0.56 | 0.36 | 0.16 | |
| | C | 1.00 | 0.84 | 0.56 | 0.28 | |
| | D | 1.00 | 0.84 | 0.56 | 0.28 | |

[a]For Double Width Double Inlet (DWDI) fans, determine the K value from the above table for SWSI fans and calculate the system effect factor (SEF) using Equation 2, then apply the multiplier from the table below:

**Multipliers for DWDI Fans**
Elbow position A = 1.00 *SEF*
Elbow position B = 1.25 *SEF*
Elbow position C = 1.00 *SEF*
Elbow position D = 0.85 *SEF*

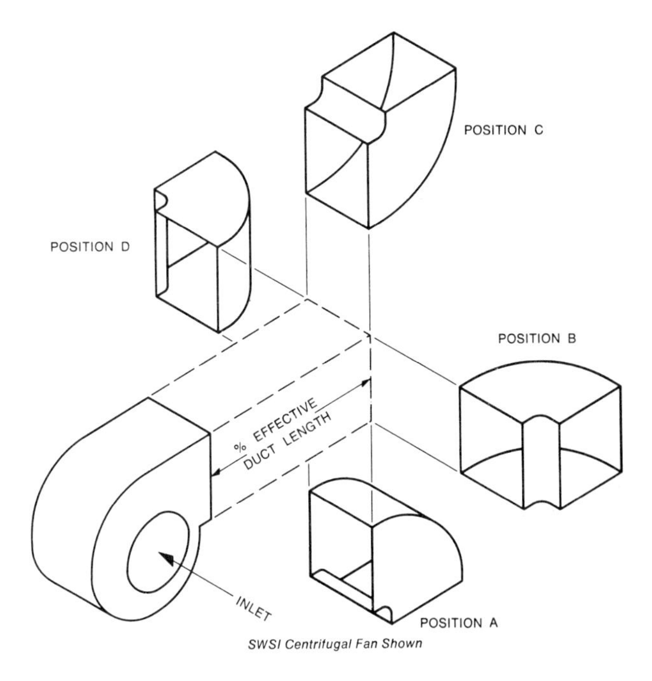


Note: Fan inlet and elbow positions must be oriented as shown for proper application of System Effect. Table 4 shows the K Values.

**Fig. 15 Outlet Duct Elbows**
(Reprinted by permission from AMCA Publication 201)

## Fan Inlet Conditions

For rated performance, the air must enter the fan uniformly over the inlet area in an axial direction without prerotation. Nonuniform flow into the inlet is the most common cause of reduced fan performance.

Such inlet conditions are not equivalent to a simple increase in the system resistance; therefore, they cannot be treated as a percentage decrease in the flow and pressure from the fan. A poor inlet condition results in an entirely new fan performance. Many poor inlet conditions affect the fan more at near-free delivery conditions than at peak pressure, so there is a continually varying difference between these two points. An elbow at the fan inlet, for example, causes turbulence and uneven flow into the fan impeller. Figures 16, 17, and 18 diagram several elbow arrangements and the corresponding K values for calculating the system effect factors. The system effect factor can be eliminated by including an adequate length of straight duct between the elbow and the fan inlet.

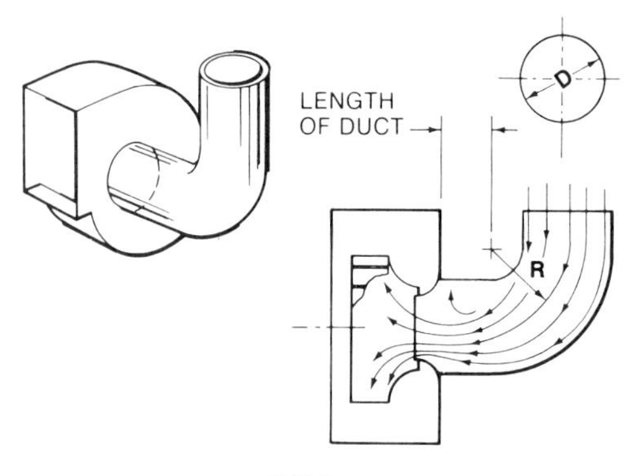


| | K Values | | |
|---|---|---|---|
| R/D | No Duct | 2D Duct | 5D Duct |
| 0.75 | 1.4 | 0.80 | 0.40 |
| 1.00 | 1.2 | 0.66 | 0.33 |
| 2.00 | 1.0 | 0.53 | 0.33 |
| 3.00 | 0.66 | 0.40 | 0.22 |

**Fig. 16 Nonuniform Flow into a Fan Inlet Induced by a 90° Round Section Elbow Without Turning Vanes**
(Adapted by permission from AMCA Publication 201)

The ideal inlet condition allows air to enter axially and uniformly without spin. A spin in the same direction as the impeller rotation reduces the pressure-volume curve by an amount dependent upon the intensity of the vortex. A counter-rotating

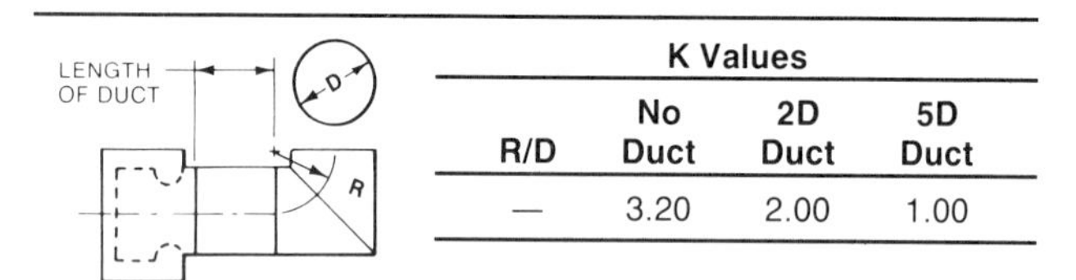


| | K Values | | |
|---|---|---|---|
| R/D | No Duct | 2D Duct | 5D Duct |
| — | 3.20 | 2.00 | 1.00 |

**Two Piece Mitered 90° Round Section Elbow—Not Vaned**

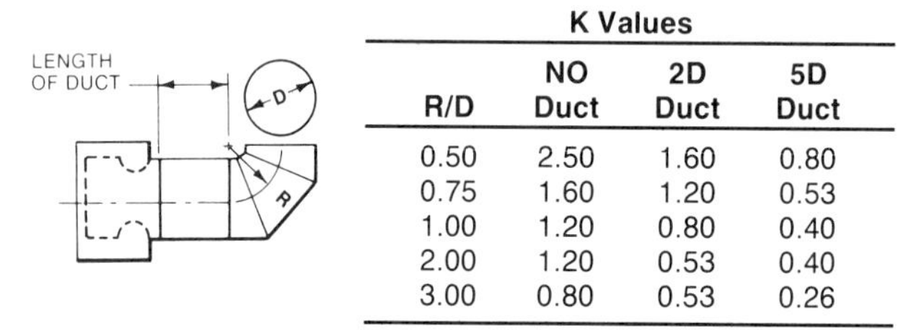


| | K Values | | |
|---|---|---|---|
| R/D | NO Duct | 2D Duct | 5D Duct |
| 0.50 | 2.50 | 1.60 | 0.80 |
| 0.75 | 1.60 | 1.20 | 0.53 |
| 1.00 | 1.20 | 0.80 | 0.40 |
| 2.00 | 1.20 | 0.53 | 0.40 |
| 3.00 | 0.80 | 0.53 | 0.26 |

**Three Piece Mitered 90° Round Section Elbow—Not Vaned**

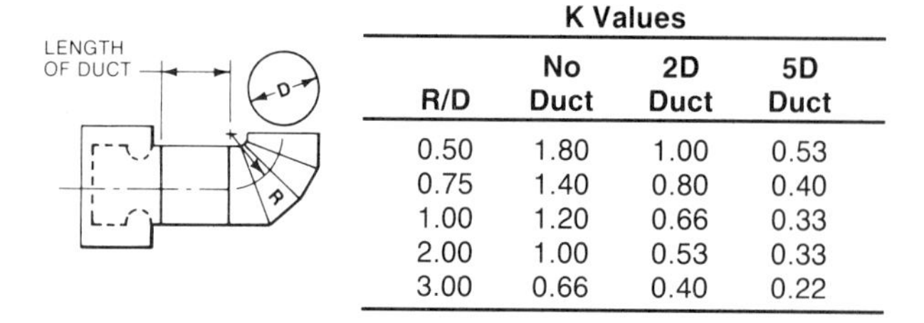


| | K Values | | |
|---|---|---|---|
| R/D | No Duct | 2D Duct | 5D Duct |
| 0.50 | 1.80 | 1.00 | 0.53 |
| 0.75 | 1.40 | 0.80 | 0.40 |
| 1.00 | 1.20 | 0.66 | 0.33 |
| 2.00 | 1.00 | 0.53 | 0.33 |
| 3.00 | 0.66 | 0.40 | 0.22 |

**Four or More Piece Mitered 90° Round Section Elbow—Not Vaned**

**Fig. 17 K Values for Various Mitered Elbows Without Turning Vanes**
(Adapted by permission from AMCA Publication 201)

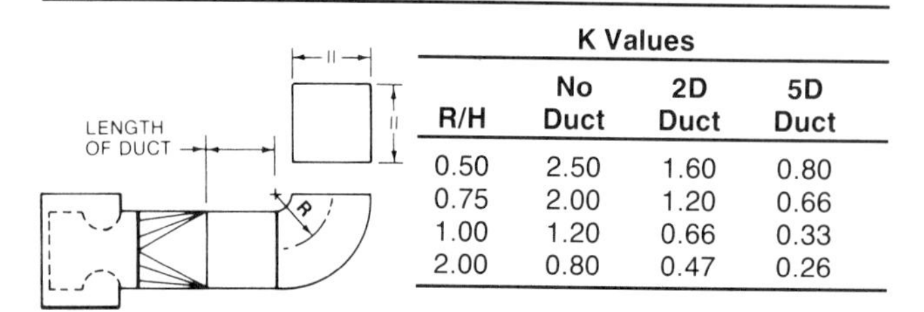


| | K Values | | |
|---|---|---|---|
| R/H | No Duct | 2D Duct | 5D Duct |
| 0.50 | 2.50 | 1.60 | 0.80 |
| 0.75 | 2.00 | 1.20 | 0.66 |
| 1.00 | 1.20 | 0.66 | 0.33 |
| 2.00 | 0.80 | 0.47 | 0.26 |

**Square Elbow With Inlet Transtion—No Turning Vanes**

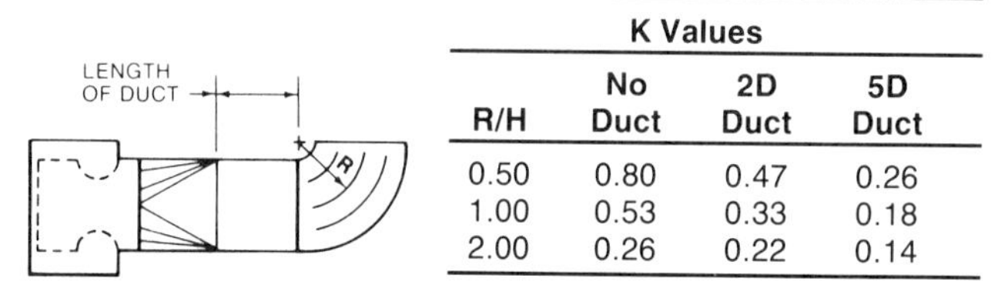


| | K Values | | |
|---|---|---|---|
| R/H | No Duct | 2D Duct | 5D Duct |
| 0.50 | 0.80 | 0.47 | 0.26 |
| 1.00 | 0.53 | 0.33 | 0.18 |
| 2.00 | 0.26 | 0.22 | 0.14 |

**Square Elbow With Inlet Transition—3 Long Turning Vanes**

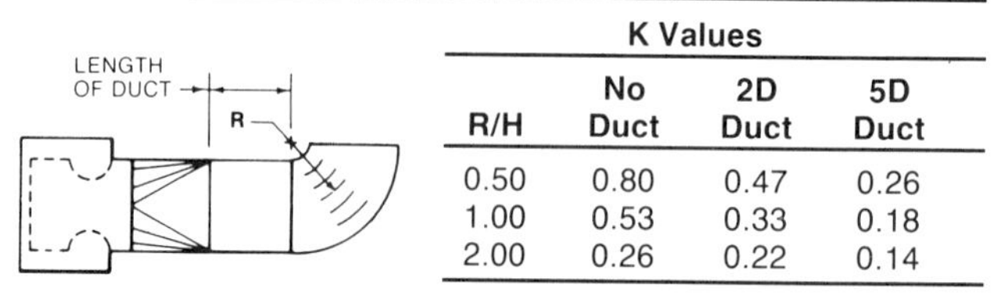


| | K Values | | |
|---|---|---|---|
| R/H | No Duct | 2D Duct | 5D Duct |
| 0.50 | 0.80 | 0.47 | 0.26 |
| 1.00 | 0.53 | 0.33 | 0.18 |
| 2.00 | 0.26 | 0.22 | 0.14 |

**Square Elbow With Inlet Transition—Short Turning Vanes**

$D = 2H/\sqrt{\pi}$

The inside area of the square duct (H × H) is equal to the inside area circumscribed by the fan inlet collar. The maximum permissible angle of any converging element of the transition is 15°, and for a diverging element 7.5°.

**Fig. 18 K Values for Various Duct Elbows**
(Adapted by permission from AMCA Publication 201)

vortex at the inlet slightly increases the pressure-volume curve, but it increases the horsepower substantially.

Inlet spin may arise from a great variety of approach conditions, and sometimes the cause is not obvious. Figure 19 illustrates some common duct connections that cause inlet spin and includes recommendations for correcting the spin. Since the variations are many, no system effect factors are tabulated.

Fans within plenums and cabinets or next to walls should be located so that air may flow unobstructed into the inlets. Fan performance is reduced if the space between the fan inlet and the enclosure is too restrictive. Figure 20 illustrates fans located in an enclosure or adjacent to walls and lists the K values for restricted inlets.

The manner in which the airstream enters an enclosure in relation to the fan inlets also affects fan performance. Plenum or enclosure inlets or walls that are not symmetrical with the fan inlets cause uneven flow and/or inlet spin. Figure 21 illustrates this condition, which must be avoided to achieve maximum performance from a fan. If this is not possible, inlet conditions can usually be improved with a splitter to break the inlet vortex, as illustrated in Figure 21.

## FAN SELECTION

After the system pressure loss curve of the air distribution system has been defined as described above, a fan can be selected to meet the requirements of the system (Graham 1966, 1977).

Fan manufacturers present performance data in either the graphic (curve) form (see Figure 22) or the tabular form (multirating tables). Multirating tables usually provide only perfor-

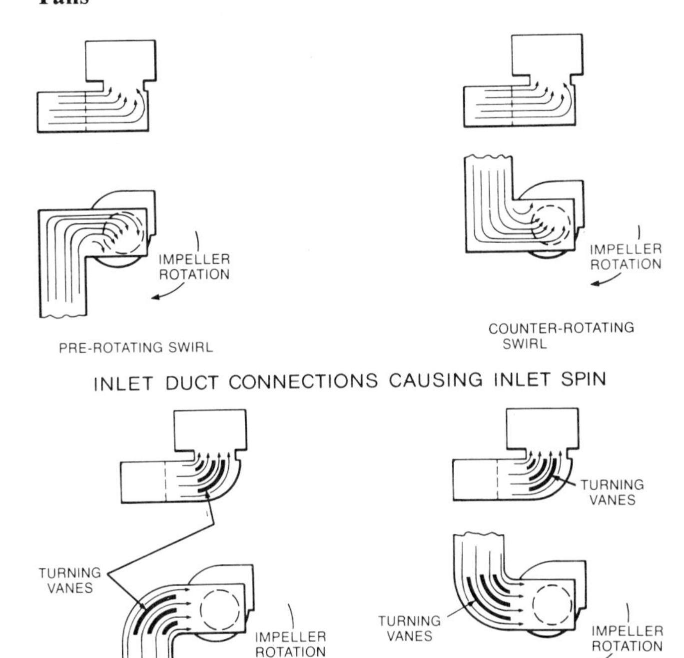


**Fig. 19 Inlet Duct Connections Causing Inlet Spin and Corrections for Inlet Spin**

(Adapted by permission from AMCA Publication 201)

mance data within the recommended operating range. The optimum selection area or peak efficiency point is identified in various ways by different manufacturers.

Performance data as tabulated in the usual fan tables are based on arbitrary increments of flow rate and pressure. In these tables, adjacent data, either horizontally or vertically, represent different points of operation (*i.e.*, different points of rating) on the fan performance curve. These points of rating depend solely on the fan's characteristics; they cannot be obtained one from the other by the Fan Laws. However, these points of operation listed in the multirating tables are usually close together, so immediate points may be interpolated arithmetically with adequate accuracy for fan selection.

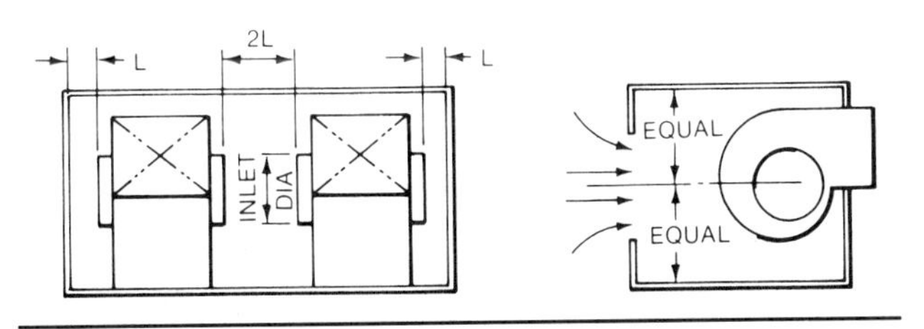


| Distance of Inlet to Wall, L | K Values |
|---|---|
| 0.75 Dia. of Inlet | 0.015 |
| 0.5 Dia. of Inlet | 0.025 |
| 0.4 Dia. of Inlet | 0.03 |
| 0.3 Dia. of Inlet | 0.05 |
| 0.2 Dia. of Inlet | 0.08 |

**Fig. 20 K Values for Fans Located in Plenums and Cabinet Enclosures**

(Adapted by permission from AMCA Publication 201)

The selection of a fan for a particular air-distribution system requires that the fan pressure characteristics fit the system pressure characteristics. Thus, the total system must be evaluated and the flow requirements, resistances, and system effect factors existing at the fan inlet and outlet must be known (see Chapter 33 of the 1985 FUNDAMENTALS Volume).

Fan speed and power requirements are then calculated, using one of the many methods available from fan manufacturers. These may consist of the multirating tables or of single or multispeed performance curves or graphs.

In using curves, it is necessary that the point of operation selected (see Figure 23) represent a desirable point on the fan curve, so maximum efficiency and resistance to stall and pulsation can be attained. On systems where more than one point of operation is encountered during operation, it is necessary to look at the range of performance and evaluate how the selected fan reacts within this complete range. This analysis is particularly necessary for variable-volume systems, where not only the fan undergoes a change in performance, but the entire system deviates from the relationships defined in Eq. (1). In these cases, it is necessary to look at actual losses in the system at performance extremes.

## PARALLEL FAN OPERATION

The combined performance curve for two fans operating in parallel may be plotted by using the appropriate pressure for the ordinates and the sum of the volumes for the abscissas. When two fans having a pressure reduction to the left of the peak pressure point are operated in parallel, a fluctuating load condition may result if one of the fans operates to the left of the peak static point on its performance curve.

The $P_t$ curves of a single fan and two identical fans operating in parallel are shown in Figure 24. Curve A-A shows the pressure characteristics of a single fan. Curve C-C is the combined performance of the two fans. The unique figure-8 shape is a plot of all possible combinations of volume flow at each pressure value for the individual fans. All points to the right of CD are the result of each fan operating at the right of its peak point of rating. Stable performance results for all systems with less obstruction to airflow than is shown on the $\Delta P$ curve D-D. At points of operation to the left of CD, it is possible to satisfy system requirements with one fan operating at one point of rating while the other fan is at a different point of rating. For example, consider $\Delta P$ E-E, which requires a pressure of 1.00 in. of water and a volume of 5000 cfm. The requirements of this system can be satisfied with each fan delivering 2500 cfm at 1.00 in. of water pressure, Point CE. The system can also be satisfied at Point CE′ by one fan operating at 1400 cfm at 0.9 in. of water, while the second fan delivers 3400 cfm at the same 0.90 in. of water.

Note that system curve E-E passes through the combined performance curve at two points. Under such conditions, unstable operation can result. Under conditions of CE′, one fan is underloaded and operating at poor efficiency. The other fan delivers most of the requirements of the system and uses substantially more power than the underloaded fan. This imbalance may reverse and shift the load from one fan to the other.

## FAN NOISE

Fan noise is a function of the fan design, volume flow rate $Q$, total pressure $P_t$, and efficiency $\eta_t$. After a decision has been made regarding the proper type of fan for a given application (keeping in mind the system effects), the best size selection of that fan must be based on efficiency, since the most efficient

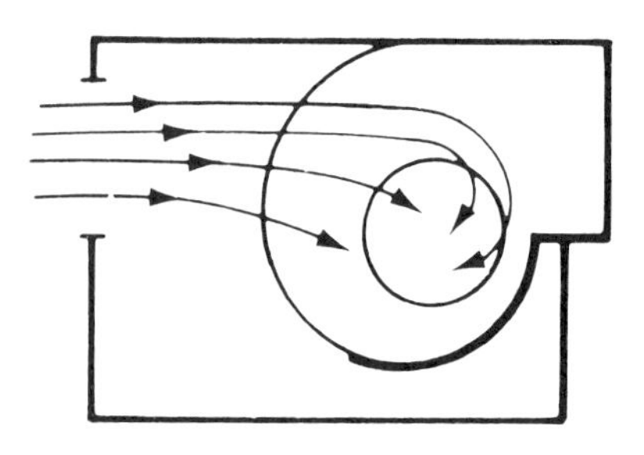

Enclosure Inlet Not Symmetrical
With Fan Inlet and
Induces a Prerotational Vortex


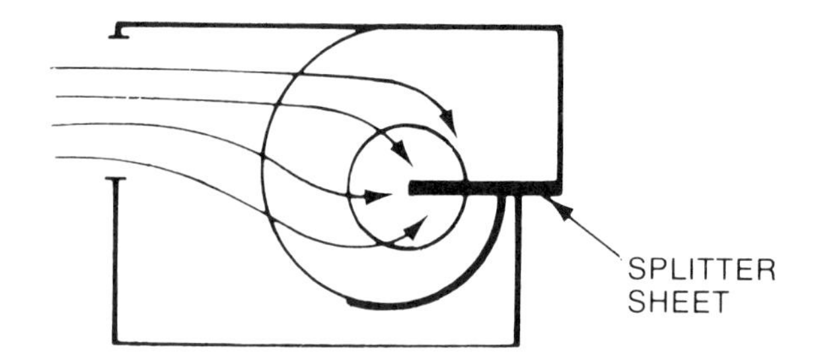


**Fig. 21 Effect of Inlet Location on Airflow**
(Adapted by permission from AMCA Publication 201)

operating range for a specific line of fans is normally the quietest. Low outlet velocity does not necessarily ensure quiet operation, so selections made on this basis alone are not appropriate. Also, noise comparisons of different types of fans, or fans offered by different manufacturers, made on the basis of rotational or tip speed are not valid. The only valid basis for comparison are the actual sound power levels generated by the different types of fans when they are all producing the required volume flow rate and total pressure.

The data are reported by fan manufacturers as *sound power levels* in eight octave bands. These levels are determined by using a reverberant room for the test facility and comparing the noise generated by the fan to the noise generated by a noise source of known sound power. The measuring technique is described in AMCA *Standard* 300-85, *Reverberant Room Method for Sound Testing of Fans*. ASHRAE *Standard* 68-1986/AMCA *Standard* 330-86, *Laboratory Method of Testing In-Duct Sound Power Measurement Procedure for Fans*, describe an alternate test method to determine the sound power a duct fan radiates into a supply and/or return duct terminated by an anechoic chamber. These standards do not fully evaluate the pure tones generated by some fans; these tones can be quite objectionable when they are radiated into occupied spaces. On critical installations, special allowance should be made by providing extra sound attenuation in the octave band containing the tone.

Typical sound power level values for various types of fans may be found in Chapter 52 of the 1987 HVAC Volume; these are typical values and significant variations can be expected as a result of differences in impeller and housing designs. Therefore, the values should be used only for general calculations and should not be used where critical noise requirements must be met, such as in auditoriums, conference rooms, and lecture halls. Where possible, sound power level data should be obtained from the fan manufacturer for the specific fan being considered.

## FAN ARRANGEMENT AND INSTALLATION

Direction of rotation is determined from the drive side of the fan. On single-inlet centrifugal fans, the drive side is usually considered the side opposite the fan inlet. (AMCA has published standard nomenclature to define positions.)

### Fan Isolation

In air-conditioning systems, ducts should be connected to fan outlets and inlets with unpainted canvas or other flexible

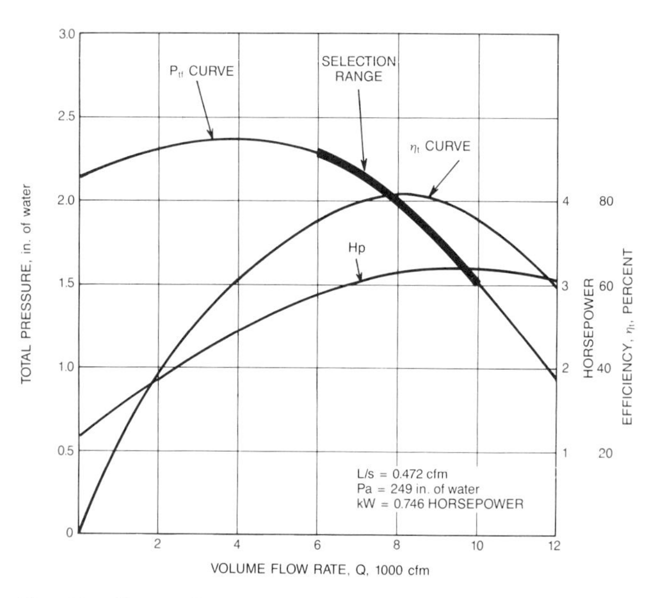


**Fig. 22 Conventional Fan Performance Curve Used by Most Manufacturers, Showing the Performance of a Fixed Fan Size Running at a Fixed Speed**

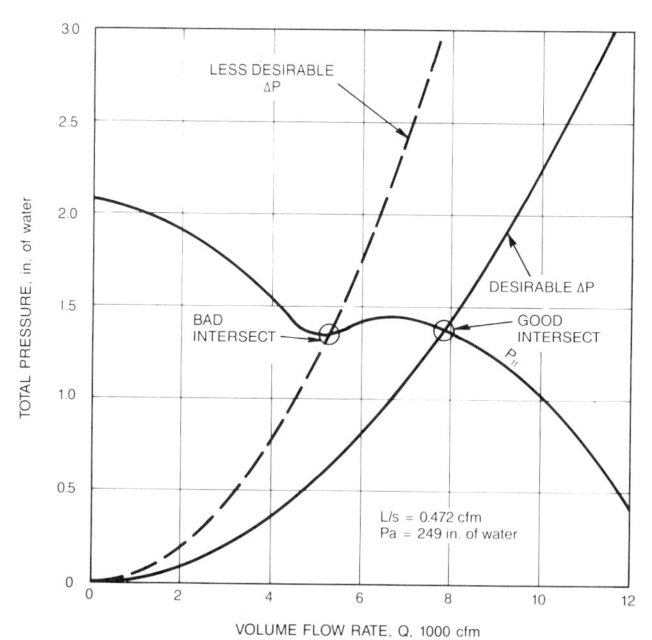


**Fig. 23 Desirable Combination of $P_{tf}$ and $\Delta P$ Curves**

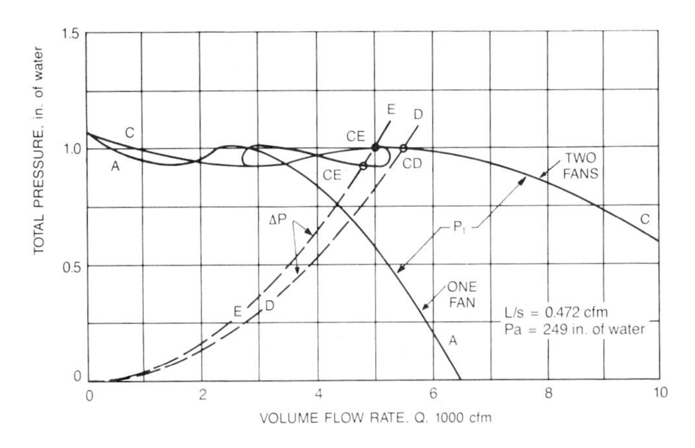


**Fig. 24 Two Forward Curve and Centrifugal Fans in Parallel Operation**

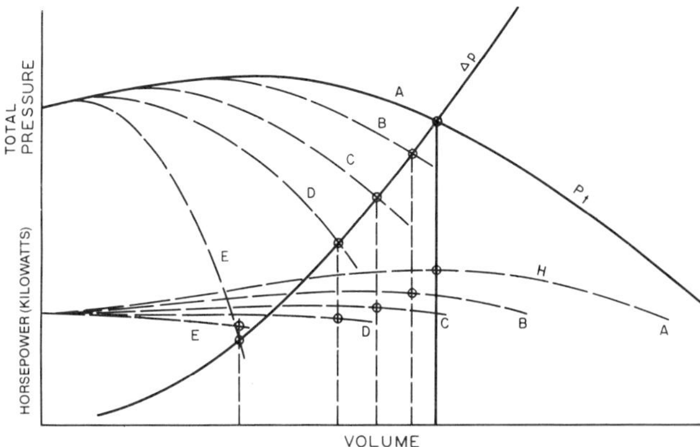


**Fig. 25 Effect of Inlet Vane Control on Backward Curve Centrifugal-Fan Performance**

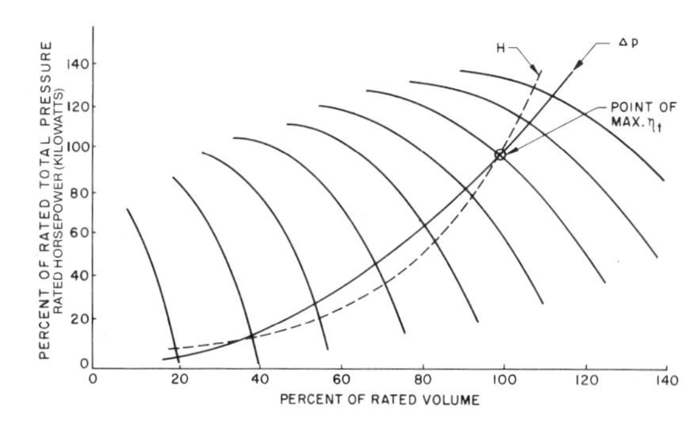


**Fig. 26 Effect of Blade Pitch on Controllable Pitch Vaneaxial-Fan Performance**

material. Access should be provided in the connections for periodic removal of any accumulations tending to unbalance the rotor. When operating against high resistance or when noise level requirements are low, it is preferable to locate the fan in a room removed from occupied areas or in one acoustically treated to prevent sound transmission. The lighter building constructions, which are common today, make it desirable to mount fans and driving motors on resilient bases designed to prevent transmission of vibrations through floors to the building structure. Conduits, pipes, and other rigid members should not be attached to fans. Noises that result from obstructions, abrupt turns, grilles, and other items not connected with the fan may be present. Treatments for such problems, as well as the design of sound and vibration absorbers, are discussed in Chapter 52 of the 1987 HVAC Volume.

## FAN CONTROL

In many heating and ventilating systems, the volume of air handled by the fan varies. The choice of the proper method for varying flow for any particular case is influenced by two basic considerations: (1) the frequency with which changes must be made and (2) the balancing of reduced power consumption against increases in first cost.

To control flow, the characteristic of either the system or the fan must be changed. The system characteristic curve may be altered by installing dampers or orifice plates. This technique reduces flow by increasing the system pressure required and, therefore, increases power consumption. Figure 10 shows three different system curves, A, B, and C, such as would be obtained by changing the damper setting or orifice diameter. Dampers are usually the lowest first cost method of achieving flow control; they can be used even in cases where essentially continuous control is needed.

Changing the fan characteristic ($P_t$ curve) for control can reduce power consumption. From the standpoint of power consumption, the most desirable method of control is to vary the fan speed to produce the desired performance. If the change is infrequent, belt-driven units may be adjusted by changing the pulley on the drive motor of the fan. Variable-speed motors or variable-speed drives, whether electrical or hydraulic, may be used when frequent or essentially continuous variations are desired. When speed control is used, the revised $P_t$ curve can be calculated by the Fan Laws.

Inlet vane control is frequently used. Figure 25 illustrates the change in fan performance with inlet vane control. Curves A, B, C, D, and E are the pressure and power curves for various vane settings between wide open (A) and nearly closed (E).

Tubeaxial and vaneaxial fans are made with adjustable pitch blades to permit balancing of the fan against the system or to make infrequent adjustments. Vaneaxial fans are also produced with controllable pitch blades (*i.e.*, pitch that can be varied while the fan is in operation) for frequent or continuous adjustment. Varying pitch angle retains high efficiencies over a wide range of conditions. Figure 26 performance is from a typical fan with variable pitch blades. From the standpoint of noise, variable speed is somewhat better than variable blade pitch; however, both of these control methods give high operating efficiency control and generate appreciably less noise than inlet vane or damper control.

## FAN MOTOR SELECTION

Whenever an electric motor is selected as the prime mover for a centrifugal fan, both the fan's maximum power and starting torque should be considered. In some cases, where larger centrifugal fans are operated with relatively small motors, the motor may not accelerate the fan/impeller within some allowable starting time. Excessive starting time raises the temperature of the motor windings to a point where circuit breakers may trip out or the motor may be damaged. The user must consider this problem when selecting fan and motor combinations.

The two main factors that control starting acceleration are the fan/impeller inertia ($WR^2$ in the fan industry) and the starting torque characteristic of the electric motor. Curves for the motor starting torque as a percentage of full load torque at any speed are available from the motor supplier.

In most normal fan applications, the fan is not run at the

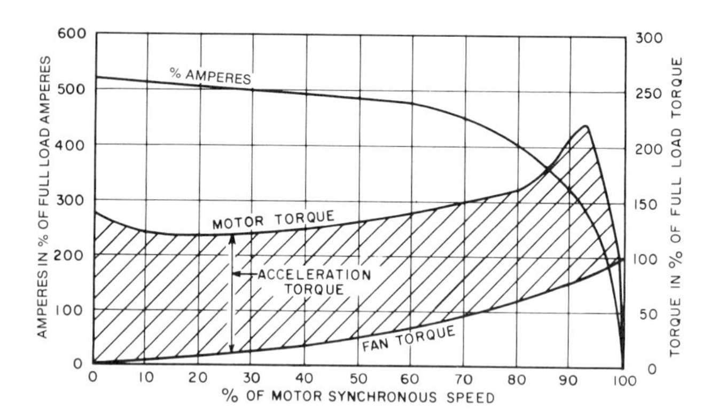


**Fig. 27 Typical Drip-proof Motor Performance Test Curve**

motor speed but through some drive combination, and the $WR^2$ must be corrected to represent the apparent $WR^2$ as seen by the motor. Along with the $WR^2$ of the fan/impeller, the additional effect of the $WR^2$ of the shaft and fan sheave and the inertia of the motor rotor must be considered.

Because the available motor torque must be divided between the fan load and the energy needed for acceleration, the acceleration torque at any point can be represented by $T_A = T_m - T_F$, where $T_m$ is motor torque and $T_F$ is fan torque. This acceleration torque is graphically represented in Figure 27 as the vertical distance between the motor torque and the fan torque curves at any speed.

Normally, a maximum allowable value for acceleration time is available from the motor manufacturer; a typical time for induction motors is around 10 seconds. Other factors, such as ambient conditions and the number of starts per day, also affect this allowable acceleration time.

## REFERENCES

AMCA. 1985. Laboratory Methods of Testing Fans for Rating. ANSI/AMCA *Standard* 210-85, ANSI/ASHRAE *Standard* 51-1985, Arlington, Heights, IL.

AMCA. 1985. Reverberant Room Method for Sound Testing of Fans. *Standard* 300-85, Arlington Heights, IL.

AMCA. 1983. Designation for Rotation & Discharge of Centrifugal Fans, *Standards Handbook* 99-2406-83, Arlington Heights, IL.

AMCA. 1982. Drive Arrangements for Tubular Centrifugal Fans, *Standards Handbook* 99-2410-82, Arlington Heights, IL.

AMCA. 1978. Drive Arrangements for Centrifugal Fans, *Standards Handbook* 99-2404-78, Arlington Heights, IL.

AMCA. 1973. Fans and Systems, *Standard* 201-73, Arlington Heights, IL.

AMCA. 1966. Motor Positions for Belt or Chain Drive Centrifugal Fans. *Standards Handbook* 99-2407-66, Arlington Heights, IL.

Buffalo Forge Co. 1983. *Fan Engineering*, 8th ed. R. Jorgensen, ed. Buffalo, NY.

Clark, M.S., J.T. Barnhart, F.J. Bubsey, and E. Neitzel. 1978. The Effects of System Connections on Fan Performance. ASHRAE *Transactions* 84(2):227.

Graham, J.B. 1971. Methods of Selecting and Rating Fans. ASHRAE *Symposium Bulletin,* Fan Application, Testing and Selection.

Graham, J.B. 1966. Fan Selection by Computer Techniques. *Heating, Piping and Air Conditioning,* April, p. 168.

CHAPTER 4

# EVAPORATIVE AIR-COOLING EQUIPMENT



THIS chapter addresses direct and indirect evaporative air-cooling equipment and air washers used for air cooling, humidification, dehumidification, and air cleaning. Maintenance and water treatment are also discussed. Residential and industrial humidification equipment are covered in Chapter 5 of this volume.

Packaged evaporative coolers, air washers, indirect air coolers, evaporative condensers, vacuum cooling apparatus, and cooling towers exchange sensible heat for latent heat. This equipment falls into two general categories: (1) apparatus for air cooling and (2) apparatus for heat rejection. This chapter primarily addresses air-cooling equipment.

Evaporative air cooling evaporates water into an airstream. Figure 1 illustrates thermodynamic changes that occur between the air and water that are in direct contact in a moving airstream. The continuously recirculated water reaches an equilibrium temperature equal to the wet-bulb temperature of the entering air. The heat and mass transfer between the air and water lowers the air dry-bulb temperature and increases the humidity ratio at a constant wet-bulb temperature.

The extent to which the leaving air temperature approaches the thermodynamic wet-bulb temperature of the entering air or the extent to which complete saturation is approached is expressed as percentage evaporative cooling or saturation effectiveness and is defined in Equation (1) as:

$$e_c = 100\,\frac{(t_1 - t_2)}{(t_1 - t')} \qquad (1)$$

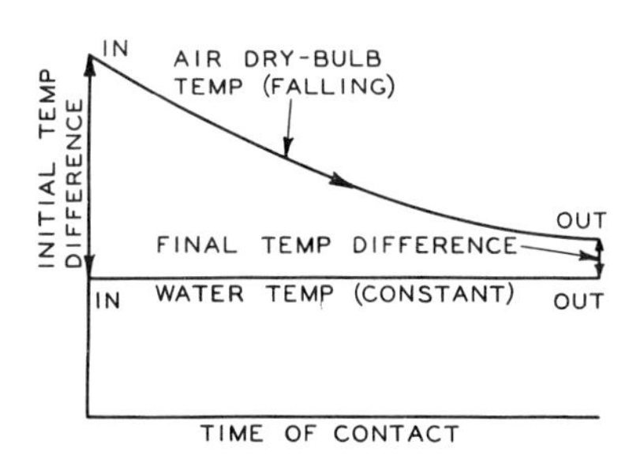


**Fig. 1 Interaction of Water and Air in Evaporative Air Cooler**

The preparation of this chapter is assigned to TC 5.7, Evaporative Cooling.

*where*

$e_c$ = evaporative cooling or saturation effectiveness (percent)
$t_1$ = dry-bulb temperature of the entering air
$t_2$ = dry-bulb temperature of the leaving air
$t'$ = thermodynamic wet-bulb temperature of the entering air
(All temperatures are expressed in consistent units)

The term *effectiveness* is used rather than *efficiency* since efficiency implies energy, power, or work. In some literature, the term efficiency (evaporative cooling, saturation, or humidifying) may be used.

If warm or cold unrecirculated water is used, the air can be heated and humidified or cooled and dehumidified. These processes are described later in the chapter.

Evaporative air-cooling equipment can be classified as either direct or indirect. Direct evaporative equipment cools air by direct contact with the water, either by an extended wetted-surface material (as in packaged air coolers) or with a series of sprays (as in an air washer). Indirect systems cool air in a heat exchanger. The heat exchanger transfers heat to either a secondary airstream that has been evaporatively cooled (air-to-air) or to water that has been evaporatively cooled (as by a cooling tower).

Combination systems can involve both direct and indirect principles, as well as heat exchangers and cooling coils. In such systems, air may exit below the initial wet-bulb temperature. While such systems can be complex, the high cost of energy may justify their use in certain geographic areas.

## PART I: DIRECT EVAPORATIVE AIR COOLERS

### WETTED-MEDIA AIR COOLERS

These coolers (Figure 2) contain evaporative pads (usually of aspen wood fibers). A water-circulating pump lifts the sump water to a distributing system, and it flows down through the pads back to the sump.

A fan within the cooler pulls the air through the evaporative pads and delivers it to the space to be cooled. The fan discharges either through the side of the cooler cabinet or through the sump bottom. Wetted-pad packaged air coolers are made in sizes ranging from 2000 to 20,000 cfm (0.9 to 9.4 m$^3$/s).

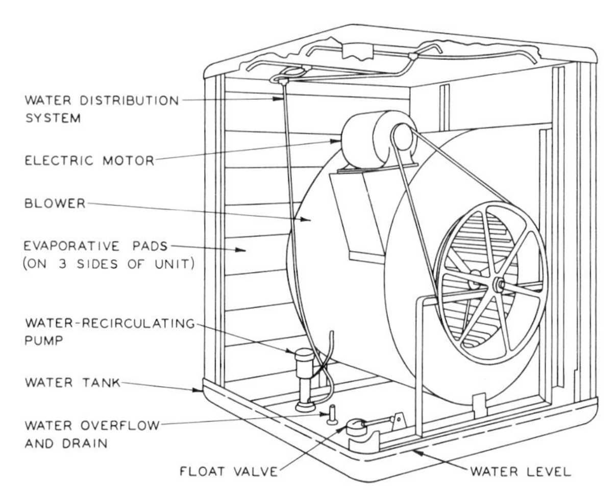


**Fig. 2 Typical Wetted-Pad Evaporative Cooler**

When clean and well maintained, commercial wetted-media air coolers operate at an effectiveness of approximately 80%. They remove particulate matter 10 microns and larger. In some units, supplementary filters ahead of or following the evaporative pads remove additional dirt. These filters prevent dirt from entering the cooling system when it is operated without water to circulate fresh air. The evaporative pads may be chemically treated to increase wettability. An additive may be included in the fibers to help them resist attack by bacteria, fungi, and other microorganisms.

The coolers are usually designed for an evaporative pad face velocity of 100 to 300 fpm (0.5 to 1.5 m/s), with a pressure drop of 0.1 in. of water (25 Pa). The aspen fibers are packed to a weight of approximately 0.3 to 0.4 lb/ft$^2$ (1.5 to 2.0 kg/m$^2$) of face area based on 2 in. (50 mm) pad thickness. Pads are mounted in removable louvered frames, which are usually made of galvanized steel or molded plastic. Troughs distribute water to the pads. A centrifugal pump with submerged inlet pumps the water through tubes that provide an equal flow of water to each trough. The sump or water tank has a water makeup connection, float valve, overflow pipe, and a drain. Provisions for the bleed-off of water to prevent the buildup of minerals and scale are usually incorporated in the design. Alternately, if the water supply is unlimited, fresh water may be used with no recirculation. This practice reduces mineral and scale buildup but wastes water.

Water usage depends on the airflow, the effectiveness of the evaporator pad, and the wet-bulb temperature of the incoming air. The humidity ratio (mass of water vapor per mass of dry air) for the entering and leaving air can be determined from a psychrometric chart. From the difference, along with the airflow rate, the actual water consumption can be calculated. An approximation of water usage is 1.3 gph of water per 1000 cfm for each 10°F reduction in dry-bulb temperature (0.52 mL/s water per m$^3$/s airflow for each °C of reduction).

The blower is usually a forward-curved, centrifugal fan, complete with motor and drive. The V-belt drive may include an adjustable pitch motor sheave to facilitate balancing air delivery against the static pressure requirements of the supply air-duct system and to permit operation of the motor below its rated ampere draw (FLA). The motor enclosure may be open, drip proof, or totally enclosed.

## RIGID-MEDIA AIR COOLERS

Another type of wetted-surface evaporative cooler design uses sheets of rigid, corrugated material as the wetted surface (Figure 3). Materials include cellulose and fiberglass that have been treated chemically with anti-rot and rigidifying salts and saturants. The sheets are laid with specific angle corrugations running in alternating directions such that the air and water flow in opposite directions. The depth of fill, in the direction of airflow, is commonly 12 in. (300 mm) but may be between 6 and 24 in. (150 and 600 mm). The media have the desirable characteristics of low resistance to airflow, high saturation effectiveness, and self cleaning by flushing.

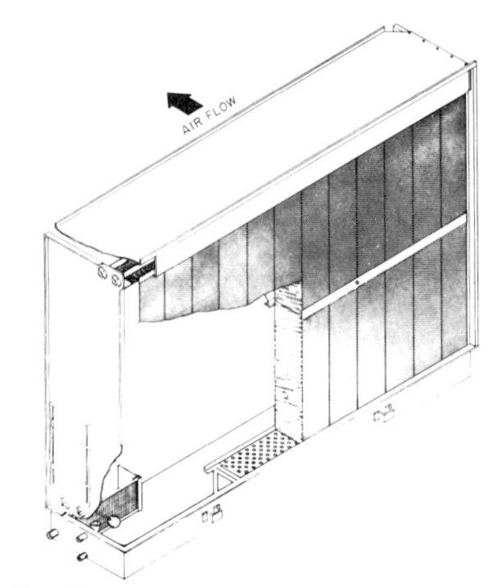


**Fig. 3 Typical Rigid-Media Air Cooler**

Air washers or evaporative coolers using this material can be built in sizes to handle as much as 200,000 cfm (94 m$^3$/s). Saturation effectiveness varies from 70% to over 95%, depending on media depth and air velocity. Air flows horizontally while the recirculating water flows vertically over the media surfaces by gravity feed from a flooding header and water distribution chamber. A pump recirculates the water from a lower reservoir, which is constructed of heavy gauge material with a corrosion-resistant coating and overflow and drain connections. The upper media enclosure is fabricated of reinforced galvanized steel or other corrosion-resistant sheet metal.

Flanges at the entering and leaving faces allow connecting the unit to ductwork. A float valve maintains proper water level in the reservoir, makes up water evaporated, and supplies fresh water for dilution to prevent over-concentration of solids and minerals. Because the water recirculation rate is low and because spray nozzles are not needed to saturate the media, pumping horsepower is low when compared with spray-filled air washers with equivalent evaporative cooling effectiveness performance.

## SLINGER PACKAGED AIR COOLERS

Slinger air coolers consist of an evaporative cooling section and a fan section (Figure 4). The fan is usually a forward curved, double inlet, double width, centrifugal fan that is V-belt driven by an electric motor. In the cooling section, outdoor air is drawn through a water spray, an evaporative filter pad, and an entrained-moisture eliminator pad. The spray is created by a motor-driven, vertical, clog proof disk that is partially immersed in the water sump. The rated evaporative cooling effectiveness may be up to 80%. Air capacities range up to 30,000 cfm (14 m$^3$/s). Higher capacities may be obtained by using multiple cooling sections with one or more fans discharging air into a distribution system. In addition to the spray apparatus and pads, the cooling section includes a housing, a bleed-off valve, a fresh water make-up float valve, overflow pipe, and drain plug. An air inlet filter, inlet louver, and rain hood would be optional accessories. The evaporator pad and moisture eliminator pad may be of latex-coated fiber, glass fiber, or coated non-ferrous metal construction; from 0.75 to 2 in. (20 to 50 mm) deep; and washable. The face air velocity across the pad surface may be

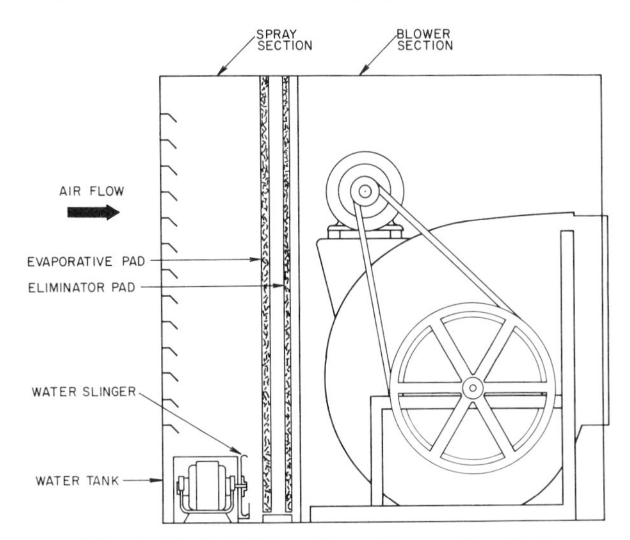


**Fig. 4 Typical Slinger-Type Evaporative Cooler**

from 300 to 600 fpm (1.5 to 3 m/s), depending on the cooling effectiveness desired. Chemical treatment may be used to increase wettability and resistance to fungi, bacteria, and fire.

The air-washing capability of the spray and filter arrangement, plus the non-clogging characteristic of the water spray disk, minimizes maintenance requirements.

## PACKAGED ROTARY AIR COOLERS

Rotary coolers (Figure 5) wet and wash the evaporative pad by rotating it through a water bath. The evaporative pad and other parts in contact with the water are made from corrosion-resistant materials. Like the slinger cooler, it consists of a cooling section and a blower compartment. The air handling part of the cooler is similar to that of the wetted-pad and slinger coolers. Packaged rotary air coolers are built in capacities ranging from 2000 to 12,000 cfm (0.9 to 5.7 $m^3/s$).

There are two types of packaged rotary air coolers. One uses a rotating drum, which is partially submerged in the water reservoir. Air passes through the thickness dimension of the drum. The second rotary cooler has a continuously looped pad. The lower end of the pad is submerged in the water reservoir; the upper end is looped around a horizontal drive shaft at the top of the cooler. Several pad mediums are used, with varying thicknesses (depending on the medium material). Pad face velocities

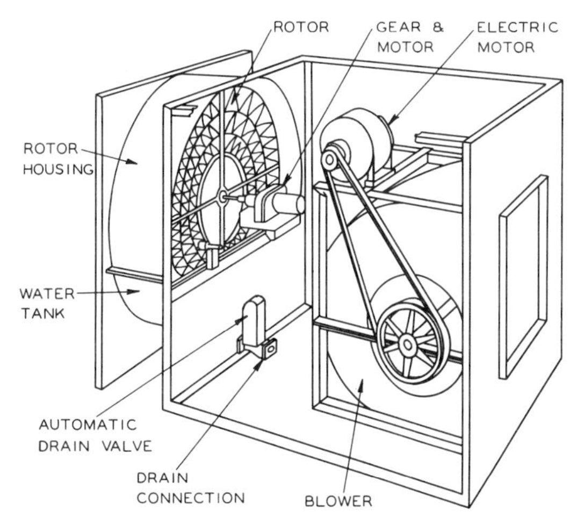


**Fig. 5 Typical Rotary-Type Evaporative Cooler**

of 100 to 600 fpm (0.5 to 3.0 m/s) are typical, with pressure drops in the range of 0.5 in. of water (125 Pa).

Both rotary cooler types may be furnished with an automatic flush valve and timer to flush the reservoir water periodically, thus minimizing any build-up of minerals, salts, and solids in the water. The timer setting is usually adjustable. An outdoor anticipation thermostat may also be connected to the fill and drain valve for freeze protection.

## REMOTE PAD EVAPORATIVE COOLING EQUIPMENT

Greenhouses, animal shelters, and similar applications use a system of exhaust fans in the wall or roof of a structure, with wetted pads placed so that outdoor air is drawn through the enclosed space. The pads are wetted from above by a perforated trough or similar device, with the excess water draining to a gutter from which it may be wasted or collected for recirculation. Water is supplied at the rate of 1.3 gph per 1000 cfm for each 10°F of reduction of the dry-bulb temperature (0.52 mL/s water per $m^3/s$ airflow for each °C of reduction). The pad should be sized for an air velocity of approximately 150 fpm (0.76 m/s).

# PART II: INDIRECT EVAPORATIVE AIR COOLERS

## INDIRECT PACKAGED AIR COOLERS

In indirect evaporative air coolers, outdoor air or exhaust air from the conditioned space passes through one side of a heat exchanger. This air (the secondary airstream) is cooled by evaporation by one of several methods: (1) direct wetting of the heat exchanger surface; (2) passing through wetted pad media; (3) atomizing spray; (4) disk evaporator, etc. The surfaces of the heat exchanger are cooled by contact with the secondary airstream. On the other side of the heat exchanger surface, the primary airstream (conditioned air to be supplied to the space) is sensibly cooled by contact with the heat exchanger surfaces.

Although the primary air is cooled by the evaporatively cooled secondary air, no moisture is added to the primary air. Hence, the process is called *indirect* evaporative air cooling. The supply (primary) air may be recirculated room air or outside air (or a mixture). The enthalpy of the primary airstream decreases because no moisture is added to it. This process contrasts with direct evaporative cooling, which is essentially adiabatic (constant enthalpy). The usefulness of indirect evaporative cooling is related to the wet-bulb depression of the secondary air below the dry-bulb temperature of the entering primary air.

Since neither the evaporative cooling nor the heat transfer effectiveness can reach 100%, the leaving dry-bulb temperature of the primary air must always be above the entering wet-bulb temperature of the secondary airstream. Dehumidifying in the primary airstream could occur only when the dew point of the primary airstream is several degrees higher than the wet-bulb temperature of the secondary airstream.

A packaged indirect evaporative air cooler includes the heat exchanger, wetting apparatus, secondary air fan assembly, secondary air inlet louver, and enclosure. The heat exchanger may be constructed with folded metal sheets, with or without a corrosion-resistant or moisture-retaining coating; or it may be constructed with tubes so that one airstream flows inside the tubes and the other flows over the exterior tube surfaces. Air filters should be placed upstream from both the primary and secondary heat exchangers to minimize fouling by dust, insects, or other airborne contaminants.

Since minerals in the water being evaporated would increase the concentration of minerals in the water being recirculated, continuous waste and dilution by addition of fresh water is necessary. Water treatment may be necessary to control corrosion of heat-exchanger surfaces and other metal parts.

The packaged indirect evaporative air cooler may either be self-contained, with its own primary air supply fan assembly, or part of a built-up or more complete packaged air-handling system. The cooling system could use a single stage of indirect evaporative cooling, or it could include indirect evaporative cooling as the first stage, with additional direct evaporative cooling and/or refrigerated (chilled water or direct expansion) cooling stages. When the indirect evaporative cooler is placed in series (upstream) with a conventional refrigerated coil, it reduces the sensible load on the coil and refrigeration system (Figure 6). Energy required for the indirect precooling stage includes the pump and secondary air fan motor, as well as some additional fan energy to overcome resistance added in the primary air. The energy consumed by the indirect evaporative cooling stage is less than the energy saved by reduction in load on the refrigeration apparatus. As a result, the overall system efficiency may increase dramatically because energy costs and demands are reduced. Another saving could result from the reduction in size of the refrigeration equipment required. Indirect evaporative cooling also may reduce the yearly hours during which the refrigeration equipment must be operated.

Evaporatively cooled air can be discharged across air-cooled refrigeration condenser coils to improve the efficiency of the condenser and refrigeration system. Chapter 56 of the 1987 HVAC Volume includes sample calculations covering energy-saving evaporative cooling systems. Manufacturers' data should be consulted for equipment selection, cooling performance, pressure drops, and space requirements.

Cooling performance of indirect evaporative cooling equipment is expressed frequently as a *Performance Factor* (P.F.), which is defined as the reduction in dry-bulb temperature in the primary airstream divided by the difference between the initial dry-bulb temperature of the primary airstream minus the entering wet-bulb temperature of the secondary air. The term *Indirect Evaporative Cooling Effectiveness* (as a percentage) is used also.

Manufacturers' ratings require careful interpretation. The *basis of ratings* should be specified, since, for the same apparatus, the performance is affected by changes in primary and secondary air velocities and mass flow rates, wet-bulb temperature, altitude, etc.

Typically, the air resistance on both primary and secondary sides ranges between 0.2 and 2.0 in. water (50 and 500 Pa). The ratio of *secondary* air to conditioned *primary* air may range from less than 0.6 to greater than 1.0 and has an effect on performance. Based on manufacturers' ratings, available equipment may be selected for performance factors (indirect evaporative cooling effectiveness) ranging from 0.6 to 0.8 (60% to 80%).

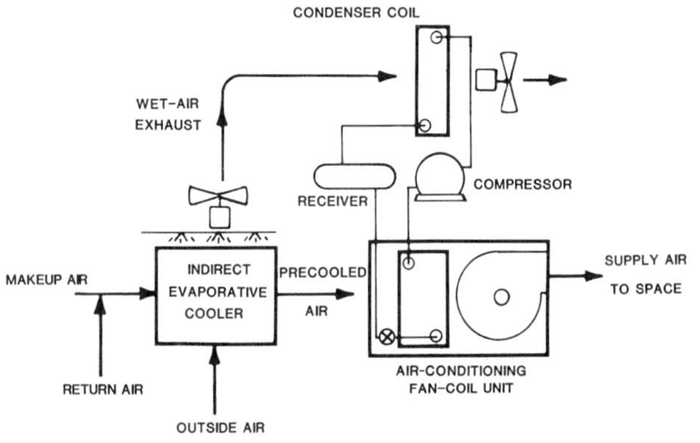


**Fig. 6 Indirect Evaporative Air Cooler Used as a Precooler**

Presently, ASHRAE has no standard for testing and rating evaporative cooling equipment.

## COOLING TOWER/COIL SYSTEMS

The combination of a cooling tower or other evaporative water cooler with a water-to-air heat exchanger coil and water circulating pump is another type of indirect evaporative cooling system. Water is pumped from the reservoir of the cooling tower to the coil and returns to the upper distribution header of the tower. Both open-water and closed-loop systems are used.

The recirculated water is evaporatively cooled to within a few degrees of the wet-bulb temperature as it flows over the wetted surfaces of the cooling tower. As the cooled water flows through the tubes of the coil in the conditioned airstream, it picks up heat from the conditioned air. The temperature of the water increases, and the primary air is cooled without the addition of moisture to the primary air. The water is again cooled as it recirculates through the cooling tower. A float valve controls the fresh water makeup, which replaces the evaporated water and prevents excessive concentration of minerals in the recirculated water. Air filtration on the air inlet of the cooling tower is also recommended.

One advantage of this system, especially for retrofit applications, is that the cooling tower may be located remote from the cooling coil. Also, the system is accessible for maintenance. Overall indirect evaporative cooling effectiveness (P.F.) may range between 0.55 and 0.75 or higher.

Evaporative water coolers designed specifically for indirect evaporative cooling applications can keep the temperature of the discharge water very close to the wet-bulb temperature of the entering air. These coolers have a wetted surface media, which has a high ratio of wetted surface area per unit of fill volume. Performance depends on depth of fill, air velocity over the fill surface, water flow to airflow ratio, wet-bulb temperature, and water-cooling range. Because of the closer approach of the water temperature to the wet-bulb temperature, the overall system P.F may be increased several percent as compared to a conventional cooling tower.

## OTHER INDIRECT EVAPORATIVE COOLING APPARATUS

Other combinations of evaporative coolers and heat exchange apparatus can accomplish indirect evaporative cooling. Heat exchangers finding application in these systems include heat pipes and rotary heat wheels, in addition to the plate, pleated media, and shell and tube types. If the conditioned (primary) air and the exhaust or outside (secondary) airstream are side by side, a heat pipe apparatus or heat wheel can transfer heat from the warmer air to the cooler air. Counter-current flow of the primary and secondary air is best for indirect evaporative cooling effectiveness. Evaporative cooling of the secondary airstream by spraying water directly on the surfaces of the heat exchanger or by an evaporative cooler upstream of the heat exchanger may cool the primary air indirectly by transferring heat from it to the secondary air.

## INDIRECT/DIRECT COMBINATIONS

In a two-stage combination, indirect/direct evaporative cooling system, a first-stage indirect evaporative cooler lowers both the dry-bulb temperature and the wet-bulb temperature of the incoming supply air. After leaving the indirect stage, the supply air passes through a second-stage direct evaporative cooler. Figure 7 shows the process on the psychrometric chart. First-stage cooling follows a line of constant humidity ratio, since no moisture is added to the primary airstream. The second stage follows the wet-bulb line at the condition of the air leaving the first stage.

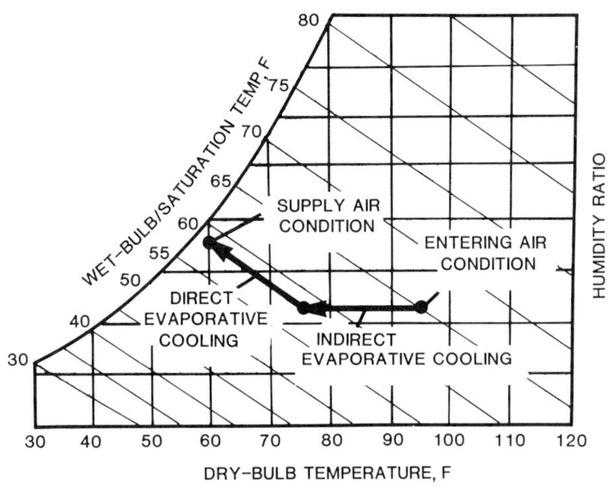


**Fig. 7 Combination Indirect/Direct Evaporative Cooling Process**

The indirect evaporative cooling stage may be any of the types described in the previous paragraphs. Figure 8 shows a system using a rotary heat wheel. The *secondary* air may be exhaust air from the conditioned space or outdoor air. When the secondary air passes through the evaporative cooler, the dry-bulb temperature is lowered by evaporative cooling. As this air passes through the heat wheel, the mass of the media is cooled to a temperature approaching the wet-bulb temperature of the secondary air. The heat wheel rotates so that its cooled mass enters the primary air and, in turn, sensibly cools the primary (supply) air. Following the heat wheel, a direct evaporative cooler further reduces the dry-bulb temperature of the primary air. This approach may achieve supply air dry-bulb temperatures of 5 °F (3 °C) or more below the secondary air wet-bulb temperature.

In areas where the 1% mean-coincident wet-bulb design temperature is 66 °F (18.9 °C) or lower (see Chapter 24, 1985 FUNDAMENTALS Volume), average annual cooling power consumption of indirect/direct systems may be as low as 0.22 kW/ton (0.06 kW/kW). When the 1% mean-coincident wet-bulb temperature is as high as 74 °F (23.3 °C), an indirect/direct cooling system can have an average annual cooling power consumption as low as 0.81 kW/ton (0.23 kW/kW). By comparison, the typical refrigeration system with an air-cooled condenser may have an average annual power consumption of greater than 1.0 kW/ton (0.28 kW/kW).

In dry environments, indirect/direct evaporative cooling systems are usually designed to supply 100% outdoor air to the conditioned spaces of a building. In these once-through applications, space latent loads and return air sensible loads are exhausted from the building rather than returned to the conditioning equipment. Consequently, the cooling capacity required from these systems may be less than that required from a conventional refrigerated cooling system.

In areas of higher wet-bulb design temperatures or when system design requires a supply air temperature lower than that attainable using indirect/direct evaporative cooling systems, a third cooling stage may be required. This third stage may be a refrigerated direct expansion or chilled water coil located either upstream or downstream from the direct evaporative cooling stage, but always downstream from the indirect evaporative stage. Refrigerated cooling is energized only when evaporative stages cannot achieve the required supply air temperature.

The psychrometric chart may be used to plot estimated system performance for single-, two-, and three-stage systems. Chapter 56 of the 1987 HVAC Volume has more details.

Figure 9 shows a schematic of a three-stage configuration (indirect/direct, with optional third-stage refrigerated cooling). The third-stage refrigerated cooling coil is located downstream from the direct evaporative cooler. In this arrangement, the conventional air-conditioning equipment operates less efficiently because a lower coil surface temperature is required to condense the water vapor introduced by the direct evaporative cooler. Locating the third-stage refrigerated coil upstream of the direct evaporative cooler improves the operating efficiency of the air conditioner but requires that the coil be larger because the coil surface will be dry.

The designer should consider options using exhaust and/or outside air as secondary air for the indirect evaporative cooling stage. If the latent load in the space is significant, the wet-bulb temperature of the exhaust air in the cooling mode may be higher than that of the outside air. In this case, outside air may be used more effectively as secondary air to the indirect evaporative cooling stage.

Custom configurations are available for indirect/direct and three-stage systems to permit many choices for locating the return, exhaust, and outside air, and mixing of airstreams, bypass of components, or variable volume control.

In direct, indirect, indirect/direct, and other staged evaporative cooling systems, the elements of the system that may be controlled include modulating outside air and return air mixing dampers; secondary air fans and recirculating pumps of an indirect evaporative stage; recirculating pumps of a direct evaporative cooling stage; face and bypass dampers for the direct stage; chilled water flow or temperature or refrigerant flow for a refrigerated stage; and system or individual terminal volume by the use of variable volume terminals, fan variable inlet vanes, adjustable pitch, or variable fan speed.

For sequential control in indirect/direct evaporative cooling systems, the indirect evaporative cooler is energized for first-stage cooling, the direct evaporative cooler for second-stage cooling, and the refrigeration coil for third-stage cooling. In some applications, reversing the sequence of the direct evaporative cooler and indirect evaporative cooler may reduce the first stage energy requirement.

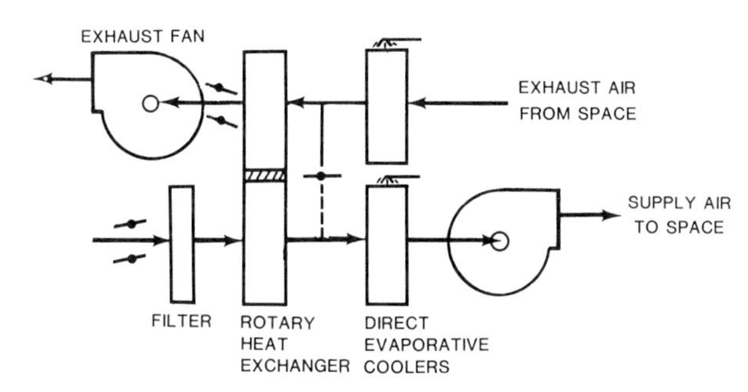


**Fig. 8 An Indirect/Direct Evaporative Cooling System Using a Rotary Heat Wheel**

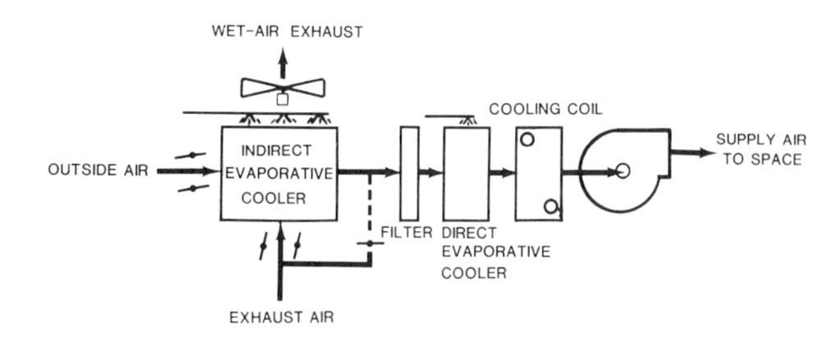


**Fig. 9 Three-Stage Indirect/Direct Evaporative Cooling System**

# PART III: AIR WASHERS

## SPRAY-TYPE AIR WASHERS

Spray-type air washers consist of a chamber or casing containing a spray-nozzle system, a tank for collection of spray water as it falls, and an eliminator section for removal of entrained drops of water from the air. A pump recirculates water at a rate higher than the evaporation rate. Intimate contact between the spray water and the airflow causes heat and mass transfer between the air and the water (see Figure 10).

Washers are commonly available from 2000 to 250,000 cfm (0.9 to 118 $m^3/s$) capacity, but no limit exists for sizes specially constructed. No standardization exists; each manufacturer publishes tables giving physical data and ratings for specific products. Therefore, air velocity, water-spray density, spray pressure, and other design factors must be considered for each application.

The simplest design has a single bank of spray nozzles with a casing usually 4 to 7 ft (1.2 to 2.1 m) long. This type of washer is applied primarily as an evaporative cooler or humidifer. It is sometimes used as an air cleaner when the dust is wettable, although the air-cleaning efficiency is relatively low. Cleaning efficiency may be increased by the addition of flooding nozzles to wash the eliminator plates. Two or more spray banks are generally used when a very high degree of saturation is necessary and for cooling and dehumidification applications that require chilled water. Two-stage washers are used for dehumidification when the quantity of chilled water is limited or when the water temperature is above that required for the single-stage design. Arranging the two stages for counterflow of the water permits a small quantity of water with a greater water temperature rise.

The lengths of washers vary considerably. Spray banks are approximately 2.5 to 4.5 ft (0.8 to 1.4 m) spaced; the first and last banks of sprays are located about 1 to 1.5 ft (0.3 to 0.45 m) from the entering or leaving end of the washer. In addition, air washers may be furnished with heating or cooling coils within the washer chamber, which may affect the overall length of the washer.

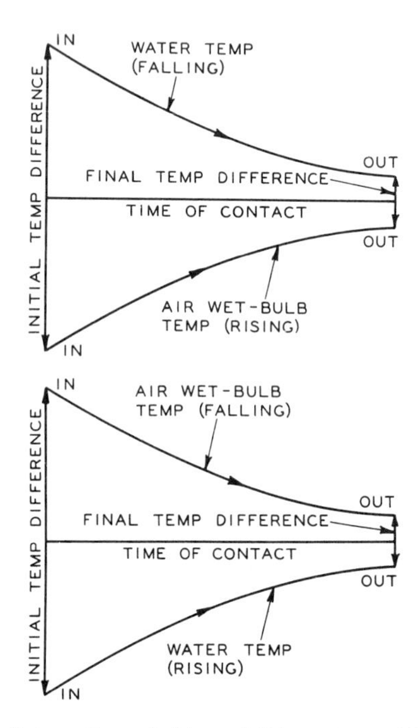


**Fig. 10 Interaction of Air and Water in an Air Washer Heat Exchanger**

Figure 11 shows the construction features of conventional spray-type air washers. Essential requirements in the air washer operations are (1) uniform distribution of the air across the spray chamber; (2) an adequate amount of spray water broken up into fine droplets; (3) good spray distribution across the airstream; (4) sufficient length of travel through the spray and wetted surfaces; and (5) the elimination of free moisture from the outlet air. The cross-sectional area is determined by the design velocity of the air through the spray chamber. The units usually have an air velocity of 300 to 600 fpm (1.5 to 3.0 m/s); however, with special eliminators, velocities as high as 1500 fpm (7.6 m/s) may be used.

Spray-water requirements for spray-type air washers that are used for washing or evaporative cooling vary from 4 gpm per 1000 cfm (0.5 L/s per $m^3/s$) with a single bank to 10 gpm per 1000 cfm (1.3 L/s per $m^3/s$) for double banks. Pumping heads usually range from 55 to 100 ft of water (160 to 300 kPa), depending on nozzle pressure, height of apparatus, pressure losses in pipe and strainers, etc. Approximately 10% of the water handled by the pump is bled off to reduce the chemical buildup from the evaporated water, thereby reducing the incidence of nozzle clogging or mineral buildup on wetted surfaces. A higher percentage of bleed-off may be required with heavily mineralized water.

Spray nozzles produce a finely atomized spray and are spaced to give uniform coverage of the chamber through which the air passes. Nozzle pressures normally vary from 20 to 40 psig (140 to 280 kPa), depending on the duty. Small orifices at pressures of up to about 40 psig (280 kPa) produce a fine spray, necessary for high saturation effectiveness, while larger orifices with pressures of about 25 psig (170 kPa) are common for dehumidification. When the water contains large amounts of chemicals that can clog the nozzles, larger orifice nozzles should be installed, even if larger pumping capacity is required. Self-cleaning nozzles are available. Spray-nozzle capacities vary from about 1 to 3.75 gpm (0.06 to 0.2 L/s) per nozzle, and spray densities are usually 1 to 5 gpm per $ft^2$ (0.7 to 3.4 L/s per $m^2$) of cross-sectional area per bank. Spacing varies from about 0.75 to 2.5 nozzles per $ft^2$ (8 to 27 per $m^2$) per bank. For lower spray densities, smaller orifices should be used to avoid bypassing air because of poor spray coverage. Flooding nozzles that are used for washing the eliminators discharge about 1.0 gpm (0.06 L/s) per nozzle in a flat stream at 3 to 5 psig (20 to 35 kPa) pressure.

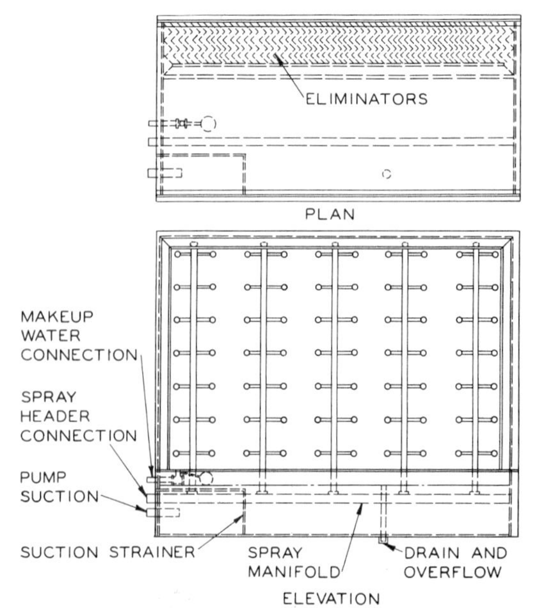


**Fig. 11 Typical Single-Bank Air Washer**

Strainers, usually made of fine mesh copper or brass screen, extend across the width of the tank. They are sometimes placed in a cylindrical shape over the pump suction connection within the tank. Strainers should be readily removable for inspection and cleaning. The openings in the screen should be smaller than the spray orifice to minimize nozzle clogging. Belt-type, automatic, and other specially designed strainers are available for use in textile mills and other industries where lint or heavy concentrations of dust are present. Accessories that should be included are (1) a float valve to maintain the minimum water level automatically, (2) a quick-fill connection, (3) a trapped overflow, (4) a tank drain opening, (5) a suction connection, and (6) a marine light on the spray chamber.

The resistance to airflow through an air washer varies with the type and number of baffles, eliminators, and wetted surfaces; the number of spray banks and their direction and air velocity; the size and type of other components, such as cooling and heating coils; and other factors such as air density. Pressure drop may be as low as 0.25 in. of water (60 Pa) or as high as 1 in. of water (250 Pa). The manufacturer should be consulted regarding the resistance of any particular washer design combination.

The casing and the tank may be constructed of various materials. One or more doors are commonly provided for inspection and access. The tank is normally at least 16 in. (400 mm) high with a 14-in. (350 mm) water level; it may extend beyond the casing on the inlet end to make the suction strainer more accessible. The tank may be partitioned by the installation of a weir, usually in the entering end, to permit recirculation of spray water for control purposes in dehumidification work. The excess then returns over the weir to the central water chilling machine.

Eliminators consist of a series of vertical plates that are spaced about 0.75 to 2 in. (20 to 50 mm) on centers at the exit of the washer. The plates are formed with a number of bends to deflect the air and obtain impingement on the wetted surfaces. Hooks on the edge of the plates improve moisture elimination. Perforated plates may be installed on the inlet end of the washer to obtain more uniform air distribution through the spray chamber. Louvers, which prevent the backlash of spray water, may also be installed for this purpose.

## HIGH-VELOCITY SPRAY-TYPE AIR WASHERS

High-velocity air washers generally operate at air velocities in the range of 1200 to 1800 fpm (6 to 9 m/s). Some have been applied as high as 2400 fpm (12 m/s), but 1200 to 1600 fpm (6 to 8 m/s) is the most accepted range for optimum application. The reduced cross-sectional area allows air washers to be used in smaller apparatus than those of lower velocities. This space reduction makes air washers the primary heat transfer means for cooling and dehumidifying (as well as humidifying) industrial air-conditioning systems. High capacities per unit of space available from high-velocity spray devices permit practical packaging of prefabricated central station units in either completely assembled and transportable form or, for large capacity units, easily handled modules. Manufacturers supply units of up to 150,000 cfm (70 $m^3/s$) capacity shipped in one piece, including spray system, eliminators, pump, fan, dampers, filters, and other functional components. Such units are self-housed, prewired, prepiped, and ready for hoisting into place (see Figure 12).

The number and arrangement of nozzles vary with different capacities and manufacturers. Adequate values of saturation effectiveness and heat transfer effectiveness are achieved by the use of higher spray densities.

Eliminator blades come in varying shapes, but most are a series of aerodynamically clean, sinusoidal shapes. Collected

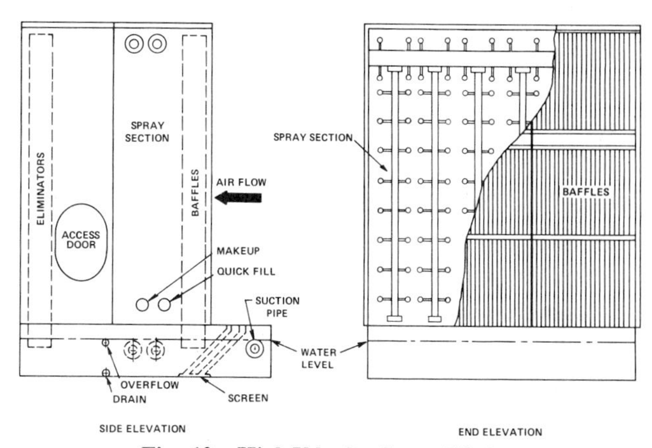


**Fig. 12 High-Velocity Spray Washer**

moisture flows down grooves or hooks designed into their profiles, then drains into the storage tank. Washers may be built with shallow drain pans and connected to a central storage tank. High-velocity washers are rectangular in cross section and, except for the eliminators, are similar in appearance and construction to conventional lower-velocity types. Pressure losses are in the 0.5 to 1.5 in. of water (120 to 370 Pa) range. These washers are available either as freestanding separate devices for incorporation into field-built central stations or in complete preassembled central station packages from the factory.

## CELL-TYPE AIR WASHERS

These washers obtain intimate air-water contact by passing the air through cells packed with glass, metal, or fiber screens (see Figure 13). Water passes over cells arranged in tiers. Behind the cells are blade-type or glass-mat eliminators. Most cell-type washers are arranged for concurrent air and water flow. They are also constructed for special duty with counter-current flow characteristics or in a combination of both arrangements. Cell washers come in many sizes, of insulated or uninsulated con-

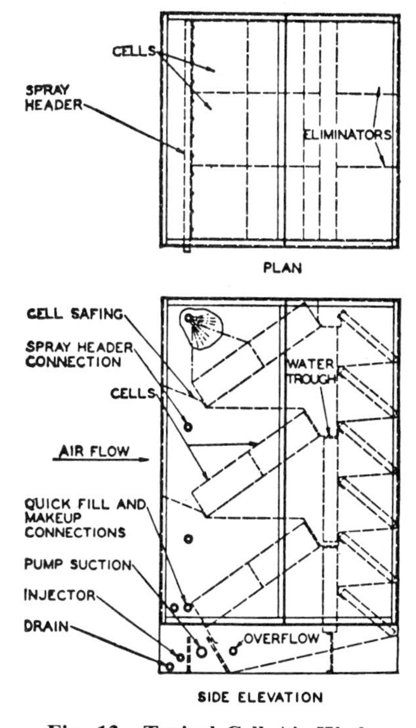


**Fig. 13 Typical Cell Air Washer**

struction. Standard washers are available up to 10 cells high by 12 cells wide with a capacity of up to 210,000 cfm (100 $m^3$/s). They are also available up to 30,000 cfm (14 $m^3$/s), complete with fan, motor, pump, and external spray piping.

Atomization of the spray water is not required in cell-type washers, but good water distribution over the face of the cell is essential. A saturation effectiveness of 90 to 97% is possible with units that have 6-in. (150-mm) deep cells using fine fibers. Units with vertically positioned cells (usually 2 in. [50 mm] deep) and coarser fibers will have 70 to 80% saturation effectiveness. Water requirements vary from 0.75 to 1.5 gpm per 1000 cfm (0.1 to 0.2 L/s per $m^3$/s) of airflow, with resistance of airflow ranging from 0.15 to 0.65 in. of water (35 to 160 Pa), depending on the water quantity circulated and the air velocity through the cells. Although classified as air washers, they should not be used as heat exchangers because of the low volume of water flowing over the cells.

Each washer consists of a number of cells, normally 20 in. square (510 mm square), arranged in tiers. A typical cell is a metal frame packed with glass fiber strands. The glass occupies only 3 to 6% of the volume of the cell but, when sprayed, presents a total wetted area of approximately 100 to 120 $ft^2$ (9.3 to 11.1 $m^2$). Wire mesh screens at both faces of the frame hold the glass pack. Each tier is independent of the others and has its own spray header, drain sheet, and, except for the lowest tier, conduit to the tank below.

Eliminators downstream from the cells remove entrained moisture from the airstream. They may be a metal-blade-type for deflection of the airstream, with hooks for trapping impinged droplets or a glass-mat-type arranged in tiers similar to the cell. The glass mat is 2 in. (50 mm) deep and is randomly packed with glass fiber in a metal frame.

Connections are provided for internal spray headers, drain, overflow, a steam injector for cleaning the cells in place, pump suction, water quick fill, and makeup water controlled by a float valve.

The Dynel/polyester mat-type of cell washer operates at a basic spray rate of 0.75 gpm of water per 1000 cfm (0.1 L/s per $m^3$/s) and airflow resistance of 0.7 in. of water (170 Pa). Where water is not recirculated, the sprays operate on a once-through basis, carrying away solids; this eliminates the need for the pump.

Saturation effectiveness may be selected from 82 down to 70% by controlling air velocity through the cell. In this shallow cell system, because moisture carry-over is prevented by the fine fiber media and mat design, moisture eliminators are not needed. Air-cleaning results are typical of comparable impingement-type filters, with some gain from wetting action and the self-flushing of particulate matter. For year-round cooling and makeup-air systems, freedom from winter freezing can be achieved by specifying self-draining headers, plus one main stop-and-drain cock in the supply main below the washer elevation.

## HUMIDIFICATION WITH AIR WASHERS

Air can be humidified with an air washer in three ways: (1) using recirculated spray water without prior treatment of the air, (2) preheating the air and washing it with recirculated spray water, and (3) using heated spray water. In any air-washing installation, the air should not enter the washer with a wet-bulb temperature of less than 39 °F (4 °C), otherwise the spray water may freeze.

### Recirculated Spray Water

Except for both the small amount of outside energy added by the recirculating pump in the form of shaft work and the small amount of heat leakage into the apparatus from outside (including the pump and its connecting piping), the process is strictly adiabatic. Evaporation from the liquid is recirculated. Its temperature should adjust to the thermodynamic wet-bulb temperature of the entering air.

The whole airstream is not brought to complete saturation, but its state point should move along a line of constant thermodynamic wet-bulb temperature. As defined in Equation 1, the extent to which the leaving air temperature approaches the thermodynamic wet-bulb temperature of the entering air is expressed by *saturation effectiveness* ratio. In humidifiers, this ratio is often referred to as the *humidifying effectiveness.*

The following is representative of the saturation or humidifying effectiveness of a spray-type air washer for these spray arrangements.

| Bank Arrangement | Length, ft (m) | Effectiveness, % |
|---|---|---|
| 1 downstream | 4 (1.2) | 50 to 60 |
| 1 downstream | 6 (1.8) | 60 to 75 |
| 1 upstream | 6 (1.8) | 65 to 80 |
| 2 downstream | 8 to 10 (2.4 to 3.0) | 80 to 90 |
| 2 opposing | 8 to 10 (2.4 to 3.0) | 85 to 95 |
| 2 upstream | 8 to 10 (2.4 to 3.0) | 90 to 98 |

The degree of saturation depends on the extent of contact between air and water. Other conditions being equal, a low-velocity airflow is conducive to higher humidifying effectiveness.

### Preheating the Air

Preheating the air increases both the dry- and wet-bulb temperatures and lowers the relative humidity, but it does not alter the humidity ratio (mass ratio, water vapor to dry air). At a higher wet-bulb temperature, but with the same humidity ratio, more water can be absorbed per unit mass of dry air in passing through the washer (if the humidifying effectiveness of the washer is not adversely affected by operation at the higher wet-bulb temperature). The analysis of the process that occurs in the washer is the same as that for recirculated spray water. The final preferred conditions are achieved by adjusting the amount of preheating to give the required wet-bulb temperature at the entrance to the washer.

### Heated Spray Water

Even if heat is added to the spray water, the mixing in the washer may still be regarded as adiabatic. The state point of the mixture should move toward the specific enthalpy of the heated spray. It is possible (by elevating the water temperature) to raise the air temperature, both dry bulb and wet bulb, above the dry-bulb temperature of the entering air.

The relative humidity of the leaving air may be controlled by (1) bypassing some of the air around the washer and remixing the two airstreams downstream or (2) automatically reducing the number of operating spray nozzles by operating valves in the different spray header branches.

## DEHUMIDIFICATION AND COOLING WITH AIR WASHERS

Air washers are also used to cool and dehumidify air. Heat and moisture removed from the air raise the water temperature. If the entering water temperature is below the entering wet-bulb temperature, both the dry-bulb and wet-bulb temperatures are lowered. Dehumidification results if the leaving water temperature is below the entering dew point temperature. Moreover, the final water temperature is determined by the sensible and latent heat pickup and the quantity of water circulated. However, this final temperature must not exceed the final required dew point, with one or two degrees below the dew point being common practice.

The air leaving a spray-type dehumidifier is substantially saturated. Usually, the spread between dry- and wet-bulb temperatures is less than 1°F (0.6°C). The spread between leaving air and leaving water depends on the difference between entering dry- and wet-bulb temperatures and on certain design features, such as the length and height of spray chamber, air velocity, quantity of water, and character of the spray pattern. The rise in water temperature is usually between 6 and 12°F (3.3 and 6.7°C), although higher rises have been used successfully. The lower rises are ordinarily selected when the water is chilled by mechanical refrigeration because of possible higher refrigerant temperatures. It is often desirable to make an economic analysis of the effect of higher refrigerant temperature compared to the benefits of a greater rise in water temperature. For systems receiving water from a well or other source at an acceptable temperature, it may be desirable to design on the basis of a high temperature rise and minimum water flow.

The most common air washer arrangement for cooling and dehumidifying air has two spray banks and is 8 to 9 ft (2.4 to 2.7 m) long. If the air washer can cool and dehumidify the entering air to a wet-bulb temperature equal to that of the leaving water temperature, it is convenient to assign such a washer a performance factor of 1.0. The actual performance factor of any washer ($F_p$) is the actual enthalpy change divided by the enthalpy change in a washer of 1.0 performance:

$$F_p = \frac{(h_1 - h_2)}{(h_1 - h_3)} \tag{2}$$

*where*

$h_1$ = enthalpy at wet-bulb temperature of entering air
$h_2$ = enthalpy at wet-bulb temperature of leaving air at actual condition
$h_3$ = enthalpy at wet-bulb temperature leaving a washer with $F_p = 1.0$
(All are expressed in consistent units.)

By knowing the performance factor of a particular air washer, the actual conditions of operation can be graphically determined (see Figure 14). Points 1 and 2 are plotted on the saturation curve representing total heat at the entering and leaving air wet-bulb temperature $t_1'$ and $t_2'$. Point 5 represents the condition at which leaving air wet-bulb and leaving water temperature would be the same. This point is determined by solving Equation 2 for $h_3$:

$$h_3 = h_1 - \frac{(h_1 - h_2)}{F_p} \tag{3}$$

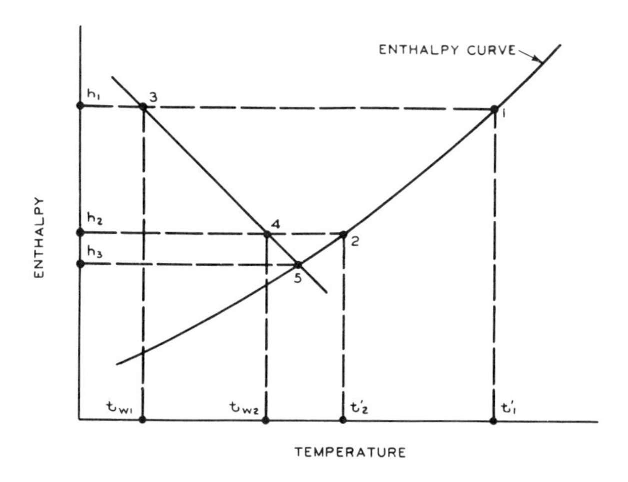


**Fig. 14 Graphical Solution of Conditions Produced by a Dehumidifying Air Washer**

A diagonal line is drawn through Point 5 with a negative slope equal to the water-to-air weight ratio. Points 3 and 4, at which the diagonal line intersects horizontal lines through Points 1 and 2, show the required entering and leaving water temperatures, $t_{w1}$ and $t_{w2}$. A check of the solution can be made from the fundamental heat balance expression: heat absorbed by the water = heat removed from the air.

The graphical method can be used to arrive quickly at solutions of air washer cooling and dehumidifying problems when there are a number of unknown factors, including quantity of water to be used and entering and leaving water temperatures. The actual performance factor of a particular washer, however, must be obtained from the manufacturer's data.

Another method expresses the performance of the dehumidifier as a relationship of the leaving to entering spread between air and water temperatures:

$$F_{p1} = 1 - \frac{(t_2' - t_{w2})}{(t'_1 - t_{w1})} \tag{4}$$

*where*

$t_1'$ = thermodynamic wet-bulb temperature of entering air
$t_2'$ = thermodynamic wet-bulb temperature of leaving air
$t_{w1}$ = entering water temperature
$t_{w2}$ = leaving water temperature

Performance factors expressed in these terms normally vary from about 70% for a 6-ft (1.8-m) long single-bank washer to nearly 100% for an 8-ft (2.4-m) long unit with two opposed spray banks.

A third method used to express the performance of a spray-type dehumidifier is given by the following formula, which, although similar to Equation 2, is numerically different because of a change in determining the value of $h_3$ as:

$$F_p = \frac{(h_1 - h_2)}{(h_1 - h_3)} \tag{5}$$

*where*

$h_1$ = enthalpy at wet-bulb temperature of entering air
$h_2$ = enthalpy at wet-bulb temperature of leaving air
$h_3$ = enthalpy of air at the leaving water temperature

Note that the performance factors determined by the second and third methods are not equal; the difference can be seen in Figure 15.

The representative curves in Figure 15 are based on entering wet-bulb temperatures of 65 and 70°F (18.3 and 21.1°C), with leaving difference between air and water of 1.5°F (0.8°C) and

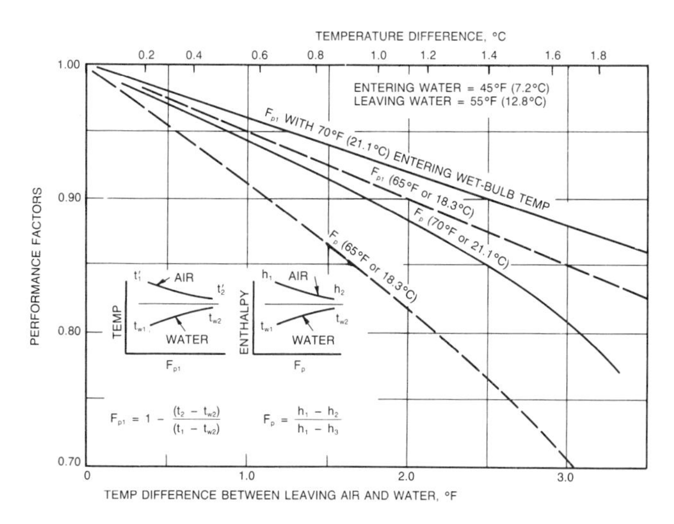


**Fig. 15 Comparison of Performance Factors**

a water temperature rise of 10°F (5.6°C). The calculation of $F_p$ is independent of rise in water temperature. However, the equation for $F_{pl}$ does involve the difference between entering and leaving water temperature. For example, the performance factor for 65°F (18.3°C) entering wet-bulb temperature and 1.5°F (0.8°C) leaving difference falls from 0.925 for 10°F (5.6°C) rise in water to 0.912 for 7°F (3.9°C) rise, assuming that the leaving difference is still 1.5°F (0.8°C).

The performance concept being applied should be stated when factors are being considered. As an added precaution, the washer requirement should also be given in terms of leaving temperature difference between air and water for a given set of conditions.

In this section, $F_p$ describes *dehumidification performance factor* as with spray-filled air washers circulating chilled water. This is historic usage of the term. As indirect evaporative cooling (described in Part II of this chapter) has gained extensive application, the term *Performance Factor* (P.F.) has been used to describe indirect evaporative cooling effectiveness. As standards for testing and rating evaporative cooling equipment develop, standard terminology should eliminate the confusion caused by these similar terms.

## AIR CLEANING WITH AIR WASHERS

The dust removal efficiency of air washers depends largely on the size, density, wettability, and solubility of the dust particle. The larger, more wettable particles are the easiest to remove. Separation is largely a result of the impingement of particles on the wetted surface of the eliminator plates. Since the force of impact increases with the size of the solid, the impact (together with the adhesive quality of the wetted surface) determines the washer's usefulness as a dust remover. The spray is relatively ineffective in removing most atmospheric dusts.

Air washers are of little use in removing soot particles because of the absence of an adhesive effect from the greasy surface. They are also ineffective in removing smoke because the inertia of the small particles (less than one micron) does not allow them to impinge and be held on the wet plates. Instead, the particles follow the air path between the plates because they are unable to pierce the water film covering the plates.

However, cell-type washers are efficient air cleaners. In practice, the air-cleaning results are typical of comparable impingement-type filters. When cell-type washers remain in the airstream without being wetted or are used where large amounts of fibrous materials are present in the air, they become plugged with airborne dust in a short time unless highly efficient filtering devices are placed upstream from the cells. The cells should be replaced if they are filled with dirt in a dry condition.

If water is not required on the cells, it may be possible to operate the water flow for a few minutes out of each hour to wash the dirt from the cell before the airflow is blocked. With some kinds of dirt, the wetted surface may increase the blockage, unless there is constant water flow.

# PART IV: MAINTENANCE AND WATER TREATMENT

Regular inspection and maintenance of evaporative coolers, air washers, and ancillary equipment ensures proper service and efficiency of the system. Users should follow currently accepted recommendations for maintenance and operational procedures. Adherence to them should minimize the involvement of HRVAC equipment in distributing possible airborne contaminants. Water lines, water-distribution troughs or pans, pumps, and pump filters must be clean and free of dirt, scale, and debris. Inadequate water flow causes dry areas on the evaporative media and a reduction in the cooling efficiency. Water and air filters should be cleaned or replaced, as required. Proper sump water level or spray pressure must be maintained. Bleed-off is the most practical means to minimize scale accumulation. The bleed-off rate should be 50 to 100% of the evaporation rate, depending on water hardness. Regular inspections should be made to ensure that the bleed-off rate is adequate and is maintained.

Pretreatment of water supply with quaternary salts or other chemicals intended to hold dissolved material in suspension is best prescribed by a local water-treatment specialist. Using water treated by a zeolite-type ion exchange softener is not recommended because the zeolite exchange of calcium for sodium results in a soft voluminous scale that is actively corrosive to galvanized steel. Any chemical agents used should not harm the cabinet, pads, or heat exchanger materials.

Periodic checks should be made for algae, slime, and bacterial growth. If required, an EPA-registered biocide should be added.

Motors and bearings should be lubricated and fan drives checked, as required.

Units that have heat exchangers with a totally wetted surface on the outside of the tubes and materials that are not harmed by chemicals can be descaled periodically with a commercial descaling agent and then flushed out. Mineral scale deposits on a wetted indirect unit heat exchanger are usually soft and allow wetting through to the tube and evaporation at the surface of the tube. Excess scale thickness causes a loss in heat transfer and should be removed.

The air washer spray system requires the most attention. Partially clogged nozzles are indicated by a rise in spray pressure, while a fall in pressure is symptomatic of eroded orifices. Strainers can minimize this problem. Continuous operation requires either a bypass around pipe-line strainers or duplex strainers. Air washer tanks should be drained and dirt deposits removed regularly. Eliminators and baffles should be periodically inspected and repainted to prevent corrosion damage.

With cell-type washers, it is also necessary to clean the glass media. A differential draft gauge can be used to measure the air resistance across the cells to determine the frequency of cleaning needed. Mat-type eliminators in these washers should be removed and cleaned with a detergent solution. Any glass mats that have been eroded by the sprays should be replaced.

## BIBLIOGRAPHY

Anderson, W.M. 1986. Three-Stage Evaporative Air Conditioning Versus Conventional Mechanical Refrigeration. ASHRAE *Transactions.* 92(1B):358.

ASHRAE. 1981. *Position Statement on Legionnaires' Disease.*

Eskra, V. 1980. Indirect/Direct Evaporative Cooling Systems. ASHRAE *Journal,* May, p. 21.

Hendrickson, H.M. 1954. How to Calculate Air Washer Performance. *Heating, Piping and Air Conditioning,* September, p. 116.

Sherwood, T.K., and R.L. Pigford. 1952. *Absorption and Extraction, 2nd Edition.* McGraw-Hill Book Company, New York.

Supple, R.G. 1982. Evaporative Cooling for Comfort. ASHRAE *Journal,* August, p.36.

Treybal, R.E. 1955. *Mass-Transfer Operations.* McGraw-Hill Book Company, New York.

Watt, J.R. 1986. *Evaporative Air Conditioning Handbook.* Chapman and Hall, London.

# CHAPTER 5

# HUMIDIFIERS



THE selection and application of humidification equipment considers (1) the environmental conditions of the occupancy or process and (2) the characteristics of the building enclosure. These may not always be compatible, so a compromise solution may be necessary, particularly in the case of existing buildings. The two factors are dealt with separately in the sections "Environmental Conditions" and "Enclosure Characteristics."

## ENVIRONMENTAL CONDITIONS

The environmental conditions for a particular occupancy or process may dictate a specific relative humidity, a required range of relative humidity, or certain limiting maximum or minimum values. The following classifications explain the effects of relative humidity and provide guidance as to requirements for most applications.

### Human Comfort

The effect of relative humidity on all aspects of human comfort has not yet been established completely. It is generally assumed that higher temperatures are necessary for thermal comfort to offset decreased relative humidity. This is exemplified by the effective temperature lines of the ASHRAE Comfort Chart (ASHRAE *Standard* 55-1981).

Low relative humidity may be undesirable for reasons other than those based on thermal comfort. Low levels increase evaporation from the membranes of the nose and throat and cause drying of the skin and hair. Some medical opinions attribute the increased incidence of respiratory complaints to the drying of mucous membranes by low indoor humidities in winter. However, no well-documented evidence indicates a serious hazard to health resulting from exposure to low humidity.

Despite the lack of complete information, extremes of humidity are undesirable and, for human comfort, relative humidity should neither exceed 60% nor be less than 20%.

### Process Control and Materials Storage

The relative humidity required by a process are usually specific and related to the control of moisture content or regain, the rate of chemical or biochemical reactions, the rate of crystallization, product accuracy or uniformity, corrosion, and static electricity. Typical conditions of temperature and relative humidity for the storage of certain commodities and the manufacturing and processing of others may be found in Chapter 28 of the 1987 HVAC Volume.

Low humidities in winter may cause drying and shrinking of furniture, wood floors, and interior trim. Winter humidification may be desirable to maintain a relative humidity closer to that experienced during manufacture or installation.

For storage of hygroscopic materials, a constant humidity is as important as humidity level. Temperature control may also be important because of the danger of condensation on products through a transient lowering of temperature.

### Static Electricity

Electrostatic charges are generated when materials of high electrical resistance move one against the other. The accumulation of such charges may result in (1) unpleasant sparks to people walking over carpets; (2) difficulties in handling sheets of paper, fibers, and fabric; (3) the objectionable clinging of dust to oppositely charged objects; and (4) dangerous situations when explosive gases are present. Increasing the relative humidity of the environment tends to prevent the accumulation of such charges, but the optimum level of humidity depends, to some extent, on the materials involved.

Relative humidities of 45% or more usually reduce or eliminate electrostatic effects with many materials, but wool and some synthetic materials may require still higher humidities (Sereda and Feldman 1964). Under some conditions, and with certain materials, maximum electrostatic charging occurs at relative humidities of 25 to 35% or higher. Increasing the humidity from a low value may increase the electrostatic problem unless humidity is carried well beyond the value where charging is a maximum.

Hospital operating rooms, where explosive mixtures of anesthetics are used, constitute a special and critical case in regard to electrostatic charges. A relative humidity of 50% or more is usually required with special grounding arrangements and restrictions on the types of clothing worn by occupants. Conditions of 72°F (22°C) and 55% relative humidity are usually recommended for comfort and safety.

### Prevention and Treatment of Disease

Relative humidity has been shown to have a significant effect on the control of airborne infection. At 50% relative humidity, the mortality rate of certain organisms is highest, and the influenza virus loses much of its virulence. The mortality rate decreases both above and below this value.

The preparation of this chapter is assigned to TC 8.7, Humidifying Equipment.

A relative humidity of 65% is regarded as optimum in nurseries for premature infants, while a 50% value is suitable for full-term and observational nurseries. In the treatment of allergic disorders, humidities well below 50% have proven satisfactory. (Detailed discussion of air-conditioning requirements in the prevention and treatment of disease and recommendations for hospital air conditioning may be found in Chapter 23 of the 1987 HVAC Volume.)

### Miscellaneous

The air absorption of sound waves is at a maximum at 15 to 20% relative humidity and increases with frequency (Harris 1963). A marked reduction in absorption is obtained at 40% relative humidity; above 50%, the effect of air absorption is negligible. Air absorption does not affect speech significantly, but may merit consideration in large halls or auditoriums where optimum acoustics are required for musical performances.

Certain microorganisms are occasionally present in poorly maintained household humidifiers. To deter the propagation and spread of such detrimental microorganisms, periodic cleaning of the humidifier and draining of the reservoir (particularly at the end of the heating season) are required. Cold-water atomizing types have been banned as room humidifiers in some hospitals because of germ propagation.

Laboratories and test chambers, in which precise control of a wide range of relative humidity is desired, require special attention. Because of the interrelation between temperature and relative humidity, precise humidity control requires equally precise temperature control. The cooling load should be reduced and the method of heat removal should be considered if high humidities are to be maintained (Solvason and Hutcheon 1965).

## ENCLOSURE CHARACTERISTICS

The extent to which a building may be humidified in winter depends on the ability of its walls, roof, and other elements to prevent or tolerate condensation. Condensed moisture or frost on surfaces exposed to the building interior (*visible condensation*) can deteriorate the surface finish, cause mold growth and subsequent indirect moisture damage and nuisance, and reduce visibility through windows. If walls and roof have not been specifically designed to prevent the entry of moist air or vapor from inside, *concealed condensation* within these constructions may occur, even at low interior humidities, and cause serious deterioration.

Chapter 21 of the 1985 FUNDAMENTALS Volume addresses the design of enclosures to prevent condensation.

### Visible Condensation

Condensation forms on any interior surface when the dew point temperature of the air in contact with it exceeds the temperature of the surface. The maximum permissible relative humidity that may be maintained without condensation is thus influenced by the thermal properties of the enclosure and the interior and exterior environment.

Average surface temperatures may be calculated by the methods outlined in Chapter 23 of the 1985 FUNDAMENTALS Volume for most insulated constructions. However, localized cold spots result from high conductivity paths such as through-the-wall framing, projected floor slabs, and metal window frames that have no thermal breaks. The vertical temperature gradient from air space and surface convection in windows and similar sections results in lower air and surface temperatures at the sill or floor. Drapes and blinds closed over windows lower surface temperature further, while heating units under windows may raise the temperature significantly.

Windows present the lowest surface temperature in most buildings and provide the best guide to permissible humidity levels for no condensation. While calculations based on overall thermal coefficients provide reasonably accurate temperature predictions at mid-height, actual minimum surface temperatures are best determined by test. One method of relating the characteristics of windows is with a temperature index (Wilson and Brown 1964). This index is defined as $(t - t_o) / (t_i - t_o)$, where $t$ is the inside window surface temperature, $t_i$ is the indoor air temperature, and $t_o$ is the outdoor air temperature.

The results of limited tests on actual windows indicate that the temperature index at the bottom of a double, residential-type window that has a full thermal break is between 0.55 and 0.57, with natural convection on the warm side. Sealed, double-glazed units exhibit an index of from 0.33 to 0.48 at the junction of glass and sash, depending on sash design. The index will likely rise to values of 0.53 or greater only 1 in. (25 mm) above the junction.

With continuous under-window heating, the minimum index for a double window with full break may be as high as 0.60 to 0.70. Under similar conditions, windows with poor thermal breaks may be increased similarly.

Figure 1 shows the relationship between temperature index and the relative humidity and temperature conditions at which condensation occurs. The limiting relative humidities for various outdoor temperatures intersect a vertical line representing the particular temperature index. A temperature index of 0.55 has been selected to represent an average for double-glazed, residential-type windows and indicates the limiting relative humidities shown in Table 1. The values for single glazing in the table are based on a calculated temperature index of 0.22.

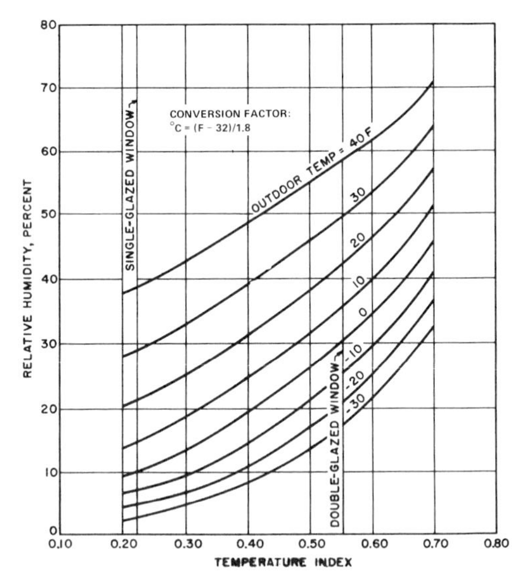


**Fig. 1 Limiting Relative Humidity for No Window Condensation**

### Concealed Condensation

The humidity level that a given building can tolerate without serious concealed condensation may be much lower than indicated by visible condensation criteria. The migration of water vapor through the inner envelope by diffusion or air leakage brings the vapor into contact with surfaces at temperatures approaching the outside temperature. Unless the building has been designed to eliminate or reduce this possibility effectively, the permissible humidity may be determined by the ability of the building enclosure to handle internal moisture rather than pre-

**Table 1 Maximum Relative Humidity Within a Space for No Condensation on Windows**

| Natural Convection, Indoor Air at 74°F (23.3°C) | | |
|---|---|---|
| **Outdoor Temp., °F (°C)** | **Single Glazing** | **Double Glazing** |
| 40 ( 4.4) | 39 | 59 |
| 30 ( −1.1) | 29 | 50 |
| 20 ( −6.7) | 21 | 43 |
| 10 (−12.2) | 15 | 36 |
| 0 (−17.8) | 10 | 30 |
| −10 (−23.3) | 7 | 26 |
| −20 (−28.9) | 5 | 21 |
| −30 (−34.4) | 3 | 17 |

vent its occurrence. Houses in cold regions of North America have been developed to a stage where adequate protection is reasonably ensured. In existing commercial buildings, such measures have not always been appreciated fully by the designers, so, for cold climates, increased humidification for such buildings should be approached with caution.

## ENERGY CONSIDERATIONS

When calculating energy requirements, the effect of the *dry air environment* should be considered for any material that gives moisture to it. This conversion of moisture (in its liquid state) from containment in a hygroscopic material to the vapor state is an evaporative process and, therefore, it requires energy. The source of energy causing this evaporation is the heat energy contained in the air. Thus, heat lost from the air to evaporate moisture equals the heat necessary to produce an equal amount of moisture vapor with an efficient humidifier. Therefore, in some cases, no new energy is required to maintain the desired humidity levels.

This apparent tradeoff has its costs in the destructive effects of moisture migration from hygroscopic materials if proper humidity levels are not maintained.

The true energy required for a humidification system must be calculated from what the actual humidity level will be in the building, not from the theoretical level. Without a humidification system, methods of calculating actual humidity have not been practically established.

A study of residential heating and cooling systems showed a correlation between infiltration and inside relative humidity, indicating a significant energy saving from increasing the inside relative humidity, which reduced infiltration of outside air by up to 50% during the heating season (Luck and Nelson 1977). This reduction is apparently due to the sealing of window cracks by the formation of frost.

## LOAD CALCULATIONS

The humidification load depends primarily on either the rate of natural ventilation of the space to be humidified or the amount of outside air introduced by mechanical means. Other sources of moisture gain or loss should also be considered, however. The humidification load can be calculated as follows:

For ventilation systems (*natural infiltration*):

$$H = \frac{VR}{B}(W_i - W_o) - S + L \qquad (1)$$

For fixed quantity of outside air systems (*mechanical ventilation*):

$$H = \frac{Q_o A}{B}(W_i - W_o) - S + L \qquad (2)$$

For variable quantity of outside air systems:

$$H = \left(\frac{T_i - T_m}{T_i - T_o}\right)\frac{Q_t A}{B}(W_i - W_o) - S + L \qquad (3)$$

*where*

$H$ = humidification load, lb of water/h (kg of water/h)
$V$ = volume of the space to be humidified, $ft^3$ ($m^3$)
$R$ = ventilation rate, air changes/h
$Q_o$ = quantity of outside air, cfm (L/s)
$Q_t$ = total quantity of air (outside air plus return air), cfm (L/s)
$T_i$ = temperature of air indoor design, °F (°C)
$T_m$ = temperature of mixed air design, °F (°C)
$T_o$ = temperature of outside air design, °F (°C)
$W_i$ = humidity ratio at indoor design conditions, lb of water/lb of dry air (kg of water/kg of dry air)
$W_o$ = humidity ratio at outdoor design conditions, lb of water/lb of dry air (kg of water/kg of dry air)
$S$ = contribution of internal moisture sources, lb of water/h (kg of water/h)
$L$ = other moisture losses, lb of water/h (kg of water/h)
$A$ = conversion factor, cfm to $ft^3$/h, 60 (L/s to $m^3$/h, 3.6)
$B$ = specific volume of air, 13.5 $ft^3$/lb of dry air, (0.843 $m^3$/kg of dry air)

### Design Conditions

Interior design conditions are dictated by the occupancy or process, as discussed under "Environmental Conditions" and "Enclosure Characteristics." Outdoor design data may be obtained from Chapter 24 of the 1985 FUNDAMENTALS Volume. Outdoor humidity can be assumed at 70 to 80% below 32°F (0°C) or 50% above 32°F (0°C) in winter for most areas. Corresponding absolute humidity values can subsequently be obtained from Chapter 6 of the 1985 FUNDAMENTALS Volume or from the ASHRAE Psychrometric Chart.

For systems handling quantities of outside air fixed by natural infiltration rates or mechanical ventilation, load calculations are based on outdoor design conditions. Equation 1 should be used for natural infiltration; Equation 2 for mechanical ventilation.

For systems that achieve a fixed mixed air temperature by varying outside air, special considerations are needed to determine the maximum humidification load. This load occurs at an outside air temperature other than the lowest design temperature, since it is a function of the amount of outside air introduced and the air's existing moisture content. Equation 3 should be used, repeated at various outside air temperatures, to determine the maximum humidification load.

In residential load calculations, the outdoor design conditions are usually taken as 20°F (−6.7°C) and 70% relative humidity, while indoor conditions are taken as 70°F (21.1°C) and 35% relative humidity. These values yield an absolute humidity difference ($W_i - W_o$) of 0.0032 lb/lb (kg/kg) of dry air for use in Equation 1. With the required lowering of interior relative humidity with outdoor temperature, indicated in Table 1, these values represent the point where maximum load occurs.

### Ventilation Rate

Ventilation of the humidified space may be the result of natural infiltration, either alone or in combination with intentional mechanical ventilation. Natural infiltration varies according to the indoor-outdoor temperature difference, wind velocity, and tightness of construction, as discussed in Chapter 22 of the 1985 FUNDAMENTALS Volume. The rate of mechanical ventilation may be determined from building design specifications or estimated from fan performance data.

In load calculations, it may be necessary to consider the water vapor removed from the air during cooling by air-conditioning or refrigeration equipment. This moisture may have to be replaced by humidification equipment to maintain the desired level of relative humidity in certain industrial projects where the load may be greater than that required for ventilation and heating.

Estimates of infiltration rate are made in determining heating and cooling load calculations for buildings; these values also apply to humidification load calculations. For residences where such data are not available, it may be assumed that a *tight house* will have an air change rate of 0.5/h; an *average house*, 1/h; and a *loose house* may have as high as 1.5 air changes/h. A tight house can be assumed to be well insulated and to have vapor barriers, tight storm doors, windows with weather stripping, and a dampered fireplace. An average house will be insulated and have vapor barriers, loose storm doors and windows, and a dampered fireplace. A loose house will probably be one constructed before 1930 and have little or no insulation, no storm doors or windows, no weather stripping, no vapor barriers, and quite often, a fireplace without an effective damper. For building construction, refer to local codes and building specifications.

### Additional Moisture Losses

Hygroscopic materials, which have a lower moisture content than materials in the humidified space, absorb moisture and place an additional load on the humidification system. An estimate of this load depends on the absorption rate of the particular material selected. Chapter 28 of the 1987 HVAC Volume lists the equilibrium moisture content of hygroscopic materials at various relative humidities.

In cases where humidity levels must be maintained regardless of condensation on exterior windows and walls, the dehumidifying effect of these surfaces constitutes a load that may need to be considered, if only on a transient basis. The loss of water vapor by diffusion through enclosing walls to the outside or areas at a lower vapor pressure may also be involved in some applications. The properties of materials and flow equations given in Chapter 20 of the 1985 FUNDAMENTALS Volume can be applied in such cases. Normally, this will constitute a small load unless openings exist between the humidified space and adjacent rooms at lower humidities.

### Internal Moisture Gains

The introduction of a hygroscopic material can cause moisture gains to the space if its moisture content is above that of the space conditions. Similarly, moisture may diffuse through walls separating the space from areas of higher vapor pressure or move by convection through openings in these walls (Brown, Solvason, and Wilson 1963).

**Table 2 Moisture Production from Various Residential Operations (Hite and Dry 1948)**

| Operation | | | Moisture lb | (kg) |
|---|---|---|---|---|
| Floor mopping—80 ft² (7.4 m²) kitchen | | | 2.40 | (1.09) |
| Clothes drying* (not vented) | | | 26.40 | (11.97) |
| Clothes washing* | | | 4.33 | (1.96) |
| Cooking (not vented)* | *From Food* | *From Gas* | | |
| Breakfast | 0.34 (0.16) | 0.56 (0.25) | 0.90 | (0.41) |
| Lunch | 0.51 (0.23) | 0.66 (0.33) | 1.17 | (0.53) |
| Dinner | 1.17 (0.53) | 1.52 (0.69) | 2.69 | (1.22) |
| Bathing—shower | | | 0.50 | (0.23) |
| Bathing—tub | | | 0.12 | (0.05) |
| Dishwashing* | | | | |
| Breakfast | | | 0.20 | (0.09) |
| Lunch | | | 0.15 | (0.07) |
| Dinner | | | 0.65 | (0.29) |
| Human Contribution—Adults | | | | |
| When resting | | Per hour | 0.2 | (0.09) |
| Working hard | | Per hour | 0.6 | (0.27) |
| Average | | Per hour | 0.4 | (0.18) |
| House plants | | Per hour | 0.04 | (0.02) |

*Based on family of four.

Moisture contributed by human occupancy is affected by the number of occupants and their degree of physical activity. Estimates of the rate of moisture gain from human sources may be made from the data of Table 2. This table also lists the moisture production from household operations and equipment (Hite and Dry 1948). As a guide to residential applications, the average rate of moisture production for a family of four may be taken as 0.7 lb/h (0.32 kg/h).

Industrial processes may constitute additional moisture sources. Single-color offset printing presses, for example, may give off 0.45 lb/h (0.20 kg/h) of water. Information on process contributions can best be obtained from the manufacturer of the particular equipment.

## EQUIPMENT

Generally, humidification equipment can be classified as either residential or industrial, although residential humidifiers can be used for small industrial applications, and small industrial units can be used in large homes. Equipment designed for use in central air systems also differs from that for space humidification, although some units are adaptable to both.

Air washers and evaporative coolers may be used as humidifiers, but they are usually selected for some additional function such as air cooling or air cleaning, as discussed in Chapter 4.

The output rates for residential humidifiers are generally based on 24 hours of operation; output rates for industrial and commercial humidifiers are based on 1 hour of operation. During normal usage, the time of operation can be considerably less, which should be considered in design. Published evaporation rates are established by equipment manufacturers based on certain test criteria that are not consistent among manufacturers. Rates and test methods should be evaluated during selection of equipment. For residential rating, the Air-Conditioning and Refrigeration Institute (ARI) has established *Standard* 610-82 for central system humidifiers. ANSI/AHAM *Standard* HU-1-1980 addresses self-contained units.

### Residential Humidifiers for Central Air Systems

This unit depends on airflow in the heating system for evaporation and distribution. General principles of operation and description of equipment are as follows:

**Pan Type.** Output varies with temperature, humidity, and airflow in the system.

1. *Basic Pan:* A shallow pan is normally installed within the furnace plenum with a flow-control device connected to the household water supply to maintain a constant water level in the pan. Humidification rate is low.
2. *Electrically Heated Pan Type:* An electric heater increases the water temperature in the pan, thus increasing the rate of evaporation. Humidification rate is positive.

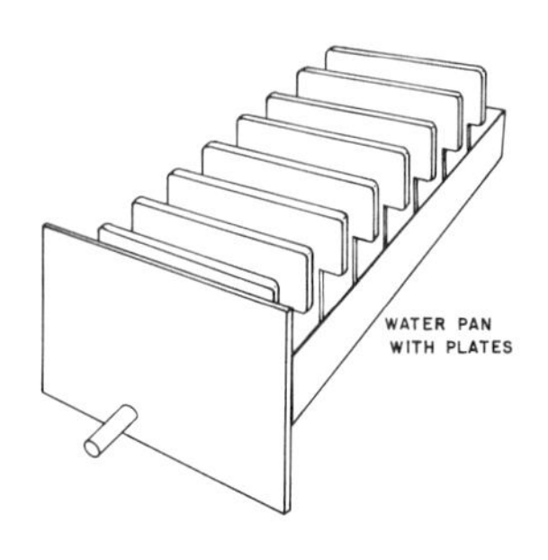


**Fig. 2 Pan Humidifier**

3. *Pans with Wicking-Type Plates:* This is similar to the basic pan, except that a number of vertical, water-absorbent plates are added to the pan to increase the wetted surface area, as in Figure 2. Humidification rate is low.

**Wetted Elements.** Output varies with temperature, humidity, and airflow in the system.

These units make use of an open-textured, wetted media through or over which air is circulated. The evaporating surface may take the form of a fixed pad, wetted by either sprays or water flowing through by gravity from a header at the top, or the pad may be a paddle wheel, drum, or belt rotating through a water reservoir.

Airflow through such units is usually accomplished in one of the three following ways:

1. *Fan Type:* These units have a small fan or blower to draw air from the furnace plenum, through the wetted pad, and back to the plenum. A fixed pad may be used as is illustrated in Figure 3, or a rotating drum-type pad as in Figure 4.

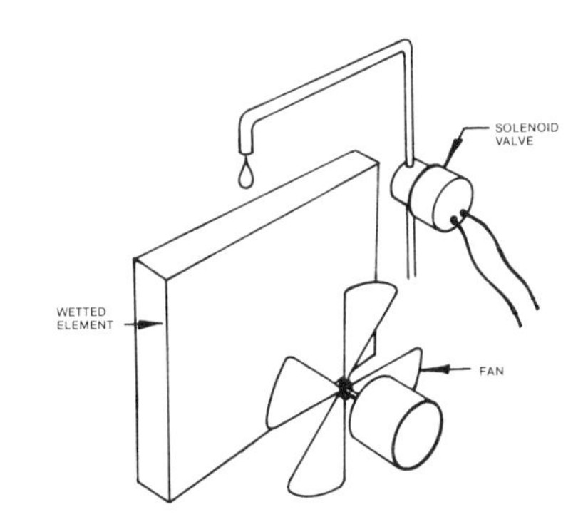


**Fig. 3 Power Wetted Element Humidifier**

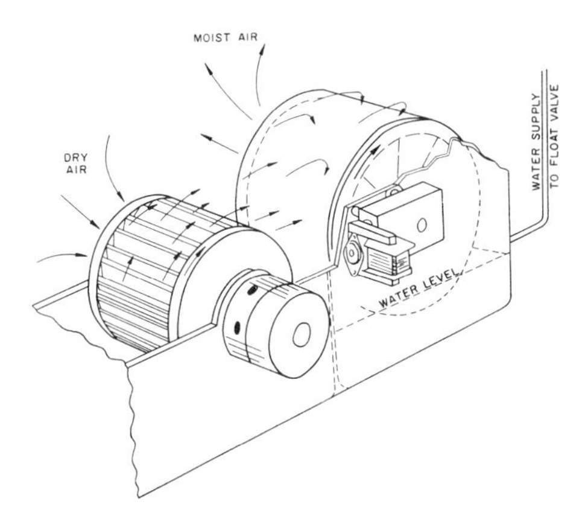


**Fig. 4 Wetted Drum Humidifier**

2. *Bypass Type:* These units do not have their own fan but rather are mounted on the supply or return plenum of the furnace with an air connection to the return plenum, as in Figure 5. The difference in static pressure created by the furnace blower serves to circulate air through the unit.
3. *Duct Mounted:* These units are designed for installation within the furnace plenum or ductwork with a drum element rotated by either the air movement within the duct or a small electric motor.

**Atomizing Types.** Output of this unit does not depend on the conditions of the air. The ability of the air to absorb the moisture depends upon temperature, airflow, and moisture content of the air moving through the system. In this type of humidifier, small particles of water are introduced directly into the airstream by the following:

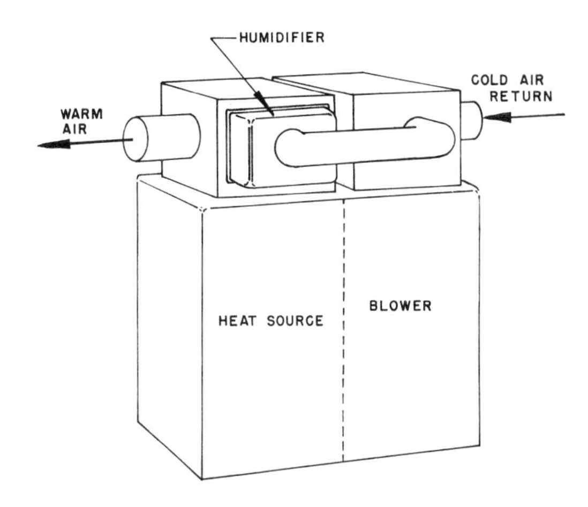


**Fig. 5 Bypass Wetted Element Humidifier**

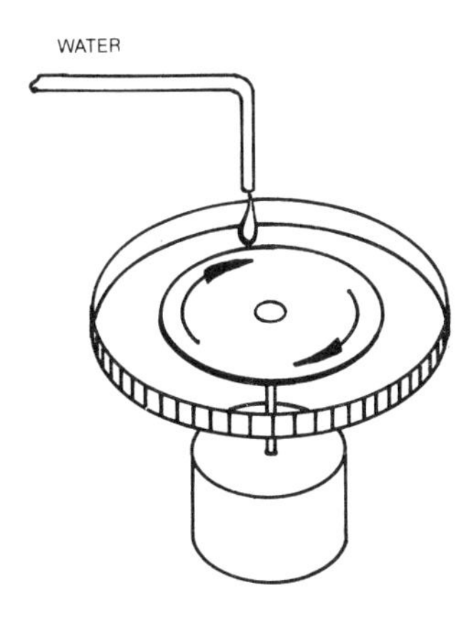


**Fig. 6 Atomizing Humidifier**

1. A spinning disk or cone, which breaks the water into a fine mist as in Figure 6
2. Spray nozzles that rely on water pressure to create fine droplets
3. A rotating disk, which slings water droplets into the airstream from a water reservoir
4. Sprays that use compressed air to create a fine mist

## Residential Humidifiers for Nonducted Applications

Many portable or room humidifiers are used in residences heated by such nonducted systems as hydronic or electric or in residences where the occupant is prevented from making a permanent installation. Most of these humidifiers are equipped with humidity controllers.

Portable units evaporate water by any of the previously described means, such as fixed or moving wetted element, atomizing spinning disk, heated pan, etc. They may be tabletop size or a larger, furniture-styled appliance, as in Figure 7. A multispeed motor on the fan or blower may be used to adjust output. Usually, the portable humidifiers require periodic filling from a bucket or filling hose.

Some portable units are offered with an auxiliary package for semipermanent water supply. This package includes such things as manual shutoff valve, float valve, copper or other tubing with

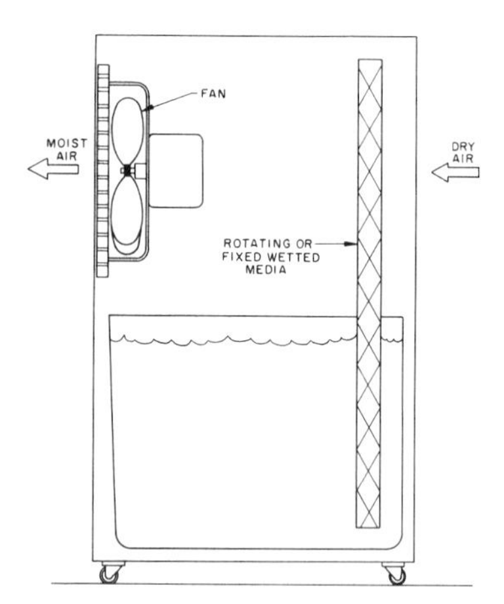


Fig. 7 Appliance Portable Humidifier

fittings, etc. The lack of provision for water overflow to drain may result in water damage.

Some units, using the same principles, may be recessed in the wall between studs, mounted on wall surfaces, or installed below floor level. These units are permanently installed in the structure and use forced-air circulation. They may have an electric element for reheat, when desired. Other types for use with hydronic systems involve a simple pan or pan-plate, either installed within a hot-water convector or using the steam from a steam radiator.

## Industrial Humidifiers for Central Air Systems

Humidifiers must be installed where the air can absorb the vapor and will not be cooled below the dew point downstream.

**Heated Pan Type.** These units offer a broad range of capacity and may be heated by an electrical element, steam, or hot-water coil. Some units can be attached directly under a system duct, as shown in Figure 8. Others have a fan, which allows them to be installed remote from the ductwork. Electric pan humidifiers are usually provided with a low water level cut-off switch as a protection device for the heating elements.

Steam coils are commonly used in pan humidifiers. At steam pressures above 15 psi (103 kPa) gauge, moisture carry-over occurs because of splashing caused by nucleate boiling. The pan humidification conditions are at optimum when nucleate boiling bubbles form on the coil surface but dissipate in the water above before reaching the surface. As these superheated bubbles dissipate, they superheat the water, which rises to the water surface, the liquid-vapor interface, where evaporation takes place. No water splashing occurs, and essentially no sensible heat is added to the airstream. If the bubbles reach the water surface at a rate sufficient to create violent surface boiling, splashing occurs and sensible heat is added to the airstream. To prevent boiling over, baffle splash eliminators should be used according to the manufacturers' instructions. Eliminators are essential where steam pressure is greater than 15 psi (103 kPa) gauge.

**Steam Type.** Direct steam-type humidifiers cover a wide range of designs and capacities. Since water vapor is steam at low pressure and temperature, the whole process can be simplified by introducing steam directly into the air to be humidified. This method is essentially an isothermal process because the temperature of the air remains constant as the moisture is added in vapor form. The steam-control valve may be modulating or two-position in response to a humidity controller. The steam may be either used from an external source with enclosed grid, cup, or jacketed dry steam humidifiers or produced within the humidifier, as in the self-contained type. When the steam is supplied from a separate source at a constant supply pressure, it responds quickly to system demand. Steam humidifiers usually prevent free water drops from being released into the conditioned air; they must be installed at a location where the air can absorb the vapor. Otherwise, condensation produces wet spots in the duct.

1. *Enclosed Steam Grid Humidifiers* (see Figure 9) should be used on low steam pressures—under 12 psi (80 kPa) gauge—to prevent splashing of condensate in the duct. The drain should be located on the side opposite the control valve. A drip leg—dimension H, a minimum of 12 in. (300mm), in Figure 9—should provide the pressure head to flow the condensate through the trap.
2. *The Cup or Pot-Type Steam Humidifier* (see Figure 10) is usually attached under a system duct. Steam is attached tangentially to the inner periphery of the cup by one or more steam inlets, depending on the capability of the unit. The steam supply line should have a suitable steam trap. There may be a tendency toward supersaturation due to stratification along the bottom of the duct. Multiple units may be required to produce satisfactory distribution. Under certain conditions, droplets of condensate may be injected into the airstream.
3. *The Jacketed Dry-Steam Humidifier* uses an integral steam valve with a steam-jacketed duct-traversing dispersing tube and condensate separator to prevent condensate from being introduced into the airstream (see Figure 11). An inverted bucket-type steam trap is required to drain the separating

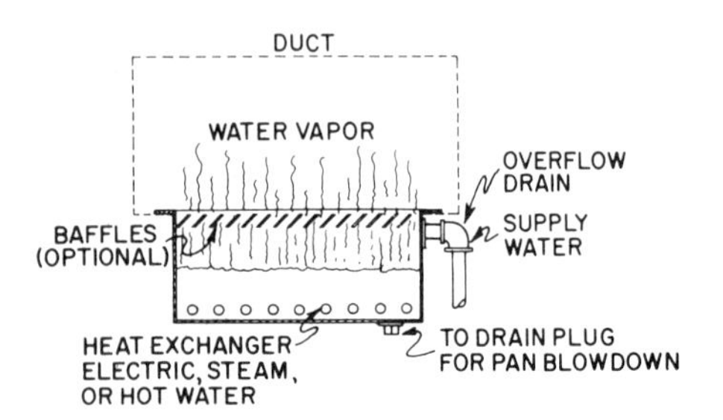


Fig. 8 Heated Pan Humidifier

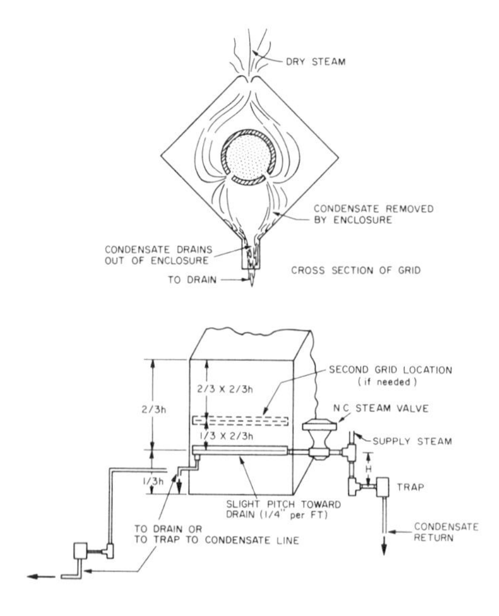


Fig. 9 Enclosed Steam Grid Humidifier

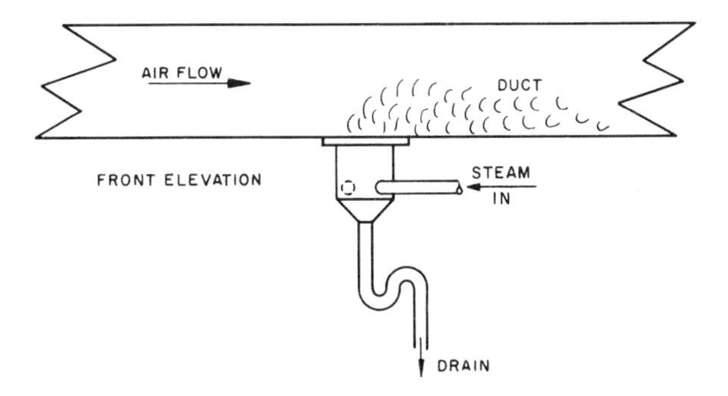


Fig. 10 Cup Steam Humidifier

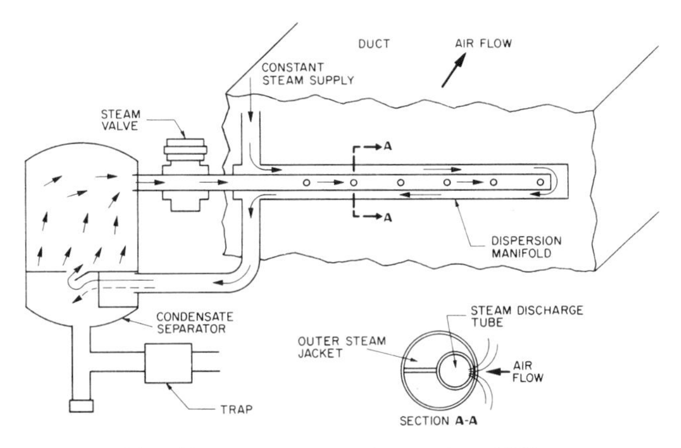


Fig. 11 Jacketed Dry-Steam Humidifier

chamber. This humidifier may be used without the jacketed tube in nonducted installations.

4. *The Self-Contained Steam Humidifier* converts tap water to steam by electrical energy (electrode boiler principle). This steam is injected into the duct system through a dispersion manifold (see Figure 12), or the humidifier may be free standing for nonducted applications.

**Atomizing Humidifiers, with Optional Filter-Eliminator** (see Figure 13). Centrifugal atomizers use a high speed disk, which slings water through a fine comb to create a fine mist that is introduced directly into the air where it is evaporated. The ability of the air to absorb the moisture depends on temperature, air velocity, and moisture content.

Where mineral fallout from hard water is a problem, optional filter-eliminators may be added to remove mineral dust from humidified air, or water demineralizers may be installed.

Two additional atomizing methods use nozzles; one uses water pressure and the other uses both air and water, as shown in Figure 14. Mixing air and water streams at combined pressures atomizes water into a fine mist, which is evaporated in the room air.

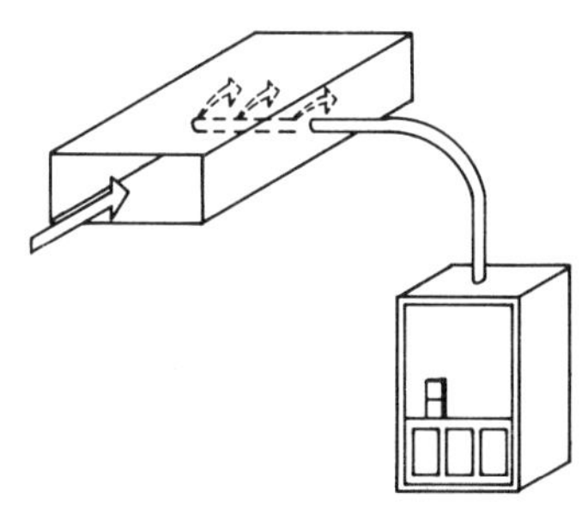

Fig. 12 Self-Contained Steam Humidifier

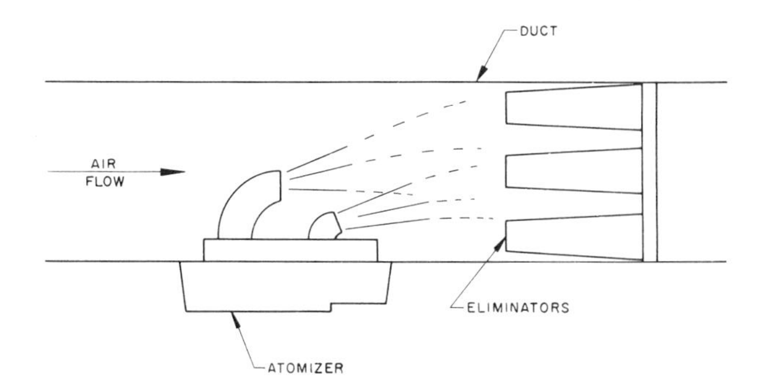


Fig. 13 Atomizing Humidifier with Optional Filter-Eliminator

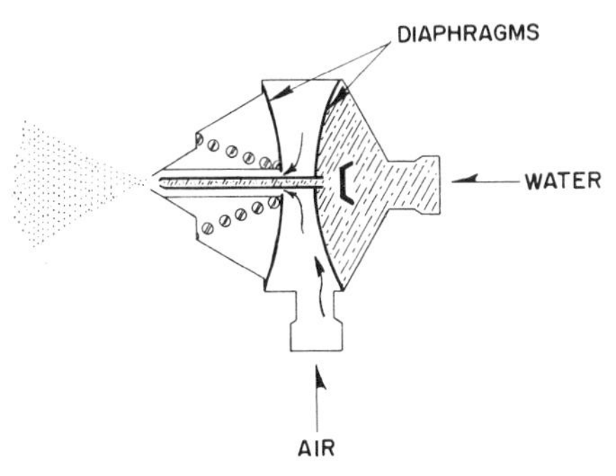


Fig. 14 Pneumatic Atomizing Humidifier

**Wetted Element Humidifier.** Wetted element humidifiers use a wetted media, sometimes in modular configurations, through or over which air is circulated to evaporate water. This unit depends on airflow for evaporation; rate varies with temperature, humidity, and velocity of air.

### Industrial Humidifiers For Nonducted Applications

All of the above methods of humidifying are available as unitary equipment. These units may include an integral air-moving device for better movement of humidified air.

## WATER SUPPLY

In areas with water of high mineral content, precipitated solids may be a problem. The precipitated solids clog nozzles, tubes, evaporative elements, and controls. In addition, solids allowed to enter the airstream leave a fine layer of white dust over furniture, floors, and carpets. If the water is softened by ion exchange, the residue increases because the magnesium or calcium salts are exchanged for those of sodium. Some wetted-element humidifiers bleed off and replace all or some of the water passing through the element to reduce the concentration of salts in the water reservoir.

Dust, scaling, biological organisms, and corrosion are all problems associated with water in humidifiers. Stagnant water provides a fertile breeding ground for algae and bacteria, which have been linked to odor and respiratory ailments. Bacterial slime reacts with sulfates in the water to produce hydrogen sulfide and its characteristic bad odor. Periodic maintenance and regular use of odorless microbiocides and other additives are needed to eliminate organisms from the water (Sriramamurty 1977).

### Scaling

Industrial pan humidifiers require water with low mineral content or nonstick coatings to prevent excessive scaling of the heating surfaces. The hardness of water is a measure of its scale-forming tendencies; hardness is indicated by the presence of calcium and magnesium ions, which are measured by the standard soap tests. Analysis results are now stated in parts per million, while the older practice states them in grains per gallon (17.1 ppm = 1 gr/gal). In SI units, 1 gr/gal = 17.1 mg/L.

Tests on steam pan humidifiers without nonstick coatings have shown that with untreated city water having a hardness of 137 ppm (8 gr/gal), the evaporating capacity of the humidifier after 80 hours running was reduced by 58 to 75% of the capacity measured with a clean coil in the steam gauge pressure range of 5 to 15 psi (35 to 103 kPa). Tests show that the nonstick coating on the heat-exchanger coil reduces the capacity loss stated above by about 60%.

Commercial water softeners may be used to eliminate scaling from hard water. However, treated water contains some additional dissolved solids, which are left in solution in the pan when water is vaporized into steam. When this concentration exceeds certain limits, the evaporating capacity is reduced. Complete purging (blowdown) of the water carrying excessive concentration of solids and replacement with fresh soft water keep the solids concentration within acceptable limits. The length of purging time depends on the hardness of the water, the evaporation rate, and the total time the unit is operating each day. Purging can be accomplished either automatically with a solenoid valve operated by a time clock or manually with a hand valve on the drain connection.

The use of demineralized water is an alternative recommended by humidifying equipment manufacturers. The construction materials of the humidifier and the piping must be capable of withstanding the corrosive effects of this water. Commercial demineralizers remove hardness and other total dissolved solids completely from the makeup water to the humidifier. They are more expensive than water softeners, but no humidifier purging is required. The sizing of demineralizers is based on the maximum required water flow to the humidifier and the amount of total dissolved solids in the makeup water. In all cases, periodic maintenance is required because of residue buildup. The conditions of the water dictate the amount of service necessary. Access for servicing should be considered during installation.

## HUMIDITY CONTROLS

Many humidity-sensitive materials are available. Most of them are organic, such as nylon, human hair, wood, and animal membranes. Sensors that change electrical resistance with humidity are also available. The most commonly used controllers and sensors may be classified as follows:

### Mechanical Controls

Mechanical sensors depend on a change in the length or size of the sensor as a function of relative humidity. The most commonly used sensors are synthetic polymers or human hair. They can be attached to a mechanical linkage to control a valve's or motor's mechanical, electrical, or pneumatic switching element. This design is suitable for most "human comfort" applications, but it may lack the necessary accuracy needed in industrial applications.

A humidity controller is normally designed to control at a set point selected by the user. Some controllers have a set-back feature that lowers the relative humidity setpoint as outdoor temperatures lower to reduce condensation within the structure.

### Electronic Controllers

As the humidity changes, electrical sensors change electrical resistance. They typically consist of two conductive materials separated by a humidity-sensitive hygroscopic insulating material (polyvinyl acetate, polyvinyl alcohol, or a solution of certain salts). Small changes are detected as air passes over the sensing surface.

The use of electronic humidity control is common in laboratory or process applications where precise control is required. It is also used to control fan speed on portable humidifiers to control humidity in the space more closely and reduce noise and draft to a minimum.

## SYSTEM DESIGN

In centrally humidified structures, the humidity controller is most commonly mounted in a space under control. Another method is to mount the humidity controller in the return air duct of an air-handling system to sense average relative humidity. Reference should be made to manufacturers' instructions regarding the controller's use on counterflow furnaces, because reverse airflow after the fan turns off can substantially shift the humidity control point in a home.

In central systems, a relatively low-temperature supply air might be required to compensate for the sensible heat gain of the room. This required temperature may be at a dew point below that necessary to maintain the desired relative humidity. This means the cooling air is saturated and cannot absorb more moisture. Operating the humidifier under these conditions causes condensation in the ducts and fogging in the room. High-capacity humidifiers installed in central air-conditioning systems should have a high-limit humidistat downstream of the humidifier to prevent condensation.

## REFERENCES

Brown, W.G.; K.R. Solvason; and A.G. Wilson. 1963. Heat and Moisture Flow Through Openings by Convection. ASHRAE *Journal*, 5(9):49, September.

Harris, C.M. 1963. Absorption of Sound in Air in the Audio-Frequency Range. *Journal of the Acoustical Society of America*, Vol. 35, January.

Hite, S.C. and J.L. Dry. 1948. Research in Home Humidity Control. *Experiment Station Research Series*, Purdue University, No. 106.

Luck, J.R. and L.W. Nelson. 1977. The Variation of Infiltration Rate with Relative Humidity in a Frame Building. ASHRAE *Transactions* 83(1).

Sereda, P.J. and R.F. Feldman. 1964. Electrostatic Charging on Fabrics at Various Humidities. *Journal of the Textile Institute,* Vol. 55, No. 5, p. T288.

Solvason, K.R. and N.B. Hutcheon. 1965. Principles in the Design of Cabinets for Controlled Environments. *Humidity and Moisture, Measurement and Control in Science and Industry.* Reinhold Publishing Corporation, Vol. 2, p. 241, New York, NY.

Sriramamurty, D.V. 1977. Application and Control of Commercial and Industrial Humidifiers. ASHRAE *Transactions* 83(1).

Wilson, A.G. and W.P. Brown. 1964. Thermal Characteristics of Double Windows. *Canadian Building Digest,* Division of Building Research, National Research Council, No. 58, Ottawa, Ontario.

## BIBLIOGRAPHY

Bender, D.O. 1973. Humidification in Pressrooms. ASHRAE *Symposium Bulletin,* NO-72-12.

Heiman, R.I. and O.E. Ulrich. 1973. Residential-Commercial-Industrial Humidifiers. ASHRAE *Symposium Bulletin,* NO-72-12.

Howery, J.F. and R.M. Pasch. 1977. Applying Residential Humidifiers to Central Systems. ASHRAE *Transactions,* 83(1).

O'Dell, L.R. 1977. Considerations for Sizing and Installing Commercial and Industrial Steam Humidifiers. ASHRAE *Transactions,* 83(1).

Smith, G.D. 1973. Considerations in Building Humidification. ASHRAE *Symposium Bulletin,* NO-72-12.

Spethmann, D.H. 1973. Humidity Control in Textile Mills. ASHRAE *Symposium Bulletin,* NO-72-12.

Wilkes, J.F. 1973. Commercial-Industrial Water and Steam Treatment—Effect on Humidification. ASHRAE *Symposium Bulletin,* NO-72-12.

## CHAPTER 6

# AIR-COOLING AND DEHUMIDIFYING COILS



COIL equipment used for cooling an airstream under forced convection may consist of a single coil section or a number of individual coil sections built up into banks and assembled in the field. Coils are also used extensively as components in central station air-handling units, room terminals, and factory-assembled self-contained air conditioners. The applications of each type of coil are limited to the field within which the coil is rated. Other limitations are imposed by code requirements, proper choice of materials for the fluids used, the condition of the air handled, and economic analysis of the possible alternates for each installation.

## USES FOR COILS

Coils are used for air cooling with or without accompanying dehumidification. Examples of cooling applications without dehumidification are precooling coils that use well water or other relatively high temperature water to reduce the load on the refrigerating machinery, and chilled water coils that remove sensible heat from chemical moisture-absorption apparatus. Most coil equipment provides both air sensible cooling and dehumidification simultaneously. The assembly usually includes the means for cleaning air to protect the coil from accumulation of dirt and to keep dust and foreign matter out of the conditioned space. Although cooling and dehumidification are their principal functions, cooling coils are wetted also with water or a hygroscopic liquid to aid in air cleaning, odor absorption, or frost prevention. Coils are also evaporatively cooled with a water spray to improve efficiency or capacity. Chapter 4 details evaporative cooling.

The preparation of this chapter is assigned to TC 8.4, Air-to-Refrigerant Heat-Transfer Equipment.

## COIL CONSTRUCTION AND ARRANGEMENT

In fin coils, the external surface of the tubes is *primary,* and the fin surface is *secondary.* The primary surface generally consists of rows of round tubes or pipes that may be staggered or placed in line with respect to the airflow. Flattened tubes or those with other non-round internal passageways are sometimes used. The inside surface of the tubes is usually smooth and plain, but some designs have various forms of internal fins or turbulence promoters (either fabricated or extruded) to enhance performance. The individual tube passes in a coil are usually interconnected by return bends (or with hairpin bend tubes) to form the serpentine arrangement of multipass tube circuits.

Cooling coils for water or halocarbon refrigerants usually have aluminum fins and copper tubes, although copper fins on copper tubes and aluminum fins on aluminum tubes are also used. Adhesives are sometimes used to bond header connections, return bends, and fin-tube joints, particularly for aluminum-to-aluminum joints. Many makes of cooling coils of the lightweight extended-surface type for both heating and cooling are available. Common outside diameters are 5/16, 3/8, 1/2, 5/8, 3/4, and 1 in. (7.9, 9.5, 12.7, 15.9, 19.1, and 25.4 mm), with fins spaced 4 to 14 per inch (1.8 to 6.4 mm apart). Tube spacing ranges from 0.6 to 2.5 in. (16 to 64 mm) on equilateral (staggered) or rectangular (in-line) centers, depending on the width of individual fins and other performance considerations. Fins should be spaced according to the job to be performed, with special attention given to air friction, possibility of lint accumulation, and frost accumulation, particularly at lower temperatures.

### Water Coils

Good performance of water coils requires the elimination of air from the water circuit and the proper distribution of water.

Unless vented, air may accumulate in the coil tube circuits, reducing thermal performance and possibly causing noise or vibration in the piping system. Air vent connections are usually provided on the coil water headers. The system piping eliminates air, as described in Chapter 13 of the 1987 HVAC Volume. A properly designed water circuit ensures a pressure drop sufficient for adequate water distribution, but does not require an excessive pumping head where large water quantities are handled.

Depending on performance requirements, the water velocity inside the tubes usually ranges from approximately 1 to 8 fps (0.3 to 2.4 m/s), and the design water pressure drop across the coils varies from about 5 to 50 ft (15 to 150 kPa) water head. The water may contain considerable sand and other foreign matter—as in precooling coils using well water, or in applications where minerals in the cooling water deposit on and foul the tube surface. Removable water headers or a removable plug for each tube allows the tube to be cleaned, which ensures a continuation of rated performance after the cooling units have been in service. Where buildup of scale deposits or fouling of the water-side surface may occur, a scale factor is sometimes included when calculating thermal performance of the coils. Cupronickel, aluminum bronze, and other tube alloys help protect against corrosion. The core tubes of properly designed and installed coils should drain completely during the *off* cycle (Figure 1). Usually, drains are installed in the water piping at the coil headers.

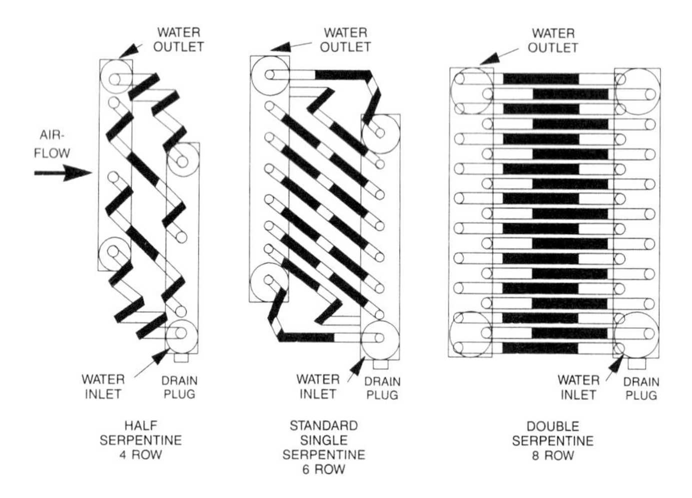


**Fig. 1 Typical Water Circuit Arrangement**

## Direct-Expansion Coils

Coils for halocarbon refrigerants present more complex cooling fluid distribution problems than do water or brine coils. The coil should be cooled effectively and uniformly throughout, with even refrigerant distribution. Direct-expansion coils are used on two types of refrigeration systems: *flooded* and *direct expansion*.

The flooded system is used mainly when a small temperature difference between the air and refrigerant is desired. Chapter 3 of the 1986 REFRIGERATION Volume describes flooded systems in more detail.

For direct-expansion systems, two of the most commonly used refrigerant liquid metering devices are the *capillary tube* and the *thermostatic expansion valve*. The capillary tube is applied in factory-assembled, self-contained air conditioners up to approximately 10-tons (35-kW) capacity and is used extensively on the smaller capacity models, such as window- or room-type units.

In this system, the bore and length of a capillary tube are sized so that at full load, under design conditions, just enough liquid refrigerant is metered from the condenser to the evaporator coil to be evaporated completely. While the capillary tube system does not operate over a wide range of conditions as efficiently as a thermostatic expansion valve system, its performance is good for design conditions. (Description and design information for capillary tube systems are given in Chapter 19 of this volume.)

The thermostatic expansion valve system is commonly used for all direct-expansion coil applications described in this chapter, particularly field-assembled coil sections, as well as those used in central air-handling units and the larger, factory-assembled hermetic air conditioners. This system depends on the thermostatic expansion valve to regulate the rate of refrigerant liquid flow automatically to the coil in direct proportion to the evaporation rate of refrigerant liquid in the coil, thereby maintaining the superheat at the coil suction outlet within the usual predetermined limits of 6 to 10°F (3.3 to 5.6°C). Because the thermostatic expansion valve responds to the superheat at the coil outlet, this superheat should be produced with the least possible sacrifice of active evaporating surface.

The length of a coil's refrigerant circuit, from the distributor feed tubes through the suction header, should be equal. The length of each circuit should be optimized to provide good heat transfer and oil return, and a reasonable pressure drop across the circuit.

To ensure reasonably uniform refrigerant distribution in multicircuit coils, a distributor placed between the thermostatic expansion valve and coil inlets divides the refrigerant equally among the coil circuits. The refrigerant distributor should distribute both liquid and vapor effectively because the entering refrigerant is usually a mixture of the two. Distributors can be placed in either the vertical or the horizontal position; however, the vertical position usually best distributes refrigerant between coil circuits.

The individual coil circuit connections from the refrigerant distributor to the coil inlet are made of small diameter tubing; they are all the same length and diameter so the same flow occurs between the refrigerant distributor and the coil. To approximate uniform refrigerant distribution, the refrigerant should flow to each refrigerant distributor circuit in proportion to the load on that circuit. The heat load must be distributed equally to each of its refrigerant circuits to obtain optimum coil performance. If the coil load cannot be uniformly distributed, the coil should be recircuited with more than one thermostatic expansion valve to feed the circuits (with separate coil suctions) so the refrigerant distribution matches the unequal circuit loading. Unequal circuit loading may be caused by such variables as uneven air velocity across the face of the coil, uneven entering air temperature, improper coil circuiting, oversized orifice size in distributor, etc.

## Control of Coils

Cooling capacity of water coils is controlled either by varying the rate of water flow or airflow. Water flow can be controlled by a three-way mixing valve or a throttling valve. For airflow control, face and bypass dampers are used. When cooling demand decreases, the coil face damper starts to close and the bypass damper opens. In some cases, the airflow is varied by controlling the fan capacity with speed controls, inlet vanes, discharge dampers, etc.

Chapter 51 of the 1987 HVAC Volume addresses controlling air-cooling coils to meet system or space requirements and factors to consider in sizing automatic valves for water coils. The selection and application of refrigerant flow-control devices—thermostatic expansion valves, capillary tubes, constant pressure expansion valves, evaporator pressure regulators, suction pressure regulators, and solenoid valves—as used with direct-expansion coils are covered in Chapter 19.

For factory-assembled self-contained systems or field-assembled systems employing direct-expansion coils equipped with thermostatic expansion valves, a single valve is sometimes used for each coil; in other cases, two or more valves are employed. The thermostatic expansion valve controls the refrigerant flow rate through the coil circuits so the refrigerant vapor at the coil outlet is superheated properly. Superheat is obtained with a suitable design of the coil and the proper valve selection. Thermostatic expansion valves alone do not control the refrigeration system's capacity or the temperature of the leaving air, nor do they maintain air conditions in specific spaces.

To match the refrigeration load requirements for the conditioned space with the cooling capacity of the coil(s), a thermostat, located in the conditioned space(s) or in the return air, temporarily interrupts refrigerant flow to the direct-expansion cooling coils by stopping the compressor(s) and/or closing the solenoid liquid valve(s). For jobs with only a single zone of conditioned space, the compressor's *on-off* control is frequently used to modulate coil capacity. Chapter 19 covers the selection and application of evaporator pressure regulators and similar regulators that are temperature operated (and respond to the temperature of the air or liquid being cooled).

Applications with multiple zones of conditioned space often use solenoid liquid valves to vary coil capacity. These valves should be used where thermostatic expansion valves feed certain types (or sections) of evaporators, which may—according to load variations—require a temporary but positive interruption of refrigerant flow. This particularly applies to multiple-section evaporators in a unit where one or more sections need to be shut off temporarily to reduce capacity. In such cases, a solenoid liquid valve should be installed directly ahead of each thermostatic expansion valve(s). More than one thermostatic expansion valve may be operated by a single solenoid valve. For coils with multiple thermostatic expansion valves, there are three arrangements: (1) face control, in which the coil is divided across its face; (2) row control; and (3) interlaced circuit (Figure 2).

Face control, the most popular arrangement, equally loads all refrigerant circuits within the coil. Face control has the disadvantage of permitting reevaporation of condensate on the coil and bypassing air into the conditioned space during partial load conditions when some of the thermostatic expansion valves are on an off cycle. Row control, seldom available as standard equipment, eliminates air bypassing during partial load operation and minimizes condensate reevaporation. Interlaced circuit control uses whole face area and depth of coil when some of the thermostatic expansion valves are shut off.

### Flow Arrangement

The relation of the fluid flow within the coil tubes to the coil depth greatly influences heat transfer performance. Air-cooling and dehumidifying coils are usually multi-row and circuited for counterflow operation. Air enters at right angles to the coil's tube face (coil height), where the refrigerant's outlet header is located. The air exits on the opposite side of the coil, where the corresponding inlet header is located. A counterflow arrangement transfers the greatest amount of heat with the shortest coil row depth.

Most direct-expansion coils operate in a counterflow manner, but proper superheat control may require a combination of parallel and counterflow. (Air flows in the same direction as the refrigerant in parallel flow operation.)

*Coil hand* refers to either the right hand (RH) or left hand (LH) arrangement of a multi-row counterflow coil. No convention of what constitutes LH or RH exists, so manufacturers usually establish a convention for their own coils. Most manufacturers designate the location of the inlet water header or refrigerant distributor as the coil hand reference point. Figure 3 illustrates the more widely accepted coil hand designation for multi-row water or refrigerant coils.

### Applications

Figure 4 shows a typical arrangement of coils in a field built-up central station system. All air should be filtered to prevent dirt, insects, and foreign matter from accumulating on the coils. The cooling coil and humidifier (when used) should include a drain pan to catch the condensate formed during the cooling cycle and the excess water from the humidifier. The drain connection should be on the downstream side of the coils, made of ample size (not less than 1¼-in [32-mm] pipe), have accessible cleanouts, and discharge to an indirect waste or storm sewer. The drain also requires a deep-seal trap so there is no possibility of sewer gas entering the system. Precautions must be taken if a possibility of freezing exists. The drain pan, unit casing, and water piping should be insulated to prevent sweating.

Factory-assembled central station air-handling units incorporate the design features outlined above. Generally, these units will accommodate various sizes, types, and row depths of cooling and heating coils to meet the job requirements and eliminate the need for a field built-up central system.

The design features of the coil (fin spacing, tube spacing, type of fins), together with the amount of moisture on the coil and the degree of surface cleanliness, determine the air velocity at which condensed moisture is blown off the coil. Condensed

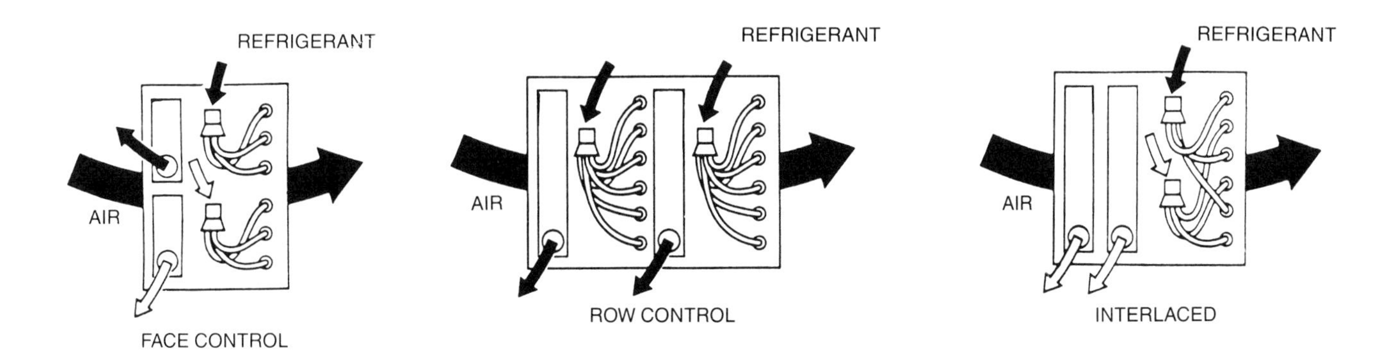


**Fig. 2 Arrangements for Coils with Multiple Thermostatic Expansion Valves**

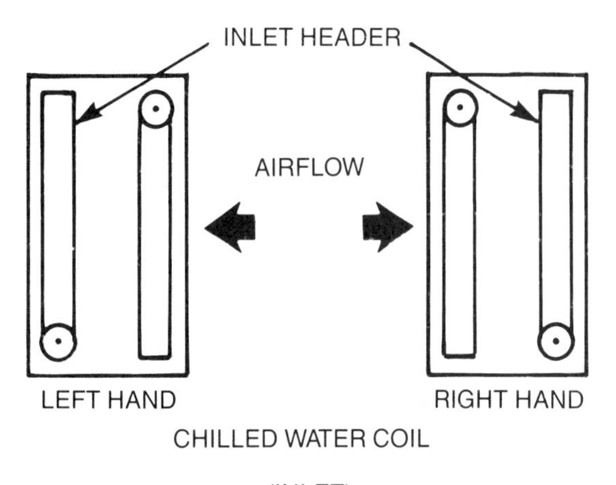


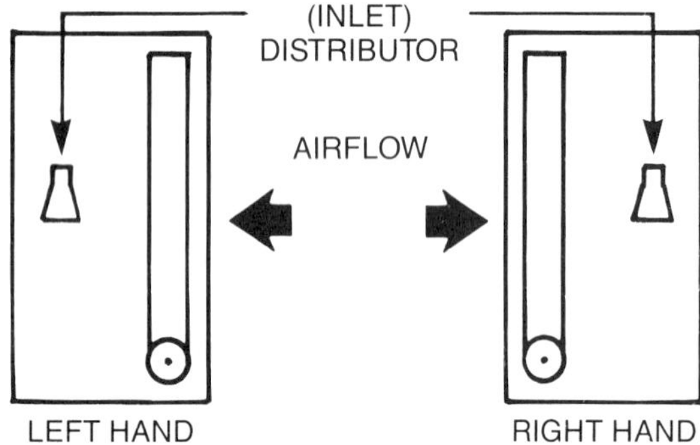


**Fig. 3 Typical Coil Hand Designation**

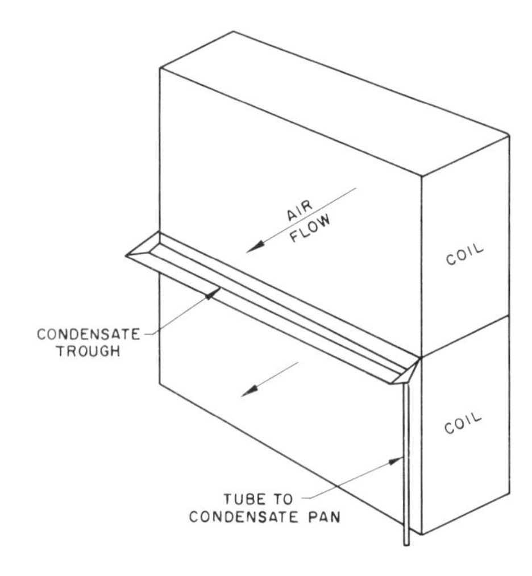


**Fig. 5 Coil Bank Arranged with Intermediate Condensate Trough**

water begins to be blown off the coil at air face velocities above 600 fpm (3 m/s). Water blown off the coils into air ductwork external to the air-conditioning unit should be avoided. However, water blowoff from the coils is not necessarily harmful if the unit is arranged to catch and dispose of the condensate. When water is likely to carry over from the air conditioning unit into external air ductwork, and no other means of prevention is provided, eliminator plates should be installed on the downstream side of the coils. Usually, eliminator plates are not included in central station air-handling units or factory-assembled self-contained air conditioners because other means of preventing carryover are included in the design. When a number of coil sections are stacked one above another, the condensate is carried into the airstream as it drips from one coil to the next. In this case, an intermediate drain pan or condensate trough (Figure 5) collects the water and conducts it directly to the main drain pan.

In field-assembled systems or factory-assembled central station air-handling units, the fans are usually positioned downstream from the coil(s) in a *draw-through* arrangement, which provides uniform airflow over the entire coil frontal area. For some multizone systems, particularly factory-assembled self-contained unit designs, the fans are located upstream from the coil in a *blow-through* arrangement that may require air baffles or diffuser plates between the fan and the cooling coil to obtain uniform air distribution over the coil face.

Where suction line risers are used for air-cooling coils in direct-expansion refrigeration systems employing a Group I volatile refrigerant, the suction line must be sized properly to ensure oil return from coil to compressor at minimum load conditions. The oil return provision is normally intrinsic with factory-assembled self-contained-type air conditioners, but must be considered for factory-assembled central station units or field-installed cooling coil banks where suction line risers are required and are assembled at the job site. Chapter 3 of the 1986 REFRIGERATION Volume describes the sizing, design, and arrangement of suction lines and their risers.

The enclosure around a coil or filter bank should be corrosion resistant and have adequate access doors for the changing of air filters, cleaning of coils, adjusting expansion valves, and oiling motors.

Cooling and dehumidifying coils are sometimes sprayed with water to increase the rate of heat transfer, provide outlet air approaching saturation, and continually wash the surface of the coil. Coil sprays require a collecting tank and recirculating pump, as shown in Figure 6. The figure also shows an air bypass, which helps a thermostat control hold the humidity ratio by returning air through the bypass, instead of mixing return and outdoor air.

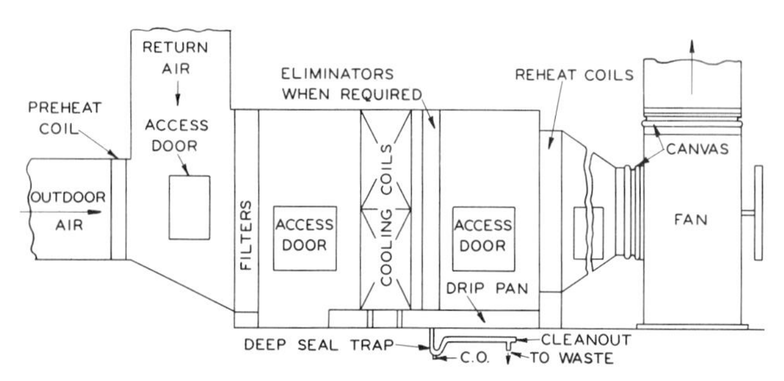


**Fig. 4 Typical Arrangement of Cooling Coils in a Built-Up Central System**

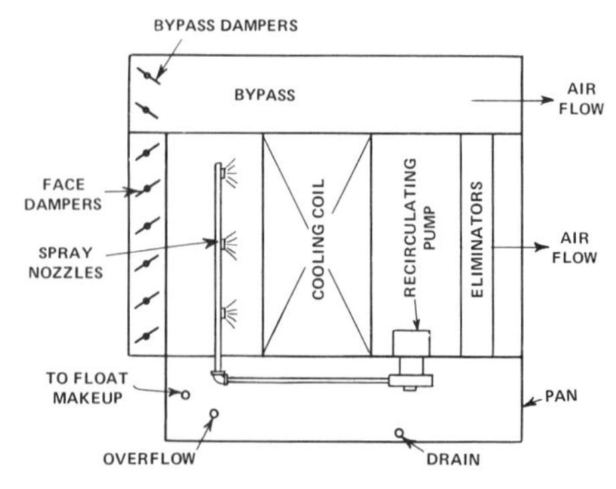


**Fig. 6 Sprayed Coil System with Air Bypass**

## COIL SELECTION

When selecting a coil, the following factors should be considered:

1. The job requirements: cooling, dehumidifying, and the capacity required to maintain balance with other system components, such as compressor equipment in the case of direct-expansion coils
2. The temperature of entering air: dry bulb only if there is no dehumidification; dry and wet bulb if moisture is to be removed
3. Available cooling media and operating temperatures
4. Space and dimensional limitations
5. Air quantity limitations
6. Allowable frictional resistances in air circuit (including coils)
7. Allowable frictional resistances in cooling media piping system (including coils)
8. Characteristics of individual coil designs
9. Individual installation requirements, such as the type of automatic control to be used or the presence of corrosive atmosphere
10. Coil air face velocity

The duties required may be determined from information in Chapter 26 of the 1985 FUNDAMENTALS Volume.

The air quantity is influenced by such factors as design parameters, codes, space, and equipment. The resistance through the air circuit influences the fan power and speed. This resistance may be limited to allow the use of a given size fan motor, to keep the operating expense low, or because of sound-level requirements. The air friction loss across the cooling coil—in summation with other series air pressure drops for such elements as air filters, water sprays, heating coils, air grilles, and ductwork—determines the static pressure requirement for the complete airway system. The static pressure requirement is used in selecting the fans and drives to obtain the design air quantity under operating conditions. Fan selection is described in Chapter 3.

The air face velocity is determined by economic evaluation of initial and operating costs for the complete installation as influenced by (1) heat transfer performance of the specific coil surface type for various combinations of face areas and row depths as a function of the air velocity and (2) air side frictional resistance for the complete air circuit (including coils), which affects fan size, power, and sound-level requirements. The allowable friction through the water or brine circuit may be dictated by the head available from a given size pump and pump motor.

Interference with uniform airflow through the coil affects performance. Interference may be caused by air entering at odd angles or by inadvertent blocking of a portion of the coil face. To obtain rated performance, the air quantity must be adjusted on the job to correspond with the amount used when selecting the coil, and it must be kept at that value. The most common causes of an air quantity reduction are fouling of the filters and collection of dirt or frost on the coils. These difficulties can be avoided by proper design.

The required total heat capacity of the cooling coil should be in balance with the capacity of other refrigerant system components, such as the compressor, water chiller, condenser, and refrigerant liquid metering device. Chapter 18 describes methods of estimating balanced system capacity under various operating conditions when using direct-expansion coils for both factory and field-assembled systems.

Dehumidifying coils must have the proper surface area to maintain the air dry-bulb and wet-bulb temperatures desired in the conditioned space. Chapter 26 of the 1985 FUNDAMENTALS Volume gives a method for calculating sensible and total heat loads. With that information, the leaving air conditions at the coil necessary for maintaining the desired sensible-to-total heat ratio may be calculated.

Different air quantities through the coil (including outside and return air) can maintain the same air conditions in the room; however, for a given total air quantity with a fixed ratio of outside-to-return air, only one set of air conditions leaving the coil will maintain the room's design conditions. Once the air quantity and leaving air conditions at the coil are selected, usually only one combination of face area, row depth, and air face velocity will maintain the desired conditions. Thus, in making the final coil selection, the initial selection should be rechecked to be sure the leaving air conditions, as calculated by the coil selection procedure, matches those determined from the cooling load estimate. A recheck may also find a more economical design.

Coil ratings and selection procedures can be obtained from manufacturers' catalogs. Most catalogs contain extensive tables giving the performance of coils at various air and water velocities and entering humidity and temperatures; most manufacturers provide free computerized coil selections to potential customers. The final choice can then be made based on all system performance and economic requirements.

### Application Range

Based on information in ARI *Coil Standard* 410-81, dry surface (sensible cooling) coils and dehumidifying coils (which accomplish both cooling and dehumidification), particularly those used for field-assembled coil banks or factory-assembled central station air conditioners using different combinations of coils, are usually rated within the following limits:

| | |
|---|---|
| Entering air dry bulb | = 65 to 100 °F (18.3 to 37.8 °C) |
| Entering air wet bulb | = 60 to 85 °F (15.6 to 29.4 °C) |
| Air face velocity | = 300 to 800 fpm (1.5 to 4.1 m/s) [sometimes as low as 200 fpm (1.0 m/s) and as high as 1500 fpm (7.6 m/s)] |
| Evaporator refrigerant saturation temperature | = 30 to 55 °F (−1.1 to 12.8 °C) at coil suction outlet [refrigerant vapor superheat at coil suction outlet is 6 °F (3.3 °C) or higher] |
| Entering chilled water temperature | = 35 to 62 °F (1.7 to 18.3 °C) |
| Water quantity | = 1.2 to 6 gpm per ton (equivalent to a water temperature rise of from 4 to 20 °F) [0.02 to 0.10 L/s per kW]; (equivalent to a water temperature rise of from 51.8 to 35.6 °F [11.1 to 2.2 °C]) |
| Water velocity | = 1.0 to 8.0 fps (0.3 to 2.4 m/s) |

For special applications, the range in the above design variables may be exceeded.

### Sensible Cooling Coils

Because air is not dehumidified for this application, sensible coils are selected on the basis of dry-bulb temperatures involving sensible-heat transfer only—the same as with heating coils.

### Dehumidifying Coils

The ratio of air-side sensible-to-total heat removed varies in practice from about 0.6 to 1.0, *i.e.*, sensible heat is from 60 to

100% of the total, depending on the application. (See Chapter 26 of the 1985 FUNDAMENTALS Volume.) For a given coil surface design and arrangement, the required sensible heat ratio may be satisfied by wide variations in and combinations of air face velocity, in-tube temperature and flow rate, entering air temperature, coil depth, etc., although the variations may be self-limiting. The maximum coil air face velocity should be limited to a value that prevents water carryover into the air ductwork. Dehumidifying coils for comfort application are frequently selected in the range of 400 to 500 fpm (2 to 3 m/s) air face velocity.

The ratings of dehumidifying coils for factory-assembled self-contained air conditioners are generally determined in conjunction with laboratory testing for the system capacity of the complete unit assembly. The current industry standards call for ratings at 33.4 cfm/1000 Btu/h or 400 cfm/ton (53.6 L/s per kW) of refrigeration. The standard ratings at 80 °F (26.7 °C) dry bulb, and 67 °F (19.4 °C) wet bulb, are representative of the entering air conditions encountered in many comfort operations. While the indoor conditions are usually lower than 67 °F (19.4 °C) wet bulb, the introduction of outdoor air usually brings the mixture of air to the cooling coil up to about 67 °F (19.4 °C) wet bulb entering air condition at design conditions.

The selection of dehumidifying coils for field-assembled projects and central station air-handling units are usually selected according to coil rating tables. The airflow volume is usually based on standard air at 70 °F (20 °C) and a barometric pressure of 14.696 psi (101.325 kPa). Selecting coils from the load division indicated by the load calculation works satisfactorily for the usual human comfort applications. Additional design precautions and refinements are necessary for more exacting industrial applications and for all types of air conditioning in more humid areas. One of these refinements uses a separate cooling coil to cool and dehumidify the ventilation air before mixing it with recirculated air. This procedure removes one of the main sources of moisture in the usual application. Reheat is required for some industrial applications and is used on some commercial and comfort applications.

When checking the operation of dehumidifying coils in various air-conditioning installations, climatic conditions must be considered. The majority of problems are encountered at light-load conditions, when the cooling requirement is considerably less than at design conditions. In hot, dry climates, where the outdoor dew points are so consistently low, dehumidifying is not generally a problem, and the light-load condition does not pose any special problems. In hot, humid climates, the light-load condition has a higher proportion of moisture and a correspondingly lower proportion of sensible heat. The result is higher dew points in the conditioned spaces during light-load conditions, unless special means for controlling the inside dew points are used.

Freezing at light loads should be avoided. Generally, freezing occurs when the coil surface temperature falls to 32 °F (0 °C). Freezing does not occur with standard coils for comfort installations, unless the evaporating temperature at the coil outlet is about 25 to 30 °F saturated (−3.9 to −1.1 °C); the exact value depends on the design of the coil and the amount of loading. Although it is not customary to choose coil and condensing units to balance at low temperatures at peak loads, freezing may occur when the load decreases. The danger of freezing is increased if a bypass is used, causing less air to be passed through the coil at light loads.

## AIRFLOW RESISTANCE

A cooling coil's airflow resistance (air friction) depends on the physical geometry (tube size and spacing, fin configuration, and number of rows), the coil face velocity, and the amount of moisture on the coil. The coil air friction may also be affected by the degree of aerodynamic cleanliness of the coil core; burrs on fin edges may increase coil friction and increase the tendency to pocket dirt or lint on the surfaces. A completely dry coil, removing only sensible heat, offers less resistance to airflow than a dehumidifying coil removing both sensible and latent heat.

For a given surface and airflow, an increase in the number of rows or fin spacing increases the airflow resistance. Therefore, the final selection becomes the economic balancing of the initial cost of the coil against the operating costs of the coil geometry combinations available.

## HEAT TRANSFER

The heat transmission rate of air passing over a clean tube (with or without extended surface) to a fluid flowing within it, is impeded principally by three thermal resistances. The first is from the air to the surface of the tube, called the *external surface* or *air-film thermal resistance.* The second is the thermal resistance to the conduction of heat through the fin and tube metal. Finally, a tube-side film thermal resistance impedes the flow of heat between the internal surface of the metal and the fluid in the tube. For some applications, an additional thermal resistance is included to account for external and/or internal surface fouling. For the applications considered here, both the resistance of the metal to heat conduction and the tube-side film resistance are usually low, as compared with the air-side surface resistance. The external surface resistance of a coil that uses space economically and has low weight approaches the value of its metal and internal film resistances. External fins, for example, decrease the external surface resistance. Water sprays applied to the surface may slightly increase the overall heat transfer, although they may serve other purposes such as air and coil cleaning.

The transfer of heat between the cooling medium and the airstream is influenced by the following variables:

1. The temperature difference between fluids
2. The design and surface arrangement of the coil
3. The velocity and character of the airstream
4. The velocity and character of the coolant in the tubes

The rating of cooling coils for combined cooling and dehumidification is addressed later in this chapter. With water coils, the water temperature rises; with volatile refrigerants, an appreciable pressure drop and a corresponding change in evaporating temperature through the refrigerant circuit often occurs. The problem for direct-expansion coils is further complicated by refrigerant evaporating in part of the circuit and superheating in the remainder. With halocarbon refrigerants, a cooling coil is tested and rated with a specific distributing and liquid-metering device, and the capacities are stated for a given superheat condition of leaving vapor.

At the same air mass velocity, varying performance can be obtained, depending on the turbulence of airflow into the coil and the uniformity of air distribution over the coil face. The latter is necessary to obtain reliable test ratings and realize rated performance in actual installations. The air resistance through the coils assists in distributing the air properly, but the effect is frequently inadequate where the inlet duct connections are brought in at sharp angles to the coil face. Reverse air currents may pass through a portion of the coils. These currents reduce the capacity, but can be avoided with proper layout or by the use of inlet vanes or baffles. Coils should be selected from curves or tables of coil performance prepared from adequate and reliable laboratory tests. For cases when available data must be extended, or for the design of a single, unique installation, the following material and illustrative examples for calculating cooling coil performance are useful.

## PERFORMANCE OF SENSIBLE COOLING COILS

The performance of sensible cooling coils depends on the following factors:

(1) The overall coefficient ($U_o$) of sensible heat transfer between airstream and coolant fluid
(2) The mean temperature difference ($\Delta t_m$) between airstream and coolant fluid
(3) The physical dimensions of and data for the coil (such as $A_a$ and $A_o$) with characteristics of the heat-transfer surface

The sensible heat cooling capacity of a given coil is expressed by the following basic equation:

$$q_{td} = U_o F_s (A_a) N_r (\Delta t_m) \quad (1a)$$

*where*

$$F_s = A_o/(A_a)N_r \quad (1b)$$

Assuming no extraneous heat losses, the same amount of sensible heat is lost from the airstream:

$$q_{td} = w_a (c_p)(t_{a1} - t_{a2}) \quad (2a)$$

*where:*

$$w_a = 1248(A_a)V_a \quad (2b)$$

The same amount of sensible heat is absorbed by the coolant, and for a nonvolatile type is:

$$q_{td} = w_r (c_r)(t_{r2} - t_{r1}) \quad (3)$$

The mean temperature difference in Eq. (1a) for a nonvolatile coolant in thermal counterflow with the air, based on the conventional logarithmic value, is expressed as:

$$\Delta t_m = \frac{(t_{a1} - t_{r2}) - (t_{a2} - t_{r1})}{\ln_e[(t_{a1} - t_{r2})/(t_{a2} - t_{r1})]} \quad (4)$$

The proper temperature differences for various crossflow situations are given in many texts, including that by Mueller (1973). These calculations are based on various assumptions, among which is that $U$ is constant throughout. While this assumption generally is not valid for multirow coils, the use of crossflow temperature differences from Mueller (1973), or other texts, should be preferable to Eq. (4), which applies only to the counterflow. However, the use of the log mean temperature difference is widespread.

The overall heat-transfer coefficient ($U_o$) for a given coil design, whether bare-pipe- or finned-type, with clean, non-fouled surfaces consists of the combined effect of three individual heat-transfer coefficients:

(1) The film coefficient of sensible heat transfer ($f_a$) between air and the external surface of the coil
(2) The unit conductance ($1/R_{md}$) of the coil material, *i.e.*, tube wall, fins, ribs, etc.
(3) The film coefficient of heat transfer ($f_r$) between the internal coil surface and the coolant fluid within the coil

These three individual coefficients acting in series form an overall coefficient of heat transfer in accordance with the material given in Chapters 2 and 23 of the 1985 FUNDAMENTALS Volume. For a bare-pipe coil, the overall coefficient of heat transfer for sensible cooling (without dehumidification) can be expressed by a simplified basic equation:

$$U_o = \frac{1}{(1/f_a) + (D_o - D_i)/2k + (B/f_r)} \quad (5a)$$

When pipe or tube walls are thin and made of material with high conductivity (as in typical heating and cooling coils), the term $(D_o - D_i)/2k$ in Eq. (5a) frequently becomes negligible and generally is disregarded. (This effect in typical bare-pipe cooling coils seldom exceeds 1 to 2% of the overall coefficient.) Thus, in its simplest form, for bare pipe:

$$U_o = \frac{1}{(1/f_a) + (B/f_r)} \quad (5b)$$

For finned coils, the equation for the overall coefficient of heat transfer can be written:

$$U_o = \frac{1}{1/\eta(f_a) + (B/f_r)} \quad (5c)$$

*where*

$\eta$, called the *fin effectiveness,* allows for the resistance to heat flow encountered in the fins. It is defined as:

$$\eta = \frac{E(A_s) + A_p}{A_o} \quad (6)$$

For typical cooling surface designs, the surface ratio $B$ ranges from about 1.03 to 1.15 for bare-pipe coils and 10 to 30 for finned coils. Chapter 3 in the 1985 FUNDAMENTALS Volume describes how to estimate fin efficiency, $E$.

The tube side heat-transfer coefficient $f_r$ for non-volatile fluids can be calculated with the usual equations for flow inside tubes, as shown in Chapter 3 of the 1985 FUNDAMENTALS Volume. Estimation of the air side heat-transfer coefficient $f_a$ is more difficult because well verified general predictive techniques are not available. Hence, direct use of experimental data is usually necessary. For plate fin coils, some correlations that satisfy several data sets are available (McQuiston 1981, Kusuda 1969). Webb (1980) has reviewed the air-side heat-transfer and pressure-drop correlations for various geometries. Mueller (1973) and Chapter 3 of the FUNDAMENTALS Volume provide guidance on this subject.

For analyzing a given heat exchanger, the concept of *effectiveness* is useful. Expressions for effectiveness have been derived for various flow configurations and can be found in many textbooks and also in Mueller (1973) and Kusuda (1969). The cooling coils covered in this chapter actually involve various forms of crossflow. However, the case of counterflow is addressed here to illustrate the value of this concept. The effectiveness for counterflow heat exchangers is given by the following equations:

$$q_{td} = w_a (c_p)(t_{a1} - t_{r1})E_a \quad (7a)$$

*where*

$$E_a = \frac{t_{a1} - t_{a2}}{t_{a1} - t_{r1}} \quad (7b)$$

*or*

$$E_a = \frac{1 - e^{-c_o(1-M)}}{1 - (M)e^{-c_o(1-M)}} \quad (7c)$$

$$c_o = \frac{A_o U_o}{w_a c_p} = \frac{F_s N_r U_o}{1248 V_a c_p} \quad (7d)$$

*and*

$$M = \frac{w_a c_p}{w_r c_r} = \frac{1248 A_a V_a c_p}{w_r c_r} \quad (7e)$$

Note the two following special conditions:

If $M = 0$, then $E_a = 1 - e^{-c_o}$
If $M = 1$, then

$$E_a = \frac{1}{(1/c_o) + 1}$$

With a given design and arrangement of heat-transfer surface used as cooling coil core material for which basic physical and heat-transfer data are available to determine $U_o$ from Eqs. (5a), (5b), and (5c), the selection, sizing, and performance calculation of sensible cooling coils for a particular application generally fall within either of two categories:

1. The heat-transfer surface area ($A_o$) or the coil row depth ($N_r$) for a specific coil size is required and initially unknown. The sensible cooling capacity ($q_{td}$), flow rates for both air and coolant, entrance and exit temperatures of both fluids, and mean temperature difference between fluids are initially known or can be assumed or determined from Eqs. (2a) to (4). The required surface area ($A_o$) or coil row depth ($N_r$) can then be calculated directly from Eq. (1a).
2. The sensible cooling capacity ($q_{td}$) is initially unknown and required for a specific coil. The face area and heat-transfer surface area are known or can be readily determined. The flow rates for air and coolant and the entering temperatures of each are also known. The mean temperature difference ($\Delta t_m$) is unknown, but its determination is unnecessary to calculate the sensible cooling capacity ($q_{td}$), which can be found directly by solving Eq. (7a). Equation 7a also provides a basic means to determine $q_{td}$ for a given coil or related family of coils, over the complete rating ranges of both air and coolant flow rates and operating temperatures.

The two categories of application problems are illustrated, respectively, in *Examples 1* and *2*:

---

**Example 1:** Standard air flowing at a mass rate equivalent to 4250 L/s is to be cooled from 29.4 to 23.9°C, using 2.5 L/s chilled water supplied at 10°C in thermal counterflow arrangement. Assuming an air face velocity of $V_a$ = 3 m/s and no air dehumidification, calculate the coil face area ($A_a$), sensible cooling capacity ($q_{td}$), required heat-transfer surface area ($A_o$), coil row depth ($N_r$), and coil air-side pressure drop ($\Delta P_{st}$), using a clean, non-fouled, thin-walled bare copper tube surface design for which the following physical and performance data have been predetermined:

$B$ = 1.07 surface ratio
$c_p$ = 1.0 kJ/kg • °C
$F_s$ = 1.41 (external surface area)/(face area)(rows deep)
$f_a$ = 85 W/m$^2$ • °C
$f_r$ = 4544 W/m$^2$ • °C
$\Delta P_{st}/N_r$ = 6.7 Pa/rows deep

*Solution:* Calculate the coil face area required:

$$A_a = \text{L/s} \bullet V_a = 4250/3(1000\ \text{L//m}^3) = 1.4\ \text{m}^2$$

Neglecting the effect of tube wall, from Eq. (5b):

$$U_o = \frac{1}{(1/85) + (1.07/4544)} = 83\ \text{W/m}^2\ (\text{external}) \bullet {}^\circ\text{C}$$

From Eqs. (2a) and (2b), the sensible cooling capacity is:

$$q_{td} = 1248 + 1.4 + 3 + 1.0(29.4 - 23.9) = 28\ 830\ \text{W}$$

From Eq. (3):

$$t_{r2} = 10 + (28\ 830)/(63)(168) = 12.7^\circ\text{C}$$

From Eq. (4):

$$\Delta t_m = \frac{(29.4 - 12.7) - (23.9 - 10)}{\ln[(29.4 - 12.7)/(23.9 - 10)]} = 14.9^\circ\text{C}$$

From Eqs. (1a) and (1b), the surface area required is:

$$A_o = 28\ 830/(83)(14.9) = 23.3\ \text{m}^2\ \text{external surface}$$

From Eq. (1b), the required row depth is:

$$N_r = 23.3/(1.41)(1.4) = 11.8\ \text{rows deep}$$

The installed 1.4 m$^2$ coil face, 12 rows deep, slightly exceeds the required capacity. The air-side pressure drop for the installed row depth is then:

$$\Delta P_{st} = (\Delta P_{st}/N_r)N_r = 6.7 + 12 = 80\ \text{Pa at } 20^\circ\text{C}$$

In *Example 1,* for some applications where such items as $V_a$, $w_r$, $t_{r1}$, f$_r$, etc., may be arbitrarily varied with a fixed design and arrangement of heat-transfer surface, a trade-off between coil face area ($A_a$) and coil row depth ($N_r$) is sometimes made to obtain alternate coil selections that would produce the same sensible cooling capacity ($q_{td}$). In the above solution, *e.g.*, an eight-row coil could be selected, but would require a larger face area ($A_a$) with lower air face velocity ($V_a$) and a lower air-side pressure drop ($\Delta P_{st}$).

**Example 2:** An air-cooling coil using a finned-tube-type heat-transfer surface has physical data as follows:

$A_a$ = 0.93 m$^2$
$A_o$ = 75 m$^2$ external
$B$ = 20 surface ratio
$F_s$ = 27 (external surface area)/(face area)(rows deep)
$N_r$ = 3 rows deep

Air at a face velocity of $V_a$ = 4 m/s and 35°C entering air temperature is to be cooled by 0.95 L/s of well water supplied at 12.8°C. Calculate the sensible cooling capacity ($q_{td}$), leaving air temperature ($t_{a2}$), leaving water temperature ($t_{r2}$), and air-side pressure drop ($\Delta P_{st}$). Assume clean and non-fouled surfaces, thermal counterflow between air and water, no air dehumidification, standard barometric air pressure, and that the following data are available or can be predetermined:

$c_p$ = 1.0 kJ/kg • °C
$f_a$ = 96.6 W/m$^2$ • °C
$f_r$ = 2840 W/m$^2$ • °C
$\eta$ = 0.9 fin effectiveness
$\Delta P_{st}/N_r$ = 55 Pa/rows deep

*Solutions:* From Eq. (5c):

$$U_o = \frac{1}{[1/(0.9 \times 96.6) + (20/2840)]} = 54\ \text{W/m}^2\ (\text{external}) \bullet {}^\circ\text{C}$$

From Eq. (7d):

$$c_o = \frac{75 \times 54}{1248 \times 0.93 \times 4 \times 1.0} = 0.87$$

From Eq. (7e):

$$M = \frac{1248 \times 0.93 \times 4 \times 1.0}{63 \times 62.9} = 1.17$$

From Eq. (7c):

$$-c_o(1 - M) = -0.87(1 - 1.17) = 0.148$$

and

$$E_a = \frac{1 - e^{0.148}}{1 - 1.17 \times e^{0.148}} = \frac{1 - 1.159}{1 - 1.36} = 0.442$$

From Eq. (7a), the sensible cooling capacity is:

$$q_{td} = 1248 \times 0.93 \times 4 \times 1.0(35 - 12.8) \times 0.442 = 45\ 555\ \text{W}$$

From Eqs. (2a) and (2b), the leaving air temperature is:

$$t_{a2} = 35 - \frac{45\ 555}{1248 \times 0.93 \times 4 \times 1.0} = 25.2^\circ\text{C}$$

From Eq. (3), the leaving water temperature is:

$$t_{r2} = 12.8 + \frac{45\ 555}{63 \times 62.9} = 24.3^\circ\text{C}$$

The air-side pressure drop is:

$$\Delta P = (\Delta P/N)N = 55 \times 3 = 165\ \text{Pa}$$

---

The preceding equations and two illustrative examples demonstrate the method for calculating thermal performance of sensible cooling coils that operate with dry surface. However, when cooling coils operate wet or act as dehumidifying coils, the performance cannot be predicted unless the effect of air-side moisture transport (or latent heat removal) is included.

## PERFORMANCE OF DEHUMIDIFYING COILS

A *dehumidifying coil* normally removes both moisture and sensible heat from entering air. In most air-conditioning processes, the air to be cooled is a mixture of water vapor and dry gases. Both lose sensible heat during contact with the first part of the cooling coil, which functions as a dry cooling coil. Moisture is removed only in the part of the coil that is below the dew point of the entering air. Figure 7 shows the assumed psychrometric conditions of this process. As the leaving dry-bulb temperature is lowered below the entering dew point temperature, the difference between the leaving dry-bulb temperature and the leaving dew point for a given coil, airflow, and entering air condition is lessened.

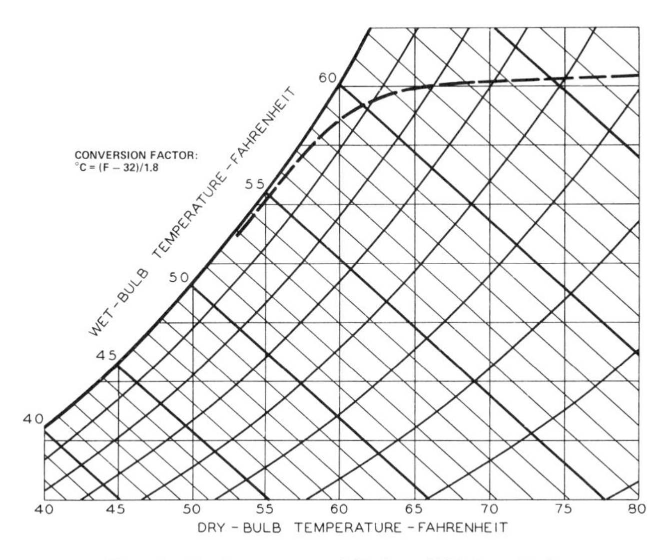


**Fig. 7 Performance of Dehumidifying Coil**

When the coil starts to remove moisture, the cooling surfaces carry both the sensible and latent heat load. As the air approaches saturation, each degree of sensible cooling is nearly matched by a corresponding degree of dew point decrease. In contrast, the latent heat removal per degree of dew point change varies considerably. The following table compares the amount of moisture removed from saturated air that is cooled from 60 to 59°F (16 to 15°) and air that is cooled from 50 to 49°F (10 to 9°C).

| Dew Point | $W_s \times 10^3$lb/lb | Dew Point | $W_s \times 10^3$lb/lb |
|---|---|---|---|
| 60°F | 11.087 | 50°F | 7.661 |
| 59°F | 10.692 | 49°F | 7.378 |
| Difference | 0.395 | Difference | 0.283 |

| Dew Point | $W_s$, g/kg | Dew Point | $W_s$, g/kg |
|---|---|---|---|
| 16°C | 11.413 | 10°C | 7.661 |
| 15°C | 10.692 | 9°C | 7.157 |
| Difference | 0.721 | Difference | 0.504 |

These numerical values conform to Table 1 in Chapter 6 of the 1985 FUNDAMENTALS Volume.

The combination of the coil with its refrigerant distributor equipment must be tested at the higher and lower capacities of its rated range. Testing at the lower capacities checks whether the refrigerant distributor provides equal distribution and if the control is able to modulate without hunting. Testing at the higher capacities checks the maximum feeding capacity of the flow control device at the greater pressure drop that occurs in the coil system.

Most manufacturers develop and produce their own performance rating tables from data obtained from suitable tests. ASHRAE *Standard* 33-78 gives one test method for rating cooling coils. ARI *Standard* 410-81 gives a method for rating thermal performance or dehumidifying coils by extending test data (determined from laboratory tests on prototypes) to other operating conditions, coil sizes, and row depths. To account for simultaneous transfer of both sensible and latent heat from the airstream to the surface, ARI *Standard* 410-81 uses essentially the same method for determining cooling and dehumidifying coil thermal performance, as described by McElgin and Wiley (1940).

The potential or driving force for transferring total heat from airstream to tube-side coolant is composed of two components in series heat flow, *i.e.*, (1) an air-to-surface air enthalpy difference and (2) a surface-to-coolant temperature difference.

Figure 8 is a typical thermal diagram for a coil in which the air and a nonvolatile coolant are arranged in counterflow. The top and bottom lines in the diagram indicate, respectively, changes across the coil in the airstream enthalpy ($h_a$) and the coolant temperature ($t_r$). To illustrate continuity, the single middle line in Figure 8 represents both surface temperature ($t_s$) and the corresponding saturated air enthalpy ($h_s$), although the temperature and air enthalpy scales do not actually coincide as shown. The differential surface area ($dA_w$) represents any specific location within the coil thermal diagram where operating conditions are such that the air-surface interface temperature ($t_s$) is lower than the local air dew point temperature. Under these conditions, both sensible and latent heat are removed from the airstream and the cooler surface actively condenses water vapor. The driving forces for the total heat ($q_t$) transferred from air-to-coolant are (1) an air enthalpy difference ($h_a - h_s$) be-

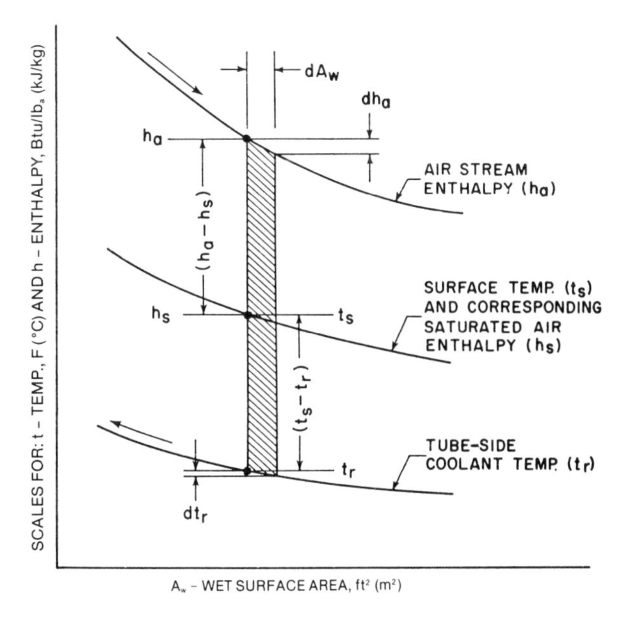


**Fig. 8 Two-Component Driving Force between Dehumidifying Air Coolant**

tween airstream and surface interface and (2) a temperature difference ($t_s - t_r$) across the surface metal and into the coolant.

Neglecting the enthalpy of the condensed water vapor leaving the surface and any radiation and convection losses, the total heat lost from the airstream in flowing over $dA_w$ is:

$$dq_t = -w_a(dh_a) \quad (8)$$

The total heat transferred from the airstream to the surface interface, according to McElgin and Wiley (1940), is:

$$dq_t = \frac{(h_a - h_s)dA_w}{(c_p)R_{aw}} \quad (9)$$

The total heat transferred from the air-surface interface across the surface elements and into the coolant is the same as in Eqs. (8) and (9):

$$dq_t = \frac{(t_s - t_r)\,dA_w}{R_{mw} + R_r} \quad (10)$$

The same quantity of total heat is also gained by the nonvolatile coolant in passing across $dA_w$:

$$dq_t = -w_r(c_r)(dt_r) \quad (11)$$

If Eqs. (9) and (10) are equated and the terms rearranged, the expression for the coil characteristic ($C$) is obtained:

$$C = \frac{R_{mw} + R_r}{c_p(R_{aw})} = \frac{t_s - t_r}{h_a - h_s} \quad (12)$$

Equation 12 shows the basic relationship of the two components of the driving force between air and coolant in terms of the three principal thermal resistances. For a given coil, these three resistances of air, metal and intube fluid ($R_{aw}$, $R_{mw}$, and $R_r$) are usually known or can be determined for the particular application, which gives a fixed value for $C$. Equation 12 can then be used to determine point conditions for the interrelated values of airstream enthalpy ($h_a$); coolant temperature ($t_r$); surface temperature ($t_s$); and the enthalpy ($h_s$) of saturated air corresponding to the surface temperature. When both $t_s$ and $h_s$ are unknown, a trial-and-error solution is necessary; however, this can be solved graphically by a surface temperature chart, such as that shown in Figure 9.

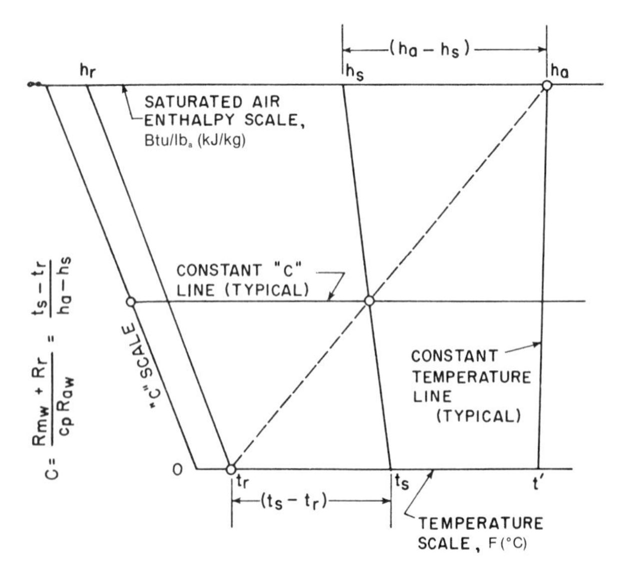


**Fig. 9 Surface Temperature Chart**

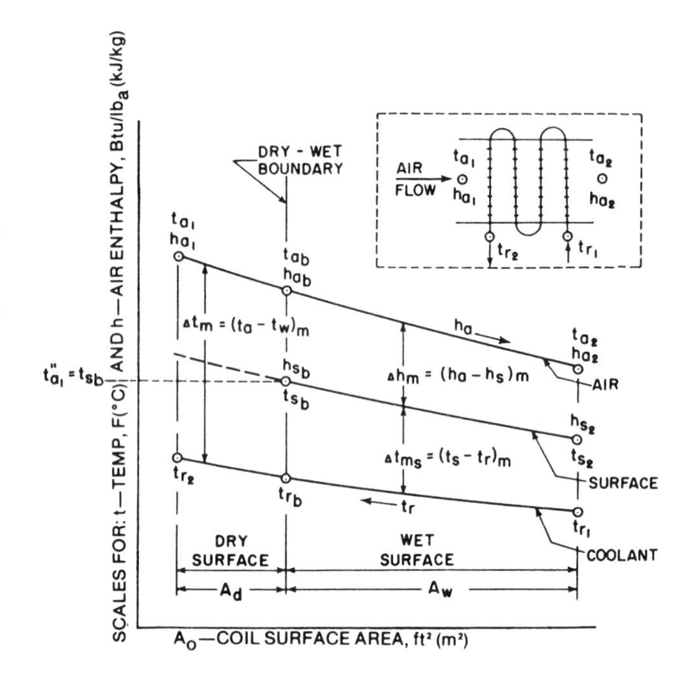


**Fig. 10 Thermal Diagram for General Case When Coil Surface Operates Partially Dry**

Figure 10 shows a typical thermal diagram for a portion of the coil surface when it is operating dry. The illustration is for counterflow, using a halocarbon refrigerant. The diagram at the top of the figure illustrates a typical coil installation in an air duct with tube passes circuited countercurrent to airflow. Locations of the entering and leaving conditions for both air and coolant are shown.

The thermal diagram in Figure 10 is of the same type as in Figure 8, showing three lines to illustrate local conditions for the air, surface, and coolant throughout a coil. The dry-wet boundary conditions are located where the coil surface temperature ($t_{sb}$) equals the entering air dew point temperature ($t''_{a1}$). Thus, the surface area ($A_d$) to the left of this boundary is dry, with the remainder of the coil surface area ($A_w$) operating wet.

When using halocarbon refrigerants in a thermal counterflow arrangement, as illustrated in Figure 10, the dry-wet boundary conditions can be determined from the following relationships:

$$y = \frac{t_{r1} - t_{r2}}{h_{a1} - h_{a2}} = \frac{w_a}{w_r(c_r)} \quad (13)$$

$$h_{ab} = \frac{t''_{a1} - t_{r2} + y(h_{a1}) + C(h''_{a1})}{C + y} \quad (14)$$

The value of $h_{ab}$ from Eq. (14) serves as an index of whether the coil surface is operating fully wetted, partially dry, or completely dry, according to the three limits below:

(1) If $h_{ab} \geqslant h_{a1}$, all surface is fully wetted.
(2) If $h_{a1} > h_{ab} > h_{a2}$, the surface is partially dry.
(3) If $h_{ab} \geqslant h_{a2}$, all surface is completely dry.

Other dry-wet boundary properties are then determined as:

$$t_{sb} = t_{a1} \quad (15)$$

$$t_{ab} = t_{a1} - (h_{a1} - h_{ab})/c_p \quad (16)$$

$$t_{rb} = t_{r2} - y(c_p)(t_{a1} - t_{ab}) \quad (17)$$

The dry surface area ($A_d$) requirements and capacity ($q_{td}$) are calculated by conventional sensible heat transfer relationships, as follows:

The overall thermal resistance ($R_o$) is comprised of three basic elements:

$$R_o = R_{ad} + R_{md} + R_r \tag{18}$$

*where*

$$R_r = B/f_r \tag{19}$$

The mean temperature difference between air dry bulb and coolant, using symbols from Figure 10, is:

$$\Delta t_m = \frac{(t_{a1} - t_{r2}) - (t_{ab} - t_{rb})}{\ln[(t_{a1} - t_{r2})/(t_{ab} - t_{rb})]} \tag{20}$$

The dry surface area required is:

$$A_d = \frac{q_{td}(R_o)}{\Delta t_m} \tag{21}$$

The air-side total heat capacity is:

$$q_{td} = w_a(c_p)(t_{a1} - t_{ab}) \tag{22a}$$

or from the coolant-side:

$$q_{td} = w_r(c_r)(t_{r2} - t_{rb}) \tag{22b}$$

The wet surface area ($A_w$) and capacity ($q_{tw}$) are determined by the following relationships, using terminology in Figure 10:

For a given coil size, design, and arrangement, the fixed value of the coil characteristic ($C$) can be determined from the ratio of the three prime thermal resistances for the job conditions:

$$C = \frac{R_{mw} + R_r}{c_p(R_{aw})} \tag{23}$$

Knowing the coil characteristic ($C$), for point conditions, the interrelations between the airstream enthalpy ($h_a$), the coolant temperature ($t_r$), the surface temperature ($t_s$), and its corresponding enthalpy of saturated air ($h_s$) can be determined by use of a surface temperature chart (Figure 9) or by a trial-and-error procedure using Eq. (24):

$$C = \frac{t_{sb} - t_{rb}}{h_{ab} - h_{sb}} = \frac{t_{s2} - t_{r1}}{h_{a2} - h_{s2}} \tag{24}$$

The mean effective difference in air enthalpy between airstream and surface from Figure 10 is:

$$\Delta h_m = \frac{(h_{ab} - h_{sb}) - (h_{a2} - h_{s2})}{\ln[(h_{ab} - h_{sb})/(h_{a2} - h_{s2})]} \tag{25}$$

Similarly, the mean temperature difference between surface and coolant is:

$$\Delta t_{ms} = \frac{(t_{sb} - t_{rb}) - (t_{s2} - t_{r1})}{\ln[(t_{sb} - t_{rb})/(t_{s2} - t_{r1})]} \tag{26}$$

The wet surface area required, calculated from air-side enthalpy difference, is:

$$A_w = \frac{q_{tw}(R_{aw})(c_p)}{\Delta h_m} \tag{27a}$$

or from the coolant-side temperature difference:

$$A_w = \frac{q_{tw}(R_{mw} + R_r)}{\Delta t_{ms}} \tag{27b}$$

The air-side total heat capacity is:

$$q_{tw} = w_a[h_{ab} - (h_{a2} + h_{fw})] \tag{28a}$$

*where*

$$h_{fw} = (W_1 - W_2)(t'_{a2} - 32) \tag{28b}$$

Note that $h_{fw}$ for normal air-conditioning applications is about 1% of the airstream enthalpy difference ($h_{a1} - h_{a2}$), and is neglected in many cases. *Example 4* illustrates this effect.

The coolant-side heat capacity is:

$$q_{tw} = w_r c_r(t_{rb} - t_{r1}) \tag{28c}$$

The total surface area requirement of the coil is:

$$A_o = A_d + A_w \tag{29}$$

The total heat capacity for the coil is:

$$q_t = q_{td} + q_{tw} \tag{30}$$

The leaving air dry-bulb temperature is found by the method illustrated in Figure 11, which represents part of a psychrometric chart showing the air saturation curve and lines of constant air enthalpy closely corresponding to constant wet-bulb temperature lines.

For a given coil and air quantity, a straight line projected through the entering and leaving air conditions intersects the air saturation curve at a point denoted as the effective coil surface temperature ($t_{\bar{s}}$). Thus, for fixed entering air conditions ($t_{a1}$, $h_{a1}$) and a given effective surface temperature ($t_{\bar{s}}$), the leaving air dry bulb ($t_{a2}$) increases but is still located on this straight line if the air quantity is increased or if the coil depth is reduced. Conversely, a decrease in air quantity, or an increase in coil depth, produces a lower $t_{a2}$ that is still located on the same straight line segment.

An index of the air-side effectiveness is the heat-transfer exponent ($c$), which is defined as:

$$c = \frac{A_o}{w_a c_p R_{ad}} \tag{31}$$

This exponent ($c$), sometimes called the *number of air-side transfer units* ($NTU_a$), is also defined as:

$$c = \frac{t_{a1} - t_{a2}}{\Delta t_{\bar{m}}} \tag{32}$$

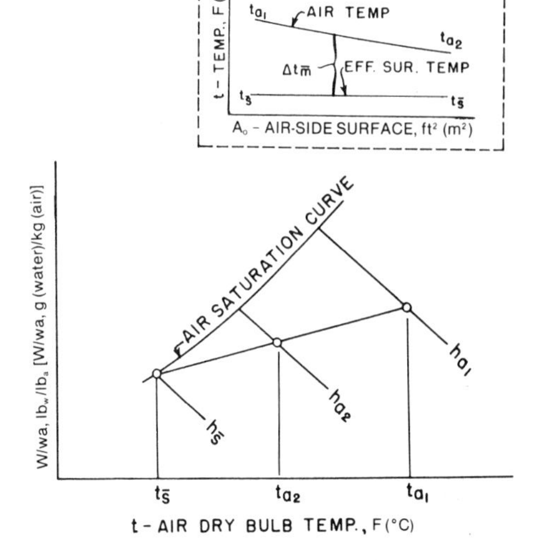


**Fig. 11 Leaving Air Dry-Bulb Temperature Determination for Air-Cooling and Dehumidifying Coils**

The temperature drop ($t_{a1} - t_{a2}$) of the airstream and the mean temperature difference ($\Delta t_{\overline{m}}$) between air and effective surface in Eq. (32) are illustrated in the small diagram at the top of Figure 11.

Knowing the exponent ($c$) and the entering ($h_{a1}$) and leaving ($h_{a2}$) enthalpies for the airstream, the enthalpy of saturated air ($h_{\bar{s}}$) corresponding to the effective surface temperature ($t_{\bar{s}}$) is calculated as follows:

$$h_{\bar{s}} = h_{a1} - \frac{(h_{a1} - h_{a2})}{1 - e^{-c}} \tag{33}$$

After finding the value of ($t_{\bar{s}}$), which corresponds to ($h_{\bar{s}}$) from the saturated air enthalpy tables, the leaving air dry-bulb temperature can then be determined as:

$$t_{a2} = t_{\bar{s}} + e^{-c}(t_{a1} - t_{\bar{s}}) \tag{34}$$

The air-side sensible heat ratio can then be calculated as:

$$\text{SHR} = \frac{c_p(t_{a1} - t_{a2})}{(h_{a1} - h_{a2})} \tag{35}$$

For the thermal performance of a coil to be determined from the foregoing relationships, values of the three principal resistances to heat flow between air and coolant (listed as follows) must be known:

(1) the metal thermal resistance across the fin and tube surface elements for both dry ($R_{md}$) and wet ($R_{mw}$) surface operation
(2) the air-film thermal resistance for dry ($R_{ad}$) and wet ($R_{aw}$) surface
(3) the tube-side coolant film thermal resistance ($R_r$)

In ARI *Standard* 410-81, the metal thermal resistance ($R_m$) is calculated, knowing the physical data, material, and arrangement of the fin and tube elements, together with the fin efficiency ($E$) for the specific fin configuration. The metal resistance ($R_m$) is variable as a weak function of the effective air-side heat-transfer coefficient ($f_a$) for a specific coil geometry, as illustrated in Figure 12.

For wetted surface application, Brown (1954), with certain simplifying assumptions, showed that the effective air-side coefficient ($f_a$) is directly proportional to the rate-of-change ($m''$) of enthalpy for saturated air ($h_s$) with the corresponding surface temperature ($t_s$). This slope ($m''$) of the air enthalpy saturation curve is illustrated in the small inset graph at the top of Figure 12.

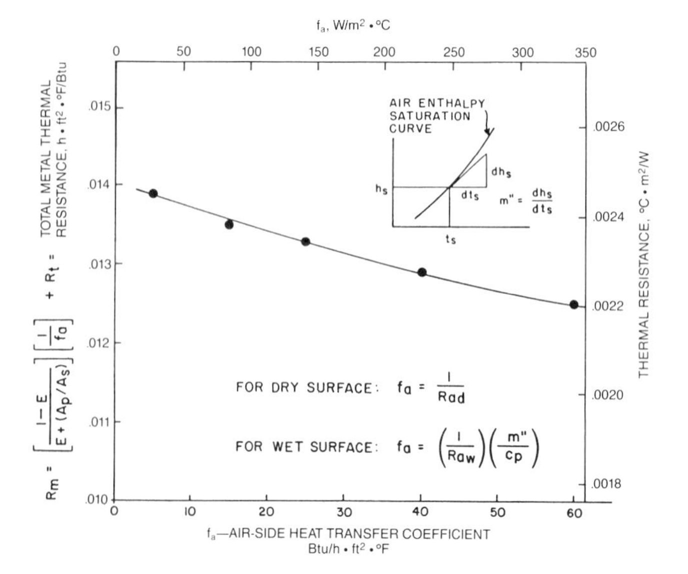


**Fig. 12 Typical Total Metal Thermal Resistance of Fin and Tube Assembly**

The abscissa for $f_a$ in the main graph of Figure 12 is an effective value, which, for dry surface, is the simple thermal resistance reciprocal ($1/R_{ad}$). For a wet surface, $f_a$ is the product of the thermal resistance reciprocal ($1/R_{aw}$) and the multiplying factor ($m''/c_p$). ARI *Standard* 410-81 outlines a method for obtaining a mean value of ($m''/c_p$) for a given coil and job condition. The total metal resistance ($R_m$) in Figure 12 includes the resistance ($R_t$) across the tube wall. For most coil designs, $R_t$ is quite small compared to the resistance through the fin metal.

The air-side thermal resistances for dry ($R_{ad}$) and wet ($R_{aw}$) surfaces, together with their respective air-side pressure drops, ($\Delta P_{st}/N_r$) and ($\Delta P_{sw}/N_r$), are determined from tests on a represented coil model over the full range in the rated airflow rate. Typical plots of the experimental data for these four performance variables versus coil air face velocity ($V_a$) at standard 70°F (21.1°C) conditions are illustrated in Figure 13.

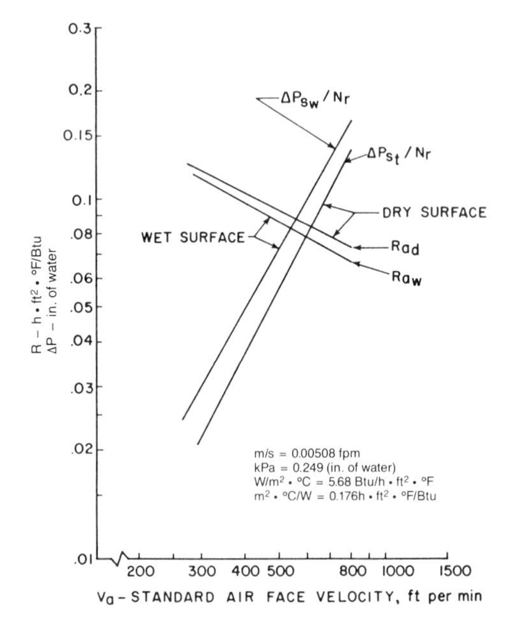


**Fig. 13 Typical Air-Side Application Rating Data Experimentally Determined for Cooling and Dehumidifying Water Coils**

The heat-transfer coefficient ($f_r$) of water used as the tube-side coolant is calculated from data in the literature (Eq. (8), ARI *Standard* 410-81). For evaporating refrigerants, many predictive techniques for calculating $f_r$ are listed in Chapter 3 of the 1985 FUNDAMENTALS Volume. The most verified predictive technique is the Shah correlation (Shah 1976, 1982). A series of tests is specified in ARI *Standard* 410-81 to obtain heat-transfer data for direct-expansion refrigerants inside tubes of a given diameter.

ASHRAE *Standard* 33-78 specifies the laboratory apparatus and instrumentation and the procedure for conducting tests on representative coil prototypes to obtain the basic performance data. Procedures are available in ARI *Standard* 410-81 for reducing this experimental data to the performance parameters necessary to rate a line or lines of various air coils. All thermodynamic properties of air and water used in solving the following equations may be found in ASHRAE Psychometric Chart and Tables of Chapter 6 of the 1985 FUNDAMENTALS Volume.

**Example 3:** This example illustrates a method for selecting a coil size, row depth, and performance data to satisfy initially specified job requirements. The application is for a typical cooling and dehumidifying coil selection under conditions in which a part of the coil surface on the entering air side operates dry, with the remaining surface wetted with condensing moisture. Figure 10 shows the thermal diagram, dry-wet boundary conditions, and terminology used in the problem solution.

*Known Initial Conditions:*

Standard air flowing at a mass rate equivalent of 3162 L/s enters a coil at 26.7 °C dry-bulb temperature ($t_{a1}$) and 19.4 °C wet-bulb temperature ($t'_{a1}$). The air is to be cooled to 13.3 °C leaving wet-bulb temperature ($t'_{a2}$) using 2.5 L/s of chilled water, supplied to the coil at an entering temperature ($t_{r1}$) of 6.7 °C, in a thermal counterflow arrangement. Assume a coil air face velocity of 2.8 m/s and a clean non-fouled, finned-tube-type heat transfer surface in the coil core, for which the following physical and performance data (such as illustrated in Figures 16 and 17) can then be predetermined:

$B$ = 25.9 surface ratio
$c_p$ = 1.0 kJ/kg • °C
$F_s$ = 32.4 (external surface area)/(face area)(rows deep)
$f_r$ = 4260 W/m² (internal) • °C
$\Delta P_{st}/N_r$ = 41 Pa/rows deep, dry surface
$\Delta P_{sw}/N_r$ = 67 Pa/rows deep, wet surface
$R_{ad}$ = 0.013 m² • °C/W
$R_{aw}$ = 0.012 m² • °C/W
$R_{md}$ = 0.004 m² • °C/W
$R_{mw}$ = 0.003 m² • °C/W

*Required Items:*

Referring to the symbols and typical diagram for applications in which only a part of the coil surface operates wet in Figure 10, determine: coil face area ($A_a$); total refrigeration load ($q_t$); leaving coolant temperature ($t_{r2}$); dry-wet boundary conditions; heat transfer surface area required for dry ($A_d$) and wet ($A_w$) sections of the coil core; leaving air dry-bulb temperature ($t_{a2}$); total number of installed coil rows deep ($N_{ri}$); and dry ($\Delta P_{st}$) and wet ($\Delta P_{sw}$) coil air friction.

*Solution:*
From the ASHRAE Psychrometric Chart and tables, first determine the following properties from the known conditions:

$h_{a1}$ = 73.4 kJ/kg
$W_1$ = 11.2 g/kg
$t''_{a1}$ = 15.7 °C
$h''_{a1}$ = 62.1 kJ/kg
*$h_{a2}$ = 55.5 kJ/kg
*$W_2$ = 9.5 g/kg

*As an approximation, assume leaving air is saturated (*i.e.*, $t_{a2} = t'_{a2}$).

Calculate coil face area required:

$$A_a = L/s \text{ per } V_a = 3162/2.8(1000\ L/m^3) = 1.12\ m^2$$

From Eq. (28b), find condensate heat rejection:

$$h_{fw} = (11.2 - 9.5)(13.3 - 0) = 22.61\ g/kg$$

Compute the total refrigeration load:

$$q_t = 1248(L/s)[h_{a1} - (h_{a2} + h_{fw})]$$

$$= 1248(3162)/(1000\ L/m^3)[73.4 - (55.5 + 22.61/1000\ g/kg)] = 70\ 000\ W$$

From Eq. (3), calculate coolant temperature leaving the coil:

$$t_{r2} = t_{r1} + q_t/(w_r \bullet c_r) = 6.7 + 70\ 000/(63 \bullet 168) = 13.2\,°C$$

From Eq. (19), determine coolant film thermal resistance:

$$R_r = 25.9/4260 = 0.006\ m^2 \bullet °C/W$$

From Eq. (23) calculate the wet coil characteristic:

$$C = \frac{0.003 + 0.006}{1.0(0.012)} = 0.75\ kg \bullet °C/kJ$$

From Eq. (13), calculate:

$$y = \frac{1248(3162)/(1000\ L/m^3)}{63 \bullet 168} = 0.372\ kg \bullet °C/kJ$$

*Determining dry-wet boundary conditions*

From Eq. (14), the boundary airstream enthalpy is:

$$h_{ab} = \frac{15.7 - 13.2 + 0.372(73.4) + 0.75(62.1)}{0.75(0.372)} = 68.2\ kJ/kg$$

According to limit (2) under Eq. (14), a part of the coil surface on the entering air side will be operating dry, as shown in Figure 14, since $h_{a1} > 68.2 > h_{a2}$.

From Eq. (16), the boundary airstream dry-bulb temperature is:

$$t_{ab} = 26.7 - (73.4 - 68.2)/1.0 = 21.5\,°C$$

The boundary surface conditions are:

$$t_{sb} = t''_{a1} = 15.7\,°C \text{ and } h_{sb} = h''_{a1} = 62.1\ kJ/kg$$

From Eq. (17), the boundary coolant temperature is:

$$t_{rb} = 13.2 - 0.372(1.0)(26.7 - 21.5) = 11.2\,°C$$

The refrigeration load for the dry surface part of the coil is now calculated from Eq. (22b):

$$q_{td} = 63(168)(13.2 - 11.2) = 21\ 160\ W$$

From Eq. (18), the overall thermal resistance for the dry surface section is:

$$R_o = 0.013 + 0.004 + 0.006 = 0.023^2 \bullet °C/W$$

From Eq. (20), the mean temperature difference between air dry bulb and coolant for the dry surface is:

$$\Delta t_m = \frac{(26.7 - 13.2) - (21.5 - 11.2)}{\ln[(26.7 - 13.2)/(21.5 - 11.2)]} = 12\,°C$$

From Eq. (21), the dry surface area required is:

$$A_d = (21\ 160)(0.023)/12.0 = 41\ m^2$$

The refrigeration load for the wet surface section of the coil, calculated from Eq. (30), is:

$$q_{tw} = 70\ 000 - 21\ 160 = 48\ 840\ W$$

Knowing $C$, $h_{a2}$, and $t_{r1}$ the surface conditions at the leaving air side of the coil is calculated from Eq. (24):

$$0.75 = \frac{t_{s2} - 6.7}{55.5 - h_{s2}}$$

The numerical values for $t_{s2}$ and $h_{s2}$ are then determined directly by use of a surface temperature chart, as shown in Figure 13, or by trial-and-error procedure, using saturated air enthalpies from Table 1, Chapter 6, of the 1985 FUNDAMENTALS Volume, thus:

$$t_{s2} = 11.1\,°C \text{ and } h_{s2} = 50\ kJ/kg$$

From Eq. (25), the mean effective difference in air enthalpy between airstream and surface is:

$$\Delta h_m = \frac{(68.2 - 62.1) - (55.5 - 50)}{\ln[(68.2 - 62.1)/(55.5 - 50)]} = 5.9\ kJ/kg$$

From Eq. (27a), the wet surface area required is:

$$A_w = (48\ 840 \bullet 0.012 \bullet 1.0)/5.9 = 99\ m^2$$

From Eq. (29), the net total surface area requirements are then:

$$A_o = 41 + 99 = 140\ m^2 \text{ external}$$

The net air-side heat transfer exponent from Eq. (31) is:

$$c = 140/(1248 \cdot 3.162 \cdot 1.0 \cdot 0.013) = 2.73$$

From Eq. (33), the enthalpy of saturated air corresponding to the effective surface temperature is:

$$h_{\bar{s}} = 73.4 - (73.4 - 55.5)/(1 - e^{-2.73}) = 54 \text{ kJ/kg}$$

The effective surface temperature, which corresponds to $h_{\bar{s}}$, is then obtained from the saturated air enthalpy tables as $t_{\bar{s}} = 12.9$°C. From Eq. (34), the leaving air dry-bulb temperature is calculated as:

$$t_{a2} = 12.9 + e^{-2.73}(26.7 - 12.9) = 13.8\text{°C}$$

The air-side sensible heat ratio is then determined from Eq. (35) as:

$$\text{SHR} = \frac{1.0(26.7 - 13.8)}{73.4 - 55.5} = 0.721$$

The calculated coil row depth ($N_{rc}$) to match job requirements is:

$$N_{rc} = A_o/A_aF_s = 140/(1.12 \cdot 32.4) = 3.86 \text{ rows deep}$$

In most coil selection problems of the type illustrated, the initial calculated row depth ($N_{rc}$) to satisfy job requirements is usually a non-integral value. In many cases, there is sufficient flexibility in fluid flow rates and operating temperature levels to recalculate the required row depth ($N_{rc}$) of a given coil size to match an available integral row depth more closely. For example, if the initial calculated row depth were $N_{rc} = 3.5$, and three or four row deep coils were commercially available, operating conditions and/or fluid flow rates could possibly be changed to recalculate a coil depth close to either three or four rows. Also, alternate coil selections for the same job are often desirable by trading off coil face size and row depth.

A coil calibrating computer program can quickly do the iteration needed to predict the operating values for the specific row depth chosen.

For Example 3, assume the initial coil selection, actually requiring 3.86 rows deep, is sufficiently refined not to need a recalculation. It is then customary to select the next highest integral row depth, thus providing some safety factor (about 5% in this case) in surface area selection. The installed row depth ($N_{rc}$) is then:

$$N_{rc} = 4 \text{ rows deep}$$

The completely wetted and completely dry air-side frictions are, respectively:

$$\Delta P_{sw} = (\Delta P_{sw}/N_{ri})N_{ri} = 67.23(4) = 269 \text{ Pa}$$

and

$$\Delta P_{st} = (\Delta P_{st}/N_{ri})N_{ri} = 41.08(4) = 164 \text{ Pa}$$

The amount of heat-transfer surface area installed is:

$$A_{oi} = A_aF_sN_{ri} = 1.12(32.4)(4) = 145 \text{ m}^2 \text{ external}$$

In summary, the final coil size selected and calculated performance data for *Example* 3 follows:

$A_a$ = 1.12 m² coil face area
$N_{ri}$ = 4 rows installed coil depth
$A_{oi}$ = 145 m² installed heat-transfer surface area
$A_o$ = 140 m² required heat-transfer surface area
$q_t$ = 70,000 W total refrigeration load
$t_{r2}$ = 13.2°C leaving coolant temperature
$t_{a2}$ = 13.8°C leaving air dry-bulb temperature
$SHR$ = 0.721 air sensible heat ratio
$\Delta P_{sw}$ = 67 Pa wet coil surface air friction
$\Delta P_{st}$ = 41 Pa dry coil surface air friction

## DETERMINING REFRIGERATION LOAD

The following determination of the refrigeration load shows a division of the true sensible and latent heat loss of the air, which is accurate within the limitations of the data. This division will not correspond to load determination obtained from approximate factors or constants.

The total refrigeration load, $q_t$, of a cooling and dehumidifying coil (or air washer) per unit mass of dry air is indicated on Figure 14 and consists of the following components:

(1) The sensible heat, $q_s$, removed from the dry air and moisture in cooling from entering temperature, $t_1$, to leaving temperature, $t_2$
(2) The latent heat, $q_e$, removed to condense the moisture at the dew point temperature, $t_4$, of the entering air
(3) The heat required, $q_w$, to cool the condensate additionally from its dew point, $t_4$, to its leaving condensate temperature, $t_3$

Items 1, 2, and 3 may be related by:

$$q_t = q_s + q_e + q_w \tag{36}$$

If only the total heat value is desired, it may be computed by:

$$q_t = (h_1 - h_2) - (W_1 - W_2)h_{f3} \tag{37}$$

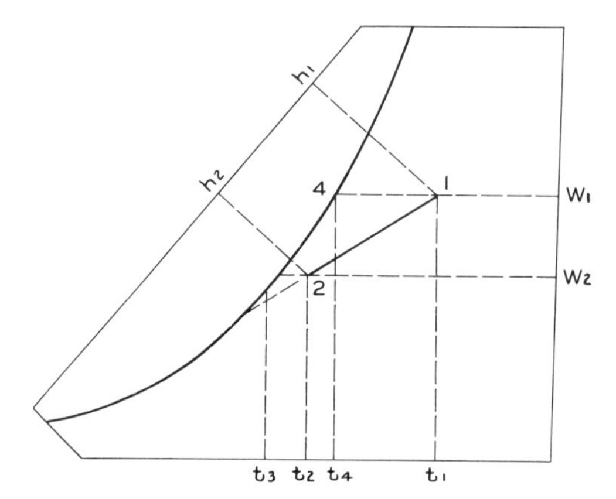


**Fig. 14 Psychrometric Performance of Cooling and Dehumidifying Coil**

*where*

$h_1$ and $h_2$ = enthalpy of air at points 1 and 2, respectively
$W_1$ and $W_2$ = humidity ratio at points 1 and 2, respectively
$h_{f3}$ = enthalpy of saturated liquid at final temperature, $t_3$

If a breakdown into latent and sensible heat components is desired, the following relations may be used.

The latent heat may be found from:

$$q_e = (W_1 - W_2)h_{fg4} \tag{38}$$

*where*

$h_{fg4}$ = latent heat of water vapor at the condensing temperature, $t_4$

The sensible heat may be shown to be:

$$q_s + q_w = (h_1 - h_2) - (W_1 - W_2)h_{g4} + (W_1 - W_2)(h_{f4} + h_{f3}) \tag{39a}$$

*or*

$$q_s + q_w = (h_1 - h_2) - (W_1 - W_2)(h_{fg4} + h_{f3}) \tag{39b}$$

*where*

$h_{g4} = h_{fg4} + h_{f4}$ = enthalpy of saturated water vapor at the condensing temperature, $t_4$
$h_{f4}$ = enthalpy of saturated liquid at the condensing temperature, $t_4$

The last term in Eq. (39a) is the heat of subcooling the condensate from the condensing temperature, $t_4$, to its final temperature, $t_3$. Then:

$$q_w = (w_1 - w_2)(h_{f4} - h_{f3}) \quad (40)$$

The final condensate temperature, $t_3$, leaving the system, is subject to substantial variations, depending on the method of coil installation as affected by coil face orientation, airflow direction, and air duct insulation. In practice, $t_3$ is frequently the same as the leaving wet-bulb temperature. Within the normal air-conditioning range, precise values of $t_3$ are not necessary as heat of the condensate $(W_1 - W_2)h_{f3}$, removed from the air, usually represents about 0.5 to 1.5% of the total refrigeration load.

---

**Example 4:** Air enters a coil at 32.2°C dry bulb, 23.9°C wet bulb; it leaves at 16.1°C dry bulb, 14.4°C wet bulb. The assumed temperature of leaving water is 12.2°C, which is between the leaving air dew point and coil surface temperature. Find the total, latent, and sensible cooling loads on the coil with the air at standard barometric pressure.

*Solution:*

From the ASHRAE *Psychrometric Chart*, find the following:

$h_1$ = 89.52 kJ/kg of dry air
$h_2$ = 58.48 kJ/kg of dry air
$t_4$ = 20.6°C dew point of entering air
$W_1$ = 15.25 g/kg of dry air
$W_2$ = 9.6 g/kg of dry air

From Table 2, find:

$h_{f4}$ = 86.37 kJ/kg
$h_{f3}$ = 51.45 kJ/kg
$h_{fg4}$ = 2456.45 kJ/kg
$h_{g4}$ = 2542.82 kJ/kg

From Eq. (37), the total heat is:

$$q_t = (89.52 - 58.48) - [(15.25 - 9.6)/(1000 \text{ g/kg})] \cdot 51.45$$
$$= 31.04 - 0.29 = 30.75 \text{ kJ per kg of dry air.}$$

From Eq. (38), the latent heat is:

$$q_e = 0.00565 \cdot 2456.45 = 13.88 \text{ kJ per kg of dry air}$$

The sensible heat, by difference, is:

$$q_s + q_w = q_t - q_e = 30.75 - 13.88 = 16.87 \text{ kJ per kg of dry air}$$

Or the sensible heat may be computed from Eq. (39a) as:

$$\begin{aligned} q_s + q_w &= (89.52 - 58.48) - (0.00565 \cdot 2542.82) \\ &+ 0.00565(86.37 - 51.45) \\ &= 31.04 - 14.37 + 0.20 \\ &= 16.87 \text{ kJ per kg of dry air} \end{aligned}$$

---

The subcooling of the condensate as a part of the sensible heat is indicated by the last term of the equation, 0.20 kJ/kg of dry air.

## MAINTENANCE

If the coil is to deliver its full cooling capacity, both its internal and external surfaces must be clean. The tubes generally stay clean in pressurized water or brine systems. Should large amounts of scale form when untreated water is used as coolant, chemical or mechanical cleaning of internal surfaces at frequent intervals is necessary. When coils use evaporating refrigerants, oil accumulation is possible and occasional checking and oil drainage is desirable. While outer tube surfaces can be cleaned in a number of ways, they are often washed with water. The surfaces can also be brushed and cleaned with a vacuum cleaner. In cases of marked neglect—especially in restaurants, where grease and dirt have accumulated—it is sometimes necessary to remove the coils and wash off the accumulation with steam, compressed air and water, or hot water. The best practice, however, is to inspect and service the filters frequently.

## LETTER SYMBOLS

$A_a$ = Coil face or frontal area, ft$^2$ (m$^2$)
$A_d$ = Dry external surface area, ft$^2$ (m$^2$)
$A_o$ = Total external surface area, ft$^2$ (m$^2$)
$A_p$ = Exposed external prime surface area, ft$^2$ (m$^2$)
$A_s$ = External secondary surface area, ft$^2$ (m$^2$)
$A_w$ = Wet external surface area, ft$^2$ (m$^2$)
$B$ = Ratio of external to internal surface area, dimensionless
$C$ = Coil characteristic as defined in Eqs. (12) and (23), lb • °F/Btu (kg • °C/kJ)
$c$ = Heat-transfer exponent, or NTU$_a$, as defined in Eqs. (31) and (32), dimensionless
$c_o$ = Heat-transfer exponent, as defined in Eq. (7d), dimensionless
$c_p$ = Air humid specific heat, Btu/lb • °F (kJ/kg • °C)
$c_r$ = Specific heat of nonvolatile coolant, Btu/lb • °F (kJ/kg • °C)
$D_i$ = Tube inside diameter, in. (mm)
$D_o$ = Tube outside diameter, in. (mm)
$E$ = Fin efficiency, dimensionless
$E_a$ = Air-side effectiveness defined in Eq. (7b), dimensionless
ln = Naperian logarithm base, dimensionless
$F_s$ = Coil core surface area parameter = [external surface area, ft$^2$ (m$^2$)] ÷ [face area, ft$^2$ (m$^2$) (row deep)]
$f$ = Convection heat-transfer coefficient, Btu/h • ft$^2$ • °F (W/m$^2$ • °C)
$h$ = Air enthalpy (actual in airstream or saturation value at surface temperature), Btu/lb (kJ/kg) of air
$\Delta h_m$ = Mean effective difference of air enthalpy, as defined in Eq. (25), Btu/lb (kJ/kg) of air
$k$ = Thermal conductivity of tube material, Btu/h • ft • °F (W/m • °C per mm thickness)
$M$ = Ratio of the nonvolatile coolant-to-air temperature changes for sensible heat cooling coils, as defined in Eq. (7e), dimensionless
$n''$ = Rate-of-change of air enthalpy at saturation with air temperature, Btu/lb • °F (kJ/kg • °C)
$N_r$ = Number of coil rows deep in airflow direction, dimensionless
$\eta$ = Fin effectiveness, as defined in Eq. (6), dimensionless
$\Delta P_{st}$ = Isothermal dry surface air-side friction at standard 70°F (20°C) conditions, in. of water (Pa)
$\Delta P_{sw}$ = Wet surface air-side friction at standard 70°F (20°C) conditions, in. of water (Pa)
$q$ = Heat-transfer capacity, W
$R$ = Thermal resistance, referred to external area, $A_o$, m$_2$ • °C/W.
SHR = Ratio of air sensible heat to air total heat, dimensionless
$t$ = Temperature, °F (°C)
$\Delta t_m$ = Mean effective temperature difference, air dry-bulb-to-coolant, °F (°C)
$\Delta t_{ms}$ = Mean effective temperature difference, surface-to-coolant, °F (°C)
$\Delta t_{\overline{m}}$ = Mean effective temperature difference, air dry-bulb to effective surface temperature, $t_{\overline{s}}$, °F (°C)

$U_o$ = Overall sensible heat-transfer coefficient, Btu/ft$^2$ • °F (W/m$^2$ • °C)
$V_a$ = Coil air face velocity at standard 70°F (21°C) conditions, fpm (m/s)
$W$ = Air humidity ratio, pounds of water per pound of air (grams of water per kilogram of air)
$w$ = Mass flow rate, lb/h (kg/s)
$y$ = Ratio of nonvolatile coolant temperature rise to air stream enthalpy drop, as defined in Eq. (13), lb$_a$ • °F/Btu (kg • °C/kJ)

Superscripts:
$'$ = Wet bulb
$''$ = Dew point

Subscripts:

$_1$ = Condition entering coil
$_2$ = Condition leaving coil
$_a$ = Airstream
$_b$ = Dry-wet surface boundary
$_m$ = Metal (with $R$) and mean (with other symbols)
$_o$ = Overall (except for $A$)
$_r$ = Coolant
$_s$ = Surface
$_{\bar{s}}$ = Effective surface
$_t$ = Tube (with $R$) and total (with $q$)
$_{ab}$ = Air, dry-wet boundary
$_{ad}$ = Dry air
$_{aw}$ = Wet air
$_{md}$ = Dry metal
$_{mw}$ = Wet metal
$_{rb}$ = Coolant dry-wet boundary
$_{sb}$ = Surface dry-wet boundary
$_{td}$ = Total heat capacity, dry surface
$_{tw}$ = Total heat capacity, wet surface

## REFERENCES


ARI. 1981. Forced-Circulation Air-Cooling and Air-Heating Coils. ARI *Standard* 410-81, Air-Conditioning and Refrigeration Institute, Arlington, VA.

ASHRAE. 1978. Method of Testing for Rating Forced-Circulation Air-Cooling and Heating Coils. ASHRAE *Standard* 33-78.

McElgin, John and D.C. Wiley. 1940. Calculation of Coil Surface Areas for Air Cooling and Dehumidification. *Heating, Piping and Air Conditioning*, March, p. 195.

Brown, Gosta. 1954. Theory of Moist Air Heat Exchangers. Royal Institute of Technology *Transactions* No. 77, Stockholm, Sweden, p. 12.

Mueller, A.C. 1973. Heat Exchangers. *Handbook of Heat Transfer*, W.M. Rohsenow and J.P. Hartnett, editors, McGraw Hill Book Company, NY, (Revised 1986).

Kusuda, T. 1969. Effectiveness Method for Predicting the Performance of Finned Tube Coils. *Heat and Mass Transfer to Extended Surfaces*, ASHRAE.

Shah, M.M. 1976. A New Correlation for Heat Transfer During Boiling Flow Through Pipes. ASHRAE *Transactions* 82 (2).

Shah, M.M. 1982. CHART Correlation for Saturated Boiling Heat Transfer, Equations and Further Study. ASHRAE *Transactions* 88 (1).

Shah, M.M. 1978. Heat Transfer, Pressure Drop, and Visual Observation Test Data for Ammonia Evaporating Inside Tubes. ASHRAE *Transactions* 84 (2).

Webb, R.L. 1980. Air-Side Heat Transfer in Finned Tube Heat Exchangers. *Heat Transfer Engineering*, Vol. 1, No. 3, p. 33.

# CHAPTER 7

# SORPTION DEHUMIDIFICATION AND PRESSURE DRYING EQUIPMENT



DEHUMIDIFICATION is the removal of water vapor from air, gases, or other fluids. Dehydration and drying of gases are synonymous terms with dehumidification. There is no limitation of pressure in this definition, and sorption dehumidification equipment has been designed and operated successfully for system pressures ranging from subatmospheric to as high as 6000 psi (40 MPa). The term dehumidification is normally limited to equipment that operates at essentially atmospheric pressures and is built to standards similar to other types of air-handling equipment. For drying gases under pressure, or liquids, the term *dryer* or *dehydrator* is normally used.

Both liquid and solid sorbent materials are used in dehumidification equipment. They either adsorb the water on the surface of the sorbent (adsorption) or chemically combine with water (absorption). In regenerative equipment, the mechanism for water removal is reversible.

Nonregenerative equipment uses hygroscopic salts such as calcium chloride, urea, or sodium chloride. Regenerative systems usually use a form of silica or alumina gel, activated alumina, molecular sieves, lithium chloride salt, lithium chloride solution, or glycol solution. The choice of sorbent depends on the requirements of the installation, the design of the equipment, and chemical compatibility with the gas to be treated or impurities in the gas. Additional information on sorbent materials and how they operate is located in Chapter 19 of the 1985 FUNDAMENTALS Volume.

Some of the more important commercial applications of dehumidification include the following:

1. lowering the relative humidity to facilitate manufacturing and handling of hygroscopic materials
2. lowering the relative humidity to prevent condensation on products manufactured in low-temperature processes
3. providing protective atmospheres for the heat treatment of metals
4. maintaining controlled humidity conditions in warehouses and caves used for storage
5. preserving ships and other surplus equipment that would otherwise deteriorate.
6. maintaining a dry atmosphere in a closed space or container for numerous static applications, such as the cargo hold of a ship
7. condensation and corrosion control
8. drying air for wind tunnels
9. drying natural gas
10. drying of gases that are to be liquefied
11. drying of instrument air and plant air
12. drying of process and industrial gases
13. dehydration of liquids

This chapter covers (1) the types of dehumidification equipment for liquid and solid sorbents, including high-pressure equipment; (2) performance curves; (3) variables of operation; and (4) some typical applications. The use of sorbents for the drying of refrigerants is addressed in Chapter 37 of the 1986 REFRIGERATION Volume.

## METHODS OF DEHUMIDIFICATION

Dehumidification can be accomplished by compression, refrigeration, liquid and solid sorption, or a combination of these systems.

Figure 1 illustrates three methods by which dehumidification with sorbent materials or sorbent equipment may be accomplished. Air in the condition at Point A is dehumidified and cooled to Point B. This is done in a liquid sorption system with intercooling directly. In a solid sorption unit, this process can

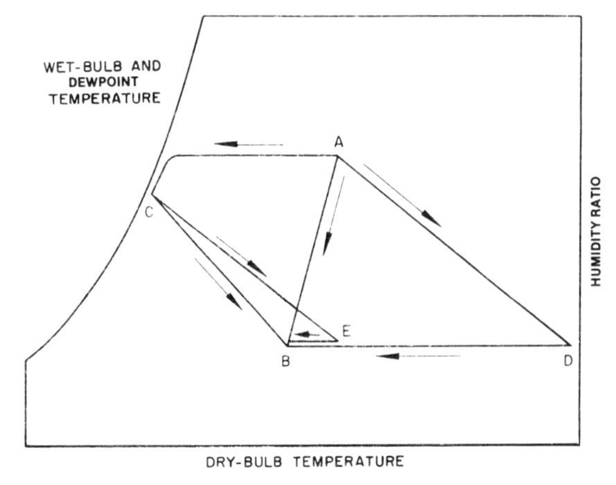


**Fig. 1 Methods of Dehumidification**

The preparation of this chapter is assigned to TC 3.5, Sorption.

be completed by precooling and dehumidifying from Point A to Point C, then desiccating from Point C to Point E and, finally, cooling to Point B. It could also be accomplished with solid sorption equipment by desiccating from Point A to Point D and then by refrigeration from Point D to Point B.

**Compression** of the gas to be dehumidified reduces its absolute moisture content but generally produces a saturated condition at the elevated pressure. This method is not economically advantageous, but is of value for pressure systems, since part of the moisture is removed by compression of the gas; further dehumidification may be accomplished by cooling alone, with sorption, or both, depending on the final dew point desired. Expansion of high-pressure gas lowers the dew point at the lower pressure because the actual volume increases. See the dew point conversion chart in Chapter 19 of the 1985 FUNDAMENTALS Volume.

**Refrigeration** of the gas below its dew point is the most common method of dehumidification. This method is advantageous when the gas is comparatively warm, has a high moisture content, and the outlet dew point desired is above 40 °F (5 °C). Frequently, refrigeration is used in combination with sorption dehumidifiers to obtain an extremely low dew point at minimum cost.

**Liquid sorption** dehumidification systems pass the gas through sprays of a liquid sorbent such as lithium chloride or a glycol solution. The sorbent in active state has a vapor pressure below that of the gas to be dehumidified and absorbs moisture from the gas stream. The sorbent solution, during the process of absorption, becomes diluted with moisture. During regeneration, this moisture is given up to an outdoor airstream in which the solution is heated. A partial bleedoff of the solution is used for continuous reconcentration of the sorbent in a closed circuit between the spraying or contactor unit and the regenerator unit.

**Solid sorption** passes the gas stream through a granular desiccant or over a fixed desiccant structure. Outdoor air passes through beds or layers of the sorbent, which, in its active state, has a vapor pressure below that of the gas to be dehumidified and absorbs moisture from the gas stream. After becoming saturated with moisture, the desiccant needs periodic reactivation to give up previously adsorbed moisture to an outdoor air or gas stream.

The solid sorption process is advantageous when the required dew point must be very low. For processes at elevated pressures (or with gases other than air) where a high dew point depression is desired, and when the entering conditions are at high temperature and moisture content, a combination of dehydration by compression, refrigeration, or both, followed by solid sorption, may be most economical. System space and power demand are reduced in comparison to systems using compression refrigeration or solid sorption alone for the entire process. A solid sorption system for the entire process may be justified when the cost of regeneration energy is unusually low. Dehumidification systems for similar requirements would normally combine refrigeration and solid sorption.

The degree of dehumidification required varies with different applications and influences the choice of the method used. Economics and availability of utilities are also important considerations.

## TYPES OF SORPTION DEHUMIDIFICATION EQUIPMENT

As defined in the preceding section, liquid sorbents or solid sorbents with granular desiccants or fixed desiccant structure may be used in the dehumidification equipment. Both types of sorbents, liquid and solid, may be used in equipment designed for the drying of air and gases at atmospheric or elevated pressures. Regardless of pressure levels, basic principles remain the same, and only the sorbent towers or chambers require special design consideration.

Sorbent capacity and actual dew point performance depend on the specific equipment used, the characteristics of the various sorbents, initial temperature and moisture content of the gas to be dried, reactivation methods, etc. Factory-assembled units are available up to a capacity of about 60,000 cfm (28 $m^2/s$). Greater capacities can be obtained with field-erected units.

Tests reveal that both liquid and solid dehumidifiers effectively remove bacteria from air passing through the system. Significant lowering of bacterial levels, combined with simultaneous dehumidification, is especially important for hospitals, laboratories, pharmaceutical manufacturing, etc.

## LIQUID ABSORPTION EQUIPMENT

Figure 2 shows a flow diagram for a typical liquid absorption system with extended-surface contactor coils. For dehumidifying, the strong absorbent solution is pumped from the sump of the unit and sprayed over the contactor coils. Air to be conditioned passes over the contactor coils and comes in intimate contact with the hygroscopic solution. Airflow can be parallel with, or counter to, the sprayed solution flow, depending on the space and application requirements.

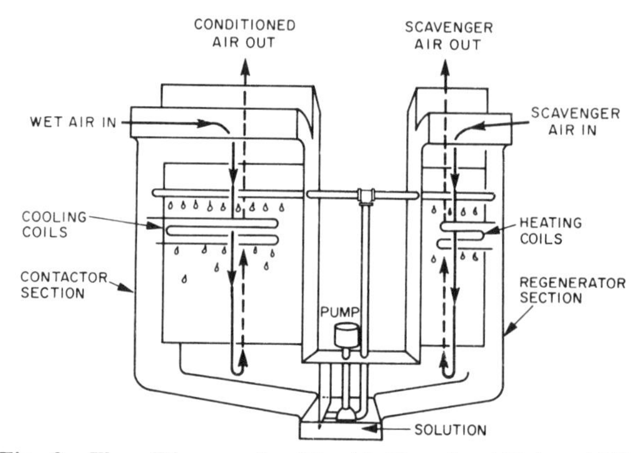


**Fig. 2 Flow Diagram for Liquid-Absorbent Dehumidifier**

The degree of dehumidification depends on the concentration, temperature, and characteristics of the hygroscopic solution. Moisture is absorbed from the air by the solution because of the vapor-pressure difference between the air and the liquid absorbent. The moisture content of the outlet air can be precisely maintained by varying the coolant flow in the coil to control the absorbent solution contact temperature. Continuous regeneration maintains the absorbent solution at the proper concentration.

The heat generated in absorbing moisture from the air consists of the latent heat of condensation of water vapor and the heat of solution, or the heat of mixing, of the water and the absorbent. The heat of mixing varies with the liquid absorbent used and with the concentration and temperature of the absorbent. The solution is maintained at the required temperature by cooling with city, well, refrigerated, or cooling-tower water, or by refrigerant flowing inside the tubes of the contactor coil.

The total heat that must be removed by the conditioner coil consists of the heat of absorption, sensible heat removed from the air, and the residual heat load added by the regeneration process. This residual heat load can be reduced substantially by using a two-sump economizer system or a liquid-to-liquid heat exchanger. In the two-sump economizer system, a small amount of the cool-dilute absorbent solution is metered to the regeneration system and replaced by a small amount of warm, highly

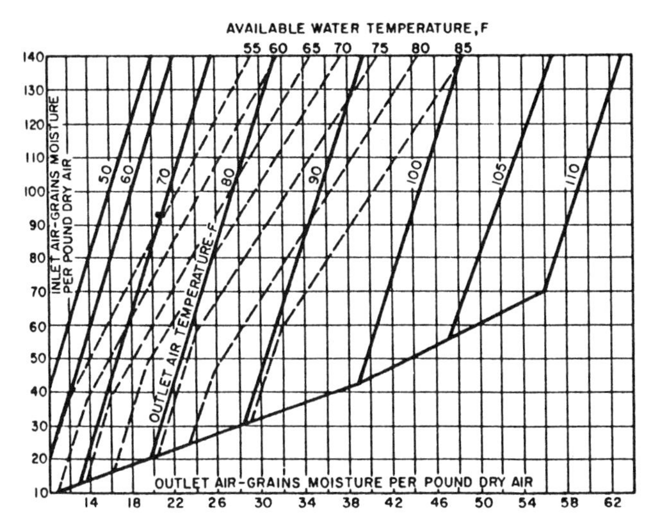


Fig. 3 (I-P) Performance Data for Liquid-Absorbent Dehumidifier Using Lithium Chloride

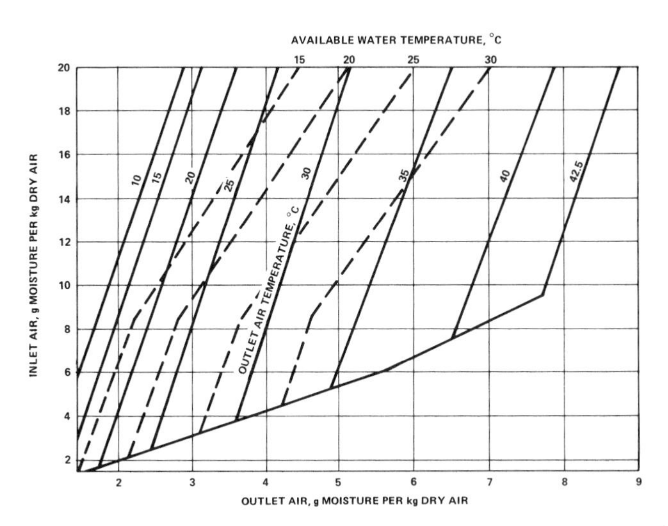


Fig. 3 (SI) Performance Data for Liquid-Absorbent Dehumidifier Using Lithium Chloride

concentrated solution. This system reduces the heat load on the conditioner cooling coil, thus reducing the amount of coolant required.

Since the contactor coil is the heat-transfer surface, proper selection of a specific absorbent solution concentration and temperature can create the desired space temperature and humidity conditions.

The dry-bulb temperature of the air leaving the liquid absorbent contactor at constant flow rate is a function of the temperature of the liquid absorbent and the amount of contact surface between the air and the solution.

The liquid absorbent is maintained at the proper concentration for moisture removal by automatically removing the water from it. A small quantity of the solution, usually 10 to 20% of the flow to the contactor coils, is passed over the regenerator coil, where the liquid is heated with steam or another heating medium. The liquid absorbents commonly used may be regenerated with steam at about 2 to 25 psig (14 to 172 kPa gauge). The vapor pressure of the liquid absorbent at temperatures corresponding to about 2 psig (14 kPa gauge) steam is considerably higher than that of the outdoor air. The hot solution, at relatively high vapor pressure, contacts outdoor air in the regenerator. The vapor-pressure difference between the outdoor air and the hot solution causes the air to absorb the water from the solution. The hot moist air from the regenerator is discharged to the outdoors, and the concentrated solution flows to the sump, where it is mixed with the dilute solution. The solution is then ready for another cycle.

The steam flow to the regenerator coil is regulated by a control that responds to the concentration of the solution circulated over the contactor coils, such as a level control, specific gravity control, or boiling-point control.

For humidifying operations, the liquid absorbent is maintained at the required temperature by adding heat in proportion to the water absorbed by the air from the solution. Water is added to the solution automatically to maintain the proper concentration. Liquid absorption dehumidification is often used to lower the moisture content of warm, moist air to a comfortable level, and it is advantageous when low-pressure steam can be used for the regeneration of the sorbent. Figure 3 shows drying performance for a liquid-absorbent dehumidifier with lithium chloride solution.

Figure 4 indicates existing applications in which liquid sorbents apply. Medium temperature range is defined as leaving dry-bulb

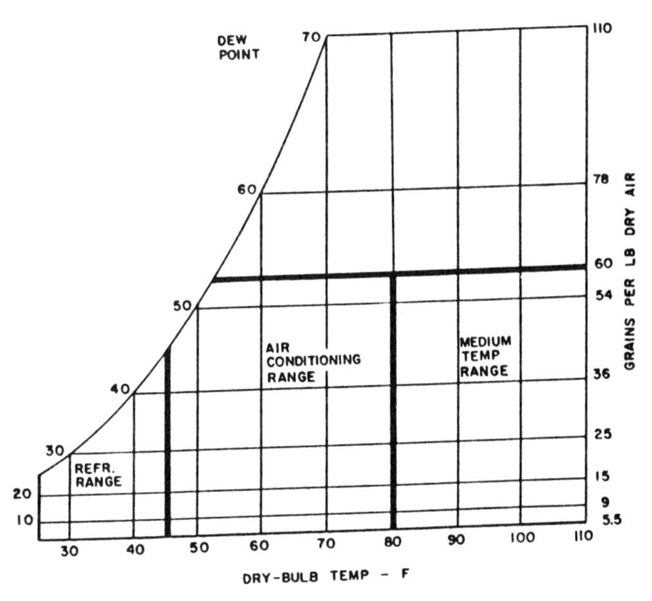


Fig. 4 (I-P) Range of Conditions for Applications Using Liquid Sorbents

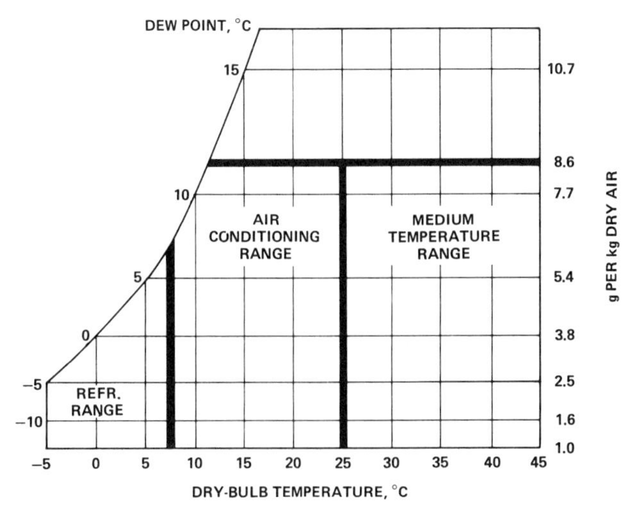


Fig. 4 (SI) Range of Conditions for Applications Using Liquid Sorbents

temperature between 80 and 110°F (27 and 43°C), with moisture content of 60 gr/lb (8.6 g/kg) of dry air or lower. Applications in this range include gelatin and glue driers, cake-icing driers or trolley rooms of bakeries, drying basic vitamins, pharmaceuticals, the application of candy coatings on gums, and the handling of many other products during manufacturing if these products are spoiled or have a shortened shelf life (because the temperature rises, during production, above a certain critical level).

The air-conditioning range is defined as leaving dry-bulb temperature between 45 and 80°F (7 and 27°C), with moisture content below 60 gr/lb (8.6 g/kg). This is the basic range for comfort and process air conditioning. In these installations, the liquid sorbents are generally used in combination with well water or refrigeration to handle the sensible heat-load requirements.

In the refrigeration range, a leaving dry-bulb temperature of 0 to 45°F (−18 to 7°C) with moisture content below 45 gr/lb (6.4 g/kg) is required of the equipment. Liquid sorbents, in combination with cold brines and direct-expansion or flooded refrigeration, are used. Typical applications include beer cellars, beer fermentation rooms, meat storage, penicillin processing, sugar cooling and conveying, and the manufacture of chewing gum.

Below the ranges shown in Figure 4, there are additional applications from −94 to 32°F (−70 to 0°C), in which the dew point of the leaving air will be at or near the dry-bulb temperature. Examples are wind tunnels, laboratories, plant-growing rooms, and tunnel-cooling of candy bars.

## SOLID SORPTION

Dehumidification by use of a solid desiccant, such as silica gel, molecular sieves, activated alumina, or hygroscopic salts, may be performed under static or dynamic conditions. In the static method, air or gas does not circulate through the desiccant. Instead, the air immediately surrounding the desiccant is dried, and, through convection and vapor diffusion, water vapor from more remote areas reaches the desiccant, where it is adsorbed. Since considerable time may be required for the air and the desiccant to establish equilibrium, this type of dehumidification is best suited for closed, sealed containers used to store or transport moisture-sensitive products.

Intermediate equipment does not incorporate an air-moving device, but the air passes through the desiccant container by thermal expansion or displacement. After some period, the desiccant is removed and is either discarded or reactivated in an oven or desiccant reactivator. These are commonly used on oil tanks or similar containers, where the evacuation of the fluid or solid in the tank is compensated for by makeup air passing through the tank *breather*, where it is dried.

In dynamic dehumidification, the air or gas being dried flows through the desiccant bed or structure. An air-moving device, such as a fan that forces the gas through the desiccant bed, and a heater or other device that periodically reactivates the desiccant are integral parts of the dehumidifier. For drying high-pressure gases or liquids, the pump, compressor, or natural pressure of the gas, and the heater used for reactivation may be entirely separated from the dryer. Following is a description of the mechanism and variables that cover any dynamic sorption system.

When moist air with a relatively high water-vapor pressure passes into activated desiccant (low vapor pressure), it surrenders a certain amount of its water vapor. The rate of moisture pickup and the humidity condition of the leaving air are functions of many variables (some of which are addressed later in this chapter). The ratio of the amount of water adsorbed by the desiccant in a given time to the amount of water vapor in the air entering the desiccant bed during that time is called *sorption efficiency*.

A characteristic of adsorbents in dynamic use is that this efficiency remains constant and at a relatively high level from the beginning of a sorption phase until some later point in the cycle, at which time the efficiency begins to drop. This point is known as the *breakpoint*, and the amount of water adsorbed on the desiccant at this point is called the *breakpoint capacity*. Ideally, the breakpoint capacity would coincide with the equilibrium capacity. In actual operation, however, breakpoint capacity can be a small proportion of the equilibrium capacity, depending on operating conditions. For isothermal operation with low flow rates and deep beds, breakpoint capacity approaches equilibrium capacity. High inlet temperature and humidity, small bed depths, and high flow rates all tend to decrease the breakpoint capacity. Although additional drying can be effected beyond the breakpoint, that the desiccant should be regenerated at or near this point. Sorption carried beyond the breakpoint continues at an increasingly slower rate until the adsorbent is saturated completely, which is known as *completion*.

When regeneration of the sorbent is desired, the heater is energized and the direction of airflow through the bed is usually reversed. The dry-bulb temperature of the effluent air rises rapidly at first and then levels off. During this period of level or slowly increasing temperature, the major portion of the heat input is being used to desorb the water. When most of the water contained in or on the desiccant is released, the heat goes into sensible heat gain and sharply increases the dry-bulb temperature of the effluent air. This period, measured from the beginning of desorption, is called *temperature-rise time*. Although additional regeneration (at a slower rate) can be attained by adding heat beyond the temperature-rise time, commercial practice ends the reactivation near this point, unless very dry effluent air is required in the following sorption phase. Regeneration past this point, until the sorbent is in moisture equilibrium with the airstream, is known as *complete desorption* or *desorption to completion*. The energy expended in the heater per unit weight of water desorbed for any given time is called *economy of desorption* and is expressed in kWh/lb (J/kg) of water desorbed.

During the process of sorption, heat is liberated, which elevates the effluent air temperature. This heat is equivalent to the latent heat of vaporization of the adsorbed liquid, plus an added quantity known as the *heat of wetting*, which is defined as the heat developed when the liquid and the solid surface contact one another. The heat of wetting is relatively large when adsorbing the first water molecules on a freshly reactivated desiccant and tapers off to a very low value as the desiccant approaches saturation. As the adsorbed vapor condenses, the latent heat is converted to sensible. All of the released heat, known cumulatively as the *heat of sorption*, is dissipated into the desiccant, the enclosure, and the passing airstream.

In comfort air conditioning, it is often necessary to cool the effluent air before its introduction into the conditioned space. But in most other dehumidification applications, this heat is not objectionable, and no provisions are made for its removal. An operation of this type is called *adiabatic*, meaning that the released heat is taken up (for the most part) by the passing airstream and that no attempt is made to cool the sorbent or the effluent air. Although ordinary air-conditioning dehumidification only approaches the adiabatic process, it is called adiabatic to differentiate it from the isothermal adsorption, in which the bed is cooled by cooling coils. Even then, only part of the heat sorption can be removed. Under true adiabatic conditions at atmospheric pressure, temperature in the desiccant bed can rise as high as 250°F (120°C) when there is drying air of high moisture content. At such elevated temperatures, the efficiency of most sorbents is reduced. Following are some of the variables that influence the results of a dynamic dehumidification operation:

*Variables concerning the desiccant bed*
1. type of desiccant
2. dry weight of desiccant
3. particle size
4. bulk density
5. shape of bed
6. area of bed normal to gas flow
7. depth of bed
8. packing of the desiccant in the bed
9. pressure drop through the bed

*Variables concerning the gas to be dried*
1. flow rate
2. temperature
3. moisture content
4. pressure
5. contact time between gas and sorbent; a function of inlet face velocity and bed depth

*Variables concerning reactivation*
1. reactivation temperature
2. rate and magnitude of heat supply
3. heat storage capacity of the bed
4. temperature gradient of the bed
5. amount of insulation
6. gas flow rate

*Miscellaneous*
1. cycle time or rotational speed of rotary equipment
2. leakage from the apparatus

**Use of cooling.** Refrigeration or other cooling means often supplement the dehumidifier. It is oten desirable to precool the entering process air to provide the desiccant bed with air having a higher relative humidity and lower temperature. Thus, some moisture is removed at high dew points, where refrigeration is more economical, and the remainder is removed by the desiccant dehumidifier at its most efficient operating condition. Since the dry-bulb temperature of the air leaving a desiccant dehumidifier is higher than the entering air temperature, the dehumidifier can also be provided with aftercooling coils to control the outlet temperature.

The source for the precooling and aftercooling coils can be self-contained refrigeration units an external cooling source (such as brine, well water, or cooling-tower water).

**Use of units in series.** Two solid sorption dehumidification units may be operated in series to provide air at an extremely low moisture content. The first unit in the system discharges sufficient process air for both the process and the reactivation airflows for the second machine. In this manner, the bed of the second machine is regenerated with extremely low dew point air; thus, the resulting drying ability of the bed is increased greatly. However, a cooling coil should be used to lower the inlet process air temperature to maintain the most efficient operation of the second machine.

## SOLID ABSORPTION EQUIPMENT

Figure 5 illustrates the principle of operation and the arrangement of major components of a typical solid absorption system. To achieve dehumidification, moist air passes through a desiccant structure, where the water vapor in the air is absorbed by the desiccant. The systems use a desiccant structure, which may be a disk, drum, or wheel, filled or impregnated with an absorbent such as lithium chloride. The desiccant structure rotates slowly through a heated stream of air, where the desiccant is desorbed or reactivated. By continuous rotation, freshly reactivated portions of the desiccant are always available for drying.

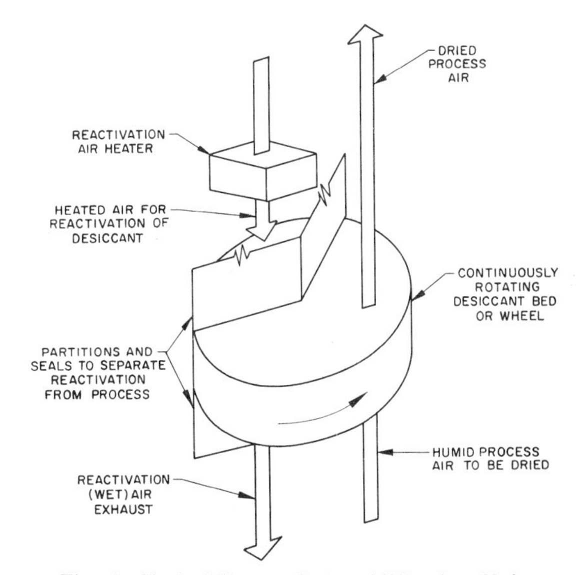


**Fig. 5 Typical Rotary Dehumidification Unit**

The amount of drying depends mainly on the temperature and absolute humidity of the air and the useful concentration of desiccant. The useful concentration of the desiccant is affected by the following:

1. The quantity of desiccant in relation to the mass flow of air and water vapor
2. The energy (amount of heat and temperature) for reactivation
3. The frequency of reactivation (speed of rotation)

The latent heat of condensation and the sensible heat conducted from the reactivated portion of the desiccant structure cause an increase in the dry-bulb temperature of the treated air. Frequently, this increase in temperature accelerates the drying process. In other instances where control of dry-bulb temperature is required, aftercoolers and/or afterheaters are installed. For a given inlet condition, the exit condition is constant and does not cycle or fluctuate. The absolute humidity of the air from the system may be readily controlled by automatically regulating the input rate of reactivating energy to produce a controlled uniform outlet condition.

The solid absorption equipment in commercial production that uses a fixed desiccant structure tends to be compact in size for a given capacity because sorption as great as 1 lb of water to 1 lb (1kg/kg) of desiccant may be used. An impregnated desiccant structure has the further advantage of tolerating large variations in the velocity of the process air without loss of desiccant to the airstream, and it reduces the need to balance pressures between process and reactivation airstreams.

Figure 6 shows typical performance of a solid sorption dehumidifier using a fixed desiccant structure.

## SOLID ADSORPTION EQUIPMENT

Solid adsorption dehumidifiers are classed in three general categories: single-bed, dual-bed, and rotary-bed units.

The single-bed unit has a bed of solid desiccant. The unit incorporates a valving system that allows an internal fan or air compressor to force air to be dried through the unit and return it to the space. Subsequently, a program timer reverses the valve mechanism and turns on a regeneration heater. Purge air is forced through the bed, and the heat drives off the moisture on the desiccant. At the completion of a designated time, the

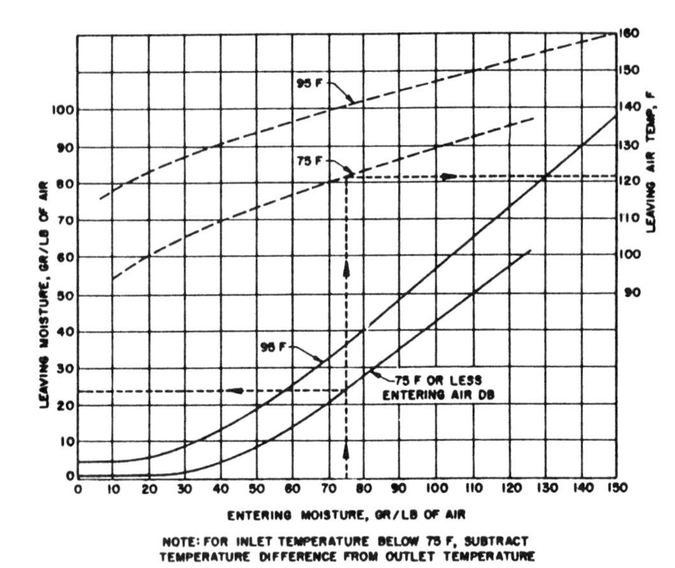


**Fig. 6 (I-P) Typical Performance Data for Rotary Solid Sorption Dehumidifier**

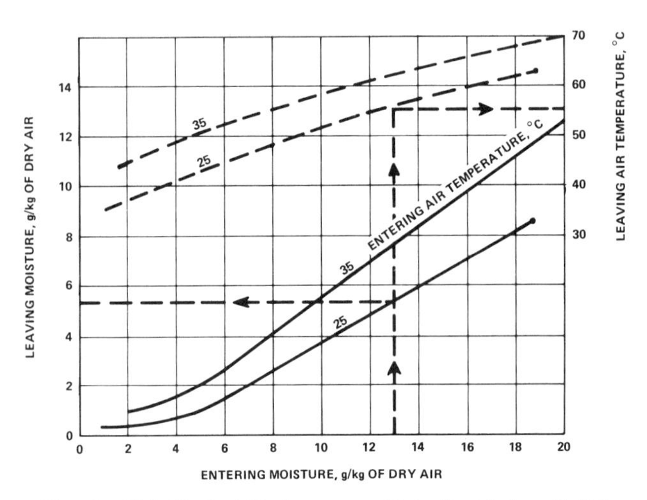


**Fig. 6 (SI) Typical Performance Data for Rotary Solid Sorption Dehumidifier**

heat shuts off, and a period of cooling by means of the purge airflow may occur. Finally, the valves reverse again, and the machine is ready to dehumidify. Delivery of dried gas from a singlebed unit is necessarily intermittent, since the desiccant bed must be alternately used for sorption and desorption. All other equipment delivers a continuous supply of dehumidified gas.

The dual-bed unit has two beds of solid desiccant and a system of valves, fans, and reactivation heaters to permit one bed to be reactivated, while the other bed removes moisture from the air or gas to be dried. After an interval (usually controlled by a timer), the valves redirect the reactivation and dry airstreams, so that the tower with reactivated desiccant receives the air to be dried and the tower with spent desiccant has the desiccant reactivated. Variations of this design replace the fans and reactivation heaters and use energy in the compressed airstream instead.

Rotary dehumidifiers (see Figure 5) have one or more beds with solid granular desiccants. The beds are usually in parallel airflow, and each bed handles air to be dried and regeneration air simultaneously. The two airstreams are separated by seals to prevent mixing, and the beds are physically rotated within the casing to expose one portion of the desiccant to the process stream and one portion to the regeneration stream. The rotational speed is usually within the range of 0.5 to 6 r/h. Rotary equipment usually gives more capacity per unit of equipment volume.

Since a rotary dehumidifier has a constantly moving bed, the effluent conditions are constant when the inlet conditions are held constant. Figure 6 shows a typical rotary dehumidifier capacity chart.

Relatively low dew points can be obtained in a closed room with a machine having a thin desiccant bed by recirculating the room air through the machine and progressively lowering the room dew point as drier air enters the machine. The machine capacity must be greater than the internal moisture load and the infiltration of moisture from external sources. Deep desiccant beds are used when extreme dryness is required in one pass through the dehumidifier and when the air is not recirculated.

Actual performance depends on the type of desiccant used, method of regeneration, and other factors. The application of solid sorbent dehumidifiers is, in general, similar to those indicated above for liquid sorbents. However, actual dew point and dry-bulb temperatures required for a specific application may favor one system over the other.

## APPLICATIONS FOR ATMOSPHERIC PRESSURE DEHUMIDIFICATION

### Preservation of Materials in Storage

The Armed Forces use, to some extent, dehumidified warehouses for special moisture-sensitive materials in long-term storage. Tests by the *Bureau of Supplies and Accounts* of the Navy Department concluded that 40% relative humidity is an adequately safe level in large warehouses, under relatively stable temperature conditions. These relative humidity maximums are selected to control deterioration of materials. Others have indicated that 60% rh is low enough to control microbiological attack. With storage at 40% rh, no undesirable effects on metals or rubber-type compounds have been noted. Some organic materials such as sisal, hemp, and paper may lose flexibility and strength, but they recover these characteristics when moisture content is regained.

Commercial storage relies on similar equipment for applications that include beer fermentation rooms, meat storage, and penicillin processing, as well as storage of machine tools, candy, food products, furs, furniture, seeds, paper stock, and chemicals. For recommended conditions of temperature and humidity, refer to Section II of the 1986 REFRIGERATION Volume.

### Process Dehumidification

The requirements for dehumidification in industrial processes are many and varied. Some of these processes are as follows:

1. Metallurgical processes, in conjunction with the controlled atmosphere annealing of metals
2. Conveying of hygroscopic materials
3. Film-drying
4. Candy, chocolate, and chewing gum manufacturing
5. Manufacturing of drugs and chemicals
6. Manufacturing of plastic materials
7. Manufacturing of laminated glass
8. Packaging of moisture-sensitive products
9. Assembly of motors and transformers
10. Solid propellent mixing
11. Manufacturing of electronic components, such as transistors and microwave components

## Condensation Prevention

Many applications require moisture control to prevent condensation. Moisture in the air will condense on cold cargo in a ship's hold when it reaches moist climates. Mositure will condense on the ship when the moist atmosphere in a cargo hold is cooled by the hull and deck plates, while the ship goes from a warm to a cold climate.

In pumping stations and sewage lift stations, the piping causes condensation, especially in the spring when the weather warms and water in the pipes is still cold. Dehumidification is also used to prevent moisture in the air from dripping into oil and gasoline tanks and into open fermentation tanks.

Electronic equipment is often cooled by refrigeration, and dehumidifiers are required to prevent internal condensation of moisture. Electronic and instrument compartments in missiles are purged with low dew point air prior to launching to prevent malfunctioning due to condensation.

Wave guides and radomes are also usually dehumidified, as are telephone exchanges and relay stations. Missile and radar sites, for the proper operation of their components, depend (to a large extent) on the prevention of condensate on interior surfaces.

## Independent Humidity Control in Air-Conditioned Spaces

Comfort air conditioning generally requires the maintenance of summer temperatures between 75 and 80°F (24 and 27°C) and relative humidities of 45 to 55%. Normally, these conditions are most economically obtained with refrigeration systems. However, if the latent heat load is larger than the cooling load, sorption systems can be used with refrigeration to give optimum results. Examples of this are high-occupancy areas and other places of assembly where the internal latent heat load is high. Large buildings and hotels located in high-humidity areas may also depend on dehumidification systems for proper comfort air conditioning.

In general, installations with panel cooling require the circulation of small quantities of highly dehumidified air for the maintenance of ventilation requirements and the prevention of condensation on the panel surfaces (see Chapter 7 of the 1987 HVAC Volume).

Some spaces must be held at a lower humidity than is available from an existing air-conditioning system. An example would be a special treatment room for extremely hygroscopic material. A dehumidifier, independent from the main air-conditioning system, should handle this load.

## Testing

Many test procedures require dehumidification with sorption equipment. Frequently, other means of dehumidification may be used in conjunction with sorbent units, but the low moisture content requirements can be obtained only by liquid or solid sorbents. Some of the typical testing applications are as follows:

1. wind tunnels
2. spectroscopy rooms
3. paper and textile testing
4. bacteriological and plant growth rooms
5. dry boxes
6. environmental rooms and chambers

# SORPTION DRIERS FOR ELEVATED PRESSURES

The same sorption principles that pertain to atmospheric dehumidification apply to drying of high-pressure air, process, or other gases. The sorbents described previously can be used with equal effectiveness.

## Absorption

Solid absorption systems use a desiccant that dissolves calcium chloride, generally in a single-tower unit that requires periodic replacement of the desiccant as it dissolves with the absorbed moisture. Normally, the inlet air or gas temperature does not exceed 90 to 100°F (32 to 38°C) saturated. The rate of replacement of the desiccant is proportional to the moisture in the inlet process flow. A dew point depression of 20 to 40°F (11 to 22°C) at pressure can be obtained when the system is operated in the range of 60 to 100°F (15 to 38°C) saturated entering temperature and 100 psig (700 kPa gauge) operating pressure. At lower pressures, the ability to remove moisture reduces proportionally as a ratio of absolute pressure. Solid-absorption units do not require a power source for operation because the desiccant is not regenerated. However, additional desiccant must be added to the system periodically.

## Adsorption

Drying with an adsorptive desiccant such as silica gel, activated alumina, or molecular sieve usually incorporates regeneration equipment, so the desiccant can be reactivated and reused. These desiccants can be readily regenerated by heat, by purging with dry gas, or by a combination of both. Depending on the desiccant selected, the dew point performance expected would be in the range of −40 to −100°F (−40 to −73°C) measured at the operating pressure with inlet conditions of 90 to 100°F (32 to 38°C) saturated and 100 psig (700 kPa gauge). Figure 7 shows typical performance using activated alumina or silica gel desiccant.

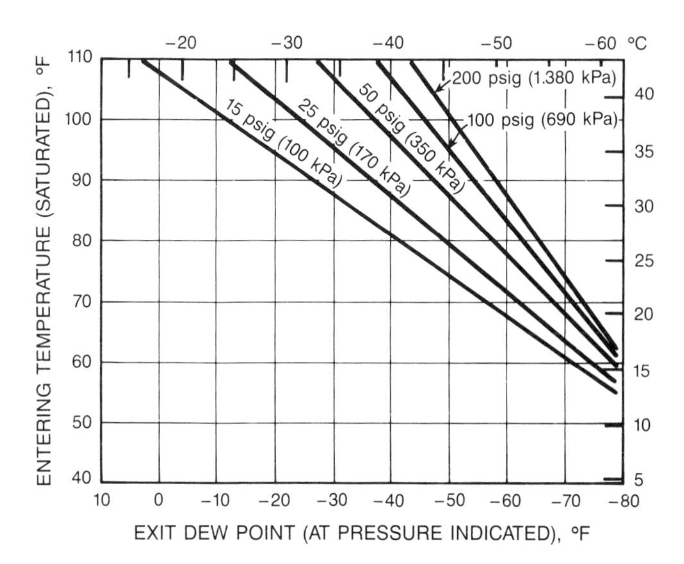


**Fig. 7 Typical Performance Data for Solid Sorption Dryers at Elevated Pressures**

Equipment design may vary considerably in detail, but most basic adsorption units use twin-tower construction for continuous operation, with an internal or external heat source, with air or process gas as the reactivation purge for liberating moisture adsorbed previously. A single adsorbent bed may be used for intermittent drying requirements. Adsorption units are generally constructed in the same manner as atmospheric-pressure units, except that the vessels are suitable for the operating pressure. Units have been operated successfully at pressures as high as 6000 psig (40 MPa).

Prior compression or cooling by water, brine, or refrigeration to below the dew point of the gas to be dried reduces the total moisture load to be handled by the sorbent, thus permitting the use of smaller drying units. An economic balance must be made to determine if removal of some water before adsorption is desirable. The cost of compression, cooling, or both must be balanced against the cost of a larger adsorption unit.

**Heat-Reactivated, Purge-Type Dryers** normally operate on four-hour (or longer) adsorption periods and are generally designed with heaters embedded in the desiccant. They use a small portion of the dried process gas as a purge to remove the moisture liberated during reactivation heating.

**Heatless-Type Dryers** operate on a short adsorption period (usually 60 to 300 s). Depressurization of the gas in the desiccant tower lowers the vapor pressure, so the adsorbed moisture is liberated from the desiccant and removed by a high purge rate of the dried process gas. The use of an ejector reduces the purge gas requirements.

**Convection-Type Dryers** usually operate on four-hour (or longer) adsorption periods and are designed with an external heater and cooler as the reactivation system. Some designs circulate the reactivation process gas through the system by a blower, while other designs divert a portion or all of the process gas flow through the reactivation system prior to adsorption. Both heating and cooling are by convection.

**Radiation-Type Dryers** operate on four-hour (or longer) adsorption periods and are designed with an external heater and blower to force heated atmospheric air through the desiccant tower for reactivation. Cooling of the desiccant tower is by radiation to atmosphere.

Figure 8 illustrates a typical adsorption-type desiccant dryer of the heat-reactivated, purge-type.

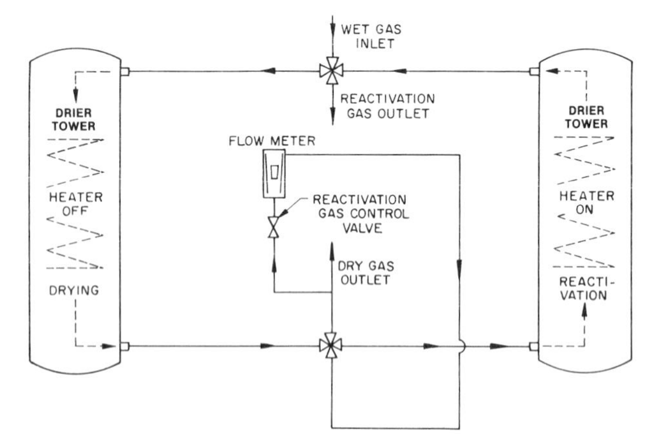


**Fig. 8 Typical Adsorption Dryer for Elevated Pressures**

## APPLICATIONS FOR DRYING AT ELEVATED PRESSURE

### Preservation of Materials

Generally, materials in storage are preserved at atmospheric pressure, but a few materials are stored at elevated pressures, especially when the dried medium is an inert gas. These materials deteriorate when they are subjected to high relative humidity or oxygen content in the surrounding media. The drying of high-pressure air, which is subsequently reduced to 3.5 to 10 psig (24 to 70 kPa gauge), has been used most effectively in pressurizing coaxial cables to eliminate electrical shorts caused by moisture infiltration. This same principle, at somewhat lower pressures, is also used in wave guides and radomes to prevent moisture film on the envelope.

### Process-Drying of Air and Other Gases

Drying of *instrument air* to a dew point of −40 °F (−40 °C), particularly in areas where the air lines are outdoors or are exposed to temperatures below the dew point of the air leaving the aftercooler, prevents condensation or freezeup in the instrument control lines.

To prevent condensation and freezing, it is necessary to dry the plant air used for pneumatically operated valves, tools, and other equipment in areas where the piping is exposed to low ambient temperatures. Additionally, dry air prevents rusting of the air lines, which produces abrasive impurities, causing excessive wear on tools.

Drying of industrial gases or fuels such as natural gas has been accepted. For example, fuels (including natural gas) are cleaned and dried before storage underground to ensure that valves and transmission lines do not freeze from condensed moisture during extraordinarily cold weather, when the gas is most needed. Propane must also be clean and dry to prevent ice accumulation. Other gases, such as bottled oxygen, nitrogen, hydrogen, and acetylene, must have a high degree of dryness. In the manufacture of liquid oxygen and ozone, the weather air supplied to the particular process must be clean and dry.

Drying of air or inert gas for conveying of hygroscopic materials in liquid or solid state ensures continuous, trouble-free plant operation. Normally, gases for this purpose are dried to a −40 °F (−40 °C) dew point. Purging and blanketing operations in the petrochemical industry depend on the use of dry inert gas for the reduction of explosive hazards, the reaction of chemicals with moisture or oxygen, and other such problems.

### Testing of Equipment

Dry, high-pressure air is used extensively for the testing of refrigeration condensing units to ensure tightness of components and to prevent moisture infiltration. Similarly, dry inert gas is used in the testing of copper tubing and coils to prevent corrosion or oxidation. The manufacture and assembly of solid-state circuits and other electronic components require exclusion of all moisture, and final testing in dry boxes must be carried out in moisture-free atmospheres. The simulation of dry high-altitude atmospheres for testing of aircraft and missile components in wind tunnels requires extremely low dew point conditions.

### Liquid Phase Drying

Solid adsorbents offer an economical method for removing water from liquid hydrocarbons where high purity is desired, and specified water content is relatively low. Among liquids that can be dried with good results are: benzene, toluene, zylene, butane, propane, trichlorethylene, R-11, R-12, and methyl chloride.

## BIBLIOGRAPHY

ASHRAE. 1975. Symposium on Sorption Dehumidification. ASHRAE *Transactions*. 81(1):606-638, 4 papers.

ASHRAE. 1980. Symposium on Energy Conservation in Air Systems Through Sorption Dehumidifier Techniques. ASHRAE *Transactions*. 86(1):1007-1036, 5 papers.

ASHRAE. 1985. Symposium on Changes in Supermarket Heating, Ventilating and Air-Conditioning Systems. ASHRAE *Transactions*. 91(1B):423-468, 5 papers.

Jones, B.W.; B.T. Beck; and J.P. Steele. 1983. Latent Loads in Low Humidity Rooms Due to Moisture. ASHRAE *Transactions*. 89(1A):35-55.

## CHAPTER 8

# FORCED-CIRCULATION AIR COOLERS



FORCED-CIRCULATION air coolers contain, within a cabinet enclosure, two basic components: a cooling coil and a motor-driven fan assembly. Coil defrost equipment is added for low-temperature applications. Primarily, the *unit cooler* and *product cooler* style are designed to operate continuously within refrigerated enclosures. They provide low-temperature cooling and freezing, as well as the equipment to deliver proper airflow.

In a broad sense, blower coils, unit coolers, product coolers, cold diffuser units, air-conditioning air handlers, etc., are essentially forced-air coolers when operated under refrigeration duty conditions. Many design and construction choices are available such as the following: (1) finned or bare tube coils; (2) electric, gas, air, or water defrosting; (3) discharge air velocity; (4) centrifugal or propeller fans, either belt- or direct-driven; (5) ducted or non-ducted, and/or (6) free standing or ceiling suspended.

## TYPES OF FORCED-CIRCULATION AIR COOLERS

### Sloped Front Unit Coolers

These units, which range from 5 to 10 in. (125 to 250 mm) high, are commonly used in back-bar and under-the-counter fixtures, as well as in vertical, self-serve, glass door reach-in enclosures. The slope fronts are designed for horizontal top mounting as a single unit, or installation as a group of parallel connected units. Direct-drive fans are sloped to fit within the restricted return air sweep rising across the enclosure access doors. These are small unit coolers, with airflows usually not exceeding 150 cfm (70 L/s) per fan.

### Low Air Velocity Unit Coolers

These units can have a *half-round* appearance, but commonly long, narrow, dual-coil units are used in meat-cutting rooms and in meat and floral walk-in coolers. The unit has an amply finned coiled surface to maintain high humidities in the room. Discharge air velocities at the coil face range from 85 to 200 fpm (0.4 to 1.0 m/s).

### Standard Air Velocity Unit Cooler

Units that are about 12 to 15 in. (300 to 380 mm) high are called *low silhouette* units. *Medium* or *mid-height* units are 18 to 48 in. (450 to 1200 mm) high or greater. Some can be classified as *high silhouette* unit coolers. The air velocity at the coil face can be as high as 600 fpm (3 m/s), but generally it is 300 to 400 fpm (1.5 to 2.0 m/s).

### High Air Velocity Product Coolers

These units are used in blast tunnel freezing and special cooling of products not affected adversely by moderate dehydration during rapid cooling. They generally draw air through the cooler at discharge velocities of about 2000 fpm (10 m/s).

Forced-circulation coolers direct air over a refrigerated coil in the enclosure. The coil lowers the air temperature below its dew point, which causes condensate or frost to form on the coil surface. However, the normal refrigeration load is a sensible heat load, *i.e.*, dry-bulb referenced, and the coil surface is considered dry. Rapid and frequent defrosting on a timed cycle can maintain this dry-surface condition.

### Sprayed Coil Product Coolers

Spray coils feature a saturated coil surface that can cool the processed air closer to the coil surface temperature than a dry coil. In addition, the spray continuously defrosts the low-temperature coil. Unlike unit coolers, spray coolers are usually floor mounted and discharge air vertically. The unit sections include a drain pan/sump, coil with spray section, moisture eliminators, and fan with drive. The eliminators remove airborne droplets of spray to prevent their discharge into the refrigerated area. Typically, belt-driven centrifugal fans draw air through the coil at 600 fpm (3 m/s) or less.

For coil surfaces above freezing, water can be used as the spray medium. Below freezing, a suitable material such as listed below must be added to the water to lower the freezing point to about 12 °F (−11 °C) or lower than the coil surface temperature. Some recirculating solutions, when mixed properly, are usable as follows:

**Sodium chloride** is limited to a room temperature of 10 °F (−12 °C) or higher. Its minimum freezing point is −6 °F (−21 °C).

**Calcium chloride** can be used for enclosure temperatures down to about −10 °F (−23 °C), but its use may be prohibited in enclosures containing food products.

**Propylene glycol** is used at higher temperatures than ethylene glycol, which may be used for temperatures as low as −35 °F (−37 °C). Because of its toxicity, sprayed ethylene glycol in other than sealed tunnels or freezers (no human access allowed during process) is usually prohibited by most jurisdictions. When a glycol mix is sprayed in food storage rooms, any carryover of the spray must be maintained within the limits prescribed by the USDA, the Bureau of Animal Industries, and all applicable local codes.

The preparation of this chapter is assigned to TC 8.4, Air-to-Refrigerant Heat-Transfer Equipment.

**Lithium chloride** is another possible spray solution, but it is no longer widely used.

All brines are hygroscopic; thus, they absorb condensate and become progressively weaker. This dilution can be corrected by continually adding salt to the solution to maintain sufficient below-freezing temperature. Salt is extremely corrosive, so it must be contained in the sprayed coil unit with suitable corrosive-resistant materials, which must receive periodic inspection and maintenance.

Sprayed coil units are usually installed in refrigerated enclosures requiring high humidity, *e.g.*, chill coolers. Paradoxically, the same sprayed coil units can be used in special applications requiring low relative humidity. For such dehydration applications, both a high brine concentrate (near its eutectic point) and a wide difference between the process air and the refrigerant temperature are maintained. Process air reheat downstream of the sprayed coil corrects the dry-bulb temperature.

## COMPONENT DESIGNS AND CONTROL METHODS

### Draw-Through and Blow-Through Airflow

Unit fans may draw air through the cooling coil and discharge it through the fan outlet into the enclosure; or the fans may blow air through the cooling coil and discharge it from the coil face into the enclosure. Blow-through units have a slightly higher thermal efficiency because heat from the fan is removed from the forced air stream by the coil. Draw-through fan energy adds to the heat load of the refrigerated enclosure, but neither load from the fractional horsepower fan motors is significant. Selection depends more on a manufacturer's design features for the unit size required, as well as the air throw required within the particular enclosure.

Blow-through design has a lower discharge air velocity stream because the entire coil face area is usually the discharge opening (grilles and diffusers not withstanding). An air throw of 33 ft (10 m) or less is common for the average standard air velocity from a blow-through unit. Greater throw, in excess of 100 ft (30 m), is normal for draw-through centrifugal fan units. The propeller fan in the high silhouette draw-through unit cooler is popular for intermediate ranges of air throw.

### Fan Assemblies

Direct-drive propeller fans (motor plus blade) are popular because they are simple, economical, and can be installed in multiple assemblies in a unit cooler housing. Additionally, they require less motor power for a given airflow rate capacity.

The centrifugal fan assembly usually includes belts, bearings, sheaves, and coupler drives along with each of their inherent problems. Yet, this design is necessary for applications having high air distribution static pressure losses. These applications include enclosures with ductwork runs, tunnel conveyors, and high-density stacking of products. Centrifugal fan-equipped units are also used in produce ripening rooms, where a large air blast is required during the ripening process.

### Casing

Casing materials are selected for compatibility with the enclosure environment in which they are installed. Construction usually features coated aluminum or galvanized steel. Stainless steel is also used in food storage or preparation enclosures where sanitation must be maintained. On larger cooler units, internal framing is fabricated of sufficiently substantial material, such as galvanized steel, and casings are usually made with similar material. Some plastic casings are used in small unit coolers, while some large, ceiling-suspended units may feature all aluminum construction (to keep their weight down).

### Coil Construction

Coil construction varies from uncoated (all) aluminum tube and fin to hot-dipped galvanized (all) steel tube and fin, depending on the type of refrigerant used and the environmental exposure to the coil surface. The most popular unit coolers have copper tube/aluminum fin construction. Ammonia refrigerant equipment is not constructed from copper tube coil. Also, sprayed coils are not constructed from aluminum fin and tube materials.

Fin spacings vary from 6 to 8 fins/inch (3.2 to 4.2 mm between fins) for coils with surfaces above 32 °F (0 °C) and between 0 and 32 °F (−18 and 0 °C) when latent loads are insignificant. Otherwise, 3 to 6 fins/inch (4.2 to 8.5 mm) is the accepted spacing for coil surfaces below 32 °F (0 °C), with a spacing of 4 per inch (6.4 mm) when latent loads exceed 15% of the total load.

### Defrosting

Coils must be defrosted when frost accumulates on their surfaces. The frost (or ice) is usually greatest at the air entry side of the coil; therefore, the required defrost cycle is determined by the inlet surface condition. A reduced secondary surface-to-primary surface ratio, in contrast, produces greater frost accumulations at the coil outlet face. In theory, the accumulation of greater frost at the coil entry air surface may improve the coil's heat transfer capacity. However, accumulated coil frost usually has two negative effects: (1) it impedes heat transfer because of its insulating effect and (2) it reduces the airflow because it restricts the free air area within the coil.

Depending on the defrost method, as much as 80% of the defrost heat load could be transferred into the enclosure. This heat load is not normally included as part of the enclosure heat gain calculation, but it is usually accounted for by the factor that estimates the hours per day of refrigeration running time.

A longer time between defrost cycles can be achieved by using more coil tube rows and a wider fin spacing. Also ice accumulation should be avoided to reduce the defrost time. For example, in low temperature applications having high latent loads, unit coolers should not be located above freezer doors.

### Controls

Electromechanical controls cycle to maintain the desired enclosure temperature. Modulating control valves such as evaporator pressure regulators are also used. In its simplest form, the temperature control is a thermostat mounted in the enclosure that either cycles the compressor *on* and *off* or a liquid line feed solenoid that opens and closes. A suction pressure switch could substitute for the wall-mounted thermostat.

An electronically based energy management system (EMS) maintains optimum energy control by equating suction pressure transducer readings with the signal from a temperature diode sensor in the enclosure. EMS controls are commonly used in large warehouses and supermarkets.

Defrost options, where required, operate either on demand or at fixed time intervals. Fixed time defrost, which is widely accepted and used, is initiated by a 24-hour clock at predetermined defrost intervals and durations. The defrost cycle terminates when the coil temperature rises sufficiently to melt the frost. A rise in evaporator pressure or an elapsed time clock are also used to terminate the defrost cycle. On many applications, a mix of these various defrost control methods is common and adapts well to the system control. Further details are addressed under "Frost Condition" in the "Ratings and Applications" section.

## AIR MOVEMENT AND DISTRIBUTION

Air distribution and velocity are two important concerns in selecting and locating a unit cooler. The direction of the air and air throw should be such that air moves where there is a heat gain; this principle applies to the enclosure walls and ceiling as well as to the product. Unit cooler(s) should be placed (1) so they do not discharge air at any doors or openings; (2) away from doors that do not incorporate an entrance vestibule or pass to another refrigerated enclosure to keep from inducing additional infiltration into the enclosure; and (3) away from the airstream of another unit to avoid defrost difficulties.

The velocity and relative humidity of air passing over an exposed product affect the amount of surface drying and weight loss. Air velocities of 500 fpm (2.5 m/s) over the product are typical for most freezer applications. Higher velocities require additional fan power and, in some cases, only slightly decrease the cooling time. For example, air velocities in excess of 500 fpm (2.5 m/s) for freezing plastic-wrapped bread reduce freezing time very little. However, increasing the air velocity from 500 to 1000 fpm (2.5 to 5.0 m/s) over unwrapped pizza reduces the freezing time and product exposure almost in half.

This variation shows the importance of product testing to design the special enclosures intended for blast freezing accurately and/or automated food processing. Sample tests should yield the following information: ideal air temperature, air velocity, product weight loss, and dwell time. With this information, the proper unit or product coolers, as well as the supporting refrigeration equipment and controls, can be selected.

## RATING AND APPLICATIONS

Currently, no industry standard exists for rating unit and product coolers. Part of the difficulty in developing a workable standard is that many variables are encountered. Cooler coil performance and capacities should be based on a fixed set of conditions and greatly depend on (1) air velocity, (2) refrigerant velocity, (3) temperature difference, (4) frost condition, and (5) superheating adjustment. The most significant items are refrigerant velocity, as related to refrigerant feed through the coil, and frost condition and defrosting in low-temperature applications. The following sections address some of the problems involved in arriving at a common rating.

### Refrigerant Velocity

Refrigerant velocity varies, depending on the commercial refrigerant feed method used as do the capacity ratings of the cooler. The following feed methods are used.

**Dry Expansion.** In this system, a thermostatically controlled, direct-expansion valve allows just enough liquid refrigerant into the cooling coil to ensure that it vaporizes at the outlet. In addition, about 5 to 15% of the coil surface is used to superheat the vapor. The direct expansion (DX) coil ratings are usually the lowest of the various feed methods.

**Recirculated Refrigerant.** This system is similar to a dry expansion feed because it has a hand expansion valve or metering device to control the flow of the entering liquid refrigerant. The coil is intentionally overfed to eliminate superheating of the refrigerant vapor. The amount of liquid refrigerant pumped through the coil may be two to six times that of a dry expansion coil. As a result, this coil's capacity is higher than that for a dry expansion feed. Chapter 2 of the 1986 REFRIGERATION Volume covers this type of feed in greater detail.

**Flooded.** This system has a liquid reservoir (surge drum) located adjacent to each unit or set of units. The surge drum is filled with a subcooled refrigerant and connected to the cooler coil. To ensure gravity flow of this refrigerant and a completely wet internal coil surface, the liquid level in the surge drum must be higher than the top of the coil. The capacity of gravity-recirculated feed is usually the highest attainable. Chapter 2 in the 1986 REFRIGERATION Volume has more detail.

**Brine.** This term encompasses any liquid or solution that absorbs heat within the coil without a change in state. Ethylene glycol and water, propylene glycol and water, and R-11 are used, in addition to solutions of calcium chloride or sodium chloride and water. The capacity ratings for brine coils depend on many variables such as flow, viscosity, specific heat, density, etc. As a result, this rating is obtained by special request from a coil manufacturer and generally is 10 to 40% less than its flooded rating application.

### Frost Condition

Frost condition and defrosting are perhaps the most indeterminate variables that affect the capacity rating of forced-air cooler coils. Any unit operating below 32 °F (0 °C) *coil* temperature accumulates frost or ice. Although a light frost accumulation improves heat transfer of the coil slightly, the continuous accumulation effects coil performance negatively. Ultimately, defrosting is the only solution.

**Enclosure Air Temperature Above 35 °F (1.7 °C).** Whenever the enclosure *air* is 35 °F (1.7 °C) or slightly warmer, it can be used to defrost the coil. However, some of the moisture on the coil surface evaporates into the air, which is undesirable for low-humidity application. The following methods of control are commonly used.

1. If the refrigeration cycle is interrupted by a *defrost timer*, the continually circulating air melts the coil frost and ice. The timer can operate either the compressor or a refrigerant solenoid valve.
2. An *oversized unit cooler* controlled by a wall thermostat defrosts during its normal *off* and *on* cycling. The thermostat can control a refrigeration solenoid in a multiple-coil system or the compressor in a unitary installation.
3. *Pressure Control* can be used for slightly oversized unitary systems. A low-pressure switch connected to the compressor suction line is set at a *cut-out* point such that the design suction pressure corresponds to the saturated temperature required to handle the maximum enclosure load. The suction pressure at the compressor drops and causes the compressor motor to stop as the enclosure load fluctuates or as the oversized compressor overcomes the maximum loading.

   A thermostatic expansion valve on the unit cooler controls the liquid refrigerant flow into the coil, which varies with the load. The *cut-in* point, which starts the compressor motor, should be set at the suction pressure, which corresponds to the equivalent saturated temperature of the desired refrigerated enclosure air. The pressure differential between the cut-in and cut-out points corresponds to the temperature difference between the enclosure air and the coil temperatures. The pressure settings should allow for the pressure drop in the suction line.

**Enclosure Air Temperature Below 35 °F (1.7 °C).** Whenever the enclosure air is below 35 °F (1.7 °C), supplementary heat must be introduced into the enclosure to defrost the coil surface and drain pan. Unfortunately, some of this defrost heat remains in the enclosure until the unit starts operation after the defrost cycle. No two defrost methods have the same input of timing.

1. *Hot gas defrost* can be the fastest and most efficient method if an adequate supply of hot gas is available. The hot refrig-

erant discharge gas internally cleans the tubes and aids in returning the lubricants back to the compressor. It can be used for small commercial units and large industrial systems. It can also be used for low-temperature applications. It can also improve capacity because it can remove some of the load from the condenser when used to defrost multiple evaporators alternately on a large, continuously operating compressor system. This method of defrost puts the least amount of heat into the enclosure air, especially when latent gas defrosting is used.

2. *Electric defrost* effectiveness depends on the location of the electric heating elements. The electric heating elements can be placed either in contact with the finned coil surface or inserted into the interior of the coil element. In this latter method, either special fin holes or *dummy tubes* are used. Electric defrost can be efficient and rapid; it is simple to operate and maintain, but it does dissipate the most heat into the enclosure.
3. *Heated Air.* Although air is the defrosting medium in these situations, it is not the enclosure air as described above. Some unit coolers are constructed to isolate the frosted coil from the cold enclosure air. Once isolated, the air around the coil is heated by hot gas or electric heating elements and is circulated to hasten the defrost. This heated air also must heat a drain pan, which is needed in any enclosures at temperatures of 34 °F (1 °C) or less. Some units have specially constructed housings and ducting to run warm air from adjoining areas.
4. *Water Defrost* is the quickest method of defrosting a unit. It is efficient, as well as effective for rapid cleaning of the complete coil surface. It can be performed manually or on an automatic timed cycle. This method becomes less desirable as the enclosure temperature decreases much below freezing, but it has been successfully used in cases as low as −40 °F (−40 °C). Water defrost is used more for industrial product cooler units than for small ceiling-suspended units.
5. *Hot Brine.* Brine-cooled coils can have a heater to heat brine for the defrost cycle. This system heats from within the coil and is as rapid as hot gas defrost. The heat source can be steam, electric resistance elements, or condensing water.

Each defrosting method is done with the fan turned off. Frequently, two methods are run simultaneously to shorten the defrost cycle. Both inadequate defrost time and overdefrosting can degrade overall performance; thus, a defrost cycle is best ended by monitoring temperature. A thermostat may be mounted within the cooler coil to sense a rise in the temperature of the finned or tubed surface. A temperature of at least 40 °F (4.5 °C) indicates the removal of frost and automatically returns the unit to the cooling cycle.

Fan operation is delayed, usually by the same thermostat, until the coil surface temperature approaches its normal operating level. This practice prevents unnecessary heating of the enclosure after defrost and also prevents drops of defrost water from being blown off the coil surface. In some applications, fan delay after defrost becomes essential to prevent a rapid buildup of air pressure, which could structurally damage the enclosure.

Initiation of defrost can be automated by time clocks, running time monitors, air pressure differential controls, or by monitoring the air temperature difference through the coil (which increases as the airflow is reduced by frost accumulation). Adequate supplementary heat for the drain pan and drain lines should be considered. Also, drain lines should be pitched properly and *trapped* outside the cold area to keep them from freezing.

Most rating tables state gross capacity and assume that the fan assembly or defrost heat is included as part of the enclosure load calculation. Some manufacturers' cooler coil ratings may appear as *sensible*, while some may be listed as *total capacity*, which includes the sensible and latent capacity. Some ratings involve *reduction factors* to account for frost accumulation in low-temperature applications or for some unusual condition. Some include multiplication factors for various refrigerant types.

The rating, known as basic cooling capacity, is based on the temperature difference between the inlet air and the refrigerant within the coil; Btu/h (W) per degree TD (temperature difference) is the dimension used. The coil inlet air temperature is considered the same as the enclosure air temperature. This practice is common for smaller enclosure applications. For larger installations, as well as heavy use process work, it is advisable to apply manufacturer's published ratings based on the average of the coil inlet-to-outlet temperatures. This is regarded as the *average enclosure temperature*. The refrigerant temperature is usually the temperature equivalent to the saturated pressure at the coil outlet. Chapter 6 gives information on how to estimate heat transfer and pressure drop in cooler coils.

The TD necessary to obtain the unit cooler capacity varies with each application. It may be as low as 8 °F (4.5 °C) for wet storage coolers and as high as 25 °F (14 °C) for workrooms. The TD can be related to the desired humidity requirements. The smaller the TD, the smaller the amount of dehumidification caused by the coil; thus, a higher humidity level can be maintained. The following information gives general guidance.

**Medium temperature applications** above 25 °F (−4 °C) saturated suction:

1. For a very high relative humidity (about 90%), a temperature difference of 8 to 10 °F (4.5 to 5.5 °C) is common.
2. For a high relative humidity (approximately 80%), a temperature difference of 10 to 12 °F (5.5 to 6.5 °C) is recommended.
3. For a medium relative humidity (approximately 75%), a temperature difference of 12 to 16 °F (6.5 to 9 °C) is recommended.
4. Temperature differences beyond these limits usually result in low enclosure humidities, which dry the product. However, for packaged products and workrooms, a TD of 25 to 30 °F (14 to 16.5 °C) is not unusual. Paper storage or similar products also require a low humidity level. Here, a TD of 20 to 30 °F (11 to 16.5 °C) may be necessary.

**Low-temperature applications** below 25 °F (−4 °C) saturated suction:

For a low-temperature application, the temperature difference is generally kept below 15 °F (8 °C) because of system economics and frequency of defrosting, rather than for humidity control.

## INSTALLATION AND OPERATION

Whenever possible, refrigerating air-cooling units should be located away from enclosure entrance doors and passageways. This practice helps reduce coil frost accumulation, as well as fan blade icing. The cooler manufacturer's *Installation, Start-Up, and Operation* instructions generally give the best information. Upon installation, the unit nameplate data (model, refrigerant type, electrical data, warning notices, certification emblems, etc.) should be recorded. This information should be compared to the job specifications, as well as the manufacturer's instructions for correctness.

## GENERAL INFORMATION

Additional information on the selection, ratings, installation, and maintenance of cooler coil units is available from the various unit manufacturers. Also, the 1986 REFRIGERATION Volume, Section II, "Food Refrigeration," has specific product cooling information.

# CHAPTER 9

# AIR-HEATING COILS



AIR-HEATING coils are used to heat air under forced convection. The total coil surface may consist of a single coil section or several coil sections assembled into a bank. The coils described in this chapter only apply to comfort heating and air conditioning using steam or hot water.

## COIL CONSTRUCTION AND DESIGN

**Extended-surface coils** consist of a primary and a secondary heat-transfer surface. The primary surface is the surface of the round tubes or pipes that are arranged in a repetitive pattern with respect to the airflow. The secondary surface (fins) consists of thin metal plates or a spiral ribbon uniformly spaced or wound along the length of the primary surface. It is in intimate contact with the primary surface for good heat transfer. This bond must be maintained permanently to ensure continuation of rated performance.

The heat-transfer bond between the fin and the tube may be achieved in numerous ways. Bonding is generally accomplished by expanding the tubes into the tube holes in the fins to obtain a permanent mechanical bond. The tube holes frequently have a formed fin collar, which provides the area of thermal contact and may serve to space the fins uniformly along the tubes.

The fins of spiral or ribbon-type fin coils are tension-wound onto the tubes. In addition, an alloy with a low melting point, such as solder, may be used to provide a metallic bond between fins and tube. Some types of spiral fins are knurled into a shallow groove on the exterior of the tube. Sometimes, the fins are formed out of the material of the tube.

The fin designs most frequently used for heating coils are flat plate fins, plate fins of special shape, and spiral or ribbon fins.

Copper and aluminum are the materials most commonly used in the fabrication of extended-surface coils. Tubing made of steel or various copper alloys is used in applications where corrosive forces might attack the coils from either inside or outside. The most common combination for low-pressure applications is aluminum fins on copper tubes. Low-pressure steam coils are usually designed to operate up to 150 to 200 psi (1.0 to 1.4 MPa) guage. Above that point, tube materials such as red brass, Admiralty, or Cupro-Nickel are selected.

The preparation of this chapter is assigned to TC 8.4, Air-to-Refrigerant Heat-Transfer Equipment.

### Steam Coils

For proper performance of steam-heating coils, condensate and air or other noncondensables must be eliminated rapidly and the steam must be distributed uniformly to the individual tubes. Noncondensable gases (such as carbon dioxide) remaining in a coil cause chemical corrosion and result in early coil failure.

Uniform steam distribution is accomplished by methods such as the following:

1. Individual orifices in the tubes
2. Distributing plates in the steam headers
3. Special perforated, small-diameter, inner steam-distributing tubes that extend into the larger diameter tubes of the primary surface

Freeze-resistant coils of the perforated inner-tube type are constructed with arrangements such as the following:

1. Supply and return on one end, with incoming steam used to heat the leaving condensate
2. Supply and return on opposite ends
3. Supply and return on one end and a supply on the opposite end

Particular care in the design and installation of piping, controls, and installation is necessary to protect the coils from freeze-up because of incomplete draining of condensate. When the entering air temperature is 32°F (0°C) or below, it is best not to modulate the steam supply to the coil. Coils located in series in the airstream, with each coil sized to be on or completely off in a specific sequence (depending on the entering air temperature), are not likely to freeze. The use of bypass dampers is also common. During part-load conditions, air is bypassed around the steam coil with full steam being kept on the coil. In this system, high velocity jets of low-temperature air must not impinge on the coil when the face dampers are in a partially closed position. See Chapter 51 of the 1987 HVAC Volume for more details.

### Water Coils

The performance of water coils for heating depends on the elimination of air from the system and the proper distribution of the water to the individual tube circuits. Air elimination in system piping is described in Chapter 13 of the 1987 HVAC Volume.

To produce the most efficient capacity without excessive water-pressure drop through the coil, various circuit arrangements are used. A single-tube serpentine circuit can be used on small booster heaters requiring small water quantities up to a maximum of approximately 4 to 5 gpm (0.25 to 0.32 L/s). With this arrangement, a single tube handles the entire water quantity, provided the tube makes a number of passes across the airstream.

The most common circuiting arrangement is often called **single-row serpentine** or **standard** circuiting. With this arrangement, all tubes in each coil row are supplied with an equal amount of water through a manifold, commonly called the **coil header.** When the water volume is small, so that water flow in the tubes is laminar, **turbulators** are sometimes installed in each tube circuit to produce turbulent water flow.

Special circuiting (serpentine) arrangements accomplish the same effect as mechanical turbulators when the water volume supplied to the coil is small. Serpentines reduce the number of circuits, thus increasing the water velocity in each circuit and creating turbulent flow.

Figure 1 illustrates commonly selected water-circuit arrangements. When hot-water coils are used with entering air temperatures below freezing (an antifreeze brine would be preferable to water), piping the coil for parallel flow rather than counterflow, as shown, should be considered. This arrangement places the highest water temperature on the entering air side. Coils piped for counterflow have the water enter the tube row on the exit air side of the coil. Coils piped for parallel flow have the water entering the tube row on the enter air side of the coil.

Water coils are usually designed to be self-venting by supplying water to the coil so that the water flows upward in the coil, forcing air out through the return water connection. This design also ensures that the coil is always completely filled with water, regardless of the water volume supplied. Water circuits must be designed to ensure complete coil drainage.

Methods for controlling water coils to produce uniform leaving air temperatures are discussed in Chapters 13, 14, 15 and 51 of the 1987 HVAC Volume.

## Flow Arrangement

The relative directions of fluid flows influence the performance of the heat-transfer surface. In air-heating coils with only one row of tubes, air flows at right angles relative to the heating medium. In coils with more than one row of tubes in the direction of the airflow, the heating medium in the tubes may be circuited in various ways.

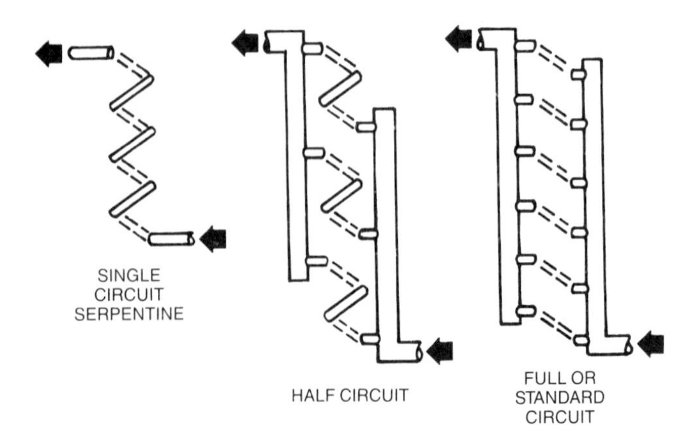


**Fig. 1 Water Circuit Arrangements- Two Row Heating Coils**

Crossflow is common in steam-heating coils. The steam temperature within the tubes remains substantially uniform, and the mean temperature difference is basically the same, whatever the direction of flow relative to the air.

Parallel flow and counterflow arrangements are common in water coils. Counterflow is the preferred arrangement to obtain the highest possible mean temperature difference. The mean temperature potential determines the *heat transfer driving force* of the coil. The greater the mean temperature difference, the greater is the heat transfer capacity of the coil.

## Electric Heating Coils

An electric heating coil consists of a length of resistance wire (commonly nickel/chromium) to which a voltage is applied. The resistance wire may be bare or sheathed. The sheathed coil is a resistance wire encased by an electrically insulating layer such as magnesium oxide, which is encased in a finned steel tube. There has been considerable debate on the relative merits of bare and sheathed coils. The sheathed coils are more expensive, have a higher air-side pressure drop, and require more space (41,200 Btu/h • ft$^2$ [130 kW/m$^2$] of face area compared to 101,000 Btu/h • ft$^2$ [320 kW/m$^2$] for bare coils). The outer surface temperature of sheathed coils is lower, the coils are mechanically stronger, and contact with body or housing is not dangerous. Sheathed coils are generally preferred for dust-laden or potentially explosive atmospheres or where there is a high probability of direct contact of personnel with coils.

## Applications

Airflow in heating coils is vertical or horizontal, although the latter arrangement is more common.

For steam heating, the coils may be installed with the tubes in a vertical or horizontal position. Horizontal-tube coils should be pitched or inclined toward the return connection to drain the condensate. Water-heating coils generally have horizontal tubes to avoid air and water pockets. Where water coils may be exposed to below-freezing temperatures, drainability must be considered, or a fluid such as glycol must be put in the coil. If a coil is to be drained and then exposed to below-freezing temperatures, it should first be flushed with an antifreeze solution.

When the leaving air temperature is controlled by modulation of the steam supply to the coil, **steam-distributing tube-type coils** are optimum for providing uniform exit air temperatures. (See section on "Steam Coils.")

Correctly designed steam-distributing tube coils limit the exit air temperature stratification to a maximum of 5 to 6°F (about 3°C) over the entire length of the coil, even when the steam supply is modulated to a small fraction of the full-load capacity.

As an added precaution, in the application of both steam- and water-heating coils, the outdoor-air inlet dampers usually close automatically when the fan is stopped (system shut down). This control arrangement minimizes the danger of freezing. In steam systems with very low temperature outdoor air conditions [such as −20°F (−29°C) or below], it is desirable to have the steam valve go to full open position when the system is shut down. If outside air is used for proportioning building makeup air, its damper should be an opposing blade design.

Heating coils are designed to allow for expansion and contraction resulting from the temperature ranges within which they operate. Care must be taken to prevent imposing strains from the piping on the coil connections. This is particularly important on high-temperature hot-water applications. Expansion loops, expansion or swing joints, or flexible connections usually provide this protection. (See Chapter 11 of the 1987 HVAC Volume.)

It is good practice to support banked coils individually in an angle-iron frame or a similar supporting structure. With this arrangement, the lowest coil is not required to support the weight of the coils stacked above. This design also facilitates the removal of individual coils in a multiple-coil bank for repair or replacement.

Low-pressure steam systems, coils controlled by modulating the steam supply, or both, should have a vacuum breaker or be drained through a vacuum-return system to ensure proper condensate drainage. It is good practice to install a closed vacuum breaker (where one is required) connected to the condensate return line through a check valve. This unit breaks the vacuum by equalizing the pressure, yet minimizes the possibility of air being bled into the system. Steam traps should be located at least 12 in. (300 mm) below the condensate outlet to enable the coil to drain properly. Also, coils supplied with low-pressure steam or controlled by modulating the steam supply should not be trapped directly to overhead return lines. Condensate can be lifted to overhead returns only when sufficient pressure is available to overcome the condensate head and any return-line pressure. If overhead returns are necessary, the condensate should be pumped to the higher elevation.

## COIL SELECTION

The following factors should be considered in coil selection:

1. The required duty or capacity, considering the other system components
2. Temperature of the air entering the coil
3. Available heating media
4. Space and dimensional limitations
5. Air quantity
6. Permissible resistances for both the air and heating media
7. Characteristics of individual coil designs
8. Individual installation requirements, such as the type of control to be used
9. Coil face velocity

The duties required may be determined from information in Chapters 23, 25, and 26 of the 1985 FUNDAMENTALS Volume. See also Chapter 1 of the 1987 HVAC Volume. There may be a choice of heating media, as well as operating temperatures, depending on whether the installation is new or is being modified. The air handled may be limited by the use of ventilating ducts for air distribution, or it may be determined by requirements for satisfactory air distribution or ventilation.

The resistance through the air circuit influences the fan's power and speed. This resistance may be limited to allow the use of a given size of fan motor or to keep the operating expense low. It may also be limited because of sound-level requirements.

The permissible water resistance of a hot-water coil may be dictated by the available pump head from a given size of pump and pump motor. This is usually controlled within limits by careful selection of the coil header size and the number of tube circuits. The performance of a heating coil depends on the correct choice of the original equipment and on proper application and maintenance. For steam coils, performance and selection of the correct type and size of steam trap is of the utmost importance.

Coil ratings are based on uniform face velocity. Nonuniform airflow through the coil will affect performance. Nonuniform airflow may be caused by air entrance at odd angles or by inadvertent blocking of a portion of the coil face. To obtain rated performance, air quantity in the field must correspond with design quantity and must always be maintained.

Complete mixing of return and outdoor air is essential to the proper operation of a coil. Mixing damper design is critical to the operation of a system. Systems in which the air passes through a fan before flowing through a coil do not ensure proper air mixing.

Coils are commonly cleaned by washing them with water. They can sometimes be brushed and cleaned with a vacuum cleaner. In extreme cases of neglect, especially in restaurants where grease and dirt have accumulated, it is sometimes necessary to remove the coils and wash off the accumulation with steam, compressed air and water, or hot water and a suitable detergent. Often, outside makeup air coils have no upstream air filters. Overall, it is best to periodically inspect and service on a regular schedule.

### Coil Ratings

Steam and hot-water coils are usually rated within the following limits, which may be exceeded for special applications:

1. **Air Face Velocity.** Between 200 to 1500 fpm (1 and 8 m/s), based on air at standard density of 0.075 lb/ft$^3$ (1.2 kg/m$^3$).
2. **Entering Air Temperature.** Minus 20°F to 100°F (−29 to 38°C) for steam coils; 0°F to 100°F (−18 to 38°C) for hot-water coils.
3. **Steam Pressures.** From 2 to 250 psi (14 to 1720 kPa) gauge at the coil steam supply connection (pressure drop through the steam control valve must be considered).
4. **Hot-Water Temperatures.** Between 120°F and 250°F (50 and 120°C).
5. **Water Velocities.** From 0.5 to 8 fps. (0.2 to 2.5 m/s).

Individual installations vary widely, but the following values can be used as a guide.

The most common air face velocities used are between 500 and 1000 fpm (2.5 and 5 m/s). Delivered air temperatures vary from about 72°F (22°C) for ventilation only to about 150°F (66°C) for complete heating. Steam pressures vary from 2 to 15 psi (14 to 103 kPa) guage, with 5 psi (35 kPa) gauge being the most common. A minimum steam pressure of 5 psi (35 kPa) gauge is recommended for systems with entering air temperatures below freezing. Water temperatures for comfort heating are commonly between 180 and 200°F (80 and 95°C), with water velocities between 4 and 6 fps (1.2 and 1.8 m/s).

Water quantity is usually based on about 20°F (11°C) temperature drop through the coil. Air resistance is usually limited to 3/8 to 5/8 in. (90 to 155 Pa) of water for commercial buildings and to about 1 in. (250 Pa) for industrial buildings. High-temperature water systems have water temperatures commonly between 300 and 400°F (150 and 200°C), with up to 100°F (56°C) drops through the coil.

The selection of heating coils is relatively simple because it involves dry-bulb temperatures and sensible heat only, without the complication of simultaneous latent heat loads, as in cooling coils. Heating coils are usually selected from charts or tables giving final air temperatures at various air velocities, entering air temperatures, and steam or water temperatures. Oversizing of modulating steam valves and steam coils in this system should be avoided because such oversizing makes control more difficult.

The selection of hot-water heating coils is more complicated because of the added water velocity variable and the fact that the water temperature decreases as it flows through the coil circuits. Therefore, in selecting hot-water heating coils, the water velocity and the mean temperature difference (between the hot water flowing in the tubes and the air passing over the fins) must be considered.

Most coil manufacturers have their own methods of producing performance rating tables from a suitable number of coil performance tests. A method of testing air heating coils is given in ASHRAE *Standard* 33-78, *Method of Testing for Rating Forced-Circulation Air Cooling and Heating Coils.* A basic method of rating to provide a fundamental means for establishing thermal performance of air-heating coils by extension of test data, as determined from laboratory tests on prototypes, to other operating conditions, coil sizes, and row depths of a particular surface design and arrangement is given in ARI *Standard* 410-81, *Forced-Circulation Air-Cooling and Air-Heating Coils.*

## HEAT TRANSFER AND PRESSURE DROP

For air-side heat transfer and pressure drop, the information given in Chapter 6 for sensible cooling coils is applicable. For hot-water coils, the information given in Chapter 6 for water-side heat transfer and pressure drop is also applicable here. For steam coils, the heat-transfer coefficient of condensing steam has to be calculated. Chapter 3 of the 1985 FUNDAMENTALS Volume lists several equations for this purpose. Shah (1981) reviewed the available information on heat transfer during condensation in horizontal, vertical, and inclined tubes and has made design recommendations. For estimation of pressure drop of condensing steam, see Chapter 5 of the 1985 FUNDAMENTALS Volume.

### Parametric Effects

The heat-transfer performance of a given coil can be changed by varying the flow rates of the air and/or the temperature of the heating medium. Understanding the interaction of these parameters is necessary for designing satisfactory coil-capacity and control. A review of manufacturers' catalogs, many of which are listed in the ARI Directory, shows the effects of varying these parameters.

## REFERENCE

Shah, M.M. 1981. Heat Transfer during Film Condensation in Tubes and Annuli: A Review of the Literature. ASHRAE *Transactions,* 87 (1): 1086-1105.

## CHAPTER 10

# AIR CLEANERS FOR PARTICULATE CONTAMINANTS



THIS chapter discusses the cleaning of both ventilation air and recirculated air for the conditioning of building interiors. Complete air cleaning may require the removal of inert particles, microorganisms, and gaseous pollutants. This chapter only addresses the removal of inert particles. The total suspended particulate concentration in this case seldom exceeds 2 mg/m$^3$ and is usually less than 0.2 mg/m$^3$ of air (NTIS 1969). This contrasts to exhaust gases from processes, flue gases, etc., where dust concentration typically is from 200 to 40,000 mg/m$^3$. Chapter 50 of the 1987 HVAC Volume covers the removal of gaseous contaminants; Chapter 11 of this volume covers exhaust-gas pollution control.

With certain exceptions, the air cleaners addressed in this chapter are not applicable for exhaust gas streams, principally because of extreme differences in dust concentration and temperatures. However, the principles of air cleaning covered are applicable to exhaust streams, and these air cleaners are used extensively in supplying gases to industrial processes that are essentially free of particles.

## ATMOSPHERIC DUST

Atmospheric dust is a complex mixture of smokes, mists, fumes, dry granular particles, and fibers. (When suspended in a gas, this mixture is called an *aerosol*.) A sample of atmospheric dust gathered at any given point generally contains materials common to that locality, together with other components that originated at a distance but were transported by air currents or diffusion. These components and their concentrations vary with the geography of the locality (urban or rural), the season of the year, weather, the direction and strength of the wind, and proximity of dust sources. A sample of atmospheric dust usually contains soot and smoke, silica, clay, decayed animal and vegetable matter, organic materials in the form of lint and plant fibers, and metallic fragments (Moore *et al.* 1954). It may also contain living organisms, such as mold spores, bacteria, and plant pollens, which may cause diseases or allergic responses. (Chapter 11 of the 1985 FUNDAMENTALS Volume contains further information on atmospheric contaminants.)

The particles in the atmosphere range in size from less than 0.01 $\mu$m to the dimensions of lint, leaves, and insects. Particulate contamination indoors is influenced by smoking, human sources, human activities, and other sources such as equipment, furnishings, and pets (McCrone *et al.* 1967). Almost all conceivable shapes and sizes are represented. This wide range of particulate size and concentration makes it impossible to design one type of cleaner that would be best for all applications.

## VENTILATION AIR CLEANING

Different fields of application require different degrees of air cleaning effectiveness. In industrial ventilation, it may only be necessary to remove the coarser dust particles from the airstream for cleanliness of the structure and protection of mechanical equipment. In other applications, surface discoloration must be prevented. Unfortunately, the smaller components of atmospheric dust are the worst offenders in smudging and discoloring building interiors. Electronic air cleaners or high efficiency dry filters are required for small particle removal, especially the respirable fraction, which often must be controlled for health reasons. In clean room applications or when radioactive or other dangerous particles are present, high efficiency filters should be selected.

The characteristics of aerosols most affecting the performance of an air filter include particle size and shape, specific gravity, concentration, and electrical properties. The most important of these is particle size. Figure 1 of Chapter 11 of the 1985 FUNDAMENTALS Volume gives data on the sizes and characteristics of airborne particulate matter and the wide range of particle size that may be encountered. Cleaning efficiency with certain media is affected to a negligible extent by the velocity of the airstream. The degree of air cleanliness required and aerosol concentration are major factors influencing filter design and selection.

In addition to criteria affecting the degree of air cleanliness, considerations of cost (initial investment and maintenance), space requirements, and airflow resistance have resulted in the development of a wide variety of air cleaners. Comparisons can be made only from data obtained by standardized test methods. Test procedures must be understood in order to use test data properly. The following discussion of current methods of rating and testing will aid in understanding the subject of air cleaners.

## RATING AIR CLEANERS

The three operating characteristics that distinguish the various types of air cleaners are *efficiency, airflow resistance,* and *dust-*

The preparation of this chapter is assigned to TC 2.4, Particulate Air Contaminants and Particulate Contaminant Removal Equipment.

*holding capacity.* Efficiency measures the ability of the air cleaner to remove particulate matter from an airstream. Average efficiency during the life of the filter is the most meaningful for most types and applications. However, because the efficiency of many dry-type filters increases with dust load, in applications with low dust concentrations, the initial (clean filter) efficiency should be considered for design.

Airflow resistance (or resistance) is the static pressure drop across the filter at a given airflow rate. The term pressure drop is used interchangeably with resistance. Dust-holding capacity defines the amount of a particular type of dust that an air cleaner can hold when it is operated at a specified airflow rate to some maximum resistance value or before its efficiency drops seriously as a result of the collected dust.

Complete evaluation of air cleaners then requires data on efficiency, resistance, dust-holding capacity, and the effect of dust loading on efficiency and resistance. When applied to automatic renewable media devices (roll filters), the rating system must evaluate the rate of media supply needed to maintain constant resistance under a specified feed rate of standard dust. As applied to electronic air cleaners, the effect of dust buildup on efficiency must be evaluated. Also, it is essential that the manufacturer supply information on the maintenance needed to keep the filter at its rated efficiency.

Air filter testing is complex and no individual test adequately describes all filters. Ideally, performance testing of equipment should simulate the operation of the device under operating conditions and furnish performance ratings of the characteristics important to the equipment user. In the case of air cleaners, the wide variations in the amount and type of particulate matter in the air being cleaned make rating difficult. Another complication is the difficulty of closely relating measurable performance to the specific requirements of users. Recirculated air tends to have a larger proportion of lint than does fresh air drawn from the outside. These difficulties should not obscure the principle that tests should simulate actual use as closely as possible.

In general, four types of tests, together with certain variations, determine air cleaner efficiency:

**Arrestance.** A standardized synthetic dust consisting of various particle sizes is fed into the air cleaner, and the weight fraction of the dust removed is determined. In the ASHRAE Test Standard summarized later in this chapter, this type of efficiency measurement is named *synthetic dust weight arrestance (arrestance)* to distinguish it from other efficiency values.

**Dust Spot Efficiency.** Atmospheric dust is passed into the air cleaner, and the discoloration effect of the cleaned air is compared with that of the incoming air. This type of measurement is called *atmospheric dust spot efficiency.*

**Fractional Efficiency** or **Penetration.** Uniform-sized particles are fed into the air cleaner and the percentage removed by the cleaner is determined, typically by a photometer or condensation nuclei counter.

**Particle Size Efficiency.** Atmospheric dust is fed to the air cleaner, and air samples taken upstream and downstream are drawn through a particle counter to obtain efficiency versus particle size.

The indicated weight arrestance of air filters, as in the *arrestance* test, depends greatly on the particle size distribution of test dust, which, in turn, must consider its state of agglomeration. Therefore, this filter test requires a high degree of standardization of the test dust, the dust dispersion apparatus, and other elements of test equipment and procedures. This test is particularly suited to low- and medium-efficiency air filters that are most commonly used on recirculating systems. It does not distinguish between filters of higher efficiency.

The *dust spot* test measures the ability of a filter to reduce the soiling of fabrics and building interior surfaces. Since these effects depend mostly on fine particles, this test is most useful for high efficiency filters. A disadvantage is the variety and variability of atmospheric dust (McCrone, *et al.* 1967; Whitby, *et al.* 1958; and Horvath 1967), which may cause the same filter to test at different efficiencies at different locations (or even at the same location at different times). This variation is most apparent in low-efficiency filters.

ASHRAE *Standard* 52 specifies both a weight test and a dust spot test and requires that values for both be reported. These results allow a comparison between air filter devices.

In *fractional efficiency* tests, use of uniform particle-size aerosols has proven to be an accurate measure of the particle size versus efficiency characteristic of filters over the entire atmospheric size spectrum. The method is time-consuming and has been used primarily in research. However, the DOP test for HEPA filters is widely used for production testing at a narrow particle size range.

Recent indoor air pollution and air pollution control strategies have focused on the need for efficiencies versus size data in the sub-micron range. To meet this need, several manufacturers publish efficiencies for a size range of atmospheric dust. The data presented may be for a clean filter or for an average over the life of the filters. No test standard currently exists for *particle size efficiency* testing.

The measurement of *pressure drop* is relatively simple, but exact measurement of true dust-loading capacity is complicated by the variability of atmospheric dust; therefore, standardized synthetic dust is normally used. Such a dust also shortens the dust-loading cycle to hours instead of weeks.

Since synthetic dusts are not exactly the same as atmospheric dusts, dust-holding capacity as measured by these accelerated tests may be different from that achieved by tests on atmospheric dust. It is impossible to determine the exact life of a filter in field use by laboratory testing. However, tests of filters under standard conditions do provide a rough guide to the relative effect of dust on performance of various units. Several types of dust-loading tests have been developed for this purpose.

Reputable laboratories perform accurate and repeatable filter tests. Differences in reported values generally lie within the variability of the test aerosols and dusts. Since most media are made of random air- or water-laid fibrous materials, the inherent media variations affect filter performance. Awareness of these variations prevents misunderstanding and specification of impossibly close performance tolerances. Caution must be exercised in interpreting published efficiency data, since the performance of two cleaners tested by different procedures generally cannot be compared. Values of air cleaner efficiency can be related only approximately to the rate of soiling of a space or of mechanical equipment or to the intensity of other objectionable effects.

## AIR CLEANER TEST METHODS

Air cleaner test methods have been developed in several areas: the heating and air-conditioning industry, the automotive industry, the atomic energy industry, and government and military agencies. The following tests have become standard in general ventilation applications in the United States. In 1968, the test techniques developed by the U.S. National Bureau of Standards (NBS) and the Air Filter Institute (AFI) were unified (with minor changes) into a single test procedure, ASHRAE *Standard* 52 (52-76). Dill (1938), Whitby *et al.* (1956), and Nutting and Logston (1953) give details of the original codes.

In general, the ASHRAE Weight Arrestance Test parallels the AFI, making use of a very similar test dust. The ASHRAE *Atmospheric Dust Spot Efficiency Test* parallels the AFI and NBS *Atmospheric Dust Spot Efficiency Tests* and specifies a dust-loading technique.

### Weight Arrestance Test

ASHRAE *Standard* 52-76 specifies a test dust as follows:

ASHRAE Synthetic Dust shall consist of 72% Standardized Air Cleaner Dust-Fine; 23% by weight Molocco Black; 5% by weight No. 7 cotton linters ground in a Wiley Mill with a 4-mm screen.

A known amount of the prepared test dust at a known and controlled rate is fed into the test unit. The concentration of dust in the air leaving the filter is determined by passing the entire airflow through a high efficiency after-filter and measuring the gain in filter weight. Arrestance (in percent) = 100 [1 − (weight gain of after-filter/weight of dust fed)].

Atmospheric dust particles range in size from a small fraction of a micrometre up to particles tens of micrometres in diameter. The artificially generated dust cloud used in the ASHRAE weight arrestance method is considerably coarser than typical atmospheric dusts. It tests the ability of a filter to remove the largest atmospheric dust particles and gives little indication of the filter performance in removing the smallest particles. But, where the weight of dust in the air is the primary concern, this is a valid test because most of the weight is contained in the larger particles. Where extremely small particles are objectionable, the weight arrestance method of rating does not differentiate between filters.

### Atmospheric Dust Spot Efficiency Test

One objectionable characteristic of the finer airborne dust particles is the capacity to soil walls and other interior surfaces. The discoloring rate of white, filter-paper targets (microfine glass fiber HEPA filter media) filtering samples of air constitutes an accelerated simulation of this effect. By measuring the rate of light change transmitted by these targets, the efficiency of the filter in preventing wall discoloration may be computed.

The accepted dust spot method is an integral part of ASHRAE *Standard* 52-76. In this procedure, untreated atmospheric dust samples are drawn upstream and downstream of the tested filter. These samples are drawn at equal flow rates through identical targets of glass fiber filter paper. The downstream sample is drawn continuously; the upstream sample is interrupted in a timed cycle so that the average rate of discoloration of the upstream and downstream targets is approximately equal. The percentage of off-time approximates the efficiency of the filter. The filter efficiency in percent is:

$$E = 100\left(1 - \frac{Q_1}{Q_2} \cdot \frac{(T_{20} - T_{21})}{(T_{10} - T_{11})} \cdot \frac{(T_{10})}{(T_{20})}\right)$$

*where*

$Q_1$ = total quantity of air drawn through upstream target
$Q_2$ = total quantity of air drawn through downstream target
$T_{10}$ = initial light transmission of upstream target
$T_{11}$ = final light transmission of downstream target
$T_{20}$ = initial light transmission of downstream target
$T_{21}$ = final light transmission of downstream target

The standard allows dust spot efficiencies to be taken at intervals during a synthetic dust-loading procedure. This characterizes the change of dust spot efficiency as dust builds up in the filter in service.

### Dust-Holding Capacity Test

In the ASHRAE *Standard* 52-76, *Dust-Holding Capacity Test*, the same synthetic test dust mentioned in the section above is fed to the filter. The pressure drop across the filter (its resistance) rises as dust is fed. The test is normally terminated when the resistance reaches the maximum operating resistance set by the manufacturer. However, not all filters retain collected dust equally well. The test, therefore, requires that arrestance be measured at least four times during the dust-loading process and that the test be terminated when two consecutive arrestance values less than 85% or one value equal to or less than 75% arrestance of the maximum arrestance have been measured. The ASHRAE *Dust-Holding Capacity* is, then, the integrated amount of dust held by the filter up to the time the dust-loading test was terminated. A typical set of curves for an ASHRAE air filter test report on a fixed cartridge-type filter is shown in Figure 1. Both Synthetic Dust Weight Arrestance and Atmospheric Dust Spot Efficiencies are shown. The standard also specifies how self-renewable devices are to be loaded with dust to establish their performance under standard conditions. Figure 2 shows the results of such a test on an automatic roll media filter.

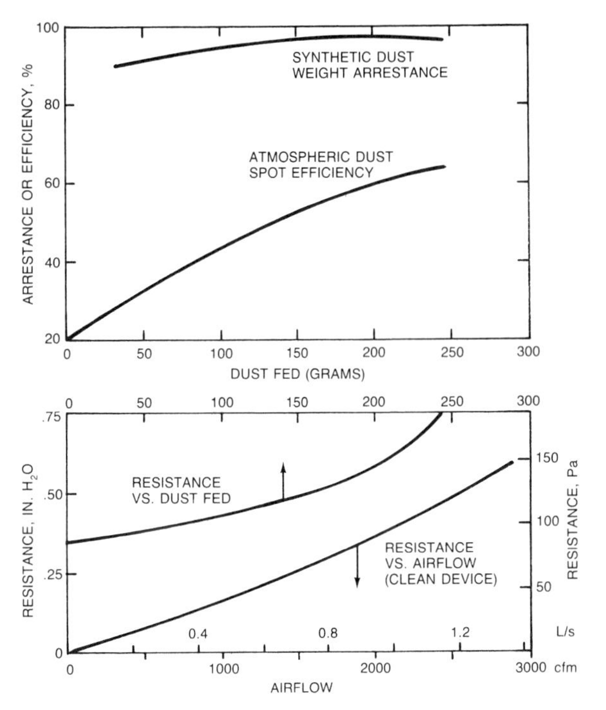


**Fig. 1 Typical Performance Curves for a Fixed Cartridge-Type Filter by ASHRAE *Standard* 52**

### DOP Penetration Test

For high efficiency filters of the type used in clean rooms and nuclear applications (HEPA filters), the normal test in the United States is the Thermal DOP method, as outlined in the U.S. Military Standard, MIL-STD-282 (1956). In this method, a smoke cloud of DOP droplets condenses from DOP vapor (DOP = Di-Octyl Phthalate, or bis-[2-ethylhexyl] phthalate, an oily, high-boiling-point liquid).

The count median diameter is about 0.18 $\mu$m, while the mass median diameter is about 0.27 $\mu$m with the cloud concentration about 80 mg/m$^3$ under properly controlled conditions. The procedure is sensitive to the mass median diameter, and DOP test results are commonly referred to as efficiency on 0.3 micrometre particles.

This smoke cloud is fed to the filter, which is held in a special test chuck. Any smoke that penetrates the body of the filter or leaks through gasket cracks passes into the downstream region where it is thoroughly mixed. The air leaving the chuck thus

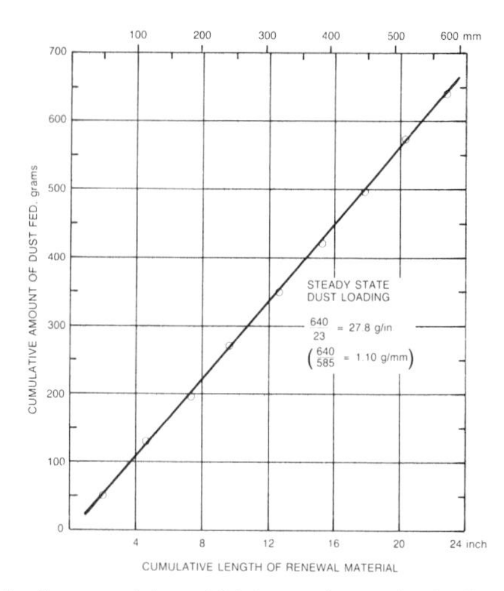


[a]Calculate loading per unit by multiplying steady-state dust loading by 144 ($10^6$) and dividing by the media width, which is normally 24 in. (610 mm). For example: $2.78 \cdot 144/24 = 167$ g/ft$^2$ ($1.10 \cdot 10^6/610 = 1800$ g/m$^2$).

**Fig. 2 Typical Dust-Loading Graph for Self-Renewable Air Filter**

contains the average concentration of penetrating smoke. This concentration, as well as the upstream concentration, is measured by a light-scattering photometer. The filter penetration, $P$, is:

$$P = 100 \left(\frac{\text{Downstream Concentration}}{\text{Upstream Concentration}}\right) \%$$

Penetration, not efficiency, is usually specified because HEPA filters have efficiencies so near 100%. The two terms are related. Thus, $E = 100 \ (1 - P)$ %.

U.S. Government specifications frequently call for the testing of HEPA filters at both rated flow and 20% of rated flow. This procedure helps reveal gasket leaks and pinholes in media that would otherwise escape notice. Such defects, however, are not located by this test. The Institute of Environmental Sciences has published a *Recommended Practice for HEPA Filters*, covering five levels of performance and two grades of construction. Reference to this document will aid the designer in clarifying specifications and intent (IES-RP-CC-001-86).

## Leakage (Scan) Tests

In the case of HEPA filters, leakage tests are sometimes desirable to show that no small *pinhole* leaks exist or to locate and eliminate any that may exist. Essentially the same technique as the DOP *Penetration Test* is performed, except that the downstream concentration is measured by scanning the face of the filter and its gasketed perimeter with a moving probe. The exact point of smoke penetration can then be located and repaired. This same test can be performed after the filter is installed; in this case, a portable aspirator-type DOP generator is used instead of the bulky thermal generator. Often called the *Cold DOP Test,* the smoke produced by such a generator is not uniform in size, but its average diameter can be approximately 0.6 $\mu$m. Particle diameter is less critical for leak location than for penetration measurement.

## Particle Size Efficiency Tests

No standard exists for determining the efficiency of air cleaners as a function of particle size. Such measurements depend heavily on the type of aerosol used as a filter challenge and, to a lesser extent, on the type of particle spectrometer being used. Figure 3 presents particle size efficiency curves for typical air filters compiled and averaged from manufacturers' data. These data are for general guidance only and must not be used to specify air filters.

One arrangement for obtaining such data is run concurrently with (or in place of) the dust spot test of ASHRAE *Standard* 52. For these measurements, sampling probes upstream and downstream of the test filter are located near the upstream and downstream dust spot target holders. With atmospheric or synthetic dust as the challenge, particles are counted both upstream and downstream to obtain an average count efficiency. The natural temporal variations in the test dust make it essential to choose up and down count cycles for statistical validity. Both *laser* and *white light* particle spectrometers are used. The major advantage of the laser spectrometer is its ability to sense particles for the size range of 0.1 to 0.3 $\mu$m, the latter being the lower limit for white light spectrometers.

The Anderson-type cascade sampler is an accepted standard for measuring particle concentration as a function of particle size. This sampler requires the selection of proper filters, the careful measuring of filter weight (mass) change, and careful control of the sampled air volume. A real time, 10-stage cascade impactor with quartz crystal microbalance mass monitors in each stage and a microprocessor is also available. The nominal aerodynamic diameters for particles are, in micrometres: 0.05, 0.1, 0.2, 0.4, 0.8, 1.6, 3.2, 6.4, 12.5, and 25.0. Near real-time devices are available to record particle mass continuously over a range of sizes using nuclei-counting or light-scattering principles.

## Specialized Performance Test

AHAM *Standard* AC-1-1986 method measures the ability of portable household air cleaners to reduce generated particulate matter suspended in the air in a room-size test chamber. The procedure measures the natural decay of three contaminants: dust, smoke, and pollen and then compares it with the air cleaner in operation.

## Miscellaneous Performance Tests

Standardized tests have been developed for air cleaners in other countries and for specialized uses. Examples are the British *Sodium-Flame* and *Methylene Blue Tests*, the German *Staubforschungs Institut Tests*, and the *U.S. Bureau of Mines Respirator Tests* (B.S. 2831; B.S. 3928; and U.S. Bureau of Mines).

## Environmental Tests

Air cleaners may be subjected to fire, high humidity, a wide range of temperatures, mechanical shock, vibration, and other environmental hazards. Several standardized tests exist for evaluating these environmental effects on air cleaners. Mil-STD-282 includes shock tests (shipment rough handling) and filter media water resistance tests. Several Atomic Energy Commission agencies (now part of the Department of Energy) specify humidity and temperature-resistance tests (USAEC 212; Peters 1962).

Underwriters Laboratories has two major standards for air cleaner flammability. The first, for commercial applications, determines flammability and smoke production. UL (UL *Standard* 900) Class 1 filters are those that, when clean, do not contribute fuel when attacked by flame and emit only negligible amounts of smoke. UL *Standard* 900 Class 2 filters are those that, when clean, burn moderately when attacked by flame or emit moderate amounts of smoke, or both. In addition, UL *Standard* 586 for flammability of HEPA filters has been established. The UL tests do not evaluate the effect of collected dust on filter flammability; depending on the dust, this effect may be drastic. Another UL Standard (UL 867) applies to electronic air cleaners.

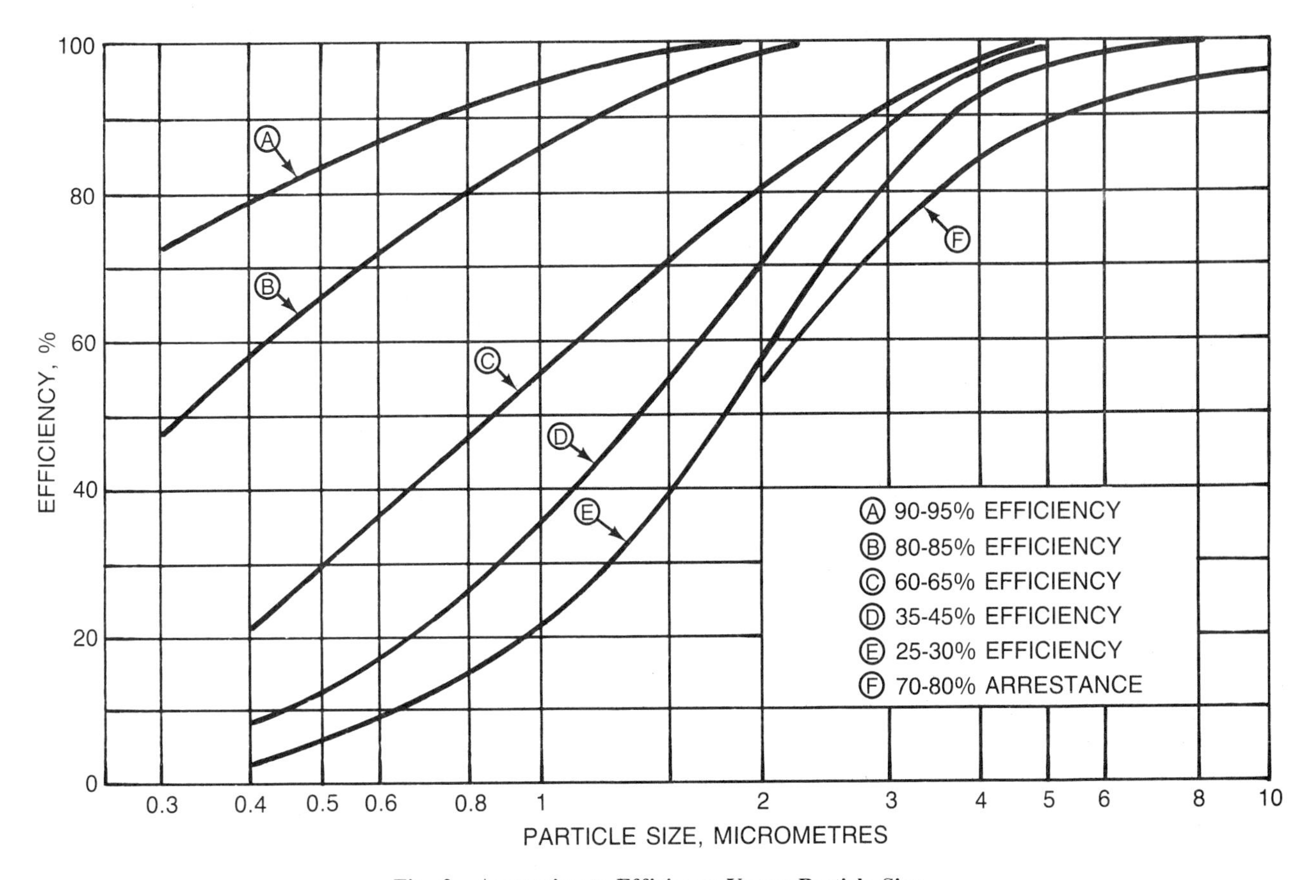


**Fig. 3 Approximate Efficiency Versus Particle Size for Typical Air Filters[a]**

[a]Compiled and averaged from manufacturers data. Efficiency and arrestance per ASHRAE *Standard* 52-76 Test Methods.

**Caution:** Curves are approximations only for general guidance. Values from them *must not* be used to specify air filters, since a generally recognized test standard does not exist.

## ARI Standards

The Air-conditioning and Refrigeration Institute has published two standards for air filter equipment (ARI 680-86; ARI 850-84). Although these standards are not widely used, they establish (1) definitions and classification; (2) requirements for testing and rating (performance test methods are per ASHRAE *Standard* 52); (3) specification of standard equipment; (4) performance and safety requirements; (5) proper marking; (6) conformance conditions; and (7) literature and advertising requirements.

## MECHANISMS OF PARTICLE COLLECTION

In the collection of particulate matter, air cleaners rely on the following four main principles or mechanisms:

**Straining.** The coarsest kind of filtration strains particles through a membrane opening smaller than the particulate being removed. It is most often observed as the collection of large particles and lint on the filter surface. The mechanism is not adequate to explain the filtration of submicron aerosols through fibrous matrices, which requires isolated fiber theory mechanisms, as follows.

**Direct Interception.** The particles follow a fluid streamline close enough to a fiber that the particle contacts the fiber and remains there. The process is nearly independent of velocity.

**Inertial Deposition.** Particles in the airstream are large enough or of large enough specific gravity that they cannot follow the fluid streamlines around a fiber; thus, they cross over stream lines, contact the fiber, and remain there. At high velocities (where these inertia effects are most pronounced), the particle may not adhere to the fiber because drag forces are so high. In this case a viscous coating applied to the fiber obtains the full benefit and is the predominant mechanism in an adhesive-coated, wire screen impingement filter.

**Diffusion.** Very small particles have random motion about their basic streamlines (Brownian motion), which contributes to deposition on the fiber. This deposition creates a concentration gradient in the region of the fiber, further enhancing filtration by diffusion. The effects increase with decreasing particle size and velocity.

**Electrostatic Effects.** Objects with opposite electrical charge are attracted to each other. The electronic air cleaner attracts particles to its collecting plates. Under certain circumstances, electrostatic charges may be created within fibrous filter media, which usually assists dust collection.

Some progress has been made in calculating theoretical filter media efficiency from the physical constants of the media by considering the effects of the collection mechanisms described previously (Fuchs 1964; Davies 1966; Lundgren and Whitby 1968; and Wilson and Cavanagh 1967).

## TYPES OF AIR CLEANERS

Common air cleaners are broadly grouped as follows:

**Fibrous Media Unit Filters,** in which the accumulating dust load causes pressure drop to increase up to some maximum permissible value. During this period, efficiency normally increases. However, at high dust loads, dust may adhere poorly to filter fibers and efficiency will drop. Filters in such condition should be replaced or reconditioned, as should filters that have reached their final (maximum permissible) pressure drop. This category includes viscous impingement and dry-type air filters, available in low efficiency to ultra-high efficiency construction.

**Renewable Media Filters,** in which fresh media is introduced into the airstream as needed to maintain essentially constant resistance, which also maintains constant efficiency.

**Electronic Air Cleaners,** which, if maintained properly by regular cleaning, have relatively constant pressure drop and efficiency.

**Combination air cleaners** of the above types are used. For example, an electronic air cleaner may be used as an agglomerator with a fibrous media downstream to catch the agglomerated particles blown off the plates. Electrode assemblies have been installed in air-handling systems, making the filtration system more effective (Frey 1985, 1986). Also, a renewable media filter may be used upstream of a high efficiency unit filter to extend its life. Charged media filters are also available that increase particle deposition on media fibers by an electrostatic field. In this case, the pressure loss increases like a fibrous media filter.

## FILTER TYPES AND THEIR PERFORMANCE

### Panel Filters

**Viscous Impingement Filters.** These are panel filters made up of coarse fibers with a high porosity. The filter media are coated with a viscous substance, such as oil, which adheres to particles that impinge on the fibers. Design air velocity through the media is usually in the range of 200 to 800 fpm (1 to 4 m/s). These filters are characterized by low pressure drop, low cost, and good efficiency on lint but low efficiency on normal atmospheric dust. They are commonly made 1/2 to 4 in. (13 to 100 mm) thick. Unit panels are available in standard and special sizes up to about 24 by 24 in. (610 by 610 mm). This filter is often used as a prefilter for higher efficiency filters.

A number of different materials have been used as the filtering medium, including coarse (15 to 60 μm diameter) glass fibers, animal hair, vegetable fibers, synthetic fibers, metallic wools, expanded metals and foils, crimped screens, random-matted wire, and synthetic open-cell foams. The arrangement of the medium in this type of filter involves three basic configurations:

1. *Sinuous Media.* The filtering medium (consisting of corrugated metal or screen strips) is held more or less parallel to the airflow. The direction of airflow is forced to change rapidly in passing through the filter, thus giving inertial impingement of dust on the metal elements.
2. *Formed-Screen Media.* Here, the filter media (screens or expanded metal) are crimped to produce high porosity media that avoid collapsing. Air flows through the media elements and dust impinges on the wires. The relatively open structure allows the filter to store substantial quantities of dust and lint without plugging.
3. *Random Fiber Media.* Fibers, either with or without bonding material, are formed into mats of high porosity. Media of this type are often designed with fibers packed more densely on the leaving air side than on the entering air side. This arrangement permits both the accumulation of larger particles and lint near the air-entering face of the filter and the filtration of finer particles on the more closely packed air-leaving face. Fiber diameters may also be graded from coarse at the air-entry face to fine at the exit face.

Although viscous-impingement filters usually operate in the range of 300 to 600 fpm (1.5 to 3 m/s), it is possible for them to operate at higher velocities. The limiting factor, other than increased flow resistance, is the danger of blowing off agglomerates of collected dust and the viscous coating on the filter.

The loading rate of a filter depends on the type and concentration of the dirt in the air being handled and the operating cycle of the system. Manometers, draft gauges, or pressure transducers are often installed to measure the pressure drop across the filter bank and thereby indicate when the filter requires servicing. The final allowable pressure drop may vary from one installation to another; but, in general, unit filters are serviced when their operating resistance reaches 0.5 in. of water (125 Pa). The decline in filter efficiency (which is caused by the absorption of the viscous coating by dust, rather than by the increased resistance because of dust load) may be the limiting factor in operating life.

The manner of servicing unit filters depends on their construction. Disposable viscous impingement, panel-type filters are constructed of inexpensive materials and are discarded after one period of use. The cell sides of this design are usually a combination of cardboard and metal stiffeners. Permanent unit filters are generally constructed of metal to withstand repeated handling. Various cleaning methods have been recommended for permanent filters; the most widely used involves washing the filter with steam or water (frequently with detergent) and then recoating it with its recommended adhesive by dipping or spraying. Unit viscous filters are also sometimes arranged for in-place washing and recoating.

The adhesive used on a viscous-impingement filter requires careful engineering. Filter efficiency and dust-holding capacity depend on the specific type and quantity of adhesive used; this information is an essential part of test data and filter specifications. Desirable adhesive characteristics, in addition to efficiency and dust-holding capacity, are (1) a low percentage of volatiles to prevent excessive evaporation; (2) a viscosity that varies only slightly within service temperature range; (3) the ability to inhibit growth of bacteria and mold spores; (4) a high capillarity or the ability to wet and retain the dust particles; (5) a high flash point and fire point; and (6) freedom from odor.

Typical performance of viscous-impingement unit filters operating within typical resistance limits is shown as Group I in Figure 4.

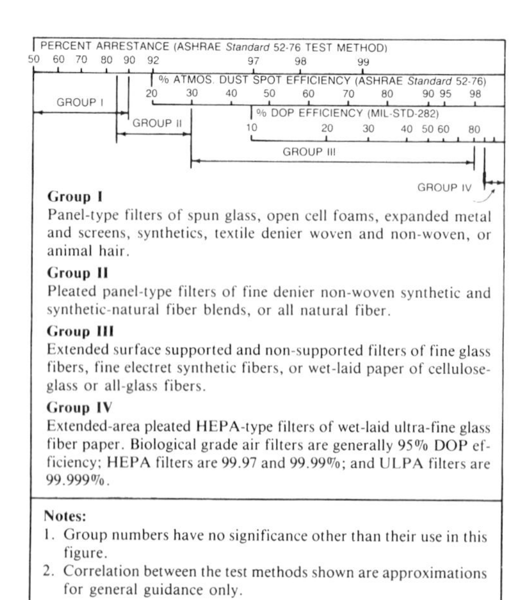


**Fig. 4 Comparative Performance of Viscous Impingement and Dry Media Filters**

**Dry-Type Extended-Surface Filters.** The media used in dry-type air filters is random fiber mats or blankets of varying thicknesses, fiber sizes, and densities. Media of bonded glass fiber, cellulose fibers, wool felt, synthetics, and other materials have been used commercially. The media in filters of this class is frequently supported by a wire frame in the form of pockets, V-shaped, or radial pleats. In other designs, the media may be self-supporting because of inherent rigidity or because airflow inflates it into extended form. Pleating of the media provides a high ratio of media area to face area, thus allowing reasonable pressure drop.

In some designs, the filter media is replaceable and is held in position in permanent wire baskets. In most designs, the entire cell is discarded after it has accumulated its dirt load.

The efficiency of dry-type air filters is usually higher than that of panel filters, and the variety of media available makes it possible to furnish almost any degree of cleaning efficiency desired. Modern dry-type filter media and filter configurations also give dust-holding capacities generally higher than panel filters.

Coarse prefilters placed ahead of extended surface filters sometimes may be economically justified by the somewhat longer life of the main filters. Economic considerations should include the prefilter material cost, changeout labor, and increased fan system power. Generally, prefilters should be considered only where they will substantially reduce the part of the dust that may plug the protected filter (usually a filter having a dust spot efficiency of 70% or more). Temporary prefilters are worthwhile during building construction to capture heavy loads of coarse dust. HEPA-type filters of 95% DOP efficiency and greater should always be protected by prefilters of 80% or greater ASHRAE atmospheric dust spot efficiency.

Typical performance of some types of filters in this group, when they are operated within typical rated resistance limits and over the life of the filters, is shown as Groups II and III in Figure 4.

The initial resistance of an extended surface filter varies with the choice of media and the filter geometry. Commercial designs typically have an initial resistance from 0.1 to 1.0 in. of water (25 to 250 Pa). It is customary to replace the media when the final resistance of 0.5 in. of water (125 Pa) is reached for the low resistance units and 2.0 in. of water (500 Pa) for the highest resistance units. Dry media providing higher orders of cleaning efficiency have a higher resistance to airflow. The operating resistance of the fully dust-loaded filter must be considered in the system design, since that is the maximum resistance against which the fan will operate. Variable air volume systems and systems with constant air volume controls prevent high airflows or possible fan motor overloading from occuring when filters are clean.

Flat panel filters with media velocity equal to duct velocity are possible only in the lowest efficiency units of the dry type (open cell foams and textile denier nonwoven media). Initial resistance of this group, at rated airflow, is mainly between 0.05 and 0.25 in. of water (12 and 63 Pa). They are usually operated to a final resistance of 0.50 to 0.70 in. of water (125 to 175 Pa).

On extended-surface filters of the intermediate efficiency ranges, the filter media area is much greater than the face area of the filter; hence, velocity through the filter media is substantially lower than the velocity approaching the filter face. Media velocities range from 6 to 90 fpm (0.03 to 0.46 m/s), although the approach velocities run to 750 fpm (3.8 m/s). Depth in direction of airflow varies from 2 to 36 in. (50 to 915 mm).

Filter media used in the intermediate efficiency range include those of (1) fine glass fibers, 0.7 to 10 μm in diameter, in mat form up to 1/2-in. (13-mm) thick; (2) thin nonwoven mats of fine glass fibers, cellulose, or cotton wadding; and (3) nonwoven mats of comparatively large diameter fibers (more than 30 μm) in greater thicknesses (up to 2 in. [50 mm]).

Electret filters are composed of electrostatically charged fibers. The charges on the fibers augment the collection of smaller particles by Brownian diffusion with Coulombic forces caused by the charges on the fibers. There are three types of these filters: resin wool, electret, and an electrostatically sprayed polymer. The charge on the resin wool fibers is produced by friction during the carding process. During production of the electret a corona discharge injects positive charges on one side of a thin polypropylene film and negative charges on the other side. These thin sheets are then shredded into fibers of rectangular cross-section. The third process spins a liquid polymer into fibers in the presence of a strong electric field, which produces the charge separation. The efficiency of the charged-fiber filters is due to both the normal collection mechanisms of a media filter and the strong local electrostatic effects. The effects induce efficient preliminary loading of the filter to enhance the caking process. However, dust collected on the media can reduce the efficiency of electret filters.

Very high efficiency dry filters, HEPA (High Efficiency Particulate Air), and ULPA (Ultra Low Penetration Air) filters, are made in an extended surface configuration of deep space folds of submicron glass fiber paper. Such filters operate at duct velocities near 250 fpm (1.3 m/s), with resistance rising from 0.5 to 2.0 in. of water (125 to 500 Pa) or more over their service life. These filters are the standard for clean room, nuclear, and toxic-particulate applications.

Membrane filters are used predominately for air sampling and specialized small-scale applications where their particular characteristics compensate for their fragility, high resistance, and high cost. They are available in many pore diameters and resistances and in flat sheet and pleated forms.

## Renewable Media Filters

**Moving-Curtain Viscous Impingement Filters.** Automatic moving-curtain viscous filters are available in two main types. In one type, random-fiber media is furnished in roll form. Fresh media is fed manually or automatically across the face of the filter, while the dirty media is rewound onto a roll at the bottom. When the roll is exhausted, the tail of the media is wound onto the takeup roll, and the entire roll is thrown away. A new roll is then installed and the cycle is repeated.

Moving-curtain filters may have the media automatically advanced by motor drives on command from a pressure switch, timer, or media light-transmission control. A pressure switch control measures the pressure drop across the media and switches *on* and *off* at chosen upper and lower set points. This control saves media, but only if the static pressure probes are located properly and unaffected by modulating outside air and return air dampers. Most pressure drop control systems do not function properly. Timers and media light-transmission controls avoid these problems; their duty cycles can be adjusted to provide satisfactory operation with acceptable media consumption.

Filters of this design are generally constructed to be failsafe by having a signal that indicates when the roll of media is about to become exhausted. At the same time, the drive motor is de-energized so that the filter cannot run out of media. The normal service requirements involve insertion of a clean roll of media at the top of the filter and disposing of the loaded dirty roll. Automatic filters of this design are not, however, limited in application to the vertical position. Horizontal arrangements are available for use with makeup air units and air-conditioning units. Adhesives must have qualities similar to those for panel-type viscous impingement filters, plus the ability to withstand media compression and long storage.

On the other type of automatic viscous-impingement filter, linked metal mesh media panels are installed on a traveling curtain, which intermittently passes through an adhesive reservoir. There, the filters give up dust load and, at the same time, take on a new coating of adhesive. The panels thus form a continuous curtain that moves up one face and down the other face. The media curtain, continually cleaned and renewed with fresh adhesive, lasts the life of the filter mechanism. The precipitated dirt must be removed periodically from the adhesive reservoir.

The resistance of an automatic filter (media roll or moving metal panels) remains approximately constant as long as proper operation is obtained. A resistance of 0.40 to 0.50 in. of water (100 to 125 Pa) at a face velocity of 500 fpm (2.5 m/s) is typical of this class.

**Moving-Curtain Dry-Media Filters.** Random-fiber (nonwoven) dry media of relatively high porosity are also used in moving-curtain (roll) filters for general ventilation service. Operating duct velocities are generally lower than for viscous-impingement filters—being in the area of 200 fpm (1 m/s).

Special automatic dry filters are also available, which are designed for the removal of lint in textile mills and dry-cleaning establishments and the collection of lint and ink mist in press rooms. The medium used is extremely thin and serves only as a base for the buildup of lint, which then acts as a filter medium. The dirt-laden media is discarded when the supply roll is used up.

Another form of filter designed specifically for dry lint removal consists of a moving curtain of wire screen, which is vacuum cleaned automatically at a position out of the airstream. Recovery of the collected lint is sometimes possible with such a device.

**Performance of Renewable Media Filters.** ASHRAE arrestance, efficiency, and dust-holding capacities for typical viscous-impingement and dry renewable media filters are listed in Table 1.

## Electronic Air Cleaners

Electronic air cleaners can be highly efficient filters by using electrostatic precipitation to remove and collect particulate contaminants such as dust, smoke, and pollen. The designation *electronic air cleaner* denotes a precipitator for HVAC air filtration. The filter consists of an ionization section and a collecting plate section.

In the ionization section, small diameter wires with a positive direct current potential of between 6 and 25 kV DC are suspended equidistant between grounded plates. The high voltage on the wires creates an ionizing field for charging the particulates. The positive ions created in the field flow across the airstream and strike and adhere to (charge) the particles, which then pass into the collecting plate section.

The collecting plate section consists of a series of parallel plates equally spaced with a positive direct current voltage of 4 to 10 kV DC applied to alternate plates. Plates that are not charged are at ground potential. As the particles pass into this section, they are forced to the plates by the electric field on the charges they carry, and thus they are removed from the airstream and collected by the plates. Particulate retention is a combination of electrical and intermolecular adhesion forces and may be augmented by special oils or adhesives. Figure 5 shows a typical electronic air cleaner cell.

In lieu of positive direct current, a negative potential also functions on the same principle, but more ozone is generated.

With voltages of 4 to 25 kV DC, safety measures are required. A typical arrangement makes the air cleaner inoperative when the doors are removed for cleaning the cells or servicing the power pack.

Electronic air cleaners typically operate from a 120- or 240-volt ac single-phase electrical service. The high voltage supplied to the air cleaner cells is normally created with solid-state power supplies. The electrical power consumption ranges from 20 to 40 watts per 1000 cfm (40 to 85 W per $m^3/s$) of air cleaner capacity.

This type of air filter can remove and collect airborne contaminants with average efficiencies of up to 98% at low airflow velocities (150 to 350 fpm [0.8 to 1.8 m/s]) when tested per the ASHRAE *Standard* 52, *Atmospheric Dust Spot Test Method.* Efficiency decreases (1) as the collecting plates become loaded with particulates, (2) with higher velocities, or (3) with non-uniform velocity.

As with most air filtration devices, it is important to arrange the duct approaches to and from the air cleaner housing so that the airflow is distributed uniformly over the face area. Panel prefilters should also be used to help distribute the airflow and to trap large particles that might short out or cause excessive arcing within the high voltage of the air cleaner cell. Electronic air cleaner design parameters of air velocity, ionizer field strength, cell plate spacing, depth, and plate voltage must match the application requirements. These include contaminant type, particle size, volume of air, and efficiency required. Many units are designed for installation into central heating and cooling systems for total air filtration. Other self-contained units are furnished with complete air movers for source control of contaminants in specific applications that need an independent system.

Electronic air cleaner cells must be cleaned periodically with detergent and hot water. Some designs incorporate automatic wash systems that clean the cell in place; in other designs, the cells must be removed for cleaning. The frequency of cleaning (washing) the cell depends on the contaminant and the concen-

**Table 1 Performance of Renewable Media Filters (Steady-State Values)**

| Description | Type of Media | ASHRAE Weight Arrestance, % | ASHRAE Atmospheric Dust Spot Efficiency, % | ASHRAE Dust-Holding Capacity, g/ft² (g/m²) | Velocity, fpm (m/s) |
|---|---|---|---|---|---|
| 20 to 40 μm glass and synthetic fibers, 2 to 2 1/2 in. (50 to 64 mm) thick | Viscous Imp. | 70 to 82 | <20 | 60 to 180 (600 to 2000) | 500 (2.5) |
| Permanent metal media cells or overlapping elements | Viscous Imp. | 70 to 80 | <20 | NA (permanent media) | 500 (2.5) |
| Coarse textile denier nonwoven mat, 1/2 to 1 in. (12 to 25 mm) thick | Dry | 60 to 80 | <20 | 15 to 70 (150 to 750) | 500 (2.5) |
| Fine textile denier nonwoven mat, 1/2 to 1 in. (12 to 25 mm) thick | Dry | 80 to 90 | <20 | 10 to 50 (100 to 550) | 200 (1) |

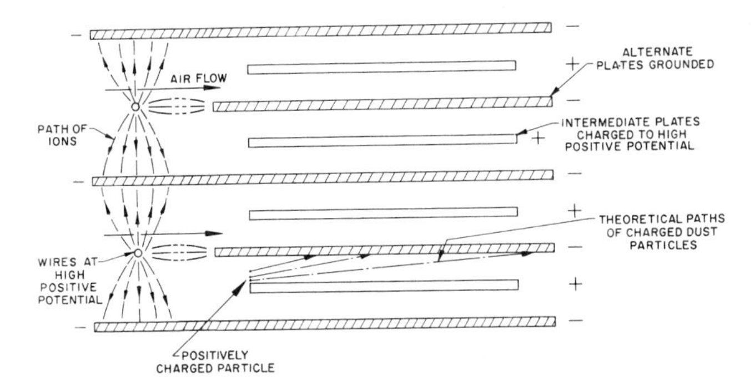


**Fig. 5 Diagrammatic Cross Section of Ionizing-Type Electronic Air Cleaner**

tration. Industrial applications may require cleaning every 8 hours, and a residential unit may only require cleaning at one to three month intervals. A good cleaning schedule is important to keep the unit performing at peak efficiency. With some contaminants, special attention must be given to cleaning the ionizing wires.

Optional features are often available for electronic air cleaners. Afterfilters such as roll filters collect particulates that agglomerate and blow off the cell plates. These are used mainly where heavy contaminant loading occurs and extension of the cleaning cycle is desired. Cell collector plates may be coated with special oils, adhesives, or detergents to improve both particle retention and removal during cleaning. High efficiency, dry-type, extended media area filters are also used as afterfilters in special designs. The electronic air cleaner used in this system improves the service life of the dry filter and collects small particles such as smoke.

Another device, a negative ionizer, uses the principle of particle charging but does not use a collecting section. Particulates enter the ionizer of the unit and receive an electrical charge. They then migrate to a grounded surface closest to the travel path. If use is continued in an area of heavy particulate concentration, a space charge can be built up, which tends to drive the charged particulate to surfaces throughout the room.

**Space Charge.** Particulates that pass through an ionizer and are charged but not removed carry the electrical charge into the space. If continued on a large scale, a space charge will be built up, which tends to drive these charged particles to the walls. Thus, a low-efficiency electronic air cleaner used in areas of high ambient dirt concentrations or a malfunctioning unit can blacken walls faster than if no cleaning device were used (Penney and Hewitt 1949; Sutton *et al.* 1964).

**Ozone.** All high voltage devices are capable of producing ozone, which is toxic and damaging to paper, rubber, and other materials. When properly designed and maintained, electronic air cleaners produce ozone concentrations that only reach a fraction of the levels acceptable for continuous human exposure and are less than those prevalent in many American cities (EPA). Continuous arcing and brush discharge in an electronic air cleaner may yield ozone levels that are annoying or mildly toxic; this will be indicated by a strong ozone odor. Although the nose is sensitive to ozone, only actual measurement of the concentration can determine that a hazardous condition exists. ASHRAE *Standard* 62, *Natural and Mechanical Ventilation*, defines acceptable concentrations of oxidants, of which ozone is the major offender. OSHA specifies allowable average exposure for an 8-hour period at 0.1 ppm (0.1 mg/kg). Indoor ozone levels have been shown to be only 30% of the outdoor level with ionizing air cleaners operating (Sutton *et al.* 1976).

## SELECTION AND MAINTENANCE

To evaluate filters and air cleaners properly for a particular application, three factors should be carefully weighed: (1) the degree of air cleanliness required, (2) the disposal of the dirt after it is removed from the air, and (3) the amount and type of dust in the air to be filtered. These factors affect initial costs, operating costs, and the extent of maintenance required. Savings—from reduction in housekeeping expenses, protection of valuable property and equipment, ability to carry on dust-free manufacturing processes, improved working conditions, and even health benefits—should be credited against the cost of installing and operating an adequate system. The capacity and physical size of the unit required may emphasize the need for low maintenance cost. Operating costs, predicted life, and efficiency are as important as initial cost because air cleaning is a continuing process.

Panel filters do not have efficiencies as high as can be expected from extended-surface filters, but their initial cost and upkeep are generally low. They require more careful attention than the moving-curtain type if the resistance is to be maintained within reasonable limits.

If higher efficiencies are required, extended-surface filters or electronic air cleaners should be considered. The use of very fine glass fiber mats or other materials in extended-surface filters has made these available in the highest efficiency ranges.

Initial costs of extended-surface filters are lower than for electronic types, but higher than for panel types. Operating and maintenance costs may be higher than for panel types and electronic air cleaners, but the efficiencies are always higher than for panel types, and the cost/benefit ratio must be considered. Pressure drop is greater and slowly increases during their useful life. The advantages include the fact that no mechanical or electrical services are required. Choice should be based on both initial and operating costs, as well as on the degree of cleaning efficiency and maintenance requirements.

While electronic air cleaners have a higher initial cost, they exhibit high initial efficiencies in cleaning atmospheric air—largely because of their ability to remove fine particulate contaminants. System resistance remains unchanged as particles are collected, and the resulting residue is disposed of directly to prepare the equipment for further duty.

Table 2 lists some applications of filters classified according to their efficiencies and type.

## FILTER INSTALLATION

Many air cleaners are available in units of convenient size for manual installation, cleaning, and replacement. A typical unit filter may be 20 to 24 in. (510 to 610 mm) square, from 1 to 40 in. (25 to 1000 mm) thick, and of either the dry or viscous-impingement types. In large systems, the frames in which these units are installed are bolted or riveted together to form a filter bank. Automatic filters are constructed in sections offering several choices of width up to 70 ft (21 m) and generally range in height from 40 to 200 in. (1 to 5 m) in 4 to 6 in. (100 to 150 mm) increments. Several sections may be bolted together to form a filter bank.

Several manufacturers provide side-loading filter sections for various types of filters. Filters are changed from outside the duct, making service areas in the duct unnecessary and thus saving cost and space.

The in-service efficiency of an air filter is, of course, sharply reduced if air leaks through either leaky bypass dampers or poorly designed frames. The higher the efficiency of a filter, the more attention must be paid to the rigidity and sealing effectiveness of the frame. In addition, high efficiency filters must be han-

**Table 2 Typical Filter Applications Classified by Filter Efficiency and Type[a]**

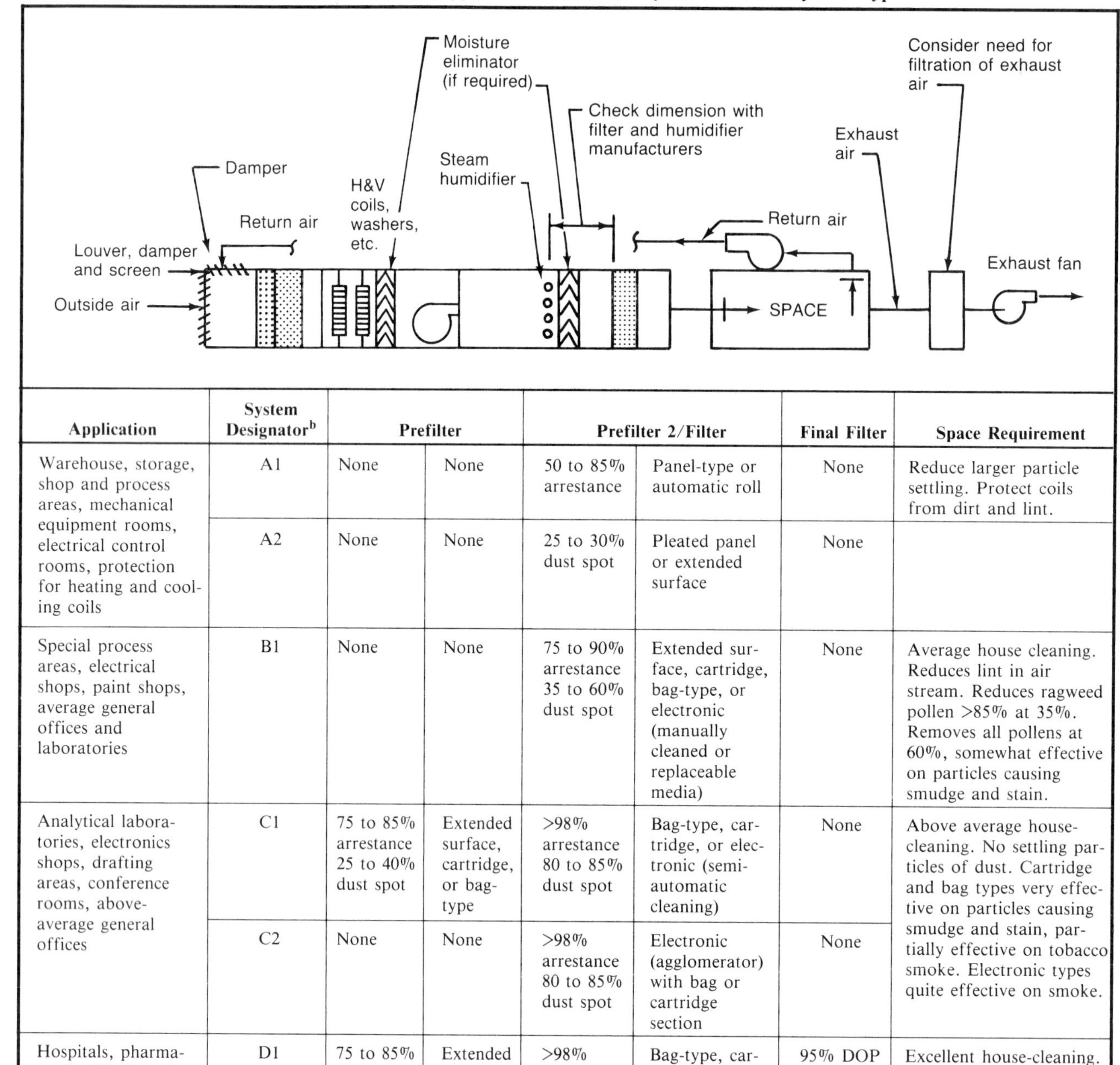


| Application | System Designator[b] | Prefilter | | Prefilter 2/Filter | | Final Filter | Space Requirement |
|---|---|---|---|---|---|---|---|
| Warehouse, storage, shop and process areas, mechanical equipment rooms, electrical control rooms, protection for heating and cooling coils | A1 | None | None | 50 to 85% arrestance | Panel-type or automatic roll | None | Reduce larger particle settling. Protect coils from dirt and lint. |
| | A2 | None | None | 25 to 30% dust spot | Pleated panel or extended surface | None | |
| Special process areas, electrical shops, paint shops, average general offices and laboratories | B1 | None | None | 75 to 90% arrestance 35 to 60% dust spot | Extended surface, cartridge, bag-type, or electronic (manually cleaned or replaceable media) | None | Average house cleaning. Reduces lint in air stream. Reduces ragweed pollen >85% at 35%. Removes all pollens at 60%, somewhat effective on particles causing smudge and stain. |
| Analytical laboratories, electronics shops, drafting areas, conference rooms, above-average general offices | C1 | 75 to 85% arrestance 25 to 40% dust spot | Extended surface, cartridge, or bag-type | >98% arrestance 80 to 85% dust spot | Bag-type, cartridge, or electronic (semi-automatic cleaning) | None | Above average house-cleaning. No settling particles of dust. Cartridge and bag types very effective on particles causing smudge and stain, partially effective on tobacco smoke. Electronic types quite effective on smoke. |
| | C2 | None | None | >98% arrestance 80 to 85% dust spot | Electronic (agglomerator) with bag or cartridge section | None | |
| Hospitals, pharmaceutical R&D and manufacturing (non-aseptic areas only), some clean ("gray") rooms | D1 | 75 to 85% arrestance 25 to 40% dust spot | Extended surface, cartridge, or bag-type | >98% arrestance 80 to 85% dust spot | Bag-type, cartridge, electronic (semi-automatic cleaning) | 95% DOP disposable cell | Excellent house-cleaning. Very effective on particles causing smudge and stain, smoke and fumes. Highly effective on bacteria. |
| | D2 | None | None | >98% arrestance 80 to 95% dust spot | Electronic (agglomerator) with bag or cartridge section | None | |
| Aseptic areas in hospital and pharmaceutical R&D and manufacturing. Clean rooms in film and electronics manufacturing, radio-active areas, etc.[c] | E1 | 75 to 85% arrestance 25 to 40% dust spot | Extended surface, cartridge, bag-type | >98% arrestance 80 to 85% dust spot | Bag-type, cartridge, electronic (semi-automatic cleaning) | ≥99.97% DOP disposable cell | Protects against bacteria, radioactive dusts, toxic dusts, smoke and fumes. |

[a]Adapted from a similar table courtesy of E.I. du Pont de Nemours & Company.
[b]System designators have no significance other than their use in this table.
[c]Electronic agglomerators and air cleaners are not usually recommended for clean room applications.

dled and installed with extreme care. Gilbert and Palmer (1965) suggest some of the precautions needed for filters of the HEPA type.

Air cleaners should be installed in the outdoor-air intake ducts of buildings and residences and in the recirculation and bypass air ducts, as well. Cleaners are logically placed ahead of heating or cooling coils and other air-conditioning equipment in the system to protect them from dust. The dust captured in an air intake duct is likely to be mostly particulate matter of a greasy nature, while lint may predominate in dust from within the building.

Where high efficiency filters protect critical areas such as clean rooms, it is important that the filters be installed as close to the room as possible to prevent the pickup of particles between the filters and the outlet. The ultimate in this trend is the so-called *laminar flow room,* in which the entire ceiling or one entire wall becomes the final filter bank.

The published performance data for all air filters are based on straight-through unrestricted airflow. Filters should be installed so that the face area is at right angles to the airflow whenever possible. Eddy currents and dead air spaces should be avoided; air should be distributed uniformly over the entire filter surface, using baffles or diffusers, if necessary. Filters are sometimes damaged if higher-than-normal air velocities impinge directly on the face of the filter.

Failure of air-filter installations to give satisfactory results can, in most cases, be traced to faulty installation or improper maintenance or both. The most important requirements of a satisfactory and efficiently operating air filter installation are as follows:

1. The filter must be of ample size for the amount of air and dust load it is expected to handle. An overload of 10 to 15% is regarded as the maximum allowable. When air volume is subject to future increase, a larger filter should be installed.
2. The filter must be suited to the operating conditions, such as degree of air cleanliness required, amount of dust in the entering air, type of duty, allowable pressure drop, operating temperatures, and maintenance facilities.
3. The filter type should be the most economical for the specific application. The initial cost of the installation should be balanced against efficiency and depreciation, as well as expense and convenience of maintenance.

The following recommendations apply to filters installed with central fan systems:

1. Duct connections to and from the filter should change size or shape gradually to ensure even air distribution over the entire filter area.
2. Sufficient space should be provided in front of or behind the filter, or both, depending on its type, to make it accessible for inspection and service. A distance of 20 to 40 in. (0.5 to 1 m) is required, depending on the filter chosen.
3. Access doors of convenient size should be provided to the filter service areas.
4. All doors on the clean-air side should be gasketed to prevent infiltration of unclean air. All connections and seams of the sheet-metal ducts on the clean-air side should be as airtight as possible. The filter bank must be caulked to prevent bypass of unfiltered air, especially when high efficiency filters are used.
5. Electric lights should be installed in the chamber in front of and behind the air filter.
6. Filters installed close to an air inlet should be protected from the weather by suitable louvers. A large mesh wire screen should be placed in front of the louvers.
7. Filters, other than electronic air cleaners, should have permanent indicators to give a warning when the filter resistance reaches too high a value or is exhausted, as with automatic roll media filters.
8. Electronic air cleaners should have an indicator or alarm system to indicate when high voltage is off or shorted out.

## SAFETY REQUIREMENTS

Safety ordinances should be investigated when the installation of an air cleaner is contemplated. Combustible filtering media may not be permitted in accordance with some existing local regulations. Combustion of dust and lint on a filtering medium is possible, although the medium itself may not burn. This may cause a substantial increase in filter combustibility. Smoke detectors and fire sprinkler systems may be considered for filter bank locations.

## REFERENCES


AHAM. 1986. Method of Measuring Performance of Portable Household Electric Cord-Connected Room Air Cleaners. AHAM *Standard* AC-1-1986. Association of Home Appliance Manufacturers, Chicago, IL.

ARI. 1984. Commercial and Industrial Air Filter Equipment. ARI *Standard* 850-84. Air-Conditioning and Refrigeration Institute, Arlington, VA.

ARI. 1986. Residential Air Filter Equipment. ARI *Standard* 680-86. Air-Conditioning and Refrigeration Institute, Arlington, VA.

ASHRAE. 1976. Methods of Testing Air Cleaning Devices Used in General Ventilation for Removing Particulate Matter. ASHRAE *Standard* 52-76.

British Standards Institution. 1965. Methods of Test for Low-Penetration Air Filters. B.S. 3928, London, W.1, England.

British Standards Institution. 1957. Methods of Test for Air Filters Used in Air Conditioning and General Ventilation. B.S. 2831, London, W.1, England.

Davies, C.N. 1966. *Aerosol Science*. Academic Press, New York.

Dill, R.S. 1938. A Test Method for Air Filters. ASHRAE *Transactions*, Vol. 44, p. 379.

EPA. *Air Quality Criteria for Photochemical Oxidants*. Environmental Protection Agency, Office of Air Noise and Radiation, Office of Air Quality Planning and Standards, Durham, NC.

Frey, A.H. 1985. Modification of Aerosol Size Distribution by Complex Electric Fields. *Bulletin of Environmental Contamination and Toxicology* 34:850-857.

Frey, A.H. 1986. The Influence of Electrostatics on Aerosol Deposition. ASHRAE *Transactions* 92(1B):55-64.

Fuchs, N.A. 1964. *The Mechanics of Aerosols*. Pergamon Press, New York.

Gilbert, H. and J. Palmer. High Efficiency Particulate Air Filter Units. USAEC, TID-7023. Available from NTIS, Springfield, VA 22151.

Horvath, H. 1967. A Comparison of Natural and Urban Aerosol Distribution Measured with the Aerosol Spectrometer. *Environmental Science and Technology*, August, p. 651.

IES. 1986. Recommended Practice for HEPA Filters. Institute of Environmental Sciences *Publication* IES-RP-CC-001-86, Mount Prospect, IL.

Lundgren, D.A. and K.T. Whitby. 1968. Effect of Particle Electrostatic Charge on Filtration by Fibrous Filters. *Industrial and Engineering Chemistry Process Design and Development*, 4(4):345, October.

McCrone, W.C.; R.G. Draftz; and J.G. Delley. 1967. *The Particle Atlas*. Ann Arbor Science Publishers, Ann Arbor, MI.

Moore, C.E.; R. McCarthy; and R.F. Logsdon. 1954. A Partial Chemical Analysis of Atmospheric Dirt Collected for Study of Soiling Properties. *Heating, Piping, and Air Conditioning*, October, p. 145.

National Air Pollution Control Administration. 1969. *Air Quality Criteria for Particulate Matter*. Obtainable from NTIS, Springfield, VA 22151.

Nutting, A. and R.F. Logsdon. 1953. New Air Filter Code. *Heatng, Piping and Air Conditioning*, June, p. 77.

Penney, G.W. and G.W. Hewitt. 1949. Electrically Charged Dust in Rooms. AIEE *Transactions*.

Peters, A.H. 1962. Application of Moisture Separators and Particulate Filters in Reactor Containment. USAEC-DP812.

Peters, A.H. 1965. Minimal Specification for the Fire-Resistant High Efficiency Filter Unit. USAEC *Health and Safety Information*, Issue No. 212, Washington, DC 20545.

Sutton, D.J. *et al*. 1964. Performance and Application of Electronic Air Cleaners in Occupied Spaces. ASHRAE *Journal*, June, p. 55.

Sutton, D.J.; K.M. Nodolf; and H.K. Makino. 1976. Predicting Ozone Concentrations in Residential Structures. ASHRAE *Journal*, September, p. 21.

U.S. Government Printing Office. 1956. Filter Units, Protective Clothing, Gas-Mask Components, and Related Products: Performance-Test Methods. MIL-STD-282.

U.S. Bureau of Mines. Filter Type Dust Fume and Mist Respirator. *Schedule* 21B, Part 14, Pittsburgh, PA 15213

UL. 1980. Electrostatic Air Cleaners. UL *Standard* 867-80. Underwriters Laboratories, Inc., Northbrook, IL.

UL. 1985. High Efficiency, Particulate, Air Filter Units. UL *Standard* 586-85. Underwriters Laboratories, Inc., Northbrook, IL.

UL. 1977. Test Performance of Air Filter Units. UL *Standard* 900-77. Underwriters Laboratories, Inc., Northbrook, IL.

Whitby, K.T.; A.B. Algren; and R.C. Jordan. 1958. Size Distribution and Concentration of Airborne Dust. ASHRAE *Transactions*, Vol. 64, p. 129.

Whitby, K.T.; A.B. Algren; and R.C. Jordan. 1956. The Dust Spot Method of Evaluating Air Cleaners. *Heating, Piping and Air Conditioning*, December, p. 151.

Wilson, L.G. and P. Cavanagh. 1967. The Relative Importance of Brownian Diffusion and Other Factors in Aerosol Filtration. *Atmospheric Environment*, Vol. 1, No. 3, p. 261.

## BIBLIOGRAPHY

ASHRAE. 1965. Symposium on Applied Air Cleaner Evaluation. Coblentz, C.W. Testing of Air Filters; McIver, S.H. Practical Aspects of an Air Cleaner Evaluation Apparatus; Peterson, C.M. and K.T. Whitby. Fractional Efficiency Characteristics of Unit Type Collectors; Margard, W.L. and R.F. Logsdon. An Evaluation of the Bacterial Filtering Efficiency of Air Filters in the Removal and Destruction of Airborne Bacteria; Sutton, D.J.; H.A. Cloud; and P.E. McNall, Jr. Methods and Results of the Evaluation of Air Filters on Ragweed Pollen; Annis, J.C. Ragweed Pollen Efficiency of Filters for Unitary (Window) Air Conditioners.

Davies, C.N. 1973. *Aerosol Filtration*. Academic Press, London.

Dennis, R. 1976. *Handbook on Aerosols*. Technical Information Center, Energy Research and Development Administration. Oak Ridge, TN.

Dorman, R.G. 1974. *Dust Control and Air Cleaning*. Pergamon Press, New York.

Drinker, P. and T. Hatch. *Industrial Dust*. McGraw-Hill Book Co., New York.

Duncan, S.F. 1963. Effect of Filter Media Microstructure on Dust Collection. ASHRAE *Journal*, Vol. 6, April, p. 37.

Engle, P.M. Jr. and C.J. Bauder. 1964. Characteristics and Applications of High Performance Dry Filters. ASHRAE *Journal*, Vol. 6, May, p. 72.

Hinds, W.C. 1982. *Aerosol Technology: Properties, Behavior, and Measurement of Airborne Particles*. John Wiley and Sons, New York.

Hunt, C.M. 1972. An Analysis of Roll Filter Operation Based on Panel Filter Measurements. ASHRAE *Transactions* 78(2):227.

Jorgensen, R. ed. 1961. *Fan Engineering*, 6th ed. Buffalo Forge Company, Buffalo, NY.

Licht, W. 1980. *Air Pollution Control Engineering, Basic Calculations for Particulate Collection*. Marcel Dekker, Inc., New York.

Lioy, P.J. and M.J.Y. Lioy, eds. 1983. *Air Sampling Instruments for Evaluation of Atmospheric Contaminants*, 6th edition. American Conference of Governmental Industrial Hygienists, Cincinnati, OH.

Liu, B.Y.H. 1976. *Fine Particles*. Academic Press, New York.

Lundgren, D.A.; F.S. Harris, Jr.; W.H. Marlow; M. Lippmann; W.E. Clark; and M.D. Durham, eds. 1979. *Aerosol Measurement*. University Presses of Florida. Gainesville, FL.

Matthews, R.A. 1963. Selection of Glass Fiber Filter Media. *Air Engineering*, Vol. 5, October, p. 30.

McDonough, J. and G. King. 1987. An In-Depth Review of Federal Standard 209C: Defining Cleanroom Performance for the VSLI Era. *Microcontamination*, Vol. 5, January, p. 18.

McNall, P.E. Jr. 1986. Indoor Air Quality—A Status Report. ASHRAE *Journal*, Vol. 28, June, p.39.

National Research Council. 1981. *Indoor Pollutants*. National Academy Press, Washington, DC.

Ogawa, A. 1984. *Separation of Particles from Air and Gases*. Volumes I and II. CRC Press, Boca Raton, FL.

Penney, G.W. and N.G. Ziesse. 1968. Soiling of Surfaces by Fine Particles. ASHRAE *Transactions* 74(1).

Rose, H.E. and A.J. Wood. 1966. *An Introduction to Electrostatic Precipitation*. Dover Publishing Company, New York.

Stern, A.C. ed. 1977. *Air Pollution*, 3rd edition. Vol IV, Engineering Control of Air Pollution. Academic Press, New York.

Stern, A.C. ed. 1960. *Air Pollution*, 2nd ed. Academic Press.

Swanton, J.R. Jr. 1971. Field Study of Air Quality in Air-Conditioned Spaces. ASHRAE *Transactions* 77(1):124.

U.S. Government. 1973. Clean Room and Work Station Requirements, Controlled Environments. Federal *Standard* 209B, amended 1976. Available from GSA, Washington, DC.

Gieske, J.A.; E.R. Blosser; and R.B. Rief. 1975. A Study of Techniques for Evaluation of Airborne Particulate Matter. Battelle Institute Report to ASHRAE, Project No. TRP-97.

USAEC/ERDA/DOE. Air Cleaning Conference Proceedings. Available from NTIS, Springfield, VA 22151.

Walsh, P.J.; C.S. Dudney; and E.D. Copenhaver, eds. 1984. *Indoor Air Quality*. CRC Press, Boca Raton, FL.

Whitby, K.T. 1965. Calculation of the Clean Fractional Efficiency of Low Media Density Filters. ASHRAE *Journal*, Vol. 7, September, p. 56.

White, P.A.F. and S.E. Smith, eds. 1964. *High Efficiency Air Filtration*. Butterworth, London, W C2, England.

Yocom, J.E. and W.A. Cote. 1971. Indoor/Outdoor Air Pollutant Relationships for Air-Conditioned Buildings. ASHRAE *Transactions* 77(1):61.

Ziesse, N.G. and G.W. Penney. 1968. The Effects of Cigarette Smoke on Space Charge Soiling of Walls When Air Is Cleaned by a Charging-Type Electrostatic Precipitator. ASHRAE *Transactions* 74(2).

# CHAPTER 11

# INDUSTRIAL GAS CLEANING AND AIR POLLUTION CONTROL



A GROWING need to maintain low levels of air pollution has resulted in stringent regulations pertaining to collection of dust and gaseous contaminates. Industrial gas-cleaning installations that exhaust to the outdoor environment are regulated by the U.S. Environmental Agency (EPA); those that exhaust to the work place are regulated by the Occupational Safety and Health Administration (OSHA) of the Department of Labor.

This chapter addresses the technology for cleaning of industrial process gas and air pollution control. Pollution from such gas streams has been classified as either *particulant* or *gaseous.* Control equipment that removes particulates from the gas stream may also remove some gaseous contaminants; on the other hand, equipment that is primarily intended for removal of gaseous pollutants also removes paticulate to some degree. However, the removal mechanisms are different, so the control equipment is described in two sections of this chapter.

As the cost of pollution control affects manufacturing costs, an early engineering evaluation of alternative processes should be made to minimize the impact pollution control may have on the total cost of a product. An alternative manufacturing process may reduce or eliminate pollution-control costs.

An industrial air-cleaning installation performs one or more of the following functions:

1. Complies with air pollution control laws or regulations.
2. Prevents nuisance or physical damage to individuals or adjacent properties.
3. Prevents reentry of contaminants to working spaces.
4. Reclaims usable materials, heat, and energy.
5. Reduces fire, explosion, or other hazards.
6. Permits recirculation of cleaned air to working spaces.
7. Uses cleaned gases for processes.

## DEGREE OF AIR CLEANING REQUIRED

The amount of material that can be discharged to the atmosphere is established by federal, state, or local regulations as *emission standards,* which are intended to accomplish certain levels of ambient air purity called *air quality standards.* The need for control within an industrial plant is often important to protect equipment, improve the product, or provide a clean working environment. Effective air-cleaning devices prevent the reentry of contaminants to working spaces, since clean effluent is discharged to the ambient atmosphere. Public complaints may occur even when the effluent concentration discharged to the atmosphere is below the permissible limits of concentration and visibility. Thus, plant location, contaminants involved, and meteorological condition of the areas must be evaluated, in addition to codes or regulations.

Combustible dusts, vapors, or gases create fire and explosion hazards. Loss of visibility, settlement, or accumulation of these materials also affects safety. Proper application of air-cleaning devices requires considering explosion hazards in the air-cleaning device. The degree of air cleaning for recovery of usable material is a matter of economic evaluation, which varies with such factors as quantity and value of material collected, investment, and operating costs. In most cases, emission standards require a higher degree of air cleaning than justified for purely economic recovery, if recovery were possible. Air cleanliness must be a high priority where toxic materials are involved and the cleaned air is recirculated to the work area.

For those industrial processes that exhaust to the outdoor environment, the U.S. EPA has established *Standards of Performance for New Stationary Sources* (NSPS) (USGPO 1986). Together with more restrictive *State Implementation Plants* (SIPs) and local codes, a regulatory basis accomplishes certain levels of ambient air purity called *Air Quality Standards.* Information on the NSPS can be obtained from the regional offices of the U.S. EPA. The Industrial Ventilation Committee of the American Conference of Governmental Hygienists (ACGIH) and National Institute of Occupational Safety and Health (NIOSH) have established criteria for recirculation of cleaned process air (ACGIH 1974, NIOSH 1978).

## SAMPLING METHODS

The U.S. EPA has developed methods to measure the particulate and gaseous components of many industrial processes and has incorporated these in the NSPS by reference. Appendix A of EPA Report No. EPA-450/2-76-028 lists the Reference Methods (EPA 1976). These are updated regularly in the issues of the *Federal Register.* Guidance for the use of the Reference Methods can be found in the *Quality Assurance Handbook for Air Pollution Measurement Systems* (EPA 1984).

The U.S. EPA has also approved other test methods; ie., ASTM Methods, when cited in the NSPSs or the applicable U.S. EPA

The preparation of this chapter is assigned to TC 5.4, Industrial Process Air Cleaning and Air Pollution Control.

Reference Methods. The experience gained by the U.S. EPA and the standards organizations is a valuable resource for designing and testing gas-cleaning equipment, even if the system is not regulated as a polluting source.

## PARTICULATE EMISSION CONTROL

Air-cleaning devices that remove particulate matter are available over a tremendous range of intended duty, from light loads (such as clean rooms and air conditioning) to industrial-process air or gas cleaners used by heavy industry. Light-duty air cleaners typically used in air-conditioning systems handle dust concentrations up to 0.002 grain/ft$^3$ (4 mg/m$^3$), whereas heavy-duty devices for process exhaust cleaning typically handle up to 35 grain/ft$^3$ (70 g/m$^3$), with no well-defined upper limit. Some process or pneumatic conveyors handle hundreds of times the above values. Light-duty equipment is often selected to control emissions that constitute a health hazard: for example, radioactive particles, beryllium particles, or biological airborne wastes, in which case the mass loading may be low. This chapter considers heavy-duty devices, termed *dust collectors,* usually applied as an air pollution control measure to process emissions having particulate concentration ranging from 0.01 to 20 grain/ft$^3$ (20 to 40 000 mg/m$^3$).

In any industrial situation in which a particulate emission problem exists, a thorough analysis of the process or system should be conducted. Cleaning devices may not be necessary if process and system control prevents the formation and subsequent emission of airborne pollutants. However, even when control equipment is required, process and system control can be used to minimize the loading of the collection device. To conserve energy, the contaminants should not be diluted with extraneous air because the volume of gas to be handled is a major factor in the owning and operating costs of control equipment.

An industrial process may be changed from *dirty* to *clean* by product substitution (switching to cleaner burning fuel or pretreating the existing fuel source). Equipment redesign, such as enclosing pneumatic conveyors or recycling noncondensable gases may also clean the process. Occasionally, additives such as chemical dust suppressants used in quarrying or liquid animal fat applied to dehydrated alfalfa prior to grinding reduce the air pollution potential.

If process control is insufficient to prevent excessive emissions, a collection device must be installed. For even a preliminary selection, a valid analysis of substances contained in the process effluent must be available. The analysis may involve careful source sampling and accurate laboratory techniques. Source sampling involves the following:

1. Astute selection of sampling site
2. Acceptable sampling points
3. Withdrawal of samples isokinetically
4. Recording of temperature, moisture content, and gas-velocity profiles in the stack and comparable data on the sample gas flow
5. Physical properties of the stack
6. Barometric pressure
7. Notation of cyclic variations in gas velocity, temperature, and dust loading
8. Availability of services and disposal facility
9. Process weight rate, consumption of raw materials, energy use, or other regulator base for emission standard

Samples and field data laboratory analysis can be used to determine the following information as it is needed:

1. Particle size distribution
2. Dust concentration, average and extreme values, a stack profile of dust concentrations
3. Aerodynamic size distribution (incorporates such effects as actual size, density, and shape)
4. Particle bulk density
5. Particle in-situ electrical resistivity (for electrostatic precipitator)
6. Particle handling characteristics—erosion, abrasion, flocculent, adhesive, or sticky qualities
7. Particle composition, recovery value
8. Flammable or explosive limits
9. Toxicity
10. Particle solubility (particularly for wet collector)
11. Computation of maximum, minimum, and normal gas-flow rates and composition at actual conditions, reduced to standard conditions, and corrected, if necessary, to basis of applicable emission standard.
12. Computation of dust loading per unit of raw material consumed.
13. Determination of degree of removal required, normally regulated by federal or local codes.

When selecting dust collectors it is important to know particle size distribution because the collection efficiency of most devices strongly depends on particle size. Since different techniques yield different results, the method of sampling and particle sizing should be selected by an expert. Reliable stack sampling and size distribution data may be difficult to obtain without special equipment and trained field or laboratory personnel. If the engineer does not have experience in this area, an air pollution consulting firm should be engaged.

## SELECTION OF GAS CLEANERS

When all information about the particular effluent problem and the process involved is known, the required collection equipment may be selected. Table 1 shows common industrial collectors and applications (Kane and Alden 1967). No one type of collector is universally applicable. Dust collectors differ in basic design and removal efficiency (Caplan 1968), first cost, energy requirements, maintenance, land use, and operating costs. The basic types of available dust collecting equipment are as follows:

*Gravity and Momentum Collectors:* settling chambers, louvers, baffle chambers
*Centrifugal Collectors:* cyclones, mechanical centrifugal collectors
*Fabric Filters:* baghouses, fabric collectors
*Electrostatic Precipitators:* tubular, plate, wet, dry
*Wet Collectors:* spray scrubbers, impingement scrubbers, wet cyclones, venturis, packed towers, mobile-bed scrubbers
*Incinerators:* afterburner, catalytic, flare
*Adsorbers*
*Electrostatic Augmented Scrubbers*

## PERFORMANCE

Dust collectors may be evaluated for overall efficiency, penetration, or fractional efficiency. The overall collection efficiency ($\eta$) is generally expressed as a percentage in terms of mass in accordance with Eq. (1).

$$\eta = 100(C_i - C_o)/C_i = 100(W_i - W_o)/W_i = 100\ W_c/W_i \quad (1)$$

*where*

$\eta$ = collection efficiency, %
$C_i$ = inlet concentration
$C_o$ = outlet concentration
$W_i$ = mass rate of contaminant in inlet gas
$W_o$ = mass rate of contaminant in outlet gas
$W_c$ = mass rate caught or removed by collector

Table 1 Collectors Used In Industry (Kane and Alden 1967)

| | | | | | Wet Collectors | | | | |
|---|---|---|---|---|---|---|---|---|---|
| Operation | Concentration | Particle Sizes | Cyclone | High Eff. Centrifugal | Medium Pressure | High Energy | Fabric Arrester | High Volt. Electrostatic | See Note No. |
| CERAMICS | | | | | | | | | |
| a. Raw product handling | Light | Fine | Rare | Seldom | Frequent | N/U | Frequent | N/U | 1 |
| b. Fettling | Light | Fine to med. | Rare | Occasional | Frequent | N/U | Frequent | N/U | 2 |
| c. Refractory sizing | Heavy | Coarse | Seldom | Occasional | Frequent | Rare | Frequent | N/U | 3 |
| d. Glaze and vitr. enamel spray | Moderate | Medium | No | No | Usual | N/U | Occasional | N/A | |
| CHEMICALS | | | | | | | | | |
| a. Material handling | Light to moderate | Fine to med. | Occasional | Frequent | Frequent | Frequent | Frequent | N/U | 4 |
| b. Crushing, grinding | Moderate to heavy | Fine to coarse | Often | Frequent | Frequent | Occasional | Frequent | N/U | 5 |
| c. Pneumatic conveying | Very heavy | Fine to coarse | Usual | Occasonal | Rare | Rare | Usual | N/U | 6 |
| d. Roasters, kilns, coolers | Heavy | Med.-coarse | Occasional | Usual | Usual | Frequent | Rare | Often | 7 |
| e. Incineration | Light to medium | Fine | No | No | No | Frequent | Rare | Frequent | 8 |
| COAL MINING AND POWER PLANT | | | | | | | | | |
| a. Material handling | Moderate | Medium | Rare | Occasional | Occasional | N/U | Usual | N/A | 9 |
| b. Bunker ventilation | Moderate | Fine | Occasional | Frequent | Occasional | N/U | Usual | N/A | 10 |
| c. Dedusting, air cleaning | Heavy | Med.-coarse | Occasional | Frequent | Occasional | N/U | Usual | N/A | 11 |
| d. Drying | Moderate | Fine | Rare | Occasional | Frequent | Occasional | N/U | N/A | 12 |
| FLY ASH | | | | | | | | | |
| a. Coal burning—chain grate | Light | Fine | N/A | Rare | N/U | N/U | Frequent | N/U | 13 |
| b. Coal burning—spreader stoker | Moderate | Fine to coarse | Rare | Rare | N/U | N/U | Frequent | Rare | 14 |
| c. Coal burning—pulverized fuel | Heavy | Fine | N/A | Frequent | N/U | N/U | Frequent | Usual | 14 |
| d. Woodburning | Varies | Coarse | Usual | Usual | N/U | N/U | Occasional | Occasional | 15 |
| FOUNDRY | | | | | | | | | |
| a. Shakeout | Light to moderate | Fine | Rare | Rare | Rare | Seldom | Usual | N/U | 16 |
| b. Sand handling | Moderate | Fine to med. | Rare | Rare | Usual | N/U | Rare | N/U | 17 |
| c. Tumbling mills | Moderate | Med.-coarse | N/A | N/A | Frequent | N/U | Usual | N/U | 18 |
| d. Abrasive cleaning | Moderate to heavy | Fine to med. | N/A | Occasional | Frequent | N/U | Usual | N/U | 19 |
| GRAIN ELEVATOR, FLOUR AND FEED MILLS | | | | | | | | | |
| a. Grain handling | Light | Medium | Usual | Occasional | Rare | N/U | Frequent | N/A | 20 |
| b. Grain driers | Light | Coarse | N/A | N/A | N/U | N/U | (See note 20) | N/A | 21 |
| c. Flour dust | Moderate | Medium | Rare | Often | Occasional | N/U | Usual | N/A | 22 |
| d. Feed mill | Moderate | Medium | Often | Often | Occasional | N/U | Frequent | N/A | 23 |
| METAL MELTING | | | | | | | | | |
| a. Steel blast furnace | Heavy | Varied | Frequent | Rare | Frequent | Frequent | N/U | Frequent | 24 |
| b. Steel open hearth, basic oxygen | Moderate | Fine to coarse | N/A | N/A | N/A | Often | Rare | Frequent | 25 |
| c. Steel electric furnace | Light | Fine | N/A | N/A | N/A | Occasional | Usual | Rare | 26 |
| d. Ferrous cupola | Moderate | Varied | N/A | N/A | Frequent | Often | Frequent | Occasional | 27 |
| e. Nonferrous reverberatory | Varied | Fine | N/A | N/A | Rare | Occasional | Usual | N/U | 28 |
| f. Nonferrous crucible | Light | Fine | N/A | N/A | Rare | Rare | Occasional | N/U | 29 |
| METAL MINING AND ROCK PRODUCTS | | | | | | | | | |
| a. Material handling | Moderate | Fine to med. | Rare | Occasional | Usual | N/U | Considerable | N/U | 30 |
| b. Driers, kilns | Moderate | Med.-coarse | Frequent | Occasional | Frequent | Occasional | N/U | Occasional | 31 |
| c. Cement rock drier | Moderate | Fine to med. | N/A | Frequent | Occasional | Rare | N/U | Occasional | 32 |
| d. Cement kiln | Heavy | Fine to med. | N/A | Frequent | Rare | N/U | Usual | Usual | 33 |
| e. Cement grinding | Moderate | Fine | N/A | Rare | N/U | N/U | Usual | Rare | 34 |
| f. Cement clinker cooler | Moderate | Coarse | N/A | Occasional | N/U | N/U | N/U | N/U | 35 |
| METAL WORKING | | | | | | | | | |
| a. Production grinding, scratch brushing, abrasive cutoff | Light | Coarse | Occasional | Frequent | Considerable | N/U | Considerable | N/U | 36 |
| b. Portable and swing frame | Light | Medium | Rare | Frequent | Frequent | N/U | Considerable | N/U | |
| c. Buffing | Light | Varied | Frequent | Rare | Frequent | N/U | Rare | N/U | 37 |
| d. Tool room | Light | Fine | Frequent | Frequent | Frequent | N/U | Frequent | N/U | 38 |
| e. Cast-iron machining | Moderate | Varied | Rare | Frequent | Considerable | N/U | Considerable | N/U | 39 |
| PHARMACEUTICAL AND FOOD PRODUCTS | | | | | | | | | |
| a. Mixers, grinders, weighing, blending, bagging, packaging | Light | Medium | Rare | Frequent | Frequent | N/U | Frequent | N/U | 40 |
| b. Coating pans | Varied | Fine to med. | Rare | Rare | Frequent | N/U | Frequent | N/U | 41 |
| PLASTICS | | | | | | | | | |
| a. Raw material processing | (See comments under Chemicals) | | | | | | | | 42 |
| b. Plastic finishing | Light to moderate | Varied | Frequent | Frequent | Frequent | N/U | Frequent | N/U | 43 |
| RUBBER PRODUCTS | | | | | | | | | |
| a. Mixers | Moderate | Fine | N/A | N/A | Frequent | N/U | Usual | N/U | 44 |
| b. Batchout rolls | Light | Fine | N/A | N/A | Usual | N/U | Frequent | N/U | 45 |
| c. Talc dusting and dedusting | Moderate | Medium | N/A | N/A | Frequent | N/U | Usual | N/U | 46 |
| d. Grinding | Moderate | Coarse | Often | Often | Frequent | N/U | Often | N/U | 47 |
| WOODWORKING | | | | | | | | | |
| a. Woodworking machines | Moderte | Varied | Usual | Occasional | Rare | N/U | Frequent | N/U | 48 |
| b. Sanding | Moderate | Fine | Frequent | Occasional | Occasional | N/U | Frequent | N/U | 49 |
| c. Waste conveying, hogs | Heavy | Varied | Usual | Rare | Occasional | N/U | Occasional | N/U | 50 |

## Notes for Table 1

[1] Dust released from bin filling, conveying, weighing, mixing, pressing, forming. Refractory products, dry pan, and screening operations more severe.
[2] Operations found in vitreous enameling, wall and floor tile, pottery.
[3] Grinding wheel or abrasive cutoff operation. Dust abrasive.
[4] Operations include conveying, elevating, mixing, screening, weighing, packaging. Category covers so many different materials that recommendation will vary widely.
[5] Cyclone and high-efficiency centrifugals often act as primary collectors, followed by fabric or wet-type.
[6] Usual setup uses cyclone as product collector followed by fabric arrester for high overall collection efficiency.
[7] Dust concentration determines need for dry centrifugal; plant location, product value determines need for final collectors. High temperatures are usual and corrosive gases not unusual.
[8] Ionizing wet scrubbers are widely used.
[9] Conveying, screening, crushing, unloading.
[10] Remote from other dust-producing points. Separate collector usually.
[11] Heavy loading suggests final high-efficiency collector for all except very remote locations.
[12] Loadings and particle sizes vary with the different drying methods.
[13] Boiler blowdown discharge is regulated, generally for temperature and in some places for pH limits; check local environmental codes on sanitary discharge.
[14] Collection for particulate and sulfer control usually requires a scrubber (dry or wet) and a fabric filter or electrostatic precipitator.
[15] Public nuisance from settled wood char indicates collectors are needed.
[16] Hot gases and steam usually involved.
[17] Steam from hot sand, adhesive clay bond involved.
[18] Concentration very heavy at start of cycle.
[19] Heaviest load from airless blasting because of higher cleaning speed. Abrasive shattering greater with sand than with grit or shot. Amounts removed greater with sand castings, less with forging scale removal, least when welding scale is removed.
[20] Operations such as car unloading, conveying, weighing, storing.
[21] Special filters are successful.
[22] In addition to grain handling, cleaning rolls, sifters, purifiers, conveyors, as well as storing, packaging operations are involved.
[23] In addition to grain handling, bins, hammer mills, mixers, feeders, conveyors, bagging operations need control.
[24] Primary dry trap and wet scrubbing usual. Electrostatic is added where maximum cleaning required.
[25] Air pollution control expensive for open hearth, accelerating the use of substitute melting equipment such as Basic Oxygen and Electric-Arc.
[26] Fabric collectors have found extensive application for this air pollution problem.
[27] Cupola control will vary with plant size, location, melt rate, and air pollution emission regulations.
[28] Corrosive gases can be a problem, especially in secondary aluminum.
[29] Zinc oxide plume can be troublesome in certain plant locations.
[30] Crushing, screening, conveying, storing involved. Wet ores often introduce water vapor in exhaust air stream.
[31] Dry centrifugals used as primary collectors, followed by final cleaner.
[32] Same as No. 30.
[33] Collectors usually permit salvage of material and also reduce nuisance from settled dust in plant area.
[34] Salvage value of collected material high. Same equipment used on raw grinding before calcining.
[35] Coarse abrasive particles readily removed in primary collector types.
[36] Roof discoloration, deposition on autos can occur with cyclones and, less frequently, with dry centrifugal. Heavy-duty air filters sometimes used as final cleaners.
[37] Linty particles and sticky buffing compounds can cause trouble in high efficiency centrifugals and fabric arresters. Fire hazard is also often present.
[38] Unit collectors extensively used, especially for isolated machine tools.
[39] Dust ranges from chips to fine floats, including graphitic carbon.
[40] Materials involved vary widely. Collector selection may depend on salvage value, toxicity, sanitation yardsticks.
[41] Controlled temperature and humidity of supply air to coating pans makes recirculation from coating pans desirable.
[42] Manufacture of plastic compounds involves operations allied to many in chemical field and varies with the basic process employed.
[43] Operations are similar to woodworking and collector selection involves similar considerations.
[44] Concentration is heavy during feed operation. Carbon black and other fine additions make collection and dust-free disposal difficult.
[45] Often, no collection equipment is used where dispersion from exhaust stack is good and stack location is favorable.
[46] Salvage of collected material often dictates type of high-efficiency collector.
[47] Fire hazard from some operations must be considered.
[48] Bulky material. Storage for collected material is considerable; bridging from splinters and chips can be a problem.
[49] Production sanding produces heavy concentration of particles too fine to be effectively caught by cyclones or dry centrifugals.
[50] Primary collector invariably indicated with concentration and partial size range involved; wet or fabric collectors, when used, are employed as final collectors.
N/A = Not applicable because of inefficiency or process incompatibility.
N/U = Not widely used.

$C_o$ and/or $C_i$ must be corrected to the same gas condition or both to dry standard condition.

Although Eq. (1) is the basic efficiency equation, it should be used with caution. Normal errors in field measurement of these parameters can lead to large errors in computed efficiency.

The mass penetration ($P$) of Eq. (1a) is more important than the collection efficiency because small changes in efficiency result in proportionately greater changes in penetration. It also offers a better way of computing efficiency through the relationship in Eq. (1b).

$$P = 100\ C_o/C_i = 100\ W_o/W_i = 100\ W_o/(W_o + W_c) \quad (1a)$$

$$\eta = 100 - P \quad (1b)$$

For example, if a collector operating at 95% efficiency with a penetration of 5% were changed to an efficiency of 90% and a penetration of 10%, the efficiency would have decreased 5%, but the penetration would have increased 100%. The concept of penetration helps in the analysis of collectors in series.

Particulate collectors generally exhibit a *fractional efficiency,* the relationship of efficiency to particle size. In general, for industrial applications, the smaller the particle the lower the collection efficiency. Efficiencies of common dust collection equipment are compared in Figure 1 and Table 2.

The U.S. EPA has proposed fine particle standards for ambient air quality (EPA 1984) and will complete final action on them during 1987. These revised standards, known as *PM*-10 *Standards,* will focus particulate abatement efforts towards fine particle control. The fine particles (with aerodynamic particle size smaller than 10 micron) are a concern because they penetrate deeply into the lungs. With the development of these standards, concern has risen over the fractional efficiency of gas cleaners.

## GRAVITY AND MOMENTUM COLLECTORS

Gravity settling removes large particles above 50 to 70 μm from a gas stream. Gravitational settling chambers, although simple in design and operation, have low efficiency and require a large space. The velocity must be low enough to prevent reentrainment of the deposited dust; a velocity below 60 fpm (0.3 m/s) is satisfactory for most materials. Gravity chambers are occasionally used in conjunction with more efficient dust-control equipment such as fabric filters or electrostatic precipitators. The efficiency of a gravity collector may be expressed as follows:

$$\eta = \frac{100 u_t L}{HV} \quad (2)$$

*where*

$\eta$ = efficiency, mass percentage of particles having settling velocity of , $u_t$
$u_t$ = settling velocity of dust, fps (m/s)
$L$ = chamber length, ft (m)
$H$ = chamber height, ft (m)
$V$ = gas velocity, fps (m/s)

Figure 1 in Chapter 11 of the 1985 FUNDAMENTALS Volume shows settling velocities for particles of various sizes.

Momentum collectors (see Figure 2) rely on sudden directional changes in the gas stream to separate particles. Because of their inertia, the particles continue in the same direction as the original gas stream, strike a target or trap, and are collected. Momen-

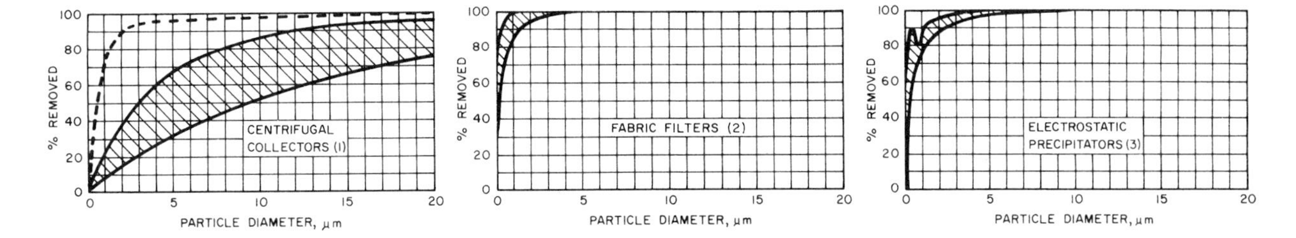


**Fig. 1 Typical Fractional-Efficiency Curves for Dry Collectors**

tum collectors need considerably less space than settling chambers and can collect 20 μm and larger particles with reasonable efficiency. Such units frequently consist of an arrangement of baffles or louvers, which cause a sudden change in direction of airflow, separating the coarser and heavier particles and some airflow into a smaller, more heavily loaded airstream. This concentrated stream then flows to a cyclone or similar separator, where the dust is removed.

## DRY CENTRIFUGAL COLLECTORS

A dry centrifugal collector is an inertial collector that removes particulates by spinning the contaminated gas. The mechanical type or skimming chamber is motor-driven and can be designed to move the gas. The cyclone has no moving parts and requires an external fan. Dry centrifugal collectors (see Figure 3) can be made from special materials (refractory-lined for high temperatures or special metal to resist corrosive gases).

In operation, a cyclone transforms the velocity of an inlet gas stream into a confined vortex, from which centrifugal forces throw suspended particles to the wall of the cyclone body. Centrifugal forces are considerably greater than gravity; therefore, a cyclone collects smaller particles than a simple gravity collector (Caplan 1968b).

Low-pressure cyclones operate at static pressure drops of 1 to 1.5 in. of water (250 to 370 Pa) between the inlet and outlet and can remove 50% of the suspended 5 to 10 μm particles. High efficiency cyclones have a drop of up to 8 in. of water (2 kPa) and can remove 70% of the particulates having about 5 μm diameter. The removal efficiency depends on true particle density, shape, and size (the aerodynamic size). Cyclone efficiency may be estimated from Figure 4, developed by Lapple (1968). The parameter $D_{pc}$, known as the *cut size,* is defined as the diameter of particles collected with 50% efficiency and may be calculated by the following:

For I-P calculations:

$$D_{pc} = \sqrt{\frac{9\,\mu b}{2\,N_e V_i (P_p - P_g)\pi}} \quad \text{(3 I-P)}$$

For SI calculations:

$$D_{pc} = \sqrt{\frac{1.435\,\mu b}{N_e V_i\,(\varrho_p - \varrho_g)}} \quad \text{(3 SI)}$$

**Table 2 (I-P) Equipment Applications (IGCI 1964)**

| Type of Dust-Collecting Equipment | Particle Diameter, μ m[a] | Max. Loading, gr/ft³[b] | Collection Efficiency, Weight, % | Pressure Loss Gas, in. of water | Pressure Loss Liquid, psi | Utilities per per 1,000 cfm | Comparative Energy Requirements | Gas Velocity fpm | Capacity Limits 1,000 cfm | Space Required, (Relative) |
|---|---|---|---|---|---|---|---|---|---|---|
| Dry inertial collectors | | | | | | | | | | |
| Settling chamber | >50 | >5 | <50 | <0.2 | — | — | 1 | 300 to 600 | None | Large |
| Baffle chamber | >50 | >5 | <50 | 0.1 to 0.5 | — | — | 1.5 | 1,000 to 2,000 | None | Medium |
| Skimming chamber | >20 | >1 | <70 | <1 | — | — | 3 | 2,000 to 4,000 | 50 | Small |
| Louver | >20 | >1 | <80 | 0.5 to 2 | — | — | 1.5 to 6 | 2,000 to 4,000 | 30 | Medium |
| Cyclone | >10 | >1 | <85 | 0.5 to 3 | — | — | 1.5 to 9 | 2,000 to 4,000 | 50 | Medium |
| Multiple cyclone | >5 | >1 | <95 | 2 to 6 | — | — | 6 to 20 | 2,000 to 4,000 | 200 | Small |
| Impingement | >10 | >1 | <90 | 1 to 2 | — | — | 3 to 6 | 2,000 to 4,000 | None | Small |
| Dynamic | >10 | >1 | <90 | Provides head | — | 1 to 2 hp. | 10 to 20 | — | 50 | — |
| Wet scrubbers | | | | | | | | | | |
| Gravity-spray | >10 | >1 | <70 | <1 | 20 to 100 | 0.5 to 2 gpm | 5 | 100 to 200 | 100 | Medium |
| Centrifugal | >5 | >1 | <90 | 2 to 6 | 20 to 100 | 1 to 10 gpm | 12 to 26 | 2,000 to 4,000 | 100 | Medium |
| Impingement | >5 | >1 | <95 | 2 to 8 | 20 to 100 | 1 to 5 gpm | 9 to 31 | 3,000 to 6,000 | 100 | Medium |
| Packed-bed | >5 | >0.1 | <90 | 1 to 10 | 5 to 30 | 5 to 15 gpm | 4 to 34 | 100 to 300 | 50 | Medium |
| Dynamic | >1 | >1 | <95 | Provides head | 5 to 30 | 1 to 5 gpm, 3 to 20 hp. | 30 to 200 | 3,000 to 4,000 | 50 | Small |
| Submerged-orifice | >2 | >0.1 | <90 | 2 to 6 | None | No pumping | 9 to 21 | 3,000 | 50 | Medium |
| Jet | 0.5 to 5 | >0.1 | <90 | Provides head | 50 to 100 | 50 to 100 gpm | 15 to 30 | 2,000 to 20,000 | 100 | Small |
| Venturi | >0.5 | >0.1 | <99 | 10 to 30 | 5 to 30 | 3 to 10 gpm | 30 to 300 | 12,000 to 42,000 | 100 | Small |
| Fabric filters | >0.2 | >0.1 | <99 | 2 to 6 | — | — | 6 to 20 | 1 to 20 | 200 | Large |
| Electrostatic precipitators | <2 | >0.1 | <99 | 0.2 to 1 | — | 0.1 to 0.6 kw | 4 to 11 | 100 to 600 | 10 to 2,000 | Large |

[a]Minimum particle diameter for which device is effective.
Particle Size can be defined as:
Fine—50% in 0.5 to 7 μm diameter range
Medium—50% in 7 to 15 μm diameter range
Coarse—50% over 15 μm in diameter

[b]Concentration (or Loading) can be defined as:
Light = under 2 gr/ft³
Moderate = 2 to 5 gr/ft³
Heavy = over 5 gr/ft³

**Table 2 (SI) Equipment Applications**

| Type of Dust-Collecting Equipment | Particle Diameter, μm[a] | Max. Loading, g/m³[b] | Collection Efficiency, Weight, % | Pressure Loss Gas, Pa | Pressure Loss Liquid, kPa | mL (water) per m³(gas) | Comparative Energy Requirements | Gas Velocity m/s | Capacity Limits, m³/s | Space Required, Relative |
|---|---|---|---|---|---|---|---|---|---|---|
| Dry inertial collectors | | | | | | | | | | |
| Settling chamber | 50 | 10 | 50 | 50 | — | — | 1 | 1.5 to 3 | None | Large |
| Baffle chamber | 50 | 10 | 50 | 25 to 125 | — | — | 1.5 | 5 to 10 | None | Medium |
| Skimming chamber | 20 | 2 | 70 | 250 | — | — | 3 | 10 to 20 | 24 | Small |
| Louver | 20 | 2 | 80 | 125 to 500 | — | — | 1.5 to 6 | 10 to 20 | 14 | Medium |
| Cyclone | 10 | 2 | 85 | 125 to 750 | — | — | 1.5 to 9 | 10 to 20 | 24 | Medium |
| Multiple cyclone | 5 | 2 | 95 | 500 to 1500 | — | — | 6 to 20 | 10 to 20 | 94 | Small |
| Impingement | 10 | 2 | 90 | 250 to 500 | — | — | 3 to 6 | 10 to 20 | None | Small |
| Dynamic | 10 | 2 | 90 | Provides head | — | — | 10 to 20 | — | 24 | — |
| Wet scrubbers | | | | | | | | | | |
| Gravity-spray | 10 | 2 | 70 | 250 | 140 to 690 | 70 to 270 | 5 | 0.5 to 1 | 47 | Medium |
| Centrifugal | 5 | 2 | 90 | 500 to 1500 | 140 to 690 | 140 to 1400 | 12 to 26 | 10 to 20 | 47 | Medium |
| Impingement | 5 | 2 | 95 | 500 to 2000 | 140 to 690 | 140 to 700 | 9 to 21 | 15 to 30 | 47 | Medium |
| Packed-bed | 5 | 0.6 | 90 | 250 to 2500 | 35 to 210 | 700 to 70 000 | 4 to 34 | 0.5 to 1.5 | 24 | Medium |
| Dynamic | 1 | 2 | 95 | Provides head | 35 to 210 | 140 to 700 | 30 to 200 | 15 to 20 | 24 | Small |
| Submerged-orifice | 2 | 0.2 | 90 | 500 to 1500 | None | — | 9 to 21 | 15 | 24 | Medium |
| Jet | 2 | 0.2 | 95 | Provides head | 345 to 610 | 7000 to 14 000 | 15 to 30 | 10 to 100 | 47 | Small |
| Venturi | 0.5 | 0.2 | 99 | 2500 to 7500 | 35 to 210 | 400 to 1400 | 30 to 300 | 60 to 210 | 47 | Small |
| Fabric filters | 0.2 | 0.2 | 99 | 500 to 1500 | — | — | 6 to 20 | 0.005 to 0.1 | 94 | Large |
| Electrostatic precipitators | 2 | 0.2 | 99 | 50 to 250 | — | — | 4 to 11 | 0.5 to 3 | 5 to 940 | Large |

[a]Minimum particle diameter for which device is effective.
Particle Size can be defined as:
Fine—50% in 0.5 to 7 μm diameter range
Medium—50% in 7 to 15 μm diameter range
Coarse—50% over 15 μm in diameter

[b]Concentration (or Loading) can be defined as:
Light = under 4 g/m³
Moderate = 4 to 10 g/m³
Heavy = over 10 g/m³

*where*

$D_{pc}$ = diameter cut-size particle collected at 50% efficiency, ft (m)
$\mu$ = absolute gas viscosity, centipoise (Pa • s)
$b$ = cyclone inlet width, ft (m)
$N_e$ = effective number of turns within cyclone. The number of turns are about five for a high-efficiency cyclone but may vary from 0.5 to 10 for other cyclones
$V_i$ = inlet gas velocity, fpm (m/s)
$\varrho_p$ = true particle density, lb/ft³ (kg/m³)
$\varrho_g$ = gas density, lb/ft³ (kg/m³)

The pressure drop through a given cyclone is proportional to the inlet velocity pressure, hence to the square of the gas flow. Efficiency of a cyclone increases at a rate less than linearly proportional to the gas flow, up to some limiting value of about 60 to 100 ft/s (25 to 30 m/s) inlet velocity. Above this limiting velocity, internal turbulence prevents further improvement in efficiency.

The mechanical centrifugal separator (see Figure 3) generally is not used for fine dust above 4.8 grain/ft³ (11 000 mg/m³) but is used in the handling of fly ash, woodworking, and plastics; it is a primary collector for pneumatic conveying where the particulates are substantially above 5 μm diameter. The mechanical centrifugal can act as prime mover for the gas stream and control larger particulates.

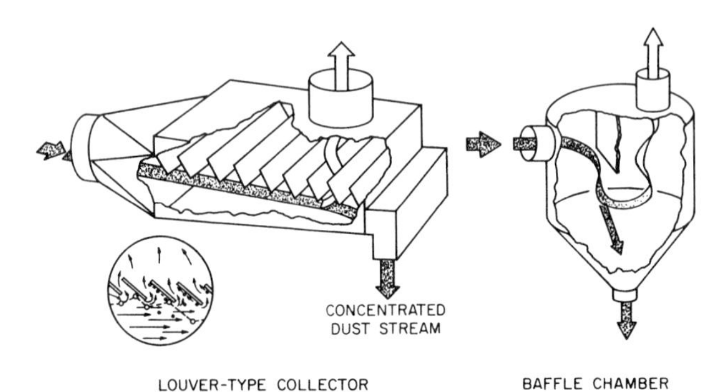


**Fig. 2 Momentum Collectors**

The inability of the centrifugal collector to control finer dust is apparent. However, these collectors work in many areas where dust is not easily friable and remains relatively large during the cleaning process. They are often used as precleaners to reduce the loading of more efficient pollution-control devices downstream. Table 1 indicates a few applications for the dry centrifugal collector.

## FABRIC FILTERS

Fabric filters offer high collection efficiency at reasonable cost if, in practical terms, they can be properly applied to the conditions of a gas stream.

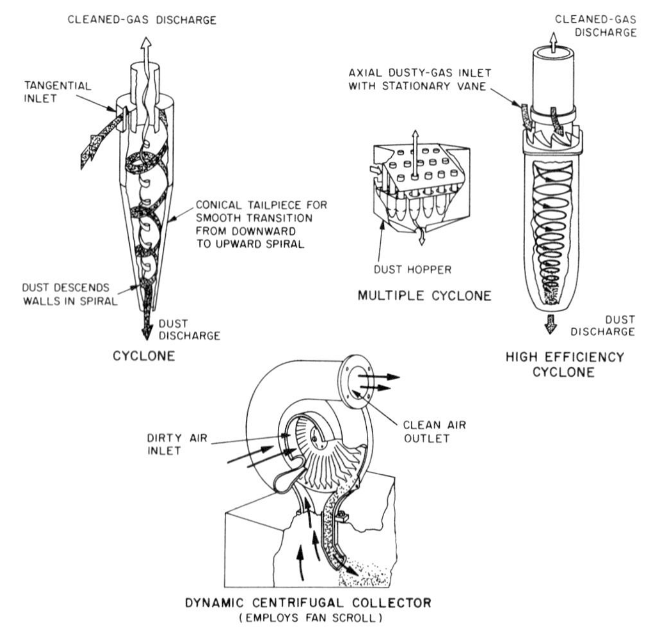


**Fig. 3 Dry Centrifugal Collectors**

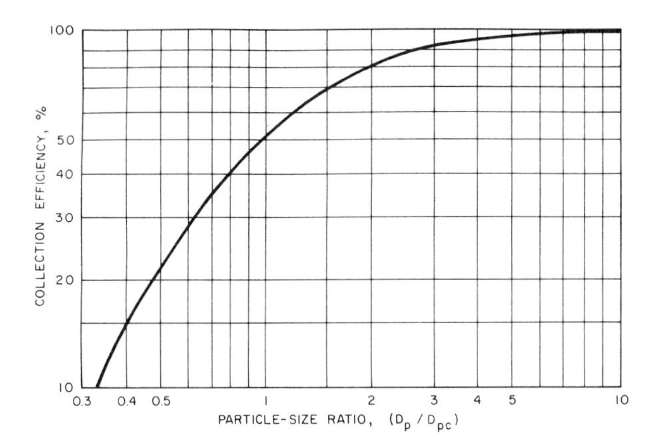


**Fig. 4 Cyclone Efficiency versus Particle-Size Ratio (Lapple 1968)**

## Principle of Operation

Fabric filters intercept a dust-laden gas stream with large areas of fabric. The gas passes through, leaving the particulates adhering to the fabric or to the dust cake on the gas entering side. Collectors must have mechanisms to dislodge the collected material frequently to prevent excessive resistance to the gas flow.

For woven cloth, particules collect on a buildup of a dust layer or cake on the cloth surfaces; the cake does the actual filtering, removing the particles from the gas stream by impaction, impingement, and diffusion. With new cloth, an initial period of poorer performance will occur as the particles bridge the openings in the cloth, which are many times the size of the particles to be removed. Once the cake is formed, the initial layers become part of the fabric and are not destroyed when the bulk of the collected material is dislodged during the cleaning cycle. Cotton and wool fibers in the woven media and most felted fabrics will build up the initial dust cake in a few minutes. Synthetic woven fabrics may require a few hours in the case of submicroscopic materials because of the smoothness of the monofilament threads. Felted fabrics contain no straight-through openings and have a reasonably good efficiency for most particulates, even when clean. The dust cake builds internally, as well as on the fabric surface; it cannot be kept porous by merely shaking.

Pleated paper filter media in cartridge form are considered fabric filters because they operate in the same general manner as high-efficiency fabrics. High initial collection efficiency is increased further by formation of a dust cake on the dirty air side of the media. A reverse pulse of compressed air is the normal method of cleaning pleated paper cartridges.

## Pressure-Volume Relationships

The major factor in fabric filter selection is the media area required to pass the desired gas volume. In this respect, the fabric filter differs from other collector mechanisms. An error in application seldom influences gas flow through a dry-mechanical, a wet-scrubber, or an electrostatic precipitator but can influence collection efficiency. With fabric, high collection efficiency is ensured, but excessive resistance and a major reduction in needed gas flow can occur with insufficient fabric surfaces. Also, unlike electrostatic precipitators and wet-scrubbers, the collection media (fabric) may be blinded by fine or sticky dust particles, and the resulting high pressue drop may force shutdown of the process. Because the potential for blinding by an internal cake cannot be predicted, media is selected by pilot test or by comparative tests of media in a full-scale installation. Thus, fabric filters must be designed on the basis of empirical data and experience.

It is difficult to obtain the significant data covering particle-size distribution and emission quantities needed for calculating pressure-volume relationships for a new installation. However, the following considerations form a supplement to the field data obtained from operating equipment. As laminar flow prevails at the gas flow rates through fabric and dust cake, resistance increase during a cycle is directly proportional to the approach velocity, as:

$$\Delta R = KV_iW \tag{4}$$

*where*

$\Delta R$ = pressure drop, in. of water (Pa)
$V_i$ = approach velocity to the fabric, fpm (m/s)
$W$ = mass of the dust cake, oz/ft$^3$ (g/m$^3$)
$K$ = permeability, in. of water/fpm/oz of dust/ft$^3$ (Pa/m/s/g of dust/m$^3$

The approach velocity becomes especially critical because it influences the fabric area over which the dust accumulation can be spread. At a constant dust concentration, for example, cutting the fabric area in half would double the thickness of the dust cake and its resistance to gas flow. When the resistance of the increased dust-mat thickness is added to the increased resistance caused by the increased flow rate, the pressure for a given concentration of dust in gas will vary approximately as the square of the approach velocity. If the total amount of dust to be collected remains relatively constant, as is frequently the case (concentration is lowered as gas volume increases), the pressure drop will vary linearly with the approach velocity.

Permeability, as defined in Eq. (4), is a function of particle-size distribution. It is high with coarser dusts and low where material is close to 1 μm in diameter. Use of a primary collector to remove the coarse fraction seldom causes a significant change in the pressure relationships because the reduction of dust-cake mass is offset by the lower permeability of the predominately small particle sizes left for dustcake formation.

Figure 5 represents pressure relationships of a single-compartment fabric filter operating through its cycle at constant volume and with a uniform dust loading and a constant size distribution. In practice, neither of these conditions occurs. Industrial dust-producing operations do not release airborne contaminants at a uniform rate and the generated material is of varying particle-size distribution. However, the pressure rise that occurs when a cycle extends for 2 to 8 hours is small, so these variations are insignificant for most systems, and a pressure-drop plot at constant volume approaches the linear slope of the illustration.

Exhaust volume through the usual single-compartment fabric filter is not constant because of the increasing resistance to gas flow as the dust cake accumulates. The reduction in flow rate

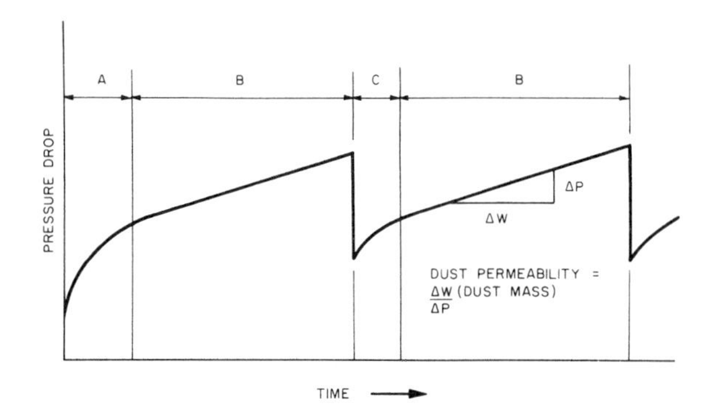


**Fig. 5 Basic Cycles**

is a function of the system pressure relationships, the exhaust fan characteristics, and point of rating. Proper fan selection can minimize the reduction in exhaust volume by increased filter resistance. A fan should operate on a steep portion of the fan pressure-volume curve. An *undersized* fan has a steeper characteristic than an *oversized* fan for the same duty and also is noisier.

When new, clean fabric is installed in a collector, the resistance is low, and the fan motor may be overloaded. This may be prevented by a throttling damper in the main duct, preferably on the clean side of the filter, to be used for start-up only. Overloading may also be prevented by using a backward-curved blade (nonoverloading fan) on the clean side of the collector.

When fabric filters using mechanical shaker or reverse-flow cleaning mechanisms are designed for continuous operation, pressure relationships become more complex, but they approach a constant value. The smaller the fraction of the elements taken out of service at any one time for dust-cake removal, the more uniform the pressure drop of the fabric filter. Figure 6 shows typical pressure diagrams for four- and six-compartment, continuous fabric filters with mechanical shakers or reverse-flow cleaning. Pressure variations are minimal in pulsejet designs because the cleaning cycle is extremely short and relatively few elements are taken out of service at any one time.

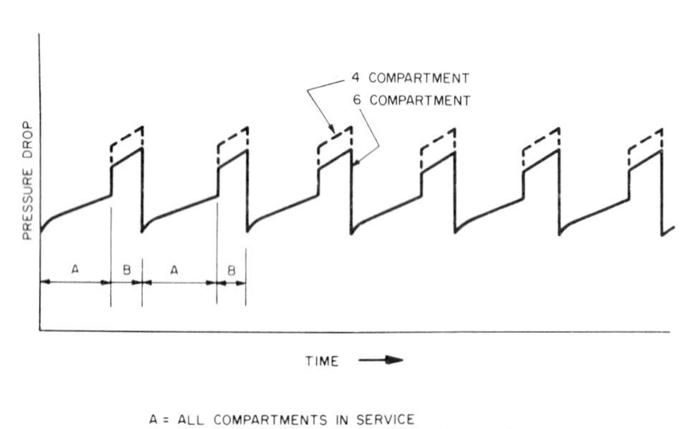


**Fig. 6 Compartmented Collector Cycles**

## Fabric Cleaning Mechanisms

For shaker-type cleaning, the fabric, formed into envelopes or cylindrical tubes, is arranged as shown in Figures 7 and 8. The tubes or envelopes are reconditioned by agitating the fabric with a shaking device to dislodge the accumulated material. The dislodged material settles in the storage hoppers before the collector compartment is placed back on stream. When operations can be interrupted on a 4-hour frequency (usually lunchtime or shift-change), and where sufficient fabric is provided so that pressure-drop increase does not produce excessive exhaust-volume variations, a single-compartment unit is the least expensive and widely used of fabric filter designs. If the system cannot be stopped for reconditioning, collectors can be divided into a number of separate sections, usually four to six. Through a system of dampers and timers, compartments are taken out of service, in sequence, for reconditioning. Because it is difficult to maintain dampers airtight, relief dampers are often included, which introduce a small volume of reverse gas to keep gas-flow relationships at the fabric suitable for cake removal.

With shaker-cleaning mechanisms, flow rates through the fabric are usually in the 2 to 4 fpm (10 to 20 mm/s) range and even lower where submicrometre particulates such as metallurgical fumes are collected. Use of compartments with their more frequent cleaning cycles does not permit substantial increase in flow rates over that of a single-compartment unit cleaned every 4 hours. The best condition for fabric reconditioning occurs with the system stopped, because even small particles will fall into the hopper.

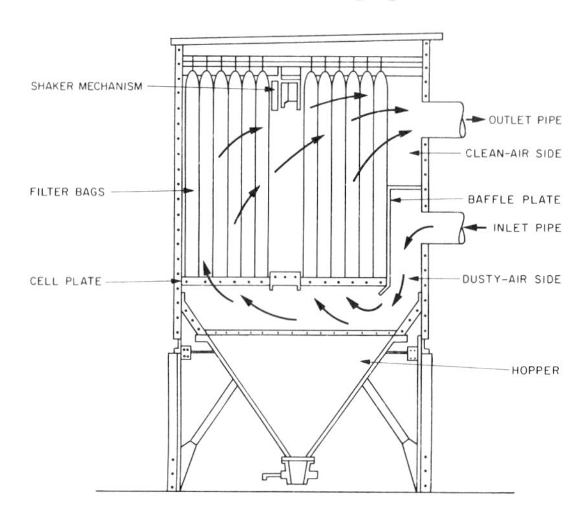


**Fig. 7 Bag-Type Fabric Collector**

**Reverse flow.** Fabric can be reconditioned by reversing the direction of flow through the fabric, assisted by the partial collapse of the bags (see Figure 9). This method reduces the number of moving parts and eliminates the cleaning-mechanism and bag supports of the mechanical shaker design, an advantage from a maintenance standpoint, especially when large volumes are cleaned. However, the cleaning or reconditioning is less vigorous and the residual drag of the reconditioned fabric is higher. Reverse-flow cleaning is usual for fabrics like glass cloth where fiber damage can occur from the flexing and snapping action of conventional shaker reconditioning.

Reverse flow requires compartmented designs because the substantial reverse-flow volumes entrain considerable collected material that must be directed to an on-stream compartment for cleaning. This reverse-flow circuit adds to the gross gas volume flowing through the collector's active filter surfaces and reduces the usual net-flow rate per unit area of filter media to below that of shaker-operated mechanisms on compartmented collectors.

With shaker mechanisms, bags are usually 4 to 8 in. (100 to 200 mm) in diameter and 10 to 20 ft (3 to 6 m) long. Reverse-flow bags are usually 8 to 12 in. (200 to 300 mm) in diameter and 20 to 33 ft (6 to 10 m) long and are generally operated at flow velocities in the 2 to 4 fpm (10 to 20 mm/s) range. Consequently, space requirements in plan dimensions are favorable for the reverse-flow unit.

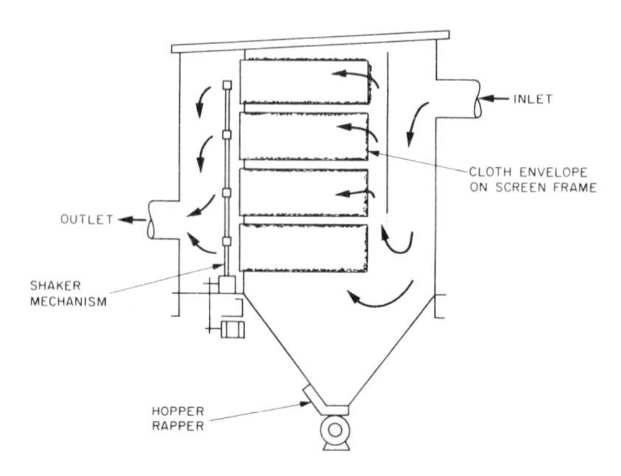


**Fig. 8 Envelope-Type Fabric Collector**

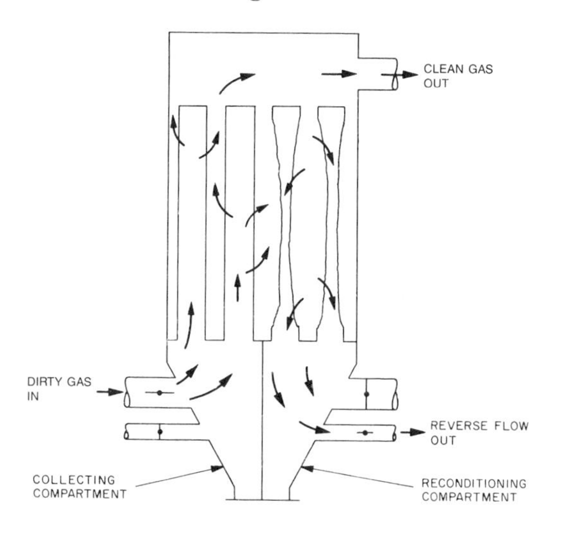


Fig. 9 Reverse-Flow Collector

**High performance designs.** Efforts to decrease fabric filter sizes by increasing the flow rates through the fabric have concentrated on methods of frequent or continuous cleaning cycles, without taking major portions of the filter surface out of service. In the pulse jet (Figure 10), a compressed-air jet operating for a fraction of a second causes a rapid vibration or ripple of the fabric, which dislodges accumulated dust cake. The pulse-jet design is predominantly used because it is easier to maintain than the reverse-jet mechanism.

Flow rates for reverse-jet or pulse-jet designs range from 6 to 10 fpm (30 to 50 mm/s) for favorable dusts, but require greater fabric area for many materials that produce a low-permeability dust cake. Felted fabrics should be used for these designs because the cleaning mechanism opens the pores of woven cloth and excessive leakage may result.

Most pulse-jet designs call for 4.5 to 6 in. (115 to 150 mm) diameter bags of 6 to 12 ft (2 to 4 m) length. The pulse jet usually requires high-pressure, compressed air, leading to higher power costs and condensation problems if the air is not dried.

A specialized pulse-jet design uses pleated media cartridges instead of bags (see Figure 11). With the pleated construction, a large media surface area is included in a relatively small housing. Because of configuration and high efficiency media, a slower surface velocity, in the range of 1 to 3 fpm (5 to 15 mm/s), is used.

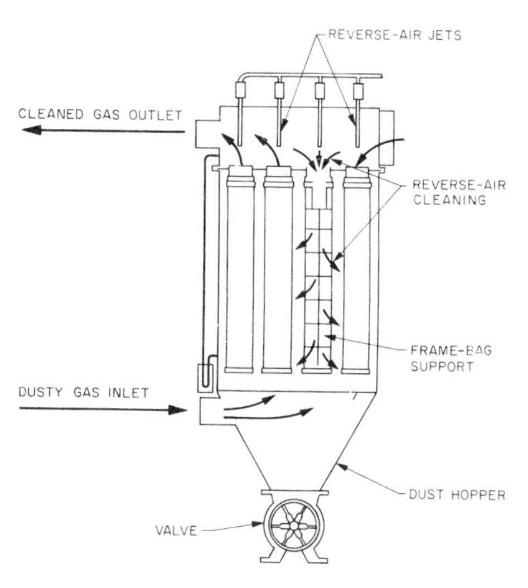


Fig. 10 Pulse-Jet Collector

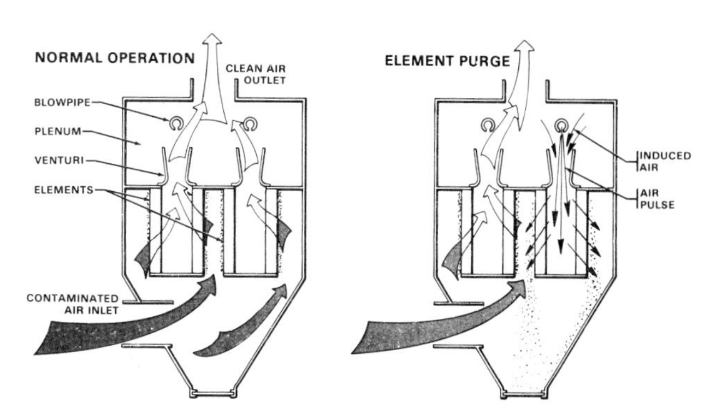


Fig. 11 Cartridge Collector

Media available include many synthetics and some cellulose-synthetic blends. These are frequently selected to control toxic dusts, fine particulates, and fumes. They are frequently used in systems where the clean air is recirculated for heat-conserving recirculation systems.

## Fabrics

A filter fabric specification considers cost, fiber diameter, type of weave, fabric weight, tensile strength, dimensional stability, chemical resistance, finish, permeability, and abrasion resistance. Often, the supplier must be relied on to choose the proper material to meet service conditions. Fabrics have become rather standardized, and there are relatively few alternatives. For ambient-air applications, a woven cotton or polypropylene fabric is the usual selection for shaker and reverse-flow cleaning mechanisms—felted synthetic for pulse-jet designs. Synthetics are used where moist air or hygroscropic materials must be exhausted through fabric collectors or where adhesive characteristics make removal of the dust cake difficult. Polypropylene, for example, has become a frequent selection. Synthetics often permit greater migration or the escape of particulates through the fabric during the cleaning cycle. Synthetic fabrics also allow fabric collectors to be applied to higher temperature particulate collection.

Table 3 shows upper temperature limits for usual collector fabrics. While higher temperatures are acceptable for short periods, reduced fabric life can be anticipated with continued exposure. Thermostatically controlled air bleed-in or collector bypass dampers may be used to protect against excessive gas temperatures.

**Table 3 Characteristics of Collector Fabrics**

| Fiber | Max. Continuous Operating Temp. °F | °C | Acid Resistance | Alkali Resistance | Flex Abrasion |
|---|---|---|---|---|---|
| Cotton | 180 | 82 | Poor | Very good | Very good |
| Wool | 200 | 93 | Good | Poor | Fair to good |
| Nylon*† | 200 | 93 | Poor | Excellent | Excellent |
| Nomex*† | 400 | 204 | Fair | Very good | Very good |
| Acrylic | 260 | 127 | Good | Fair | Good |
| Polypropylene | 180 | 82 | Excellent | Excellent | Very good |
| Polyethylene | 145 | 62 | Excellent | Excellent | Very good |
| Teflon† | 425 | 218 | Excellent | Excellent | Good |
| Glass fiber | 500 | 260 | Fair to good | Fair to good | Poor |
| Polyester*† | 275 | 135 | Good | Good | Very good |
| Cellulose | 180 | 82 | Poor | Good | Good |

* These fibers are subject to hydrolysis to varying degrees when they are exposed to hot, wet atmospheres.
† Du Pont trademark.

Because high-temperature, gas-cleaning applications often involve combustion products or other corrosive materials, it becomes necessary to anticipate chemical attack on fabric and housing if temperatures drop below their dew points (the dew points may be substantially higher than that of water). Often the relative resistance of the available media to the corrosive agent involved influences the choice of fabric.

For most high-temperature applications, gas requires cooling to meet fabric limits. Cooling through air-to-air heat transfer is recommended, if practical. Controlled evaporative cooling increases the dew point and the danger of collector fabric blinding from condensation but can be used for temperature reductions to 400°F (204°C) in most applications. Cold-air bleed is expensive because gas volume and, consequently, collector size are increased substantially; this method is most often used in the critical range of 332°F (167°C) to 260°F (127°C).

### Exhauster Location

In the preferred arrangement, the exhauster is located on the clean air side of the fabric. Advantages include the following:

1. Fan on clean air side handles clean air and has no abrasive exposure from the collected product.
2. High efficiency backward blade and air-foil designs can be selected because accumulation on the fan wheel is not a factor.
3. Escape of hazardous materials due to leaks is minimized.
4. Collector can be installed inside the plant, even near the process, because leakage is minimal.

For economic reasons, an exhauster may be located on the dust-laden gas side of the collector if particulates are of light concentration, are relatively nonabrasive, and the equipment can be located outdoors. Although this arrangement should be avoided because of the nuisance leakage potential, the following factors may be considered in the engineering evaluation:

1. Airtight collector construction on the clean side is not considered essential.
2. Collector can be bypassed in case of malfunction.
3. The collector housing can be designed to withstand only the collector pressure drop, neglecting that of the total system.
4. Slight energy savings may be realized because velocity pressure and acceleration losses between clean-air plenum and fan inlet are eliminated.
5. Less ductwork is needed.

### Efficiency

For correctly applied fabric filters equipped with the proper fabric, an efficiency of 99.9% or better is reliably attained on almost all industrial and process dust applications. In the use of a very large baghouse for fine metallurgical fumes, efficiencies in the range of 98.5 to 99.0% are attained. Most manufacturers guarantee such efficiencies on applications in which they have prior experience. Lower efficiency is caused by poor maintenance (torn fabric seams, loose connections, etc.) or improper selection of lighter/higher permeability fabrics in an effort to reduce cost.

Filtration theory based on separate fibers and individual dust particles predicts a much lower efficiency for fine particles in the 0.1 to 1.0 or 2.0 $\mu$m range. Laboratory experimentation confirms the theory for the same assumed conditions. The theory and corroborating experimental results do not apply to industrial-type fabric filters because the particulate loading is heavy, covers a wide range of sizes, and the filtering medium is a dust cake or a fiber-dust matrix. That is, particle size is not the major factor influencing efficiency attained from a fabric filter.

### Strengths and Weaknesses of Fabric Filter Technology

Fabric filters have several advantages: (1) collection efficiency, even on submicroscopic particulates (such as smoke and metallic fumes), meets the most stringent emission standards; (2) material is collected dry, often a significant factor if product recovery is desired; and (3) process upsets seldom violate emission standards.

Inherent weaknesses of a fabric filter include (1) potential fire or explosion hazard; (2) large space requirement; (3) moist, sticky materials or condensation that causes *blinding* of the fabric or dust mat reduces or stops the gas flow; and (4) secondary dust problem during collected material removal, transport, and disposal. Collected material handling should be an integral part of initial system design to avoid a later in-plant or external air pollution problem. Fabric collectors are rarely used for cleaning moisture-laden gases. When a large volume of gas is to be cleaned, consideration should be given to the tremendous number of fabric tubes (envelopes) required and the maintenance problem of detecting, locating, and replacing a damaged element.

### Application of Fabric Filters

Fabric filters offer relatively trouble-free service for most applications in which exhaust air is at ambient-air temperature and where dust is generated mechanically—crushed, ground, screened, and transported.

In the high-temperature gas-cleaning fields, fabric filters are (1) usual for electric arc furnace fumes in basic steel and foundry; (2) an alternate solution to electrostatic precipitators for cement kiln control, especially dry-process kilns; (3) the alternate solution to high-energy wet scrubbing for foundry cupola; and (4) the usual solution for furnace fumes from nonferrous melting and carbon black manufacturing.

Table 1 compares the use of fabric filters and collectors of other designs—mechanical, wet, electrostatic—in major fields of collector application.

### Recirculation

In some cases where the air has *not* been affected by combustion, solvent vapors, or toxic materials, it may be desirable to recirculate the air to reduce energy costs or balance static pressue in a space. The application of high-efficiency filters typically used in general ventilation systems to reduce particle concentrations to levels acceptable for recirculated air requires special consideration.

Guidelines for such applications are being established (NIOSH 1978); however, in many cases, field experience is limited. Manufacturers' recommendations should help avoid misapplication of such systems. Efficiency and dust concentration ranges for which ventilation filters are designed are vastly different than those for dust collectors and are measured in different terms.

## ELECTROSTATIC PRECIPITATORS

Electrostatic precipitators electrically charge particulates suspended in a gas stream and then separate the charged particulate from the gas stream by forces in an electrical field. The charged particles move to the grounded collecting surface where they agglomerate to form a deposit. The deposit may be removed by mechanical vibration or by continuous or periodic washing. The deposits fall to the hoppers where they may be removed from the electrostatic precipitator.

There are two general classes of precipitators: (1) the electronic air cleaner used commonly for ventilation and air-conditioning applications (see Chapter 10) and (2) the industrial electrostatic precipitator used primarily for the heavier particulate emissions of industry. Electronic air cleaners are generally low voltage (less than 12 kV), positive polarity for ozone reduction, and of two-stage design. Industrial electrostatic precipitators are designed for higher voltages (greater than 20 kV) and can use negative or positive polarity, depending on electrode design.

Industrial precipitators are normally custom-engineered for a particular application and can be designed to operate at collection efficiencies above 99.5% for closely specified conditions. Properties of the dust (such as particle-size distribution and electrical resistivity) can affect performance significantly, as can variations in gas composition and flow rate.

The earliest precipitators were tubular in shape and operated dry or wet with a continuous or intermittent flow of liquid down the inner wall of the pipe. Although still used for certain applications, the tube-type has given way to the plate-type precipitator to reduce space requirements.

Figure 12 shows various types of industrial plate-type precipitators. For each precipitator, the charging electrodes are located between parallel collecting plates. The gas flow through these precipitators is horizontal. Full-scale precipitators based on these designs ordinarily operate with gas velocity between 60 and 400 fpm (0.3 and 2.0 m/s) and with treatment time in the range of 2 to 10 seconds. Operating voltages from 20 to 60 kV (RMS) are applied between the charging electrode and the grounded plates. Temperatures up to 850°F (450°C) are handled routinely by precipitators consisting almost entirely of carbon steel. Special design techniques are required for temperatures above 975°F (525°C).

## Principle of Operation

The particulate is charged in an electrical corona surrounding the charging electrode (Oglesby, *et al.* 1970). The voltage at which corona begins is proportional to a power function of the radius of curvature of the surface. Therefore, high curvature regions—such as small diameter wires, projection points, or needles—are used as charging electrodes. The precipitator must operate between the voltage levels at which corona is initiated and the arc-over voltage, at which point the direct arc between the high voltage and grounded electrode occurs.

A common empirical equation for the efficiency of electrostatic precipitation is the *Deutsch Equation*:

$$\eta = 1 - e^{-t/\tau} \quad (5)$$

*where*

$\eta$ = collection efficiencies, expressed as a fraction
$t/\tau$ = precipitation rate parameter
$t$ = residence time, s

In practice, the Deutsch Equation is expressed empirically using $A/Q$ (specific collection area):

$$\eta = 1 - e^{-\omega A/Q} \quad (6)$$

*where*

$\omega$ = effective migration velocity, fpm (m/s), given in Table 4
$A$ = plate area, $ft^2$ ($m^2$)
$Q$ = gas flow, cfm ($m^3/s$)

Theoretically derived values for drift velocity provide less accurate estimates of efficiency when used with the Deutsch Equation than do empirical data because of the following conditions:

1. Reentrainment of particles
2. Variable distance at which a particle approaching the plates is captured
3. Nonuniformity of electric field
4. Temperature and humidity of gas
5. Nonuniformity of gas flow

Empirical models, which describe the performance of an electrostatic precipitator, rely on a consistent application of an electrode design. It is generally impossible to compare the parameters of empirical models for this equipment. Each supplier of electrostatic precipitators has experience with their electrode design to aid in sizing decisions.

Because the characteristics of emissions vary from location to location, even for similar processes, on-site pilot plant testing

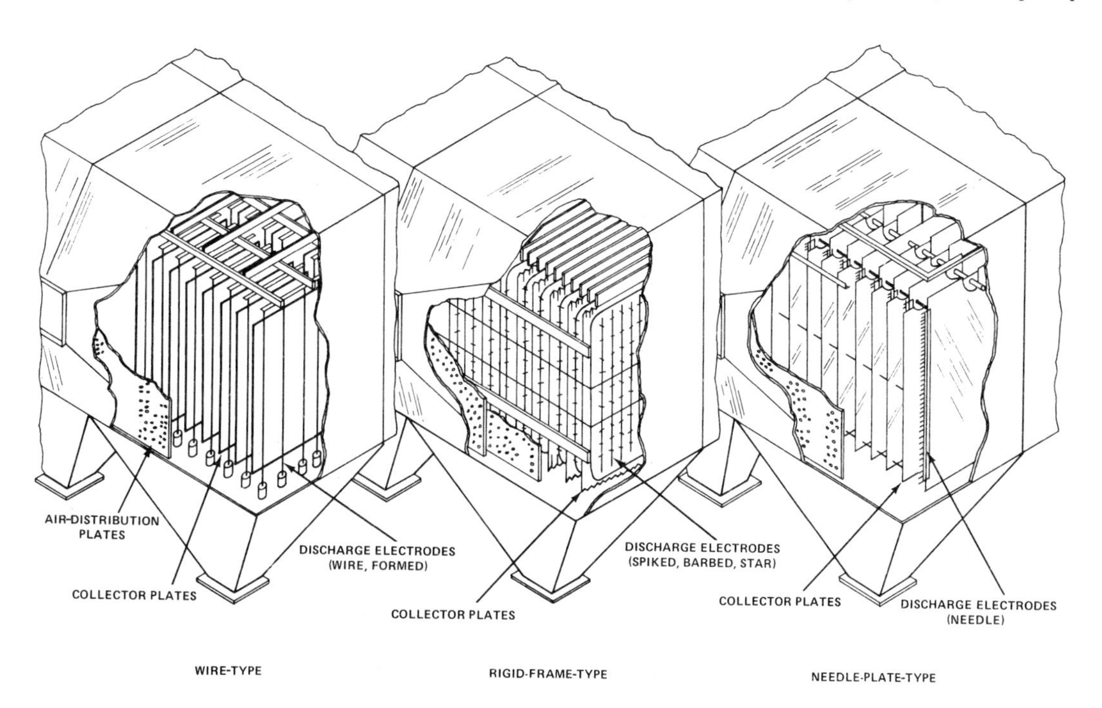


**Fig. 12 Various Types of Plate Precipitators (Cutaways Show Discharge Electrode Methods)**

**Table 4 Effective Migration Velocity, ω**

| Application | ω, fpm | ω, m/s |
|---|---|---|
| Pulverized coal (fly ash) | 4 to 30 | 0.02 to 0.15 |
| Paper mills | 16 | 0.08 |
| Open-hearth furnace | 12 | 0.06 |
| Secondary blast furnace (80% foundry iron) | 26 | 0.13 |
| Gypsum | 32 to 40 | 0.16 to 0.2 |
| Hot phosphorus | 6 | 0.03 |
| Acid mist ($H_2SO_4$) | 12 to 16 | 0.06 to 0.08 |
| Acid mist ($TiO_2$) | 12 to 16 | 0.06 to 0.08 |
| Flash roaster | 16 | 0.08 |
| Multiple-hearth roaster | 16 | 0.08 |
| Portland cement manufacturing (wet process) | 20 to 22 | 0.1 to 0.11 |
| Portland cement manufacturing (dry process) | 12 to 14 | 0.06 to 0.07 |
| Catalyst dust | 4 to 16 | 0.02 to 0.08 |
| Gray iron cupola (iron-coke ratio = 10) | 6 to 8 | 0.03 to 0.04 |

From *Air Pollution Engineering Manual* (HEW 1967).

is often necessary to determine the proper system configuration and EP size (Noll 1984).

## Design of Electrostatic Precipitators

Uniformity of gas velocity through the precipitator is of primary importance to achieve good efficiency. With the large gas flows commonly encountered, space allocations do not normally permit streamlined ducting; instead, elbow splitters, baffles, etc., are used. Uniformity of gas flow is so important that vendors frequently conduct flow-model experiments as a part of the design process.

The electrical resistivity of the collected dust layer is another characteristic that influences the sustained collectibility of particulate in a given application (White 1974). Resistivity is a measure of the electrical resistance of a dust layer 1 cm in cross-sectional area and 1 cm thick. To obtain good performance, electrical resistivities should be below $5 \times 10^{10}$ Ω • cm. Resistivity can vary with temperature and moisture and can be reduced significantly by absorption of various electrolytic materials added to the gas stream. As an example, precipitators have served utility power-plant operations effectively because traces of sulfur trioxide are present in the flue gas. As little as 5 to 10 ppm (5 to 10 mg/kg) of sulfur trioxide at 300°F (150°C) reduces the resistivity of fly ash, which makes the difference between poor performance and very good performance.

Electrostatic precipitators have been used successfully in collecting particulate emissions from utility boilers, wet-process cement kilns, glass production furnaces, sulfuric acid mist separators, blast furnaces, open hearth furnaces, and basic oxygen furnaces. Precipitators have not served as effectively for carbon black plants, organic fume collection, dry-process cement plants, and lime kiln operations. Electrostatic filters have a low pressure loss compared to fabric filters or high energy scrubbers and result in significant energy savings.

## Electrical Equipment

Electrical power supplies are generally designed for 55 kV (RMS) service with current ratings from 100 to 1500 mA (Hall 1975). Each power supply is equipped with manual or automatic voltage controls to maintain the voltage as high as possible. Silicon-rectifier controls maintain a high average voltage. They prevent mechanical damage when an arc occurs by reducing the voltage and reenergizing the precipitator at a slightly lower voltage. When the arc ceases, the voltage is raised on each successive cycle until another spark occurs. By setting the rate of rise of voltage between arcs, the sparking rate is set to give a good balance between maintaining the voltage as high as possible and keeping the fields energized. Sparking rates of about two sparks per second are typical.

*Safety* is an important consideration in precipitator installations. The internal, high-voltage equipment must not be accessible to personnel until the power supply is disconnected and grounded. Key interlock systems, which prevent opening hatchways, for example, while the high-voltage equipment is energized, are normally provided. Also, care should be exercised to ensure that sample connections and cleanout points are not located so as to allow pitot tubes or rodding tools to contact the high-voltage equipment.

## Rapping

To maintain a continuous process of precipitation, the precipitated material must be removed continuously or periodically from the collecting electrodes. In the case of dry dust, the collecting electrodes are vibrated or rapped, causing the material to dislodge and fall into a hopper. Since the material must fall through the gas stream to reach the hopper, steps must be taken to minimize reentrainment. Examples of such measures include (1) designing the collecting electrodes to minimize surface velocity; (2) minimizing rapping frequency; (3) rapping only a small section of the precipitator at any one time; and (4) providing more than one precipitating field in the direction of flow when high values of collection efficiency are required.

In some applications, the electrodes are cleaned by irrigation with water or other liquid. Typically, the method for removing the deposits from the electrodes must be *tuned* in the field after start-up to get best performance from the system.

## Dust-Handling Equipment

Although the precipitator collects and drops particulate into hoppers beneath the collecting plates, the job is still incomplete. Many precipitators fail because of poor design of the hoppers or material-handling equipment. Insufficient hopper capacity or an inadequate rate of removal causes collected material to rise above safe design levels. The plates and charging electrodes can be shorted electrically, resulting in failure of a complete electrical section. This common cause of precipitator failure occurs most frequently when dust loadings are increased because of higher ash content of fuel or raw material or because the furnace or reactor is overloaded.

Other common problems with dust-handling equipment include (1) plugging of hoppers, (2) blockage of dust valves with solid objects (such as wire weights), and (3) improper or insufficient maintenance.

## Fields of Application

Table 4 lists industrial operations in which precipitators have been extensively used. For these applications, a great quantity of data are available. Most manufacturers of precipitator equipment can provide expert assistance in selecting the proper equipment. The number of applications for precipitators continues to grow, and several areas deserve comment.

Low-sulfur coal, mined often in the western states, is being substituted for high-sulfur coal in many areas to avoid excessive $SO_2$ emissions. This coal creates a difficult situation for conventional fly ash precipitators because the sulfur-trioxide content of the flue gas is too low to yield acceptable resistivities of the fly ash. To lower the range of resistivity for more adequate particulate collection, the fly ash is either *preconditioned* by adding sulfur trioxide or the precipitator is designed for operation at high temperature (650 to 900°F [340 to 475°C]). Either of these operations involves new problems and operating techniques that require special care in mechanical design.

*Oil-fired boilers* require particulate abatement if the ash or ash-plus-sulfur-trioxide content of the flue gas is too high. This is usually the case when Venezuelan residual oil is burned in a large power station, causing a relatively dense gray or white plume whose major component is sulfur trioxide or sulfate compounds. Some of the special problems involved in the application of precipitators to oil-fired boilers are as follows:

1. The ash may have a high enough carbon content to be combustible, requiring special sensing equipment for safety or the use of combustion additives.
2. The fly ash may be very fine and cause high emissions during rapping. However, rapping can be much gentler and less frequent than on coal-fired applications.
3. If the precipitator operates at a relatively high temperature (350 F [175°C] or higher), a substantial part of the sulfur trioxide may pass through the precipitator as a gas and condense in the atmosphere to form a white fume.

*Municipal incinerators* have long operated without treatment of the flue gas or with wet scrubbers of simple design. The increasing demands for recovery of useful energy from refuse burning have led to the construction of many new refuse-fired boilers, as well as conversions to partial refuse firing. Wet scrubbers can meet relatively high efficiency standards, but the cost of fan operation is substantial and the HCl generated by the combustion of vinyl chloride plastics causes a serious corrosion problem. Precipitators have neither of these drawbacks, but do require good temperature control and preconditioning of the gas to produce acceptable resistivity of the ash.

In the *glass industry*, despite the submicrometre particulate size and the high resistivities common to glass melting, most glass-melting furnaces are now under control through the use of electrostatic precipitators.

Other application of the ESP include bio-mass combustors, lime and calcining kilns, fiberglass forming and curing, frit smelters, recovery boilers (direct contact and low odor), woodchip dryers, etc. Several emerging applications are coal slurry combusiton, atmospheric fluidized bed combustion, hazardous waster incineration, flue gas desulfurization and dry scrubbing, steam enhanced oil recovery, and regeneration of scrubber sludge.

## WET COLLECTORS

Wet-type dust-collecting devices apply some type of liquid (usually water) to capture and remove particulates (dust, mist, and fumes). Particle diameter sizes can range from 0.3 to 50 μm or larger. Wet collectors can be classified into three categories: (1) low-energy-type (up to 1 W/cfm [2 J/L]); (2) medium-energy-type (1 to 3 W/cfm [2 to 6 J/L]); and (3) high-energy-type (over 3 W/cfm [6 J/L]).

### Principle of Operation

The more important mechanisms involved in the capture and removal of the particulates are inertial impaction, Brownian motion diffusion, and condensation.

**Inertial impaction** (see Figure 13) occurs when a dust particle and a liquid droplet collide, resulting in the particle being captured. The resulting relatively large, liquid dust-particle droplet may be easily removed from the carrier gas stream by gravitation or impingement on separators.

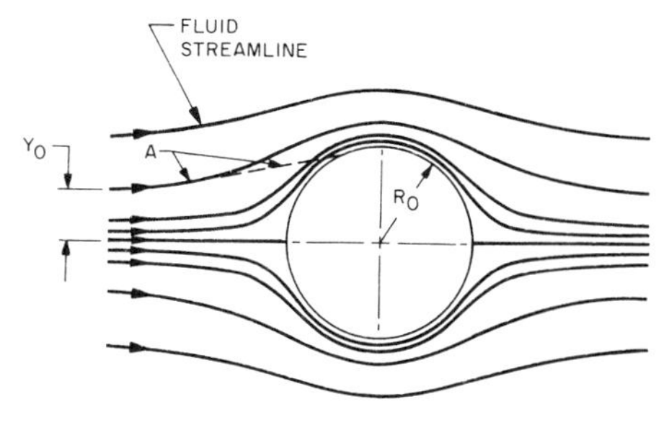


**Fig. 13 Inertial Impaction**

**Brownian motion diffusion** (see Figure 14) occurs when the dust particles are extremely small and have motion independent of the carrier gas stream. These small particles collide with one another, making larger particles, or collide with a liquid droplet and are captured.

**Condensation** occurs when the gas or air is cooled below its dew point. When moisture is condensed from the gas stream, fogging occurs, and the dust particles serve as condensation nuclei. The dust particle becomes larger as a result of the condensed liquid, and the probability of removal by impaction is increased.

Wet collectors perform two individual operations. The first operation occurs in the *contact zone* where the dirty gas comes in contact with the liquid. The second is in the *separation zone* where the liquid that has captured the particulate is removed from the gas stream. All well-designed wet collectors use one or more of the following principles: (1) high liquid to gas ratio, (2) intimate contact between the liquid and dust particles, and (3) formation of larger numbers of small liquid droplets to increase the chances of impaction.

For a given type of wet collector, the greater the power applied to the system, the higher will be the collection efficiency (Lapple and Kamack 1955). This is described as the *Contacting Power Theory.* Figure 15 compares fractional efficiencies of wet collectors. The relationship of venturi scrubber pressure drop to the removal of particulates is indicated in Figure 16.

### Types of Wet Collectors

Spray towers (see Figure 17) and *impingement scrubbers* (see Figure 18) are available in many different arrangements. The gas stream may be subjected to a single spray or a series of sprays, or the gas may be forced to impinge on a series of baffles. Except for packed towers, these types of scrubbers are of the low energy category; thus, they have a relatively low degree of par-

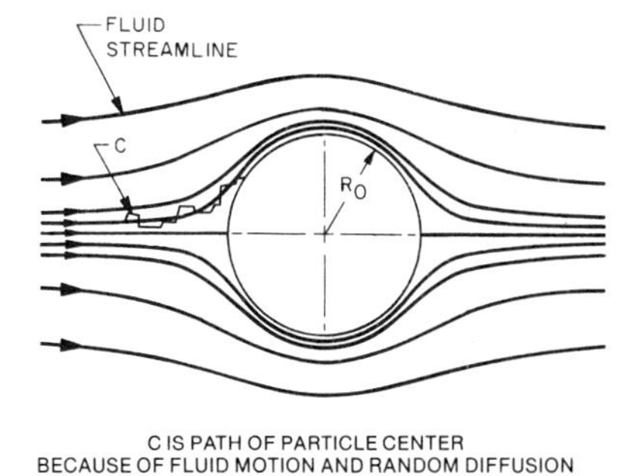


**Fig. 14 Diffusion**

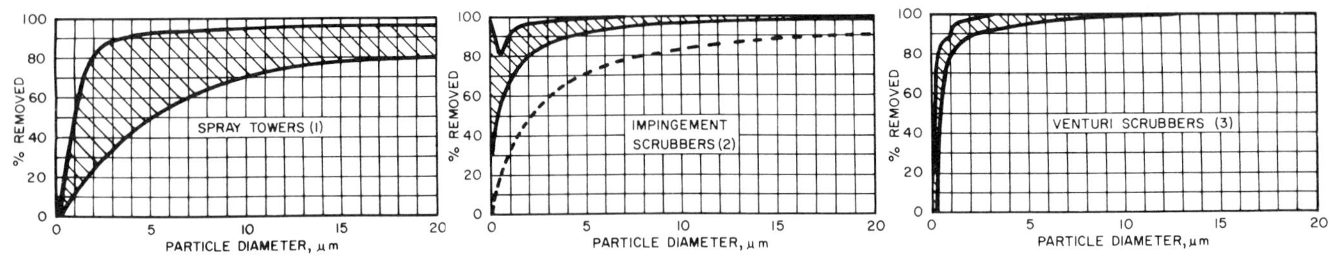


*Notes:*
1. Efficiency depends on liquid distribution. High efficiency is congruent with high fluid rate and pressure.
2. Upper curve is packed tower; lower curve is orifice-type wet collector. Dashed lines indicate less efficient irrigated baffles or rods.
3. High efficiency is congruent to high pressure drop.

**Fig. 15 Typical Fractional-Efficiency Curves for Wet Collectors**

ticulate removal. Efficiency ranges from 50 to 75% on 2-μm dust. Water consumption ranges from 2.8 to 11 gpm/1000 $cfm_{gas}$ (700 to 2500 $L_{gas}/L_{water}$). The efficiency of a spray tower can be improved by adding high-pressure sprays (see Figure 17). The spray pressure ranges from 30 to 100 psig (300 to 800 kPa abs.). Depending on the spray pressure used, the efficiency can range from 95 to 99% on 2-μm dust particles.

*Centrifugal-type collectors* (see Figure 18) are characterized by the tangential entry of the gas stream and are classified as medium energy devices. The efficiency ranges from 90 to 95% for 2-μm particulates. Water consumption ranges from 2.5 to 5 gpm/1000 $cfm_{gas}$ (1400 to 2800 $L_{gas}/L_{liquid}$).

*Orifice-type collectors* (see Figure 19) are also classified in the medium energy category. Usually, the gas stream is made to impinge on the surface of the scrubbing liquid and is forced through constrictions where the gas velocity is increased and where the liquid-gaseous-particulate interaction occurs. Collection efficiency is about 90 to 94% for 2-μm particles. Water usage for orifice collectors is limited to evaporation loss and removal of collected pollutants.

A *high energy venturi scrubber* (see Figure 20) passes the gas through a venturi-shaped orifice where the gas is accelerated to 12,000 fpm (60 m/s) or greater. Depending on the design, the scrubbing liquid is added at, or ahead of, the throat. The rapid acceleration of the air shears the liquid into many millions of liquid droplets, increasing the chance of liquid-particulate impaction. Yung developed a mathematical performance and design model for venturi scrubber (Semrau 1958-59). Subsequent validation experiments (Rudnick *et al.* 1986) have demonstrated that this model yields the most representative prediction of actual venturi scrubber performance in comparison with other performance models that have been developed.

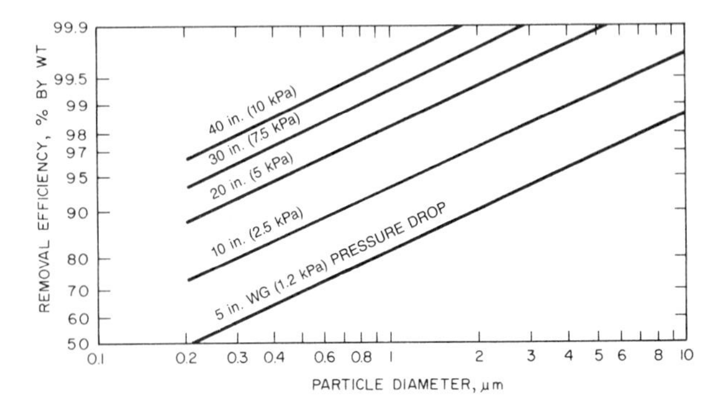


**Fig. 16 Relationship between Fractional Efficiency and Particle Size in Venturi Scrubbers**

In typical applications, venturi pressure drops range from 0.36 to 2 psi (2.5 to 15 kPa). Water circulation is high, requiring 0.5 to 10 $gpm_{liquid}$/1000 $cfm_{gas}$ (700 to 14,000 $L_{gas}/L_{liquid}$; thus, venturi systems use water-reclamation systems.

## Applications

Wet collectors may be used for the collection of most particulates from industrial process gas streams where economics allow for collection of the material in a wet state. Some disadvantages of wet collectors include (1) high susceptibility to corrosion, (2) high humidity in the gas stream, (3) large pressure drops, and (4) high power requirement. Some important advantages are (1) no secondary dust sources, (2) small spare-parts requirement, (3) ability to collect both gases and particulates, (4) low cost, and (5) ability to handle high temperature and high humidity gas streams, as well as reduce the possibility of fire or explosion.

## Auxiliary Equipment

When slurry from wet collectors cannot be returned directly

**Fig. 17 Spray Tower**

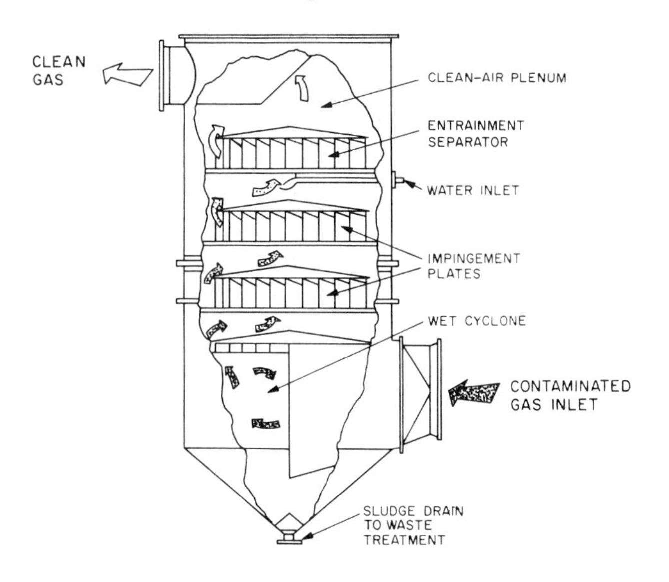


**Fig. 18 Impingement Scrubber**

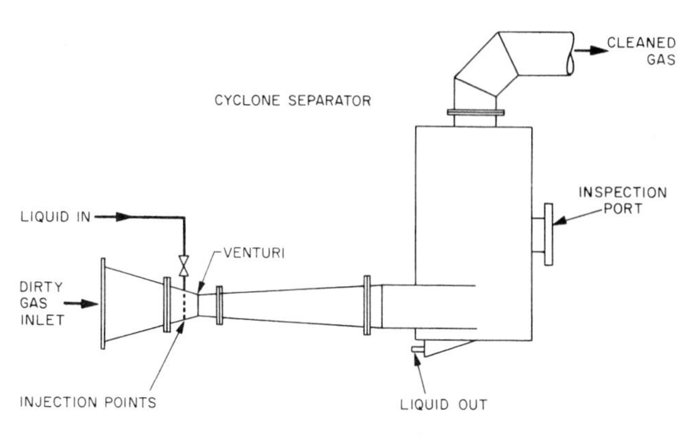


**Fig. 20 Venturi Scrubber**

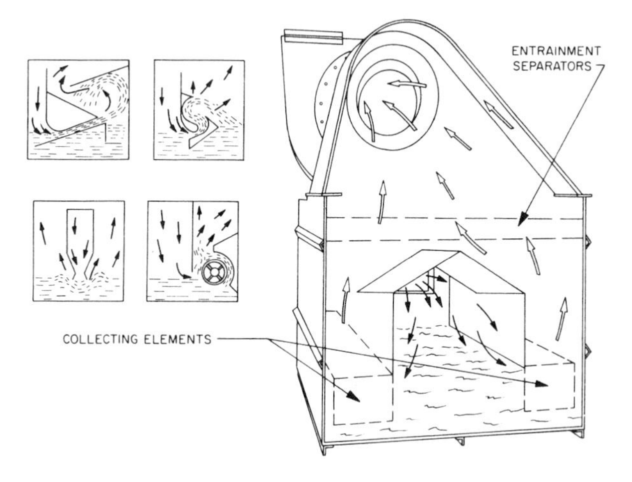


**Fig. 19 Typical Wet-Orifice-Type Collector**

to the process or tailing pond, liquid clarification and treatment is needed to permit recycling of the water and prevents stream pollution. Stringent stream pollution regulations make even a small stream of bleed water a problem. Clarification equipment may include settling tanks, sludge-handling facilities, and possibly centrifuges or vacuum filters. Provisions must be provided for handling and disposal of dewatered sludge so that secondary pollution problems do not develop.

## GASEOUS CONTAMINANT REMOVAL

### Wet-Packed Gas Scrubbers

Packed scrubbers are used to remove objectionable constituents from a gas stream. To effect this removal, the scrubbers are irrigated with some liquid (usually water). Packed scrubbers present a large wetted surface over which gas must flow. Scrubbing is accomplished by (1) impingement of particulate matter contained in the gas on the wetted surface of the packing, and/or (2) absorption of soluble gas or vapor molecules mixed with the air by contact with the liquid wetted surface. Removal of a solid or liquid particulate involves physical capture by wetting after impingement on the liquid surface. There is no limit to the amount of particulate capture so long as the properties of the liquid film are unchanged.

Gas or vapor removal is more complex, because the constituent removed from the air stream is dissolved in the liquid, which changes its properties. A solution of gas or vapor in the scrubbing liquid possesses a vapor pressure of the solute above the liquid. This vapor pressure typically increases with increasing concentration of solute in the liquid and/or with increasing liquid temperature. Scrubbing of the undesirable constituent from the air continues as long as the partial pressure of that constituent in the gas exceeds the vapor pressure of that same constituent above the liquid. The rate of contaminant removal is a function of the difference between the partial pressure and vapor pressure, as well as the rate of diffusion of the contaminant.

The efficiency of a packed scrubber is expressed as the fraction of the particulate or gas contaminant, which is removed per unit of packed depth or by a given device of fixed-bed depth. This efficiency usually is reported as a percentage of a particulate removed by weight or of a gas removed by volume. Packings are designed to present a large surface area that will wet evenly with liquid. Further, they should have a high void fraction so that pressure drop will be low.

High efficiency packings promote turbulent mixing of the gas and liquid. In most common fume-scrubbing operations, increasing the gas rate only slightly decreases the removal efficiency; therefore, these devices are operated at high gas velocities. Usually, increasing the liquid rate has little effect on efficiency; therefore, liquid flow is kept near the minimum required for satisfactory operation. As the gas rate is increased above a certain value, there is a tendency to strip liquid from the surface of the packing and entrain the liquid out of the scrubber with the exit gas. If pressure drop is not limiting, maximum scrubber capacity will occur at a gas rate just below the rate that causes excessive liquid entrainment.

Most scrubbers are operated on the suction side of the fan. This not only eliminates the leakage of contaminants into the atmosphere, but also allows for servicing the unit while it is in operation. Additionally, such an arrangement minimizes corrosion of the fan. Because the exit gas from a wet-packed scrubber is saturated with water vapor, the discharge stack should be arranged to drain condensate rather than allow it to reenter the fan.

### Scrubber Packings

Figure 21 illustrates six types of randomly dumped packings. Packings are available in ceramic, metal, and thermoplastic materials. Plastic packings have found extensive use in scrubbers because of their light weight and resistance to mechanical

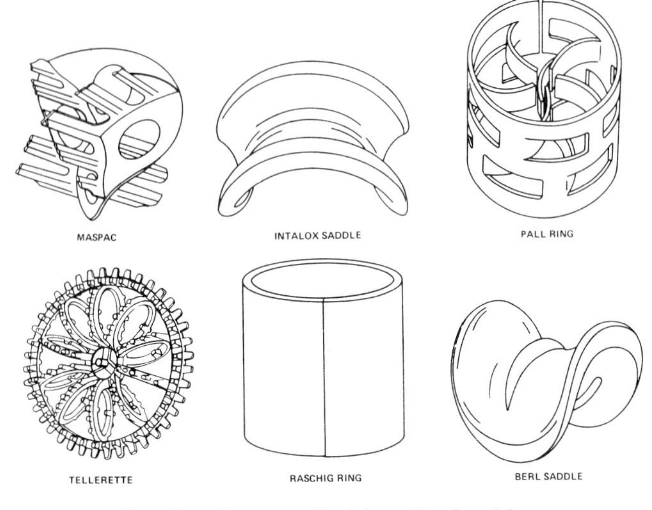


**Fig. 21 Common Packings for Scrubbers**

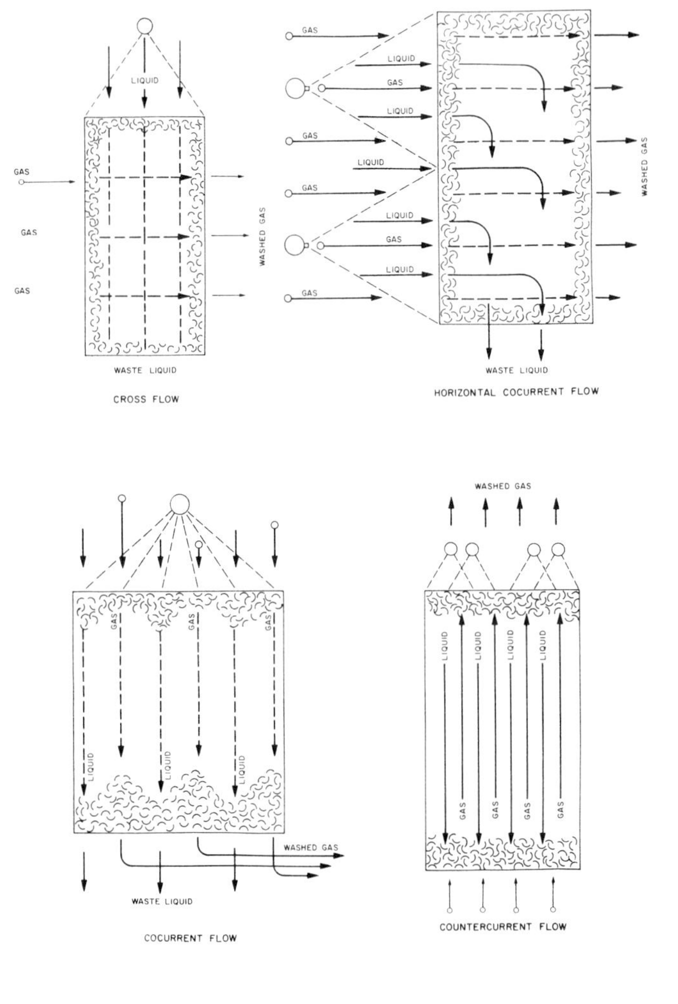


**Fig. 22 Flow Arrangement through Packed Beds**

damage. They offer a wide range of chemical resistance to acids, alkalies, and many organic compounds; however, plastic packing can be deformed by excessive temperatures or by solvent attack.

The relative capacity of tower packings at constant pressure drop can be obtained by calculation from the packing factor shown in Table 5. The smaller the packing factor of a given packing, the greater is its gas handling capacity. The gas-handling capacity of a packing is inversely proportional to the square root of the packing factor, as illustrated in Eq. (7).

$$G \propto 1/\sqrt{F} \quad (7)$$

## Arrangements of Packed Scrubbers

Wet-packed scrubbers operate in four general arrangements (Figure 22) classified by the way the liquid contacts with the contaminated gas stream.

**Cocurrent-flow** scrubbers can be operated horizontally or vertically. A horizontal cocurrent scrubber depends on the gas velocity to carry the liquid into the packed bed. Thus, this device operates as a wetted entrainment separator with limited gas and liquid contact time. Gas velocity is limited to about 650 fpm (3.3 m/s) by liquid reentrainment. A vertical cocurrent scrubber can be operated at very high pressure drop (1 to 3 in. of water/ft of depth [800 to 2400 Pa/m of depth]), since there is no flooding limit for the gas velocity. Contact time is a function of bed depth. Absorption driving forces are reduced, because the liquid-containing contaminate is in contact with the exit gas stream.

**Cross-flow** scrubbers use vertically downflowing liquid and a horizontally moving gas stream. The cross-sectional area for gas flow is thus different from the liquid-irrigated area. This device has absorption driving forces intermediate between vertical cocurrent scrubbers and countercurrent scrubbers.

**Countercurrent** scrubbers move the liquid flows vertically downward while the gas stream moves vertically upward. The gas-handling capacity of this device is limited by pressure drop or liquid entrainment. Contact time can be controlled by the depth of packing used. Absorption driving forces are at a maximum, since the exit gas is in contact with fresh scrubbing liquid.

## Other Packed Scrubbers

For removal of gaseous contaminants by absorption, the countercurrent packed scrubber (Figure 23) is most commonly used. Because driving forces are high, the exit gas is reduced to the lowest contaminant concentration. Further, liquid consumption is kept to a minimum, since the effluent liquid can have the highest contaminant concentration.

Extended surface packings have been used successfully for the absorption of highly soluble gases such as HCl, since the required contact time is minimal. This type of packing consists of a woven mat of plastic fibers. Because of the small diameter of the fibers, they must be formed from a plastic material that is not affected by chemical exposure. Figure 24 illustrates such a scrubber consisting of three wetted stages in series. A final dry mat is used as an entrainment eliminator. A wetted impingement stage precedes the wetted mats if solids are present in the inlet gas stream to prevent plugging of the woven mats.

Figure 25 shows a vertical arrangement of an extended surface packed scrubber. This design uses three complete stages in series. The horizontal mat at the bottom of each stage operates as a flooded-bed scrubber. The two inclined upper mats operate as entrainment eliminators. Because the scrubbing bed operates in a flooded mode, water consumption is minimized.

## Pressure Drop

The pressure drop through a particular packing in countercurrent scrubbers can be calculated from the airflow and water flow per unit area. Charts, such as Figure 26, are available from the manufacturer for each type and size of packing. However,

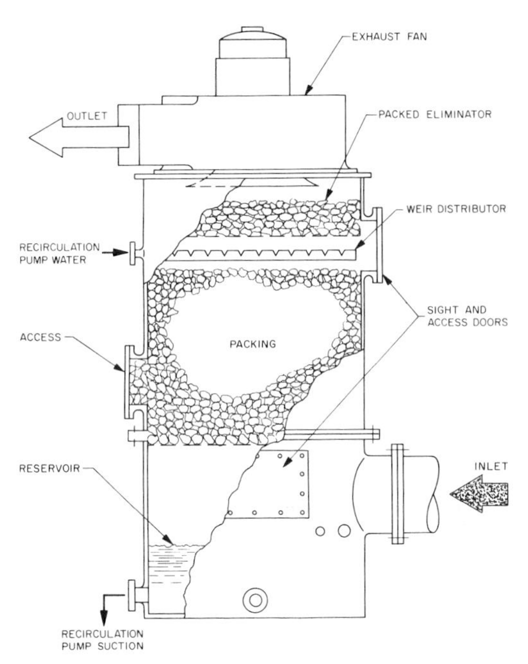


**Fig. 23 Countercurrent Packed Tower**

the pressure drop for any packing can be easily determined by using the packing factor from Table 5 and the modified generalized pressure-drop correlation shown in Figure 27. This correlation has been developed specifically for a gas stream substantially of air, with water as the scrubbing liquid. This correlation should not be used if the properties of the gas or liquid vary significantly from air or water, respectively. Countercurrent scrubbers are generally designed to operate at pressure drops between 0.25 and 0.65 in. of water/ft of depth (200 and 530 Pa/m of depth). Liquid irrigation rates typically vary between 5 and 20 gpm/ft$^2$ (3.4 and 13.6 L/s • m$^2$) of bed area.

## Absorption Efficiency

Calculating the efficiency of a packed bed is much more complex than determining capacity because it involves the mechanics of absorption. Some of the factors affecting efficiency are gas rate, liquid rate, packing size, type of packing, gas concentration, liquid concentration, temperature, and reaction rate.

Practically all commercial packings have been tested for absorption rate (mass transfer coefficient) using a standard of $CO_2$ in air and a solution of NaOH in water. This system was selected because the interaction of the variables was known through extensive study. Further, the mass-transfer coefficients for this system are low so they can be determined accurately by experiment. The values of mass transfer coefficients ($K_Ga$) for various packings under standard test conditions with this system are given in Table 6. Over 95% of gas pollution control problems involve low concentrations of contaminants in air (less than 0.005 mol fraction). The vast majority of scrubbers use water or dilute caustic solution as the scrubbing liquid. The concentration of contaminant in the effluent liquid is less than 0.005 mol fraction; thus, a very small vapor pressure of contaminant exists above this liquid. Because the gas and liquid flow at a constant rate through the scrubber and there is negligible vapor pressure of contaminant above the liquid, it is possible to simplify the design approach.

Mass transfer from the gas to the liquid is explained by the *Two Film Theory.* First, the gaseous contaminant travels by diffusion from the main gas stream through the gas film, then through the liquid film and finally into the main liquid stream. Whether the gas film or the liquid film limits (controls) the absorption rate depends on the solubility of the contaminant in

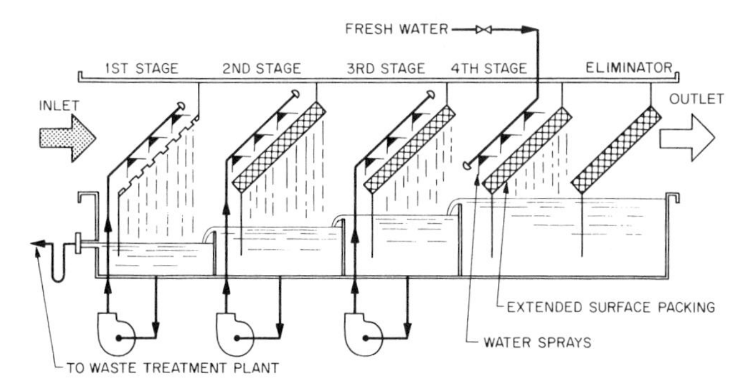


**Fig. 24 Horizontal Multistage Extended Surface Scrubber**

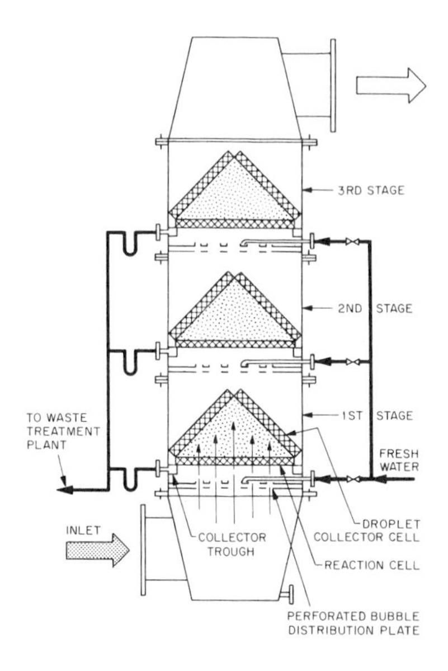


**Fig. 25 Vertical Extended Surface Scrubber**

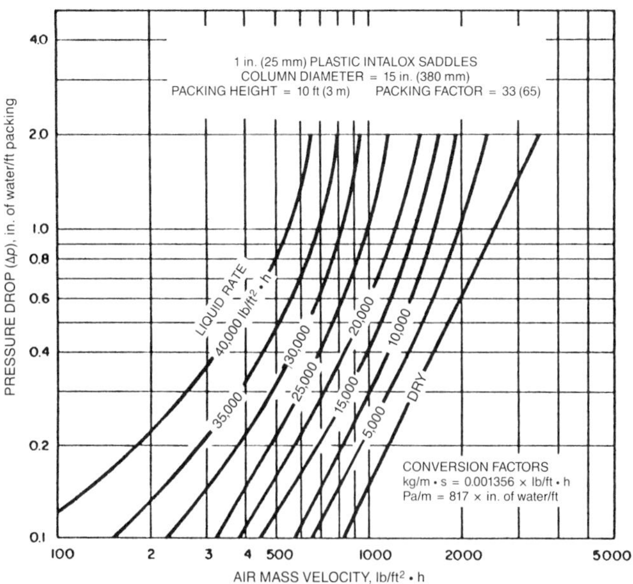


**Fig. 26 Pressure Drop versus Gas Rate for a Typical Packing**

**Table 5 Packing Factors, $F$ (Dumped Packings)**

| Type of Packing | Material | Nominal Size | | | | |
|---|---|---|---|---|---|---|
| | inch / mm | 3/4 / 20 | 1 / 25 | 1.5 / 40 | 2 / 50 | 3 or 3.5 / 80 or 90 |
| Super Intalox | Plastic | | 40 | | 21 | 16 |
| Super Intalox | Ceramic | | 60 | | 30 | |
| Intalox Saddles | Ceramic | 145 | 92 | 52 | 40 | 22 |
| Berl Saddles | Ceramic | 170 | 110 | 65 | 45 | |
| Raschig Rings | Ceramic | 255 | 155 | 95 | 65 | 37 |
| Hy-Pak | Metal | | 43 | 26 | 18 | 15 |
| Pall Rings | Metal | | 48 | 33 | 20 | 16 |
| Pall Rings | Plastic | | 52 | 40 | 24 | 16 |
| Tellerettes | Plastic | | 36 | | 18 | 16 |
| Maspac | Plastic | | | | 32 | 21 |

the liquid. Sparingly soluble gases, like $H_2S$ and $CO_2$, are said to be liquid-film controlled. Highly soluble gases, such as HCl and $NH_3$, are called gas-film controlled. In liquid-film controlled systems, the mass transfer coefficient varies with the liquid rate, but is only slightly affected by the gas rate. In gas-film controlled systems, the mass transfer coefficient is a function of gas and liquid rates.

The percent removal of the contaminant (by volume) from the air is a function of the inlet and outlet concentration of contaminant in the air stream, where $Y$ is the mol fraction of contaminant, as:

$$\% \text{ removed} = 100\,[1 - (Y_{out}/Y_{in})] \tag{8}$$

The driving force for absorption (assuming negligible vapor pressure above the liquid) is the logarithmic mean average of inlet and outlet concentrations of the contaminant, where $P_{ln}$ is the pressure in atm (kPa), as:

$$\Delta P_{ln} = P\left[\frac{Y_{in} - Y_{out}}{\ln(Y_{in}/Y_{out})}\right] \tag{9}$$

The rate of absorption of contaminant (mass-transfer coefficient) is related to the depth of packing as:

$$K_G a = N/HA\Delta P_{ln} \tag{10}$$

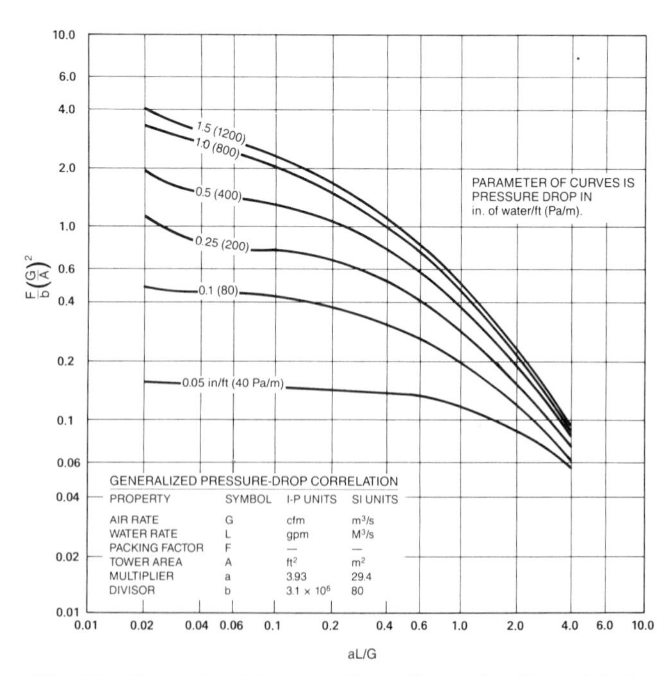


**Fig. 27 Generalized Pressure-Drop Curves for Packed Beds Operating with Air and Water Near 75°F (24°C)**

**Table 6 (I-P) Relative $K_G a$ for Various Packings lb • mol/hr • ft³ • atm***

| Type of Packing | Material | Nominal Size, inches | | | |
|---|---|---|---|---|---|
| | | 1 | 1.5 | 2 | 3 or 3.5 |
| Super Intalox | Plastic | 2.19 | | 1.44 | 0.887 |
| Intalox Saddles | Ceramic | 1.96 | 1.71 | 1.44 | 0.820 |
| Raschig Rings | Ceramic | 1.73 | 1.50 | 1.21 | |
| Hy-Pak | Metal | 2.20 | 1.87 | 1.69 | 1.09 |
| Pall Rings | Metal | 2.32 | 1.87 | 1.62 | 0.91 |
| Pall Rings | Plastic | 1.98 | 1.73 | 1.46 | 0.89 |
| Tellerettes | Plastic | 2.19 | | 1.98 | |
| Maspac | Plastic | | | 1.44 | 0.89 |

*System: $CO_2$ and NaOH; Liquid rate: 4 gpm/ft²; Gas rate: 110 cfm/ft²

**Table 6 (SI) Relative $K_G a$ for Various Packings g • mol/s • m³ • kPa***

| Type of Packing | Material | Nominal Size, mm | | | |
|---|---|---|---|---|---|
| | | 25 | 40 | 50 | 80 or 90 |
| Super Intalox | Plastic | 0.096 | | 0.063 | 0.039 |
| Intalox Saddles | Ceramic | 0.086 | 0.075 | 0.063 | 0.036 |
| Raschig Rings | Ceramic | 0.076 | 0.066 | 0.053 | |
| Hy-Pak | Metal | 0.097 | 0.082 | 0.074 | 0.048 |
| Pall Rings | Metal | 0.102 | 0.082 | 0.071 | 0.040 |
| Pall Rings | Plastic | 0.087 | 0.076 | 0.064 | 0.039 |
| Tellerettes | Plastic | 0.096 | | 0.087 | |
| Maspac | Plastic | | | 0.063 | 0.039 |

*System: $CO_2$ and NaOH
Liquid rate: 9780 kg/m² • h; Gas rate: 2445 kg/m² • h
22.7 lb • mol/hr • ft³ • atm = g • mol/s • m³ • kPa

*where*

$H$ = packed depth, ft (m)
$A$ = scrubber cross-sectional area, ft² (m²)
$N$ = solute absorbed, lb • mol/h (g • mol/s)

The value of $N$ can be determined from Eq. (11), where $G$ is the airflow rate in lb • mol/h (g • mol/s).

$$N = G\,(Y_{in} - Y_{out}) \tag{11}$$

The air velocity is a function of the unit airflow rate and the gas density, as:

$$V = T\,G\,M_v\,/\,C_1\,A \tag{12}$$

*where*

$V$ = air velocity, fpm (m/s)
$M_v$ = molar volume, ft³/lb mol (m³/g • mol)
$T$ = exit gas temperature, °R (K)
$C_1$ = 492 (273)

By combining these equations and assuming atmospheric operating pressure, a graphical solution can be derived for both liquid-film controlled and gas-film controlled systems. Figures 28, 29, and 30 show the height of packing required versus percent removal for various mass transfer coefficients at air velocities of 120, 240, and 360 fpm (0.61, 1.22, and 1.83 m/s), respectively, with liquid-film controlled systems.

Figures 31, 32, and 33 show the height of packing versus percent removal for various mass transfer coefficients at the same air velocities with gas-film controlled systems. These graphs can be used to determine the height of 2.0-in. (50-mm) plastic Intalox Saddles required to give the desired percentage contaminant removal required. The height for any other type or size of packing is inversely proportional to the ratio of standard $K_G a$ taken from Table 6. Thus, if a 13.0-ft (4.0-m) depth were required

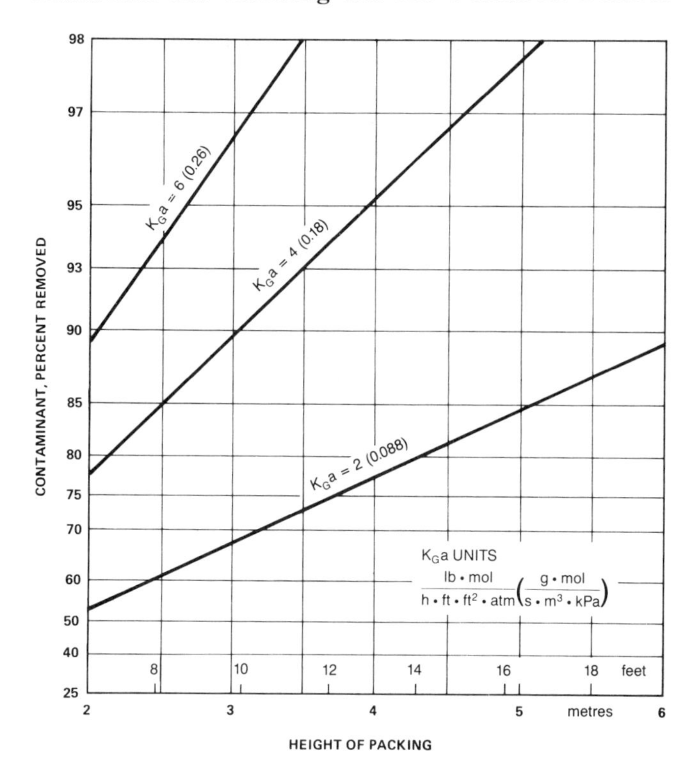


Fig. 28 Height of Packing versus Percent of Contaminant Removed for Liquid-Film-Controlled Systems 120 fpm (0.61 m/s) Air Velocity

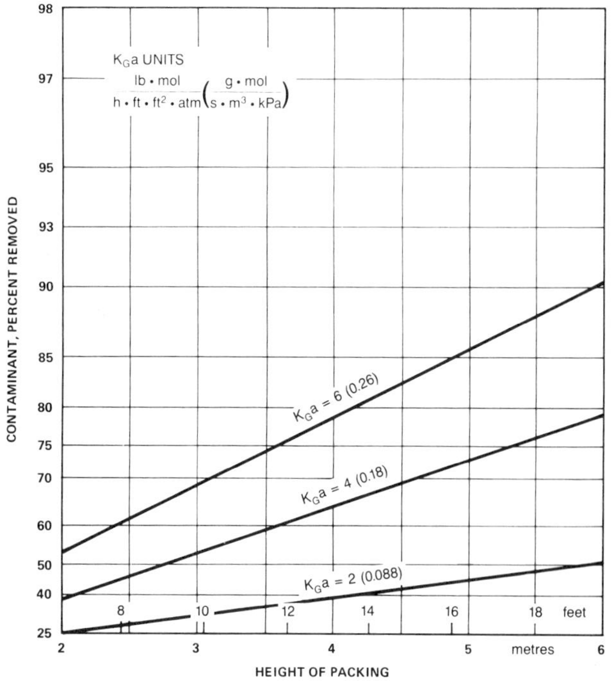

Fig. 30 Height of Packing versus Percent of Contaminant Removed for Liquid-Film-Controlled Systems 360 fpm (1.83 m/s) Air Velocity

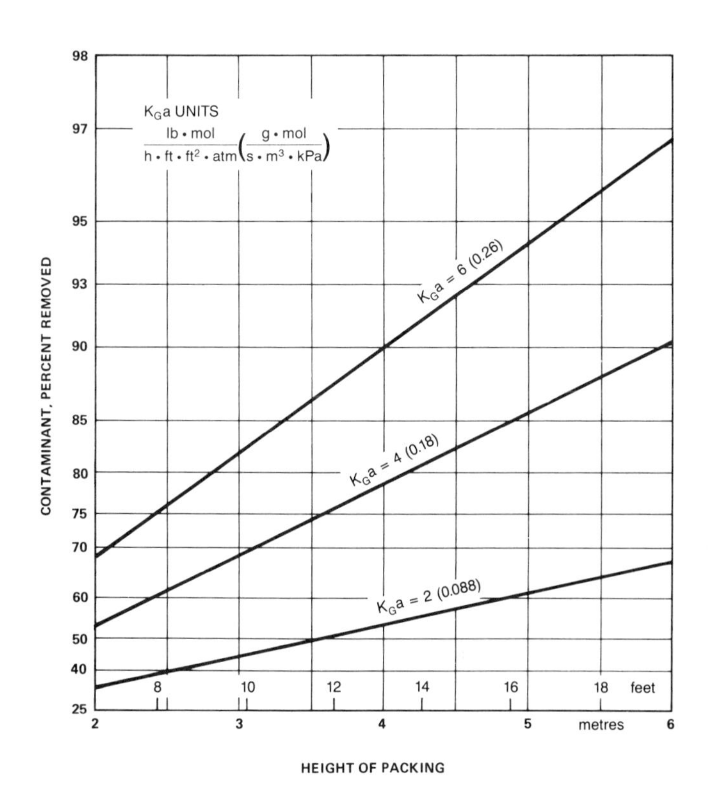


Fig. 29 Height of Packing versus Percent of Contaminant Removed for Liquid-Film-Controlled Systems 240 fpm (1.22 m/s) Air Velocity

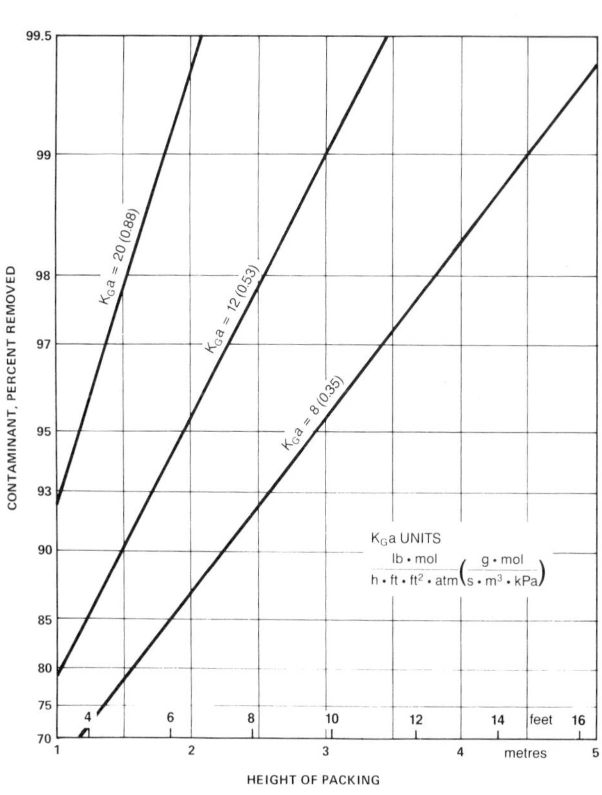


Fig. 31 Height of Packing versus Percent of Contaminant Removed for Gas-Film-Controlled Systems 120 fpm (0.61 m/s) Air Velocity

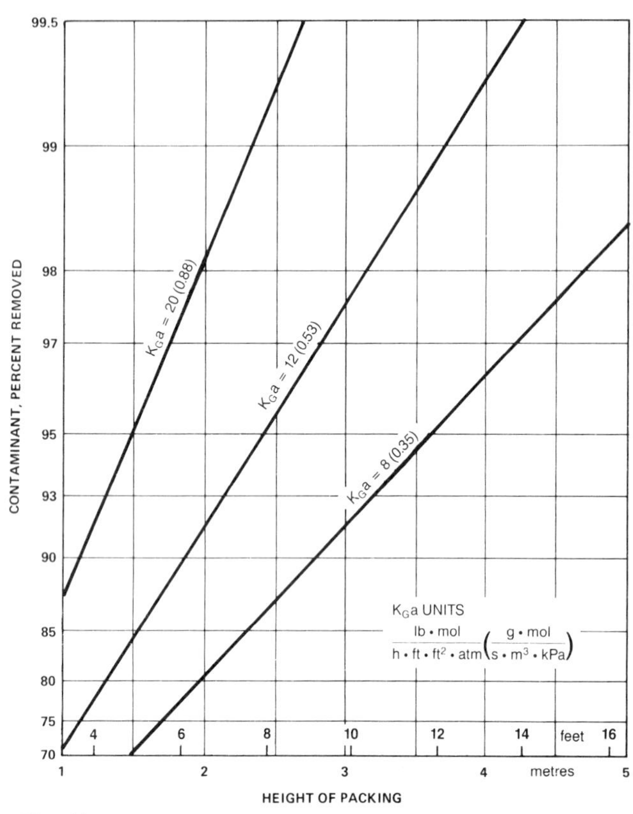


**Fig. 32 Height of Packing versus Percent of Contaminant Removed for Gas-Film-Controlled Systems 240 fpm (1.22 m/s) Air Velocity**

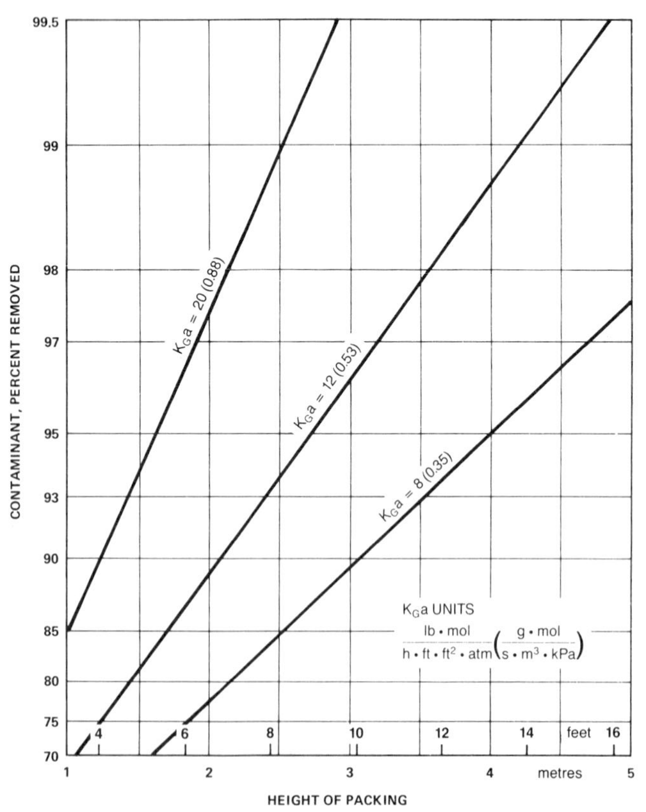


**Fig. 33 Height of Packing versus Percent of Contaminant Removed for Gas-Film-Controlled Systems 360 fpm (1.83 m/s) Air Velocity**

**Table 7 Relative $K_Ga$ for Various Packings***

| Gas Contaminant | Scrubbing Liquid | $K_Ga$, lb • mol / h • ft³ • atm | $K_Ga$, g • mol / s • m³ • kPa |
|---|---|---|---|
| $CO_2$ | 4 wt % NaOH | 2.0 | 0.088 |
| $H_2S$ | 4 wt % NaOH | 5.92 | 0.26 |
| $SO_2$ | Water | 2.96 | 0.13 |
| HCN | Water | 5.92 | 0.26 |
| HCHO | Water | 5.92 | 0.26 |
| $Cl_2$ | Water | 4.55 | 0.20 |

*Liquid-film controlled systems, 2-in. (50-mm) plastic Super Intalox. Temperatures: from 60 to 75°F (16 to 24°C); liquid rate: 10 gpm/ft² (6.8 L/s • m²); gas rate: 215 cfm/ft² (1.1 m/s).

**Table 8 Relative $K_Ga$ for Various Packings***

| Gas Contaminant | Scrubbing Liquid | $K_Ga$, lb • mol / h • ft³ • atm | $K_Ga$, g • mol / s • m³ • kPa |
|---|---|---|---|
| HCl | Water | 18.66 | 0.82 |
| HBr | Water | 5.92 | 0.26 |
| HF | Water | 7.96 | 0.35 |
| $NH_3$ | Water | 17.30 | 0.76 |
| $Cl_2$ | 8 wt % NaOH | 14.33 | 0.63 |
| $SO_2$ | 11 wt % $Na_2CO_3$ | 11.83 | 0.52 |
| $Br_2$ | 5 wt % NaOH | 5.01 | 0.22 |

*Gas-film controlled systems, 2-in. (50-mm) plastic Super Intalox. Temperatures: from 60 to 75°F (16 to 24°C); liquid rate: 10 gpm/ft² (6.8 L/s • m²); gas rate: 215 cfm/ft² (1.1 m/s).

for 95% removal of contaminants, the same removal could be obtained with a 9.5-ft (2.9-m) depth of 1-in. (25-mm) plastic Pall Rings at the same gas and liquid rates. However, the pressure drop would be higher for the smaller diameter packing.

To use these charts, the value of the masstransfer coefficient typical for the particular contaminant to be removed must be known. Table 7 gives such values for 2.0-in. (50-mm) plastic Intalox Saddles in typical liquid-film controlled systems. Likewise, Table 8 gives mass transfer coefficients for 2.0-in. (50-mm) plastic Intalox Saddles in common gas-film controlled systems.

In Tables 7 and 8, where the scrubbing liquid is not water, the mass transfer coefficient listed applies only when at least 33% excess reagent is used over that required to combine with the gaseous contaminant. When HCl is dissolved in water, there is little vapor pressure of HCl above solutions of less than 8% (by weight) concentration. When $NH_3$ is dissolved in water, however, there is an appreciable vapor pressure of $NH_3$ above solutions even of low concentrations. The packed height for $NH_3$ removal obtained from Figures 31, 32, and 33 is based on the use of dilute acid to maintain the pH of the solution below 7.

## Typical Scrubbing Problem

The problem is to remove 600 ppm (0.0006 mol fraction) of HF from the air at 90°F. The exit gas concentration of HF should not exceed 30 ppm (0.00003 mol fraction); or from Eq. (8), the percent HF removed = 100 [1 − (0.00003/0.0006)] = 95%.

The design factors are as follows:

2 in. polypropylene Intalox saddles
F = 21, from Table 5.
$K_Ga$ = 7.96, from Table 8.
G = Total Gas flow = 4600 cfm
Water flow = 3.75 gpm/ft²

Water temperature = 68°F
Packed tower diameter = 4 ft or
A = Area = $\pi(4/2)^2$ = 12.57 $ft^2$
L = Total water flow rate = 3.75 • 12.57 = 47.1 gpm

Figure 27 may be used to calculate the pressure drop through the packed tower. The abscissa and ordinate values for Figure 27 are as follows:

$$\text{x axis (abscissa)} = 3.93\ (47.1/4600) = 0.40$$

$$\text{y axis (ordinate)} = \frac{21}{3.1 \times 10^6}\left(\frac{4600}{12.57}\right)^2 = 0.90$$

From Figure 27, the pressure drop through the packed tower is approximately 0.28 in. of water/ft.

From Figure 33, the depth of packing required for 95% removal is 13 ft. Thus, the total pressure drop is (13)(0.28) = 3.64 in. of water.

### General Efficiency Comparisons

Figure 33 shows that 90% removal efficiency in the preceding example requires only 10 ft (3.1 m) of packed depth or 23% less than 95% removal efficiency. At the same gas velocity, both liquid-film and gas-film controlled systems require a bed-depth increase of 43% to increase the removal efficiency from 80% to 90%. In a gas-film controlled system, increasing the gas rate by 50% requires only a 12% increase in bed depth to maintain equal removal efficiency. In a liquid-film controlled system, however, increasing the gas rate by 50% requires a 50% increase in bed depth to maintain equal removal efficiency.

Thus, in a gas-film controlled system, the gas rate can be increased significantly with only a small increase in bed depth required, since the mass transfer coefficient increases with increasing gas velocity. In such a system, a scrubber of fixed depth can handle an overload condition with only a minor loss of removal efficiency. In liquid-film controlled systems, the depth of bed is a direct function of gas velocity because the mass transfer coefficient does not change appreciably with gas velocity. In this system, an overload on the scrubber decreases removal efficiency significantly.

### Liquid Effects

Some liquids tend to foam when they are contaminated with particulates or soluble salts, so the pressure drop should be kept in the lower half of the normal range—0.25 to 0.40 in. of water/ft (200 to 330 Pa/m) of packed height.

When water is used as the scrubbing liquid, the effluent from the scrubber will contain suspended particulate or dissolved solute. These contaminants may alter the pH or the suspended solids such that water treatment is required before discharge of scrubber effluent from the plant. Most scrubbers that use chemical solutions for scrubbing the air must treat these spent solutions to neutralize them and/or remove toxic substances before the solutions can be discharged. In the control of gaseous pollution, most systems do not destroy the pollutant but merely remove it from the air.

## INCINERATION OF GASES AND VAPORS

Incineration is the process which converts volatile organic compounds (VOCs), organic aerosols, and most odorous materials in a contaminated airstream to innocuous carbon dioxide and water vapor using heat energy. Incineration is the most effective means for totally eliminating VOCs (EPA 1976). The types of incineration commonly used are *thermal* and *catalytic*.

Thermal incinerators, often called afterburners or direct flame incinerators, consist of an insulated oxidation chamber in which gas and/or oil burners are typically located. The contaminated airstream enters the chamber and comes into direct contact with the flame, which provides the heat energy necessary to promote oxidation. Proper conditions of time, temperature, and turbulence oxidize the airstream contaminants effectively. The contaminated airstream enters the incinerator near the burner where turbulence-inducing devices are usually installed. The final contaminant conversion efficiency largely depends on good mixing within the contaminated airstream and the temperature of the oxidation chamber.

Supplemental fuel is used for start-up to raise the temperature of the contaminated airstream enough to initiate contaminant oxidation. Once oxidation begins, the temperature rises further due to the energy released by the contaminant. The supplementary fuel feed rate is then modulated to maintain the desired incinerator operating temperature. Most organic gases oxidize to approximatley 90% conversion efficiency if a temperature of at least 1200°F (650°C) and a residence time of 0.3 to 0.5 seconds is achieved within the oxidation chamber. However, incinerator temperatures are typically maintained in the range of 1400 to 1500° F (760 to 820°C) to ensure conversion efficiencies of 90% or greater.

In the past, thermal incineration was not recommended due to the high operating cost associated with the supplementary fuel requirement for maintaining the high operating temperatures. However, with the advent of widespread air pollution control requirements for VOCs, reliable heat-recovery methods for use with thermal incinerators were developed. Incineration systems now incorporate primary heat recovery to preheat the incoming contaminated airstream and, in some cases, secondary heat recovery for process heat or building heat. Primary heat recovery is almost always achieved using air-to-air heat exchangers. Use of a regenerable, ceramic media for heat recovery has increased due to superior heat recovery efficiency. Secondary heat recovery may incorporate an air-to-air heat exchanger or a waste heat boiler (DOE 1979).

Incineration systems using conventional air-to-air heat exchangers can achieve 80% primary efficiency. Regenerative heat exchanger units have claimed as high as 95% heat recovery, and are routinely operated with 85 to 90% heat recovery efficiency. When operated at these high heat-recovery efficiencies and with inlet VOC concentrations of 15 to 25% of the lower explosive limit (LEL), the incineration process approaches a self-sustaining condition, thereby requiring almost no support fuel.

Catalytic incinerators operate under the same principle as thermal incinerators, except that they use a catalyst to promote oxidation. The catalyst allows oxidation to occur at lower temperatures than a thermal incinerator for the same VOC concentration. Therefore, catalytic incinerators require less supplemental fuel to preheat the contaminated airstream and have lower overall operating temperatures.

A catalytic incinerator generally consists of a preheat chamber followed by the catalyst bed. Residence time and turbulence are not as important as with thermal incinerators, but it is essential that the contaminated airstream be heated uniformly to the required catalytic reaction temperature. The required temperature varies, depending upon the catalyst material and configuration.

The contaminated gas stream temperature is raised in the preheat chamber by a conventional burner. Although the contaminated airstream contacts the burner flame, the heat input is significantly less than that for a thermal incinerator and only a small degree of direct contaminant oxidation occurs. Natural gas is preferred to prevent catalyst contamination, which could occur with sulfur-bearing fuel oils. However, No. 2 fuel oil units have been operated successfully. The most effective catalysts contain the noble metals such as platinum or palladium.

Catalysis occurs at the molecular level. Therefore, an available, active catalyst surface area is an important factor for maintaining high conversion efficiencies. If particulate materials contact the catalyst either as discrete or partially oxidized aerosols, they can ash on the catalyst surface. Then the catalyst surface eventually becomes coated with the ash, which reduces the available, active catalyst surface and results in significantly reduced oxidation efficiency. This problem is usually evidenced by a secondary pollution problem, i.e., odorous emissions caused by the partially oxidized organic compounds.

The greatest concern to catalytic incinerator operation is catalyst poisoning or deactivation. Poisoning is caused by specific airstream contaminants that chemically combine or alloy with the active catalyst material. The list of poisons frequently cited includes phosphorous, bismuth, arsenic, antimony, lead, tin, and zinc. Some organic compounds such as polyester amides and imides have also proven to be poisons. The first five materials are considered to be fast-acting poisons and must be excluded from the contaminated airstream. Even trace quantities of the fast-acting poisons can cause rapid catalyst deactivation. The last two materials are slow-acting poisons; catalysts are somewhat tolerant of these materials, particulary at temperatures less than 1000 °F (540 °C). However, even the slow poisons should be excluded from the contaminated airstream to ensure continuous, reliable performance. Therefore, galvanized steel, another possible source of the slow poisons, should not be used for the ductwork leading to the incinerator.

Sulfur and halogens are also regarded as catalyst poisons. In most cases, their chemical interaction with the active catalyst material is reversible. That is, catalyst activity can be restored by operating the catalyst without the halogen or sulfur-bearing compound in the airstream. The potential problem of greater concern with respect to the halogen-bearing compounds is the formation of hydrogen chloride or fluoride gas, or hydrochloric or hydrofluoric acid emissions.

Catalytic incinerators generally cost less to operate than thermal incinerators because of the lower fuel consumption. With the exception of regenerative heat-recovery techniques, primary and secondary heat recovery can also be incorporated into a catalytic incineration system to reduce operating costs further. Maintenance costs are usually higher for catalytic units, particularly if frequent catalyst cleaning or replacement is necessary. The concern regarding catalyst life has been the major factor limiting more widespread application of catalytic incinerators.

Exhaust gases are incinerated for the following reasons:

**Odor Control.** All highly odorous pollutant gases are combustible or are chemically changed to less odorous pollutants when they are sufficiently heated. Often, the concentration of odorous materials in the waste gas is extremely low and the only feasible method of control is by incineration. Odors from rendering plants and mercaptans and organic sulfides from kraft pulping operations are examples of effluents that can be so controlled. Other forms of oxidation such as chlorination or ozonation can achieve the same ends (see Chapter 50 of the 1987 HVAC Volume).

**Reduction in Opacity of Plumes.** Flame afterburning is often used to destroy organic aerosols that cause visible plumes. Examples are coffee roasters, smoke houses, and enamel baking ovens. Such burners can also be used for heating wet stack gases, which might otherwise show a steam plume.

**Reduction in Emissions of Reactive Hydrocarbons.** Some air pollution control agencies regulate the emission of organic gases and vapors because of their involvement in photochemical smog reactions. Flame afterburning is one effective way of destroying these materials.

**Reduction in Explosion Hazard.** Refineries and chemical plants are among the industries that must dispose of highly combustible or otherwise dangerous organic materials. The safest method of disposal is usually by burning in flares or specially designed furnaces. However, special precautions and equipment design must be used in the handling of potentially explosive mixtures.

## ADSORPTION OF VAPOR EMISSIONS

All solids have weak attractive forces that extend outward from their surfaces and can attract nearby molecules in an adjacent gas or liquid. The captured molecules form a thin layer, one to a few molecules thick, on the surface of the solid. Commercial adsorbents are porous solids whose enormous internal surface area enables them to capture and hold large numbers of molecules. A typical activated carbon for solvent recovery contains about 5,000,000 $ft^2/lb$ (1000 $m^2/g$) of internal surface area. This property has led to the use of adsorbents for removing organic vapors, water vapor, odors, and hazardous pollutants from gas streams.

Adsorbents in large-scale use for processing gas streams include activated carbons, activated alumina, silica gel, and molecular sieves. Activated carbon derived from coal, petroleum, or coconut shells is the most widely used adsorbent for air pollution control applications because it adsorbs organic compounds in preference to water. The other three major gas phase adsorbents have a great affinity for water and will adsorb it to the exclusion of any organic molecules also present in a gas stream. They are used primarily as gas-drying agents. Molecular sieves, however, find use in several specialized pollution-control applications, including removal of mercury vapor, sulfur dioxide, or nitrogen oxides from gas streams.

The capacity of a particular activated carbon to adsorb any organic vapor from an exhaust airstream is related to the concentration and molecular weight of the organic compound and the temperature of the airstream. Higher molecular mass compounds are usually more strongly adsorbed than lower ones. The capacity of carbon to adsorb any given organic is greater at high concentrations of the organic than at low. Adsorption is favored by low temperatures and it decreases as the temperature is raised. These points are illustrated in Figure 34, which shows adsorption capacities of a solvent-recovery carbon for toluene (molecular mass 92) and acetone (molecular mass 58) at various temperatures and concentrations.

Adsorption is a reversible phenomenon. An increase in temperature causes some or all of an adsorbed vapor to desorb. The temperature of low-pressure steam is sufficient to drive off most of a low-boiling solvent previously adsorbed at ambient temperature. Higher boiling organics may require high-pressure steam or hot inert gas to secure good desorption. Very high molecular mass compounds can require reactivation of the carbon in a furnace at 1350 °F (750 °C) to drive off all of the adsorbed material. Regeneration of the carbon can also be accomplished, in some instances, by washing it with an aqueous solution of a chemical, which will react with the adsorbed organic and make it water soluble. An example is caustic soda washing of carbon containing adsorbed sulfur compounds.

The difference between an adsorbent's capacity under adsorbing and desorbing conditions in any application is called its *working capacity.* A major use of activated carbon for air pollution control is in canisters under the hoods of most automobiles, which capture gasoline vapors escaping from the carburetor

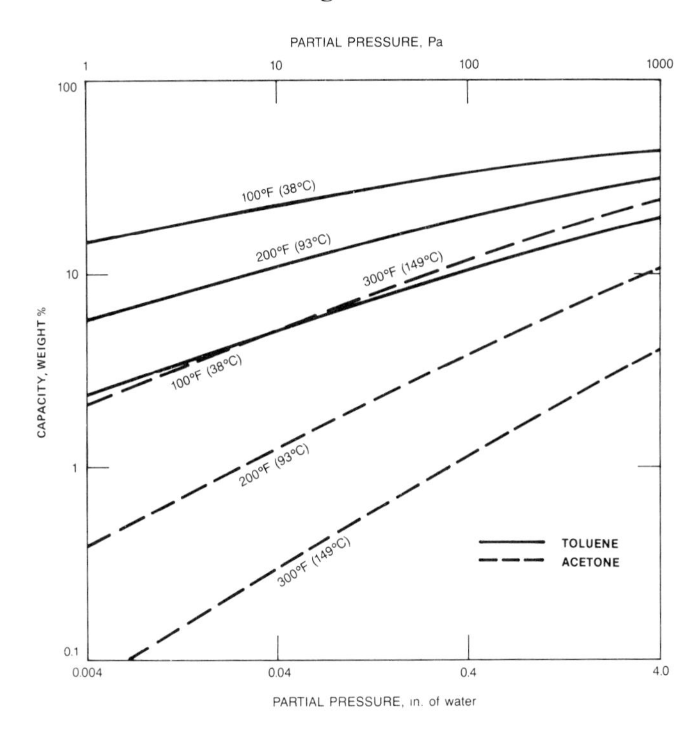


**Fig. 34 Adsorption Isotherms on Activated Carbon**

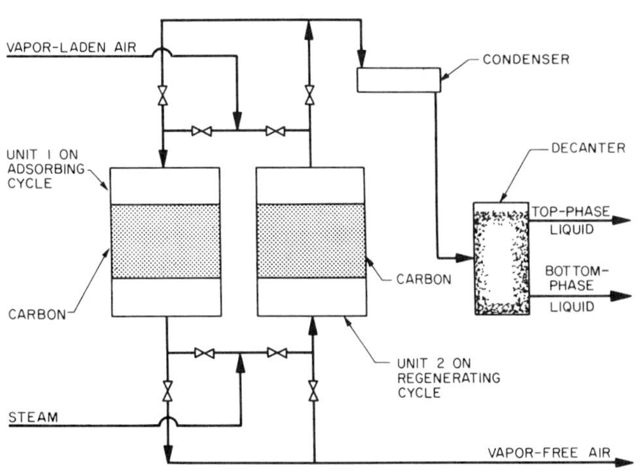


**Fig. 35 Schematic of Two-Unit Fixed-Bed Adsorber**

(when the engine is stopped) and from the fuel tank breather vent. Desorption of gasoline vapors is accomplished by pulling fresh air through the carbon canister and into the carburetor when the engine is running. Although there is no temperature differential between adsorbing and desorbing conditions in this case, the outside airflow desorbs enough gasoline vapor to give the carbon a substantial working capacity and the carbon remains effective for many years.

For applications where only traces of an odorous pollutant must be removed from exhaust air, the life of a carbon bed is often so long that it is more economical to discard the carbon when it is spent and replace it with new carbon, rather than to invest in the additional facilities required for regenerating the carbon.

## Solvent Recovery

The largest use of adsorption for controlling air pollution from stationary sources is in recovering solvent vapors emitted by various manufacturing and cleaning processes. Operations in which such recovery is practiced include solvent degreasing, rotogravure printing, dry cleaning, and the manufacture of products such as synthetic fibers, adhesive labels, tapes, coated copying paper, rubber goods, and coated fabrics.

Figure 35 shows a schematic of a typical solvent-recovery system using two carbon beds. One bed adsorbs while the other is regenerated with low-pressure steam. Desorbed solvent vapor and steam go to a water-cooled condenser. If the solvent is immiscible with water, an automatic decanter separates the solvent for re-use. A distillation column is used for water-miscible solvents.

Adsorption time per cycle typically runs from one-half to several hours. The adsorbing carbon bed is switched to regeneration by an automatic timer shortly before solvent vapor breakthrough from the bed, or immediately thereafter by an organic vapor-sensing control device in the exhaust airstream.

Low-pressure steam consumption for regeneration is generally about 3.5 lb per lb (3.5 kg/kg) of solvent recovered (Boll 1976), but can range from 2 to over 5 lb (2 to over 5 kg), depending on the specific solvent and its concentration in the exhaust airstream being stripped. Steam with only a slight superheat is normally used so that it will condense quickly and give rapid heat transfer.

After steaming, the hot moist carbon bed is usually cooled and partially dried before being put back on stream. Heat for drying is supplied by the cooling down of the carbon and adsorber and sometimes by an external air heater. In most cases, it is desirable to leave some moisture in the bed. When solvent vapors are adsorbed, there is an evolution of heat. For most common solvents, the heat of adsorption is 40 to 60 Btu/mol (kJ/mol). When high concentration vapors are adsorbed in a dry carbon bed, this heat can cause a substantial temperature rise and can even ignite the bed, unless it is controlled. If the bed contains moisture, the water consumes heat and helps to prevent an undue rise in bed temperature; also, certain applications may require heat sensors and automatic sprinklers.

As the adsorptive capacity of activated carbon depends on temperature, it is important that solvent-laden air going to a recovery unit be as cool as practicable. The exhaust from many solvent-emitting processes (such as drying ovens) is at elevated temperatures. A water- or air-cooled heat exchanger needs to be installed to reduce the gas temperature going to the adsorbers.

Very low solvent-vapor concentrations can be recovered in an activated carbon system. The size and cost of the recovery unit, however, depend on the volume of air to be handled, so it is important to minimize the volume of an exhaust stream and keep the solvent-vapor concentration as high as possible, consistent with safety requirements. Insurance carriers specify that solvent-vapor concentrations must not exceed 25% of the lower explosive limit when intermittent monitoring is used. With continuous monitoring, a concentration up to 50% of the LEL (Low Explosive Limit) is permissible.

Standard solvent recovery systems are available as a skid-mounted package with air-handling capacities up to about 11,000 cfm (5000 L/s). Multiples of these package units can be used for larger airflows. Custom-designed systems are built in all sizes, up to large ones handling 200,000 cfm (100 $m^3/s$) or more. Materials of construction may be painted carbon steel, stainless steel, Monel, or even titanium, depending on the nature of the gas mixture. The activated carbon is usually placed in horizontal flat beds or vertical cylindrical beds. The latter design minimizes ground space required for the system. Other alternatives are possible; one manufacturer uses a segmented horizontal rotating cylinder of carbon in which one segment is adsorbing while others are being steamed and cooled.

Recovery of over 99% of the solvent contained in the air processed by solvent-recovery systems is often obtained in commercial plants. The efficiency of the collecting hoods at the solvent-emitting sources is the most important factor in determining the percentage of total solvent emissions recovered.

Dust filters are generally placed ahead of the carbon beds to prevent a buildup of dust, which could blind the carbon bed. Occasionally, the carbon is removed for screening to eliminate accumulated dust and fine particles of carbon.

The working capacity of the activated carbon decreases with time if the solvent mixture contains high boiling components, which are only partially removed by low-pressure steam. In this situation, two alternatives should be evaluated:

1. Periodic removal of the carbon and return to its manufacturer for high-temperature furnace reactivation to near virgin carbon activity.
2. Use of more rigorous solvent-desorbing conditions in the solvent recovery system. Either high-temperature steam, hot inert gas, or a combination of electrical-heating elements and application of a vacuum may be used. The latter method is selected, for example, to recover high boiling lithography ink solvents.

### Adsorption and Incineration

Alternate cycles of adsorption and desorption in an activated carbon bed are used to concentrate solvent or odor vapors prior to incineration, thus greatly reducing the fuel required for burning organic vapor emissions. Fuel savings of 98% are possible versus direct incineration. The process is particularly useful in cases where emission levels vary from hour to hour.

Contaminated gas is passed through a carbon bed until saturation occurs. The gas stream is then switched to another carbon bed, and the exhausted bed is shut down for desorption. A hot inert gas, usually burner flue gas, is introduced to the adsorber to drive off concentrated organic vapors and convey them to an incinerator. The volume of this desorbing gas stream is much smaller than the original contaminated gas volume so that only a small incinerator, operating intermittently, is required (Grandjacques 1977).

### Odor Control

Where odorous compounds are emitted in high concentrations, such as from cookers in rendering plants, incineration or scrubbing are usually the most economical methods of controlling them. Many odorous emissions, however, contain low concentrations of vapors, which are still offensive. The odor threshold of acrolein in air, for example, is only 0.21 ppm (0.21 mg/kg) by volume for 100% response, while that of ethyl mercaptan is 0.001 ppm (0.001 mg/kg), and hydrogen sulfide is 0.0005 ppm (0.0005 mg/kg) (MCA 1968). Activated carbon beds effectively overcome many odor emission problems. Carbon applications for odor control include chemical and pharmaceutical manufacturing operations, foundries, sewage treating plants, oil and chemical storage tanks, lacquer drying ovens, food-processing plants, and rendering plants. In some of these applications, carbon is the sole odor-control method; in others, it is applied to the exhaust from a scrubber.

The hardware used for activated carbon odor control varies from simple steel drums fitted with appropriate gas inlet and outlet piping to large vertical moving bed installations in which carbon is contained between louvered side panels, as shown in Figure 36. In this arrangement, fresh carbon can be added from

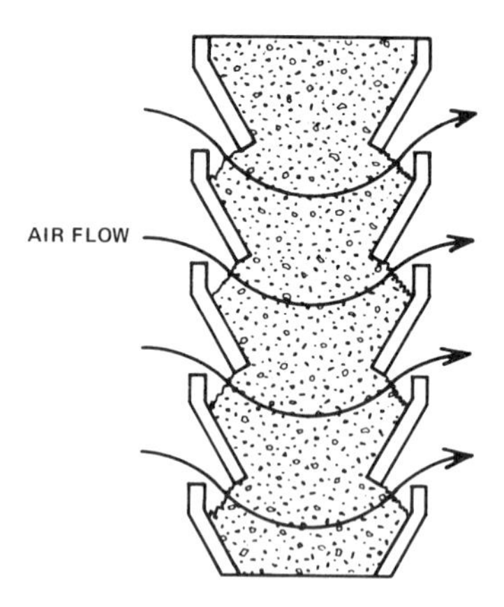


Fig. 36 Moving-Bed Adsorber

the top while spent or dust-laden carbon is periodically removed from the bottom.

Figure 37 shows another type of odor adsorption equipment. Adsorbers of this general configuration are available as standard packaged systems complete with motor and blower. Air-handling capacities range from 500 to 12,000 cfm (250 to 5500 L/s).

Activated carbon life in odor-control service ranges from a few weeks to a year or more, depending on the concentration of the odorous emission. The volume of carbon required is often so small and its life so long that it is simply discarded and replaced with new carbon when spent. Larger quantities can be returned to the carbon manufacturer for high-temperature thermal reactivation. Regeneration in place by steam, hot inert gas, or washing with a solution of alkali is sometimes practiced.

### Impregnated Adsorbents

When plain physical adsorption is too weak to remove a particular contaminant from an exhaust gas stream, adsorbents impregnated with a reactive chemical are often used. When a reactive chemical is spread over the immense internal surface area of an adsorbent, it maximizes the contaminant-reactive chemical contact so that relatively shallow beds are effective. A reduction of the bed depth can reduce the size of equipment. But,

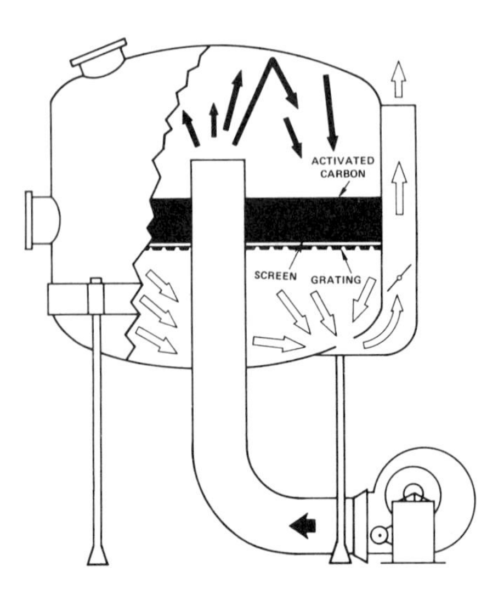


Fig. 37 Odor Adsorber

more importantly, it may reduce the operating pressure drop, thus causing a net reduction in energy for the operation of the equipment.

Typical applications of impregnated adsorbents are listed as follows:

1. Sulfur- or iodine-impregnated carbon removes mercury vapor from air, hydrogen, or other gases by forming mercuric sulfide or iodine (see Figure 38).
2. Metal-oxide impregnated carbons remove hydrogen sulfide.
3. Amine- or iodine-impregnated carbons and silver-exchanged zeolites remove radioactive methyl iodide from nuclear power plant work areas and exhaust gases.
4. Alkali-impregnated carbons remove acid gases.
5. Activated alumina, impregnated with potassium permanganate, removes acrolein and formaldehyde.
6. Other impregnated adsorbents are available for specific applications.

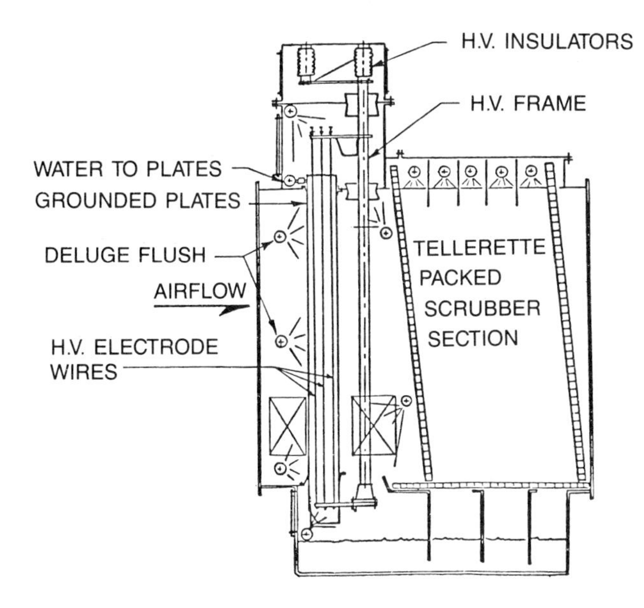


**Fig. 38 Electrostatic Augmented Scrubber**

## ELECTROSTATICALLY AUGMENTED SCRUBBERS

Several air-cleaning devices combine high-voltage particle charging and wet scrubbing. Because these devices can be constructed of corrosion-resistant plastics, they are especially useful where the dirty airstream is highly corrosive and contains submicron aerosols in the form of condensed fumes or mists and soluble or reactive gases such as HCl, HF, and $SO_2$. Compared to the conventional venturi scrubber, these devices remove submicron particulates at a much lower pressure drop. Thus, energy consumption for operation is less costly and is usually about 2.0 to 2.5 bhp/1000 cfm (3 to 4 J/L) per stage.

The three designs of electrostatically augmented scrubbers most used are as follows:

1. Aerosols charged at 28 kV and higher pass through a contact chamber containing randomly oriented packing elements of dielectric material (see Figure 38).
2. Aerosols charged at 12 kV pass through a low-energy venturi scrubber.
3. Aerosols charged negatively flow into a spray chamber where the scrubbing liquid droplets are positively charged.

Depending on the loading of the submicron aerosols and the velocity through the particle-charging section, collection efficiency per stage (one charging section followed by one scrubber section) on aerosols is normally 50 to 90%. Higher collection efficiencies can be obtained by using two or more stages. Removal efficiency of gaseous pollutants depends on the mass transfer and absorption design of the scrubber section.

In most applications of electrostatically augmented scrubbers, the dirty airstream is quenched by adiabatic cooling with liquid sprays and thus contains a large amount of water vapor that wets the solid aerosol particles. For this reason, collection efficiency of these devices is not affected by particle resistivity as much as with the conventional electrostatic precipitator.

Electrical equipment for particle charging is similar to that for electrostatic precipitators. The scrubber section is usually equipped with a liquid recycle pump, recycle piping, and a liquid distribution system.

Electrostatically augmented scrubbers have been applied successfully to clean and reduce the opacity of airstreams from industrial incinerators burning hazardous liquid and solid wastes, from plywood veneer driers, refractory tunnel kilns, phosphate rock defluorination reactors, coke calciners, and municipal incinerators.

## REFERENCES

ACGIH. 1974. *A Manual of Recommended Practice,* 13th ed. Committee on Industrial Ventilation, American Conference of Governmental Industrial Hygienists, Lansing, MI.

Boll, C.H. 1976. Recovering Solvents by Adsorption. *Plant Engineering,* January 8.

Caplan, K.J., ed. 1968a. Control Equipment. *Air Pollution Control Manual, Part II.* American Industrial Hygiene Association, Akron, OH.

Caplan, K.J. 1968b. Source Control by Centrifugal Force and Gravity. *Air Pollution,* A.C. Stern, ed., Vol. III. Academic Press, NY.

DOE. 1979. The Coating Industry: Energy Savings with Volatile Organic Compound Emission Control. *Report No.* TID-28706, Department of Energy, Washington, DC.

EPA. 1976. Control Methods for Surface Coating Operations. *Report No.* EPA-450/2-76-028, Control of Volatile Organic Emissions from Existing Stationary Sources—Vol. 1. Environmental Protection Agency, Washington, DC.

EPA. 1984. Proposed Fine Particle Standards for Ambient Air Quality. *Federal Register* 48, 10408, March 30, Environmental Protection Agency, Washington, DC.

EPA. 1984. Quality Assurance Handbook for Air Pollution Measurement Systems. *Report No.* EPA-600/4-77-027b, Environmental Protection Agency, Washington, DC.

GPA. 1986. *Code of Federal Regulations*, Vol. 40, Part 60. pp. 182-786. U.S. Government Printing Office, Washington, DC.

Grandjacques, B. 1977. Carbon Adsorption Can Provide Air Pollution Control with Savings. *Pollution Engineering,* August.

Hall, H.J. 1975. Design and Application of High Voltage Power Supplies in Electrostatic Precipitation. Air Pollution Control Association *Journal*, Vol. 25, No. 2.

HEW. 1967. *Air Pollution Engineering Manual.* HEW *Publication* No. 999-AP-40. Department of Health and Human Services, Washington, DC. (formerly Dept. of Health, Education, and Welfare).

IGCI. 1964. Determination of Particulate Collection Efficiency of Gas Scrubbers. *Publication No.* 1, Industrial Gas Cleaning Institute.

Kane, J.M. and J.L. Alden. *Design of Industrial Exhaust Systems,* 4th ed. Industrial Press, New York.

Lapple, C.E. and H.J. Kamack. 1955. Performance of Wet Scrubbers. *Chemical Engineering Progress*, March.

MCA. 1968. *Odor Thresholds for 53 Commercial Chemicals.* Manufacturing Chemists Association, Washington DC. October.

NIOSH. 1978. A Recommended Approach to Recirculation of Exhaust Air. *Publication No.* 78-124, National Institute of Occupational Safety and Health, Washington DC.

Noll, C.G. 1984. Demonstration of a Two-Stage Electrostatic Precipitator for Application to Industrial Processes. *Proceedings of 2nd International Conference on Electrostatic Precipitation,* Kyoto, Japan. November. pp. 428-434.

Oglesby, S. *et al.* 1970. A Manual of Electrostatic Precipitator Technology, Part I—Fundamentals. NTIS *Publication* PB-196 380, U.S. National Air Pollution Control Administration, Washington DC.

Rudnick, S.N. *et al.* 1986. Particle Collection Efficiency in a Venturi Scrubber: Comparison of Experiments with Theory. *Environmental Science & Technology*, Vol. 20, No. 3, pp. 237-242.

White, H.J. 1974. Resistivity Problems in Electrostatic Precipitation. Air Pollution Control Association *Journal*, Vol. 24, No. 9.

# CHAPTER 12

# COMPRESSORS



THE compressor is one of the six essential parts of the compression refrigeration system; the others are the condenser, the expansion device, the evaporator, the controls, and the interconnecting piping. A knowledge of the design and operation of all six components is fundamental for the design and operation of a refrigeration system.

This chapter covers the design features and performance of commercially available refrigeration and air-conditioning compressors, with reference to application and selection. Theoretical and actual cycles are also treated in Chapter 1 of the 1985 FUNDAMENTALS Volume.

## PART I: POSITIVE DISPLACEMENT COMPRESSORS

A positive displacement compressor is a machine that increases the pressure of the refrigerant vapor by reducing the volume of the compression chamber by a fixed amount through work applied to the mechanism. Such compressors include reciprocating, screw, vane, rolling piston, scroll, and trochoidal types.

The preparation of this chapter is assigned to TC 8.1, Positive Displacement Compressors and TC 8.2, Centrifugal Machines.

### PERFORMANCE

Compressor performance is the result of design compromises involving physical limitations of the refrigerant, compressor, and motor, while attempting to provide the following:

1. The greatest trouble-free life expectancy
2. The most refrigeration effect for the least power input
3. The lowest applied cost
4. A wide range of operating conditions
5. An acceptable vibration and sound level

Two useful measures of compressor performance are capacity (which is related to compressor volume displacement) and efficiency.

**Capacity** is the rate of heat removal by the refrigerant pumped by the compressor in a refrigerating system. Capacity equals the product of the mass flow of refrigerant produced by the compressor and the difference in specific enthalpies of the refrigerant vapor at its thermodynamic state when it enters the compressor and the refrigerant liquid at the saturation temperature corresponding to the pressure of the vapor leaving the compressor. It is measured in Btu/h (watts).

The coefficient of performance (COP) for a hermetic compressor includes the combined operating efficiencies of the motor and the compressor.

$$\text{COP (hermetic)} = \frac{\text{Capacity, Btu/h (W)}}{\text{Input power to motor, Btu/h (W)}}$$

The COP for an open compressor does not include motor efficiency.

$$\text{COP (open)} = \frac{\text{Capacity, Btu/h (W)}}{\text{Input power to shaft, Btu/h (W)}}$$

## The Ideal Compressor

The capacity of a compressor at a given operating condition is a function of the mass of gas compressed per unit time. Ideally, the mass flow rate is equal to the product of the compressor displacement per unit time and the gas density, as shown in Eq. (1):

$$\omega = \varrho V_d \tag{1}$$

*where*

$\omega$ = ideal mass rate of gas compressed, lb/h (kg/s)
$\varrho$ = density of gas entering the compressor, lb/ft$^3$ (kg/m$^3$)
$V_d$ = geometric displacement of the compressor, ft$^3$/h (m$^3$/s)

The ideal refrigeration cycle is addressed in Chapter 1 of the 1985 FUNDAMENTALS Volume; the following quantities can be determined from the pressure-enthalpy diagram in Figure 6 of that chapter:

$$Q_{\text{refrigeration effect}} = (h_1 - h_4) \quad \text{Btu/lb (J/kg)}$$

$$Q_{\text{work of compression}} = (h_2 - h_1) \quad \text{Btu/lb (J/kg)}$$

Using $\omega$, the mass flow rate of gas as determined by Eq. (1):

$$\text{Ideal Capacity} = \omega\, Q_{\text{refrigeration effect}} \quad \text{Btu/h (W)}$$

$$\text{Ideal Power Input} = \omega\, Q_{\text{work}} \quad \text{Btu/h (W)}$$

## Actual Compressor Performance

Deviations from ideal performance of positive displacement compressors occur because of various losses, with a resulting decrease in capacity and an increase in power input. Depending on the type of compressor, some or all of the following factors have a major effect on compressor performance.

1. *Pressure drops within the compressor:*
   *a.* Through the shutoff valves (suction, discharge, or both)
   *b.* Across suction strainer
   *c.* Across motor (hermetic compressor)
   *d.* In manifolds (suction and discharge)
   *e.* Through valves and valve ports (suction and discharge)
   *f.* In internal muffler
2. *Heat gain to refrigerant from:*
   *a.* Hermetic motor
   *b.* Friction
   *c.* Heat of compression; heat exchange within compressor
3. *Valve inefficiencies resulting from imperfect mechanical action*
4. *Internal gas leakage*
5. *Oil circulation*
6. *Re-expansion*—The volume of gas remaining in the compression chamber after discharge, which re-expands into the compression chamber during the suction cycle and limits the mass of fresh gas that can be brought into the compression chamber.
7. *Deviation from isentropic compression*—When considering the ideal compressor, an isentropic compression cycle is assumed. In the actual compressor, the compression cycle deviates from isentropic compression primarily because of fluid and mechanical friction and heat transfer within the compression chamber. The actual compression cycle and the work of compression must be determined from measurements.
8. *Over and under compression*

All of these losses are difficult to consider individually. They can, however, be grouped together and considered by category. Their effect on ideal compressor performance is measured by the following efficiencies:

**Compression Efficiency** considers only what occurs within the compression volume and is a measure of the deviation of the actual compression from isentropic compression. It is defined as the ratio of the work required for isentropic compression of the gas to the work delivered to the gas within the compression volume (as obtained by measurement).

**Mechanical Efficiency** is the ratio of the work delivered to the gas (as obtained by measurement) to the work input to the compressor shaft.

**Volumetric Efficiency** is the ratio of actual volume of gas entering the compressor to the theoretical displacement of the compressor.

**Isentropic (Adiabatic) efficiency** is the ratio of the work required for isentropic compression of the gas to work input to the compressor shaft.

**Actual Capacity** is a function of the ideal capacity and the overall volumetric efficiency, $e_o$, of the actual compressor.

$$\text{Actual capacity} = e_o \omega\, Q_{\text{ref effect}} \quad \text{Btu/h (W)} \tag{2}$$

Actual shaft power is a function of the power input to the ideal compressor and the compression, mechanical, and volumetric efficiencies of the compressor, as shown in Eq. (3).

$$bp = \frac{ip \cdot e_o}{e_2 \cdot e_3} = \frac{i_p \cdot e_o}{e_4} \tag{3}$$

*where*

$bp$ = actual shaft power, Btu/h (W)
$ip$ = ideal power input = $\omega \cdot Q_{\text{work}}$, Btu/h (W)
$e_o$ = volumetric efficiency
$e_2$ = compression efficiency
$e_3$ = mechanical efficiency
$e_4$ = isentropic efficiency

**Liquid Subcooling** is not accomplished by the compressor. However, the effect of liquid subcooling is included in the ratings for compressors by most manufacturers of air-conditioning halocarbon compressors, since liquid subcooling is usually designed into the system. Ammonia compressors and industrial refrigeration compressors are not normally rated with subcooling, however, since liquid subcooling is usually not designed into the system.

## Heat Rejection

The heat rejected by the condenser is the sum of the refrigeration effect and the heat equivalent of the power input to the compressor minus the heat removed by oil cooling or other means, if applicable. In a hermetic compressor, this heat rejection includes the inefficiency of the motor. Compressor heat rejection factors are necessary for sizing condensers.

# MOTORS AND PROTECTION DEVICES

## Hermetic Motors

Motors for positive-displacement compressors range from fractional horsepower to several thousand horsepower. When selecting a motor for driving a compressor, the following factors should be considered:

a. horsepower and RPM
b. starting and pull-up torques
c. ambient and maximum rise temperatures
d. cost and availability
e. insulation system
f. efficiency and performance
g. starting currents
h. voltage and phase
i. type of protection required
j. Multi-speed or variable speed

Large, industrial open compressors can be run with motors running from 900 to 3600 rpm and voltages ranging from 230 to 4160 volts. Due to local codes, requirements of local utilities, or end-user requirements, motors can be started across-the-line, part winding, Wye-Delta, autotransformer, or solid state to limit the starting currents. Caution must be taken that the starting method will supply enough torque to accelerate the motor and overcome the torque required for compression.

Hermetic motors can be more highly loaded than comparable open motors because of the gas cooling used.

To use a hermetic motor effectively, the maximum design load should be as close as possible to the breakdown torque at the lowest voltage used. This design operates better at lighter loads and higher voltage conditions. The limiting factor at high loads is normally the motor temperature, while at light loads the limiting factor is the discharge gas temperature. Overdesign of the motor increases discharge gas temperature at light loads.

The compressor operates most efficiently at high compression ratios, so it is important to keep compression losses to a minimum. The single-phase motor presents more design problems than the polyphase because the relationship between main and auxiliary windings becomes critical together with the necessary starting equipment.

The locked-rotor rate of temperature rise must be kept low enough to prevent excessive motor temperatures with the motor protection available. The maximum temperature under these conditions should be held to 300°F (150°C). With better protection, a higher rate of rise can be tolerated, and a less expensive motor can be used.

The materials selected for these motors must have high dielectric strength, resist fluid and mechanical abrasion, and be compatible with an atmosphere of refrigerant and oil.

The types of hermetic motors selected for various applications are as follows:

Refrigeration compressors—1 phase
- Low to medium torque—split phase or PSC (Permanent Split Capacitor)
- High torque—CSCR (Capacitor Start—Capacitor Run) and CSIR (Capacitor Start—Induction Run)

Room air-conditioner compressors—1 phase
- PSC or CSCR
- 2 speed, pole switching
- variable speed

Central air-conditioning and commercial refrigeration
- 1 phase, PSC and CSCR to 6 hp (4500 W)
- 3 phase, 2 hp (1500 W) and above, across-the-line start
- 10 hp (7500 W) and above; part winding; Wye-Delta and across-the-line start
- 2 speed, pole switching
- variable speed

For further information on motors and motor protection, see Chapter 31 of this volume.

## Protective Devices

Compressors are also provided with one or more of the following devices for protection against abnormal conditions and to comply with various codes.

1. *High pressure protection* as required by Underwriters Laboratories and per ARI *Standards* and ANSI/ASHRAE 15-1978, *Safety Code for Mechanical Refrigeration.* This may include the following:
   *a.* A high pressure cutout.
   *b.* A high to low side internal relief valve or rupture member. To comply with the ANSI/ASHRAE *Standard* 15-1978, all compressors 50 cfm (24 L/s) displacement and larger must be equipped with either an internal or external high- to low-side relief valve to prevent rupture of the crankcase or housing. The differential pressure setting depends on the refrigerant used and the operating conditions. Care must be taken to ensure that the relief valve will not accidentally blow on a fast pulldown. Many welded hermetic compressors have an internal high- to low-pressure relief valve to limit maximum system pressure in units not equipped with other high-pressure control devices.
   *c.* A relief valve or rupture member on the compressor casing similar to those used on other pressure vessels.
2. *High temperature control devices* to protect against overheating and oil breakdown.
   *a.* Motor overtemperature protective devices, which are addressed in the section on "Hermetic Motor Protection" in Chapter 31 of this volume.
   *b.* To protect against lubricant and refrigerant breakdown, a temperature sensor is sometimes used to stop the compressor when discharge temperatures exceed safe values. The switch may be placed internally (near the compression chamber) or externally (on the discharge line). Discharge temperatures may reach unsafe values at high pressure ratios or at partial compressor loads.
   *c.* On larger compressors, lubricant temperatures are controlled by cooling with either a heat exchanger or direct liquid injection, or the compressor may shut down on high lubricant temperature.
   *d.* Where lubricant sump heaters are used to maintain a minimum lubricant sump temperature, a thermostat is usually used to limit the maximum lubricant temperature.
3. *Low pressure protection* may be provided for:
   *a.* Suction gas. Many compressors or systems are limited to a minimum suction pressure by a protective switch. Motor cooling, freeze-up, or pressure ratio usually determine the pressure setting.
   *b.* The compressor. Lubricant pressure protectors are used with forced-feed lubrication systems to prevent the compressor from operating with insufficient pressure. Low lubricant pressure safety switches stop the compressor on the loss of lubricant pressure after a preset time delay (10 to 120 sec).
4. *Time delay,* or lockouts with manual resets, prevent damage to both compressor motor and contactors from repetitive rapid-starting cycles.
5. *Low voltage and phase loss* or reversal protection is used on some systems. Phase reversal protection is used with multiphase devices to ensure the proper direction of rotation. See Chapter 31 of this volume.

## APPLICATION

### Liquid Hazard

A gas compressor is not designed to handle large quantities of liquid. The damage that may occur depends on the quantity of liquid, the frequency of occurrence, and the design of the compressor. Slugging, floodback, and flooded starts are three ways in which liquid can damage the valves or dilute and wash away the lubricating oil.

**Slugging** is the short-term pumping of a large quantity of liquid refrigerant or oil. It can occur just after startup if refrigerant accumulated in the evaporator during shutdown returns to the compressor. It can also occur when system operating conditions change radically, such as during a defrost cycle. When the pumping rate of an unloading compressor is suddenly increased after prolonged light load operation, a slug of oil may be returned.

**Floodback** is the continuous return of liquid refrigerant mixed with the suction gas. It occurs because of improper selection, setting, or operation of the device feeding the evaporator; because of loss of evaporator load; or because of an excessive charge of refrigerant.

A **flooded start** occurs when a large volume of refrigerant accumulates in the crankcase or oil sump during shutdown, diluting the lubricating oil and even entering the compression spaces. The refrigerant accumulates because the compressor is at the lowest temperature or elevation in the system or because of the affinity of the halocarbon refrigerants for oil. In addition to the possible valve and bearing damage, low-side oil sump machines may have a serious oil loss because of the violent foaming that takes place at start-up. High-side sump machines can blow out oil due to high levels or violent agitation.

Compressors can be protected from damage caused by liquid by combinations of the following:

1. Design the system for the minimum practical refrigerant charge and with no refrigerant or oil traps.
2. Properly choose and set the evaporator liquid-feed device.
3. Provide a suction accumulator to catch any returning liquid and evaporate it or meter it back to the compressor. Provide a method for returning or removing the trapped oil. A high level shutdown should be installed if a liquid charge could fill the accumulator.
4. Apply a crankcase heater to operate during shutdown. A sump heater guards against flooded starts only, and a small elevation of compressor temperature over the rest of the system is effective in limiting refrigerant accumulation.
5. Isolate the compressor from the system at shutdown. A check valve in the discharge line and a solenoid valve in the liquid line will accomplish this.
6. Pump the refrigerant out of the low side of the system at shutdown. A single pump-out by closing a liquid line solenoid valve at shutdown may be used or the pumpdown may be repeated automatically during shutdown as low-side pressure rises.

### Sound Level

An acceptable sound level is a basic requirement of good design and application. Chapter 52 in the 1987 HVAC Volume covers design criteria in more detail. Noises with high-peaked discrete frequencies are particularly annoying. Whenever possible, final acceptance should be based on performance in the unit application. This is especially true of small-size equipment.

Compressor noise comes from mechanical components, from valves and rubbing surfaces, and from the response of other members to generated frequencies. Electric motor noise comes from generation and excitation. Gas noise comes because of pulses, windage, and turbulence.

A program to decrease the sound level involves the following:

1. Modification and refinement of the members generating the energy to reduce the driving forces.
2. Reducing response to the driving forces by designing components so that their natural frequencies do not coincide with the running frequency, twice line frequency, or the higher harmonics of these. This is particularly true of hermetics where no isolation of the motor from the running gear and housing of the compressor can be made.
3. Internal and external isolation, wherever possible.
4. Muffling suction and discharge gases.
5. Use of materials that dampen or absorb sound energy.
6. Balance of running gear.

### Vibration

Compressor vibration results from gas-pressure pulses and inertia forces associated with the moving parts. The problems of vibration can be handled in the following ways:

1. **Isolation.** With this common method the compressor is resiliently mounted in the unit by springs, synthetic rubber mounts, etc. In hermetics, the internal compressor assembly is usually spring-mounted within the welded shell, and the entire unit is externally isolated. Transmission to the base can be reduced to 5% by these means. The desired degree of isolation dictates the flexibility of the system and therefore the motion during shipment, starting, and stopping. Spring lateral stiffness and snubbers and stops or both are commonly used to limit this motion.
2. **Reduction of Amplitude.** The amount of movement can be reduced by adding mass to the compressor. Mass is added either by rigidly attaching the compressor to a base, condenser, or chiller, or by providing a solid foundation. When structural transmission is a problem, particularly with large machines, the entire assembly is then resiliently mounted.
3. **Elimination.** Vibrations can be reduced to acceptable levels by following good design principles, maintaining close balancing tolerances, maintaining good alignment, or using properly selected isolators.

On large units, generally field installations, the problems must be solved as they appear. On installations in which the compressor is well balanced and runs well, vibration often occurs at light loads or when independent of the system. The problem is then one of resonance of the adjoining members or the result of gas pulsations in the low- or high-side circuits.

Resonance of a member can be detected either by the effect of clamping added mass or by determining natural frequencies. Changing the stiffness of the member, changing its shape, and/or changing the mass will correct it. A word of caution: a line has many modes of vibration and many natural frequencies.

Vibrations from gas pulses cannot be readily dampened; they are more pronounced at bends in the lines. Pulsations are best handled by a good muffler, which eliminates the need to match impedances of lines and coils to the compressor. The muffler should be placed as close to the compressor as possible. Within the compressor, at the time of design and testing, port size and length, manifold volumes, and internal line sizes and configuration should be carefully selected to smooth gas pulses, reflect waves, and eliminate conditions of resonance.

Chapter 52 in the 1987 HVAC Volume has further information.

### Shock

In designing for shock, three types of dynamic loads are recognized:

1. Suddenly applied loads of short duration
2. Suddenly applied loads of long duration
3. Sustained periodic varying loads

Since the forces are primarily inertial, the basic approach is to maintain low equipment mass and make the strength of the carrying structure as great as possible. The degree to which this practice is followed is a function of the amount of shock loading.

**Commercial Units.** The major shock loading to these units occurs during shipment or when they operate on commercial carriers. Train service provides a severe test because of low forcing frequencies and high shock load. Shock loads as high as 10 g have been recorded; 5 g can be expected.

Trucking service results in higher forcing frequencies, but shock loads can be equal to, or greater than, those for rail transportation.

Aircraft service forcing frequencies generally fall in the range of 20 to 60 Hz with shocks to 3 g.

**Military Units.** The requirements are given in detail in specifications that exceed anything expected of commercial units. In severe applications, deformation of the supporting members and shock isolators may be tolerated, provided that the unit performs its function.

Basically, the compressor must be made of components rigid enough to avoid misalignment or deformation during shock loading. Therefore, structures with low natural frequencies should be avoided.

### Testing

Testing for ratings must be in accordance with the ASHRAE *Standard* 23-78, *Methods of Testing for Rating Positive Displacement Refrigerant Compressors.* Compressor tests are of two types: the first determines capacity, efficiency, sound level, motor temperatures, etc.; the second determines the probable life of the machine. Life testing should be conducted under conditions simulating those under which the compressor must operate; it generally goes on for years. A minimum set of conditions should include maximum discharge, maximum suction pressure operation, a medium condition with wet return gas, minimum suction, and a high discharge pressure condition with maximum suction gas temperature. In addition, most manufacturers run field trials and special tests (excessive floodback, flooded startups, starts and stops, and reversals) to prove new designs.

Where good correlation with actual field experience and long-term life tests can be shown, accelerated life tests can shorten the required test time.

### Standard Rating Conditions

To establish a uniform industry-wide basis for rating compressors, the Air-Conditioning and Refrigeration Institute (ARI) has established standard rating conditions.

# PART II: RECIPROCATING COMPRESSORS

Most reciprocating compressors are single-acting, using pistons that are driven directly through a pin and connecting rod from the crankshaft. Double-acting compressors that use piston rods, crossheads, stuffing boxes, and oil injection are not used extensively and, therefore, are not covered here.

**Single-Stage compressors** can achieve saturated suction temperatures to −50°F (−46°C) at 95°F (35°C) saturated condensing temperature by using R-502. Chapters 3 and 4 of the 1986 REFRIGERATION Volume have additional information on other halocarbon and ammonia systems.

**Booster compressors** are typically used for low-temperature applications with R-22 or ammonia. Minus 85°F (−65°C) saturated suction can be achieved by using R-22, and −65°F (−54°C) saturated suction is possible by using ammonia.

The booster raises the refrigerant pressure to the level where further compression can be achieved with a high-stage compressor without exceeding the compression-ratio limitations of the respective machines.

Since superheat is generated as a result of compression in the booster, intercooling is normally required to reduce the refrigerant stream temperature to the practical level required at the inlet to the high-stage unit. Intercooling methods include controlled liquid injection into the intermediate stream, gas bubbling through a liquid reservoir, and use of a liquid-to-gas heat exchanger where no fluid mixing occurs.

**Integral Two-Stage compressors** achieve low temperature (−80°F [−62°C], using R-22 or ammonia) within the frame of a single compressor. The cylinders within the compressor are divided into respective groups so that the combination of volumetric flow and pressure ratios are balanced to achieve booster and high-stage performance effectively. Refrigerant connections between the high-pressure suction and low-pressure discharge stages allow an interstage gas cooling system to be connected to remove superheat between stages. This interconnection is similar to the methods used for individual high-stage and booster compressors.

Capacity reduction is typically achieved by cylinder unloading, as in the case of single-stage compressors. Special consideration must be given to maintaining the correct relationship between high-pressure and low-pressure stages.

The most widely used compressor is the halocarbon compressor, which is manufactured in three types of design: (1) open, (2) semihermetic or bolted hermetic, and (3) the welded-shell hermetic.

Ammonia compressors are manufactured only in the open design because of the incompatibility of the refrigerant and hermetic motor materials.

**Open-Type compressors** are those in which the shaft extends through a seal in the crankcase for an external drive.

**Hermetic compressors** are those in which the motor and compressor are contained within the same pressure vessel, with the motor shaft integral with the compressor crankshaft and the motor in contact with the refrigerant.

A **semihermetic compressor** (bolted, accessible, or serviceable) is a hermetic compressor of bolted construction amenable to field repair.

In **welded-shell hermetic compressors** (sealed) the motor-compressor is mounted inside a steel shell, which, in turn is sealed by welding. (Combinations of design features used are shown in Table 1. Typical performance values for halocarbon compressors are given in Table 2.)

### Table 1 Typical Design Features of Reciprocating Compressors

| Item | Halocarbon Compressor Open | Halocarbon Compressor Semi-hermetic | Halocarbon Compressor Welded Hermetic | Ammonia Compressor Open |
|---|---|---|---|---|
| 1. Number of cylinders—one to: | 16 | 12 | 6 | 16 |
| 2. Power range | 0.167 hp (0.12 kW) up | 0.5 to 150 hp (0.37 to 112 kW) | 0.167 to 25 hp (0.12 to 18.7 kW) | 10 hp (7.5 kW) up |
| 3. Cylinder arrangement | | | | |
| a. Vertical, V or W, radial | x | x | | |
| b. Radial, horizontal opposed | | | x | |
| c. Horizontal, vertical V or W | | | | x |
| 4. Drive | | | | |
| a. Hermetic compressors, induction electric motor | | x | x | |
| b. Open compressors—direct drive, V belt chain, gear, by electric motor or engine | x | | | x |
| 5. Lubrication—splash or force feed, flood | x | x | x | x |
| 6. Suction and discharge valves—ring plate or ring or reed flexing | x | x | x | x |
| 7. Suction and discharge valve arrangement | | | | |
| a. Suction and discharge valves in head | x | x | x | x |
| b. Uniflow—suction valves in top of piston, suction gas entering through cylinder walls. Discharge valves in head | x | | | x |
| 8. Cylinder cooling | | | | |
| a. Suction gas cooled | x | x | x | x |
| b. Water jacket cylinder wall, head, or cylinder wall and head | x | | | x |
| c. Air cooled | x | x | x | x |
| d. Refrigerant cooled heads | x | | | x |
| 9. Cylinder head | | | | |
| a. Spring loaded | x | x | x | x |
| b. Bolted head | x | x | x | x |
| 10. Bearings | | | | |
| a. sleeve, anti-friction | x | x | x | x |
| b. tapered roller | x | | | x |

| Item | Halocarbon Compressor Open | Halocarbon Compressor Semi-hermetic | Halocarbon Compressor Welded Hermetic | Ammonia Compressor Open |
|---|---|---|---|---|
| 11. Capacity control, if provided—manual or automatic | | | | |
| a. Suction valve lifting | x | x | x | x |
| b. Bypass—cylinder heads to suction | x | x | x | x |
| c. Closing inlet | x | x | | x |
| d. Adjustable clearance | x | x | | x |
| e. Variable speed | x | x | x | x |
| 12. Materials | | | | |
| Motor insulations and rubber materials must be compatible with refrigerant and oil mixtures. Otherwise, no restrictions | x | x | x | |
| No copper or brass | | | | x |
| 13. Oil return | | | | |
| a. Crankcase separated from suction manifolds, oil return check valves, equalizers, spinners, foam breakers | x | x | | x |
| b. Crankcase common with suction manifold | | | x | |
| 14. Synchronous speeds (50 and 60 Hz) | 250 to 3600 | 1500 to 3600 | 1500 to 3600 | 250 to 3600 |
| 15. Pistons | | | | |
| a. Aluminum or cast iron | x | x | x | x |
| b. Ringless | x | x | x | x |
| c. Compression and oil control rings | x | x | x | x |
| 16. Connecting rod | | | | |
| Split rod with removable cap or solid eccentric strap | x | x | x | x |
| 17. Mounting | | | | |
| Internal spring mount | | x | x | |
| External spring mount | | x | x | |
| Rigidly mounted on base | x | x | | x |

### Table 2 Typical Performance Values

| Compressor Size and Type | | Operating Conditions and Refrigerants: Evap. Temp. −40°F (−40°C) Cond. Temp. 105°F (40.5°C) Suction Gas 65°F (18.3°C) Subcooling 0°F (0°C) R-12, 500, 502 | Evap. Temp. 0°F (−17.8°C) Cond. Temp. 110°F (43.3°C) Suction Gas 65°F (18.3°C) Subcooling 0°F (0°C) R-12, 500, 502 | Evap. Temp. 40°F (4.4°C) Cond. Temp. 105°F (40.5°C) Suction Gas 55°F (12.8°C) Subcooling 0°F (0°C) R-12, 500, 502, 22 | Evap. Temp. 45°F (7.2°C) Cond. Temp. 130°F (54.4°C) Suction Gas 65°F (18.3°C) Subcooling 0°F (0°C) R-12, 500, 502, 22 |
|---|---|---|---|---|---|
| Large, over 25 hp (19 kW) | Open | 0.21 tons/hp (0.99 W/W) | 0.40 tons/hp (1.89 W/W) | 0.91 tons/hp (4.29 W/W) | 0.74 tons/hp (3.49 W/W) |
| | Hermetic | 3.15 Btu/h per W (0.92 W/W) | 6.00 Btu/h per W (1.76 W/W) | 13.12 Btu/h per W (3.85 W/W) | 9.90 Btu/h per W (2.90 W/W) |
| Medium, 5 to 25 hp (4 to 19 kW) | Open | 0.19 tons/hp (0.90 W/W) | 0.37 tons/hp (1.74 W/W) | 0.83 ton/hp (3.91 W/W) | 0.65 tons/hp (3.06 W/W) |
| | Hermetic | 2.89 Btu/h per W (0.85 W/W) | 5.60 Btu/h per W (1.64 W/W) | 12.04 Btu/h per W (3.53 W/W) | 9.15 Btu/h per W (2.68 W/W) |
| Small, under 5 hp (4 kW) | Open | — | — | — | — |
| | Hermetic | — | 3.80 Btu/h per W (1.11 W/W) | 10.14 Btu/h per W (2.97 W/W) | 7.76 Btu/h per W (2.27 W/W) |

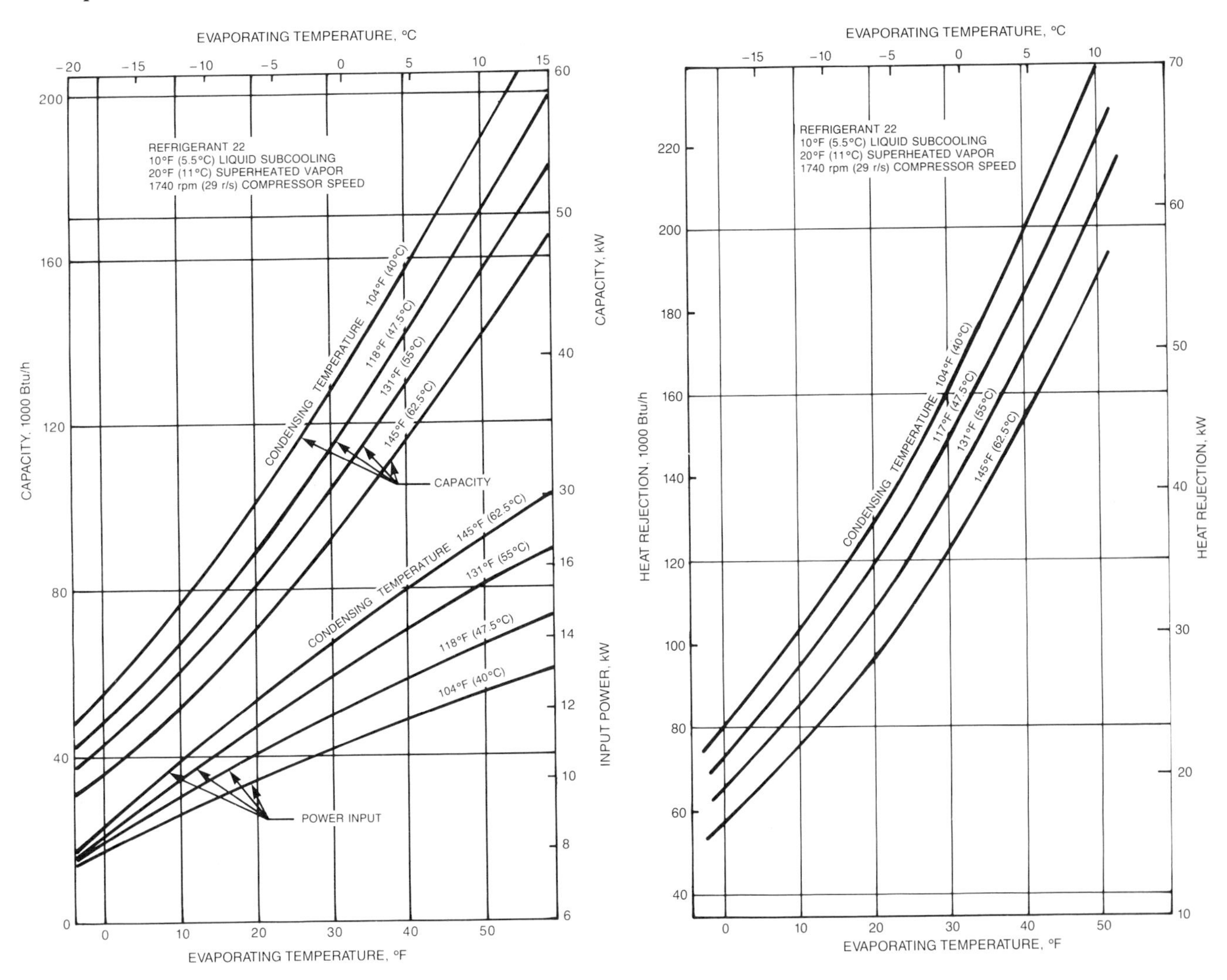


**Fig. 1 Capacity and Power-Input Curves for a Typical Hermetic Reciprocating Compressor**

**Fig. 2 Heat-Rejection Curves for a Typical Hermetic Reciprocating Compressor**

## PRESENTATION OF PERFORMANCE DATA

Figure 1 presents a typical set of capacity and power curves for a four-cylinder semihermetic compressor, 2.38-in. (60.3-mm) bore, 1.75-in. (44.4-mm) stroke, 1720 rpm, operating with Refrigerant 22. Figure 2 shows the heat rejection curves for the same compressor. Compressor curves should be labeled with the following information:

1. Compressor identification
2. Degrees subcooling or statement that data have been corrected to zero degrees subcooling
3. Compressor speed
4. Type refrigerant
5. Suction gas superheat
6. Compressor ambient
7. External cooling requirements (if any)
8. Maximum power or maximum operating conditions
9. Minimum operating conditions at fully loaded and fully unloaded operation

## MOTOR PERFORMANCE

The motor efficiency is usually the result of a compromise between cost and size. Generally, the larger a motor is for a given rating, the more efficient it can be made. Accepted efficiencies range from approximately 80% for a 3 hp (2.25 kW) motor to 92% for a 100 hp (75 kW) motor. Uneven loading has a marked effect on motor efficiency. It is important that cylinders be spaced evenly. Also, the more cylinders there are, the smaller the impulses become. Greater moments of inertia of moving parts and higher speeds reduce the impulse effect. Small and evenly spaced impulses also help reduce noise and vibration.

Since many compressors start against load, it is desirable to estimate starting torque requirements. The following equation is for a single cylinder compressor. It neglects friction, the additional torque required to force discharge gas out of the cylinder, and the fact that the tangential force at the crankpin is not always equal to the normal force at the piston. This equation also assumes considerable gas leakage at the discharge valves but little or no leakage past the piston rings or suction valves. It yields a conservative estimate.

$$T_s = \frac{(P_2 - P_1)As}{2N_2/N_1} \quad (4)$$

*where*

$T_s$ = starting torque, lb in. (N • m)
$P_2$ = discharge pressure, psi (Pa), absolute or gauge
$P_1$ = suction pressure, psi (Pa), absolute or gauge
$A$ = area of cylinder, in.$^2$ (m$^2$)
$s$ = stroke of compressor, in. (m)
$N_2$ = motor speed, rpm
$N_1$ = compressor speed, rpm

Equation 4 shows that when pressures in the system are balanced or almost equal ($P_2 = P_1$), torque requirements are considerably reduced. Thus, a pressure balancing device on an expansion valve or a capillary tube that equalizes pressures at shutdown allows the compressor to be started without excessive effort. For multicylinder compressors, an analysis must be made of both the number of cylinders that might be on a compression stroke and the position of the rods at start. Since the force needed to push the piston to the top dead center is a function of how far the rod is away from the cylinder centerline, the worst possible angles these might assume need to be graphically determined by torque-effort diagrams. The torques for some arrangements are shown below:

| No. Cylinders | Arrangement of Cranks | Angle between Cylinders | Approx. Torque from Equation 4 |
|---|---|---|---|
| 1 | Single | | $T_s$ |
| 2 | Single | 90° | $1.025T_s$ |
| 2 | 180° apart | 0° or 180° | $T_s$ |
| 3 | Single | 60° | $1.225T_s$ |
| 3 | 120° apart | 120° | $T_s$ |
| 4 | 180°, 2 rods/crank | 90° | $1.025T_s$ |
| 6 | 180°, 3 rods/crank | 60° | $1.23T_s$ |

**Pull-up torque** is as important as starting torque. It is required to accelerate the compressor from rest, overcoming both inertia and gas forces to bring itself to operating speed. The greatest pull-up torque requirement comes when starting a compressor at a pressure ratio of about 2:1.

## FEATURES

### Crankcases

The crankcase, or in a welded hermetic compressor, the cylinder block, is usually of cast iron. Aluminum is also used, particularly in small open and welded hermetic compressors. Open and semihermetic crankcases enclose the running gear, oil sump and, in the latter case, the hermetic motor. Access openings with removable covers are provided for assembly and service purposes. Welded hermetic cylinder blocks are often just skeletons, consisting of the cylinders, the main bearings, and either a barrel into which the hermetic motor stator is inserted or a surface to which the stator can be bolted.

The cylinders can be integral with the crankcase or cylinder block, in which case a material that provides a good sealing surface and resists wear must be provided. In aluminum crankcases, cast-in liners of iron or steel are usual. In large compressors, premachined cylinder sleeves inserted in the crankcase are common. With halocarbon refrigerants, excessive cylinder wear or scoring is not much of a problem and the choice of integral cylinders or inserted sleeves is often based on manufacturing considerations.

### Crankshafts

Crankshafts are made of either forged steel with hardened bearing surfaces finished to 8 microinches (0.203 μm) or iron castings. Grade 25 to 40 (25,000 to 40,000 psi or 170 to 275 MPa) tensile gray iron can be used where the lower modulus of elasticity can be tolerated. Nodular iron shafts approach the stiffness, strength, and ductility of steel and should be polished in both directions of rotation to 16 microinches (0.406 μm) maximum for best results. Crankshafts often include counterweights and should be dynamically and/or statically balanced.

While a safe maximum stress is important in shaft design, it is equally important to prevent excessive deflection that can edge-load bearings to failure. In hermetics, deflection can permit motor air gap to become eccentric, which affects starting, reduces efficiency, produces noise, and further increases bearing edge-loading.

Generally, the harder the bearing material used, the harder the shaft. With bronze bearings, a journal hardness of 350 Brinell is usual, while unhardened shafts at 200 Brinell in babbitt bearings are typical. Other combinations of materials and hardnesses have been used successfully.

### Main Bearings

It is possible to overhang both the crank and drive means with the bearings between; however, usual practice is to place the cylinders between the main bearings and, in a hermetic, to overhang the motor. Main bearings are made of steel-backed babbitt, steel-backed or solid bronze, or aluminum. In an aluminum crankcase, the bearings are usually integral. By automotive standards, unit loadings are low; however, the oil-refrigerant mixture frequently provides only marginal lubrication and 8000 h/yr operation in commercial refrigeration service is quite possible. For conventional shaft diameters and speeds, 600 psi (4.1 MPa), main bearing loading based on projected area is not unusual. Running clearances average 0.001 in./in. (1 mm/m) of diameter with steel-backed babbitt bearings and a steel or iron shaft. Bearing oil grooves placed in the unloaded area are usual. Feeding oil to the bearing is only one requirement, another is the venting of evolved refrigerant gas and oil escape from the bearing to carry away heat.

In most compressors, crankshaft thrust surfaces (with or without thrust washers) must be provided in addition to main bearings. Thrust washers may be steel-backed babbitt, bronze, aluminum, hardened steel, or polymer and are usually stationary. Oil grooves are often included in the thrust face.

### Connecting Rods and Eccentric Straps

Connecting rods have the large end split and a bolted cap for assembly. Unsplit eccentric straps require the crankshaft to be passed through the big bore at assembly. Rods or straps are of steel, aluminum, bronze, nodular iron, or gray iron. Steel or iron rods often require inserts of such bearing material as steel-backed babbitt or bronze, while aluminum and bronze rods can bear directly on the crankpin and piston pin. Refrigerant compressor service limits unit loadings to 3000 psi (20 MPa) based on projected area with a bronze bushing in the rod small bore and a hardened steel piston pin. Aluminum rod loadings at the piston pin of 2000 psi (14 MPa) have been used. Large end unit loadings are usually under 1000 psi (7 MPa).

The Scotch yoke-type of piston-rod assembly has also been used. In small compressors, it has been fabricated by hydrogen brazing steel components. Machined aluminum components have been used in large hermetic designs.

## Piston, Piston Ring, and Piston Pin

Pistons are usually made of cast iron or aluminum. Cast-iron pistons with a running clearance of 0.0004 in./in. (400 μm/m) of diameter in the cylinder will seal adequately without piston rings. With aluminum pistons, rings are required because a running clearance in the cylinder of 0.002 in./in. (2 mm/m) or more of diameter may be necessary, as determined by tests at extreme conditions. A second or third compression ring may add to power consumption with little increase in capacity; however, it may help oil control, particularly if drained. Oil scraping rings with vented grooves may also be used. Cylinder finishes are usually obtained by honing, and a 12 to 40 μin. (0.3 to 1.0 μm) range will give good ring seating. An effective oil scraper can often be obtained with a sharp corner on the piston skirt.

The minimum piston length is determined by the side thrust and is also a function of running clearance. Where clearance is large, pistons should be longer to prevent slap. An aluminum piston (with ring) having a length equal to 0.75 times the diameter, with a running clearance of 0.002 in./in. (2 mm/m) of diameter, and a rod length to crank arm ratio of 4.5, has been used successfully.

Piston pins are steel, case-hardened to Rockwell *C* 50 to 60 and ground to a 8 μin. (0.2 μm) finish or better. Pins can be restrained against rotation in either the piston bosses or the rod small end, be free in both, or be full-floating, which is usually the case with aluminum pistons and rods. Retaining rings prevent the pin from moving endwise and abrading the cylinder wall.

There is no well-defined limit to piston speed; average velocities of 1200 fpm (6 m/s), determined by multiplying twice the stroke in feet by the rpm (in metres by the r/s), have been used successfully.

## Suction and Discharge Valves

The most important components in the reciprocating compressor are the suction and discharge valves. Successful designs provide long life and low pressure losses. The life of a properly made and correctly applied valve is determined by the motion and stress it undergoes in performing its function. Excessive pressure losses across the valve result from high gas velocities, poor mechanical action, or both.

For design purposes, gas velocity is defined as being equal to the bore area multiplied by the average piston speed and divided by the valve area. Permissible gas velocities through the restricted areas of the valve are left to the discretion of the designer and depend on the level of volumetric efficiency and performance desired. In general, designs with velocities up to 12,000 fpm (60 m/s) with ammonia and up to 9000 fpm (46 m/s) with Refrigerants 12 and 22 have been successful.

An ideal valve system would meet the following requirements:

1. Large flow areas with shortest possible path
2. Straight gas-flow path, no directional changes
3. Low valve mass combined with low lift for quick action
4. Symmetry of design with minimum pressure imbalance
5. No increase in clearance volume
6. Durability
7. Low cost
8. Tight sealing at ports
9. Minimum valve flutter

Most valves in use today fall in one of the following groups:

1. A **free-floating reed valve,** with backing to limit movement, seats against a flat surface with circular or elongated ports. It is simple, and stresses can be readily determined, but it is limited to relatively small ports; therefore, multiples are often used. Totally backed with a curved stop, it is a valve that can stand considerable abuse.
2. A **reed, clamped at one end,** with full backstop support or a stop at the tip to limit movement, has a more complex motion than a free-floating reed; the resulting stresses are far greater than those calculated from the curvature of the stop. Considerable care must be taken in the design to ensure adequate life.
3. A **ring valve** usually has a spring return. A free-floating ring is seldom used because of its high-leakage losses. Improved performance is obtained by using spring return, in the form of coil springs or flexing backup springs, with each valve. Ring-type valves are particularly adaptable to compressors using cylinder sleeves.
4. A **valve formed as a ring** has part of the valve structure clamped. Generally, full rings are used with one or more sets of slots arranged in circles. By clamping the center, alignment is ensured and a force is obtained that closes the valve. To limit stresses, the valve proportions, valve stops, and supports are designed to control and limit valve motion.

## Lubrication

Lubrication systems range from a simple splash system to the elaborate forced-feed systems with filters, vents, and equalizers. The type of lubrication required depends largely on bearing loads and application.

For low to medium bearing loads and factory assembled systems where cleanliness can be controlled, the **splash system** gives excellent service. Bearing clearances must be larger, however; otherwise, oil does not enter the bearing readily. Thus, the splashing effect of the dippers in the oil and the freer bearings cause the compressor to operate somewhat noisily. Furthermore, the splash at high speed encourages frothing and oil pumping; this is no problem in package-type equipment, but may be in remote systems where gas lines are long.

A **flooded system** includes disks, screws, grooves, oil-ring gears, or other devices that lift the oil to the shaft or bearing level. These devices flood the bearing and are not much better than splash systems, except that the oil is not agitated as violently, so that quieter operation results. Since little or no pressure is developed by this method, it is not considered forced-feed.

In **forced-feed lubrication**, a pump-gear, vane, or plunger develops pressure, which forces oil into the bearing. Smaller bearing clearances can be used because adequate pressure feeds oil in sufficient quantity for proper bearing cooling. As a result, the compressor may be quieter in operation.

Gear pumps are used to a large extent. Spur gears are simple but tend to promote flashing of the refrigerant dissolved in the oil because of the sudden opening of the tooth volume as two teeth disengage. This disadvantage is not apparent in internal-type eccentric gear or vane pumps where a gradual opening of the suction volume takes place. The eccentric gear pump, the vane pump, or the piston pump therefore give better performance than simple gear pumps when the pump is not submerged in the oil.

Oil pumps must be made with minimum clearances to pump a mixture of gas and oil. The discharge of the pump should have provision to bleed a small quantity of oil into the crankcase. A bleed vents the pumps, prevents excess pressure, and ensures faster priming.

A strainer should be inserted in the suction line to keep foreign substances from the pump and bearings. If large quantities of very fine particles are present and bearing loadings are high, it may be necessary to add an oil filter to the discharge side of the pump.

Oil must return from the suction gas into the compressor crankcase. A flow of gas from piston leakage opposes this oil flow, so the velocity of the leakage gas must be low to permit oil to separate from the gas. A separating chamber may be built as part of the compressor to help separate oil from the gas.

In many designs, a check valve is inserted at the bottom of the oil return port to prevent a surge of crankcase oil from entering the suction. This check valve must have a bypass, which is always open, to permit the check valve to open wide after the oil surge has passed. When a separating chamber is used, the oil surge is trapped before it can enter the suction port, thus making a check valve less essential.

### Seals

Stationary and rotary seals have been used extensively on open-type reciprocating compressors. Older stationary seals usually used metallic bellows and a hardened shaft for a wearing surface. Their use has diminished because of high cost.

The rotary seal costs less and is troublefree. A synthetic seal tightly fitted to the shaft prevents leakage and seals against the back face of a carbon nose. The front face of this carbon nose seals against a stationary cover plate. This design has been used on shafts up to 4 in. (100 mm) in diameter. The rotary seal should be designed so that the carbon nose is never subjected to the full thrust of the shaft; the spring should be designed for minimum cocking force; and materials should be such that a minimum of swelling and shrinking is encountered.

## SPECIAL DEVICES

### Capacity Control

An ideal capacity-control system would have the following operating characteristics (not all of these benefits can occur simultaneously):

1. Continuous adjustment to load
2. Full-load efficiency unaffected by the mechanism
3. No loss in efficiency at part load
4. Reduction of starting torque requirement
5. No reduction in compressor reliability
6. No reduction of compressor operating range
7. No increase in compressor vibration and sound level at part load

Capacity control may be obtained by (1) controlling suction pressure by throttling; (2) controlling discharge pressure; (3) returning discharge gas to suction; (4) adding re-expansion volume; (5) changing the stroke; (6) opening a cylinder discharge port to suction while closing the port to discharge manifold; (7) changing compressor speed; (8) closing off cylinder inlet, and (9) varying suction valve opening.

The most commonly used methods are the opening of the suction valves by some external force, gas bypassing within the compressor, and gas bypassing outside the compressor.

When capacity control compressors are used, system design becomes more important and the following must be considered:

1. Possible increase in compressor vibration and sound level at unloaded conditions
2. Minimum operating conditions as limited by discharge or motor temperatures (or both) at part-load conditions
3. Good oil return at minimum operating conditions when fully unloaded
4. Rapid cycling of unloaders
5. Refrigerant feed device capable of controlling at minimum capacity

### Crankcase Heaters

During shutdown, refrigerants tend to migrate to the coldest part of the refrigeration system. In cold weather, the compressor oil sump is usually the coldest area. When the system refrigerant charge is large enough to dilute the oil excessively and cause flooded starts, a crankcase heater should be used. The heater should maintain the oil at a minimum of 20°F (11°C) above the rest of the system at shutdown and well below the breakdown temperature of the oil at any time.

### Internal Centrifugal Separators

Some compressors are equipped with **anti-slug** devices in the gas path to the cylinders. This device centrifugally separates oil and liquid refrigerant from the flow of foam during a flooded start and thus protects the cylinders. It does not eliminate the other hazards caused by liquid refrigerant in gas compressors.

## APPLICATION

To operate through the entire range of conditions for which the compressor was designed and to obtain the desired service life, it is important that the mating components in the system be correctly designed and selected. Suction superheat must be controlled, oil must return to the compressor, and adequate protection must be provided against abnormal conditions.

Chapters 21 through 25 of the 1986 REFRIGERATION Volume cover system design. Chapter 7 in that volume gives details of system cleanup in the event of a hermetic motor burnout.

### Suction Superheat

No liquid refrigerant should be present in the suction gas entering the compressor because it causes oil dilution and gas formation in the lubrication system. If the liquid carry-over is severe enough to reach the cylinders, excessive wear of valves, stops, pistons, and rings can occur; liquid slugging may actually break valves.

Suction gas without superheat is not harmful to the compressor as long as no liquid refrigerant is entrained; some systems are even designed to operate this way, although suction superheat is intended in the design of most systems. Measuring suction superheat can be difficult, and the indication of a small amount does not necessarily mean that liquid is not present. An effective suction separator may be necessary to remove all liquid with either type of system.

High superheat may result in dangerously high discharge temperatures and, in hermetics, high motor temperatures.

Normal values of superheat are usually 10 to 20°F (5 to 11°C) in the air-conditioning range. For R-12 and R-22, an increase of about 0.5% in compressor capacity and a decrease of 0.5%

in power can be expected for every 10°F (5°C) increase in superheat obtained in the evaporator.

### Automatic Oil Separators

Oil separators are used most often to reduce the amount of oil discharged into the system by the compressor and to return oil to the crankcase. They are recommended when the oil is not miscible with the refrigerant (ammonia and R-13, in particular). Because oil readily mixes with halocarbon refrigerants, oil separators are usually unnecessary, except in special cases such as low temperature or flooded systems.

### Parallel Operation

Where multiple compressors are used, the trend is toward completely independent refrigerant circuits. This has an obvious advantage in the case of a hermetic motor burnout.

Parallel operation of compressors in a single system has some operational advantage at part load. Careful attention must be given to apportioning returned oil to the multiple compressors so that each always has an adequate quantity. Figure 3 shows the method most widely used. Line A connects the tops of the crankcases and equalizes the pressure above the oil, while line B permits oil equalization at the normal level. Lines of generous size must be used. In addition to lines A and B, the high-pressure sides of the compressor oil pumps can be cross-connected so that if one pump ceases to function the other will supply lubricant to both machines. Generally, line A is a large diameter, while line B is a small diameter, which limits the possible blowing of oil from one crankcase to the other.

A central reservoir for returned oil may also be used with means (such as crankcase float valves) for maintaining the proper levels in the various compressors. With staged systems, the low-stage compressor oil pump can sometimes deliver a measured amount of oil to the high-stage crankcase. The high-stage oil return is then sized and located to return a slightly greater quantity of oil to the low-stage crankcase. Where compressors are at different elevations and/or staged, the use of pumps in each oil line is necessary to maintain adequate crankcase oil levels. In both cases, proper gas equalization must be provided.

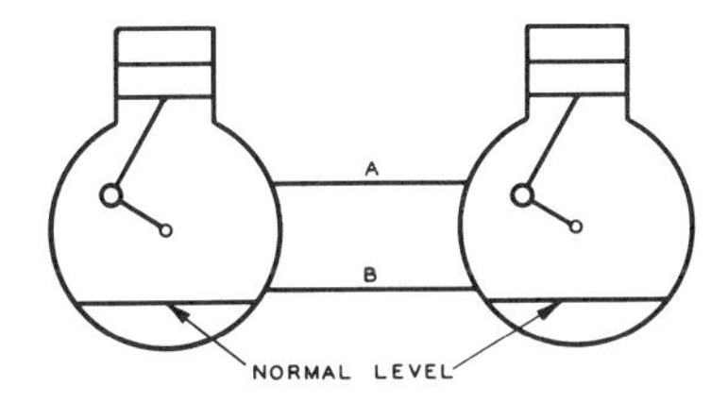


**Fig. 3 Modified Oil-Equalizing System**

# PART III: ROTARY COMPRESSORS

This section considers small rotary compressors used in household refrigerators and small air-conditioning units of up to 5 hp (4 kW) capacity.

## SMALL ROTARY COMPRESSORS

Rotary compressors are characterized by circular, or *rotary*, motion, as opposed to reciprocating motion. Their positive displacement compression process is nonreversing and is either continuous or cyclical, depending on the mechanism used. Most are direct-drive machines.

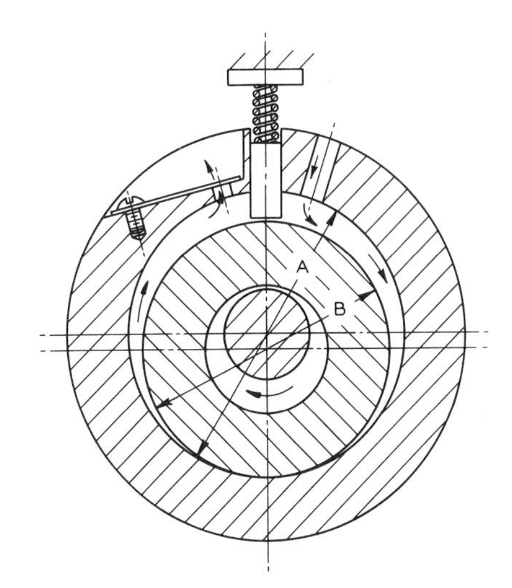


**Fig. 4 Fixed-Vane, Rolling-Piston Rotary Compressor**

Figures 4 and 5 show two common types of rotary compressors—the fixed vane-type (Figure 4) and the rotating vane-type (Figure 5). These machines are similar in size and weight.

The fixed vane uses a roller mounted on an eccentric shaft with a single vane or blade suitably positioned in the nonrotating cylindrical housing. This blade can reciprocate with the eccentrically moving roller.

The rotating vane-type has a rotor concentric with the shaft, with vanes in the rotor; this assembly is offcenter with respect to the cylindrical housing.

Displacements for the two types of rotary compressors are:

**Fixed Vane**

$$V_d = \pi H (A^2 - B^2)/4 \qquad (5)$$

**Rotating vane** (a simple approximation for a two-bladed circular cylinder compressor)

$$V_d = H(A - B)\left[\frac{\pi}{4}(A + B) - t\right] \qquad (6)$$

*where*

$V_d$ = displacement
$H$ = cylinder height
$A$ = cylinder diameter
$B$ = rolling piston or rotor diameter
$t$ = blade thickness

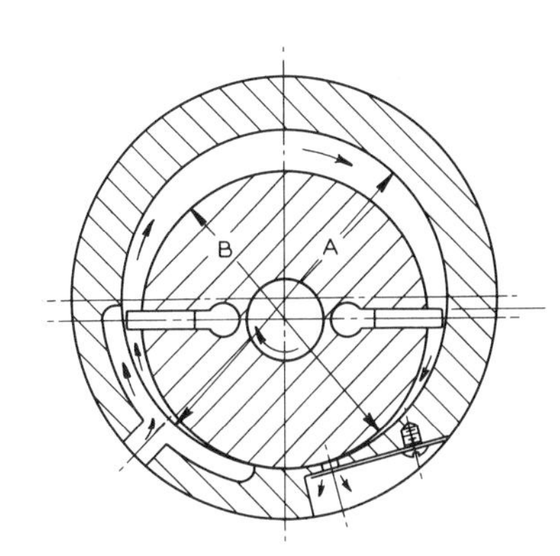


**Fig. 5 Rotating-Vane Rotary Compressor**

The above formula can also be used for a single vane rotary with a contoured bore. In this case, the cylinder diameter, $A$, is the average of the major and minor diameters.

For all rotary vane compressors, the gas-flow mechanics do not change if the blades are offset or multiple blades, multiple cylinders, or if both are used. An oval shaped bore produces a double lobe, effectively a two-cylinder rotary compressor.

The high-side crankcase is used because of the simplicity of its lubrication system and the lack of oiling and compressor cooling problems. Minimizing both the heat transfer and gas leakage area improves the performance.

Internal leakage is controlled through hydrodynamic sealing and selection of mating parts for optimum clearances. Hydrodynamic sealing depends on clearance, surface speed, oil viscosity, and surface finish of the parts. Smoother finishes and close tolerances generally support the hydrodynamic sealing and thus reduce gas losses.

Re-expansion losses may be made small, but some re-expansion is necessary to retain in the cylinder some oil held in the leading edge of the vane. A notch on the cylinder by the discharge port will supply oil for the cylinder wall, relieve hydraulic pressure, and provide a larger exit to the discharge port, which minimizes overcompression losses.

## PERFORMANCE

Rotary compressors have a high volumetric efficiency because of the small clearance volume and correspondingly low re-expansion loss inherent in its design. Figure 6 and Table 3 show performance typical of rotating vane compressors now in production.

An acceptable sound level is important in the design of any small compressor, particularly when it is intended for use in the home. Figure 7 illustrates a convenient method for analyzing and evaluating compressor sound level that is typical for rotary compressors. Since gas flow is continuous and no suction valve is required, rotary compressors can be relatively quiet. Suction and discharge mufflers are generally used to reduce sound pulses, however. One of the most important benefits of a discharge muffler is that it prevents the pulsating discharge gas flow from vibrating the connecting tubing and the condenser. The sound level for a rotary compressor is directly related to horsepower and speed in the same design. Increasing the speed increases the sound level, although not uniformly in all the frequency bands. The sound level is most prominent in the high-frequency bands.

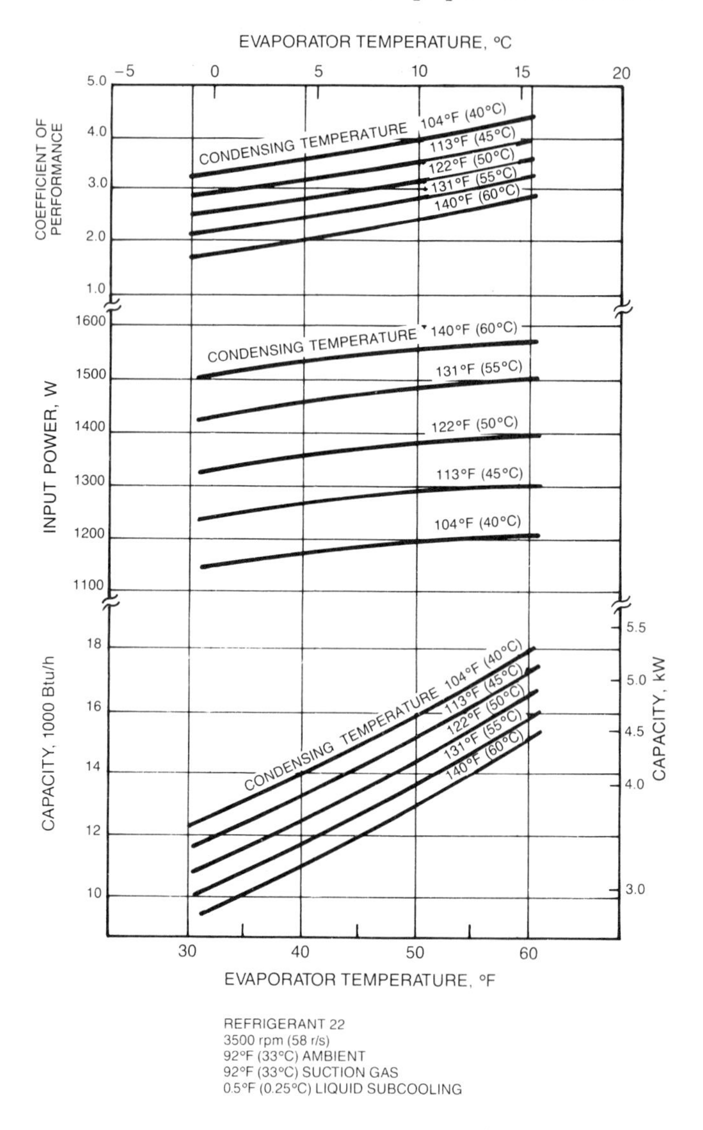


**Fig. 6 Performance Curves for a Typical Rotating-Vane Compressor**

**Table 3 Typical Rotating Vane Compressor Performance**

| | | |
|---|---|---|
| Compressor speed, rpm (r/s) | 3450 (58) | 3450 (58) |
| Refrigerant | 12 | 22 |
| Operating conditions | | |
| Condensing temp, °F (°C) | 130 (54.4) | 130 (54.4) |
| Liquid temp, °F (°C) | 90 (32.2) | 115 (46.1) |
| Evaporator temp, °F (°C) | −10 (−23.3) | 45 (7.2) |
| Suction pressure, psia (kPa) | 19.2 (132) | 90.7 (625) |
| Suction gas temp, °F (°C) | 90 (32.2) | 95 (35) |
| Capacity (cooling effect at the evaporator measured on the calorimeter), Btu/h (W) | 1140 (330) | 21,800 (6390) |
| Coefficient of performance factor, Btu/Wh (W/W) | 3.6 (1.06) | 9.6 (2.81) |
| Power, watts | 315 | 2274 |
| Power distribution, watts | | |
| Work of compression | 200 | 1580 |
| Motor losses | 80 | 450 |
| Compressor losses | 35 | 245 |

## FEATURES

### Shafts

The shaft should allow the minimum oil film thickness in the bearing under maximum loading conditions. Shaft deflection under load is caused primarily by compression-gas loading on the projected surface of the roller and by the torsional and side pull loading of the motor rotor. Strength design criteria must allow for the minimum oil film under maximum run and starting conditions. The motor rotor should have minimal deflection to eliminate motor starting problems under extreme conditions of torque.

A relative hardness, depending on the materials chosen, should be maintained between the mating bearing and journal. Journals are ground round as accurately as possible and honed or polished to a finish of 10AA or better. The finish required is dictated by the minimum oil thickness required.

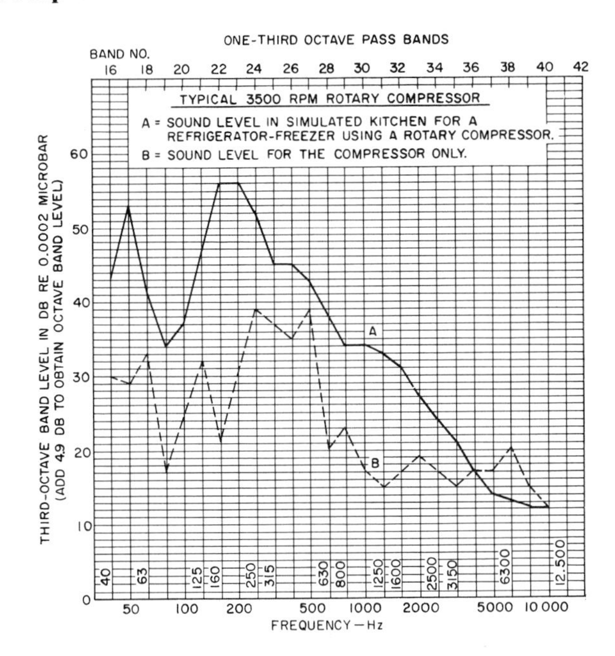


**Fig. 7 Sound Level of a Combination Refrigerator-Freezer with a Typical Rotary Compressor**

## Journals and Bearings

The bearing must support the rotating member under all conditions of use. It must also be reliable and resistant to possible damage during run-in, must require no maintenance, and must provide a minimum of wear. Normally, a well designed sleeve bearing meets these requirements.

Powdered metal has gained popularity as a bearing material due to the porous properties that help it wick oil. Powdered metal can also be formed into complex bearing shapes with little machining required. Powdered metal ports can be made from various materials, depending on the particular application required. Apparent hardness and particle hardness are important parameters from a strength and wear standpoint.

The gray cast-iron bearing is another good choice for a bearing material. It must be a high quality, fully pearlitic, non-porous casting. It should also be ASTM Type A graphite, contain no ferrite, and have a hardness range of 187 to 220 BHN (*Brinnell Hardness Number*).

## Vanes

Vanes are designed for maximum reliability by the choice of materials and lubrication. The vanes are hardened, ground, and polished to the best finish obtainable. Various materials are suitable, including powdered metal materials of full density. Particle hardness ranges typically in the 65 to 80 $R_c$ range, and apparent hardness ranges in the 35 to 55 $R_c$ range.

Usually, rotating blade compressors have two vanes, although three or more may be used where analysis of the compression cycle and machine dimensions show that they are needed. Performance and economics do not favor more than two blades for small compressors.

Vane thickness is a function of the material selected and the maximum allowable deflection. The maximum amount of deflection is relative to the vane slot clearance and the vane extension into the cylinder bore. Deflection should be based on the maximum section/discharge differential pressure.

To obtain good sealing and proper lubrication, the shape of the vane tip must conform to the surface of generation along the cylinder wall in accordance with hydrodynamic theory.

Springs or mechanical linkages to control vane movement are not used in the rotating vane machine. The under-vane pressure is the most important of several forces acting on the vane against the force of compression, which tends to raise the vane tip from the cylinder wall. The under-vane pressure may be as high as the condensing pressure.

## Vane Springs

The fixed vane design incorporates a vane spring to aid the compressor at start-up. The spring rate is dependent upon the inertia of the vane. The vane spring should be designed with adequate torsional and deflective stresses to ensure long life.

## Valves

Suction valves are not required by rotary compressors because suction flow is continuous. However, it is common practice to equip small machines with a check valve at the suction inlet that closes automatically when the compressor stops. This practice prevents high pressure gas from migrating into the evaporator.

The discharge valve is usually a simple reed valve made of high grade steel. For compressors below 5 hp (4 kW), thickness is in the range of 0.005 to 0.012 in. (0.12 to 0.3 mm). In the design of these valves, the material is not stressed beyond 50,000 psi (345 MPa). In most compressors, 15,000 to 35,000 psi (103 to 240 MPa) is more typical. The lower values apply to small machines and the higher values to larger machines. In making the valve, care is taken to avoid notched or rough edges along the stressed areas to avoid stress. All valves are tumbled to obtain the desired smoothness of valve edges.

## Lubrication

A good lubricating system circulates an ample supply of clean oil to all working surfaces, bearings, blades, blade slots, and seal faces. When the compressor shell is at a high-side pressure and there is a sufficient pressure differential across the bearings, passageways are needed to distribute oil to the bearing surfaces. Larger machines may include a simple impeller or centrifugal oil pump to urge the oil along and further ensure the reliability of the system.

Oil grooves must have outlets to permit free flow of the excess lubricant and to vent the bearing during start-up.

# MECHANICAL EFFICIENCY

High mechanical efficiency depends on minimizing other sources of friction, as well as those in the bearings. Friction losses occur between the vanes and the slot walls, at the vane tip on the cylinder wall, and between the rotor faces and the top and bottom bearing faces for rotating vane compressors. For fixed-vane compressors, high mechanical efficiency depends on minimizing losses around the vane and the vane slot, the roller to the top and bottom bearing faces, and the vane tip on the roller's outside diameter. The amount and distribution of the friction losses vary based on the geometry required to obtain the necessary displacement.

# MOTOR SELECTION

Breakdown torque requirements depend on the displacement of the compressor, the refrigerant, and the operating conditions.

A small compressor for home refrigeration (home freezers or household refrigerators) with R-12 uses about 36 to 38 oz-ft per $in^3$ of compressor displacement per revolution (186 to 196 $kN \cdot m/m^3$). Similarly, larger compressors using R-22 for window air conditioners use about 67 to 69 oz-ft per $in^3$ (346 to 357 $kN \cdot m/m^3$) of breakdown torque.

Motor torque required for a given condensing range is almost constant over a wide range of suction pressure conditions. For example, at 10°F (−12°C) evaporator temperature, the cooling capacity of a given compressor is about half the cooling capacity of the same machine at 45°F (7°C) evaporator temperature.

Rotary machines do not usually require complete unloading for successful starting. The starting torque available in standard split-phase motors is ample for smaller sizes. Start-capacitor motors are used on 0.33, 0.25, and 0.20 hp (250, 190, and 150 W) sizes to limit the starting current as recommended by Underwriters Laboratories. Permanent split-capacitor motors for the air-conditioner sizes provide sufficient starting torque and improve the power factor to the required range. Motor manufacturers' test data and recommendations should be consulted when selecting a suitable motor for any specific application.

## LARGE ROTARY COMPRESSORS

Larger rotary, blade-type compressors are used in the low-temperature field as high-volume, low-stage, or booster compressors. These booster compressors are applied at saturated suction conditions ranging from −125 to −5°F (−87 to −20°C) with Refrigerants 12, 22, and 717. Available units range in power from 10 to 600 hp (7.5 to 450 kW) and in displacement from 60 to 3600 cfm (30 to 1700 L/s) in a single unit.

These compressors are compact and lightweight, yet rugged, and are designed for use in industrial processing plants, cold storage plants, freezing plants, chemical processing plants, or for any application requiring a low-temperature system.

The basic design of the compressor is similar to that shown in Figure 8. The number of blades shown is typical of compressors 10 to 50 hp (7.5 to 37 kW) in size with the number of blades increasing as barrel or cylinder diameters are increased. Gas trapped between successive blades is compressed by the volume reduction resulting as the rotor rotates.

The discharge port is located to operate at the desired design point during compression. Although the compressor operates at compression ratios above or below this design point, either undercompression (backflow) or overcompression losses occur, which increases power requirements. In practice these compressors are limited to compression ratios not exceeding 7:1 with the actual point determined by the specific application. They are also limited to operate at relatively low differential pressures to limit stresses in the vanes and bearings and rotor deflection.

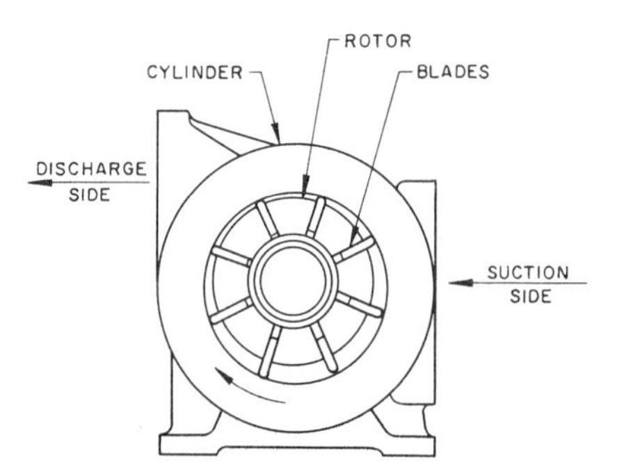


**Fig. 8 Interior View of a Large Rotary Compressor**

Figure 8 illustrates a nonjacketed type in which cooling is obtained by injection of excess lubricating oil. This lubricant is separated from the discharge gas, cooled in a heat exchanger, and returned to the compressor in a continuous flow. Internal oil cooling gives the following advantages:

1. The cylinder temperature is maintained at close to the rotor temperature, thus with less danger of rubbing at the close running seal. Rubbing can be caused by overcooling of the cylinder.
2. A complete lubricant seal of the vane and rotor end clearances is maintained.
3. The excess oil provides an oil film and direct cooling of the vane rubbing edge, preventing damage from overheating (usually, the prime cause of vane failure).
4. Improved performance results from compression at temperatures lower than that of constant entropy.

Full unloading for starting and capacity control from 100% down to 20% with proportional power reduction is available for this type of compressor. (Inability to provide full unloading for starting was an undesirable characteristic in large vane rotary compressors.) Valves in both end covers accomplish the unloading. These valves are normally open until oil pressure has been generated when the compressor is up to speed. The same valves are used for capacity control and as a safety relief against excessive pressures.

Porting slots are helical and cover the full length of the rotor, permitting a complete venting of hydraulic pressures from the compressor without vane damage. Inspection plugs provide a means to check rotor end clearance without disassembly. The desired rotor end clearance is maintained by a thrust bearing independently assembled and adjusted without reference to the load-carrying bearings. The oil cooling lubricant is also the shaft and bearing lubricant.

## SCROLL COMPRESSOR

The scroll compressor is a rotary, positive-displacement compressor used in automotive air-conditioning, as well as residential air-conditioning systems and heat pumps. Production units in some sizes are now available in Japan and the United States. To date, the scroll compressor has been considered for systems in the 1.5 through 10 ton (5 through 35 kW) range.

A scroll is a free-standing, involute spiral bounded on one side by a flat plate or base. The scroll set, which is the basic compression unit, consists of two scrolls, one is fixed in space and the other moves in a controlled orbit around a fixed point on the fixed scroll. The two involutes are phased 180° apart. Figure 9 shows the principle of operation. The suction gas enters the scroll set on the outer periphery. The meshing of the involutes forms cresent-shaped pockets, which, starting from the outside, reduce in-volume as the orbiting scroll translates to increase the pressure of the trapped gas. The closed pockets move radially inward until a discharge port is uncovered, resulting in the discharge of high-pressure gas. The scroll is unidirectional—it functions as a compressor when rotated in one direction and as an expander when rotated in the opposite direction.

The moving or orbiting scroll is driven by a short-throw crank mechanism. The proper indexing of the orbiting scroll relative to the fixed scroll is maintained by a coupling that forces the orbiting scroll to translate rather than rotate as a result of the action of the crank. Internal leakage effects must be controlled

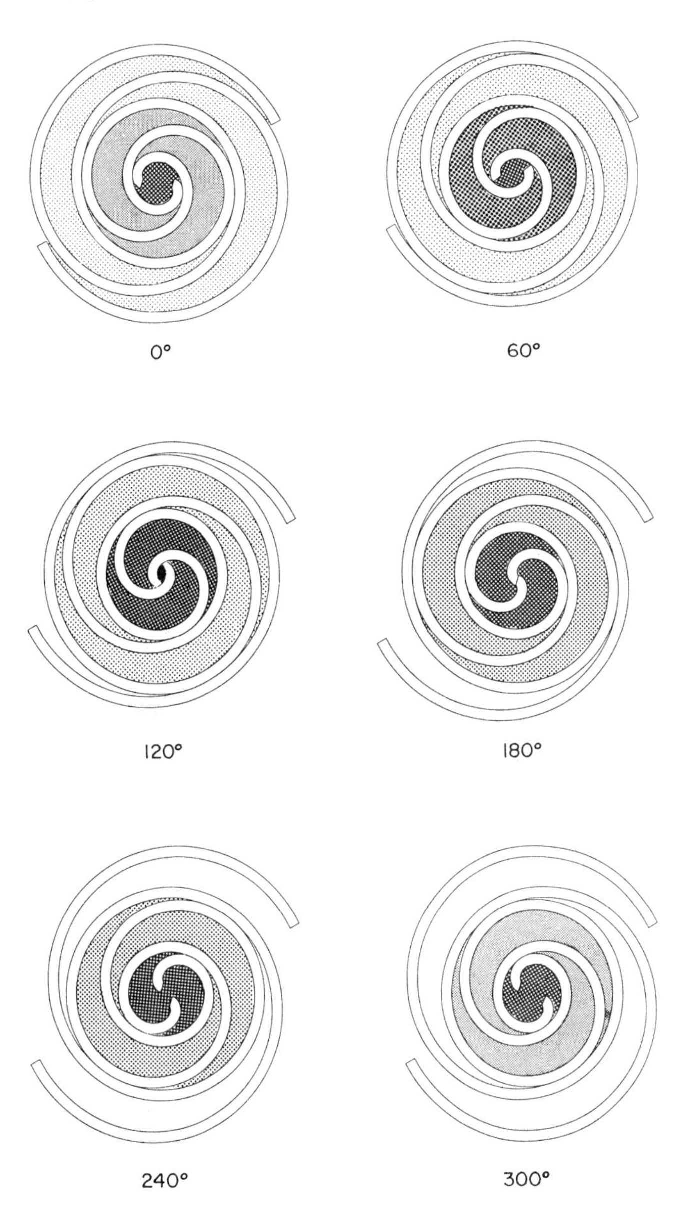


**Fig. 9 Sequence of Operation for the Scroll Compressor**

for good performance. Leakage sites include the gaps on the flanks of the involutes and between the tips of the involute and the opposing scroll base plate. Presently, flank leakage is controlled through the use of very accurately machined scrolls or a linkage mechanism that compliantly holds the involute of the orbiting scroll against the involute of the fixed scroll. Tip leakage is controlled by a pressure balance that forces the scrolls together axially or by including a sealing element at the tip of the involute.

The scroll compressor is a constant volume ratio machine. As such, it has no valves to control the gas flow through the scroll set during normal operation. By controlling the number of wraps of the involute and the location of the discharge port, the optimum pressure ratio is established for the compressor. Operating losses occur when the compressor is forced to operate at pressure ratios other than the optimum. These losses are small, however, for normal operating conditions. The performance levels for scroll compressors are generally high. Performance factors at the ARI rating point (45°F [7.2°C] saturated suction temperature and 130°F [54.4°C] compressor outlet temperature with 15°F [8.3°C] subcooling and 20°F [11.1°C] superheat) can range from 10 to 11 Btu/watt hour (2.9 to 3.2 W/W).

### TROCHOIDAL COMPRESSOR

The trochoidal compressor is a rotary, positive-displacement compressor. Because it is able to run at high speeds (up to 9000 rpm) and is small compared to many other compressors of similar capacity, it is being used for automobile air conditioning. The rotor and casing can, theoretically, take many forms. In the Wankel compressor, a housing with a two-node paritrochoid curve and a rotor that revolves around it form the compression and intake chambers (Figure 10). The epitrochoid, with a three-sided rotor and two envelope casing trochoidal unit, is in production in Japan. The Wankel compressor is built in capacities of up to 2 tons (7 kW).

## PART IV: HELICAL ROTARY COMPRESSORS

The helical rotary or screw compressor belongs to the broad class of positive-displacement compressors. Screw compressors currently in production for refrigeration and air-conditioning applications comprise two distinct types: single screw and twin screw.

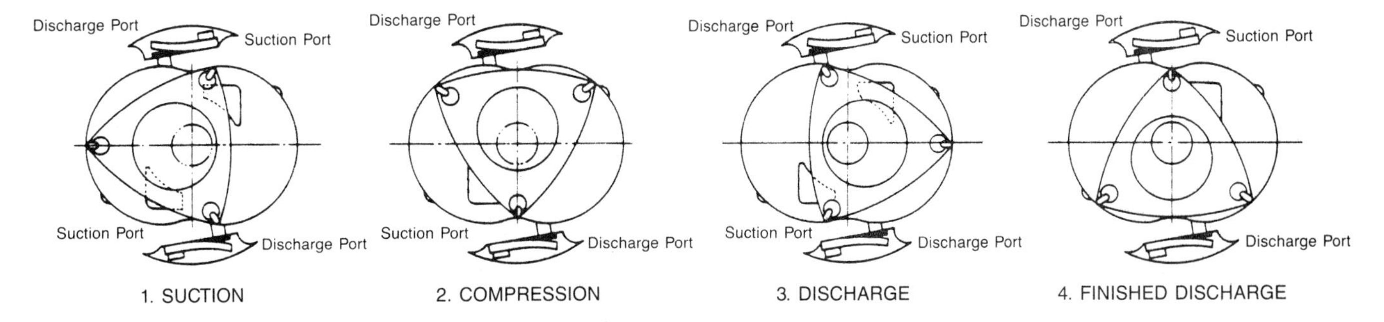


**Fig. 10 Sequence of Operation of a Wankel Rotary Compressor**

Many principles of screw compressor design and operation are common to both types. Both are conventionally used in the fluid injection mode where sufficient fluid cools and seals the compressor. Several features such as volume ratio, capacity control, and economizer facilities also apply to both types and are covered in this section.

## SINGLE-SCREW COMPRESSORS

Today's single-screw compressor is the result of developments started in the early 1960's. During the first ten years, its application was in the air compressor market, but since then it has been progressively introduced in the refrigeration, air-conditioning, heat-pump, and process markets in a wide range of capacities. It has the capability to operate at pressure ratios above 20:1 single stage. The capacity range currently available is from 20 to 1300 tons (70 to 4600 kW).

### Description

The single-screw (or globoid) compressor consists of a single cylindrical main rotor that works with a pair of gate rotors. Both the main rotor and gate rotors can vary within wide limits in terms of form and mutual geometry. Figure 11 shows the design normally encountered in refrigeration.

The main rotor has six helical flutes with a cylindrical periphery and a globoid (or hour-glass shape) root profile. The two identical gate rotors each have 11 teeth and are located on opposite sides of the main rotor. The tooth profile is identical to the flute section, and the casing enclosing the main rotor has two slots, which allow the teeth of the star wheels to pass through. Two diametrically opposed discharge ports of profiled shape, which use a common discharge chamber, are in the casing.

The compressor is driven through the main rotor shaft, and the gate rotors follow by direct meshing action at 6/11 of the main rotor speed. The geometry of the single-screw compressor is such that 100% of the gas compression power is transferred directly from the main rotor to the gas. No power (other than small frictional losses) needs to be transferred across the meshing points to the gate rotors.

### The Compression Process

The operation of the single-screw compressor can be divided into three distinct phases: suction, compression, and discharge. With reference to Figure 12, the process is as follows:

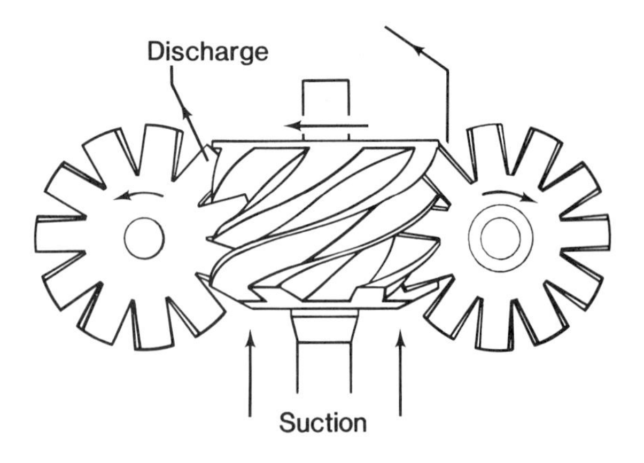


**Fig. 11 Schematic of a Single-Screw Compressor**

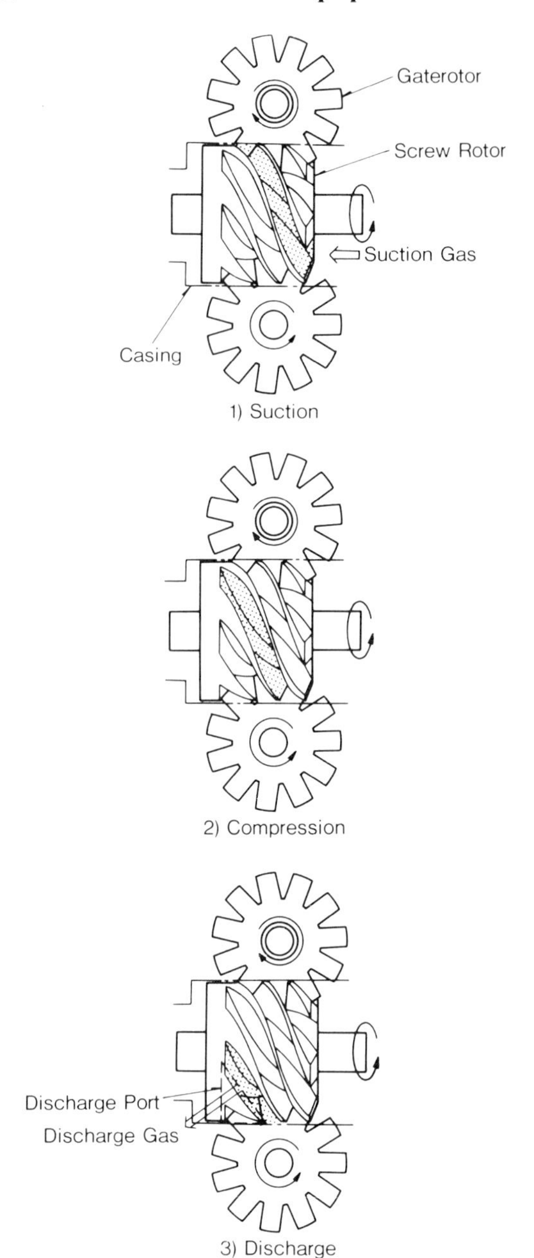


**Fig. 12 Sequence of Operation of a Single-Screw Compressor**

**Suction.** During rotation of the main rotor, a typical flute in open communication with the suction chamber gradually fills with suction gas. The tooth of the gate rotor in mesh with the flute acts as an aspirating piston.

**Compression.** As the main rotor turns, the flute engages a tooth on the other gate rotor and is covered simultaneously by the cylindrical main rotor casing. The gas is trapped in the space formed by the three sides of the flutes, the casing, and the gate rotor tooth. As rotation continues, the flute volume decreases and compression occurs.

**Discharge.** At the geometrically fixed point where the leading edge of the flute and the edge of the discharge port coincide, compression ceases, and the gas discharges into the delivery line until the flute volume has been reduced to zero.

## MECHANICAL FEATURES

### Rotors

The screw rotor is normally made of cast iron, and the mating gate rooters are made from an engineered plastic. The inherent lubricating quality of the plastic, as well as its compliant nature, allows the single-screw compressor to achieve close clearances with conventional manufacturing practices.

The gate rotors are mounted on a metal support that carries the differential pressure from discharge to suction pressure. The gate rotor acts as a rotating seal. The gate rotor, having a low moment of inertia, is attached to its support by a simple spring and dashpot mechanism. This method of attachment allows the gate rotor assemblies to be true idlers by giving them an angular degree of freedom to dampen out transients without damage or wear.

### Bearings

In a typical open or semi-hermetic single-screw compressor, the main screw shaft contains one pair of angular contact ball bearings (an additional angular contact or roller bearing is used for some heat pump semi-hermetics). On the opposite side of the screw, one roller or needle bearing is used.

It is important to note that the compression process takes place simultaneously on each side of the main rotor of the single-screw compressor. This balanced gas pressure results in zero loads on the rotor bearings during full load and while symmetrically unloaded. Should the compressor be unloaded asymmetrically (see economizer operation below 50% capacity), the designer is not restricted by the rotor geometry and can easily add bearings with a long design life to handle the load. Axial loads are also low because the flutes terminate on the outer cylindrical surface of the rotor and suction pressure is vented to both ends of the rotor.

Each gate rotor shaft has one pair of angular contact ball bearings and one roller bearing. The gate rotor bearings must overcome a small moment of force due to the gas acting on the compression surface of the gate rotor. Since the single-screw compressor's physical geometry places no constraints on bearing size, $B_{10}$ lives of 200,000 hours are typical.

### Volume Ratio

The degree of compression within the rotor flutes is predetermined for a particular port configuration on screw compressors having fixed suction and discharge ports. A characteristic of the compressor is the **volume ratio,** $V_i$, which is defined as the ratio of volume of the flute at the start of compression to the volume of the same flute when it first begins to open to the discharge port. Hence, the volume ratio is determined by the size and shape of the discharge port.

For maximum efficiency, the pressure generated within the flutes during compression should exactly equal the pressure in the discharge line at the moment when the flute opens to it. If this is not the case, then either overcompression or undercompression occurs, both resulting in internal losses. Although such losses cause no harm to the compressor, they increase power consumption and reduce efficiency.

Screw compressors equipped with economizers and properly selected volume ratios can reduce and, in some cases (such as air-conditioning and heat-pump applications with pressure ratio ranges from 2 to 6), eliminate these power increases due to an increase in pressure ratio. This is because as the pressure ratio increases, the flash gas supercharging the closed flute also increases, thus reducing any under-compression.

Volume ratios can be specified on most compressors. Selection should be made according to operating conditions. The built-in pressure ratio of a screw compressor is a function of the volume ratio, as follows:

$$P_i = V_i^n$$

*where*

n is the isentropic exponent for the refrigerant being used.

Usually, the alternative port configurations to give the required volume ratio are designed into the capacity control components, thus providing ease of interchangeability, both during construction and after installation (although partial disassembly is required).

### Capacity Control

As with all positive displacement compressors, both speed modulation and suction throttling can be used. Ideal capacity modulation for any compressor includes (1) continuous modulation from 100% to less than 10%, (2) good part-load efficiencies, (3) unloaded starting, (4) unchanged reliability.

Variable compressor displacement, the most common means for meeting these criteria, usually takes the form of two movable valves in the compressor casing (the single-screw compressor has two gate rotors forming two compression areas). At part load, each slide valve produces a slot that retards the point at which compression begins. This causes a reduction in flute volume and, hence, in compressor throughput. As the intake charge is displaced before compression, little or no thermodynamic losses occur. However, if no other steps were taken, this mechanism would result in an undesirable drop in the effective volume ratio, and would result in undercompression and inefficient part-load operation. This problem is avoided by arranging the valve mechanism to reduce the discharge port area at the same time as the bypass slot is created. A fully modulating mechanism is provided in most large single-screw compressors, while two-position slide valves fit the requirements of the air-conditioning field. The specific part-load performance will be affected by compressor built-in volume ratio, $V_i$, evaporator temperature, and condenser temperature.

Detail design of the valve mechanism differs between makes of compressors but usually consists of what is shown in Figure 13 and 14. Figure 13 shows an axial sliding valve along each side of the rotor casing. This mechanism is usually operated by a hydraulic or gas piston and cylinder assembly located within the compressor itself. The piston is actuated either by oil, discharge gas, high-pressure liquid refrigerant, or a positioning motor at system discharge pressure being driven in either direction according to the operation of a four-way solenoid valve.

Figure 14 shows a rotatable ring built into the casting around the main rotor. This ring incorporates the bypass slot and discharge port and is turned by an external gear motor.

### Cooling, Sealing, and Bearing Lubrication

A major function of injecting a fluid into the compression area is the removal of the heat of compression. Also, since the single-screw compressor (like all screw compressors) has fixed leakage areas, the fluid also helps seal potential leakage paths. Fluid is normally injected into a closed flute through ports either in the casing or through ports in the moving capacity control

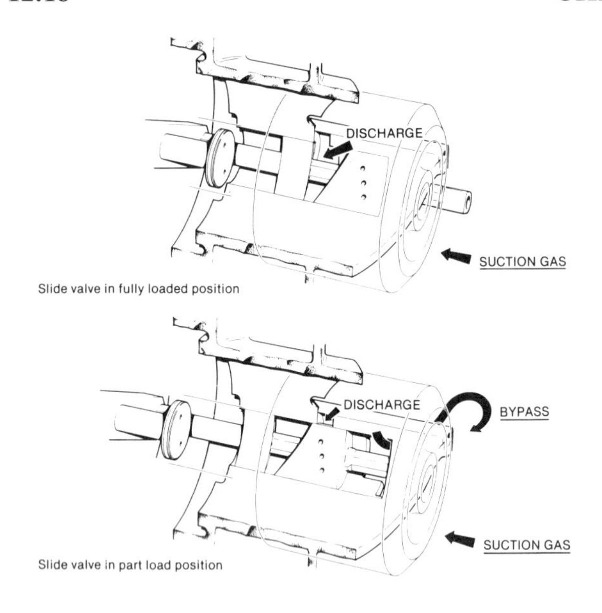


**Fig. 13 Single-Screw Axial-Type Capacity Control Mechanism**

slide. Most single-screw compressors allow the use of many different injection fluids to suit the gas being compressed.

## Oil-Injected Compressors

Oil is used in single-screw compressors to seal, cool, lubricate, and actuate capacity control. It gives a flat efficiency curve over a wide compression ratio and speed range, thus decreasing discharge temperature and reducing noise levels. In addition, the compressor can handle some liquid flood-back because it tolerates oil.

Oil-injected single-screw compressors operate at high head pressures using the common high-pressure refrigerants such as R-12, R-22, R-502, R-114, and R-717. They also operate effectively at high-pressure ratios because the injected oil cools the compression process. Currently, compressors with capacities in the 20 to 1500 hp (15 to 1125 kW) range are manufactured.

Oil injection requires an oil separator to remove the oil from the high-pressure refrigerant. For those applications with exacting demands for low oil carry-over, separation equipment is available to leave less than 5 ppm (5mg/kg) oil in the circulated refrigerant.

With most compressors, oil can be injected automatically without a pump because of the pressure difference between the oil reservoir (discharge pressure) and the reduced pressure in a

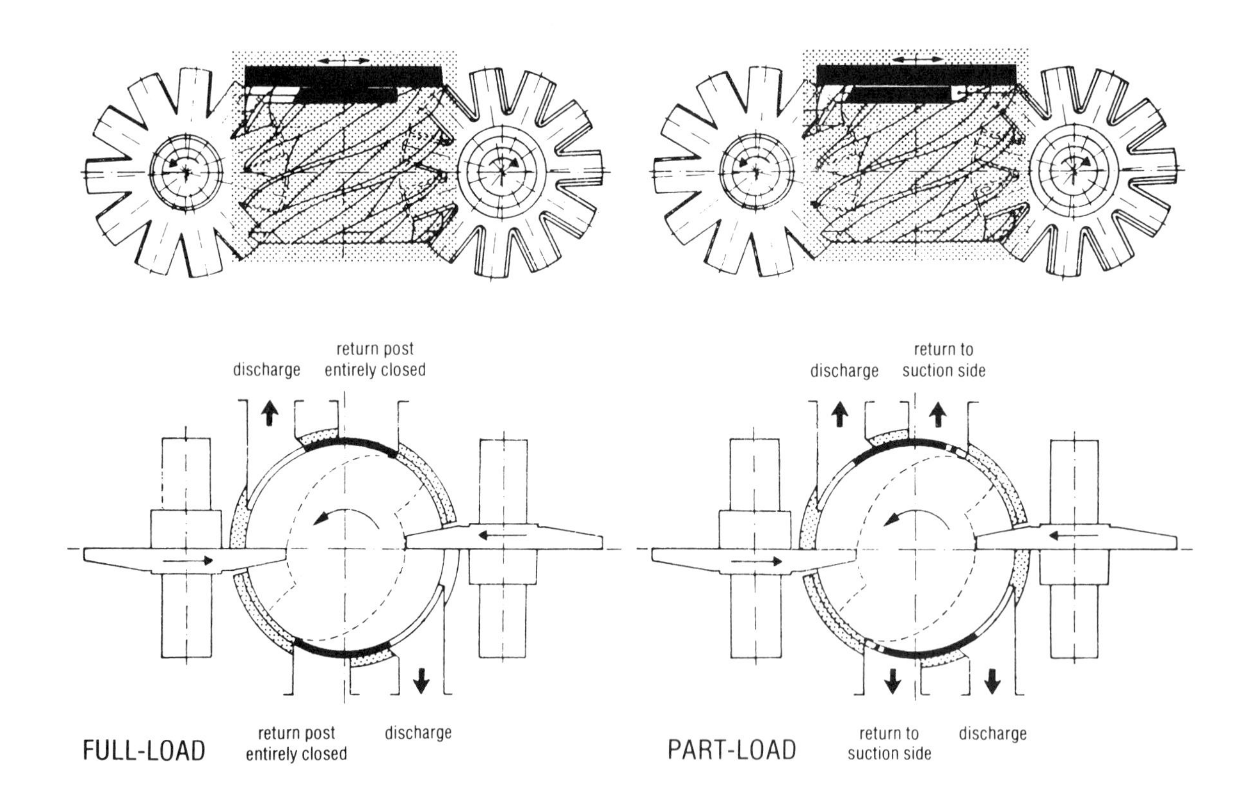


**Variable Return Port**

The groove volume, V1, which is determined by the rotor diameter, determines the capacity. Gradually reducing V1 controls this capacity at constant speed and delays the beginning of compression. This delay is caused by a variable return port, which allows the groove (after it has been sealed off by a tooth) to communicate with the suction chamber.

**Rotating Control Ring**

A rotating control ring allows capacity control adjustment. The ring, located in a recess in the cylindrical rotor housing, forms a return port with an infinitely variable size. A ring gear positions the control externally, so it achieves loss-free and stepless capacity through a range from 20 to 100%.

**Fig. 14 Single-Screw Capacity Control (Rotating Ring-Type)**

flute or bearing assembly during compression. A continuously running oil pump is used in some compressors to generate an oil pressure of 30 to 45 psi (200 to 300 kPa) over compressor discharge pressure. This pump requires 0.3 to 1.0% of the compressor's motor power.

Several methods of oil cooling are listed below:

1. Direct injection of liquid refrigerant into the compression process. Injection is controlled directly from the compressor discharge temperature, and loss of compressor capacity is minimized as injection takes place in a closed flute just before discharge occurs. This method requires very little power (typically less than 5% of compressor power).
2. A small refrigerant pump draws liquid from the receiver and injects it directly into the compressor discharge line. The injection rate is controlled by sensing discharge temperature and modulating the pump motor speed. The power penalty in this method is the pump power (about 1 hp for compressors up to 1000 hp, which can result in energy savings over refrigerant injected into the compression chamber.
3. External oil cooling between the oil reservoir and the point of injection is possible. Various heat exchangers are available to cool the oil: (1) separate water supply, (2) chiller water on a package unit, (3) condenser water on a package unit, (4) water from an evaporative condenser sump, (5) forced air-cooled oil cooler, and (6) high pressure liquid recirculation (thermosyphon).

   The heat added to the oil during compression is the amount usually removed in the oil cooler.

## Oil-Injection-Free Compressors

While the single-screw compressor operates well with oil injection, it also operates with equal or better efficiencies in an *oil-injection-free* (OIF) mode with many common halocarbon refrigerants. When pressure ratios are in the range of 2 to 8, oil normally injected into the casing may be replaced by liquid refrigerant. No lubricaiton is required because the only power transmitted from the screw to the gate rotors is that needed to overcome small frictional losses. Thus, the refrigerant need only cool and seal the compressor. The liquid refrigerant may still contain a small amount of oil to lubricate the bearings (0.1 to 1%, depending on compressor design).

OIF operation has the following advantages:

1. Single-screw compressors require no discharge oil separators.
2. Semi-hermetic compressors require no oil or refrigerant pumps.
3. External coolers are not required.
4. The compressor accepts liquid refrigerant entering the suction.
5. Higher condensing temperatures than conventional oil-injected compressors are permitted because the compressor discharges at saturation temperature.

## Performance Characteristics

Figures 15 and 16 show typical efficiencies of all designs of single-screw compressors. High isentropic and volumetric efficiencies are the result of internal compression, the absence of suction or discharge valves and their losses, and extremely small clearance volumes. The curves show the importance of selecting the correct volume ratio. Although volumetric efficiency is affected little by volume ratio, isentropic efficiency depends greatly on the correct ratio.

Manufacturer's data for operating conditions or speed should not be extrapolated. Screw compressor performance at reduced

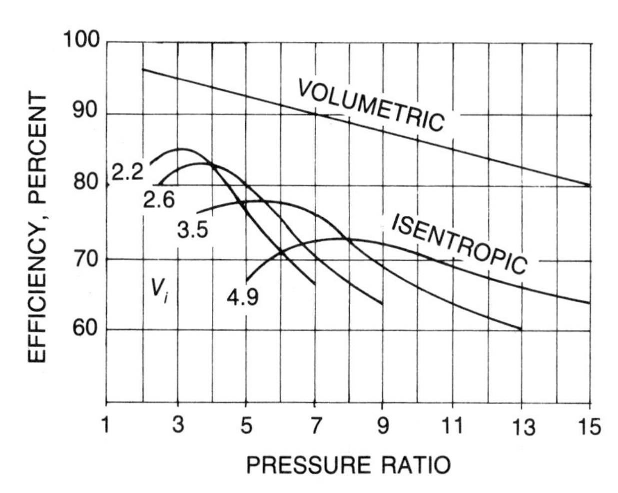


**Fig. 15 Efficiency of a Typical Single-Screw Compressor Operating with R-717 (Ammonia)**

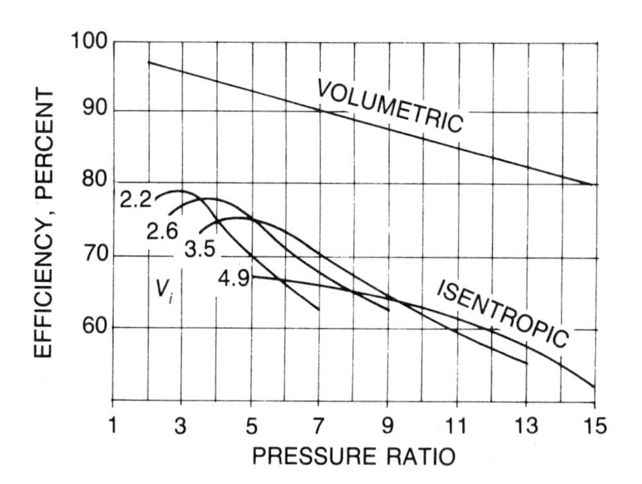


**Fig. 16 Efficiency of Typical Single-Screw Compressor Operating with R-22**

speed is usually significantly different from that specified at the normally rated point. Performance data normally include information about the degree of liquid subcooling and suction superheating assumed in the data. In the case of economizer operation (Figure 19), the liquid temperature approach to the economizer will be specified. Figure 17 gives typical characteristics during part-load power consumption.

## Economizers

Screw compressors are available with a secondary suction port between the primary compressor suction and discharge port. This port, when used with an economizer, improves compressor-useful refrigeration and improves compressor efficiency (Figure 18).

In operation, gas is drawn into the rotor flutes in the normal way from the suction line. The flutes are then sealed off in sequence and compression begins. An additional charge is added to the closed flute through a suitably placed port in the casing by an intermediate gas source at a slightly higher pressure than that reached in the compression process at that time. The original

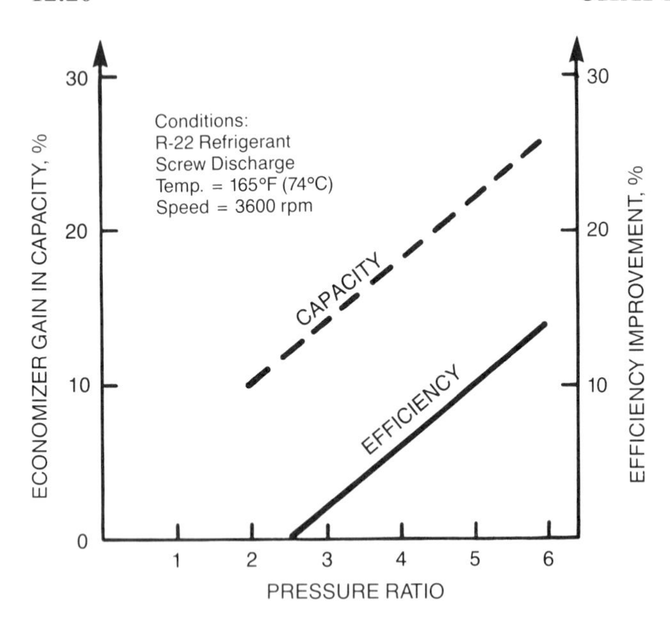


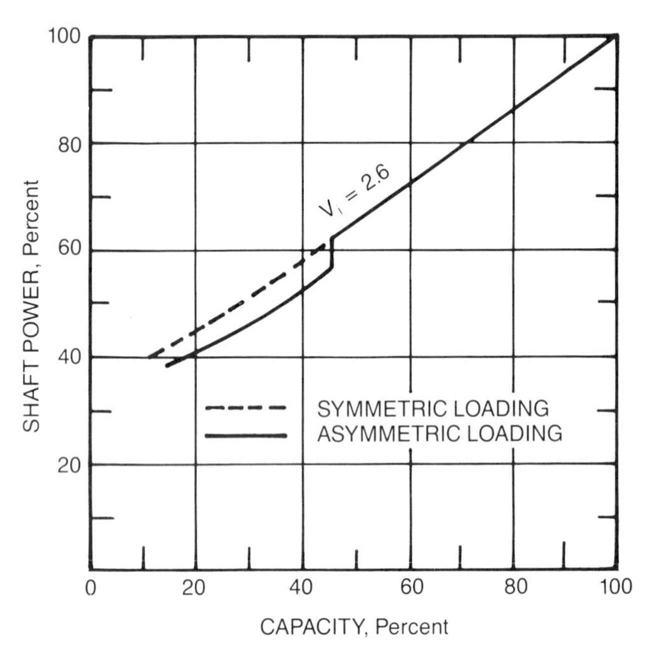


Fig. 17 Single-Screw Compressor Part-Load Characteristics at Low-Pressure Ratios

and the additional charge are then compressed together to discharge conditions. The pumping capacity of the compressor at suction conditions is not affected by this additional flow through the economizer port.

When the port is used with an economizer, most of the flash gas generated by the expansion of the high-pressure liquid to low-pressure liquid enters a closed flute. Since this does not affect the suction capacity of the compressor, the effective refrigerating capacity of the economized compressor is increased over the non-economized compressor by the amount of the flash gas (volume) plus the increased heat absorption capability ($\Delta H$) of the liquid entering the evaporator. Furthermore, the only additional mass flow the compressor must handle is the flash gas entering a closed flute, which is above suction pressure. Thus, under most conditions, the capacity improvement is accompanied by an efficiency improvement (Figure 18). Economizers become effective when system pressure ratios are equal to 4 and above.

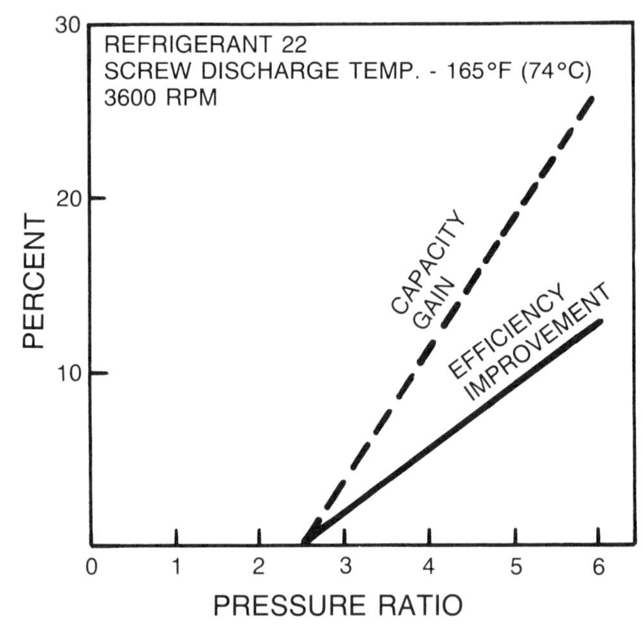


Fig. 18 Typical Improvement in Screw-Compressor Efficiency and Capacity with an Economizer on R-22

Figure 19 shows a pressure-enthalpy diagram for a flash-tank economizer system. In it, high-pressure liquid passes through

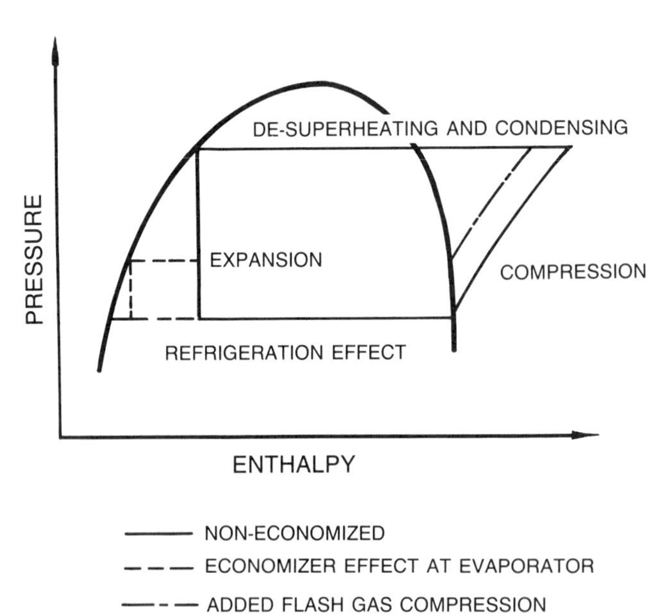


Fig. 19 Theoretical Economizer Cycle on a Pressure-Enthalpy Diagram

an expansion device and enters a tank at an intermediate pressure between suction and discharge. This pressure is maintained by the pressure in the compressor's closed flute (closed from suction). The gas generated from the expansion enters the compressor through the economizer port. The liquid (which is now saturated at the intermediate pressure), when passed to the evaporator, gives a larger refrigeration capacity per pound (kg). In addition, the compressor has more useful volume capacity at suction because the flash is eliminated at the suction, and the increase in power input is a lower percentage than the capacity increase (due to the flash gas entering the closed flute and being compressed to discharge).

As screw compressors are unloaded, the economizer pressure falls toward suction pressure. As a result, the additional capacity and improved efficiency of the economizer system falls to zero at between 70 to 80% of full-load capacity.

The single-screw compressor has two compression chambers, each having its own slide valve. Each slide valve can be operated independently, which gives the opportunity to introduce economizer gas into one side of the compressor. By operating the slide valves independently, the chamber without the economizer gas can be unloaded to 0% capacity (50% capacity of the compressor). The other chamber remains at full capacity and retains the full economizer effect, making the economizer effective below 50% compressor capacity.

The secondary suction port may also be used for (1) a system side load or (2) second system evaporator that operates at a temperature level above the system evaporator.

### Semi-hermetic Design

The semi-hermetic single-screw compressor was introduced for air-conditioning and heat-pump services in the early 1980's. Figure 20 shows a semi-hermetic single-screw compressor.

## DOUBLE HELICAL ROTARY SCREW COMPRESSORS

The twin-screw compressor is a positive-displacement compressor. It consists of two mating helically grooved rotors—a male (lobes) and a female (flutes or gullies) in a stationary housing with inlet and outlet gas ports (Figure 21). The flow of gas in the rotors is mainly in an axial direction. Frequently used lobe combinations are 4+6, 5+6, and 5+7 (male + female). For instance, with a four-lobe male rotor, the driver, rotates at 3600 rpm; the six-lobe female rotor follows at 2400 rpm. The female rotor can be driven through synchronized timing gears or directly driven by the male rotor on a light oil film. In some applications, it is practical to drive the female rotor, which results in a 50% speed and displacement increase over the male-driven compressor, assuming 4+6 lobe combination. Geared speed increasers are also used on some applications to increase the capacity delivered by a particular compressor size.

Double helical screws find application in many air-conditioning, refrigeration, and heat-pump applications—typically in the industrial and commercial market. Machines can be designed to operate at high- or low-pressure levels and are often applied below 2:1 and above 20:1 compression ratios single stage. Commercially available compressors are suitable for application on all normally used high-pressure refrigerants.

### The Compression Process

Compression is obtained by direct volume reduction with pure rotary motion. For clarity, the following description of the three basic compression phases is limited to one male rotor lobe and one female rotor interlobe space (Figure 22).

**Suction.** As the rotors begin to unmesh, a void is created on both the male side (male thread) and the female side (female

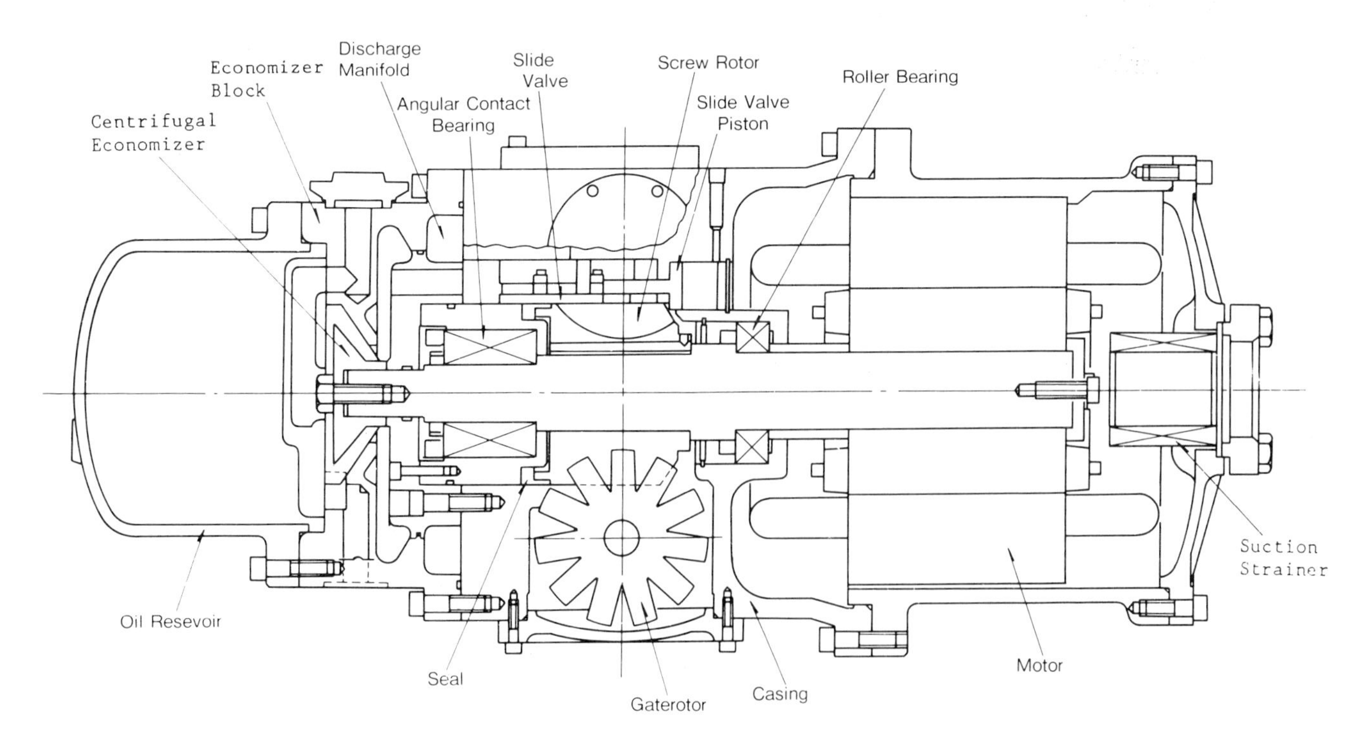


**Fig. 20 OIF Semi-Hermetic Single-Screw Compressor**

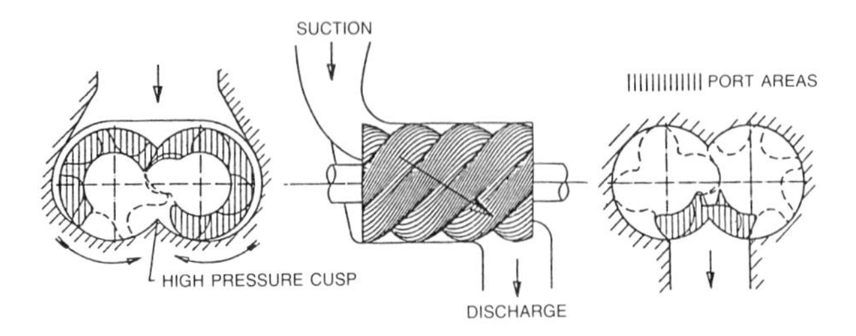


Fig. 21 Twin-Screw Compressor

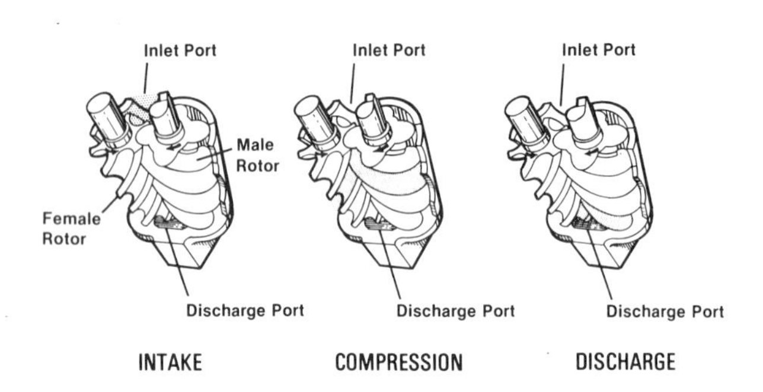


Fig. 22 Compression Process

thread), and gas is drawn in through the inlet port. As the rotors continue to turn, the interlobe space increases in size and gas flows continuously into the compressor. Just prior to the point at which the interlobe space leaves the inlet port, the entire length of the interlobe space is completely filled with gas.

**Compression.** Further rotation starts the meshing of another male lobe with another female interlobe space on the suction end and progressively compresses the gas in the direction of the discharge port. Thus, the occupied volume of the trapped gas within the interlobe space is decreased and the gas pressure consequently increased.

**Discharge.** At a point determined by the designed built-in volume ratio, the discharge port is uncovered and the compressed gas is discharged by further meshing of the lobe and interlobe space.

During the remeshing period of compression and discharge, a fresh charge is drawn through the inlet on the opposite side of the meshing point. With four male lobes rotating at 3600 rpm (60 r/s), four interlobe volumes are filled and are giving 14,400 discharges per minute (240 discharges per second). Since the intake and discharge cycles overlap effectively, a smooth continuous flow of gas results.

## MECHANICAL FEATURES

### Rotor Profiles

Helical rotor design has evolved over the years, starting with an asymmetric point-generated rotor profile in 1935. This profile was only used in compressors with timing gears (dry compression). The symmetric, *circular* rotor profile was introduced in 1947 because it was easier to manufacture than the preceding profile and it could be used without timing gears for wet compression.

The asymmetric, point-, and line-generated profile was introduced in 1967, giving higher performance due to better rotor dynamics and decreased leakage areas. This design allowed the possibility for female rotor drive, as well as the conventional male drive.

During the 1980's, helical rotor design has spread to a large family of rotor profiles and lobe combinations, designated for different applications, manufacturing methods, and rotor materials. These newer rotor profiles are normally asymmetric and line-generated. Rotor profile, blowhole, length of sealing line, quality of sealing line, torque transmission between rotors, rotor-housing clearances, interlobe clearances, and lobe combinations are optimized for specific pressures, temperatures, speeds, and wet or dry operation. Optimal rotor tip speed is 3000 to 8000 fpm (15 to 40 m/s) for wet operation (oil flooded) and 12,000 to 24,000 fpm (60 to 120 m/s) for dry operation.

### Rotor Contact and Loading

Contact between the male and female rotors is mainly rolling and primarily situated at a contact bank on each rotor's pitch circle. Rolling at this contact band means virtually no rotor wear occurs.

**Gas Forces.** On the driven rotor, the internal gas forces always create a torque in a direction opposite to the direction of rotation. This has been called *positive* or braking torque.

On the undriven rotor, the design can be made so that the torque is positive, negative, or zero, except on female drive designs where zero or negative torque does not occur. Negative torque occurs when internal gas forces tend to drive the rotor. If the average torque on the undriven rotor is near zero, this rotor is subjected to torque reversals as it goes through its phase angles. Under certain conditions, this can cause instability problems. Torque transmitted between the rotors does not create problems because they are mainly in rolling contact.

**Male Drive.** The transmitted torque from male rotor to female rotor is normally 5 to 25% of input torque.

**Female Drive.** The transmitted torque from female rotor to male rotor is normally 50 to 60% of input torque.

**Rotor Loads.** The rotors in an operating compressor are subjected to radial, axial, and tilting loads. Tilting loads are radial loads caused by axial loads outside of the rotor center line. The axial load is normally balanced with a balancing piston for larger high-pressure machines (rotor diameter above 6 in. (150 mm) and discharge pressures above 160 psi (1.1 MPa). Balancing pistons are typically close tolerance, labyrinth-type devices with high-pressure oil or gas on one side and low pressure on the other side. They are used to produce a thrust load to offset some of the primary gas loading on the rotors, thus reducing the amount of thrust load the bearings support.

### Bearings

Twin-screw compressors normally have either four or six bearings, depending on whether one or two bearings are used for the radial and axial loads. Some designs incorporate multiple rows of smaller bearings per shaft to share the loads. Sleeve bearings have been used historically to support radial loads in machines with male rotor diameters larger than 6 in. (150 mm), while anti-friction bearings are generally applied to smaller machines. However, improvements in anti-friction designs and materials have led to compressors with up to 12 in. (300 mm) rotor diameter with full anti-friction bearing designs. Cylindrical and tapered roller bearings, and various types of ball bearings are used in screw compressors for carrying radial loads. The most common thrust or axial load-carrying bearings are angular contact ball bearings, though tapered rollers or tilting pad bearings are used in some machines.

### General Design

Screw compressors are often designed for particular pressure ranges. Low-pressure compressors have long, high-displacement rotors and adequate space to accommodate bearings to handle the relatively light loads. They are frequently designed without thrust balance pistons, since the bearings alone can handle the low thrust loads and still maintain good life.

High-pressure compressors have short and strong rotors (shallow groves) and, therefore, have space for large bearings. They are normally designed with balancing pistons for high thrust bearing life.

**Rotor Materials.** Rotors are normally made of steel, though aluminum, cast iron, and nodular iron are used in some applications.

## CAPACITY CONTROL

As with all positive-displacement compressors, both speed modulation and suction throttling can reduce the volume of gas drawn into a screw compressor. Ideal capacity modulation for any compressor would be (1) continuous modulation from 100% to less than 10%, (2) good part-load efficiency, (3) unloaded starting, and (4) unchanged reliability. However, not all applications need ideal capacity modulation. Variable compressor displacement and variable speed are the best means for meeting these criteria. Variable compressor displacement is the most common capacity-control method used. Various mechanisms achieve variable displacement, depending on the requirements of a particular application.

### Capacity Slide Valve

A slide valve for capacity control is a valve with sliding action parallel to the rotor bores. They are placed within or close to the high-pressure cusp region, face one or both rotor bores, and bypass a variable portion of the trapped gas charge back to suction, depending on their position. Within this definition, there are two types of capacity slide valves.

1. **Capacity slide valve regulating discharge port.** This type of slide valve is located within the high-pressure cusp region. It controls capacity, as well as the location of the radial discharge port at part load. The axial discharge port is designed for a volume ratio giving good part-load performance without losing full-load performance. Figure 23 shows a schematic view of the most common arrangement.
2. **Capacity slide valve that does not regulate the discharge port.** A slide valve outside the high-pressure cusp region controls only capacity and not the radial discharge port.

   The first type is the most common arrangement. It is, in general, the most efficient of the available volume reduction methods, due to its indirect correction of built-in volume ratio at part load and its ability to give large capacity reductions without large movement of the slide valve.

### Capacity Slot Valve

A capacity slot valve consists of a number of slots that follow the rotor helix and face one or both rotor bores. The slots are gradually opened or closed with a plunger or turn valve. These recesses in the casing wall increase the volume of the compression space and also create leakage paths over the lobe tips. The result is somewhat lower full-load performance when compared to a design without slots.

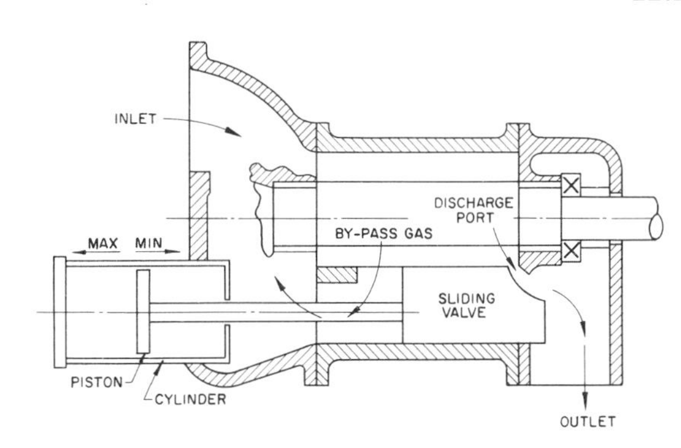


**Fig. 23 Slide-Valve Unloading Mechanism**

### Capacity Lift Valve

Capacity lift valves or plug valves are movable plugs in one or both rotor bores (with radial or axial lifting action) that regulate the actual start of compression. These valves control capacity in a finite number of steps, rather than by the infinite control of a conventional slide valve. (Figure 24).

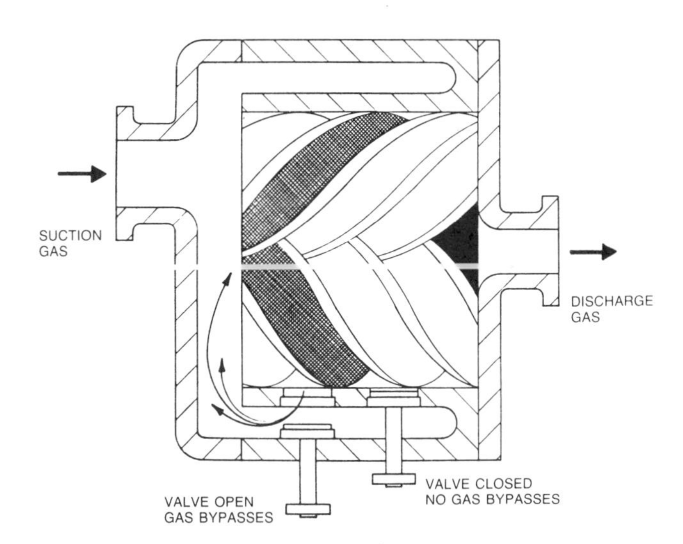


**Fig. 24 Lift-Valve Unloading Mechanism**

Neither slot valves nor lift valves offer quite as good efficiency at part load as a slide valve, because they do not relocate the radial discharge port. Thus, under-compression losses at part load can be expected if the machines have the correct volume ratio for full-load operation and the compression ratio at part load does not reduce.

## VOLUME RATIO

In all positive-displacement rotary compressors with fixed-port location, the degree of compression within the rotor thread is determined by the location of the suction and discharge ports. The built-in volume ratio, $V_i$, of screw compressors is defined as the ratio of volume of the thread at the start of the compression process to the volume of the same thread when it first begins to open to the discharge port. The suction port must be located to trap the maximum suction charge; hence, the volume ratio is determined by the location of the discharge port.

Only the suction pressure and volume ratio of the compressor determine the internal pressure achieved before opening to discharge. However, the condensing and evaporating temperatures determine the system discharge pressure and the compression ratio in the piping that leads to the compressor. Any mismatch between the internal and system discharge pressures results in under- or over-compression losses and poorer efficiency.

If the operating conditions of the system seldom change, it is possible to specify a fixed-volume-ratio compressor that will give good efficiency. Compressor manufacturers normally make compressors with three or four possible discharge port locations that correspond to system conditions encountered frequently. Generally, the system designer is responsible for specifying a compressor that most closely matches expected pressure conditions.

The required volume ratio for a particular application can be calculated as follows:

$$CR = P_s/P_d \quad (7)$$

*where*

$CR$ = system compression ratio
$P_s$ = expected suction pressure in absolute units
$P_d$ = expected discharge pressure in absolute units

Then, compression as an isentropic process is approximated as follows:

$$P_i = V_i^k \quad (8)$$

*where*

$P_i$ = internal pressure ratio
$V_i$ = compressor volume ratio
$k$ = ratio of specific heats for the refrigerant used

And finally, the selected compressor, as nearly as possible, has the following internal pressure ratio:

$$P_i = CR \quad (9)$$

Usually, in slide-valve-equipped compressors, the radial discharge port is located in the discharge end of the slide valve. A short slide valve gives a low volume ratio, and a long slide valve gives a higher volume ratio. The difference in length basically locates the discharge port earlier or later in the compression process. Different length slide valves allow changing the volume ratio of a given compressor, though disassembly is required.

### Variable Volume Ratio

Some newer twin-screw compressors adjust the volume ratio of the compressor, while operating, to the most efficient ratio for whatever system pressures are encountered.

In fixed-volume ratio compressors, the motion of the slide valve toward the inlet end of the machine is stopped when it comes in contact with the rotor housing in that area. In the most common of the variable volume ratio machines, this portion of the rotor housing has been replaced with a second slide, the *moveable slide stop*, which can be actuated to different locations in the slide valve bore (Figure 25).

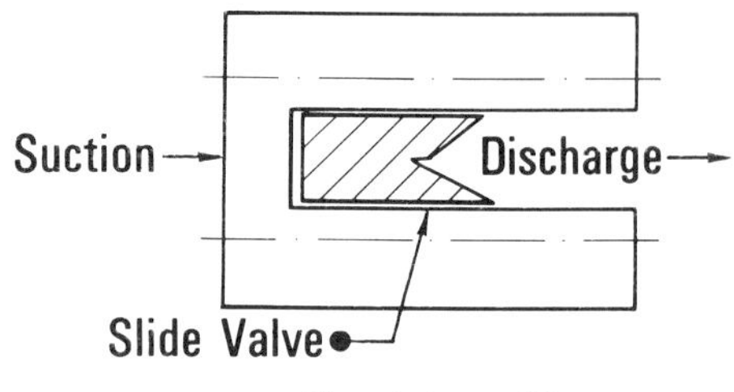


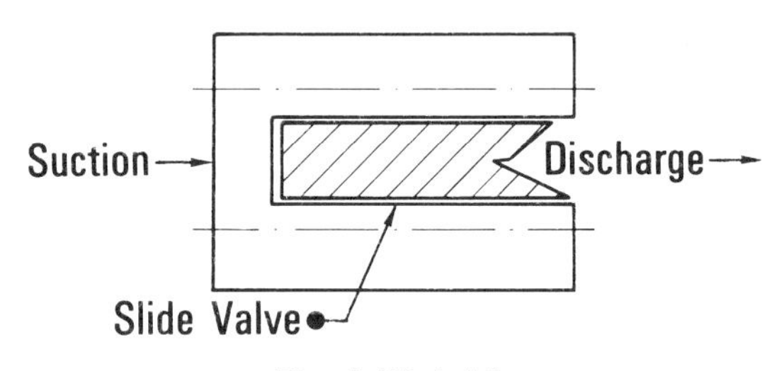


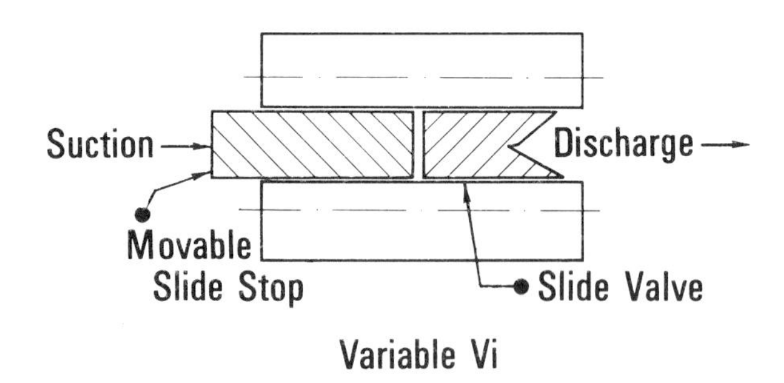


**Fig. 25 View from Above of Fixed- and Variable-Volume Ratio (Vi) Slide Valves**

By moving the slides back and forth, the radial discharge port can be relocated during operation to match the compressor volume ratio to the optimum. This added flexibility allows operation at different suction and discharge pressure levels while still maintaining maximum efficiency. The comparative efficiencies of fixed and variable volume ratio screw compressors are shown in Figure 26 for full-load operation on ammonia and R-22 refrigerants. The figure shows that a variable volume ratio compressor efficiency curve encompasses the peak efficiencies of compressors with fixed-volume ratios over a wide range of system pressure ratios.

Following are other secondary effects of variable volume ratio:

- less oil foam in oil separator (no over-compression)
- less oil carried over into the refrigeration system (because of less oil foam in oil separator)
- extended bearing life—minimized load on bearings
- extended efficient operating range with economizer discharge port corrected for flash gas from economizer, as well as gas coming from suction
- lower noise levels
- lower discharge temperatures and oil cooler heat rejection

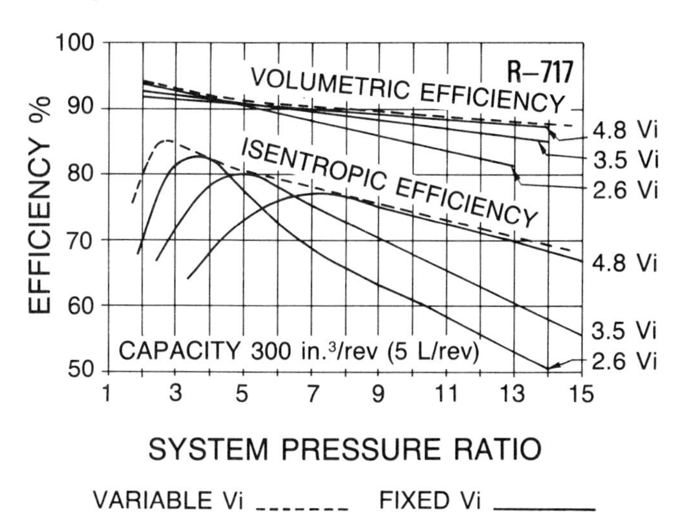


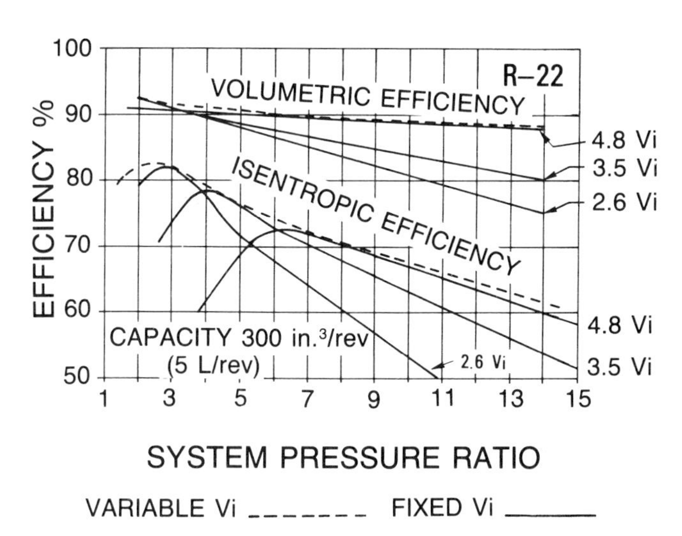


**Fig. 26 Twin-Screw Compressor Efficiency Curves**

The greater the change in either suction or condensing pressure that a given system experiences, the more benefits are possible with variable volume ratio. Efficiency improvements as high as 30% are possible, depending on the application, refrigerant, and system operating range.

## OIL INJECTION

Two primary types of compressor lubrication systems are employed in twin-screw compressors: dry and oil flooded.

### Dry Operation (No Rotor Contact)

Since the two rotors in twin-screw compressors are parallel, timing gears are a practical means of synchronizing the rotors so they do not touch each other. Eliminating rotor contact eliminates the need for lubrication in the compression area. Initial screw compressor designs were based on this approach, and dry screws still find application in the gas process industry.

Synchronized twin-screw compressors once required high rotor tip speed and were, therefore, rather noisy. However, with today's profile technology, the synchronized compressor can run at lower tip speeds and higher pressure ratios, giving quieter operation. The added cost of timing gears and internal seals generally make the dry screw more expensive than an oil-flooded screw for normal refrigeration or air-conditioning duty.

### Oil-Flooded Operation

The oil-flooded twin-screw compressor is the most common type of screw used in refrigeration and air conditioning. Compressor capacities range from 6 to 6000 cfm (3 to 2800 L/s). Oil-flooded compressors typically have oil supplied to the compression area at a volume rate of about 0.5% of the displacement volume. Part of this oil is used for lubrication of the bearings and shaft seal prior to injection. Typically, paraffinic- or napthenic-based mineral oils are used, though synthetics are used on some applications. The oil is normally injected into a closed thread through ports in the moving slide valve or through stationary ports in the casing.

The oil fulfills three primary purposes: sealing, cooling, and lubrication. It also tends to fill any leakage paths between and around the rotors. This keeps volumetric efficiency (VE) high, even at high compression ratios. Normal compressor VE exceeds 85% even at 25:1 single stage (ammonia, 7.6 in. or 193 mm rotor diameter). It also gives flat efficiency curves with decreasing speeds, where desired, for quiet operation. Oil transfers much of the heat of compression from the gas to the oil, keeping typical discharge temperature below 190°F (88°C), which allows high compression ratios without danger of breaking down the refrigerant or the oil. The lubrication function of the oil protects bearings, seals, and the rotor contact areas.

The ability of a screw compressor to tolerate oil also permits the compressor to handle a certain amount of liquid floodback, as long as the liquid quantity is not large enough to lock the rotors hydraulically.

### Oil Separation and Cooling

Oil injection requires an oil separator to remove oil from the high-pressure refrigerant. Separation equipment routinely gives less than 5 ppm (5 mg/kg) oil in the circulated refrigerant.

Oil injection is normally achieved by one of two methods: (1) with a continuously running oil pump, capable of generating an oil pressure of 30 to 45 psi (0.2 to 0.3 MPa) over compressor discharge pressure, representing 0.3 to 1.0% of compressor motor power; or (2) with some compressors, oil can be injected automatically without a pump because of the pressure difference between the oil reservoir (discharge pressure) and the reduced pressure in a thread during the compression process.

Since the oil absorbs a significant amount of the heat of compression in oil-flooded operation, it must be cooled to maintain low discharge temperatures. One cooling method is by direct injection of liquid refrigerant into the compression process. The injected liquid refrigerant amount corresponds to 0.02% of displacement volume. The amount of liquid injected is normally controlled by sensing the discharge temperature and injecting enough liquid to maintain a constant temperature level. Some of the injected liquid mixes with the oil and leaks to lower pressure threads, where it tends to raise pressures and reduce the amount of gas the compressor can draw in. Also, any of the liquid that has time to absorb heat and expand to vapor must be recompressed, tending to raise absorbed power levels. Compressors are normally designed with the liquid injection ports as late as possible in the compression to minimize capacity and power penalties. Typical penalties for liquid injection are in the 1 to 10% range, depending on compression ratio.

Another method of oil cooling draws liquid from the receiver with a small refrigerant pump and injects it directly into the compressor discharge line. The power penalty in this method is the pump power (about 1 hp for compressors up to 1000 hp).

In the third method, the oil can be cooled outside the compressor between the oil reservoir and the point of injection. Various configurations of heat exchangers are available for this purpose, and the oil cooler heat rejection can be accomplished by (1) separate water supply, (2) chiller water on a package unit, (3) condenser water on a package unit, (4) water from an evaporative condenser sump, (5) forced air-cooled oil cooler, (6) liquid refrigerant, and (7) high-pressure liquid recirculation (thermosyphon).

External oil coolers using water or other means from a source independent of the condenser allow for the condenser to be reduced in size by an amount corresponding to the oil cooler capacity. Where oil cooling is carried out from within the refrigerant system by means such as (1) direct injection of liquid refrigerant into the compression process or the discharge line, (2) direct expansion of liquid in an external heat exchanger, (3) using chiller water on a package unit, (4) recirculation of high-pressure liquid from the condenser, or (5) water from an evaporative condenser sump, the condenser must be sized for the total heat rejection, *i.e.*, evaporator load plus shaft power.

With an external oil cooler, the mass flow rate of oil injected into the compressor is usually determined by the oil cooler performance rather than by the compressor sealing requirements, since the oil is acting predominately as a heat-transfer medium. Conversely, with direct liquid injection cooling, the oil requirement is dictated by the compressor lubrication and sealing needs.

## ECONOMIZERS

Twin-screw compressors are available with a secondary suction port between the primary compressor suction and discharge ports. This port can accept a second suction load at a pressure above the primary evaporator, or flash gas from a liquid subcooler vessel, called an economizer.

In operation, gas is drawn into the rotor thread in the normal way from the suction line. The thread is then sealed in sequence and compression begins. An additional charge may be added to the closed thread through a suitably placed port in the casing. The port is connected to an intermediate gas source at a pressure slightly higher than that reached in the compression process at that time. Both original and additional charges are then compressed to discharge conditions.

When the port is used as an economizer, a portion of the high-pressure liquid is vaporized at the side-port pressure and subcools the remaining high-pressure liquid nearly to the saturation temperature at the operating side port. Since this has little effect on the suction capacity of the compressor, the effective refrigerating capacity of the compressor is increased by the amount of the flash gas (volume) plus the increased heat absorption capacity ($\Delta H$) of the liquid entering the evaporator. Furthermore, the only additional mass flow the compressor must handle is the flash gas entering a closed thread, which is above suction pressure. Thus, under most conditions, the capacity improvement is accompanied by an efficiency improvement.

Economizers become effective when system pressure ratios are equal to about 2 and above (dependent of volume ratio). The subcooling can be made with a heat exchanger, flash tank, or other separation means.

One concern occurs as twin-screw compressors are unloaded. As the economizer pressure falls toward suction pressure, the additional capacity and improved efficiency of the economizer system is no longer available below a certain percentage of capacity, depending on design.

## HERMETIC COMPRESSORS

Hermetic screw compressors are common, particularly in small sizes. The hermetic motors can operate under discharge, suction, or intermediate pressure. Motor cooling can be with gas, oil, and/or liquid refrigerant. The oil separator in oil-flooded operation is also normally integrated with the compressor. Figure 27 shows one type of hermetic twin-screw design.

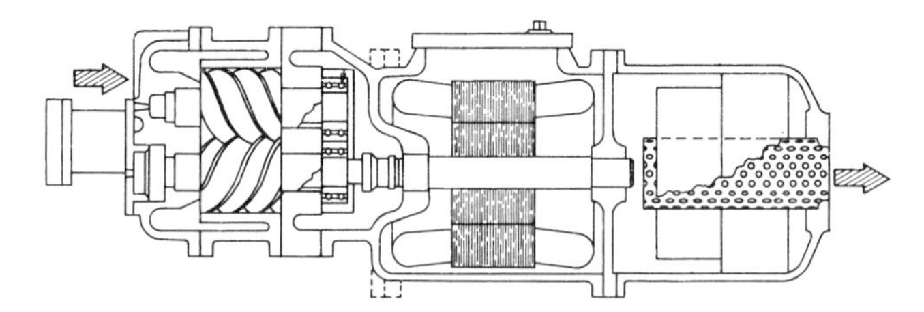

**Fig. 27 Hermetic Twin-Screw Compressor**

## PERFORMANCE CHARACTERISTICS

Figure 26 gives full-load efficiencies of a modern twin-screw compressor. Both fixed- and variable-volume ratio compressors without economizers are shown. High isentropic and volumetric efficiencies are the result of internal compression, the absence of suction or discharge valves, and small clearance volumes. The curves show that volumetric efficiency depends little on the choice of volume ratio, but isentropic efficiency depends strongly on it.

Performance data normally note the degree of liquid subcooling and suction superheating assumed. If an economizer is used, the liquid temperature approach and pressure drop to the economizer should be specified.

# PART V: CENTRIFUGAL COMPRESSORS

Centrifugal compressors, sometimes called **turbocompressors,** are members of a family of turbomachines that includes fans, propellers, and turbines. These machines continuously exchange angular momentum between a rotating mechanical element and a steadily flowing fluid. Because their flows are continuous, turbomachines have greater volumetric capacities, size-for-size, than do positive displacement devices. For effective momentum exchange, their rotative speeds must be higher, but little vibration or wear results because of the steadiness of the motion and the absence of contacting parts.

Centrifugal compressors are used in a variety of refrigeration and air-conditioning installations. Suction flow rates of compressors range between 60 and 30,000 cfm (0.03 and 15 $m^3/s$), with rotational speeds between 1800 and 90,000 rpm (30 and 1500 r/s). However, the high rotative speed associated with a low volumetric flow tends to establish a minimum practical capacity for most centrifugal applications. The upper capacity limit is determined by physical size, a 30,000 cfm (15 $m^3/s$) compressor being about 6 or 7 ft (1.8 or 2.1 m) in diameter.

Suction temperatures are usually between 50 and −150°F (10 and −100°C), with suction pressures between 2 and 100 psia (14 and 700 kPa) and discharge pressures up to 300 psia (2.1 MPa). Pressure ratios range between 2 and 30. Almost any refrigerant can be used.

As many as ten stages can be installed in a single casing. Sideloads can be introduced between stages so that one compressor performs several functions at several temperature levels. Multiple casings can be connected in tandem to a single driver. These can be operated in series, in parallel, or even with different refrigerants.

## THEORY

### Refrigeration Cycles

Figure 28 illustrates a simple vapor compression cycle in which a centrifugal compressor operates between states 1 and 2. Typical applications might involve a single- or two-stage halocarbon compressor or a seven-stage ammonia compressor.

Figure 29 shows a more complex cycle with interstage liquid flash cooling. This cycle has a higher coefficient of performance than the simple cycle. It is frequently used with two- and four-stage halocarbon and hydrocarbon compressors.

More than one stage of flash cooling can be applied to compressors that have more than two impellers. Liquid subcooling and interstage desuperheating can also be used to advantage. For information on refrigeration cycles, see Chapter 1 of the 1985 FUNDAMENTALS Volume.

### Angular Momentum

The momentum exchange between a centrifugal impeller and a flowing refrigerant is expressed by:

$$W_i = u_i c_u \qquad (10)$$

*where*

$W_i$ = impeller work input per lb (kg) of refrigerant
$u_i$ = impeller blade tip speed
$c_u$ = tangential component of the refrigerant velocity leaving the impeller blades

These velocities are shown in Figure 30, where refrigerant flows out from between the impeller blades with relative velocity $b$ and absolute velocity $c$. The relative velocity angle $\beta$ is a few degrees less than the blade angle because of a phenomenon called *slip.*

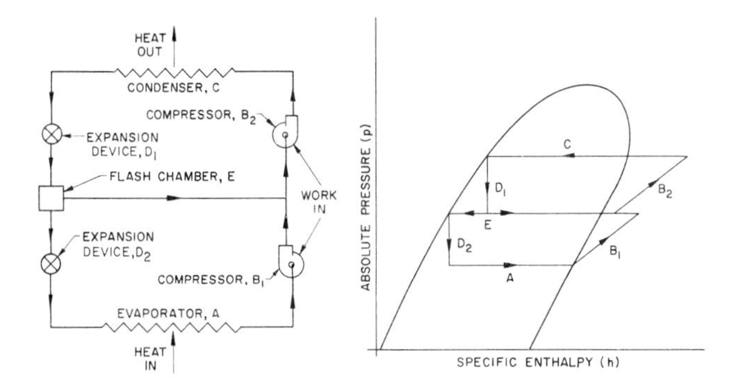


**Fig. 28 Simple Vapor Compression Cycle**

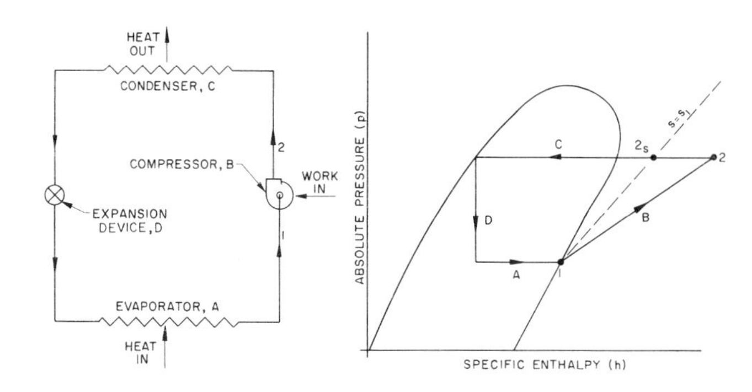


**Fig. 29 Compression Cycle with Flash Cooling**

Equation (10) assumes that the refrigerant enters the impeller without any tangential velocity component or swirl. This is generally the case at design flow conditions. If the incoming refrigerant was already swirling in the direction of rotation, the impeller's ability to impart angular momentum to the flow would be reduced. A subtractive term would then be required in the equation.

Some of the work done by the impeller increases the refrigerant pressure, while the remainder only increases its kinetic energy. The ratio of pressure-producing work to total work is called the **impeller reaction.** Since this varies from about 0.4 to about 0.7, an appreciable amount of kinetic energy leaves the impeller with magnitude $c^2/2000$.

To convert this kinetic energy into additional pressure, a diffuser must follow the impeller. Radial vaneless diffusers are most common, but vaned and scroll diffusers are also used.

In a multistage compressor, the flow leaving the first diffuser is guided to the inlet of the second impeller and so on through the machine, as can be seen in Figure 31. The *total compression work input per* lb (kg) *of refrigerant* is the sum of the individual stage inputs:

$$W = \Sigma W_i \qquad (11)$$

provided that the mass flow rate is constant throughout the compressor.

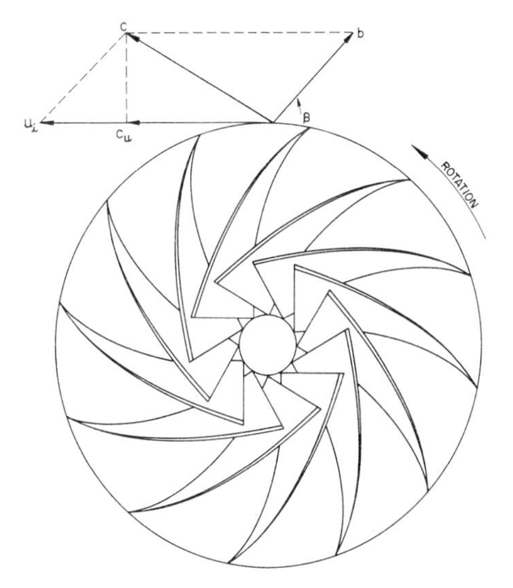


**Fig. 30 Impeller Exit Velocity Diagram**

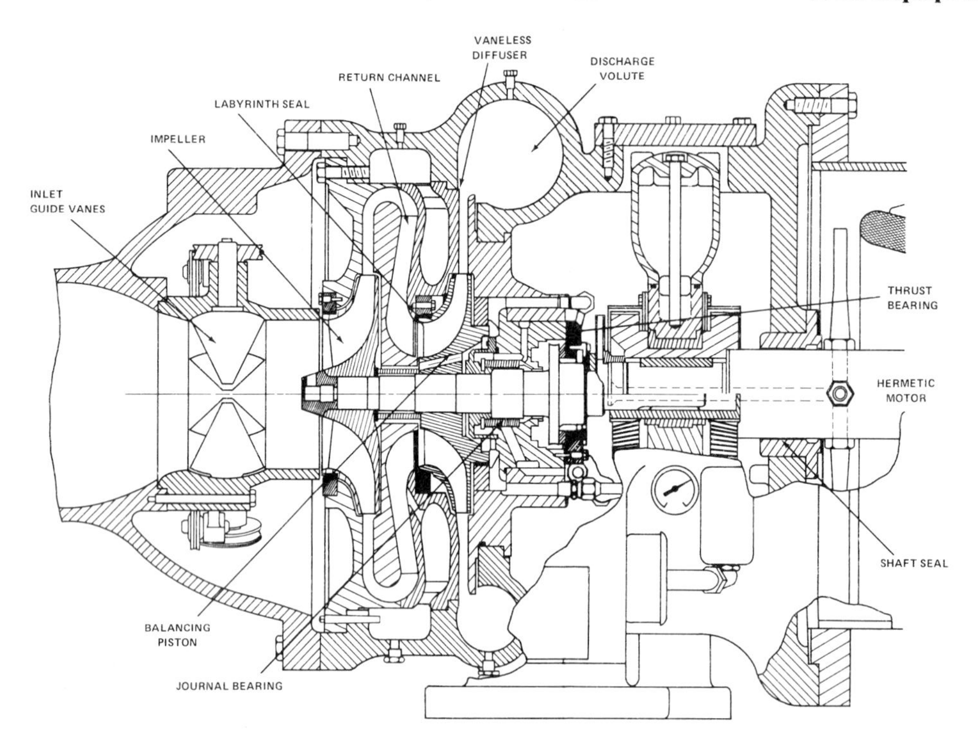


**Fig. 31 Centrifugal Refrigeration Unit Cross-Section**

## ISENTROPIC ANALYSIS

The static pressure that results from a compressor's work input or, conversely, the amount of work required to produce a given pressure rise, depends on the efficiency of the compression and the thermodynamic properties of the refrigerant. For an adiabatic process, the work input required is minimum if the compression is isentropic. Therefore, actual compression is often compared to an isentropic process and the performance thus evaluated is based on an isentropic analysis.

The reversible work required by an isentropic compression between states 1 and $2_s$ in Figure 28 is called the **adiabatic work:**

$$W_s = h_{2_s} - h_1 \tag{12}$$

The irreversible work done by the actual compressor is

$$W = h_2 - h_1 \tag{13}$$

assuming negligible cooling occurs. Flash-cooled compressors cannot be analyzed by this procedure unless they are subdivided into uncooled segments with the cooling effects evaluated by other means. Compressors with side-flows must also be subdivided. In Figure 29 the two compression processes must be analyzed individually.

Equation (13) also assumes a negligible difference in the kinetic energies of the refrigerant at states 1 and 2. If this is not the case, a kinetic energy term must be added to the equation. All of the thermodynamic properties throughout the section "Centrifugal Compressors" are *static* properties as opposed to *stagnation* properties; the latter artificially includes kinetic energy.

The ratio of isentropic work to actual work is the **adiabatic efficiency:**

$$\eta_s = \frac{W_s}{h_2 - h_1} \tag{14}$$

This varies from about 0.62 to about 0.83, depending on the application. Because of the thermodynamic properties of gases, a compressor's overall adiabatic efficiency does not completely indicate its individual stage performance. The same compressor produces different adiabatic results with different refrigerants and also with the same refrigerant at different suction conditions.

Despite its shortcomings, isentropic analysis has a definite advantage in that adiabatic work can be read directly from thermodynamic tables and charts similar to those presented in Chapter 17 of the 1985 FUNDAMENTALS Volume. Where these are unavailable for the particular gas or gas mixture, they can be accurately calculated and plotted using thermodynamic relationships and a computer.

## POLYTROPIC ANALYSIS

An alternate and preferred approach to compressor evaluation is to imagine a reversible polytropic process which duplicates the actual compression between states 1 and 2 in Figure 28. The path equation for this reversible process is:

$$\eta = v\,(dp/dh) \tag{15}$$

where $\eta$ is the **polytropic efficiency.** The reversible work done along the polytropic path is called the **polytropic work** and is given by:

$$W_p = \int_{p_1}^{p_2} v\,dp \tag{16}$$

It follows from Eqs. (13), (15), and (16) that the polytropic efficiency is the ratio of reversible work to actual work:

$$\eta = \frac{W_p}{h_2 - h_1} \quad (17)$$

Thermodynamic relations for an ideal gas:

$$pv^n = p_1 v_1^n = p_2 v_2^n \quad (18)$$

can be used to permit integration so that Eq. (16) can be written:

$$W_p = \frac{n}{n-1} p_1 v_1 \left[\left(\frac{p_2}{p_1}\right)^{(n-1)/n} - 1\right] \quad (19)$$

Further manipulation eliminates the exponent:

$$W_p = \left[\frac{p_2 v_2 - p_1 v_1}{\ln(p_2 v_2 / p_1 v_1)}\right] \ln (p_2/p_1) \quad (20)$$

For greater accuracy in handling gases with properties known to deviate substantially from those of a perfect gas, a more complicated procedure is required:

Equation (15) can be approximated by using Eq. (18) and:

$$\frac{p^m}{T} = \frac{p_1^m}{T_1} = \frac{p_2^m}{T_2} \quad (21)$$

*where*

$$m = \frac{ZR}{c_p}\left(\frac{1}{\eta} + X\right) = \frac{(k-1/k)(1/\eta + X)Y}{(1+X)^2} \quad (21a)$$

$$n = \frac{1}{Y - (ZR/c_p)(1/\eta + X)(1 + X)}$$

$$= \frac{1 + X}{Y[(1/k)(1/\eta + X) - (1/\eta - 1)]} \quad (21b)$$

*and:*

$$X = \frac{T}{v}\left(\frac{\partial v}{\partial p}\right)_p - 1 \quad (22a)$$

$$Y = -\frac{p}{v}\left(\frac{\partial v}{\partial p}\right)_T \quad (22b)$$

$$Z = \frac{pv}{RT} \quad (22c)$$

The accuracy with which Eqs. (18) and (21) represent Eq. (15) depends on the constancy of $m$ and $n$ along the polytropic path. Because these exponents usually vary, mean values between states 1 and 2 should be used.

**Compressibility functions** $X$ and $Y$ have been generalized for gases in corresponding states by Schultz (1962) and their equivalents are listed by Edminster (1961). For usual conditions of refrigeration interest, *i.e.,* for $p < 0.9\, p_c$, $T < 1.5\, T_c$, and $0.6 < Z$, these functions can be approximated by:

$$X = 0.1846\, (8.36)^{1/Z} - 1.539 \quad (23a)$$

$$Y = 0.074\, (6.65)^{1/Z} + 0.509 \quad (23b)$$

The compressibility factor $Z$ has been generalized by many authors, including those of Edminster (1961) and Hougen et al. (1959). Generalized corrections for the specific heat at constant pressure $c_p$ can also be found in these references.

Equations (18) and (21) make possible the integration of Eq. (16):

$$W_p = f\left(\frac{n}{n-1}\right) p_1 v_1 \left[\left(\frac{p_2}{p_1}\right)^{(n-1)/n} - 1\right] \quad (24)$$

In Eq. (24), the polytropic work factor $f$ corrects for whatever error may result from the approximate nature of Eqs. (18) and (21). Since the value of $f$ is between 1.00 and 1.02 in most refrigeration applications, it is generally neglected.

Once the polytropic work has been found, the efficiency follows from Eq. (17). Polytropic efficiencies range from about 0.70 to about 0.84, with a typical value being 0.76.

The highest efficiencies are obtained with the largest compressors and the densest refrigerants because of a Reynolds number effect discussed by Davis *et al.* (1951). A small number of stages is also advantageous because of parasitic losses associated with each stage.

Overall polytropic work and efficiencies are more consistent from one application to another because they represent an average stage aerodynamic performance. Therefore, values calculated by the polytropic analysis have greater utility than those of the isentropic analysis.

Instead of using Eqs. (15) through (24), it is easier and often more desirable to determine the adiabatic work by means of isentropic analysis and then convert to polytropic work by

$$\frac{W_p}{W_s} \cong \eta \left[\frac{(p_2/p_1)^{(k-1)/k\eta} - 1}{(p_2/p_1)^{(k-1)/k} - 1}\right] \quad (25)$$

Equation (25) is strictly correct only for a perfect gas, but because it is a ratio involving comparable errors in both numerator and denominator, it is of more general utility. Equation (25) is plotted in Figure 32 for $\eta = 0.76$. To obtain maximum accuracy, the ratio of specific heats $k$ must be a mean value for states 1, 2, and $2_s$. If $c_p$ is known, $k$ can be determined by:

$$k = \frac{1}{1 - (ZR/c_p)(1 + X)^2/Y} \quad (26)$$

Chapter 17 of the 1985 FUNDAMENTALS Volume lists properties of refrigerants.

The **gas compression power** is:

$$P = wW \quad (27)$$

where w is the mass flow rate. To this must be added the mechanical friction losses to obtain total shaft power. Friction losses vary from less than 1% of the gas power to more than 10%. A typical estimate is 3%.

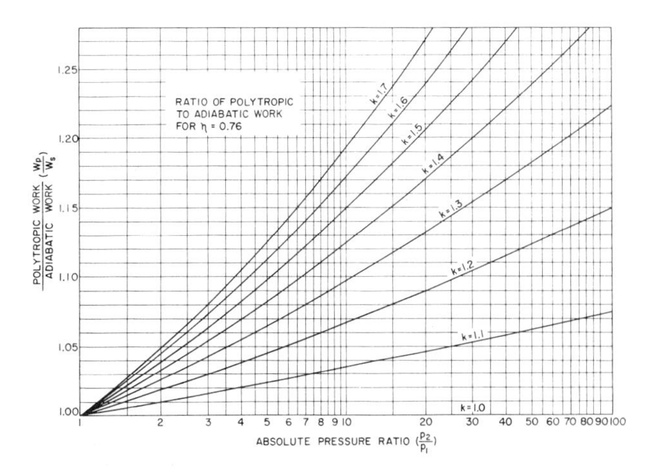


**Fig. 32 Ratio of Polytropic to Adiabatic Work**

## Mach Number

Two different Mach numbers are used:

The **flow Mach number** is the ratio of flow velocity to acoustical velocity at a particular point in the fluid stream:

$$a = v\sqrt{-(\partial p/\partial v)_s}\ \sqrt{n_s pv} \quad (28)$$

Values of acoustical velocity for a number of saturated vapors at various temperatures are presented in Table 4.

The flow Mach numbers in a typical compressor vary from about 0.3 at the stage inlets and outlets to about 1.0 at the impeller exits. With increasing flow Mach numbers, the losses increase because of separation, secondary flow, and shock waves.

The **Impeller Mach number,** which is a pseudo Mach number, is the ratio of impeller tip speed to acoustical velocity at the stage inlet:

$$M_i = u_i/a_i \quad (29)$$

## Impellers

Three key parameters are used to describe impellers, namely *flow coefficient, polytropic work coefficient,* and *specific speed.*

**Flow Coefficient.** Desirable impeller diameters and rotational speeds are determined from blade tip velocity by a dimensionless flow coefficient ($Q/ND^3$) in which $Q$ is the volumetric flow rate. Practical values for this coefficient range from 0.02 to 0.35, with good performance falling between 0.11 and 0.21. Optimum results occur between 0.15 and 0.18. Impeller diameter $D_i$ and rotational speed $N$ follow from:

$$(Q/ND^3) = \pi(Q_i/u_i D_i^2) = \pi^3(Q_i N^2/u_i^3) \quad (30)$$

The maximum flow coefficient in multistage compressors is found in the first stage and the minimum in the last stage (unless large side loads are involved). For high-pressure ratios, special measures may be necessary to increase the last stage ($Q/ND^3$) to a practical level. Side loads are beneficial in this respect, but interstage flash cooling is not.

**Polytropic Work Coefficient.** Because of the different base unit mass in SI (versus weight in the customary I-P system) *polytropic work* and *polytropic head* cannot be used interchangeably. Since this chapter is concerned with polytropic work, that term will be used exclusively. Polytropic head is related to it by:

$$H_p = W_p/g \quad (31)$$

Besides the power requirement, polytropic work also determines impeller blade tip speed and number of stages. For an individual stage, the stage work is related to speed by:

$$W_{pi} = \mu_i u_i^2 \quad (32)$$

*where* $\mu_i$ is the stage work coefficient

The overall polytropic work is the sum of the stage works:

$$W_p = \Sigma W_{pi} \quad (33)$$

and the overall work coefficient is:

$$\mu = gW_p/\Sigma u_i^2 \quad (34)$$

Values for $\mu$ (and $\mu_i$) range from about 0.42 to about 0.74, with 0.55 representative for estimating purposes. Compressors designed for modest work coefficients have backward-curved impeller blades. These tend to have greater part-load ranges and higher efficiencies than do radial-bladed designs.

Maximum tip speeds are limited by strength considerations to about 84,600 fpm (430 m/s). For reasons of cost and reliability, a more common limitation is 59,000 fpm (300 m/s). The maximum polytropic work capability of a typical stage on this basis is about 15,000 ft-lb$_f$/lb (50 kJ/kg).

A greater restriction on stage work capability is often imposed by the impeller Mach number $M_i$. For adequate performance, $M_i$ must be limited to about 1.8 for stages with impellers overhung from the ends of shafts and to about 1.5 for impellers with shafts passing through their inlets. For good performance, these values must be even lower. Such considerations limit maximum stage work to about 1.5 $a_i^2$.

**Specific Speed.** This nondimensional index of optimum performance characteristic of geometrically similar stages is defined by:

$$N_s = N\sqrt{Q_i}/W_{pi}^{0.75} = (1/\pi^3\mu_i^{0.75})\sqrt{Q_i/ND_i^3} \quad (35)$$

**Table 4 (I-P) Acoustical Velocities of Saturated Vapors in ft/s**

| Refrigerant | Evaporator Temperature, °F | | | | | | | | |
|---|---|---|---|---|---|---|---|---|---|
| | −300 | −250 | −200 | −150 | −100 | −50 | 0 | 50 | 100 |
| 11 | | | | | | | 430 | 446 | 456 |
| 12 | | | | | 413 | 432 | 444 | 446 | 436 |
| 13 | | | | 417 | 433 | 434 | 416 | 371 | |
| 13B1 | | | | 349 | 369 | 381 | 381 | 368 | 333 |
| 22 | | | | | 500 | 523 | 535 | 534 | 516 |
| 50[a] | 808 | 898 | 927 | 883 | | | | | |
| 113 | | | | | | | 362 | 378 | 389 |
| 114 | | | | | | 358 | 374 | 383 | 383 |
| 170[b] | | | 738 | 791 | 823 | 829 | 801 | 729 | |
| 290[c] | | | | | 683 | 711 | 724 | 717 | 685 |
| 500 | | | | | 455 | 476 | 489 | 490 | 475 |
| 502 | | | | | 425 | 442 | 448 | 439 | 409 |
| 717[d] | | | | | 1180 | 1250 | 1310 | 1340 | 1350 |
| 718[e] | | | | | | | | 1370 | 1430 |
| 1150[f] | | | 791 | 839 | 859 | 850 | 793 | | |
| 1270[g] | | | | | 704 | 733 | 745 | 736 | 699 |

a = Methane; b = Ethane; c = Propane; d = Ammonia; e = Water; f = Ethylene; g = Propylene.

**Table 4 (SI) Acoustical Velocities of Saturated Vapors in m/s**

| Refrigerant | Evaporator Temperature, °C | | | | | | | | |
|---|---|---|---|---|---|---|---|---|---|
| | −200 | −170 | −140 | −110 | −80 | −50 | −20 | 10 | 40 |
| 11 | | | | | | | 131 | 136 | 139 |
| 12 | | | | | 124 | 130 | 135 | 136 | 133 |
| 13 | | | | 124 | 130 | 131 | 126 | 113 | |
| 13B1 | | | | 104 | 111 | 115 | 116 | 112 | 102 |
| 22 | | | | | 150 | 158 | 162 | 163 | 158 |
| 50[a] | 224 | 258 | 271 | 262 | | | | | |
| 113 | | | | | | | 110 | 115 | 119 |
| 114 | | | | | | 108 | 114 | 117 | 117 |
| 170[b] | | | 216 | 235 | 247 | 250 | 243 | 222 | |
| 290[c] | | | | | 205 | 215 | 220 | 218 | 210 |
| 500 | | | | | 136 | 144 | 148 | 149 | 145 |
| 502 | | | | | 127 | 133 | 136 | 134 | 125 |
| 717[d] | | | | | 354 | 377 | 398 | 408 | 413 |
| 718[e] | | | | | | | | 418 | 437 |
| 1150[f] | | | 232 | 249 | 257 | 257 | 241 | | |
| 1270[g] | | | | | 211 | 221 | 226 | 224 | 214 |

a = Methane; b = Ethane; c = Propane; d = Ammonia; e = Water; f = Ethylene; g = Propylene.

## PERFORMANCE

From an applications standpoint, more useful parameters than $\mu$ and $(Q/ND^3)$ are $\Omega$ and $\Theta$ (Sheets 1952):

$$\Omega = gW_p/a_i^2 = m(\Sigma u_i^2/a_i^2) \quad (36a)$$

$$\Theta = Q_1/a_1D_1^2 = (M_1/\pi)(Q_1/ND_1^3) \quad (36b)$$

They are quite as general as the customary test coefficients and produce performance maps like Figure 33, with speed expressed in terms of first-stage impeller Mach number.

A compressor user with a particular installation in mind may prefer more explicit curves, such as pressure ratio and power versus volumetric flow rate at constant rotational speeds. Plots of this sort require fixed suction conditions to be entirely accurate. This is even more the case if discharge pressure and power are plotted against mass flow rate or refrigeration effect.

A typical compressor performance map is shown in Figure 33 where percent of rated work is plotted with efficiency contours against percent of rated volumetric flow at various speeds. Point A is the *design point* at which the compressor operates with maximum efficiency. Point B is the *selection* or *rating point* at which the compressor is being applied to a particular system. From the application or user's point of view, $\Omega$ and $\Theta$ have their 100% values at Point B.

To save money, refrigeration compressors are selected for heads and capacities above their peak efficiency regions, as in Figure 33. The opposite selection would require larger impellers and additional stages. Refrigerant acoustical velocity and the ability to operate at a high enough Mach number are also of concern. If the compressor of Figure 33 were a multistage design, $M_1$ would be about 1.2. For a single-stage compressor, it would be about 1.5.

Another acoustical effect is seen on the right side of the performance map, where increasing speeds do not produce corresponding increases in capacity. The maximum flow rates at $M_1$ and 1.1 $M_1$ approach a limit determined by the relative velocity of the refrigerant entering the first impeller. As this velocity approaches a sonic value, the flow becomes choked and further increases become impossible.

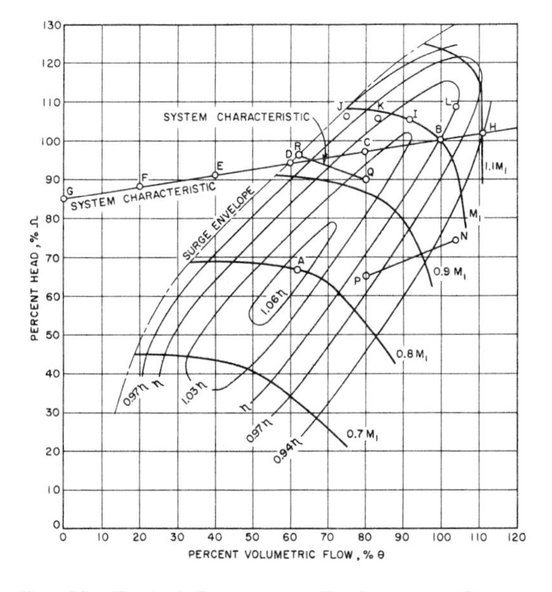


Fig. 33 Typical Compressor Performance Curves

### Testing

When a centrifugal compressor is tested, overall $\mu$ and $\eta$ versus $Q_1/ND_1^3$ at constant $M_1$ are plotted. Because they are needed to convert test results with one gas to field performance with another. When side-flows and cooling are involved, the overall work coefficient is found from Eqs. (33) and (34) by evaluating the mixing and cooling effects between stages separately. The *overall efficiency* in such cases is:

$$\eta = \frac{\Sigma w_i W_{pi}}{\Sigma w_i W_i} \quad (37)$$

Testing with a fluid other than the design refrigerant is a common practice called *equivalent performance testing.* Its need arises from the impracticability of providing test facilities for the complete range of refrigerants and powers for which centrifugal compressors are designed. Equivalent testing is possible because a given compressor produces the same $\mu$ and $\eta$ at the same $(Q/ND^3)$ and $M_i$ with any fluids whose volume ratios $(v_1/v_2)$ and Reynolds numbers are the same.

The thermodynamic performance of a compressor can be evaluated according to either the stagnation or the static properties of the refrigerant, and it is important to distinguish between these concepts. The *stagnation efficiency,* for example, may be higher than the *static efficiency.* The safest procedure is to use static properties and evaluate kinetic energy changes separately.

### Surging

Part-load range is limited on the other side of the performance map by a *surge envelope.* Satisfactory compressor operation to the left of this line is prevented by unstable *surging* or *hunting,* in which the refrigerant alternately surges backward and forward through the machine, accompanied by much noise, vibration, and heat. Prolonged operation under these conditions can damage the system, as well as the compressor.

The flow reverses during surging about once every two seconds. Small systems surge at higher frequencies and large ones at lower. Surging can be distinguished from other kinds of noise and vibration by the fact that its flow reversals alternately unload and load the driver. Motor current varies markedly during surging, and turbines alternately speed up and slow down.

Another kind of instability, called *incipient surge* or *stall,* may occur slightly to the right of the true surge envelope. This phenomenon involves the formation of rotating stall pockets or cells in the diffuser. It produces a roaring noise at a frequency determined by the number of cells formed and the impeller running speed. The driver load is steady during incipient surge, which is harmless to the compressor but may still vibrate system components excessively.

### System Balance

If a refrigeration system characteristic is superimposed on a compressor performance map, it shows the speeds and efficiencies at which the compressor operates in that particular application. A typical brine cooling system curve is plotted in Figure 33, passing through Points B, C, D, E, F, G, and H. With increased speed, the compressor at Point H will produce more than its rated capacity; with decreased speeds at Points C and D, it will produce less. Because of surging, the compressor cannot be operated satisfactorily at Points E, F, or G.

The system can be operated at these capacities, however, by using a hot-gas bypass. The volume flow at the compressor suction must be at least that for Point D in Figure 33; this volume flow is reached by adding recirculated desuperheated gas from the compressor discharge to the actual system volume flow.

## Capacity Control

When the driver speed is constant, a common method of altering capacity is to swirl the refrigerant entering one or more impellers. Adjustable inlet guide vanes called *prerotation vanes,* as shown in Figure 31, produce the swirl. Setting these vanes to swirl the flow in the direction of rotation produces a new compressor performance curve without any change in speed. Controlled positioning of the vanes can be accomplished by pneumatic, electrical, or hydraulic means.

Typical curves for five different vane positions are shown in Figure 34 for the compressor of Figure 33 at the constant speed $M_1$. With the prerotation vanes wide open, the performance curve is identical to the $M_1$ curve in Figure 33. The other curves are different, as are the efficiency contours and the surge envelope.

The same system characteristic has been superimposed on this performance map, as in Figure 33, to provide a comparison of these two modes of operation. In Figure 34, Point E can be reached with pre-rotation vanes, Point H cannot. Theoretically, turning the vanes against rotation would produce a performance line passing through Point H, but sonic relative inlet velocities prevent this, except at low Mach numbers. Hot-gas bypass is still necessary at Points F and G with pre-rotation vane control, but to a lesser extent than with variable speed.

The gas compression powers for both control methods are listed in Table 5. For the compressor and system assumed in this example, the table shows that speed control requires less gas compression power down to about 55% of rated capacity. Prerotation vane control requires less power below 55%. For a complete analysis, friction losses and driver efficiencies must also be considered.

For both control schemes, a system characteristic that requires decreasing polytropic work with decreasing flow is most compatible. In Figure 34, for example, the compressor could operate at 15% of its rated capacity with 28% of its rated power if the system polytropic work requirement could be reduced by 29%.

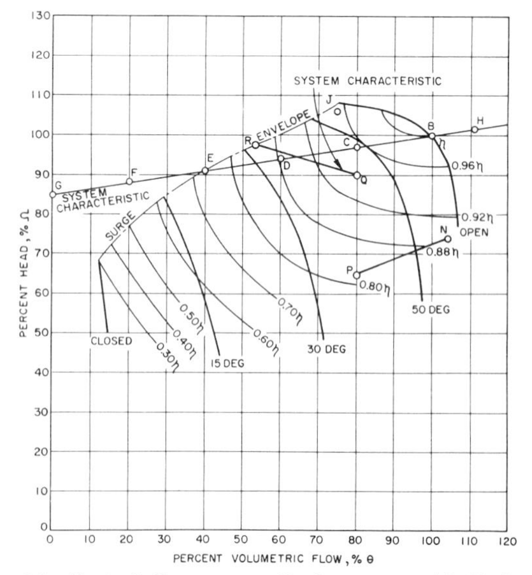


**Fig. 34 Typical Compressor Performance with Various Prerotation Vane Settings**

**Table 5 Typical Part-Load Gas Compression Power Input for Speed and Vane Controls**

| System Volumetric Flow, % | Power Input, % Speed Control | Power Input, % Vane Control |
|---|---|---|
| 111 | 120 | — |
| 100 | 100 | 100 |
| 80 | 76 | 81 |
| 60 | 59 | 64 |
| 40 | 55 | 50 |
| 20 | 51 | 46 |
| 0 | 47 | 43 |

Since fixed-speed motors are the most common drivers of centrifugal compressors, pre-rotation vane control is more prevalent than speed variation. Less common control methods are (1) suction throttling; (2) adjustable diffuser vanes; (3) movable diffuser walls; (4) impeller throttling sleeve; and (5) combinations of these with pre-rotation vanes and variable speed. Each has its advantages and disadvantages in terms of performance, complexity, and cost.

# APPLICATION

## Critical Speeds

Centrifugal compressors are designed so that the first lateral critical speed is either well above or well below the operating speed. Operation at speeds between 0.8 and 1.1 times the first lateral speed is generally unacceptable from a reliability standpoint. The second lateral critical speed should be at least 25% above the operating speed of the machine.

The operating speeds of hermetic compressors are fixed, and each manufacturer has full responsibility for making sure the critical speeds are not too close to the operating speeds. For open-drive compressors, however, the speed depends on the application. Thus, the designer must make sure that the critical speeds are sufficiently far from the operating speeds.

In applying open-drive machines, it is also necessary to consider torsional critical speeds. These are a function of the designs of the compressor, the drive turbine or motor, and the coupling(s). In geared systems, the gearbox design is also involved. Manufacturers of centrifugal compressors have computer programs for calculating the torsional natural frequencies of the entire system, including the driver, the coupling(s), and the gears, if any. Responsibility for performing this calculation and ascertaining that the torsional natural frequencies are sufficiently far away from torsional exciting frequencies should be shared between the compressor manufacturer and the systems designer.

For engine drives, it may be desirable to use a fluid coupling to isolate the compressor (and gear set) from engine torque pulsations. Depending on compressor bearing design, there may be other speed ranges that should be avoided to prevent the nonsynchronous shaft vibration commonly called *oil whip* or *oil whirl.* Responsibilities for avoiding this should also be clearly established among the parties concerned.

## Vibration

Excessive vibration of a centrifugal compressor is an indication of malfunction, which may lead to failure. Periodic checks of the vibration spectrum at suitable locations or continuous monitoring of vibration at such locations are, therefore, useful in ascertaining the operational health of the machine. The relationship between internal displacements and stresses and external vibration is different for each compressor design. In a given design, this relationship also differs for the various causes of internal displacements and stresses, such as imbalance of the rotating parts (either inherent or caused by deposits, erosion, corrosion, looseness, or thermal distortion), bearing instabilities, misalignments, distortion because of piping loads, broken motor rotor bars, or cracked impeller blades. It is, therefore, impossible to establish any universal rules for what vibration is considered excessive.

To establish meaningful criteria for a given machine or design, it is necessary to have baseline data indicative of proper operation. Significant increases of any of the frequency components of the vibration spectrum above the baseline will then indicate a deterioration in the machine's operation; the frequency component for which this increase occurs is a good indication of the part of the machine that is deteriorating. Increases in the component at the fundamental running frequency, for instance, are usually because of deterioration of balance. Increases at approximately half the fundamental running frequency are because of bearing instabilities, and increases at twice the running frequency are usually because of deterioration of alignment, particularly coupling alignment.

As a general guide to establishing satisfactory vibration levels, a constant velocity criterion is sometimes used. In many cases, a velocity amplitude of 0.2 in./s (5.0 mm/s) constitutes a reasonable criterion for vibration measured on the bearing housing.

Although measurement of the vibration amplitude on the bearing housing is convenient, the value of such measurements is limited because the stiffness of the bearing housing in typical centrifugal compressors is generally considerably larger than that of the oil film. Thus, vibration monitoring systems often use non-contacting sensors, which measure the displacement of the shaft relative to the bearing housing, either instead of, or in addition to, monitoring the vibration of the bearing housing (Mitchell 1977). Such sensors are also useful for monitoring the axial displacement of the shaft relative to the thrust bearing.

In some applications, compressor vibration levels, which are perfectly acceptable from a reliability standpoint, can cause noise problems if the machine is not isolated properly from the building.

## Noise

The satisfactory application of centrifugal compressors requires careful consideration of noise control, especially if compressors are to be located near a noise-sensitive part of a building. The noise of centrifugal compressors is primarily of aerodynamic origin. In geared compressors, gear-mesh noise also contributes. Most of the predominant noise sources are of a sufficiently high frequency (above 1000 Hz) so that significant noise reductions can be achieved by carefully designed acoustical and structural isolation of the machine. While the noise originates within the compressor proper, most of it is usually radiated from the discharge line and the condenser shell. Reductions of equipment-room noise by up to 10 dB can be obtained by lagging the discharge line and the condenser shell. This is addressed in Chapter 52 of the 1987 HVAC Volume.

The equipment room, however, may not be the most important location. Noise problems with centrifugal refrigeration equipment usually occur in noise-sensitive parts of the building, such as a nearby office or conference room. The cost of controlling the transmission of compressor noise to such areas should be considered in the building layout and weighed against cost factors for alternative locations of the equipment in the building.

If the equipment room must be located close to noise sensitive building areas, it is usually cost effective to have the noise and vibration isolation designed by an experienced acoustical consultant, since small errors in design or execution can make the results unsatisfactory (Hoover 1960).

Blazier (1972) covers general information on typical noise levels near centrifugal refrigeration machines. Data on the noise output of a specific machine should be obtained from the manufacturer; the request should specify that the measurements are to be in accordance with ARI *Standard* 575-79, *Method of Measuring Machinery Sound within Equipment Rooms.*

## Heat Recovery

The primary objective of heat recovery is to conserve energy (usually rejected as thermal waste to a condenser cooling medium) by transferring it to a temperature level suitable for heating purposes. A typical application for a heat-recovery unit would be in a building in which excess heat developed in the core zones is used to heat the perimeter zones. The heat-recovery unit must work as a straight water chiller during the summer months and as a combined water chiller/heat pump when heating is required. The simplest designs (double bundle condenser and radiator systems) use one common condensing temperature level for both heat transfer into the heating loop and heat rejection to the ambient. This operation results in a typical system characteristic similar to Line N-P-Q-R in Figures 33 and 34. The condenser saturation pressure has to be raised at P when heating starts. Then, the entire mass flow must be brought to that high-pressure level, and hot-gas bypass must be used for volume flows smaller than at R. To avoid these drawbacks, more sophisticated systems—such as multiple machine installations, cascade systems, and booster systems—can be used.

Energy savings can only be determined by analyzing the operation of the heat-recovery unit during the course of the entire year. Only the end result is important, not how much energy is saved during just one mode of operation (cooling or heating). Additional information concerning centrifugal heat pumps can be found in Chapter 17 of this volume.

## Drivers

Centrifugal compressors are driven by almost any prime mover, be it motor, turbine, or engine. Power requirements range from 33 to 12,000 hp (25 to 9000 kW). Sometimes the driver is coupled directly to the compressor; often, however, there is a gear set between them, usually because of low driver speed. Flexible couplings are required to accommodate the angular, axial, and lateral misalignments that may arise within a drive train. (Additional information concerning prime movers is contained in Chapters 30 and 31 of this volume.)

Many specialized applications are made of centrifugal refrigeration compressors, an outstanding example being their use in hermetic water chilling systems of 85 to 2000 tons (300 to 7000 kW) capacity. These units use standardized single- and two-stage compressors driven by integral motors operating in refrigerant atmospheres. Liquid or gaseous refrigerants cool the motors, making them quieter and less costly than conventional open designs and eliminating the need for any mechanical shaft-seal.

Hermetic compressors operating at rotative speeds higher than two-pole motor synchronous speed (3600 rpm) are driven by internal speed-increasing gears. These also operate in refrigerant atmospheres. A discussion of centrifugal water chilling systems is contained in Chapter 17 of this volume.

Standardized single- and two-stage compressors with non-hermetic drivers are also used in water chilling systems for 85 to 10,000 tons (300 to 35 000 kW). Internal gears, which are quieter, less costly, and more compact than external gearboxes, are available in compressors for 85 to 1400 tons (300 to 5000 kW).

Starting torque must be considered in selecting a driver, particularly if it is to be a motor or single-shaft gas turbine. Compressor torque is roughly proportional to both speed squared and to the refrigerant density. The latter is often much higher at startup than at rated operating conditions. If pre-rotation vanes or suction throttling cannot provide sufficient torque reduction for starting, the system standby pressure must be lowered by some auxiliary means.

In certain applications, a centrifugal compressor drives its prime mover backward at shutdown. The compressor is driven backward by refrigerant equalizing through the machine. The extent to which reverse rotation occurs depends on the kinetic energy of the drive train relative to the expansive energy in the system. Large installations with dense refrigerants are most susceptible to running backward, a modest amount of which is harmless if suitable provisions have been made. Reverse rotation can be minimized or eliminated by closing discharge valves, side-load valves, and pre-rotation vanes at shutdown and opening hot-gas bypass valves and liquid refrigerant drains.

## Paralleling

The problems associated with paralleling turbine-driven centrifugal compressors at reduced load are illustrated by points I and J in Figure 33. These represent two identical compressors connected to common suction and discharge headers and driven by identical turbines. A single controller sends a common signal to both turbine governors so that both compressors should be operating at part-load Point K (full load is at Point L). But the I machine is running 1% faster than its twin because of their respective governor adjustments, while the J compressor is working against 1% more pressure difference because of the piping arrangement. The result is a 20% discrepancy between the two compressor loadings.

One remedy is to readjust the turbine governors so that the J compressor runs 0.5% faster than the other unit. A more permanent solution, however, is to eliminate one of the common headers and provide either separate evaporators or separate condensers. This increases the compression ratio of whichever machine has the greater capacity, decreases the compression ratio of the other, and shifts both toward Point K.

The best solution of all is to install a flow meter in the discharge line of each compressor and use a *master-slave* control system in which the original controller signals only one turbine, the *master,* while a second controller causes the *slave* unit to match the master's discharge flow.

The problem of imbalance, associated with turbine-driven centrifugal compressors, is minimal in fixed-speed compressors with vane controls. A loading discrepancy comparable to the example given above would require a 25% difference in vane positions.

The paralleling of centrifugal compressors offers advantages in redundancy and improved part-load operation. This arrangement provides the capability of efficiently unloading to a lower percentage of total system load. When the unit requirement reduces to 50%, one compressor can carry the complete load, and it will be operating at a higher percent volumetric flow and efficiency than a single large compressor.

## Other Special Applications

Other specialized applications of centrifugal compressors are found in petroleum refineries and in the chemical industry, as discussed in Chapter 36 of the 1986 REFRIGERATION Volume. Marine requirements are detailed in ASHRAE *Standard* 26-1985, *Recommended Practice for Mechanical Refrigeration Installations on Shipboard.*

# MECHANICAL DESIGN

## Impellers

Impellers without covers, such as the one in Figure 30, are known as *open* or *unshrouded* designs. Those with covered blades, as in Figure 31, are called *shrouded* impellers. Open models must operate in close proximity to contoured stationary surfaces to avoid excessive leakage around their vanes. Shrouded designs must be fitted with labyrinth seals around their inlets for a similar purpose. Labyrinth seals behind each stage are required in multistage compressors.

Impellers must be shrunk, clamped, or bolted to their shafts to prevent loosening from thermal and centrifugal expansions. Generally, they are made of cast or brazed aluminum or of cast, brazed, riveted, or welded steel. Aluminum has a higher strength-weight ratio than steel, up to about 300 °F (150 °C), which permits higher rotating speeds with lighter rotors. Steel impellers retain their strength at higher temperatures and are more resistant to erosion. Lead-coated and stainless steels can be selected in corrosive applications.

## Casings

Centrifugal compressor casings are about twice as large as their largest impellers, with suction and discharge connections sized for flow Mach numbers between 0.1 and 0.3. They are designed for the pressure requirements of ASHRAE *Standard* 15-1978, *Safety Code for Mechanical Refrigeration.* A hydrostatic test pressure 50% greater than the maximum design working pressure is customary.

Cast iron is the most common casing material, having been used for temperatures as low as −150 °F (−100 °C) and pressures as high as 300 psia (2100 kPa). Nodular iron and cast or fabricated steel are also used for low temperatures, high pressures, high shock, and hazardous applications. Multistage casings are usually split horizontally, although unsplit *barrel* designs can also be used.

## Lubrication

Like motors and gears, the bearings and lubrication systems of centrifugal compressors can be considered *internal* or *external,* depending on whether or not they operate in refrigerant atmospheres. For reasons of simplicity, size, and cost, most air-conditioning and refrigeration compressors have internal bearings, as shown in Figure 31. They often have internal oil pumps as well, driven either by an internal motor or the compressor shaft; the latter arrangement requires an auxiliary oil pump for starting.

Most refrigerants are soluble in lubricating oils, the extent increasing with refrigerant pressure and decreasing with oil temperature. A compressor's oil may typically contain 20% refrigerant (by weight) during idle periods of high system pressure and 5% during normal operation. Thus, refrigerant will come out of solution and *foam* the oil when such a compressor is started.

To prevent excessive foaming from cavitating the oil pump and starving the bearings, oil heaters minimize refrigerant solubility during idle periods. Standby oil temperatures between 130 and 150°F (55 and 65°C) are required, depending on system pressures. Once a compressor has started, its oil should be cooled to increase oil viscosity and maximize refrigerant retention during the pulldown period.

A sharp reduction in system pressure before starting tends to *supersaturate* the oil. This produces more foaming at startup than would the same pressure reduction after the compressor has started. Machines designed for pressure ratios of 20 or more may reduce system pressures so rapidly that excessive oil foaming cannot be avoided, except by maintaining a low standby pressure. Additional information on the solubility of refrigerants in oil can be found in Chapters 2 and 8 of the 1986 REFRIGERATION Volume.

*External* bearings avoid the complications of refrigerant-oil solubility at the expense of some oil-recovery problems. Any nonhermetic compressor must have at least one shaft-seal. Mechanical seals are commonly used in refrigeration machines because they are leak-tight during idle periods. These seals do require some lubricating oil leakage when operating, however. The amounts range up to 20 drops per minute, depending on seal face velocities and refrigerant pressures. Shaft-seals leak oil out of compressors with internal bearings and into compressors with external bearings. Means for recovering seal oil leakage with a minimum loss of refrigerant must be provided in external lubrication systems.

## Bearings

Centrifugal compressor bearings are generally of hydrodynamic design, with sleeve bearings being the most common for radial loads; tilting pad, tapered land, and pocket bearings are customary for thrust. The usual materials are aluminum, babbit, and bronze.

Thrust bearings tend to be the most important in turbomachines, and centrifugal compressors are no exception. Thrust comes from the pressure behind an impeller exceeding the pressure at its inlet. In multistage designs, each impeller adds to the total unless some are mounted backward to achieve the opposite effect. In the absence of this opposing balance, it is customary to provide a *balancing piston* behind the last impeller, with pressures on the piston thrusting oppositely to the stages. To avoid axial rotor vibration, some net thrust must be retained in either balancing arrangement.

## Accessories

The minimum accessories required by a centrifugal compressor are an oil filter, an oil cooler, and three safety controls. Oil filters are usually rated for 15 μm or less. They may be built into the compressor but are more often externally mounted. Dual filters can be provided so that one can be serviced while the other is operating.

Oil coolers, single or dual, usually use condenser water, chilled water, refrigerant, or air as their cooling medium. Water- and refrigerant-cooled models may be built into the compressor, and refrigerant-cooled oil coolers may be built into a system heat exchanger. Many oil coolers are mounted externally for maximum serviceability.

Safety controls, with or without anticipatory alarms, must include a low oil pressure cutout, a high oil temperature switch, and high discharge and low suction pressure (or temperature) cutouts. A high motor temperature device is necessary in a hermetic compressor. Other common safety controls and alarms sense discharge temperature, bearing temperature, oil filter pressure differential, oil level, low oil temperature, shaft-seal pressure, balancing piston pressure, surging, vibration, and thrust bearing wear.

Pressure gauges and thermometers are useful indicators of the critical items monitored by the controls. Suction, discharge, and oil pressure gauges are the most important, followed by suction, discharge, and oil thermometers. Suction and discharge instruments are often attached to system components rather than to the compressor itself, but they should be provided. Interstage pressures and temperatures can also be helpful, either on the compressor or on the system.

## OPERATION AND MAINTENANCE

Reference should be made to the compressor manufacturer's operating and maintenance instructions for recommended procedures. A planned maintenance program, as described in Chapter 59 of the 1987 HVAC Volume, should be established. As part of this program, an operating log should be kept, tabulating pertinent unit temperatures, pressures, flows, fluid levels, and electrical data. These can be compared periodically with values recorded for the new unit. Gradual changes in data can be used to signify the need for routine maintenance; abrupt changes can be clues of system or component difficulty. A successful maintenance program requires that the operating engineer be able to recognize and identify the reason for these data trends. In addition, by having a knowledge of the system component parts and their operational interaction, he or she will be able to use these symptoms to prescribe the proper maintenance procedures.

The following items deserve attention in the establishment of a planned compressor maintenance program:

1. *A tight system* is important. Leaks on compressors operating at subatmospheric pressures allow noncondensables and moisture to enter the system, adversely affecting system operation and component life. Leakage in higher-pressure systems allows oil and refrigerant loss. The existence of vacuum leaks can be detected by a change in operational pressures not supported by either corresponding refrigerant temperature data or the frequency of purge unit operation. Pressure leaks are characterized by those symptoms related to refrigerant charge loss such as low suction pressures and high suction superheat. Such leaks should be located and fixed to prevent component deterioration.
2. *Compliance with the manufacturer's recommended oil filter inspection and replacement schedule* allows visual indication of the condition of the compressor lubrication system. Repetitive clogging of filters can mean system contamination. Periodic oil sample analysis can monitor acid, moisture, and particulate levels to assist in problem detection.
3. *Operating and safety controls* should be checked periodically and calibrated to ensure system reliability.
4. *The electrical resistance of hermetic motor windings* between phases and to ground should be checked (megged) regularly, following the manufacturer's outlined procedure. This will

assist detection of any internal electrical insulation deterioration or the formation of electrical leakage paths before a failure occurs.

5. *Water-cooled oil coolers* should be systematically cleaned on the water side (depending on water conditions), and the operation of any automatic water control valves should be checked.
6. *For some compressors, periodic maintenance is required,* such as manual lubrication of couplings and other external components, and shaft seal replacement. Prime movers and their associated auxiliaries all require routine maintenance. Such items should be made part of the planned compressor maintenance schedule.
7. *The necessary steps for preparing the unit for prolonged shutdown, i.e.,* winter, and a specified instruction for starting after this standby period, should both be part of the program. Compressors that have internal lubrication systems should have provisions to have their oil heaters energized continuously throughout this period or have their oil charges replaced just prior to putting them back into operation.

## REFERENCES


API. 1979. Centrifugal Compressors for General Refinery Services. API *Standard* 617-79. American Petroleum Institute, Washington, DC.

ARI. 1983. Centrifugal Water-Chilling Packages. ARI *Standard* 550-83.

ASHRAE. 1973. Centrifugal Chiller Noise in Buildings. ASHRAE *Symposium* LO-73-9.

ASHRAE. 1969. Centrifugal Heat Pump Systems. ASHRAE *Symposium* DV-69-4.

ASHRAE. 1968. The Operation and Maintenance of Centrifugal Units. ASHRAE *Symposium* LP-68-2.

ASME. 1983. Pressure Vessels Division 1. SI Sec 8-D-1-83. American Society of Mechanical Engineers, New York.

ASME. 1965. Compressors and Exhausters. ASME *Performance Test Code* 10-65. American Society of Mechanical Engineers, New York.

Blazier, W.E., Jr. 1972. Chiller noise: its impact on building design. ASHRAE *Transactions*, Vol. 78, Part I, p. 268.

Davis, H.; H. Kottas; and A.M.G. Moody. 1951. The influence of Reynolds number on the performance of turbomachinery. ASME *Transactions,* July, p. 499.

Edmister, W.C. 1961. *Applied Hydrocarbon Thermodynamics.* Gulf Publishing Company, Houston, TX, pp. 22 and 52.

Hoover, R.M. 1960. Noise Levels Due to a Centrifugal Compressor Installed in an Office Building Penthouse. *Noise Control,* No. 6, p. 136.

Hougen, O.A.; K.M. Watson; and R.A. Ragatz. 1959. Chemical Process Principles, Part II—Thermodynamics. John Wiley and Sons, Inc., New York, pp. 579 and 611.

Kerschbaumer, H.G. 1975. An investigation into the parameters determining energy consumption of centrifugal compressors operating in heat recovery systems. *International Congress of Refrigeration,* Moscow, Paper B.2.61.

Military Specification, Refrigerating Unit, Centrifugal for Air Conditioning. 1968. MIL-R-24085A, *Ships*, March.

Mitchell, J.S. 1977. Monitoring machinery health. *Power,* Vol. 121, Part I, March p. 46; Part II, May, p. 87; Part III, July, p. 38.

Schultz, J.M. 1962. The polytropic analysis of centrifugal compressors. ASME *Transactions,* January and April, pp. 69 and 222.

Sessler, S.M. 1973. Acoustical and mechanical considerations for the evaluation of chiller noise. ASHRAE *Journal*, Vol. 15, No. 10, October, p. 39.

Sheets, H.E. 1952. Nondimensional compressor performance for a range of Mach numbers and molecular weights. ASME *Transactions,* January, p. 93.

# CHAPTER 13

# ABSORPTION COOLING, HEATING, AND REFRIGERATION EQUIPMENT



ABSORPTION machines are heat-operated refrigeration machines that operate on one of the earliest known principles of refrigeration. The cycle uses an absorbent as a secondary fluid to absorb the primary fluid, which is a gaseous refrigerant that has been vaporized in the evaporator. The evaporation process absorbs heat, thus providing the needed refrigeration.

Both the absorption cycle and the mechanical compression cycle evaporate and condense a refrigerant liquid at two or more pressure levels within the unit. The domestic absorption refrigerator cycle is an exception because these processes occur at the same total pressure (but at differing partial pressures). The absorption cycle uses a heat-operated generator, a heat-rejecting absorber, and a liquid solution pump to produce the pressure differential. The mechanical compression cycle differs in that it uses a compressor. The absorption cycle substitutes a physiochemical process (and a pump) for a purely mechanical compressor. Both cycles require energy for operation—heat and a small amount of mechanical energy for the absorption cycle and mechanical energy for the compression cycle.

Chapter 1 of the 1985 FUNDAMENTALS Volume describes the operating principles and thermodynamics of refrigerant-absorbent equipment currently manufactured and predominately used for airconditioning and process cooling. This equipment is generally water cooled and uses water and lithium bromide (with water as the refrigerant) or ammonia and water (with ammonia as the refrigerant). Additives inhibit corrosion and promote wetting. Chapter 17 of the 1985 FUNDAMENTALS Volume lists properties of these substances. Bogart (1981) has further information on absorption cycles.

Ammonia-water absorption equipment has been used in large capacity industrial applications requiring low temperatures for process cooling. The air-cooled ammonia-water-hydrogen absorption cycle, which uses ammonia as the refrigerant and water as the absorbent, is used in the domestic refrigerator to keep food and liquid cold. The water-lithium bromide absorption equipment is used to produce hot water for comfort heating, process heating, and domestic purposes, as well as for cooling. This cycle also can be used to deliver heat at a temperature higher than that of the driving heat source. These heat pumps are called *heat transformers* or *temperature boosters.*

The following terms should be used for all refrigerant-absorbent mixtures. NOTE: Industry convention is to use these terms for the water-lithium bromide cycle, but the opposite convention is used for the ammonia-water cycle.

**Weak absorbent** is the solution that has picked up refrigerant in the absorber and has the least affinity for refrigerant.
**Strong absorbent** is the solution that has had refrigerant driven from it in the generator and, therefore, has a strong affinity for refrigerant.

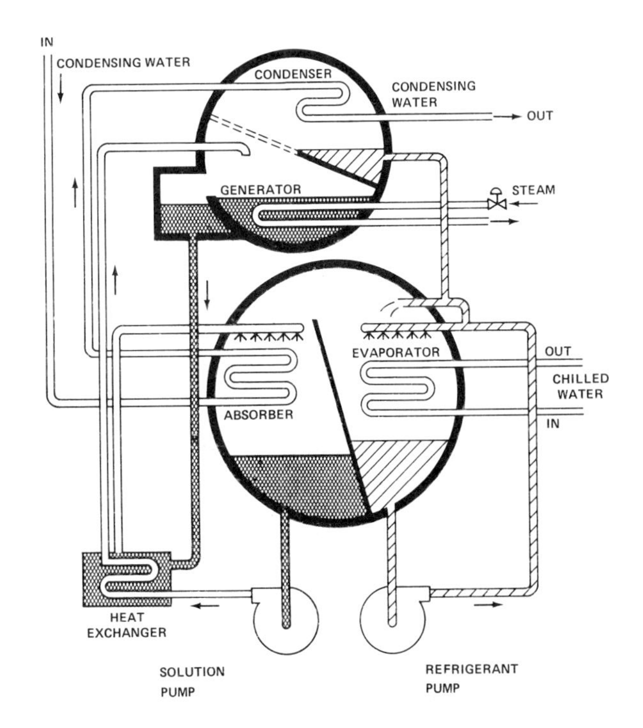


**Fig. 1 Diagram of Two-Shell, Lithium Bromide Cycle Water Chiller**


The preparation of this chapter is assigned to TC 8.3, Absorption and Heat-Operated Machines.

## LARGE CAPACITY WATER-LITHIUM BROMIDE EQUIPMENT

This equipment is classified by its method of firing and whether it has a single-stage or two-stage generator. Machines using a steam or hot fluid heat source are *indirect fired,* and those using a flame source are *direct fired.* Those machines using clean, hot, waste gases as a heat source can also be classified as indirect fired, but they are more often called *heat-recovery units.* Units with two-stage generators are called *dual-* or *double-effect units.*

### Indirect Fired Units

Figures 1 and 2 are typical schematic diagrams of machines that are available as single-stage indirect-fired liquid chillers with capacities ranging from 50 to 1500 tons (175 to 5280 kW). Figure 3 is a schematic of a dual- or double-effect unit, i.e. one with

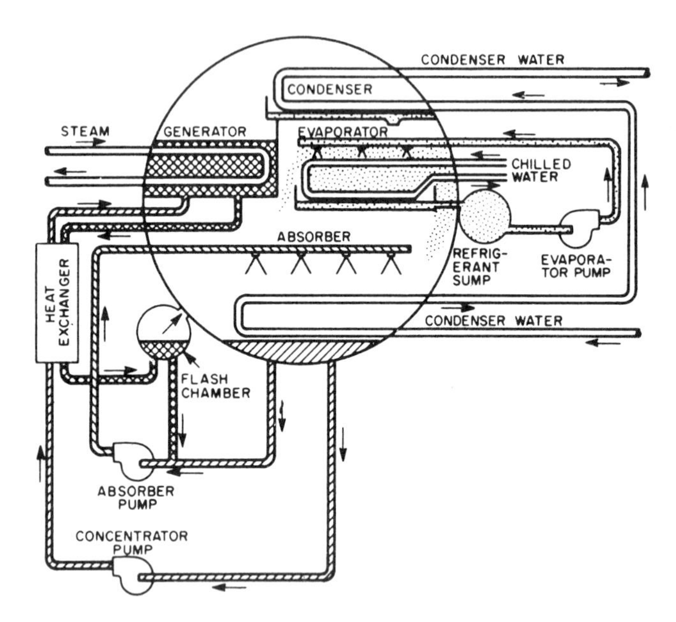


**Fig. 2 Diagram of One-Shell, Lithium Bromide Cycle Water Chiller**

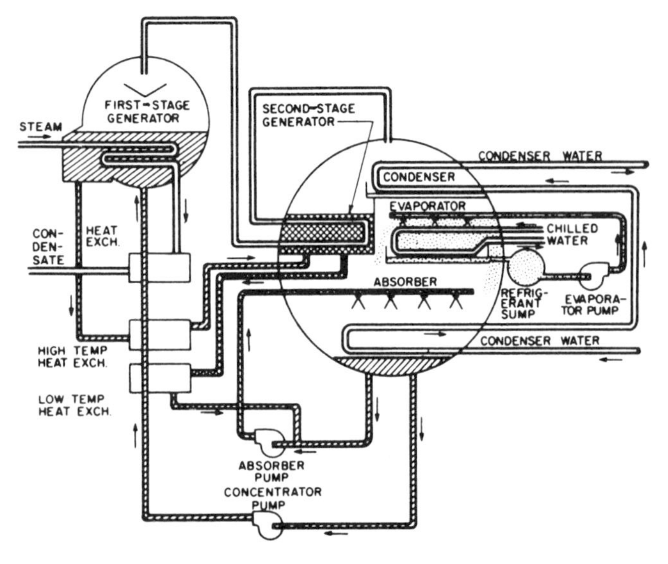


**Fig. 3 Diagram of a Double Effect Lithium Bromide Cycle Used as a Water Chiller**

two generation stages. The first-effect generator receives the external heat that boils the refrigerant from the weak absorbent. This hot refrigerant vapor goes to a second generator, where, on condensing, it supplies heat for further refrigerant vaporization from the absorbent of intermediate concentration that flows from the first generator and is cooled by passing through a first-stage heat economizer. Other than the added generator and economizer, all components of the double-effect water-lithium bromide absorber units are common to the single-effect units. The advantage of the double-effect unit is higher performance. Steam rates are typically about 60% of those of single-stage machines. Heat source temperature is at least 122°F (50°C) above that required for a single-stage unit.

### Direct-Fired Units

At present, direct-fired, large capacity, water-lithium bromide machines of single- and dual-effect design are manufactured only in Japan. The most common of these machines is the double-effect, direct-fired, dual-fuel type. These machines can produce both chilled water for summer air conditioning and hot water for winter heating. Some of them can produce both chilled and hot water simultaneously. Figures 4 and 5 show the chilling and heating cycles of one of these machines.

### Heat Recovery

Some manufacturers also build two-stage machines that use the heat rejected from such sources as gas turbines, diesel engines, or process operations that emit hot, clean gases with temperatures above 550°F (290°C). These machines can accept this heat energy, which would otherwise be wasted, to generate chilled or hot water for space conditioning or process use. In addition, they can accomplish this at higher efficiencies and with less auxiliary equipment than other systems. A more refined version of the heat-recovery type chiller-heater is capable of using both the hot exhaust gases and hot water from such sources as diesel or gas reciprocating engines. In these machines, both forms of heat energy are used to produce chilled water. Figure 6 shows a typical application of this type of machine.

Heat Transformer, Temperature Amplifier, or Type II are synonymous terms applied to an absorption system that elevates the temperature of a fluid stream above the temperature level of any other fluid stream supplied to the cycle. In most such applications, one stream heats the generator and evaporator, while the absorber heats a second, higher temperature stream. An alternative is that a portion of a single stream can heat the generator and evaporator, while the absorber heats the remainder of that stream to a significantly higher temperature level. These systems can be classified as heat-recovery devices.

The thermal lift, above that of the source, is achieved by the absorption of a vapor into the absorbing solution and a resultant rise in the solution's temperature. A heat exchanger transfers the energy from the solution to the process steam.

Figure 7 shows a heat transformer using a water-lithium bromide solution. This unit is similar to the absorption chiller cycle used for air conditioning and process cooling. However, the function of all components are interchanged, and the flow direction of all fluid and heat streams are reversed. Consequently, the generator of the chiller becomes the absorber of the heat transformer, the chiller evaporator a condenser, and vice versa. Also, the pumps replace expansion valves, and the expansion valves replace pumps. Also in the heat transformer, the generator and condenser operate at lower pressures than the absorber and evaporator; that is, they operate opposite that of the chiller cycle.

While the useful cooling capacity of the evaporator is the primary interest of a chiller, the absorber capacity and tempera-

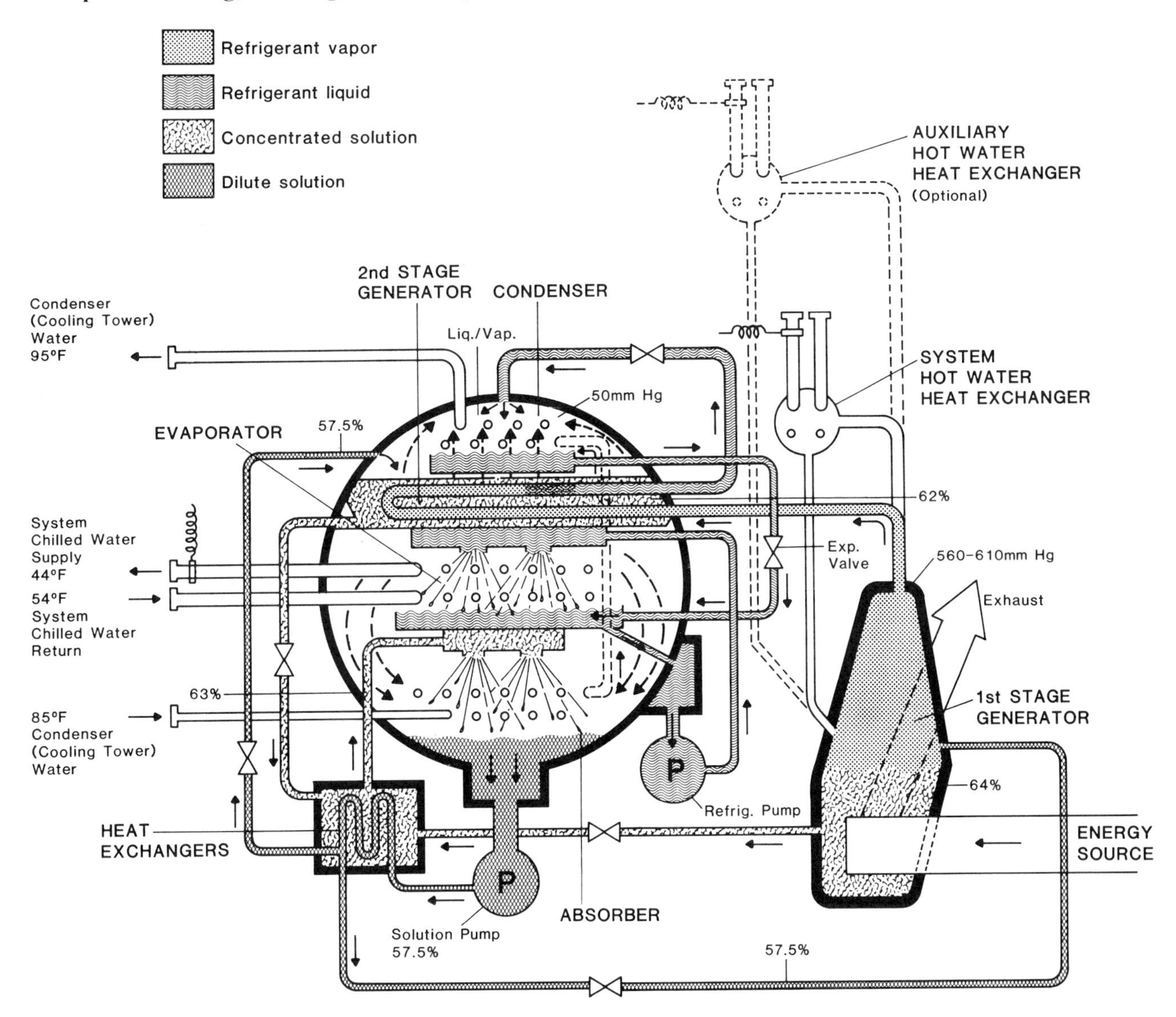


Fig. 4 Chilling Cycle

ture lift are of similar importance in the heat transformer. As with the chiller, the dual-effect cycle has been operated for the water-lithium bromide versions of heat transformers. Single-effect units can produce low pressure steam, if they are given the necessary condenser water and waste heat source temperatures. Other working fluids may also be used, and the various component streams may be in series as opposed to being in parallel, as shown in the figure.

## Components

**Generators** (concentrators, desorbers) are heat exchangers, usually of the tube bundle type. They are submerged in solution or arranged to accommodate a falling film of solution and heated by steam, hot liquids, or hot gases. The heat then evaporates the refrigerant from the solution.

**Evaporators** (coolers) are tube bundles over which liquid refrigerant is sprayed or dripped and evaporated. The liquid to be cooled passes inside the tubes.

**Condensers** are internally water-cooled tube bundles located in the refrigerant vapor space over or near the generator. Droplet eliminators shield the condenser from solution carry-over from the generator. In double-effect machines, the highstage condenser is incorporated in the tube-side of the lower stage generator.

**Absorbers** are internally water-cooled tube bundles over which strong absorbent is sprayed or dripped in the presence of refrigerant vapor. The vapor is absorbed by the solution flowing over the outside of the tubes, thus releasing heat, which is removed by the cooling water.

**Solution Heat Exchangers** or **Economizers** are usually of the tube and shell or plate and frame designs and are made with *ferrous* metal. They recover heat from liquid streams to increase cycle performance.

**Solution and Refrigerant Pumps** are generally electric-motor-driven centrifugal pumps of hermetic design. They are cooled and lubricated by the cycle fluids or a captive coolant circuit.

**Purgers** are devices that remove noncondensable gases. These gases, even when present in small quantities, raise the total pressure in the absorber, which increases the evaporator pressure enough to cause an appreciable increase in evaporator temperature.

**Throttling Devices** are most commonly orifices or traps. They throttle either the refrigerant liquid between the condenser and evaporator or the strong absorbent leaving an economizer.

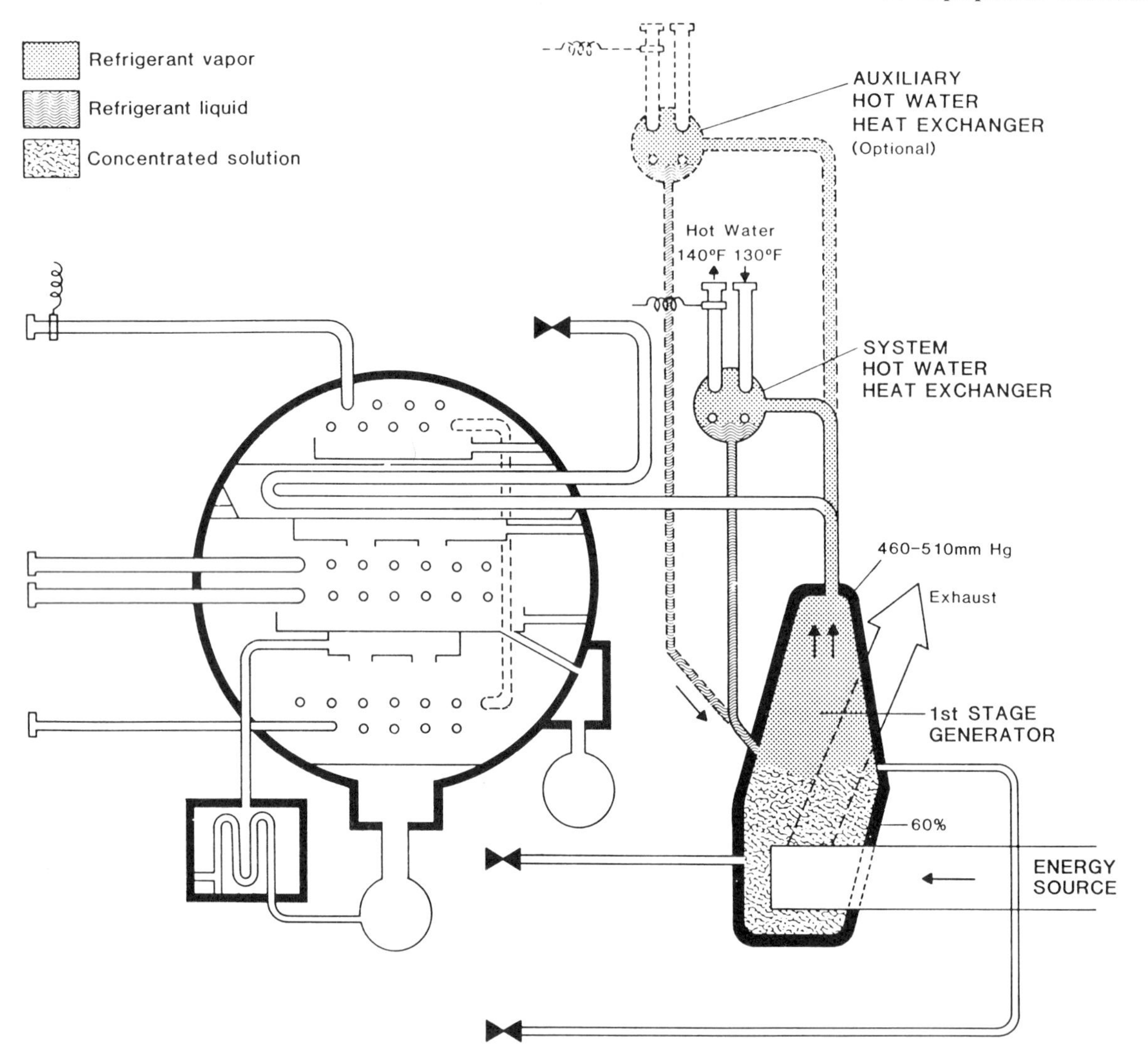


Fig. 5 Heating Cycle

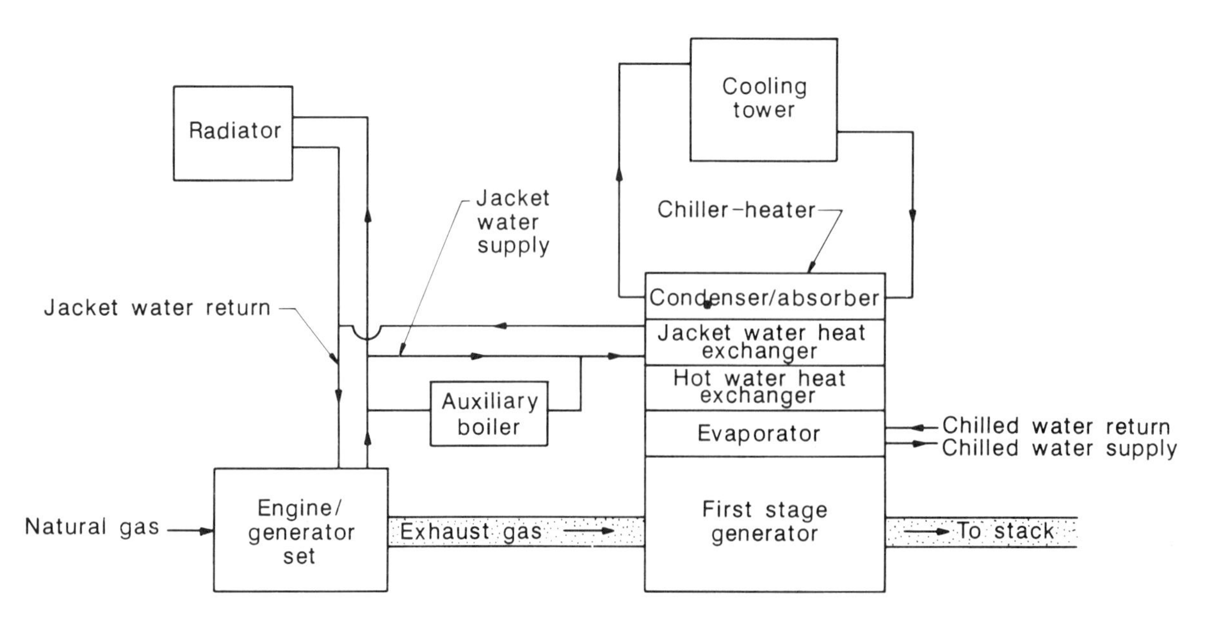


Fig. 6 Heat Recovery Chiller-Heater with Prime Mover in a Cogeneration Plant

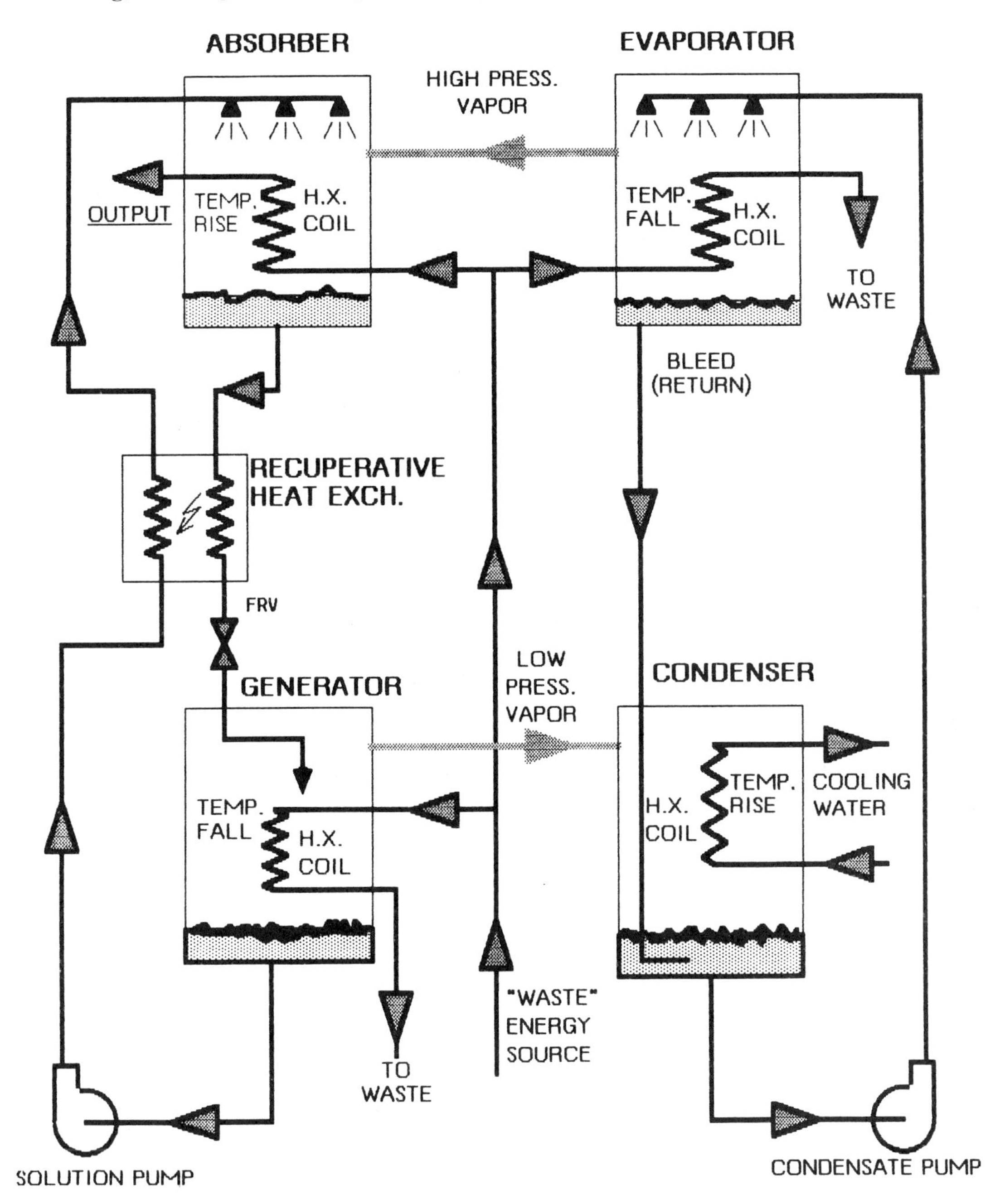


**Fig. 7 Heat Transformer**

## CAPACITY CONTROL

Most water-lithium bromide absorption machines meet load variation and maintain chilled-water temperature control by varying the reconcentration rate of the absorbent solution. At a given constant load, the chilled water temperature is maintained by a temperature difference between the refrigerant and the chilled water. The refrigerant temperature is maintained, in turn, both by the absorber being supplied with a flow rate and concentration of absorbent and by the absorber cooling-water temperature.

Load changes are reflected by corresponding changes in chilled water temperature. A load reduction, for example, causes a smaller temperature difference in the evaporator and a reduced requirement for solution flow or concentration. Most units have a sensor, usually located at the chilled water exit, to sense the chilled-water temperature changes caused by a load change. In most indirect units available, the chilled-water thermostat controls an automatic valve that regulates the energy input of the heat source. Energy management control systems monitor many system and chiller temperatures and reset the chilled-water temperature to maintain maximum efficiency.

## CONDENSING-WATER TEMPERATURE CONTROL

Some absorption machines and some applications (e.g., process applications) require the condensing (cooling) water temperature to vary no more than ±5°F (±3°C) from design. A three-way valve, installed between the cooling tower and the machine and controlled from the water temperature to the absorber, generally permits the tower water and return water to recirculate and mix in order to control the temperature within the specified limits.

The nominal cooling water design temperatures are 85°F (29°C) entering the unit and about 100°F (38°C) leaving it; the typical lower limit is about 60°F (16°C). Low cooling water temperatures can be considered as transients or as design temperatures. One or a combination of the following methods are usually used to accommodate these low temperatures:

1. Dilution of solution—A control permits stored liquid refrigerant to transfer directly to the absorbent circuit.
2. Control heat input—An override control limits the amount of input to the generator.
3. Lower freezing point of refrigerant—A control permits measured amounts of absorbent to transfer to the refrigerant circuit.

Dilution and heat input control reduce the concentration of the solution, while a lowered freezing point prevents freezing of the refrigerant. At the lowest limit of water temperature and refrigeration load, the unit may be permitted to cycle on and off. Since colder cooling water enhances efficiency, the ability of machines to use colder cooling water, when available, means lower energy demand and improved seasonal economy.

## PROTECTIVE DEVICES

In addition to capacity controls, water-lithium bromide machines require several protective devices. Some controls keep the units operating within safe limits, while others stop the unit before damage occurs from a malfunction that would cause the safe limits to be exceeded. The following typical protective devices are used.

**Low Temperature Cut-Out.** This low-limit thermostat stops the unit when the evaporator temperature falls too low. When the refrigerant temperature warms to the cut-in point, the switch resets to start the machine again.

**Cooling-Water Switch.** This pressure- or flow-sensitive switch stops the unit when the cooling water supply fails. The switch resets when the cooling water supply is restored.

**Chilled-Water Switch.** This pressure- or flow-sensitive switch stops the unit when the chilledwater flow drops below design limits. The switch resets when the proper chilled-water flow is reestablished.

**Concentration Limiters.** Some machines have controls to limit either the capacity of the generator, the capacity of the absorber, or both. They respond to sensors that indicate limiting conditions of absorbent concentration and reset when the limiting conditions no longer exist. Crystallization, which is caused by high solution concentration and low cooling water temperatures, can be prevented by these devices.

**Overflow Piping (or "J" Loop).** This important protective device allows hot solution from the generator to overflow to the absorber in case another control fails. This device also prevents crystallization in the economizer.

**Generator Shell Overpressure Relief.** On hot-water, high pressure steam, direct-fired, and heat-recovery units, a rupture disk may be placed in the generator shell.

**Protective Devices for Direct-Fired Machines.** The following four controls are used in the high temperature generator to provide safe operations:

1. Low Level Switch (ensures adequate solution level)
2. High Pressure Switch
3. High Temperature Control
4. Flame Ignition and Monitoring Control

**Hot Water Temperature Cut-Out.** For machines with heating capabilities, a high temperature control limits the temperature and restarts the machine when an acceptable level is reached.

## OTHER CONTROLS

Most absorption units have a control panel containing electric motor starters, fuses, and other electric and pneumatic controls that can be grouped in a panel. The panel may also have a disconnect switch and a control voltage transformer. The panel, motors, and controls are factory installed and wired.

If not furnished with the unit, the following controls are generally available:

1. Disconnect Switch
2. Control Transformer
3. Condensing (cooling) Water-Flow Switch
4. Controls for series or parallel unit operation
5. Solution Flow Control
6. Automatic Decrystallization Control
7. Steam Demand Limiter

## WATER-LITHIUM BROMIDE CYCLE SYSTEM PERFORMANCE

Figure 8 shows the typical performance characteristics of a single-stage water-lithium bromide absorption machine with an indirect heat source generator. Figure 9 shows the effect of steam pressure on the maximum available capacity.

The coefficient of performance (COP), when defined as cooling effect divided by heat input, typically ranges from 0.65 to 0.75 for a single-stage water-lithium bromide absorption machine operating at the nominal conditions listed in Table 1. A COP

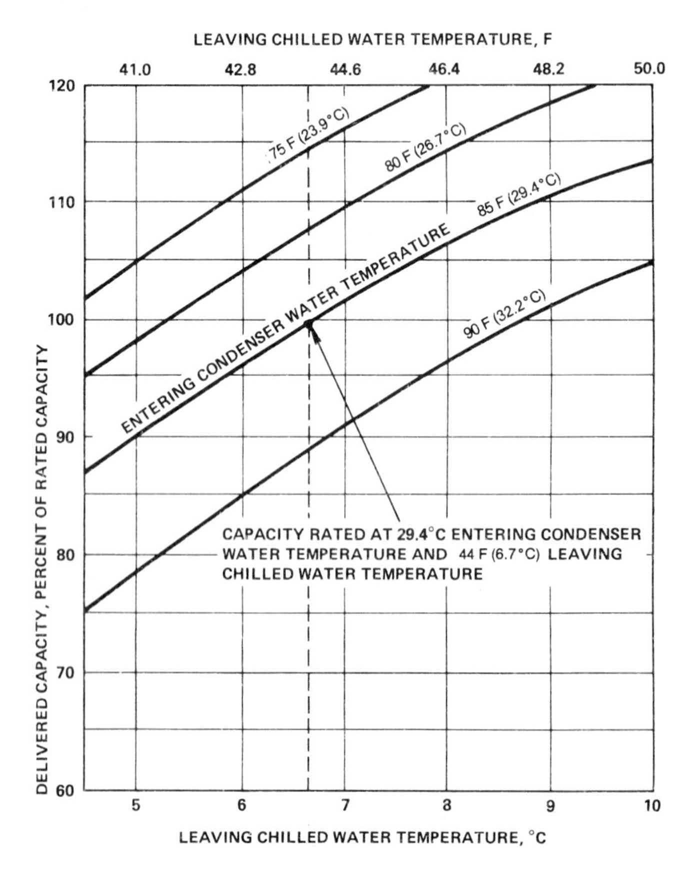


**Fig. 8 Performance Characteristics of Lithium Bromide Cycle Water Chiller**

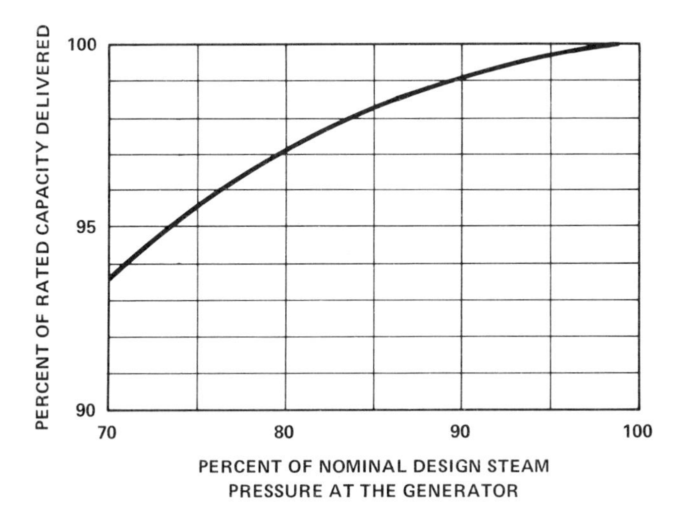


**Fig. 9 Effect of Steam Pressure on Capacity**

higher that 0.75 can be reached whenever chilled water temperatures exceed the nominal or condensing water temperatures drop below the nominal. Reversing these temperature conditions reduces the COP to below 0.60. A COP of 0.68 corresponds to a steam rate of 18 lb/h per ton of refrigeration (0.68 kW/kW).

Table 2 shows the nominal rating conditions of a double-effect, indirect-fired unit. Figure 10 shows the effect of steam pressure on maximum available capacity. These machines can perform on steam rates as low as 9.3 lb/h per ton of refrigeration (0.75 kW/kW) with steam inlet pressure as low as 43 psi (296 kPa) gauge.

## EQUIPMENT SELECTION AND APPLICATION

The size of an absorption machine chosen is based on the flow rate and the inlet and exit temperatures of the liquid to be chilled (usually water). Available steam pressure may also be a constraint, particularly if waste steam is to be used. The condensing water flow is adjusted to produce the required refrigeration capacity. Generally, the selected machine is the smallest size that will produce the required refrigeration capacity.

The manufacturer's data should be used, when available, to establish the unit size. These data are available in tabular, graphical, or computer format. Computer selection is more accurate and detailed and can consider special energy requirements of a particular application. Such data helps determine the following information, which is useful for economic analyses:

1. Nominal unit size
2. Required cooling-water rate and temperature rise
3. Cooling-water pressure loss
4. Chilled-water pressure loss
5. Steam rate at full and reduced loads
6. Physical dimensions and weight

**Table 1 Nominal Rating Conditions for Single-Stage Absorption**

| | | |
|---|---|---|
| Leaving chilled water temperature | 44 °F | 6.7 °C |
| Chilled water temperature differential | 10 °F | 5.5 °C |
| Entering condenser water temperature | 85 °F | 29.4 °C |
| Steam pressure at control valve inlet, gauge pressure dry and saturated | 9 to 12 psig | 62 to 83 kPa |
| Scale factor for evaporator, condenser, and absorber | 0.0005 h • ft² • °F/Btu | 90 mm² • K/W |

**Table 2 Nominal Rating Conditions for Double-Effect Absorption**

| | | |
|---|---|---|
| Leaving chilled water temperature | 44 °F | 6.7 °C |
| Chilled water temperature differential | 10 °F | 5.5 °C |
| Entering condenser water temperature | 85 °F | 29.4 °C |
| Steam pressure at control valve inlet, gauge pressure dry and saturated | 43 to 130 psig | 296 to 896 kPa |
| Scale factor for evaporator, condenser, and absorber | 0.0005 h • ft² • °F/Btu | 90 mm² • K/W |

If a manufacturer's selection data are not available, a first estimate for preliminary comparisons may be based on the following:

### Single Stage

1. Cooling water rate = 3.6 gpm/ton (0.065 L/s per kW)
2. Cooling water temperature rise = 16 °F (9 °C)
3. Steam rate = 18.5 lb/hr per ton (1.5 kW/kW)

### Dual Effect

1. Cooling water rate = 3.0 gpm/ton (0.054 L/s per kW)
2. Cooling water temperature rise = 15 °F (8 °C)
3. Steam rate = 12 lb/hr per ton (0.43 g/s per kW)
4. Direct fired (with natural gas) = 11.7 ft³/h per ton (26 mL/s per kW or 1.0 kW/kW)

For flexibility and economy of operation, a multiple-unit installation may be designed, even though single units of sufficient size are available. Also, multiple units must be installed when the total refrigeration requirement exceeds the capacity of the largest available machine. Multiple absorption units may be arranged for series (Figure 11) or parallel (Figure 12) operation. In either case, a return water temperature sensor (RWT) may be used to cycle either the lead or the lag unit (number one or number two in parallel operation) when one unit can carry the load. In parallel operation, when one unit is shut down, chilled water flowing through the downed unit bypasses the operating unit. This situation creates a mixed chilled-water temperature that exceeds the control point setting. To avoid this problem, an override thermostat (OTH) may be used to operate the *on* unit at a lower temperature than its own chilled-water controller setting. This thermostat then keeps the mixed chilled-water temperature at an acceptable level.

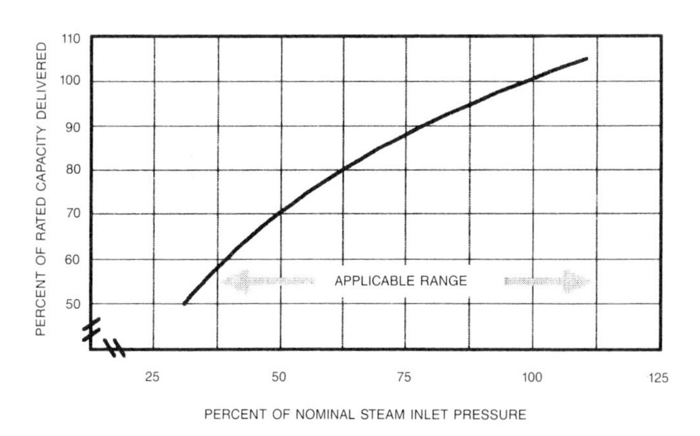


**Fig. 10 Effect of Steam Pressure on Capacity of Double Effect Absorption Unit**

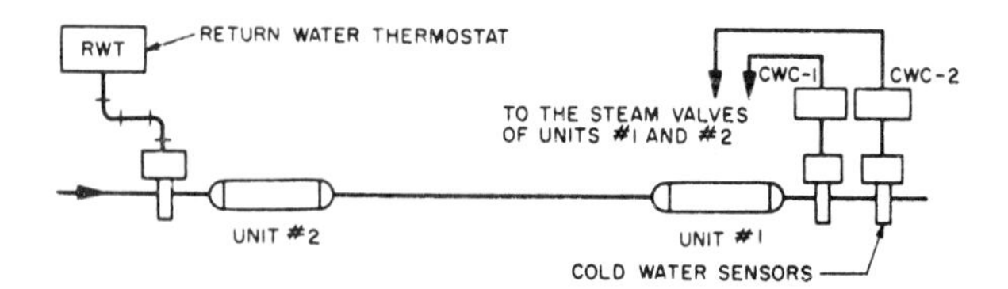


**Fig. 11 Multiple Absorption Units—Series Operation**

High temperature hot water is commonly used as an energy source for single-effect systems. Cooling capacity falls off steeply at inlet water temperatures below 190°F (88°C). Low-temperature applications, such as solar and engine heat recovery, must account for this performance reduction either by (1) oversizing the chillers and setting a lower limit on supply temperature or (2) selecting a machine designed for low temperature firing.

Low-temperature cooling water sources may allow lower hot water supply temperatures. Cooling-water piping can be arranged for parallel flow to the condenser and absorber instead of the normal series flow. This change requires more cooling-water flow, but higher flows are desirable in either case. Higher hot-water flow can reduce the approach between the generator temperature and the entering water temperature. Also, chilled-water temperature should be raised as much as possible by designing for a minimum approach in the air handlers.

Specially designed chillers with improved performance for the lower hot-water temperatures from solar collectors are available. The operating characteristics are similar to conventional machines, however, and the same remedies apply.

Machines are available that provide "free cooling" when condensing water below 50°F (10°C) is available. Capacity levels up to 60% of nominal are available by storing solution out of circulation and condensing refrigerant vapor directly on cold absorber tubes.

Applications for hot gas machines have become more numerous over the past few years for two reasons. First, the heat rejected by gas turbines and reciprocating engines is recovered with a COP of 1.14. And second, industrial processes using thermal or catalytic incinerators and also having chilled water requirements benefit from the availability of low cost chilled water.

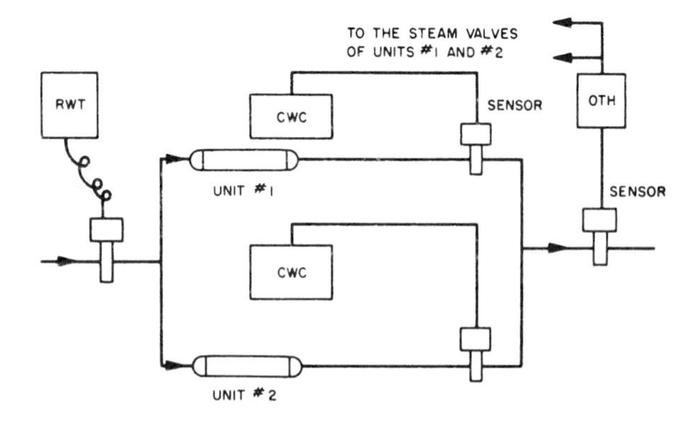


**Fig. 12 Multiple Absorption Units—Parallel Operation**

## OPERATION AND MAINTENANCE PROCEDURES

### Operation

Large capacity, water-lithium bromide absorption machines are comparatively trouble-free and simple to operate. Internal cleanliness and leaktightness that last the life of the unit are the most important aspects of operation and maintenance. Units are manufactured to rigid standards of vacuum integrity and internal cleanliness. Equipment such as electronic halide leak detectors and helium mass spectrometers is used to ensure that the equipment has no leaks prior to shipment from the factory. The length of a machine's useful life and the extent to which it gives trouble-free operation is directly related to the care taken in maintaining the unit's cleanliness and tightness.

Because these machines operate at low pressure (relatively high vacuum), atmospheric air can leak into units that are improperly operated or maintained. Units must be purged regularly by either automatic or manual purgers furnished with the machines, to prevent air and other noncondensable gases from accumulating in the unit. Such accumulation adversely affects refrigeration capacity, promotes crystallization of salt within the unit, and, over time, permits internal parts to deteriorate through corrosive attack.

*Crystallization* describes the precipitation of salt crystals from absorbent. It results in a slush-like mixture that can plug pipelines and other fluid passages within the machine and make it inoperable. Crystallization is not common in later-model machines that are operated and maintained properly.

Many manufacturers provide concentration control to avoid crystallization. In those rare instances where crystallization occurs through equipment malfunction, automatic devices, which are standard equipment, reduce the need for applying external heat. However, should crystallization occur, it may be necessary to apply heat and perhaps add water to the affected portions of the machine (usually the heat exchanger) to dissolve the crystals. Crystallization does not harm the equipment, but it is a nuisance. It is a symptom of trouble, and its cause should be found and corrected. Commonly, the causes are controls that are improperly set or malfunctioning, sudden drops in cooling-water temperature, atmospheric air leaks into the machine, and electrical power failure.

Fouling of external or machine-side tube surfaces of an absorption machine is not a problem because there is not a continuous source of scale-forming substances such as algae. However, as with all heat-transfer devices involving the heating of water inside tubes, it is necessary to use good water treatment practices on the internal or water side of the absorber and condenser tubes.

Improper performance may be experienced with steam that is not sufficiently dry. However, a more common steam deficiency is a supply pressure lower that the design specification. Under this condition, the machine does not produce its design refrigeration capacity.

Unit refrigeration capacity is also adversely affected if the condensing or cooling water is supplied to the absorber at a temperature higher than or a flow less than that specified in the design. Of course, higher water temperatures also contribute to higher rates of fouling.

All manufacturers of large capacity water-lithium bromide absorption machines recommend adding a heat-transfer additive to the unit. Particularly at initial start-up, this addition noticeably improves a new machine's refrigeration capacity. The additive does not deplete rapidly during normal operation; however, the manufacturer's additive replenishment instructions should be followed.

### Maintenance

Units built today use corrosion inhibitors to help protect the internal parts from corrosive attack. The manufacturer's instructions for the proper inhibitors required and the schedule for replenishment should be followed. The use of corrosion inhibitors still requires proper maintenance with regard to purging and leaktightness.

Whenever a machine is opened to the atmosphere for repair and maintenance, nitrogen should be used to break the vacuum. Also, as long as the machine is open, a nitrogen atmosphere with a small amount of bleed should be sustained in the unit. Nitrogen, because it is an inert gas, will prevent corrosive attack on internal parts while they are open to atmosphere.

Because purging is so important, the manufacturer's instructions on purge system maintenance should be followed, and the effectiveness of the purge system should be verified periodically. As an example of maintenance where mechanical purge pumps are used, the oil in the pump should be changed at prescribed intervals. Also, leaktightness should be evaluated periodically. All manufacturers describe procedures for measuring the leak rate, bubble count, or noncondensable accumulation rate of their machines. If the measured leak rate is excessive, it is important to find and repair the leak as soon as possible.

The internal cooling-water surfaces of the absorber usually require periodic cleaning by mechanical, chemical, or by both means, regardless of the effectiveness of the water treatment practices followed. The more effective the water treatment, however, the longer the allowable period between tube cleaning.

## SMALL CAPACITY WATER-LITHIUM BROMIDE EQUIPMENT

Small units of 3 to 30 tons (10 to 105 kW) capacity are designed for residential or small commercial use. Indirect- and direct-fired liquid chiller, chiller-heater, and air-conditioning equipment in both single-stage and dual-effect configurations have been produced. Currently available equipment uses one or more of the following unique capabilities of the water-cooled cycle:

1. The cycle can be fired by a flat-plate solar collector heat source with good efficiency, but with lower capacity.
2. Heating can be derived from the cooling cycle by stopping the cooling-water flow. When this water flow stops in some designs, a solution trap between the high and low sides opens and allows refrigerant vapor to flow to and condense on the evaporator coil and heat the water flowing inside the tubes.
3. Solution can be circulated either thermally by a vapor-lift action in a pump tube or by a mechanical pump.

Figure 13 illustrates a gas-fired heater/chiller with vapor-lift solution circulation. The solution returns from the generator to the absorber by gravity flow. A one-pass flow of liquid delivered through capillary drippers wets the absorber and evaporator tubes.

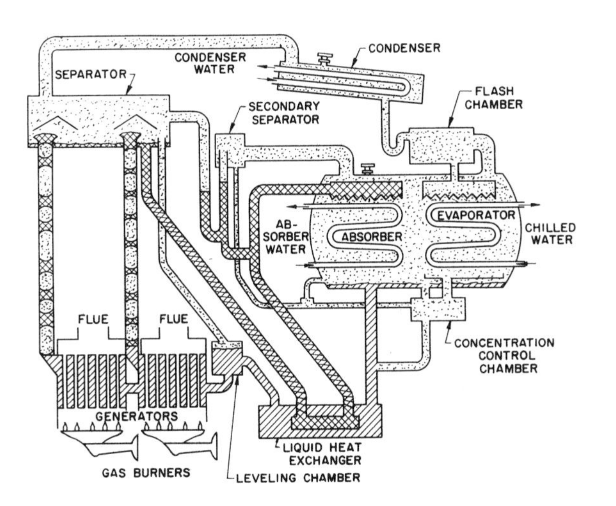


**Fig. 13 Diagram of Direct-Fired Lithium Bromide Cycle Water Chiller/Heater**

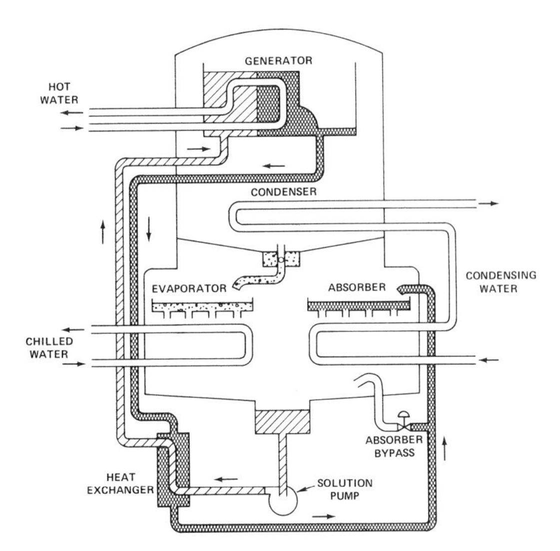


**Fig. 14 Diagram of Vertical Shell, Small Capacity, Lithium Bromide Cycle Water Chiller for Solar Cooling**

Generators are of steel fire-tube construction, usually with atmospheric gas burners. Power gas burners, oil burners, and hot water for solar applications are used as well.

Solution heat exchangers are formed of closely spaced steel plates to combine low pressure drop and good heat transfer.

Figure 14 shows a residential chiller with a mechanical solution pump designed for optimum solar cooling. The pump prevents crystallization at low inputs and reduces submergence in the generator, which is a factor in minimizing the firing temperature.

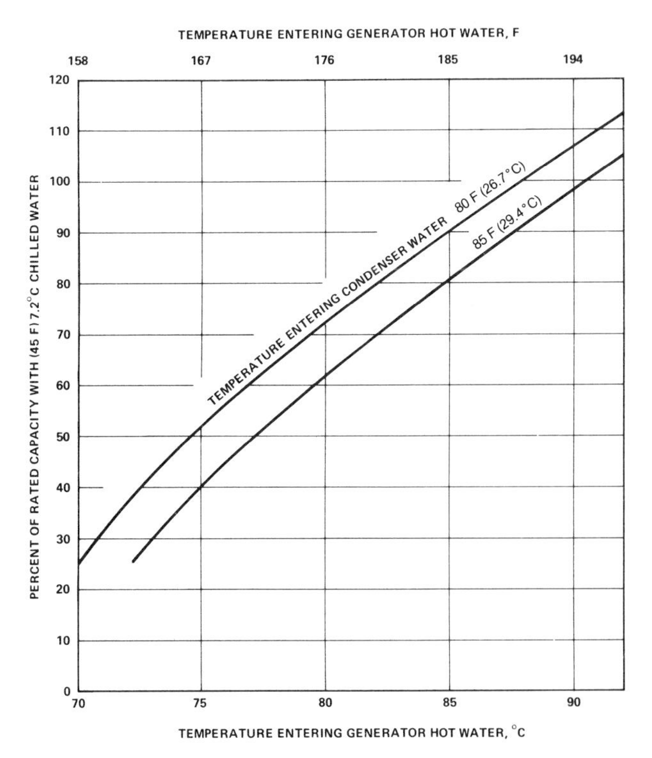


**Fig. 15 Generator Hot Water Temperature Requirements for Solar Optimized Absorption Chiller**

Much of the information about large capacity units applies to small units. To keep costs low, small-capacity units generally have no pumps, controls, or a need for periodic maintenance. These cost factors have lead to units with one-pass refrigerant and solution flow from capillary drippers onto capillary surfaces, on/off control by room thermostat in the smallest units, and a palladium-cell purger.

## Performance

Direct-fired units operate at a single input with a rating-point COP of 0.5 based on the higher heating value of the fuel. Indirect-fired unit values range from 0.65 to 0.72.

Figure 15 plots the generator hot water temperature versus percent rated capacity of chillers designed for solar cooling. Rated flows in this case are as follows:

Hot water = 3.6 gpm/ton (65 mL/s per kW)
Chilled water = 2.4 gpm/ton (43 mL/s per kW)
Condensing water = 3.6 gpm/ton (65 mL/s per kW)

## Solar Application

The water-lithium bromide absorption cycle uniquely cools at a high COP with the moderate temperatures supplied by flat-plate solar collectors. Figure 16 shows a typical solar heating/cooling application. Solar energy is collected by the collector loop and stored in the thermal storage tank. When the storage temperature is high enough to operate the absorption generator, the absorption cycle can be energized upon demand for cooling. If cooling is required but solar temperatures are not high enough, an auxiliary boiler can supply the heat.

Many variations are possible, including firing the generator directly from the collector to improve energy availability, cold storage instead of hot storage, or a combination of the two.

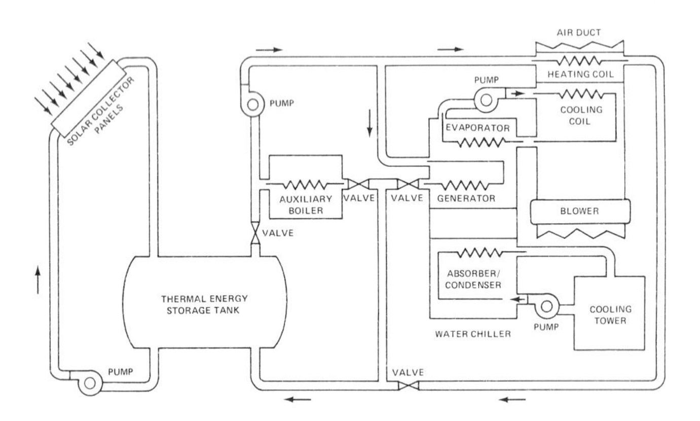


**Fig. 16 Typical Solar Heating/Absorption Application Schematic**

## Cogeneration Application

Small commercial or residential units in the 30- to 40-ton (105- to 140-kW) capacity range have been developed to work with small cogeneration units (100 to 125 kW). This combination uses the heat of both the exhaust gas and jacket water from reciprocating engines (Figure 6). A feature of these packages is that one piece of equipment can supply electricity, as well as chilled and hot water, to meet all building requirements with no additional energy cost.

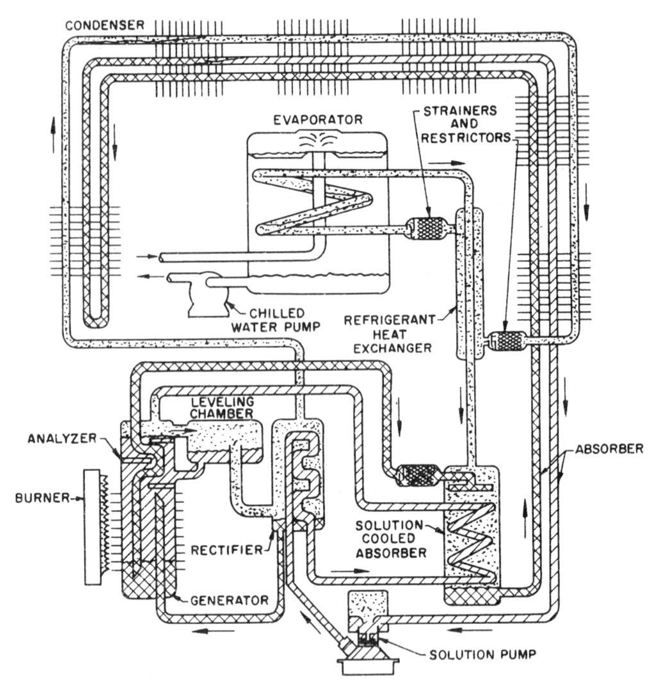


**Fig. 17 Diagram of Ammonia-Water, Direct-Fired, Air-Cooled Chiller**

# AMMONIA-WATER EQUIPMENT

Figure 17 is a typical schematic diagram of an ammonia-water machine, which is available as a direct-fired, air-cooled liquid chiller in capacities of 3 to 5 tons (10 to 18 kW).

Ammonia-water equipment varies from water-lithium bromide equipment to accommodate these three major differences:

1. Water (the absorbent) is also volatile, so the regeneration of weak absorbent to strong absorbent is a fractional distillation process.
2. Ammonia (the refrigerant) causes the cycle to operate at condenser pressures around 300 psia (2070 kPa) and at evaporator pressures of about 70 psia (480 kPa). As a result, vessel sizes are held to a diameter of 6 in. (150 mm) or less, and positive displacement solution pumps are used.
3. Air cooling requires condensation and absorption to occur inside the tubes so that the outside can be finned for greater air contact.

## Components

**Generator**. These vertical tanks are finned on the outside to extract heat from the combustion products. Internally, a system of analyzer plates create intimate counterflow contact between the vapor generated, which rises, and the absorbent, which descends. Atmospheric gas burners depend on the draft of the condenser air fan to sustain adequate combustion airflow to fire the generator. The exiting flue products mix with the air that has passed over the condenser and absorber.

**Heat Exchangers**. Heat exchange between strong and weak absorbents takes place partially within the generator-analyzer. A tube bearing strong absorbent spirals through the analyzer plates and in the solution-cooled absorber, where strong absorbent metered from the generator through the solution capillary passes over a helical coil bearing weak absorbent. In the solution-cooled absorber, the strong absorbent absorbs some of the vapor from the evaporator, thus retaining its heat of absorption within the cycle to improve the cycle COP. The strong absorbent and unabsorbed vapor continue from the solution-cooled absorber into the air-cooled absorber, where absorption is completed and the heat of absorption is rejected to the air.

**Rectifier**. Vapor issuing from the generator is still partially laden with water vapor, which can be reduced to a negligible amount by cooling. The rectifier comprises a spiral coil through which weak absorbent from the solution pump passes on its way to the absorber and generator. Some type of packing is included to assist counterflow contact between condensate from the coil (which is refluxed to the generator) and the vapor (which continues onto the air-cooled condenser).

**Absorber and Condenser**. These finned tubes are arranged so most of the incoming air flows over the condenser tubes and most of the exit air flows over the absorber tubes.

**Evaporators**. The liquid to be chilled drips over a coil bearing evaporating ammonia, which absorbs the refrigeration load. On the chilledwater side, which is at atmospheric pressure, a pump circulates the chilled liquid to the load source. Refrigerant to the evaporator is metered from the condenser through restrictors. A *tube-in-tube heat exchanger* provides the maximum refrigeration effect per unit mass of refrigerant. The tube-in-tube design is particularly effective in this cycle because water present in the ammonia produces a liquid residue that evaporates at increasing temperatures as the amount of residue decreases.

**Solution Pumps**. Solution is moved through suction and discharge valves by the reciprocating motion of a flexible sealing diaphragm. This motion is imparted by hydraulic fluid pulses delivered to the opposite side of the diaphragm by a hermetic vane or piston pump at atmospheric suction pressure.

**Capacity Control**. A thermostat usually cycles the machine *on* and *off*. A chilled-water switch is included to shut the burners off if the water temperature drops close to freezing. Units may also be underfired by 20% to derate to a lower load.

**Protective Devices**. Typical protective devices included are (1) a sail switch that verifies airflow before allowing gas burners to come on, (2) a pressure relief valve, and (3) a generator high-temperature switch.

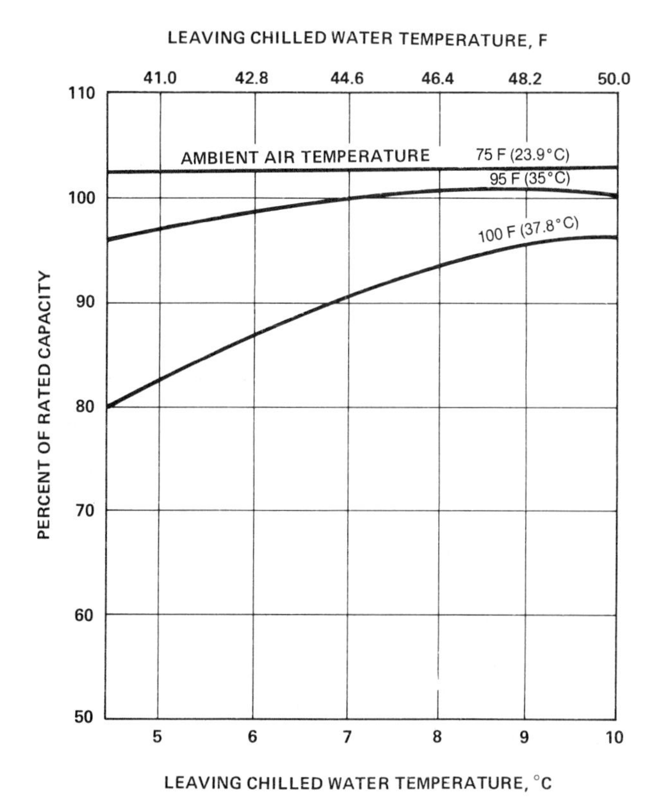


**Fig. 18 Performance Characteristics of Direct-Fired, Air-Cooled, Ammonia-Water Chiller**

## EQUIPMENT PERFORMANCE AND SELECTION

Absorption equipment is built and rated to meet ANSI *Standard* Z21.40.1 for outdoor installation. The rating conditions are ambient air at 95°F (35°C) dry bulb and 75°F (24°C) wet bulb and chilled water delivered at the manufacturer's specified flow at 45°F (7°C). A COP of about 0.5 is realized, based on the higher heating value of the gas. Figure 18 shows a typical performance curve.

Although most units are piped to a single furnace, duct, or fan coil and operated as air conditioners, multiple units supplying a multi-coil chilled-water system are common. If a boiler is part of the system, positive shutoff valving must be installed between the chiller and boiler to prevent loss of water during heating to the atmospheric evaporator tank in the chiller.

Chillers are also packaged with outdoor furnaces and coils to comprise a self-contained cooling and heating air conditioner for roof-top or slab mounting.

## DOMESTIC REFRIGERATOR

### Ammonia-Water-Hydrogen Cycle

Domestic refrigerators use an absorption cycle with three fluids. These units are popular for recreational vehicles because they can be dualfired by gas of electric heaters. The refrigeration unit is hermetically sealed. All spaces within the system are open to each other and, hence, are at the same total pressure, except for minor variations caused by fluid columns used to circulate the fluids.

The elements of the system shown in Figure 19 include (1) a generator, (2) a condenser, (3) an evaporator, (4) an absorber, and (5) a hydrogen reserve vessel. The following three distinct fluid circuits exist in the system: (1) an ammonia circuit, which includes the generator, condenser, evaporator, and absorber; (2) a hydrogen circuit, which includes the evaporator and absorber; and (3) a solution circuit, which includes the generator and absorber.

Starting with the generator, a gas burner or other heat source applies heat to expel ammonia from the solution. The ammonia vapor generated then flows through an analyzer (6) and a rectifier (7) to the condenser. The small amount of residual water vapor in the ammonia is condensed by atmospheric cooling in the rectifier and drains to the generator (1) through the analyzer (6).

The ammonia vapor passes onto the section (2a) of the condenser (2), where it is liquified by air cooling. Fins on the condenser increase the cooling surface. The liquified ammonia then flows into an intermediate point of the evaporator (3). A liquid trap between the condenser section (2a) and the evaporator prevents hydrogen from entering the condenser. Ammonia vapor that does not condense in the condenser section (2a) passes to the other section (2b) of the condenser and is liquified. It then flows through another trap into the top of the evaporator.

The evaporator has two sections. The upper section (3a) has fins and cools the food space directly, while the lower section (3b) cools the freezing compartment directly.

Hydrogen gas enters the lower evaporator section (3b) and, after passing through a precooler, flows upward and counterflow to the downward flowing liquid ammonia. The hydrogen atmosphere, because it is above the liquid ammonia, reduces the partial pressure of the ammonia vapor in accordance with Dalton's law. While the total pressures in the evaporator and the con-

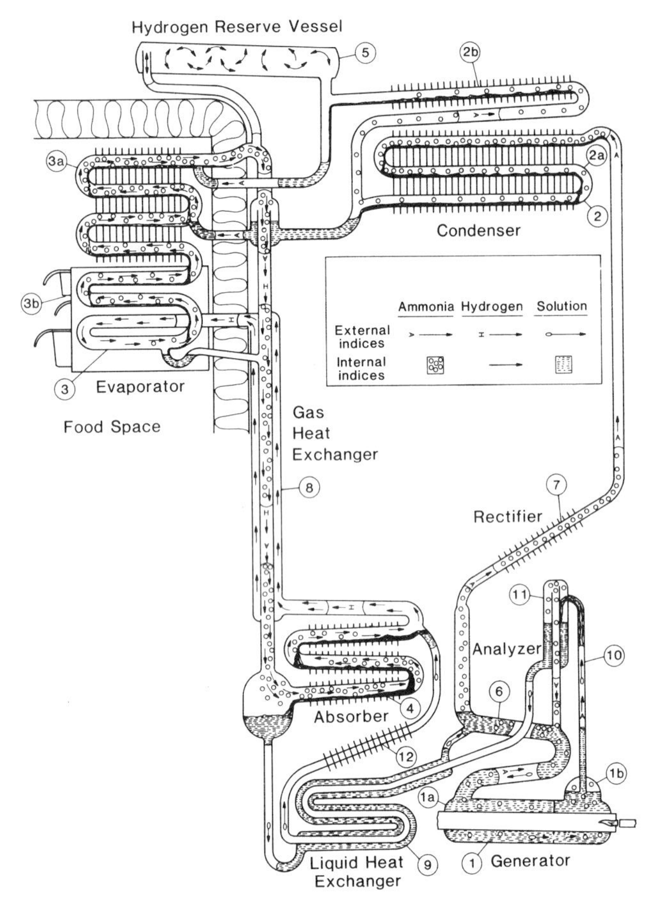


**Fig. 19 Schematic of Domestic Absorption Refrigeration Cycle**

denser are the same, substantially pure ammonia is in the space where condensation takes place, and the vapor pressure of the ammonia essentially equals the total pressure. As a result, the ammonia evaporates because its partial vapor pressure in the evaporator is less than the total pressure by the amount of partial pressure of the hydrogen.

The gas mixture of hydrogen and ammonia leaves the top of the evaporator and passes down through the center of the gas heat exchanger (8) to the absorber (4). Here, ammonia is absorbed by water, and the hydrogen, which is almost insoluble, passes up from the top of the absorber, through the external chamber of the gas heat exchanger (8), and into the evaporator. Some ammonia vapor passes with the hydrogen from the absorber to the evaporator. Because of the difference in molecular mass of ammonia and hydrogen, the gas circulation is maintained between the evaporator and absorber.

Countercurrent flow in the evaporator permits placing the box cooling section of the evaporator at the top of the food space, which is the most effective location. Also, the gas leaving the lower temperature evaporator section (3b) can pick up more ammonia at the higher temperature in the box cooling evaporator section (3a), thus increasing capacity and efficiency. In addition, the liquid ammonia flowing to the lower temperature evaporator section is precooled in the upper evaporator section. The dual liquid connection between the condenser and the evaporator permits extending the condenser below the top of the evaporator to provide more surface, while maintaining gravity flow of liquid ammonia to the evaporator. The two-temperature evaporator partially segregates the freezing function from the box cooling function, thus giving better humidity control.

In the absorber, the strong absorbent flows counter to and is diluted by direct contact with the gas. From the absorber, the weak absorbent flows through the liquid heat exchanger (9) to the analyzer (6) and then to the weak absorbent chamber (1a) of the generator (1). Heat applied to this chamber causes vapor to pass up through the analyzer (6) and to the condenser. The solution passes through an aperture in the generator partition into the strong absorbent chamber (1b). Heat applied to this chamber causes vapor and liquid to pass up through the small diameter pipe (10), as in an "air lift," to the separation vessel (11). While liberated ammonia vapor passes through the analyzer (6) to the condenser, the strong absorbent flows through the liquid heat exchanger (9) to the absorber. The finned air cooled loop (12) between the liquid heat exchanger and the absorber precools the solution further. The heat of absorption is rejected to the surrounding air.

The hydrogen reserve vessel (5), which is connected between the condenser outlet and the hydrogen circuit, is a reservoir for the hydrogen gas and compensates for changes in load and environmental conditions.

## Controls

**Burner Ignition and Monitoring Control.** These controls are either electronic or thermomechanical. Electronic controls ignite, monitor, and shut off the main burner as required by the thermostat. For thermomechanical control, a thermocouple monitors the main flame. The low temperature thermostat then changes the input to the main burner in a two-step mode. A pilot is not required because the main burner acts as the pilot on low fire.

**Low Temperature Thermostat.** This thermostat monitors the temperature in the cabinet and controls the gas input.

**Safety Device.** Each unit has a fuse plug to relieve pressure in the event of fire. Nominal operating conditions are as follows:

Ambient temperature: 95°F (35°C)
Freezer temperature: 10°F (−12°C)
Cabinet temperature: 30°F (−1°C)
Input: 200 Btu/h per $ft^3$ of cabinet interior (2.1 $kW/m^3$)

## BIBLIOGRAPHY

Auh, P.C. 1977. A Survey of Absorption Cooling Technology in Solar Applications. Brookhaven National Laboratory *Report* BNL-50704.

Bogart, M. 1981. *Ammonia Absorption Refrigeration in Industrial Processes*. Gulf Publishing Co.

Grossman, G. and K.W. Childs. 1983. Computer Simulation of a Lithium Bromide-Water Absorption Heat Pump For Temperature Boosting. ASHRAE *Transactions* 89(1).

Huntley, W.R. 1984. Performance Test Results of a Lithium Bromide-Water Absorption Heat Pump that Uses Low Temperature Waste Heat. Oak Ridge National Laboratory *Report* No. ORNL/TM9702, Oak Ridge, TN.

Kreider, J.F. and F. Kreith. 1975. *Solar Heating And Cooling*. Scripta Book Company, Hemisphere Publishing Corporation, Washington, DC.

Merrick, R.H. and R.A. English. 1960. An Air-Cooled Absorption Cycle. ASHRAE *Journal*, August, p. 39.

Miller, D.K. 1976. The Performance of Water Cooled Lithium Bromide Absorption Units for Solar Energy Applications. Heating/Piping/Air Conditioning, January.

Niebergall, W. 1981. *Handbuch der Kaltetechnik*, Volume 7: Sorptionsmaschinen. R. Plank, ed. Springer Verlag, Berlin.

Stoecker, W.F. and J.W. Jones. 1982. *Refrigeration and Air Conditioning*. McGraw Hill Book Company, New York.

Threlkeld. J.L. 1970. *Thermal Environmental Engineering*. Prentice-Hall, Englewood Cliffs, NJ.

G.C. Vliet, M.B. Lawson and R.A. Lithgow. 1982. Water Lithium Bromide Double-Effect Cooling Cycle Analysis. ASHRAE *Transactions* 88(1):811-823.

# CHAPTER 14

# AIR-CYCLE EQUIPMENT



AIR-CYCLE refrigeration systems are more commonly used in the air conditioning of aircraft than in surface and stationary applications because their lightweight, compact nature usually offsets their inherently low efficiency. Air-cycle air conditioning is not economical for residential and commercial buildings because of the high power required. However, it is used in specialized applications where efficiency is not the primary factor. For example, these units are used with portable gas turbine power plants to control the environment in remotely located, temporary military bases.

Air-cycle refrigeration may be designed and operated as an open or a closed system. Open air-cycle systems have been widely used in both military and commercial airplanes. In the open air-cycle principle, air is the refrigerant, and it directly cools the space requiring refrigeration. Unlike the refrigerant of a vapor-cycle system, which continuously changes phase from liquid to gas and back to liquid again, the refrigerant of an air-cycle system remains in the gaseous phase throughout the cycle.

Refrigeration in a simple open air-cycle system is obtained by (1) compression, (2) heat transfer, and (3) expansion accompanied by work extraction. A compressor first compresses air to a pressure higher than the pressure of the space to be cooled. A heat exchanger then rejects the heat of compression to a suitable heat sink such as ambient air. In the third step, a turbine cools the air by extracting work from it as it expands. A simple expansion or throttling device, such as an orifice or valve, does not noticeably lower the air temperature. The turbine, in contrast, extracts work from the expanding air to cool it. The load may be a fan that draws cooling air through a heat exchanger located upstream from the turbine. The cold air then discharges from the turbine and is used directly for cooling. Air-cycle units have been built in various sizes ranging in weight up to 250 lb (110 kg) and turbine flow capacity up to 235 lb/min (1.8 kg/s).

In the closed or dense-air system, the air refrigerant remains within the piping or component parts of the system at all times and usually remains at pressures above atmospheric. The term *dense-air system* is derived from the higher pressures maintained in comparison with the open system. Closed air-cycle systems have not been used in aircraft to date, but they are being considered for future aircraft. Both the open and closed air-cycle systems, as originally applied to cold-storage and heater-cooling systems, are almost obsolete. Their original advantage of using a safe and cheap refrigerant was nullified by the introduction of halocarbon refrigerants. It is not possible to overcome the inherently low efficiency; however, one manufacturer has developed a small reciprocating compressor-expander that operates on an open air cycle to produce cryogenic refrigeration.

Another type of air-cycle cooling is the Ranque-Hilsch vortex tube. It has found application in supplying small amounts of refrigeration, such as for small drinking-water coolers, spot cooling of a hard-to-reach critical component in an electronic control system, and air conditioning of helmets and suits for workers in hot, humid, or toxic locations. This device converts compressed air into hot and cold air. It is a simplified radial-inflow turbine. The air is injected tangentially at sonic velocity into a chamber, creating a cyclonic spinning. The air at the center of the vortex is cooled as it expands and transfers its angular momentum into heating the air at the periphery. Cold air is drawn out of the center, and the heated air is ejected from a relatively long tube. Vortex tube efficiency is low (about 10% of a comparable vapor refrigeration cycle), but its simplicity, light weight, and reliability make it attractive for some applications.

The preparation of this chapter is assigned to TC 9.3, Transportation Air Conditioning.

## AIR-CYCLE REFRIGERATION SYSTEMS

The common air-cycle systems addressed here are (1) simple, (2) bootstrap, and (3) simple/bootstrap. Other air-cycle systems are variations or combinations of these systems.

### Simple Air-Cycle Refrigeration

The simple air-cycle refrigeration system, shown in Figure 1, consists of an air-to-air heat exchanger and a cooling turbine.

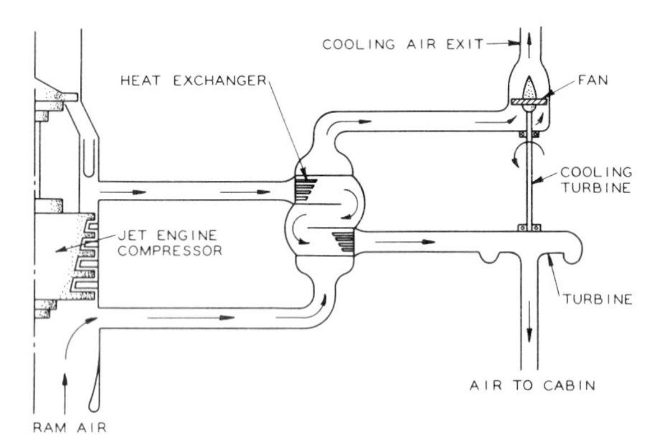


**Fig. 1 Simple Air-Cycle Refrigeration System**

High-pressure, high-temperature air from the high stage compressor of a jet engine is cooled initially in the heat exchanger and then cooled further in the cooling turbine by the process of expansion with work extraction. The heat sink for this system is *ram air* (the ambient air rammed into an airplane through a scoop, such as an engine inlet or auxiliary air inlet, as the airplane moves through the air). A fan connected to the turbine draws cooling air over the heat exchanger at low aircraft speeds or during static conditions while the aircraft is on the ground.

### Bootstrap Air-Cycle Equipment

The term *bootstrap,* as used in air-cycle refrigeration systems, indicates a system in which the pressure of the working fluid (high-pressure air) is raised to a higher level in the compressor section of the cooling turbine unit before expanding in the turbine section (Figure 2). The compressor is driven in bootstrap fashion by the turbine, which is on the same shaft.

The bootstrap air-cycle refrigeration system, shown in Figure 2, consists of a primary heat exchanger, a secondary heat exchanger, and a cooling turbine. High-pressure air is cooled first in the primary heat exchanger. The air is then compressed to a higher pressure and temperature in the compressor of the cooling turbine. A substantial amount of the heat of compression is removed in the secondary heat exchanger, and the air is cooled further as it expands through the turbine section of the cooling turbine. Ram air is used as a heat sink in the primary and secondary heat exchangers. Other heat sinks are sometimes used.

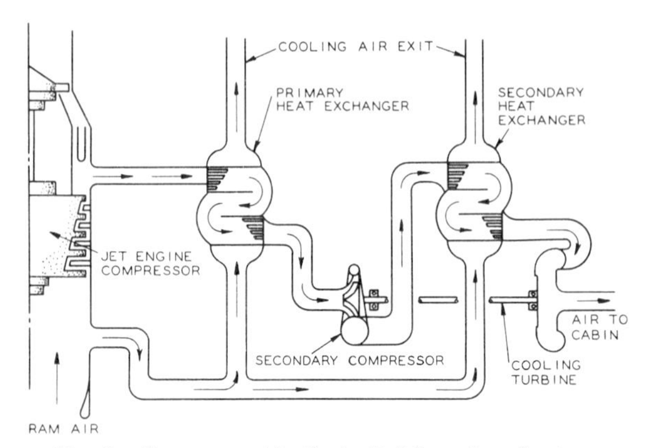


**Fig. 2 Bootstrap Air-Cycle Refrigeration System**

### Simple/Bootstrap Air-Cycle Equipment

Simple/bootstrap is a combination of the simple and bootstrap types in which the power from the turbine, usually through a single-shaft machine, drives a fan and compressor.

In the combination system, shown in Figure 3, the fan blows cooling air across the two heat exchangers during ground static operation, and the check valve prevents recirculating flow. In flight, the ram-pressured airflow is higher than the fan capacity, and a portion of the flow bypasses the fan, passing through the check valve.

Generally, the fan absorbs only a small portion of the turbine power (10 to 15%), and the compressor absorbs the remainder of the power. In some installations, the fan is located downstream of the heat exchangers and pulls the cooling air through the system.

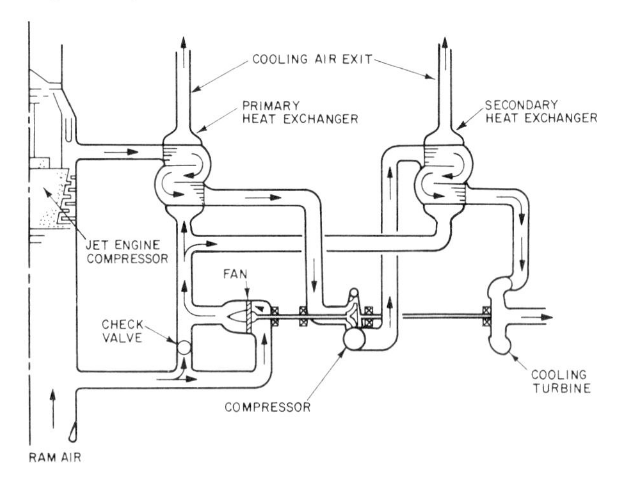


**Fig. 3 Simple/Bootstrap Air-Cycle Refrigeration System**

## CONTROL SYSTEMS

Typical control systems for air-cycle air-conditioning systems are those that provide (1) automatic temperature control of the compartment being cooled, (2) automatic protection against the supply of too much or too little air to the compartment, and (3) automatic protection against ice collection in the water separator for systems with a low-pressure water separator, or the condenser in a system using high-pressure water separation.

Figure 4 shows schematically the control system for a bootstrap air-cycle air-conditioning system.

### Compartment Air Temperature Control

The electronic system for compartment air temperature control, shown in Figure 4, consists of the following items:

*Temperature controller*, located in the compartment.
*Temperature selector*, located in the compartment.

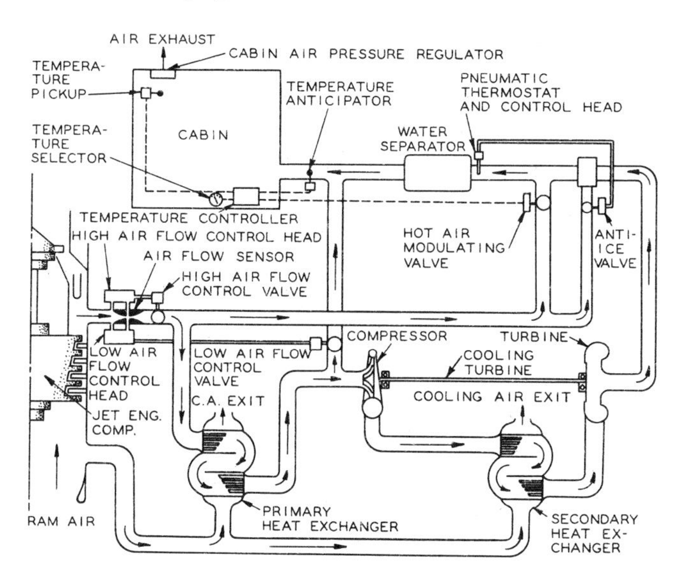


**Fig. 4 Bootstrap Air-Cycle Air-Conditioning System**

*Duct air temperature anticipator,* located in the duct connecting the water separator and the compartment.
*Compartment air temperature pickup,* located in the compartment.
*Hot-air modulating valve,* located in the duct connecting the hot-air supply duct with the cold-air duct downstream of the turbine discharge.

The temperature controller converts signals from the temperature anticipator, temperature pickup, and temperature selector into movement of the actuator of the hot-air modulating valve. Positioning of the valve permits hot air to bypass the heat exchangers and the cooling turbine. This hot air is mixed with cold air from the turbine discharge in the proper proportions to satisfy the compartment temperature for which the selector is set.

An all-pneumatic compartment temperature-control system consists of the same number of components as the electronic temperature-control system. Each component performs the same function as its corresponding part in the electronic system.

## Airflow Control

One type of airflow control system is a simple venturi. Flow is limited only when the throat of the venturi is choked, *i.e.,* when sonic velocity exists in the nozzle throat.

If the pressure of the high-pressure air source varies over a wide range without an airflow-control system, more air is sometimes forced into the compartment than desired; at other times, less air is delivered to the compartment than required for proper ventilation. To eliminate these undesirable conditions, a high airflow-control system and a low airflow-control system that are independent of each other are sometimes incorporated in the air-conditioning system. A typical airflow-control system is the *constant* $\Delta P$-type control. The $\Delta P$ is the differential between venturi upstream pressure and venturi throat pressure.

Two independent airflow-control systems are shown in Figure 4 for the bootstrap air-cycle air-conditioning system. One is a high airflow-control system, which limits the airflow to the compartment to a maximum value; the other is a low airflow-control system, which limits airflow to the compartment to a minimum value. Both systems consist of the following:

*Airflow sensor* (venturi), located in the high pressure air duct upstream from the primary heat exchanger and the high airflow-control valve. A single venturi is used.
*High airflow-control head,* mounted on the venturi.
*Low airflow-control head,* mounted on the venturi.
*High airflow-control valve,* located in the high-pressure air duct upstream of the primary heat exchanger.
*Low airflow-control valve,* located in the duct that bypasses the secondary heat exchanger, cooling turbine, and water separator.

When the high airflow control is regulating system airflow, a constant venturi $\Delta P$ is maintained by the automatic positioning of the high airflow-control valve, which throttles the supply air pressure. When the high airflow-control valve is throttling, the low airflow-control valve is always closed.

When the low airflow control is regulating system airflow, a constant venturi $\Delta P$ is maintained by the automatic positioning of the low airflow-control valve. The venturi $\Delta P$ maintained by the low airflow control is a value less than the venturi $\Delta P$ maintained by the high airflow-control system when that system is controlling. Modulation of the low airflow-control valve permits cool air from the discharge of the primary heat exchanger to bypass the secondary heat exchanger, the cooling turbine, and the water separator, and then mix with the cold discharge air from the water separator. When the low airflow-control valve is modulating, the high airflow-control valve is always open.

Actuation of the high and low airflow-control valves is usually accomplished by pneumatic actuators because high-pressure air is available.

## Water Separator and Turbine Anti-Ice Control

Under humid ambient conditions, condensation occurs in the expansion process of the air discharged from an aircraft air-cycle refrigeration unit, and the air often contains water in the form of mist, fog, or even snow. The most severe fogging conditions occur during low-altitude flight, such as in landing approaches, where good visibility is essential. In addition, the presence of moisture in the aircraft cabin causes deterioration of equipment and reduces reliability of operation. Therefore, it is important to install provisions in an air-cycle refrigeration unit for the removal of entrained moisture.

In many air-cycle systems, entrained moisture is removed by a water separator located in the low-pressure portion of the cycle downstream of the turbine, as shown in Figure 4. In this location, free moisture that condenses in the expansion cooling process can be agglomerated by a coalescer bag and removed from the airstream by inertial means.

A typical low-pressure water separator is shown in Figure 5. Its main parts are the housing, a fabric coalescer, a vortex generator, a collector-drain section, and a pressure relief valve. Water-laden air enters the water separator as a fog or mist. Finely dispersed water particles are agglomerated into larger droplets as they pass through the coalescer. The air containing these droplets is then swirled as it passes through the vanes of the vortex generator. This swirling action centrifuges the water droplets into a stagnant area of the collector section, where they are drained from the water separator. The water drained from the separator may be used to supplement heat-exchanger cooling. The low-pressure water separator is only 80 to 85% efficient so that the conditioned air has some entrained moisture. Traps in the downstream ducting and drains remove the entrained mositure before in enters the cabin.

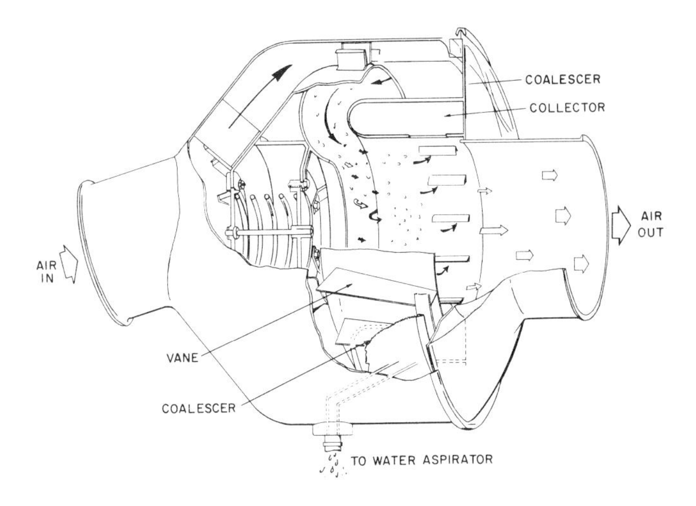


**Fig. 5 Water Separator, Low Pressure**

The pneumatic water separator anti-ice control system (Figure 4) consists of the following items:

*Pneumatic thermostat and control head,* located in the duct upstream from the water separator.

*Anti-ice valve,* located in the duct connecting the hot-air supply duct with the cold-air duct between the turbine discharge and the water separator inlet.

The anti-ice control system limits the dry-bulb temperature of the air entering the water separator to a minimum nominal value of 35°F (1.7°C). The thermostat senses the temperature of the air entering the water separator and maintains that air at a temperature of 35°F (1.7°C) by positioning the anti-ice valve so that the proper amount of hot air is mixed with cold turbine discharge air, which prevents freezing of water condensed in the expansion process.

Another water separator anti-ice control system is the $\Delta P$-type. This device consists of two pressure-sensing lines, a control head, and an anti-ice valve. The pressure differential across the water separator is sensed and the control differential pressure is set at a value higher than the maximum pressure drop existing across the water separator under normal operating conditions. Thus, if icing conditions are not prevalent, the anti-ice valve remains closed. If ice begins to collect in the water separator, the water separator pressure drop begins to increase until the control differential pressure is reached. The anti-ice valve is then modulated to maintain the control differential pressure constant. The water separator inlet-air temperature will be maintained at approximately 32°F (0°C) dry bulb or at the dew point of the air, if the dew point is less than 32°F (0°C).

An alternate sensing device for controlling dew point is the anti-ice screen. The screen is located between the turbine discharge and the water separator. Ice collecting on the grids of the screen increases the $\Delta P$ across the screen, which is sensed by the anti-ice valve. This type of sensing eliminates several disadvantages of the $\Delta P$ sensing across the water separator: (1) erroneous signal because of a dirty water separator coalescer bag and (2) potential instability problems caused by slower ice buildup or control response.

Under certain operating conditions, ice may form on the turbine wheel. If the ice accumulates enough to cause wheel unbalance and damaging vibration or excessive noise levels, it may be necessary to add turbine anti-ice control. This may be accomplished by using a back-pressure device at the turbine discharge, by variable-area turbine nozzles, by moving the water separator anti-ice valve thermostat to the turbine discharge, by modulating cooling air, or by ducting the higher temperature bypass flow through a shroud around the exducer of the turbine wheel, which would then mix with the turbine discharge air.

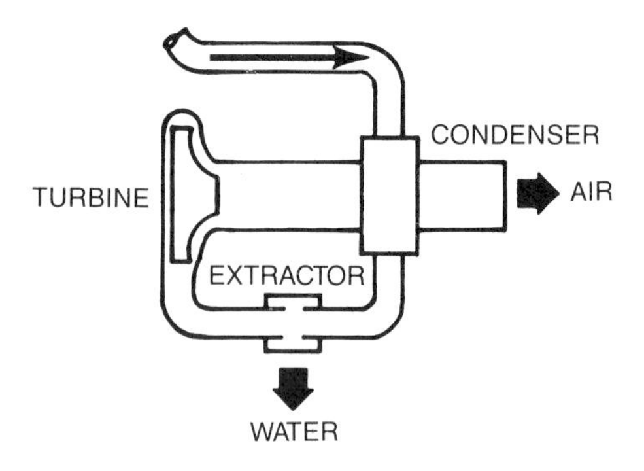


**Fig. 6 High-Pressure Water Removal Schematic**

Some air-cycle systems are designed to remove the entrained moisture in the high-pressure portion of the cycle upstream from the turbine, as shown by the schematic of Figure 6. Air at high pressure cannot hold as much water vapor, and, as a result, condensation occurs at a higher temperature. The condensation takes place on the finned surfaces of a condenser heat exchanger, and relatively large droplets are blown out of the condenser with the airstream. The air then enters the water extractor where the droplets are removed by inertial means.

## SOURCES OF HIGH-PRESSURE AIR

The following common sources of high-pressure air are used in air-cycle refrigeration systems for aircraft:

1. Jet engine and prop-jet engine compressors
2. Auxiliary air compressors driven by main engine, either through an air turbine or by means of a shaft
3. Auxiliary gas turbine compressor

A common source of high-pressure air for air-cycle air-conditioning systems in jet-propelled airplanes is the compressor of the turbojet engine.

In some instances, a gas turbine compressor supplies high-pressure air for air conditioning of airplanes on the ground. Some airplanes carry an auxiliary gas turbine and use it during flight for cabin pressurization and air conditioning. To justify the weight penalty of an on-board auxiliary gas turbine, the unit must supply additional services, such as ground electrical power and engine starting.

## ESTIMATING PERFORMANCE OF FIXED-SIZE AIR-CYCLE REFRIGERATION UNITS

The method of determining the quantity and temperature of refrigerated air delivered by a fixed-size air-cycle refrigeration system is illustrated by the following examples. Example 1 deals with a simple system of fixed size, and Example 2 deals with a bootstrap system of fixed size.

The symbols used in the examples are defined below.

$A$ = geometric area of turbine nozzle, in.$^2$
$A_n$ = effective area of turbine nozzle, in.$^2$
$c_p$ = specific heat at constant pressure. For air, $c_p$ = 0.24 Btu/lb • °F
$D$ = tip diameter of compressor impeller, fan wheel, or turbine wheel, in.
$E$ = effectiveness of heat exchanger (simple system), dimensionless
$E_1$ = effectiveness of primary heat exchanger (bootstrap system), dimensionless
$E_2$ = effectiveness of secondary heat exchanger (bootstrap system), dimensionless
$F_{CF}$ = compressor flow factor, dimensionless
$F_{TF}$ = turbine flow factor, dimensionless
$F_{TP}$ = turbine power factor, dimensionless
$F_V$ = turbine velocity factor = $N/(K\sqrt{\Delta T_i})$

where $K = \sqrt{J_{eq} g c_p \left(\frac{12 \times 60}{\pi D}\right)^2}$, dimensionless

$g$ = 32.2 ft/s$^2$
$J_{eq}$ = mechanical equivalent of heat, 778 ft-lb/Btu
$M$ = Mach number, dimensionless
$N$ = rotational speed of cooling turbine shaft, rpm
$P$ = total pressure of air, in. Hg absolute
$P_{avg}$ = average total pressure of air between two points in the system, in. Hg absolute
$R$ = gas constant for air = 53.3 ft-lb/lb • °R
$r_c$ = compressor (or fan) pressure ratio (greater than 1.0), dimensionless
$r_t$ = turbine pressure ratio (greater than 1.0), dimensionless
$T$ = total temperature of air, °R
$T_{avg}$ = average total temperature of air between two points in the system, °R
$W_B$ = flow rate of bleed air, lb/min
$W_C$ = flow rate of cooling air (simple system), lb/min
$W_{C1}$ = flow rate of cooling air in primary heat exchanger (bootstrap system), lb/min
$W_{C2}$ = flow rate of cooling air in secondary heat exchanger (bootstrap system), lb/min
$Y_c$ = compression factor = $r_c^{(\gamma-1)/\gamma} - 1$, dimensionless
$Y_e$ = expansion factor = $1 - \frac{1}{r_t^{(\gamma-1)/\gamma}}$, dimensionless
$\Delta P$ = pressure difference, in. Hg
$\Delta T$ = temperature difference, °F
$\Delta T_i$ = isentropic temperature difference of air across turbine, °F
$\gamma$ = ratio of specific heat at constant pressure to specific heat at constant volume, dimensionless. For air, $\gamma$ = 1.395
$\eta_c$ = isentropic efficiency of compressor, dimensionless
$\eta_m$ = mechanical efficiency of cooling turbine = $\frac{\text{fan or compressor power}}{\text{turbine power}}$, dimensionless
$\eta_t$ = isentropic efficiency of turbine, dimensionless
$\sigma$ = ratio of any air density to standard air density, dimensionless

Numerical subscripts denote locations in the flow schematics for each example.

**Example 1:** For the simple system and conditions shown in Figure 7, determine (*a*) the flow rate of bleed air $W_B$; (*b*) the flow rate of cooling air $W_C$; and (*c*) turbine discharge air temperature $T_3$. The effective area of the turbine nozzle is 0.328 in.$^2$ All temperatures and pressures are total values unless otherwise stated. Dry air is assumed throughout the analysis.

*Solution:* A trial-and-error solution is necessary. Values are assumed for the bleed and cooling airflow rates $W_B$ and $W_C$, and the turbine speed $N$.

Assume $W_B$ = 20 lb/min, $W_C$ = 56 lb/min, and $N$ = 67,000 rpm.

*Bleed Air Circuit*

From Figure 8, using the assumed values for $W_B$ and $W_C$, read $E$ = 0.932.

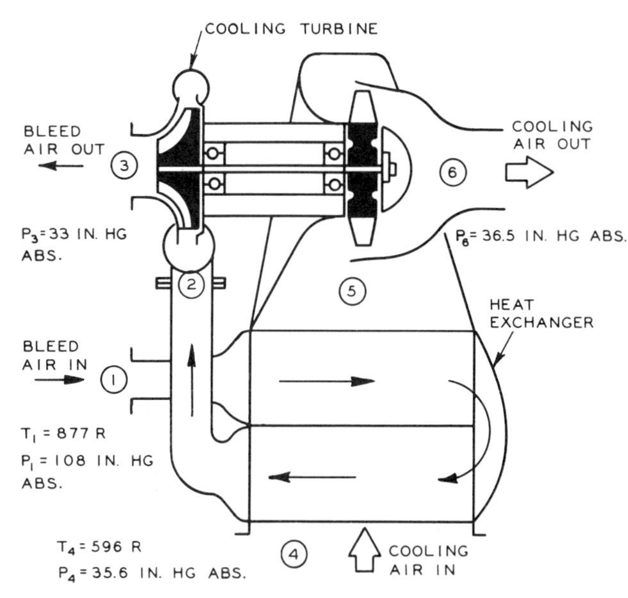


**Fig. 7 Simple Air-Cycle Refrigeration System for Example 1**

$$E = \frac{T_1 - T_2}{T_1 - T_4} = \frac{\Delta T_{1-2}}{T_1 - T_4}$$

$$\Delta T_{1-2} = E(T_1 - T_4) = 0.932\ (877 - 596) = 262\text{°F}$$
$$T_2 = T_1 - DT_{1-2} = 877 - 262 = 615\text{°R}$$

The average temperature between points 1 and 2 is

$$(T_{avg})_{1-2} = \frac{T_1 + T_2}{2} = \frac{877 + 615}{2} = 746\text{°R}$$

From Figure 9, for $W_B$ = 20 lb/min, read $\varrho \Delta P_{1-2}$ = 12.3 in. Hg, since

$$\varrho = 17.35 \frac{(P_{avg})_{1-2}}{(T_{avg})_{1-2}}$$

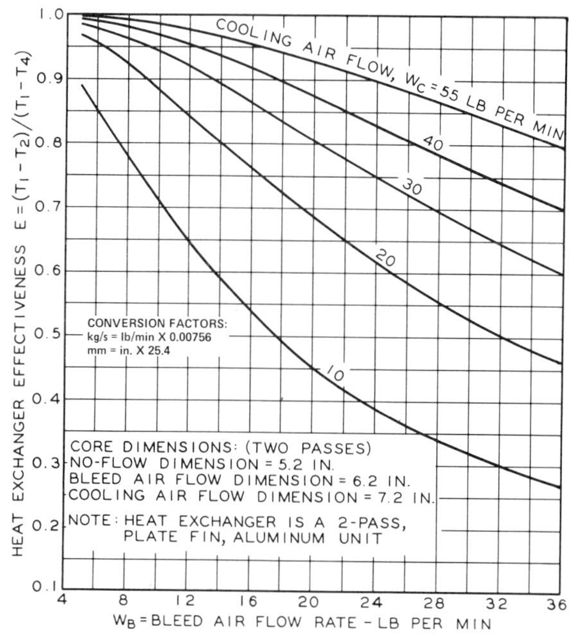


[a]For air-to-air heat exchanger in a simple air-cycle refrigeration unit

**Fig 8 Test Performance Curve—Effectiveness versus Bleed Airflow Rate[a]**

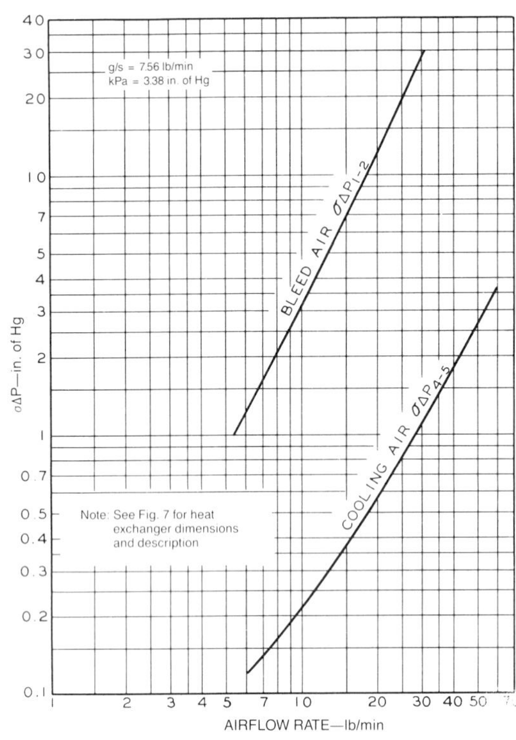


[a]For air-to-air heat exchanger in a simple air-cycle refrigeration unit.

**Fig. 9 Test Performance Curve—Pressure Drop versus Airflow Rate[a]**

*where* 17.35 is the ratio of standard temperature to standard pressure (519/29.92).

$$\Delta P_{1-2} = \frac{(\varrho \Delta P_{1-2})(T_{avg})_{1-2}}{17.35\ [P_1 - (\Delta P_{1-2}/2)]} = \frac{(12.3)(746)}{17.35\ [108 - (\Delta P_{1-2}/2)]}$$

$$= 5 \text{ in. Hg}$$

(Note that this equation is arranged in a form convenient for solution on a calculator but requires a trial-and-error process. It can be rearranged in quadratic form and solved directly for $\Delta P_{1-2}$.)

$$P_2 = P_1 - \Delta P_{1-2} = 108 - 5 = 103 \text{ in. Hg abs.}$$

The turbine pressure ratio $r_t = P_2/P_3 = 103/33 = 3.12$

The expansion factor is

$$Y_e = 1 - \frac{1}{r_t^{(\gamma-1)/\gamma}} = 1 - \frac{1}{(3.12)^{0.395/1.395}} = 0.275$$

Isentropic temperature drop

$$(\Delta T_i)_{2-3} = T_2 Y_e = 615(0.275) = 169\,°\text{F}$$

For the assumed turbine speed $N$ of 67,000 rpm

$$N/\sqrt{T_2} = 67{,}000/\sqrt{615} = 2700$$

From Figure 10, read the turbine flow factor $F_{TF} = 0.94$.

$$F_{TF} = \frac{W_B\sqrt{T_2}}{15.63 P_2 A_n}$$

*where*

$$15.63 = \left[\left(\frac{r^2}{\gamma - 1}\right)\left(\frac{2}{\gamma + 1}\right)^{(\gamma+1)/\gamma-1}\right]^{0.5}\sqrt{\frac{g}{Jc_p}\left(\frac{60}{2.036}\right)}$$

$$W_B = \frac{15.63 P_2 A_n}{\sqrt{T_2}}(F_{TF}) = \frac{(15.63)(103)(0.328)(0.94)}{\sqrt{615}}$$

$$= 20.01 \text{ lb/min}$$

The calculated value of $W_B$ checks with the original assumed value. If the calculated value had not checked, a second trial solution would have been necessary.

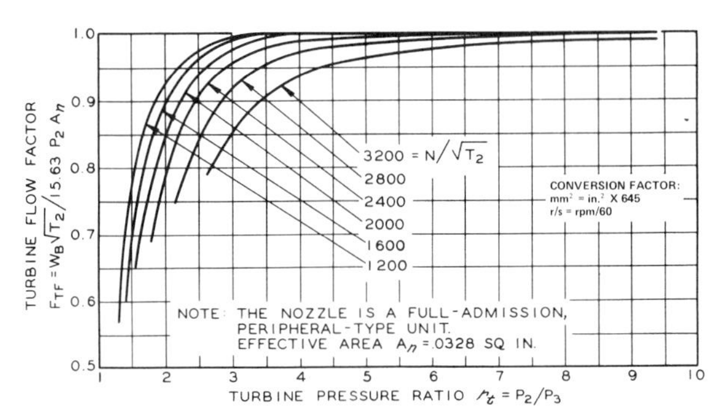


[a]For turbine nozzle of cooling turbine in a simple air-cycle refrigeration unit.

**Fig. 10 Test Performance Curve—Flow Factor versus Pressure Ratio[a]**

The turbine velocity factor

$$F_V = \frac{N}{K\sqrt{(\Delta T_i)_{2-3}}}$$

*where*

$$K = \sqrt{Jgc_p/[(12 \times 60)/(\pi D)]^2}$$

$$= \sqrt{(778)(32.2)(0.24)[(720)/(3.14 \times 3.5)]^2}$$

$$= 5111 \text{ rpm}/\sqrt{°\text{R}}$$

Therefore,

$$F_V = \frac{67{,}000}{5111\sqrt{169}} = 1.007$$

From Figure 11, read the turbine efficiency $\eta_t = 79.6\%$.

$$\Delta T_{2-3} = \frac{\eta_t}{100}(\Delta T_i)_{2-3} = \frac{79.6}{100}(169) = 134.5\,°\text{R}$$

The turbine discharge air temperature is

$$T_3 = T_2 - DT_{2-3} = 615 - 134.5 = 480.5\,°\text{R}$$

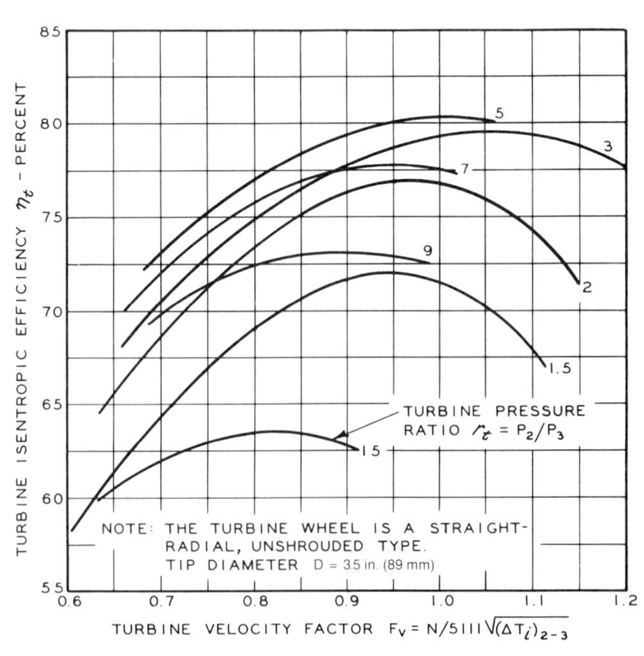


[a]For turbine wheel of cooling turbine in a simple air-cycle refrigeration unit.

**Fig. 11 Test Performance Curve—Efficiency versus Velocity Factor[a]**

*Cooling Air Circuit*

$$\Delta T_{4-5} = \left(\frac{W_B}{W_C}\right)\Delta T_{1-2} = \frac{20}{56}(262) = 94°F$$

$$T_5 = T_4 + DT_{4-5} = 596 + 94 = 690°R$$

$$(T_{avg})_{4-5} = \frac{T_4 + T_5}{2} = \frac{596 + 690}{2} = 643°R$$

From Figure 9, read $\sigma\Delta P_{4-5} = 3.27$ in. Hg.

$$\Delta P_{4-5} = \frac{(\sigma\Delta P_{4-5})(T_{avg})_{4-5}}{17.35\,[P_4 - (\Delta P_{4-5}/2)]} = \frac{(3.27)(643)}{17.35[(35.6 - (\Delta P_{4-5}/2)]}$$

$$= 3.58 \text{ in. Hg.}$$

$$P_5 = P_4 - \Delta P_{4-5} = 35.6 - 3.58 = 32.02 \text{ in. Hg abs.}$$

The turbine power factor $F_{TP}$ is

$$\frac{W_B\Delta T_{2-3}}{P_5\sqrt{T_5}} = \frac{(20)(134.5)}{32.02\sqrt{690}} = 3.20$$

$$\frac{P_6}{P_5} = \frac{36.5}{32.02} = 1.14$$

From Figure 12, read

$$\frac{N}{\sqrt{T_5}} = 2550, \text{ and } \frac{W_C\sqrt{T_5}}{P_5} = 46$$

$$N = \sqrt{T_5}\left(\frac{N}{\sqrt{T_5}}\right) = 690(2550) = 67{,}000 \text{ rpm}$$

$$W_C = \frac{P_5}{\sqrt{T_5}}\left(\frac{W_C\sqrt{T_5}}{P_5}\right) = \frac{32.02}{690}(46) = 56 \text{ lb/min}$$

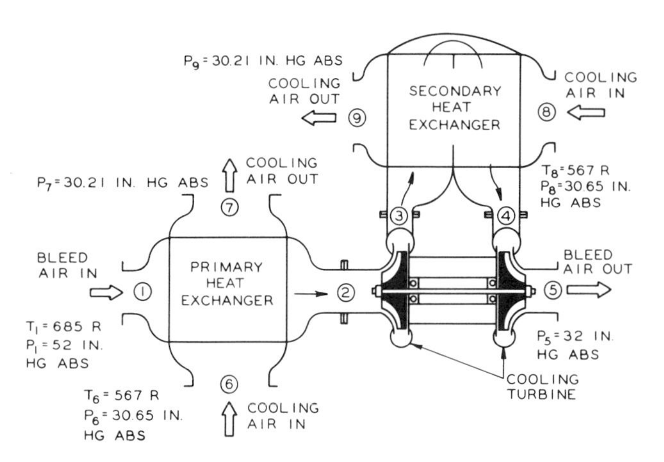


**Fig. 13 Bootstrap Air-Cycle Refrigeration System for Example 2**

The calculated values of $N$ and $W_C$ check with the original assumptions.

To summarize, the required values obtained by the trial-and-error solution are:

$W_B$ = 20 lb/min, $W_C$ = 56 lb/min, and $T_3$ = 480.5°R.

**Example 2:** For the bootstrap system shown in Figure 13, determine: (*a*) the flow rate of bleed air $W_B$; (*b*) the flow rate of cooling air through the primary heat exchanger $W_{C1}$; (*c*) the flow rate of cooling air through the secondary heat exchanger $W_{C2}$; and (*d*) the turbine discharge air temperature $T_5$. The geometric area of the turbine nozzle is 1.497 in$^2$.

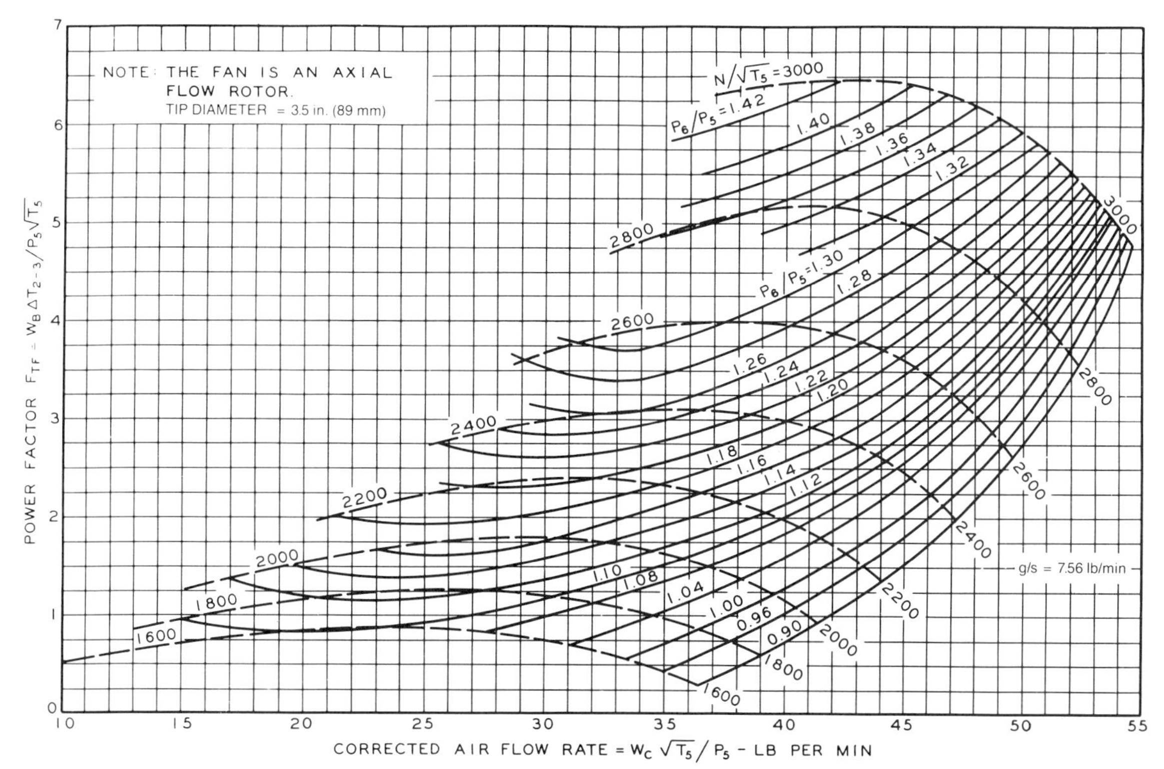


[a]For fan of cooling turbine in a simple air-cycle refrigeration unit.

**Fig. 12 Test Performance Curve— Power Factor versus Corrected Airflow Rate[a]**

*Solution:* As in Example 1, a trial-and-error solution is required. Values are assumed for $W_B$, $W_{C1}$, $W_{C2}$, and the compressor pressure ratio $P_3/P_2$.

Assume $W_B$ = 70 lb/min, $W_{C1}$ = 200 lb/min, $W_{C2}$ = 138 lb/min, and $P_3/P_2$ = 1.62.

*Bleed Air Circuit*

From Figure 14, read $E_1$ = 0.86. From Figure 15, read $E_2$ = 0.86.

$$E_1 = \frac{T_1 - T_2}{T_1 - T_6} = \frac{\Delta T_{1-2}}{T_1 - T_6}$$

$$\Delta T_{1-2} = E_1(T_1 - T_6) = 0.86(685 - 567) = 101\text{°F}$$

$$T_2 = T_1 - \Delta T_{1-2} = 685 - 101 = 584\text{°R}$$

$$(T_{avg})_{1-2} = \frac{T_1 + T_2}{2} = \frac{685 + 584}{2} = 634.5\text{°R}$$

From Figure 16, read $\sigma\Delta P_{1-2}$ = 0.193 in. Hg

$$\Delta P_{1-2} = \frac{(\sigma\Delta P_{1-2})(T_{avg})_{1-2}}{17.35\ [P_1 - (\Delta P_{1-2}/2)]}$$

$$= \frac{(0.193)\ (634.5)}{17.35\ [53 - (\Delta P_{1-2}/2)]} = 0.137 \text{ in. Hg}$$

(Note that this equation is arranged in a form convenient for solution on a calculator but requires a trial-and-error process. It can be rearranged in quadratic form and solved directly for $\Delta P_{1-2}$.)

$$P_2 = P_1 - \Delta P_{1-2} = 52 - 0.14 = 51.86 \text{ in. Hg abs.}$$

The compressor flow factor is

$$F_{CF} = \frac{W_B\sqrt{T_2}}{D^2P_2}$$

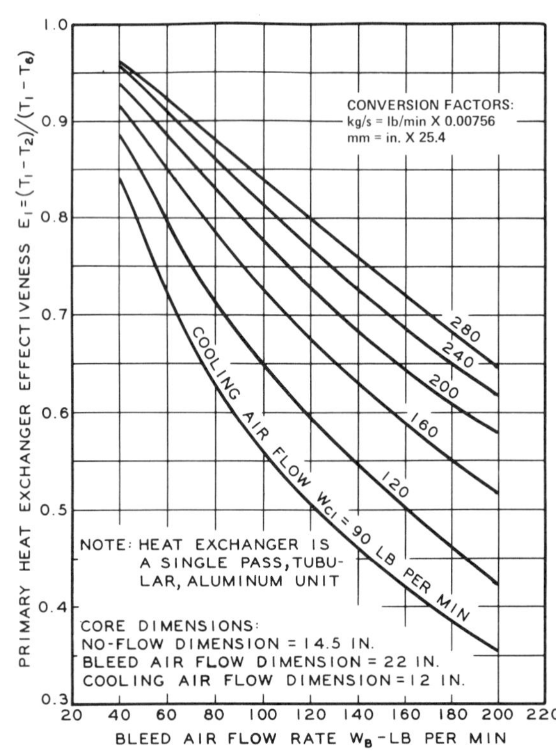


[a]For primary air-to-air heat exchanger in a bootstrap air-cycle refrigeration system.

**Fig. 14 Test Performance Curve—Effectiveness versus Bleed Airflow Rate[a]**

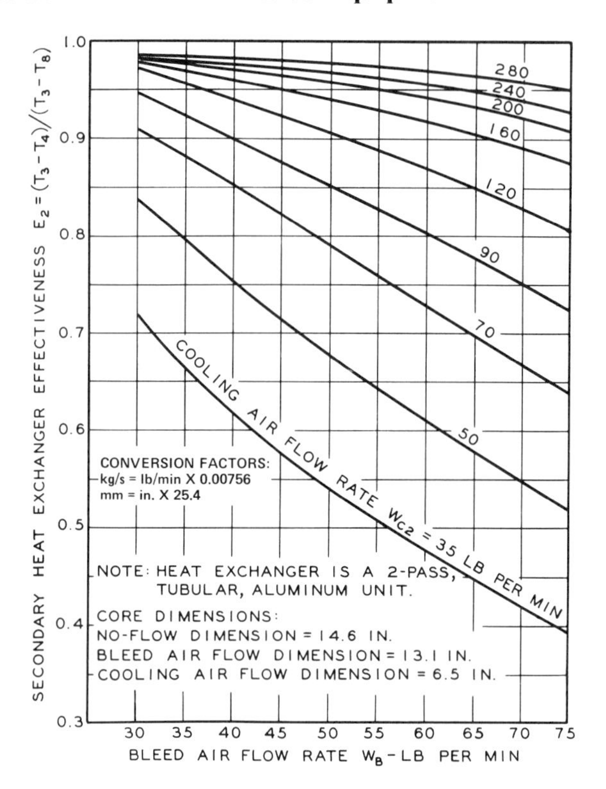


[a]For secondary air-to-air heat exchanger in a bootstrap air-cycle refrigeration system.

**Fig. 15 Test Performance Curve—Effectiveness versus Bleed Airflow Rate[a]**

Because the compressor tip diameter $D$ is 6.2 in. (Figure 17),

$$F_{CF} = \frac{70\sqrt{584}}{(6.2)^2(51.86)} = 0.85$$

From Figure 17, read the compressor Mach number $M$ = 0.85.

$$M = \frac{DN}{11{,}235\sqrt{T_2}}$$

*where*

$$11{,}235 = \frac{12 \cdot 60\sqrt{\gamma\, g\, R}}{\pi}, \text{ in.} \cdot \text{rpm}/\sqrt{\text{°R}}$$

$R$ = gas constant for air = 53.3 ft-lb/lb • °R

$\pi$ = 3.1416

The compressor speed is

$$N = \frac{11{,}235\sqrt{T_2}\ (M)}{D} = \frac{11{,}235\sqrt{584}\ (0.845)}{6.2}$$

From Figure 17, read the compressor efficiency $\eta_c$ = 77%. The compression factor is

$$Y_c = (r_c)^{(\gamma-1)/\gamma} - 1 = (P_3/P_2)^{(\gamma-1)/\gamma} - 1$$

$$= (1.62)^{0.395/1.395} - 1 = 0.1463$$

$$\Delta T_{2-3} = \frac{Y_cT_2}{\eta_c} = \frac{(0.1463)(584)}{0.77} = 111\text{°F}$$

$$T_3 = T_2 + \Delta T_{2-3} = 584 + 111 = 695\text{°R}$$

$$P_3 = P_2\left(\frac{P_3}{P_2}\right) = 51.86\ (1.62) = 84 \text{ in. Hg abs.}$$

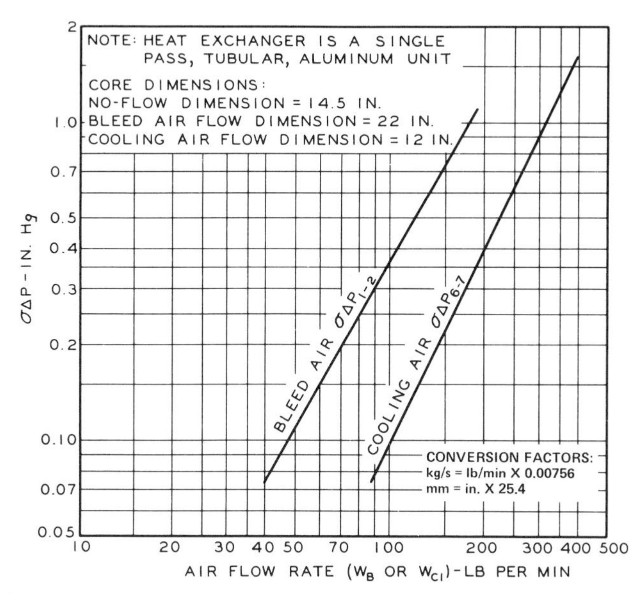


[a]For primary air-to-air heat exchanger in a bootstrap air-cycle refrigeration system.

**Fig. 16 Test Performance Curve—Pressure Drop versus Airflow Rate[a]**

$$E_2 = \frac{T_3 - T_4}{T_3 - T_8} = \frac{\Delta T_{3-4}}{T_3 - T_8}$$

$$\Delta T_{3-4} = E_2(T_3 - T_8) = 0.86\ (695 - 567) = 110\,°\text{F}$$

$$T_4 = T_3 - \Delta T_{3-4} = 695 - 110 = 585\,°\text{R}$$

$$(T_{avg})_{3-4} = \frac{T_3 + T_4}{2} = \frac{695 + 585}{2} = 640\,°\text{R}$$

From Figure 18, read $\sigma\Delta P_{3-4} = 3.22$ in. Hg.

$$\Delta P_{3-4} = \frac{(\varrho\Delta P_{3-4})(T_{avg})_{3-4}}{17.35\ [P_3 - (\Delta P_{3-4}/2)]} = \frac{(3.22)\ (640)}{17.35\ [84 - (\Delta P_{3-4}/2)]}$$
$$= 1.43 \text{ in. Hg.}$$

$$P_4 = P_3 - \Delta P_{3-4} = 84 - 1.43 = 82.57 \text{ in. Hg abs.}$$

The turbine pressure ratio $r_t$ is $P_4/P_5 = 82.57/32 = 2.58$.

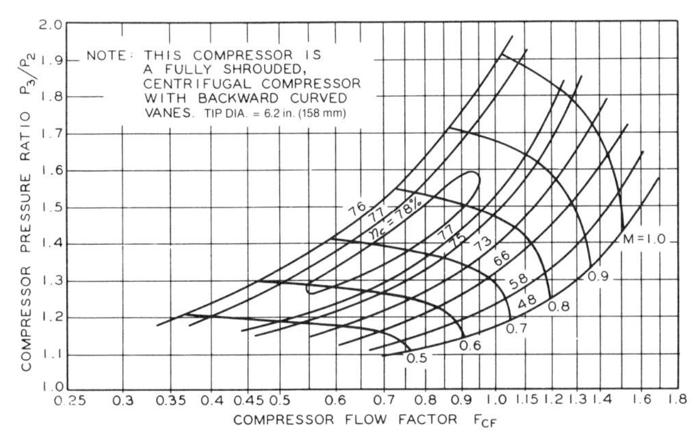


[a]For compressor of cooling turbine in a bootstrap air-cycle refrigeration system.

**Fig. 17 Test Performance Curve—Pressure Ratio versus Flow Rate Factor[a]**

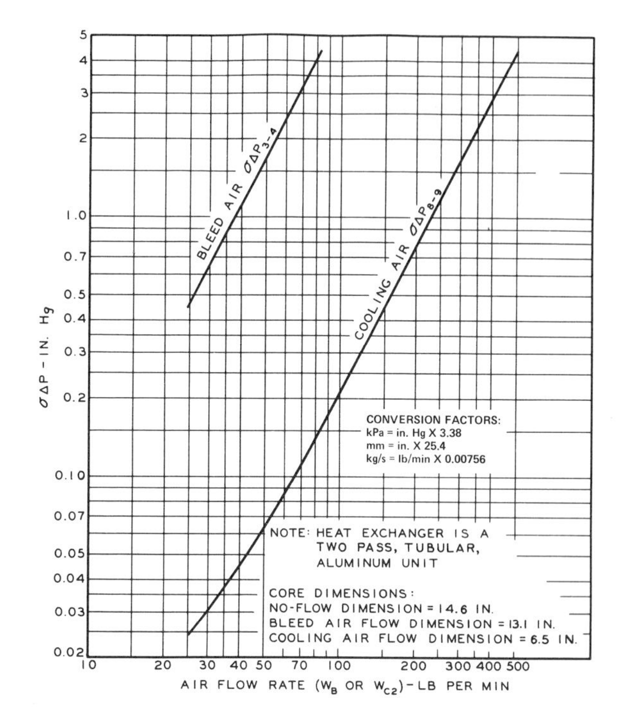


[a]For secondary air-to-air heat exchanger in a bootstrap air-cycle refrigeration system.

**Fig. 18 Test Performance Curve—Pressure Drop versus Airflow Rate[a]**

The turbine Mach number is

$$M = \frac{N}{1725\sqrt{T_4}} = \frac{37{,}300}{1725\sqrt{585}} = 0.894$$

From Figure 19, read the turbine flow factor $F_{TF} = 13.7$.

$$F_{TF} = \frac{W_B\sqrt{T_4}}{P_4 A}$$

$$W_B = \frac{P_4 A}{\sqrt{T_4}}(F_{TF}) = \frac{(82.57)(1.497)}{\sqrt{585}}(13.7)$$
$$= 70 \text{ lb/min.}$$

This checks with the assumed value of $W_B$. The expansion factor $Y_c$ is

$$Y_c = 1 - \frac{1}{(r_t)^{(\gamma-1)/\gamma}} = 1 - \frac{1}{(2.58)^{0.395/1.395}} = 0.235$$

The isentropic temperature drop between Points 4 and 5 is

$$(\Delta T_i)_{4-5} = T_4 Y_c = (585)(0.235) = 137.5\,°\text{F}$$

The turbine velocity factor is

$$F_V = \frac{N}{2750\sqrt{(\Delta T_i)_{4-5}}}$$

*where*

$$2750 = \sqrt{Jgc_p\ [(12 \bullet 60)/(\pi \bullet 6.5)]^2}, \text{ rpm}/\sqrt{°\text{R}}$$

$$F_V = \frac{37{,}300}{2750\sqrt{137.5}} = 1.157$$

From Figure 20, read the turbine efficiency $\eta_t = 83.3\%$.

$$\Delta T_{4-5} = \eta_t(\Delta T_i)_{4-5} = 0.833(137.5) = 114.5\,°\text{F}$$

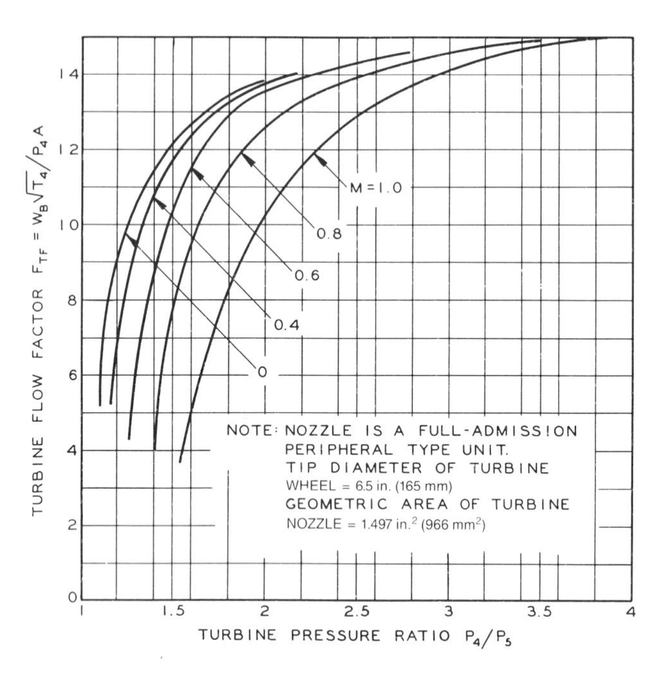


[a]For turbine nozzle of cooling turbine in a bootstrap air-cycle refrigeration system.

**Fig. 19 Test Performance Curve—Flow Factor versus Pressure Drop[a]**

$$\text{Turbine Power} = \frac{\text{Compressor Power}}{\eta_m}$$

$$W_B c_p \Delta T_{4-5} = \frac{W_B c_p \Delta T_{2-3}}{\eta_m}$$

$$\Delta T_{4-5} = \frac{\Delta T_{2-3}}{\eta_m} = \frac{111}{0.97} = 114.5\,°\text{F}$$

This value of $\Delta T_{4-5}$ checks with the value calculated for the equation $\Delta T_{4-5} = \eta_t(\Delta T_i)_{4-5}$, and indicates that a balance exists between turbine power and compressor power. This also shows the assumed value of compressor pressure ratio $r_c$ is correct.

$$T_5 = T_4 - \Delta T_{4-5} = 585 - 114.5 = 470.5\,°\text{R}$$

*Primary Heat Exchanger Cooling Air Circuit*

$$\Delta P_{6-7} = 30.65 - 30.21 = 0.44 \text{ in. Hg}$$

$$(P_{avg})_{6-7} = \frac{P_6 + P_7}{2} = \frac{30.65 + 30.21}{2} = 30.43 \text{ in. Hg abs.}$$

$$\Delta T_{6-7} = \frac{W_B}{W_{C1}}(\Delta T_{1-2}) = \frac{70}{200}(101) = 35\,°\text{F}$$

$$T_7 = T_6 + \Delta T_{6-7} = 567 + 35 = 602\,°\text{R}$$

$$(T_{avg})_{6-7} = \frac{T_6 + T_7}{2} = \frac{567 + 602}{2} = 584.5\,°\text{R}$$

$$\sigma \Delta P_{6-7} = 17.35 \frac{(P_{avg})_{6-7}}{(T_{avg})_{6-7}} (\Delta P_{4-5})$$

$$= 17.35 \left(\frac{30.43}{584.5}\right)(0.44) = 0.397 \text{ in. Hg}$$

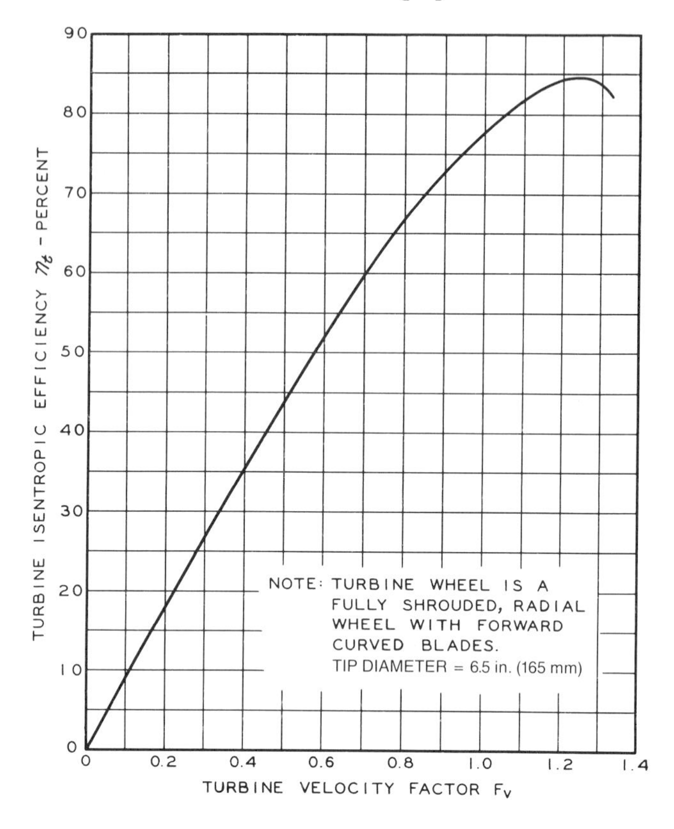


[a]For turbine wheel of cooling turbine in a bootstrap air-cycle refrigeration system.

**Fig. 20 Test Performance Curve—Efficiency versus Velocity Factor[a]**

From Figure 16, read $W_{C1}$ = 200 lb/min. This value checks with the assumed value for $W_{C1}$.

*Secondary Heat Exchanger Cooling Air Circuit*

$$\Delta P_{8-9} = 30.65 - 30.21 = 0.44 \text{ in. Hg}$$

$$(P_{avg})_{8-9} = \frac{P_8 + P_9}{2} = 30.43 \text{ in. Hg abs.}$$

$$\Delta T_{8-9} = \frac{W_B}{W_{C2}}(\Delta T_{3-4}) = \frac{70}{138}(110) = 56\,°\text{F}$$

$$T_9 = T_8 + \Delta T_{8-9} = 567 + 56 = 623\,°\text{R}$$

$$(T_{avg})_{8-9} = \frac{T_8 + T_9}{2} = 595\,°\text{R}$$

$$\sigma \Delta P_{8-9} = 17.35 \frac{(P_{avg})_{8-9}}{(T_{avg})_{8-9}} (\Delta P_{8-9})$$

$$= 17.35 \left(\frac{30.43}{595}\right)(0.44) = 0.391 \text{ in. Hg}$$

From Figure 18, read $W_{C2}$ = 138 lb/min. This value checks with the originally assumed value for $W_{C2}$.

## DESIGN OF SYSTEMS FOR OTHER AIRFLOW CAPACITY

The test performance curves for the fan, compressor, turbines, and heat exchangers in fixed-size systems may be used to design systems with larger or smaller airflow capacity. The methods used for each of the components are described below.

### Fan

Figure 12 presents the performance of a 3.5-in. (89 mm) tip diameter axial flow fan. The performance of any geometrically similar fan with a tip diameter, $D$, between 3 and 10 in. (76 and 254 mm) may be predicted, with reasonable accuracy, by using Figure 12 and the following equations (the value 3.5 becomes 89 in SI units):

$$\left(\frac{W_C\sqrt{T_5}}{P_5}\right)_{3.5} = \left(\frac{W_C\sqrt{T_5}}{P_5}\right)_D \left(\frac{3.5}{D}\right)^2$$

$$\left(\frac{W_B\Delta T_{2-3}}{P_5\sqrt{T_5}}\right)_{3.5} = \left(\frac{W_B\Delta T_{2-3}}{P_5\sqrt{T_5}}\right)_D \left(\frac{3.5}{D}\right)^2$$

$$\left(\frac{P_6}{P_5}\right)_{3.5} = \left(\frac{P_6}{P_5}\right)_D$$

### Compressor

Figure 17 presents the performance of a 6.2-in. (158 mm) tip diameter centrifugal compressor with backward-curved impeller vanes. The curve is actually presented in dimensionless form so it is directly applicable, with reasonable accuracy, to any geometrically similar compressor with an impeller tip diameter, $D$, between approximately 4 and 12 in. (100 and 300 mm).

$$(F_{CF})_{6.2} = (F_{CF})_D$$

or

$$\left(\frac{W_B\sqrt{T_2}}{D^2P_2}\right)_{6.2} = \left(\frac{W_B\sqrt{T_2}}{D^2P_2}\right)_D$$

$$(M)_{6.2} = (M)_D$$

or

$$\left(\frac{DN}{11235\sqrt{T_2}}\right)_{6.2} = \left(\frac{DN}{11235\sqrt{T_2}}\right)_D$$

$$\left(\frac{P_3}{P_2}\right)_{6.2} = \left(\frac{P_3}{P_2}\right)_D$$

$$(\eta_c)_{6.2} = (\eta_c)_D$$

### Turbine

Figures 10 and 11 present performance of a straight radial inward-flow turbine wheel with a full admission nozzle. The wheel tip diameter is 3.5 in. (89 mm). These curves may be used to predict performance, with reasonable accuracy, for any geometrically similar turbine wheel with a tip diameter, $D$, between 3 and 10 in. (75 and 250 mm).

$$(F_V)_{3.5} = (F_V)_D\left(\frac{D}{3.5}\right)$$

or

$$\left(\frac{N}{5111\sqrt{(\Delta T_i)_{2-3}}}\right)_{3.5} = \left(\frac{N}{5111\sqrt{(\Delta T_i)_{2-3}}}\right)_D \left(\frac{D}{3.5}\right)$$

$$\left(\frac{W_B\sqrt{T_2}}{15.63P_2A_n}\right)_{3.5} = \left(\frac{W_B\sqrt{T_2}}{15.63P_2A_n}\right)_D$$

$$\left(\frac{P_2}{P_3}\right)_{3.5} = \left(\frac{P_2}{P_3}\right)_D$$

$$\left(\frac{N}{\sqrt{T_2}}\right)_{3.5} = \left(\frac{N}{\sqrt{T_2}}\right)_D \left(\frac{D}{3.5}\right)$$

$$(A_n)_D = (A_n)_{3.5}\left(\frac{D}{3.5}\right)^2$$

Figures 19 and 20 present performance of an inward-flow, radial-turbine wheel with forward-curved vanes and a full admission nozzle. The wheel tip diameter is 6.5 in. (165 mm). These two curves may be used, with reasonable accuracy, to predict the performance of any geometrically similar turbine wheel with a tip diameter $D$ between approximately 3.5 and 12 in. (90 and 300 mm).

$$(F_V)_{6.5} = (F_V)_D\left(\frac{D}{6.5}\right)$$

or

$$\left(\frac{N}{2750\sqrt{(\Delta T_i)_{4-5}}}\right)_{6.5} = \left(\frac{N}{2750\sqrt{(\Delta T_i)_{4-5}}}\right)_D \left(\frac{D}{6.5}\right)$$

$$(F_{TF})_{6.5} = (F_{TF})_D$$

$$\left(\frac{W_B\sqrt{T_4}}{P_4A}\right)_{6.5} = \left(\frac{W_B\sqrt{T_4}}{P_4A}\right)_D$$

$$\left(\frac{P_4}{P_5}\right)_{6.5} = \left(\frac{P_4}{P_5}\right)_D$$

$$(M)_{6.5} = (M)_D$$

or

$$\left(\frac{N}{1725\sqrt{T_4}}\right)_{6.5} = \left(\frac{N}{1725\sqrt{T_4}}\right)_D \left(\frac{D}{6.5}\right)$$

$$(A)_D = (A)_{6.5}\left(\frac{D}{6.5}\right)^2$$

### Heat Exchangers

Figures 8 and 9 present the performance of a two-pass, combination cross-counterflow, air-to-air, aluminum heat exchanger with a plate-fin-type heat-transfer surface. The overall core dimensions of this unit are 5.2 in. (132 mm) in the no-flow direction ($L_n$), 6.2 in. (158 mm) in the bleed airflow direction ($L_B$), and 7.2 in. (183 mm) in the cooling airflow direction ($L_c$). The weight of this core proper is approximately 6.7 lb (3 kg).

Figures 14 and 16 present the performance of a single-pass, crossflow, air-to-air, aluminum heat exchanger with a flattened and dimpled tubular type heat-transfer surface. The overall core dimensions of this unit are 14.5 in. (368 mm) in the no-flow direction ($L_n$), 22 in. (559 mm) in the bleed airflow direction ($L_B$), and 12 in. (305 mm) in the cooling airflow direction ($L_c$). The weight of the core is approximately 37 lb (16.8 kg).

Figures 15 and 18 present the performance of a two-pass cross-counterflow, air-to-air, aluminum heat exchanger with a flattened and dimpled tubular-type heat-transfer surface. The overall core dimensions of this unit are 14.6 in. (371 mm) in the no-flow direction ($L_n$), 13.1 in. (333 mm) in the bleed airflow direction ($L_B$), and 6.5 in. (165 mm) in the cooling airflow direction ($L_c$). The weight of the core is approximately 48 lb (21.8 kg).

By increasing or decreasing the length of any of the three heat exchangers in the no-flow direction only, the performance of the new heat exchangers can be determined from the given performance by the following formulas:

$$(W_B)_{new} = (W_B)_{old}\frac{(L_n)_{new}}{(L_n)_{old}}$$

$$(W_C)_{new} = (W_C)_{old}\frac{(L_n)_{new}}{(L_n)_{old}}$$

$$(\sigma \Delta P)_{new} = (\sigma \Delta P)_{old}$$

$$(\text{Core Weight})_{new} = (\text{Core Weight})_{old}\frac{(L_n)_{new}}{(L_n)_{old}}$$

## Correction for Moisture

As previously noted, dry air was assumed throughout the sample analysis. The dry-bulb discharge air temperature of an air-cycle cooling turbine operating with completely dry air, or with inlet and discharge conditions such that no moisture condensation occurs, is the *dry air rated temperature.* With humid air where condensation occurs in the expansion process, the dry-bulb discharge temperature of the cooling turbine is the *wet air rated temperature.* With the anti-ice valve described previously, this *wet air rated temperature* would be nominally about 35 °F (2 °C) dry bulb; with the pneumatic thermostat, it would be 32 °F (0 °C) dry bulb, or the dew point if the dew point is less than 32 °F (0 °C). The low-pressure water separator is less than 100% efficient, and some condensate will be entrained in the airstream. Some of this entrained moisture may be reevaporated by heat transfer into the air as it passes through ducting to the airplane cabin, essentially doing further cooling of the air in the process.

Depending on such factors as duct insulation, duct bends, and air velocity, part of the water entrained as it leaves the separator settles out of the airstream and does not reevaporate. These factors are difficult to evaluate, except by testing. However, the theoretical dry-bulb discharge temperature, which assumes that the entrained moisture is reevaporated adiabatically from a point on the psychrometric chart saturation curve, is the cooling turbine *effective discharge temperature.* Where inlet moisture conditions and water separator efficiency are known, the preceding calculations can be corrected for the reevaporation effect. If low-pressure water separator efficiency is not known, a good representative value to assume is 85%, *i.e.*, approximately 15% of condensed moisture may reevaporate adiabatically, thus adding to effective cooling of the air supply.

## BIBLIOGRAPHY

Crabtree, R.E.; M.P. Saba; and J.E. Strang. 1980. The Cabin Air Conditioning and Temperature Control System for the Boeing 767 and 757 Airplanes. ASME-80, ENAS-55, American Society of Mechanical Engineers, New York.

Dieckmann, R.R.; O.R. Kosfeld; and L.C. Jenkins. 1986. Increased Avionics Cooling Capacity for F-15 Aircraft. SAE 860910, Society of Automotive Engineers, Warrendale, PA.

SAE. 1969. Aerospace Applied Thermodynamics Manual. AIR 1168, Society of Automotive Engineers, Warrendale, PA.

Vortex Tubes. 1972. Vortex Corporation *Bulletin*, Cincinnati, OH.

# CHAPTER 15

# CONDENSERS



THE condenser in a refrigeration system is a heat exchanger that usually rejects all the heat from the system. This heat consists of heat absorbed by the evaporator plus the heat equivalent of the energy input to the compressor. The compressor discharges hot, high-pressure refrigerant gas into the condenser, which rejects heat from the gas to some cooler medium. Thus, the cool refrigerant condenses back to the liquid state and drains from the condenser to continue in the refrigeration cycle.

The common forms of condensers may be classified on the basis of the cooling medium as (1) water-cooled, (2) air-cooled, and (3) evaporative (air- and water-cooled).

## PART I: WATER-COOLED CONDENSERS

### HEAT REMOVED

The heat rejection rate in a condenser for each unit of heat removal produced in the cooler may be estimated from a graph (Figure 1). The theoretical values shown are based on Refrigerant 22 with 10 °F (5.6 °C) suction superheat, 10 °F (5.6 °C) liquid subcooling, and compressor efficiency of 80%. Actually, the heat removed is slightly higher or lower than these values, depending on compressor efficiency.

Usually, the heat rejection requirement ($q_o$) can be accurately determined from known values of evaporator load ($q_i$) and the heat equivalent of the actual power required ($q_w$) for compression (obtained from the compressor manufacturer's catalog):

$$q_o = q_i + q_w \tag{1}$$

*Note:* $q_w$ is reduced by any independent heat-rejection processes (oil cooling, motor cooling, etc.).

The volumetric flow rate of condensing water required may be found from Eq. (2):

$$Q = \frac{q_o}{\varrho(t_2 - t_1)\, c_p} \tag{2}$$

*where*

$Q$ = volumetric flow rate of water, ft³/h (L/s) (multiply ft³/h by 0.125 to obtain gpm)
$q_o$ = heat-rejection rate, Btu/h (W)
$\varrho$ = density of water, lb/ft³ (kg/m³)
$t_1$ = temperature of water entering condenser, °F (°C)
$t_2$ = temperature of water leaving condenser, °F (°C)
$c_p$ = specific heat of water, Btu/lb • °F (kJ/kg • °C)

---

**Example 1.** Estimate the volumetric flow rate of condensing water required for the condenser of a Refrigerant 22 water-cooled unit operating at a condensing temperature of 105 °F, a cooler temperature of 40 °F, 10 °F liquid subcooling, and 10 °F suction superheat. Water enters the condenser at 86 °F and leaves at 95 °F. The refrigeration load is 100 tons.

*Solution.* From Figure 1, the heat rejection factor for these conditions is about 1.19.

$q_o$ = 100(1.19) = 119 ton

$\varrho$ = 62.1 lb/ft³ @90.5 °F

$c_p$ = 1.0 Btu/lb • °F

---

The preparation of this chapter is assigned to TC 8.4, Air-to-Refrigerant Heat-Transfer Equipment; TC 8.5, Liquid-to-Refrigerant Heat Exchangers; and TC 8.6, Cooling Towers and Evaporative Condensers.

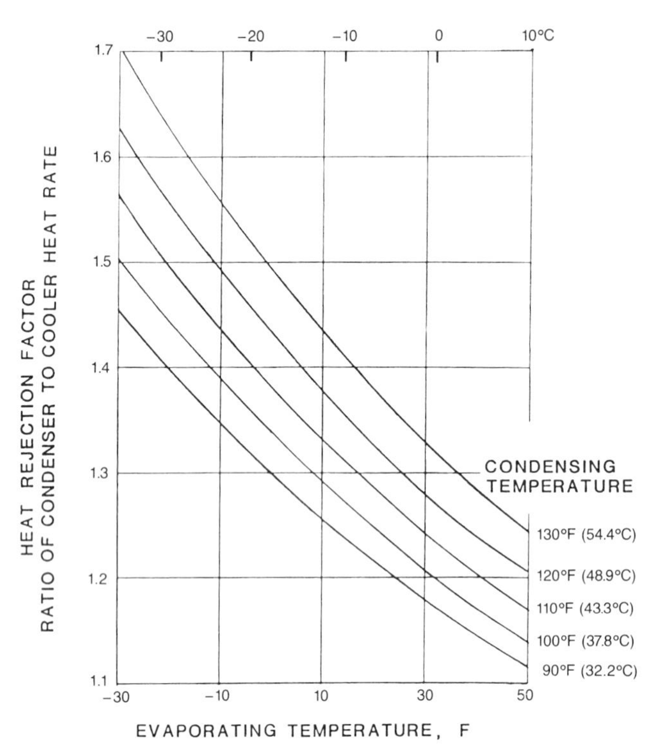


Refrigerant-22, 10°F (−5.6°C) liquid subcooling, 10°F (−5.6°C) suction superheat, 80% compressor efficiency

**Fig. 1 Heat Removed in a Condenser**

From Eq. (2):

$$Q = \frac{1496(119)}{62.1(95 - 86)1.0} = 319 \text{ gpm}$$

*Note:* 1496 is a conversion factor with units, Btu • gal/ton • min • ft$^3$.

## HEAT TRANSFER

A water-cooled condenser transfers heat in three stages: sensible cooling in the gas desuperheating and condensate subcooling stages, and transfer of latent heat in the condensing stage. Condensing is by far the dominant process in normal refrigeration system application, accounting for 83% of the heat rejection in the previous example. Because the tube wall temperature is normally lower than the condensing temperature at all locations in the condenser, the condensation process takes place throughout the condenser.

The heat transfer coefficient of a condenser varies almost inversely as the temperature difference between entering superheated gas and the ambient temperature. As a result, an average overall heat-transfer coefficient and the mean temperature difference (calculated from the condensing temperature corresponding to the saturated condensing pressure and the entering and leaving water temperatures) give reasonably accurate predictions of performance.

Subcooling has an effect on the average overall heat-transfer coefficient when tubes are submerged in liquid. The heat rejection rate is then determined as:

$$q = UA\Delta t_m \tag{3}$$

*where*

$q$ = total heat transfer, Btu/h (W)
$U$ = overall heat-transfer coefficient, Btu/h • ft$^2$ • °F (W/m$^2$ • °C)
$A$ = heat transfer surface area associated with $U$, ft$^2$ (m$^2$)
$\Delta t_m$ = mean temperature difference, °F (°C)

Chapter 3 of the 1985 FUNDAMENTALS Volume shows how to calculate $\Delta t_m$. When a detailed analysis of integral desuperheating and/or subcooling processes in condenser design is desired, computational schemes have been developed. (Bell 1972).

### Overall Heat Transfer ($U_o$)

The overall heat transfer in a water-cooled condenser with water inside the tubes may be computed from calculated or test-derived heat-transfer coefficients of the water and refrigerant sides, from physical measurements of the condenser tubes, and from a fouling factor on the water side, by using Eq. (4).

$$U_o = \frac{1}{(A_o/A_i)/h_w + S_R \cdot r_{fw} + (t/k)(A_o/A_m) + 1/(h_r \phi_w)} \tag{4}$$

*where*

$U_o$ = overall heat-transfer coefficient, based on the external surface and the mean temperature difference, between the external and internal fluids, Btu/h • ft$^2$ • °F (W/m$^2$ • °C)
$A_o/A_i$ = ratio of external to internal surface area
$h_w$ = internal or water-side film coefficient, Btu/h • ft$^2$ • °F (W/m$^2$ • °C)
$r_{fw}$ = fouling resistance on water side, ft$^2$ • h • °F/Btu (m$^2$ • °C/W)
$t$ = thickness of tube wall, ft, (m)
$k$ = thermal conductivity of tube material, Btu/h • ft • °F (W/m • °C)
$A_o/A_m$ = ratio of external to mean heat-transfer surface areas of metal wall
$h_r$ = external, or refrigerant side, coefficient, Btu/h • ft$^2$ • °F (W/m$^2$ • °C)
$\phi_w$ = weighted fin efficiency (100% for bare tubes)

For tube-in-tube condensers, or other condensers where the refrigerant flows inside the tubes, the equation for $U_o$, in terms of water side surface, becomes:

$$U_o = \frac{1}{(A_o/A_i)/h_r + r_{fw} + (t/k) + 1/h_w} \tag{5}$$

*where*

$h_r$ = internal or refrigerant-side coefficient, Btu/h • ft$^2$ • °F (W/m$^2$ • °C)
$h_w$ = external or water-side coefficient, Btu/h • ft$^2$ • °F (W/m$^2$ • °C)

### Water-Side Film Coefficient ($h_w$)

Values of the water-side coefficient may be calculated from equations in Chapter 3 of the 1985 FUNDAMENTALS Volume. For turbulent flow, at Reynolds numbers exceeding 10,000 in horizontal tubes and using average water temperatures, the general equation (McAdams 1954) is:

$$h_w D/k = 0.023\,(DG/\mu)^{0.8}(c_p\mu/k)^{0.4} \tag{6}$$

*where*

$D$ = inside tube diameter, ft (m)
$k$ = thermal conductivity of water, Btu/h • ft • °F (W/m • °C)
$G$ = mass velocity of water, lb/h • ft$^2$ (kg/h • m$^2$)
$\mu$ = viscosity of water, lb/ft • h (mPa • s)
$c_p$ = specific heat of water at constant pressure, Btu/lb • °F (kJ/kg • °C)

The constant (0.023) in Eq. (6) reflects plain ID tubes. Bergles (1973 and 1976) discusses numerous water-side enhancement methods, which increase the value of the constant in Eq. (6).

Because of its strong influence on the value of $h_w$, water velocity generally should be maintained as high as possible without excessive pressure drop. Maximum velocities with clean water of 10 to 13 fps (3 to 4 m/s) are commonly used. Experiments by Sturley (1975) at velocities up to approximately 26 fps (8 m/s) show no damage to copper tubes after prolonged operation. A minimum velocity of 3 fps (1 m/s) is good practice when the water quality is such that noticeable fouling or corrosion could result. With clean water, the velocity may be lower if it must be conserved or if it has a low temperature.

### Refrigerant-Side Film Coefficient ($h_r$)

Factors influencing the value of $h_r$ are listed below:

1. Type of refrigerant being condensed
2. Geometry of condensing surface (plain tube OD and finned-tube fin spacing, height, and cross-section profile)
3. Condensing temperature
4. Condensing rate in terms of mass velocity or rate of heat transferred
5. Arrangement of tubes in bundle
6. Vapor distribution and rate of flow
7. Condensate drainage
8. Liquid subcooling

Values of the refrigerant-side coefficients may be estimated from correlations shown in Chapter 4 of the 1985 FUNDAMENTALS Volume. Information on the effects of the type of refrigerant, condensing temperature, and loading (temperature drop across the condensate film) on the condensing film coefficient can be found in the section on "Condensing" in the same chapter. (Also Katz *et al.* 1947; Katz and Robinson. 1947; and Kratz, MacIntire, and Gould 1930 give further information.)

Actual values of $h_r$ for a given physical condenser design can be determined from test data by use of a *Wilson Plot* (McAdams 1954 and Briggs and Young 1969).

The type of condensing surface has considerable effect on the condensing coefficient. Most halocarbon refrigerant condensers use finned tubes where the fins are integral with the tube. Water velocities normally used are large enough for the resulting high water-side film coefficient to justify using extended external surface to balance the heat-transfer resistances of the two surfaces. Peason and Withers (1969) compared the refrigerant condensing performance of some integral finned tubes with different fin spacings.

Physical aspects of a given condenser design—such as tube spacing and orientation, shell-side baffle arrangement, orientation of multiple water-pass arrangements, and the number of tubes high in the bundle—affect the refrigerant-side coefficient by influencing vapor distribution and flow rate through the tube bundle and condensate drainage from the bundle. Butterworth (1977) reviewed correlations accounting for these variables in predicting the heat-transfer coefficient for shell-side condensation. Kistler *et al.* (1976) developed analytical procedures for design use within these parameters.

As refrigerant condenses on the tubes, it falls upon the tubes in lower rows. Due to the added resistance of this liquid film, the effective film coefficient for lower rows should be less than that for upper rows. Therefore, the average overall refrigerant film coefficient should decrease as the number of tube rows increase. However, the additional compensating effects of added film turbulence and direct contact condensation on the subcooled liquid film make actual row effect uncertain.

When the liquid is subcooled by raising the condensate level to submerge a desired number of tubes, the refrigerant film coefficient associated with the submerged tubes is less than the condensing coefficient. If the refrigerant film coefficient in Eq. (4) is an an average based on all tubes in the condenser, its value decreases as a greater portion of the tubes is submerged.

### Tube-Wall Resistance ($t/k$)

Most refrigeration condensers, with the exception of ammonia, use relatively thin-walled copper tubes. Where these are used, the temperature drop or gradient across the tube wall is not significant. If the tube metal has a high thermal resistance, as does 70/30 cupro-nickel, considerable temperature drop will occur or, conversely, an increase in the *mean temperature difference* ($\Delta t_m$) or in surface area is required to transfer the same amount of heat, compared to copper. This temperature difference penalty is accounted for in Eqs. (4) and (5) when calculating $t/k$. In the case of a heavy-tube wall of low-conductivity material, see Chapter 3 of the 1985 FUNDAMENTALS Volume for a correlation to improve the accuracy of the tube-wall resistance calculation.

### Weighted-Fin Efficiency ($\phi_w$)

For a finned tube, a temperature gradient exists from the root of a fin to its tip because of the thermal resistance of the fin material. A factor called *fin efficiency* accounts for this effect. (See Chapter 3 of the 1985 FUNDAMENTALS Volume.) For tubes with low conductivity material, high fins, or high values of fin pitch, the fin efficiency becomes increasingly significant. Young and Ward (1957), along with the UOP *Engineering Data Book II* (1984) describe methods of evaluating these effects.

### Fouling Factor ($r_{fw}$)

Manufacturers' ratings are based on commercially clean equipment with an allowance for the possibility of water-side fouling. This allowance often takes the form of a *fouling factor*, which is a thermal resistance referenced to the waterside area of the heat-transfer surface. Thus, the temperature penalty imposed on the condenser is equal to the heat flux at the water-side area, multiplied by the fouling factor. Increased fouling increases overall heat transfer resistance, due to the parameter $(A_o/A_i)r_{fw}$ in Eq. (4). Fouling increases the $\Delta t_m$ required to obtain the same capacity—with a corresponding increase in condenser pressure and system power—or lowers system capacity.

Allowance for a given fouling factor has a greater influence on equipment selection than simply increasing the overall resistance (Starner 1976). Required increases in surface area result in lower water velocities. Consequently, the increase in heat transfer surface required for the same performance is due to both the fouling resistance and the additional resistance which results from lower water velocities.

For a given tube surface, load, and water temperature range, the tube length can be optimized to give a desired condensing temperature and water-side pressure drop. The solid curves in Figure 2 show the effect on water velocity, tube length, and overall surface required due to increased fouling. A fouling factor of 0.00072 $ft^2 \cdot h \cdot °F/Btu$ (130 $mm^2 \cdot °C/W$) doubles the required surface area compared with that with no fouling allowance.

A worse case occurs when an oversized condenser must be selected to meet increased fouling requirements but without the flexibility of increasing tube length. As shown by the dashed lines in Figure 2, the water velocities decrease more rapidly as the total surface increases to meet the required performance. Here, the required surface area doubles with a fouling factor of only 0.00049 $ft^2 \cdot h \cdot °F/Btu$ (86 $mm^2 \cdot °C/W$). If the application can afford more pumping power, the water flow rate may be increased to obtain a higher velocity, which increases the water

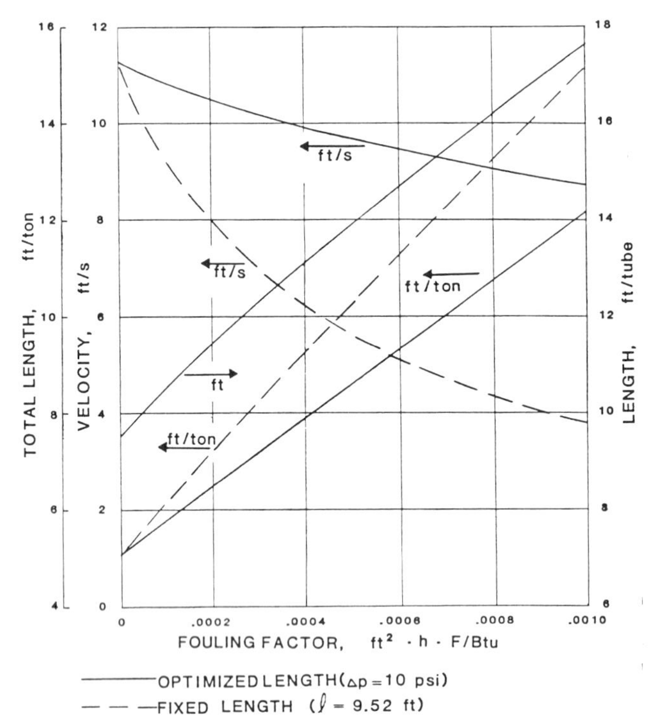


$T_{wi}$ = 85°F, $T_{wo}$ = 95°F, $T_{c_{sat}}$ = 103°F, 2 pass, externally finned tube (0.75 in. OD, 26 fins/in.), $h_{r_o}$ = 620 Btu/hr • ft² • °F, plain ID

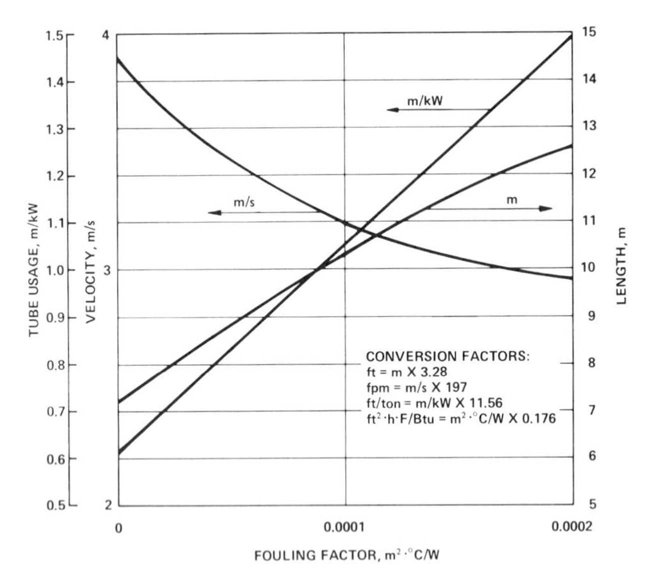


[a]Externally finned tube (1.023 fins/mm, 19-mm OD × 0.71-mm wall); $h_{r_o}$ = 2839 W/m²·°C shell-side; water range, 5.56°C; 6.5 LMTD; $T_{H_2O}$ = 32.2°C; $\Delta P_{f,\ H_2O}$ = 74.7 kPa.

**Fig. 2 Effect of Fouling on Condenser**

film heat transfer coefficient. This factor, plus the lower leaving water temperature, reduces the condensing temperature.

Fouling is a major unresolved problem in heat exchanger design (Taborek, *et al.* 1972). The major uncertainty is which fouling factor to choose for a given application or water condition to obtain expected performance from the condenser; use of too low a fouling factor wastes compressor power, while too high a factor wastes heat exchanger material.

Fouling may result from sediment, biological growths, and corrosion products. Scale results from the deposition of chemicals from the cooling water onto the warmer surface of the condenser tube. Chapter 53 in the 1987 HVAC Volume discusses water chemistry and water treatment factors that are important in corrosion and scale control of condenser cooling water.

Tables of fouling factors are available; however, in many cases, the values are larger than necessary (TEMA 1978). Water velocities above 3 fps (1 m/s) are often recommended to minimize fouling.

Extensive research has been conducted into the causes, rates, and values of water-side fouling. Numerous models have been proposed. These studies generally found fouling resistance reaches an asymptotic value with time (Suitor *et al.* 1976). Many fouling research data are based on surface temperatures that are considerably higher than those found in air-conditioning and refrigeration condensers. In the absence of suspended solids or biological fouling, long-term fouling of condenser tubes does not exceed 0.0002 ft² • h • °F/Btu (35 mm² • °C/W), and short-term fouling does not exceed 0.0001 ft² • h • °F/Btu (18 mm² • °C/W) (ASHRAE 1982). Periodic cleaning of condenser tubes (mechanically or chemically) usually maintains satisfactory performance, except in severe environments. The appropriate edition of ARI *Standard* 450 for water-cooled condensers should be referred to when reviewing manufacturers' ratings. This standard gives methods for correcting ratings for different values of fouling.

## WATER-PRESSURE DROP

The water (or other fluid) pressure drop is important for designing or selecting condensers. Where a cooling tower cools the condensing water, the water-pressure drop through the condenser is generally limited to about 10 psi (70 kPa). If the condenser water comes from another source, the pressure drop through the condenser should be lower than the available pressure to allow for pressure fluctuations and additional flow resistance caused by fouling.

Pressure drop through horizontal condensers includes the loss through the tubes, tube entrance and exit losses, and losses through the heads or return bends (or both). The effect of the coiling of the tubes must be considered in shell-and-coil condensers.

Expected pressure drop through tubes can be calculated from a modified Darcy-Weisbach equation:

$$\Delta p = N_p[K_H + f(L/D)](\varrho V^2/2g_c) \tag{7}$$

*where*

$\Delta p$ = pressure drop, psi (kPa)
$N_p$ = number of tube passes
$K_H$ = entrance and exit flow resistance coefficient, number of velocity heads
$f$ = friction factor
$L$ = length of tube, ft (m)
$D$ = inside tube diameter, ft (m)
$\varrho$ = fluid density, lb/ft³ (kg/m³)
$V$ = fluid velocity, fps (m/s)
$g_c$ = gravitational constant, 32.17 $lb_m \cdot ft/lb_f \cdot s^2$ (1 kg • m/N • s²)

For tubes with smooth ID's, the friction factor may be determined from a Moody chart or various relations, depending on the flow regime and wall roughness (see 1985 FUNDAMENTALS Volume, Chapter 2). For tubes with internal enhancement, the friction should be obtained from the tube manufacturer.

The value of $K_H$ depends on tube entry and exit conditions and the flow path between passes. As a minimum, a value of 1.5 is recommended. This factor is more critical with short tubes.

Predicting pressure drop for shell-and-coil condensers is more difficult than it is for shell-and-tube condensers because of design curvature of the coil and tube flattening or kinking during coil bending. Seban and McLaughlin (1963) discuss the effect of curvature or bending of pipe and tubes on the pressure drop.

## LIQUID SUBCOOLING

The amount of condensate subcooling provided by the condensing surface in a shell-and-tube condenser is small, generally less than 2°F (1°C). When a specific amount of subcooling is required, it may be obtained by submerging tubes in the condensate. Tubes in the lower portion of the bundle are used for this purpose. If the condenser is multipass, then the subcooling tubes should be included in the first pass to gain exposure to the coolest water.

Subcooling benefits are achieved at the expense of both condensing surface and refrigerant inventory. An optimum design should be based on these opposing factors.

When means are provided to elevate the condensate level for the desired submergence of the subcooler tubes, the heat is transferred principally by natural convection.

Subcooling performance can be improved by enclosing tubes in a separate compartment within the condenser to obtain the benefits of forced convection of the liquid condensate.

Segmental baffles may be provided to produce flow across the tube bundle. Kern and Kraus (1972) describe how heat-transfer performance can be estimated analytically by use of longitudinal or cross-flow correlations, but it is more practically determined by test because of the large number of variables. It is important to keep the refrigerant pressure drop incurred along the flow path from exceeding the pressure difference permitted by the saturation pressure of the subcooled liquid.

## CIRCUITING

Varying the number of water side passes in a condenser can affect the saturated condensing temperature significantly, which affects system performance. Figure 3 shows the change in condensing temperature for one, two, or three passes in a particular condenser. As an example, at a loading of 22,000 Btu/h • tube (6450 W/tube) a two-pass condenser with a 10°F (5.6°C) range would have a condensing temperature of 102.3°F (39.1°C). At the same loading with one-pass, 5°F (2.8°C) range, this unit would have a condensing temperature of 99.3°F (37.4°C). The one-pass option does, however, require twice the water flow rate with an associated increase in pumping power. A three-pass design may be favorable when the costs associated with water flow outweigh the system gains from lower condensing temperature. Hence, different numbers of passes (if an option) and ranges should be weighed against other parameters (water source, pumping power, cooling tower design, etc.) to optimize overall system performance and cost.

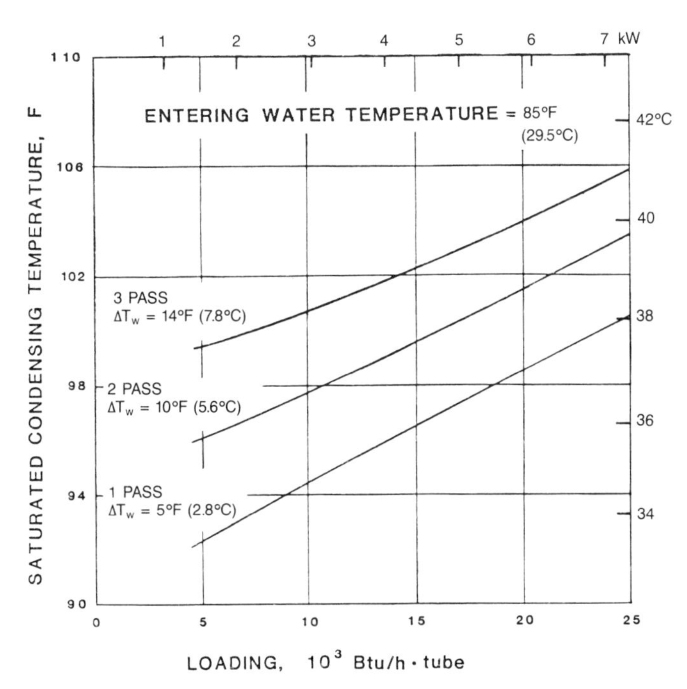


**Fig. 3 Effects of Condenser Circuiting**

## CONDENSER TYPES

The most common types of water cooled refrigerant condensers are (1) shell-and-tube; (2) shell-and-coil; and (3) tube-in-tube. The type selected depends on the size of the cooling load, the refrigerant used, the quality and temperature of the available cooling water, the amount of water that can be circulated, the location and space allotment, the required operating pressures (water and refrigerant sides), cost, and maintenance considerations.

**Shell-and-Tube Condensers** are built in sizes from 1 to 10,000 tons (3.5 to 35 000 kW). The refrigerant condenses outside the tubes and the cooling water circulates through the tubes in a single or multipass circuit. Fixed tube sheet, straight tube construction is usually used, although, U-tubes that terminate in a single tube sheet are sometimes used. Typically, shell-and-tube condenser tubes run horizontally. Where floor installation area is limited, the condenser tubes may be oriented vertically. However, vertical tubes have poor condensate draining, which reduces the refrigerant film coefficient. Vertical condensers with open water systems have been used with ammonia.

Gas inlet and liquid outlet nozzles should be located carefully. The proximity of these nozzles may adversely affect condenser performance by requiring excessive amounts of liquid refrigerant to seal the outlet nozzle from inlet gas flow. This effect can be diminished by the addition of baffles at the inlet and/or outlet connection.

Halocarbon refrigerant condensers have been made with many types of material specifications, including all prime surface or finned, ferrous, or non-ferrous tubes. Common tubes are nominal 0.75 and 1.0 in. (20 and 25 mm) OD copper tubes with integral fins on the outside. These tubes are often available with fin heights from 0.035 to 0.061 in. (0.9 to 1.5 mm) and fin spacings of 19, 26, and 40 fins/in. (1.33, 1.02, and 0.64 mm spacing). For ammonia condensers, prime surface steel tubes, 1.25 in. (32 mm) OD, with 0.095 in. (2.4 mm) average wall thickness, are commonly used.

An increased number of tubes designed for enhanced heat transfer are available (Bergles 1976). On the inside of the tube, common enhancemments include longitudinal or spiral grooves and ridges, internal fins, and other devices to promote turbulence and augment heat transfer. On the refrigerant side, condensate surface tension and drainage are important in design of the tube outer surface. Tubes are available with the outsides machined or formed specifically to enhance the condensation and promote drainage. Heat transfer design equations should be obtained from the manufacturer.

Because the water and refrigerant film resistances with enhanced tubes are reduced, the effect of fouling becomes relatively large. Where high levels of fouling may occur, the fouling resistance may easily account for over 50% of the total. In such cases, the advantages of enhancement may diminish. On the other hand, water side augmentation, which creates turbulence, may reduce fouling. The actual value of fouling resistances depends on the particular type of enhancement and the service conditions. (Starner 1976 and Watkinson *et al.* 1974).

Similarly, refrigerant side enhancements may not show as much benefit in very large tube bundles as in smaller bundles. This is due to the *row effect* addressed previously in the section on refrigerant film coefficient.

The tubes are either brazed into thin copper, copper alloy, or steel tube sheets or they are rolled into heavier non-ferrous or steel tube sheets. Straight tubes with a maximum OD less than the tube hole diameter and rolled into tube sheets are removable. This construction facilitates field repair in the event of tube failure.

The required heat transfer area for a shell-and-tube condenser can be found by solving Eqs. (2), (3), and (4). The mean temperature difference is the logarithmic mean temperature difference, with the entering and leaving refrigerant temperatures taken as the saturated condensing temperature. Depending on the parameters fixed, an iterative solution may be required.

Shell-and-U-tube condenser design principles are the same as those outlined for horizontal shell-and-tube units, with one exception: the water pressure drop through the U-bend portion of the U-tube is generally less than that through the compartments in the water head where the direction of water flow is reversed. The pressure loss is a function of the inside tube diameter and the ratio of the inside tube diameter to bending centers. Pressure loss should be determined by test.

**Shell-and-Coil Condensers** have the cooling water circulated through one or more continuous or assembled coils contained within the shell. The refrigerant condenses outside the tubes. Capacities range from 0.5 to 15 tons (1.8 to 53 kW). Due to the type of construction, the tubes are neither replacable nor mechanically cleanable.

Again, Eq. (2), (3), and (4) may be used for performance calculations, with the saturated condensing temperature used for the entering and leaving refrigerant temperatures in the logarithmic mean temperature difference. The value of $h_w$ (the water side film coefficient) and, especially, the pressure loss on the water side requires close attention; laminar flow can exist at considerably higher Reynolds numbers in coils than in straight tubes. Because the film coefficient for turbulent flow is greater than that for laminar flow, values of $h_w$, as calculated from Eq. (6), will be too high if the flow is not turbulent. Once the flow has become turbulent, the film coefficient will be greater than that for a straight tube (Eckert 1963). Pressure drop through helical coils can be much greater than that through smooth straight tubes for the same length of travel. The section on "Water Pressure Drop" in this chapter outlines the variables that make the accurate determination of the pressure loss difficult. The pressure loss and heat-transfer rates should be determined by test because of the large number of variables inherent in this condenser.

**Tube-in-Tube Condensers** consist of one or more assemblies of two tubes, one within the other, in which the refrigerant vapor is condensed in either the annular space or the inner tube. These units are built in sizes from 0.3 to 50 tons (1 to 180 kW).

Equations (2) and (3) would be used to size a tube-in-tube condenser. Because the refrigerant may have a significant pressure loss through its flow path, the refrigerant temperatures used in calculating the mean temperature difference should be selected carefully. The refrigerant temperatures should be consistent with the model used for the refrigerant film coefficient. Logarithmic mean temperature difference for either counterflow or parallel flow should be used, depending on the piping connections. Eq. (4) can be used to find overall heat transfer coefficient when the water flows in the tubes and Eq. (5) when the water flows in the annulus.

Tube-in-tube condenser design differs from those outlined previously, depending on whether the water flows through the inner tube or through the annulus. Condensing coefficients are more difficult to predict when condensation occurs within a tube or annulus because the mechanism differs considerably from condensation on the outside of a horizontal tube. Where the water flows through the annulus, disagreement exists regarding the appropriate method used to calculate the water-side film coefficient and the water-pressure drop. The problem is further complicated if the tubes are also formed in a spiral.

The water side is mechanically cleanable only when the water flows inside straight tubes and cleanout access is provided. Tubes are not replaceable.

## NONCONDENSABLE GASES

When first assembled, most refrigeration systems contain gases: usually air and water vapor. As addressed later in this chapter, these gases are detrimental to condenser performance. Therefore, it is important to evacuate the entire refrigeration system before operation.

For low-pressure refrigerants, where the operating pressure of the evaporator is less than ambient pressure, even slight leaks can be a continuing source of noncondensables. In such cases, a purge system, which automatically expels noncondensable gases, is recommended. Figure 4 shows the refrigerant loss associated with the use of purging devices at various operating conditions.

When present, noncondensable gases collect on the high pressure side of the system and raise the condensing pressure above that corresponding to the temperature at which the refrigerant is actually condensing. The increased condensing pressure increases power consumption and reduces capacity. Also, if oxygen is present at a point of high discharge temperature, the oil may oxidize.

The excess pressure is caused by the partial-pressure of the noncondensable gas. These gases form a resistance film over some of the condensing surface, thus lowering the heat-transfer

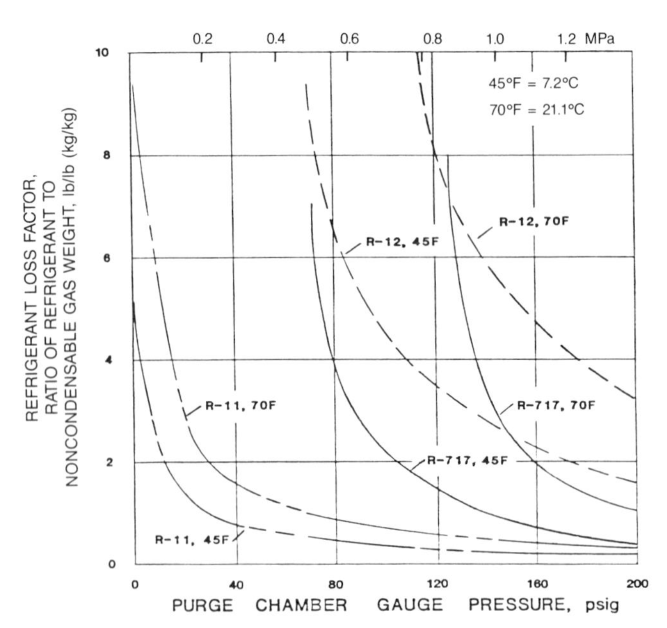


**Fig. 4 Loss of Refrigerant during Purging at Various Gas Temperatures and Pressures**

coefficient. Henderson and Marchello (1969), Webb *et al.* (1980), Wanniarachchi (1981), and Chapter 4 of the 1985 FUNDAMENTALS Volume describe how a small percentage of noncondensables can cause major decreases in the refrigerant film coefficient in shell-and-tube condensers. The noncondensable situation of a given condenser is difficult to characterize because such gases tend to accumulate in the coldest and least agitated part of the condenser or in the receiver. Thus, a fairly high percentage of noncondensables can be tolerated if the gases are confined to areas far from the heat-transfer surface. One way to account for noncondensables is to treat them as a refrigerant or gas-side fouling resistance in Eq. (4). Some predictions are presented by Wanniarachchi and Webb (1982). Webb (1982) also defined the effect of noncondensables on refrigeration systems.

As an example of the effect on system performance, experiments performed on a 250-ton R-11 chiller condenser reveal that 2% noncondensables by volume caused a 15% reduction in condensing coefficient. Also, 3 and 8% noncondensables by volume caused power increases of 2.6 and 5%, respectively.

The presence of noncondensable gases can be tested by shutting down the refrigeration system, while allowing the condenser water to flow long enough for the refrigerant to reach the same temperature as the water. If the condenser pressure is higher than the pressure corresponding to the refrigerant temperature, noncondensable gases are present. This test may not be sensitive enough to detect the presence of small amounts of noncondensables, which can, nevertheless, decrease shell-side condensing coefficients.

## CONSTRUCTION AND TEST CODES

Pressure vessels must be constructed and tested under the rules of national, state, and local codes. The introduction of the current ASME *Boiler and Pressure Vessel Code*, Section VIII, gives guidance on rules and exemptions.

The more common applicable codes and standards are as follows:

ARI *Standard* 450-87, *Standard for Water-Cooled Refrigerant Condensers, Remote Type*. This standard covers industry criteria for standard equipment, standard safety provisions, marking, and recommended procedure for testing and rating water-cooled refrigerant condensers.

ASHRAE *Standard* 22-78, *Methods of Testing for Rating Water-Cooled Refrigerant Condensers*. This standard covers the recommended rating point and fouling factors, and testing and rating methods.

ANSI/ASHRAE *Standard* 15-78, *Safety Code for Mechanical Refrigeration*. This code specifies design criteria, use of materials, and testing. It refers to the ASME *Boiler and Pressure Vessel Code*, Section VIII, for refrigerant-containing sides of pressure vessels, where applicable. Factory test pressures are specified, and minimum design working pressures are given by this code. This code requires pressure-limiting and pressure-relief devices on refrigerant-containing systems, as applicable, and defines the setting and capacity requirements for these devices.

ASME *Boiler and Pressure Vessel Code, Unfired Pressure Vessels*, Section VIII. This code covers the safety aspects of design and constructions. Most states require condensers to meet the requirements of the ASME if they fall within scope fo the ASME code. Some of the exceptions from meeting the ASME requirements listed in the ASME code are as follows:

1. Condenser shell ID is 6 in. (152 mm) or less.
2. 15 psig (103 kPa gage) or less.
3. The fluid (water) portion of the condenser need not be built to the requirements of the ASME code if the fluid is water and the volume is 120 gal (454 L) or less.

Condensers meeting the requirements of the ASME code will have a ASME stamp. The ASME stamp is a *U* or *UM* inside a three-leaf clover. The *U* can be used for all condensers; the *UM* can be used for those with net refrigerant side volume less than 1.5 ft$^3$ (0.042 m$^3$).

UL *Standard* 207-86, *Refrigerant-Containing Components and Accessories*. This standard covers specific design criteria, use of materials, testing, and initial approval by Underwriters Laboratories. A condenser with the ASME *U* stamp does not require UL approval.

### Design Pressure

1. Refrigerant side, as a minimum, should be saturated pressure for the refrigerant used at 105°F (40.6°C), or 15 psig (103 kPa gage), whichever is greater. Standby temperatures and temperatures encountered during shipping of units with a refrigerant charge should also be considered.
2. Required fluid (water) side pressure varies, depending largely on the following conditions: static head, pump head, transients due to pump start-up, and valve closing. A common water-side design pressure is 150 psig (1.03 MPa gage), although with taller building construction, requirements for 300 psig (2.07 MPa gage) are not uncommon.

## OPERATION AND MAINTENANCE

When a water-cooled condenser is selected, anticipated operating conditions, including water and refrigerant temperatures, have usually been determined. Standard practice allows for a fouling factor in the selection procedure. A new condenser, therefore, operates at a condensing temperature lower than the design point because it has not yet fouled. Once a condenser starts to foul or scale, economic considerations determine how frequently the condenser should be cleaned. As the scale builds up in a condenser, the condensing temperature and subsequent power consumption increase, while the unit capacity decreases. This effect can be seen in Figure 5 for a condenser with a design

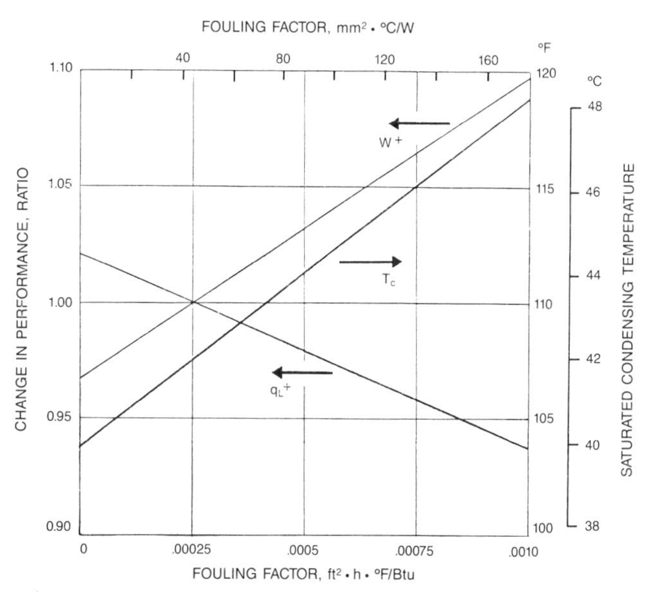


$q_L^+$ = chiller actual capacity/chiller design capacity
$W^+$ = compressor actual kW/compressor design kW
$T_c$ = saturated condensing temperature
Design condenser fouling factor = 0.00025 ft$^2$·h·°F/Btu (44 mm$^2$·°C/W)
Cooler leaving water temperature = 44°F (6.7°C)
Condenser entering water temperature = 85°F (29.4°C)

**Fig. 5 Effect of Fouling on Chiller Performance**

fouling factor of 0.00025 ft$^2$ • h • °F/Btu (44 mm$^2$ • °C/W). At some point, the increased cost of power can be offset by the labor cost of cleaning.

Local water conditions, as well as the effectiveness of chemical water treatment, if used, make the use of any specific maintenance schedule difficult.

Cleaning can be done either mechanically with a brush or chemically with an acid solution. In applications where water-side fouling may be severe, on-line cleaning can be accomplished by brushes installed in cages in the water heads. Through the use of valves, flow is reversed at set intervals, propelling the brushes through the tubes (Kragh 1975). The most effective method depends on the type of scale formed. Competent advice in selecting the particular method of tube cleaning is advisable.

Occasionally, one or more tubes may develop leaks because of corrosive impurities in the water or through improper cleaning procedures. These leaks must be found and repaired as soon as possible; this can normally be done by replacing the leaky tubes, a procedure requiring tools and skills that are best found through the original condenser manufacturer. In large condensers, where the contribution of a single tube is relatively insignificant, a simpler approach may be to seal the ends of the leaking tube.

If the condenser is located where the water can freeze during the winter, special precautions should be taken when it is idle. Opening all vents and drains may be sufficient, but water heads should be removed and tubes blown free of water.

If refrigerant vapor is to be released from the condenser, and there is water in the tubes, the pumps should be on and the water flowing. Otherwise, freezing can easily occur.

Finally, the condenser manufacturer's installation recommendations on orientation, piping connections, space requirements for tube cleaning or removal, and other important factors should be followed.

# PART II: AIR-COOLED CONDENSERS

Heat is transferred in an air-cooled condenser in three main phases: (1) desuperheating, (2) condensing, and (3) subcooling. Figure 6 shows the changes of state of Refrigerant 12 passing through the condenser coil and the corresponding temperature change of the cooling air as it passes through the coil. Desuperheating, condensing, and subcooling zones vary 5 to 10%, depending on the entering gas temperature and the leaving liquid temperature, but Figure 6 is typical for most common refrigerants.

Condensing takes place in approximately 85% of the condenser area at a substantially constant temperature. The indicated drop in condensing temperature is the result of the friction loss through the condenser coil.

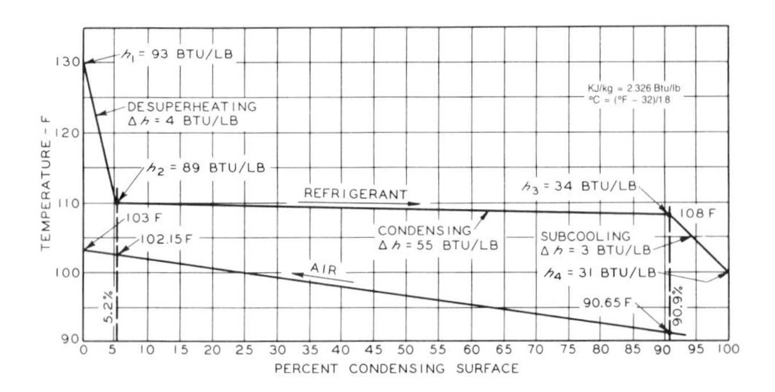


**Fig. 6 Temperature and Enthalpy Changes in an Air-Cooled Condenser**

An air-cooled condenser may be classified as (1) condenser remote from the compressor or (2) condenser as part of a condensing unit. A further subdivision is forced airflow or free convection airflow. Indoor and outdoor applications may also be used as classifications, but many units are used for both.

## COIL CONSTRUCTION

A condenser coil with optimum circuiting requires the least heat transfer surface and has the least operational problems, such as trapping oil and refrigerant, high discharge pressure, etc. Coils are commonly constructed of copper, aluminum, or steel tubes, ranging from 0.25 to 0.75 in. (6 to 20 mm) in diameter. Copper is easy to use in manufacturing and requires no protection against corrosion. Aluminum requires exact manufacturing methods and special protection if aluminum to copper joints are made. Steel tubing requires weather protection.

Tube diameter is chosen as a compromise between factors, as manufacturing facilities, cost, header difficulties, air resistance, and refrigerant flow resistance. Where a choice exists, the smaller diameter gives more flexibility in coil circuit design and results in lower system refrigerant charge.

Fins improve the air-side heat transfer. Most fins are made of aluminum, but copper and steel are also used. The most common forms are plate fins that make a coil bank, plate fins individually fastened to the tube, and fins spirally wound onto the tube. Other forms, such as plain tube-fin extrusions or tube extrusions with accordion-type fins are also used. The most common fin spacings range from 8 to 18 fins per in. (3.2 to 1.4 mm spacing).

A refrigeration system with a receiver, such as a remote split system, usually has a section designed as an integral part of the condenser that subcools the liquid refrigerant as it flows from the receiver. A small system, such as a residential or closed coupled system, usually subcools the refrigerant by flooding part of the condenser with refrigerant.

The amount of subcooling depends on the heat transfer characteristics of the condenser and the temperature difference between the condenser and the entering air. Compressor capacity increases about 0.5% per 1°F subcooling (0.9% per °C) at the same suction and discharge pressure. In systems without a receiver, an increase in subcooling increases the discharge pressure, which decreases capacity. In effect, systems without a receiver have optimum subcooling at design conditions. Increasing subcooling within limits keeps the liquid refrigerant temperature at the expansion valve below saturation temperature and eliminates refrigerant flashing at the expansion valve.

## FANS AND AIR REQUIREMENTS

Condenser coils can be cooled by free convection, wind, or fans of various designs such as propeller, centrifugal, and vaneaxial. Because efficiency of material usage is sharply increased (up to a point) by increased air velocity across the coil, fans are predominantly used, rather than free convection or wind.

For a given condensing temperature, the lower limit of air quantity occurs when the leaving air temperature approaches the entering saturated refrigerant condensing temperature. The upper limit is usually determined by air resistance in the coil and casing, noise, fan power, or fan size. Here, as with the coil, a system optimum must be reached by balancing operating cost and first cost with size and sound requirements. Commonly used values are 600 to 1200 cfm/ton (80 to 160 L/s per kW). Temperature differences (TD) at the normal cooling condition between condensing refrigerant and entering air range from 15 to 40°F (8 to 22°C). Fan power requirements generally range from 0.1 to 0.2 hp/ton (21 to 42 W/kW).

The type of fan depends primarily on static pressure and unit shape requirements. Propeller fans are well-suited for units with low internal pressure drop and free air discharge. Speeds are selected in the range 850 to 1750 rpm (14.2 to 29.2 r/s), depending on size and sound requirements. The airflow pattern has a marked effect on the sound level. A partially obstructed fan inlet or outlet can drastically increase noise level. Support brackets, etc., close to the fan blade, have the same effect. Propeller fans are often mounted directly on the motor shaft for simplicity, but care must be taken that the motor bearings and mountings are suited for this. The belt drive offers more flexibility but usually requires more maintenance. Propeller fans operating at low static pressures are more efficient than centrifugal fans.

The centrifugal fan performs better at higher static pressures. It is used almost exclusively where any but the simplest duct is required. From a standpoint of sound, the centrifugal fan is less sensitive to inlet obstructions than the propeller fan, but it is more sensitive to outlet obstructions. Considerations of direct drive versus belt drive are the same for centrifugal fans, but because of the static capabilities, a wide speed range is more desirable. Consequently, belt drives are used more on centrifugal fans than on propeller fans. Vaneaxial fans are often more efficient than centrifugal fans. They also offer the possibility of efficient flow modulation by automatic blade pitch adjustment.

## HEAT TRANSFER AND PRESSURE DROP

The overall heat-transfer coefficient of air-cooled condensers can be expressed by Eq. (5c) in Chapter 6. For estimation of air-side heat-transfer coefficients and fin effectiveness, the information in Chapter 6 (regarding sensible cooling coils) also applies to condensers. Heat transfer on the refrigerant side is complex. Refrigerant enters the condenser as superheated vapor and cools by single-phase convection to saturation temperature, after which condensation starts. If the tube-wall temperature is lower than the saturation temperature, condensation can occur while the vapor core is superheated. Eventually, all the vapor condenses to liquid. In most condensers, the condensate liquid is subcooled before leaving the condenser in a single-phase convection process. Chapter 3 in the 1985 FUNDAMENTALS Volume lists equations for heat-transfer coefficients in the single-phase flow sections. For condensation from superheated vapors, there is no verified general predictive technique. For condensation from saturated vapors, one of the best methods is the Shah correlation (Shaw 1979 and Shaw 1981). For estimation of two-phase pressure drop, see Chapter 4 of the 1985 FUNDAMENTALS Volume.

## CONDENSERS REMOTE FROM COMPRESSOR

Remote air-cooled condensers are used for refrigeration systems from one-half ton to over 500 tons (1.8 to over 1800 kW). Forced-air condensers are the most common. The main components consist of (1) a finned condensing coil, (2) one or more fans and motors, and (3) an enclosure or frame.

The coil-fan arrangement can take almost any form ranging from the horizontal coil with upflow air to the vertical coil with horizontal airflow. The choice of design depends primarily on the intended application and the manufacturing facilities available.

## CONDENSERS AS PART OF CONDENSING UNIT

Condensing units are produced in such a number of shapes and sizes that detailing their form is not included in this chapter. Also, while some units are particularly suited to either outdoor or indoor use, a large number of units are being used for both applications, and is no longer justifiable to use indoor-outdoor as major classifications. Basic design considerations are the same as those for remote condensers, but the inclusion of the compressor makes it desirable to enclose the components in a cabinet, in most cases. This cabinet also contains controls, precharged line fittings or valves, and liquid receiver, if separate from the coil. When open-type compressors are used, the compressor motor often drives the fan, as well. The airflow created in the cabinet is often directed over the fan motor for improved cooling, and sometimes this effect is used for compressor and controls, as well. In very small systems, the fan is sometimes omitted and free convection is used. Service access is obtained through one or more panels.

The following related factors should be considered sound when the unit is designed:

1. Avoid a straight-line path from the compressor to the listener.
2. Acoustical material may be used inside the cabinet and to line the air outlets; streamline air passages as much as possible.
3. Mount the compressor carefully on the base.
4. A light gauge base is often superior to a heavier, stiffer base, which readily transmits vibration to other panels.
5. Natural frequencies of panels and refrigerant lines must be different from basic compressor and fan frequencies.
6. Refrigerant lines with many bends have many pulsation forces and natural frequencies.
7. Sometimes, a condenser with a very low refrigerant pressure drop has one or two passes resonant with the compressor discharge pulsations.
8. Fan and motor supports can be made to isolate noise and vibration.
9. Fan selection is a major factor; steep propeller blades and unstable centrifugal fans can be serious noise producers.
10. Reduced speed fan operation is often used at night when noise is more objectionable.

Comparisons between water-cooled and air-cooled equipment are often made. Where plenty of low-cost water is available, both first cost and operating cost may be lower for water-cooled equipment. A 20% larger compressor is generally used for air-cooled operation, so operating cost is higher. With water shortages common in many areas, however, the factors change. When cooling towers are used to produce cooling water, the initial cost of a water-cooled system may be higher. Also, the lower operating cost of the water-cooled system may be offset by cooling tower pump and fan power and maintenance costs. When making a comparison, the complete system should be considered, including the cost of cleaning fouled water-cooled condensers and clogged air-cooled condenser fins.

Noncondensable gases should be purged from the condenser for proper condenser operation (see the section "Noncondensable Gases").

## APPLICATION AND RATING OF CONDENSERS

Condensers are rated in terms of *Total Heat Rejection* (THR), which is the total heat removed in desuperheating, condensing, and subcooling of the refrigerant. This value is the product of the weight rate of refrigerant flow and the difference in enthalpy of the refrigerant vapor entering the condenser coil and the enthalpy of the leaving refrigerant liquid.

A condenser may also be rated in terms of *Net Refrigeration Effect* (NRE), which is the total heat rejection less the heat of compression added to the refrigerant in the compressor. The NRE is a practical expression of the capacity of a refrigeration system.

Table 1 Net Refrigeration Effect Factors

| Open-Type Compressors[a] | Air-Cooled and Evaporative Condensers | | | | | | | | | | |
|---|---|---|---|---|---|---|---|---|---|---|---|
| Saturated Suction Temperature, °F (°C) | Condensing Temperature, °F (°C) | | | | | | | | | | |
| | 85 (29.4) | 90 (32.2) | 95 (35) | 100 (37.8) | 105 (40.6) | 110 (43.3) | 115 (46.1) | 120 (46.9) | 125 (51.7) | 130 (54.4) | 135 (57.2) |
| −40 (−40) | 0.71 | 0.70 | 0.69 | 0.68 | 0.67 | 0.65 | 0.64 | 0.63 | 0.62 | 0.60 | — |
| −20 (−28.9) | 0.77 | 0.76 | 0.74 | 0.73 | 0.72 | 0.71 | 0.70 | 0.69 | 0.67 | 0.66 | — |
| 0 (−17.8) | 0.82 | 0.80 | 0.79 | 0.78 | 0.77 | 0.76 | 0.75 | 0.74 | 0.73 | 0.71 | — |
| +20 (−6.7) | 0.86 | 0.85 | 0.84 | 0.83 | 0.82 | 0.81 | 0.79 | 0.78 | 0.77 | 0.76 | 0.75 |
| +40 (+4.4) | 0.91 | 0.90 | 0.89 | 0.87 | 0.86 | 0.85 | 0.84 | 0.83 | 0.82 | 0.81 | 0.80 |

[a]Factors based on 15°F (8.3°C) of superheat entering the compressor for Refrigerant 22; actual suction gas temperature of 65°F (18.3°C) for saturated suction temperature down to −10°F (−23.3°C), 55°F (12.8°C) at −20°F (−28.9°C), 35°F (1.7°C) at −40°F (−40°C), for Refrigerants 12,500, and 502; 10°F (5.6°C) for ammonia (R-717).

| Sealed Compressors[b] | Air-Cooled and Evaporative Condensers | | | | | | | | | | |
|---|---|---|---|---|---|---|---|---|---|---|---|
| Saturated Suction Temperature, °F (°C) | Condensing Temperature °F (°C) | | | | | | | | | | |
| | 85 (29.4) | 90 (32.2) | 95 (35) | 100 (37.8) | 105 (40.6) | 110 (43.3) | 115 (46.1) | 120 (48.9) | 125 (51.7) | 130 (54.4) | 135 (57.2) |
| −40 (−40) | 0.55 | 0.54 | 0.53 | 0.52 | 0.51 | 0.50 | 0.49 | 0.47 | 0.46 | 0.44 | — |
| −20 (−28.9) | 0.65 | 0.64 | 0.62 | 0.61 | 0.60 | 0.59 | 0.58 | 0.55 | 0.53 | 0.51 | — |
| 0 (−17.8) | 0.72 | 0.71 | 0.70 | 0.69 | 0.67 | 0.66 | 0.64 | 0.62 | 0.60 | 0.58 | — |
| +20 (−6.7) | 0.77 | 0.76 | 0.75 | 0.74 | 0.72 | 0.71 | 0.69 | 0.68 | 0.66 | 0.64 | 0.62 |
| +40 (+4.4) | 0.81 | 0.80 | 0.79 | 0.78 | 0.77 | 0.75 | 0.74 | 0.72 | 0.71 | 0.70 | 0.68 |
| +50 (+10) | 0.83 | 0.82 | 0.81 | 0.80 | 0.79 | 0.78 | 0.76 | 0.75 | 0.74 | 0.73 | 0.72 |

[b]Factors based on 15°F (8.3°C) of superheat entering the compressor for Refrigerant 22; actual suction gas temperature of 65°F (18.3°C) for saturated suction temperature down to −10°F (−23.3°), 55°F (12.9°C) at −20°F (−28.9°C), 35°F (1.7°C) at −40°F (−40°C), for Refrigerants 12,500, and 502.

NOTES:
1. These factors should be used only for air-cooled and evaporative condensers connected to reciprocating compressors.
2. Condensing temperature is that temperature corresponding to the saturation pressure as measured at the discharge of the compressor.
3. Net Refrigeration Effect Factors are an aproximation only. When used in published ratings, the net refrigeration capacity must be represented as *Approximate Only*.
4. For more accurate condenser selection, and for condensers connected to centrifugal compressors, use the total heat-rejection effect of the condenser and the compressor manufacturer's total rating, which includes the heat of compression.

For open compressors, the THR is the sum of the actual power input to the compressor and the NRE. For hermetic compressors, the THR is obtained by adding the NRE to the total motor power input and substracting the heat losses from the surface of the compressor and discharge line. The surface heat losses are generally 0 to 10% of the power consumed by the motor. All quantities must be expressed in consistent units. Table 1 recommends factors for converting condenser THR ratings to NRE for both open and hermetic reciprocating compressors.

Ratings of air-cooled condensers are based on Temperature Difference (TD) between the dry-bulb temperature of the air entering the coil and the saturated condensing temperature corresponding to the pressure at the inlet. Typical TD values are 10 to 15°F (5.6 to 8.3°C) for low temperature systems at a −20°F (−28.9°C) evaporator temperature, 15 to 20°F (8.3 to 11.1°C) for medium temperature systems at a 20°F (−6.7°C) evaporator temperature, and 25 to 30°F (13.9 to 16.7°C) for air-conditioning systems at a 45°F (7.2°C) evaporator temperature. The THR capacity of the condenser is approximately proportional to the TD. The capacity at 30°F (16.7°C) TD is about 50% greater than the same condenser selected for 20°F (11.1°C) TD.

In determining air temperature entering the condenser coil, weather data, as published in Chapter 24 of the 1985 FUNDAMENTALS Volume are widely used. The 2.5% value is suggested for summer operation.

The specifying engineer must choose the specific design dry-bulb temperature carefully, especially for refrigeration serving process cooling. Entering air temperature that is higher than expected quickly causes higher than design head pressure and higher compressor power. Both of these factors may cause unexpected system shutdown, usually when it can least be tolerated. Congested or unusual locations may create entering air temperatures that are higher than general ambient conditions.

Capacity of an air-cooled condenser equipped with an integral subcooling circuit varies, depending on the refrigerant charge. The charge is greater when the subcooling circuit is full of liquid, which increases the subcooling. When the subcooling circuit is used for condensing, the refrigerant charge is lower, the condensing capacity is greater, and the liquid subcooling is reduced.

Publication of ratings should be made as a standard rating and application range in accordance with ARI *Standard* 460-80 and ASHRAE *Standard* 20-70, *Methods of Testing for Rating Remote Mechanical Draft Air-Cooled Refrigerant Condensers.*

Condenser ratings are based on THR capability in Btu/h (W) for a given TD and the specified refrigerant. When application ratings are given in terms of NRE, they must always be defined with respect to the suction temperature, TD, type of compressor (either open or hermetic), and the refrigerant. Condenser selections for hermetic compressors should be made on the basis of THR ratings, if these are published by the particular compressor manufacturer, and the THR effect of the air-cooled condenser.

Compressors and condensers are available in various increments of capacity and the capacity of the two will seldom be the same at a specified set of conditions. The capacity balance point of these two components can be determined by either trial and error or graphically by cross-plotting condenser THR against compressor THR, as shown in Figure 7.

The compressor THR ($Q_d$) is plotted for a given saturated suction temperature at various saturated discharge temperatures.

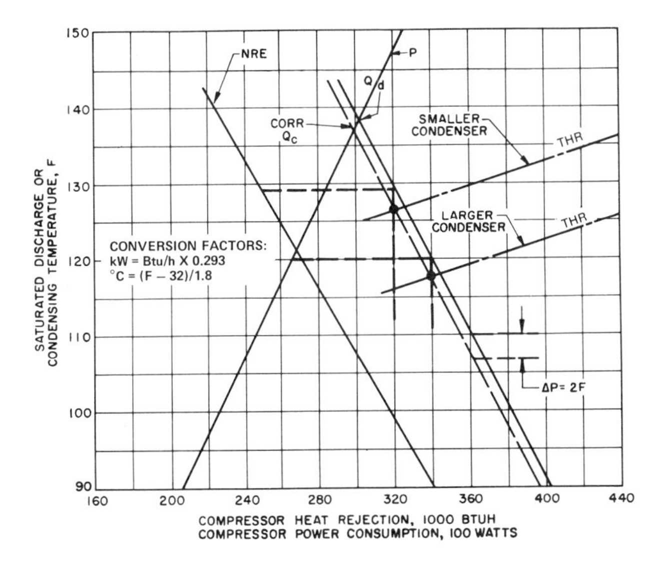


**Fig. 7 Performance of Condenser-Compressor Combination**

The saturated discharge temperature at the compressor must be corrected for hot-gas line pressure drop to obtain the saturated condensing temperature. The pressure drop is usually assumed to equal a 2°F (1.1°C) drop in saturated temperature. The corrected curve ($Q_c$) is represented by a dotted line. Compressor NRE and Power Input ($P$) are also plotted on the same coordinates.

The condenser THR is plotted for various saturated condensing temperatures for a given entering air temperature. The balanced compressor and condenser THR and the saturated condensing temperature are read at the intersection of the corrected compressor curve ($Q_c$) and the condenser THR curve. The compressor saturated discharge is read from the compressor curve ($Q_d$) for the same THR. The net refrigeration capacity and power input are read at the saturated discharge temperature.

Figure 7 is a plot of a compressor operating at a 90°F (32.2°C) ambient with two sizes of an air-cooled condenser. The compressor will operate at a saturated discharge temperature of 129°F (53.9°C) with a saturated condensing temperature of 127°F (52.8°C) when using the smaller condenser. The THR is 320,000 Btu/h (94 kW). The NRE is 250,000 Btu/h (73 kW) and the power consumption is 28 kW. With the larger condenser, the compressor operates at a saturated discharge temperature of 120°F (48.9°C) and a saturated condensing temperature of 118°F (47.8°C). The THR is 340,000 Btu/h (100 kW) and the power consumption is 26.5 kW.

The condenser selected must satisfy the anticipated cooling requirements at design ambient conditions at the lowest possible total compressor-condenser system operating cost, including an evaluation of the cost of control equipment.

## CONTROL OF AIR-COOLED CONDENSERS

For a refrigeration system to function properly, the condensing pressure and temperature must be maintained within certain limits. An increase in condensing temperature causes a loss in capacity, requires extra power, and may overload the compressor motor. Low condensing pressures hinder flow through the conventional liquid feed devices; this hindrance starves the evaporator and causes loss of capacity, possible freezing of the coil, and tripout on low pressure. Some systems have low-pressure drop thermostatic expansion valves. These systems require a means of controlling the liquid temperature so that it is subcooled at the expansion valve.

To prevent excessively low head pressure during winter operation, two basic control methods are used with conventional expansion valves: (1) refrigerant-side control, and (2) air-side control. Many methods accomplish head pressure control in each of these two categories. Only the basic method is addressed here.

**Refrigerant-side control** is accomplished by modulating the amount of active condensing surface available for condensing by flooding the coil with liquid refrigerant. This method requires a receiver and an extra charge of refrigerant to flood the coil. Several valving arrangements give the required amount of flooding to meet the variable needs. Both temperature and pressure actuation are used.

**Air-side control** may be accomplished by any of three methods, or a combination of two of them: (1) cycling of fans, (2) modulating dampers, and (3) fan speed control. Any method of air-side control must consider the direction and force of the prevailing wind. The unit should be oriented so that the wind does not create adverse operating conditions.

Fan cycling in response to head pressure at certain times of the year often results in rapid cycling of the fan motor, causing expansion valve hunt and possibly a burnout of the fan motor.

Fan cycling in response to outdoor ambient temperature eliminates rapid cycling but is limited to use with multiple fan units or is supplemental to other control methods. A common method for control of a two-fan unit is to cycle only one fan. A three-fan unit may cycle two fans. The capacity of two-fan condensers will be reduced to approximately 55% capacity by cycling one fan, while the capacity of a three-fan condenser will be reduced to approximately 40% by cycling two fans. Further reduction in capacity to approximately 15% is possible by modulating the airflow through the uncontrolled fan section with either speed control or dampers on the air intake or discharge side (Figure 8). In multiple fan arrangements, idle fans should not be driven backward by the short-circuiting of air. Such low-speed operation may result in the failure of certain bearings; a simple fan separator baffle prevents this.

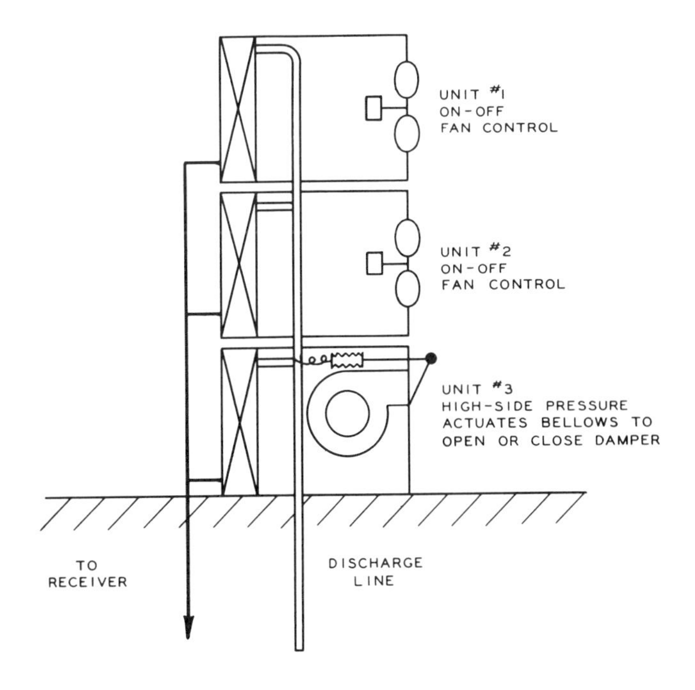


**Fig. 8 Unit Condensers Installed in Parallel with Combined Fan Cycling and Damper Control**

Damper operators, powered directly by refrigerant pressure or conventionally controlled in response to either receiver pressure or ambient temperature, are being used to control head pressure. These devices throttle the airflow through the condenser coil from 100% to zero (Figure 9). Propeller fans for such units should have flat power characteristics so that the fan motor does not overload when the damper is nearly closed.

Solid-state controls can vary the speed of any alternating current motor through frequency modulation. Head pressure can be controlled automatically by varying the fan speed. As fan horsepower is proportional to the cube of its speed, large savings in power consumption are achieved at low ambients and part-load operation.

Parallel operation of condensers, especially with capacity-control devices, requires careful design. Connecting only identical condensers in parallel reduces operational problems. The "Multiple Condenser Installation" section of this chapter has further information. Figures 14 and 15 also apply to air-cooled condensers.

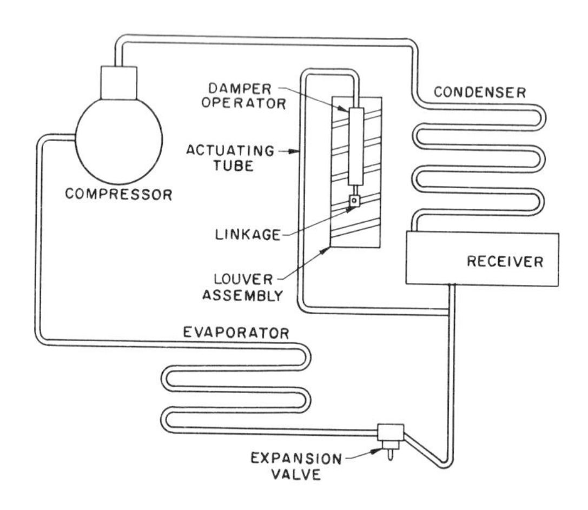


**Fig. 9 Control of Head Pressure by Use of Damper Operators**

The control methods described maintain sufficient head pressure for proper expansion valve operation. On *off* cycles, when the outdoor temperature is lower than the temperature of the indoor space to be conditioned or refrigerated, the refrigerant migrates from the evaporator and receiver and condenses in the cold condenser. The system pressure drops and may correspond to the outdoor temperature. On start-up, insufficient feeding of liquid refrigerant to the evaporator because of low head pressure causes cycling of the compressor until the suction pressure reaches an operating level above the setting of the low pressurestat. At extremely low outdoor temperatures, the pressure in the system may be below the cut-in point of the low pressurestat and the compressor will not start. This difficulty is solved by either of two methods: (1) bypassing the low-pressure switch on start-up or (2) using the condenser isolation method of control.

1. On system start-up, a time-delay relay may bypass the low-pressure switch for approximately 180 seconds to permit the head pressure to build up and allow compressor operation.
2. The condenser isolation method uses a valve arrangement to isolate the condenser from the rest of the system to prevent the refrigerant from migrating to the condenser coil during the *off* cycle.

When the following are used, a noncondensable charged thermostatic expansion valve must be provided: (1) head pressure control of the flooding type or (2) air-side control where the liquid temperature from the condenser may be lower than the evaporator temperature. This type of valve prevents erratic operation of the expansion valve because of condensation of the thermal bulb charge on the head of the expansion valve.

## INSTALLATION AND MAINTENANCE

The installation and maintenance of remote condensers require little labor because of their relatively simple design. Remote condensers are located as close as possible to the compressor, either indoors or outdoors. They may be located above or below the level of the compressor. Various considerations relating to installation and maintenance follow:

**Indoor condensers.** When condensers are located indoors, the warm discharge air must be conducted to the outdoors. An outdoor air intake opening near the condenser is provided and may be equipped with shutters. Indoor condensers can be used for space heating during the winter and ventilation during the summer (Figure 10).

**Outdoor installations.** In outdoor installations of vertical face condensers, prevailing winds should blow toward the air intake, or discharge shields should deflect opposing winds.

**Piping.** Piping practice with condensers is identical to that established by experience with other remote condensers. Discharge piping should be sized for a total pressure drop equivalent to 2 °F (1.1 °C) saturated temperature drop. Standard piping procedures should be followed.

**Receiver.** A condenser may be equipped with a built-in receiver, or it may be separate and installed at a distance from the condenser. Most water- and air-cooled systems function without receivers. If a receiver is used and located in a comparatively warm ambient temperature, the liquid drain line from the condenser to the receiver should be sized for a liquid velocity of 100 fpm (0.5 m/s) for gravity drainage and venting to the condenser. The receiver should be equipped with a purge valve and sight glass.

On a system using an air-cooled condenser with integral subcooling circuits, a receiver must not be used as a part of the operating system located between the subcooling coil outlet and the metering device. Liquid will drain from the subcooling circuit to the receiver, and it is impossible to maintain a liquid seal in the subcooling circuit unless the receiver is filled with liquid refrigerant, in which case the receiver would be superfluous. Furthermore, since it is impossible to keep a subcooled liquid and vapor in the same vessel, the liquid and vapor temperature soon equalizes at saturated conditions in the receiver.

If a receiver must be used, there are two possibilities: either the receiver serves as a storage tank, in which case it must be valved off from condensing and subcooling circuits during operation, or the receiver is located between the condensing circuit and the subcooling circuit. The latter method maintains a liquid-

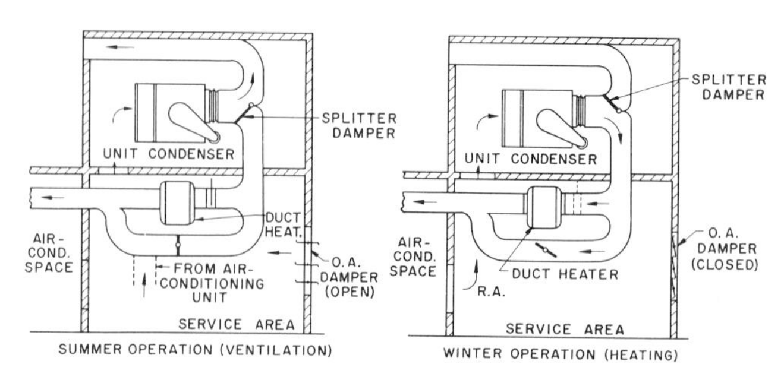


**Fig. 10 Air-Cooled Unit Condenser for Winter Heating and Summer Ventilation**

filled subcooling circuit if a liquid seal is maintained in the receiver, but usually requires cutting into the internal refrigerant piping of the condenser.

**Maintenance.** Relubrication of fan motor and fan bearings should be scheduled. Adjustment of belt tension may be necessary after installation and at yearly intervals. *Industrial refrigeration experience indicates that the maintenance cost of air-cooled condensing equipment is only 25% of that of water-cooled equipment.* In addition, the capacity loss resulting from fouling with increased age is less with air-cooled condensing equipment than with wet condensing equipment.

Condensers are usually placed in locations that are relatively free of dust or dirt. In dirty locations, the condenser coil may require periodic removal of lint or leaves, either with a long bristle brush, by washing with a hose, or with compressed air. A cleaning solution is required to remove grease. Indoor condensers in dusty locations, such as textile mills or supermarkets, should be equipped with washable filters.

# PART III: EVAPORATIVE CONDENSERS

As with the water- and air-cooled condensers, the evaporative condenser rejects heat from a condensing vapor into the environment. In fact, the evaporative condenser is a combination of a water-cooled condenser and an air-cooled condenser that rejects heat through the evaporation of water into an airstream traveling across a condenser coil.

Evaporative condensers reduce the water pumping and chemical treatment requirements associated with cooling tower/refrigerant condenser systems. In addition, they require substantially less fan horsepower than air-cooled condensers of comparable capacity. Most importantly, however, systems using evaporative condensers can be designed for lower condensing temperatures and, subsequently, lower compressor energy input than systems with conventional air- or water-cooled condensers.

The evaporative condenser can operate at a lower condensing temperature than an air-cooled condenser because the air-cooled condenser is limited by the ambient dry-bulb temperature. In the evaporative condenser, the heat rejection is limited by the ambient wet-bulb temperature, which is normally 14 to 25 °F (8 to 14 °C) lower than the ambient dry bulb. The evaporative condenser also provides lower condensing temperatures than the cooling tower/water-cooled condenser because the heat-transfer/mass-transfer steps are reduced from two (between the refrigerant and the cooling water and between the water and ambient air) to one step (refrigerant directly to ambient air).

## PRINCIPLE OF OPERATION

In an evaporative condenser, vapor circulates through a condensing coil that is continually wetted on the outside by a recirculating water system. As seen in Figure 11, air is simultaneously directed over the coil, causing a small portion of the recirculated water to evaporate. This evaporation removes heat from the coil, thus cooling and condensing the vapor. Principal components of an evaporative condenser include the condensing coil, the fan(s), the spray water pump, the water distribution system, cold water sump, drift eliminators, and water makeup assembly.

## COILS

Evaporative condensers generally use bare pipe or tubing without fins. The high rate of energy transfer from the wetted external surface to the air eliminates the need for extended surface. Furthermore, bare coils sustain performance better because they are less susceptible to fouling and they are easier to clean. Coils are usually fabricated from steel tubing, copper tubing, iron pipe, or stainless-steel tubing. Ferrous materials are generally hot-dip galvanized for exterior protection.

## METHOD OF COIL WETTING

The pump moves water from the cold water sump to the distribution system located above the coil. The water descends through the air circulated by the fan(s), over the coil surface, and eventually returns to the pan sump. Water distribution systems are designed for complete and continuous wetting of the full coil surface. This ensures the high rate of heat transfer achieved with wet tubes and prevents excessive scaling, which is more likely to occur on intermittently or partially wetted sur-

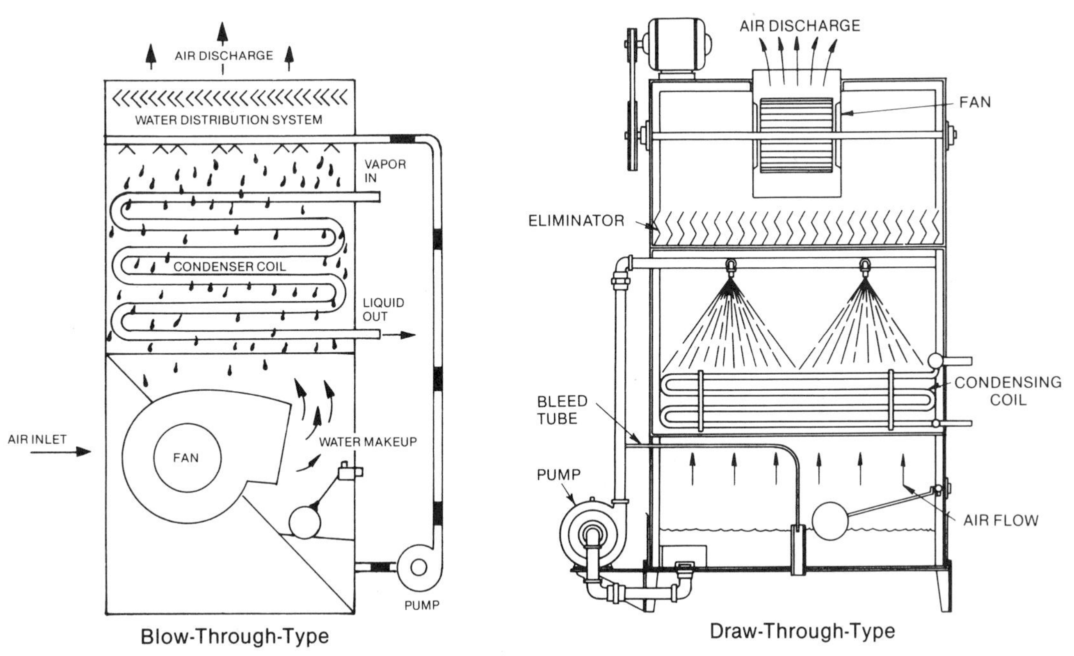


**Fig. 11 Functional View of Evaporative Condenser**

faces. Such scaling is undesirable because it decreases the heat-transfer efficiency (which tends to raise the condensing temperature) of the unit. Water lost through evaporation and blowdown from the cold-water sump (see Chapter 21) is replaced through an assembly which typically consists of a mechanical float valve or solenoid valve and float switch combination.

## AIRFLOW

Most evaporative condensers have fan(s) to either blow or draw air through the unit (Figure 11). Typically, the fans are either the centrifugal or propeller type, depending on the external pressure needs, permissible sound levels, and energy usage requirements.

Drift eliminators recover entrained moisture from the airstream. These eliminators strip most of the water from the discharge airstream; however, a certain amount is discharged as drift. The rate of drift loss from an evaporative condenser is a function of the unit configuration, eliminator design, airflow through the evaporative condenser, and the water flow rate. Generally, an efficient eliminator design can reduce drift loss to a range of 0.002 to 0.2% of the water circulation rate.

If the air inlet is near the sump, louvers or deflectors may be installed to prevent water from splashing out of the unit.

## HEAT-TRANSFER PROCESS

In an evaporative condenser, heat flows from the condensing refrigerant vapor inside the tubes, through the tube wall, to the water film outside the tubes, and then from the water film to the air. Figure 12 shows temperature trends in an evaporative condenser. The driving potential in the first step of heat transfer is the temperature difference between the condensing refrigerant and the surface of the water film, whereas the driving potential in the second step is a combination of temperature and concentration difference between the water surface and the air. Sensible heat transfer between the water stream and the airstream at the water-air interface occurs because of the temperature gradient, while mass transfer (evaporation) of water vapor from the water-air interface to the airstream occurs because of the concentration gradient. Commonly, a single enthalpy driving force between the air saturated into the temperature of the water-film surface and the enthalpy of air in contact with that surface is applied to simplify an analytical approach. The *exact* formulation of the heat- and mass-transfer process requires consideration of the two driving forces simultaneously.

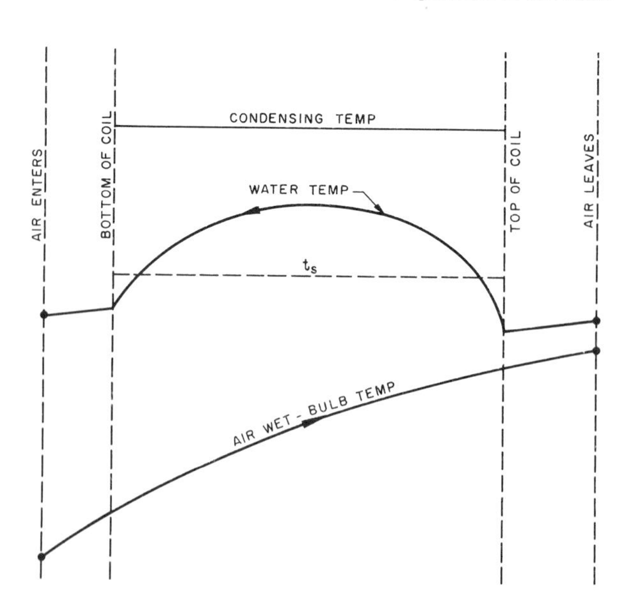


**Fig. 12 Heat-Transfer Diagram for an Evaporative Condenser**

Because the performance of an evaporative condenser cannot be simply represented solely by a temperature difference or an enthalpy difference, simplified predictive methods can only be used for interpolation of data between test points or between tests of different size units, provided that the air velocity, water flow rate, refrigerant velocity, and tube bundle configuration are comparable.

The rate of heat flow from the refrigerant through the tube wall and to the water film can be expressed as:

$$q = U_s A(t_c - t_s) \tag{6}$$

*where*

$q$ = rate of heat flow, Btu/h (W)
$U_s$ = heat-transfer coefficient, Btu/h • ft$^2$ • °F (W/m$^2$ • °C)
$t_c$ = saturation temperature at the pressure of refrigerant entering condenser, °F (°C)
$t_s$ = temperature of water-film surface, °F (°C)
$A$ = outside surface area of condenser tubes, ft$^2$ (m$^2$)

The rate of heat flow from the water-air interface to the airstream can be expressed as:

$$q = U_c A(h_s - h_e) \tag{7}$$

*where*

$q$ = heat input to condenser, Btu/h (W)
$U_c$ = heat-transfer coefficient from the water-air interface to the airstream, Btu/h • ft$^2$ • enthalphy difference in Btu/lb (W/m$^2$ • enthalpy difference in kJ/kg)
$h_s$ = enthalpy of air saturated at $t_c$, Btu/lb (kJ/kg)
$h_e$ = enthalpy of air entering condenser, Btu/lb (kJ/kg)

Equations (6) and (7) have three unknowns: $U_s$, $U_c$, and $t_s$ ($h_s$ is a function of $t_s$). Consequently, it is not possible to solve these equations directly. Rather, the solution requires an iterative procedure that estimates one of the three unknowns and then solves for the remaining two unknowns. A comparison is then made of the $q$ predicted by Eqs. (6) and (7), and if these are not equal (within acceptable convergence limits), a new estimate of the same unknown is made and the process is repeated until the convergence criterion is satisfied. Such procedures are well-suited to computerized techniques.

Leidenfrost and Korenic (1979, 1982), Korenic (1980), and Leidenfrost *et al.* (1980) have evaluated heat-transfer performance of evaporative condensers by in-depth analysis of internal conditions within a coil. This one-dimensional steady-state analysis uses a finite element technique to calculate the change in the state of the air and water along their respective passages of the coil. The analytical model is based on an exact graphical presentation of the heat- and mass-transfer processes between the air and the water at their interface (Bosnjakovic, 1965).

## COMPARISON OF ATMOSPHERIC HEAT-REJECTION EQUIPMENT

Heat may be rejected from a condensing refrigerant to the atmosphere by (1) an evaporative condenser, (2) a combination of water-cooled condenser and cooling tower, or (3) an air-cooled condenser. All three methods involve air; however, in the case of both the evaporative condenser and the cooling tower, the air has a higher heat-absorbing capacity. Hence, in comparison with an air-cooled condenser, an evaporative condenser requires less coil surface and airflow to reject the same heat. Or alternately, greater operating efficiencies can be achieved by operating at a lower condensing temperature.

While both the water-cooled condenser/cooling tower combination and the evaporative condenser use evaporative heat rejection, the former has added nonevaporative heat transfer from the condensing refrigerant to the circulating water, which requires more surface area. Evaporative condensers are, therefore, the most compact for a given capacity.

## CONDENSER LOCATION

Most evaporative condensers are located outdoors, frequently on the roofs of machine rooms. They may also be located indoors and ducted to the outdoors. Generally, centrifugal fan models must be used for indoor applications to overcome the external static resistance of the duct system.

Evaporative condensers installed outdoors can be protected from freezing in cold weather by a remote sump arrangement in which the water and pump are located in a heated space that is remote to the condensers (Figure 13). Piping is arranged so that whenever the pump stops, all of the water drains back into

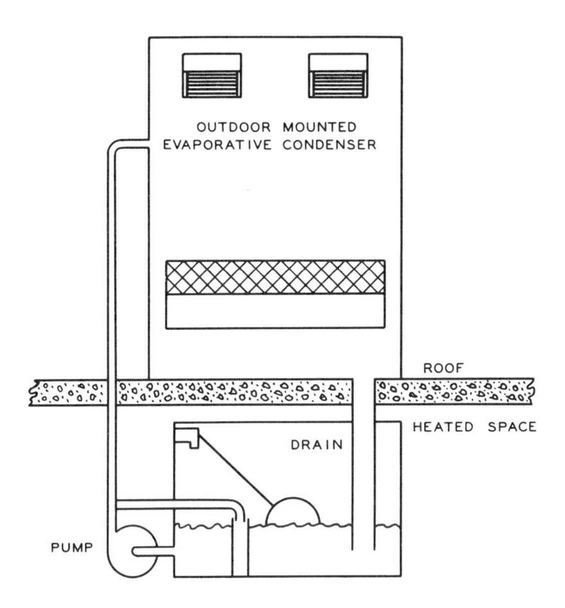


**Fig. 13 Evaporative Condenser Arranged for Year-Round Operation**

the sump to prevent freezing. Where remote sumps are not practical, reasonable protection can be provided by sump heaters, such as electric immersion heaters, steam coils, or hot water coils. Water pumps and lines must also be protected such as with electric heat tracing tape and insulation.

Where the evaporative condenser is ducted to the outdoors, moisture from the warm saturated air can condense in condenser discharge ducts, especially if the ducts pass through a cool space. Some condensation may be unavoidable even with short, insulated ducts. In such cases, the condensate must be drained. Also, in these ducted applications, the drift eliminators must be highly effective.

## MULTIPLE CONDENSER INSTALLATIONS

Large refrigeration plants may have several evaporative condensers connected in parallel or evaporative condensers in parallel with shell-and-tube condensers. In such systems, unless all condensers have the same refrigerant-side pressure loss, refrigerant liquid will load those condensers with the highest pressure loss. Also, in periods of light load where some condenser fans are off, liquid will load the active condensers.

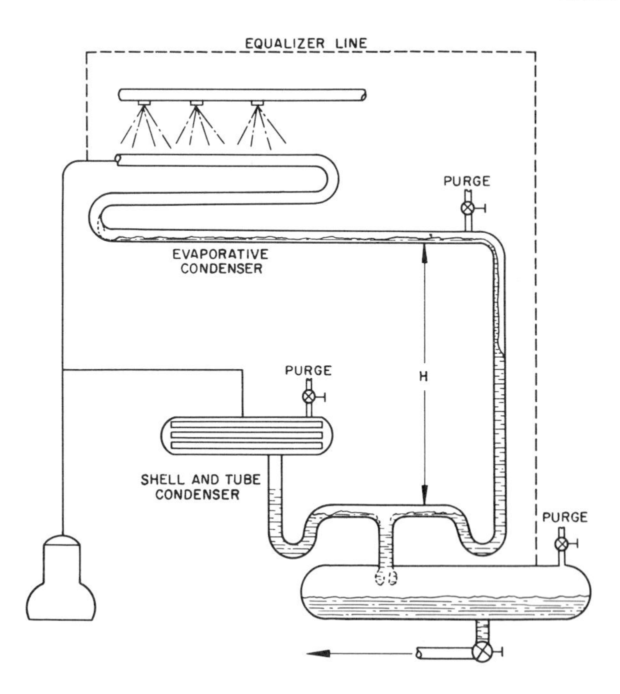


Head (H) above the trap must be greater than the internal resistance of the condenser. Note purge connections at both condensers and receiver.

**Fig. 14 Parallel Operation of Evaporative and Shell-and-Tube Condensers**

Trapped drop legs, as illustrated in Figures 14 and 15, provide proper control of such multiple installations. Effective height of drop legs (H) must equal the head loss through the condenser at maximum loading. Particularly in cold weather, when condensing pressure is controlled by shutting down some condenser fans, active condensers may be loaded considerably above nominal rating, with higher head losses. The drop leg should be high enough to anticipate this condition.

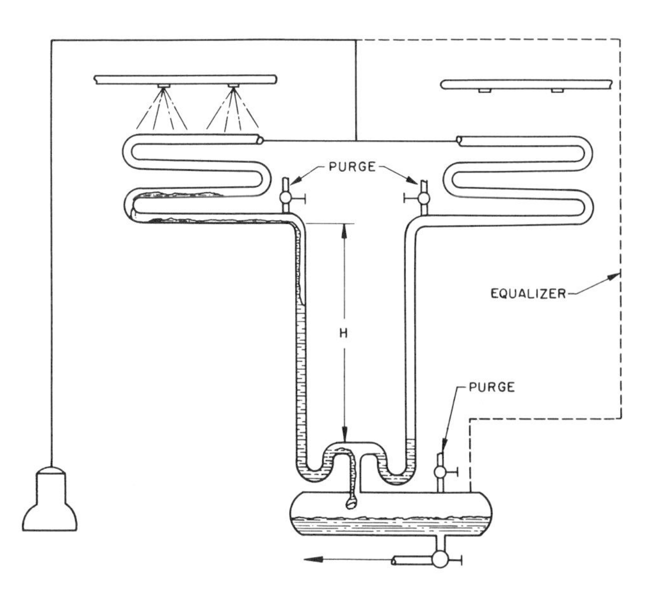


Either condenser can operate independent of the other. Head (H) above the trap, for either condenser, must be not less than the internal resistance of the condenser. Note purge connections at both condensers and the receiver.

**Fig. 15 Parallel Operation of Two Evaporative Condensers**

## RATINGS

Heat rejected from an evaporative condenser is generally expressed as a function of the saturated condensing temperature and entering air wet bulb. The type of refrigerant has considerable effect on ratings; this effect is handled by separate tables (or curves) for each refrigerant or by a correction factor when the difference is small. Superheat of refrigerant entering the condenser may also affect the rating; test standards such as ASHRAE *Standard* 64-74, *Methods of Testing Remote Mechanical-Draft Evaporative Refrigerant Condensers*, establish superheat at a typical value.

It is common to select evaporative condensers in terms of system refrigeration capacity. Manufacturers' ratings include evaporative temperatures and generally assume a nominal heat of compression. Suction gas-cooled hermetic and semihermetic compressors add heat. In such cases, the total wattage input to such compressors should be added to the refrigeration load to obtain the total heat rejected by the condenser.

Rotary screw compressors use oil to lubricate the moving parts and provide a seal between the rotors and the compressor housing. This oil is subsequently cooled either in a separate heat exchanger or by refrigerant injection. In the former case, the amount of heat removed from the oil in the heat exchanger is subtracted from the sum of the refrigeration load and the compressor brake horsepower to obtain the total heat rejected by the evaporative condenser. When liquid injection is used, the total heat rejection is the sum of the refrigeration load and the brake horsepower. Heat rejection rating data together with any ratings based on refrigeration capacity should be included.

## DESUPERHEATING COILS

A desuperheater is an air-cooled finned coil usually installed in the discharge airstream of an evaporative condenser (Figure 16). The primary function of the desuperheater is to increase the condenser capacity by removing some of the superheat from the discharge vapor before the vapor enters the wetted condensing coil. The amount of superheat removed is a function of the desuperheater surface, condenser airflow, and the temperature difference between the refrigerant and air. In practice, a desuperheater is limited to reciprocating compressor ammonia installations where discharge temperatures are relatively high (250 to 300°F [120 to 150°C]).

## REFRIGERANT LIQUID SUBCOOLERS

The refrigerant pressure at the expansion device feeding the evaporator(s) can be lower than the receiver pressure because of liquid line pressure losses. If the evaporator is located above the receiver, the static head difference further reduces the pressure at the expansion device. To avoid liquid line flashing where these conditions exist, it is necessary to subcool the liquid refrigerant after it leaves the receiver. The minimum amount of subcooling required is the temperature difference between the condensing temperature and the saturation temperature corresponding to the saturation pressure at the expansion device. Subcoolers are often used with halocarbon systems but are seldom used with ammonia systems for the following reasons:

1. Because ammonia has a relatively low liquid density, liquid line static head losses are small.
2. Ammonia has a very high latent heat; hence, the amount of flash gas resulting from typical pressure losses in the liquid line is extremely small.
3. Ammonia is seldom used in a direct-expansion feed system where subcooling is critical to proper expansion valve performance.

One method commonly used to supply subcooled liquid for halocarbon systems places a subcooling coil section in the evaporative condenser below the condensing coil (Figure 17).

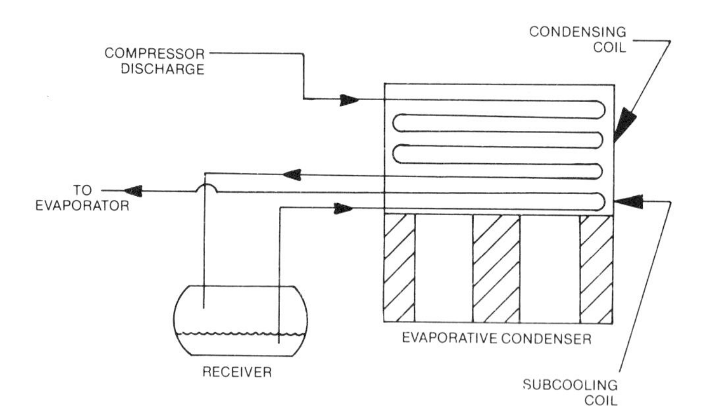


**Fig. 17 Evaporative Condenser with Liquid Subcooling Coil**

Depending on the design wet-bulb temperature, condensing temperature, and subcooling coil surface area, the subcooling coil section normally furnishes 10 to 15°F (5.5 to 8.3°C) of liquid subcooling. As shown in Figure 17, a receiver must be installed between the condensing coil and subcooling coil to provide a liquid seal for the subcooling circuit.

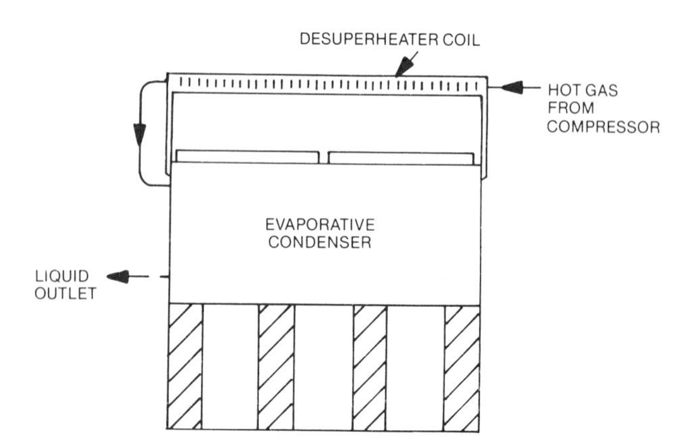


**Fig. 16 Evaporative Condenser with Desuperheater Coil**

## MULTICIRCUIT CONDENSERS AND COOLERS

Evaporative condensers and evaporative fluid coolers, which are essentially the same, may be multicircuited to condense different refrigerants or cool different fluids simultaneously, providing the condensing temperatures or leaving temperatures are the same. Typical multicircuits (1) condense different refrigerants found in food market units where R-12 and R-502 are condensed in separate circuits, (2) condense and cool different fluids in separate circuits such as condensing the refrigerant from a screw compressor in one circuit and cooling the water or glycol for its oil cooler in another circuit, and (3) cool different fluids in separate circuits such as in many industrial applications where it is necessary to keep the fluids from different heat exchangers in separate circuits.

A multicircuit unit is usually controlled by sensing sump water temperature and modulating an air volume damper or by cycling a fan motor. Two-speed motors or separate high-speed and low-speed motors are also used. Using sump water temperature has an averaging effect on control. The total sump water volume is recirculated once every minute or two, making it a good indicator of changes in load.

## WATER TREATMENT

As the recirculated water evaporates in an evaporative condenser, the dissolved solids originally present in the water remain in the system. Continued concentration of these dissolved solids can lead to scaling and/or corrosion problems. In addition, airborne impurities and biological contaminants are often introduced into the recirculated water. If these impurities are not controlled, they can cause sludge or biological fouling. Accordingly, a water treatment program is needed to control all potential contaminants. While simple bleedoff may be adequate for control of scale and corrosion, it will not control biological contamination, which must be addressed in any treatment program. (Chapter 53 of the 1987 HVAC Volume has additional information.) Specific recommendations on water treatment can be obtained from any competent water treatment supplier.

## WATER CONSUMPTION

For the purpose of sizing the makeup water piping, all of the heat rejected by an evaporative condenser is assumed to be latent heat (approximately 1050 Btu/lb [2.44 MJ/kg] of water evaporated). The heat rejected depends on operating conditions, but it can range from 14,000 Btu/h per ton of air conditioning (1.17 W/W) to 17,000 Btu/h per ton of freezer storage (1.42 W/W). The evaporated water ranges from about 1.6 to 2 gph/ton (0.48 to 0.60 mL/s per kW) of refrigeration. In addition, a small amount of water can be lost in the form of drift through the eliminators. With good, quality makeup water, the bleed rates may be as low as half the evaporation rate, and the total water consumption would range from 2.4 gph/ton (0.72 mL/s per kW) for air conditioning to 3 gph/ton (0.90 mL/s per kW) for refrigeration.

## CAPACITY MODULATION

To ensure operation of expansion valves and other refrigeration system components, extremely low condensing pressures must be avoided. Capacity reduction, controlled by a pressure switch, is obtained in several ways: (1) intermittent operation of the fan(s), (2) a modulating damper in the airstream to reduce the airflow (centrifugal fan models only), and (3) multispeed fan motors. Multi-speed fan motors are available as two-speed motors or variable-frequency motors. Two-speed fan motors usually operate at 100% and 50% fan speed, which provides 100% and approximately 60% condenser capacity, respectively. With the fans off and the water pump operating, condenser capacity is approximately 10%.

Often, two-speed fan motors provide sufficient capacity control, because it is seldom necessary to hold condensing pressure to a very tight tolerance other than to maintain a certain minimum condensing pressure to ensure refrigerant liquid feed pressure for the system low side and/or sufficient pressure for hot gas defrost requirements. Variable speed motors are more expensive but offer a full range of speed control for those applications requiring close control on condensing pressure. Careful evaluation of possible critical speed problems should be reviewed with the manufacturer on all variable speed motor applications.

Modulating air dampers also offer closer control on condensing pressure, but they do not offer as much fan horsepower reduction at part load as fan speed control. Water pump cycling for capacity control is not recommended because the periodic drying of the tube surface promotes a buildup of scale.

## PURGING

Refrigeration systems that are operating below atmospheric pressure and systems that are opened for service may require purging to remove air that causes high condensing pressures. With the system operating, purging should be done from the top of the outlet connection. On multiple-coil condensers or multiple-condenser installations, one coil at a time should be purged. Purging two or more coils at one time equalizes coil outlet pressures and can cause refrigerant liquid to back up in one or more of the coils, thus reducing the system operating efficiency. Purging may also be done from the high point of the evaporative condenser refrigerant feed but is only effective when the system is not operating. During normal operation, noncondensables are dispersed throughout the high-velocity vapor, and excessive refrigerant would be lost if purging were done from this location.

## MAINTENANCE

Evaporative condensers are often installed in remote locations and may not receive the routine attention of operating and maintenance personnel. Therefore, programmed maintenance is essential; where manufacturers' recommendations are not available, the following list may be used as a guide:

| Maintenance Item | Frequency |
|---|---|
| 1. Check fan and motor bearings and lubricate, if necessary. Check tightness and adjustment of thrust collars on sleeve bearing units and locking collars on ball bearing units. | Q |
| 2. Check belt tension. | M |
| 3. Clean strainer. If air is extremely dirty, strainer may need frequent cleaning. | W |
| 4. Clean and flush sump. | S |
| 5. Check operating water level in sump, and adjust makeup valve, if required. | W |
| 6. Check water distribution, and clean as necessary | W |
| 7. Check bleed water line to ensure it is operative and adequate as recommended by manufacturer. | W |
| 8. Check fans and air inlet screens and remove any dirt or debris. | D |
| 9. Inspect unit carefully for general preservation and cleanliness, and make any needed repairs immediately. | R |
| 10. Check operation of controls such as modulating capacity control dampers. | M |
| 11. Check operation of freeze control items such as pan heaters and their controls. | Y |
| 12. Check the water treatment system for proper operation. | W |
| 13. Inspect entire evaporative condenser for spot corrosion. Treat and refinish any corroded spot. | Y |

D = Daily; W = Weekly; M = Monthly; Q = Quarterly; S = Semiannually; Y = Yearly; R = As required.

## CONSTRUCTION CODES

If state or municipal codes do not take precedence, design pressures, materials, welding, tests, and relief devices should be in accordance with the ASME *Boiler and Pressure Vessel Code,* Section VIII, Div. 1, and ANSI/ASHRAE *Standard* 15-1978, *Safety Code for Mechanical Refrigeration.* Evaporative condensers are exempt, however, from the ASME Code on the basis of Item (c) of the Scope of the Code, which states if the ID of the condenser shell is 6 in. (150 mm) or less, it is not governed by the Code.

## REFERENCES


ARI. 1987. Standard for Water-Cooled Refrigerant Condensers, Remote Type. ARI *Standard* 450-87, Air-Conditioning and Refrigeration Institute, Arlington, VA.

ASHRAE. 1982. Waterside Fouling Resistance Inside Condenser Tubes. Reasearch Note 31 (RP 106), ASHRAE *Journal*, Vol. 24, No. 6 (June), p. 61.

Bell, K.J. 1972. Temperature Profiles in Pure Component Condensers with Desuperheating and/or Subcooling. AICHE *Symposium Paper* 14a. February, American Institute of Chemical Engineers, New York.

Bergles, A.E. 1973. Recent Developments in Convective Heat Transfer Augmentation. *Applied Mechanics Reviews*, Vol. 26, No. 6 (June), p. 675.

Bergles, A.E. 1976. Survey of Augmentation of Two-Phase Heat Transfer. ASHRAE *Transactions* 82(1): 881.

Bosnjakovic, F. 1965. *Technical Thermodynamics*. Holt, Rinehart, and Winston, New York.

Briggs, D.E. and E.H. Young. 1969. Modified Wilson Plot Techniques for Obtaining Heat Transfer Correlations for Shell and Tube Heat Exchangers. *Chemical Engineering Progress Symposium Series*, Vol. 65, No. 92, p. 35.

Butterworth, D. 1977. Developments in the Design of Shell-and-Tube Condensers. ASME *Paper No.* 77-WA/HT-24. American Society of Mechanical Engineers, New York.

Eckert, E.R.G. 1963. *Heat and Mass Transfer*. McGraw-Hill, New York, p. 143.

Henderson, C.L. and J.M. Marchello. 1969. Film Condensation in the Presence of a Noncondensable Gas. ASME *Journal of Heat Transfer, Transactions*, August, p. 447.

Katz, D.L.; R.E. Hope; S.C. Datsko; and D.B. Robinson. 1947. Condensation of Freon-12 with Finned Tubes. *Refrigerating Engineering*, March, p. 211 and April, p. 315.

Katz, D.L. and D.B. Robinson. 1947. Condensation of Refrigerants and Finned Tubes. *Heating and Ventilating*, Reference Section, November.

Kern, D.Q. and A.D. Kraus. 1972. *Extended Surface Heat Transfer*, Chapter 10. McGraw-Hill, New York.

Kistler, R.S.; A.E. Kassem; and J.M. Chenoweth. 1976. Rating Shell-and-Tube Condensers by Stepwise Calculations. ASME *Paper No.* 76-WA/HT-5. American Society of Mechanical Engineers, New York.

Korenic, B. 1980. *Augmentation of Heat Transfer by Evaporative Coolings to Reduce Condensing Temperatures*. PhD Thesis, Purdue University, W. Lafayette, IN.

Kragh, R.W. 1975. Brush Cleaning of Condenser Tubes Saves Power Costs. *Heating, Piping, Air Conditioning*, September.

Kratz, A.P.; H.J. MacIntire; and R.E. Gould. 1930. Heat Transfer in Ammonia Condensers. *Bulletin No.* 209, Part III, June. University of Illinois—Engineering Experiment Station.

Leidenfrost, W. and B. Korenic. 1979. Analysis of Evaporative cooling and Enhancement of Condenser Efficiency and of Coefficient of Performance. *Warme- und Stoffubertragung* 12.

Leidenfrost, W. and B. Korenic. 1982. Experimental Verification of a Calculation Method for the Performance of Evaporatively Cooled Condensers. *Brennstoff-Warme-Kraft* 34, No. 1, p. 9. VDI Association of German Engineers, Dusseldorf.

Leidenfrost, W.; K.H. Lee; and B. Korenic. 1980. Conservation of Energy Estimated by Second Law Analysis of a Power-Consuming Process. *Energy*, Vol. 5, p. 47.

McAdams, W.H. 1954. *Heat Transmission*, 3rd ed., McGraw-Hill Book Company, New York, pp. 219 and 343.

Pearson, J.F. and J.G. Withers. 1969. New Finned Tube Configuration Improves Refrigerant Condensing. ASHRAE *Journal*, June, p. 77.

Seban, R.A. and E.F. McLaughlin. 1963. Heat Transfer in Tube Coils with Laminar and Turbulent Flow. *International Journal of Heat Mass Transfer*, Vol. 6, p. 387.

Shah, M.M. 1979. A General Correlation for Heat Transfer During Film Condensation in Tubes. *International Journal of Heat and Mass Transfer*, Vol. 22, No. 4 (April), p. 547.

Shah, M.M. 1981. Heat Transfer During Film Condensation in Tubes and Annuli: A Review of the Literature. ASHRAE *Transactions* 87(1).

Starner, K.E. 1976. Effect of Fouling Factors on Heat Exchanger Design. ASHRAE *Journal*, May, p. 39.

Sturley, R.A. 1975. Increasing the Design Velocity of Water and Its Effect in Copper Tube Heat Exchangers. Paper No. 58, The International Corrosion Forum, Toronto, Ontario. April.

Suitor, J.W.; W.J. Marner; and R.B. Ritter. 1976. The History and Status of Research in Fouling of Heat Exchangers in Cooling Water Service. Paper No. 76-CSME/CS Ch E-19, National Heat Transfer Conference, St. Louis, MO, August 8-11.

Taborek, J.; F. Voki; R. Ritter; J. Pallen; and J. Knudsen. 1972. Fouling—The Major Unresolved Problem in Heat Transfer. *Chemical Engineering Progress Symposium Series* 68, Parts I and II, Nos. 2 and 7.

TEMA. 1978. *Standards of the Tubular Exchanger Manufacturers' Association*, 6th ed. Tubular Exchanger Manufacturers' Association, New York.

UOP, Wolverine Division. 1984. *Engineering Data Book II*, Section 1, p. 30.

Wanniarachchi, A.S. 1981. *The Effect of Noncondensable Gases on the Performance of Condensers*. PhD Thesis, Pennsylvania State University, University Park.

Wanniarachchi, A.S. and R.L. Webb. 1982. Noncondensible Gases in Shell-Side Refrigerant Condensers. ASHRAE *Transactions* 88(2): 170-184.

Watkinson, A.P.; L. Louis; and R. Brent. 1974. Scaling of Enhanced Heat Exchanger Tubes. *The Canadian Journal of Chemical Engineering*, Vol. 52, p. 558.

Webb, R.L.; A.S. Wanniarachchi; and T.M. Ruby. 1980. The Effect of Noncondensible Gases on the Performance of an R-11 Centrifugal Water Chiller Condenser. ASHRAE *Transactions* 86(2):57.

Webb, R.L. 1982. Gas in Refrigerant Condensers. Research Note 55 (RP 225). ASHRAE *Journal* Vol. 28, No. 5 (May), p. 52.

Young, E.H. and D.J. Ward. 1957. Fundamentals of Finned Tube Heat Transfer. *Refining Engineer*, Part I, November.

# CHAPTER 16

# LIQUID COOLERS



A *liquid cooler* (hereafter called a cooler) is a component of a refrigeration system in which the refrigerant is evaporated, thereby producing a cooling effect on a fluid (usually water or brine). This chapter addresses the performance, design, and application of coolers. It briefly describes various types of liquid coolers and the refrigerants commonly used as listed in Table 1.

## TYPES OF LIQUID COOLERS

### Direct-Expansion

Refrigerant evaporates inside tubes of a direct-expansion cooler. These coolers are usually used with positive-displacement compressors, such as reciprocating, rotary, or rotary screw compressors, to cool water or brine. Shell-and-tube is the most common arrangement, although tube-in-tube coolers are also available.

Figure 1 shows a typical shell-and-tube cooler. A series of baffles channels the fluid throughout the shell side. The baffles increase the velocity of the fluid, thereby increasing its heat transfer coefficient. The velocity of the fluid flowing perpendicular to the tubes should be at least 2 ft/s (0.6 m/s) to clean the tubes and less than 10 ft/s (3 m/s) to prevent erosion.

Distribution is critical in direct-expansion coolers. If some tubes are fed more refrigerant than others, they tend to bleed liquid refrigerant into the suction line. Since most direct-expansion coolers are controlled to a given suction superheat, the remaining tubes must produce a higher superheat to evaporate the liquid bleeding through. This unbalance causes poor heat transfer. Uniform distribution is usually achieved by a spray distributor or by keeping the volume of the refrigerant inlet head to a minimum. Both methods create sufficient turbulence to mix the liquid and vapor refrigerant entering the cooler so each tube gets the same mixture of liquid and vapor.

The number of refrigerant passes is another important item in the performance of a direct-expansion cooler. A single pass cooler must evaporate all the refrigerant before it reaches the end of the tubes, which requires long tubes or enhanced inside tube surfaces. A multiple-pass cooler can have less or no surface enhancement, but after the first pass, good distribution is difficult to obtain.

A tube-in-tube cooler is similar to a shell-and-tube design, except that it consists of one or more pairs of coaxial tubes. The fluid usually flows inside the inner tube while the refrigerant flows in the annular space between the tubes. By running the fluid through the inner tube, the fluid side can be mechanically cleaned if access to the header is provided.

Most direct-expansion coolers are designed for horizontal mounting. If mounted vertically, performance may vary considerably from that predicted.

### Flooded Shell-and-Tube

In a flooded cooler, the refrigerant vaporizes on the outside of tubes, which are submerged in liquid refrigerant within a closed shell. The fluid flows through the tubes as shown in Figure 2. Flooded coolers are usually used with rotary screw or centrifugal compressors to cool water or brine.

Refrigerant liquid/vapor mixture usually feeds into the bottom of the shell through a distributor that distributes the refrigerant vapor equally under the tubes. The relatively warm fluid in the tubes heats the refrigerant liquid surrounding the tubes, causing it to boil. As bubbles rise up through the space


The preparation of this chapter is assigned to TC 8.5, Liquid-to-Refrigerant Heat Exchangers.

**Table 1 Types of Coolers**

| Type of Cooler | Usual Refrigerant Feed Device | Usual Range of Capacity Tons (kW) | Commonly Used with Refrigerant Nos. |
|---|---|---|---|
| Flooded Shell-and-Tube | Low-pressure float,<br>High-pressure float,<br>Fixed orifice(s), Weir | 25 to 2000<br>(88 to 7000) | 11, 12, 22, 113,<br>114, 500, 502<br>717 |
| Spray type Shell-and-Tube[a] | Low-pressure float,<br>High-pressure float | 50 to 10,000<br>(175 to 35 000) | 11, 12, 13B1, 22<br>113, 114 |
| Direct-Expansion Shell-and-Tube | Thermal expansion valve | 2 to 350<br>(7 to 1220) | 12, 22, 500, 502<br>717 |
| Baudelot (Flooded) | Low-pressure float | 10 to 100<br>(35 to 350) | 717 |
| Baudelot (Direct-Expansion) | Thermal expansion valve | 5 to 25<br>(17 to 88) | 12, 22, 717 |
| Tube-in-Tube[b] | Thermal expansion valve | 5 to 25<br>(17 to 88) | 12, 22, 717 |
| Shell-and-Coil | Thermal expansion valve | 2 to 10<br>(7 to 35) | 12, 22, 717 |

[a]See "Flooded Cooler" section of text.
[b]See "Direct-Expansion Cooler" section of text.

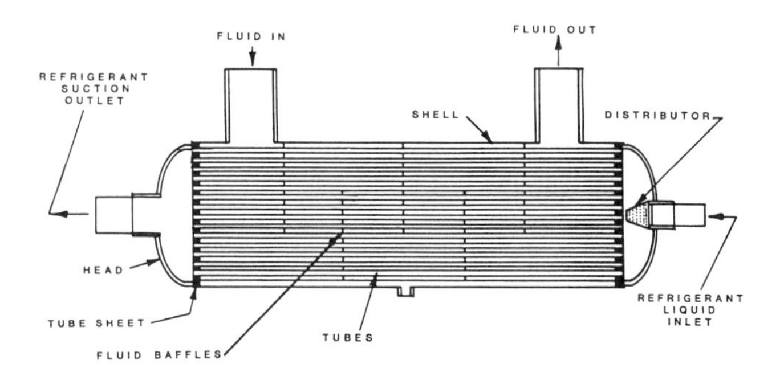


**Fig. 1 Direct-Expansion Shell-and-Tube Cooler**

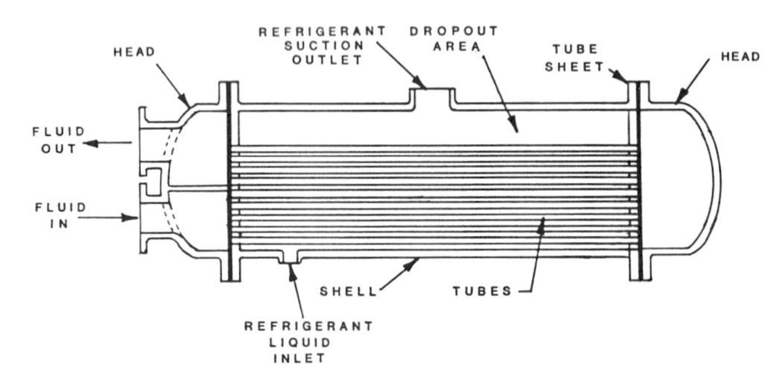


**Fig. 2 Flooded Shell-and-Tube Cooler**

between tubes, the liquid surrounding the tubes becomes increasingly bubbly (or foamy, if much oil is present).

The refrigerant vapor must be separated from the mist generated by the boiling refrigerant. The simplest separation method is provided by a drop-out area between the top row of tubes and the suction connectors. If this drop-out area is insufficient, a coalescing filter may be required between the tubes and connectors. Perry and Green (1984) give additional information on mist elimination.

The size of tubes, number of tubes, and number of passes should be determined to maintain the fluid velocity between 3 and 12 ft/s (1 and 4 m/s). A velocity below or above these limits may be used if the fluid is free of suspended abrasives and fouling substances (Sturley 1975).

One variation of this cooler is the *Spray-Type Shell-and-Tube Cooler.* In large diameter coolers with a refrigerant that has a heat transfer coefficient that is adversely affected by the head of the refrigerant, a spray can be used to cover the tubes with liquid rather than flooding them. A mechanical pump circulates liquid from the bottom of the cooler to the spray heads.

Flooded shell-and-tube coolers are generally unsuitable for other than horizontal orientation.

## Baudelot

Baudelot coolers (Figure 3) are used to cool a fluid to near its freezing point in industrial, food, and dairy applications. In this type of cooler, the fluid is circulated over the outside of a number of columns of horizontal tubes or vertical plates, making them easy to clean. The inside surface of the tubes or plates is cooled by evaporating the refrigerant. The fluid to be cooled is distributed uniformly along the top of the heat exchanger and then flows by gravity to a collection pan below. The cooler may be enclosed by insulated walls to avoid unnecessary loss of refrigeration effect.

Refrigerant 717 (ammonia) is commonly used with the Baudelot cooler and arranged for flooded operation, using a conventional gravity-feed system with a surge drum. A low pressure float valve maintains a suitable refrigerant liquid level

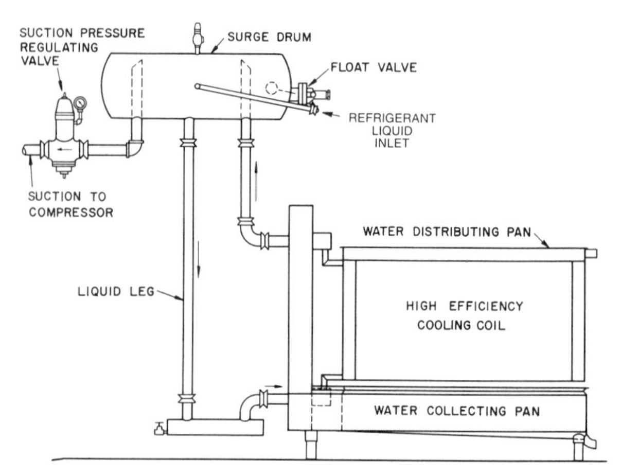


**Fig. 3 Baudelot Cooler**

in the surge drum. Baudelot coolers using other common refrigerants are generally of the direct-expansion type, with thermostatic expansion valves.

### Shell-and-Coil

A shell-and-coil cooler is a tank containing the fluid to be cooled with a simple coiled tube used to cool the fluid. This type of cooler has the advantage of cold water storage to offset peak loads. In some models, the tank can be opened for cleaning. Most applications are at low capacities for bakeries, photographic labs, and to cool drinking water.

The coiled tube containing the refrigerant can be either inside the tank (Figure 4) or attached to the outside of the tank in such a manner to permit heat transfer.

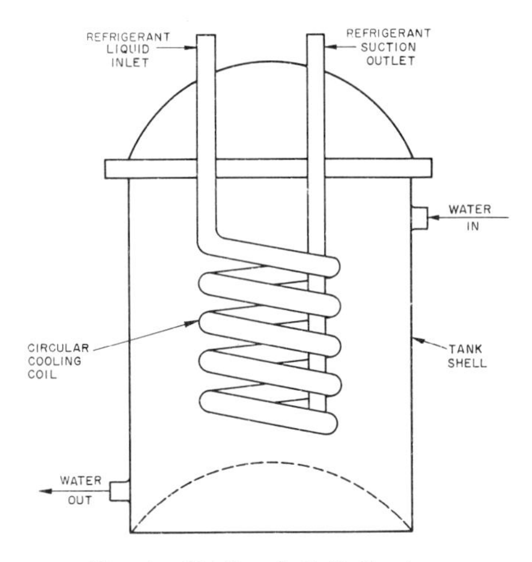


**Fig. 4 Shell-and-Coil Cooler**

## HEAT TRANSFER

Heat transfer for liquid coolers can be expressed by the following steady-state heat transfer equation:

$$q = U \cdot A \cdot \Delta t_m \quad (1)$$

*where*

$q$ = total heat transfer rate, Btu/h (W)
$\Delta t_m$ = mean temperature difference, °F, (°C)
$A$ = area of heat transfer surface, ft$^2$ (m$^2$)
$U$ = overall heat transfer coefficient Btu/h • ft$^2$ • °F (W/m$^2$ • °C)

The area $A$ can be calculated if the geometry of the cooler is known. Chapter 3 of the 1985 FUNDAMENTALS Volume describes the calculation of the mean temperature difference.

This chapter only discusses the components of $U$ and does not discuss it in depth. $U$ may be calculated by one of the following equations.

**Based on Inside Surface Area**

$$U = \frac{1}{1/h_i + (A_i/(A_o h_o)) + (t/k)(A_i/A_m) + r_{fi}} \quad (2)$$

**Based on Outside Surface Area**

$$U = \frac{1}{(A_o/(A_i h_i)) + 1/h_o + (t/k)(A_o/A_m) + r_{fo}} \quad (3)$$

*where*

$h_i$ = inside heat transfer coefficient based on inside surface area, Btu/h • ft$^2$ • °F (W/m$^2$ • °C)
$h_o$ = outside heat transfer coefficient based on outside surface area, Btu/h • ft$^2$ • °F (W/m$^2$ • °C)
$A_o$ = outside heat transfer surface area, ft$^2$ (m$^2$)
$A_i$ = inside heat transfer surface area, ft$^2$ (m$^2$)
$A_m$ = mean heat transfer area of metal wall, ft$^2$ (m$^2$)
$k$ = thermal conductivity of heat transfer surface, Btu/h • ft • °F (W/m • °C)
$t$ = thickness of heat transfer surface, ft (m)
$r_{fi}$ = fouling factor of fluid side based on inside surface area, ft$^2$ • h • °F/Btu (m$^2$ • °C/W)
$r_{fo}$ = fouling factor of fluid side based on outside surface area, ft$^2$ • h • °F/Btu (m$^2$ • °C/W)

Note: If fluid is on inside, multiply $r_{fi}$ by $A_o/A_i$ to find $r_{fo}$.
If fluid is on outside, multiply $r_{fo}$ by $A_i/A_o$ to find $r_{fi}$.

### Heat Transfer Coefficients

The refrigerant side coefficient usually increases under the following conditions: (1) increase in cooler load, (2) decrease in suction superheat, (3) decrease in oil concentration, and (4) increase in saturated suction temperature. The amount of increase or decrease varies, depending on the type of cooler. Flooded coolers have a relatively small change in heat transfer coefficient as a result of a change in load, whereas a direct expansion cooler shows a significant increase in heat transfer coefficient with an increase in load. A Wilson Plot of test data (Jacob 1949, 1957; Briggs and Young 1969) can show actual values for the refrigerant side coefficient of a given cooler design. Young and Ward (1957) give additional information on refrigerant side heat transfer coefficients.

The fluid side coefficient is determined by cooler geometry, fluid flow rate, and fluid properties (viscosity, specific heat, thermal conductivity, and density). For a given fluid, the fluid side coefficient increases with an increase in fluid flow rate due to increased turbulence, and an increase in fluid temperature due to improvement of fluid properties as temperature increases.

Direct-expansion and flooded coolers show a significant increase in heat transfer coefficient with an increase in fluid flow. The effect of flow is smaller for Baudelot and shell-and-coil coolers. Many of the listed references give additional information on fluid side heat transfer coefficients.

To increase the heat transfer coefficients of coolers, an enhanced heat transfer surface can help in the following ways:

1. It increases heat transfer area, thereby increasing overall heat transfer rate, even if refrigerant side heat transfer coefficient is unchanged.
2. Where the flow of fluid or refrigerant is low, it improves heat transfer coefficients by increasing turbulence at the surface and mixing the fluid at the surface with fluid away from the surface.
3. In flooded coolers, an enhanced surface provides more and better nucleation points to promote boiling of refrigerant on the outside of tubes.

### Fouling Factors

Most fluids over time foul the fluid side heat transfer surface, thus reducing the overall heat transfer coefficient of the cooler. If fouling is expected to be a problem, a mechanically cleanable cooler should be used, such as a flooded, Baudelot, or cleanable direct-expansion tube-in-tube cooler. Direct-expansion shell-and-tube and shell-and-coil coolers can be cleaned chemically. Flooded coolers and direct-expansion tube-in-tube coolers with enhanced fluid-side heat transfer surfaces have a tendency to be self-cleaning due to high fluid turbulence, and a smaller fouling factor can probably be used for these coolers. ARI *Standard* 480 discusses fouling calculations.

The refrigerant side of the cooler is not subject to fouling, and a fouling factor need not be included for the that side.

### Heat Exchanger Properties

Conductivity is a constant for any given material and can be obtained from thermal conductivity tables. The thickness and heat transfer areas are determined by the geometry of the cooler.

## PRESSURE DROP

### Fluid Side

Pressure drop is usually minimal in Baudelot and shell-and-coil coolers, but must be considered in direct-expansion and flooded coolers. Both the direct-expansion and flooded coolers rely on turbulent fluid flow to improve heat transfer. This turbulence is obtained at the expense of pressure drop. Increasing the pressure drop increases heat transfer, and decreasing it decreases heat transfer.

For air-conditioning service, the pressure drop is commonly limited to 10 psi (70 kPa) to keep pump size and energy costs reasonable. The following equation projects the change in pressure drop due to a change in flow.

$$\text{New Pressure Drop} = \text{Original Pressure Drop} \cdot \left[\frac{\text{New Flow}}{\text{Original Flow}}\right]^{1.8} \quad (4)$$

### Refrigerant Side

The refrigerant side pressure drop must be considered for the following coolers: direct-expansion, shell-and-coil, and sometimes Baudelot. When there is a pressure drop on the refrigerant side, the refrigerant inlet and outlet pressures and corresponding saturated temperature are different. This difference causes a change in the mean temperature difference, which effects the total heat transfer rate. If the pressure drop is high, operation of the expansion valve may be affected due to reduced pressure drop across the valve. This pressure drop varies, depending on the refrigerant used, operating temperature, and type of tubing (Wallis 1969, Martinelli and Nelson 1948).

## VESSEL DESIGN

### Mechanical Requirements

Pressure vessels must be constructed and tested under the rules of national, state, and local codes. The introduction of the current ASME Boiler and Pressure Vessel Code, Section VIII, gives guidance on rules and exemptions.

The more common applicable codes and standards are as follows:

1. ARI *Standard* 480-87, *Remote Type Refrigerant Cooled Liquid Coolers*, covers industry criteria for standard equipment, standard safety provisions, marking, and recommended procedure for testing and rating refrigerant liquid coolers.
2. ASHRAE *Standard* 24-78, *Methods of Testing for Rating Liquid Coolers*, covers the recommended rating point and fouling factors, and testing and rating methods.
3. ANSI/ASHRAE *Standard* 15-1978, *Safety Code for Mechanical Refrigeration*, involves specific design criteria, use of materials, and testing. It refers to the ASME *Boiler Code*, Section VIII, for refrigerant-containing sides of pressure vessels, where applicable. Factory test pressures are specified, and minimum design working pressures are given. (Table 2 of this chapter gives minimum design pressures for evaporators, as taken from Table 5, Section 8 of ANSI/ASHRAE 15-78.) This code requires pressure limiting and pressure-relief devices on refrigerant-containing systems, as applicable, and defines the setting and capacity requirements for these devices.
4. ASME *Boiler and Pressure Vessel Code, Unfired Pressure Vessels*, Section VIII, covers the safety aspects of design and construction. Most states require coolers to meet the requirements of the ASME if they fall within scope of the ASME code. Some of the exceptions from meeting the ASME requirements listed in the ASME code are as follows:
   - Cooler Shell ID of 6 in. (150 mm) or less.
   - 15 psi (103 kPa) gauge or less.
   - The fluid (water) portion of the cooler need not be built to the requirements of the ASME code if the fluid is water and the volume is 120 gal. (454 L) or less.

   Coolers meeting the requirements of the ASME code will have an ASME stamp, which is a *U* or *UM* inside a three-leaf clover. The *U* can be used for all coolers and the *UM* can be used for small coolers.
5. Underwriters Laboratories *Standard* UL-207-82, *Refrigerant-Containing Components and Accessories*, involves specific design criteria, use of materials, testing, and initial approval by Underwriters Laboratories. A cooler with the ASME *U* stamp does not require UL approval.

**Table 2 Minimum Design Pressure Based on Standby Conditions for Common Refrigerants**

| Refrigerant Number | Standby Temperature 80°F (26.7°C) | 90°F (32.2°C) | 100°F (37.8°C) | 110°F (43.3°C) |
|---|---|---|---|---|
| | Minimum Design Pressure, psi (kPa) gauge | | | |
| R-11 | 15 (103)* | 15 (103)* | 15 (103)* | 15 (103)* |
| R-12 | 85 (586)* | 100 (690) | 118 (814) | 137 (945) |
| R-22 | 144 (993)* | 169 (1165) | 196 (1351) | 227 (1565) |
| R-114 | 18 (124)* | 25 (172) | 32 (221) | 40 (276) |
| R-500 | 102 (703)* | 121 (834) | 142 (979) | 165 (1137) |
| R-502 | 162 (1117)* | 188 (1296) | 217 (1496) | 248 (1710) |
| R-717 | 139 (958)* | 166 (1145) | 198 (1365) | 233 (1607) |

*Minimum per ANSI/ASHRAE *Standard* 15-1978

**Design Pressure** on the refrigerant side as a minimum should be the saturated pressure for the refrigerant used at 80°F (27°C) as per ANSI/ASHRAE *Standard* 15-1978. Standby temperatures and temperatures encountered during shipping of chillers with a refrigerant charge should also be considered. Table 2 gives saturation pressures for various temperatures of common refrigerants.

Required fluid (water) side pressure varies depending largely on the following conditions (1) static head, (2) pump head, (3) transients due to pump start-up, and (4) valve closing.

### Chemical Requirements (Perry and Green 1984, NACE 1974)

**Refrigerant 717 (Ammonia).** Carbon steel and cast iron are the most widely used materials for ammonia systems. Stainless steel alloys are satisfactory but more costly. Copper and high copper alloys are avoided because they are attacked by ammonia when moisture is present. Aluminum and aluminum alloys may be used with caution with ammonia.

**Halocarbon Refrigerants.** Almost all the common metals and alloys are used satisfactorily with these refrigerants. Exceptions include magnesium and aluminum alloys containing more than 2% magnesium, where water may be present. Zinc is not recommended for use with Refrigerant 113; it is more chemically reactive than other common construction metals and, therefore, it is usually avoided when other halogenated hydrocarbons are used as a refrigerant. Under some conditions with moisture present, halocarbon refrigerants form acids that attack steel and even nonferrous metals. This problem does not commonly occur in properly cleaned and dehydrated systems. ANSI/ASHRAE *Standard* 15-1978, paragraph 8.1.2, states that aluminum, zinc, or magnesium shall not be used in contact with methyl chloride nor magnesium alloys with any halogenated refrigerant.

**Water**. Relatively pure water is satisfactory with both ferrous and nonferrous metals. Brackish or sea water, and some river waters, are quite corrosive to iron and steel and also with copper, aluminum, and many alloys of these metals. A reputable water consultant, who knows the local water condition, should be consulted. Chemical treatment by pH control, inhibitor applications, or both, may be required. Where this is not feasible, more noble construction materials or special coatings must be used.

**Brines**. Ferrous metals and a few nonferrous alloys are almost universally used with sodium chloride and calcium chloride brines. Copper base alloys can be used, if adequate quantities of sodium dichromate are added and caustic soda is used to neutralize the solution. Even with ferrous metals, these brines should be treated periodically to hold the pH value near the neutral point.

Ethylene glycol and propylene glycol are stable compounds that are less corrosive than chloride brines.

### Electrical Requirements

When fluid being cooled is electrically conductive, the system must be grounded to prevent electrical chemical corrosion.

## APPLICATION CONSIDERATIONS

### Refrigerant Flow Control

**Direct-expansion coolers**. The constant superheat thermal expansion valve is the most common control used. It is located directly upstream of the cooler. A thermal bulb strapped to the suction line leaving the cooler senses refrigerant temperature. The valve can be adjusted to produce a constant suction superheat during steady operation.

The thermal expansion valve adjustment is commonly set at a suction superheat of 10°F (5.6°C), which is sufficient to ensure that liquid is not carried into the compressor. Direct expansion cooler performance is affected greatly by superheat setting. Reduced superheat improves cooler performance, so suction superheat should be set as low as possible while avoiding liquid carryover to the compressor.

**Flooded coolers**, as the name implies, must have good liquid refrigerant coverage of the tubes to achieve good performance. Liquid level control in a flooded cooler becomes the principal issue in flow control. Some systems are designed *critically charged* so when all the liquid refrigerant is delivered to the cooler, it is just enough for good tube coverage. In these systems, an orifice is often used as the throttling device between condenser and cooler.

**Float Valve**. This device is used as another method of flooded cooler control. A high-side float valve can be used, with the float sensing condenser liquid level, to just drain the condenser. For more exact control of liquid level in the cooler, a low side float valve is used, where the valve senses cooler liquid level and controls the flow of entering refrigerant.

### Freeze Prevention

Freeze prevention must be considered for coolers operating near the freezing point of the fluid. Freezing of the fluid in some coolers causes extensive damage. Two methods can be used for freeze protection: (1) hold the saturated suction pressure above the fluid freezing point or (2) shut the system off if the temperature of the fluid gets near its freezing point.

A suction pressure regulator can hold the saturated suction pressure above the freezing point of the fluid. A low-pressure cutout can shut the system off before the saturated suction pressure goes below the freezing point of the fluid. The leaving fluid temperature can be monitored to cut the system off before a danger of freezing, usually about 10°F (5.6°C) above the fluid freezing temperature. It is recommended that both methods be used.

Baudelot, shell-and-coil and direct-expansion shell-and-tube coolers are all resistant to damage caused by freezing, and ideal for applications where freezing may be a problem.

If a cooler is installed in an unconditioned area, possible freezing due to low ambient temperature must be considered. If the cooler is used only when the ambient temperature is above freezing, the fluid should be drained from the cooler for cold weather. As an alternate to draining, if the cooler is in use year-round, the following methods can be used:

1. A heat tape or other heating device can be used to keep the cooler above freezing.
2. For water, adding an appropriate amount of ethylene glycol will prevent freezing.
3. Continuous pump operation may also prevent freezing.

### Oil Return

Most compressors discharge a small percentage of oil in the discharge gas. This oil mixes with the condensed refrigerant in the condenser and flows to the cooler. Since the oil is nonvolatile, it does not evaporate and may tend to collect in the cooler.

In direct-expansion coolers, the gas velocity in the tubes and the suction gas header is usually sufficient to carry the oil from the cooler and into the suction line. From there, with proper piping design, it can be carried back to the compressor. At light loads and low temperature, oil may tend to gather in the superheat section of the cooler, detracting from performance. For this reason, operation of refrigerant circuits at light load for long periods should be avoided, especially at low-temperature conditions.

In flooded coolers, vapor velocity above the tube bundle is usually insufficient to return oil up the suction line, so oil tends to accumulate in the cooler. With time, depending on compressor oil loss rate, the oil concentration in the cooler may become large. When concentration exceeds about 5%, heat transfer performance will be adversely affected.

It is common in flooded coolers to take some oil-rich liquid and return it to the compressor on a continuing basis, in order to establish a rate of return equal to compressor oil loss rate.

### Maintenance

Maintenance of coolers centers around two areas: (1) safety and (2) cleaning of the fluid side. The cooler should be inspected periodically for any weakening of the pressure boundaries of the cooler. The inspection should include visual inspection for corrosion, erosion, and any deformities. Any pressure relief device should also be inspected. The insurer of the cooler may require regular inspection of the cooler. If the fluid side is subjected to fouling, it may require periodic cleaning. Cleaning may be by either mechanical or chemical means. The manufacturer or service organization experienced in cooler maintenance should have details for cleaning.

### Insulation

A cooler operating at a saturated suction temperature lower than the dew point of the surrounding air should be insulated to prevent condensation. Chapter 20 of the 1985 FUNDAMENTALS Volume describes insulation in more detail.

## REFERENCES

Bergelin, O.P.; K.J. Bell and M.D. Leighton. 1958. Heat Transfer and Fluid Friction During Flow Across Banks of Tubes. University of Delaware Engineering Experiment Station *Bulletin* No. 4.

Briggs, D.E. and E.H. Young. 1969. Modified Wilson Plot Techniques for Obtaining Heat Transfer Correlations for Shell and Tube Heat Exchangers. Chemical Engineering Progress *Symposium* Series, Vol. 65, No.92, p. 25.

Donohue, D.A. 1949. Heat Transfer and Pressure Drop in Heat Exchangers. International Engineering Chemistry, Vol. 41, No. 11, November.

Jacob, M. 1949 and 1957. *Heat Transfer.* John Wiley and Sons, New York, Vols. I and II.

Martinelli, R.C. and D.B. Nelson. 1948. Prediction of Pressure Drop During Forced Circulation Boiling of Water. ASME *Transactions*, August, p. 695.

McAdams, W.H. 1954. *Heat Transmission*, 3rd ed. McGraw-Hill Book Company, New York.

NACE. 1974. *Corrosion Data Survey*, 5th ed. Compiled by N.E. Hamner for the National Association of Corrosion Engineers, Houston, TX.

Palen, J.W. and J. Taborek. 1969. Solution of Shell Side Pressure Drop and Heat Transfer by Stream Analysis Method. Chemical Engineering Progress *Symposium* Series, Vol. 65. No. 92.

Perry, J.H. and R.H. Green. 1984. *Chemical Engineers Handbook*, 6th ed. McGraw Hill Book Company, New York.

Scovill Manufacturing Company, 1957. *Heat Exchanger Tube Manual*, 3rd ed.

Sturley, R.A. 1975. Increasing the Design Velocity of Water and Its Effect on Copper Tube Heat Exchangers. Paper No. 58, The International Corrosion Forum, Toronto, Canada.

Tinker, T. 1956. Shell Side Characteristics of Shell-and-Tube Heat Exchangers. ASME Paper No. 56-A-123. American Society of Mechanical Engineers, New York.

Wallis, G.B. 1969. *One Dimensional Two Phase Flow.* McGraw Hill Book Company, New York.

Wolverine Division of UOP, Inc. 1984. *Engineering Data Book II.*

Young, E.H. and D.J. Ward. 1957. Fundamentals of Finned Tube Heat Transfer, Part I. Refining Engineer, November.

# CHAPTER 17

# LIQUID CHILLING SYSTEMS



A liquid chilling system cools water, brine, or other secondary coolant for air conditioning or refrigeration. The system may be either factory assembled and wired or shipped in sections for erection in the field. The most frequent application is water chilling for air conditioning, although both brine cooling for low-temperature refrigeration and chilling of fluids in industrial processes are also common.

The basic components of a liquid chilling system include a compressor, a liquid cooler (evaporator), a condenser, a compressor drive, a refrigerant flow-control device, and a control center; the system may also include a receiver, an intercooler, and/or a subcooler. In addition, certain auxiliary components may be used, such as an oil cooler, an oil separator, an oil-return device, a purge unit, an oil pump, a refrigerant transfer unit, and additional control valves.

## PART I: GENERAL CHARACTERISTICS

### PRINCIPLES OF OPERATION

Liquid (usually water) enters the cooler, where it is chilled by liquid refrigerant evaporating at a lower temperature. The refrigerant vaporizes and is drawn into the compressor, which increases the pressure of the gas so that it may be condensed at a higher temperature in the condenser. The condenser cooling medium is warmed in the process. The condensed liquid then flows to the evaporator through a metering device. A fraction of the liquid refrigerant changes to vapor (*flashes*) as the pressure drops between the condenser and the evaporator. This flashed vapor produces no refrigeration effect, but the compressor requires energy to pump it through the cycle. The following modifications minimize the problem of flashing.

**Subcooling.** Condensed refrigerant is often subcooled to a temperature below its saturation temperature in either the subcooler section of a water-cooled condenser or a separate heat exchanger.

**Intercooling** (with multiple compression stage compressors [centrifugal and reciprocating] or compressors with intermediate pressure ports [screws]). In the case of two-stage compression, the condensed liquid is throttled to a pressure between condensing and evaporating pressures. The resulting flash gas goes to the high-stage suction of the compressor; the remaining liquid is throttled further to evaporator pressure. In the case of intermediate ported compressors, flash gas from a subcooler is injected into the compressor at a pressure level between suction and discharge pressure.

Both of these methods (sometimes combined for maximum effect) reduce flash gas and increase the net refrigeration effect per unit power consumption.

**Liquid Injection.** Condensed liquid is throttled to the intermediate pressure and injected into the second-stage suction of

---

The preparation of this chapter is assigned to TC 8.1, Positive Displacement Compresors and TC 8.2, Centrifugal Machines.

the compressor to prevent excessively high discharge temperatures and, in the case of centrifugal machines, to reduce noise. In the case of screw compressors, condensed liquid is injected into a port fixed at slightly below discharge pressure to provide oil cooling.

## COMMON LIQUID CHILLING SYSTEMS

### Simple System

The refrigeration cycle of a simple system is shown in Figure 1. Chilled water enters the cooler at 54 °F (12 °C), for example, and leaves at 44 °F (7 °C). Condenser water leaves a cooling tower at 85 °F (30 °C), enters the condenser, and returns to the cooling tower at 95 °F (35 °C). Condensers may also be cooled by air or through evaporation.

This system, with a single compressor and one refrigerant circuit employing a water-cooled condenser, is extensively used to chill water for air conditioning. It is relatively simple and compact.

Both hermetic and external drive liquid chilling machines are available. An *external drive* machine uses a compressor driven by a turbine, an engine, or an external electric motor. The compressor driver is easily accessible for repair or replacement. A drive shaft seal is necessary to isolate the refrigerant and oil from the atmosphere.

A *hermetic* unit employs a hermetic compressor with the electric motor totally enclosed in the refrigerant atmosphere. The possibility of refrigerant leakage to the outside through a shaft seal is eliminated, and motor operating noise is subdued by the housing. Since forced refrigerant cooling of the motor is very effective, smaller, less expensive motors are used. The need for a heavy external base to preserve motor-compressor shaft alignment is eliminated. Hermetic machines are less expensive than external drive machines, have slightly greater power consumption (than do otherwise identical external drive models), and are quieter. Should motor failure occur, however, the repair cost is higher.

Some type of reduced voltage starting (wye-delta, autotransformer) is often used with centrifugal and screw compressor chillers to reduce starting current because the starting torque required by these compressors is relatively low. Solid-state starters, which are stepless and compact enough to be mounted on the package, are also available.

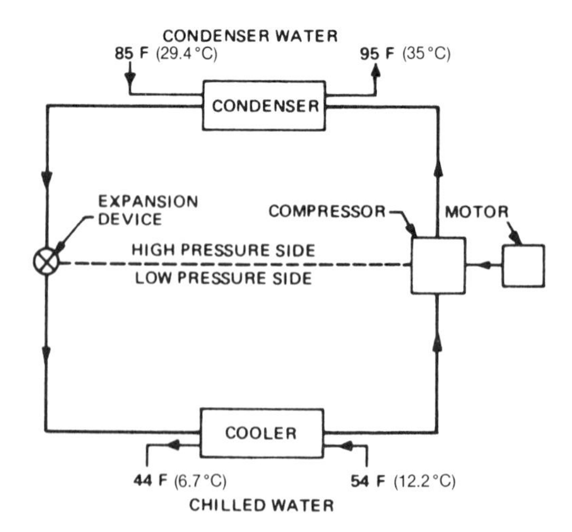


**Fig. 1 Equipment Diagram for Simple Liquid Chiller**

### Multiple Chiller System

Multiple chiller systems offer some standby capacity if repair work must be done on one chilling machine. Starting in-rush current and power costs at partial-load conditions are reduced. Maintenance can be scheduled for one chilling machine during part-load times, and sufficient cooling can still be provided by the remaining unit(s). These advantages require a significant increase in installed cost, however.

Two basic multiple chiller systems are used: *parallel* and *series* chilled water flow. In the parallel arrangement, liquid to be chilled is divided among the liquid chillers; the multiple chilled streams are combined again in a common line after chilling. As the cooling load decreases, one unit may be shut down, but the remaining unit(s) must then provide colder-than-design chilled liquid so that when all streams combine (including one from the idle machine), design chilled liquid supply temperature is provided in the common line.

In the case of water chilling, when the design chilled water temperature is above approximately 45 °F (7 °C), all units should be controlled by the combined exit water temperature (or by the return water temperature, as is often used with reciprocating chillers), since overchilling will not cause dangerously low water temperature in the operating machine(s). Return chilled water temperature can be used to cycle one unit off when it drops below a value corresponding to a capacity that can be matched by the remaining units.

When the design chilled water temperature is below about 45 °F (7 °C), each machine should be controlled by its own chilled water temperature, both to prevent dangerously low evaporator temperatures and to avoid frequent shutdowns by the low-temperature cutout. In this case, the temperature differential setting of the RWT (*return water temperature*) must be adjusted carefully to prevent short cycling caused by the step increase in chilled water temperature when one chiller is cycled off. These control arrangements are shown in Figures 2 and 3.

The series arrangement is better than the parallel in most respects, except that chilled liquid pressure drop may be higher if shells with fewer liquid-side passes or baffles are not available. No overchilling by either unit is ever required, and compressor power consumption is lower than it is for the parallel arrangement at partial loads. Since the evaporator temperature never drops below the design value (because no overchilling is necessary), the chances of evaporator freezeup are minimized. However, the chiller should still be protected by a low-temperature safety control.

When condensers are water-cooled, they are best piped in series counterflow so that the lead machine is provided with warmer condenser and chilled water and the lag machine ex-

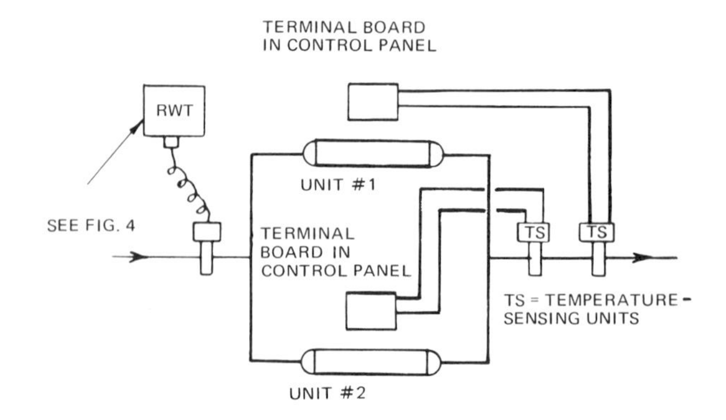


**Fig. 2 Parallel Operation High Design Water Leaving Coolers (Approximately 45 °F [7.2 °C] and above)**

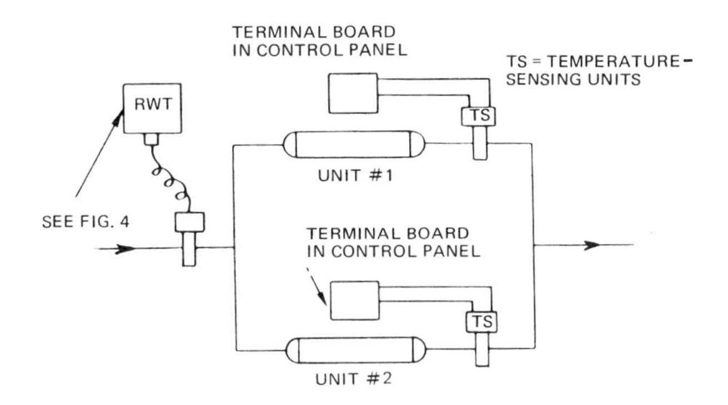


**Fig. 3 Parallel Operation Low Design Water Leaving Coolers (Approximately 44°F [6.7°C] or below)**

periences colder entering condenser and chilled water. Refrigerant compression for each unit is thus nearly the same. If about 55% of design cooling capacity is assigned to the lead machine and about 45% to the lag machine, identical units can be used. In this way, either machine can provide the same standby capacity if the other is down, and lead and lag machines may be interchanged to equalize the number of operating hours on each. (The controls may be arranged to allow shutdown of either the lead or lag unit at part-load times as shown in Figure 4.

A control system for two machines in series is shown in Figure 5. (On reciprocating chillers, RWT sensing is usually used instead of leaving water sensing because it allows closer temperature control.) Both units are modulated to minimum capacity, at which time one unit shuts down, leaving 100% load on the operating machine.

One machine should be shut down as soon as possible, with the remaining unit carrying the full load. This not only reduces the number of operating hours on a unit, but also leads to less total power consumption because the COP tends to decrease below the full load value when unit load drops much below 50%.

## Heat-Recovery Systems

Any building or plant requiring the simultaneous operation of heat-producing and cooling equipment is a potential heat-

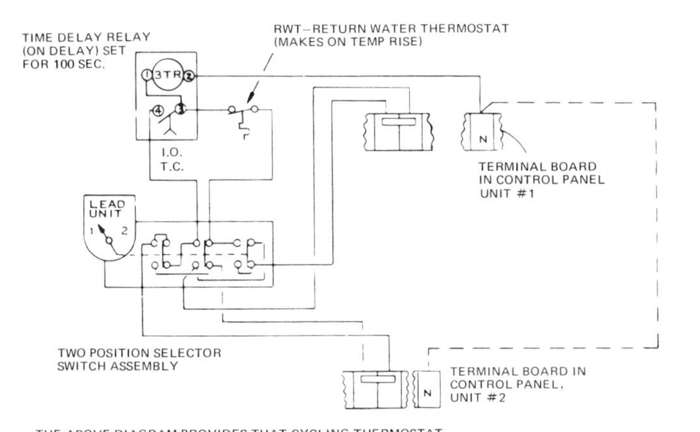


**Fig. 4 Parallel or Series-Lead Selector and Cycling Thermostat (Two Units)**

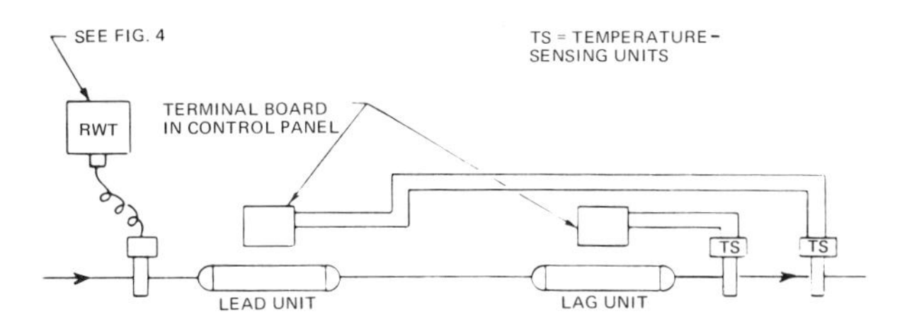


**Fig. 5 Series Operation**

recovery installation. Heat-recovery equipment should be considered for all new or retrofit installations. In some cases, the installed cost may be less because of the elimination or reduction of both heating equipment and the space required for it. Special packaged heat-recovery units are available that produce water temperatures above 200°F (93°C); standard packages are available to produce water temperatures to 158°F (70°C).

Heat-recovery systems extract heat from chilled liquid and reject some of that heat, plus the energy of compression, to a warm-water circuit for reheat or heating. Air-conditioned spaces thus furnish heating for other spaces in the same building. During the full-cooling season, all heat must be rejected outdoors, usually by a cooling tower. During spring or fall, some heat is required inside, while a portion of the heat extracted from the air-conditioned spaces must be rejected outside simultaneously.

Heat recovery offers the user low heating costs and reduces space requirements for mechanical equipment. A control system must be designed carefully, however, to take the greatest advantage of the recovered heat and to maintain proper temperature and humidity in all parts of the building. Chapter 6 of the 1987 HVAC Volume covers balanced heat-recovery systems.

Since cooling tower water is not satisfactory for heating coils, a separate, closed warm-water circuit with another condenser bundle or auxiliary condenser, in addition to the main water chiller condenser, must be provided. In some cases, it is economically feasible to use a standard condenser and a closed-circuit water cooler.

A recommended control scheme is shown in Figure 6. The heating water temperature is controlled by a cooling tower bypass valve, which modulates the flow of condenser cooling water to the tower. An outside air thermostat resets the hot water control point upward as outdoor temperature drops. Also, the thermostat should reset the chilled water temperature control point upward on colder days. In this way, extra power is not consumed unnecessarily by the compressor in attempting to maintain summer design coil temperatures during dry, cold outdoor conditions.

# EQUIPMENT SELECTION

The largest factor in total liquid chiller owning cost is the cooling load size; therefore, an accurate calculation of total needed liquid chiller capacity should be made. The practice of adding a 10 to 20% safety factor to load estimates is not only unnecessary, because of the availability of accurate load estimating methods, but it also proportionately increases costs related to equipment purchase, installation, and poor efficiency resulting from wasted power. Oversized equipment can also cause operational difficulties such as frequent on-off cycling or surging of centrifugal machines at low loads. The penalty for a small underestimation of cooling load, however, is not serious. On the few design load days of the year, an increase in chilled liquid temperature is often acceptable.

The primary criterion for equipment selection is total owning cost, which is composed of the following:

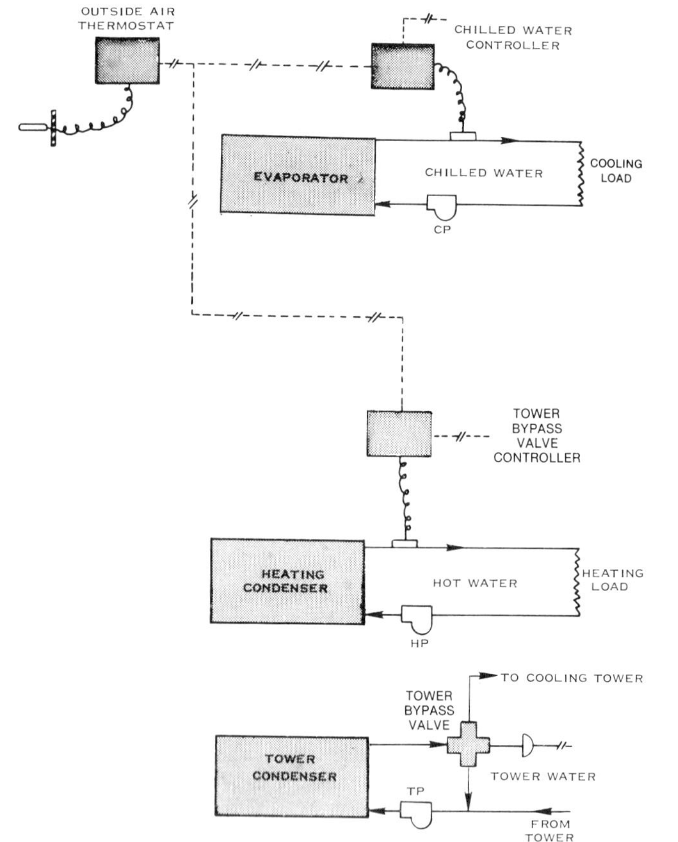


**Fig. 6 Heat-Recovery Control System**

1. **Purchase Price.** Each machine type and/or manufacturer's model should include all the necessary auxiliaries such as starters and vibration mounts. If these are not included, their price should be added to the base price.
2. **Installation Cost.** Factory-packaged machines are both less expensive to install and usually considerably more compact, resulting in space savings. The cost of field assembly of field-erected chillers must also be evaluated.
3. **Energy Cost.** Using an estimated load schedule and part-load power consumption curves furnished by the manufacturer, a year's energy cost should be calculated.
4. **Maintenance Cost.** Each bidder may be asked to quote on a maintenance contract on a competitive basis.
5. **Insurance and Taxes.** Hermetic units often require higher insurance premiums.

The most used criterion is purchase price. However, the cost of energy is causing the operating cost to be carefully scrutinized in certain sectors. Package arrangements and accessories, which offer increased operating economy, are increasing in use and should continue to do so.

When heat-recovery package chillers are considered, it is necessary to investigate beyond the specific piece of equipment and its accessories and compare system cost and performance. For instance, the heat-recovery chiller installed cost should be compared with the installed cost of a chiller plus a separate heating system. The energy costs of the alternative methods should be calculated. Maintenance, standby arrangement, relationship of heating to cooling loads, effect of package selection on sizing, and type of peripheral equipment must be factored into a reasonable decision.

Condensers and coolers are often available with either *liquid heads,* which require disconnection of water piping for tube access and maintenance, or *marine-type water boxes,* which permit tube access with water piping intact. The former is considerably lower in price. The cost of disconnecting piping must be greater than the additional cost of marine-type water boxes to justify their use. However, it is only necessary to install a union or flange connection in the piping to facilitate the removal of heads.

The above approaches are helpful in choosing between equipment types and makes. Chapter 49 of the 1987 HVAC Volume is recommended for further information. As a guide, the following is suggested for determining the types of liquid chillers that are generally used for air conditioning:

| | |
|---|---|
| Up to 25 tons (88 kW) | —Reciprocating |
| 25 to 80 tons (88 to 280 kW) | —Screw or Reciprocating |
| 80 to 200 tons (280 to 700 kW) | —Screw, Reciprocating, or Centrifugal |
| 200 to 800 tons (700 to 2800 kW) | —Screw or Centrifugal |
| Above 800 tons (2800 kW) | —Centrifugal |

For air-cooled condenser duty, brine chilling, or other high head applications from 80 to about 200 tons (280 to about 700 kW), reciprocating and screw liquid chillers are more frequently installed than centrifugals. Centrifugal liquid chillers (particularly multistage machines), however, may be applied quite satisfactorily at high head conditions.

Factory packages are available to about 2400 tons (8400 kW), and field-assembled machines to about 10,000 tons (35 MW).

## CONTROL

### Liquid Chiller Controls

The *chilled liquid temperature sensor* sends an air pressure (pneumatic control system) or electrical (electronic control system) signal to the control circuit, which modulates compressor capacity in response to leaving or return chilled liquid temperature change with load.

Compressor capacity adjustment is accomplished differently on the following liquid chillers:

*Reciprocating chillers* use combinations of cylinder unloading and on-off compressor cycling of single or multiple compressors.

*Centrifugal liquid chillers,* driven by electric motors, commonly use adjustable prerotation vanes. Turbine drives allow the use of speed control in addition to prerotation vane modulation, reducing power consumption at partial loads.

*Screw compressor liquid chillers* use a slide valve to adjust the length of the compression path.

In air-conditioning applications, most centrifugal and screw compressor chillers modulate from 100% to approximately 10% load.

Reciprocating chillers are available with simple on-off cycling control in small capacities and with multiple steps of unloading down to 12.5% in the largest multiple compressor units. Most intermediate sizes provide unloading to 50, 33, or 25% capacity. When continuous operation below the minimum possible capacity is desired, hot-gas bypass can reduce capacity to as low as 0%.

The *water temperature controller* is a thermostatic device that unloads or cycles the compressor(s) when the cooling load drops below minimum unit capacity. An *anti-recycle timer* is sometimes used to limit starting frequency.

On centrifugal or screw compressor chillers, a *current limiter* or *demand limiter* limits compressor capacity during periods of possible high power consumption (such as pulldown) to prevent current draw from exceeding the design value; such a limiter can be set to limit demand, as described in the section on "Centrifugal Liquid Chillers."

### Controls That Influence the Liquid Chiller

Condenser coolant control is required to regulate head pressure. Normally, a control senses the temperature of the water leaving a cooling tower and accomplishes this control by cycling fans, dampers, or a water bypass around the tower.

A flow-regulating valve is another common means of control. The orifice of this valve modulates in response to condenser pressure. For example, a reduction in pressure decreases the coolant flow, which, in turn, raises the condenser pressure to the desired level.

A reciprocating chiller usually has a thermal expansion valve, which requires a restricted range of head pressure to avoid starving the evaporator (at low head).

An expansion valve(s) usually controls a screw compressor chiller. Cooling tower water temperature can be allowed to fall with decreasing load from the design condition to the chiller manufacturer's recommended minimum limit.

A thermal expansion valve may control a centrifugal chiller at low capacities, while higher tonnage machines employ a high pressure float, orifice(s), or even a low-side float valve to control refrigerant liquid flow to the cooler. These latter types of controls allow relatively low condenser pressures, particularly at partial loads. Also, a centrifugal machine may surge if head pressure is not reduced when cooling load decreases. In addition, low head pressure reduces compressor power consumption and operating noise. For these reasons, in a centrifugal installation, cooling tower water temperature should be allowed to fall naturally with decreasing load and wet-bulb temperature, except that the liquid chiller manufacturer's recommended minimum limit must be observed.

### Safety Controls

Some or all of the cutouts listed below may be provided in a liquid chilling package to stop the compressor(s) automatically. Cutouts may be manual or automatic reset.

1. **High Condenser Pressure.** This pressure switch opens if the compressor discharge pressure reaches the value prescribed in the ANSI/ASHRAE *Standard* 15-1978, *Safety Code for Mechanical Refrigeration.*
2. **Low Refrigerant Pressure (or Temperature).** This device opens when evaporator pressure (temperature) reaches a minimum safe limit.
3. **High Oil Temperature.** This device protects the compressor if loss of oil cooling occurs or if a bearing failure causes excessive heat generation.
4. **High Motor Temperature.** If loss of motor cooling or overload because of a failure of operating controls occurs, this device shuts down the machine. It may consist of direct-operating bimetallic thermostats, thermistors, or Balco wire sensors embedded in the stator windings; it may be located in the discharge gas stream of the compressor.
5. **Motor Overload.** Some small reciprocating compressor hermetic motors may use a directly operated overload in the power wiring to the motor. Some larger motors use pilot-operated overloads. Centrifugal and screw compressor motors generally use starter overloads to protect against overcurrent.
6. **Low Oil Sump Temperature.** This switch is used either to protect against an oil heater failure or prevent starting after a prolonged shutdown before the oil heaters have had time to drive off refrigerant dissolved in the oil.
7. **Low Oil Pressure.** To protect against clogged oil filters, blocked oil passageways, loss of oil, or an oil pump failure, a switch shuts down the compressor when oil pressure drops below a minimum safe value or if sufficient oil pressure is not developed shortly after compressor startup.
8. **Chilled Liquid Flow Interlock.** This device may not be furnished with the liquid chilling package but is needed in the external piping to protect against a cooler freezeup in the event of a liquid flow stoppage.
9. **Condenser Water Flow Interlock.** This device is sometimes used in the external piping.
10. **Low Chilled Liquid Temperature.** Sometimes called *freeze protection* on reciprocating chillers, this cutout operates at a minimum safe value of leaving chilled liquid temperature to prevent cooler freezeup in the case of an operating control malfunction.
11. **Relief Valves.** In accordance with the ANSI/ASHRAE *Standard* 15-1978, *Safety Code for Mechanical Refrigeration,* relief valves or rupture disks, set to relieve at the shell design working pressure, must be provided on most pressure vessels. Relief valves are connected to the high-pressure side of the chilling machine, while rupture disks may be connected to the high or low side. Fusible plugs may also be used in some locations. Pressure relief devices should be vented outdoors in accordance with ANSI/ASHRAE *Standard* 15-1978.

## STANDARDS

ARI *Standards* 550-83, *Centrifugal or Rotary Water-Chilling Packages*, and 590-81, *Reciprocating Water-Chilling Packages*, provide guidelines for the rating of centrifugal and reciprocating liquid chilling machines, respectively.

The design and construction of refrigerant pressure vessels are governed by the ASME *Boiler and Pressure Vessel Code,* Section VIII, except when design working pressure is 15 psig (103 kPa ) or less (as is usually the case for R-11 and R-113 liquid chilling machines). When the water side of a condenser or cooler contains more than 120 gal (0.45 $m^3$) in a single compartment, the ASME code also governs its construction.

The ANSI/ASHRAE *Standard* 15-1978, *Safety Code for Mechanical Refrigeration,* applies to all liquid chillers. Methods for the measurement of unit sound levels are described in ANSI *Standard* S1.2.

## METHODS OF TESTING

All tests of reciprocating liquid chillers for rating or verification of rating should be conducted in accordance with ASHRAE *Standard* 30-78, *Methods of Testing Liquid Chilling Packages.*

Centrifugal or screw liquid chiller ratings should be derived and verified by test in accordance with ARI *Standard* 550-83.

## GENERAL MAINTENANCE

Listed below are some of the general maintenance specifications that apply equally to reciprocating, centrifugal, and screw chillers. The equipment should be neither overmaintained nor neglected. A preventive maintenance schedule should be established; the items covered can vary with the nature of the application.

The following list is intended as a guide; in all cases, the manufacturer's specific recommendation should be followed.

#### Continual Monitoring

Condenser water treatment—treatment is determined specifically for the condenser water used.

Operating conditions—daily log sheets are recommended.

**Periodic Checks**

Leak check
System dryness
Oil level
Oil filter pressure drop
Refrigerant quantity or level
System pressures and temperatures
Water flows
Expansion valves operation

**Regularly Scheduled Maintenance**

Condenser and oil cooler cleaning
Calibrating pressure, temperature, and flow controls
Tightening wires and power connections
Inspection of starter contacts and action
Dielectric checking of hermetic and open motors
Oil filter and drier change
Analysis of oil and refrigerant
Seal inspection
Partial or complete valve or bearing inspection, as per manufacturer's recommendations

# PART II: RECIPROCATING LIQUID CHILLERS

## EQUIPMENT DESCRIPTION

### Components and Their Function

The **Reciprocating Compressor,** described in Chapter 12, is a positive displacement machine that maintains fairly constant volume flow rate over a wide range of pressure ratios. The following three types of compressors are commonly used in liquid chilling machines:

1. Welded hermetic, to about 25 tons (90 kW) chiller capacity
2. Semihermetic, to about 200 tons (700 kW) chiller capacity
3. Direct-drive open, to about 200 tons (700 kW) chiller capacity

**Open Liquid Chillers** are usually more expensive than *hermetics* and are declining in use for this reason. *Hermetic motors* are generally suction gas cooled; the rotor is mounted on the compressor crankshaft.

**Condensers** may be evaporative, air- or watercooled. Water-cooled versions may be either tube-in-tube or shell-and-coil for low cost, or shell-and-tube for compactness. Most shell-and-tube condensers can be repaired, while others must be replaced if a refrigerant-side leak occurs.

Air-cooled condensers are much more common than evaporative condensers. Less maintenance is needed for air-cooled heat exchangers than for the evaporative type. Remote condensers can be applied with condenserless packages. (Information on condensers can be found in Chapter 15.)

**Coolers** are usually direct-expansion, in which refrigerant evaporates while it is flowing inside tubes and chilled liquid is cooled as it is guided several times over the outside of the tubes by shell-side baffles. Tube-in-tube coolers are sometimes used for low cost with small machines; they offer low cost when repairability and required space are not important criteria. (A detailed description of coolers can be found in Chapter 16.) The *thermal expansion valve* modulates refrigerant flow from the condenser to the cooler to maintain enough suction superheat to prevent any unevaporated refrigerant liquid from reaching the compressor. Excessively high values of superheat are avoided so that unit capacity is not reduced. (For additional information, see Chapter 19.)

Oil cooling is not usually required for air conditioning. However, oil cooling may be obtained by a refrigerant-cooled coil in the crankcase or by a water-cooled oil cooler. Oil coolers are often used in conjunction with low suction temperatures or high-pressure ratio applications when extra oil cooling is needed.

### Capacities and Types Available

Available capacities range from about 2 to 200 tons (7 to 700 kW). Multiple reciprocating compressor units have become popular for the following reasons:

1. The number of capacity increments is greater, resulting in closer liquid temperature control, lower power consumption, less current in-rush during starting, and extra standby capacity.
2. Multiple refrigerant circuits are employed, resulting in the potential for limited servicing or maintenance of some components while maintaining cooling.

### Selection of Refrigerant

If compressor displacement required for a given capacity is minimized, then compressor size can be reduced and total chiller cost is low. For this reason, R-22 is most popular for air conditioning. R-717 has the same advantage, but its odor and incompatibility with copper makes it unsuitable for some applications.

R-22, R-502, and R-717 are commonly used in low temperature applications.

Although R-22 is generally used in reciprocating liquid chillers, R-12 (or R-500) is suited to high condensing temperature applications, such as heat pumps and heat recovery units, because of lower condensing pressures and discharge temperatures when compared to R-22. Condensing temperature can be extended to provide elevated outlet water temperatures without exceeding compressor design working pressures, which are generally based on R-22 pressures.

R-12 and R-502 are suited to high condensing and low evaporating temperature applications because discharge gas temperature does not increase rapidly with compression ratio.

## PERFORMANCE CHARACTERISTICS AND OPERATING PROBLEMS

A distinguishing feature of the reciprocating compressor is its pressure rise versus capacity characteristic. Pressure rise has only a slight influence on the volume flow rate of the compressor, and, therefore, a reciprocating liquid chiller retains nearly full cooling capacity, even on above-design wet-bulb days. It is well suited for air-cooled condenser application and low-temperature refrigeration.

A typical characteristic is shown in Figure 7 and compared with the centrifugal and screw compressors. Methods of capacity control are furnished by the following:

1. Unloading of compressor cylinders (one at a time or in pairs)
2. On-off cycling of compressors
3. Hot-gas bypass
4. Compressor speed control
5. A combination of the above methods

Figure 8 illustrates the relationship between system demand and performance of a compressor with three steps of unloading. As cooling load drops to the left of the fully loaded compressor

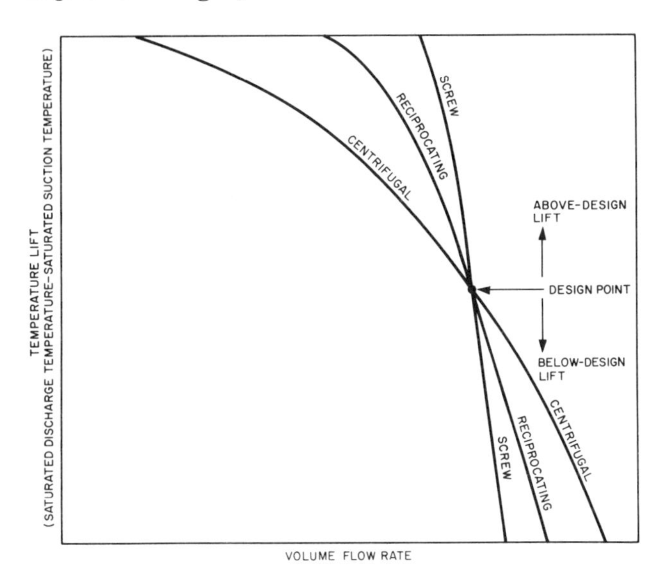


**Fig. 7 Comparison of Single-Stage Centrifugal, Reciprocating, and Screw Compressor Performance**

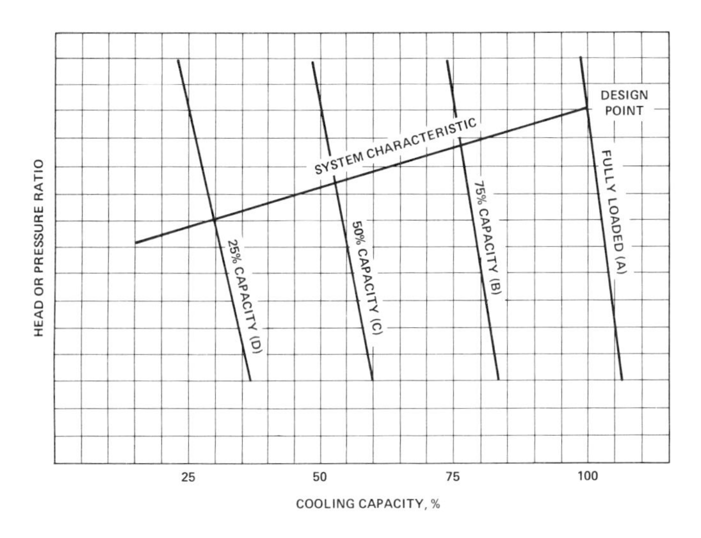


**Fig. 8 Reciprocating Liquid Chiller Performance with Three Equal Steps of Unloading**

characteristic line A, compressor capacity is reduced to that represented by line B, which produces the required refrigerant flow. Since cooling load varies continuously while machine capacity is available in fixed increments, some compressor on-off cycling or successive loading and unloading of cylinders is required to maintain fairly constant liquid temperature. In practice, a good control system minimizes the load-unload or on-off cycling frequency while maintaining satisfactory temperature control.

## METHOD OF SELECTION

### Ratings

Two types of ratings are published: The first, for a packaged liquid chiller, lists values of capacity and power consumption for many combinations of leaving condenser water and chilled water temperatures (ambient dry-bulb temperatures for air-cooled models). The second shows capacity and power consumption for different condensing temperatures and chilled water temperatures. This type of rating permits selection with a remote condenser—evaporative, watercooled, or air-cooled. Sometimes the required rate of heat rejection is also listed to aid in selection of a separate condenser.

### Power Consumption

With all liquid chilling systems, power consumption increases as condensing temperature rises. Therefore, the smallest package, with the lowest ratio of input to cooling capacity, can be used when condenser water temperature is low, the remote air-cooled condenser is relatively large, or when leaving chilled water temperature is high. The cost of the total system, however, may not be low when liquid chiller cost is minimized. Increases in cooling tower or fan coil cost will reduce or offset the benefits of reduced compression ratio.

### Fouling

A fouling allowance of 0.0005 ft$^2$ • °F • h/Btu (0.088 m$^2$ • °C/kW) is included in manufacturers' ratings in accordance with ARI *Standard* 590-81. However, fouling factors greater than 0.0005 should be considered in the selection if water conditions are other than ideal.

## SPECIFIC CONTROL CONSIDERATIONS

A reciprocating chiller is distinguished from centrifugal and screw compressor-operated chillers by its use of increments of capacity reduction rather than continuous modulation. Therefore, unique arrangements must be used to establish precise chilled liquid temperature control while maintaining stable operation free from excessive on-off cycling of compressors or unnecessary loading and unloading of cylinders.

To help provide good temperature control, return chilled liquid temperature sensing is normally used by units with steps of capacity control. The resulting flywheel effect in the chilled liquid circuit damps out excessive cycling. Leaving chilled liquid temperature sensing has the advantage of preventing excessively low leaving chilled liquid temperatures if chilled liquid flow falls significantly below the design value. It may not provide stable operation, however, if rapid changes in load are encountered.

An example of a basic control circuit for a single compressor-packaged reciprocating chiller with three steps of unloading is shown in Figure 9. The unit is started by moving the on-off switch to the *on* position. The programmed timer will start operating. Assuming that the flow switch, field interlocks, and chiller safety devices are closed, pressing the momentarily closed reset button will energize control relay C1, locking in the safety circuit and the motor starting circuit. When the timer completes its program, timer switch 1 closes and timer switch 2 opens. Timer relay TR energizes, stopping the timer motor. When timer switch 1 closes, the motor starting circuit is completed and the motor contactor holding coil is energized, starting the compressor.

The four-stage thermostat controls the capacity of the compressor in response to the demand of the system. Cylinders are loaded and unloaded by de-energizing and energizing the unloader solenoids. If the load is reduced so that the return water temperature drops to a predetermined setting, the unit shuts down until the demand for cooling increases.

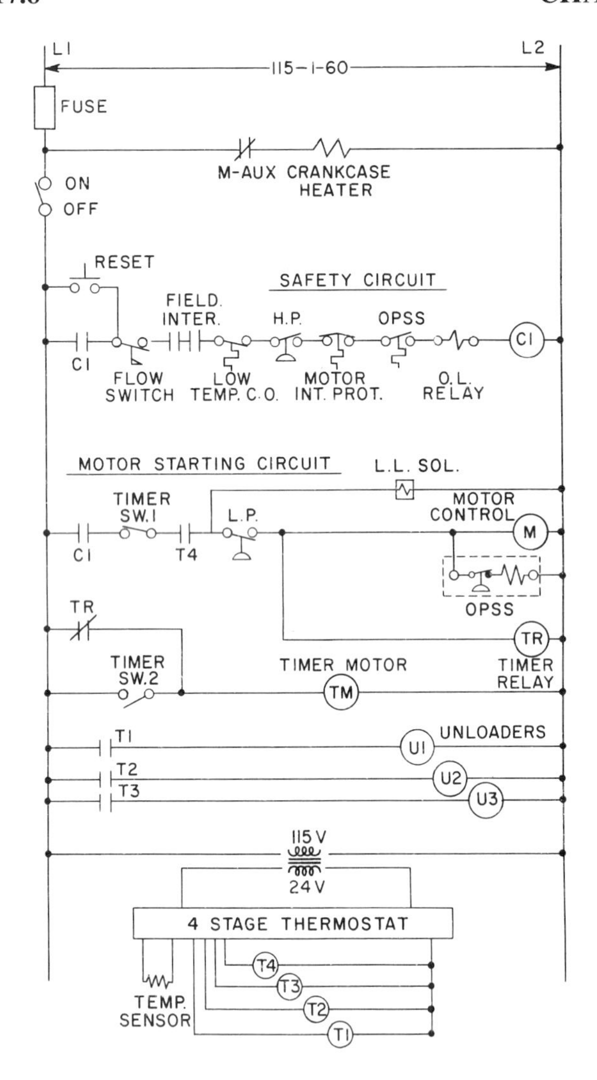


**Fig. 9 Reciprocating Liquid Chiller Control System**

Opening a device in the safety circuit will de-energize control relay C1 and shut down the compressor. The liquid line solenoid is also de-energized. Manual reset is required to restart. The crankcase heater is energized whenever the compressor is shut down.

If the automatic reset, low-pressure cutout opens, the compressor shuts down, but the liquid line solenoid remains energized. The timer relay (TR) is de-energized, causing the timer to start and complete its program before the compressor can be restarted. This prevents rapid cycling of the compressor under low head pressure conditions. A time delay low-pressure switch can also be used for this purpose with the proper circuitry.

## SPECIAL APPLICATIONS

For multiple chiller applications and a 10°F (5.6°C) chilled liquid temperature range, the use of a parallel chilled liquid arrangement is common because of the high cooler pressure drop resulting from the series arrangement. For a large (18°F [10°C]) range, however, the series arrangement eliminates the need for overcooling during operation of one unit only. Special coolers, with low water pressure drop, may also be used to reduce total chilled water pressure drop in the series arrangement.

# PART III: CENTRIFUGAL LIQUID CHILLERS

## EQUIPMENT DESCRIPTION

### Components and Their Function

The *centrifugal compressor* is described in Chapter 12 of this volume. Since it is not of the constant displacement type, it offers a wide range of capacities continuously modulated over a limited range of pressure ratios. By altering built-in design items (including number of stages, compressor speed, impeller diameters, and choice of refrigerant), it can be used in liquid chillers having a wide range of design chilled liquid temperatures and design cooling fluid temperatures. Its ability to vary capacity continuously to match a wide range of load conditions with nearly proportionate changes in power consumption makes it desirable for both close temperature control and energy conservation. Its ability to operate at greatly reduced capacity makes for more on-the-line time with infrequent starting.

The hour of the day for starting an electric-drive centrifugal liquid chiller can often be chosen by the building manager to minimize peak power demands. It has a minimum of bearing and other types of contacting surfaces that can wear; this wear is minimized by providing forced lubrication to those surfaces prior to startup and during shutdown. Bearing wear is usually more dependent on the number of startups than the actual hours of operation. Thus, by reducing the number of startups, the life of the system is extended and its maintenance cost is reduced.

Both open and hermetic compressors are selected. Open compressors may be driven by steam turbines, gas turbines or engines, or electric motors, with or without speed-changing gears. (Engine and turbine drives are covered in detail in Chapter 32, electric motor drives are covered in Chapter 31.)

Packaged electric-drive chillers may be of the open or hermetic type and use two-pole, 50 Hz, or 60 Hz polyphase electric motors, with or without speed-increasing gears. Hermetic units use only polyphase induction motors. Speed-increasing gears and their bearings, in both open and hermetic-type packaged chillers, operate in a refrigerant atmosphere, and the lubrication of their contacting surfaces is incorporated in the compressor lubrication system.

Magnetic and SCR (silicon controlled rectifier) motor controllers are used with packaged chillers. When purchased separately, the controller must meet the specifications of the chiller manufacturer to ensure adequate equipment safety. When timed step starting methods are used, the time between steps should be long enough for the motor to overcome the relatively high inertia of the compressor and attain sufficient speed to minimize the electric current drawn immediately after transition.

Flooded coolers are commonly used, although direct-expansion coolers are employed by some manufacturers in the lower capacity ranges. The typical flooded cooler uses copper tubes that are mechanically expanded into the tube sheets, and, in some cases, into intermediate tube supports, as well.

Since refrigerant liquid flow into the compressor increases power consumption, mist eliminators or baffles are often used in flooded coolers to minimize refrigerant liquid entrainment in the suction gas. (Additional information on coolers for liquid chillers can be found in Chapter 16.)

The condenser is generally water-cooled, with refrigerant condensing on the outside of copper tubes. Large condensers may have refrigerant drain baffles, which direct the condensate from within the tube bundle directly to the liquid drains, reducing the thickness of the liquid film on the lower tubes.

Air-cooled condensers can be used with units that use higher pressure refrigerants, but with considerable increase in power consumption. Operating costs should be compared with systems using cooling towers and condenser water circulating pumps.

System modifications, including subcooling and intercooling, as described in Part I of this chapter, are often used to conserve energy. (Additional information concerning thermodynamic cycles can be found in Chapter 1 of the 1985 FUNDAMENTALS Volume. For information on condensers and subcoolers, see Chapter 15 of this volume.)

Some units combine the condenser, cooler, and refrigerant flow control in one vessel; a subcooler may also be incorporated.

## Capacities and Types Available

Centrifugal packages are currently available from about 80 to 2400 tons (280 kW to 8.4 MW) at nominal conditions of 44°F (6.7°C) leaving chilled water temperature and 95°F (35°C) leaving condenser water temperature. This upper limit is continually increasing. Field-assembled machines extend to about 10,000 tons (35 MW). Single-stage, two-stage internally geared machines, and two-stage direct-drive machines are commonly used in packaged units. Electric motor-driven machines constitute the majority of units sold.

Units with hermetic motors, cooled by refrigerant gas or liquid, are offered from about 80 to 2000 tons (280 kW to 7 MW). Open-drive units are not offered by all hermetic manufacturers in the same size increments but are generally available from 80 to 10,000 tons (280 kW to 35 MW).

## Selection of Refrigerant

The selection of refrigerant is an important consideration in determining equipment and operating costs. In choosing refrigerant, consider coefficient of performance, operating pressures, flow rate, heat-transfer properties, stability, and toxicity or flammability. Halogenated hydrocarbons are normally used as refrigerants because they offer reasonable characteristics with respect to the application.

The centrifugal compressor is particularly suitable for handling relatively high flow rates of suction vapor. As the volumetric flow of suction vapor increases with higher capacities and lower suction temperatures, the higher pressure refrigerants, for example R-12 and R-22, become more popular. The physical size and weight of the refrigerant piping, and, often, other components of the refrigeration system, are reduced by the use of higher pressure refrigerants. In order of decreasing volumetric flow and increasing pressures, the commonly used refrigerants are R-113, R-11, R-114, R-12 or R-500, and R-22.

Pressure vessels for use with R-113 and R-11 usually have a design working pressure of 15 psi (103.4 kPa) on the refrigerant side. The vessel shells are usually stronger than necessary for this requirement to ensure sufficient rigidity and prevent collapse under vacuum.

The thermal stability of the refrigerant and its compatibility with materials it contacts are also important. Special attention is given to the selection of elastomers and electrical insulating materials because many common materials of this sort are affected by the refrigerants. (Additional information concerning refrigerants can be found in Chapters 16 and 17 of the 1985 FUNDAMENTALS Volume.)

## PERFORMANCE AND OPERATING CHARACTERISTICS

Figures 10 and 11 assume isentropic conditions to show reversible or specific work available from a compressor, and required

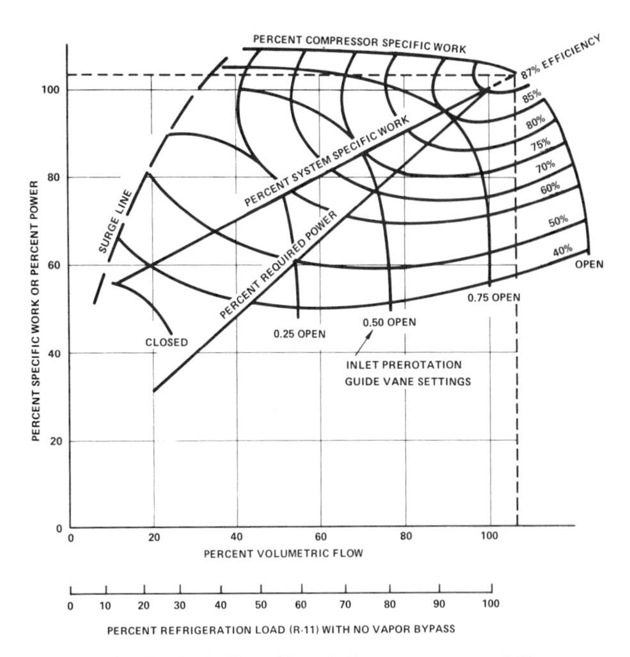


**Fig. 10 Typical Centrifugal Compressor and System at Constant Speed**

for a system, as functions of volumetric flow. Figure 10 shows a compressor's performance at constant speed with various inlet guide vane settings. Figure 11 shows that compressor's performance at various speeds with open inlet guide vanes. The required system specific work can be obtained from refrigerant tables and charts for particular suction and condensing temperatures obtained from cooler and condenser performance data.

However, for comparative purposes, the *temperature lift* (con-

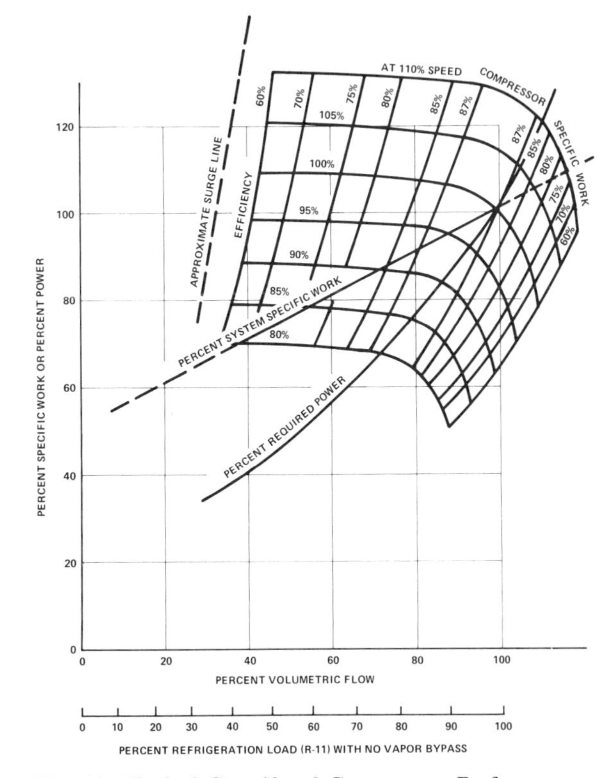


**Fig. 11 Typical Centrifugal Compressor Performance at Various Speeds**

densing temperature minus evaporating temperature) closely represents the specific work when a particular refrigerant is considered. At less than full load, the percent volumetric flow will usually be a little less than the percent load, unless some vapor passes through or bypasses the refrigerant expansion device. The use of single or series orifices as expansion devices can provide some bypass vapor at reduced loads, which makes the percent volumetric flow equal to or greater than the percent load.

A *hot-gas bypass* allows the compressor to operate at zero load. This feature is a particular advantage for such intermittent industrial applications as the cooling of quenching tanks. Bypass vapor obtained by either method increases the power consumption. At light loads, some bypass vapor, if introduced into the cooler below the tube bundle, may increase the evaporating temperature by agitating the liquid refrigerant and thereby more thoroughly wetting the tube surfaces.

Capacity is modulated at constant speed by automatic adjustment of prerotation vanes that whirl the refrigerant gas at the impeller eye. This effect matches system demand by shifting the compressor performance curve downward and to the left (as shown in Figure 10). Compressor efficiency, when unloaded in this manner, is superior to suction throttling. Some manufacturers also automatically reduce diffuser width or throttle the impeller outlet with decreasing load.

Speed control for a turbine-driven centrifugal compressor offers even lower power consumption, down to about 50% capacity, at which time speed may be reduced gradually to prevent surging, and control is transferred to the prerotation vanes for operation below 50% load.

Figure 12 shows how temperature lift varies with load. A typical reduction in entering condenser water temperature of 10°F (5.6°C) helps to reduce temperature lift at low load. Other factors producing lower lift at reduced loads are (1) the reduction in condenser cooling water range (the difference between entering and leaving temperatures, resulting from decreasing heat rejection); (2) the decrease in temperature difference between

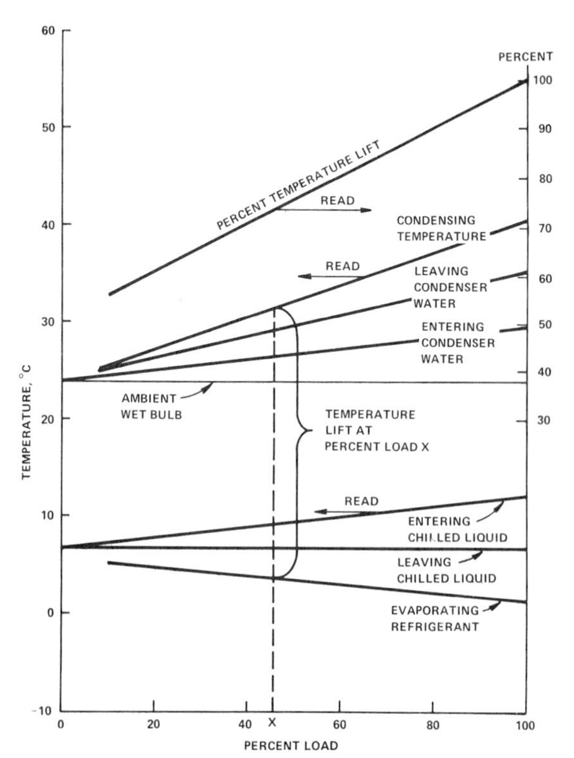


**Fig. 12 Temperature Relations in a Typical Centrifugal Liquid Chiller**

condensing refrigerant and leaving condenser water; and (3) a similar decrease between evaporating refrigerant and leaving chilled liquid temperature.

In many cases, the actual reduction in temperature lift is even greater because the wet-bulb temperature usually drops with the cooling load, producing a greater decrease in entering condenser water temperature.

In Figure 10, the temperature lift versus percent load of Figure 12 is translated into percent specific work versus percent volumetric suction vapor flow and superimposed on a constant speed centrifugal performance map. Below the horizontal scale for percent volumetric flow, corresponding percentages of the refrigeration load are indicated. The compressor map has curves for several inlet guide vane settings from *closed* to *open,* showing the developed specific work versus percent volumetric flow with the 100% values of both specific work and volumetric flow corresponding to design load conditions. At any other load, some specific fractional vane opening compressor specific work curve will intersect the system specific work curve. Note that if the system specific work curve is extended, it intersects the open compressor curve at 106% system volumetric flow and 103% system specific work. The compressor has some extra capacity.

The compressor map also has a number of contour lines of constant efficiency. At each point along the percent system specific work curve, the efficiency can be interpolated. Thus, for any specific percent load, the percent volumetric suction vapor flow, the specific work, and the compressor efficiency can be obtained from Figure 10. Using the evaporating temperature shown in Figure 12 and the refrigerant tables and/or chart, the specific volume of the suction vapor can be obtained.

*Specific work,* by definition, is the reversible work per unit mass. The total work per unit mass is therefore the specific work divided by the efficiency, neglecting the mechanical losses in bearings, gears, etc., because such losses are usually not included in the compressor efficiency. Power (the work done per unit time) is thus the product of the work per unit mass and the mass flow rate, the latter being the volumetric flow, divided by the specific volume, *i.e.*:

$$P_{th} = W_s Q_1 / \eta_s V_1$$

*where*

$P_{th}$ = theoretical power
$W_s$ = isentropic specific work
$\eta_s$ = isentropic efficiency
$Q_1$ = suction volumetric flow
$V_1$ = specific volume of suction vapor

Figure 10, from such calculations, shows the percent power curve. Mechanical losses can be added or theoretical power may be divided by the mechanical efficiency to obtain the total power required.

The above analysis is for single-stage compressors or for multistage units without interstage cooling or introduction of vapor between stages. The specific work shown for multistage units is then the total specific work developed and the volumetric flow of the suction vapor to the first stage. Where interstage flash coolers are used, each stage must be handled separately.

Figure 11 shows the effects resulting from changes in compressor speed for a compressor with similar performance characteristics as the one shown in Figure 10. The load curve of Figure 10 is also indicated in Figure 11. Note that the compressor's design parameters, including impeller diameters, flow areas, and 100% speed, are selected to match the system specific work curve at a volumetric flow slightly greater than for obtaining peak efficiency. As the load decreases from 100% and the speed progressively lessens, the operating efficiency remains high for a considerable range in capacity. Further reductions in capacity and speed show progressively lower efficiencies and a progression toward instability as indicated by the approximate surge line. The curved nature of the Percent Required Power Curve shows the effects resulting from the changes in efficiency along the load line as the specific work and capacity vary.

As stated earlier, speed control is usually used from 100% down to about 50% load; below 50%, inlet vane control is used with performance similar to that shown in Figure 10. Power consumption is reduced when the coldest possible condenser water is used, consistent with the chiller manufacturer's recommended minimum condenser water temperature. In cooling tower applications, minimum water temperatures should be controlled by a cooling tower bypass and/or by cooling tower fan control, not by reducing the water flow through the condenser. Maintaining a high flow rate at lower temperatures minimizes fouling and power requirements.

Surging occurs when the system specific work becomes greater than the compressor developed specific work or above the surge line indicated in Figures 10 and 11.

Excessively high temperature lift and corresponding specific work commonly originate from (1) excessive condenser or evaporator water-side fouling beyond the specified allowance; (2) inadequate cooling tower performance and higher-than-design condenser water temperature; and (3) noncondensables in the condenser, which increase condenser pressure. When surging occurs, correct the above conditions that exist.

## METHOD OF SELECTION

### Ratings

A refrigeration system with specified details is chosen from selection tables for given capacities and operating conditions. Rating tables differ from selection tables in that they list the capacities and operating data for each refrigeration system under various operating conditions, often with specific details for the listed conditions. The details specified for centrifugal systems include the number of passes in each of the heat exchangers and may include changes in rated motor kilowatt capacity or turbine size, code indication for driving gear ratio, and code indication for impeller diameters.

The maximum number of condenser and cooler water passes should be used, without producing excessive water pressure drop. The greater the number of water-side passes, the less the power consumption. Sometimes a slight reduction in condenser water flow (and slightly higher leaving water temperature) allows a better selection (lower power consumption or smaller model) than will the choice of fewer water passes when a rigid pressure-drop limit exists.

### Fouling

In accordance with ARI *Standard* 550-83, a fouling allowance of 0.0005 $h \cdot ft^2 \cdot {}^\circ F/Btu$ ($0.088\ m^2 \cdot {}^\circ C/kW$) is included in manufacturers' ratings. (Chapter 15, Part I, has further information about fouling factors.)

To reduce fouling, a minimum water velocity of about 3.3 ft/s (1 m/s) is recommended in coolers or condensers. Maximum water velocities exceeding 7 ft/s (2.1 m/s) are not recommended because of potential erosion problems.

Proper water treatment and regular tube cleaning are recommended for all liquid chillers to reduce power consumption and operating problems. See Chapter 53 in the 1987 HVAC Volume for water treatment information.

Continuous or daily monitoring of the quality of the condenser water is desirable. Checking the quality of the chilled liquid is also desirable. The intervals between checks become greater as the possibilities for fouling contamination become less—for example, closed-loop water-circulating systems for air conditioning, where an annual check should be sufficient. Corrective treatment is required, and periodic, usually annual, cleaning of the condenser tubes usually keeps fouling within the above specified allowance. In applications where more frequent cleaning is desirable, an on-line cleaning system may be economically viable.

### Noise and Vibration

An important consideration in the application of large equipment is the noise generated and transmitted to adjoining areas. A standard for sound measurement and rating has been issued by the Air-Conditioning and Refrigeration Institute in ARI *Standard* 575-87. Additional information on chiller noise, measurement, and control is given in Chapter 7 of the 1985 FUNDAMENTALS Volume and Chapter 52 of the 1987 HVAC Volume.

The chiller manufacturer's recommendations for mounting should be followed to prevent transmission or amplification of vibration to adjacent equipment or structures. Auxiliary pumps, if not connected with flexible fittings, can induce vibration of the centrifugal unit, especially if the rotational speed of the pump is nearly the same as either the compressor prime mover or the compressor. Flexible tubing becomes less flexible when it is filled with liquid under pressure and some vibration can still be transmitted.

## SPECIFIC CONTROL CONSIDERATIONS

The section on "Control" in Part I describes the chilled liquid temperature sensor. In centrifugal systems, it is usually placed in thermal contact with the leaving chilled water. In electrical control systems, the electrical signal is transmitted to an electronic control module, which controls the operation of an electric motor or motors positioning the capacity controlling inlet guide vanes. A control limiter is usually included on electric motor-driven machines. An electrical signal from a current transformer in the compressor motor controller is sent to the electronic control module. The module receives indications of both the leaving chilled water temperature and the compressor motor current. The portion of the electronic control module responsive to motor current is called the control limiter. It overrides the demands of the temperature sensor.

The inlet guide vanes, independent of the demands for cooling, do not open more than the position that results in the current setting of the control limiter. Pneumatic capacity controls operate in a similar manner. The chilled liquid temperature sensor provides a pneumatic signal. The controlling module receives both that signal and the motor current electrical signal and controls the operation of a pneumatic motor or motors positioning the inlet guide vanes. Both controlling systems have sensitivity adjustments.

The control limiter on most machines can limit current draw during periods of high electrical demand charges. This control can be set from about 40 to 100% of full load amperes. Whenever power consumption is limited, cooling capacity is correspondingly reduced. If cooling load is only 50% of the full value, the current (or demand) limiter can be set at 50% without loss of cooling. By setting the limiter at 50% of full current draw, any subsequent high demand charges are prevented during pulldown after startup. Even during periods of high cooling load, it may be desirable to limit electrical demand if a small increase in chiller liquid temperature is acceptable. If the temperature continues to decrease after the capacity control has reached its minimum position, a low-temperature control stops the compressor and restarts it when a rise in temperature indicates the need for cooling. Manual controls may also be provided to bypass the temperature control. Provision is included to ensure that the capacity control is at its minimum position when the compressor starts to provide an unloaded starting condition.

Additional operating controls are needed for appropriate operation of oil pumps, oil heaters, purge units, and refrigerant transfer units. An antirecycle timer should also be included to prevent frequent motor starts. Multiple unit applications require additional controls for capacity modulation and proper sequencing of units. (See the "Multiple Chiller System" section in Part I.)

Safety controls protect the unit under abnormal conditions. Safety cutouts that may be required are high condenser pressure, low evaporator refrigerant temperature or pressure, low oil pressure, high oil temperature, high motor temperature, and high discharge temperature. Auxiliary safety circuits are usually provided on packaged chillers. At installation, the circuits are field wired to field-installed safety devices, including auxiliary contacts on the pump motor controllers and flow switches in the chilled water and condenser water circuits. Safety controls are usually provided in a lockout circuit, which will trip out the compressor motor controller and prevent automatic restart. The controls reset automatically, but the circuit cannot be completed until a manual reset switch is operated and the safety controls return to their safe positions.

## AUXILIARIES AND SPECIAL APPLICATIONS

### Auxiliaries

Purge units are required for centrifugal liquid chilling machines using R-11, R-113, or R-114 because evaporator pressure is below atmospheric pressure. If a purge unit were not used, air and moisture would accumulate in the refrigerant side over a period of time. Noncondensables collect in the condenser during operation, reducing the heat-transfer coefficient and increasing condenser pressure as a result of both their insulating effect and the partial pressure of the noncondensables. Compressor power consumption increases, capacity is reduced, and surging may occur.

Moisture may build up until saturation of refrigerant occurs and free moisture is present. Acids will then be produced from reaction with the refrigerant and internal corrosion will begin. A purge unit is designed to prevent the accumulation of noncondensables and ensure internal cleanliness of the liquid chilling machine. It is not intended, however, to reduce the need for proper leak checking and repairing of leaks, which is required maintenance for any liquid chiller. Purge units may be manual or automatic, compressor-operated, or compressorless.

Oil coolers may be water-cooled, using condenser water when the quality is satisfactory, or chilled water when a small loss in net cooling capacity is acceptable. Oil coolers may also be refrigerant- or air-cooled, eliminating the need for water piping to the cooler.

**Refrigerant Transfer Unit.** A refrigerant transfer unit may be provided for centrifugal liquid chillers using refrigerants with a boiling point below ambient temperature at atmospheric pressure (R-12, R-114, R-22, R-500). The unit consists of a small reciprocating compressor with electric motor drive, a condenser (air-cooled or water-cooled), an oil reservoir and oil separator, and valves and interconnecting piping. Refrigerant transfer has three steps:

1. *Gravity Drain.* When the receiver is at the same level as or below the cooler, some liquid refrigerant may be transferred to the receiver by opening valves in the interconnecting piping.
2. *Pressure Transfer.* By resetting valves and operating the compressor, refrigerant gas is pulled from the receiver and is used to pressurize the cooler, forcing refrigerant liquid from the cooler to the storage receiver. If the chilled liquid and condenser water pumps can be operated to establish temperature differences, the migration of refrigerant from the warmer vessel to the colder vessel can also be used to assist in the transfer of refrigerant.
3. *Pump-Out.* After the refrigerant liquid has been transferred, valve positions are changed and the compressor is operated to pump refrigerant gas from the cooler to the transfer unit condenser, which sends condensed liquid to the storage receiver. If any secondary refrigerant liquid (water, brine, etc.) remains in the cooler, pump-out must be stopped before cooler pressure drops below the saturation condition corresponding to the freezing point of the secondary refrigerant.

When recharging, if the saturation temperature corresponding to cooler pressure is below the secondary refrigerant freezing point, refrigerant gas from the storage receiver must be introduced until the cooler pressure is above this condition. The compressor can then be operated to pressurize the receiver and move refrigerant liquid into the cooler without danger of freezeup.

Water-cooled transfer unit condensers provide fast refrigerant transfer. Air-cooled condensers eliminate the need for water, but they are slower and more expensive.

### Special Applications

**Heat-Recovery System.** One special application of centrifugal liquid chillers is heat recovery. Instead of rejecting all heat extracted from the chilled liquid to a cooling tower, a separate closed condenser cooling water circuit is heated by the condensing refrigerant for such purposes as comfort heating, preheating, or reheating. Factory packages are now available that include an extra condenser water circuit, either in the form of a double-bundle condenser or an auxiliary condenser.

The control requirements for a centrifugal heat-recovery package are as follows:

1. *Control of Chilled Liquid Temperature* is accomplished by a sensor in the leaving chilled liquid line signaling the capacity control device.
2. *Control of Hot Water Temperature* is accomplished by a sensor in the hot water line that modulates a cooling tower bypass valve. As the heating requirement increases, hot water temperature drops, opening the tower bypass slightly. Less heat is rejected to the tower, condensing temperature increases, and hot water temperature is restored as more heat is rejected to the hot water circuit.

The hot water temperature selected has a bearing on the installed cost of the centrifugal package, as well as on the power consumption while heating. Lower hot water temperatures of 95 to 105 °F (35 to 40 °C) result in a less expensive machine that uses less power. Higher temperatures require a greater compressor motor output, perhaps higher pressure condenser shells, sometimes extra compression stages, or a cascade arrangement. Installed cost of the centrifugal heat-recovery machine is increased as a result.

Another consideration in the design of a central chilled water plant with heat-recovery centrifugals is the relative size of the cooling and heating loads seen by the liquid chilling machine. It is best to equalize the heating and cooling loads on each machine so that the compressor may operate at optimum efficiency during both the full cooling and full heating seasons. When the heating requirement is considerably smaller than the cooling requirement, multiple packages will lower operating costs and allow standard air-conditioning centrifugal packages of lower cost to be used for the remainder of the cooling requirement (ASHRAE Symposium Bulletin DV-69-4). In multiple packages, only one unit is designed for heat recovery and carries the full heating load.

**Free Cooling.** The use of centrifugal liquid chillers to cool without operation of the compressor is called *free cooling.* When a supply of condenser water is available at a temperature below that of the needed chilled water temperature, the chiller can operate as a thermal siphon. Low-temperature condenser water condenses refrigerant, which is either drained by gravity or pumped into the evaporator. Higher-temperature chilled water causes the refrigerant to evaporate, and vapor flows back to the condenser because of the pressure difference between the evaporator and the condenser. Free cooling is limited to about 10 to 30% of the chiller design capacity. The actual free cooling capacity depends on the chiller design and the temperature difference between the desired chilled water temperature and the condenser water temperature.

**Air-Cooled System.** Two types of air-cooled centrifugal systems are prevalent. One consists of a water-cooled centrifugal package with a closed-loop condenser water circuit. The condenser water is cooled in a water/air heat exchanger. This arrangement results in higher condensing temperature and increased power consumption. In addition, winter operation requires the use of glycol in the condenser water circuit, reducing the heat-transfer coefficient.

In the other type, systems that are directly air-cooled eliminate the intermediate heat exchanger and condenser water pumps, resulting in lower power requirements. However, attention must be given to keeping the condenser and refrigerant piping leaktight. Air-cooled centrifugal packages are now on the market.

Because a centrifugal will surge if it is subjected to a head appreciably higher than the design value, it is important that the air-cooled condenser be designed for the required heat rejection. It is common practice to select a reciprocating air-cooled machine based on an outside dry-bulb temperature that will be exceeded 5% of the time. A centrifugal may be unable to operate during such times because of surging, unless the chilled water temperature is raised proportionately. Thus, the compressor impeller(s) and/or speed should be selected for the maximum dry-bulb temperature to ensure that the desired chilled water temperature will be maintained at all times.

An air-cooled centrifugal chiller should allow the condensing temperature to fall naturally to about 70°F (21°C) during colder weather. An important decrease in compressor power consumption results, which is greater than that for reciprocating systems controlled by thermal expansion valves.

During winter shutdown, precautions must be taken to prevent freezing of the cooler liquid resulting from a free cooling effect from the air-cooled condenser. A thermostatically controlled heater in the cooler, in conjunction with a low refrigerant pressure switch to start the chilled liquid pumps, will protect the system.

**Brine Cooling.** The most frequent use of centrifugal liquid chilling units is for water-chilling applications. Centrifugals are also applied for brine cooling duty with brines such as calcium chloride, methylene chloride, ethylene glycol, and propylene glycol. (For a complete description of brines, see Chapter 18 of the 1985 FUNDAMENTALS Volume.) Brine properties must be considered in calculating heat-transfer performance and pressure drop. Because of the greater temperature lift, higher compressor speeds and possibly more stages may be required for brine cooling duty. Compound and/or cascade systems are required for low-temperature applications. (See Chapter 1 of the 1986 REFRIGERATION Volume for additional information.)

## OPERATION AND MAINTENANCE

Proper operation and maintenance are essential for reliability, longevity, and safety. See Chapter 59 of the 1987 HVAC Volume for general information on principles, procedures, and programs for effective maintenance. The manufacturer's operation and maintenance instructions should also be consulted for specific procedures.

Normal operation conditions should be established and recorded at initial startup. Changes from these conditions can be used to signal the need for maintenance. One of the most important items is to maintain a leaktight unit. Leaks on units operating at subatmospheric pressures will result in air and moisture entering the unit and increasing the condenser pressure.

While the purge unit can remove noncondensables sufficiently to prevent an increase in condenser pressure, continuous entry of air and attendant moisture into the system promotes refrigerant and oil breakdown and corrosion. If frequent or continuous purging is required, the leaks should be repaired as soon as possible.

Periodic analysis of the oil and refrigerant charge can also be used to identify system contamination problems. High condenser pressure or frequent purge unit operation indicate leaks that should be corrected as soon as possible. With positive operating pressures, leaks result in loss of refrigerant and such operating problems as low evaporator pressure. A leak check should also be included in preparations for a long-term shutdown. (See Chapter 7 of the 1986 REFRIGERATION Volume for a detailed discussion of the harmful effects of air and moisture.)

Normal maintenance should include periodic oil and refrigerant filter changes as recommended by the manufacturer. All safety controls should be checked periodically to ensure that the unit is protected properly.

Cleaning of inside tube surfaces may be required at various intervals, depending on the water condition. Condenser tubes need only annual cleaning if proper water treatment is maintained. Cooler tubes need less frequent cleaning if the chilled water circuit is a closed loop.

If it is necessary to remove the refrigerant charge and open the unit for service, the unit should be leak-checked, dehydrated, and evacuated properly before recharging. (See Chapter 21 for dehydrating, charging, and testing and ASHRAE Symposium Bulletin LP-68-2.)

# PART IV: SCREW LIQUID CHILLERS

## EQUIPMENT DESCRIPTION

### Components and Their Function

The single- and twin-screw compressors are both positive displacement machines with nearly constant flow performance. Compressors for liquid chillers can be both *oil-injected* and *oil-injection-free.* (Chapter 12 describes screws compressors in greater detail.)

The **cooler** may be *flooded* or *direct-expansion.* There is no particular cost advantage to one design over the other. The flooded cooler is more sensitive to freezeup, requires more refrigerant, and requires closer evaporator pressure control, but its performance is easier to predict and it can be cleaned. The direct-expansion cooler requires closer mass flow control, is less likely to freeze, and returns oil to the oil system rapidly. The decision to use one or the other is based on the relative importance of these factors on a given application.

Screw coolers have the following characteristics: (1) high maximum working pressure, (2) continuous oil scavenging, (3) no mist eliminators (flooded coolers), and (4) distributors designed for high turndown ratios (direct-expansion coolers). A suction gas, high-pressure liquid heat exchanger is sometimes incorporated into the system to provide subcooling for increased thermal expansion valve flow and reduced power consumption. (For further information on coolers, see Chapter 16.)

Flooded coolers were once used in units with a capacity larger than about 400 tons (1400 kW). Today, direct-expansion coolers are also used in larger units up to 800 tons (2800 kW) because of the development of a new type of servo-operated expansion valve with an electronic controller that measures evaporating pressure, leaving brine temperature, and suction gas superheat.

The condenser may be included as part of the liquid chilling package when water-cooled, or it may be remote. Air-cooled liquid chilling packages are also available. When remote air-cooled or evaporative-cooled condensers are applied to liquid chilling packages, a liquid receiver generally replaces the water-cooled condenser on the package structure. Water-cooled condensers are the cleanable shell-and-tube type (see Chapter 15).

Oil cooler loads vary widely, depending on the refrigerant and application, but they are substantial because oil injected into the compressor absorbs a portion of the heat of compression. Oil cooling is by (1) a water-cooled oil cooler using condenser water, evaporative condenser sump water, chilled water, or a separate water- or glycol-to-air cooling loop; (2) an air-cooled oil cooler using an oil-to-air heat exchanger; (3) a refrigerant-cooled oil cooler (where oil cooling load is low); (4) liquid injection into the compressor; or (5) condensed refrigerant liquid thermal recirculation (where appropriate head is available). The latter two are the most economical both in first cost and overall operating cost because cooler maintenance and special water treatment are eliminated.

Efficient oil separators are required. The types and efficiencies of these separators vary according to refrigerant and application. Field built-up systems require better separation than complete factory-built systems. Ammonia applications are most stringent because there is no appreciable oil return with the suction gas from the flooded coolers normally used in ammonia applications. However, separators are available for ammonia packages, which do not require the periodic addition of oil that is customary on other R-717 systems. The types of separators in use are centrifugal, de-mister, gravity, coalescer, and combinations of these.

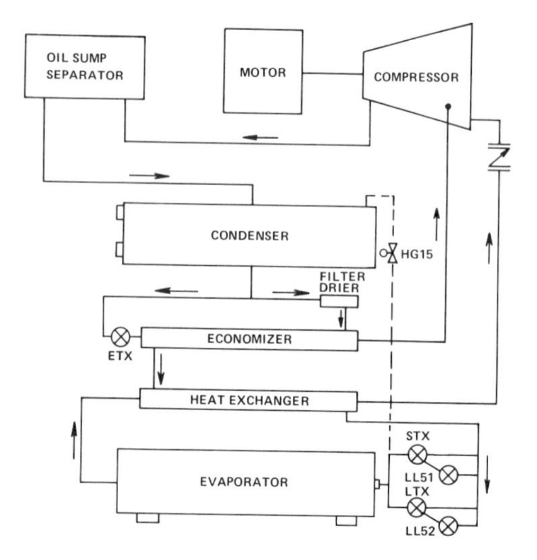


Fig. 13 Refrigeration System Schematic

Hermetic compressor units may use a centrifugal separator as an integral part of the hermetic motor while cooling the motor with discharge gas and oil simultaneously. A schematic of a typical refrigeration system is shown in Figure 13.

## Capacities and Types Available

Screw compressor liquid chillers are available as factory-packaged units from about 40 to 850 tons (140 to 3000 kW). Both open and hermetic styles are manufactured. Packages without water-cooled condensers, with receivers, are made for use with air-cooled or evaporative-cooled condensers. Most factory-assembled liquid chilling packages use R-22.

Additionally, compressor units, comprised of a compressor, hermetic or open motor, oil separator, and oil system, are available from 20 to 2000 tons (70 to 7000 kW). These are used with remote evaporators and condensers for low, medium, and high evaporating temperature applications. Condensing units, similar to compressor units in range and capacity but with water-cooled condensers, are also built. All of the above are suitable for R-12, R-502, and R-22. Similar open motor-drive units are available for R-717, as are booster units.

## Selection of Refrigerant

Refrigerants used are R-12, R-22, R-500, R-502, and R-717. R-22 and R-717 are popular because the compressor size required is relatively small. Since screw compressors are now available with intermediate pressure porting, which allows subcooling and intercooling, the merits of each refrigerant can be fully exploited in each application. R-500 and R-12 are typically used in heat-recovery installations where temperatures of the heat-transfer media up to 158°F (70°C) are required.

The screw compressor operating characteristic shown in Figure 7 is compared with reciprocating and centrifugal performance. Additionally, since the screw compressor is a positive displacement compressor, it does not surge. Since it has no clearance volume in the compression chamber, it pumps high volumetric flows at high heads. Because of this, screw compressor chillers suffer the least capacity reduction at high condensing temperatures.

The screw compressor provides stable operation over the whole working range because it is a positive-displacement machine. The working range is wide because the discharge temperature is kept low and is not a limiting factor because of oil injection into the compression chamber. Consequently, the compressor is able to operate single-stage at high pressure ratios. An economizer system can be used to improve the capacity and lower the power consumption at full-load operation.

An example of such an economizer arrangement is shown in Figure 13, where the main refrigerant liquid flow is subcooled in a heat exchanger connected to the intermediate pressure port in the compressor. The evaporating pressure in this heat exchanger is higher than the suction pressure of the compressor. Oil separators must be sized for the size of compressor, type of system (factory assembled or field connected), refrigerant, and type of cooler. Direct-expansion coolers have less stringent separation requirements than do flooded coolers.

In a direct-expansion cooler, the refrigerant is evaporated in the tubes, which means that the velocity is kept so high that the oil rapidly returns to the compressor. In a flooded evaporator, the refrigerant is outside the tubes, and some type of external oil-return device must be used to minimize the oil concentration in the cooler. Suction or discharge check valves are used to minimize spinback and oil loss during shutdown.

Since the oil system is on the high-pressure side of the unit, precautions must be taken against oil dilution. Dilution can also be caused by excessive floodback through the suction or intermediate ports; it may go unnoticed until serious operating or mechanical problems are experienced.

## METHOD OF SELECTION

### Ratings

Screw liquid-chiller ratings are generally presented similarly to those for centrifugal-chiller ratings. Tabular values of capacity and power consumption at various chilled water and condenser water temperatures are shown.

In addition, ratings are given for packages minus the condenser, listing capacity and power versus chilled water temperature and condensing temperature. Ratings for compressors alone are also common, showing capacity and power consumption versus suction temperature and condensing temperature for a given refrigerant.

### Power Consumption

Typical part-load power consumption is shown in Figure 14. Power consumption of screw chillers benefits from reduction of condensing water temperature as the load decreases, as well as operating at the lowest practical head pressure at full load. However, because direct-expansion systems require a pressure differential, the power consumption saving is not as great at part load as shown.

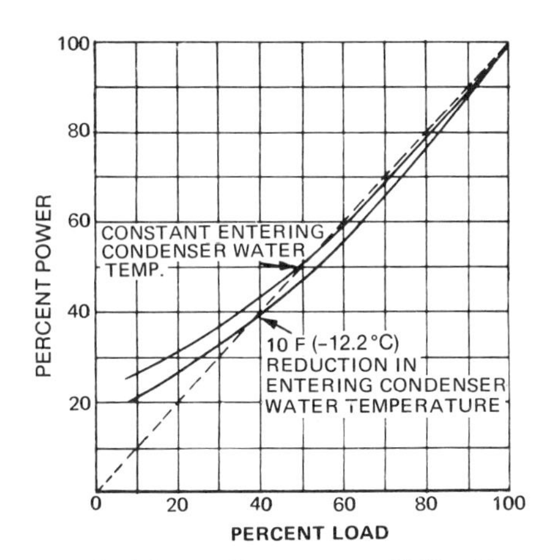


**Fig. 14 Typical Screw Compressor Chiller Part-Load Power Consumption**

### Fouling

A fouling allowance of 0.0005 $ft^2 \cdot {}^\circ F \cdot h/Btu$ (0.088 $m^2 \cdot {}^\circ C/kW$) is incorporated in screw compressor chiller ratings. Excessive fouling (above the design value) increases power consumption and reduces capacity. Fouling of water-cooled oil coolers results in higher than desirable oil temperatures.

## SPECIFIC CONTROL CONSIDERATIONS

Screw chillers provide continuous capacity modulation, from 100% capacity down to 10% or less. The leaving chilled liquid temperature is sensed for capacity control. Safety controls commonly required are (1) oil failure switch, (2) high-low pressure cutout, (3) cooler flow switch, (4) high oil or discharge temperature cutout, (5) hermetic motor inherent protection, and (6) oil pump and compressor motor overloads. The compressor is unloaded automatically (slide valve driven to minimum position) before starting. Once it starts operating, the slide valve is controlled hydraulically by a temperature-load controller that energizes the load and unload solenoid valves.

The temperature-load controller protects against motor overload from higher than normal condensing temperatures or low-voltage conditions and also allows demand to be set, if desired. An antirecycle timer is used to prevent overly frequent recycling. Oil sump heaters are energized during the off cycle. A hot gas capacity control is optionally available and prevents automatic recycling at no-load conditions such as is often required in process liquid chilling. A suction to discharge starting bypass sometimes aids starting and allows the use of standard starting torque motors.

Some units are equipped with electronic regulators specially developed for the screw compressor characteristics. These regulators include PI-control (Proportional-Integrating) of the leaving brine temperature and such functions as automatic/manual control, capacity indication, time circuits to prevent frequent recycling and to bypass the oil pressure cutout during startup, switch for unloaded starting, etc. (Typical external connections are shown in Figure 15.)

## AUXILIARIES AND SPECIAL APPLICATIONS

### Auxiliaries

A refrigerant transfer unit is similar to the unit described in the section on "Auxiliaries" in Part III. It is designed for R-22

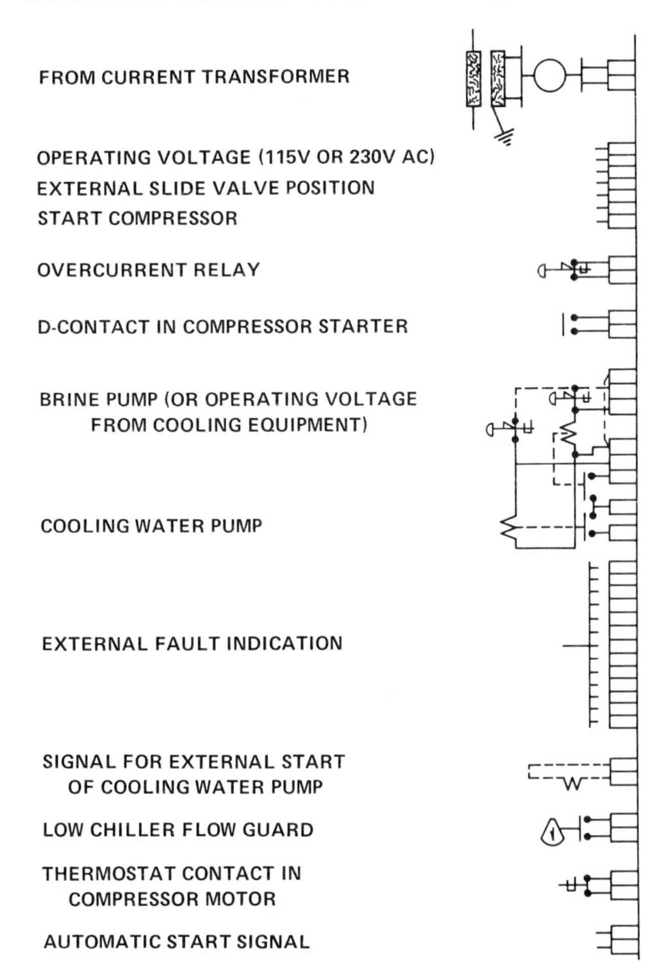


**Fig. 15 Typical External Connections for Screw Compressor Chiller**

operating pressure. Its flexibility is increased by including a reversible liquid pump on the unit. It is available as a portable unit or mounted on a storage receiver.

An oil-charging pump is useful for adding oil to the pressurized oil sump. Two types are used: a manual pump and an electric motor-driven positive-displacement pump.

Acoustical enclosures are available for installations where low noise levels are required.

### Special Applications

Because of the screw compressor's positive displacement characteristic and oil-injected cooling, its use for high differential applications is limited only by power considerations and maximum design working pressures. Therefore, it is being used for a number of special applications because of reasonable compressor cost and no surge characteristic. Some of the fastest growing areas are listed below:

1. Heat-recovery installations
2. Air-cooled
   a. Split packages with field-installed interconnecting piping
   b. Factory-built rooftop packages
3. Low-temperature brine chillers for process cooling
4. Ice rink chillers
5. Power transmission line oil cooling

High temperature compressor and condensing units are being used increasingly for air conditioning because of the higher efficiency of direct air-to-refrigerant heat exchange resulting in higher evaporating temperatures (see Chapter 6). Many of these installations have air-cooled condensers (see Chapter 15, Part II).

## MAINTENANCE

Periodic maintenance of screw liquid chillers is important; it is essential that the manufacturer's instructions be followed, especially since there are some items that differ substantially from reciprocating or centrifugal units.

Water-cooled condensers must be cleaned of scale periodically (see Part I). If the condenser water is also used for the oil cooler, this should be considered in the treatment program. Oil coolers operate at higher temperatures and lower flows than condensers, so it is possible that the oil cooler may have to be serviced more often than the condenser.

Since large oil flows are a part of the screw compressor system, the oil filter pressure drop should be monitored carefully and the elements changed periodically. This is particularly important in the first month or so after startup of any factory-built package and is absolutely essential on field-erected systems. Since the oil and refrigeration systems merge at the compressor, much of the loose dirt and fine contaminants in the system eventually find their way to the oil sump, where they are removed by the oil filter. Similarly, the filter-drier cartridges should be monitored for pressure drop and moisture during initial start and regularly thereafter. Generally, if a system reaches an acceptable dryness level, it stays that way unless it is opened.

It is good practice to check the oil for acidity periodically, using commercially available acid test kits. Oil does not have to be changed unless it is contaminated by water, acid, or metallic particles. Also, a refrigerant sample should be analyzed yearly to determine its condition.

There are certain procedures that should be followed on a yearly basis or during a regularly scheduled shutdown. These include checking and calibrating all operation and safety controls, tightening all electrical connections, inspecting power contacts in starters, dielectric checking of hermetic and open motors, and checking the alignment of open motors.

Leak testing of the unit should be performed regularly. For a water-cooled package used for summer cooling, this should be performed yearly. A flooded unit with proportionately more refrigerant in it, used for year-round cooling, should be tested every four to six months. A process air-cooled chiller designed for year-round operation 24 hours per day should be checked every one to three months.

The screw compressor is a simple, rugged machine. The decision concerning whether to tear down, and how far to go, depends, in part, on the nature of its application, *i.e.*, process, computer cooling, air conditioning, the design of the rest of the system, whether there is standby equipment, and the cost of a breakdown.

Based on 6000 operating hours per year and depending on the above considerations, a reasonable inspection or changeout timetable is shown below:

| | | |
|---|---|---|
| Shaft seals | 1.5 to 4 years | Replace |
| Hydraulic cylinder seals | 1.5 to 4 years | Replace |
| Thrust bearings | 4 to 6 years | Check preload and/or replace |
| Shaft bearings | 7 to 10 years | Inspect |

## BIBLIOGRAPHY


ASHRAE. 1973. Centrifugal Chiller Noise in Buildings. ASHRAE *Symposium Bulletin* LO-73-9.

ASHRAE. 1969. Centrifugal Heat Pump Systems. ASHRAE *Symposium Bulletin* DV-69-4.

ASHRAE. 1968. The Operation and Maintenance of Centrifugal Units. ASHRAE *Symposium Bulletin* LP-68-2.

Soumerai, H. 1967. Large screw compressors for refrigeration. ASHRAE *Journal,* March, p. 38.

# CHAPTER 18

# COMPONENT BALANCING IN REFRIGERATION SYSTEMS



ONE principal design aspect of an air-to-air refrigeration system, whether factory-built or field-assembled, is the selection of the components, including the compressor, condenser, evaporator, refrigerant flow control device, fan, motor, and controls.

The characteristics of each of the components are related to the other components, and the system formed by them must perform properly at design conditions and every other condition that may be expected during operation. Most refrigeration systems have varying conditions of both the fluid from which heat is removed and the fluid to which heat is discharged.

## REFRIGERATION SYSTEMS

A system that cools and dehumidifies air may have a varying rate of airflow. Even with a constant flow rate and a constant entering air dry-bulb temperature, however, the load may fluctuate because of a varying wet-bulb temperature.

Air dry-bulb temperature and air wet-bulb temperature are the important factors for air-cooled and evaporative condensers.

Chapter 1 of the 1985 FUNDAMENTALS Volume provides complete information on the basic theory of refrigeration. Also, Chapters 2, 3, 4, and 5 of the 1986 REFRIGERATION Volume contain more information on this subject. Components are discussed in that volume and in Chapters 6, 12, 15, 16, and 19 of this volume.

Selection of the best and most economical components for a particular system requires a familiarity with the effects of changing operating conditions on the performance of each component.

The preparation of this chapter is assigned to TC 10.1, Custom Engineered Refrigeration Systems.

### Evaporator

The evaporator transfers heat from the fluid being cooled to the boiling refrigerant. If the fluid does not change state, the effect of change is relatively simple. The evaporator capacity decreases, together with a decrease in either fluid flow rate or temperature difference between refrigerant and fluid.

If the fluid (or some part of it) changes state as it is cooled, then the transfer rate is affected. In a mixture of air and water vapor, there will be no change of state, and the performance will be as mentioned, if none of the surface temperature over which the mixture passes is below the dew point of the water vapor. If any part of the surface is below the dew point of the water vapor, some condensation occurs and additional heat transfer results. The extent of the heat transfer that results from condensation of water vapor depends on the proportion of the surface that is below the dew point, as well as the difference between the initial vapor pressure and the vapor pressure corresponding to the weighted average surface temperature.

Chapter 6 of this volume addresses this subject in detail and shows that there is no easy method of indicating which part of a coil surface will be dry or wet. Plots based on entering wet-bulb temperature give reasonable results and can be used, in standard cases, to find the system balance point at various conditions of operation.

### Compressor

The compressor raises the pressure of the refrigerant gas from the evaporator pressure (minus any pressure loss between evaporator and compressor suction) to the condensing pressure (plus any pressure loss between the compressor discharge and the condenser).

With constant discharge pressure, a positive-displacement compressor gas-handling rate is almost directly proportional to the absolute pressure at the compressor suction. The compression ratio change causes some variation, but this difference is small unless the ratio is great enough to result in a low volumetric efficiency. At ratios of five or less, with reciprocating compressors (higher with screw compressors), the effect of a small change in compression ratio is not significant.

Changes in the gas temperature at the compressor suction have a small effect on capacity for a few degrees change in suction temperatures above −50°F (−46°C).

The relationship between capacity and absolute pressure at the compressor suction helps judge the effect of suction line pressure loss or lower evaporating pressure.

If the evaporating pressure is 75 psia (520 kPa), a 2 psi (14 kPa) loss in the suction line will reduce the compressor capacity by about 2.7%. If the evaporating pressure is 20 psia (140 kPa), a 2 psi (14 kPa) reduction will result in about a 10% capacity reduction.

Small increases in compressor discharge pressure, with no increase in condensing temperature (such as results from discharge line loss), have little effect on compressor capacity. The principal result is an increase in power.

An increase in discharge pressure (the result of higher condensing temperature causing a higher temperature of liquid to the expansion valve) reduces the capacity because it reduces the heat-removal capacity of the refrigerant handled.

## Condenser

If the condenser is air-cooled, the difference between the condensing temperature and the available air dry-bulb temperature depends on the rate of airflow over the condenser and the rate of heat rejection. Various methods are available for the control of condensing temperature. The method or combination of methods chosen depends on personal preference, availability, practicality, and cost.

Controlling the flow rate of air over the condenser is simple and low in cost and is usually the method selected if available in the proper size.

Controlling the temperature of the air by variation of partial recirculation is used when one or more condensers are installed in an enclosed space and ventilation maintains the desired space temperature. This arrangement permits use of condenser heat to maintain the space temperature.

Partially flooding the condenser to reduce the effective condensing surface requires special condenser circuiting or a separate vessel as a receiver. When this receiver is arranged so that the liquid flows through it and the receiver is equalized, the liquid temperature rises between the condenser outlet and the receiver outlet, slightly reducing system capacity and practically eliminating subcooling at the entrance to the liquid line.

If the condenser is the evaporative-type, the difference between the condensing temperature and the wet-bulb temperature of the available air depends on the airflow rate over the condenser and the rate of heat rejection. Condensing temperature can be controlled by partial recirculation of air from the condenser, operation of the condenser without sprays, or reduced airflow rate. Reduced airflow rate is impractical at wet-bulb temperatures that result in surface temperatures below 32°F (0°C).

## Refrigerant Flow-Control Device

A refrigerant flow-control device controls the flow rate of refrigerant liquid between condenser and evaporator and regulates the flow rate of refrigerant into the evaporator to match the evaporation rate, which corresponds to the actual load (or heat removal) at any time.

The flow-control device must permit the highest flow rate required at the lowest pressure difference available. It must also function with the existing pressure difference at the lowest rate used.

The pressure difference available is that at the condenser liquid outlet, less the pressure at the evaporator entrance, plus (or minus) the static head because of the refrigerant liquid column, minus other possible losses that result from flow resistance or pressure difference required to open pilot-operated valves (any but the smallest solenoid valves) and some expansion valves (particularly the larger ones). Pressure loss or gain values, except for liquid column static and valve opening pressures, vary with the 1.8 to 2.0 power of the refrigerant flow rate. Losses directly related to flow rate include the liquid line with manual valves, filters, driers, and distributor. Also included are solenoid valve and expansion valve losses above those required to open any that are pilot-operated.

All the losses are based on liquid (no vapor) reaching the expansion valve. The temperature of the liquid in the line must be no higher than that corresponding to saturation at the pressure existing at that point. Sometimes this requires subcooling at the start of the liquid line. Receivers through which the liquid passes should be avoided because liquid leaving such a receiver has little or no subcooling. The system must be designed so that none of the refrigerant entering the expansion valve is in vapor form. A separate subcooler may be desirable, especially after the liquid has passed through a receiver.

Some pressure difference is required to move refrigerant liquid from the condenser to the evaporator; this difference must be maintained at the value necessary for the refrigerant flow rate at any particular time. Power per unit of refrigeration is reduced by keeping this pressure difference as small as possible. If condenser pressure must be controlled at a higher pressure than is required by the sink temperature, it should be controlled at the lowest pressure consistent with operation. At many periods, the load may be below design maximum and thus require a smaller pressure difference for the lower rate of refrigerant liquid transfer from condenser to evaporator.

Many factors determine the need for condenser pressure control. In some cases, balance of the system at very low evaporating and even surface temperatures is acceptable. In other cases, cooling surface temperatures that are too low will result in freezing or in suction temperatures that are too low. These low temperatures reduce refrigeration capacity below an acceptable point.

The pressure difference for transfer of liquid refrigerant from condenser to evaporator is affected by the relative elevations of condenser and evaporator. Each system must be considered individually.

In the interest of lower operating cost and better operation, condensing pressure should be controlled with reference to the *critical parameter.* If this value is the difference in pressure between the inlet and outlet of the expansion valve, then the control should sense that difference and regulate the condenser capacity to maintain the minimum difference required. More frequently, the evaporating pressure must be kept above a predetermined value; the control should sense this and regulate condenser capacity to maintain this value.

Either of these, particularly the latter, may require a control that senses condensing pressure, which overrides the main control to avoid high condensing pressure at startup.

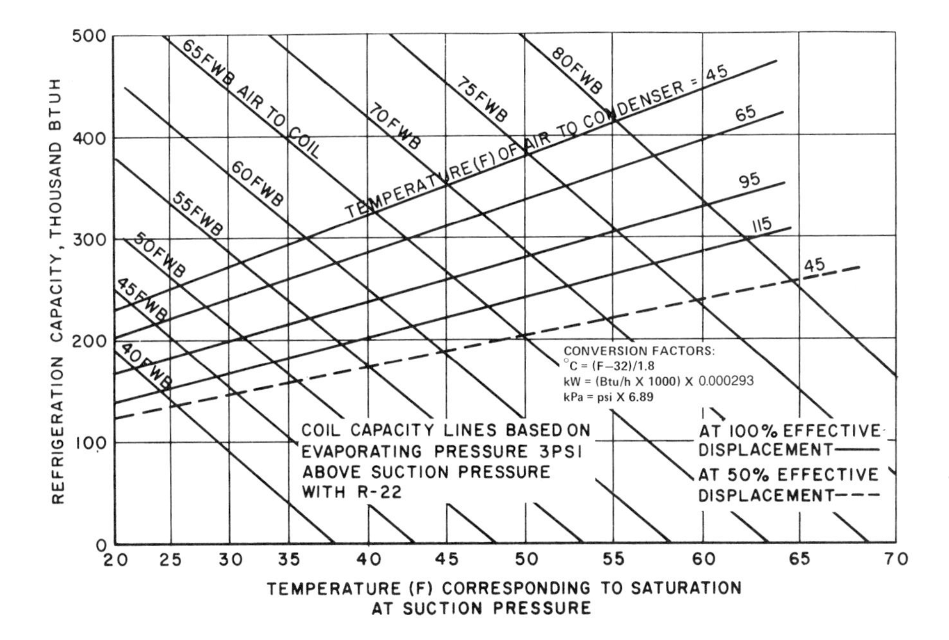


**Fig. 1 Performance Diagram for Air-Cooling Coil and Condensing Unit with Various Temperatures of Air to Condenser**

## SYSTEM PERFORMANCE

Charts similar to Figures 1 and 2 help check the performance of a system. These charts are drawn with the temperature scale spaced on the basis of the refrigerant pressure. This spacing results in easier plotting because the compressor (or condensing unit) capacity lines and the wet-bulb coil capacity lines are practically straight.

To illustrate the analysis of a system, assume that an air-cooling coil has been selected to satisfy an actual air-conditioning load in which it must remove 250,000 Btu/h (73 kW) when air at 65 °F (18 °C) wet bulb enters and the evaporating temperature is about 48 °F (9 °C). Coil capacity lines for various entering wet-bulb temperatures have been plotted in Figure 1. These lines are drawn about 3 psi (21 kPa) to the left of the evaporating temperature to allow for that pressure loss between the evaporator outlet and the compressor inlet.

It is required that this 250,000 Btu/h (73 kW) be removed by the condensing unit when air is entering it at 95 °F (35 °C). The condensing unit capacity curves in Figure 1 are for a unit that meets the requirement. Note that the 95 °F (35 °C) condensing unit line intersects the 65 °F (18 °C) coil line at about 260,000 Btu/h (76 kW) and just under 46 °F (8 °C)—about 48 °F (9 °C) at the 3 psi (21 kPa) higher evaporator pressure. At design conditions, performance is almost as desired; however, design conditions do not always occur.

If the wet-bulb temperature entering the coil is very much higher, the 75 °F (24 °C) wet-bulb line intersects the 95 °F (35 °C) line at over 300,000 Btu/h (88 kW) and 55 °F (13 °C) suction. This may overload the compressor motor. If air enters the condenser at 45 °F (7 °C) and the air entering the coil is at 50 °F (10 °C) wet bulb, the intersection is at 250,000 Btu/h (73 kW) and 26 °F (−3 °C) suction. This new balance point causes several changes.

First, the condensing pressure will be lower. The actual condensing temperature (and corresponding pressure) at this suction and air temperature can be obtained from the condenser manufacturer's data. The difference available for transfer of liquid to the evaporator can then be compared to the value required for the new flow rate and the actual liquid line, valves, etc., for the particular installation. Figure 1 assumes that the available pressure difference is adequate. If not, the suction pressure (and temperature) capacity fall to the point where the pressure difference available matches the value required.

The evaporating pressure—at a little above 26 °F (−3 °C), or lower, as a result of liquid pressure loss—may result in a coil surface temperature that is below freezing. The lowest surface temperature can be calculated from the coil characteristics and the temperature of the air leaving the coil. If the surface temperature is below freezing, it may still be acceptable if it is not below the dew point of the air passing over it.

Reducing condenser capacity, so that the condensing pressure is maintained at about the same value as if 95 °F (35 °C) air entered it, results in approximately 200,000 Btu/h (59 kW) capacity at 31 °F (−0.6 °C) suction.

If the compressor can be controlled to operate at 50% capacity, the balance with 50 °F (10 °C) wet bulb air to the coil and 45 °F (7 °C) air to the condenser will be at about 150,000 Btu/h (44 kW) and 36 °F (2 °C) suction. This latter arrangement results in less power per unit of refrigeration.

To illustrate other possibilities, assume that the compressor can operate at 25, 50, 75, or 100% capacity. The condensing temperature is held constant at 105 °F (41 °C). With 65 °F (18 °C) wet bulb air over the coil at 43 °F (6 °C) suction, 280,000 Btu/h (82 kW) is required. Figure 2 shows coil and compressor capacity lines that meet this requirement.

Assume it is necessary to reduce the rate of airflow over the coil to maintain conditions when the load is less and to maintain the suction temperature near 43 °F (6 °C) to keep relative humidity nearly constant. In Figure 2, the 100% compressor capacity line intersects the 65 °F (18 °C) wet bulb and full airflow capacity line at about 280,000 Btu/h (82 kW) and 43 °F (6 °C). If the load is reduced, the rate of airflow might be reduced to 50%; if no change of compressor capacity occurs, however, the suction temperature would be about 30 °F (−1 °C). When the airflow is about 50% and compressor capacity is also about 50%, the intersection is close to 43 °F (6 °C) suction and 140,000 Btu/h (41 kW).

This change can be automatic and can be controlled from room temperature in one of two ways. The room thermostat can control the four steps of compressor capacity, and the dampers controlling the airflow over the coil can be controlled from the suction pressure or the temperature of the air leaving the coil; or, the room thermostat can regulate the dampers controlling the airflow over the coil, and the compressor capacity can be varied in response to suction pressure or the temperature of air leaving the coil.

When the compressor has a limited number of steps of capacity, the first method is preferable because the compressor capacity does not oscillate between two capacities in response to rapidly changing suction pressure or air temperature leaving the coil. Instead, the compressor capacity changes infrequently, in response to the relatively slowly changing room temperature.

Figure 2 does not show anything with reference to the condenser. The compressor capacity lines are based on the assumption that the condenser and heat removal medium are adequate to remove the heat at the highest rate without permitting the condensing temperature to rise above 105 °F (41 °C). It is also assumed that a control exists to keep this temperature from falling below 105 °F (41 °C).

Figure 1 includes the condenser, but only in combination with a particular compressor. The curves show the performance of the combination without any indication of the condensing temperature at various operating conditions. If a particular combination of compressor and condenser is used that performs at all desired operating conditions, there is no need to consider the condensing temperature, except to check for adequate pressure difference to transfer liquid refrigerant from condenser to evaporator.

When the condenser and compressor are selected separately or the performance is not available for all the operating conditions expected or to be considered, separate compressor and condenser performance curves should be plotted. Figure 3 is an example of such curves (the compressor heat rejection curves are based on a hermetic compressor with the motor waste heat also transferred to the refrigerant). The compressor performance requires two curves for each suction saturation temperature—the solid line indicating the heat from the evaporator and the broken line indicating the heat to the condenser. The difference between the two, at any condensing and evaporating temperature, represents the energy to lift the heat from the evaporating temperature to the condensing temperature. The larger the difference in temperatures, the larger the heat energy added by the compressor for the lift. The heat added is almost directly proportional to the temperature difference for ranges from 0 to 150 °F (−18 to 66 °C), and differences up to 100 °F (56 °C), assuming there are no other changes.

Since the heat rejected is the amount that enters the condenser, the intersections of the broken lines with the condenser performance lines are the only intersections with any significance. If the suction saturation temperature is 40 °F (4 °C) and the temperature of air to the condenser is 105 °F (41 °C), the intersection of the broken line for 40 °F (4 °C) with the condenser line for 105 °F (41 °C), as shown in Figure 3, is at about 129 °F (54 °C) condensing temperature and 290,000 Btu/h (85 kW) heat rejected. Moving vertically down to the solid line for 40 °F (4 °C), the heat removed from the evaporator at this condition is about 213,000 Btu/h (62 kW).

A chart such as Figure 3 facilitates consideration of the effects of various condensers and air rates over the condenser. The capacity lines for a condenser of twice the capacity would have angles with the vertical, the tangents of which would be half of those shown.

Reducing the airflow rate over the condenser increases the angle with the vertical. Of course, the capacity of the condenser does not change proportionately with a change in air rate. At 50% air rate, the capacity would be more than 50%. The actual value depends on the characteristics of the particular condenser.

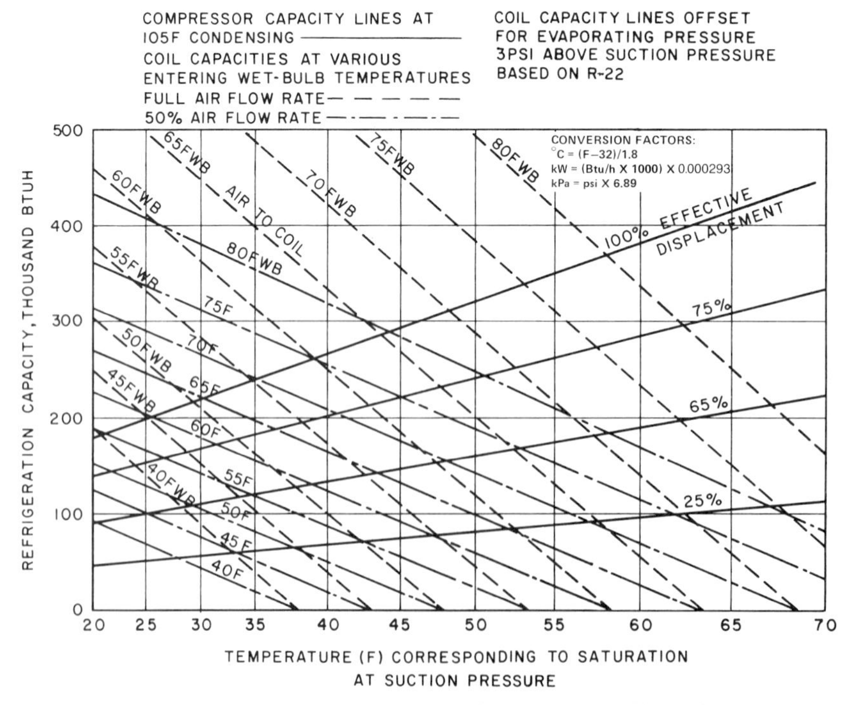


**Fig. 2 Performance Diagram for Air-Cooling Coil and Compressor at Fixed Condensing Temperature**

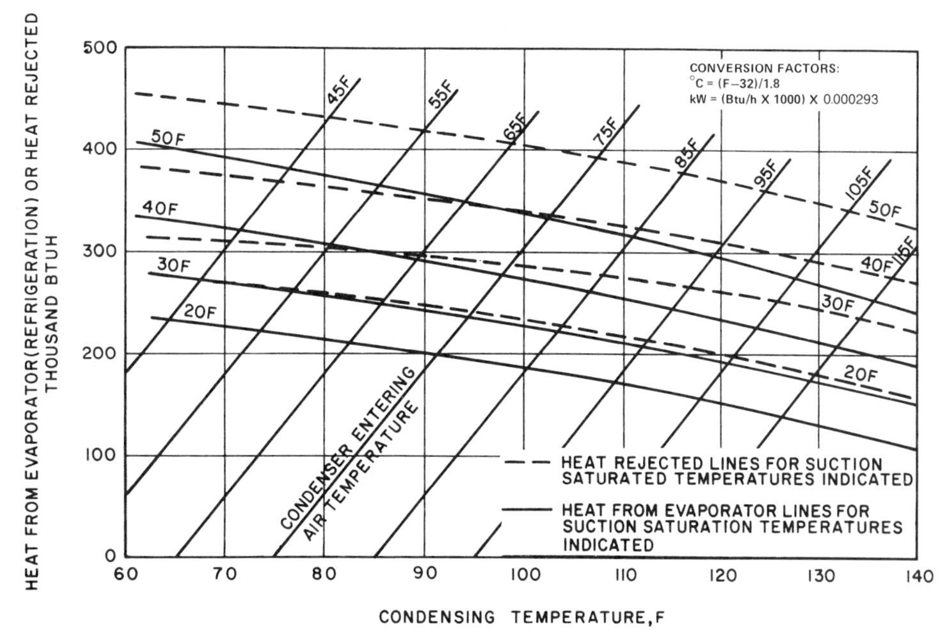


**Fig. 3 Performance Diagram for Air-Cooled Condenser with Compressor Refrigeration Capacity and Heat Rejection**

Figure 3 is based on air-cooled condensers. With the necessary data available, performance lines can be drawn for evaporative condensers for various wet-bulb temperatures. Such a chart might be more easily drawn with the condensing temperature scale spaced on the basis of saturation pressure because the condenser capacity lines would be more nearly straight.

Lines representing condenser performance with various air or water flow rates can be drawn if the actual condenser performance at such conditions is known or calculated.

## LIQUID FROM CONDENSER TO EVAPORATOR

Many variables influence the pressure difference required to transfer refrigerant liquid from the condenser to the evaporator. When all the factors are known, this difference can be plotted to predict performance.

In Figure 4 (based on R-22), the capacity of the expansion valve is plotted against the pressure loss through the valve. The plot includes correction for the greater Btu/lb (kJ/kg) available with the lower temperature liquid when the condensing temperature is lower.

To allow for losses in addition to those through the expansion valve, the suction and condensing temperature lines for an air-cooled condenser have been plotted so that the difference in pressure at the intersections is about 35 to 40 psi (240 to 280 kPa) less than the actual difference.

The unit capacity curve for constant load assumes air entering the evaporator at a constant rate and nearly constant enthalpy. As the condensing temperature is reduced, the capacity of the combined system components increases, even at the lower suction pressure, until the expansion valve capacity is reached. As condensing temperature is reduced further, the suction temperature and capacity are limited by the expansion valve capacity. The coil surface temperature falls below freezing because of reduction in pressure difference between condenser and evaporator.

As this capacity curve at constant load is followed from the right to the left, the condensing temperature decreases. Assuming the average difference of 30 °F (16.7 °C) between entering air temperature and condensing temperature, the outdoor air would be at about 95 °F (35 °C) for the 125 °F (52 °C) condensing line and 40 to 45 °F (4 to 7 °C) at the 70 °F (21 °C) line. The outdoor dew point would be about 40 °F (4 °C) or less, compared to the 57.5 °F (14 °C) dew point at 78 °F (26 °C) dry bulb with 50% relative humidity. Unless considerable moisture is added to the conditioned space, the interior dew point and wet bulb will be less, and the load will not remain constant as the outside temperature falls. The unit capacity with the lower load might be similar to the curve for decreasing wet bulb and load. In this case, the expansion valve capacity is not the limiting factor until the evaporating or suction temperature is below 20 °F (−7 °C) and the surface temperature is probably below 32 °F (0 °C). The low coil temperature is not caused by too little pressure difference between condenser and evaporator. It results from increased system capacity at lower condensing temperature, combined with reduced load at lower inside wet-bulb temperature.

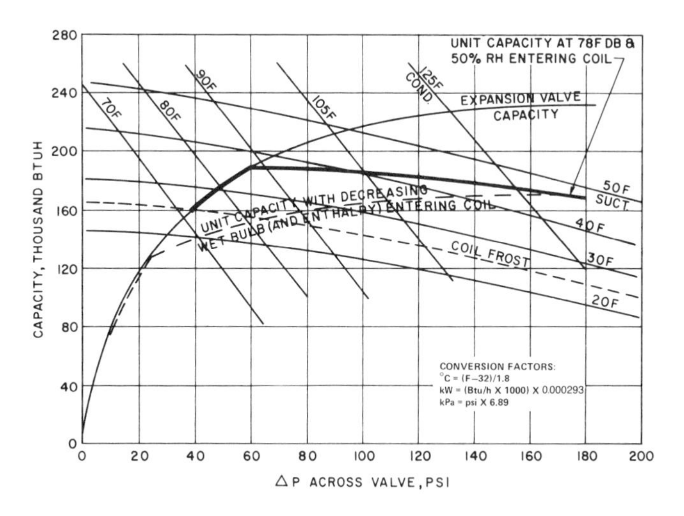


**Fig. 4 Refrigeration Capacity versus Expansion Valve Capacity**

Under this latter condition, control to maintain a higher condensing temperature helps to increase coil surface temperature because it reduces system capacity and thus increases the suction temperature balance point. This is particularly true if the condenser pressure control system increases the liquid temperature approximately as it increases the condensing temperature. If the control is the type that results in a large increase in subcooling, the capacity will not decrease as much.

At times, the reduction in inside dew point is such that it is about as low as the coil surface temperature. Then, although the coil surface temperature is below freezing, little or no frost accumulates.

With a condition like this, operation may be satisfactory if the compressor can operate at less than 100% effective displacement. This means further reduction of refrigeration capacity. Depending on how the heat gain varies with outside temperature, this control may or may not satisfy the requirements. Full analysis requires not only system performance, but a comparison of that performance with the possibly varying refrigeration requirements.

## OTHER CONSIDERATIONS

The charts shown in Figures 1, 2, 3, and 4 are examples of the many possibilities for graphically illustrating the performance of each item to show the actual result at design conditions, as well as the effect of various changes. Performance of equipment at conditions far from that of design, or the usual use, may not be easy to obtain. However, most of it can be calculated with data and methods provided in this volume and the 1986 REFRIGERATION Volume.

Performance of cooling and dehumidifying coils and air-cooled or evaporative condensers at higher altitudes is not easily obtained but may be calculated (Ramsey 1968).

When air-cooled or evaporative condensers operate with reduced airflow or spray so that the difference between the condensing and entering air temperatures is large, the difference between the condensing and liquid temperatures will also be large. With low entering air temperature and a low airflow rate that is controlled to maintain a relatively high condensing temperature, the actual capacity will be higher than that indicated by the capacity line with full condenser airflow rate. This is because of the lower-temperature liquid (assuming that the liquid does not go through a receiver before reaching the expansion valve). The charts in this chapter do not have curves to show this factor, but the values can be calculated, and such capacity lines can be added when needed.

Low liquid temperature is desirable, but the liquid temperature entering the expansion valve should never be less than that corresponding to saturation at the evaporating pressure. This is particularly so if full evaporator capacity is desired because the refrigerant-side heat-transfer rate for the non-boiling liquid is considerably less than that for evaporation. Such low liquid temperatures are not usual, but with 0°F (−18°C) air to an air-cooled condenser, at a flow rate to maintain 70°F (21°C) condensing temperature, and an evaporating pressure corresponding to 50°F (10°C), the liquid temperature is likely to be below the evaporating temperature.

Depending on the objectives of the analysis, performance diagrams may be prepared using the refrigerant flow rate instead of the refrigerating capacity. If the system uses water or brine as an intermediate fluid between the refrigerant and the air or other cooled fluid, the performance diagrams must include the characteristics of the additional heat exchangers involved.

## TROUBLE DIAGNOSIS

Performance diagrams often help in diagnosing operating difficulties. Comparing actual performance and values of such variables as refrigerant temperatures and pressures with the values predicted by diagrams often suggests the difficulty. It may only be necessary to obtain an indication of the direction in which certain measurable variables should move with changes in operating conditions. For this purpose, free-hand sketches may often be used without plotting the performance characteristics quantitatively to scale.

## SYSTEM PERFORMANCE

Although this chapter only addresses the balancing of air-to-air refrigeration system components, the basic balancing process may be applied to other component systems or may be enlarged to include the characteristics of the load on the refrigeration system.

An example of the use of the procedure in other component systems is the combination of fan, motor, and system resistance to airflow, as discussed in Chapter 3 of this volume.

In air-conditioning system design, superposition of the heat gain characteristics of the conditioned space on the capacity characteristic of the cooling system often will be of value in determining the degree of complexity actually required to meet the objectives of the system.

While the component balancing process is essential to the technical process of selecting components, it also helps developing an overall appraisal of performance. The designer may consider the system as merely an assembly of components that has a known sensitivity to the variables that affect its performance, overlooking the fact that the sensitivity of the system as a whole to the external variables is generally lower. This can result in unwarranted system and control complexity, which can have an important influence on the cost and reliability of the overall system. A study of the way in which components balance when combined yields valuable information on the overall response of the system.

Figures 1, 2, and 3 can all be placed on a single cross-plot with the lines to show interrelations between components under all operating conditions. For widely fluctuating temperatures, loads, and pulldown conditions, multiple coil circuiting, dampers, and expansion valves may be appropriate. MOP (Maximum Operating Pressure) and low ambient compensating features of expansion valves can enhance reliability and performance. Evaporator performance with pump and liquid overfeed systems can also be represented in a graphical analysis. System controls should assume minimal cycling for stability.

## BIBLIOGRAPHY

Ramsey, M.A. 1968. Effect of Altitude on Equipment Performance. *Air Conditioning, Heating and Ventilating,* March, p. 41 and April, p. 57.

Ramsey, M.A. 1966. Tested Solutions to Problems in Air Conditioning and Refrigeration. *Industrial Press, Inc.,* New York.

# CHAPTER 19

# REFRIGERANT-CONTROL DEVICES



THE control of refrigerant flow is essential in any refrigeration system. Part I of this chapter details control switches, including (1) pressure control switches, (2) temperature control switches, (3) differential control switches, and (4) float switches. Part II addresses the operation, selection, and application of control valves, including (1) thermostatic expansion valves, (2) electronic expansion valves, (3) constant pressure expansion valves, (4) evaporator pressure and temperature regulators, (5) suction pressure regulators, (6) condenser pressure regulators, (7) high-side float valves, (8) low-side float valves, (9) solenoid valves, (10) refrigerant-reversing valves, (11) condensing water regulators, (12) check valves, and (13) relief devices. Part III covers oil separators, and Part IV deals with capillary tubes. For further information on automatic control, see Chapter 51 of the 1987 HVAC Volume.

## PART I: CONTROL SWITCHES

The *control switch* operates one or more sets of electrical contacts and opens or closes water or refrigerant valves in response to changes in pressure, temperature, or liquid level. Pressure- and temperature-responsive controls have one or more power elements, which may use bellows, diaphragms, or bourdon tubes to produce the force needed to operate the mechanism. Level-responsive controls use either a float or an electronic probe to operate (directly or indirectly) one or more sets of electrical contacts.

### PRESSURE CONTROL SWITCHES

The refrigerant pressure is applied directly to the power element, which moves against a spring that can be adjusted to control any operation at the desired pressure. If the control is to operate in the subatmospheric (or *vacuum*) range, the bellows or diaphragm force is sometimes reversed to act in the same direction as the adjusting spring. To counteract this, an additional spring may be needed to overcome the reversed bellows force; consequently, the force of both the bellows and the vacuum spring oppose the adjusting spring. However, the controls are sometimes built without any spring and use the spring force of the bellows or diaphragm in place of the adjusting spring.

The force available for doing work in this control switch depends on the pressure in the system and on the area of the bellows or diaphragm. With proper area, a sufficient force can be produced to operate heavy switches, water valves, or refrigerant valves. In heavy-duty controls, the minimum differential is large because of the massiveness of the bellows and the opposing spring system.

### TEMPERATURE CONTROL SWITCHES

Temperature control switches for refrigeration are pressure control switches in which the pressure-responsive power element

The preparation of this chapter is assigned to TC 8.8, Refrigerant System Controls and Accessories.

is replaced by a temperature-responsive power element. The exact temperature-pressure or temperature-volume relationship of the fluid used in the power element allows the temperature of the bulb to control the switch accurately. Operation of the control switch results from these changes in pressure or volume.

## DIFFERENTIAL CONTROL SWITCHES

Control switches that maintain a given difference in pressure or temperature between two pipe lines or spaces are called *differential control switches.* An example of this type of control is the oil safety switch used with reciprocating compressors that have forced-feed pressure lubrication. These controls must have two elements (either pressure- or temperature-sensitive) because they must sense conditions in two different locations. Figure 1 is a schematic diagram of a differential pressure control switch that uses bellows as power elements.

As shown, the two elements are rigidly connected by a rod so that motion of one causes motion of the other. On the connecting rod, a power takeoff operates either single-pole double-throw contacts (as shown) or valves. A compression spring permits the setting of the differential pressure at which the device operates. The sum of the forces developed by the low-pressure bellows and the scale spring equals the force developed by the high-pressure bellows at the control point.

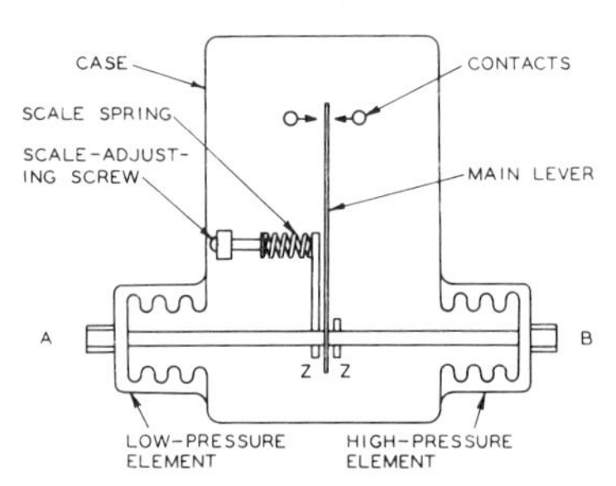


**Fig. 1 Differential Pressure Control Switch**

*Instrumental differential* is the difference in pressure between the low- and the high-pressure elements for which the instrument is adjusted. In the case of a temperature element, this difference is expressed in degrees. *Operating differential* is the difference in pressure or temperature required to open or close the switch contacts. It is actually the change in instrument differential from *cut-in* to *cut-out* for any setting. Operating differential can be varied by a second spring that acts in the same direction as the first and takes effect only at the *cut-in* or *cut-out* point, without affecting the other spring. A second method is the adjustment of the distance between collars *Z-Z* on the connecting rod. The greater the distance between them, the greater the operating differential.

Where a constant instrument differential is required on a temperature-sensitive differential control switch throughout a large temperature range, it is usually necessary to use a different fill in one element than in the other, if both are of the vapor-charged type. The alternative is to use liquid-filled power elements.

A second type of differential-temperature control uses two sensing bulbs and capillaries connected to one bellows with a liquid fill. This is known as a *constant-volume* fill, because the operating point corresponds to a constant volume of the two bulbs, capillaries, and bellows. If the two bulbs have equal volume, a rise in the temperature of one bulb requires an equivalent fall in the temperature of the other to maintain the operating point.

## FLOAT SWITCHES

The *float switch* has a float that operates one or more sets of electrical contacts through variation in the level of a liquid. It is connected by equalizing lines to the vessel in which the liquid level is to be maintained or indicated.

### Operation and Selection

Some float switches (see Figure 2) operate from the movement of a magnetic armature located in the field of a permanent magnet. Others use true solid-state circuits in which a variable signal is generated by liquid contact with a probe that replaces the float. The latter methods are adapted to remote-controlled applications and are preferred for ultralow temperature applications. Switches that have mercury-tube contacts are usually not recommended for installation in an ambient temperature lower than −25°F (−32°C), since mercury freezes solid at temperatures of approximately −38°F (−39°C).

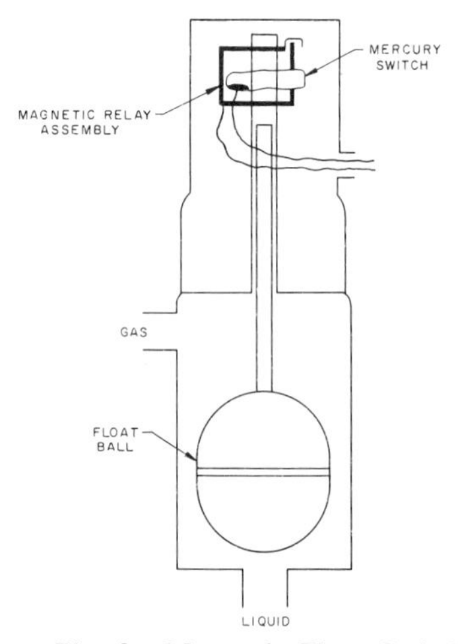


**Fig. 2 Magnetic Float Switch**

### Application

The float switch can maintain or indicate the level of a liquid, operate an alarm, control the operation of a pump, or perform other functions. A float switch, solenoid liquid valve, and hand expansion valve combination can control the refrigerant level on the high- or low-pressure side of the refrigeration system in the same way that high- or low-side float valves are used. The hand expansion valve, located in the refrigerant liquid line immediately downstream of the solenoid valve, is initially adjusted to provide a refrigerant flow rate at maximum load to keep the solenoid liquid valve in the open position 80 to 90% of the time; it need not be adjusted thereafter. From the outlet side of the hand expansion valve, the refrigerant passes through a line and enters either the evaporator or the surge drum, depending on the unit design.

When the float switch is applied for low-side level control, proper precaution must be taken to provide a quiet liquid level that falls in response to an increase in evaporator load and rises with a decrease in evaporator load. The same recommendations for insulation of the body and liquid leg of the lowside float valve apply to the float switch when it is used for refrigerant-level control on the low-pressure side of the refrigeration system. To avoid floodback in this application, controls should be wired to prevent the opening of the solenoid liquid valve when the solenoid suction valve closes or the compressor stops.

# PART II: CONTROL VALVES

Valves are used to start, stop, direct, and modulate the flow of refrigerant to satisfy system requirements in accordance with

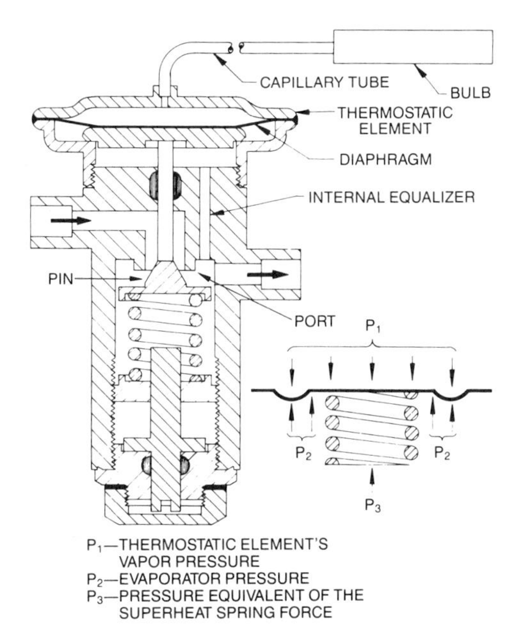


**Fig. 3 Typical Thermostatic Expansion Valve**

load requirements. To ensure satisfactory performance, valves should be protected adequately from foreign material, excessive moisture, and corrosion in refrigeration systems by the installation of properly sized strainers, filters, and filter-driers. Other valve designs and constructions are available. Either a diaphragm or a bellows can be used in various types of refrigerant flow-control valves.

## THERMOSTATIC EXPANSION VALVES

The thermostatic expansion valve controls the flow rate of liquid refrigerant entering the evaporator in response to the superheat of the refrigerant gas leaving it. Its basic function is to keep the evaporator active without permitting unevaporated refrigerant liquid to be returned through the suction line to the compressor. This is done by controlling the mass flow rate of refrigerant entering the evaporator so it equals the rate at which the refrigerant can be completely vaporized in the evaporator by the absorption of heat. Since the thermostatic expansion valve is operated by the superheated refrigerant gas leaving the evaporator and is responsive to changes in superheat of this gas, a portion of the evaporator must be used to superheat the refrigerant gas.

Unlike the constant pressure-type, the thermostatic expansion valve is not limited to constant load applications. It is used successfully for controlling the refrigerant flow to all types of direct-expansion evaporators in air-conditioning and commercial, low-temperature, and ultralow temperature refrigeration systems.

### Operation

A schematic cross section of the typical thermostatic expansion valve, with the principal components identified, is shown in Figure 3. The forces below govern thermostatic expansion valve operation:

$P_1$ = the vapor pressure of the thermostatic element (a function of the bulb's refrigerant charge and the bulb temperature), which is applied to the top of the diaphragm and acts to open the valve.

$P_2$ = the evaporator pressure, which is applied under the diaphragm through the equalizer passage and acts in a closing direction.

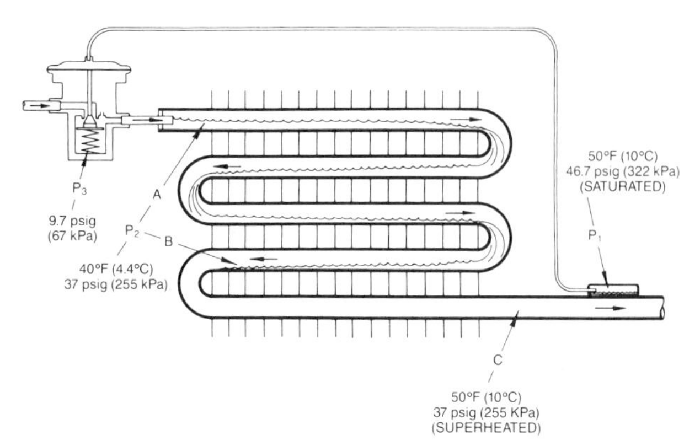


**Fig. 4 Thermostatic Expansion Valve Controlling Flow of Liquid Refrigerant 12 Entering Evaporator, Assuming a Refrigerant 12 Charge in the Bulb**

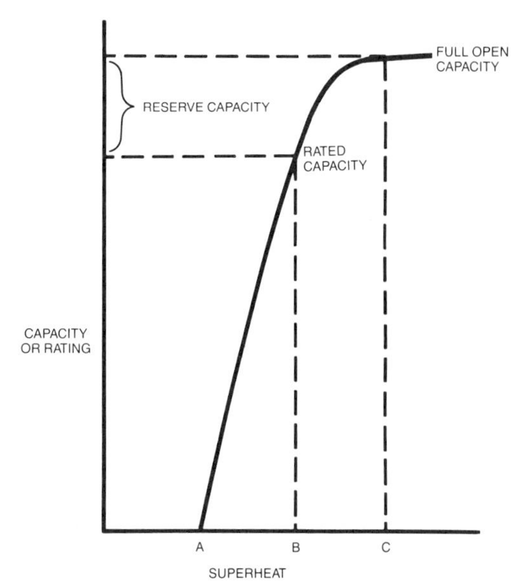


**Fig. 5 Typical Gradient Curve for Thermostatic Expansion Valves**

$P_3$ = the pressure equivalent of the superheat spring force, which is applied underneath the diaphragm and is also a closing force.

At any constant operating condition, these forces are balanced and $P_1 = P_2 + P_3$.

An additional force, which is small and not considered fundamental, arises from the unbalanced pressure across the valve port. To a degree, it can affect thermostatic expansion valve operation. For the configuration shown in Figure 3, the force resulting from port imbalance is the product of the pressure drop across the port and the difference in area of the port and the stem; it is an opening force. In other designs, depending on the direction of flow through the valve, the port imbalance may result in a closing force.

The principal effect of port imbalance is on the stability of the valve control. As with any modulating control, if the ratio of the diaphragm area to the port is kept large, the imbalanced port effect is minor. However, depending on this ratio or system operating conditions, valves are made with balanced port construction.

Figure 4 shows an evaporator that uses R-12 and operates at a saturation temperature of 40°F (4.4°C) (37 psig [255 kPa]).

Liquid refrigerant enters the expansion valve, is reduced in pressure and temperature at the valve port, and enters the evaporator at Point A as a mixture of saturated liquid and vapor. As flow continues through the evaporator, more and more of the boiling refrigerant is evaporated. Assuming there is no pressure drop, the refrigerant temperature remains at 40°F (4.4°C) until the liquid portion is evaporated by the absorption of heat at Point B. From this point, additional heat absorption increases the temperature and superheats the refrigerant gas, while the pressure remains constant at 37 psig (255 kPa), until at Point C (the outlet of the evaporator) the refrigerant gas temperature is 50°F (10°C). At this point, the superheat is 10°F (5.6°C) (50 to 40°F [10 to 4.4°C]).

An increase in the heat load on the evaporator increases the temperature of the refrigerant gas leaving the evaporator. The bulb of the thermostatic expansion valve senses this increase; the thermostatic charge pressure, $P_1$, increases and causes the valve to open wider. The increased flow rate results in a higher evaporator pressure, $P_2$, and a balanced control point is again established. Conversely, a decrease in the heat load on the evaporator decreases the temperature of the refrigerant gas leaving the evaporator and causes the thermostatic expansion valve pin to move in a closing direction.

The new control point, following an increase in valve opening, is at a slightly higher superheat because of the spring rate of the diaphragm and superheat spring. Conversely, a decrease in load results in a superheat slightly lower than the original control point.

These superheat changes in response to load changes are illustrated by the gradient curve of Figure 5. Superheat at no load A, *static superheat,* ensures sufficient spring force to keep the valve closed during equipment shutdown. An increase in valve capacity or load is roughly proportional to superheat increase until the valve is open fully. The *opening superheat,* represented by the distance AB, may be defined as the increase in superheat required to open the valve to the capacity at which it is intended to operate or at which it is rated. *Operating superheat* is the sum of the static superheat and opening superheat.

## Capacity

The *factory superheat setting* (static superheat setting) of thermostatic expansion valves is made when the valve pin starts to move away from the seat. Valve manufacturers establish capacity ratings on the basis of opening superheat from 4 to 8°F (2.2 to 4.4°C), depending on valve design, valve size, and application. Full-open capacities usually exceed rated capacities by 10 to 40% to allow a reserve, represented by the distance BC in Figure 5, for manufacturing tolerances and application contingencies.

System design should not be based on use of the reserve capacity of the thermostatic expansion valve, which is obtained only at the expense of higher superheat. The added superheat gradient may have an adverse effect on valve performance. Because valve gradients used for rating purposes generally produce optimum modulation for a given valve design, manufacturers' recommendations should be followed.

Thermostatic expansion valve capacities are normally published for various evaporator temperatures and valve pressure drops. (See ASHRAE *Standard* 17-82 and ARI *Standard* 750-81 for testing and rating methods.) Nominal capacities apply at 40°F (4.4°C) evaporator temperature. Capacities are reduced at lower evaporator temperatures. These reductions in capacity are the result of the change in the refrigerant pressure-temperature relationship at lower temperatures. For example, if Refrigerant 12 is used, the change in saturated pressure between 40 and 45°F (4.4 and 7.2°C) is 4.7 psi (32.4 kPa), whereas between −20 and −15°F (−28.9 and −26.1°C) the change is 1.9 psi (13.1 kPa). Although the valve responds to pressure changes, capacities are based on superheat gradients. Thus, the valve opening and, consequently, the valve capacity, is less for a given superheat change at the lower evaporator temperatures.

Pressure drop across the valve port is always the net pressure drop available at the valve, rather than the difference between compressor discharge and compressor suction pressures.

Allowances must be made for the following:

1. *Friction loss* through condenser, liquid lines, and fittings.
2. *Liquid line accessories,* such as filters, driers, solenoid valves, etc.
3. *Static head* in a vertical liquid line. If the thermostatic expansion valve is at a higher level than the receiver, there will be a pressure loss in the liquid line because of the static head of liquid.
4. *Distributor pressure drop.*
5. *Evaporator pressure drop.*
6. *Suction line accessories,* such as evaporator regulators, solenoid valves, etc.
7. *Suction line friction losses.*

Although the mass flow rate through the valve is greater for higher pressure drops, the capacity increase is not proportional. The mass flow rate through the valve varies according to the square root of the pressure drop but because of the greater enthalpy of the entering liquid refrigerant, a greater mass flow rate is required to obtain a given refrigerating effect. At high liquid temperatures, the capacity increase resulting from higher pressure drops may be offset by the loss in net refrigerating effect, resulting in no change in capacity.

If liquid refrigerant subcooling is increased along with the liquid pressure, the capacity changes differ from those addressed previously. Subcooling, in addition to increasing the net refrigerating effect per unit of mass flow, reduces the amount of flashing in the valve port and also adds to valve capacity. Most manufacturers publish liquid temperature correction factors to be applied to valve capacity ratings.

Thermostatic expansion valve ratings are based on vaporfree saturated liquid entering the valve. If flash gas is present in the entering liquid, the valve capacity is reduced substantially because the gas must be handled along with the liquid. Flashing of the liquid refrigerant may be caused by pressure drop in the liquid line because of friction losses through the line, the filters, the driers, the vertical lift, or a combination of these. If the refrigerant subcooling at the receiver outlet is not adequate to prevent the formation of flash gas, additional subcooling means must be used to remove it. Liquid-to-suction heat exchange provides a moderate degree of subcooling, but for extreme requirements, a separate liquid-cooling coil may be necessary.

## Thermostatic Charges

Each type of thermostatic charge has certain advantages and limitations. The principal types of thermostatic charges and their characteristics are described below.

**Gas Charge.** Conventional gas charges are limited liquid charges that use the same refrigerant in the thermostatic element that is used in the refrigeration system. The amount of charge is such that, at a predetermined temperature, all of the liquid has vaporized and any temperature increase above this point results in virtually no increase in element pressure. Figure 6 shows the pressure-temperature relationship of the Refrigerant 12 gas charge in the thermostatic element. Because of the characteristic pressure-limiting feature of its thermostatic element, the gas-charged valve can provide compressor motor

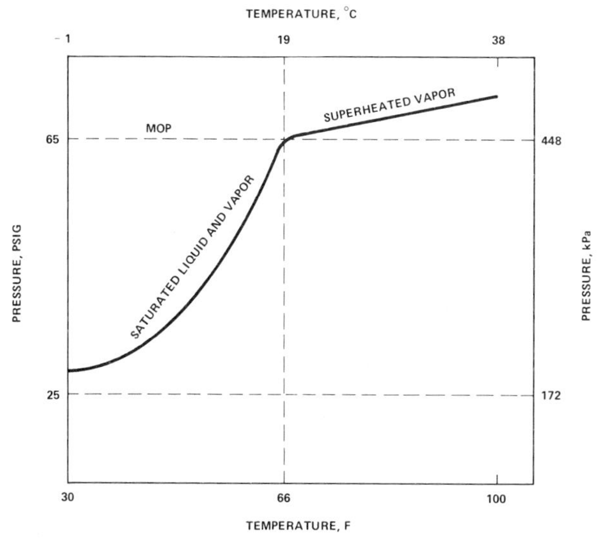


**Fig. 6 Pressure-Temperature Relationship of Refrigerant 12 Gas Charge in Thermostatic Element**

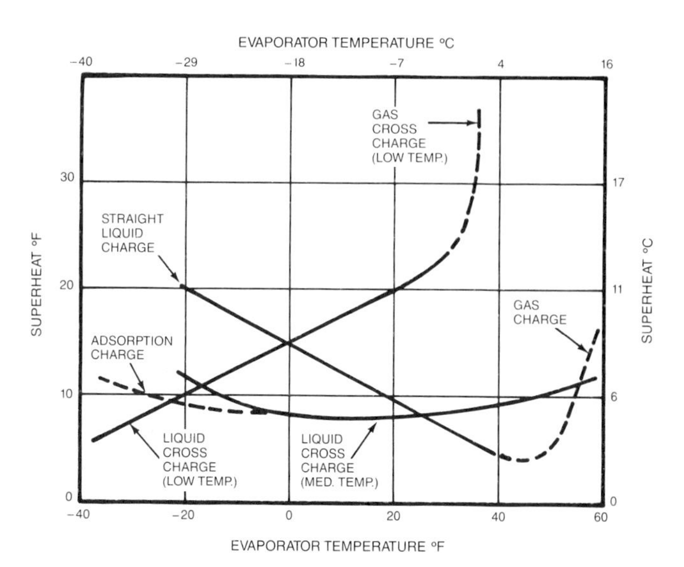


**Fig. 7 Typical Superheat Characteristics of Common Thermostatic Charges**

overload protection on some systems by limiting the maximum operating suction pressure. It also helps to prevent *floodback* (the return of refrigerant liquid to the compressor through the suction line) on starting. Increasing the superheat setting lowers the maximum operating suction pressure; decreasing the superheat setting raises it because the superheat spring, together with the evaporator pressure, acts directly on the element pressure through the diaphragm.

Gas-charged valves must be carefully applied to avoid loss of control from the bulb. If either the diaphragm chamber or the capillary tube becomes colder than the bulb, the small amount of charge in the thermostatic element condenses. Control at the bulb is lost and the valve throttles or closes. This matter is detailed in the section on "Application."

**Liquid Charges.** Straight liquid charges use the same refrigerant in the thermostatic element that is used in the refrigeration system. The volumes of the bulb, capillary tubing, and diaphragm chamber are so proportioned that the bulb contains some liquid under all temperature conditions. Therefore, the bulb always controls the valve operation, even with a colder diaphragm chamber or capillary tube.

The characteristics of the straight liquid charge (see Figure 7) result in an increase in operating superheat as the evaporator temperature decreases. This usually limits the use of the straight liquid charge to moderately high evaporator temperatures. The valve setting required for a reasonable operating superheat at a low evaporator temperature may cause floodback during pulldown from normal ambient temperatures.

**Liquid Cross Charges.** The liquid cross charges, unlike the conventional liquid charges, use a liquid in the thermostatic element that is different from the refrigerant in the system. Cross charges have flatter pressure-temperature curves than the system refrigerants with which they are used. Consequently, their superheat characteristics differ considerably from those of the straight liquid or gas charges.

Cross charges in the commercial temperature range generally have superheat characteristics that are nearly constant or that deviate only moderately through the evaporator temperature range. This charge, also illustrated in Figure 7, is generally used in the evaporator temperature range of 40 to 0°F (4.4 to −17.8°C) or slightly below.

For evaporator temperatures substantially below 0°F (−17.8°C), a more extreme cross charge may be used. At high evaporator temperatures, the valve controls at a high superheat. As the evaporator temperature is reduced to the normal operating range, the operating superheat is also reduced to normal. This characteristic prevents floodback on starting, reduces the load on the compressor motor after start-up, and permits a rapid pulldown of suction pressure. To avoid floodback, valves with this type of charge must be set for the optimum operating superheat at the lowest evaporator temperature expected.

**Gas Cross Charges.** The gas cross charges combine the features of the gas charge and the liquid cross charge. They use a limited amount of liquid, thereby providing a maximum operating pressure. The liquid used in the charge is different from the refrigerant in the system and is chosen to provide superheat characteristics similar to those of the liquid cross charges (low temperature). Consequently, they provide both the superheat characteristics of a cross charge and the maximum operating pressure of a gas charge (Figure 7). While a commercial (medium temperature) gas cross charge is possible, its uses are limited.

**Adsorption Charge.** Typical adsorption charges depend on the property of an adsorbent, such as silica gel or activated charcoal, that is used in an element bulb to adsorb and desorb a gas such as carbon dioxide, with accompanying changes in temperature. The amount of adsorption or desorption changes the pressure in the thermostatic element to produce a variable valve motion. Since adsorption charges respond only to the temperature of the adsorbent material, they are essentially unaffected by cross-ambient conditions. The comparatively slow response time of the adsorbent results in a charge characterized by its stability. Superheat characteristics can be varied by using different charge materials, adsorbents, and/or charge pressures. The pressure-limiting feature of the gas or gas cross charges is not available with the adsorption-type element.

**Mechanical Pressure-Limiting Means.** While the low-temperature liquid cross charge offers compressor-motor overload protection, and the gas and gas cross charges offer more positive protection, each has limitations. The liquid cross charges partially limit suction pressure. The gas and gas cross charges can suffer from charge condensation or migration, if the element becomes colder than the bulb.

Thermostatic expansion valves, constructed with a mechanical pressure-limiting means to restrict suction pressure, may use either the liquid or the liquid cross charge without any charge migration problems. The valve is throttled by a collapsible member, which operates only when the evaporator pressure exceeds a specified value. At lower evaporator pressures, there is no interference with normal valve operation.

## Type of Equalization

**Internal Equalizer.** When the refrigerant pressure drop through a single-circuit evaporator is equivalent to a 2°F (1.1°C) or less drop in evaporator temperature, a thermostatic expansion valve that has an internal equalizer may be used. Internal equalization describes valve outlet pressure transmitted through an internal passage to the underside of the diaphragm (see Figure 3).

The friction loss in many evaporators is greater than the 2°F (1.1°C) equivalent. Where a pressure-drop refrigerant distributor is used, the pressure drop across the distributor may be as high as 25 to 35 psi (170 to 240 kPa), and the pressure at the outlet of the expansion valve is considerably higher than the pressure at the evaporator outlet. As a result, an internally equalized valve controls at an abnormally high superheat because $P_2$ is a false pressure and does not reflect the true pressure in the evaporator. Under these conditions, the evaporator does not perform efficiently because it is starved for refrigerant. Furthermore, the false pressure is not constant but varies with evaporator load and cannot be compensated for by adjusting the superheat setting of the valve.

**External Equalizer.** Because evaporator and/or refrigerant distributor pressure drop produces poor system performance with an internal equalizer valve, a valve that has an external equalizer is used. Instead of the internal communicating passage shown in Figure 3, an external connection to the underside of the diaphragm is provided. The external equalizer line is connected either to the suction line, as shown in Figure 8, or into the evaporator at a point downstream from the major pressure drop.

## Alternate Types of Construction

**Pilot-operated thermostatic expansion valves** are used on large systems where the required capacity per valve is beyond the range of direct-operated valves. The pilot-operated valve consists of a piston-type pilot-operated regulator, which is used as the main expansion valve, and a low-capacity thermostatic expansion valve, which serves as an external pilot valve. The small pilot thermostatic expansion valve supplies pressure to the piston chamber or, depending on the regulator design, bleeds pressure from the piston chamber in response to a change in the operating superheat. Pilot operation permits the use of a characterized port in the main expansion valve to provide good modulation over a wide loading range. Therefore, the pilot-operated valve performs well on refrigerating systems that have some form of compressor capacity reduction, such as cylinder unloading. Figure 9 illustrates such a control valve applied to a large capacity, direct-expansion chiller.

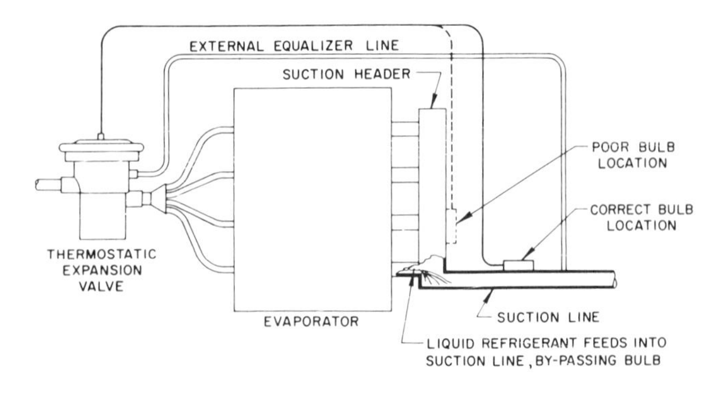


**Fig. 8 Bulb Location for Thermostatic Expansion Valve**

The auxiliary pilot controls should be sized to handle only the pilot circuit flow; this permits the use of smaller controls than would be needed if the full system capacity were to be handled. For example, in Figure 9, a small solenoid valve in the pilot circuit, installed ahead of the thermostatic expansion valve, converts the pilot-operated valve into a stop valve when the solenoid valve is closed.

**Equalization Features.** When the compressor stops, a thermostatic expansion valve usually moves to the *closed* position. This movement sustains the difference in refrigerant pressures in the evaporator and the condenser. Low-starting torque motors require that these pressures be equalized to reduce the torque needed to restart the compressor. One way to lower starting torque is to add, parallel to the main valve port, a small fixed auxiliary passageway, such as a slot or drilled hole in the valve seat or valve pin. This opening permits a limited fluid flow through the control, even when the valve is closed and allows the system pressures to equalize on the *off* cycle. The size of such a fixed auxiliary passageway must be limited so its flow capacity is not greater than the smallest flow that must be controlled in normal system operation.

Another more complex control is available for systems requiring shorter equalizing times than can be achieved with the fixed auxiliary passageway. This control incorporates an auxiliary valve port, which bypasses the primary port and is opened by the element diaphragm as it moves toward and beyond the position at which the primary valve port is closed. The flow capacity of such an auxiliary valve can be considerably larger than that of the fixed auxiliary passageway, so that pressures can equalize more rapidly.

**Flooded System.** Thermostatic expansion valves are seldom applied to flooded evaporators because superheat is necessary for proper valve control; only a few degrees of suction vapor superheat in a flooded evaporator incurs a substantial loss in capacity. If the bulb is installed downstream from a liquid-to-suction heat exchanger, a thermostatic expansion valve can be made to operate at this point on a higher superheat. Valve control is apt to be poor because of the variable rate of heat exchange as flow rates change (see "Application" section).

Expansion valves with modified thermostatic elements are available in which electrical heat is supplied to the bulb. The bulb is inserted in direct contact with refrigerant liquid in a low-side accumulator. The contact of cold refrigerant liquid with the bulb overrides the artificial heat source and throttles the expansion valve. As the liquid falls away from the bulb, the valve

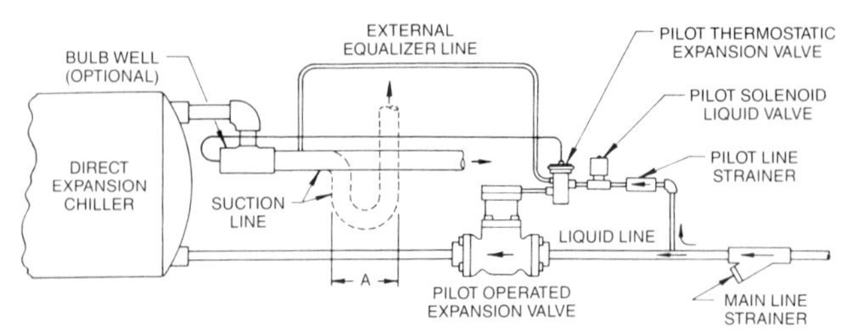


**Fig. 9 Pilot-Operated Thermostatic Expansion Valve Controlling Liquid Refrigerant Flow to a Direct-Expansion Chiller**

feed increases again. Although similar in construction to a thermostatic expansion valve, it is essentially a modulating liquid level control valve.

**Desuperheating Valves.** Thermostatic expansion valves with special thermostatic charges are used to reduce gas temperatures (superheat) on various air conditioning and refrigeration systems. Suction gas in a single-stage system can be desuperheated by injecting liquid directly into the suction line. This cooling may be required with or without discharge gas bypass used for compressor capacity control. The line upstream of the valve bulb location must be long enough so the injected refrigerant can mix adequately with the gas being desuperheated. On compound, multistage systems, liquid is injected directly into the interstage line upstream of the valve bulb.

## Application

**Hunting** is alternate overfeeding and starving of the refrigerant feed to the evaporator, which produces sustained cyclic changes in the pressure and temperature of the refrigerant gas leaving the evaporator. Extreme hunting reduces the capacity of the refrigeration system because the mean evaporator pressure and temperature are lowered, and the compressor capacity is reduced. If overfeeding of the expansion valve causes intermittent flooding of liquid into the suction line, the compressor may be damaged.

Although hunting is commonly attributed to the thermostatic expansion valve, it is seldom solely responsible. One reason for hunting is that all evaporators have a time lag. When the bulb signals for a change in refrigerant flow, the refrigerant must traverse the entire evaporator before a new signal reaches the bulb. This lag or time lapse may cause continuous overshooting of the valve both opening and closing. In addition, the thermostatic element, because of its mass, has a time lag that may be in phase with the evaporator lag and amplify the original overshooting.

It is possible to alter the response rate of the thermostatic element by either using thermal ballast or changing the mass or heat capacity of the bulb, thereby damping or even eliminating the hunting. A change in valve gradient may produce the same result.

Extremely high refrigerant velocity in the evaporator can also cause hunting. Liquid refrigerant under these conditions moves in waves, called *slugs,* that fill a portion of the evaporator tube and erupt into the suction line. These unevaporated slugs chill the bulb and temporarily reduce the feed of the valve, resulting in intermittent starving of the evaporator.

On multiple-circuit evaporators, a lightly loaded or overfed circuit will also flood into the suction line, chill the bulb, and throttle the valve. Again, the effect is intermittent; when the valve feed is reduced, the flooding ceases, and the valve reopens.

Hunting can be minimized or avoided by the following:

1. Select the proper valve size from the valve capacity ratings rather than nominal valve capacity; oversized valves aggravate hunting.
2. Throttle the valve by changing the valve adjustment. A lower superheat setting usually (but not always) increases hunting (Huelle 1972; Wedeking and Stoecker 1966; Stoecker 1966).
3. Select the correct thermostatic element charge. Cross-charged elements have inherent anti-hunt characteristics.
4. Design the evaporator section for even refrigerant and airflow. Uniform heat transfer from the evaporator is only possible if refrigerant is distributed by a properly selected and applied refrigerant distributor and air distribution is controlled by a properly designed housing. (Air-cooling and dehumidifying coils, including refrigerant distributors, are detailed in Chapter 6.)
5. Size and arrange suction piping correctly.
6. Locate and apply the bulb correctly.
7. Select the best location for the external equalizer line connection.

**Bulb Location.** Most installation requirements are met by strapping the bulb to the suction line to obtain good thermal contact between them. Normally, the bulb is attached to a horizontal line upstream of the external equalizer connection (if used) at a 3 or 9 o'clock position as close to the evaporator as possible. While the bulb is not normally placed near or after suction line traps, some system designers test and prove locations that differ from these recommendations. A good moisture-resistant insulation over the bulb and suction line eliminates any adverse effect of varying ambient temperatures at the bulb location.

Occasionally, the bulb of the thermostatic expansion valve is installed downstream from a liquid-suction heat exchanger to compensate for a capacity shortage due to an undersized evaporator. While this procedure seems to be a simple method of obtaining maximum evaporator capacity, installing the bulb downstream of the heat exchanger is undesirable from a control standpoint. As the valve modulates, the liquid flow rate through the heat exchanger changes, causing the rate of heat transfer to the suction vapor to change. An exaggerated valve response follows, resulting in hunting. If may be possible to find a bulb location downstream from the heat exchanger that reduces the hunt considerably. However, the danger of floodback to the compressor normally overshadows the need to attempt this method.

Certain installations require increased bulb sensitivity as a protection against floodback. The bulb, if located properly in a well in the suction line, has a rapid response feature because of its intimate contact with the refrigerant stream. The bulb sensitivity can be increased by the use of a bulb smaller than is normally supplied. However, the use of the smaller bulb is limited to gas-charged valves. Good piping practice also effects expansion valve performance.

Figure 10 illustrates the proper piping arrangement when the suction line runs above the evaporator. An oil trap as short as possible is located downstream from the bulb. The vertical riser or risers must be sized to produce a refrigerant velocity that ensures continuous return of oil to the compressor. The terminal end of the riser or risers enters the horizontal run at the top of the suction line; this avoids both interference from the overfeeding of any other expansion valve or any drainback during the *off* cycle.

If circulated with oil-miscible refrigerants, a heavy concentration of oil elevates the refrigerants' boiling temperatures. The response of the thermostatic charge of the expansion valve is related to the saturation pressure and the temperature of pure refrigerant. In an operating system, the false pressure-temperature signals of oil-rich refrigerants cause both floodback or operating superheats considerably lower than indicated and quite often cause erratic valve operations. To keep the oil concentration at an acceptable level, either the oil pumping rate must be reduced or an effective oil separator must be used.

The external equalizer line is ordinarily connected at the evaporator outlet, as shown in Figure 8. It may also be connected at the evaporator inlet or at any other point in the evaporator downstream of the major pressure drop. On evaporators with long refrigerant circuits that have inherent lag, hunting may be minimized by changing the connection point of the external equalizer line.

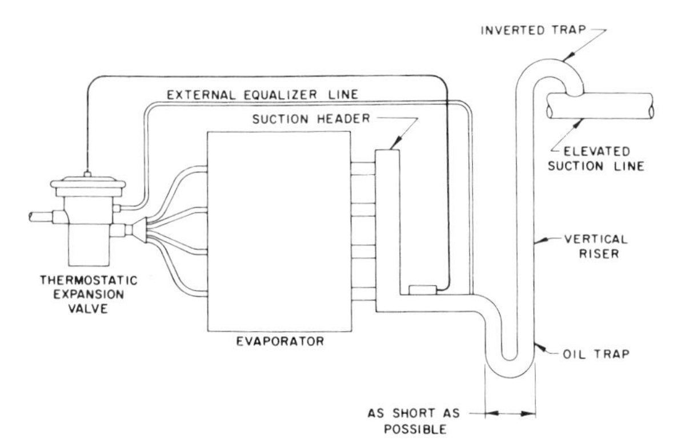


**Fig. 10 Bulb Location When Suction Main Is Above Evaporator**

The ambient temperature at the valve does not ensure a corresponding temperature of the diaphragm chamber because a certain amount of heat is conducted to the cooler valve body. It is not unusual for the diaphragm-chamber temperature to fall below the bulb temperature. When this fall occurs with a gas-charged valve, the entire thermostatic charge can condense in the diaphragm chamber, and the refrigerant feed can then be controlled from that point. Extreme starving of the evaporator, to the point of complete cessation of feed, is characteristic of this condition. For this reason, gas-charged valves are normally used only when sufficient pressure drop exists between the outlet of the valve and the bulb location. On multiple-circuit evaporators, the refrigerant distributor serves this purpose. The pressure drop through the distributor elevates the valve body temperature above that of the evaporator and assists in maintaining control at the valve bulb. Straight liquid and liquid cross-charged valves operate regardless of the temperature of the space in which they are located due to sufficient charge in the diaphragm housing, cap tube, and bulb to ensure control at the bulb.

Direct expansion chillers located in a heated environment are among the few applications in which gas charged valves operate successfully without refrigerant distributors. This is possible because operation occurs at considerably lower superheats (4 to 6°F [2 to 3°C]) and narrower ranges of temperatures than other types of direct expansion evaporators. When this type of chiller is subjected to cold outdoor ambients, special considerations to prevent charge migration to the diaphragm housing of the valve element are needed. The absorption charge can be used for these applications because it is unaffected by the colder ambients. If a pressure-limiting feature is necessary, special hydraulic elements can use the gas cross charge in such a way as to prevent charge migration.

## ELECTRONIC EXPANSION VALVES

Electronically controlled, solid-state expansion valves have become widely available. The electronic expansion valve, similar to the thermostatic expansion valve, is a liquid refrigerant flow-control device. These valves may be categorized into the following four types:

1. Heat motor operated
2. Magnetically modulated
3. Pulse width modulated; *on-off* type
4. Step motor driven

*Heat motor valves* may be either of two types. In the first type, one or more bimetallic elements are heated electrically, causing them to deflect. The bimetals are linked mechanically to a valve pin or poppet. In a second type of heat motor expansion valve, a volatile material charge is contained within an electrically heated chamber so that the charge temperature (and pressure) is controlled by electrical power input to the heater. The charge pressure is made to act on a diaphragm or bellows, which is balanced against either ambient air pressure or refrigeration system suction pressure. The diaphragm is linked to a pin or poppet.

In *magnetically modulated valves*, a direct current electromagnet modulates smoothly, while an armature, or plunger, compresses a spring progressively as a function of coil current. The modulating electromagnet plunger may be connected to a valve pin or poppet directly or may be used as the pilot element in a servo loop to operate a much larger valve. When the modulating plunger operates a pin or poppet directly, the valve may be of a pressure-balanced port design so that pressure differential has little or no influence on valve opening.

The *pulse width modulated valve* is an *on-off* solenoid valve with special design features that allow it to function as an expansion valve through a life of millions of cycles. Even though the valve is either fully opened or closed, it operates as an infinitely variable metering device by pulsing the valve open regularly. The duration of each opening, or pulse, is regulated by the electronics. For example, a valve may be pulsed every six seconds. If 50% flow is needed, the valve would be held open three seconds and closed for three seconds.

A *step motor* is an electronically commutated multi-phase motor capable of running continuously in forward or reverse, or it can be discreetly positioned in increments of a small fraction of a revolution. Step motors are used in instrument drives, plotters, and other applications where accurate positioning is required. Step motors require electronics, such as an integrated circuit, to switch windings in proper sequence. When used to drive expansion valves, a lead screw changes the rotary motion of the rotor to a linear motion suitable for stroking a valve pin or poppet. The lead screw may be driven directly from the rotor, or a reduction gearbox may be placed between the motor and lead screw. The motor may be hermetically sealed within the refrigerant environment, or the motor and gearbox can be sealed to operate in air.

Electric expansion valves may be controlled by either digital or analog electronic circuits. Electronic control gives the additional flexibility to consider control schemes that are impossible with conventional valves.

## CONSTANT PRESSURE EXPANSION VALVES

The constant pressure expansion valve is operated by the evaporator or valve-outlet pressure; it regulates the mass flow rate of liquid refrigerant entering the evaporator and maintains this pressure at a constant value. Although this valve was first used as a liquid refrigerant expansion valve, other applications are also addressed.

### Operation

Figure 11 shows a schematic cross section of a constant pressure expansion valve. The valve has both an adjustable spring, which exerts its force on top of the diaphragm in an opening direction, and a spring beneath the diaphragm, which exerts its force in a closing direction. Evaporator pressure is admitted beneath the diaphragm, through either the internal or external equalizer passage, and the combined forces of the evaporator pressure and the closing spring act to counterbalance the opening spring pressure.

With the valve set and refrigerant flowing at a given pressure, a small increase in the evaporator pressure forces the diaphragm

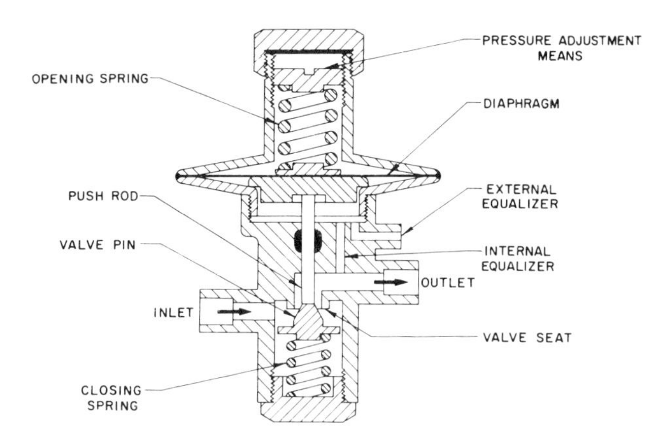


Valve is used with either internal or external equalizer, but not with both.

**Fig. 11 Constant Pressure Expansion Valve**

up, which restricts the refrigerant flow and limits the evaporator pressure. When the evaporator pressure drops below the valve setting due to a decrease in load, the top spring pressure moves the valve pin in an open direction. As a result, the refrigerant flow increases and raises the evaporator pressure to the balanced valve setting. This valve controls the evaporation of the liquid refrigerant in the evaporator at a constant pressure and temperature.

Constant pressure expansion valves automatically adjust the flow of liquid refrigerant to the evaporator to balance compressor pumping capacity. When this balance is established, evaporator pressure remains constant during the remainder of the running phase of the refrigeration cycle. This low-side pressure balancing point is selected by turning the valve adjuster to the desired pressure setting. To avoid floodback to the compressor during low heat load periods, the valve setting should be selected to use evaporator surface effectively when there is a minimum heat load on the system. When there is a maximum heat load on the system, the increase in evaporator pressure causes the valve to close partially. This prevents the pressure from rising any further and overloading the compressor.

For reasons of operation and adjustment, constant pressure expansion valves are most effective on applications with a constant heat load. When the compressor stops at the end of the running phase of the cycle, standard constant pressure expansion valves close to prevent off-cycle refrigerant flow. The rapid equalization features described in the "Thermostatic Expansion Valves" section can also be built into constant pressure expansion valves to provide off- cycle pressure equalization for use with low-starting torque compressor motors.

## Selection

The constant pressure expansion valve, for liquid refrigerant expansion service, should be selected for the required capacity and system refrigerant at the lowest expected pressure drop across the valve and should have an adjustable pressure range to provide the required evaporator (valve outlet) pressure. Deciding whether the valve should be a bleedtype or a standard expansion is up to the designer.

## Application

The constant pressure expansion valve, when applied as a liquid refrigerant expansion valve, is suitable only for constant load applications; therefore, its use is limited. Ordinarily, only one such liquid expansion valve is used on each system. When applied to a variable load, this valve starves the evaporator at the high load and overfeeds it at the low load, with possible compressor damage resulting in the latter case. Because of the difference in the operating principles of the constant pressure expansion valve and the thermostatic expansion valve, both cannot be used in the same system as liquid-refrigerant expansion valves.

The constant pressure expansion valve is used for the pilot valve in some of the large suction pressure regulating valves.

The constant pressure expansion valve may be used alone as the pilot valve on a modified suction pressure regulator, or it may be used to bypass compressor discharge gas in refrigeration systems. Such applications may arise from either the requirement for a reduction in compressor capacity or the requirement for a low-limit control of either the evaporator, the suction pressure, or both. After passing through the valve, the hot gas may be introduced directly into the suction line, in which case some additional desuperheating means may be required, or it may be introduced at the evaporator inlet, where it mixes with the cold refrigerant and is desuperheated before it reaches the compressor. Some applications suitable for constant pressure expansion valves are drink dispensers, food dispensers, water coolers, ice cream freezers, and self-contained room air-conditioning units.

# EVAPORATOR PRESSURE AND TEMPERATURE REGULATORS

The evaporator pressure regulator (back pressure regulator) regulates the evaporator pressure (pressure entering the regulator) at a constant value. It is used in the evaporator outlet or suction line wherever low-limit control of the evaporator pressure or temperature is required.

## Operation

As illustrated in Figure 12, the inlet pressure acts on the bottom of the seat disk and opposes the adjusting spring. The outlet pressure acts on the underside of the bellows and the top of the seating disk and, because the effective areas of the bellows and the port are equal, the two forces cancel each other, and the valve responds to inlet pressure only. When the evaporator pressure rises above the force exerted by the spring, the valve begins to open. When the evaporator pressure drops below the force exerted by the spring, the valve closes. In actual operation, the valve assumes a throttling position to balance system load.

This change in the pressure, which acts on and operates the diaphragm or bellows, is called *differential.* The *total valve differential* is the difference between the pressure at which the valve operates at rated stroke and the pressure at which the valve starts to open.

*Pressure drop* is the difference between the pressure at the valve inlet and the pressure at the valve outlet. Understanding the difference between differential and pressure drop when determining the proper valve size for a given set of load conditions is essential.

Pilot-operated evaporator pressure regulators are either self-contained or high-pressure driven. The self-contained regulator (Figure 13) starts to operate when the evaporator pressure rises above the pressure setting of the diaphragm spring; the diaphragm moves in an opening direction, increasing the pressure above the piston. This increase moves the piston down, thereby moving the main valve in an opening direction and causing the evaporator pressure to drop back to the pressure setting of the pilot.

When this happens, the pilot valve moves in a closing direction, allowing the pressure above the piston to decrease. Then,

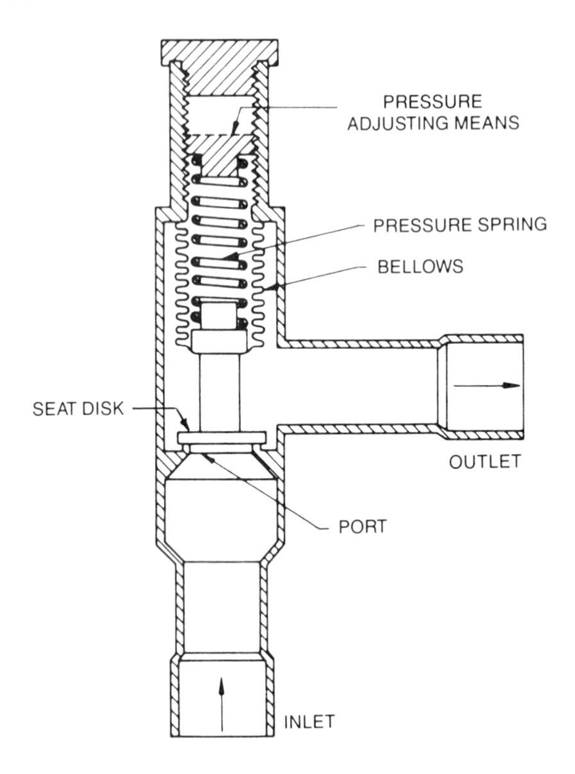


**Fig. 12 Direct-Acting Evaporator Pressure Regulator**

the main spring moves the main valve in the closing direction, preventing the evaporator pressure from falling below the pressure setting of the pilot. In operation, the pilot valve and the main valve assume either intermediate or throttling positions, depending on the load.

Many pilot-operated regulators are of a normally open design and require high-pressure liquid or gas to provide a closing force. The advantage of this regulator is that it requires no minimum operating suction pressure drop to operate. When the valve inlet pressure increases above set point, the pilot valve diaphragm lifts against the adjustment spring load (Figure 14). When the diaphragm moves up in response to the increasing inlet pressure,

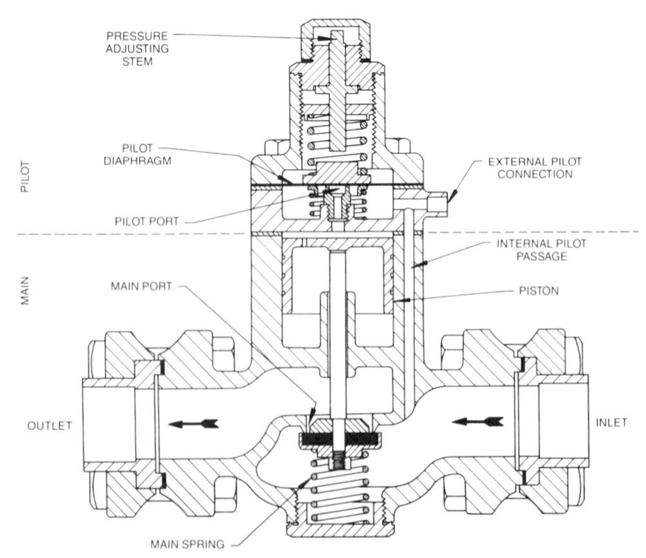


Valve is used with either internal or external equalizer, but not with both.

**Fig. 13 Pilot-Operated Evaporator Pressure Regulator (Self-Contained)**

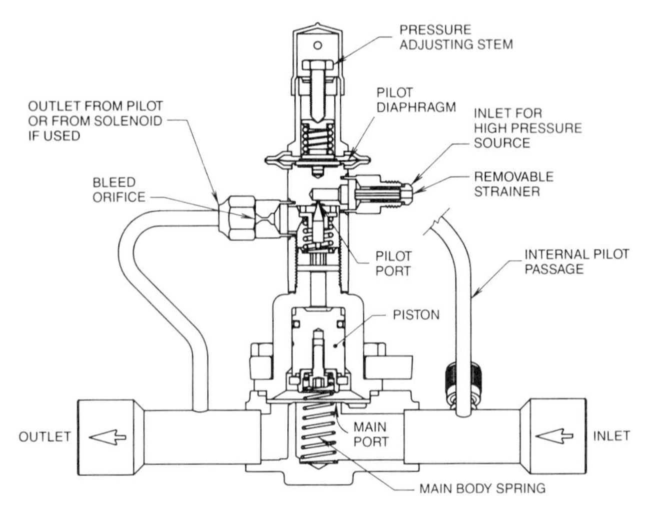


**Fig. 14 Pilot-Operated Evaporator Pressure Regulator (High Pressure Driven)**

the pilot valve pin closes the pilot valve port. Gas or liquid from the high-pressure side of the system is throttled by the pilot, allowing the main piston spring to move the valve to a more open position. Because the top of the piston chamber bleeds to the downstream side of the valve through a bleed orifice, a continuous flow of high-pressure liquid or gas through the pilot valve drives the piston down to a closed or partically closed position. A solenoid valve may be used to drive the piston down, closing the valve for defrost operation.

## Selection

Unless otherwise dictated by special system requirements, the evaporator pressure regulator should be selected for the required capacity and system refrigerant, at the required evaporator (regulator inlet) pressure and the lowest pressure drop across the regulator, consistent with good regulator performance and economical compressor operation.

## Application

Evaporator pressure regulators are used on finned-coil evaporators where frosting should be prevented or evaporator pressure must be maintained constant and higher than suction pressure to prevent dehumidification. These regulators are used on water and brine chiller applications to prevent freezeup of the chillers under low-load conditions.

On multievaporator installations, as shown in Figure 15, evaporator pressure regulators (direct-acting or pilot) can be installed to control evaporator pressure and temperature in each unit. The regulators maintain the desired evaporator temperature in the warmer units, while the compressor continues to operate to satisfy the coldest unit. With such a system, the compressor may be controlled by either a low-pressure switch or by thermostats installed in the individual units.

The evaporator pressure regulator, with internal pilot passage, receives its source of pressure for pilot operation at the regulator inlet connection, whereas the regulator with the external pilot connection not only permits a choice of pressure source for pilot operation, but also permits the use of a remote pressure pilot or pilot valves for other purposes.

If the regulator inlet pressure is unsteady and adversely affects the regulator performance, a source of pressure steadier than that available at the regulator inlet may be obtained for the pilot by connecting the external pilot line to a surge drum or enlarged section of suction line upstream from the regulator.

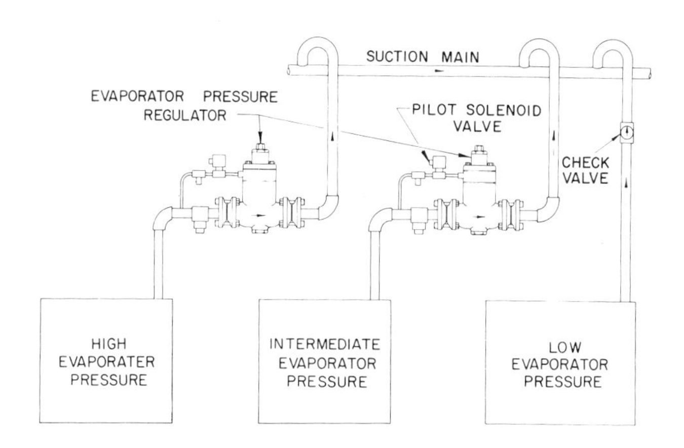


**Fig. 15 Evaporator Pressure Regulators in a Multiple System**

A remote pressure pilot installed in the external pilot line facilitates adjustment of the pressure setting when the regulator must be installed in an inaccessible location. A pressure pilot on the self-contained design with a pneumatic connection permits resetting and controlling of the evaporator pressure and temperature by a pneumatic control system.

In Figure 15 a pilot solenoid valve installed in the external pilot line enables the regulator to function as a suction stop valve, as well as an evaporator pressure regulator. This feature is particularly useful on a flooded evaporator to prevent *pumping-out* of the evaporator, which occurs when the thermostat is satisfied and the compressor continues to operate to cool other evaporators.

A dual-pressure regulator may be attained by using two pressure pilots. The lower pressure pilot is piped through a pilot solenoid valve and, when energized, controls the lower pressure. The higher pressure, which occurs when the pilot solenoid valve is deenergized, is generally used for a warmer evaporator requirement or for pressure above 32 °F (0 °C) saturation for defrosting.

In some instances, it is desirable to control, at a constant temperature, the air or liquid entering or leaving the evaporator by modulating the evaporator pressure and temperature in accordance with the load demand so the evaporator temperature is decreased during heavy loads and increased during lighter loads. This can be accomplished with an evaporator pressure regulator modified in one of the following ways:

1. By using a temperature-actuated pilot in the external pilot line in place of the pressure pilot and by placing the bulb of the temperature pilot where it will respond to the temperature of the air or liquid being cooled. The temperature pilot modulates the position of the regulator according to the load requirement and maintains the air or liquid at a constant temperature.
2. By using a pneumatic thermostat and a pressure pilot with a pneumatic connection so that the evaporator pressure and temperature are raised by an increase in air pressure. This increase is supplied to the pressure pilot at reduced loads and is lowered by a decrease in air pressure supplied to the pressure pilot at increased loads, as required to maintain the temperature of the pneumatic thermostat bulb. This bulb is placed to respond to the temperature of the air or liquid being cooled. Thus, the air or liquid being cooled is maintained at a constant temperature.
3. By using a pressure pilot that is adaptable to a reversible electric motor drive, which resets both the pressure pilot and a potentiometertype thermostat to control the reversible electric motor. The evaporator pressure and temperature are raised at reduced loads and lowered at increased loads, as required to maintain the temperature of the thermostat bulb, which is placed to respond to the temperature of the air or liquid being cooled. The air or liquid being cooled is maintained at a constant temperature.

In addition to the normal applications, these regulators, when equipped with either high-pressure pilots and suitable valve-seat material or a higher range pressure spring, such as in the direct-acting regulator, are used in several methods of application in refrigeration systems that have air-cooled condensers. The regulators are used for maintaining low limit control during cold weather operation of the compressor discharge, the liquid line pressure, or both.

Electronically piloted, suction throttling valves have been developed to control temperature in a food merchandising refrigerator or other refrigerated space (Figure 16). This valve is a form of evaporator pressure regulator, although it responds only to temperature in the space, rather than pressure in the evaporator or suction line. The system consists of a temperature sensor, an electronic control circuit, and a suction throttling valve. A temperature setting is made by turning a calibrated potentiometer or rotary switch normally located on the control circuit. The valve responds to the difference between set point temperature and the prevailing temperature. A temperature above the set point drives the valve further open, while a temperature below set point modulates the valve in the closing direction. During defrost, the control circuit drives the valve tightly closed.

By modulating suction gas flow in response to the measured temperature, the refrigerated space may be held close to the set point regardless of variations in heat load or compressor suction pressure. An electronic regulator is particularly useful on refrigerators containing fresh meat or other products where close control or temperature is important. In some instances, an energy savings may be realized, because a temperature regulating control reduces refrigeration during low heat load conditions more effectively than an evaporator pressure regulator.

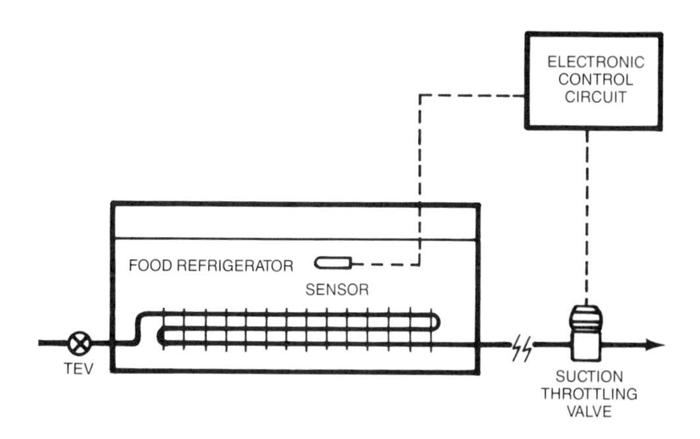


**Fig. 16 Electronic Temperature Controlling Suction Throttling Regulator**

## SUCTION PRESSURE REGULATORS

The suction pressure regulator (holdback valve or crankcase pressure regulator) limits the compressor suction pressure (regulator outlet pressure) to a maximum value. This type of regulator should be used in the suction line at the compressor on any refrigeration installation in which the liquid expansion valve cannot limit the suction pressure and compressor motor overload would otherwise exist because of the following:

1. Excessive starting load

2. Excessive suction pressure following the defrost cycle
3. Prolonged operation at excessive suction pressure
4. Low voltage and high suction pressure conditions

### Operation

Direct-acting suction pressure regulating valves sense only their outlet or downstream pressure (compressor crankcase or suction pressure). As illustrated in Figure 17, the inlet pressure acts on the underside of the bellows and the top of the seating disk. Since the effective areas of the bellows and port are equal, these forces cancel each other and do not affect valve operation. The valve outlet pressure acts on the bottom of the disk and exerts a closing force, which is opposed by the adjustable spring force. When the outlet pressure drops below the equivalent force exerted by the spring, the valve moves in an opening direction. If the outlet pressure rises, the valve moves in a closing direction and throttles the refrigerant flow to maintain the set point of the valve.

Externally and internally pilot-operated suction pressure regulators are available for larger system applications. Although their design is more complex, due to pilot operation, their method of operation is similar to that described above.

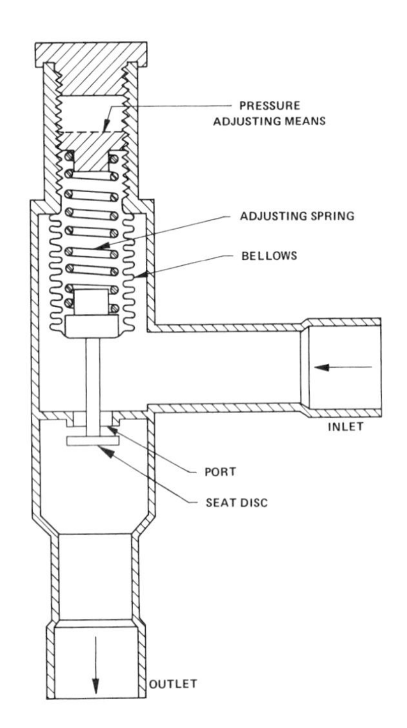


**Fig. 17 Direct-Acting Suction Pressure Regulator**

### Selection

The suction pressure regulator should be selected for the required capacity and system refrigerant, the required regulator inlet pressure, and the lowest practical pressure drop across the regulator to minimize any loss in system capacity.

### Application

In addition to the normal applications, these regulators—when equipped with high-pressure pilots, suitable valve-seat material, or a higher range pressure spring such as in the direct-acting regulator—are used in several methods of application in refrigeration systems with air-cooled condensers. They are used for maintaining low-limit control of the compressor discharge, liquid line pressure, or both, during cold weather operation. In addition, such modified regulators are used to bypass compressor discharge gas in refrigeration systems, as described in the "Application" section for the constant pressure expansion valve.

## CONDENSER PRESSURE REGULATORS

Various condenser pressure regulating valves are used to maintain sufficient condensing pressure to allow air-cooled condensers to operate properly during the winter. Both single- and two-valve arrangements have been used for this purpose. See Chapter 3 of the 1986 REFRIGERATION Volume and Chapter 15 of this volume for more information.

The two-valve arrangement often uses a valve that is constructed and operates similarly to the evaporator pressure-regulating valve shown in Figures 12 and 13. This control is installed either in the liquid line between the condenser and receiver or in the discharge line. It throttles when the condenser or discharge pressure falls as a result of a low ambient condition.

The second valve in the two-valve arrangement bypasses discharge gas around the condenser to the receiver to mix with cold liquid and maintain adequate high-side pressure. Several bypass valves are available, some of which are similar to the suction pressure-regulating valve shown in Figure 17. This valve responds to outlet pressure (receiver pressure). When receiver pressure decreases as a result of a decrease in ambient temperature, the valve opens and bypasses discharge bypass gas to the receiver. Figure 18 shows another device that responds to changes in pressure between its inlet and outlet. As the differential pressure increase, the valve opens. Thus, when the other valve in this two-valve arrangement throttles and restricts liquid flow, a differential is created, and this bypass device opens.

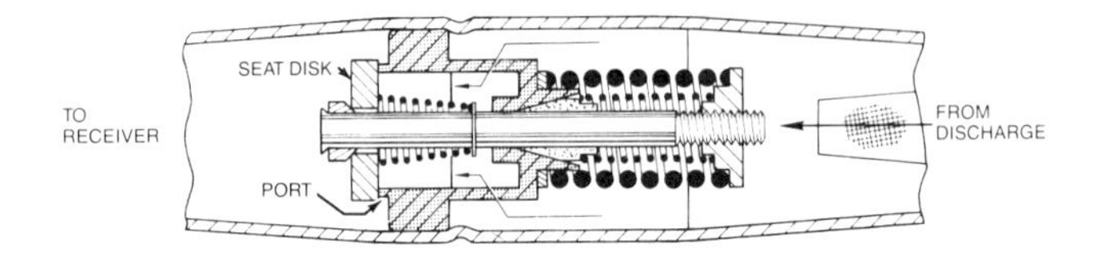


**Fig. 18 Condenser Bypass Valve**

It is sometimes an advantage to substitute a single three-way condenser pressure-regulating valve for the two-valve arrangement described previously. The three-way valve (Figure 19) simultaneously holds back liquid in the condenser and passes compressor discharge into the receiver to maintain pressure in the liquid line. The lower side of a metal diaphragm is exposed to system high-side pressure, while the upper side is exposed to a noncondensable gas charge (usually dry nitrogen). A pushrod connects the diaphragm to the valve poppet, which seats on either the upper or lower port and throttles either the discharge gas or the liquid from the condenser, respectively. During system start-up in extremely cold weather, the poppet may be tight against the lower seat, stopping all liquid flow from the condenser and bypassing discharge gas into the receiver until adequate system head pressure is developed. During stable operation in cold weather, the poppet modulates at an intermediate position, with liquid flow from the condensing coil mixing with compressor discharge gas within the valve and flowing to the receiver. During warm weather, the poppet seats tightly against the upper port, allowing free flow of liquid from the condenser but preventing flow of discharge gas.

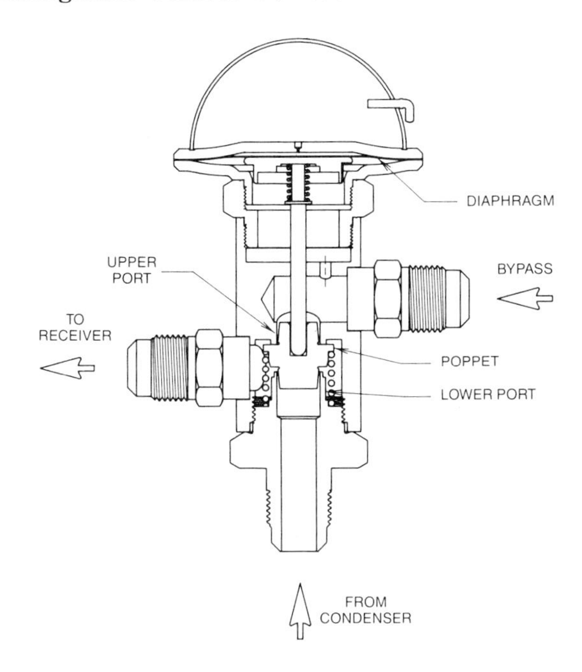


**Fig. 19 Three-Way Condenser Pressure-Regulating Valve**

Three-way condenser pressure-regulating valves are not usually adjustable by the user. The pressure setting is established by the pressure of the gas charge placed in the dome above the diaphragm during manufacture.

## HIGH-SIDE FLOAT VALVES

### Operation

A *high-side float valve* controls the mass flow rate of refrigerant liquid entering the evaporator so it equals the rate at which the refrigerant gas is pumped from the evaporator by the compressor. Figure 20 shows a cross section of a typical valve. The refrigerant liquid flows from the condenser into the high-side float valve body, where it raises the float and moves the valve pin in an opening direction, permitting the liquid to pass through the valve port, expand, and flow into the evaporator. Most of the system refrigerant charge is contained in the evaporator at all times. The highside float system is a flooded system.

### Selection

For acceptable performance, the high-side float valve is selected for the system refrigerant and a rated capacity neither excessively large nor too small. The orifice is sized for the maximum required capacity with the minimum pressure drop across the valve. The valve operated by the float may be a pin-and-port construction (Figure 20), a butterfly valve, a balanced double-ported valve, or a sliding gate or spool valve. The internal bypass vent tube allows installation of the highside float valve near the evaporator and above the condenser without danger of the float valve becoming gas bound. Some large-capacity valves use a high-side float valve for pilot operation of a diaphragm or piston-type spring-loaded expansion valve. This arrangement can provide improved modulation over a wide range of load and pressure-drop conditions.

### Application

A refrigeration system in which a high-side float valve is used

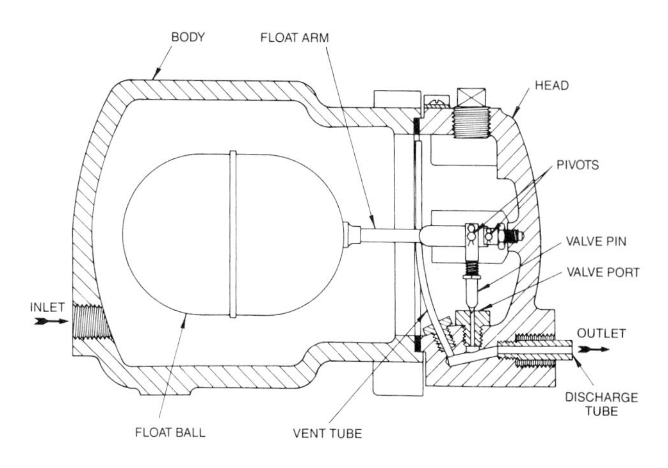


**Fig. 20 High-Side Float Valve**

consists ordinarily of a single evaporator, compressor, and condenser. The operating receiver or a liquid sump at the condenser outlet can be quite small. A full sized receiver is required for pumping out the flooded evaporator. Under certain conditions, the high-side float valve may be used to feed more than one evaporator in a system; consequently, additional control valves are required. One of the disadvantages of the high-side float valve system is that the amount of system refrigerant charge is critical. The use of an excessive amount of system charge causes floodback, while an insufficient amount of system charge causes a reduction in system capacity.

## LOW-SIDE FLOAT VALVES

### Operation

The low-side float valve performs the same function as the high-side float valve, but it is connected to the low-pressure side of the system. When the evaporator liquid level drops, the float opens the valve, which allows refrigerant liquid from the liquid line to flow through the valve port and directly enter the evaporator or surge drum. In another type of valve design, the refrigerant liquid flows through the valve port, passes through a remote feedline, and enters the evaporator through a separate connection. (A typical direct-feed valve construction is shown in Figure 21.) The low-side float system is a flooded system.

### Selection

Low-side float valves are selected in the same manner as the high-side float valves discussed previously.

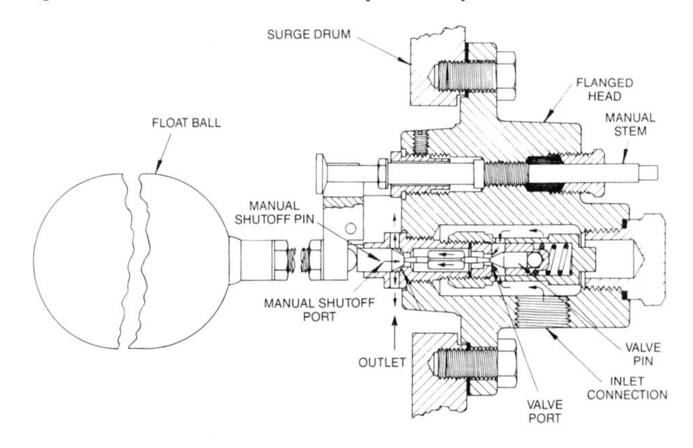


**Fig. 21 Low-Side Float Valve**

### Application

In the low-side float valve system, the refrigerant charge is not critical. The low-side float valve can be used in multiple evaporator systems in which some of the evaporators may be controlled by other low-side float valves and some by thermostatic expansion valves.

Depending on its design, the float valve is mounted either directly in the evaporator or surge drum or in an external chamber connected to the evaporator or surge chamber by equalizing lines, *i.e.*, a gas line at the top and a liquid line at the bottom. In the externally mounted type, the float valve is separated from the float chamber by a gland that maintains a quiet level of liquid in the float chamber for steady actuation of the valve.

In evaporators with high boiling rates or restricted liquid and gas passages, the boiling action of the liquid raises the refrigerant level during operation. When the compressor stops or the solenoid suction valve closes, the boiling action of the refrigerant liquid ceases, and the refrigerant level in the evaporator drops. Under these conditions, the high-pressure liquid line supplying the low-side float valve should be shut off by a solenoid liquid valve to prevent overfilling of the evaporator. Otherwise, excess refrigerant will enter the evaporator on the *off* cycle, which can cause floodback when the compressor starts or the solenoid suction valve opens.

When a low-side float valve is used, precautions must be taken that the float is in a quiet liquid level that falls properly in response to an increase in evaporator load and rises with a decrease in evaporator load. In low-temperature systems particularly, it is important that the equalizer lines between the evaporator and either the float chamber or the surge drum be generously sized to eliminate any reverse response of the refrigerant liquid level in the vicinity of the float. Where the low-side float valve is located in a nonrefrigerated room, the equalizing liquid and gas lines and the float chamber must be insulated to provide a quiet liquid level for the float.

## SOLENOID VALVES

A *solenoid valve* is closed by gravity, pressure, or spring action and opened by a plunger actuated by the magnetic action of an electrically energized coil, or vice versa. Figures 22 and 23 show cross sections of solenoid valves with their principal components identified.

Because solenoid valves are actuated electrically, they may be conveniently operated in remote locations by any suitable electric switch. These valves are always fully open or fully closed, in contrast to motorized valves, which may operate in a modulating position. Solenoid valves may be used to control the flow of many different fluids, if the pressures and temperatures involved, the viscosity of the fluid, and the suitability of the materials used in the valve construction are carefully considered.

Solenoid valves can be divided into the following general types:

1. **Normally closed solenoid valves,** in which the closure member moves away from the port to open the valve when the coil is energized, *e.g.*, a two-way solenoid valve.
2. **Normally open solenoid valves,** in which the closure member moves to the port to close the valve when the coil is energized, *e.g.*, a two-way solenoid valve.
3. **Multiaction solenoid valves,** which combine (in one body) the action of one or more normally open and one or more normally closed solenoid valves, *e.g.*, a three-way solenoid valve or a four-way solenoid valve.

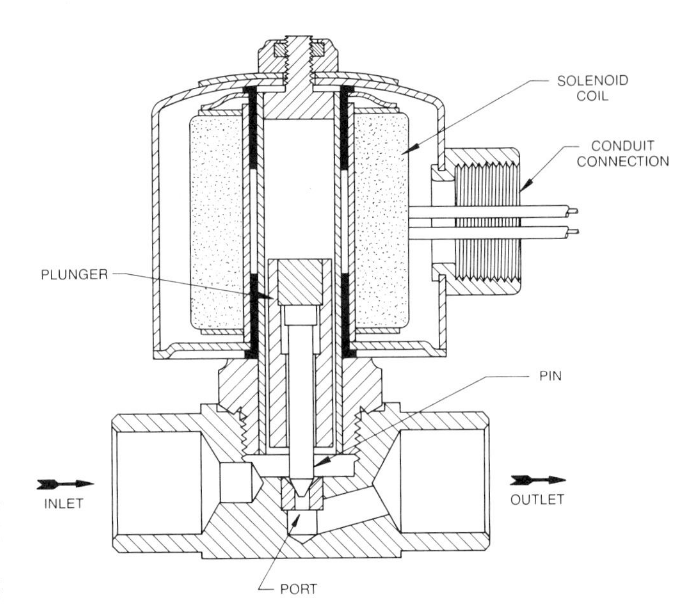


**Fig. 22 Normally Closed Direct-Acting Solenoid Valve with the Hammer-Blow Feature**

### Operation

While all types of solenoid valves are used, the normally closed type is used far more extensively. In the normally closed direct-acting solenoid valve shown in Figure 22, the solenoid coil, acting on the plunger, pulls the valve pin away from and off the valve port, thereby opening it directly. Because this valve depends on the power of the solenoid coil for operation, its port size for a given operating pressure differential is limited by the limitations of solenoid coil size.

Figure 23 shows a medium-sized, normally closed pilot-operated solenoid valve. In this valve, the solenoid coil, acting on the plunger, does not open the main port directly but opens

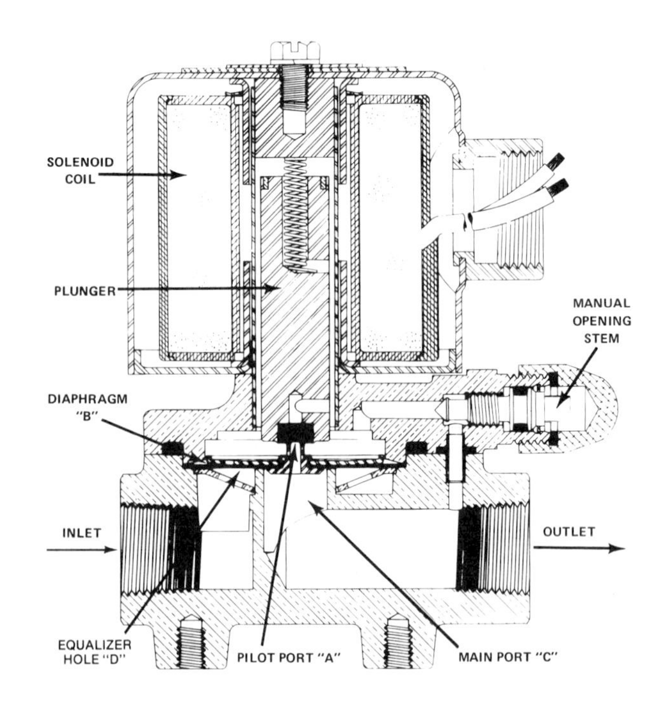


**Fig. 23 Normally Closed Pilot-Operated Solenoid Valve with the Direct-Lift Feature**

the pilot Port A. Pressure trapped on top of Diaphragm B is released through the pilot port, thus creating a pressure imbalance across the diaphragm, forcing it upward and opening the main Port C. When the solenoid coil is de-energized, the plunger drops and closes pilot Port A. Then the pressures above and below the diaphragm equalize again through the equalizer Hole D, and the diaphragm drops and closes the main port. In some pilot-operated solenoid valve designs, a piston is used for the main closing member instead of a diaphragm. In medium-sized valves, the pilot port is usually located in the main closing member.

Such pilot-operated valves depend on a certain minimum pressure drop across the valve (approximately 0.5 psi [3.5 kPa] or more) to hold the piston or diaphragm in the *open* position. If a valve is oversized such that inadequate pressure drop is developed, the valve may chatter or fail to open fully. Valves should be sized by the capacity tables provided by the manufacturer, rather than by pipe or tube size or port diameter. Where it is desirable to keep the valve open without this pressure drop penalty, such as on refrigeration suction lines, the piston or diaphragm may be linked mechanically to the solenoid valve pin and plunger, as shown in Figure 24. The opening and closing actions are the same as before. However, the increased pulling force of the plunger as it approaches its stop position in the coil is used to hold the piston in the open position, without requiring valve pressure drop.

To obtain the maximum operating pressure differential for a given solenoid pulling power, many valves leave the plunger free to gain momentum before it knocks the valve pin out of the valve port or pilot port with an *impact* or a *hammer-blow* effect (see Figures 22 and 24). Solenoid valves with this hammer-blow feature must have the full rated voltage, within the customary tolerance of +10 to −15%, applied instantaneously to their coils so that the valves open under rated conditions. Otherwise, they will fail. If, after the valve is in the open position, the line voltage drops below the hold-in voltage value but not to zero, the plunger drops and the valve closes. After the line voltage builds up again, the valve will not reopen, whether it is used on alternating or direct current, because the hammer-blow effect is lost. However, in the case of an alternating current valve, the coil will overheat and may burn out under this condition because the inherent high inrush current continues to flow through the coil when the plunger is not pulled all the way into the coil to close the air gap.

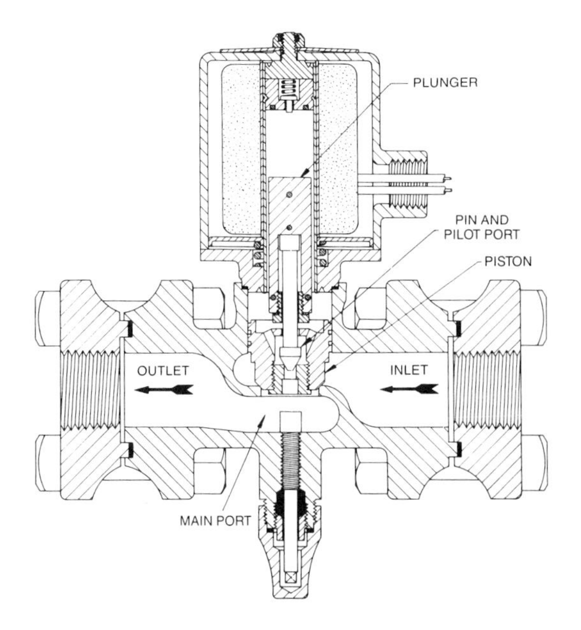


**Fig. 24 Normally Closed Pilot-Operated Solenoid Valve with the Hammer-Blow and Mechanically Linked Piston-Pin Plunger Features**

Direct lift solenoid valves (see Figure 23), in which the valve pin is an integral part of the plunger, can be designed to open fully at rated voltage, +10 to −15%, whether the voltage is applied gradually or instantaneously. Normally open solenoid valves are usually held in the *open* position by gravity or a spring force. When energized, the power of the solenoid coil, acting on the plunger, pulls the valve pin on the valve port or pilot port to close it and, therefore, reverses the valve action.

Multiaction solenoid valves are available to accommodate many different flow configurations, such as (1) a common inlet three-way valve, which directs flow from a common inlet connection to one of two outlet connections; or (2) the fourway valve described in the section "Refrigerant-Reversing Valves." Design compromises usually result in the ports of these direct-acting valves being smaller (or the maximum operating pressure differential lower) than the equivalent normally closed two-way solenoid valves. As a result, multiaction pilot-operated solenoid valves are used in most cases. Their use warrants careful application analysis because the pressure differences required to shift and hold the valves may not exist under all operating conditions.

## Selection

When solenoid valves are selected, the following factors should be considered:

1. Basic flow configuration, such as two-way and three-way normally closed
2. Type of fluid to be handled
3. Temperature and pressure conditions of the entering fluid
4. Allowable fluid flow pressure drop across the valve needed to establish the port size for the required capacity
5. Capacity in appropriate terms; do not size for less pressure drop than is needed to open a piloted valve. Use capacity tables for sizing to ensure that pressure drop at least equals the minimum pressure drop required for operation as specified by the manufacturer. If a piloted valve is too far oversized, it may chatter or fail to open fully. If no minimum pressure drop is specified, 0.5 psi (3.5 kPa) is normally safe. If the valve is advertised to have a *zero pressure drop* opening feature, minimum pressure drop need not be considered.
6. Maximum operating pressure differential under which the valve will be required: (*a*) to open, for the normally closed valve; and (*b*) to close, for the normally open valve. Three- and four-way valves require additional information on the operating conditions for which they are intended.
7. Safe working pressure. This should not be confused with the pressure difference under which the valve is required to open
8. Type and size of line connections
9. Electrical characteristics for the solenoid coil. Voltage and frequency must be specified for alternating current, but only voltage is specified for direct current
10. Ambient temperature in which the valve will be located
11. Cycling rate of valve
12. Hazard of location, which may make explosion-proof coil housings necessary

## Application

Spring loaded solenoid valves (see Figure 23) usually can be installed in vertical lines or any other position. The solenoid valves in Figure 22 should be installed upright in horizontal lines.

Solenoid valves must each be used with the correct individual current characteristics. Momentary overvoltage is not harmful, but sustained overvoltage of more than 10% may cause solenoid coils to burn out under unfavorable conditions. Undervoltage is harmful to alternating current operated valves if it reduces operating power enough to prevent the valve from opening when the coil is energized. This condition may cause burnout of an alternating current coil, as discussed previously.

When the solenoid valve is energized by a control transformer of limited capacity, the transformer must be able to provide proper voltage during the inrush load. As the inrush alternating current may be several times the holding current, it is useless to check the voltage at the coil leads when only holding current is being supplied. For such applications, the inrush current in amperes, multiplied by the rated coil voltage, gives the necessary volt-ampere capacity, which must be provided by the transformer for each solenoid valve simultaneously actuated. The inrush and holding currents of a direct-current solenoid valve are equal.

Fuses protecting electrical supply lines for solenoid valves should be sized according to holding current and should preferably be of the slow blowing-type. To protect the coil insulation and the controlling switch, a capacitor or other device may be wired across the coil leads of a high voltage, directcurrent solenoid valves when installed to absorb or destroy the counter-voltage surge generated by the coil when the circuit is broken.

Whenever it is necessary to reassemble the solenoid valve after installation, the magnetic coil sleeves (if required) must be replaced in their correct respective positions to operate the valve properly. Leaving the coil sleeves out of an alternating current valve may result in coil burnout.

To avoid valve failure because of low voltage, the solenoid valve coil should not be energized by the same contacts or at the same instant that a heavy motor load is connected to the electrical supply line. The solenoid coil can be energized immediately before or after the heavy motor load is connected to the line.

Solenoid valves are used for the following applications:

1. **Refrigerant Liquid.** To prevent flow of refrigerant liquid to the evaporator, a solenoid valve is installed in the liquid line just ahead of the expansion valve to (1) prevent flow of refrigerant liquid to the evaporator when the compressor is idle; (2) provide individual temperature control in each room of a multiple system; or (3) control the number of evaporator sections used as the load varies on a central air-conditioning installation.
2. **Refrigerant Suction Gas.** In commercial multiple systems, especially those having evaporators containing a large amount of refrigerant, solenoid valves are often provided in the suction line from each unit or room, as well as in the liquid line, to isolate each evaporator completely. Otherwise, refrigerant gas migrates from one evaporator to another through the suction line during the *off* cycle, which causes uneven performance and possible floodback when the compressor starts again.

   In applications 1 and 2, the solenoid valves are operated by thermostats. The compressor may be operated by a low-pressure switch or directly by a thermostat. The compressor may or may not be operated on a pumpdown cycle, either a continuous one or a pumpdown and lockout cycle.
3. **Refrigerant Discharge Gas.** In many hot-gas defrost applications, a solenoid valve, installed in a line connected to the discharge line between the compressor and the condenser, feeds the evaporator with hot gas, which provides heat for the defrosting operation. The solenoid valve remains closed, except during the defrosting operation. A solenoid valve installed in a bypass around one or more compressor cylinders provides compressor-capacity control. This valve may be used to bypass the entire compressor output and reduce the compressor starting load where required.
4. **Water and Other Liquids.** Solenoid valves are used to control the flow of water and many other liquids. Water is one of the more harmful liquids because it deposits solids on the internal solenoid valve surfaces, causing corrosion; therefore, solenoid valves for water service should be easy to dismantle and clean.
5. **Air.** Many air systems rely on solenoid valves to operate controls or actuators. Since rapid cycling is often required, solenoid valves for air service should be selected with consideration for endurance.
6. **Steam.** The application of solenoid valves in industrial steam systems is quite varied. Because of the continuous high steam temperatures involved, special high-temperature solenoid coils are usually required. The ambient temperatures in which the valves are located also need consideration.

## Pilot Solenoid Valve Application

*Pilot solenoid valves* are often used in industry. A few applications pertaining to refrigerating systems follow.

*Two-way, normally closed, direct-acting pilot solenoid valves* are used with (1) a large, piston-type, spring-loaded expansion valve to provide refrigerant liquid shutoff service, as shown in Figure 9; (2) either an evaporator-pressure regulator to provide shutoff service or a pressure pilot selector; (3) a large piston-type, spring-loaded regulator to provide refrigerant gas shutoff service; and (4) four-way refrigerant switching valves on heat-pump systems to accomplish cooling, heating, and defrosting.

*Three-way, direct-acting, pilot solenoid valves* are used to operate (1) a cylinder unloading mechanism for compressor capacity reduction; and (2) three- and four-way reversing valves on heat-pump systems to accomplish cooling, heating, and defrosting.

*Refrigerant-Reversing Valves.* These are three- or four-way two-position valves that are usually operated by pilot solenoid valves and designed for reversing or changing the direction of refrigerant flow through certain parts of a refrigeration system. They are used on refrigerating and year-round air-conditioning (heat-pump) systems to control cooling, heating, and defrosting operations.

## Operation

Valves may be operated by either a two-, three-, or four-way pilot solenoid valve, which may be an integral part of the reversing valve or a separate valve connected to the reversing valve by tubes. The reversing valves can be divided into the poppet valve and the slide valve group. Although design modifications can be found within each of these groups, the following explanations presents the general principles of operation.

The *four-way reversing valve* may be connected so the refrigeration system is *fail-safe* on either the heating or cooling cycle if the solenoid valve coil fails. The valve is connected by interchanging the two lines on the reversing valve, which connect to the inside and outside heat-exchange coils. Figure 25 shows refrigeration flow through a four-way slide reversing valve in the cooling (or defrosting) cycle.

During the cooling cycle shown in Figure 25, the pilot valve is energized, which opens pilot Port A to bleed high-pressure refrigerant into Chamber C. Simultaneously, pilot Port B is connected to low pressure and bleeds refrigerant from Chamber D. The pressure difference across Piston G develops enough force on Piston G to cause Pistons G and H and the connecting struc-

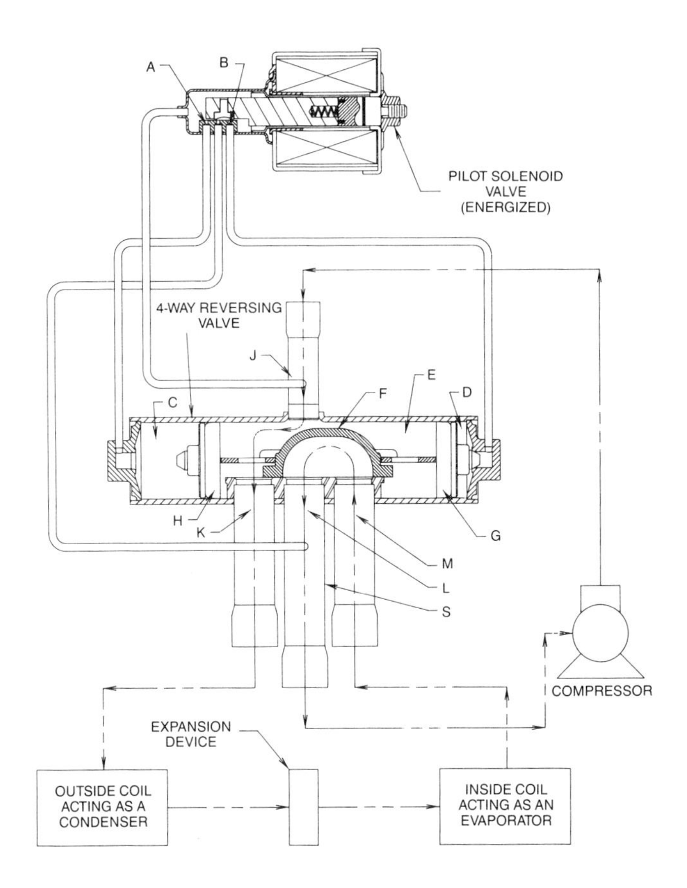


**Fig. 25 Four-Way Slide-Type Refrigerant-Reversing Valve Used in Cooling (or Defrosting) Cycle of Refrigeration System**

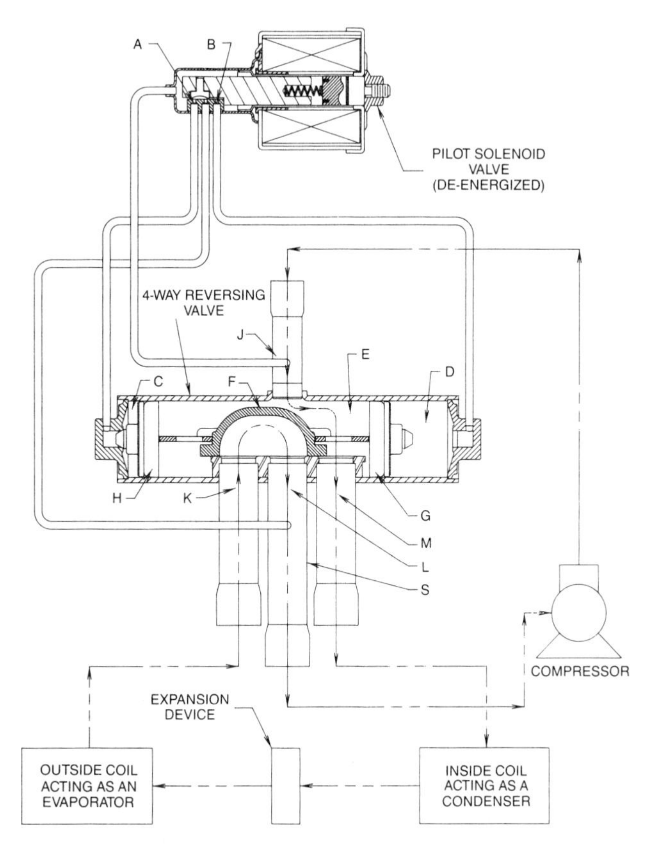


**Fig. 26 Four-Way Slide-Type Refrigerant-Reversing Valve Used in Heating Cycle of Refrigeration System**

tural member to move Slide F from encompassing Ports K and L to encompassing Ports M and L. When G reaches the end of its stroke, the valve has reversed, and the inside coil is connected to low-pressure cool refrigerant, while the outside coil is connected to high-pressure hot refrigerant.

To reverse to the heating cycle (Figure 26) the pilot valve is de-energized. This opens pilot Port B, which bleeds high-pressure refrigerant into Chamber D and connects pilot Port A to low pressure, which bleeds refrigerant from Chamber C. The valve slide moves in the same manner as before, and the inside coil is connected to high-pressure hot refrigerant, while the outside coil is connected to low-pressure cool refrigerant.

A similar reversing action of Pistons G and H can be produced by a three-way pilot valve. The connection J to the high-pressure tube of the main valve is omitted, and the pilot valve operates to connect either Chamber C or D to the low-pressure source. The high pressure in Chamber E acts on the Pistons G and H and moves the Slide F towards the chamber with the low pressure. In this case, a bleed hole in each piston allows refrigerant to fill the Chambers C and/or D when these chambers expand in volume. Also, the piston must seal the main valve end-cap ports to prevent leakage through the pilot valve to the suction Tube S of the main valve.

Reversing valves operate well with most fluorinated refrigerants, but the type of refrigerant used affects the size of the valve selected for a given capacity. The valve should have minimum pressure drops through the valve passages in both the discharge and suction gas paths so that the compressor capacity is not reduced.

## Application

In addition to their year-round air-conditioning (heat-pump) application on residential or commercial systems, four-way reversing valves are used in refrigeration systems in motor trucks, trailers, and railway refrigerator cars for the transportation of perishable cargo. The transport systems can be arranged to operate automatically and to provide cooling, heating, and defrosting, as required.

While the four-way valves are sometimes adapted for other applications, three-way reversing valves are designed specifically for commercial refrigeration systems to defrost the evaporator or for heat reclaim using an auxiliary condenser.

Regardless of the type of installation to which the reversing valve is added (*e.g.*, window unit, residential unit, commercial unit, transportation unit, or customer-built systems), the system may be made to operate automatically. If a dual-action thermostat is used, cooling automatically occurs when the temperature drops below a present value.

# CONDENSING WATER REGULATORS

## Two-Way Regulators

The condensing water regulator modulates the quantity of water passing through a water-cooled refrigerant condenser in response to the condensing pressure. This regulator is used on a vapor-cycle refrigeration system to maintain a condensing pressure that loads but does not overload the compressor motor. The regulator automatically modulates to correct for both variations in temperature or pressure of the water supply and variations in the quantity of refrigerant gas, which the compressor is sending to the condenser.

### Operation

The condensing water regulator consists of a valve and an actuator that are linked together, as shown in Figure 27. The actuator consists of a metallic bellows and adjustable spring combination connected to the system condensing pressure. For large water-flow capacities, a small condensing water regulator is used for pilot-operation of a diaphragm-type main valve.

After a compressor starts, the condensing pressure begins to rise. When the opening pressure setting of the regulator spring is reached, the bellows moves to open the valve disk gradually or slide from its seat. The regulator continues to open as the condensing pressure rises until a balance point is reached between the water flow and the heat rejection requirement, at which point the condensing pressure is stabilized. When the compressor stops, the continuing water flow through the regulator causes the condensing pressure to drop gradually, closing the regulator, which becomes fully closed when the opening pressure setting of the regulator is reached.

### Selection

Selection of a condensing water regulator depends on the system refrigerant used, the water-flow rate required, and the available water-pressure drop across the regulator. While one standard bellows operator can sometimes handle any of several refrigerants, special springs or bellows may be required for very high- or low-pressure refrigerants, as in the case of Refrigerant 717 (ammonia), where a stainless steel bellows must be used in place of a brass bellows.

The water-flow rate required depends on condenser performance data, the temperature of available water, the quantity of heat that must be rejected to this water, and the allowable leaving water temperature. For a given opening of the valve seat, which corresponds to a given pressure rise above the regulator opening point, the flow rate handled by a given size water regulator is a function of the available waterpressure drop across the valve seat. Available water-pressure drop is determined for the required flow rate by deducting condenser water pressure drop, pipeline pressure drop, and static head losses from the pressure of the water at its supply point.

The condensing water regulator should be selected from the manufacturer's data on the basis of maximum required flow rate, minimum available pressure drop, and water temperature.

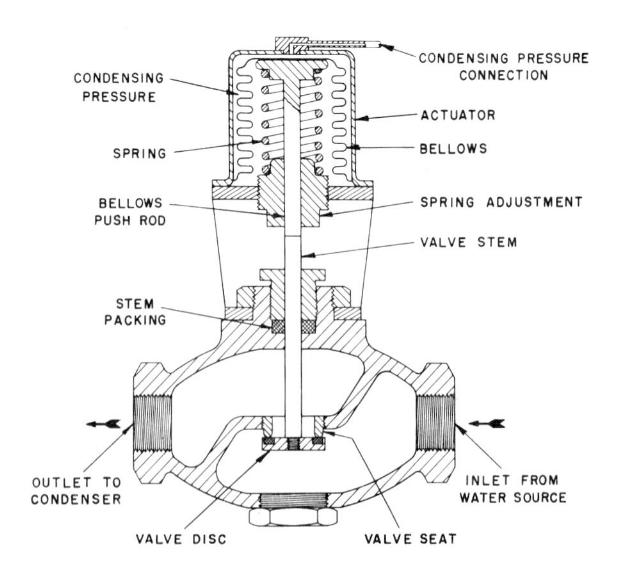


**Fig. 27 Two-Way Condensing Water Regulator**

### Application

Oversizing of a regulator should be avoided because this encourages hunting. When two-way condensing water regulators (see Figure 27) are used on recirculating cooling tower systems, the throttling action of the valves during cold weather causes a reduction in pump and tower circulation, which is undesirable for that equipment.

### Three-Way Regulators

On tower systems requiring individual condensing pressure control, three-way condensing water regulators should be used (see Figure 28). These are similar in construction to twoway regulators, but they have an additional port, which opens to bypass water around the condenser as the port controlling water flow to the condenser closes. Thus, the tower decking or sprays and the circulating pump receive a constant supply of water, although the water supply to individual condensers is modulated for control.

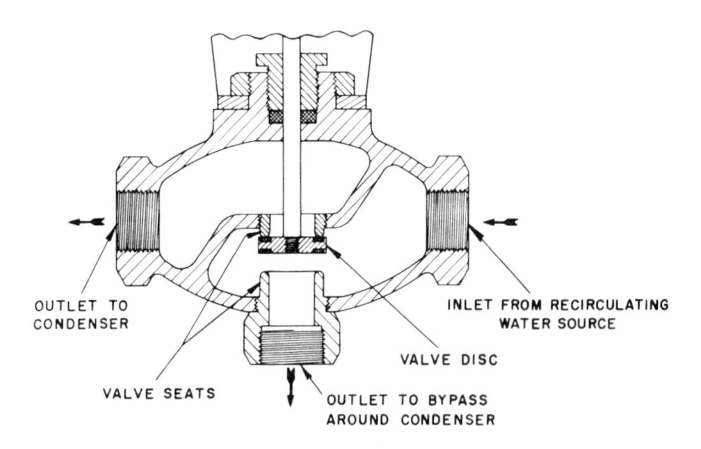


**Fig. 28 Three-Way Condensing Water Regulator**

Three-way condensing water regulators must be supplemented by other means if cooling tower systems are to be operated in freezing weather. An indoor sump is usually required, and a temperature-actuated three-way water control valve is used to divert periodically all of the condenser leaving water directly to the sump whenever the water becomes too cold.

Unlike the recommended and accepted practice pertaining to the application of refrigerant-control valves, a strainer is not ordinarily used with a water regulator because a strainer usually requires more cleaning and servicing than a regulator without a strainer.

## CHECK VALVES

Refrigerant check valves are normally used in refrigerant lines where pressure reversals can cause undesirable reverse flows. A check valve is usually opened by a portion of the pressure drop, which causes flow in the pipeline. Closing usually occurs either when a reversal of pressure takes place or when the pressure drop across the check valve is less than the minimum opening pressure drop in the normal flow direction.

Conventional check valve designs frequently use piston construction and use the globe pattern in large sizes, while in-line designs are commonly smaller than the 2-in. (51mm) size. Either design may be used with closing springs; the heavier springs give more reliable and tighter closing but require greater pressure drop for the check to open. The in-line check design does not permit a manual opening stem. Although conventional check valves may be designed to open at less than 1 psi (7 kPa) drop, they may not be reliable at temperatures below −25 °F (−31.7 °C) because the light closing springs may not overcome the viscous oil.

Other check valve designs are used for various special functions not covered here, such as *excess flow checks*, which close only when flow exceeds the maximum desired rate; *electrically lifted checks*, which require no pressure drop to remain open; *remote pressure operated checks*, which are normally open but close when they are supplied with a higher pressure source of refrigerant; and *pressure differential valves*, which maintain a uniform pressure differential between system components for special functions in a refrigeration system.

### Seat Materials

Although precision metal seats may be manufactured nearly bubble-tight, they are not economically practical for refrigerant check valves. Seats made of synthetic rubbers provide excellent tightness at medium and high temperatures but may leak at low temperatures due to lack of resilience. Because high temperatures deteriorate most rubbers suitable for refrigerants, the use of plastic seat materials has become increasingly successful, despite the possibility of damage by large pieces of foreign matter in the systems.

### Applications

In *compressor discharge lines,* check valves are used to prevent flow from the condenser to the compressor during the *off* cycle or to prevent flow from an operating compressor to an idle compressor. While a 2 to 6 psi (14 to 41 kPa) pressure drop is tolerable, the check-valve design must resist pulsations of the compressor, the temperature of discharge gas, and must be bubble-tight to prevent accumulation of liquid refrigerant at the compressor discharge valves or in the crankcase.

In liquid lines, check valves prevent reverse flow through the unused expansion device on heat-pump systems or prevent backup into the low-pressure liquid line of a recirculating system during a defrost period. While a 2 to 6 psi (14 to 41 kPa) pressure drop is usually acceptable, the check-valve seat must be bubble-tight.

In the suction line of a low-temperature evaporator, a check valve may be used to prevent the transfer of refrigerant vapor to a lower temperature evaporator on the same suction main. In this case, the pressure drop must be less than 2 psi (14 kPa), the valve seating must be reasonably tight, and the check valve must be reliable at low temperatures.

Normally, open pressure-operated check valves are used to close suction lines, gas, or liquid legs in gravity recirculating systems during defrost.

In hot-gas defrost lines, check valves may be used in the branch hot-gas lines connecting the individual evaporators to prevent crossfeed of refrigerant during the cooling cycle when the defrost operation is not taking place. In addition, check valves are used in the hot-gas line between the hotgas heating coil in the drain pan and the evaporator, to prevent pan coil sweating during the refrigeration cycle. Tolerable pressure drop is typically 2 to 6 psi (14 to 41 kPa), seating must be nearly bubble-tight, and seat materials must withstand high temperatures.

To prevent chatter or pulsation, check valves should be sized for the particular pressure drop that ensures they are in the wide-open position at the desired flow rate.

## RELIEF DEVICES

Refrigerant relief devices have either safety or functional uses. A safety relief device is designed to relieve positively at its set pressure for one crucial occasion without prior leakage. This relief may be to the atmosphere or to the low side.

A *functional* relief device is a control valve that may be called on to open, modulate, and reclose with repeatedly accurate performance. For reasons of system control, relief is usually from a portion of the system at higher pressure to a portion at lower pressure. Design refinements of the functional relief valve usually make it unsuitable or uneconomical as a safety relief device.

### Safety Relief Valves

These are most commonly *pop-type designs,* which abruptly open when the inlet pressure exceeds the outlet pressure by the valve setting pressure (see Figure 29). Seat configuration is such that once lift begins, the resulting increased active seat area causes the valve seat to pop wide open against the force of the setting spring. Because the flow rate is measured at a pressure of 10% above the setting, the valve must open within this 10% increase in pressure.

This relief valve operates on a fixed pressure differential from inlet to outlet. Because the valve is affected by back pressure, the installation of a rupture member at the valve outlet is not permissible.

Relief valve seats are made of metal, plastic, lead alloy, or synthetic rubber. The last is commonly used because it has greater resilience and, consequently, probable reseating tightness. For valves that have lead-alloy seats, an emergency manual reseating stem is occasionally provided to permit reforming of the seating surface by tapping the stem lightly with a hammer.

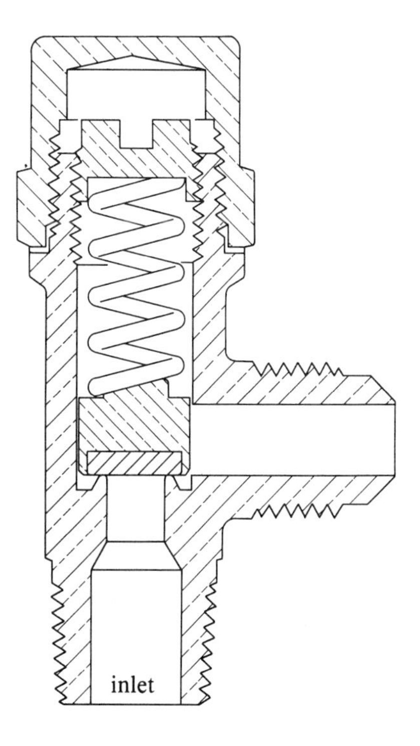


**Fig. 29 Pop-Type Safety Relief Valve**

The advantages of the pop-type relief valve are simplicity of design, low initial cost, and high discharge capacity.

### Other Safety Relief Devices

Two devices performing similar safety relief operations are the *fusible plug* and the *rupture member* (Figure 30). The former contains a fusible member that melts at a predetermined temperature corresponding to the safe saturation pressure of the refrigerant, but is limited in application to pressure vessels with internal gross volumes of 3 $ft^3$ (0.085 $m^3$) or less and internal diameters of 6 in. (152.4 mm) or less. The rupture member contains a frangible disk designed to rupture at a predetermined pressure.

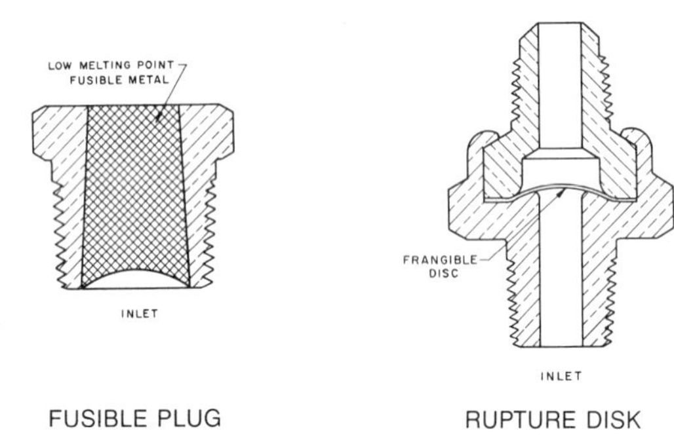


**Fig. 30 Safety Relief Devices**

### Discharge Capacity

The minimum required discharge capacity of the pressure relief device or fusible plug for each pressure vessel is determined by the following formula, specified by the ASHRAE *Standard* 15-78, *Safety Code for Mechanical Refrigeration*:

$$C = kfDL$$

*where*

$C$ = minimum required discharge capacity of the relief device, lb air/min (kg air/min)
$D$ = outside diameter of vessel, ft (m)
$L$ = length of the vessel, ft (m)
$k$ = factor dependent on units used
(k = 1 for I-P units, k = 4.88, for SI units)
$f$ = factor dependent on the kind of refrigerant, as follows:

| Kind of Refrigerant | Value of $f$ |
|---|---|
| Ammonia (Refrigerant 717) | 0.5 |
| Refrigerants 12, 22, and 500 | 1.6 |
| Refrigerants 502, 13, 13 B1, and 14 when on cascade systems | 2.5 |
| All other refrigerants | 1.0 |

Capacities of pressure relief valves are determined by test in accordance with the provisions of the ASME *Boiler and Pressure Vessel Code*. Relief valves approved by the National Board of Boiler and Pressure Vessel Inspectors are stamped with the code symbol, which consists of the letters UV in a clover leaf design with the letters NB stamped directly below this symbol. In addition, the pressure setting and capacity are stamped on the valve.

When relief valves are used on pressure vessels of 10 $ft^3$ (0.283 $m^3$) internal gross volume or more, a relief system consisting of a three-way valve and two relief valves in parallel is required.

The rated discharge capacity of a rupture member or fusible plug that discharges to the atmosphere under critical flow conditions is determined by calculation, using the formula provided in the ASHRAE *Standard* 15-78.

### Pressure Setting

The maximum pressure setting for a relief device is limited by the design working pressure of the vessel to be protected. Pressure vessels normally have a safety factor of 5. Therefore, the minimum bursting pressure is five times the rated design working pressure. The relief device must have enough discharge capacity to prevent the pressure in the vessel from rising more than 10% above its design pressure. Since the capacity of a relief device is measured at 10% above its stamped setting, the setting cannot exceed the design pressure of the vessel.

To prevent loss of refrigerant through pressure relief devices during normal operating conditions, the relief device setting must be substantially higher than the system operating pressure. For rupture members, the setting should be 50% above a static system pressure and 100% above a maximum pulsating system pressure. Failure to provide this margin of safety causes fatigue of the frangible member and rupture well below the stamped setting.

For relief valves, the setting should be 25% above the maximum system pressure. This safety factor will provide spring force on the valve seat sufficient to maintain a tight seal and still allow for setting tolerances and other factors that cause settings to vary. Although relief valves are set at the factory within a few lb (kg) of the stamped setting, the variation may be as much as 10% after the valves have been stored or placed in service for a period of time.

### Discharge Piping

The size of the discharge pipe from the pressure relief device or fusible plug should not be less than the size of the pressure relief device or fusible plug outlet. The maximum length of the discharge piping is provided in a table or may be calculated from the formula in the ASHRAE *Standard* 15-78.

### Relief Device Summary

1. A relief device with sufficient capacity for code requirements and one suitable for the type of refrigerant used should be selected.
2. The proper size and length of discharge tube or pipe should be used.
3. The relief device should not be discharged prior to installation or when pressure testing the system.
4. For systems containing large quantities of refrigerant, a three-way valve and two relief valves should be used.
5. A pressure vessel that permits the relief valve to be set at least 25% above the maximum system pressure should be provided.

### Functional Relief Valves

Functional relief valves are usually diaphragm types in which the system pressure acts on a diaphragm that lifts the valve disk from the seat (Figure 31). The other side of the diaphragm is exposed to both the adjusting spring and atmospheric pressure. The ratio of effective diaphragm area to seat area is high, so the outlet pressure has little effect on the operating point of the valve.

Because the lift of the diaphragm is not great, the diaphragm valve is frequently built as the pilot or servo of a larger piston-operated main valve, thereby providing fine sensitivity and high flow capacity. Construction and performance are similar to the previously described pilot-operated evaporator pressure regulator, except that the diaphragm valves are constructed for higher pressures. Thus, the valves are suitable for use as defrost relief from evaporator to suction pressure, as large-capacity relief from

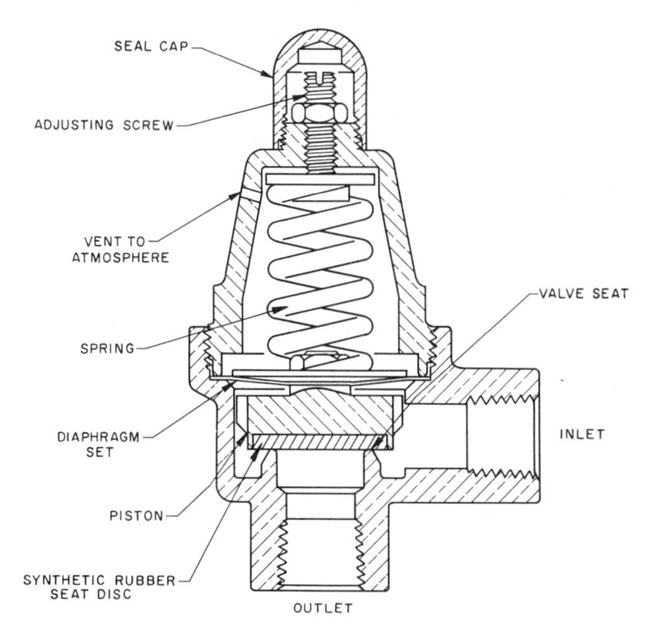


**Fig. 31 Diaphragm-Type Relief Valve**

a pressure vessel to the low side, or as a liquid refrigerant pump relief from pump discharge to the accumulator to prevent excessive pump pressures when some evaporators are valved closed.

# PART III: DISCHARGE-LINE OIL SEPARATORS

The discharge-line oil separator removes oil from the discharge gas of lubricated rotary and reciprocating compressors. Oil is separated by (1) reducing gas velocity, (2) changing direction of flow, (3) impingement on baffles, (4) mesh pads or screens, and (5) centrifugal force. The separator reduces the amount of oil reaching the low side, helps maintain the oil charge in the compressor oil sump, and muffles the sound of the gas flow.

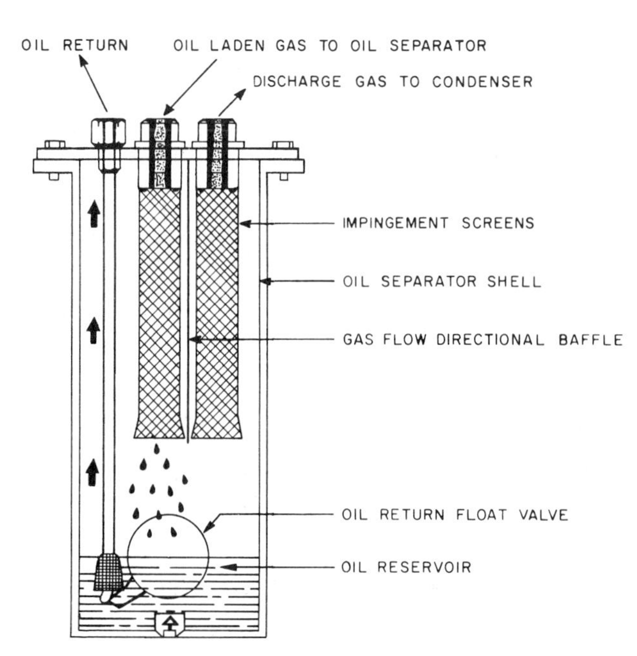


**Fig. 32 Discharge-Line Oil Separator**

Figure 32 shows a small separator incorporating inlet and outlet screens and a high-side float valve. A space below the float valve provides for dirt or carbon sludge. When oil accumulates to raise the float ball, oil passes through a needle valve and returns to the low-pressure crankcase. When the level falls, the needle valve closes, preventing the release of hot gas into the crankcase. Insulation and electric heaters may be added to prevent the refrigerant from condensing when the separator is exposed to low temperatures. A wide variety of horizontal and vertical flow separators is manufactured with one or more of such elements as centrifuges, baffles, wire mesh pads, or cylindrical filters.

## SELECTION

Separators are usually given system capacity ratings for several refrigerants at several suction and condensing temperatures. Another rating method for selection purposes gives the capacity in terms of the compressor displacement volume. Some separators also show a marked reduction in separation efficiency at some stated minimum capacity.

Because the compressor capacity increases when the suction pressure is raised or the condensing pressure lowered, the system capacity at its lowest compression ratio should be the criterion for selecting the capacity of the separator.

## APPLICATION

A discharge-line oil separator is best for ammonia or hydrocarbon refrigerants to reduce oil fouling in the evaporator. With oil soluble halocarbon refrigerants, only certain flooded systems, low-temperature systems, or systems with long suction lines or other oil return problems need oil separators. (See Chapter 3 of the 1986 REFRIGERATION Volume for more information about oil separator applications.)

# PART IV: CAPILLARY TUBES

Every refrigerating unit requires a pressure-reducing device to meter the flow of refrigerant to the low side in accordance with the demands placed on the system. The capillary tube rapidly achieved popularity, especially with the smaller unitary hermetic equipment such as household refrigerators and freezers, dehumidifiers, and room air conditioners. Capillary tube use has been extended to include larger units such as unitary air conditioners in sizes up to 10 tons (35 kW) capacity.

The capillary operates on the principle that liquid passes through it much more readily than gas. It consists of a small diameter line, which, when used for controlling a system's refrigerant flow, connects the outlet of the condenser to the inlet of the evaporator. It is sometimes soldered to the outer surface of the suction line for heat-exchange purposes.

A high-side liquid receiver is not normally used with a capillary; consequently, a corresponding reduction in refrigerant charge could result. In a few instances, such as with household refrigerators and freezers, it is common to select a small low-side accumulator. The pressure-equalizing characteristic of a capillary makes the use of a low starting torque motor compressor possible. Inherently, a capillary does not operate as efficiently over a wide range of conditions as does a thermostatic expansion valve; however, due to counter-balancing factors in most applications, its performance is generally good. The simplicity of the capillary gives it the advantage of reduced cost.

## THEORY

Because the capillary passes liquid much more readily than

gas, it is a practical metering device. When a condenser-to-evaporator capillary is sized to permit the desired flow of refrigerant, liquid will seal its inlet. If a system imbalance occurs and some gas (uncondensed refrigerant) enters the capillary, this gas considerably reduces the mass flow of refrigerant with little or no change in system pressures. If the opposite type of imbalance occurs, liquid refrigerant backs up in the condenser.

This condition causes subcooling and increases the mass flow of refrigerant. If properly sized for the application, the capillary compensates automatically for load and system variations and gives acceptable performance over a wide range of operating conditions.

A common flow condition subcools liquid at the entrance to the capillary. Bolstad and Jordan (1948) demonstrated the flow behavior from temperature and pressure measurements along the tube as follows:

With subcooled liquid entering the capillary tube, the pressure distribution along the tube is similar to that shown in the graph (see Figure 33). At the entrance to the tube, section 0-1, a slight pressure drop occurs, usually unreadable on the gauges. From point 1 to point 2, the pressure drop is linear. In the portion of the tube 0-1-2, the refrigerant is entirely in the liquid state, and at point 2, the first bubble of vapor forms. From point 2 to the end of the tube, the pressure drop is not linear, and the pressure drop per unit length increases as the end of the tube is approached. For this portion of the tube, both the saturated liquid and saturated vapor phases are present, with the percent and volume of vapor increasing in the direction of flow. In most of the runs, a significant pressure drop occured from the end of the tube into the evaporator space.

With a saturation temperature scale corresponding to the pressure scale superimposed along the vertical axis, the observed temperatures may be plotted in a more efficient way than if a uniform temperature scale were used. The temperature is constant for the first portion of the tube 0-1-2. At point 2, the pressure has dropped to the saturation pressure corresponding to this temperature. Further pressure drop beyond point 2 is accompanied by a corresponding drop in temperature, the temperature being the saturation temperature corresponding to the pressure. As a consequence, the pressure and temperature lines coincide from point 2 to the end of the tube.

The rate of refrigerant flow through a capillary always increases with an increase in inlet pressure. Flow rate also increases with a decrease in external outlet pressure down to a certain critical value, below which the flow does not change. Figure 33 illustrates a case in which the outlet pressure inside the capillary has reached the critical value, which is higher than the external pressure. Point 2 in Figure 33, where the first gas bubble appears, is called the *bubble point.* The preceding portion of capillary is called the *liquid length,* and that following is called the *two-phase length.*

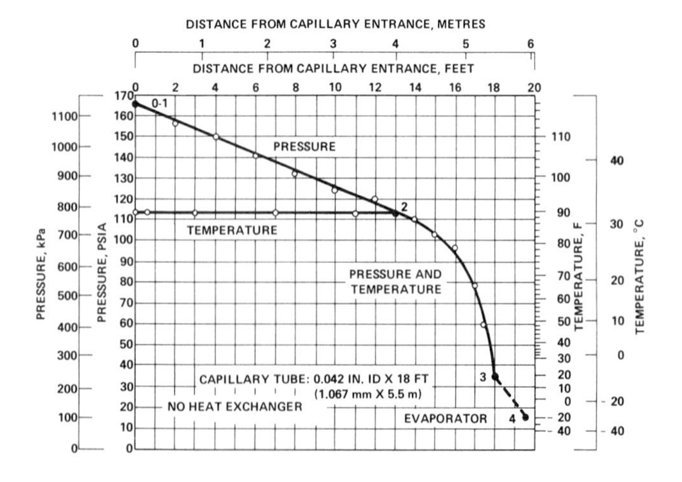


**Fig. 33 Pressure and Temperature Distribution along Typical Capillary Tube**

## SYSTEM DESIGN FACTORS

The high side must be designed carefully for use with a capillary. If liquid backs up in the condenser, enough refrigerant may be removed from the evaporator to cause an undercharged condition. This can be improved by decreasing the high-side volume so that less liquid refrigerant is needed to increase the discharge pressure to achieve balanced conditions. The total volume of the high side is usually a compromise, as it may also be necessary to provide sufficient refrigerant storage volume to protect against excessive discharge pressures during high-load conditions. To prevent hydrostatic rupture failure in case of capillary stoppage, the high-side volume should be sufficient to contain the entire refrigerant charge.

Another consideration in high-side design, where cyclic operation is involved, is unloading during the *off* period. When unit operation ceases, the capillary continues to pass refrigerant from the high side to the low side until pressures are equalized. Good drainage of the liquid into the capillary during this unloading interval should be provided. If liquid is trapped in the high side, it will evaporate there during the *off* cycle, pass to the low side as a warm gas, condense, and add latent heat to the evaporator. Liquid trapping may also increase the time required for the pressures to equalize after the compressor stops operating. If this interval is too long, the compressor may not be sufficiently unloaded at cutin to permit easy starting.

The important design parameters in low-side design are total refrigerant charge and charge tolerance. The total refrigerant charge should be held to a minimum, consistent with satisfactory performance. However, a high tolerance may be necessary to minimize the effect of a varying distribution of refrigerant in the system during operation. A good low-side design achieves both objectives.

The amount of refrigerant in the evaporator is at its maximum value during the *off* cycle and its minimum during the *running* cycle. The suction piping should be arranged to reduce the adverse effects of the variable-charge distribution. A suitable liquid accumulator is sometimes necessary.

When a suction line heat exchanger is used, the excess capillary may be coiled and located at either end of the exchanger. Although more efficient heat exchange may be obtained if the excess capillary is coiled at the evaporator, system stability is enhanced if a portion of the capillary is located at the condenser. Forming the bends and coils to avoid local restrictions should be done carefully.

All of the capillary should not be located at the evaporator or preceded by a conventional large-bore liquid line in heat exchange with the suction line. With this particular arrangement, an overcharge into the suction line places low side and high side in a heat-exchange relationship in the heat exchanger and causes serious instability. The cooling of the refrigerant entering the capillary is increased due to liquid in the suction line; the cooling increases the capillary capacity and reduces the liquid stored in the condenser.

## CAPACITY BALANCE CHARACTERISTIC

The selection of a capillary tube depends on the application and anticipated range of operating conditions. One approach to the problem involves the concept of *capacity balance.* A refrigerating system may be operating at the condition of capacity balance when the resistance of the capillary is sufficient to maintain a liquid seal at its entrance without excess liquid accumulating in the high side of the system (see Figure 34). Only one such *capacity balance point* exists for any given compressor discharge pressure. A curve through the capacity balance points for a range of compressor discharge pressures is called the *capacity balance characteristic* of the system. Such a curve is shown in Figure 35. Ambient temperatures are drawn on the chart for a typical air-cooled system. A given set of compressor discharge and suction pressures is associated with fixed condenser and evaporator pressure drops; these pressures establish the capillary inlet and outlet pressures.

The capacity balance characteristic curve for any combination of compressor and capillary may be determined experimentally by the arrangement shown in Figure 36, which makes it possible to vary independently the suction and discharge pressures until capacity balance is obtained. The desired suction pressure may be obtained by regulating the heat input to the low side, usually by electric heaters; the desired discharge pressure may be obtained by a suitable controlled water-cooled condenser. A liquid indicator is located at the entrance to the capillary. The usual test procedure is to hold the high-side pressure constant and, with gas bubbling through the sight glass, slowly increase the suction pressure until a liquid seal forms at the capillary entrance. Repeating this procedure at various discharge pressures determines the capacity balance characteristic curve similar to that shown in Figure 35. This equipment may also be used as a calorimeter to determine simultaneously the capacity of the refrigerating system.

## OPTIMUM SELECTION AND REFRIGERANT CHARGE

Whether the initial capillary selection and charge are optimum for the unit is always questioned, even in such simple applications as a condenser-to-evaporator capillary for a room air-conditioning unit. The refrigerant charge can be varied for a given capillary selection applied to a unit by locating a small refrigerant bottle (valved off and sitting on a scale) in the circuit. The interconnecting line must be flexible and arranged so it is filled with vapor instead of liquid. The charge is brought into the unit or removed from it by heating or cooling the bottle.

The only test for varying the capillary restriction is to remove

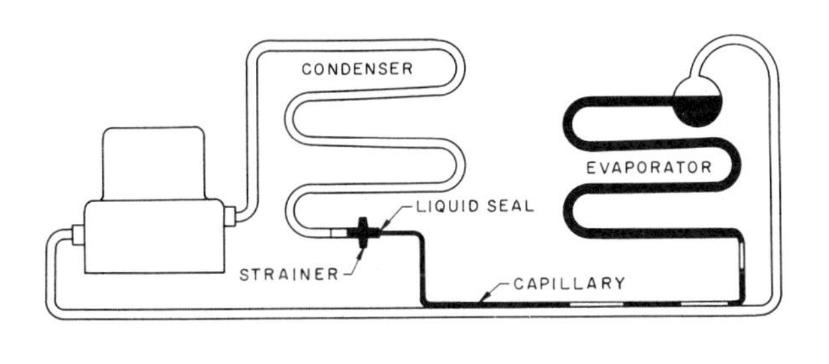


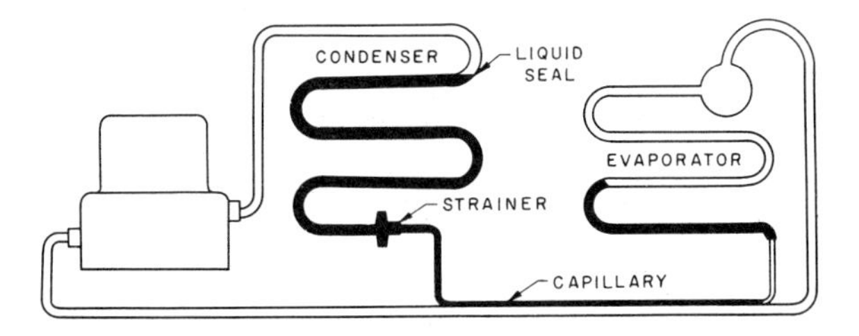


[a]Capillary selected for capacity balance conditions. Liquid seal at capillary inlet but no excess liquid in condenser. Compressor discharge and suction pressures normal. Evaporator properly charged.

[b] Too much capillary resistance—liquid refrigerant backs up in condenser and causes evaporator to be undercharged. Compressor discharge pressure may be abnormally high. Suction pressure below normal. Bottom of condenser subcooled.

**Fig. 34 Effect of Capillary Tube Selection on Refrigerant Distribution [a,b]**

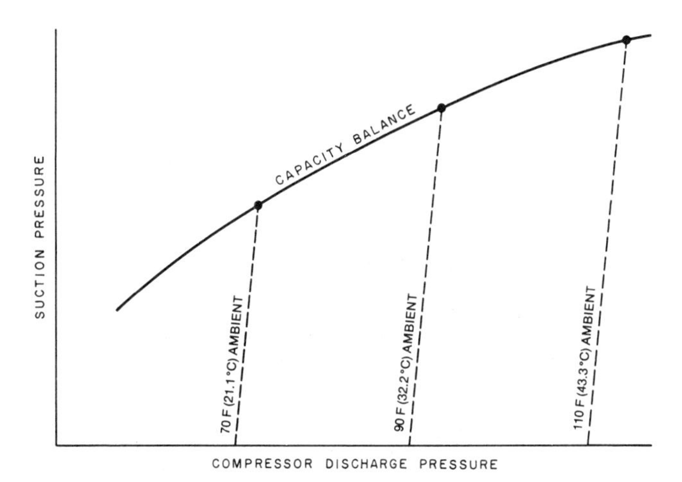


[a]Operation below this curve results in a mixture of liquid and vapor entering the capillary. Operation above the capacity balance points causes liquid to back up in the condenser and elevate its pressure.

**Fig. 35 Capacity Balance Characteristic of a Capillary System[a]**

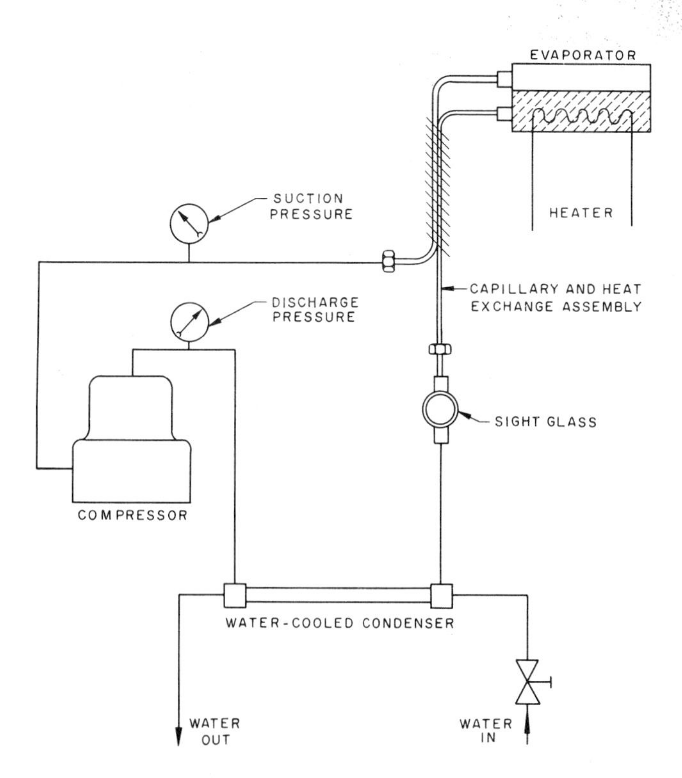


**Fig. 36 Test Setup for Determining the Capacity Balance Characteristic of a Given Combination of Compressor and Capillary and Heat Exchanger Assembly**

the element, install a new selection, and determine the optimum charge, as outlined above. A method occasionally used is to pinch the capillary to determine whether or not increased resistance is needed.

It is necessary to operate the unit through the expected range of operation to determine power and cooling capacity for any given selection and charge combination.

## SPECIAL APPLICATIONS

The scope of capillary tubes has been extended beyond the basic expansion process of flow-metering application. The mass flow may be affected through certain limits by varying the heat content of the refrigerant prior to its entrance into or along the length of the restrictor. Several important effects made possible through the use of the above principles include temperature differential control, selective flow control, and modulated flow.

A two-temperature evaporator is readily obtained by connecting a heat exchanger and a length of capillary tube between two coil sections. Refrigerant from the colder evaporator passes into the heat exchanger, where it condenses any vapor and subcools the refrigerant slightly, leaving the first coil prior to its entrance into the capillary. Consequently, for the low-pressure drops involved, very little flashing in the capillary occurs. The compressor refrigerant and mass circulation may be held approximately constant by thermostatically controlling one evaporator to a constant temperature. These two conditions of constant flow and complete liquefication throughout most of the restrictor ensure practically constant pressure or temperature differential.

Another method for realizing a temperature differential between two evaporators is to control the refrigerating effect of one of the evaporators selectively. A short section of capillary is attached to the inlet of the high-temperature coil, which is directly connected at its outlet to the lowtemperature unit. The short capillary and the high-temperature evaporator are bypassed by a large bore tube containing a solenoid valve. The latter is connected to a thermostat to control the refrigerant flow around the high-temperature evaporator and control its average temperature with respect to that of the other cooling unit.

Refrigerant flow may be modulated by attaching an electric heater to the circuit ahead of the capillary. When the heater is energized, it causes gas bubbles to form in the liquid and retards its flow. The latter results in a partially refrigerated evaporator.

In a reversible system designed for both cooling and heating cycles, the latter mode usually demands more capillary resistance and less refrigerant for optimum performance. A common method is a check valve in a branch of the capillary system that automatically changes its overall resistance with a reversal of the refrigerant flow. It is possible in heat pump systems to arrange a reservoir that is in communication with the indoor coil and in heat exchange with the outdoor coil. This arrangement allows the total refrigerant charge to be active during the cooling cycle and reduced by the differential amount stored in the reservoir during the heating cycle.

### Processing and Inspection

To prevent mechanical clogging caused by foreign particles, a strainer should precede the capillary. Also, all parts of the system must be evacuated adequately to eliminate water vapor and noncondensable gases, which may cause clogging by corrosion. The oil should be free from wax separation at the minimum operating temperature.

The interior capillary surface should be smooth and uniform in diameter. Although plug-drawn copper is more common, wire-drawn or sunk tubes are also available. Life tests should be conducted at low evaporator temperatures and high condensing temperatures to check on the possibility of corrosion and plugging. Material specifications for seamless copper tube are given in ASTM *Standard* B75-80. Similar information on hand-drawn copper tubes can be found in ASTM *Standard* B360-76.

A procedure should be established to ensure uniform flow capacities, within reasonable tolerances, for all capillaries used in product manufacture. This procedure may be conducted as follows:

The final capillary, determined from tests, is removed from the unit and given an airflow capacity rating, using the wettest meter method described in ASHRAE *Standard* 28-78. Master capillaries are then produced, by using the wet-test meter airflow equipment, to provide the maximum and minimum flow capacities for the particular unit. The maximum flow capillary has a flow capacity equal to that of the test capillary, plus a specified tolerance. The minimum flow capillary has a flow capacity equal to that of the test capillary less a specified tolerance. One sample of the maximum and minimum capillaries is sent to the manufacturer of capillary tubes to be used as tolerance guides for elements supplied for a particular unit. Samples are also sent to the inspection group for quality control.

### Preliminary Selection

The preliminary selection of a condenser-to-evaporator capillary for a given compressor rating may be determined by referring to Figure 37. Note that the compressor rating is based on the refrigeration per unit mass specified. The curves of Figure 37 apply to typical operating conditions, which are indicated on the charts. If desired, the ratings may be converted to mass flow by dividing the compressor capacity by the refrigeration per unit mass given on the chart. For the preliminary selection, no distinction is made between units with and without heat exchangers. The subcooling referred to in Figure 37 pertains to the total subcooling obtained in the unit, whether this is in the condenser or in the condenser and a heat exchanger.

In the selection of a capillary for a specific application, practical considerations influence the length. For example, the minimum length will be determined by such geometric considerations as the physical distance between the high side and low side and the length of capillary tube required for optimum heat exchange. It may also be dictated by considerations of exit velocity and noise and the possibility of plugging with foreign materials. The maximum length may be determined primarily by considerations of cost. It is fortunate, therefore, that the flow characteristics of a capillary can be adjusted independently by varying either its bore or its length. Thus, it is feasible to select the most convenient length independently and then (within certain limits) select a bore to give the desired flow. An alternate procedure is to select a standard bore and then adjust the length, as required.

Standard diameters and wall thickness for capillary tubes are given in ASTM *Standard* B360-76. Many nonstandard tubes are also used, resulting in nonuniform interior surfaces and variations in flow.

### Simplified Calculation Procedure

The optimum capillary size corresponding to any given set of system operating conditions can be calculated. These conditions of pressure, mass flow, and inlet subcooling or quality are a function of the unit design and a choice of the service operating conditions.

The capillary on room air conditioners is adjusted to give maximum unit capacity at ASHRAE rating conditions of 80°F

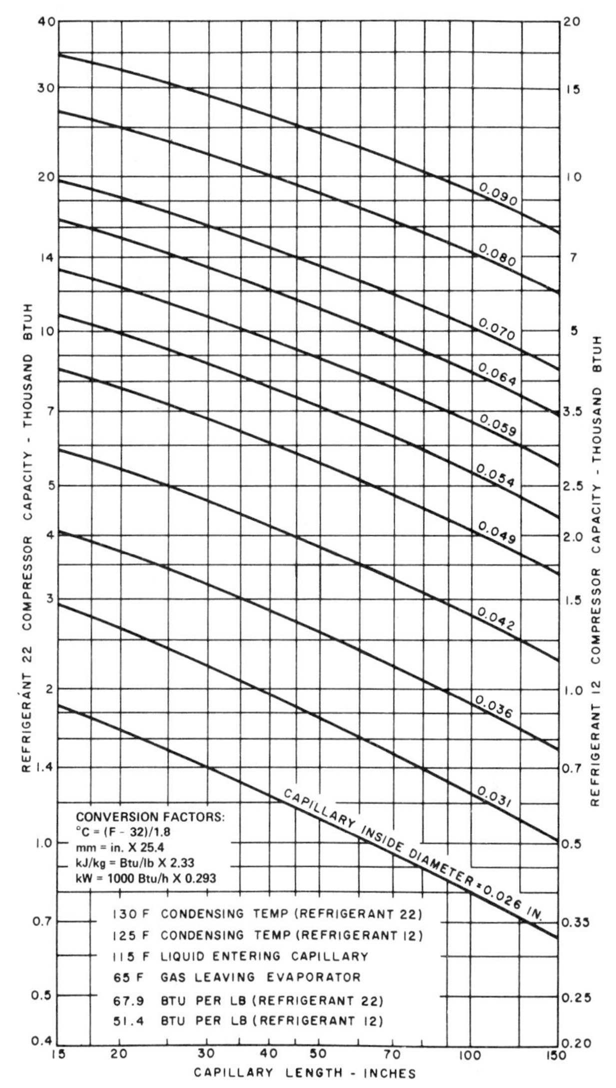


**Fig. 37 Preliminary Selection Chart for Refrigerants 12 and 22 Condenser-to-Evaporator Capillary**

(26.7 °C) dry bulb, 67 °F (19.4 °C) wet bulb indoors and 95 °F (35 °C) dry bulb, 75 °F (23.9 °C) wet bulb outdoors. Optimum performance may be obtained when there is considerable subcooling in the refrigerant prior to its entrance into the capillary.

The unit must be tested to ensure the capillary performs properly under various limiting conditions. For example, the capillary of a household refrigerator, which may have been sized to give optimum performance on no-load cycles at a particular ambient temperature, should be tested to ensure it operates properly during pulldown under maximum and minimum ambient and loading conditions.

The ratings given in Figures 38 and 39 are general application ratings for Refrigerants 12 and 22. They are presented in this form to provide easier usage, even though some accuracy is sacrificed in the process. The rating curves are quite versatile, as they can be used to calculate flow rate directly from a given capillary selection and flow condition or used to determine a capillary selection from a flow condition and flow rate.

These ratings, while presented in the form shown originally by Hopkins (1950), show different values because they were recalculated from the later work of Whitesel (1957). The difference is recognized in the gas flow-through portion of the ratings.

The ratings based only on Figures 38 and 39 yield approximate results because suction pressure is not shown as a parameter. This is not serious because Bolstad and Jordan (1948) demonstrated that drop in back pressure over a wide range causes a negligible increase in refrigerant flow rate. This is understandable when the back pressure is below the critical pressure. The curve for increase in flow rate flattens considerably above the critical pressure. Hopkins (1950) showed that using the flow factor to correct tube diameter and length results in some additional inaccuracy.

The effect of the suction pressure on capillary capacity may be calculated. First, the critical pressure is determined from Figure 40 and the correction factor is obtained from Figure 41. The capacity obtained from Figures 38 and 39 is multiplied by the correction factor to obtain the final value of flow. Relatively high back pressures cause substantial reductions in flow below the figure given by the basic rating chart for critical outlet pressure.

The effect of a suction line heat exchanger can normally be considered by subtracting the liquid temperature drop through this element from the liquid temperature at capillary inlet. This method holds rigidly when the refrigerant bubble point comes after the heat exchanger. Figure 42, which was derived mathematically for the latter condition, may be used to determine the heat-exchanger liquid subcooling.

Figure 43 may be used to determine liquid length, which indicates liquid pressure drop. Although in preparing this chart the Refrigerant 12 pressure drop values were calculated to be about 5.5% less than those for Refrigerant 22 at a given mass flow, the two sets of values were averaged for simplification. Average values of densities between −40 and 140 °F (−40 and 60 °C) were used.

Calculation procedures are illustrated in **Examples 1** to **6**:

---

**Example 1:** Determine the flow rate of Refrigerant 22 for a restrictor of 0.054 in. (1.37 mm) ID, 111-in. (2.82-m) long, operating without heat exchange at 100 °F (37.8 °C) saturated condensing temperature with 10 °F (5.6 °C) subcooling.

*Solution:* From Figure 38, at 100 °F (37.8 °C) saturated condensing temperature, or 210.6 psi (1451 kPa), and 10 °F (5.6 °C) subcooling, the flow rate for an 80 in. (2 m) length of 0.064 in. (1.63 mm) ID restrictor is found to be 97.0 lb/h (12.2 g/s).

To correct for actual diameter and length, the flow factor $\phi_1$ is found from Figure 39. For 0.054 in. (1.37 mm) ID and 111 in. (2.82 m) length, $\phi_1 = 0.55$. Corrected flow rate is $97.0 \cdot 0.55 = 53.4$ lb/h (6.7 g/s).

**Example 2:** Determine the condition of flow for the resistor of Example 1 if the flow rate falls to 40 lb/h (5 g/s) at 120 °F (48.9 °C) saturated condensing temperature, or 274.3 psi (1890 kPa) pressure.

*Solution:* The flow factor $\phi_1 = 0.55$ (from Example 1). With a flow rate of 40 lb/h (5 g/s), the uncorrected flow rate to be used with Figure 38 is $40/0.55 = 72.7$ lb/h (9.2 g/s).

From Figure 38, with a flow rate of 72.7 lb/h (9.2 g/s) and 274.3 psi (1890 kPa) condensing pressure, the flow condition is read as 12% inlet quality at the resistor entrance.

**Example 3:** Determine the length of 0.042 in. (1.07 mm) ID tube, which is equivalent to 30 in. (0.76 m) of 0.036 in. (0.91 mm) ID tube.

*Solution:* From Figure 39, the flow factor for 30 in. (0.76 m) of 0.036 in. (0.91 mm) of ID tube is read as 0.35. For this value of flow factor, it is found from Figure 39 that 73 in. (1.85 mm) of 0.042 in. (1.07 mm) ID tube are required.

**Example 4:** Determine the length of 0.054 in. (1.372 mm) ID capillary that will pass 50 lb/h (6.3 g/s) of Refrigerant 12, with no subcooling and no gas at the inlet, when the inlet pressure outside the capillary is 150 psig (1034 kPa).

*Solution:* The absolute inlet pressure is 164.7 psi (1135 kPa). From Figure 38, the flow rate is 65 lb/h (8.2 g/s). The flow factor $\phi_1$ is $50/65 = 0.77$. From Figure 42: for $\phi_1 = 0.77$ and 0.054 in. (1.37 mm) ID, the length of capillary required is 53 in. (1.35 m).

**Example 5:** What will be the flow rate for either Refrigerant 12 or Refrigerant 22 for a 0.070 in. (1.78 mm) ID capillary 131-in. (3.33-m)

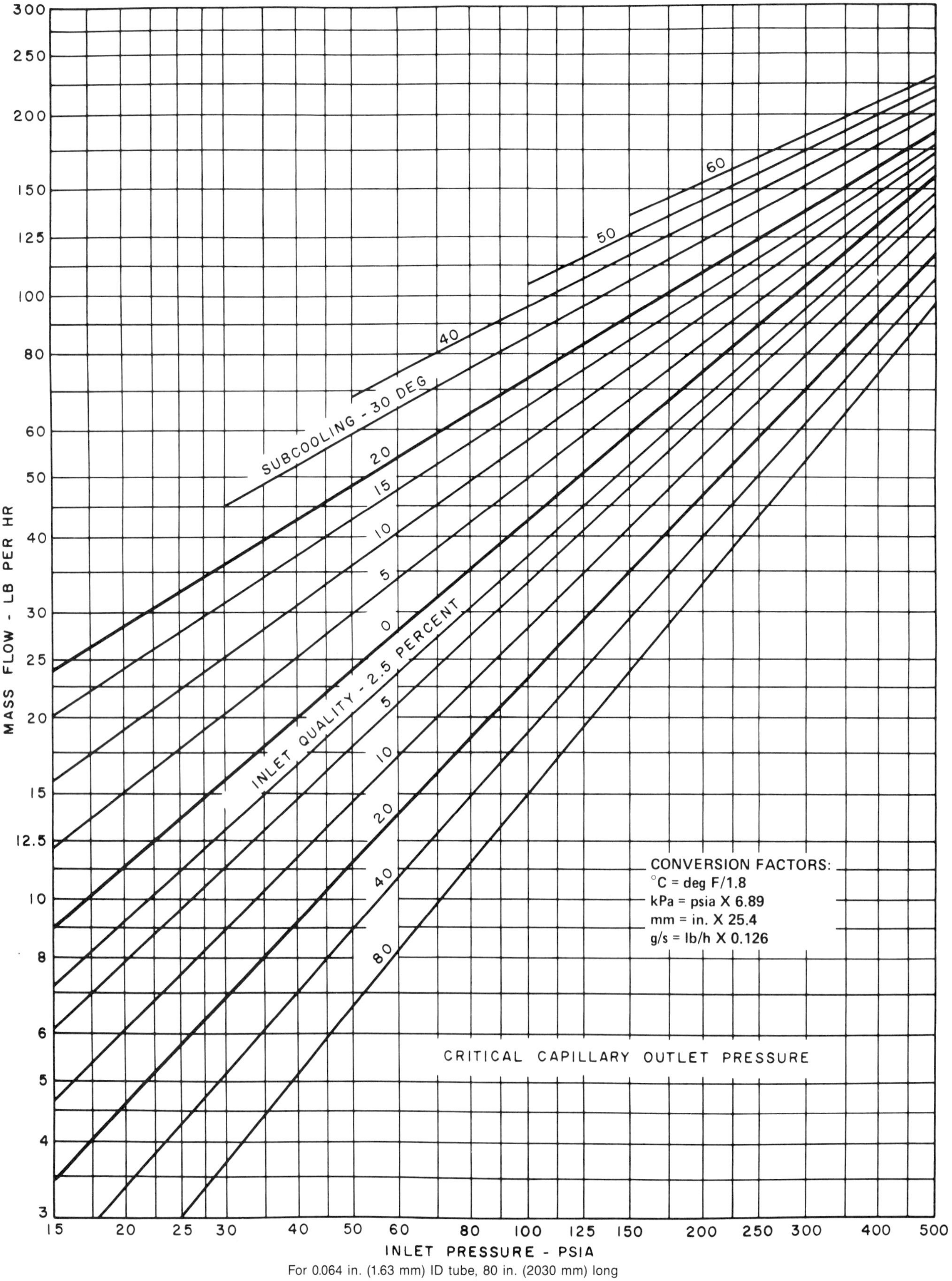

MASS FLOW - LB PER HR

INLET PRESSURE - PSIA

SUBCOOLING - 30 DEG, 40, 50, 60, 20, 15, 10, 5, 0

INLET QUALITY - 2.5 PERCENT, 5, 10, 20, 40, 80

CRITICAL CAPILLARY OUTLET PRESSURE

CONVERSION FACTORS:
°C = deg F/1.8
kPa = psia X 6.89
mm = in. X 25.4
g/s = lb/h X 0.126

For 0.064 in. (1.63 mm) ID tube, 80 in. (2030 mm) long

**Fig. 38 Basic Rating Curves for Condenser-to-Evaporator Capillary (Refrigerants 12 and 22)**

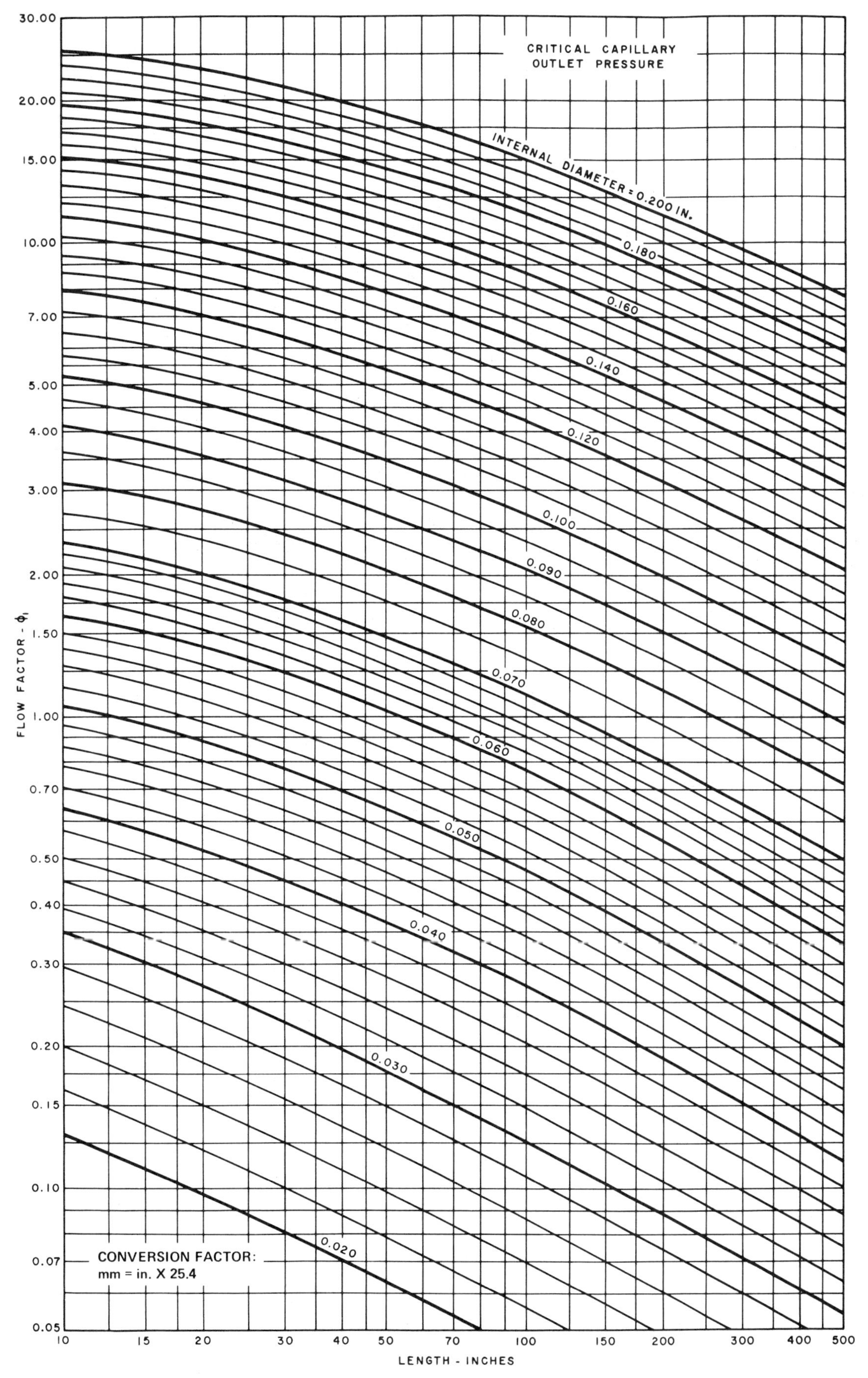


Fig. 39 Capillary Flow Factors (Refrigerants 12 and 22)

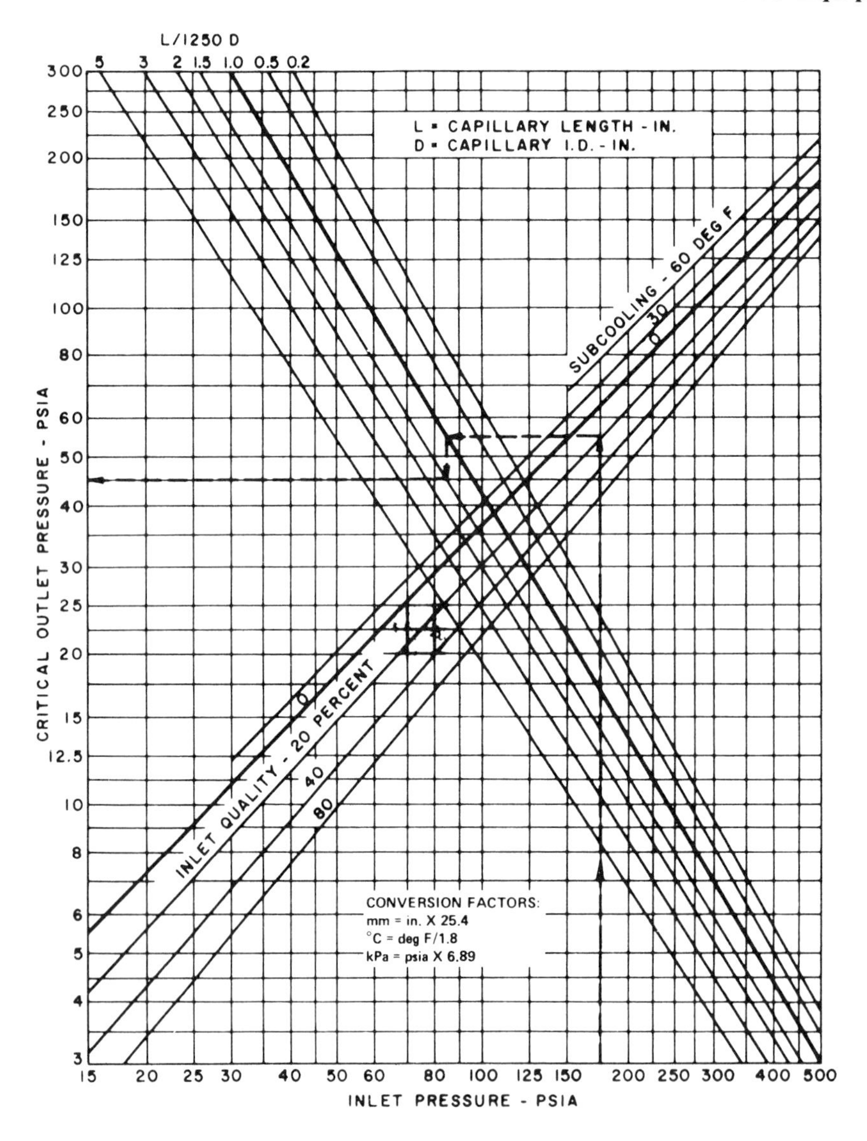


**Fig. 40 Capillary Critical Pressure Chart (Refrigerants 12 and 22)**

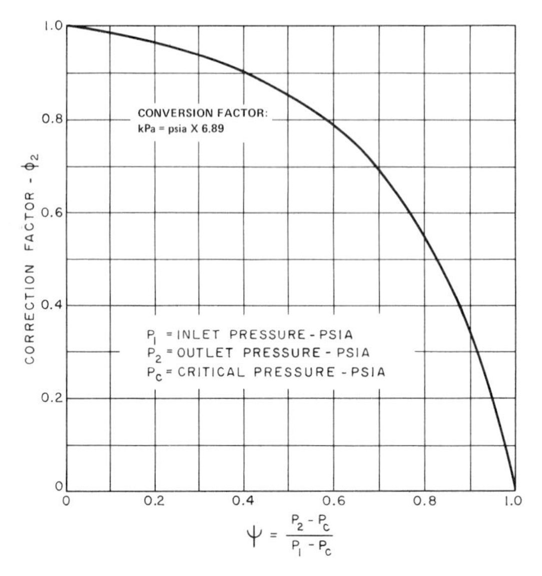


**Fig. 41 Critical Correction Factor (Refrigerants 12 and 22)**

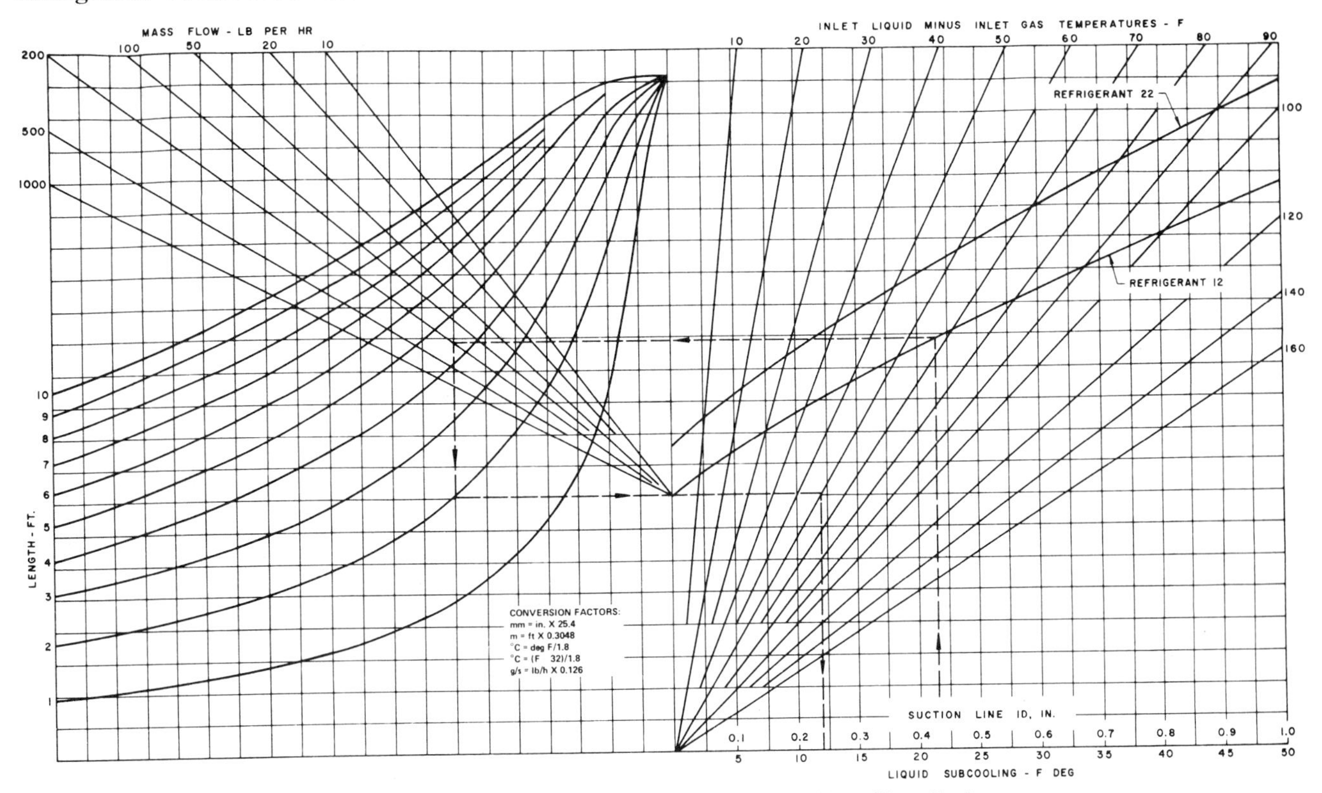


Fig. 42 Performance Chart for Suction Line and Capillary Heat Exchanger (Liquid-to-Gas Counterflow

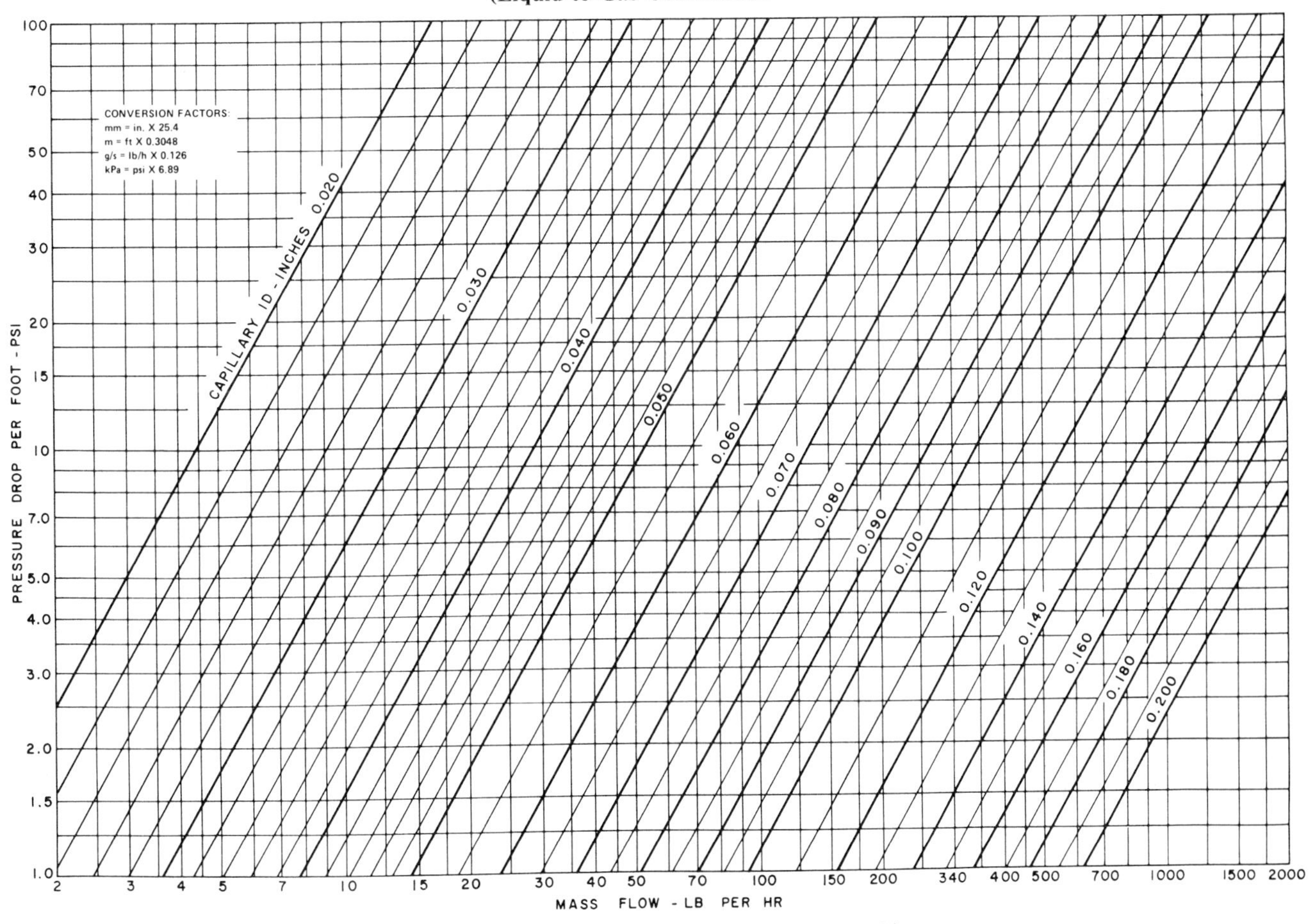


Fig. 43 Pressure-Drop Chart for Capillary Liquid (Refrigerants 12 and 22)

long with an inlet pressure of 175 psi (1206 kPa), 20% inlet quality, and an outlet pressure of 150 psi (1034 kPa).

*Solution:* From Figure 38, the basic rating is 41 lb/h (5.2 g/s). From Figure 38, $\phi_1$ = 0.98. The basic flow is then 0.98 • 41 = 40.1 lb/h (5.1 g/s). $L/1250D$ = 131/(1250)(0.070)[3330/(1250)(1.78)] = 1.5.

Enter Figure 40 at 175 psi (1206 kPa) inlet pressure and proceed vertically to 20% inlet quality. Proceed horizontally to the $L/1250D$ line of unity. Proceed vertically to the $L/1250D$ line of 1.5. Proceed horizontally to a critical outlet pressure of 45 psi (310 kPa). The value of $\pi$ for Figure 41 is (150 − 45)/(175 − 45) = 0.808. From Figure 39, $\phi_2$ = 0.53. The corrected flow is 40.1 • 0.53 = 21.3 lb/h (2.68 g/s).

**Example 6:** Determine the amount of subcooling obtained in a counterflow heat exchanger having a 0.43 in. (10.9 mm) ID suction tube 2-ft (0.61-m) long. The flow rate of Refrigerant 12 is 200 lb/h (25.2 g/s) and the temperature difference between the inlet liquid and inlet gas is 60°F (33.3°C).

*Solution:* Enter Figure 42 at 0.43 in. (10.9 mm) ID suction line and proceed vertically to the diagonal Refrigerant 12 line. Proceed horizontally from the intersection to the 200 lb/h (25.2 g/s) flow rate. From this point, drop vertically to the 2 ft (0.61 m) length curve, and proceed horizontally to the 60°F (33.3°C) temperature difference line. The liquid subcooling is read on the horizontal scale at the bottom of Figure 42 at 12°F (6.67°C).

## REFERENCES


Bolstad, M.M. and R.C. Jordan. 1948. Theory and Use of the Capillary Tube Expansion Device. *Refrigerating Engineering*, December, p. 519.

Hopkins, N.E. 1950. Rating the Restrictor Tube. *Refrigerating Engineering*, November, p. 1087.

Huelle, Z.R. 1972. The MSS Line—A New Approach to the Hunting Problem. ASHRAE *Journal*, October, p. 43.

Stoecker, W.F. 1966. Stability of an Evaporator-Expansion Valve Control Loop. ASHRAE *Transactions*, Vol. 72, Part II, p. IV.3.1-8.

Wedeking, G.L. and W.F. Stoecker. 1966. Transient Response of the Mixture-Vapor Transition Point in Horizontal Evaporating Flow. ASHRAE *Transactions* Vol. 72, Part II, p. IV.2.1-15.

Whitesel, H.A. 1957. Capillary Two-Phase Flow. *Refrigerating Engineering*, April, p. 42.

Whitesel, H.A. 1957. Capillary Two-Phase Flow—Part II. *Refrigerating Engineering*, September, p. 35.


## BIBLIOGRAPHY


ASHRAE. 1978. Safety Code for Mechanical Refrigeration. ASHRAE *Standard* 15-78.

Bolstad, M.M. and R.C. Jordan. 1949. Theory and Use of the Capillary Expansion Device—Part II, Nonadiabatic Flow. *Refrigerating Engineering*, June, p. 577.

Cooper, L.; W.R. Brisken; and C.K. Chu. 1957. Simple Selection Method for Capillaries Derived from Physical Flow Conditions. *Refrigerating Engineering*, July, p. 37.

Halter, E.J. 1975. Droplet and Drop Size and Distribution Estimation. ASHRAE *Transactions* 81(2): 459-470.

Lathrop, H.F. 1948. Application and Characteristics of Capillary Tubes. *Refrigerating Engineering*, August, p. 129.

Marcy, G.P. 1949. Pressure Drop with Change of Phase in a Capillary Tube. *Refrigerating Engineering*, January, p. 53.

Prosek, J.R. 1953. A Practical Method of Selecting Capillary Tubes. *Refrigerating Engineering*, June, p. 644.

Smith, F.G. 1956. Turbulent Flow of Air Through Capillary Tubes. *Refrigerating Engineering*, October, p. 48.

Staebler, L.A. 1948. Theory and Use of a Capillary Tube for Liquid Refrigerant Control. *Refrigerating Engineering*, January, p. 54.

Staebler, L.A. 1950. The Capillary Tube and Its Applications to Small Refrigerating Systems. *Refrigeration Service Engineers Society Service Manual*, Section 18.

Swart, R.H. 1946. Capillary Tube Exchangers. *Refrigerating Engineering*, September, p. 221.

## CHAPTER 20

# COOLING TOWERS



MOST air-conditioning systems and industrial processes generate heat that must be removed and dissipated. Water is commonly used as a heat-transfer medium to remove heat from refrigerant condensers or industrial process heat exchangers.

In the past, this was accomplished by drawing a continuous stream of water from a natural body of water or a utility water supply, heating it as it passed through the process, and then discharging the water directly to a sewer or returning it to the body of water. Water purchased from utilities for this purpose has now become prohibitively expensive because of increased water supply and disposal costs. Similarly, cooling water drawn from natural sources is relatively unavailable because the increased temperature of the discharge water disturbs the ecology of the water source.

Air-cooled heat exchangers may be used to cool the water by rejecting heat directly to the atmosphere, but the first cost and the fan energy consumption of these devices are high. They are economically capable of cooling the water to within approximately 20°F (11°C) of the ambient dry-bulb temperature. Such temperature levels are often too high for the cooling water requirements of most refrigeration systems and many industrial processes.

Cooling towers overcome most of these problems and, as such, are commonly used to dissipate heat from water-cooled refrigeration, air-conditioning, and industrial process systems. The water consumption rate of a cooling tower system is only about 5% of that of a once-through system, making it the least expensive system to operate with purchased water supplies. Additionally, the amount of heated water discharged (blowdown) is very small, so the ecological effect is reduced greatly. Lastly, cooling towers can cool water to within 5 to 10°F (3 to 6°C) of the ambient wet-bulb temperature or about 35°F (19°C) lower than air-cooled systems of reasonable size.

The preparation of this chapter is assigned to TC 8.6, Cooling Towers and Evaporative Condensers.

## PRINCIPLE OF OPERATION

A cooling tower cools water by using a combination of heat and mass transfer. The water to be cooled is distributed in the tower by spray nozzles, splash bars, or filming-type fill, which exposes a very large water surface area to atmospheric air. Atmospheric air is circulated by (1) fans, (2) convective currents, (3) natural wind currents, or (4) induction effect from sprays. A portion of the water absorbs heat to change from a liquid to a vapor at constant pressure. This heat of vaporization at atmospheric pressure is transferred from the water remaining in the liquid state into the airstream.

Figure 1 shows the temperature relationship between water and air as they pass through a counterflow cooling tower. The curves indicate the drop in water temperature (Point A to Point B) and the rise in the air wet-bulb temperature (Point C to Point D) in their respective passages through the tower. The temperature difference between the water entering and leaving the cooling tower (A minus B) is the *range*. For a system operating in a steady state, the range is the same as the water temperature rise through the load heat exchanger. Accordingly, the range is determined by the heat load and water flow rate, not by the size or capability of the cooling tower.

The difference between the leaving water temperature and the entering air wet-bulb temperature (B minus C) in Figure 1 is the *approach to the wet bulb* or simply the *approach* of the cooling tower. The approach is a function of cooling tower capability,

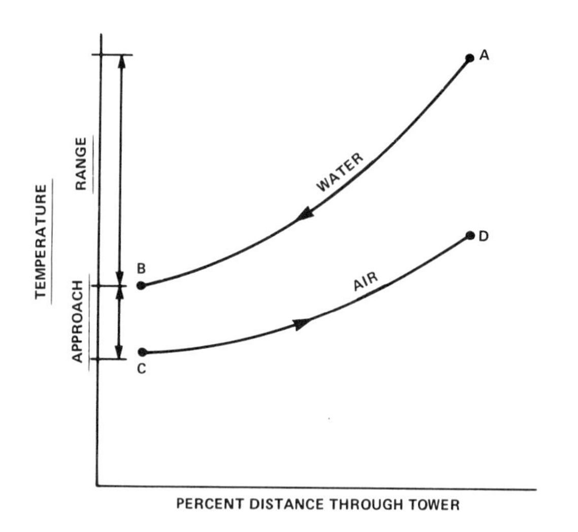


**Fig. 1 Temperature Relationship Between Water and Air in a Counterflow Cooling Tower**

and a larger cooling tower produces a closer approach (colder leaving water) for a given heat load, flow rate, and entering air condition. Thus, the amount of heat transferred to the atmosphere by the cooling tower is always equal to the heat load imposed on the tower, while the temperature level at which the heat is transferred is determined by the thermal capability of the cooling tower.

The thermal performance of a cooling tower is affected by the entering air wet-bulb temperature. Entering air dry-bulb temperature and relative humidity have an insignificant effect on thermal performance, but they do affect the rate of water evaporation. A psychrometric analysis of the air passing through a cooling tower illustrates this effect (Figure 2). Air enters at the ambient condition Point A, absorbs heat and mass (moisture) from the water, and exits at Point B in a saturated condition (at very light loads, the discharge air may not be saturated). The amount of heat transferred from the water to the air is proportional to the difference in enthalpy of the air between the entering and leaving conditions ($h_B - h_A$). Because lines of constant enthalpy correspond almost exactly to lines of constant wet-bulb temperature, the change in enthalpy of the air may be determined solely by the change in wet-bulb temperature of the air.

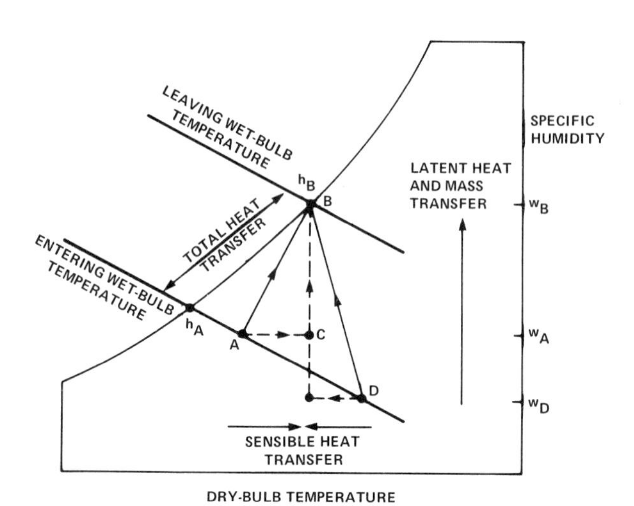


**Fig. 2 Psychrometric Analysis of Air Passing through a Cooling Tower**

Vector AB in Figure 2 may be separated into component AC, which represents sensible air heating (sensible water cooling), and component CB, which represents latent air heating (latent water cooling). If the entering air condition is changed to Point D at the same wet-bulb temperature but at a higher dry-bulb temperature, the total heat transfer remains the same, but the sensible and latent components have changed. AB represents sensible cooling of the water by evaporation and sensible and latent heating of the air. Case BD still represents sensible cooling of the water by evaporation, but sensible cooling and latent heating of the air. Thus, for the same water-cooling load, the amount of evaporation depends on the amount of sensible heating or cooling of the air.

The ratio of latent to sensible heat is important in analyzing the water usage of a cooling tower. Mass transfer (evaporation) occurs only in the latent portion of the heat-transfer process and is proportional to the change in specific humidity. Because the entering air dry-bulb temperature or relative humidity affects the latent to sensible heat-transfer ratio, it also affects the rate of evaporation. In Figure 2, the rate of evaporation is less in Case AB ($W_B - W_A$) than in Case DB ($W_B - W_D$) because the latent heat transfer (mass transfer) represents a smaller portion of the total.

The evaporation rate at typical design conditions is approximately 1% of the water flow rate for each 12.6°F (7°C) of water temperature range. The actual annual evaporation rate is less than the design rate because the sensible component of total heat transfer increases as the entering air temperature decreases.

In addition to water loss from evaporation, losses also occur because of liquid carryover into the discharge airstream and blowdown to maintain acceptable water quality. Both of these factors are addressed later in this chapter.

## DESIGN CONDITIONS

The thermal capability of any cooling tower may be defined by the following parameters:

1. Entering and leaving water temperatures
2. Entering air wet-bulb or entering air wet-bulb and drybulb temperatures
3. Water flow rate

The entering air dry-bulb temperature affects the amount of water evaporated from the water cooled in any evaporative-type cooling tower. It also affects airflow through hyperbolic towers and directly establishes thermal capability within any indirect-contact cooling tower component operating in a dry mode. Variations in tower performance associated with changes in the remaining parameters are covered in the "Performance Curves" section of this chapter.

The thermal capability of cooling towers for air-conditioning is identified in *nominal* tonnage, based on heat dissipation of 15,000 Btu/h (1.25 kW) per condenser ton (kW) and a water circulation rate of 3 gpm per ton (54 mL/s per kW) cooled from 95 to 85°F (35 to 29.4°C) at 78°F (25.6°C) wet-bulb temperature. For specific applications, however, nominal tonnage ratings are not used, and the thermal performance capability is usually stated in terms of flow rate at specified operating conditions (entering and leaving water temperatures and entering air wet-bulb and/or dry-bulb temperatures).

## TYPES OF COOLING TOWERS

Two basic types of evaporative cooling devices are used. The first type involves direct contact between heated water and atmosphere (see Figure 3). The direct-contact device (*cooling*

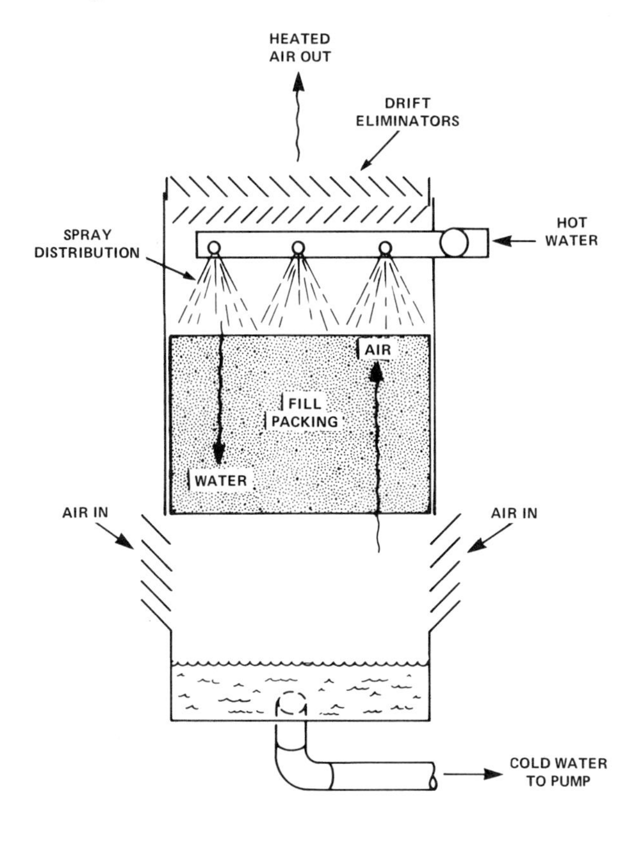


**Fig. 3 Direct-Contact Evaporative Cooling Tower, Showing Counterflow Water-Air Relationship**

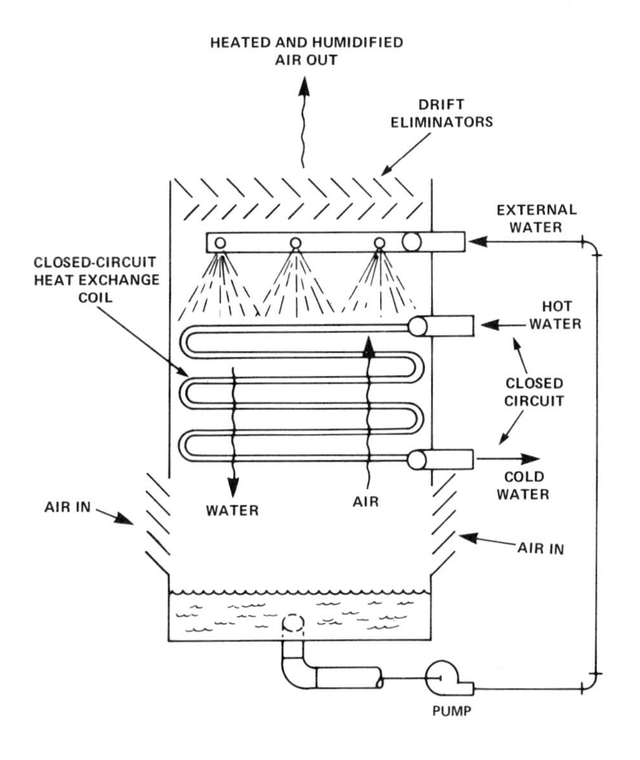


**Fig. 4 Indirect-Contact Evaporative Cooling Tower, Showing Counterflow Water-Air Relationship**

*tower*) exposes water directly to the cooling atmosphere, thereby transferring the source heat load directly to the air. The second type involves indirect contact between heated fluid and atmosphere (see Figure 4).

Indirect-contact towers (*closed-circuit fluid coolers*) contain two separate fluid circuits: (1) the external circuit in which water is exposed to the atmosphere as it cascades over the tubes of a coil bundle, and (2) an internal circuit in which the fluid to be cooled circulates inside the tubes of the coil bundle. In operation, heat flows from the internal fluid circuit, through the tube walls of the coil, to the external water circuit, which is cooled evaporatively. As the internal fluid circuit never contacts the atmosphere, this unit can be used to cool fluids other than water and/or to prevent contamination of the primary cooling circuit with airborne dirt and impurities.

Spray-filled towers expose water to air without using a heat-transfer medium, which is the most rudimentary method of exposing water to air in direct-contact devices. The amount of water surface exposed to the air depends on the efficiency of the sprays, and the time of contact depends on the elevation and pressure of the water-distribution system.

To increase contact surfaces, as well as time of exposure, a heat-transfer medium, or *fill*, is installed below the water-distribution system, in the path of the air. The two types of fill in use are splash-type and film-type (Figure 5). Splash-type fill

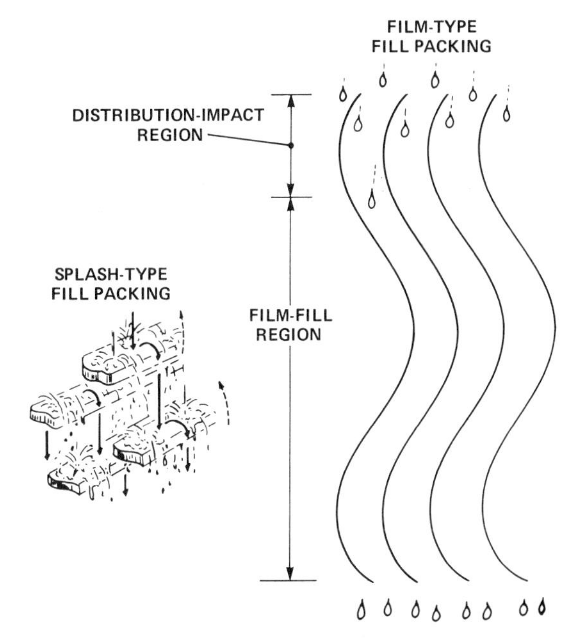


**Fig. 5 Types of Fill**

maximizes contact area and time by forcing the water to cascade through successive elevations of splash bars arranged in staggered rows. Film-type fill achieves the same effect by causing the water to flow in a thin layer over closed-spaced sheets (principally PVC) that are arranged vertically.

Either type of fill is applicable to both counterflow and crossflow towers. For thermal performance levels typically encountered in air conditioning and refrigeration, the tower with film-type fill is usually more compact. However, splash-type fill is less sensitive to initial air and water distribution and is usually the fill of choice for water qualities conducive to plugging.

## Types of Direct-Contact Cooling Towers

**Nonmechanical Draft Towers,** aspirated by sprays or density differential, do not contain fill and do not use a mechanical

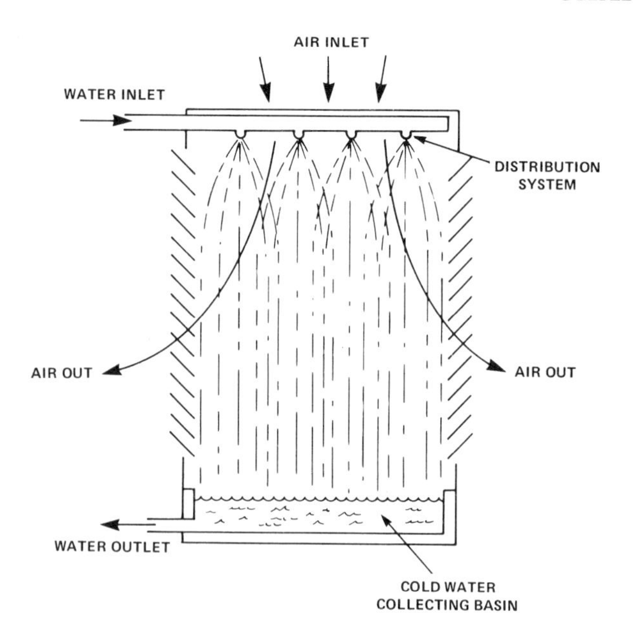


Fig. 6 Vertical Spray Tower

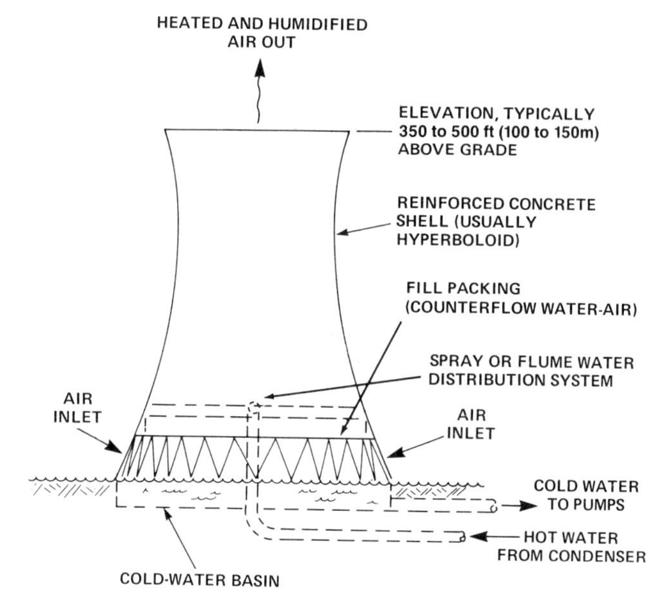


Fig. 8 Hyperbolic Tower

device (fan). The aspirating effect of the water spray, either vertically or horizontally (Figures 6 and 7), induces airflow through the tower in a parallel-flow pattern.

Because air velocities (both entering and leaving) are relatively low, such towers are susceptible to adverse wind effects and, therefore, are normally used to satisfy a low cost requirement when operating temperatures are not critical to the system. Some horizontal spray towers use high-pressure sprays to induce large air quantities and improve air/water contact. Multispeed or staged pumping systems are normally recommended to reduce energy use in periods of reduced load and ambient conditions.

*Chimney Towers (Hyperbolic)* have been used primarily for larger power installations, but may be of generic interest (Figure 8). The heat-transfer mode may be either counterflow, crossflow, or parallel flow. Air is induced through the tower by the air density differentials, which exist between the lighter heat-humidified chimney air and the outside atmosphere. Fills are typically the splash- or film-type.

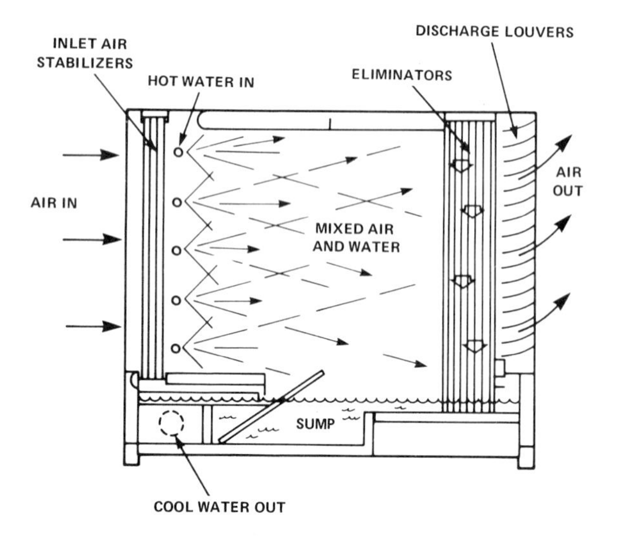


Fig. 7 Horizontal Spray Tower

Primary justification of these high first cost products comes through reduction in auxiliary power requirements (elimination of fan energy), reduced property acreage, and elimination of recirculation and/or vapor plume interference. Materials used in chimney construction have been primarily steel-reinforced concrete, while early-day timber structures had limitations of size.

**Mechanical Draft Towers.** *Conventional Towers* (Figure 9). The fans on mechanical draft towers may be on the inlet air side (forced-draft) or exit air side (induced-draft). The centrifugal or propeller fan is chosen, depending on external pressure needs, permissible sound levels, and energy usage requirements. Water is downflow, while the air may be upflow (counterflow heat transfer) or horizontal-flow (crossflow heat transfer). Air may be single-entry (in one side of tower) or double-entry (in two sides of tower). All four combinations have been produced in various sizes, for example, (1) forced-draft counterflow, (2) induced-draft counterflow, (3) forced-draft crossflow, and (4) induced-draft crossflow.

Towers are typically classified as either *factory-assembled* (Figure 10), when the entire tower or a few large components are factory-assembled and shipped to the site for installation, or *field-erected* (Figure 11), where the tower is completely constructed on site.

Most factory-assembled towers are of metal construction, usually galvanized steel. Other constructions include treated wood, stainless steel, and plastic towers or components. Field-erected towers are predominantly framed of preservative-treated redwood or treated fir with fiberglass-reinforced plastic used for special components and the casing. Coated metals, primarily steel, are used for complete towers or components. Concrete or ceramic materials are usually restricted to the largest sizes (see the "Materials of Construction" section of this chapter).

*Special Purpose Towers.* Towers containing a conventional mechanical draft unit in combination with an air-cooled (finned-tube) heat exchanger (Figure 12) are *wet/dry towers.* They are used either for vapor plume reduction or water conservation. The hot, moist plumes discharged from cooling towers are especially dense in cooler weather. On some installations, limited

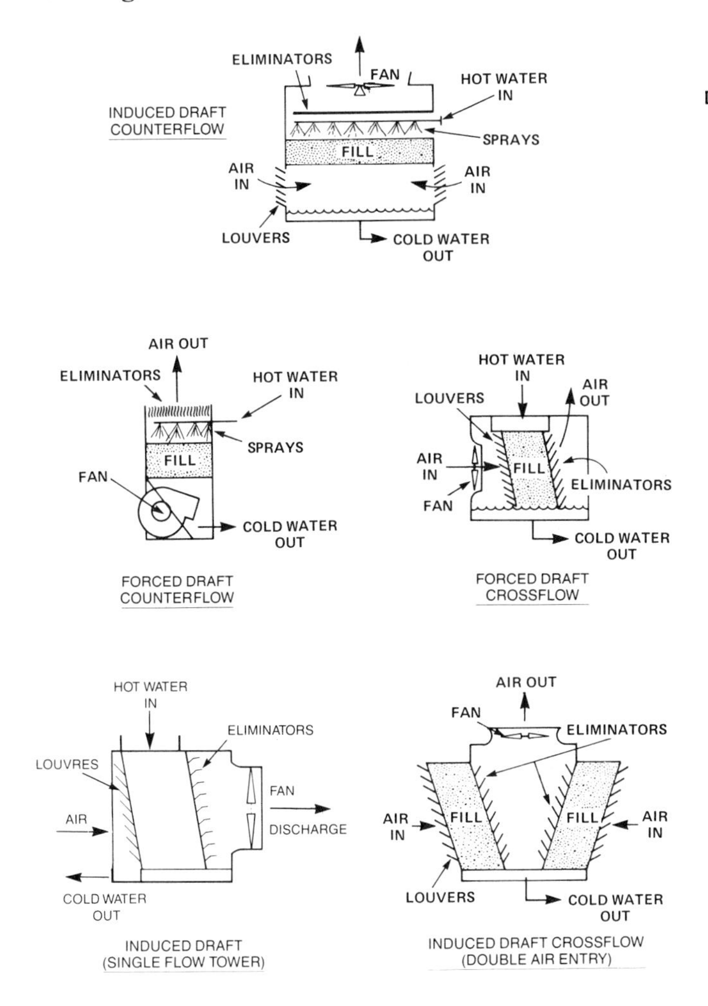


**Fig. 9 Conventional Mechanical Draft Cooling Towers**

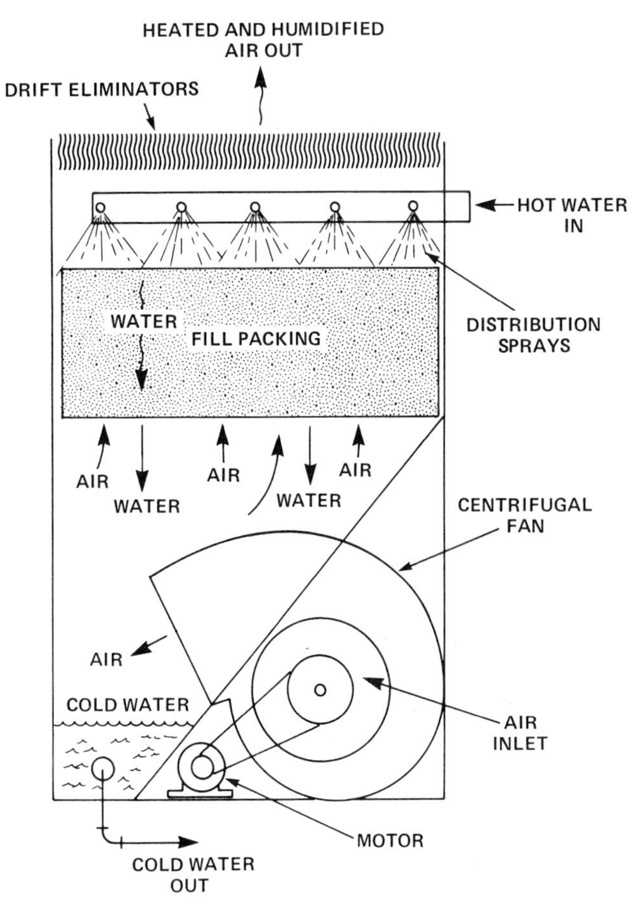


**Fig. 10 Counterflow Forced-Draft Tower**

abatement of these plumes is required to avoid restricted visibility on roadways, bridges, buildings, etc.

A vapor plume abatement tower usually has a relatively small air-cooled component, which tempers the leaving airstream to reduce the relative humidity and thereby minimize the fog-generating potential of the tower. Conversely, a water conservation tower usually requires a large, air-cooled component to provide significant savings in water consumption. Some designs can handle heat loads entirely by the nonevaporative air-cooled heat exchangers during reduced ambient conditions.

A variant of the wet/dry tower is an *evaporatively precooled/air-cooled heat exchanger.* It uses an adiabatic saturator (air precooler/humidifier) to enhance the summer performance of an air-cooled exchanger, thus conserving water in comparison to conventional cooling towers (annualized) (Figure 13). Evaporative fill sections usually operate only during specified summer periods, while full dry operation is expected below 50 to 70°F (10 to 20°C)) ambient conditions. Integral water pumps return the lower basin water to upper distribution systems of the adiabatic saturators in a manner similar to the closed-circuit fluid cooler and evaporative condenser products.

**Ponds, Spray Ponds, Spray Module Ponds, and Channels.** Heat dissipates from the surface of a body of water by evaporation, radiation, and convection. Captive lakes or ponds (man-made or natural) are sometimes used to dissipate heat by natural air currents and wind. This system is usually used in large plants where real estate is not limited.

A pump-spray system above the pond surface improves heat transfer by spraying the water in small droplets, thereby extending the water surface and bringing it into intimate contact with the air. The heat transfer is largely the result of evaporative cooling (see the "Cooling Tower Theory" section of this chapter). The system is a piping arrangement using branch arms and nozzles to spray the circulated water into the air. The pond acts largely as a collecting basin. Control of temperatures, real estate demands, limited approach to the wet-bulb temperature, and winter operational difficulties have ruled out the spray pond in favor of more compact and more controllable mechanical or natural draft towers.

Equation (1) can be used to estimate cooling pond area. Because of variations in wind velocity and solar radiation, a substantial margin of safety should be added to the results.

$$w_p = \frac{A(95 + 0.425v)}{h_{fg}} [p_w - p_a] \qquad \text{(1 I-P)}$$

$$w_p = \frac{A(0.0887 + 0.07815v)}{h_{fg}} [p_w - p_a] \qquad \text{(1 SI)}$$

*where*

$w_p$ = evaporation of water lb/hr (kg/s)
$A$ = area of pool surface, ft$^2$ (m$^2$)
$v$ = air velocity over water surface, fpm (m/s)
$h_{fg}$ = latent heat required to change water to vapor at surface water temperature, Btu/lb (kJ/kg)
$p_a$ = saturation pressure at room air dew point, in. Hg/(kPa)
$p_w$ = saturation vapor pressure taken at the surface water temperature, in. Hg/(kPa)

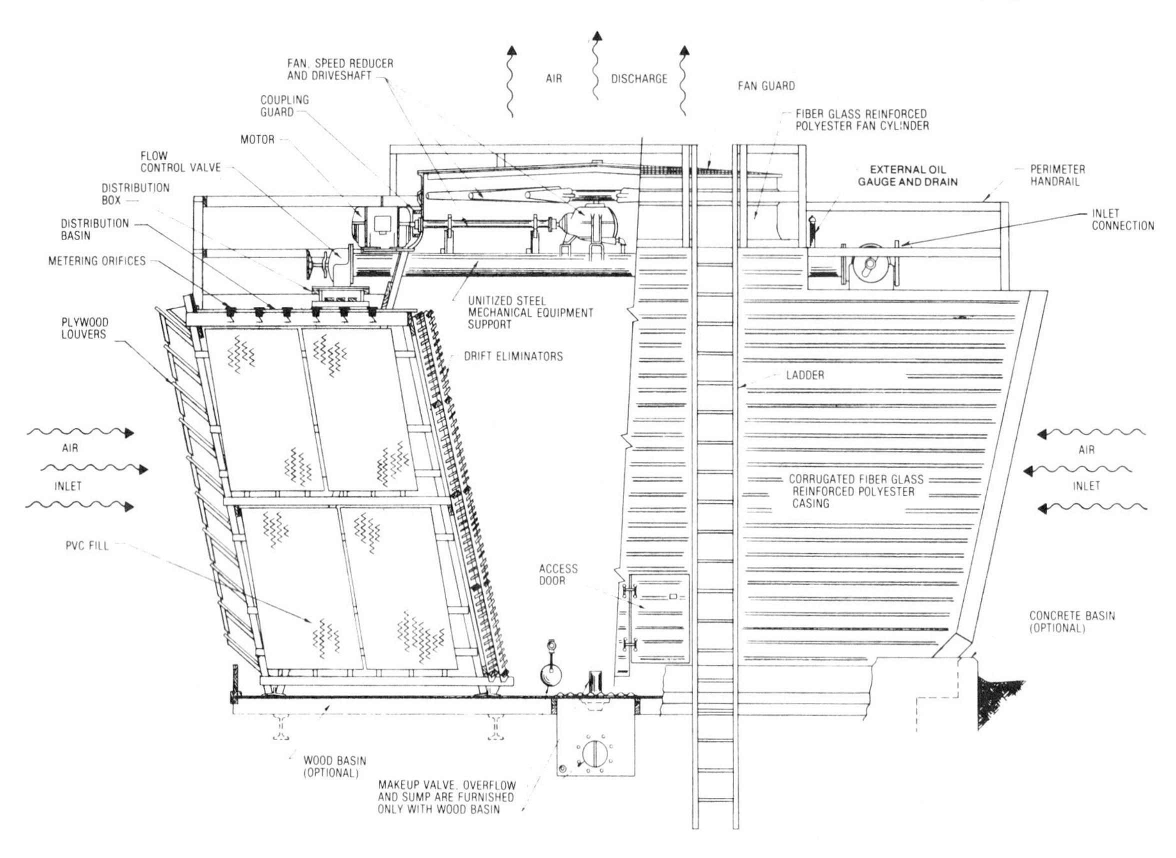


**Fig. 11 Crossflow Mechanical Draft Tower**

### Types of Indirect-Contact Towers

**Closed-Circuit Fluid Coolers (Mechanical Draft).** Both counterflow- and crossflow-types are used in forced and induced fan arrangements (Figure 4). The tubular heat exchangers are typically serpentine bundles, usually arranged for free-gravity internal drainage. Pumps are integrated in the product to transport water from the lower collection basin to the upper distribution basins or sprays. The internal coils can be fabricated from any of several materials, but galvanized steel or copper are predominent. Closed circuit fluid coolers, which are similar to evaporative condensers, are used increasingly on heat-pump systems and screw compressor oil-pump systems.

The indirect-contact towers require a closed-circuit heat exchanger (usually tubular serpentine coil bundles) which is exposed to air/water cascades similar to the fill of a cooling tower (Figure 4). Some types include supplemental film or splash fill sections to augment the external heat-exchange surface area.

**Coil Shed Towers (Mechanical Draft).** Coil shed towers usually consist of isolated coil sections (nonventilated), which are located beneath a conventional cooling tower (Figure 14). Counterflow- and crossflow-types are available with either forced- or induced-fan arrangements. Redistribution water pans, located at the tower's base, feed cooled water by gravity-flow to the tubular heat exchange bundles (coils). These units are similar in function to closed-circuit fluid coolers, except that supplemental fill is always required, and the airstream is directed only through the fill regions of the tower. Typically, these units are arranged as a field-erected, multifan cell towers and are used primarily in the process cooling industry.

## MATERIALS OF CONSTRUCTION

Materials found in cooling tower construction are usually selected to resist the corrosive water and atmospheric conditions.

**Woods** have been used extensively for all static components except hardware. Redwood and fir predominate, usually with post-fabrication pressure treatment of waterborne preservative chemicals, typically chromated-copper-arsenate (CCA) or acid-copper-chromate (ACC). These microbiocide chemicals prevent the attack of wood-destructive organisms, such as termites or fungi.

**Metals.** Steel with galvanized zinc is used for small- and medium-size installations. Hot-dip galvanizing after fabrication is used for larger weldments. Hot-dip galvanizing and cadmium and zinc plating are used for hardware. Brasses and bronzes are selected for special hardware, fittings, and tubing material. Stainless steels (principally 302, 304, and 316) are often used for sheet metal, drive shafts, and hardware in exceptionally corrosive atmospheres. Cast iron is a common choice for base castings, fan hubs, motor or gear reduction housings, and piping-valve components. Metals coated with polyurethane and polyvinyl-chloride are used selectively for special components. Epoxy-coal tar compounds and epoxy-powdered coatings are also used for key components or entire cooling towers.

**Plastics.** Fiberglass-reinforced polyester materials are used for components such as piping, fan cylinders, fan blades, casing, louvers, and structural connecting components. Polypropylene and ABS are specified for injection-molded components, such as fill bars and flow orifices. PVC is increasingly used as fill, eliminator, and louver materials. Reinforced plastic mortar is used in larger piping systems, coupled by neoprene O-ring-gasketed ball and socket joints.

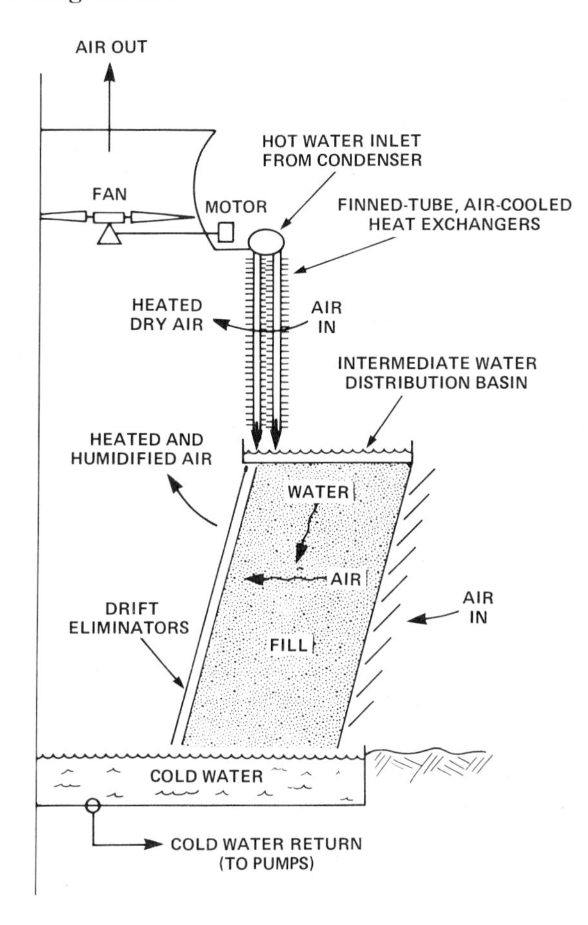


Fig. 12 Combination Wet-Dry Tower

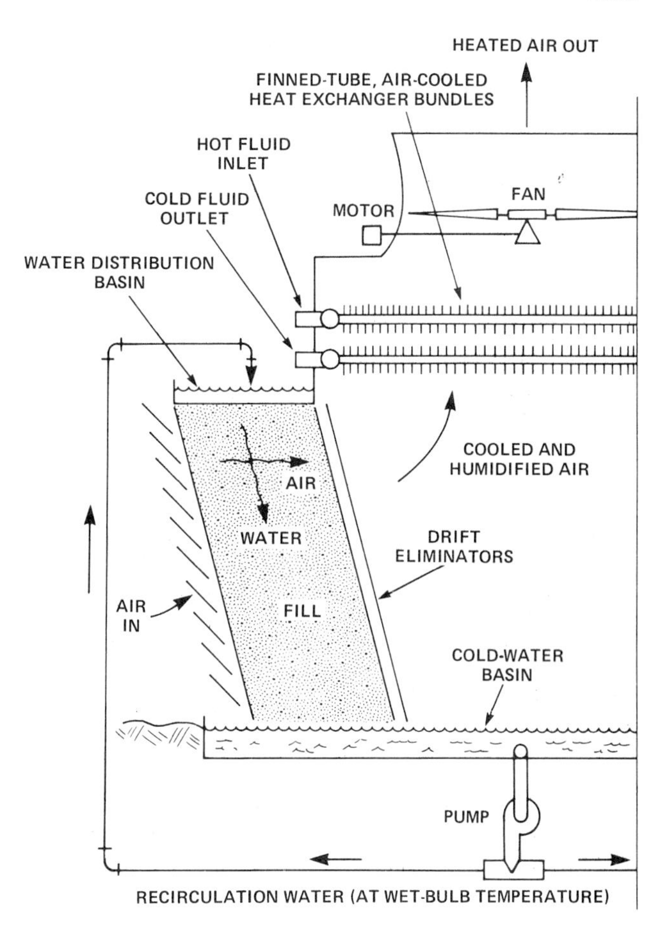


Fig. 13 Adiabatically Saturated Air-Cooled Heat Exchanger

**Concrete, Masonry, and Tile.** Concrete is typically specified for cold-water basins of field-erected cooling towers and is used in piping, casing, and structural systems of the largest towers, primarily in the power industry. Special tiles and masonry are used when aesthetic considerations are important.

## SELECTION CONSIDERATIONS

Selecting the proper water-cooling equipment for a specific application requires consideration of (1) cooling duty, (2) economics, (3) required services, and (4) environmental conditions. Many of these factors are interrelated, but should be evaluated individually.

Because a wide variety of water-cooling equipment may meet the required cooling duty, such items as size, height, length, width, plan area, volume of airflow, fan and pump energy consumption, materials of construction, water quality, and availability influence the final equipment selection.

The optimum choice is generally made after an economic evaluation. Chapter 49 of the 1987 HVAC Volume describes two common methods of economic evaluation—life-cycle costing and payback analysis. Each of these procedures compares equipment on the basis of total owning, operating, and maintenance costs.

*Initial cost* comparisons consider the following factors:

1. Erected cost of equipment
2. Costs of interface with other subsystems, which include items such as:
   *a.* Basin grilleage and value of the space occupied
   *b.* Cost of pumps and prime movers
   *c.* Electrical wiring to pump and fan motors
   *d.* Electrical controls and switchgear
   *e.* Cost of piping to and from the tower (Some designs require more inlet and discharge connections than others, thus affecting the cost of piping.)

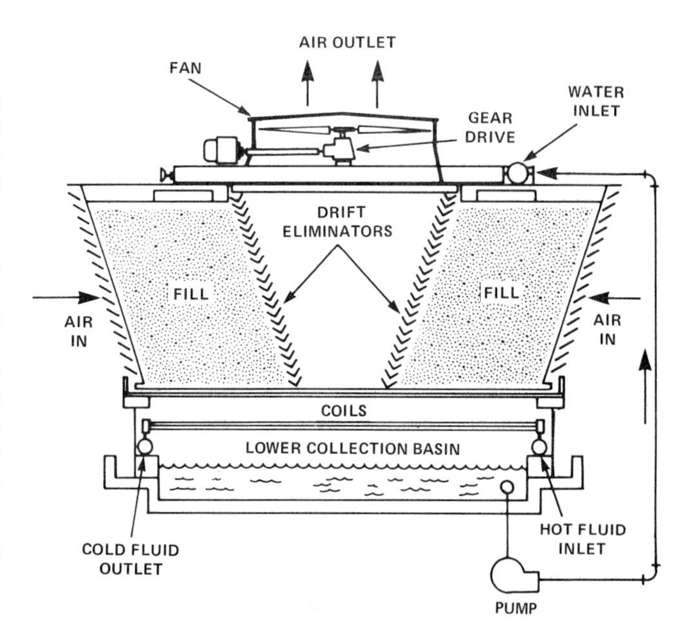


Fig. 14 Coil Shed Cooling Tower

*f.* The tower basin, sump screens, overflow piping, and makeup lines, when they are not furnished by the manufacturer
*g.* Shutoff and control valves, when they are not furnished by the manufacturer
*h.* Walkways, ladders, etc., providing access to the tower

In evaluating *owning and maintenance costs*, major items to consider are as follows:

1. System energy costs (fans, pumps, etc.) on the basis of operating hours per year
2. Demand charges
3. Expected equipment life
4. Maintenance and repair costs
5. Money costs

Other factors are (1) safety features and safety codes; (2) conformity to building codes; (3) general design and rigidity of structures; (4) relative effects of corrosion, scale, or deterioration on service life; (5) availability of spare parts; (6) experience and reliability of manufacturers; and (7) operating flexibility for economical operation at varying loads or during seasonal changes. In addition, equipment vibration, sound levels, acoustical attenuation, and compatibility with the architectural design are important. The following section details many of these more important considerations.

## APPLICATION

This chapter describes some of the major design considerations, but the manufacturer of the cooling tower should be consulted for more detailed recommendations.

### Siting

When a cooling tower can be located in an open space with free air motion and unimpeded air supply, siting is normally not a problem in obtaining a satisfactory installation. However, towers are often situated indoors, against walls, or in enclosures. In such cases, the following factors must be considered:

1. Sufficient free and unobstructed space should be provided around the unit to ensure an adequate air supply to the fans and to allow proper servicing.
2. The tower discharge air should not be deflected in any way that might promote recirculation (a portion of the warm, moist discharge air reentering the tower [Figure 15]). Recirculation raises the entering wet-bulb temperature, causing increased hot- and cold-water temperatures and, during cold weather operation, can promote the icing of air intake areas. The possibility of air recirculation should be particularly considered on multiple tower installations.

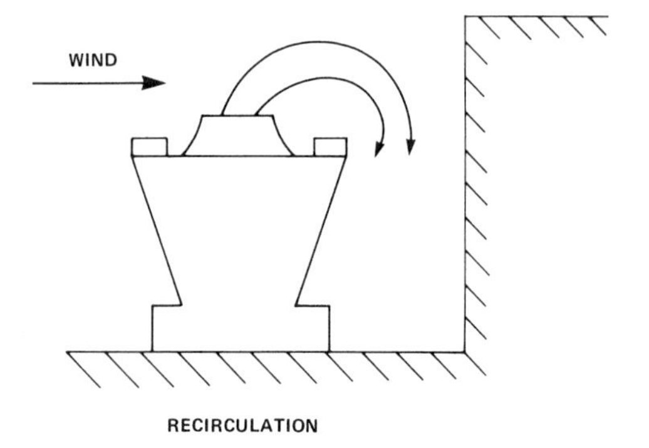


**Fig. 15 Discharge Air Reentering Tower**

Location of the cooling tower is usually a result of one or more of the following: (1) structural support requirements, (2) rigging limitations, (3) local codes and ordinances, (4) cost of bringing auxiliary services to the cooling tower, and (5) architectural compatibility. Sound, fog, and drift considerations are also best handled by proper site selection during the planning stage.

### Piping

Piping should be adequately sized according to standard commercial practice. Chapter 34 of the 1985 FUNDAMENTALS Volume, Chapter 14 of the 1987 HVAC Volume, and Chapter 33 of this volume discuss pipe sizing in detail.

All piping should be designed to allow expansion and contraction. If the tower has more than one inlet connection, balancing valves should be installed to balance the flow to each cell properly. Positive shutoff valves should be used, if necessary, to isolate individual cells for servicing.

When two or more towers are operated in parallel, an equalizer line between the tower sumps handles imbalances in the piping to and from the units and changing flow rates that arise from such obstructions as clogged orifices and strainers. All heat exchangers, and as much tower piping as possible, should be installed below the operating water level in the cooling tower to prevent overflowing of the cooling tower at shutdown and to ensure satisfactory pump operation during start-up. Tower basins need to carry the proper amount of water during operation to prevent air entrainment into the water suction line. Tower basins should also have enough reserve volume between the operating and overflow levels to fill riser and water-distribution lines on start-up and to fulfill the *water-in-suspension* requirement of the tower.

### Capacity Control

Most cooling towers encounter substantial changes in ambient wet-bulb temperature and load during the normal operating season. Accordingly, some form of capacity control may be required to maintain prescribed condensing temperatures or process conditions.

Fan cycling is the simplest method of capacity control on cooling towers and is often used on multiple-unit or multiple-cell installations. In nonfreezing climates, where close control of the exit water temperature is not essential, fan cycling is an adequate and inexpensive method of capacity control. However, motor burnout from too-freqent cycling is a concern.

Two-speed fan motors, in conjunction with fan cycling, can double the number of steps of capacity control, when compared to fan-cycling alone. This is particularly useful on single-fan motor units, which would have only one step of capacity control by fan cycling. Two-speed fan motors are commonly used on cooling towers as the primary method of capacity control, and they provide the added advantage of reduced energy consumption at reduced load conditions.

Modulating dampers in the discharge of centrifugal blower fans are used for cooling tower capacity control, as well as for energy management. In many cases, modulating dampers are used in conjunction with two speed motors. The frequency-modulating controls currently being developed permit a multiplicity of fan speeds and give promise of finite capacity control and energy management, as do the newer automatic-variable-pitch propeller fans.

Cooling towers that inject water to induce the airflow through the cooling tower have various pumping arrangements for capacity control. Multiple pumps in series or two-speed pumping provide capacity control and also reduce energy consumption.

Modulating water bypasses for capacity control should be used *only* after consultation with the cooling tower manufacturer. This is particularly important at low ambient conditions in which the reduced water flows can promote freezing within the tower.

## Free Cooling

With an appropriately equipped and piped system, reduced load and/or reduced ambient conditions can significantly reduce energy consumption by using the tower for *free cooling.* Because the tower's cold water temperature drops as the load and ambient temperature drop, the water temperature will eventually be low enough to serve the load directly, allowing the energy-intensive chiller to be shut off. Figures 16, 17, and 18 outline three (of several) methods of free cooling, but do not show all of the piping, valving, and controls that may be necessary for the functioning of a specific system.

**Indirect Free Cooling** keeps the condenser water and chilled-water circuits separate and may be accomplished in the following ways:

1. In the *vapor migration* system (Figure 16), bypasses between the evaporator and the condenser permit the migratory flow of refrigerant vapor to the condenser; they also permit gravity-flow of liquid refrigerant back to the evaporator, without operation of the compressor. Not all chiller systems are adaptable to this arrangement, and those that are may offer limited load capability under this mode of operation. In some cases, auxiliary pumps enhance refrigerant flow and, therefore, load capability.
2. A separate heat exchanger (Figure 17) in the system (usually of the plate-and-frame-type) allows heat to transfer from the chilled-water circuit to the condenser water circuit by total bypass of the chiller system.
3. An indirect-contact evaporative cooling tower (Figure 4) also permits indirect free cooling. Its use is covered under "Direct Free Cooling."

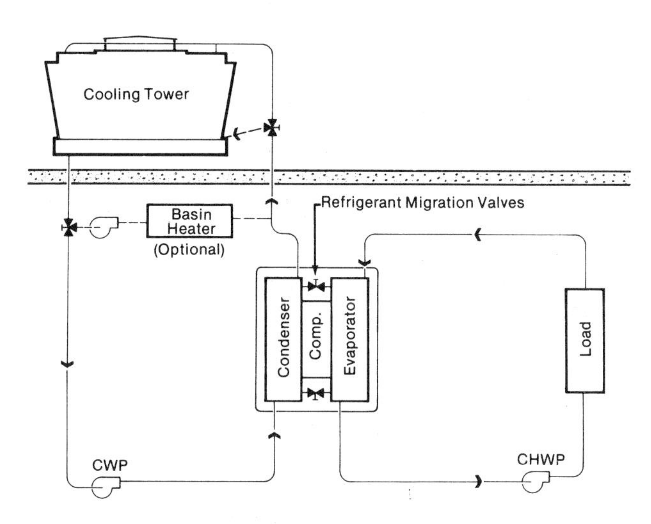


**Fig. 16 Free Cooling by Use of Refrigerant Vapor Migration**

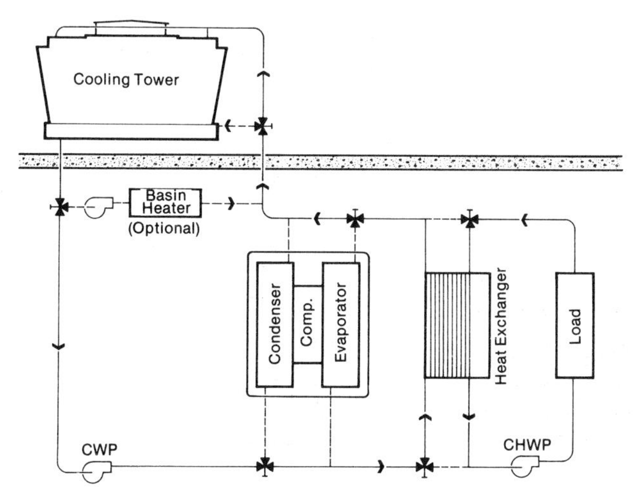


**Fig. 17 Free Cooling by Use of Auxiliary Heat Exchanger**

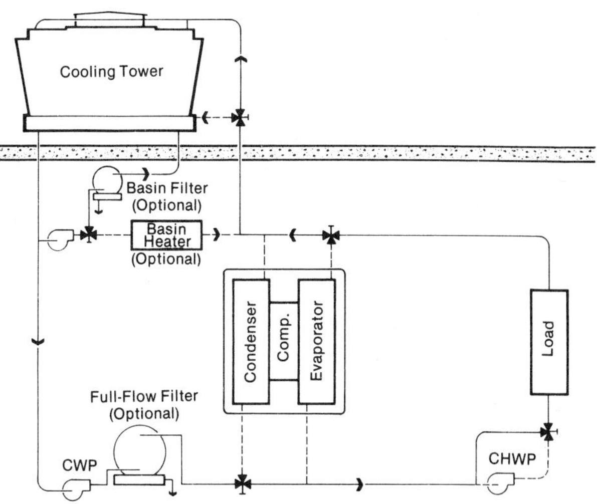


**Fig. 18 Free Cooling by Interconnection of Water Circuits**

**Direct Free Cooling** (Figure 18) involves interconnecting the condenser water and chilled-water circuits so the cooling-tower water serves the load directly. In this case, the chilled water pump is normally bypassed so design water flow can be maintained to the cooling tower. The primary disadvantage of the direct free cooling system is that it allows the relatively *dirty* condenser water to contaminate the *clean* chilled-water system. Although filtration systems (either side-stream or full-flow) minimize this contamination, many specifiers consider this contamination to be an overriding concern. As mentioned previously, the use of a closed-circuit cooling tower (Figure 4) eliminates this concern for contamination. During summer, the water from the tower is circulated in a closed loop through the condenser. During winter, the water from the tower is circulated in a closed loop directly through the chilled-water circuit.

Maximum use of the free-cooling mode of system operation occurs when a reduction in ambient reduces the need for dehumidification. Therefore, higher temperatures in the chilled-water circuit can normally be tolerated during the free-cooling season and are beneficial to the system's heating-cooling balance. In many cases, typical 45 °F (7.2 °C) chilled-water temperatures are allowed to rise to 55 °F (12.8 °C) or higher in the free-cooling mode of operation. This maximizes tower usage and minimizes system energy consumption.

## Winter Operation

When a cooling tower is to be used in subfreezing climates, the following design and operating considerations are necessary:

1. The open circulating water in a cooling tower
2. The closed circulating water in a closed-circuit evaporative fluid cooler
3. Sump water in both a cooling tower or closed-circuit evaporative cooler

**Open Circulating Water.** Cooling towers that operate in freezing climates can be *winterized* by a suitable method of capacity control. This capacity control maintains the temperature of the water leaving the tower well above freezing. In addition, during cold weather, regular visual inspections of the cooling tower should be made to ensure all controls are operating properly.

On induced-draft propeller fan towers, fans may be periodically operated in reverse to de-ice the air-intake areas. Forced-draft centrifugal fan towers should be equipped with capacity control dampers to minimize the possibility of icing.

**Closed Circulating Water.** In addition to the previously mentioned protection for the open circulating water, precautions must be taken to protect the fluid inside the heat exchanger of a closed-circuit fluid cooler. When system design permits, the best protection is to use an antifreeze solution. When this is not possible, the system must be designed to provide supplemental heat to the heat exchanger, and the manufacturer should be consulted concerning the amount of heat input required.

All exposed piping to and from the cooler should be insulated and heat traced. In case of a power failure during subfreezing weather, the heat exchanger should include an emergency draining system.

**Sump Water.** Freeze protection for the sump water in an idle tower or closed-circuit fluid cooler can be obtained by various means. A good method for protecting the sump water is to use an auxiliary sump tank located within a heated space (see Chapter 14 of the 1987 HVAC Volume). When a remote sump is not practical, auxiliary heat must be supplied to the tower sump to prevent freezing. Common sources are electric immersion heaters and steam and hot-water coils. The tower manufacturer should be consulted for the exact heat requirements to prevent freezing at design winter temperatures.

All exposed water lines susceptible to freezing should be protected by electric tape or cable and insulation. This precaution applies to all lines or portions of lines that have water in them when the tower is shut down.

## Sound

Sound has become an important consideration in the selection and siting of outdoor equipment, such as cooling towers and other evaporative cooling devices. Communities are enacting legislation that limits allowable sound levels of outdoor equipment. Even if legislation does not exist, people who live and work near a tower installation may object if the sound intrudes on their environment. Because the cost of remedying a sound problem may exceed the original cost of the cooling tower, such a problem should be considered in the early stages of system design.

To determine the acceptability of tower sound in a given environment, the first step is to establish a noise criterion for the area of concern. This may be an existing or pending code or an estimate of sound levels that will be acceptable to those living or working in the area. The second step is to estimate the sound levels generated by the tower at the critical area, taking into account the effects of geometry of the tower installation and the distance from the tower to the critical area. Often, the tower manufacturer can supply sound rating data on a specific unit, which serves as the basis for this estimate. Lastly, the noise criterion is compared to the estimated tower sound levels to determine the acceptability of the installation.

In cases where the installation may present a sound problem, several potential solutions are available. It is good practice to situate the tower as far as possible from any sound-sensitive areas. Two-speed fan motors should be considered to reduce tower sound levels (by a nominal 12 dB) during light load periods, such as at night, if these correspond to critical sound-sensitive periods. However, fan motor cycling should be held to a minimum, since a fluctuating sound usually is more objectionable than a constant sound level.

In critical situations, effective solutions may include barrier walls between the tower and the sound-sensitive area or acoustical treatment of the tower. Attenuators specifically designed for the tower are available from most manufacturers. It may be practical to install a tower larger than would normally be required and lower the sound levels by operating the unit at reduced fan speed. For additional information on sound control, see Chapter 52 of the 1987 HVAC Volume.

## Drift

Water droplets become entrained in the airstream as it passes through the tower. While eliminators strip most of this water from the discharge airstream, a certain amount discharges from the tower as drift. The rate of drift loss from a tower is a function of tower configuration, eliminator design, airflow rate through the tower, and water loading. Generally, an efficient eliminator design reduces drift loss to a range of 0.002 to 0.2% of the water circulation rate.

Because drift contains the minerals of the makeup water (which may be concentrated three to five times) and, often, water treatment chemicals, cooling towers should not be placed near parking areas, large windowed areas, or architectural surfaces sensitive to staining or scale deposits.

## Fogging (Cooling Tower Plume)

The warm air discharged from a cooling tower is essentially saturated. Under certain operating conditions, the ambient air surrounding the tower cannot absorb all of the moisture in the tower discharge airstream, and the excess condenses as fog.

Fogging may be readily predicted by projecting a straight line on a psychrometric chart (Figure 19) from the tower entering air conditions to a point representing the discharge conditions.

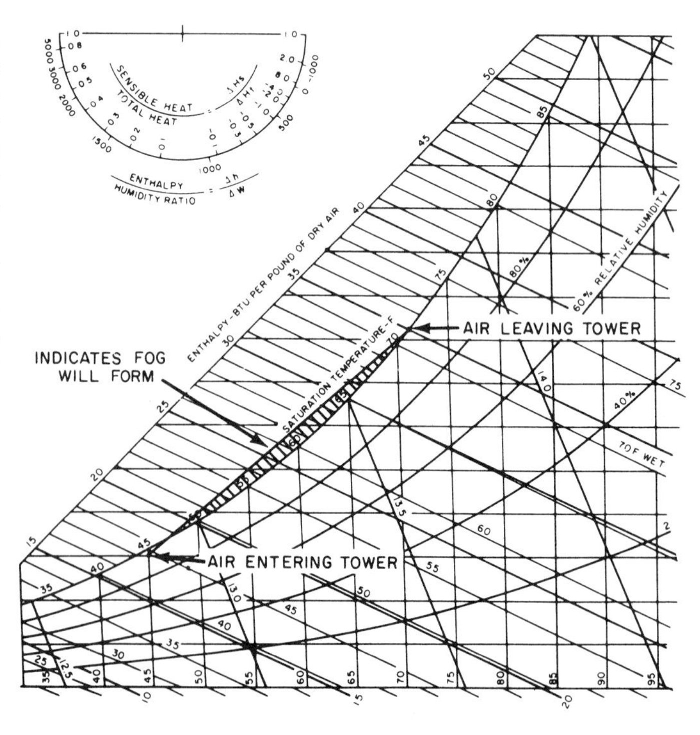


**Fig. 19 Fog Prediction Using Psychrometric Chart**

A line crossing the saturation curve indicates fog generation; the greater the area of intersection to the left of the saturation curve, the more intense the plume. Fog persistence depends on the original intensity and the degree of mechanical and convective mixing with ambient air that dissipates the fog.

Methods of reducing or preventing fogging have taken many forms, including heating the tower exhaust with natural gas burners or steam coils, installing precipitators, or spraying chemicals at the tower exhaust. However, such solutions generally are costly to operate and not always effective.

On larger, field-erected installations, combination wet-dry cooling towers, which combine the normal evaporative portion of a tower with a finned-tube, dry surface heat exchanger section, in series or in parallel, afford a more practical means of plume control. In such units, the saturated discharge air leaving the evaporative section is mixed within the tower with the warm, relatively dry air off the finned coil section to produce a subsaturated air mixture leaving the tower.

Often, however, the most practical solution to tower fogging is to locate the tower where visible plumes, should they form, will not be objectionable. Accordingly, when selecting cooling tower sites, the potential for fogging and its effect on tower surroundings, such as large windowed areas or traffic arteries, should be considered.

### Maintenance

Usually, the tower manufacturer furnishes operating and maintenance manuals that include parts lists for a specific unit; manufacturer's instructions should be followed. When these are not available, the schedule of services in Table 1 can guide the operator to establish a reasonable program of inspection and maintenance.

When a cooling tower or other evaporative cooling device is operated in subfreezing ambient temperatures, the unit should be winterized (see section on "Winter Operation"). During cold weather operation, frequent visual inspections and routine maintenance services ensure all controls are operating properly and can discover any icing conditions before they become serious.

The efficient performance of a cooling tower depends not only on mechanical maintenance but on cleanliness. Accordingly, cooling tower owners should incorporate the following as a basic part of their maintenance program.

1. Periodic inspection of mechanical equipment, fill, and both hot- and cold-water basins to ensure they are maintained in a good state of repair.
2. Periodic draining and cleaning of wetted surfaces and areas of alternate wetting and drying to prevent the accumulation of dirt, scale, or biological organisms such as algae and slime, in which bacteria may develop.
3. Proper treatment of the circulating water for biological control and corrosion, in accordance with accepted industry practice.
4. Systematic documentation of operating and maintenance functions. This is extremely important because without it, no policing can be done to determine whether an individual has actually adhered to a maintenance policy.

**Table 1 Typical Inspection and Maintenance Schedule[a]**

| | Fan | Motor | Gear Reducer | V-Belt Drives | Fan Shaft Bearings | Drift Eliminators | Fill | Cold-Water Basin | Distribution System | Structural Members | Casing | Float Valve | Bleed Rate | Drive Shaft | Flow Control Valves | Suction Screen |
|---|---|---|---|---|---|---|---|---|---|---|---|---|---|---|---|---|
| 1. Inspect for clogging | | | | | | W | W | | W | | | | | | | W |
| 2. Check for unusual noise or vibration | D | D | D | | | | | | | Y | | | | D | | |
| 3. Inspect keys and set screws | | S | S | S | S | | | | | | | | | S | | |
| 4. Lubricate | | Q | | | Q | | | | | | | | | | S | |
| 5. Check oil seals | | | S | | | | | | | | | | | | | |
| 6. Check oil level | | | W | | | | | | | | | | | | | |
| 7. Check oil for water and dirt | | | M | | | | | | | | | | | | | |
| 8. Change oil (at least) | | | S | | | | | | | | | | | | | |
| 9. Adjust tension | | | | M | | | | | | | | | | | | |
| 10. Check water level | | | | | | | | W | W | | | | | | | |
| 11. Check flow rate | | | | | | | | | | | | | M | | | |
| 12. Check for leakage | | | | | | | | S | S | | | S | | | | |
| 13. Inspect general condition | | | | M | | Y | Y | Y | | S | Y | Y | | S | S | |
| 14. Tighten loose bolts | S | S | S | | S | | | | | Y | R | | | S | | |
| 15. Clean | R | S | R | | | R | R | S | R | | | R | | R | R | W |
| 16. Repaint | R | R | R | | | | | R | R | | R | | | R | | |
| 17. Completely open and close | | | | | | | | | | | | | | | S | |
| 18. Make sure vents are open | | | M | | | | | | | | | | | | | |

D—daily; W—weekly; M—monthly; Q—quarterly; S—semiannually; Y—yearly; R—as required.
[a]More frequent inspections and maintenance may be desirable.

### Water Treatment

The quality of water circulating through an evaporative cooling system has a significant effect on the overall system efficiency, the degree of maintenance required, and the useful life of system components. Because the water is cooled primarily by evaporation of a portion of the circulating water, the concentration of dissolved solids and other impurities in the water can increase rapidly. Also, appreciable quantities of airborne impurities, such as dust and gases, may enter during operation. Depending on the nature of the impurities, they can cause scaling, corrosion, and/or silt deposits.

To limit the concentrations of impurities, a small percentage of the circulating water is wasted (called *blowdown* or *bleedoff*). The number of concentrations thus obtained in the circulating water can be calculated from the equation:

$$\text{No. of Concentrations} = \frac{\text{Evaporation} + \text{Drift} + \text{Blowdown}}{\text{Drift} + \text{Blowdown}}$$

The entries in this equation may be expressed as quantities or as percentages of the circulating rate. The evaporation loss averages approximately 1% for each 12.5°F (7°C) of cooling range, while drift loss on a mechanical draft tower is usually less than 0.2% of the circulating rate. Accordingly, for a tower operating with a 12.5°F (7°C) range and using a figure of 0.1% for drift, a blowdown rate of 0.9% of the circulating rate would be required to maintain a level of two concentrations.

In addition to blowdown, evaporative cooling systems are often treated chemically to control scale, inhibit corrosion, restrict biological growth, and control the collection of silt. Scale formation occurs whenever the dissolved solids and gases in the circulating water reach their limit of solubility and precipitate out onto piping, heat-transfer surfaces, and other parts of the system. Simple blowdown can often control scale formation, but where this is inadequate, or it is desirable to reduce the rate of blowdown, chemical scale inhibitors can be added to increase the level of concentrations at which precipitation occurs. Typical scale inhibitors include acids, inorganic phosphates, and similar compounds.

Corrosion control is generally accomplished by adding chemical compounds of chromates, phosphates, or polyphosphonates with zinc, either singly or in combination, to form a protective, insoluble film on metal surfaces throughout the system. Many compounds, such as chromates, are toxic, and their use is limited or banned by local codes and ordinances. The phosphate compounds, while nontoxic, tend to promote algae growth; their use, therefore, may be limited in the future.

Algae, slimes, fungi, and other microorganisms grow readily in evaporative cooling systems and can (1) form an insulating coating on heat-transfer surfaces, (2) restrict fluid flow, (3) promote corrosion, and/or (4) attack organic components within the system (such as wood). The common method of control is to shock treat the system on a periodic basis with a toxic agent such as chlorine or other biocide. Normally, two different biocides are added on an alternating basis to ensure microorganisms do not develop a resistance to any one compound.

During normal cooling tower operation, large quantities of airborne dirt can enter and subsequently settle out as silt deposits. Such deposits can promote corrosion, harbor microorganisms detrimental to the system, and obstruct fluid flow. Silt is normally controlled by adding polymers, which keep the silt in suspension while it flows through the system. Eventually, it settles in the tower basin where deposits can be more easily removed during periodic maintenance.

In most cases where chemical water treatment is practiced, a competent water treatment specialist can be invaluable because system requirements can vary widely. Chapter 53 of the 1987 HVAC Volume has more information on water treatment.

## PERFORMANCE CURVES

The combination of flow rate and heat load dictates the range a cooling tower must accommodate. The entering wet-bulb temperature and required system temperature level combine with cooling tower size to balance the heat rejected at a specified approach. The performance curves in this section are typical and may vary from project to project.

Cooling towers can accommodate a wide diversity of temperature levels, ranging as high as 150 to 160 °F (65 to 70 °C) hot-water temperature in the hydrocarbon processing industry. In the air-conditioning and refrigeration industry, towers are generally applied in the range of 90 to 115 °F (32 to 46 °C) hot-water temperature. A typical standard design condition for such cooling towers is 95 °F (35 °C) hot-water to 85 °F (29.4 °C) cold-water temperature, and 78 °F (25.6 °C) wet-bulb.

A means of evaluating the relative performance of a cooling tower used for a typical air-conditioning system is shown in the accompanying performance curves. The example tower was selected for a flow rate of 3 gpm per nominal ton (54 mL/s per kW) when cooling water from 95 °F (35 °C) to 85 °F (29.4 °C) at 78 °F (25.6 °C) entering wet-bulb temperature (Figure 20).

When operating at other wet bulbs or ranges, the curves may be interpolated to find the resulting temperature level (hot- and cold-water) of the system. When operating at other flow rates, this same tower performs at the levels described by the following performance curve titles (Figures 21 through 23): 2, 4, and 5 gpm per nominal ton (36, 72, and 90 mL/s per kW). Intermediate flow rates may be interpolated between charts to find resulting operating temperature levels.

The format of these curves is similar to the predicted performance curves supplied by manufacturers of cooling towers; the exception is predicted performance curves usually are families of only three specific ranges—80%, 100%, and 120% of design range—and only three charts are provided, covering 90%, 100%, and 110% of design flow. Such performance curves, therefore, bracket the acceptable tolerance range of test conditions and may be interpolated for any specific test condition within the scope of the curve families and chart flow rates.

The charts may also be used to identify the feasibility of varying the parameters to meet specific applications. For example, the subject tower can handle a greater heat load (flow rate) when operating in a lower ambient wet-bulb region. This may be seen by comparing the intersection of the 10 °F (5.6 °C) range curve with 73 °F (22.8 °C) wet bulb at 85 °F (29.4 °C) cold water to show the tower is capable of rejecting 33% more heat load at this lower ambient temperature (Figure 22).

Similar comparisons and crossplots identify relative tower capacity or degree of difficulty for a wide range of variables. The curves produce accurate comparisons within the scope of the information presented but should not be extrapolated outside the field of data given. Also, the curves are based on a typical mechanical draft, film filled, crossflow, medium size, air-conditioning cooling tower. Other types and sizes of towers produce somewhat different balance points of temperature level. However, the curves may be used to evaluate a tower for year-round or seasonal use, if they are restricted to the general operating characteristics described. (See specific manufacturer's data for maximum accuracy when planning for test or critical temperature needs.)

As stated, the water cooling tower, when selected for a specified design condition, operates at other temperature levels when the ambient temperature is off-design or when heat load or flow rate vary from the design condition. When flow rate is held constant, range falls as heat load falls, causing temperature levels to fall to a closer approach. Hot- and cold-water temperature levels fall when the ambient wet bulb falls at constant heat load, range, and flow rate. As water loading to a particular tower falls, while holding the ambient wet bulb and range constant, the tower cools the water to a lower temperature level or closer approach to the wet bulb.

## FIELD TESTING

ASME *Standard* PTC 23 and Cooling Tower Institute *Bulletin* ATC-105 are commonly followed to field test cooling towers. ASME *Standard* PTC 23, *Atmospheric Water Cooling Equipment,* considers both entering and ambient wet-bulb temperatures, but emphasizes the entering wet-bulb temperature. Hot water temperatures are normally measured in the distribution basin or inlet piping instead of near the heat exchanger (heat source), which generally has large temperature gradients. Test data is evaluated by interpolating the manufacturer's performance curves.

Cooling Tower Institute ATC-105 (1975) dictates inlet wet-bulb temperature measurement be used with the wet-bulb stations located within 4 ft (1200 mm) of the air intakes. A prior agreement between test participants is required to define the acceptable number and location of measurement stations to ensure the test average temperature is an accurate representation of the true weighted average inlet wet-bulb temperature.

Data is evaluated by either the characteristic curve or performance curve method. Tower capability is established by comparing the actual flow rate of water measured during the test (adjusted for fan horsepower) to the flow rate predicted from the performance curves at the test conditions.

The following standards and tests should assure that the cooling tower performance meets the desired cooling capacity.

1. Cooling Tower Institute *Standard* 201, *Certification Standard for Water Cooling Towers.* Because cooling towers in enclosures are not covered by this standard, they require field testing to verify performance.
2. A performance bond issued by an independent bonding firm, which quarantees the thermal performance.
3. Field testing by an independent testing agency.

Several factors must be considered if valid tower performance data are to be gathered. The conditions required to run a valid performance test include a steady heat load combined with

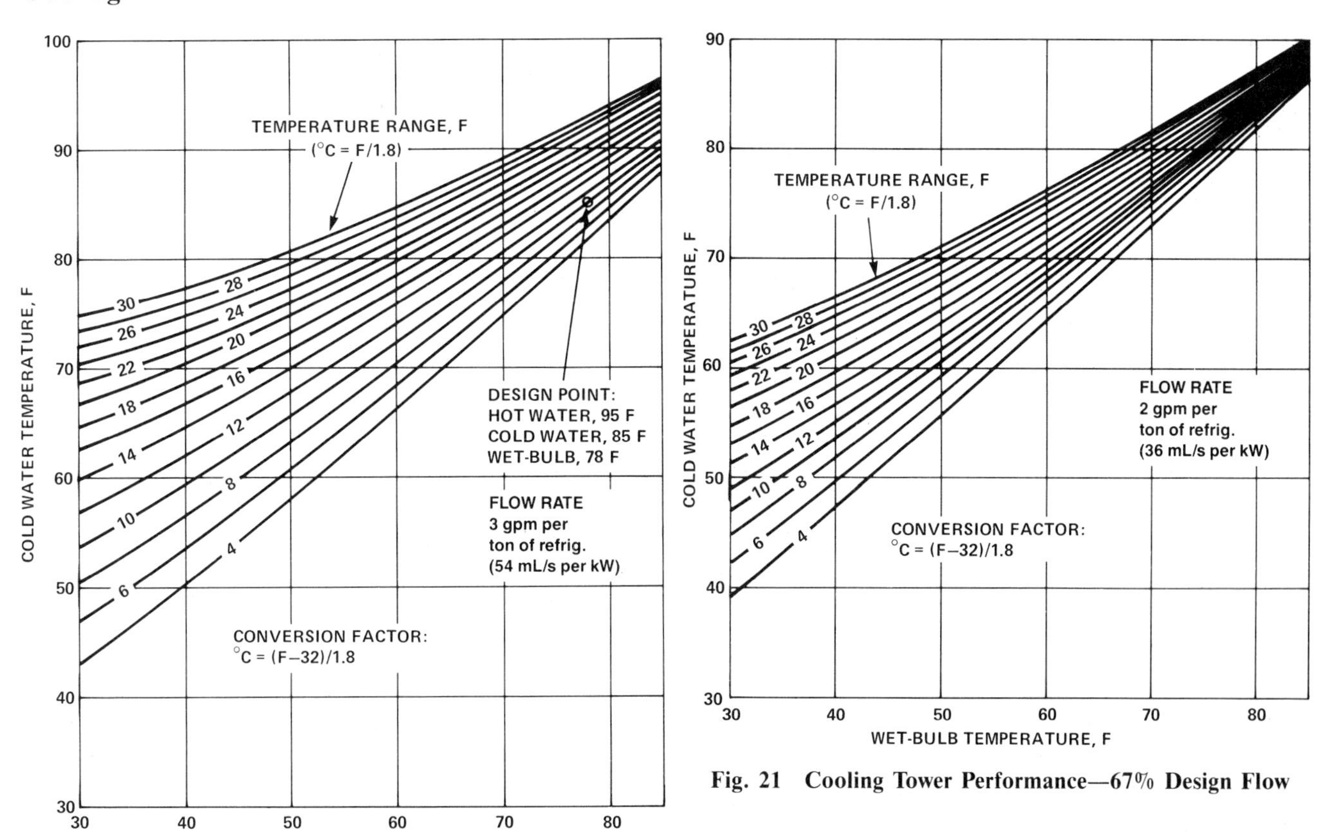


**Fig. 20 Cooling Tower Performance—100% Design Flow**

**Fig. 21 Cooling Tower Performance—67% Design Flow**

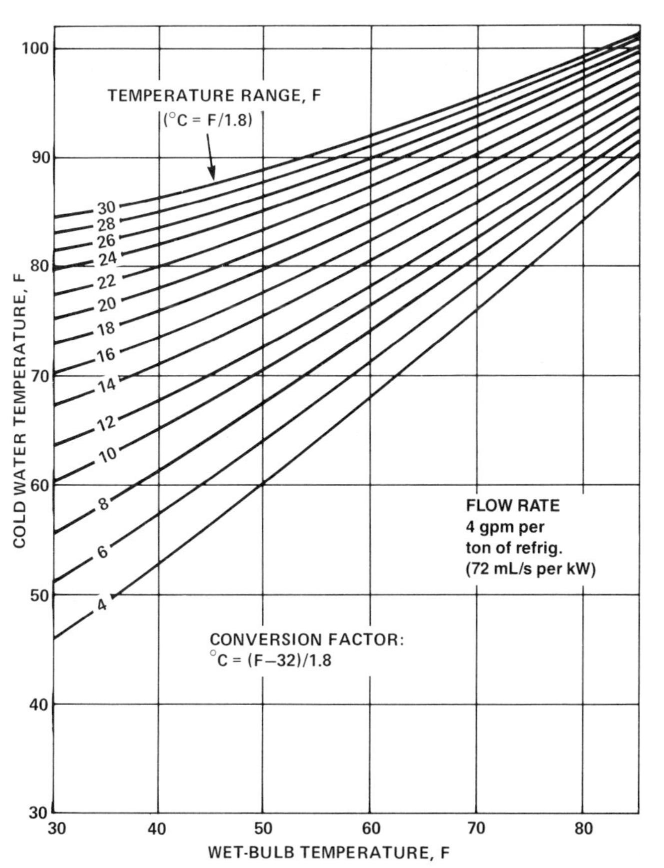


**Fig. 22 Cooling Tower Performance—133% Design Flow**

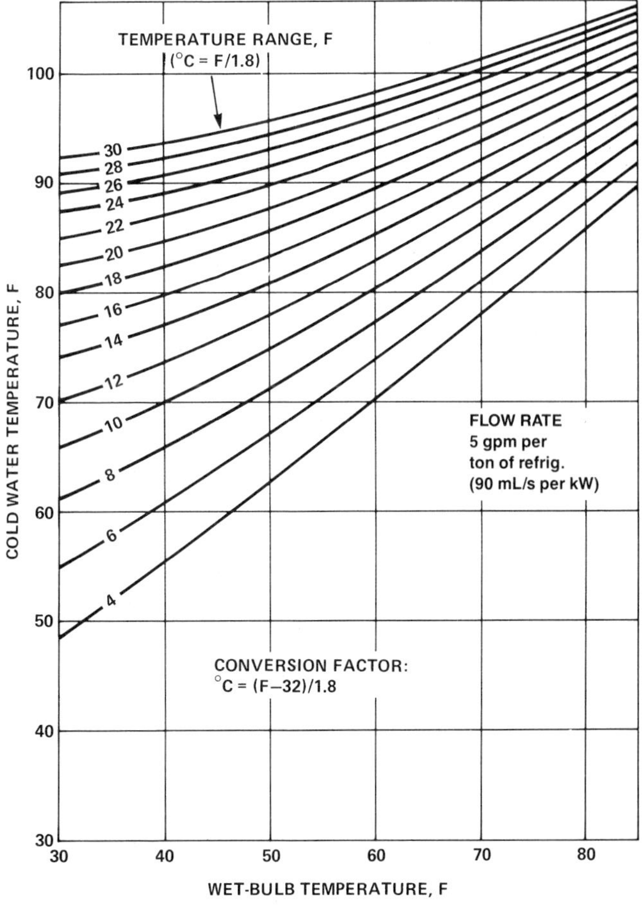


**Fig. 23 Cooling Tower Performance—167% Design Flow**

steady circulating water flow, both preferably as near design as possible. Weather conditions should be reasonably stable, with prevailing winds of 10 mph (4.5 m/s) or less. The tower should be clean and should be adjusted for proper water distribution, with all fans operating at design speed. Both the CTI and ASME codes indicate exact maximum allowable deviations from design for temperature, flow, and heat load for an acceptable test.

Accurate data with calibrated instrumentation at representative locations must be obtained. Water temperatures to the tower are normally measured in the distribution basin or inlet piping. To assure complete mixing, water temperatures from the tower are usually measured at the discharge of the circulating pumps. Wet-bulb temperatures are determined by a mechanically aspirated psychrometer. Finally, water flow to the tower should be measured by calibrated pitot tube traverse of the tower piping or by calibrated in-line flow meters.

To reduce the complexity and expense of the test, the participants may agree to (1) use sling psychrometers or static wetand dry-bulb thermometers, which are positioned in the entering airstream in lieu of mechanically aspirated psychrometers; and (2) measure circulating water rate by comparing head above gravity flow orifices in hot-water basins to calibrated head/flow data curves or by comparing the hot-water manifold spray pressure to calibrated spray distribution nozzle data curves.

## COOLING TOWER THEORY (Baker and Shryock 1961)

Consider a cooling tower having one square foot (one square meter) of plan area; cooling volume $V$ containing a ft$^2$ (m$^2$) of extended water surface per ft$^3$ (m$^3$); water rate $L$ and air rate $G$, both in lb/h (kg/s). Figure 24 schematically shows the processes of mass and energy transfer. The bulk water at temperature $t$ is surrounded by the bulk air at dry-bulb temperature $t_a$, having enthalpy $h_a$ and humidity ratio $W_a$. The interface is assumed to be a film of saturated air with an intermediate temperature $t''$, enthalpy $h''$, and humidity ratio $W''$. Assuming a constant value of 1 Btu/lb • °F (4.19 kJ/kg • °C) for the specific heat of water $c_p$, the total energy transfer from the water to the interface is:

$$dq_w = Lc_p dt = K_L a\,(t - t'')\,dV \tag{2}$$

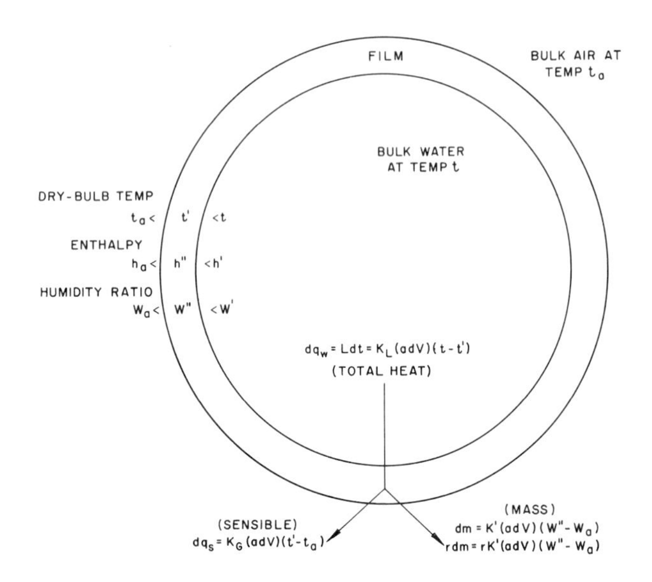


**Fig. 24 Heat and Mass Transfer Relationships between Water, Interfacial Film, and Air (Baker and Shryock 1961)**

*where*

$q_w$ = rate of heat transfer, bulk water to interface, Btu/h (W)
$K_L$ = unit conductance, heat transfer, bulk water to interface, Btu/h • ft$^2$ • °F (W/m$^2$ • °C)

The heat transfer from interface to air is:

$$dq_s = K_G a\,(t'' - t_a)\,dV \tag{3}$$

*where*

$q_s$ = rate of sensible heat transfer, interface to air stream, Btu/h (W)
$K_G$ = overall unit conductance, sensible heat transfer between interface and main air stream, Btu/h • ft$^2$ • °F (W/m$^2$ • °C)

The diffusion of water vapor from film to air is:

$$dm = K'a\,(W' - W_a)\,dV \tag{4}$$

*where*

$m$ = mass transfer rate, interface to air stream, lb/h (kg/s)
$K'$ = unit conductance, mass transfer, interface to main airstream, lb/h • ft$^2$ (lb/lb) [kg/s • m$^2$ (kg/kg)]

Considering the latent heat of evaporation a constant $r$, the heat rate is:

$$rdm = dq_L = rK'a\,(W' - W_a)\,dV \tag{5}$$

The process will reach equilibrium when $t_a = t$ and the air becomes saturated with moisture at that temperature. Under adiabatic conditions, equilibrium is reached at the temperature of adiabatic saturation or at the thermodynamic wet-bulb temperature of the air. This is the lowest attainable temperature in a cooling tower. The circulating water rapidly approaches this temperature when a tower operates without heat load. The process is the same when a heat load is applied, but the air enthalpy increases as it moves through the tower so the equilibrium temperature increases progressively. The approach of the cooled water to the entering wet-bulb temperature is a function of the capability of the tower.

Merkel (1925) assumed the Lewis relationship to be equal to one (1) in combining the transfer of mass and sensible heat into an overall coefficient based on enthalpy difference as the driving force:

$$K_G/(K'\,c_{pm}) = 1 \tag{6}$$

*where*

$c_{pm}$ = humid specific heat of moist air, Btu/lb (dry air • °F) [kJ/kg (dry air • °C)].

The relationship also explains why the wet-bulb thermometer closely approximates the temperature of adiabatic saturation in an air-water vapor mixture. Simplification yields:

$$Lc_L dt = Gdh = K'a\,(h'' - h_a)\,dV \tag{7}$$

The equation considers the transfer from the interface to the airstream, but the interfacial conditions are indeterminate. If the film resistance is neglected and an overall coefficient $K$ is postulated, based on the driving force of enthalpy $h'$ at the bulk water temperature $t$, the equation becomes:

$$Lc_L dt = Gdh = Ka\,(h' - h_a)\,dV \tag{8}$$

*or*

$$KaV/L = \int_{t_1}^{t_2} c_L dt/(h' - h_a) \tag{9}$$

*and*

$$KaV/G = \int_{h_1}^{h_2} dh/(h' - h_a) \tag{10}$$

In cooling tower practice, the integrated value of Eq. (9) is commonly referred to as the Number of Transfer Units or NTU.

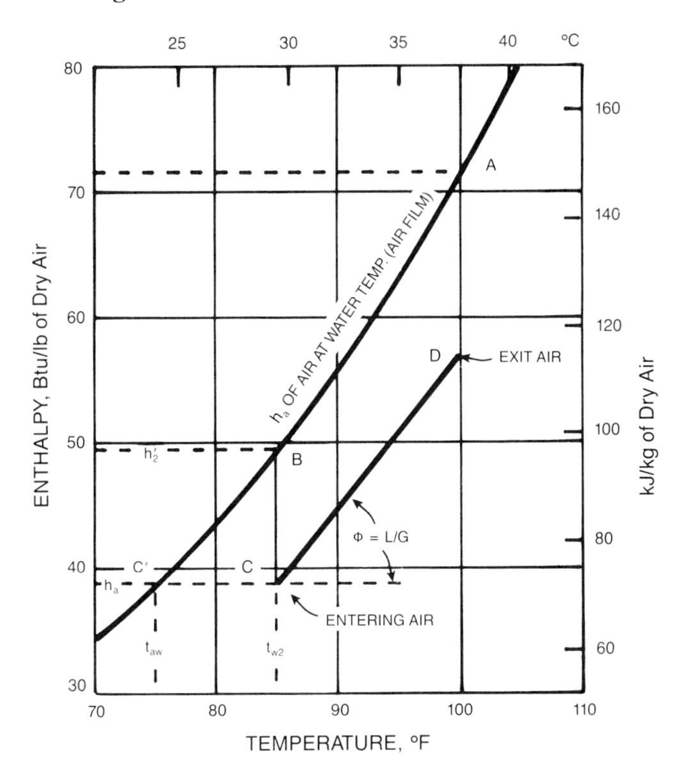


**Fig. 25 Counterflow Cooling Diagram (Baker and Shryock 1961)[a]**

This value gives the number of times the average enthalpy potential $(h' - h_a)$ goes into the temperature change of the water $(\Delta t)$ and is a measure of the difficulty of the task. Thus, one transfer unit has the definition of $c_L \Delta t/(h' - h_a)_{avg.} = 1$.

The equations are not self-sufficient and are not subject to direct mathematical solution. They reflect mass and energy balance at any point in a tower and are independent of relative motion of the two fluid streams. Mechanical integration is required to apply the equations, and the procedure must account for relative motion. The integration of Eq. (9) gives the NTU for a given set of conditions.

## Counterflow Integration

The counterflow cooling diagram (Figure 25) is based on the saturation curve for air-water vapor. As water is cooled from $t_{w1}$ to $t_{w2}$, the air film enthalpy follows the saturation curve from A to B. Air entering at wet-bulb temperature $t_{aw}$ has an enthalpy $h_a$ corresponding to C′. The initial driving force is the vertical distance BC. Heat removed from the water is added to the air so the enthalpy increase is proportional to water temperature. The slope of the air operating line CD equals $L/G$.

Counterflow calculations start at the bottom of a tower, the only point where the air and water conditions are known. The NTU is calculated for a series of incremental steps, and the summation is the integral of the process.

**Example 1:** Air enters the base of a tower at 75 °F wet-bulb temperature, water leaves at 85 °F, and $L/G$ (water-to-air) ratio is 1.2, so $dh = 1.2 \times 1 \times dt$, where 1 is the specific heat, $c_p$, of water. Calculate the NTU for various cooling ranges.

**Solution:** The calculation is as in Table 2. Water temperatures are shown in Column 1 for 1 °F increments from 85 to 90 °F and 2 °F increments from 90 to 100 °F. The corresponding film enthalpies are obtained from psychrometric tables and are shown in Column 2.

The upward air path is in Column 3. The initial air enthalpy is 38.6, corresponding to 75 °F wet-bulb and increases by the relationship $\Delta h = 1.2 \times 1 \times \Delta t$.

**Table 2 Counterflow Integration Calculations**

| 1 Water Temp $t$ | | 2 Enthalpy of Film $h'$ | | 3 Enthalpy of Air $h_a$ | | 4 Enthalpy Difference $h' - h_a$ | | 5 $1/(h' - h_a)$ | | 6 $\Delta t/(h' - h_a)$ Average | | 7 $\Sigma[\Delta t/(h' - h_a)]$ | | 8 Cooling Range | |
|---|---|---|---|---|---|---|---|---|---|---|---|---|---|---|---|
| °F | (°C) | Btu/lb | (kJ/kg) | Btu/lb | (kJ/kg) | Btu/lb | (kJ/kg) | | | | | | | °F | (°C) |
| 85 | (29.4) | 49.4 | (97.0) | 38.6 | (71.9) | 10.8 | (25.1) | 0.0926 | (0.0398) | 0.0921 | (0.0220) | 0.0921 | (0.0220) | 1 | (0.6) |
| 86 | (30.0) | 50.7 | (100.0) | 39.8 | (74.7) | 10.9 | (25.3) | 0.0917 | (0.0395) | 0.0917 | (0.0219) | 0.1838 | (0.0439) | 2 | (1.1) |
| 87 | (30.6) | 51.9 | (102.8) | 41.0 | (77.5) | 10.9 | (25.3) | 0.0917 | (0.0395) | 0.0913 | (0.0218) | 0.2751 | (0.0657) | 3 | (1.7) |
| 88 | (31.1) | 53.2 | (105.9) | 42.2 | (80.3) | 11.0 | (25.6) | 0.0909 | (0.0391) | 0.0901 | (0.0216) | 0.3652 | (0.0873) | 4 | (2.2) |
| 89 | (31.7) | 54.6 | (109.1) | 43.4 | (83.1) | 11.2 | (26.0) | 0.0893 | (0.0385) | 0.0889 | (0.0213) | 0.4541 | (0.1086) | 5 | (2.8) |
| 90 | (32.2) | 55.9 | (112.1) | 44.6 | (85.8) | 11.3 | (26.3) | 0.0885 | (0.0380) | 0.1732 | (0.0413) | 0.6273 | (0.1499) | 7 | (3.9) |
| 92 | (33.3) | 58.8 | (118.9) | 47.0 | (91.4) | 11.8 | (27.5) | 0.0847 | (0.0364) | 0.1653 | (0.0394) | 0.7925 | (0.1893) | 9 | (5.0) |
| 94 | (34.4) | 61.8 | (125.9) | 49.9 | (97.0) | 12.4 | (28.9) | 0.0806 | (0.0346) | 0.1569 | (0.0374) | 0.9493 | (0.2267) | 11 | (6.1) |
| 96 | (35.6) | 64.9 | (133.1) | 51.8 | (102.6) | 13.1 | (30.5) | 0.0763 | (0.0328) | 0.1477 | (0.0353) | 1.097 | (0.2620) | 13 | (7.2) |
| 98 | (36.7) | 68.2 | (140.7) | 54.2 | (108.2) | 14.0 | (32.5) | 0.0714 | (0.0308) | 0.1376 | (0.0329) | 1.2346 | (0.2949) | 15 | (8.3) |
| 100 | (37.8) | 71.7 | (148.9) | 56.6 | (113.8) | 15.1 | (35.1) | 0.0662 | (0.0285) | | | | | | |

SI enthalpy values were taken from an SI psychrometric chart. Values are based on 0 °C for both moisture & air.

The driving force, $h' - h_a$, at the inlet and outlet of each increment is found by subtraction and is listed in Column 4. The reciprocals $1/(h' - h_a)$ are calculated (Column 5), and the average for each increment is multiplied by $\Delta t$ to obtain the NTU for each increment (Column 6). The summation of the incremental values (Column 7) represents the NTU for the summation of the incremental temperature changes, which is the cooling range given in Column 8.

---

Because of the slope and position of CD relative to the saturation curve, the potential difference increases progressively from the bottom to the top of the tower in this example. The degree of difficulty decreases as this driving force increases, which is reflected as a reduction in the incremental NTU proportional to a variation in incremental height. This procedure determines the temperature gradient with respect to tower height.

The procedure of Example 1 considers increments of temperature change and calculates the coincident values of NTU, which correspond to increments of height. The unit-volume concept (Baker and Mart 1952) is an alternate procedure that considers increments of NTU (representing increments of height) with corresponding temperature changes calculated by iteration. The unit-volume procedure is more cumbersome but is necessary in crossflow integration because it accounts for temperature and enthalpy change, both horizontally and vertically.

## Crossflow Integration

In a crossflow tower (Figure 26) water enters at the top, and the solid lines of constant water temperature show its temperature distribution. Air enters from the left and the dotted lines show constant enthalpies. The cross section is divided into unit-volumes in which $dV$ becomes $dxdy$ and Eq. (8) becomes:

$$c_L L dt dx = G dh dy = Ka\,(h' - h_a)\,dxdy \tag{11}$$

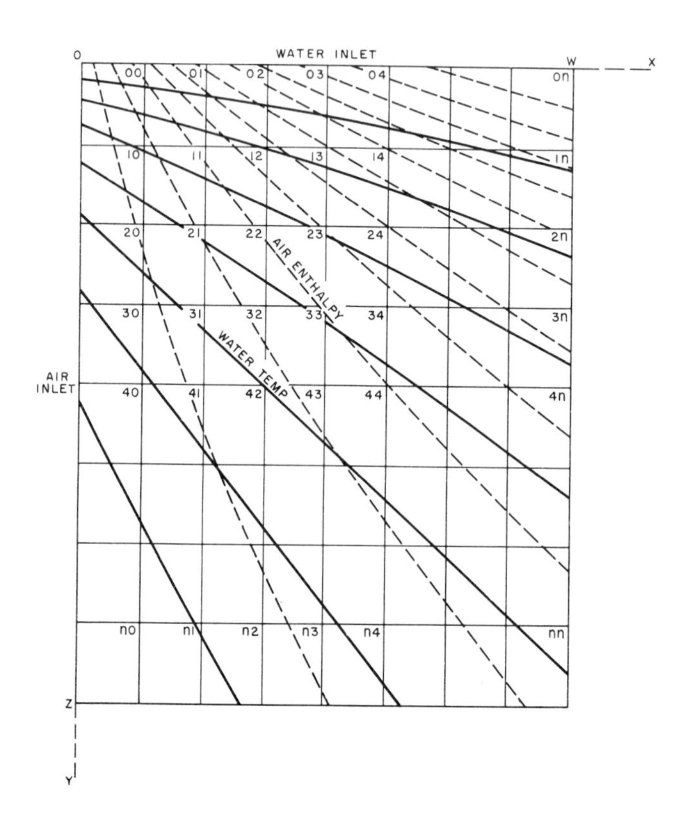


**Fig. 26 Water Temperature and Air Enthalpy Variation through a Crossflow Cooling Tower (Baker and Shryock 1961)**

L/G = 1.0
$\Delta t$(Vertically) = $0.1(h' - h)_{avg}$
$\Delta h$(Horizontally) = $\Delta t$

Conversion Factors:
°C = (°F − 32)/1.8
kJ/kg = 0.430 Btu/lb

100°F HOT WATER TEMPERATURE

| 75°F WET-BULB TEMPERATURE | 71.73 h′ | 71.73 h′ | 71.73 h′ | 71.73 h′ | 71.73 h′ |
|---|---|---|---|---|---|
| 38.59 h | 1,1<br>2.92 Δt<br>97.08°F<br>66.68 h′<br>2.92 Δh<br>41.51 h | 1,2<br>2.66 Δt<br>97.37°F<br>67.11 h′<br>2.66 Δh<br>44.17 h | 1,3<br>2.42 Δt<br>97.58°F<br>67.52 h′<br>2.42 Δh<br>46.59 h | 1,4<br>2.21 Δt<br>97.79°F<br>67.87 h′<br>2.21 Δh<br>48.80 h | 1,5<br>2.02 Δt<br>97.98°F<br>68.20 h′<br>2.02 Δh<br>50.82 h |
| 38.59 h | 2,1<br>2.49 Δt<br>94.59°F<br>62.67 h′<br>2.49 Δh<br>41.08 h | 2,2<br>2.30 Δt<br>95.07°F<br>63.42 h′<br>2.30 Δh<br>43.38 h | 2,3<br>2.13 Δt<br>95.45°F<br>64.03 h′<br>2.13 Δh<br>45.51 h | 2,4<br>1.97 Δt<br>95.82°F<br>64.62 h′<br>1.97 Δh<br>47.48 h | 2,5<br>1.83 Δt<br>96.15°F<br>65.15 h′<br>1.83 Δh<br>49.32 h |
| 38.59 h | 3,1<br>2.14 Δt<br>92.45°F<br>59.43 h′<br>2.14 Δh<br>40.72 h | 3,2<br>2.01 Δt<br>93.06°F<br>60.33 h′<br>2.01 Δh<br>42.73 h | 3,3<br>1.89 Δt<br>93.56°F<br>61.09 h′<br>1.89 Δh<br>44.62 h | 3,4<br>1.77 Δt<br>94.05°F<br>61.84 h′<br>1.77 Δh<br>46.39 h | 3,5<br>1.67 Δt<br>94.48°F<br>62.50 h′<br>1.67 Δh<br>48.06 h |
| 38.59 h | 4,1<br>1.85 Δt<br>90.55°F<br>56.69 h′<br>1.85 Δh<br>40.44 h | 4,2<br>1.77 Δt<br>91.29°F<br>57.74 h′<br>1.77 Δh<br>42.21 h | 4,3<br>1.68 Δt<br>91.88°F<br>58.59 h′<br>1.68 Δh<br>43.89 h | 4,4<br>1.60 Δt<br>92.45°F<br>59.43 h′<br>1.60 Δh<br>45.59 h | 4,5<br>1.51 Δt<br>92.97°F<br>60.20 h′<br>1.51 Δh<br>47.00 h |
| 38.59 h | 5,1<br>1.62 Δt<br>88.93°F<br>54.46 h′<br>1.62 Δh<br>40.21 h | 5,2<br>1.56 Δt<br>89.73°F<br>55.55 h′<br>1.56 Δh<br>41.77 h | 5,3<br>1.50 Δt<br>90.38°F<br>56.45 h′<br>1.50 Δh<br>43.27 h | 5,4<br>1.45 Δt<br>91.00°F<br>57.33 h′<br>1.45 Δh<br>44.72 h | 5,5<br>1.38 Δt<br>91.59°F<br>58.17 h′<br>1.38 Δh<br>46.10 h |

90.53°F AVERAGE WATER TEMPERATURE

**Fig. 27 Crossflow Calculations (Baker and Shryock 1961)**

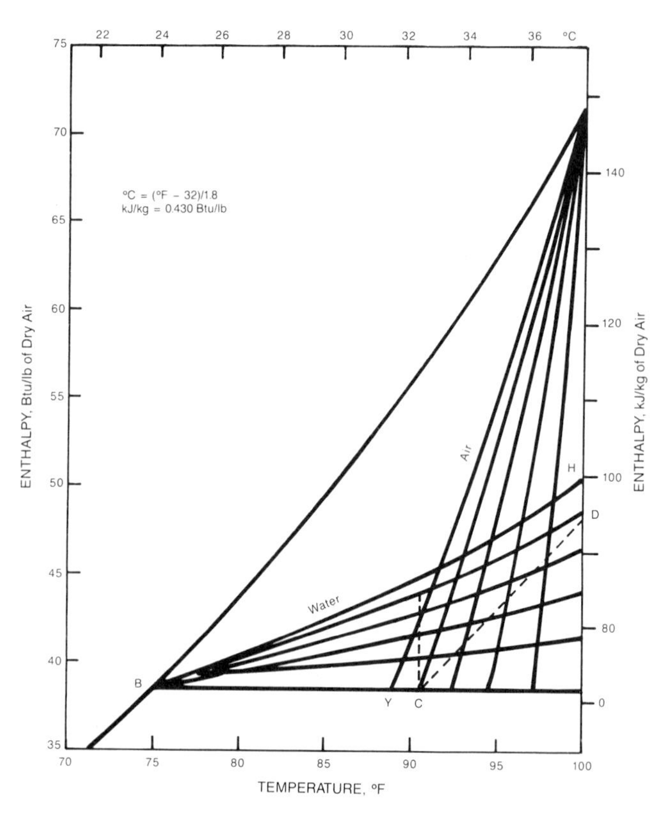


**Fig. 28 Crossflow Cooling Diagram (Baker and Shryock 1961)**

The overall $L/G$ ratio applies to each unit-volume by considering $dx/dy = w/z$. The cross-sectional shape is automatically considered when an equal number of horizontal and vertical increments are used. Calculations start at the top of the air inlet and proceed down and across. Typical calculations are shown in Figure 27 for water entering at 120°F (49°C), air entering at 75°F (24°C) wet-bulb temperature, and $L/G = 1.0$. Each unit-volume represents 0.1 NTU. Temperature change vertically in each unit is determined by iteration from:

$$\Delta t = 0.1\,(h' - h_a)_{av} \qquad (12)$$

$c_L(L/G)dt = dh$ determines the horizontal change in air enthalpy. With each step representing 0.1 NTU, two steps down and across equal 0.2 NTU, etc., for conditions corresponding to the average leaving water temperature.

Figure 26 shows air flowing across any horizontal plane is moving toward progressively hotter water, with entering hot water temperature as a limit. Water falling through any vertical section is moving toward progressively colder air that has the entering wet-bulb temperature as a limit. This is shown in Figure 28, which is a plot of the data in Figure 27. Air enthalpy follows the family of curves radiating from Point A. Air moving across the top of the tower tends to coincide with OA. Air flowing across the bottom of a tower of infinite height follows a curve that coincides with the saturation curve AB.

Water temperatures follow the family of curves radiating from Point B, between the limits of BO at the air inlet and BA at the outlet of a tower of infinite width. The single operating line CD of the counterflow diagram in Figure 25 is replaced in the crossflow diagram by a zone represented by the area intersected by the two families of curves.

## TOWER COEFFICIENTS

Calculations can reduce a set of conditions to a numerical value representing degree of difficulty. The NTU corresponding to a set of hypothetical conditions is called the required coefficient and evaluates degree of difficulty. When test results are being considered, the NTU represents the available coefficient and becomes an evaluation of the equipment tested.

The calculations consider temperatures and the $L/G$ ratio. The minimum required coefficient for a given set of temperatures occurs at $L/G = 0$, corresponding to an infinite air rate. No increase in air enthalpy occurs, so the driving force is maximum and the degree of difficulty is minimum. Decreased air rate (increase in $L/G$) decreases the driving force, and the greater degree of difficulty shows as an increase in NTU. This situation is shown for counterflow in Figure 29. Maximum $L/G$ (minimum air rate) occurs when CD intersects the saturation curve. Driving force becomes zero, and NTU is infinite. The point of zero driving force may occur at the air outlet or at an intermediate point because of the curvature of the saturation curve.

Similar variations occur in crossflow cooling. Variations in $L/G$ vary the shape of the operating area. At $L/G = 0$, the operating area becomes a horizontal line, which is identical to the counterflow diagram, and both coefficients are the same. An increase in $L/G$ causes an increase in the height of the operating area and a decrease in the width. This continues as the areas extend to Point A as a limit. This maximum $L/G$ always occurs when the wet-bulb temperature of the air equals the hot-water temperature and not at an intermediate point, as may occur in counterflow.

Both types of flow have the same minimum coefficient, at $L/G = 0$, and both increase to infinity at a maximum $L/G$. The maximums are the same if the counterflow potential reaches zero at the air outlet, but the counterflow tower will have a lower maximum $L/G$ when the potential reaches zero at an intermediate point, as in Figure 29. A cooling tower can be designed to operate at any point within the two limits, but most applications limit the design to much narrower limits determined by air velocity.

A low air rate requires a large tower, while a high air rate in a smaller tower requires greater fan power. Typical limits in air velocity are about 300 to 700 fpm (1.5 to 3.6 m/s) in counterflow and 350 to 800+ fpm (1.8 to 4.1 m/s) in crossflow.

### Available Coefficients

A cooling tower can operate over a wide range of water rates, air rates, and heat loads, with variation in the approach of the cold water to the wet-bulb temperature. Analysis of a series of test points shows the available coefficient is not a constant, but varies with the operating conditions, as shown in Figure 30.

Figure 30 is a typical correlation of a tower characteristic showing the variation of available $KaV/L$ with $L/G$ for parameters

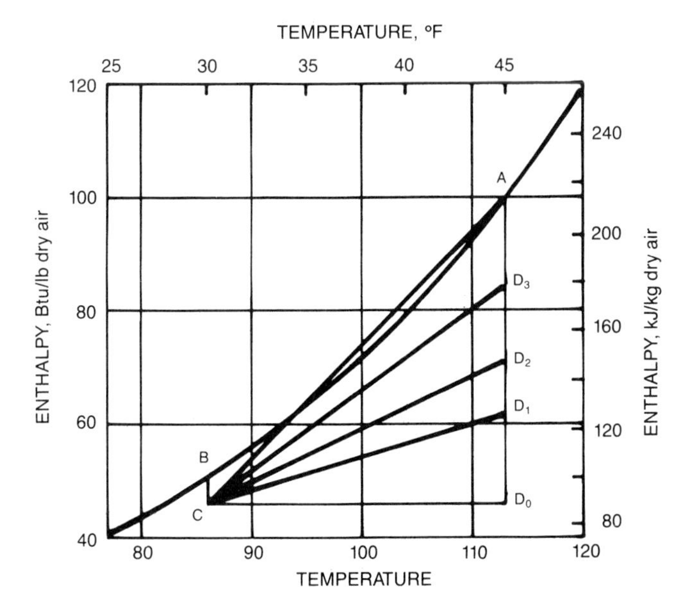


**Fig. 29 Counterflow Cooling Diagram for Constant Conditions, Variable L/G Ratios (Baker and Shryock 1961)**

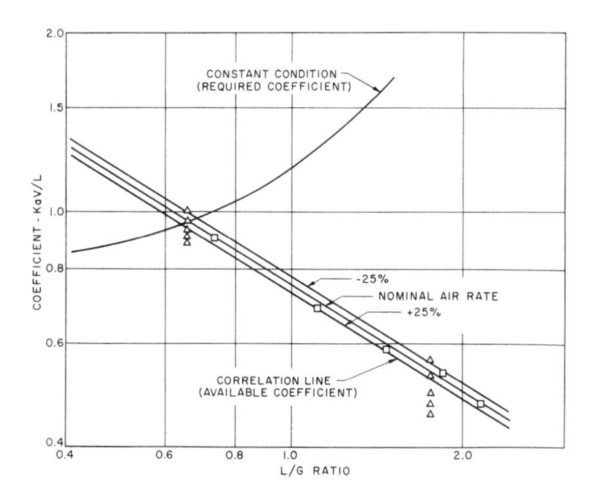


[a] Plotted points at nominal air rate. Square points at 100°F (38°C) hot-water temperature. Triangular points vary from 80 to 160°F (27 to 54°C) hot-water temperature.

**Fig. 30 Tower Characteristic, *KaV/L* vs *L/G* (Baker and Shryock 1961)[a]**

of constant air velocity. Recent fill developments and more accurate test methods have shown that some of the *characteristic lines* are curves, rather than a series of straight, parallel lines on logarithmic coordinates.

Ignoring the minor effect of air velocity, a single average curve may be considered, which corresponds to:

$$KaV/L \sim (L/G)^n \quad (13)$$

The exponent $n$ varies within a range of about $-0.35$ to $-1.1$ but averages between $-0.55$ and $-0.65$. Within the range of testing, $-0.6$ has been considered sufficiently accurate.

The family of curves corresponds to the relation:

$$KaV/L \sim (L)^n(G)^m \quad (14)$$

*where* $n$ is as above and $m$ varies numerically slightly from $n$ and is a positive exponent.

The triangular points in Figure 30 show the effect of varying temperature at nominal air rate. The deviations result from simplifying assumptions and may be overcome by modifying the integration procedure. Usual practice, as shown in Eq. (10), ignores evaporation and assumes that:

$$Gdh = c_L dt \quad (15)$$

The exact enthalpy rise is greater than this because a portion of the heat in the water stream leaves as vapor in the airstream. The correct heat balance in Inch-Pound units is (Baker and Shryock 1961):

$$Gdh = c_L dt + L_E(t_{w2} - 32) \quad (16)$$

This reduces the driving force and increases the NTU.

Evaporation causes the water rate to decrease from $L$ at the inlet to $L - L_E$ at the outlet. The water-to-air ratio varies from $L/G$ at the water inlet to $(L - L_E)/G$ at the outlet. This results in an increase in NTU.

Basic theory considers the transfer from the interface to the airstream. As the film conditions are indeterminate, the film resistance is neglected as assumed in Eq. (8). The resulting coefficients show deviations closely associated with hot-water temperature and may be modified by an empirical hot-water correction factor (Baker and Mart 1952).

The effect of film resistance (Mickley 1949) is shown in Figure 31. Water at temperature $t$ is assumed to be surrounded by a film of saturated air at the same temperature, Point B, at enthalpy $h'$, on the saturation curve. The film is actually at a lower temperature, Point B$'$, at enthalpy $h''$. The surrounding air at enthalpy $h_a$ corresponds to Point C. The apparent potential difference is commonly considered $(h' - h_a)$, but the true potential difference (Mickley 1949) is $(h'' - h_a)$. From Eqs. (2) and (7):

$$h'' - h_a/c_L(t' - t) = K_L/K' \quad (17)$$

The slope of CB$'$ is the ratio of the two coefficients. No means to evaluate the coefficients has been proposed, but a slope of $-11.1$ for crossflow towers has been reported (Baker and Shryock 1961).

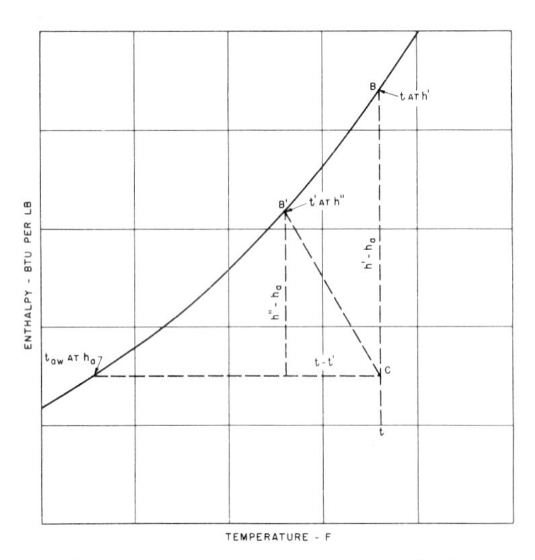


**Fig. 31 True vs Apparent Potential Difference (Baker and Shryock 1961)**

## Establishing Tower Characteristics

Maximum performance in a given volume of fill is obtained with uniform water distribution and constant air velocity throughout. Water distribution changes have a significant effect on the characteristic. Air is channeled by the cell structure, more severely in a counterflow tower, because of restricted air inlet and change in airflow direction.

Some cooling occurs in the spray chamber above the filling and also in the open space below the filling, but it is all credited to the filled volume. The indicated available coefficient reflects the performance of the entire assembly. The characteristic of a fill pattern varies widely, particularly in counterflow towers. A true tower characteristic must be developed from full-scale manufacturer's tests of the actual assembly.

## REFERENCES

ASME. 1958. *Atmospheric Water Cooling Equipment*, PTC 23-58, p. 11.

Baker, D.R. 1962. Use Charts to Evaluate Cooling Towers. *Petroleum Refiner*, November.

Baker, D.R. and L.T. Mart. 1952. Analyzing Cooling Tower Performance by the Unit-Volume Coefficient. *Chemical Engineering*, December, p. 196.

Baker, D.R. and H.A. Shryock. 1961. A Comprehensive Approach to the Analysis of Cooling Tower Performance. ASME *Transactions Journal of Heat Transfer*, August, p. 339.

Cooling Tower Institute. 1975. Acceptance Test Code for Water Cooling Towers. CTI *Bulletin*, ATC-105, Houston, TX.

Fluor Products Company. 1958. *Evaluated Weather Data for Cooling Equipment Design*.

Kohloss, F.H. 1970. Cooling Tower Application. ASHRAE *Journal*, August.

Landon, R.D. and J.R. Houx, Jr. 1973. *Plume Abatement and Water Conservation with the Wet-Dry Cooling Tower*, The Marley Company.

Merkel, F. 1925. Verduftungskuhlung. *Forschungarbeiten*, No. 275.

Mickley, H.S. 1949. Design of Forced-Draft Air Conditioning Equipment. *Chemical Engineering Progress*, Vol. 45, p. 739.

# CHAPTER 21

# FACTORY DEHYDRATING, CHARGING, AND TESTING



PROPER testing, dehydration, and charging help ensure the proper performance and extended life of the refrigeration system. This chapter covers the methods used to perform these functions. The chapter does not address criteria such as allowable moisture content, refrigerant quantity, and performance, which are specific to each machine.

## DEHYDRATION

Factory dehydration may only be feasible for certain sizes of equipment. On large equipment, which will be open to the atmosphere when it is connected in the field, factory treatment is usually limited to purge and backfill, with an inert holding charge of nitrogen. In most instances, this equipment is only stored for short periods, so this method suffices until total system evacuation and charging can be done at the time of installation.

Excess moisture in refrigeration systems may lead to the freezeup of the capillary or expansion valve. Moisture levels that are too low, however, can destabilize the oil-refrigerant mixture and cause metal corrosion, which creates acid and sludge. These contaminants lead to valve breakage, motor burnout, and bearing and seal failure. Chapter 7 of the 1986 REFRIGERATION Volume has more information on moisture and other contaminants in refrigerant systems.

Normally, these effects, with the exception of freezeup, are not detected by a standard factory test. Therefore, it is important to use a dehydration technique that results in a safe moisture level without adding foreign elements or solvents. In conjunction with this technique, an accurate method of moisture measurement must be established. Many factors such as the size of unit, its application, and the type of refrigerant determine the acceptable moisture content. Table 1 shows moisture limits specified by several manufacturers.

### Sources of Moisture

The main sources of moisture include that (1) retained on the surfaces of metals; (2) produced as a product of the combustion of a gas flame; (3) contained in liquid fluxes, oil, and refrigerant; (4) absorbed in the hermetic motor insulating materials; (5) a part of the factory ambient at the point of unit assembly; and (6) provided by free water. The moisture contained in the refrigerant has no effect on the dehydration of the component or unit at the factory. However, because the refrigerant is added after the dehydration process, it must be considered in determining the overall moisture content of the completed unit. The moisture in the oil may or may not be removed in the dehydration process, depending on when the oil is added to the component or system.

Bulk oils, as received, have 20 to 30 ppm (20 to 30 mg/kg) of moisture. Refrigerants have an accepted commercial tolerance on bulk shipments 10 to 15 ppm (10 to 15 mg/kg). Therefore, controls at the factory are needed to ensure the maintenance of these moisture levels in the oils and refrigerant.

The newer insulating materials in hermetic motors retain much less moisture when compared to the old rag paper and cotton-insulated motors. However, tests by several manufacturers have shown that the stator, with its insulation, is still the major source of moisture in compressors. Also, with these materials, the motor winding heat used in dehydration is been reduced.

### Dehydration by Heat, Vacuum, or Dry Air

Heat may be applied by placing components in an oven or by using infrared heaters. Oven temperatures of 180 to 340°F

The preparation of this chapter is assigned to TC 8.1, Positive Displacement Compressors.

Table 1 Factory-Dehydrating and Moisture-Measuring Methods for Refrigeration Systems

| Manufacturer | Component or System | Method of Dehydration | Method of Moisture Determination | Moisture Limit |
|---|---|---|---|---|
| A | Evaporator coil, small<br>Evaporator coil, large<br>0.25 hp (0.2 kW) condensing unit<br>7.5 hp (5.6 kW) condensing unit | −70 °F (−57 °C) dew point dry-air sweep 240 s<br><br>Purchased dry | $P_2O_5$ dry-air sweep 24 h<br><br>$P_2O_5$ sample refrigerant | 25 mg<br>65 mg<br>25 mg<br>80 mg |
| B | Compressor<br>Condenser and evaporator<br>Air conditioner from above components | Evacuate 4 h in 250 °F (121 °C) oven<br>Oven, hot dry-air purge<br>Vacuum at room ambient to 240 μm of mercury | Cold trap<br>Cold trap<br>$P_2O_5$ | 180 mg<br>15 mg<br>35 ppm (mg/kg) |
| C | Evaporator and condenser<br>Compressor | 200 °F (93 °C) air at −20 °F (−29 °C) dew point<br>240 °F (116 °C) oven 10 h sweeping 0.2 cfm (94.4 mL/s) −20 °F (−29 °C) dew point air | None<br>Repeat dehydration process | None<br>1 g |
| D | Coils and tubing<br>Compressor | 250 °F (121 °C) oven, sweep with −70 °F (−57 °C) dew point air<br>300 °F (149 °C) oven, sweep with −70 °F (−57 °C) dew point air | Sweep 1 $ft^3$/h (7.9 mL/s) dry nitrogen over Alnor Dew Point Recorder | 10 mg<br>100 mg |
| E | Freezer<br>Drier | Purchase dry components −40 °F (−40 °C) dew point, sweep with −70 °F (−57 °C) dew point air in room ambient<br>350 °F (177 °C) oven sweeping with −40 °F (−40 °C) dew point air for 3 h | $P_2O_5$ test on refrigerant in system | 10 ppm (mg/kg) |
| F | Compressor<br>Condenser and evaporator<br>Drier<br>Refrigerator from above components | 0.5-h dc winding heat to 356 °F (180 °C)<br>0.25-h vacuum repeat<br>1 h in oven 302 °F (150 °C) with dry-air sweep<br>6-h bake at 347 °F (175 °C)<br>Vacuum, dc winding heat, oven 230 °F (110 °C) | Cold trap 4 h at 302 °F (150 °C)<br>Cold trap 4 h at 302 °F (150 °C)<br><br>Cold trap 4 h at 302 °F (150 °C) | 0.2 g<br>0.2 g<br><br>0.2 g |
| G | Compressor<br>Air conditioner using above compressor | 0.5-h vacuum, winding heat to 190 °F (88 °C)<br>0.5-h vacuum, winding heat 3 h oven 250 °F (121 °C), 0.5-h vacuum | Cold trap 6 h at 200 °F (93 °)<br>Aminco Weaver check of refrigerant in system | 1.2 g<br>15 mg/kg R-12<br>25 mg/kg R-22 |
| H | Compressor | Before or after welding, pass through 280 °F (138 °C) oven maintained at −60 °F (−51 °C) dew point, 5.5 h | Cold trap 6 h | 0.2 g |
| I | Compressors:<br>3 to 5 ton (10.6 to 17.6 kW)<br>7.5 to 15 ton (26 to 53 kW)<br>20 to 40 ton (70 to 141 kW)<br><br>50 to 100 ton (176 to 352 kW) | <br>Dry air at 275 ±5 °F (135 ± 3 °C) for 3 h<br>0.5 h, evacuation at 275 ±15 °F (135 ± 8 °C)<br>Dry-nitrogen sweep at 275 ±15 °F (135 ± 8 °C) for 3.5 h evacuate to 200 μm of mercury<br>Evacuate at 270 ±5 °F (132 ± 3 °C) for 4 h to 1000 μm of mercury | <br>Cold trap<br><br>Cold trap<br><br>Cold trap | <br>0.25 g<br><br>0.75 g<br><br>0.75 g<br>1.00 g |
| J | Compressors:<br>1.5 to 5 ton (5.3 to 17.6 kW)<br><br>2 to 40 ton (7 to 141 kW) semihermetics<br><br>5 to 150 ton (18 to 528 kW) open compressors | <br>−100 °F (−73 °C) dew point dry air 340 °F (171 °C) oven for 1.5 h<br><br>−100 °F (−73 °C) dew point dry air 250 °F (121 °C) oven for 3.5 h<br><br>1-mm of mercury vacuum in 175 °F (79 °C) oven for 1.5 h | <br>Alnor Moisture Monitor checked daily; audited by cold trap<br>Electrolytic moisture monitor audited by cold trap<br><br>Audited by cold trap | <br>0.1 to 0.5 g<br><br>0.1 to 1.1 g<br><br>0.4 to 2.7 g |
| K | Refrigerants 12, 500, and 22;<br>3 and 4 G oils | As purchased | Karl Fischer or equivalent | 10 ppm (mg/kg)<br>30 ppm (mg/kg) |

(82 to 170 °C) are usually maintained. The oven temperature should be selected carefully to prevent damage to the synthetics used and to avoid breakdown of any residual run-in oil that may be present in compressors. The air in the oven must be maintained at a low humidity level. When using heat alone, the time and escape area are critical; therefore, the size of the parts that can be economically dehydrated by this method is restricted.

The vacuum method reduces the boiling point of water below that of the ambient temperature. The moisture then changes to vapor, which is pumped out by the vacuum pump. Table 2 in Chapter 6 of the 1985 FUNDAMENTALS Volume shows the relationship of temperature and pressure.

| | |
|---|---|
| **Low Vacuum** | 29.92 to 1.0 in. Hg. (101.325 to 3.38 kPa) |
| **Medium Vacuum** | 1.0 to $3.94 \times 10^{-5}$ in. Hg. (3.38 kPa to 133 mPa) |
| **High Vacuum** | $3.94 \times 10^{-5}$ to $3.94 \times 10^{-8}$ in. Hg. (133 to 0.13 mPa) |
| **Very High Vacuum** | $3.94 \times 10^{-8}$ to $3.94 \times 10^{-11}$ in. Hg. (133 to 0.013 μPa) |
| **Ultra High Vacuum** | $1.0 \times 10^{-11}$ Micron and beyond (1.3 pPa) |

The degree of vacuum achieved and the time required to obtain the specified moisture level is a function of (1) the type and size of vacuum pump used, (2) the internal volume of the component or system, (3) the size and composition of water-holding materials in the system, (4) the initial amount of moisture in the volume, (5) piping and fitting sizes, (6) the shape of the gas passages, and (7) the maintained external temperatures. The pumping rate of the vacuum pump is critical only if the unit is not being evacuated through a conductance-limiting orifice such as a purge valve.

If dry air or nitrogen is drawn or blown through the equipment for dehydration, it removes moisture by becoming totally or partially saturated. In systems with several passages or blind passages, flow may not be sufficient to dehydrate. The flow rate should obtain optimum moisture removal, and its success depends on the overall system design and temperature.

### Combination Methods

Each of the following methods can be effective if controlled carefully, but a combination of two or even three of the methods is preferred because of the faster and more uniform dryness of the treated system.

**Heat and Vacuum Method.** The heat of this combination method drives the deeply sorbed moisture to the surfaces of materials and removes it from walls; the vacuum lowers the boiling point, making the pumping rate more effective. The heat source can be an oven, infrared lamps, or an ac (or dc) current circulating through the internal motor windings of semi-hermetic and hermetic compressors. Combinations of vacuum, heat, and then vacuum can also be used.

**Heat and Dry-Air Method.** The heat drives the moisture from the materials. The dry air picks up this moisture and removes it from the system or component. The dry air used should have a dew point between −40 and −100°F (−40 and −73°C). The sources of heat are the same as those mentioned previously. Heat can be combined with a vacuum to accelerate the process. The heat and dry-air method is effective with open, hermetic, and semi-hermetic compressors. The heating temperature should be selected carefully to prevent damage to the compressor parts or breakdown of any residual oil that may be present.

The advantages and limitations of the various methods depend greatly on the system or component designs and the results expected. Goddard (1945) considers double evacuation with an air sweep between vacuum applications the most effective method, while Larsen and Elliot (1953) believe the dry-air method, if controlled carefully, is just as effective as the vacuum method and much less expensive, although it incorporates an evacuation of one and a half hours after the hot-air purge. Tests by one manufacturer show that a 280°F (138°C) oven bake for one and a half hours, followed by a twenty-minute evacuation, effectively dehydrates compressors that use the newer insulating materials. For additional discussions on moisture measurements, see Chapter 7 of the 1986 REFRIGERATION Volume.

## MOISTURE MEASUREMENT

Measuring the correct moisture level in a dehydrated system or part is important but not always easy. Table 1 shows measuring methods used by various manufacturers, and others are described in the literature. Few standards are available, however, and the moisture limits accepted by various manufacturers vary.

**Cold-Trap Method.** This common method of determining residual moisture generally monitors the production dehydration system to ensure that it produces equipment that meets the required moisture specifications. An equipment sample is selected after completion of the dehydration process, placed in an oven, and heated in the ranges of 150 to 275°F (65 to 135°C) (depending on the limitations of the sample) for a period of four to six hours. During this time, a vacuum is drawn through a cold-trap bottle immersed in an acetone and dry-ice solution, or equivalent, which is generally held at about −100°F (−73°C). The vacuum levels are between 10 to 100 μm of mercury (1.33 to 13.3 Pa), with the lower levels preferred. Important factors in this method of moisture determination are leaktightness of the vacuum system and cleanliness and dryness of the cold-trap bottle.

**Vacuum Leakback.** The rate of vacuum leakback is another means of checking components or systems to see that no water vapor is present. This method is used primarily in conjunction with a unit or system evacuation that removes the noncondensables prior to final charging. This test allows a check of each unit, but too rapid a pressure buildup may signify a leak, as well as incomplete dehydration. The time factor may be critical in this method and must be examined carefully. Blair and Calhoun (1946) show that a small surface area in connection with a relatively large volume of water may only build up vapor pressure slowly. This method also does not give the actual condition of the charged system.

**Dew Point.** When dry air is used, a reasonably satisfactory check for dryness is a dew point reading of the air as it leaves the part being dried. If the airflow is relatively slow, there should be a marked difference between the dew point of air entering and air leaving the part, followed by a decrease in dew point of the leaving air until it eventually equals the dew point of the entering air. As is the case with all systems and methods described in this chapter, the values considered acceptable depend on the size, usage, and moisture limits desired. Different manufacturers use different limits.

**Gravimetric Method.** In this method, described by ASHRAE *Standard* 35-1983, a controlled amount of refrigerant is passed through a train of flasks containing phosphorous pentoxide ($P_2O_5$), and the weight increase of the chemical caused by moisture is measured. Although this method is satisfactory when the refrigerant is pure, any oil contamination produces inaccurate results. This method must be used only in a laboratory or under carefully controlled conditions. Also, it consumes considerable time and cannot be used when production quantities are high. Furthermore, the method is not effective in systems containing only small charges of refrigerant because it requires from 200 to 300 grams of refrigerant for accurate results. If it is used on systems where withdrawal of any amount of refrigerant changes the performance, recharging is required.

**Aluminum Oxide Hygrometer.** This sensor consists of an aluminum strip that is anodized by a special process to provide a porous oxide layer. A very thin coating of gold is evaporated over this structure. The aluminum base and the gold layer form two electrodes that essentially form an aluminum oxide capacitor.

In the sensor, the water vapor passes through the gold layer and comes to equilibrium on the pore walls in direct relation to the vapor pressure of water in the ambient surrounding the sensor. The number of water molecules absorbed in the oxide structure determines the sensor's electrical impedance, which modulates an electrical current output that is directly proportional to the water vapor pressure. This device is suitable for both gases and liquids over a temperature range of 158 to −166°F (70 to −110° C) and a pressure range of about 10 μm of mercury absolute (1 Pa) to 5000 psig (34.5 MPa). The *Henry's Law* constant must be determined for use with each fluid. This constant is the saturation parts per million by mass of water for the fluid divided by the saturated vapor pressure of water at a constant temperature. For many fluids, this constant must be corrected for the operating temperature at the sensor.

**Christensen Moisture Detector.** For a quick check of uncharged components or units, the Christensen Moisture Detector is used on the production line. In this method, dry air is blown first through the dehydrated part and then over a measured amount of $CaSO_4$. The temperature of the $CaSO_4$ rises in proportion to the quantity of water absorbed by the $CaSO_4$, and desired limits can be set and monitored. One manufacturer reports that coils were checked in 10 seconds with this method. Moisture limits between 2 and 60 mg by this detector. Corrections must be made for variations in the desiccant grain size, quantity of air passed through the desiccant, and difference in instrument and component temperatures.

**Karl Fischer Method.** In systems containing refrigerant and oil, moisture may be determined by (1) measuring the dielectric strength or (2) the Karl Fischer Method (Reed 1954). In this method, a sample is condensed and cooled in a mixture of chloroform, methyl alcohol, and Karl Fischer reagent. The refrigerant is then allowed to evaporate as the solution warms to room temperature. When the refrigerant has evaporated, the remaining solution is titrated immediately to a *dead stop* electrometric end point, and the amount of moisture is determined. This method requires a sample of 50 to 60 g and takes about 1 hour to perform. It is generally considered inaccurate below 15 ppm (15 mg/kg); however, as the method does not require that oil be boiled off the refrigerant, it can be used for checking complete systems. Reed points out that additives in the oil, if any, must be checked to ensure that they do not interfere with the reactions of the method. The Karl Fischer Method may also be used for determining moisture in oil alone (Reed 1954, ASTM 1985, Morton and Fuchs 1960).

**Instrument Described by Taylor.** Taylor (1956) describes an electrolytic water analyzer designed specifically to analyze moisture levels in a continuous process, as well as discrete samples. The device passes the refrigerant sample, in a vapor form, through a sensitive element consisting of a phosphoric acid film surrounding two platinum electrodes—with the acid film absorbing the moisture. When a dc voltage is applied across the electrodes, the water absorbed in the film is electrolyzed into hydrogen and oxygen, and the resulting dc current, in accordance with *Faraday's First Law of Electrolysis,* flows in proportion to the weight of the products electrolyzed. Liquids and vapor may be analyzed because the device has an internal vaporizer. This device handles all of the popular hydrocarbon refrigerants, but the samples must be free of oils and other contaminants. In tests on desiccants, this method is quick and accurate with R-22.

**Sight-Glass Indicator.** In fully charged systems, a sight-glass indicator can be used in the refrigerant lines. This device consists of a colored chemical button visible through the sight glass, which indicates excessive moisture by a change in color. This method requires that the system be run for a reasonable time to allow moisture to circulate over the button. This method compares moisture only qualitatively to a fixed standard. It has been used on factory-dehydrated split systems to ensure that they are dry after field installation and charging.

**Special Considerations.** Although all the methods described in this section can effectively measure moisture, their use in the factory requires certain precautions. Operators must be trained in the use of the equipment or, if the analysis is made in the laboratory, the proper method of securing samples must be understood. Sample flasks must be dry and free of contaminants; lines must be clean, dry, and properly purged. The procedures for weighing the sample, the time during the cycle, and the location of the sample part should be clearly defined and followed carefully. Checks and calibrations of the equipment must be made on a regular basis if consistent readings are to be obtained.

Brisken (1955) maintains that the true moisture levels can be determined only after (1) the system has been running long enough to establish equilibrium between the moisture content of the vapor and the liquid and (2) the cellulose of the motor has given up to the system the amount of water that it cannot hold at operating temperature. Each refrigerant has its own moisture-equilibrium curve between liquid and vapor phases, which must be used to determine the moisture content in a static system.

## CHARGING

The accuracy needed when charging refrigerant or oil into a unit depends on its size and application. Charging equipment must also be adapted to the particular conditions of the plant: manual or automatic charging devices are used. *Standard-type charging* is used where extreme accuracy is not necessary or the production rate is not high. Fully automatic charging boards check the vacuum in the units, evacuate the charging line, and meter the desired amount of oil and refrigerant into the system. These devices are accurate and suitable for high production.

Refrigerant and oil must be handled carefully during charging; the place and time to charge the oil and refrigerant have a great bearing on the life of a system. If a complete unit is charged prior to performance testing, the presence of liquid refrigerant in the crankcase can cause damage because of slugging. If the oil is added after the refrigerant is already in the crankcase, excessive foaming and oil vapor lock may cause bearing damage. Refrigerant lines must be dry and clean, and all charging lines must be kept free of moisture and noncondensable gases. Also, new containers must be connected with proper purging devices. Carelessness in observing these precautions may lead to excess moisture and noncondensables in the refrigeration system.

Oil drums must lie on their sides in storage to prevent accumulation of dirt and water on top, which might get into the oil as soon as a drum is opened. Drums of oil should be opened in the last moment before charging, even on applications requiring that the oil be degasified. Regular checks for moisture or contamination must be made at the charging station to ensure that the oil and refrigerant delivered to the unit meet the required specifications. Compressors charged with oil for storage or shipment must be charged with dry nitrogen. Compressors without oil may be charged with dry air.

## TESTING FOR LEAKS

It is important to detect leaks prior to charging. Extended warranties and critical refrigerant charges add to the importance of proper leak detection.

Basically, the *allowable leakage rate* depends entirely on the system or component characteristics. Any leak on the low pressure side of a system operating below atmospheric pressure is dangerous, no matter how large the refrigerant charges. A system that has 4 to 6 oz (0.1 to 0.2 kg) of refrigerant and a 5-year warranty must have virtually no leak (1 oz [30 g] in ten years or more), whereas in a system that has 10 to 20 lb (4.5 to 9 kg) of refrigerant, the loss of 1 oz (30 g) of refrigerant in a year is not critical. Before any leak testing is done, the component or system should be strength tested at a pressure considerably higher than the leak test pressure. This test ensures safety when the unit is being tested under pressure in an exposed condition. Applicable design test pressures for high- and low-side components have been established by UL, ASME, ANSI, and ASHRAE. Units or components using composition gaskets as joint seals should have the final leak test after dehydration. Many have found that a final torquing of this type of joint after dehydration is beneficial in reducing leaks.

## Leak-Detection Methods

**Water Submersion Test.** The most popular method of leak and strength testing used is the water submersion test. The unit or component is pressurized to the specified positive pressure and submerged in a well-lighted tank filled with clean water. A long time may be needed to obtain the leak test sensitivity desired.

**Pressure Testing.** The unit is sealed off under pressure or vacuum, and a decrease or rise in pressure versus time is noted. The disadvantages of this test are the time involved, the lack of sensitivity, and the inability to determine the exact location of the leak.

**Halide Leak Testing.** The halide torch is used on systems charged with a halogenated refrigerant. The sensitivity of this test is approximately 1 oz (30 g) per year ($10^{-7}$ standard L/s). The gas, drawn across a faintly bluish flame, turns the flame greenish blue and varies in intensity with the size of the leak. Each joint or area can easily be probed, thus locating the leak. The sensitivity of halide torches is reduced by refrigerant contamination; therefore, testing should be done in well-ventilated areas or chambers. Large leaks, even in well-ventilated areas, may cause contamination levels so high that small leaks are not detected.

**Electronic Leak Testing.** The electronic leak detector consists of a probe that draws air over a platinum diode, whose positive ion emission is greatly increased in the presence of a halogen gas. This increased emission is translated into a visible or audible signal. Electronic leak testing shares with halide torches the disadvantages that every suspect area must be explored and that contamination makes the instrument less sensitive; however, it does have some advantages. The main advantage is increased sensitivity. With a well-maintained detector, it is possible to identify leakage at a rate of $10^{-9}$ standard L/s, which is roughly equivalent to the loss of 1 oz (30 g) of refrigerant in 100 years. Another advantage is that the detector can be calibrated in many ways so that a leak can be measured quantitatively. The instrument also can be desensitized to the point that leaks below a predetermined rate are not found. Some models have an automatic compensating feature to accomplish this.

With increased sensitivity, the problem of contamination becomes more critical. To use this improved sensitivity, the unit under test is placed in a chamber slightly pressurized with outside air, which keeps contaminants out of the production area and carries contaminating gas from leaky units. An audible signal allows the probe operator to concentrate on probing, without having to watch a flame or dial. Equipment maintenance presents a problem because the sensitivity of the probe must be checked at short intervals. Any exposure to a large amount of refrigerant causes loss of probe sensitivity. A rough check (such as air underwater testing), prior to use of the electronic device, is frequently used to find large leaks.

**Mass Spectrometer.** The most sensitive leak detection method is probably the mass spectrometer. The unit to be tested is evacuated and then surrounded by a helium and air mixture. The vacuum is then sampled through a mass spectrometer, and any trace of helium indicates one or more leaks. The sensitivity of the mass spectrometer is extremely high, as leaks of $10^{-13}$ standard L/s can be detected. Effective test levels in the manufacturing environment, though, are closer to $10^{-8}$ standard L/s. The helium for testing is normally kept inside a chamber completely closed except at the bottom. The unit to be tested is simply raised into the lighter-than-air helium atmosphere.

This method, in addition to being extremely sensitive, has the advantage of measuring all leaks on all joints simultaneously. Therefore, a quick test is possible. However, the cost of equipment is high, helium is expensive, the instrument must be maintained carefully, and a method of locating individual leaks must be developed.

The concentration of helium needed depends on the maximum leak permissible, the configuration of the system under test, the time it can be left in the helium atmosphere, and the vacuum level in the system being tested; the lower the vacuum level, the higher the helium readings. The longer a unit is exposed to the helium atmosphere, the lower a concentration is necessary to maintain the required sensitivity. If, because of the shape of the test unit, a leak occurs at a point distant from the point of sampling, a good vacuum must be drawn, and sufficient time must be allowed for traces of helium to appear on the mass spectrometer.

In general, a helium concentration of more than 10% is costly. The inherently high diffusion rate causes it to disperse to the atmosphere, no matter how effectively the chamber is designed. As in the case of other methods described in this chapter, the best testing procedure in using the spectrometer is to locate calibrated leaks at extreme points of the test unit and to adjust exposure time and helium concentration in the most economical manner. One manufacturer of refrigeration equipment found leaks at a rate of 0.05 oz (1.4 g) of refrigerant per year by using a 10% concentration of helium and exposing the tested system for ten minutes.

Although the mass spectrometer method is extremely sensitive, the sensitivity that can be used may be limited by the characteristics of the tested system. Since only the *total leak rate* is found with this method, it is impossible to tell whether a leakage rate of, for example, 1 oz (30 g) per year, is the result of one fairly large leak or a number of small leaks. If a sensitivity is desired that rejects units outside of the sensitivity range of tests listed earlier in this chapter, it is necessary to use a helium probe for the location of leaks. In this method, the component or system to be probed is evacuated fully to clear it of helium; then, while it is connected to the mass spectrometer, a fine jet of helium is sprayed over each joint or suspect area. If a large system is tested, a waiting period is necessary because some time is required for the helium to pass from the leak point to the mass spectrometer. Isolated areas, such as return bends on one end of a coil, may be hooded and sprayed with helium to determine whether the leak is in this region, thus saving time.

## Special Considerations

Two general categories of leak detection may be selected: one group furnishes a leak check before refrigerant is introduced into the system, and the other group requires the use of refrigerant. The methods that do not use refrigerant have the advantage that heat applied to repair a joint has no harmful effects. Repairs that require heat on units that contain refrigerant require that all gas be removed and vented before welding. This practice prevents breakdown of the refrigerant and pressure buildup, which would prevent the successful completion of a sound joint.

All leak-testing equipment must be calibrated frequently to ensure maximum sensitivity. The electronic leak detector or the mass spectrometer is usually calibrated with equipment furnished by the manufacturer. Mass spectrometers, for example, are usually checked by a flask containing helium. A glass orifice in the flash allows the helium to escape at a known rate, and the operator maintains the desired sensitivity by comparing the noted escape rate with a known standard.

The effectiveness of the detection system can best be checked with calibrated leaks made of glass, which can be bought commercially. These leaks can be built into a test unit and sent through the normal leak-detection cycles to evaluate the effectiveness of the detection method. Care must be taken that the test leakhole does not become closed. To check against closing, the leakage rate of the test leak must be determined before and after each system audit.

From a manufacturing standpoint, the use of any leak- detection method should be secondary to the prevention of leaks. Improper brazing and welding techniques, unclean parts, untested sealing compounds or improper fluxes and brazing materials, and poor workmanship result in leaks that occur in transit or later. Careful control and analysis of each joint or leak point make it possible to concentrate tests on areas where leaks are most likely to occur. If operators must scan hundreds of joints on each unit, the probability of finding all leaks is rather small, whereas concentration on a few suspect areas reduces field failures considerably.

## PERFORMANCE TESTING

Since there are many types and designs of refrigeration systems, this section only presents specific information on reciprocating compressor testing and covers some important aspects of performance testing of other components and complete systems.

### Compressor Testing

The two prime considerations in compressor testing are power and capacity. Secondary considerations are leakback rate, low-voltage starting, noise, and vibration.

**Testing Without Refrigerant.** A number of tests measure compressor power and capacity before the unit is exposed to refrigerant. In cases where excessive power is caused by friction of running gear, *low-voltage tests* spot defective units early in assembly. In these tests, voltage is increased from a *low* or *zero* value to the value that causes the compressor to *break away,* and this value is compared with an established standard. When valve-plates are accessible, performance can be tested by using an air pump for *leakback tests.* Air at fixed pressure is put through the unit to determine the flow rate at which proper valve opening occurs. The air pressure exerted against the closing side of the valve indicates its efficiency. This method is effective only when the valves are reasonably tight, and its use is difficult when a valve must be run in before it seats properly.

Extreme care should be taken when a compressor is used to pump air because the combination of oil, air, and high temperatures caused by compression can cause a diesel effect or result in an explosion.

In a common test using the compressor as an air pump, the discharge airflow is measured through a flow meter, orifice, or other flow-measuring device. When the volumetric efficiency of the compressor with refrigerant is known, the flow rate that can be expected with air at a given pressure may be calculated. Since this test adiabatically compresses the air, the head pressure must be low to prevent overheating of discharge lines and oil oxidation if the test lasts longer than a few minutes. (The temperature of adiabatic compression is 280°F [138°C] at 35 psig [240 kPa], but 540°F [282°C] at 125 psig [860 kPa].) When the compressor is run long enough to stabilize temperatures, both power and flow can be compared with established limits. Temperature readings at discharge and rpm measurements will aid in analyzing defective units. If a considerable amount of air is discharged or trapped, the air used in the test must be dry enough to prevent condensation from causing rust or corrosion on the discharge side.

Another method of determining compressor performance requires the compressor to pump from a free air inlet into a fixed volume. The time required to reach a given pressure is compared against a maximum standard acceptable value. The pressure used in this test is approximately 125 psig (860 kPa), so that a reasonable time spread can be obtained. The time needed for measuring the capacity of the compressor must be sufficient for accurate readings but short enough to prevent overheating. Power readings can be recorded at any time in the cycle. By shutting off the compressor, the leakback rate can be measured as an additional check. In addition to the pump-up and leakback tests noted above, a vacuum test should also be performed.

The vacuum test should be performed by closing off the suction side with the discharge open to the atmosphere. The normal vacuum obtained under these conditions is 27 in. of mercury vacuum (10 kPa absolute). Abrupt closing of the suction side also allows the oil to serve as a check on the priming capabilities of the pump because of the suppression of the oil and attempt to deaerate. This test also checks for porosity and leaking gaskets. To establish reasonable pumpup times, leakback rates, and suctions, a large number of production units must be tested to allow for production variations.

In any capacity test using air, to prevent compressor contamination, only clean, dry air should be used.

The acceptance test limits described above are best established by taking compressors of known capacity and power and observing their performance during the test. Precautions should be taken to prevent oil used repeatedly for the lubrication of many compressors from becoming acid or contaminated.

**Testing With Refrigerant.** The most common test method is the *run around* cycle. Successful variations and modifications of this cycle are described in ASHRAE *Standard* 23-78. This method requires a condenser large enough to handle the heat of compression, and an expansion device. The gas compressed by the compressor is flash-cooled until its enthalpy is the same as that at the suction conditions, and it is then expanded back to the suction state. This method eliminates the need for an evaporator and uses a condenser about one-fifth the size normally used with the compressor. On compressors of small capacity, a piece of tubing that connects discharge to suction and has a hand expansion valve can be used effectively. The measure of performance is usually the relationship of suction and discharge pressures to power. When a water-cooled condenser is used, the head pressure is usually known, and the water temperature rise and flow are used as capacity indicators.

As a further refinement, flow-measuring devices can be installed in the refrigerant lines. This system is charge-sensitive if predetermined head and suction pressures and temperatures are to be obtained. This is satisfactory when all units have the same capacity and one test point is acceptable, since the charge desired can be determined with little experimentation. When a variety of sizes is to be tested, however, or more than one test point is desired, a liquid sump or receiver after the condenser can be used for full-liquid expansion.

Refrigerant must be free of contamination, inert gases, and moisture; the tubing and all other components should be clean and sealed when they are not in use. In the case of hermetic and semi-hermetic systems, a motor burnout on the test stand makes it imperative not to use the stand until it has been thoroughly flushed and is absolutely acid-free. In all tests, oil migration must be observed carefully, and the oil must be returned to the crankcase.

The length of a compressor performance test depends on various factors. Stabilization of conditions is a prerequisite if accuracy is to be obtained. If oil-pump or oil-charging problems are inherent, the compressor should be run long enough to ensure that all defective units are detected. Most manufacturers use test periods from approximately 30 minutes to one hour. A check of the test system is usually made by running a sample of the production units on a calorimeter under controlled conditions.

### Testing of Complete Systems

In a factory, testing of any system may be done in a controlled ambient temperature or in an existing shop ambient temperature. In both cases, tests must be run carefully, and proper corrections must be made, if necessary. Since measuring air temperature and flow is difficult, production-line tests are usually more reliable when secondary conditions are used as capacity indicators. Measurements of water temperature and flow, power, cycle time, refrigerant pressures, and refrigerant temperatures are reliable capacity indicators.

When testing self-contained air conditioners, for example, a fixed load may be applied to the evaporator, using any air source and either a controlled ambient or shop ambient temperature. As long as the load is relatively constant, its absolute value is not important. For water-cooled units, in which water flow can be absolutely controlled, capacity is best measured by the heat rejected from the condenser. Suction and discharge pressures can be measured for the analysis.

Suction and discharge pressures can be used as a direct measure of capacity in units that have air-cooled condensers. As long as the load is relatively constant, the absolute value is not important. Air distribution, velocity, or temperature over the coil of the test unit must be kept constant during the test, and the performance of the test unit must then be correlated with the performance of a standard unit. Power measurements supplement the suction and discharge pressure readings. As a rule, suction and discharge temperatures are useful in determining unsatisfactory operation of the unit and are particularly important when the evaporator or condenser loads are not reasonably stable. In such cases, simultaneous readings of suction temperature and pressure throughout an entire cycle permit the experienced observer to judge the performance of the unit.

The length of the test run depends on the test used, but, in general, stabilization should be achieved in approximately 30 minutes. For units in which air flows over an evaporator, the conditions should load the test unit with a dry coil. This reduces the time necessary to balance the test system. The test requires preheating of the air temperature to a level that maintains the test unit evaporator saturated temperature above the dew point of the supplied air. For units with air-cooled condensers, the air leaving the condenser can be recirculated back to the evaporator for use as the load. This second arrangement is simple and inexpensive, but it may cause wide variation in performance if the system is not controlled carefully.

The primary function of the factory performance test is to ensure that a unit is constructed and assembled properly. Therefore, all equipment must be compared to a standard unit. This standard unit should be typical of the unit used to pass the ARI and AHAM certification programs for compressors and other units. ARI and AHAM provide rating standards with applicable maximum and minimum tolerances. Several ASHRAE *Standards* specify applicable rating tests.

Normal causes of malfunction in a complete refrigeration system are overcharging, undercharging, presence of noncondensable gases in the system, blocked capillaries or tubes, and excessive power. To determine the validity and sensitivity of any test procedure, it is best to use a unit with known characteristics and then establish limits for deviations from the test standard. If the established limits for charging are $\pm 1$ oz ($\pm 30$ g) of refrigerant, for example, the test unit is charged first with the correct amount of refrigerant and then with 1 oz (30 g) more and 1 oz (30 g) less. If this procedure does not establish clearly defined limits, it cannot be considered satisfactory, and new values must be established. This same procedure should be followed regarding all variables that influence performance and result in deviations from established limits. All equipment must be maintained carefully and calibrated if tests are to have any significance. Gauges must be checked at regular intervals and protected from vibration. Capillary test lines must be kept clean and free of contamination. Power leads must be kept in good repair to eliminate high resistance connection, and electrical meters must be calibrated and protected to obtain consistent data.

In plants where component testing and manufacturing control have been so well managed that the average unit performs satisfactorily, units are tested only long enough to find major flaws. Sample lot testing is sufficient to ensure product reliability. This approach is sound and economical because complete testing taxes power and plant capacity and is not necessary.

When the evaporator load is static, as in the case of refrigerators or freezers, time, temperature, and power measurements are used to measure performance. The time elapsed between start and first compressor shutoff or the average on-and-off period during a predetermined number of cycles in a controlled or known ambient temperature determines performance. Also, concurrent suction and discharge temperatures in connection with power readings are used to establish conformity to standards. On units where the necessary connections are available, pressure readings may be taken. Such readings are usually possible only on units where refrigerant loss is not critical because some loss is caused by gauges.

Units with complicated control circuits usually undergo an operational test to ensure that controls function within design specifications and operate in the proper sequence.

### Testing of Components

Component testing must be based on a thorough understanding of the use and purpose of the component. Pressure switches may be calibrated and adjusted with air in a bench test and need not be checked again if there is no danger of blocked passages or pulldown tripout during the operation of the switch. However, if the switch is brazed into the final assembly, precautions are needed to prevent blocking of the switch capillary.

Capillaries for refrigeration systems are checked by air testing. When the capillary limits are known, it is relatively easy to establish a flow-rate and pressure-drop test for eliminating crimped or improperly sized tubing. When several capillaries are used in a distributor, a series of water manometers check for unbalanced flow and can find damaged or incorrectly sized tubes.

In plants with good manufacturing control, only sample testing of evaporators and condensers is necessary. Close control of coils during manufacture leads to the detection of improper expansion, poor bonding, split fins, or uneven spacing. Proper inspection eliminates the need for costly test equipment. In testing the sample, either a complete evaporator or condenser or a section of the heat-transfer surface is tested. Since liquid-to-liquid furnishes the most easily and accurately measurable heat-transfer medium, a tube or coil can be tested by passing water through it while it is immersed in a bath of water. The temperature of the bath is kept constant, and the capacity is calculated by measuring the coil flow rate and the temperature differential between water entering and leaving the coil.

## REFERENCES

ASHRAE. 1983. Method of Testing Dessicants for Refrigerant Drying. ASHRAE *Standard* 35-1983.

ASTM. 1985. Guide to Test Methods and Specifications for Electrical Insulating Oils of Petroleum Origin. ASTM *Standard* D117-85. American Society for Testing and Materials, Philadelphia.

Blair, H.A. and J. Calhoun. 1946. Evaucation and Dehydration of Field Installations. *Refrigerating Engineering,* August, p. 125.

Brisken, W.R. 1955. Moisture Migration in Hermetic Refrigeration Systems. *Refrigerating Engineering,* July, p. 42.

Goddard, M.B. 1945. Moisture in Freon Refrigerating Systems. *Refrigerating Engineering,* September, p. 215.

Larsen, L.W. and J. Elliot. 1953. Factory Methods for Dehydrating Refrigeration Compressors. *Refrigerating Engineering,* December, p. 1325.

Morton, J.D. and L.K. Fuchs. 1960. Determination of Moisture in Fluorocarbons. ASHRAE *Transactions,* 66:434.

Reed, F.T. 1954. Moisture Determination in Refrigerant Oil Solutions by the Karl Fischer Method. *Refrigerating Engineering,* July, p. 65.

Taylor, E.S. 1956. New Instrument for Moisture Analysis of "Freon" Fluorinated Hydrocarbons. *Refrigerating Engineering,* July, p. 41.

CHAPTER 22

# AUTOMATIC FUEL-BURNING EQUIPMENT



## PART I: GAS-BURNING EQUIPMENT

A **gas burner** is the final device that conveys gas (or a mixture of gas and air) to the combustion zone. Burners are of the atmospheric injection, luminous flame, or power burner types.

### RESIDENTIAL EQUIPMENT

Residential gas burners are those designed for central heating plants or those designed for unit application. Gas-designed units and conversion burners are available for the several kinds of central systems and for other applications where the units are installed in the heated space.

**Central heating appliances** include warm-air furnaces and steam or hot-water boilers.

**Warm-air furnaces** come in different designs, depending on the force used to move combustion products, the force used to move heated supply and return air, the location within a building, and the efficiency required.

The force to move supply and return air can be supplied by the natural buoyancy of heated air in a gravity furnace (if the space to be heated is close to and/or above the furnace) or by a blower in a forced-air furnace.

The force to move the combustion products can be supplied by the natural buoyancy of hot combustion products in a natural draft furnace, by a blower in a forced draft or induced draft furnace, or by the thermal expansion forces in a pulse combustion furnace.

Furnaces also are available in upflow, downflow, horizontal, and other heated air directions to fit the application requirements.

The efficiency required determines, to a great extent, some of the above characteristics and the need for other characteristics such as the source of combustion air, the use of vent dampers and draft hoods, the need to recover latent heat from the combustion products, and the design of the heat-transfer components.

**Conversion burners** are not normally used in modern furnace design because the burner design is integrated with the furnace design for safety and efficiency. Older gravity furnaces usually use conversion burners more readily than can modern gravity and forced-air furnaces.

The preparation of this chapter is assigned to TC 3.7, Fuels and Combustion.

**Steam or hot water boilers** are available in cast iron, steel, and nonferrous metals. In addition to supplying space heating, many boilers are designed to provide domestic hot water, using tankless integral or external heat exchangers.

Some gas furnaces and boilers are available with essentially sealed combustion chambers. These units have no draft hood and are called **direct vent appliances**. Combustion air is piped from outdoors directly to the combustion chamber.

In some instances, the combustion air intake is in the same location as the flue gas outlet. The air and flue pipes are sometimes constructed as concentric pipes, with an appropriate terminal that exposes the air intake and flue gas outlet to a common pressure condition. No chimney or vertical vent is needed with such units. Some induced-draft combustion systems are designed to operate safely when common-vented with other appliances that have natural draft combustion systems.

For appliances not covered by U.S. Department of Energy regulations, the required minimum or listed output at the boiler nozzle or at the forced-air furnace bonnet is respectively 75 and 80% of the approved input to the burners, as set by ANSI *Standards,* although actual tests may show a higher value. The listed output for gravity furnaces is 75% (ANSI/AGA Z21.13, Z21.47). No minimum output is required for appliances covered by the U.S. Department of Energy regulations.

**Conversion burners** are complete burner and control units designed for installation in existing boilers and furnaces. Atmospheric conversion burners may have drilled-port, slotted-port, or single-port burner heads. These burners are either upshot or inshot types. Figure 1 shows a typical upshot gas conversion burner.

Several power burners are available in residential sizes. These are of gun-burner design and are desirable for furnaces or boilers with restricted flue passages or with downdraft passages.

Conversion burners for domestic application are available in sizes ranging from 40,000 to 400,000 Btu/h (12 to 120 kW) input, the maximum rate being set by ANSI/AGA *Standard* Z21.17. However, many such gas conversion burners installed in apartment buildings have input rates up to 900,000 Btu/h (260 kW) or more.

As the successful and safe performance of a gas conversion burner depends on numerous factors other than those incorporated in such equipment, installations of this kind must be made in strict accordance with current ANSI/AGA *Standard* Z2.8. Draft hoods conforming to current ANSI/AGA *Standard*

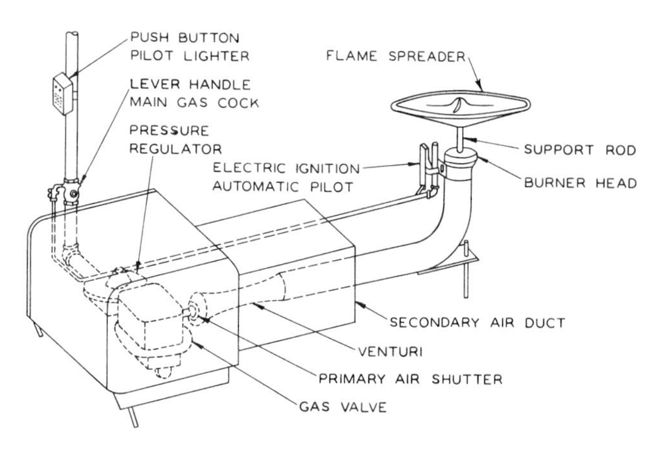


**Fig. 1 Typical Single-Port Gas Conversion Burner**

Z21.12 should also be installed (in place of the dampers used with a solid fuel) on all boilers and furnaces converted to burn gas. Due to space limitations, a converted appliance with a breeching over 12 in. (300 mm) in size is often fitted with a double-swing barometric regulator instead of a draft hood.

## COMMERCIAL-INDUSTRIAL EQUIPMENT

Many basic types of industrial gas burners are available, but only those used commonly for heating service are covered here. These burners may be of the atmospheric or power type. In addition to these types of burners for use in central heating systems, various gas-designed units such as unit heaters, duct furnaces, infrared heaters, and makeup heaters are available for space heating.

The installation of conversion burners larger than 400,000 Btu/h (120 kW) for use in large boilers is usually engineered by the burner manufacturer, the manufacturer's representative, or a local utility company. They are available in several sizes and types. In some cases, the burner may be an assembly of multiple burner heads filling the entire firebox. For conversion burner installation in boilers requiring more than 400,000 Btu/h (120 kW) input, reference should be made to current ANSI/AGA *Standard* Z83.3. Conversion power burners above 400,000 Btu/h (120 kW) are available and should conform to UL 795.

**Atmospheric burners** generally include an air shutter, a venturi tube, a gas orifice, and outlet ports. These burners are of inshot or upshot design. Inshot burners are placed horizontally, making them adaptable to firing Scotch-type boilers. Upshot burners are arranged vertically, making them more adaptable in firebox-type boilers.

**Power-type burners** use a fan to supply and control combustion air. These burners can be of the natural-draft or forced-draft type. In natural draft installations, a chimney is required to draw the products of combustion through the boiler or furnace; the burner fan supplies only enough power to move the air through the burner. Many natural draft power burners have a configuration similar to that of an inshot atmospheric burner to which a fan and windbox have been added. More complex gas-air mixing patterns are possible, and combustion capabilities are thereby improved.

The fan size and speed on power burners have gradually increased, and the combustion process has been modified so that the fan not only moves air through the burner but also forces it through the boiler. Combustion occurs under pressure in controlled airflow. While a vent of only limited height may be needed to convey the combustion products to the outdoors, higher chimneys and vents are usually required to elevate the effluent further. These burners are forced-draft burners. In a forced-draft power burner, the air and gas flows can be modulated by suitable burner controls provided by the manufacturer. Gas input is usually controlled by an appliance pressure regulator and a firing rate valve, both piped in series in the gas train. If the utility also provides a regulator, it is installed at the gas service entry at the gas meter location.

The gas is introduced into a controlled airstream designed to produce thorough gas-air mixing but still capable of maintaining a stable flame front. In a ring burner, the gas is introduced into the combustion airstream through a gas-filled ring just ahead of the combustion zone. In a premix burner, gas and primary air are mixed together, and the mixture is then introduced into secondary air in the combustion zone.

The power burner has superior combustion control, particularly in restricted furnaces and under forced draft. Commercial and industrial gas burners frequently operate with higher gas pressures than are intended for residential equipment. It is often necessary to determine both maximum and minimum gas pressures to be applied to the gas regulator and control trains. If the maximum gas pressure exceeds the 150 to 500 inches of water pressure (5.4 to 18.0 psig or 1.0 to 3.5 kPa) standard for domestic equipment, selecting appropriate gas controls is necessary. All gas-control trains must be rated for the maximum expected gas pressure. Gas pressures in densely populated areas may be significantly lower than in rural areas. Gas-control trains must be sized accordingly.

Gas-fired systems for packaged firetube or watertube boilers in heating plants consist of specially engineered and integrated combustion and control systems. The burners have forced and/or induced draft fans. The burner fan horsepower requirements of these systems and the forced-draft equipment mentioned previously differ. The system pressure, up to and including the stack outlet, may be positive. If the boiler flue breeching has positive flue gas pressure, it must be gastight.

The integrated design of burner and boiler makes it possible, through close control of air-fuel ratios and by matching of flame patterns to boiler furnace configurations, to maintain high combustion efficiencies over a wide range of loads. Combustion space and heat-transfer areas are designed for maximum heat transfer when the specific fuel for which the unit is offered is used. Most packaged units are fire-tested as complete packages prior to shipment, allowing for inspection of all components of the units to ensure that burner equipment, automatic controls, etc., function properly. These units generally bear the Underwriters Laboratories label.

**Gas-fired air heaters** are generally designed for use in airplane hangars, public garages, or similar large spaces. They are self-contained, are automatically controlled with integral means for air circulation, and equipped with automatic electric ignition of pilots, induced or forced-draft, prepurge, and fast-acting combustion safeguards. They also are used with ducts, discharge nozzles, grilles or louvers, and filters.

**Unit heaters** are used extensively for heating large spaces such as stores, garages, and factories. These heaters consist of a burner, heat exchanger, fan for distributing the air, draft hood, automatic pilot, and controls for burners and fan. They are usually mounted in an elevated position from which the heated air is directed downward by louvers. Some unit heaters are suspended from the ceiling, while others are freestanding floor units of the heat tower type. Unit heaters are classified for use with or without ducts, depending on the applicable ANSI/AGA *Standard* under which their design is certified. When connected to ducts, they must have sufficient blower capacity to deliver an adequate air quantity against duct resistance.

**Duct furnaces** are usually like unit heaters without the fan and are used for heating air in a duct system with blowers provided to move the air through the system. Duct furnaces are tested for operation at much higher static pressures than are obtained in unit heaters (ANSI/AGA 83.9).

**Infrared heaters**, vented or unvented, are used extensively for heating factories, foundries, sports arenas, loading docks, garages, and other installations where convection heating is difficult to apply. Following are two general gas-fired types of infrared heaters.

1. Surface combustion radiant heaters, which have a ported refractory or metallic screen burner face through which a self-sufficient mixture of gas and air flows. The gas-air mixture burns on the surface of the burner face heating it to incandescence and releasing heat by radiation. These units operate at surface temperatures of about 1600 °F (870 °C) and generally are unvented. Buildings containing unvented heaters discharging the combustion gases into the space should be adequately ventilated.
2. Internally fired heaters, which consist of a heat exchanger with the exposed surface radiating heat at a surface temperature of approximately 180 °F (80 °C). The exposed surface can be equipped with reflecting louvers or a single large reflector to direct the radiated energy. These units are usually vented.

These infrared heaters are usually mounted in elevated positions and radiate heat downward. (See Chapter 28.)

**Direct-fired makeup air heaters** (see Chapter 26) are used to temper the outside air supply, which replaces contaminated exhaust air. The combustion gases of the heater are mixed directly with large volumes of outside air. Such mixing is considered safe because of the high dilution ratio.

## ENGINEERING CONSIDERATIONS

With gas-burning equipment, the principal engineering considerations are sizing of gas piping, adjustment of the primary air supply, and, in some instances, adjustment of secondary aeration, as with power burners and conversion burners. Consideration of input rating may be necessary to compensate for high altitude conditions. Chapter 34 of the 1985 FUNDAMENTALS Volume has piping and sizing details.

### Combustion Process and Adjustments

Gas burner adjustment is mainly an adjustment of air supply. Most residential gas burners are of the atmospheric injection (Bunsen) type in which primary air is introduced and mixed with the gas in the throat of the mixing tube. For normal operation of most atmospheric-type burners, 40 to 60% of the theoretical air as primary air will give best operation. Slotted-port and ribbon burners may require from 50 to 80% primary air for proper operation. The amount of excess air required depends on several factors, including uniformity of air distribution and mixing, direction of gas travel from the burner, and the height and temperature of the combustion chamber. With power burners using motor-driven blowers to provide both primary and secondary air, the excess air can be closely controlled while proper combustion is secured.

Secondary air is drawn into gas appliances by natural draft. Yellow flame burners depend on secondary air for combustion.

Air shutter adjustments should be made by closing the air shutter until yellow flame tips appear and then by opening the air shutter to a final position at which the yellow tips just disappear. This type of flame obtains ready ignition from port to port and also favors quiet flame extinction.

Gas-designed equipment usually does not incorporate any means for varying the secondary air supply (and hence the $CO_2$). The amount of effective opening and baffling is determined by compliance with ANSI *Standards*. Gas conversion burners, however, do incorporate means for controlling secondary air to permit adjustment over a wide range of inputs. It is desirable, through the use of suitable indicators, to determine whether or not carbon monoxide is present in flue gases. For safe operation, carbon monoxide should not exceed 0.04% (air-free basis).

### Compensation for Altitude

Compensation for altitude must be made for altitudes higher than 2000 ft (610 m) above sea level. The typical gas-fired central heating appliance using atmospheric burners, multiple flue ways, and an effective draft diverter must be derated at high altitudes.

All the air for combustion is supplied by the chimney effect of the flue ways. The geometry of these flue ways allows a given volume of air to pass through the appliance, regardless of its weight. Each burner has a gas orifice and a venturi tube designed to permit a given volume of gas to be introduced and burned during a given period. The entire system is essentially a constant volume device and must, therefore, be derated in accordance with the change in weight of these constant volumes at higher altitudes. The de-rating factor recommended is 4% per 1000 ft (305 m) of altitude (Gas Engineers Handbook 1965).

Many commercial and industrial applications use a forced-draft gas-fired packaged system, including a burner with a forced-draft fan and a fuel-handling system. The burner head, the heat exchanger, and the vent act as a series of orifices downstream of the forced-draft fan. To compensate for the increased volume of air and flue products that must be forced through these orifices at higher altitudes, it is necessary to increase burner fan capacity until the required volume of air is delivered at a pressure high enough to overcome the fixed restrictions. These oversized fans and fan motors are usually offered as options.

One problem that frequently occurs in the selection of commercial gas-fired equipment, and particularly for forced-draft gas-combustion equipment, relates to the heat content of the gas as delivered at elevated locations. Natural gas usually has a heating value of just over 1000 Btu/ft$^3$ (37.3 kJ/L) at standard conditions. At the lower elevations, the heating value of the gas, as delivered, is close to the heating value at the standard conditions and selection of controls is relatively simple. However, at 5000 ft (1524 m) elevation, 1000 Btu/ft$^3$ (37.3 kJ/L) natural gas has a heat content of 850 Btu/ft$^3$ (31.7 kJ/L), as delivered to the burner-control train. However, some gas supplies in the mountains are enriched, and the energy content at standard conditions is higher than 1000 Btu/ft$^3$ (37.3 kJ/L). Problems frequently occur in the selection of controls sized to permit the required volume of gas to be delivered for combustion. Gas specifications should indicate if the energy value and specific gravity shown are for standard conditions or for the gas as furnished at a higher elevation.

The problem of gas supply heat content is more significant for the proper application of commercial forced-draft combustion equipment than it is for smaller atmospheric burner units because of the larger gas volumes handled and the higher design pressure drops common to this type of equipment.

As an example, a gas-control train at sea level has 14 in. of water (3.5 kPa) of gas pressure delivered to the meter inlet. Allowing 6 in. of water (1.5 kPa) pressure drop through the meter and piping to the burner-control train at full gas flow, 8 in. of water (2.0 kPa) pressure remains for delivery of gas for combustion. To force the gas through the gas pressure regulator, shutoff valve, and associated manifold piping, 4 in. of water (1.0 kPa) pressure is typically applied. The last 4 in. of water (1.0 kPa) overcomes the resistance of the gas flow control valve, the gas ring, and the furnace pressure, which may be about 2 in. of water (0.5 kPa).

Consider this same system at 5000-ft (1524-m) altitude, starting from the furnace back. Because of the larger combustion gas volume handled, the furnace pressure will have increased to about 2.5 in. of water (0.6 kPa). Some changes can be made to the gas ring to allow for a larger gas flow, but there is usually some increase in the pressure drop, approximately to 2.5 in. of water (0.6 kPa), so that 5 in. of water (1.2 kPa) of pressure is now required at the gas volume control valve. Because gas pressure drop varies with the square of the volume flow rate, the manifold pressure drop will increase to about 5.5 in. of water (1.4 kPa), and the piping pressure drop will increase to 8.5 in. of water (2.1 kPa). A meter inlet pressure of 19 in. of water (4.7 kPa) would now be required. Since higher gas pressures frequently are not available, it becomes necessary to increase piping and control manifold sizes.

# PART II: OIL-BURNING EQUIPMENT

An **oil burner** is defined by the National Oil Fuel Institute as a mechanical device for preparing fuel oil to combine with air under controlled conditions for combustion. Two methods (atomization and vaporization) are used to prepare fuel oil for combustion. Air for combustion is supplied by natural or mechanical draft. Ignition is generally accomplished by an electric spark, gas pilot flame, or oil pilot flame. Burners of different types operate with luminous or nonluminous flame. Operation may be continuous, intermittent, modulating, or high-low flame.

While most oil burners operate from automatic temperature- or pressure-sensing controls, some of the simpler types are operated manually.

Oil burners may be classified in several ways: by application, type of atomizer, or firing rate. They can be divided into two major groups: residential and commercial-industrial. Further breakdown is made by type of design and operation, such as pressure atomizing, air or steam atomizing, rotary, vaporizing, and mechanical atomizing. Unvented, portable kerosene heaters are not classified as residential oil burners or as oil heat appliances.

## RESIDENTIAL OIL BURNERS

Residential oil burners are ordinarily used in the range of 1/2 to 3 1/2 gph (0.5 to 3.7 mL/s) fuel consumption rate. However, burners up to 7 gph (7.4 mL/s) sometimes fall in the residential classification because of basic similarities in controls and standards. (Burner capacity of 7 gph (7.4 mL/s) and above is classified as commercial-industrial.) No. 2 fuel oil is generally used, although burners in the residential size range can also operate on No. 1 fuel oil. In addition to applications to boilers and furnaces for space heating, burners in the 1/2 to 1 gph (0.5 to 1.0 mL/s) size range are also used for separate tank-type residential hot water heaters, infrared heaters, space heaters, and other commercial equipment.

The majority of residential burner production (over 95%) is of the high-pressure atomizing gun burner type. Substantial numbers of other types of burners (described below) are still in operation, but only a few of these types are currently in production.

The **high-pressure atomizing gun burner** illustrated in Figure 2 supplies oil to the atomizing nozzle at 100 to 300 psi (700 to 2100 kPa). A blower supplies air for combustion, and a damper or other adjustable device regulates the air supply at the burner. Ignition is usually established by a high-voltage electric spark, which may be intermittent. This is sometimes called constant ignition (on when the burner motor is on) or interrupted ignition (on only to start combustion). Typically, these burners fire into a combustion chamber in which negative draft is maintained.

Use of retention heads and residential burner motors operating at 3500 rpm (58 r/s) instead of 1750 rpm (29 r/s) (see Figure 3) has become widespread. The retention head assists combus-

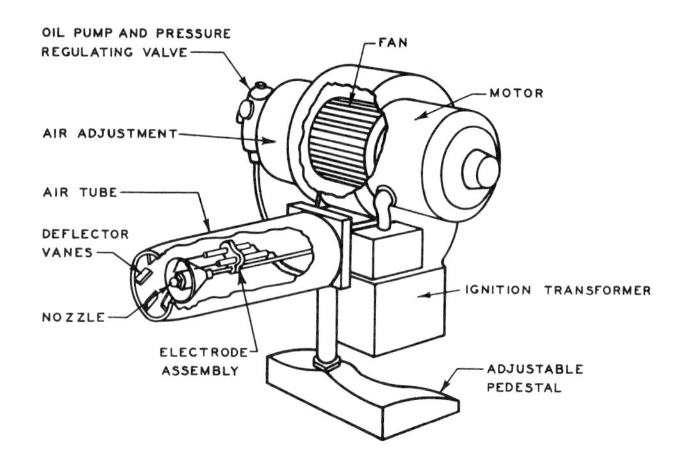


Fig. 2 High-Pressure Atomizing Oil Burner

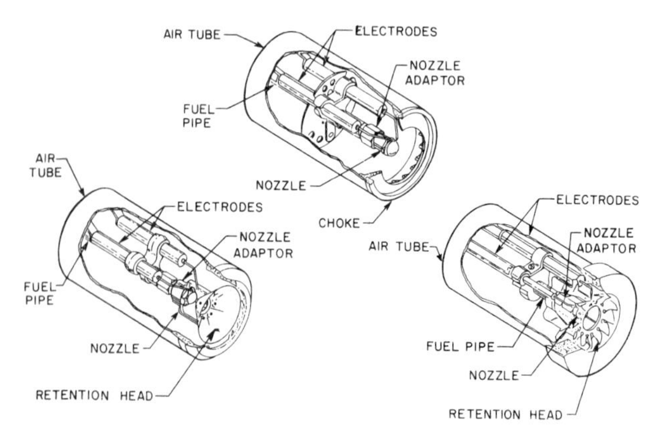


Fig. 3 Details of a High-Pressure Atomizing Oil Burner

tion by providing better air-oil mixing, turbulence, and shear. The use of 58 r/s motors (often combined with a retention head design) has resulted in a more compact burner with equal capacity and, reportedly, more tolerance for varying draft conditions.

The **pressure atomizing forced-draft gun burner** has a construction similar to that of the gun burner but is designed to fire into a combustion chamber operated under positive pressure. The combustion process is independent of chimney draft, and a draft regulator is not required unless excessively high stack drafts are encountered.

Other designs are still in operation but are not a significant part of the residential market. They include the following:

The *low-pressure atomizing gun burner* differs from the high-pressure type because it uses air at a low pressure to atomize the oil.

The *pressure atomizing induced draft burner* uses the same type of oil pump, nozzle, and ignition system as the high-pressure atomizing gun burner.

*Vaporizing burners* are designed for use with No. 1 fuel oil. Fuel is ignited by manual pilot or electrically.

*Rotary burners,* usually of the vertical wall flame type.

NFPA *Standard* 31 (ANSI *Standard* Z95.1), *Oil Burning Equipment,* prescribes correct installation practices for fuel-oil-burning appliances.

## COMMERCIAL-INDUSTRIAL OIL BURNERS

Commercial and industrial oil burners are designed for use with distillate grades or residual grades of fuel oil. With slight modifications, burners designed for residual grades can use the distillate fuel oils. Table 1, Chapter 15, of the 1985 FUNDAMENTALS Volume gives requirements for the various grades of distillate and residual fuel oils.

The commercial-industrial burners covered here are atomizers, which inject the fuel oil into the combustion space as a fine, conical spray with the apex at the burner atomizer. The burner also forces combustion air into the oil spray, causing an intimate and turbulent mixing of air and oil. An electrical spark or spark-ignited gas or oil ignitor, applied for a predetermined time, ignites the mixture, and sustained combustion takes place.

All of these burners are capable of almost complete burning of the fuel oil, without visible smoke, when they are operated with excess air as low as 20% (approximately 12% $CO_2$ in the flue gases). Atomizing oil burners are generally classified according to the method used for atomizing the oil, such as pressure atomizing, return flow pressure atomizing, air atomizing, rotary cup atomizing, steam atomizing, or mechanical atomizing burners. Descriptions of these burners are given in the following sections, together with usual capacities and applications. Table 1 lists approximate size range, fuel grade, and usual applications. All burners described are available as gas-oil (dual-fuel) burners.

### Pressure Atomizing Oil Burners

This type of burner is used in most installations where No. 2 grade fuel oil is burned. The oil is pumped at pressures of 100 to 300 psi (700 to 2100 kPa) through a suitable burner nozzle orifice that breaks it into a fine mist and swirls it into the combustion space as a cone-shaped spray. Combustion air from a fan is forced through the burner air-handling parts surrounding the oil nozzle and is directed into the oil spray.

For smaller capacity burners, ignition is usually started by an electric spark applied near the discharge of the burner nozzle. For burner capacities above 20 gph (21 mL/s), a spark-ignited gas or oil igniter is used.

Pressure atomizing burners are designated commercially as forced-draft or natural- (or induced) draft burners. The forced-draft burner has a fan and motor with capacity to supply all the air for combustion to the combustion chamber or furnace at a pressure high enough to force the gases through the heat-exchange equipment without the assistance of an induced-draft fan or chimney draft. Mixing of the fuel and air is such that a minimum of refractory material is required in the combustion space or furnace to support combustion. The natural-draft (induced-draft) burner requires a negative draft in the combustion space.

Burner range, or variation in burning rate, is changed by simultaneously varying the oil pressure to the burner nozzle and regulating the airflow by a damper. This range is limited to about 1.6 to 1 for any given nozzle orifice. Burner firing mode controls for various capacity burners vary among manufacturers. Usually, larger burners are equipped with controls that provide variable heat inputs. If the burner capacity is up to 15 gph (16 mL/s), an *on-off* control is used; if it is up to 25 gph (26 mL/s), a modulation control is used. In the latter two cases, the low burning rate is about 60% of the full load capacity of the burner.

For these burners, no preheating is required for burning No. 2 oil. No. 4 oil must be preheated to about 100°F (38°C) for

**Table 1 Classification of Atomizing Oil Burners**

| Type of Oil Burner | Heat Range 1000 Btu/h (kW or MW) | Flow Volume gph (mL/s) | Fuel Grade | Usual Application |
|---|---|---|---|---|
| Pressure atomizing | 70 to 7000 (20 to 2000 kW) | 0.5 to 50 (0.5 to 53) | No. 2 (less than 25 gph or 26 mL/s)<br>No. 4 (higher) | Boilers<br>Warm-air furnaces<br>Appliances |
| Return-flow pressure atomizing or modulating pressure atomizer | 3500 (1 MW) and above | 25 (26) and above | No. 2 and heavier | Boilers<br>Warm-air furnaces |
| Air atomizing | 70 to 1000 (20 to 290 kW) | 0.5 to 70 (0.5 to 74) | No. 2 and heavier | Boilers<br>Warm-air furnaces |
| Horizontal/rotary cup | 750 to 37,000 (0.2 to 11 MW) | 5 to 300 (5 to 315) | No. 2 for small sizes<br>No. 4 to 6 for larger sizes | Boilers<br>Large warm-air furnaces |
| Steam atomizing (register-type) | 12,000 (3.5 MW) and up | 80 (84) and up | No. 2 and heavier | Boilers |
| Mechanical atomizing (register-type) | 12,000 (3.5 MW) and up | 80 (84) and up | No. 2 to 6 | Boilers<br>Industrial furnaces |
| Return-flow mechanical atomizing | 45,000 to 180,000 (13 to 53 MW) | 300 to 1200 (315 to 1260) | No. 2 and heavier | Boilers |

proper burning. When properly adjusted, these burners operate well with 20% excess air (approximately 12% $CO_2$), no visible smoke (approximately No. 2 smoke spot number, as determined by ASTM *Standard* D2156), and only a trace of carbon monoxide in the flue gas in commercial applications.

Burners with lower firing rates used to power appliances, residential heating units, or warm-air furnaces are usually set up to operate to about 50% excess air (approximately 10% $CO_2$).

Good operation of these burners calls for a relatively constant draft—either in the furnace or at the breeching connection, depending on the burner selected.

### Return-Flow Pressure Atomizing Oil Burner

This burner is a modification of the pressure atomizing burner; it is also called a modulating pressure atomizer. It has the advantage of wide load range for any given atomizer, about 3 to 1 turndown as against the 1.6 to 1 for the straight pressure atomizing burner (see the section "Mechanical Atomizer Burners").

This wide range is accomplished through a return-flow nozzle, which has an atomizing swirl chamber just ahead of the orifice. Good atomization throughout the load range is attained by maintaining a high rate of oil flow and high pressure drop through the swirl chamber. The excess oil above the load demand is returned from the swirl chamber to the oil storage tank or to the suction of the oil pump.

Control of the burning rate is effected by varying the oil pressure in both the oil inlet and oil return lines. Except for the atomizer, load range, and method of control, the information given for the straight pressure atomizing burner applies to this burner.

### Air Atomizing Oil Burners

Except for the nozzle, this burner is similar in construction to the pressure atomizing burner. Atomizing air and oil are supplied to individual parts within the nozzle. The nozzle design allows the oil to break up into small droplet form as a result of the shear forces created by the atomizing air. The atomized oil is carried from the nozzle through the outlet orifice by the airflow into the furnace.

The main combustion air from a draft fan is forced through the burner throat and mixes intimately with the oil spray inside the combustion space. The burner ignitor is similar to that used on pressure atomizing burners.

This burner is well suited for heavy fuel oils, including grade No. 6, and has a wide load range, or turndown, without changing nozzles. Turndown ranges of 3 to 1 for the smaller sizes and about 6 or 8 to 1 for the larger sizes may be expected. Load range variation is accomplished by simultaneously varying the oil pressure, the atomizing air pressure, and the combustion air entering the burner. Some designs use relatively low atomizing air pressure (5 psi [35 kPa] and lower); other designs use air pressures up to 75 psi (520 kPa). The burner uses from 2.2 to 7.7 $ft^3$ of compressed air per gallon of fuel oil (on an air-free basis) (15 to 52 L/L).

Because of its wide load range, this burner operates well on modulating control.

No preheating is required for No. 2 fuel oil. The heavier grades of oil must be preheated to maintain proper viscosity for atomization. When properly adjusted, these burners operate well with 15 to 25% excess air (approximately 14 to 12% $CO_2$, respectively, at full load); no visible smoke (approximately No. 2 smoke spot number); and only a trace of carbon monoxide in the flue gas.

### Horizontal Rotary Cup Oil Burners

This burner atomizes the oil by spinning it in a thin film from a horizontal rotating cup and injecting high velocity primary air into the oil film through an annular nozzle that surrounds the rim of the atomizing cup.

The atomizing cup and frequently the primary air fan are mounted on a horizontal main shaft that is motor driven and rotates at constant speed—58 to 100 r/s, depending on the size and make of the burner. The oil is fed to the atomizing cup at controlled rates from an oil pump that is usually driven from the main shaft through a worm and gear.

A separately mounted fan forces secondary air through the burner windbox. The use of secondary air should not be introduced by natural draft. The oil is ignited by a spark-ignited gas or oil-burning ignitor (pilot). The load range or turndown for this burner is about 4 to 1, making it well suited for operation with modulating control. Automatic combustion controls are electrically operated.

When properly adjusted, these burners operate well with 20 to 25% excess air (approximately 12.5 to 12% $CO_2$ at full load); no visible smoke (approximately No. 2 smoke spot number); and only a trace of carbon monoxide in the flue gas.

This burner is available from several manufacturers as a package comprised of burner, primary air fan, secondary air fan with separate motor, fuel oil pump, motor, motor starter, ignition system (including transformer), automatic combustion controls, flame safety equipment, and control panel.

Good operation of these burners requires relatively constant draft in the combustion space. The main assembly of the burner with motor, main shaft, primary air fan, and oil pump is arranged for mounting on the boiler front and is hinged so that the assembly can be swung away from the firing position for easy access.

Rotary burners require some refractory in the combustion space to help support combustion. This refractory may be in the form of throat cones as well as minimum combustion chambers.

### Steam Atomizing Oil Burners (Register Type)

Atomization is accomplished in this burner by the impact and expansion of steam. Oil and steam flow in separate channels through the burner gun to the burner nozzle. There, they mix before discharging through an orifice, or series of orifices, into the combustion chamber.

Combustion air, supplied by a forced-draft fan, passes through the directing vanes of the burner register, through the burner throat, and into the combustion space. The vanes give the air a spinning motion, and the burner throat directs it into the cone-shaped oil spray, where intimate mixing of air and oil takes place.

Full-load oil pressure at the burner inlet is generally some 100 to 150 psi (0.7 to 1.0 MPa), and the steam pressure is usually kept higher than the oil pressure by about 25 psi (170 kPa). Load range is accomplished by varying these pressures. Some designs operate with oil pressure ranging from 150 psi (1.0 MPa) at full load to 10 psi (70 kPa) at minimum load, resulting in a range of turndown of about 8 to 1. This wide load range makes the steam atomizing burner suited to modulating control. Some manufacturers provide dual atomizers within a single register so that one can be cleaned without dropping load.

Depending on the burner design, steam atomizing burners use from 1 to 5 lbs of steam to atomize a gallon of oil (0.5 to 2.3 kg/L). This corresponds to 0.5 to 3.0% of the steam generated by the boiler. Where no steam is available for startup, compressed air from the plant air supply may be used for atomizing. Some designs permit the use of a pressure atomizing nozzle tip for startup when neither steam nor compressed air is available.

This burner is used mainly on watertube boilers, which generate steam at 150 psi (700 kPa) or higher and at capacities above 12,000,000 Btu/h (3.5 MW) input.

Oils heavier than grade No. 2 must be preheated to the proper viscosity for good atomization. When properly adjusted, these burners operate well with 15% excess air (14% $CO_2$) at full load, without visible smoke (approximately No. 2 smoke spot number), and with only a trace of carbon monoxide in the flue gas.

### Mechanical Atomizing Oil Burners (Register Type)

The term **mechanical atomizing,** as generally used, describes a technique synonymous with **pressure atomizing.** Both terms designate atomization of the oil by forcing it at high pressure through a suitable stationary atomizer.

The mechanical atomizing burner has a windbox with an assembly of adjustable internal air vanes called an **air register.** Usually the forced-draft fan is mounted separately, with a duct connection between fan outlet and burner wind-box.

Oil pressure of some 90 to 900 psi (0.6 to 6.2 MPa) is used, and load range is obtained by varying the pressure between these limits. The operating range or turndown for any given atomizer can be as high as 3 to 1. Because of its limited load range, this type of burner is seldom selected for new installations.

### Return Flow Mechanical Atomizing Oil Burners

This burner is a modification of a mechanical atomizing burner; atomization is accomplished by oil pressure alone. Load ranges up to 6 or 8 to 1 are obtained on a single burner nozzle by varying the oil pressure between 100 and 1000 psi (0.7 to 6.9 MPa).

The burner was developed for use in large installations such as on shipboard and in electric generating stations where wide load range is required and water loss from the system makes the use of atomizing steam undesirable. It is also used for firing large hot-water boilers. Compressed air is too expensive for atomizing oil in large burners.

This is a register burner similar to the mechanical atomizing burners. Wide range is accomplished by use of a return-flow nozzle, which has a swirl chamber just ahead of the orifice or sprayer plate. Good atomization is attained by maintaining a high rate of oil flow and high pressure drop through the swirl chamber. The excess oil above the load demand is returned from the swirl chamber to the oil storage tank or to the oil pump suction. Control of burning rate is accomplished by varying the oil pressure in both the oil inlet and the oil return lines.

### Dual-Fuel Gas/Oil Burners

Dual-fuel, combination gas/oil burners are forced-draft burners that incorporate, in a single assembly, the features of the commercial-industrial grade gas and oil burners described in the preceding sections. These burners have a three-position switch that permits the manual selection of gas, oil, or a *center-off* position. This switch contains a positive center stop or delay to ensure that the burner flame relay or programmer cycles the burner through a postpurge and prepurge cycle before starting again on the other fuel. The burner manufacturers of larger boilers design the special mechanical linkages needed to deliver the correct air-fuel ratios at full-fire, low-fire, or any intermediate rate. Smaller burners may be straight *on-off* firing. Larger burners may have low-fire starts on both fuels and use a common flame scanner. Smaller dual-fuel burners usually include pressure atomization of the oil. Air-atomization systems are included in large oil burners.

Automatic changeover dual-fuel burners are available for use with gas and No. 2 oil. A special temperature control mounted on an outside wall senses outdoor temperature. It is electrically interlocked with the dual-fuel burner control system. When the outdoor temperature drops to the outdoor control set point, it changes fuels automatically after putting the burner through a postpurge and a prepurge cycle. The *minimum* additional controls and wiring needed to operate automatically with an outdoor temperature control are (1) burner fuel changeover relay(s) and (2) time delay devices to ensure interruption of the burner control circuit at the moment of fuel changeover.

The fuel changeover relays replace the three-position manual fuel selection switch. A manual fuel selection switch can be retained as a manual override on the automatic feature. These control systems require special design and are generally provided by the burner manufacturer.

The dual-fuel burner is fitted with a gas train and oil piping that is connected to a two-pipe oil system following the principles of the preceding sections. A reserve of oil must be maintained at all times for automatic fuel changeover.

Boiler flue chimney connectors are equipped with a double-swing barometric draft regulator or, if required, sequential furnace draft control to operate an automatic flue damper. Dual-fuel burners and their accessories should be installed by experienced contractors to ensure satisfactory operation.

## EQUIPMENT SELECTION

Economic and practical factors (such as the degree of operating supervision required by the installation) generally dictate that fuel oil must be selected by considering the maximum heat input of the oil-burning unit. For heating loads and where only one oil-burning unit is operated at any given time, the relationship is as shown in Table 2 (this table is only a guide). In many cases, a detailed analysis of operating parameters results in burning the lighter grades of fuel oils at capacities far above those indicated. Process application, especially with multiple units, may require different criteria.

### Fuel Oil Storage Systems

All fuel oil storage tanks should be constructed and installed in accordance with National Fire Protection *Standard* 31 and with local ordinances.

**Storage Capacity.** Dependable and economical operation of oil-burning equipment requires ample and safe storage of fuel

**Table 2 Guide for Fuel Oil Grades versus Firing Rate**

| Maximum Heat Input of Unit, 1000 Btu/h (MW) | Volume Flow Rate, gph (mL/s) | Grade of Fuel |
|---|---|---|
| Up to 3500 (1) | Up to 25 (26) | 2 |
| 3500 to 7000 (1 to 2) | 35 to 50 (37 to 53) | 2, 4, 5 |
| 7000 to 15,000 (2 to 4.4) | 50 to 100 (53 to 105) | 5 and 6 |
| Over 15,000 (4.4) | Over 100 (105) | 6 |

oil at the site. Design responsibility should include analysis of specific storage requirements as follows:

1. Rate of oil consumption.
2. Dependability of oil deliveries.
3. Economical delivery lots. The cost of installing larger storage capacity should be balanced against the savings indicated by accommodating larger delivery lots. Truck lots and rail car lots vary with various suppliers, but the quantities are approximated as follows:

| | |
|---|---|
| Small truck lots in metropolitan areas | 500 to 2000 gal (1900 to 7600 L) |
| Normal truck lots | 3000 to 5000 gal (11 to 20 $m^3$) |
| Transport truck lots | 5000 to 9000 gal (20 to 34 $m^3$) |
| Rail tanker lots | 8000 to 12,000 gal (30 to 45 $m^3$) |

**Tank size and location.** Standard oil storage tanks range in size from 55 to 50,000 gal (0.2 to 190 $m^3$) and larger. Tanks are usually built of steel, but those for heavy oil *only* may be of concrete construction. Unenclosed tanks located in the lowest story, cellar, or basement should not exceed 660 gal (2500 L) capacity each, and the aggregate capacity of such tanks should not exceed 1320 gal (5000 L), unless each 660 gal (2500 L) tank is insulated in an approved fireproof room having a fire resistance rating of at least 2 hours.

.The storage tanks with the storage capacity at a given location exceeding about 1000 gal (3800 L) should be underground whenever practical and accessible for truck or rail delivery with gravity flow from the delivering carrier into storage. If the oil is to be burned in a central plant such as a boiler house, the storage tanks should be located, if possible, so that the oil burner pump or pumps can pump directly from storage to the burners. In case of a year-round operation, except for storage or supply capacities below 2000 gal (7580 L), at least two tanks should be installed to facilitate tank inspection, cleaning, repairs, and clearing of plugged suction lines.

When the main oil storage tank is not close enough to the oil-burning units for the burner pumps to take suction from storage, a supply tank must be installed near the oil-burning units and oil must be pumped periodically from storage to the supply tank by a transport pump at the storage location. Supply tanks should be treated the same as storage tanks regarding location within buildings, tank design, etc. On large installations, it is recommended that standby pumps be installed as a protection against heat loss in case of pump failure.

Since all piping connections to underground tanks must be at the top, such tanks should not be more than 10 ft 6 in. (3.2 m) in height from top to bottom to avoid pump suction difficulties. (This dimension may have to be less for installations at high altitudes.) The total suction head for the oil pump must not exceed 14 ft (4.3 m) at sea level.

**Connections to storage tank.** All piping connections for tanks over 275-gal (1040-L) capacity should be through the top of the tank. Figure 4 shows a typical arrangement for a cylindrical storage tank with heating coil as required for No. 5 or No. 6 fuel oils. The heating coil and oil suction lines should be located near one end of the tank.

The maximum allowable steam pressure in such a heating coil is 15 psi (100 kPa). The heating coil is unnecessary for oils lighter than No. 5, unless a combination of high pour point and low outdoor temperature makes heating necessary.

A watertight manhole with internal ladder provides access to the inside of the tank. If the tank is equipped with an internal heating coil, a second manhole is required and arranged to permit withdrawal of the coil.

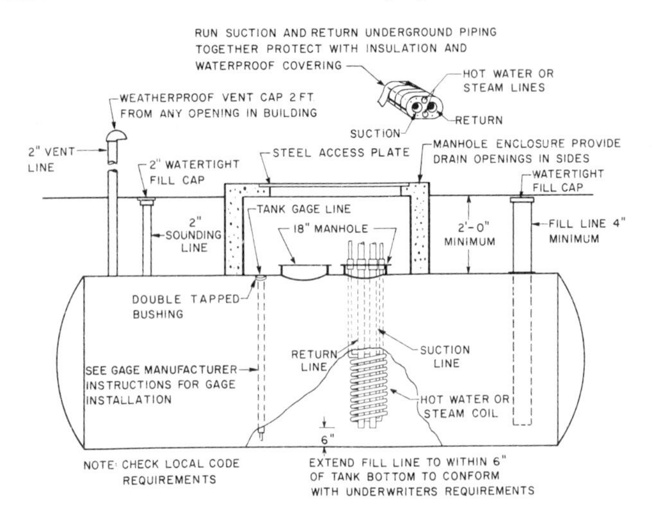


**Fig. 4 Typical Oil Storage Tank (No. 6 Oil)**

The fill line should be vertical and should discharge near the end of the tank away from the oil suction line. The inlet of the fill line must be outside the building and accessible to the oil delivery vehicle, unless an oil transfer pump is used to fill the tank. When possible, the inlet of the fill line should be at or near grade level where filling may be accomplished by gravity. The fill line should be at least 2 in. (50 mm) in diameter for No. 2 oil and 6 in. (150 mm) in diameter for No. 4, 5, or 6 oils for gravity filling. Where filling is done by pump, the fill line for No. 4, 5, or 6 oils may be 4 in. (100 mm) in diameter.

An oil return line bringing recirculated oil from the burner line to the tank should discharge near the oil suction line inlet. Each storage tank should be equipped with a vent line sized and arranged in accordance with National Fire Protection Association *Standard* 31.

Each storage tank must have a device for determining oil level. For tanks inside buildings, the gauging device should be designed and installed so that oil or vapor will not discharge into a building from the fuel supply system. No storage tank should be equipped with a glass gauge or any gauge which, when broken, would permit the escape of oil from the tank. Gauging by a measuring stick is permissible for outside tanks or for underground tanks.

## Fuel-Handling Systems

The fuel-handling system consists of the pumps, valves, and fittings for moving fuel oil from the delivery truck or car into the storage tanks and from the storage tanks to the oil burners. Depending on the type and arrangement of the oil-burning equipment and the grade of fuel oil burned, fuel-handling systems vary from simple to quite complicated arrangements.

The simplest handling system would apply to a single burner and small storage tank for No. 2 fuel oil similar to a residential heating installation. The storage tank is filled through a hose from the oil delivery truck, and the fuel-handling system consists of a supply pipe between the storage tank and the burner pump. Equipment should be installed on light oil tanks to indicate visibly or audibly when the tank is full; on heavy oil tanks, a remote-reading liquid-level gauge should be installed.

Figure 5 shows a complex oil supply arrangement for two burners on one oil-burning unit. For a unit with a single burner, the change in piping is obvious. For supplying oil to two or more

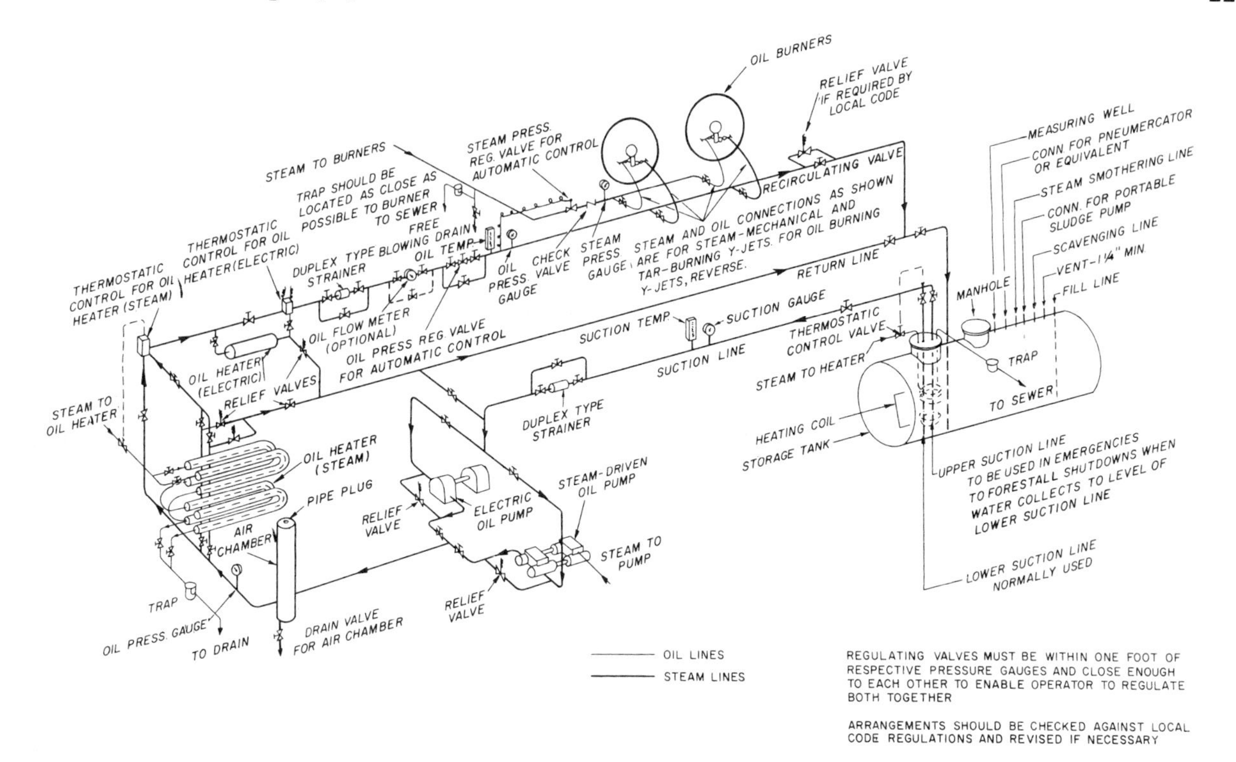


**Fig. 5 Industrial Burner Auxiliary Equipment**

units, the oil line downstream of the oil discharge strainer becomes a main supply header, and the branch supply line to each unit would include oil flow meter, automatic control valve, etc. For light oils requiring no heating, all oil heating equipment shown in Figure 5 would be omitted. Both a suction and return line should be used, except for gravity flow in residential installations.

Oil pumps (steam or electric driven) should deliver oil at the maximum rate required by the burners (this includes the maximum firing rate, the oil required for recirculating, plus a 10% margin).

The calculated suction head at the entrance of any burner pump should not exceed 10 in. Hg (34 kPa) for installations at sea level. Higher elevations require that the suction head be reduced in direct proportion to the reduction in barometric pressure.

The oil temperature at the pump inlet should not exceed 120 °F (49 °C). Where oil burners with integral oil pumps (and oil heaters) are used and where the suction lift from the storage tank is within the capacity of the burner pump, each burner may take oil directly from the storage tank through an individual suction line, except where No. 6 oil is used.

Where two or more tanks are used, the piping arrangement into the top of each tank should be the same as for a single tank so that any tank may be used at any time; any tank can be inspected, cleaned, or repaired while the system is in operation.

The length of suction line between storage tank and burner pumps should not exceed 100 ft (30 m). Where the main storage tank(s) is located more than 100 ft (30 m) from the pumps, in addition to the storage tanks, a supply tank should be installed near the pumps, and a transfer pump should be installed at the storage tanks for delivery of oil to the supply tank.

Central oil distribution systems comprising a central storage facility, distribution pumps or provision for gravity delivery, distribution piping, and individual fuel meters are used for residential communities—notably mobile home parks. The provisions of NFPA 31, *Installation of Oil Burning Equipment,* should be followed in making a central oil distribution system installation.

## Fuel Preparation System

Fuel oil preparation systems consist of oil heater, oil temperature controls, strainers, and associated valves and piping required to maintain fuel oil at the temperatures necessary to control the oil viscosity, to facilitate oil flow and burning, and to remove suspended matter.

Preparation of fuel oil for handling and burning requires heating the oil if it is grade 5 and 6. This decreases its viscosity so it will flow properly through the oil system piping and can be atomized properly by the oil burner. Grade 4 occasionally requires heating to facilitate burning. Grade 2 requires heating only under unusual conditions.

For handling residual oil from the delivering carrier into storage tanks, the viscosity should be about $156 \times 10^6$ cst (156 $m^2/s$). For satisfactory pumping, the viscosity of the oil surrounding the inlet of the suction pipe must be $444 \times 10^6$ cst (444 $m^2/s$) or lower; in the case of high pour point oil, the temperature of the entire oil content of the tank must be above the pour point.

These storage tank heaters are usually made of pipe coils or grids using steam or hot water at not over 15 psi (100 kPa) pressure as the heating medium. Electric heaters are sometimes used. For control of viscosity for pumping, the heated oil surrounds the oil suction line inlet. When heating high-pour-point oils, the heater should extend the entire length of the tank. All heaters have suitable thermostatic controls. In some cases, storage tank heating may be accomplished satisfactorily by returning or recirculating sufficient amounts of oil to the tank after it has passed through heaters located between the oil pump and oil burner.

Heaters to regulate viscosity at the burners are installed between the oil pumps and the burners. For small packaged burners, the heaters, when required, are either assembled integrally with the individual burners or mounted separately. The source of heat may be electricity, steam, or hot water. For larger installations, the heater is mounted separately and is often arranged in combination with central oil pumps forming a central oil pumping and heating set.

The separate or central oil pumping and heating set is recommended for those installations burning heavy oils where the load demand is periodical, and continuous circulation of hot oil is necessary during down periods.

Another system of oil heating to maintain pumping viscosity that is occasionally used for small- or medium-sized installations consists of an electrically heated section of oil piping. Low-voltage current is passed through the pipe section, which is isolated by nonconducting flanges.

The oil heating capacity for any given installation should be approximately 10% greater than the maximum oil flow. Maximum oil flow is the maximum oil burning rate plus the rate of oil recirculation.

Controls for oil heaters must be dependable to ensure proper oil atomization and avoid overheating of oil, which results in depositing coke inside the heaters. In the cases of steam or electric heating, an interlock should be included with a solenoid valve or switch to shut off the steam or electricity. During periods when the oil pump is not operating, the oil in the heater can become overheated and deposit carbon. This also can be a problem with high-temperature hot water; in such a case, provisions must be made to avoid overheating the oil when the oil pump is not operating.

Oil heaters with low- or medium-temperature hot water are not generally subject to coke deposits. Where steam or hot water is used in oil heaters located after the oil pumps, the pressure of the steam or water in the heaters is usually lower than the oil pressure. Consequently, heater leakage between oil and steam causes oil to flow into the water or condensing steam. To avoid such oil entering the boilers, the condensed steam or the water from such heaters should be discarded from the system, or special equipment should be provided for oil removal.

Hot-water oil heaters of double-tube-and-shell construction with inert heat-transfer oil and sight glass between are available. With this type of heater, oil leaks through an oil-side tube show up in the sight glass, and repairs can be made to the oil-side before a water-side tube leaks.

This discussion of oil-burning equipment applies to oil-fired boilers and furnaces. Chapters 23 and 24 address boilers and furnaces in more detail.

# PART III: SOLID FUEL-BURNING EQUIPMENT

A mechanical stoker is a device that feeds a solid fuel into a combustion chamber. It supplies air for burning the fuel under automatic control and, in some cases, incorporates automatic ash and refuse removal.

## CAPACITY CLASSIFICATION OF STOKERS

Stokers are classified according to their coal feeding rates. The following classification has been made by the U.S. Department of Commerce, in cooperation with the Stoker Manufacturers Association:

Class 1: capacity under 60 lb (28 kg) of coal per hour.
Class 2: capacity 60 to 100 lb (28 to 45 kg) of coal per hour.
Class 3: capacity 100 to 300 lb (46 to 136) kg of coal per hour.
Class 4: capacity 300 to 1200 lb (136 to 545 kg) of coal per hour.
Class 5: capacity 1200 lb (545 kg) of coal per hour and over.

**Class 1 Stokers** are used primarily for residential heating and are designed for quiet, automatic operation. Stokers for residential application are usually underfeed types and are similar to those shown in Figure 6, except they are usually screw feed. Although some stokers for residential application are still used, the primary thrust of stoker application is in the commercial and industrial areas. Stokers in this class feed coal to the furnace intermittently, in accordance with temperature or pressure demands. A special control is needed to ensure stoker operation in order to maintain a fire during periods when no heat is required.

**Class 2 and 3 Stokers** are usually of the screw-feed type, without auxiliary plungers or other means of distributing the coal. They are used extensively for heating plants in apartments, hotels, and industrial plants. They are of the underfeed type and are available in both the hopper type and the bin-feed type. These units are also built in a plunger-feed type with an electric motor or a steam or hydraulic cylinder coal-feed drive.

Stokers in this class are available for burning all types of anthracite, bituminous, and lignite coals. The tuyere and retort design varies according to the fuel and load conditions. Sta-

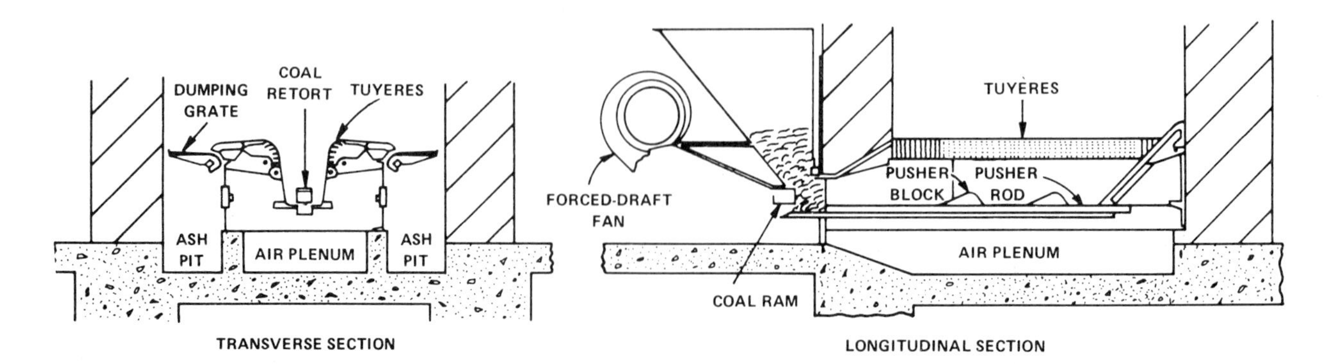


Fig. 6 Horizontal Underfeed Stoker with Single Retort

tionary grates are used on bituminous models, and the clinkers formed from the ash accumulate on the grates surrounding the retort.

Anthracite stokers in this class are equipped with moving grates that discharge the ash into a pit below the grate. This ash pit may be located on one or both sides of the grate and, in some installations, is of sufficient capacity to hold the ash for several weeks of operation.

**Class 4 Stokers** vary in details of design, and several methods of feeding coal are practiced. The underfeed stoker is widely used, although overfeed types are used in the larger sizes. Bin feed and hopper models are available in underfeed and overfeed types.

**Class 5 Stokers** are underfeed, spreader, chain grate or traveling grate, and vibrating grate. Various subcategories reflect types of grates and methods of ash discharge.

## STOKER TYPES BY FUEL-FEED METHODS

Stokers are classified according to the method of feeding fuel to the furnace, such as (1) spreader, (2) underfeed, (3) chain grate or traveling grate, and (4) vibrating grate. The type of stoker used in a given installation depends on the general system design, the capacity required, and the type of fuel burned. In general, the spreader stoker is the most widely used in the capacity range of 75,000 to 400,000 lb/h (9.5 to 50 kg/s) because it responds quickly to load changes and can burn a wide range of coals. The underfeed stokers are principally used with small industrial boilers of less than 30,000 lb/h (3.8 kg/s). In the intermediate range, the large underfeed units, as well as the chain- and traveling-grate stokers, are being displaced by spreader and vibrating-grate stokers. (Table 3 summarizes their major features.)

### Spreader Stokers

Spreader stokers use a combination of suspension burning and grate burning. As illustrated in Figure 7, coal is continually projected into the furnace above an ignited fuel bed. The coal fines are partially burned in suspension. Large particles fall to the grate and are burned in a thin, fast-burning fuel bed. Because this firing method provides extreme flexibility to load fluctuations and because ignition is almost instantaneous on in-

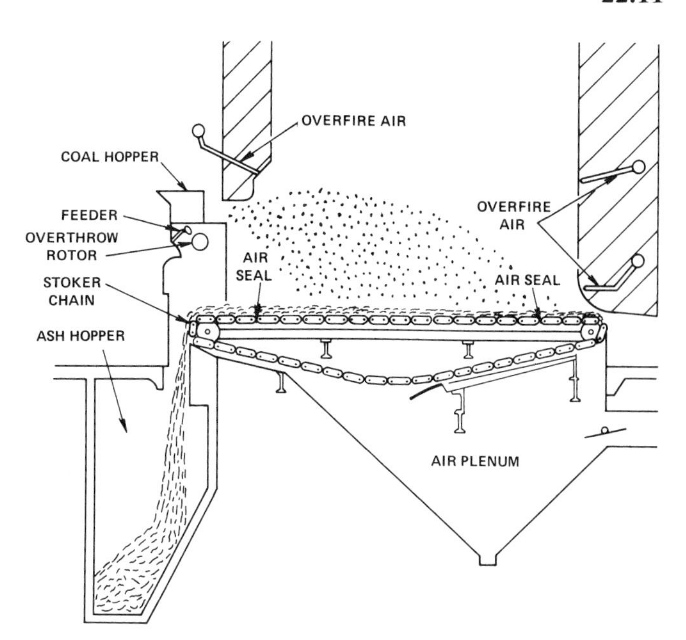


**Fig. 7 Spreader Stoker, Traveling-Grate-Type**

creased firing rate, the spreader stoker is favored over other stokers in many industrial applications.

The spreader stoker is designed to burn about 50% of the fuel in suspension. Thus, it generates much higher particulate loadings than other types of stokers and requires dust collectors to trap particulate material in the flue gas before discharge to the stack. To minimize carbon loss, fly-carbon reinjection systems are sometimes used to return this carbon into the furnace for complete burnout. This process does increase furnace dust emissions and, hence, can be used only with highly efficient dust collectors.

**Grates** for spreader stokers may be of several types. All grates are designed with high airflow resistance to avoid formation of blowholes through the thin fuel bed. The earliest designs were simple stationary grates from which ash was removed manually. Later designs allowed intermittent dumping of the grate manually or by a power cylinder. Both types of dumping grates

**Table 3 Characteristics of Various Types of Stokers (Class 5)**

| Stoker Type and Subclass | Typical Capacity Range, lb/h (kg/s) | Maximum Burning Rate, Btu/h • ft² (kW/m²) | Characteristics |
|---|---|---|---|
| *Spreader* | | | Capable of burning a wide range of coals; best to follow fluctuating loads; high fly ash carry-over, low-load smoke |
| Stationary- and dumping-grate | 20,000 to 80,000 (2.5 to 10) | 450,000 (1420) | |
| Traveling-grate | 100,000 to 400,000 (13 to 50) | 750,000 (2360) | |
| Vibrating-grate | 20,000 to 100,000 (2.5 to 13) | 400,000 (1260) | |
| *Underfeed* | | | Capable of burning caking coals and a wide range of coals (including anthracite); high maintenance, low fly ash carry-over; suitable for continuous-load operation |
| Single or double retort | 20,000 to 30,000 (2.5 to 3.8) | 400,000 (1260) | |
| Multiple retort | 30,000 to 500,000 (4 to 63) | 600,000 (1890) | |
| *Chain-grate and traveling-grate* | 20,000 to 100,000 (2.5 to 13) | 500,000 (1580) | Characteristics similar to vibrating-grate stokers, except that these stokers have difficulty in burning strongly caking coals |
| *Vibrating-grate* | 1400 to 150,000 (0.2 to 19) | 400,000 (1260) | Low maintenance, low fly ash carry-over; capable of burning wide variety of weakly caking coals; smokeless operation over entire range |

are frequently used for the small- and medium-sized boilers (see Table 3). Also, both types are sectionalized and there is a separate undergrate air chamber for each grate section and a grate section for each spreader unit. Consequently, both the air and fuel supply to one section can be temporarily discontinued for cleaning and maintenance without affecting the operation of other sections of the stoker.

For high efficiency operation, a continuous ash-discharging grate, such as the traveling grate, is necessary. The introduction of the spreader stoker with the traveling grate increased burning rates by about 70% over the stationary-grate and dumping-grate types. Although continuous ash-discharge grates of reciprocating and vibrating types have been developed, the traveling-grate stoker is preferred because of its higher burning rates.

**Fuels and fuel bed.** All spreader stokers (in particular, ones with traveling grates) are able to use fuels with a wide range of burning characteristics, including coals with caking tendencies, because the rapid surface heating of the coal in suspension destroys the caking tendency. High moisture, free-burning bituminous and lignite coals are commonly burned, while coke breeze can be burned in a mixture with a high volatile coal. However, anthracite, because of its low volatile content, is not a suitable fuel for spreader-stoker firing. Ideally, the fuel bed of a coal-fired spreader stoker is from 2 to 4 in. (50 to 100 mm) thick.

**Burning rates.** The maximum heat release rates range from 400,000 Btu/h • ft$^2$ (1.3 MW/m$^2$) (a coal consumption of approximately 40 lb/h or 5 g/s) on stationary-, dumping-, and vibrating-grate designs to 750,000 Btu/h • ft$^2$ (2.4 MW/m$^2$) on traveling-grate spreader stokers. Higher heat-release rates are practical with certain waste fuels in which a greater portion of fuel can be burned in suspension than is possible with coal.

## Underfeed Stokers

Underfeed stokers introduce the raw coal into a retort beneath the burning fuel bed. They are classified as horizontal feed and gravity feed. In the horizontal type, coal travels within the furnace in a retort parallel with the floor; in the gravity-feed type, the retort is inclined by 25 deg. Most horizontal feed stokers are designed with single or double retorts (rarely, with triple retorts), while gravity-feed stokers are designed with multiple retorts.

In the **horizontal stoker,** as shown in Figure 6, coal is fed to the retort by a screw (for the smaller stokers) or a ram (for the larger units). Once the retort is filled, the coal is forced upward and spills over the retort to form and feed the fuel bed. Air is supplied through tuyeres at each side of the retort and through air ports in the side grates. Over-fire air is used to provide additional combustion air to the flame zone directly above the bed to prevent smoking, especially at low loads.

The **gravity-feed** units are similar in operating principle. These underfeed stokers consist of sloping multiple retorts and have rear ash discharge. Coal is fed into each retort, where it is moved slowly to the rear while simultaneously being forced upward over the retorts.

**Fuels and fuel bed.** Either type of underfeed stoker can burn a wide range of coal, although the horizontal type is better suited for free-burning bituminous coal. These units can burn caking coal, provided there is not an excess amount of fines. The ash-softening temperature is an important factor in selecting coals because the possibility of excessive clinkering increases at lower ash-softening temperatures. Because combustion occurs in the fuel bed, these stokers respond slowly to load change. Fuel-bed thickness in underfeed stokers is extremely nonuniform, ranging from 8 to 24 in. (200 to over 600 mm). The fuel bed often contains large fissures separating masses of coke.

**Burning rates.** The single-retort or double-retort horizontal stokers are generally used to service boilers with capacities up to 30,000 lb/h (3.8 kg/s). These units are designed for heat-release rates of 400,000 Btu/h • ft$^2$ (1.3 MW/m$^2$).

## Chain-Grate and Traveling-Grate Stokers

Figure 8 shows a typical chain-grate or traveling-grate stoker. These stokers are often used interchangeably because they are fundamentally the same, except for grate construction. The essential difference is that the links of chain-grate stokers are assembled so that they move with a scissors-like action at the

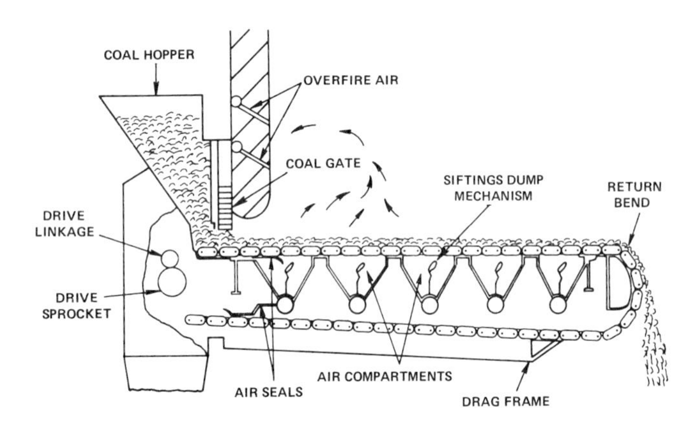


**Fig. 8 Chain-Grate Stoker**

return bend of the stoker, while in most traveling grates there is no relative movement between adjacent grate sections. Accordingly, the chain grate is more suitable for handling coals with clinkering-ash characteristics than is the traveling-grate unit.

The operation of each type is similar. Coal, fed from a hopper onto the moving grate, enters the furnace after passing under an adjustable gate to regulate the thickness of the fuel bed. The layer of coal on the grate entering the furnace is heated by radiation from the furnace gases or from a hot refractory arch. As volatile matter is driven off by this rapid radiative heating, ignition occurs. The fuel continues to burn as it moves along the fuel bed and the layer becomes progressively thinner. At the far end of the grate, where the combustion of the coal is completed, the ash is discharged into the pit as the grates pass downward over a return bend.

Often, furnace arches (front and/or rear) are included with these stokers to improve combustion by reflecting heat to the fuel bed. The front arch also serves as a bluff body to mix rich streams of volatile gases with air to reduce unburned hydrocarbons. A chain-grate stoker with overfire air jets eliminates the need of a front arch for burning volatiles. As shown in Figure 9, the stoker was zoned or sectionalized and equipped with individual zone dampers to control the pressure and quantity of air delivered to the various sections.

**Fuels and fuel bed.** The chain-grate and traveling-grate stokers can burn a variety of fuels that include peat, lignite, subbituminous coal, free-burning bituminous coal, anthracite coal, and coke, provided that the fuel is sized properly. However, strongly caking bituminous coals have a tendency to mat and prevent proper air distribution to the fuel bed. Also, a bed of strongly caking coal may not be responsive to rapidly changing loads. Fuel-bed thickness varies with the type and size of coal

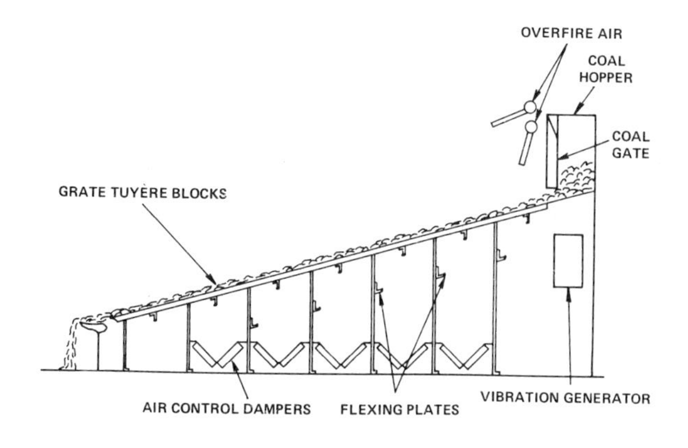


**Fig. 9 Vibrating-Grate Stoker**

burned. For bituminous coal, a 5 to 7 in. (125 to 175 mm) bed is common; for small-sized anthracite, the fuel bed is reduced to 3 to 5 in. (85 to 125 mm).

**Burning rates.** Chain-grate and traveling-grate stokers are offered for a maximum continuous burning rate of between 350,000 to 500,000 Btu/h • ft$^2$ (1100 to 1600 kW/m$^2$), depending on the type of fuel and its ash and moisture content.

### Vibrating-Grate Stokers

The vibrating-grate stoker, as shown in Figure 9, is similar to the chain-grate stoker in that both are overfeed, mass-burning, continuous ash-discharge units. However, in the vibrating stoker, the sloping grate is supported on equally spaced vertical plates that oscillate back and forth in a rectilinear direction, causing the fuel to move from the hopper through an adjustable gate into the active combustion zone. Air is supplied to the stoker through laterally exposed areas beneath the stoker formed by the individual flexing of the grate support plates. Ash is automatically discharged into a shallow or basement ash pit. The grates are water-cooled and are connected to the boiler circulating system.

The rates of coal feed and fuel-bed movement are controlled by the frequency and duration of the vibrating cycles and regulated by automatic combustion controls that proportion the air supply to optimize heat-release rates. Typically, the grate is vibrated about every 90 seconds for durations of 2 to 3 seconds, but this depends on the type of coal and boiler operation. The vibrating-grate stoker has found increasing acceptance since its introduction because of simplicity, inherently low fly ash carry-over, low maintenance, wide turndown (10 to 1), and adaptability to multiple fuel firing.

**Fuels and fuel bed.** The water-cooled vibrating-grate stoker is suitable for burning a wide range of bituminous and lignite coals. Because of the gentle agitation and compaction caused by the vibratory actions, coal having a high free-swelling index can be burned, and a uniform fuel bed without blowholes and thin spots can be maintained. The uniformity of air distribution and resultant fuel-bed conditions produce good response to load swings and smokeless operation over the entire load range. Fly ash emission is probably greater than from the traveling grate because of the slight intermittent agitation of the fuel bed. The fuel bed is similar to that of a traveling-grate stoker.

**Burning rates** of vibrating-grate stokers vary with the type of fuel. In general, however, the maximum heat-release rates should not exceed 400,000 Btu/h • ft$^2$ (1.3 MW/m$^2$) (a coal use of approximately 40 lb/h or 5 g/s) to minimize fly ash carry-over.

# PART IV: CONTROLS FOR AUTOMATIC FUEL-BURNING EQUIPMENT

This section only covers the automatic controls necessary for automatic fuel-burning equipment. Chapter 51 of the 1987 HVAC Volume addresses basic automatic control.

Automatic fuel-burning equipment requires a control system that provides a prescribed sequence of operating events and takes proper corrective action in the event of any failure in the equipment or its operation. The basic requirements for oil burners, gas burners, and coal burners (stokers) are the same. The term *burner,* unless otherwise specified, refers to all three types of fuel-burning equipment. The details of operation and the components used differ for the three types of burners, however. The controls can be classified as operating controls, limit controls, and interlocks. Operating controls include a primary sensor, secondary sensors, actuators, ignition system, firing-rate controls, draft controls, and programmers. Limit controls include flame safeguard, temperature limit, pressure limit, and water level limit controls. Several control functions are frequently included in a single component. Control systems for domestic burners and the smaller commercial burners are generally operated on low voltage (frequently 24 V). Some domestic gas burner control systems obtain their electrical power from the direct conversion of heat to low-voltage electrical energy with a thermopile. These are called self-generating systems. Line voltage and pneumatic controls are common in the larger commercial and industrial areas.

## OPERATING CONTROLS

Operating controls initiate the normal starting and stopping of the burner in response to the primary sensor acting through appropriate actuators. Secondary sensors, secondary actuators, and the ignition system are all part of the operating controls systems.

A **primary sensor** is required to monitor the effect desired by the furnace. Examples of primary sensors are: a room thermostat for a residential furnace; a pressure-actuated switch for a steam boiler; or a thermostat for a hot-water heater. Several sensors are sometimes connected together to average the measurement (temperature or pressure) at several points. Control on the basis of a temperature or pressure difference between two points can also be achieved. An example is the outside temperature-sensing element sometimes used in conjunction with the room thermostat to reset room temperature as a function of the outdoor temperature.

**Secondary sensors** are generally required also. An example is the fan control used on warm-air furnaces to control the fan as a function of plenum temperature. In some cases, the primary sensor operates a heat-control device such as a damper motor or a circulating pump. The burner is then operated from a secondary control such as an immersion thermostat in a hot-water system or a thermostat in the bonnet of the warm-air furnace. The primary sensor in a steam system with individual room control actuates a steam valve supplying that room or zone. The burner is then operated by a secondary control consisting of a pressure sensor located on a boiler.

An **actuator** is a device that converts the control system signal into a useful function. Actuators generally consist of valves, dampers, or relays.

**Valves** are required for final shutoff of the gas pilot and main gas (or oil) valves in the supply line. The type of valve used depends on the service, fuel, and characteristics of the burner. Valves are classified by types as follows:

1. **Solenoid gas or oil valves** provide quick opening and quick closing.
2. **Motor operated gas and oil valves** may be of gear train, pneumatic, or hydraulically operated types. They provide relatively slow opening and quick closing. Opening time is not adjustable. They may have provision for direct or reverse acting operating levers and, in some cases, have position indicators.
3. **Diaphragm valves for gas** are usually operated by gas pressure, although some types are steam- or air-operated. These valves may have provision for adjustable opening time, which may be adjusted to the desired burner operating characteristics. They may also have provision for direct- or reverse-acting lever arm.
4. **Manually opened safety shutoff valves for gas or oil** can be opened manually only when power is available. When power is interrupted or any associated interlock is activated, they trip free for fast closure. They are normally used on semiautomatic and manually fired installations.
5. **Burner input control valves** are required for gas or oil by some regulatory agencies in addition to safety shutoff valves. These control valves may be of the slow-opening type or may provide for high-low or modulating operation. These valves may or may not have provision for final shutoff. Requirements of local codes and approval bodies should be followed to ensure that valves selected are approved for the type of burner used and that maximum overall efficiency of the burner is obtained.

## Ignition

An automatic burner ignitor is necessary for safe operation and is, in most applications, an essential part of the automatic control system. The burner ignitor is an electric spark that directly ignites the main fuel supply or a relatively small fuel burner that ignites the main fuel supply. Most ignitors for oil burners are continuous (function continuously while burner is in operation) or intermittent (function only long enough to establish main flame). On small input gas-fired appliances, ignition of the main fuel is facilitated by a small standing gas pilot, which burns continuously.

Electric spark ignitors are generally used only for ignition of distillate grades of fuel oil. The spark is generated by a transformer, which supplies up to 10,000 V to the sparking electrodes. This ignition system is used frequently on oil burners having capacities of up to 20 gph (21 mL/s).

When pilot burners are used on larger equipment, the fuel burned may be gas, LP, or distillate fuel oil. Ignition of the pilot burner is by electric spark.

## Safety Controls and Interlocks

Safety controls and interlocks are provided to ensure against furnace explosions and other hazards, such as overheating of boilers resulting from low water, depending on the type of oil-burning units. These controls shut off the fuel or prevent firing in case of the following:

1. Flame failure—main flame or pilot flame
2. Combustion air fan failure
3. Overheating of warm-air furnace parts
4. Low water level in boilers
5. Low fuel-oil pressure
6. Low fuel-oil temperature, when burning heavy oils
7. High fuel-oil temperature, when burning heavy oils
8. Low atomizing air or atomizing steam pressure
9. High water temperature in water heaters and hot-water boilers
10. High steam pressure
11. Burner out of firing position (burner position interlock)
12. Failure of rotary burner motor (rotary cup interlock)

These interlocks do not affect the normal operation of the unit. The various types of safety devices in normal use and their application and operation are described in the following sections.

## Limit Controls

The safety features, which must be incorporated into the control system, function only when the system exceeds prescribed safe operating conditions. Such safety features are embodied in the limit controls. They actuate electrical switches that are normally closed but that will close the fuel valve in the event of an unsafe condition such as (1) excessive temperature in the combustion chamber or heat exchanger; (2) excessive pressure in a boiler or hot water heater; (3) low water level in a boiler and in larger commercial and industrial burners; (4) high or low gas pressure; (5) low oil pressure; (6) low atomizing media pressure; and (7) low oil temperature when firing residual fuel oil. Separate limit and operating controls are always recommended. The operating control is adjusted to function within the operating range, while the limit is set at a point above the operating range to stop the burner in the event of failure of the operating control.

## Flame Safeguard Controls

Loss of ignition, or flame failure while fuel is being injected into a furnace, can easily lead to a serious furnace explosion. Therefore, a flame sensor is used to shut off the fuel supply in case of flame failure. The flame detector of this device senses the presence or absence of flame through temperature, flame conductance of an electric current, or flame radiation. Through suitable electrical devices, the flame detector shuts down the burner when there is loss of flame.

Modern flame sensors of the flame conducting type and flame scanner type supervise the ignitor (pilot burner) and the main burner. Thus, at light-off, a safe ignitor flame must be established before fuel is admitted to the main burner.

# PROGRAMMING

Flame safety systems are interlocked and equipped with timing devices so that automatic lighting occurs in proper sequence with proper time intervals. In the larger units (size dependent on local authorities), these controls comprise a programmer and a flame-sensing device or flame detector.

The programmer maintains the proper safety sequence for lighting the burner, subject to supervision by the flame detector. When lightoff is called for, by push button or by the combustion control, the programmer controls the sequence of operation as follows:

1. Starts the burner fan so that the furnace, or combustion space, the gas passages of the heat exchanger, and the chimney connector are purged of any unburned combustible gases. This

flow of air is maintained until the entire volume of combustion space, gas passages, and chimney connector has been changed at least several times. It is commonly called *prepurge.*

2. After the prepurge, the controller initiates operation of the burner ignitor and allows a short time for the flame detector to prove satisfactory operation of the ignitor. If the flame detector is not satisfied, the program controller switches back to its initial start position, and an integral switch opens to prevent further operation until the electrical circuit is manually reset.
3. If the ignitor operation satisfies the flame detector, the program controller starts fuel flow to the burner. The flame detector is in control from this point on until the burner is shut off. If the flame detector is not satisfied with established stable burning during the start-up, it shuts down the burner and the program controller switches back to its initial position.
4. When the burner is shut off for any reason, the program controller causes the forced-draft fan to continue delivering air for a specific time to postpurge the unit of any atomized oil and combustible gases. This postpurge should amount to at least two air changes in the combustion area, gas passages, and chimney connector. (Some authorities disagree about the necessity of this function.)

Electronic programmers are available that, in addition to programming the entire boiler operating sequence, have data acquisition and storage capability to record operating conditions and flag the cause of burner shutdowns.

### Other Safety Devices

Safety devices to shut off the burner in case of hazardous conditions other than flame failures consist of simple sensors such as temperature elements, water level detectors, and pressure devices. These are all designed, installed, and interlocked to shut off the fuel oil and combustion air to the burner if any of the conditions listed occurs. All safety interlocks are arranged to prevent lighting the burner until satisfactory conditions have been established.

## COMBUSTION CONTROL SYSTEMS

Some modern gas- and oil-burning units are equipped with integrated devices that automatically regulate the flow of fuel and combustion air to satisfy the demand from the units. These systems, called **combustion control systems**, consist of a master controller, which senses the heat demands through room temperature or steam pressure, etc., and controlled devices, which receive impulses from the master controller. The signals change the rate of heat input by starting or stopping motors, driving oil pumps and fans, or adjusting air dampers and oil valves. The signal may be electrical or, in the case of some larger boilers, compressed air. Descriptions of various types of combustion control systems follow.

An **on-off control system** simply starts and stops the fuel burner to satisfy the heat demand and does not control a combustion system's *rate* of heat input. Until the controlled temperature of a room, the outlet temperature of a water heater, or the steam boiler pressure reaches a predetermined point, the burner remains on. At a set maximum condition of temperature or steam pressure, the burner shuts off.

The burner again ignites at a predetermined condition of temperature, pressure, etc. This control is normal for small units with burners using No. 2 fuel oil at capacities up to about 2,120,000 Btu/h or 15 gph (620 kW or 16 mL/s). These small burners have gas controls or a single motor to drive the burner fan or the fuel oil pump. Proportioning of combustion air to fuel is accomplished by the fan design and/or fixed position inlet air damper. For gas burners, pressure-atomizing oil burners, and other types of burners burning No. 2 or No. 4 oil at capacities between 2,120,000 to 4,240,000 Btu/h at 15 to 30 gph (620 to 1240 kW at 16 to 32 mL/s), the on-off control system usually has a low-start feature. This device starts the burner at less than full load (perhaps 60%) and then increases to full burning rate after several seconds. This type of control is called **low-high-off.**

A **low-high-low-off** control system (often referred to as two-position firing) also provides a low-fire start. In addition, it cycles the burner back to low-fire when the maximum control set point is reached.

In operation, the two-position control cycle has considerable advantage on a rapidly fluctuating load. However, on oil-firing, combustion conditions are poor at the low-load rate, so this cycle is seldom recommended for cases of sustained low loads, such as heating loads. The two-position control cycle provides somewhat steadier conditions. A positioning motor, through a mechanical linkage, positions the oil flow control valve and the burner damper properly. This permits damper settings for proportioning air to fuel.

A **modulating control** regulates the firing rate to follow the load demands more closely than the on-off or low-high-low-off controls. The fuel oil control valve and the burner damper respond over a range of positions within the operating range capability of the burner. The simplest and least expensive modulating control systems, as frequently used for burner capacities between 4,240,000 to 70,000,000 Btu/h or 30 to 500 gph (1.2 to 20 MW or 32 to 525 mL/s), use a positioning motor and mechanical linkage, which permit a large number of positions of the fuel control valve and the forced-draft damper. Most dampers larger than 70,000,000 Btu/h or 500 gph (20 MW or 525 mL/s) require more sophisticated controls that permit a continuous adjustment of valve and damper positions within the range of burner operation.

Modulating control systems that position the fuel oil control valve and the forced-draft damper in a predetermined relationship are called **parallel controls** or **positioning controls.** These valves maintain the fuel and air in proper proportion throughout the load range. Modulating controls for single-burner boilers usually include an on-off control for loads below the operating range of the burner.

A **full-metering and proportioning control system** is a modulating control system that maintains the proper ratio of air to fuel throughout the burner range with a fuel-air ratio controller. This system meters the airflow, usually by sensing the air pressure drop through the burner throat, and positions the air controller to suit momentary load demands. The fuel control valve immediately adjusts to match fuel flow to airflow. For some requirements, the ratio controller is reversed to match airflow to fuel flow. In unusual cases, the ratio controller is set to operate one way on load increase and the opposite way during load decrease. This control maintains optimum efficiency throughout the entire burner range, usually holding close to 15% excess air, as compared to 25% or greater with on-off controls.

### Draft Controls

Combustion control systems function properly with a reasonably constant furnace draft or pressure. The main concern is the varying draft effect of high chimneys. To control this draft, various regulators are used in the flue connections at the gas outlets of the units. Draft hoods or balanced draft dampers are used for smaller units and draft damper controllers for large boilers. These regulators should be supplied as part of the combustion control equipment. Chapter 26 has further details.

## REFERENCES

AGA. 1984. Installation of Domestic Gas Conversion Burners. ANSI/AGA *Standard* Z21.8-84.American Gas Association, Cleveland, OH.

AGA. 1981. Draft Hoods. ANSI/AGA *Standard* Z21.12-81.

AGA. 1982. Gas-Fired Low-Pressure Steam and Hot-Water Boilers. ANSI/ AGA *Standard* Z21.13-82.

AGA. 1984. Domestic Gas Conversion Burners. ANSI/AGA *Standard* Z21.17-84.

AGA. 1983. Gas-Fired Central Furnaces (Except Direct-Vent and Separated Combustion System Central Furnaces). ANSI/AGA *Standard* Z21.47-83.

AGA. 1971. Gas Utilization Equipment in Large Boilers. Revised 1983. ANSI/AGA *Standard* Z83.3-71.

AGA. 1982. Gas-Fired Duct Furnaces. ANSI/AGA *Standard* Z83.9-82.

*Gas Engineers Handbook*, Section 12, Chapter 2. The Industrial Press, New York. 1965.

UL. 1973. Commercial-Industrial Gas-Heating Equipment. UL *Standard* 795-73. Underwriters Laboratories, Northbrook, IL.

UL. 1980. Oil Burners. UL *Standard* 296, 7th ed.

UL. 1973. Oil-Fired Boiler Assemblies, UL *Standard* 726, 3rd ed.

UL. 1980. Oil-Fired Central Furnaces, UL *Standard* 727, 6th ed.

NFPA. 1987. NFPA *Standard* 85A, Prevention of Furnace Explosions in Fuel Oil- and Natural Gas-Fired Single Burner-Furnaces. National Fire Protection Association, Quincy, MA

# CHAPTER 23

# BOILERS



A *boiler* is a pressure vessel designed to transfer heat (produced by combustion) to a fluid. The definition has been expanded to include transfer of heat from electrical resistance elements to the fluid or by direct action of electrodes on the fluid. In boilers of interest to ASHRAE, the fluid is usually water, in the form of liquid or steam.

If the fluid being heated is air, the heat-exchange device is called a *furnace,* not a boiler. The firebox, or combustion space, of some boilers is also called the *furnace.*

For the purposes of this chapter, excluding special and unusual fluids, materials, and methods, a boiler is a cast-iron, steel, or copper pressure vessel heat exchanger, designed with and for fuel-burning devices and other equipment to (1) burn fossil fuels (or use electric current) and (2) transfer the released heat to water (in water boilers) or to water and steam (in steam boilers). Boiler heating surface is the area of fluid-backed surface exposed to the products of combustion, or the *fire-side* surface. Various codes and standards define allowable heat transfer rates in terms of heating surface. Boiler design provides for connections to a piping system, which delivers heated fluid to the point of use and returns the cooled fluid to the boiler.

Chapters 11 and 15 in the 1987 HVAC Volume cover most applications of heating boilers.

## BOILER CLASSIFICATIONS

Boilers may be grouped into classes based on working pressure and temperature, fuel used, shape and size, usage (such as heating or process), steam or water, and in other ways. Most classifications are of little importance to a specifying engineer, except as they affect performance. Excluding designed-to-order boilers, significant class descriptions are given in boiler catalogs or are available from the boiler manufacturer. The following basic classifications may be helpful.

### Working Temperature/Pressure

With few exceptions, all boilers are constructed to meet ASME *Boiler and Pressure Vessel Code,* Section 4, Rules for Construction of Heating Boilers (low-pressure boilers) and the ASME *Code,* Section 1, Rules for Construction of Power Boilers (medium- and high-pressure boilers).

**Low-pressure boilers** are constructed for maximum working pressures of 15 psi (103 kPa) steam and up to 160 psi (1100 kPa) hot water. Hot water boilers are limited to 250°F (120°C) operating temperature. The controls and relief valves, which limit temperature and pressure, are not part of the boiler but must be installed to protect the boiler.

**Medium- and high-pressure boilers** are designed to operate at above 15 psi (103 kPa) steam or above 160 psi water (1100 kPa) or 250°F (120°C) water boilers.

**Steam boilers** are available in standard sizes of up to 50,000 lb steam/h ($60 \cdot 10^3$ to $50 \cdot 10^6$ Btu/h or 17 to 15 000 kW), many of which are used for space heating in both new and existing systems. On larger installations, they may also provide steam for auxiliary uses, such as hot-water heat exchangers, absorption cooling, laundry, sterilizers, etc. In addition, many steam boilers provide steam (for which there is no viable substitute) at various temperatures and pressures for a wide variety of industrial processes.

**Water boilers** are available in standard sizes of up to $50 \cdot 10^6$ Btu/h (50 MBtu/h to 50,000 MBtu/h [15 to 15 000 kW]), many of which are in the low-pressure class and are used for space heating in both new and existing systems. Some water boilers may be equipped with either internal or external heat exchangers to supply domestic (service) hot water.

Every steam or water boiler is rated at the maximum working pressure determined by the ASME Boiler Code Section (or other code) under which it is constructed and tested. It also must be equipped, when installed, with safety controls and pressure-relief devices mandated by such code provisions.

### Fuel Used

Boilers may be designed to burn coal, wood, various grades of fuel oil, various types of fuel gas, or to operate as electric boilers. A boiler designed for one specific fuel type may not be convertible to another type of fuel. Some boiler designs can be adapted to burn coal, oil, or gas. Several designs allow firing with oil or gas by burner conversion or by using a dual fuel

The preparation of this chapter is assigned to TC 6.1, Hot-Water and Steam-Heating Equipment and Systems.

burner. The variations in boiler design according to fuel used are primarily of interest to boiler manufacturers, who can furnish details to a specifying engineer. The manufacturer is responsible for performance and rating according to the code or standard for the fuel used (see the section "Performance Codes and Standards").

## Materials of Construction

Most boilers, other than special or unusual models, are made of cast iron or steel. Some small boilers are made of copper or copper-clad steel.

**Cast-iron boilers** are constructed of individually cast sections, assembled into blocks (assemblies) of sections. Push or screw nipples, gaskets, or an external header join the sections pressure-tight and provide passages for the water, steam, and products of combustion. The number of sections assembled determines boiler size and energy rating. Sections may be vertical or horizontal, the vertical design being the most common.

The boiler may be **dry-base** (the firebox is beneath the fluid-backed sections), **wet-leg** (the firebox top and sides are enclosed by fluid-backed sections), or **wet-base** (the firebox is surrounded by fluid-backed sections, except for necessary openings).

At the manufacturer's option, the three boiler types can be designed to be equally efficient. Testing and rating standards apply equally to all three types. The wet-base design is easiest to adapt for combustible floor installations (applicable codes usually demand a floor temperature under the boiler no higher than 90°F [32°C], plus room temperature. A wet-base steam boiler at 215°F [102°C] or water boiler at 240°F [116°C] cannot meet this requirement without floor insulation).

**Steel boilers** are fabricated into one assembly of a given size and rating, usually by welding. The heat-exchange surface past the firebox usually is an assembly of vertical, horizontal, or slanted tubes. The tubes may be **firetube** (flue gas inside, heated fluid outside) or **watertube** (fluid inside). The tubes may be in one or more passes. As with cast-iron boilers, drybase, wet-leg, or wet-base design may be used. Most small steel heating boilers are of dry-base, vertical firetube design. Larger boilers usually have horizontal or slanted tubes; both firetube and watertube designs are used. A popular design for medium and large steel boilers is the **Scotch**, or **Scotch Marine**, which is characterized by a central fluid-backed cylindrical firebox, surrounded by firetubes in one or more passes, all within the outer shell.

Cast-iron boilers range in size from $35 \cdot 10^3$ to $10 \cdot 10^6$ Btu/h (10 to 2900 kW) gross output. Steel boilers range in size from $50 \cdot 10^3$ Btu/h (15 kW) to the largest boilers made.

## Condensing or Non-condensing

Until recently, boilers were designed to operate without condensing flue gases in the boiler. This precaution was necessary to prevent corrosion of cast-iron or steel parts. Hot-water units were often operated at 140°F (60°C) minimum return temperature to prevent rusting when operated with natural gas.

Because higher boiler efficiencies can be achieved with lower water temperatures, condensing boilers purposely allow the flue gas water vapor in the boiler to condense and drain. Figure 1 shows a typical relationship of overall boiler efficiency to return water temperature. The dew point of 130°F (55°C) shown varies with the percent of hydrogen in the fuel and the $CO_2$ (or excess air) in the flue gas.

Low return water temperatures and condensing boilers are particularly important because they are so efficient at part-load operation when high water temperatures are not required. For example, a hot water heating system that operates under light load conditions at 80°F (27°C) return water temperature has

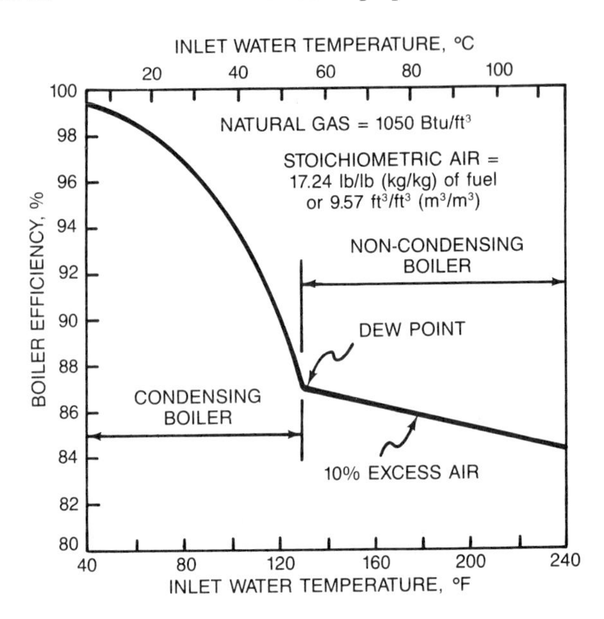


**Fig. 1 Effect of Inlet Water Temperature on Efficiency of Condensing and Non-Condensing Boilers**

a potential overall boiler efficiency of 97% when operated with natural gas of the specifications shown in Figure 1.

Figure 2 shows the effect of $CO_2$ on dew point of the flue gas. Designs with flue gas temperatures between the dew point and dew point +140°F (60°C) should be avoided unless the venting system is appropriate for condensing. Chapter 26 gives further details.

The condensing medium can be (1) return heating system water, (2) service water, or (3) other water or fluid sources at temperatures in the 70 to 130°F (20 to 55°C) range. The medium can also be a source of heat recovery in HVAC systems.

## Electric Boilers

Electric boilers are in a separate class. Since no combustion occurs, no boiler heating surface and no flue openings are necessary. Heating surface is the surface of the electric elements or electrodes immersed in the boiler water. The design of elec-

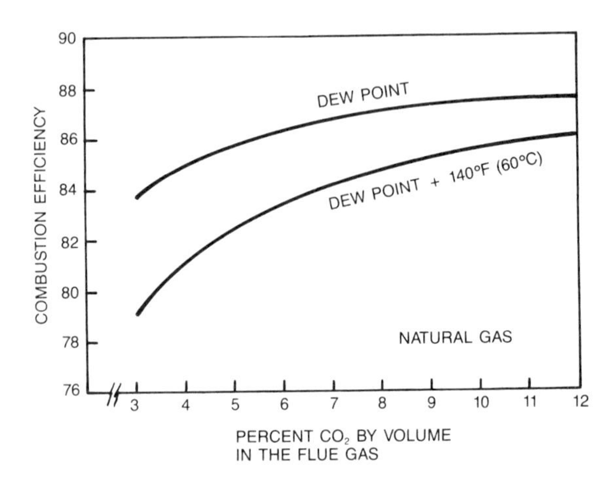


**Fig. 2 Effect of $CO_2$ on the Dew Point of Flue Gas**

tric boilers is largely determined by the shape and heat release rate of the electric heating elements used. Electric boiler manufacturers' literature describes available sizes, shapes, voltages, ratings, and methods of control.

### Selection Parameters

Boiler selection should be based on a competent review of the following parameters:

**All Boilers**

1. ASME (or other) Code Section, under which the boiler is constructed and tested
2. Net boiler output capacity, in Btu/h (kW)
3. Total heat-transfer surface, in $ft^2$ ($m^2$)
4. Water content in lb (kg)
5. Auxiliary power requirements, in kWh (MJ)
6. Internal water-flow patterns
7. Cleaning provisions for all heat-transfer surfaces
8. Operational efficiency
9. Space requirements and piping arrangement
10. Water treatment requirements

**Fuel-Fired Boilers**

1. Combustion space (furnace volume) in $ft^3$ ($m^3$)
2. Internal flow patterns of combustion products
3. Combustion air and venting requirements

**Steam Boilers**

1. Steam space, in $ft^3$ ($m^3$)
2. Steam disengaging area, in $ft^2$ ($m^2$)

The codes and standards outlined in the section "Performance Codes and Standards" include requirements for minimum efficiency, maximum temperature, burner operating characteristics, and safety control. For most purposes, test agency certification and labeling, published in boiler catalogs and shown on boiler rating plates, are sufficient. Some boilers not tested and rated by a recognized agency are produced and do not bear the label of any such agency. Nonrated boilers (rated and warranted only by the manufacturer) are used when codes do not demand a rating agency label. Almost without exception, both rated and nonrated boilers are of ASME *Code* construction and are marked accordingly.

## EFFICIENCY: INPUT AND OUTPUT RATINGS

Efficiency of fuel-burning boilers is defined in two ways: combustion efficiency and overall efficiency. Overall efficiency of electric boilers is in the 92 to 96% range.

**Combustion efficiency** is input minus stack (chimney) loss, divided by input, and ranges from 75 to 86% for most noncondensing mechanically fired boilers. Condensing boilers operate in the range of 88% to over 95% efficiency.

**Overall efficiency** is gross output divided by input. Gross output is measured in the steam or water leaving the boiler and depends on individual installation characteristics. Overall efficiency is lower than combustion efficiency by the percentage of heat lost from the outside surface of the boiler (this loss is usually termed *radiation loss*). Overall efficiency can be precisely determined only by laboratory tests under fixed test conditions. Approximate combustion efficiency can be determined under any operating condition by measuring operating flue-gas temperature and percentage $CO_2$ or $O_2$ and by consulting a chart or table for the fuel being used. The following section covers input and output measurement, efficiency, and testing and rating methods.

## PERFORMANCE CODES AND STANDARDS

Heating boilers are usually rated according to standards developed by (1) The Hydronics Institute (formerly the Institute of Boiler and Radiator Manufacturers [IBR] and The Steel Boiler Institute [SBI]); (2) The American Gas Association (AGA); and (3) The American Boiler Manufacturers Association (ABMA).

The Hydronics Institute has adopted a standard for rating cast-iron and steel heating boilers ("Testing and Rating Standard for Cast-Iron and Steel Heating Boilers") based on performance obtained under controlled test conditions. The gross output obtained by test is limited by certain factors, such as flue gas temperature, draft, $CO_2$ in flue gas, and minimum overall efficiency. This standard applies primarily to oil-fired equipment but is also used for power gas ratings for dual-fueled units.

Gas designed boilers are design-certified by AGA based on tests conducted in accordance with the *American National Standard* Z21.13 ("Gas Fired Low Pressure Steam and Hot Water Heating Boilers").

ABMA has adopted test procedures for commercial-industrial and packaged firetube boilers based on the ASME *Performance Test Code, Steam-Generating Equipment* (PTC 4.1). The units are tested for performance under controlled test conditions with minimum levels of efficiency required.

In 1978 the Department of Energy issued a test procedure applying to all gas- and oil-fired boilers up to 300,000 Btu/h (90 kW) input. This procedure supersedes industry rating methods used previously. The DOE test procedure determined both on-cycle and off-cycle losses based on a laboratory test involving cyclic conditions. The test results are applied to a computer program, which simulates an actual installation and results in an Annual Fuel Utilization Efficiency (AFUE). The steady-state efficiency developed during the test is similar to combustion efficiency described previously and is the basis for determining heating capacity, a term equivalent to gross output.

## BOILER-SIZING

Boiler-sizing is the selection of boiler output capacity to meet connected load. The boiler **gross output** is the rate of heat delivered by the boiler to the system under continuous firing at rated input. **Net rating** (IBR rating) is gross output minus a fixed percentage to allow for an estimated average piping heat loss, plus an added load for initially heating up the water in a system (sometimes called **pickup**).

**Piping loss** is variable. If all piping is within the space defined as load, loss is zero. If piping runs through unheated spaces, heat loss from the piping may be much higher than accounted for by the fixed **net rating** factor. Pickup is also variable. Pickup factor may be unnecessary when actual connected load is less than design load. On the design's coldest day, extra system output (boiler and radiation) is needed to pick up the load from a shutdown or low night setback. If night setback is not used, or if no extended shutdown occurs, no pickup load exists. Standby system input capacity for pickup, if needed, can be in the form of excess capacity in base-load boilers, or in a standby boiler.

If piping and pickup losses are negligible, boiler-sizing can be boiler gross output = design load. If piping loss and pickup load are large or variable, those loads should be calculated and equivalent gross boiler capacity added. Unless terminal units can deliver full boiler output under pickup conditions, at an inlet water temperature lower than boiler high limit setting, rated boiler capacity cannot be delivered. System input (boiler output) will be greater than system output, water temperature will rise, and the boiler will cycle on the high limit control, delivering an average input to the system that is much lower than the boiler gross output. Boiler capacity must be matched by terminal unit and system delivery capacity.

## CONTROL OF BOILER INPUT AND OUTPUT

Boiler controls regulate the rate of fuel input (on-off, step-firing, or modulating) in response to a control signal representing load change, so that average boiler output equals load within some accepted control tolerance. Boiler controls include safety controls that shut off fuel flow when unsafe conditions develop.

**Operating controls.** Steam boilers are operated by boiler-mounted, pressure-actuated controls, which vary the input of fuel to the boiler. Common examples of controls are *on-off, high-low-off,* and *modulating.* Modulating controls infinitely vary the fuel input from 100% down to a *selected minimum* point. The ratio of maximum to minimum is called the **turndown ratio.** The minimum input is usually between 5% and 25%, *i.e.,* 20 to 1, to 4 to 1, and depends on the size and type of fuel-burning apparatus.

Hot-water boilers are operated by temperature-actuated controls that are usually mounted on the boiler. Controls are the same as for steam boilers, i.e., on-off, high-low-off, and modulating.

Anytime (either during startup or in normal operation) the fire-side surface temperature in a steam boiler or water boiler is below the dew point of the flue gas, condensation of flue gas occurs on the fire side of the boiler. This condensate is corrosive to most boiler surfaces, unless the boiler is of a material that is resistant to this corrosion. Condensation can also occur in the venting system (see Chapter 26). The boiler manufacturer should be consulted if low operating temperatures or frequent cold starts are anticipated. Thermal shock should be avoided with steel firetube boilers.

Detailed descriptions of specific control systems for heating systems or for boilers are beyond the scope of this chapter. Basic control system designs and diagrams are furnished by boiler manufacturers. Control diagrams or specifications to meet special safety requirements are available from insurance agencies (such as IRI, FM, or IRM), governmental bodies, and ASME. Such special requirements usually apply to boilers with inputs of $400 \cdot 10^3$ Btu/h (120 kW) or greater, but may also apply to smaller sizes. The boiler manufacturer may furnish installed and pre-wired boiler control systems, with diagrams, when the requirements are known.

When a boiler installation code applies, the details of control system requirements may be specified by the inspector only after boiler installation. Special controls must then be supplied and the control design completed per the inspector's instructions. It is essential that the heating-system designer/specifying engineer determine the applicable codes, establish sources of necessary controls, and provide the skills needed to complete the control system. Local organizations are often available to provide these services.

## REFERENCES

ABMA. 1981. Packaged Firetube Ratings. American Boiler Manufacturers Association.

ANSI. 1987. Gas-Fired Low-Pressure Steam and Hot Water Boilers. ANSI *Standard* Z21.13-87. New York: American National Standards Institute.

ASME. 1964. Steam Generating Units. *Performance Test Code* 4.1-64. Reaffirmed 1985. New York: American Society of Mechanical Engineers.

ASME. 1982. Controls and Safety Devices for Automatically Fired Boilers. ASME *Standard* CSD-1-82.

ASME. 1986. Rules for Construction of Power Boilers. *Boiler and Pressure Vessel Code,* Section 1-86.

ASME. 1986. Rules for Construction of Heating Boilers. *Boiler and Pressure Vessel Code,* Section 4-86.

ASME. 1986. Recommended Rules for Care and Operation of Heating Boilers. *Boiler and Pressure Vessel Code,* Section 6-86.

ASME. 1986. Recommended Rules for Care of Power Boilers. *Boiler and Pressure Vessel Code,* Section 7-86.

de Lorenzi, Otto. 1967. Combustion Engineering: a Reference Book on Fuel Burning and Steam Generation. New York: Combustion Engineering, Inc.

Hydronics Institute. 1986. Testing and Rating Standard for Cast-Iron and Steel Heating Boilers.

Strehlow, R.A. 1984. Combustion Fundamentals. New York: McGraw-Hill.

Woodruff, E.B., H.B. Lammers, and T.F. Lammers. 1984. Steam-Plant Operation, 5th Ed. New York: McGraw-Hill.

# CHAPTER 24

# FURNACES



## PART I: RESIDENTIAL FURNACES

### EQUIPMENT VARIATIONS

Residential furnaces are available in a variety of self-enclosed appliances, which provide heated air through a system of ductwork into the space being heated. There are two types.

**Fuel-Burning Furnaces.** Combustion takes place within a combustion chamber. Circulating air passes over the outside surfaces of the heat exchanger so that it does not contact the fuel or the products of combustion. The products of combustion are passed to the outside atmosphere through a vent.

**Electric Furnaces.** A resistance-type heating element either heats the circulating air directly or through a metal sheath enclosing the resistance element.

Residential furnaces may be further categorized by: (1) fuel type, (2) mounting arrangement, (3) air flow direction, (4) combustion system, and (5) installation location. The following sections cover typical furnace variations according to these categories.

### NATURAL GAS FURNACES

Natural gas is the most common fuel supplied for residential heating, and the central system forced-air furnace is the most common way of heating with natural gas. This type of furnace is equipped with a blower to circulate the air through the furnace enclosure, over the heat exchanger, and through the ductwork distribution system. A typical furnace consists of the following basic components: (1) cabinet or casing; (2) heat exchangers; (3) combustion system including burners and controls; (4) draft hood or flue gas collection box; (5) circulating air blower and motor; (6) air filter; and (7) other accessories such as a vent damper, humidifier, electronic air cleaner, or a combination of these things.

#### Casing or Cabinet

The furnace casing is most commonly formed from cold-rolled steel and is painted. Doors on one side of the furnace allow access to those sections requiring service. The inside of the casing adjacent to the heat exchanger is lined with a foil-faced blanket insulation or a metal radiation shield to reduce heat losses through the casing and to limit the outside surface temperature of the furnace. On some furnaces the inside of the blower compartment is lined with insulation to acoustically dampen the blower noise.

#### Heat Exchangers

Heat exchangers are normally made of mirror-image formed parts and are joined together to form a *clam shell*. The material is generally cold-rolled steel for standard indoor furnaces. If the furnace is exposed to clean air and the heat exchanger remains

The preparation of this chapter is assigned to TC 6.3, Central Warm Air Heating and Cooling.

dry, this material has a long life and does not easily corrode. Some problems of heat-exchanger corrosion and failure have been encountered in the past because of exposure to halogens in the flue gas. These problems were caused by combustion air that was contaminated by such things as laundry bleach, cleaning solvents, and halogenated hydrocarbon refrigerants.

Coated or alloy steel are used in top-of-the-line models and furnaces for special applications. Common corrosion-resistant materials include aluminized steel, ceramic-coated cold-rolled steel, and stainless steel. Furnaces certified for use downstream of a cooling coil must have corrosion-resistant heat exchangers.

A great deal of research has been done on corrosion-resistant materials for use in condensing heat exchangers (Stickford 1985). The presence of chloride compounds in the condensate can cause the failure of a condensing heat exchanger unless a corrosion-resistant material is used.

Several manufacturers now produce liquid-to-air heat exchangers in which a liquid is heated in a finned tube and is either evaporated or pumped to a condenser section or fan-coil, which heats circulating air. Materials of construction range from copper to special steels for condensing furnace applications.

## Burners and Internal Controls

Burners are most frequently made from stamped sheet metal, though cast iron is also used. Fabricated sheet metal burners may be made from cold-rolled steel coated with high temperature paint or from a corrosion-resistant material such as stainless or aluminized steel. Burner material must be compatible with that of the heat exchangers to meet the corrosion-protection requirements of the specific application. Gas furnace burners may be either of the monoport or multiport types; the type used with each particular furnace depends on compatibility with the heat exchanger.

Furnace controls include an ignition device, gas valve, fan switch, limit switch, and other components specified by the manufacturer. These controls allow gas to flow to the burners when heat is required. The three most common ignition systems are (1) standing pilot, (2) intermittent, and (3) direct spark or hot surface ignition (ignites main burners directly). The "Technical Data" section of this chapter has more details on the functions and performance of these individual control components.

## Venting Components

Atmospheric vent indoor furnaces are equipped with a draft hood connecting the heat exchanger flue gas exit to the vent pipe or chimney. The draft hood has a relief air opening large enough to ensure that the exit of the heat exchanger is always at atmospheric pressure. One purpose of the draft hood is to make certain that the natural-draft furnace continues to operate safely without generating carbon monoxide if the chimney is blocked, if there is a downdraft, or if there is excessive updraft for any reason. Another purpose is to maintain constant pressure on the combustion system. Direct-vent furnaces do not use a draft hood of this type; in this case, a control system shuts the furnace down if the chimney or vent becomes blocked.

Power-vent furnaces use a blower to force or induce the flue products through the furnace. These furnaces may or may not have a relief air opening, but, in either case, the same safety requirements are met.

Recent research into common venting of atmospheric draft appliances (water heaters) and power vent furnaces shows that non-positive vent pressure systems may operate on a common vent. (See manufacturer's instructions for specific information.) A venturi-aspirator approach is under development for positive vent pressure systems (Deppish and DeWerth 1986).

ANSI *Standard* Z21.47A-1985 classifies venting systems. Central furnaces are categorized by temperatures and pressures attained in the vent as follows:

**Category I**—A central furnace that operates with a non-positive vent pressure and with a vent gas temperature at least 140°F (60°C) above its dew point.

**Category II**—A central furnace that operates with a non-positive vent pressure and with a vent gas temperature less than 140°F (60°C) above its dew point.

**Category III**—A central furnace that operates with a positive vent pressure and with a vent gas temperature at least 140°F (60°C) above its dew point.

**Category IV**—A central furnace that operates with a positive vent pressure and with a vent gas temperature less than 140°F (60°C) above its dew point.

Furnaces rated in accordance with ANSI *Standard* Z21.47 are marked to show they are one of the four venting categories listed above.

## Blowers and Motors

Centrifugal blowers with forward-curved blades of the double-inlet type are used in most forced-air furnaces. These blowers overcome the resistance of the furnace air passageways, filters, and ductwork. They are usually sized to provide the additional air requirement for cooling and the static pressure required for the cooling coil. The blower may be a direct-drive type with the blower wheel attached directly to the motor shaft, or it may be a belt-drive type with a pulley and V-belt used to drive the blower wheel.

Electric motors used to drive furnace blowers are usually designed for that purpose. Direct-drive motors may be of the shaded pole or permanent split-capacitor type. Speed variation may be obtained by taps connected to extra windings in the motor. Belt-drive blower motors are normally split-phase or capacitor-start. The speed of belt-drive blowers is controlled by adjustment of a variable-pitch drive pulley.

## Air Filters

An air filter in a forced-air furnace removes dust from the air that could reduce the effectiveness of the blower and heat exchanger(s). Filters supplied with a forced-air furnace are often disposable. Permanent filters that may be washed or vacuum cleaned and reinstalled are also used. The filter is *always* located in the circulating airstream ahead of the blower and heat exchanger.

## Accessories

**Humidifiers.** These are not included as a standard part of the furnace package. However, one of the advantages of a forced-air heating system is that it offers the opportunity to control the relative humidity of the heated space at a comfortable level. Several types of humidifiers are used with forced-air furnaces. Chapter 5 addresses various types.

**Electronic Air Cleaners.** These air cleaners are much more effective than the air filter provided with the furnace, and they filter out much finer particles, including smoke. Electronic air cleaners create an electric field of high voltage direct current in which dust particles are given a charge and collected on a plate having the opposite charge. The collected material is then cleaned periodically from the collector plate by the homeowner.

Electronic air cleaners are mounted in the airstream entering the furnace. Chapter 10 contains detailed information on filters.

**Automatic Vent Damper.** This device closes the furnace vent opening when the furnace is not in use, thus, reducing off-cycle losses. More information about the energy-saving potential of this accessory is included in the "Technical Data" section of this chapter.

## Airflow Variations

The components of a gas-fired, forced-air furnace can be arranged in a variety of configurations to suit a residential heating system. The relative positions of the components in the different types of furnaces are as follows:

1. The **upflow or "highboy" furnace** (see Figure 1) has the blower beneath the heat exchanger discharging vertically upward. Air enters through the bottom or the side of the blower compartment and leaves at the top. This furnace may be used in closets and utility rooms on the first floor or in basements with the return air ducted down to the blower compartment entrance.

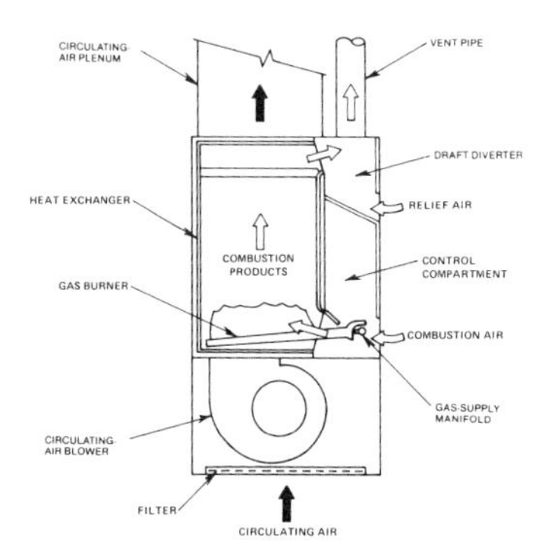


**Fig. 1 Upflow Forced-Warm-Air Furnace**

2. The **downflow furnace** (see Figure 2) has the blower located above the heat exchanger discharging downward. Air enters at the top and is discharged vertically at the bottom. This furnace is normally used with a perimeter heating system in a house without a basement. It is also used in upstairs furnace closets and utility rooms supplying conditioned air to both levels of a two-story house.
3. The **horizontal furnace** (see Figure 3) has the blower located beside the heat exchanger. The air enters at one end, travels horizontally through the blower, over the heat exchanger, and is discharged at the opposite end. This furnace is used for locations with limited head room such as attics, crawl spaces, or is suspended under the ceiling. These units are often designed so that the components may be rearranged to allow installation with airflow from left to right or from right to left.

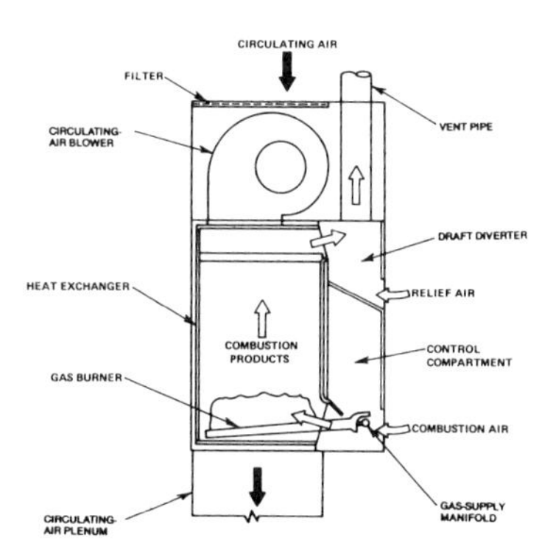


**Fig. 2 Downflow (Counterflow) Forced-Warm-Air Furnace**

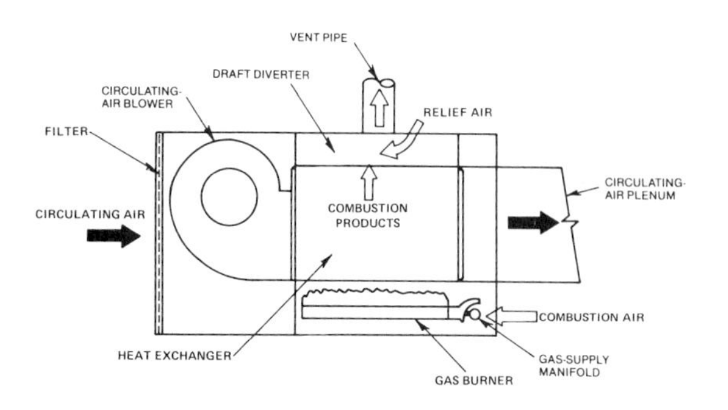


**Fig. 3 Horizontal Forced-Warm-Air Furnace**

4. The **basement or "lowboy" furnace** (see Figure 4) is a variation of the upflow furnace and requires less head room. The blower is located beside the heat exchanger at the bottom. Air enters the top of the cabinet, is drawn down through the blower, discharged over the heat exchanger, and leaves vertically at the top.
5. The **gravity furnace** is available, but not common. This furnace has larger air passages through the casing and over the heat exchanger so that buoyancy force created by the air being warmed circulates the air through the ducts.

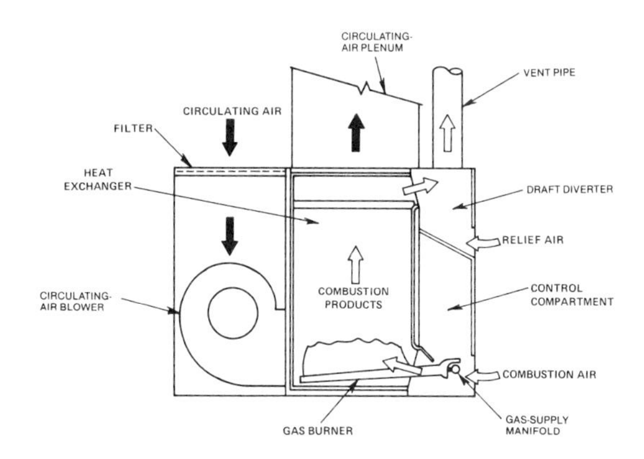


**Fig. 4 Basement (Low-Boy) Forced-Warm-Air Furnace**

## Combustion System Variations

Gas-fired furnaces use an atmospheric vent or a power vent combustion system. The atmospheric vent is more common. The buoyancy of the hot combustion products carry these products through the heat exchanger, into the draft hood, and up the chimney.

Power combustion furnaces have a combustion blower, which may be located either upstream or downstream from the heat exchangers. If the blower is located upstream, blowing the combustion air into the heat exchangers, the system is called a *forced-draft system.* If the blower is downstream, the arrangement is called an *induced-draft system.* Power combustion systems have been commonly used with outdoor furnaces in the past; however, more indoor furnaces are being designed using this concept. Power combustion furnaces do not require a draft hood, resulting in reduced off-cycle losses and improved efficiency.

Direct-vent furnaces may be either the natural-draft or power-draft type. They do not have a draft hood and obtain combustion air from outside the structure. Mobile home furnaces must be the direct-vent type.

### Indoor-Outdoor Furnace Variations

Central system residential furnaces are designed and certified for either indoor or outdoor use. Indoor furnaces come in either upflow, downflow, or horizontal arrangements. Outdoor furnaces normally are horizontal flow and are available in two styles.

One style is the heating-only outdoor furnace. This furnace is similar to the more common indoor horizontal furnace. The primary difference is that the outdoor furnace is weatherized. The motors and controls are sealed. The exposed components are made from corrosion-resistant materials such as galvanized or aluminized steel.

A common style outdoor furnace is the combination package unit. This unit is a combination of an air conditioner and gas furnace built into a single casing. The design varies, but the most common combination consists of an electric air conditioner coupled with a horizontal gas furnace. The advantage is that much of the interconnecting piping and wiring is included in the unit.

## LPG FURNACES

Natural gas furnaces are also available as liquefied petroleum gas (LPG) furnaces. Most manufacturers have their furnaces certified for both natural gas and LPG. The major difference between the two furnaces is the pressure at which the gas is injected from the manifold into the burners. For natural gas, the manifold pressure usually is controlled at 3 to 4 in. of water (0.7 to 1.0 kPa); for LPG, the pressure is usually 10 to 11 in. of water (2.5 to 2.7 kPa). The increased pressure for LPG is usually required to inject sufficient primary air into the burners.

Because of higher injection pressure and the greater heat content per volume of gas, there are certain physical differences between a natural gas furnace and an LPG furnace. One difference is that the pilot and burner orifices must be smaller for LPG furnaces. The gas valve regulator spring is also different. Sometimes it is necessary to change burners, but this is not required normally. Manufacturers sell conversion kits containing both the required parts and instructions to convert furnace operation from one gas to the other.

## OIL FURNACES

Indoor oil furnaces come in the same configuration as gas furnaces. They are available in upflow, downflow, horizontal, and low-boy configurations for ducted systems. Oil-fired outdoor furnaces and combination units are seldom available.

The major differences between oil and gas furnaces are in the combustion system, heat exchanger, and the barometric draft regulator used in lieu of a draft hood. The ducted system, oil-fired, forced-air furnaces are usually forced draft and equipped with pressure-atomizing burners. The pump pressure and the orifice size of the injection nozzle regulates the firing rate of the furnace. Electric ignition lights the burners. Other furnace controls, such as the blower switch and the limit switch, are similar to those used on gas furnaces.

Heat exchangers of oil-fired furnaces are normally heavy gauge steel formed into a welded assembly. The hot flue products flow through the inside of the heat exchanger into the chimney. The conditioned air flows over the outside of the heat exchanger and into the air supply plenum.

## ELECTRIC FURNACES

Electric-powered furnaces come in a variety of configurations and have some similarities to gas- and oil-fired furnaces. However, when used with an air conditioner, the cooling coil may be upstream from the blower and heaters. On gas- and oil-fired furnaces, the cooling coil is normally mounted downstream from the blower and heat exchangers.

Figure 5 shows a typical arrangement for an electric forced-air furnace. Air enters the bottom of the furnace and passes through the filter, then flows up through the cooling coil section into the blower. The electric heating elements are immediately above the blower so that the high-velocity air discharging from the blower passes directly through the heating elements.

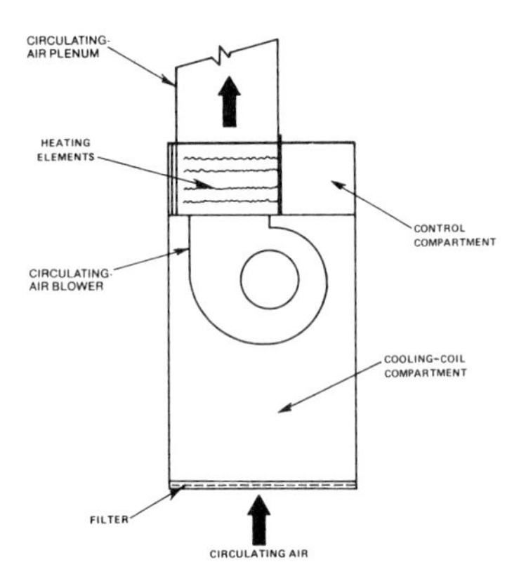


**Fig. 5 Electric Forced-Air Furnace**

The furnace casing, air filter, and blower are similar to equivalent gas furnace components. The heating elements are made in modular form with 5 kW capacity being typical for each module. Electric furnace controls include electric overload protection, contactor, limit switches, and a fan switch. The overload protection may be either fuses or circuit breakers. The contactor brings the electric heat modules on. The fan switch and limit switch functions are similar to those of the gas furnace, but one limit switch is usually used for each heating element.

Frequently, electric furnaces are made from modular sections; for example, the coil box, blower section, and electric heat section are made separately and then assembled in the field. Regardless of whether the furnace is made from a single piece casing or a modular casing, it is generally a multiposition unit. Thus, the same unit may be used for either upflow, downflow, or horizontal installations.

When the electric heating appliance is sold without a cooling coil, it is called an electric furnace. The same appliance is called a fan-coil unit when it is sold with an air-conditioning coil already installed. When the unit is used as the indoor section of a split heat pump, it is called a heat-pump fan coil. For detailed information on heat pumps, see Chapter 43.

Electric forced-air furnaces are also used with outdoor package units. These units are called packaged heat pumps or packaged air conditioners, and not electric furnaces.

## SYSTEM DESIGN AND EQUIPMENT SELECTION

### Warm-Air Furnaces

Two steps are required in selecting a warm-air furnace: (1) determining the required heating capacity of the furnace and (2) selecting a specific furnace to satisfy this requirement.

**Heating capacity** depends on several variables that may operate alone or in combination. The first variable is the design heating requirement of the residence. The heat loss of the structure can be calculated by using the procedures in the 1985 FUNDAMENTALS Volume.

The second variable to consider when selecting the capacity of a furnace is the additional heating required if the furnace is operating on a night setback cycle. During the morning recovery period, additional capacity is required to bring the conditioned space temperature up to the desired level. The magnitude of this recovery capacity depends on weather conditions, the magnitude of the night setback, and the time allowed for the furnace to return room air temperature to the desired level. Another consideration similar to night setback concerns structures that only require intermittent heating, such as churches and auditoriums. Chapter 19 of the 1987 HVAC Volume has further information.

A third variable is the influence of internal loads. Normally, the heat gain from internal loads is neglected when selecting a furnace, but if the internal loads are constant, they should be used to reduce the required capacity of the furnace. This would be particularly applicable in nonresidential applications.

The energy required for humidification is a fourth variable. The humidification energy depends on the desired level of relative humidity and the rate at which the moisture must be supplied to maintain the specified level. Net moisture requirements must take into account internal gains as a result of people, equipment, and appliances, and losses through migration in exterior surfaces, plus air infiltration. Chapter 5 gives details on how to determine the humidification requirements.

A fifth variable is the influence of off-peak storage devices. A storage device, when used in conjunction with a furnace, decreases the required capacity of the furnace. The storage device can supply the additional capacity required during the morning recovery of a night setback cycle or reduce the daily peak loads to assist in load shedding. Detailed calculations can determine the contribution of storage devices.

A sixth variable is the influence of backup systems. A furnace can exist as a backup to a solar system, a heat recovery system, or a structure requiring multiple units for uninterrupted service. Oversizing results in higher initial costs, possible increased operating costs (see "Size Selection" section below), and decreased comfort control. Undersizing produces unacceptable comfort control near the design conditions.

The seventh variable is a check on the furnace's capacity to accommodate air-conditioning, even if air-conditioning is not planned initially. The cabinet should be large enough to accept a cooling coil that satisfies the cooling load. The blower and motor should have sufficient capacity to provide increased airflow rates typically required in air-conditioning applications. Chapter 10 of the 1987 HVAC Volume includes specific design considerations.

Determining the required heating capacity is the first step in selecting a warm-air furnace and is independent of the fuel type and size of the equipment.

**Specific Furnace Selection.** The second step in the selection of a warm-air furnace is to choose a specific furnace that satisfies the required design capacity. The final decision depends on numerous parameters, which may individually or collectively influence the selection. The most significant parameter is the fuel type. The second step of the furnace selection process is subdivided by fuel types.

## Natural Gas Furnaces

**Size Selection.** Historically, furnaces have been oversized for several reasons: the calculational procedure was not exact, especially the estimate of air infiltration; weather conditions are occasionally more severe than the design conditions used to calculate the required furnace capacity; and the additional first cost of a slightly larger furnace was considered a good value in view of possible undersizing, which would be expensive to correct. Adequate airflow for cooling was another consideration. The price of natural gas was relatively inexpensive so that possible inefficiencies because of oversizing were not considered detrimental. The net result led to oversizing by one or two units.

Present studies indicate that oversizing may increase overall energy use for new houses where vent and ducts are sized to furnace capacity. However, in retrofits (where fixed vent and duct sizes are assumed), oversizing has little effect on overall energy use. In either situation, oversizing may reduce the comfort level due to wide temperature variations in the conditioned space.

Chapter 25 of the 1985 FUNDAMENTALS Volume recommends oversizing by 40% to raise the night setback 10°F (6°C) in one hour.

**Performance Criteria** or a consistent definition of *efficiency* must be used throughout. Some typical efficiencies encountered are (1) steady-state efficiency, (2) utilization efficiency, (3) annual fuel utilization efficiency, and (4) seasonal efficiency (used in California).

These efficiencies are generally used by the furnace industry in the following manner:

1. *Steady-State Efficiency* is the efficiency of a furnace when it is operated under equilibrium conditions based on an ANSI *Standard.* It is calculated by measuring the energy input, subtracting the losses for exhaust gases, and then dividing by the fuel input (cabinet loss not included), *i.e.,*

$$SS(\%) = \frac{\text{Fuel Input} - \text{Flue Loss}}{\text{Fuel Input}} \cdot 100$$

2. *Utilization Efficiency* is obtained from an empirical equation developed by the *National Bureau of Standards* (*NBSIR 78-1543*) by starting with a 100% efficiency and deducting losses for exhausted latent and sensible heat, cyclic effects, infiltration, and pilot-burner effect.
3. *Annual Fuel Utilization Efficiency* (AFUE) is the same as Utilization Efficiency, except that losses from a standing pilot during nonheating season time are deducted. This equation can also be found in *NBSIR 78-1543* and is used in the Federal Trade Commission's (FTC) *Energy Guide Fact Sheets.*

   The AFUE is determined for residential fan-type furnaces by the use of the ANSI/ASHRAE *Standard* 103-1982R, *Methods of Testing for Heating Seasonal Efficiency of Central Furnaces and Boilers,* in conjunction with the amendments issued by the U.S. Department of Energy in the March 28, 1984, *Federal Register.* This version of the test method allows the rating of non-weatherized furnaces as indoor, isolated combustion systems (ICS), or both. Weatherized furnaces are rated as outdoor.
4. *Seasonal Efficiency* (SE) is a term used by the State of California and is unique to its regulations. It contains electrical energy factors, which account for losses in generating electricity. SE is about three percentage points less than the AFUE for the identical unit. The annual fuel consumed and AFUE are calculated by using *NBSIR 78-1543,* and the formula for SE is found in the *California Energy Commission Regulations for Appliance Efficiency Standards.*

Federal law requires manufacturers of furnaces to use AFUE to rate efficiency. Table 1 gives typical values for different efficiencies on furnaces. "Typical" does not mean maximum or minimum; for any model furnace, the AFUE rating can be found at the point of sale on *Fact Sheets* required by the FTC.

Annual energy savings may be compared with the following formula:

**Table 1 Typical Values of AFUE (Furnaces Located in Conditioned Space)**

| Type of Gas Furnace | AFUE, % |
|---|---|
| 1. Atmospheric with standing pilot | 64.5 |
| 2. Atmospheric with intermittent ignition | 69.0 |
| 3. Atmospheric with intermittent ignition and auto. vent damper | 78.0 |
| 4. Same basic furnace as 2, except with power vent | 78.0 |
| 5. Same as 4, except with improved heat transfer | 81.5 |
| 6. Direct vent with standing pilot, preheat | 66.0 |
| 7. Direct vent, power vent, and intermittent ignition | 78.0 |
| 8. Power burner (forced-draft) | 75.0 |
| 9. Condensing | 92.5 |
| **Type of Oil Furnace** | **AFUE, %** |
| 1. Standard | 71.0 |
| 2. Same as 1, with improved heat transfer | 76.0 |
| 3. Same as 2, with auto. vent damper | 83.0 |
| 4. Condensing | 91.0 |

$$\text{Annual Energy Reduction} = \frac{AFUE_2 - AFUE_1}{AFUE_2}$$

where $AFUE_2$ is greater than $AFUE_1$. For example compare items 4 and 2 of Table 1.

$$\text{A.E.R.} = \frac{78 - 69}{78} = \frac{9}{78} = 11.5\%$$

The SP43 work (see "System Performance" section) confirms that this is a reasonable estimate.

**Construction Features and Limitations.** Most indoor furnaces have cold-rolled steel heat exchangers. If the furnace is exposed to clean air and the heat exchanger remains dry, this material has a long life and does not corrode easily. Many deluxe, noncondensing furnaces have a coated heat exchanger to provide extra protection against corrosion. Research by Stickford *et al.* (1985) indicated that chloride compounds in the condensate of condensing furnaces can cause the heat exchanger to fail unless it is made of stainless or specialty steel. A corrosion-resistant heat exchanger must also be used in a furnace certified for downstream use of a cooling coil.

**Design Life.** Typically, the heat exchangers are cold-rolled steel and have a design life of approximately 15 years. Special coated heat exchangers, when used for standard applications, increase the design life to as much as 20 years. Coated heat exchangers are recommended for furnace applications where corrosive atmospheres exist.

**Sound Level** must be considered in most applications. Chapter 52 in the 1987 HVAC Volume outlines the procedures to follow in determining acceptable noise levels.

**Safety.** Because of open-flame combustion, the following safety items need to be considered: (1) the surrounding atmosphere should be free of dust or chemical concentrations; (2) a path for combustion air must be provided for both sealed and open combustion chambers; and (3) the gas piping and vent pipes must be installed according to the *National Fuel Gas Code* (ANSI Z223.1), local codes, and the manufacturer's instructions.

**Applications.** Gas furnaces are primarily applied to residential heating. The majority are used in single-family dwellings but are also applicable to apartments, condominiums, and mobile homes.

**Performance versus Cost** must be considered in selection. The costs are divided between initial and life-cycle costs. Included in life-cycle costs are maintenance, energy consumption, design life, and the price escalation of the fuel. Procedures for establishing operating costs for use in product labeling and audits are available from the Department of Energy (DOE). For residential furnaces, *Fact Sheets,* provided by manufacturers, are available at the point of sale.

### Other Fuels

The system design and selection criteria for LPG furnaces are identical to those for natural gas furnaces.

The design criteria for oil furnaces are also identical to those for natural gas furnaces, except that oil-fired furnaces equipped with pressure-atomizing or rotary burners should be rated in accordance with ANSI Z91.1 test methods. This requires a minimum steady-state efficiency of 75% for forced-air furnaces.

### Electric Furnaces

The design criteria for electric furnaces are identical to those for natural gas furnaces. The selection criteria are similar, except that an electric furnace does not have the flue losses and combustion air losses of a gas furnace. For this reason, the seasonal efficiency is usually equal to the 100% steady-state efficiency less jacket losses.

The design life of electric furnaces is related to the durability of the contactors and the heating elements. The typical design life is approximately 15 years.

Safety primarily considers proper wiring techniques. Wiring should comply with the *National Electrical Code* (NEC) (ANSI/NFPA 70) and local codes, if they apply.

## TECHNICAL DATA

Detailed technical data on furnaces are available from manufacturers, wholesalers, and dealers. The data is generally tabulated in product specification bulletins printed by the manufacturer for each furnace line. These bulletins normally include performance information, electrical data, blower and air delivery data, control system information, optional equipment information, and dimensional data.

### Natural Gas Furnaces

**Capacity Ratings.** ANSI *Standard* Z21.47 requires the heating capacity to be marked on the rating plate of commercial furnaces. Heating capacity of residential furnaces, less than 225,000 Btu/h (66 kW) input, is shown on the furnace *Fact Sheet* required by the FTC. Capacity is calculated by multiplying the input by the steady-state efficiency.

Residential gas furnaces are readily available with heating capacities ranging from 35,000 to 175,000 Btu/h (10 to 51 kW). Some smaller furnaces are manufactured for special-purpose installations such as mobile homes. Smaller capacity furnaces should be more common in the future because new homes are better insulated than older homes and have a lower heat load. Larger furnaces are also available, but these are normally considered for commercial use.

Because of the overwhelming popularity of the upflow furnace, it is available in the greatest number of models and sizes. The downflow, horizontal, and combination furnaces are all readily available but are generally limited in models and sizes.

Residential gas furnaces are available as heating-only and as heat-cool models. The difference is that the heat-cool model is designed to operate as the air-handling section of a split-system air conditioner. The heating-only models typically operate with enough airflow to allow a 60 to 100°F (33 to 56°C) air temperature rise through the furnace. This rise provides good comfort conditions for the heating system with a low noise level blower and low electrical energy consumption. Condensing furnaces may be designed for a lower temperature rise—as low as 45°F (25°C).

The heat-cool model furnaces have multispeed blowers with a more powerful motor capable of delivering about 400 cfm (190 L/s) per ton of air-conditioning. Models are generally available in the 2, 3, 4, and 5 ton (7, 10.6, 14, and 17.6 kW) sizes, but all cooling sizes are not available for every furnace size. For example, a 60,000 Btu/h (18 kW) furnace would be available in models with blowers capable of handling 2 or 3 tons (7 or 10.6 kW) of air-conditioning; 120,000 Btu/h (35 kW) models would be matched to 4 or 5 tons (14 or 17.6 kW) of air-conditioning. In addition to the blower differences, controls of the heat-cool furnace models are generally designed to operate the multispeed blower motor on low speed during heating operation and on a higher speed for cooling operation. This feature provides optimum comfort conditions for year-round system operation.

**Efficiency Ratings.** Currently, gas furnaces have steady-state efficiencies that vary from 75 to 95%. Atmospheric and power vent furnaces typically range from 75 to 80% efficiency, while condensing furnaces have over 90% steady-state efficiency. Koenig (1978), Gable and Koenig (1977), Hise and Holman (1977), and Bonne, *et al.* (1977) found that oversizing of residential gas furnaces with standing pilots reduced the seasonal efficiency of heating systems in new installations with vents and ducts sized according to furnace capacity.

The AFUE of a furnace may be improved by ways other than changing the steady-state efficiency. These improvements generally add more components to the furnace. One method replaces the standing pilot with an intermittent ignition device. Gable and Koenig (1977) and Bonne *et al.* (1976) indicate that this feature can save as much as $5.6 \times 10^6$ Btu/yr (5.9 GJ/yr) per furnace. For this reason, some jurisdictions require the use of intermittent ignition devices.

Another method of improving AFUE is to take all combustion air from outside the heated space (direct-vent) and preheat it. A combustion air preheater incorporated into the vent system draws combustion air through an outer pipe that surrounds the flue pipe. Such systems have been used for many years on mobile home furnaces and outdoor furnaces. Annual energy consumption of a direct-vent furnace with combustion air preheat may be as much as 9% less than a standard furnace of the same design (Bonne *et al.* 1976). Direct-vent without combustion air preheat is not inherently more efficient, because the reduction in combustion-induced infiltration is offset by the use of colder combustion air.

An automatic vent damper (thermal or electro-mechanical) is another device that saves energy on indoor furnaces. This device, which is placed after the draft hood outlet, closes the vent when the furnace is not in operation. It saves energy during the off cycle of the furnace by (1) reducing exfiltration from the house and (2) trapping residual heat from the heat exchanger within the house rather than allowing it to flow up the chimney. These savings approach 11% under ideal conditions where combustion air is taken from the heated space that is under thermostat control. However, these savings are much less (estimates vary from 0 to 4%) if combustion air is taken from outside the heated space.

The AFUE of power-draft and power vent furnaces is higher than for standard natural-draft furnaces. These furnaces normally have such a high internal flow resistance that combustion airflow stops when the combusiton blower is off. This characteristic results in greater energy savings than those from a vent damper. Computer studies by Gable and Koenig (1977), Bonne *et al.* (1976), and Chi (1977) have estimated annual energy savings up to 16% for power-vent furnaces with electric ignition as compared to natural draft furnaces with standing pilot.

**Controls.** Externally, the furnace is controlled by a low-voltage room thermostat. Controls can be simple heating only, combination heating-cooling, multistage, or night setback. Chapter 51 of the 1987 HVAC Volume addresses thermostats in more detail. A night setback thermostat can reduce the annual energy consumption of a gas furnace. Dual setback (setting the temperature back during the night and during unoccupied periods in the day) can save even more energy. Gable and Koenig (1977) and Nelson and MacArthur (1978) estimate that energy savings of up to 30% are possible, depending on the degree and length of setback and the geographical location. The percentage of energy saving is greater in mild-climate regions; however, the total energy saving is greatest in cold regions.

Several types of gas valves provide various operating functions within the furnace. The type of valve available relates closely to the type of ignition device used. Two-stage valves, available on some furnaces, operate at full gas input or reduced rate and must be controlled with a two-stage thermostat. Their primary advantage is to provide less heat and, therefore, less temperature variation and greater comfort during mild weather conditions when full heat output is not required. Fuel savings with two-stage firing rate systems may not be realized unless both the gas and combustion air are controlled.

The fan switch controls the circulating air blower. This switch may be a temperature-sensitive switch exposed to the circulating airstream within the furnace cabinet, or it may be an electrically operated relay. Blower startup is typically delayed about one minute after the startup of the burners. This delay gives the heat exchangers time to warm up and eliminates the excessive flow of cold air when the blower comes on. Blower shutdown is also delayed several minutes after burner shutdown to remove residual heat from the heat exchangers and improve the annual efficiency of the furnace. Continuous blower operation throughout the heating season was encouraged in the past to improve air circulation and provide even temperature distribution throughout the house. However, this practice is emphasized less now because the total energy use and operating cost is greater with continuous blower operation. Continuous blower operation may be needed for good particulate control when an electronic air cleaner is used.

The limit switch prevents overheating in the event of severe reduction to circulating airflow. This temperature-sensitive switch is exposed to the circulating airstream and shuts off the gas valve if the temperature of the air leaving the furnace is excessive. The fan and limit switches are often incorporated in the same housing and are sometimes operated by the same thermostatic element.

## LPG Furnaces

Most residential natural gas furnaces are also available in a liquefied petroleum gas version with identical ratings. The technical data for these two furnaces are identical, except for the gas controls and the burner and pilot orifice sizes. Orifice sizes on LPG furnaces are much smaller because LPG has a higher density and may be supplied at a higher manifold pressure. The heating value and specific gravity of typical gases are listed in the following table:

| Gas | Heating Value | | Specific Gravity |
|---|---|---|---|
| Type | Btu/ft$^3$ | MJ/m$^3$ | Air = 1.0 |
| Natural | 1030 | 38.4 | 0.60 |
| Propane | 2500 | 93.1 | 1.53 |
| Butane | 3175 | 118.3 | 2.00 |

Gas controls may be different because of the higher manifold pressure and the requirement on indoor furnaces for pilot gas shutoff if pilot ignition should fail. Pilot gas leakage is more critical with propane or butane gas because both are heavier than air and can accumulate to create an explosive mixture within the furnace or furnace enclosure.

Since 1978, ANSI *Standard* Z21.47 requires a gas pressure regulator as part of the LPG furnace. Prior to that, the pressure regulator was supplied only with the LP supply system.

Beside natural and LPG, a furnace may be certified for manufactured gas, mixed gas, or LPG-air mixtures; however, furnaces with these certifications are not commonly available. Mobile home furnaces are certified as convertible from natural gas to LPG, and this type of furnace is readily available.

### Oil Furnaces

Oil furnaces are similar to gas furnaces in size, shape, and function, but the heat exhanger, burner, and combustion control systems are significantly different. Technical data are similar in some respects and different in others.

Input ratings are based on the oil flow rate (gal/h or L/s), and the heating capacity is calculated by the same method as for gas furnaces. The heating value of oil is 140,000 Btu/gal (39 MJ/L). There are not as many models and sizes available for oil as there are for gas, but residential furnaces in the 64,000 to 150,000 Btu/h (19 to 44 kW) heating capacity are common. Air delivery ratings are similar to gas furnaces, and both heating-only and heat-cool models are available.

The efficiency of an oil furnace can drop during normal operation if the burner is not maintained and kept clean. In this case, the oil does not atomize sufficiently to allow complete combustion, and energy is lost up the chimney in the form of unburned hydrocarbons. Because most oil furnaces use power burners and electric ignition systems, the annual efficiency is relatively high.

Oil furnaces are available in upflow, downflow, and horizontal-flow models. The thermostat, fan switch, and limit switch are similar to those of a gas furnace. Oil flow is controlled by a pump and burner nozzle, which sprays the oil-air mixture into a single chamber drum-type heat exchanger. The heat exchangers are normally heavy gauge cold-rolled steel. Humidifiers, electronic air cleaners, and night setback thermostats are available as accessories.

### Electric Furnaces

Residential electric resistance furnaces are available in heating capacities of 5 to 35 kW. Air-handling capabilities are selected to provide sufficient air to meet the requirement of an air conditioner of reasonable size to match the furnace. Small furnaces supply about 800 cfm (380 L/s) and large furnaces about 2000 cfm (940 L/s).

The only loss associated with an electric resistance furnace is in the cabinet—about 2% of input. Both the steady-state efficiency and the annual efficiency of an electric furnace is greater than 98%, and if the furnace is located within the heated space, the seasonal efficiency is 100%.

Although the efficiency of an electric furnace is high, electricity is a relatively high priced form of energy. The operating cost may be reduced substantially by using an electric heat pump in place of a straight electric resistance furnace. Heat pumps are discussed in Chapter 9 of the 1987 HVAC Volume.

Humidifiers and electronic air cleaners are available as accessories for both electric resistance furnaces and heat pumps.

Conventional setback thermostats are recommended to save energy. The electric demand used to recover from the setback, however, may be quite significant. Bullock (1977) and Schade (1978) addressed this problem by using (1) a conventional two-stage setback thermostat with *staged* supplemental electric heat or (2) solid-state thermostats, with programmed logic to inhibit supplemental electric heat from operating during morning recovery. Benton (1983) reported energy savings of up to 30% for these controls, although the recovery time may be extended up to several hours.

Electric furnaces are available in upflow, downflow, or horizontal models. Internal controls include overload fuses or circuit breakers, overheat limit switches, a fan switch, and a contactor to bring on the heating elements at timed intervals.

## SYSTEM PERFORMANCE

Both the performance and the interaction of the furnance and the distribution system with the building determine how much fuel energy input to the furnace beneficially heats the conditioned space. *System performance* is defined by the space in which the performance applies. *Conditioned space* is defined as that space whose temperature is actively controlled by a thermostat. A building can contain other space, like an attic, basement, or crawl space, which can influence the thermal performance of the conditioned space, but it is not defined as part of the conditioned space.

For houses with basements, it is important to decide whether the basement is part of the conditioned space, since it typically receives some fraction of the HVAC system output. The basement *is* part of the conditioned space only if it is under active thermostat control and warm air registers are provided to maintain comfort; otherwise it is not part of the conditioned space. The following system performance examples show designs for improving system efficiency, along with their effect on temperature in the unconditioned basement.

**The SP43 Dynamic Simulation Model.** The dynamic response and interactions between components of central forced-warm-air systems are sufficiently complex that the effects of system options on annual fuel use are not easily evaluated. ASHRAE Special Project 43 (SP43) was initiated in 1982 to assess the effects of system component and control mode options. The resulting simulation model accounts for the dynamic and thermal interactions of equipment and loads in response to varying weather patterns.

Fisher *et al.* (1984) describe the SP43 simulation model. Jakob, Fisher, *et al.* (1986), Herold, Flanigan *et al.* (1987), and Jakob *et al.* (1987) describe the validation model for the heating mode through field experiments in two houses. Herold *et al.* (1986) summarize both the project and the model.

Jakob, Locklin *et al.* (1986), and Locklin *et al.* (1987) presented the model's predictions of the effect of system and control variables on the overall performance of the forced warm-air heating system. These variables include furnace and venting types, furnace installation location and combustion air source, furnace sizing, night setback, thermostat cycling rate, blower operating strategy, basement insulation levels, duct sealing and insulation levels, house and foundation type, and climate.

## SYSTEM PERFORMANCE FACTORS

A series of *System Performance Factors*, comprising both efficiency factors and dimensionless energy factors, describe dynamic performance of the individual components and the overall system over any period of interest. Jakob, Locklin *et al.* (1986) and Locklin, Herold, *et al.* (1987) describe the factors in detail.

Table 2 identifies the performance factors and their mathematical definitions in four main categories: (1) equipment-component efficiency factors, (2) equipment-system performance factors, (3) equipment-load interaction factors, and (4) energy cost factors. Key aspects of the factors are summarized in the following sections.

**Table 2 Definitions of System Performance Factors—Annual Basis[a]**

**Equipment-Component Efficiency Factors**

$$E_F = \text{FE} = \text{Furnace Efficiency} = 100\,\frac{\text{Furnace Output}}{\text{Total Energy Input}} = 100\,\frac{\text{Duct Input}}{\text{Total Energy Input}}$$

$$E_D = \text{DE} = \text{Duct Efficiency} = \text{Duct Output} / \text{Duct Input}$$

**Equipment-System Performance Factors**

$$E_{HD} = \text{HDE} = \text{Heat Delivery Efficiency} = \frac{\text{FE} \times \text{DE}}{100} = 100\,\frac{\text{Duct Output}}{\text{Total Energy Input}}$$

$$F_{MG} = \text{MGF} = \text{Miscellaneous Gain Factor} = \frac{\text{Total Heat Delivered}^b}{\text{Duct Output}}$$

$$E_S = \text{SE} = \text{System Efficiency} = \text{HDE} \times \text{MGF} = 100\,\frac{\text{Total Heat Delivered}}{\text{Total Energy Input}}$$

**Equipment/Load-Interaction Factors**

$$F_{IL} = \text{ILF} = \text{Induced Load Factor}^c = \frac{\text{System Induced Load}^d}{\text{Total Heat Delivered}}$$

$$F_{LM} = \text{LMF} = \text{Load Modification Factor} = 1.0 - \text{ILF} = \frac{\text{Total Heat Delivered} - \text{System Induced Load}}{\text{Total Heat Delivered}}$$

$$I_S = \text{SI} = \text{System Index} = \text{SE} \times \text{LMF} = \frac{\text{Total Heat Delivered} - \text{System Induced Load}}{\text{Total Energy Input}}$$

**Energy Cost Factors**

$$R_{AE} = \text{AER} = \text{Auxiliary Energy Ratio} = \frac{\text{Auxiliary Energy Input}}{\text{Primary Energy Input}} = \frac{\text{Electrical Energy Input}}{\text{Fuel Energy Input}}$$

$$R_{CL} = \text{LCR} = \text{Local Energy Cost Ratio} = \frac{\text{Electrical Cost Per Energy Unit}}{\text{Reference Fuel Cost Per Energy Unit}} \text{ (in common units)}$$

$$F_{CR} = CRF = \text{Cost Ratio Factor} = \frac{\text{Primary Energy Input} + \text{Auxiliary Energy Input}}{\text{Primary Energy Input} + \text{LCR (Auxiliary Energy Input)}} = \frac{1.0 + \text{AER}}{1 + \text{LCR} \times \text{AER}}$$

Special Case (Electric Furnace): $F_{CR} = 1/L_{CR}$

$$I_{SCM} = \text{CMSI} = \text{Cost Modified System Index} = \text{SI} \times \text{CRF} = \frac{\text{Net Load} - \text{System Induced Load}}{\text{Primary Energy Input} + \text{LCR (Auxiliary Energy Input)}}$$

**Annual Energy Use**

AEU = Annual Energy Use (fuel and electricity) predicted by SP43 Model, in total energy units.

Annual Fuel Used = AEU / (1.0 + AER)

Annual Electricity Used = AEU / (1.0 + 1/AER)

**Percent Savings**

% Energy Savings = $100\,(I_S - (I_S)_{BC}) / I_S$ *where* $(I_S)_{BC}$ = the System Index for the base case

% Cost Savings = $100\,(I_{SCM} - (I_{SCM})_{BC}) / I_{SCM}$ *where* $(I_{SCM})_{BC}$ = the cost modified System Index for both cases

**Other Factors for Dynamic Performance**

AFUE = Annual Fuel Utilization Efficiency by DOE/ASHRAE efficiency rating, applicable to specific furnace. Values in this chapter are for generic furnaces.

SSE = Steady State Efficiency value for a given furnace by ANSI test procedure.

[a]Energy inputs and outputs are integrated over an annual period. Efficiencies (*E*) are expressed as percents. Indicies (*I*), factors (*F*), and ratios (*R*) are expressed as fractions.

[b]The *Total Heat Delivered* is the integration over time of all the energy supplied to the conditioned space by the HVAC equipment. By definition, it is exactly equal to the space-heating load.

[c]The *Induced Load Factor* may be positive or negative, depending on the value of the load relative to the selected base case.

[d]The *System Induced Load* is the difference between the space-heating load for a particular case and the space-heating load for the base case. For the base case, the System Induced Load is, by definition, zero.

### Equipment-Component Efficiency Factors

**Furnace Efficiency,** $E_F$, is the ratio of the energy that is delivered to the plenum during cyclic operation of the furnace to the total input energy on the annual basis. $E_F$ includes summertime pilot losses and blower energy.

This factor is similar to the Annual Fuel Utilization Efficiency (AFUE) in ASHRAE and DOE standards (ASHRAE 1982, DOE 1978), which provides an estimate of annual energy, taking into account assumed system dynamics. However, $E_F$ differs from AFUE in the following aspects.

1. The AFUE for a given furnace is defined by a single predetermined cyclic condition with standard dynamics; $E_F$ is based on the integrated cyclic performance over the year.
2. The AFUE does not include auxiliary electric input and it gives credit for jacket losses, except when the furnace is an outdoor unit; $E_F$ and the other efficiency factors defined here include auxiliary electric input.
3. Effects of system-induced infiltration are handled differently in the two concepts; they are covered in the factors for equipment/load interaction in the SP43 model, whereas AFUE includes an adjustment factor to account for these effects.
4. The AFUE rating is a fixed value for a given furnace model and is independent of the system or dynamics over a full heating season.

**Duct Efficiency,** $E_D$, is the ratio of the energy intentionally delivered to the conditioned space through the supply registers to the energy delivered to the furnace plenum, on an annual basis.

### Equipment-System Performance Factors

**Heat Delivery Efficiency,** $E_{HD}$, is the product of Furnace Efficiency, $F_E$ and the Duct Efficiency, $E_D$. Heat delivery efficiency is the ratio of the energy intentionally delivered to the conditioned space to the total input energy. It is a measure of how effectively the HVAC system delivers heat directly to the conditioned space on an annual basis.

**Miscellaneous Gain Factor,** $F_{MG}$, is the ratio of the total heat delivered to the conditioned space divided by the energy intentionally delivered to the conditioned space through the duct registers. This factor accounts for unintentional, but beneficial, heat that directly reaches the conditioned space from equipment components like the vent or ducts. Jacket heat loss is also included as a miscellaneous gain if the furnace equipment is located within the conditioned space. For example, under certain circumstances, the conditioned space may lose heat to the vent during the off-cycle. This loss would be included in $F_{MG}$. Because the net heat delivered to the conditioned space by the heating system may be less than the heat delivered to the conditioned space through the duct registers implies that $F_{MG}$ can be less than one.

**System Efficiency,** $E_S$, is the product of $E_{HD}$ and $F_{MG}$. System efficiency indicates how much of the HVAC system's energy input is delivered to the conditioned space. It is the ratio of the total energy delivered to the conditioned space divided by the total energy input to the furnace. Thus, it includes intentional and unintentional energy gains.

### Equipment/Load-Interaction Factors

**Load Modification Factor,** $F_{LM}$, is the ratio of the total heat delivered minus the System Induced Load to the Total Heat Delivered to the conditioned space for the case of interest. The System Induced Load is the difference between the Space-Heating Load for a particular case and the Space-Heat Load for a base case. Note that the Total Heat Delivered *equals* the Space-Heating Load. The Load Modification Factor may also be considered as the ratio of the total heat delivered for a base case to the total heat delivered for a case of interest. $F_{LM}$ adjusts the system efficiency, $E_S$, to account for equipment operation on the heating load. It accounts for the effect of the combustion-induced infiltration and off-period infiltration due to draft hood flow, as well as the effects of changes in temperature of unconditioned spaces adjacent to the conditioned space.

$F_{LM}$ credits the system efficiency for a heating load reduction, or debits the system efficiency if system operation increases the heating load with respect to the base case. $F_{LM}$ is greater than 1.0 (a credit) if the specific conditioned-space load is less than the base case load, as for a system with a vent damper that reduces off-cycle infiltration. $F_{LM}$ is less than 1.0 (a debit) if the conditioned-space load is greater than the base case load, such as occurs when additional infiltration is induced for combustion air during burner operation and for vent dilution during off periods.

**Miscellaneous heat gains** are losses from the heating systems that go *directly* to the conditioned space. For example, jacket losses become a miscellaneous gain if the equipment is located in the conditioned space.

**Load-equipment interactions** can increase or decrease the load. Examples include a vent damper, which can reduce off-cycle infiltration and, therefore, the unconditioned space load, or additional infiltration that is induced for combustion air during burner operation and for ventilation during off periods. Heating system losses to unconditioned spaces like the basement, which subsequently affect that space either by positive heat flow or by reducing heat loss, are also considered load-equipment interactions.

**Duct losses.** The magnitude of duct air leakage and convective heat losses to (1) the conditioned space, (2) to spaces inside the structure but outside the conditioned space, like the basement or attic, or (3) to the outside has a substantial effect on system performance.

**System Index,** $I_S$, is the Total Heat Delivered minus the System Induced Load divided by the total energy input. It is the product of the system efficiency, $E_S$, and the Load Modification Factor, $F_{LM}$. It is an energy-based *figure of merit* that adjusts the System Efficiency for any credits or debits due to system-induced loads relative to a base case load. The System Index is a powerful tool for comparing alternative systems. However, high values of $I_S$ are sometimes associated with lower basement temperatures—caused by designs that deliver a greater portion of the furnace output directly to the conditional space. The ratio of the System Indexes for two systems being compared is the inverse of the ratio of their Annual Energy Use, $A_{EU}$.

### Energy Cost Factors

Table 2 also defines factors that can modify the System Index to account for relative costs of fuel and electrical energy and form a Cost Modified System Index, $I_{SCM}$. The Auxiliary Energy Ratio, $A_{ER}$ is the ratio of electrical energy input to fuel energy input. The Local Energy Cost Ratio, $R_{CL}$, is the ratio of electrical and fuel energy costs in common units of energy. (A value of $R_{CL} = 4$ is used in comparsions presented here.)

### Key Implications

The following important implications apply to the definitions for the various System Performance Factors.

1. The defined conditioned space is important to the comparisons of System Index. Because the Load Modification Factor, $F_{LM}$, and the System Index, $I_S$, are based upon the same reference equipment and house configuration, the performance of various furnaces installed in basements may be compared or furnaces installed within the conditioned space (*i.e.*, closet installations) may be compared. However, performance of furnaces installed in basements or crawl spaces cannot be com-

pared with heating systems installed only within the conditioned space.

Since the System Index depends upon a reference equipment and house configuration, it may only be used as a ranking index from which the relative benefits of different system features can be derived. That is, it can be used to compare the savings of various system features in specific applications to a base case.

2. The Miscellaneous Gain Factor, $F_{MG}$, includes only those heating system losses that go *directly* to the conditioned space.
3. The Equipment-System Performance Factors relate strictly to the subject equipment whereas the Equipment/Load-Interaction Factors draw comparisons between the subject equipment and an explicitly defined *base case*. This base case is a specific load and equipment configuration to which all alternative systems are compared.

Systems with the best total energy economy have the highest System Index. Systems with leaky and uninsulated ducts could have a higher System Efficiency, even though fuel use would be higher, if basement duct losses that become gains to the conditioned space are included in the Miscellaneous Gain Factor. The foregoing definitions prevent this possibility.

## System Performance Examples

This section summarizes two phases of ASHRAE Special Project 43, which evaluated the annual energy savings attained by certain forced warm-air-heating systems. The study was limited to certain house configurations and climates, and to gas-fired equipment. Electric heat pump and zoned baseboard systems were not studied. For this reason, these data should not be used to compare systems or select a heating fuel. Several of the factors addressed are not references to performance. Instead, they are *figures of merit*, which represent the effect various components have on a system. As such, these factors should not be applied outside the scope of these examples.

The following examples of overall system thermal performance illustrate how the furnace, vent, duct system, and building can interact. Table 5 summarizes SP43 simulation model predictions of the annual system performance for a *Base Case* (a conventional, natural-draft gas furnace with an Intermittent Ignition Device, and an *Example Case* (noncondensing power vent furnace). Each are installed in a typical three-bedroom, ranch-style house of frame construction, located in Pittsburgh. Table 3 shows the assumptions for the thermal envelope of the house, and Table 4 shows other assumptions for these predictions.

Table 7 shows the predicted annual system performance factors

**Table 3 Assumptions for Thermal Envelope of House in Pittsburgh**

| | U-Values of Thermal Envelope, Btu/h • ft² • °F | (W/m² • °C) |
|---|---|---|
| **Per HUD-MPS 1980** | | |
| Attic | 0.03 | (0.17) |
| Walls | 0.07 | (0.40) |
| Windows | 0.69 | (3.92) |
| Sliding Glass Doors | 0.69 | (3.92) |
| Storm Doors | None | None |
| Floors[a] | 0.07 | (0.40) |
| **Basement Conditions** | | |
| Basement Ceiling Insulation | None | None |
| Basement Wall Insulation | 0.13 | (0.74) |
| **Design Heat Loss[b]** | 47,000 Btu/h | (13.8 kW) |

[a] Recommended value of floor insulation over an unheated space.

[b] Calculated by method described in ASHRAE GRP158, *Cooling and Heating Load Calculation Manual*, for a 1400 ft² (130 m²) ranch-style house with basement. Outdoor design temperature = 5 °F (−15 °C).

**Table 4 Other Assumptions for Simulation Predictions**

| | Base Case |
|---|---|
| **Furnace, Adjustments, & Controls:** | |
| Furnace Size | 1.4DHL |
| Circulating Air Temp. Rise | 60 °F (33 °C) |
| Thermostat Set Point | 68 °F (20 °C) |
| Thermostat Cycling Rate at 50% On-Time | 6 cycles/hour |
| Night Setback, 8 hours | None |
| Blower Control - On | 80 second |
| - Off | 90 °F (32 °C) |
| **Duct-Related Factors:** | |
| Insulation | None |
| Leakage, relative to duct flow | 10% |
| Location | Basement |
| **Load-Related Factors:** | |
| Nominal Infiltration[a] | |
| - Conditioned Space | 0.75 ACH |
| - Basement | 0.25 ACH |
| Occupancy, Persons | 3 Evening & Night, 1 Daytime |
| Internal Loads | Typical Appliances, Day & Evening Only (20 kWh/Day) |
| Shading by Adjacent Trees or Houses | None |

[a] Model runs used variable infiltration, as driven by indoor-outdoor temperature differences, wind, and burner operation. Values shown above are nominal. (ACH = Air changes per hour.)

**Table 5 System Performance Examples**

| Performance Factor | Base Case: Typical Conventional, Natural-Draft Furnace With IID | Alternative Case: Typical Non-Condensing Power Vent Furnace |
|---|---|---|
| DOE/ASHRAE AFUE, % | 69 | 81.5 |
| Furanace Efficiency, $E_F$, % | 75.5 | 85.5 |
| Duct Efficiency, $E_D$, % | 60.9 | 59.3 |
| Heat Delivery Efficiency, $E_{HD}$, % | 46.0 | 50.7 |
| Miscellaneous Gain Factor, $F_{MG}$ | 1.004 | 0.983 |
| System Efficiency, $E_S$, % | 46.1 | 49.8 |
| Load Modification Factor, $F_{LM}$ | 1.000 (Base case) | 1.099 |
| System Index, $I_S$ | 0.461 | 0.548 |
| Annual Energy Use, AEU, $10^6$ Btu (GJ) | 73.0 (77.0) | 61.5 (64.9) |
| Auxiliary Energy Ratio, AER | 0.027 | 0.028 |
| Energy savings from base case, % | — | 15.8 |
| Cost savings from base case, % (with RCL = 4) | — | 15.6 |

The factors in this table should not be extrapolated to cases other than those covered in this chapter. That is, electric heat pump, radiant heating, or zoned baseboard systems were not studied in the SP43 project and should not be compared by this model.

The values presented here do not represent only this class of equipment—electric furnaces and heat pumps in a similar installation and under similar conditions would incur similar losses. The System Index for any central air system can be improved, in comparison to the above examples, by insulating the ducts or locating the ductwork inside the conditioned space or both.

for the different furnaces installed in the same house in the same climate.

**Base Case.** The annual Furnace Efficiency of the conventional natural draft furnace is predicted to be 75.5%. Air leakage and heat loss from the uninsulated duct system result in a Duct Efficiency of 60.9%. The Heat Delivery Efficiency, which is ratio of the energy intentionally delivered to the conditioned space through the supply registers to the total input energy, is 46.0%. The Miscellaneous Gain Factor, which is 1.004, accounts for the small heat gain to the conditioned space from the heated masonry chimney passing through the conditioned space. The System Efficiency, the ratio of the total heat delivered or the space heating load, to the total energy input, is 46.1%. Because this case is designated as the base case, the Load Modification Factor is 1.0. Thus, the System Index is 0.461.

The duct losses and jacket losses are accounted for in the energy balance on the basement air and in the energy flow between the basement and the conditioned space. The increase in infiltration caused by the need for combustion air and vent dilution air is also accounted for in the energy balances on the living space air and basement air. In the base case the temperature in the unconditioned basement is nearly the same (68 °F or 20 °C) as in the first floor where the thermostat is located. This condition is caused by heat loss of the exposed ducts in the basement and by the low outdoor infiltration into the basement that was achieved by sealing normal cracks associated with construction.

**Example Case.** The Furnace Efficiency for the noncondensing power vent furnace being compared is 85.5%. The Duct Efficiency is 59.3%, slightly lower than that for the duct system with the conventional natural draft furnace. Therefore, the Heat Delivery Efficiency is 50.7%, which reflects the higher furnace efficiency. For the noncondensing power vent furnace, the Miscellaneous Gain Factor is 0.983, reflecting the small heat loss from the conditioned space to the colder masonry chimney (due to reduced off cycle vent flow). The System Efficiency, which is the Heat Delivery Efficiency times the Miscellaneous Gain Factor, is 49.8%. For this furnace system, *compared to the base case system of the conventional, natural draft furnace*, the Load Modification Factor is 1.099. Therefore, the space heating load for the house with the noncondensing power vent furnace is (1/1.099) or 91% of the space heating load for the house with the conventional, natural draft furnace. This reduction in heating load is mainly due to the reduction in off-cycle vent flow.

For the noncondensing power vent furnace system, the System Index is 0.548. Note that the ratio of System Indexes for these two cases (0.548/0.461 = 1.189) is the inverse of the ratio of their Annual Energy Use ($61.5 \times 10^6 / 73 \times 10^6 = 0.842$).

Table 6 lists the energy flows shown on Figure 6 for the two furnace systems. Table 5 also contains the combinations of these energy flows. Table 2 refines these energy flow combinations. Table 7 shows how the values of the Performance Factors listed in Table 5 are calculated from the energy flows.

## Effect of Furnace Types

Table 8 summarizes the energy impacts of several furnaces. Note that the system indexes, $I_S$, for both thermal and electric vent dampers are similar, although thermal vent dampers are slower reacting and less effective at blocking the vent. Also, the ratio of furnace AFUE to the base-case AFUE closely matches the ratio of $I_S$ to $(I_S)_{BC}$ for the corresponding furnaces. The exceptions are the vent damper cases, where the improvement in $I_S$ suggest a smaller AFUE credit. In general, the study found that the furnace AFUE is a good indication of relative annual performance of furnaces in typical systems.

The results reported in Table 8 are for homes that do not include the basement in the conditioned space. That is, energy lost to the basement contributes only indirectly to the useful heating effect. If the basement is assumed to be a part of the conditioned space, the miscellaneous gain factor and subsequent calculated efficiencies and indexes should be multipled by 1.66 to account for the beneficial effects of equipment (furnace jacket and duct system) heat losses that contribute to heating the basement. The 1.66 multiplier is the reciprocal of the electric furnace System Efficiency as shown in the following table, assuming that all electric heat is recovered for useful heating of the raw expanded conditioned space.

The following table shows values of System Efficiency and System Index that could be inferred by including the basement as a conditioned space.

| | Conditioned Space Does Not Include Basement | | Conditioned Space Includes Basement | |
|---|---|---|---|---|
| | $E_S$ | $I_S$ | $E_S$ | $I_S$ |
| Base Case | 46.0 | 0.461 | 76.4 | 0.765 |
| Direct Vent Gas | 59.2 | 0.622 | 98.3 | 1.033 |
| Electric Furnace | 60.3 | 0.651 | 100.0 | 1.081 |

The reported values in Table 8 adhere to the definitions presented earlier. Other equipment types must be analyzed on the same common basis to draw similar conclusions. For instance an electric baseboard heating system would have performance factors that are essentially the same as those for the electric furnace located in the Conditioned Space.

## Effect of Climate and Night Setback

Table 9 covers the effects of climate (insulation levels change by location) and night setback on system performance for two furnaces. The improvements in System Index with higher percent on-time (colder climates) follow improvements in duct efficiency. Furnace Efficiency appears to be relatively uniform in houses representative of typical construction practices in each city and where the furnace is sized at 1.4 times the design heat loss. Also, percent savings due to night setback increases in magnitude with warmer climates. The percent energy saved in the three climates varies with the magnitude of energy use (from 10% to 16% for the natural draft cases).

## Effect of Furnace Sizing

Furnace sizing affects the System Index depending on how the vent duct system is designed. As Table 10a indicates, where the ducts and vent are sized according to the furnace size (referred to as the *new* case), performance drops about 10%, as the furnace capacity is varied between 1.0 times DHL and 2.5 times DHL for a given application. In the *retrofit* case, where the vent and duct system are sized at a furnace capacity of 1.4 times the design heat loss, the System Index changes little with increased furnace capacity, indicating little energy savings. In a *new* case, where the duct system is designed for cooling and the vent size does not change between furnace capacities, the SP43 study indicates there is essentially no effect on System Index.

Table 10b shows similar results in condensing furnaces for the *new* case of ducts and vent sized according to the furnace capacity. In this case the decrease in System Index is smaller, about 2% over the range of 1.15 to 2.5 times DHL.

Finally, both Table 10a and 10b show that duct efficiency increases with furnace capacity, because higher capacity furnaces are on less of the time.

**Table 6 Energy Flow Descriptions and Values**

| Energy Flow Name | Description | Energy Flow, 1000 Btu (kJ = 1.055 × Btu) | |
|---|---|---|---|
| | | Conventional, Natural-Draft Furnace (Base Case) | Noncondensing Power-Vent Furnace (Example Alternative Case) |
| QWALLS | Sum of convective heat flows from conditioned space to the exterior walls | 8,333 | 8,396 |
| QCEILC | Convective heat flow from the conditioned space to the ceiling | 4,587 | 4,688 |
| QFLSPB | Convective heat flow from conditioned space floor to conditioned space air | 3,833 | 3,366 |
| QROOM | Convective heat flow from shaded section of partition to conditioned space | 1,239 | 1,164 |
| QPART | Convective heat flow from unshaded section of partition to conditioned space | 3,726 | 3,651 |
| QWINST | Sum of solar energy transmitted through windows and absorbed by surroundings and heat conducted through window glazing | 12,470 | 12,529 |
| QINF | Infiltration heat flow from the conditioned space to the environment | 25,196 | 21,356 |
| QINC | Convective portion of internal heat gain | 8,092 | 8,092 |
| QSTACS | Direct heat flow from furnace vent to the conditioned space | 118 | −530 |
| QNET | Net heat delivered to conditioned space through duct registers (supply-return) | 33,588 | 31,209 |
| QRAD | Radiative portion of internal heat gain | 8,092 | 8,092 |
| QBWAG | Convective heat flow from the basement air to the above-ground section of the basement wall | 5,022 | 4,991 |
| QBWBG | Convective heat flow from the basement air to the below-ground section of the basement wall | 5,597 | 5,535 |
| GBASF | Convective heat flow from basement air to basement floor | 8 | 60 |
| QFLBAS | Convective heat flow from conditioned space floor to basement air | 8,206 | 8,756 |
| QBASVN | Infiltration heat flow from basement to environment | 6,561 | 6,512 |
| QDUCTB | Total direct heat loss from duct system to basement (leakage + convection) | 20,790 | 20,605 |
| QSTACB | Direct heat flow from furnace vent to basement | 2,627 | 1,300 |
| QPLEN | Convective heat flow from plenum surface to basement air | 781 | 812 |
| QADINFE | Induced infiltration heat flow for combustion makeup air from basement to environment | 498 | 39 |
| QJACK | Convective heat flow from the furnace jacket to the basement air | 393 | 410 |
| QFSLOS | Sensible heat loss to furnace vent | 11,483 | 3,469 |
| QFLLOS | Latent heat loss to furnace vent | 7,045 | 5,926 |
| QEXT | Primary energy input to furnace | 71,133 | 59,841 |
| EFAN | Fan electrical consumption | 1,912 | 1,655 |
| QSTACA | Direct heat loss from furnace vent to attic air | 807 | 250 |
| QCEILA | Convective heat flow from the attic floor (conditioned space ceiling) to the attic air | 2,120 | 2,323 |
| QROOF | Convective heat flow from roof to attic air | 122 | 111 |
| QAVENT | Infiltration heat flow from attic to environment | 2,796 | 2,676 |
| QP | Pilot energy consumption | 0 | 0 |
| QDRAFT | Heat Lost by spillage from draft hood into basement | 0 | 0 |
| FO1 | Furnace Output = Enthalphy change across furnace | 55,182 | 52,605 |
| FO2 | Furnace Output = Total energy input − losses<br>FO2 = QEXT + QP + EFAN − QFSLOS − QFLLOS | 54,517 | 52,101 |
| FO3 | Furnace Output = Net heat delivered through registers + Duct loss + Plenum heat loss<br>FO3 = QNET + QDUCTB + QPLEN | 55,159 | 52,625 |
| PEI | Primary Energy Input = QEXT + QP | 71,133 | 59,841 |
| AEI | Auxiliary Energy Input = EFAN | 1,912 | 1,655 |
| TEI | Total Energy Input = PEI + AEI | 73,045 | 61,496 |
| AEU | Annual Energy Use = TEI | 73,045 | 61,496 |
| DI | Duct Input = Furnace Output = FO1 | 55,182 | 52,605 |
| DO | Duct Output = QNET | 33,588 | 31,209 |
| THD | Total Heat Delivered = QNET + QSTACS | 33,706 | 30,679 |
| RNL | Reference Net Load = Total Heat Delivered for the base case | 33,706 | 33,706 |
| SIL | System Induced Load = THD − RNL | 0 | −3,028 |

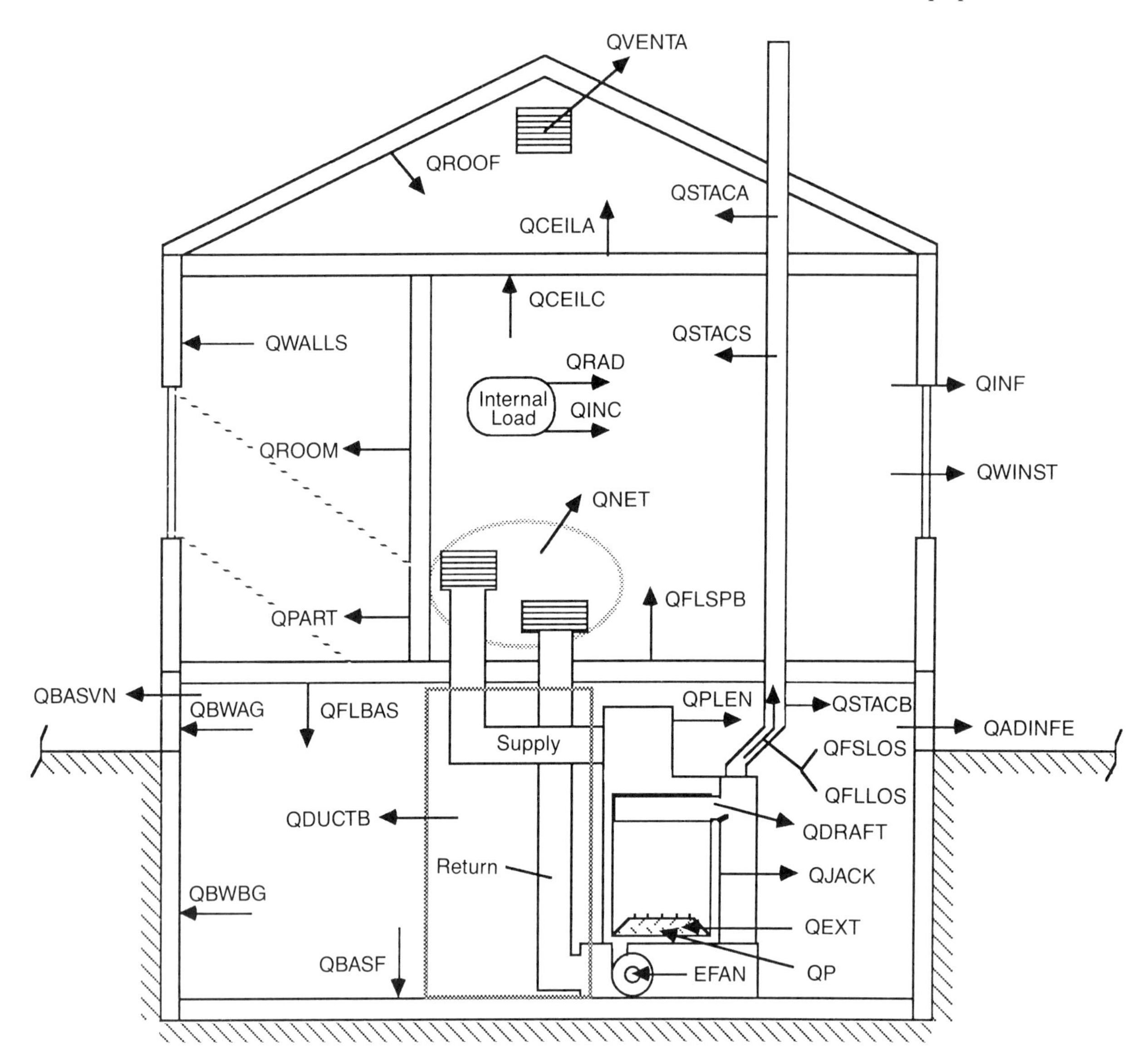


**Fig. 6 Summary of Energy Flow-Paths in SP43 Model**

## Effects of Furnace Sizing and Night Setback

Tables 10a and 10b show the relationship between furnace sizing and night setback for the retrofit case. The energy savings due to night setback, 8 hours per day at 10°F (6°C), is nearly constant at 15% and independent of furnace size. Table 9 covers the effect of climate variations on energy savings due to night setback.

**Duct Treatment.** Table 11 shows the effect of duct treatment on

**Table 7 Performance Factor Values**

| Factor | Description | Conventional, Natural-Draft Furnace (Base Case) | Noncondensing Power-Vent Furnace (Example Alternative Case) |
|---|---|---|---|
| FE | Furnace Efficiency = FOI/TEI = DI/TEI | 75.5% | 85.5% |
| DE | Duct Efficiency = DO/DI | 60.9% | 59.3% |
| HDE | Heat Delivery Efficiency = DO/TEI | 46.0% | 50.7% |
| MGF | Miscellaneous Gain Factor = THD/DO | 1.004 | 0.983 |
| SE | System Efficiency = THD/TEI | 46.1% | 49.9% |
| LMF | Load Modification Factor = (THD − SIL)/THD | 1.0 | 1.099 |
| SI | System Index = (THD − SIL)/TEI | 0.461 | 0.548 |
| AER | Auxiliary Energy Ratio = AEI/PEI | 0.0269 | 0.0277 |

Table 8 Effect of Furnace Type on Annual Heating Performance[a]

| | Furnace Characterization Typical Values[c] AFUE/SS | Predicted by SP43 Model | | | | | | | | | |
|---|---|---|---|---|---|---|---|---|---|---|---|
| | | Annual Performance Factors | | | | | | | | Auxiliary Energy Ratio AER | Average Basement °F (°C) |
| | | $E_F$ | $E_D$ | $F_{MG}$ | $F_{LM}$ | $I_S$ | $I_{SCM}$ ($R_{CL}=4$) | $I_{SCM}{}^d$ $(I_{SCM})_{BC}$ | AEU, $10^6$ Btu (GJ) | | |
| **Conventional, Natural-Draft** | | | | | | | | | | | |
| Pilot | 64.5/77 | 72.9 | 60.9 | 1.006 | 1.000 | 0.447 | 0.416 | 0.971 | 75.5 (79.7) | 0.026 | 67.9 (19.9) |
| IID (Base Case)[b] | 69/77 | 75.5 | 60.9 | 1.004 | 1.000 | 0.461 | 0.428 | 1.000 | 73.0 (77.0) | 0.027 | 67.8 (19.9) |
| IID + Thermal Vent Damper | 78/77 | 75.4 | 61.0 | 1.002 | 1.086 | 0.501 | 0.464 | 1.085 | 67.3 (71.0) | 0.027 | 68.2 (20.1) |
| IID + Electric Vent Damper | 78/77 | 75.4 | 61.2 | 0.988 | 1.105 | 0.504 | 0.467 | 1.091 | 66.9 (70.6) | 0.027 | 68.3 (20.2) |
| **Power-Vent Types** | | | | | | | | | | | |
| Non-condensing | 81.5/82.5 | 85.5 | 59.3 | 0.983 | 1.099 | 0.548 | 0.507 | 1.185 | 61.5 (64.9) | 0.028 | 67.6 (19.8) |
| Condensing[e] | 92.5/93.1 | 95.5 | 62.0 | 1.000 | 1.050 | 0.622 | 0.566 | 1.322 | 54.2 (57.2) | 0.034 | 66.9 (19.4) |
| Electric Furnace | na | 99.5 | 60.6 | 1.000 | 1.079 | 0.651 | 0.163 | 0.380 | 51.8 (54.6) | 0.020[f] | 67.1 (19.5) |

[a] The values in this table are *figures of merit* to be considered within the confines of the SP43 project, and they should not be applied outside the scope of these examples.
[b] Ranch-type house with basement in Pittsburgh climate and base conditions of 60 °F (33 °C) circulating air temperature rise, 6 cycles/hour, no setback, 10% duct air leakage.
[c] AFUE = Annual Fuel Utilization Efficiency by ASHRAE/DOE standard
SS = Steady-state efficiency by ANSI standard
[d] $R_{CL}$ = 4.0
[e] Direct vent uses outdoor air for combustion (includes preheat).
[f] Blower energy is treated as auxiliary energy.

furnace performance. Duct treatment includes sealants to reduce leaks and interior or exterior insulation to reduce heat loss due to conduction. Sealing and insulation improves the system performances as indicated by the System Index. For cases with no duct insulation, reducing duct leakage from 10% to zero increases the System Index 2.6%. Insulating the ducts also improves system performance. R5 (0.88 °C • m²/W) insulation on the exterior of the ducts increases $I_S$ by 4.4%, and R5 insulation on the interior of the duct increases $I_S$ by 8.5%.

**Basement Configuration.** Table 12 covers the effect of basement configuration and duct treatment on system performance. Insulating and sealing the ducts reduces the basement temperature. More heat is then required in the conditioned space to make up for losses to the colder basement. Where ducts pass through the attic or ventilated crawl space, insulation and sealing improves duct performance, although the total system performance is poorer. Installing the ducts within the conditioned space, on the other hand, significantly improves the Miscellaneous Gain Factor, $F_{MG}$, in which the duct losses are added directly to the conditioned space. In this case the System Index, $I_S$, would also improve.

Table 9 Effect of Climates and Night Setback on Annual Heating Performance[a,b]

| Furnace Type and Location | Setback, °F (°C) | Avg. % On-Time | Avg. Bsmt. °F (°C) | Furnace Eff., $E_F$, % | Duct Eff., $E_D$, % | System Index, $I_S$ | AEU[c] $10^6$ Btu (GJ) | % Energy Saved By Setback |
|---|---|---|---|---|---|---|---|---|
| **Conventional Natural-Draft** | | | | | | | | |
| Nashville | 0 | 12.7 | 64.6 (18.1) | 73.9 | 56.2 | 0.417 | 55.6 (58.7) | |
| | 10 (5.6) | 10.7 | 63.1 (17.3) | 74.8 | 58.9 | 0.497 | 46.7 (49.3) | 16.0 |
| Pittsburgh (Base City) | 0 | 18.0 | 67.8 (19.9) | 75.5 | 60.9 | 0.461[d] | 73.0 (77.0) | |
| | 10 (5.6) | 15.6 | 65.7 (18.7) | 75.9 | 62.4 | 0.532 | 63.4 (66.9) | 13.2 |
| Minneapolis | 0 | 20.9 | 68.0 (20.0) | 76.8 | 63.2 | 0.487 | 99.1 (104.6) | |
| | 10 (5.6) | 18.7 | 65.7 (18.7) | 77.0 | 62.4 | 0.546 | 88.4 (93.3) | 10.7 |
| **Direct, Condensing** | | | | | | | | |
| Nashville | 0 | 11.1 | 63.9 (17.7) | 94.0 | 57.0 | 0.564 | 41.0 (43.3) | |
| | 10 (5.6) | 9.5 | 62.6 (17.0) | 93.7 | 59.5 | 0.662 | 35.0 (36.9) | 14.8 |
| Pittsburgh | 0 | 16.2 | 67.8 (19.9) | 93.3 | 61.7 | 0.612 | 55.1 (58.1) | |
| | 10 (5.6) | 14.2 | 65.0 (18.3) | 93.0 | 63.2 | 0.700 | 48.1 (50.7) | 12.6 |
| Minneapolis | 0 | 18.2 | 66.8 (19.3) | 95.1 | 64.5 | 0.654 | 73.7 (77.8) | |
| | 10 (5.6) | 16.4 | 64.7 (18.2) | 94.8 | 65.6 | 0.729 | 66.2 (69.8) | 10.2 |

[a] Ranch-type house, basement, and base conditions: 60 °F (33 °C) circulating air temperature rise, 6 cycles/hr, 10% duct air leakage. Thermal envelope typical of each city; for example, no basement insulation in Nashville.
[b] The values in this table are *figures of merit* to be considered within the confines of the SP43 project, and they should not be applied outside the scope of these examples.
[c] AEU = Annual Energy Use.
[d] For a corresponding electric furnace, $I_S$ = 0.651.

**Table 10a Effect of Sizing, Setback, and Design Parameters on Annual Heating Performance—Conventional Natural-Draft Furnace[a,b]**

| Furnace Multiplier[c] | Duct Design | Setback °F (°C) | Annual Performance Factors $E_F$ | $E_D$ | $F_{MG}$ | $F_{LM}$ | $I_S/(I_S)_{BC}$ | Temperature Swing, °F (°C) | Avg. Room °F (°C) | Recovery Time, h[d] | Avg. Bsmt. °F (°C) |
|---|---|---|---|---|---|---|---|---|---|---|---|
| 1.00 | New[e] | 0 | 76.1 | 59.2 | 1.016 | 1.104 | 1.095 | 3.4 | 67.7 | n.a.[f] | 67.9 (19.9) |
| 1.15 | New | 0 | 75.6 | 59.0 | 1.009 | 1.059 | 1.032 | 4.0 (2.2) | 67.8 (19.9) | n.a. | 67.9 (19.9) |
| | Retrofit | 0 | 75.5 | 59.0 | 1.002 | 1.032 | 0.998 | 4.0 (2.2) | 67.8 (19.9) | n.a. | 67.9 (19.9) |
| | Retrofit | 10 (5.6) | 76.1 | 60.7 | 0.999 | 1.155 | 1.155 | 4.1 (2.3) | 65.8 (18.8) | 2.02 | 65.7 (18.7) |
| 1.40 | New | 0 | 75.5 | 60.8 | 1.010 | 1.029 | 1.034 | 4.8 (2.7) | 68.1 (20.1) | n.a. | 67.9 (19.9) |
| | Retrofit | 0 | 75.5 | 60.9 | 1.004 | 1.000 | 1.000[g] | 4.9 (2.7) | 68.0 (20.0) | n.a. | 67.8 (19.9) |
| | Retrofit | 10 (5.6) | 75.9 | 62.4 | 1.001 | 1.120 | 1.152 | 5.2 (2.9) | 66.0 (18.9) | 1.02 | 65.7 (18.7) |
| 1.70 | New | 0 | 74.8 | 61.3 | 1.013 | 1.017 | 1.023 | 5.9 (3.3) | 68.2 (20.1) | n.a. | 68.1 (20.1) |
| | Retrofit | 0 | 75.0 | 61.7 | 1.006 | 0.983 | 0.992 | 5.9 (3.3) | 68.2 (20.1) | n.a. | 67.9 (19.9) |
| | Retrofit | 10 (5.6) | 75.6 | 63.5 | 1.003 | 1.095 | 1.143 | 6.3 (3.5) | 66.2 (19.0) | 0.54 | 65.8 (18.8) |
| 2.50 | New | 0 | 74.9 | 63.4 | 1.008 | 0.943 | 0.978 | 8.2 (4.6) | 68.6 (20.3) | n.a. | 68.1 (20.1) |
| | Retrofit | 0 | 75.2 | 64.9 | 1.009 | 0.937 | 1.000 | 8.6 (4.8) | 68.6 (20.3) | n.a. | 67.8 (19.9) |
| | Retrofit | 10 (5.6) | 75.7 | 66.5 | 1.006 | 1.042 | 1.144 | 9.3 (5.2) | 66.6 (19.2) | 0.24 | 65.8 (18.8) |

[a] Ranch house with basement in Pittsburgh climate with base conditions of 60°F (33.3°C) circulating air temperature rise, 6 cycles/hour, 10% duct air leakage
[b] The values in this table are *figures of merit* to be considered within the confines of the SP43 project, and they should not be applied outside the scope of these examples.
[c] Furnace Output Rating or Heating Capacity = (Furnace Multiplier)(Design Heat Loss)
[d] Longest recovery time during winter (lowest outdoor temperature = 5°F or −15°C)
[e] Retrofit case was not run for furnace multiplier = 1.00
[f] n.a. indicates not applicable
[g] $(I_S)_{BC}$ = 0.4614; for the corresponding electric furnace, $I_S/(I_S)_{BC}$ = 1.410.

**Table 10b Effect of Furnace Sizing on Annual Heating Performance—Condensing Furnace with Preheat[a,b]**

| Furnace Multiplier[c] | Annual Performance Factors $E_F$ | $E_D$ | $F_{MG}$ | $F_{LM}$ | $I_S$ | $I_S/(I_S)_{BC}$ | Temperature Swing, °F (°C) | Avg. Room °F (°C) | Avg. Bsmt. °F (°C) |
|---|---|---|---|---|---|---|---|---|---|
| 1.15 | 95.1 | 60.9 | 1.000 | 1.076 | 0.623 | 1.351[d] | 3.5 (1.9) | 67.9 (19.3) | 66.7 (19.3) |
| 1.40 | 95.5 | 62.0 | 1.000 | 1.050 | 0.622 | 1.347 | 4.3 (2.4) | 68.1 (20.1) | 66.7 (19.3) |
| 2.50 | 95.4 | 64.8 | 1.000 | 0.990 | 0.611 | 1.325 | 7.3 (4.1) | 68.7 (20.4) | 66.7 (19.3) |

[a] Ranch-type house with basement in Pittsburgh climate with base conditions of 60°F (33.3°C) circulating air temperature rise, 6 cycles/hour, 10% duct air leakage
[b] The values in this table are *figures of merit* to be considered within the confines of the SP43 project, and they should not be applied outside the scope of these examples.
[c] Furnace Output Rating or Heating Capacity = (Furnace Multiplier)(Design Heat Loss)
[d] $(I_S)_{BC}$ = 0.4614; for the corresponding electric furnace, $I_S$ = 0.651.

**Table 11 Effect of Duct Treatment on System Performance**

| Case | Duct Configuration 1 | Base Case 2 | 3 | 4 | 5 | 6 | 7 |
|---|---|---|---|---|---|---|---|
| **Condition** | | | | | | | |
| Duct Insulation | None | None | None | R5 | R5 | R5 | R5[a] |
| Duct Leakage, % | 0 | 10 | 20 | 0 | 10 | 20 | 10 |
| Basement Insulation | | | | | | | |
| Ceiling | None | None | None | None | None | None | None |
| Wall | R8 | R8 | R8 | R8 | R8 | R8 | R8 |
| **Performance** | | | | | | | |
| Burner On-Time, % | 17.5 | 18.0 | 18.6 | 16.8 | 17.2 | 17.8 | 16.6 |
| Blower On-Time, % | 23.8 | 24.3 | 24.9 | 23.0 | 23.4 | 24.1 | 22.9 |
| Avg. Basement Temp., °F | 66.8 | 67.8 | 68.8 | 65.2 | 66.3 | 67.4 | 65.1 |
| (°C) | (19.3) | (19.9) | (20.4) | (18.4) | (19.1) | (19.7) | (18.4) |
| Furnace Efficiency, $E_F$, % | 75.4 | 75.5 | 75.7 | 75.0 | 75.2 | 75.4 | 75.0 |
| Duct Efficiency, $E_D$, % | 66.8 | 60.9 | 54.8 | 77.4 | 70.4 | 63.2 | 79.6 |
| Load Modification Factor | 0.94 | 1.00 | 1.07 | 0.85 | 0.91 | 0.97 | 0.84 |
| $I_S/(I_S)_{BC}$[b] | 1.026 | 1.000 | 0.970 | 1.070 | 1.044 | 1.010 | 1.085 |

[a] Case 7 is interior insulation (liner); cases 1-6 are exterior insulation (wrap).
[b] $(I_S)_{BC}$ = 0.4614; for the corresponding electric furnace, $I_S$ = 0.651.

**Table 12 Effect of Duct Treatment and Basement Configuration on System Performance[a]**

| Case | Basement Configuration 1 | 2 | 3 | 4 | 5 |
|---|---|---|---|---|---|
| **Condition** | | | | | |
| Duct Insulation | None | None | None | None | R5 |
| Duct Leakage, % | 20 | 10 | 10 | 10 | 0 |
| Basement Insulation | | | | | |
| Ceiling | None | None | R11 | R11 | R11 |
| Wall | None | None | None | R8 | R8 |
| **Performance** | | | | | |
| Burner On-Time, % | 23.6 | 22.7 | 21.8 | 18.0 | 16.3 |
| Blower On-Time, % | 29.8 | 29.2 | 28.0 | 24.2 | 22.2 |
| Avg. Basement Temp., °F | 64.4 | 63.3 | 62.4 | 67.9 | 64.3 |
| (°C) | (18.0) | (17.4) | (16.7) | (19.9) | (17.9) |
| Furnace Efficiency, $E_F$, % | 75.3 | 75.2 | 75.2 | 75.7 | 75.0 |
| Duct Efficiency, $E_D$, % | 50.9 | 56.9 | 56.7 | 61.7 | 77.0 |
| Load Modification Factor | 0.91 | 0.85 | 0.89 | 0.99 | 0.88 |
| $I_S/(I_S)_{BC}$[b] | 0.765 | 0.794 | 0.825 | 1.001 | 1.103 |

[a]See Table 7 for Base Case
[b]$(I_S)_{BC}$ = 0.4614

# PART II: COMMERCIAL FURNACES

The basic difference between residential and commercial furnaces is the size and heating capacity of the equipment. The heating capacity of a commercial furnace may range from 150,000 Btu/h (44 kW) to over 1,000,000 Btu/h (300 kW). Generally, furnaces with output capacities less than 320,000 Btu/h (93 kW) are classified as light commercial, and those above 320,000 Btu/h (93 kW) are large commercial equipment. In addition to the difference in capacity, commercial equipment is constructed from material having increased structural strength and more sophisticated control systems.

## EQUIPMENT VARIATIONS

Light commercial heating equipment comes in almost as many flow arrangements and design variations as residential equipment. Some are identical to residential equipment, while others are unique to commercial applications. Some commercial units function as a part of a ducted system, and others operate as unducted space heaters.

### Ducted Equipment

Upflow gas-fired commercial furnaces are available up to 300,000 Btu/h (88 kW) and supply enough airflow to handle up to 10 tons (35 kW) of air-conditioning. These furnaces may have high static pressure, belt-driven blowers, and frequently they consist of two standard upflow furnaces tied together in a side-by-side arrangement. They are normally incorporated into a system in conjunction with a commercial split-system air-conditioning unit and are available in either LP or natural gas. Oil-fired units may be available on a limited basis.

**Horizontal gas-fired duct furnaces** are also available for built-up light commercial systems. This type of furnace is not equipped with its own blower but is designed for uniform airflow across the entire furnace. Duct furnaces are normally certified for operation either upstream or downstream of an air-conditioner cooling coil. If a combination blower and duct furnace is desired, a package called a *blower unit heater* is available. Duct furnaces and blower unit heaters are available in natural gas, LPG, oil, and electric models.

**Electric duct furnaces** are available in a great range of sizes and are suitable for operation in upflow, downflow, or horizontal positions. These units are also used to supply auxiliary heat with the indoor section of a split-heat pump.

The most common commercial furnace is the **combination package unit**, which is sometimes called a combination rooftop unit. They are available as air-conditioning units with LPG and natural gas furnaces, electric resistance heaters, or heat pumps. Combination oil heat-electric cool units are not commonly available. Combination units come in a full range of sizes covering air-conditioning ratings from 5 to 50 tons (18 to 176 kW) with matched furnaces supplying heat-to-cool ratios of approximately 1.5 to 1.

Combination units of 15 tons (53 kW) and under are available as single-zone units. The entire unit must be in either the heating or cooling mode. All air delivered by the unit is at the same temperature. Frequently, the heating function is staged so the system operates at reduced heat output when the load is small.

Large combination units in the 15- to 50-ton (53 to 176 kW) range are available as single-zoned units, as are small units; however, they are also available as multizoned units. A multizone unit supplies conditioned air to several different zones of a building in response to individual thermostats that control those zones. These units are capable of supplying heating to one or more zones at the same time cooling is being supplied to other zones.

Large combination units are normally available only in a curbed configuration; that is, the units are mounted on a rooftop over a curbed opening in the roof. The supply and return air enters through the bottom of the unit. Smaller units may be available for either curbed or uncurbed mounting. In either case, the unit is always connected to ductwork within the building to distribute the conditioned air.

### Unducted Heaters

Three types of commercial heating equipment are used as unducted space heaters. One is the *unit heater*, which is available from about 25,000 to 320,000 Btu/h (7 to 94 kW). They are normally mounted from ceiling hangers and blow air across the heat exchanger into the heated space. Natural gas, LPG, and electric unit heaters are available. The second unducted heater that is used in commercial heating is the *infrared heater*. These units are mounted from ceiling hangers and transmit heat downward by radiation. Both gas and electric infrared heaters are available.

Finally, *floor furnaces* are used as large area heaters. They are available in capacities ranging from 200,000 to 2,000,000 Btu/h (59 to 586 kW). Floor furnaces direct heated air through nozzles for task heating or use air circulators to heat large industrial spaces.

## SYSTEM DESIGN AND EQUIPMENT SELECTION

The procedure for design and selection of a commercial furnace is similar to that for a residential furnace. First, the design capacity of the heating system must be determined, considering structure heat loss, recovery load, internal load, humidification, off-peak storage, waste heat recovery, and backup capacity. Since most commercial buildings use setback during weekends, evenings, or other long periods of inactivity, the recovery load is important, as are internal loads and waste heat recovery.

Selection criteria differ from a residential furnace in some respects and are identical in others. Sizing criteria are essentially the same, and it is recommended that the furnace be oversized 30% above total load. Since combination units must be sized accurately for the cooling load, it is possible that the smallest gas-fired capacity available will be larger than the 30% value. This will be especially true for the warmer climates of the United States.

Efficiency of commercial units is about the same as for residential units. Two-stage gas valves are frequently used with commercial furnaces, but the efficiency of a two-stage system *may* be lower than for a single-stage system. At a reduced firing rate, the excess combustion airflow through the burners increases, decreasing the steady-state operating efficiency of the furnace. Multistage furnaces with multistage thermostats and controls are commonly used to provide more uniform distribution of heat within the building.

The design life of commercial heating and cooling equipment is about 20 years. Most gas furnace heat exchangers are either coated steel or stainless steel. Since most commercial furnaces are made for outdoor application, the cabinets are made from corrosion-resistant coated steel, such as galvanized or aluminized. Blowers are usually belt-driven and are capable of delivering air at high static pressures.

The noise level of commercial heating equipment is important with some products and less important with others. Unit heaters, for example, are used primarily in industrial applications where noise is less important. Most other commercial equipment is used in schools, office buildings, and other commercial buildings where noise level is important. In general, the larger the furnace, the more air it handles, and the more noise it generates. However, commercial systems with longer and larger ductwork result in more sound attenuation. The net result is that quality commercial heating systems produce about the same noise level in the heated space as do residential systems.

Safety requirements are the same for light commercial systems as they are for residential systems. Above 400,000 Btu/h (117 kW) gas input, the ANSI Z21.47 requirements for gas controls are considerably more stringent. A large percentage of commercial heating systems are located on the rooftop or some other location outside the building. Outdoor furnaces always provide a margin of safety beyond that of an indoor furnace.

### TECHNICAL DATA

Technical data for commercial furnaces are supplied by the manufacturer. Furnaces are available with heat outputs ranging from 150,000 Btu/h (44 kW) to more than 1,000,000 Btu/h (300 kW) output. For heating-only commercial heaters, the airflow is set to supply air with a 85 °F (47 °C) temperature rise. Heat/cool combination units supply air equal to about 400 cfm per ton (54 L/s per kW) of cooling capacity. Heat-to-cool ratios are generally held at about 1.5 to 1.

The steady-state efficiency for commercial furnaces is about the same as that for residential furnaces. However, little information has been published about the annual efficiency of commercial systems. Some efficiency improvement components, such as intermittent ignition devices, are common in commercial furnaces.

## PART III: GENERAL CONSIDERATIONS

### INSTALLATION PRACTICES

Installation requirements call for the forced-air heating system to meet two basic criteria: (1) the system must be safe, and (2) it must provide comfort for the occupants of the conditioned space.

Indoor furnaces are sometimes installed as isolated combustion systems. An isolated combustion system is one in which a furnace is installed indoors and all combustion and ventilation air is admitted through grilles or is ducted from outdoors and does not interact with air in the conditioned space; for example: interior enclosures with air from attic or ducted from outdoors, exterior enclosures with air from outdoors through grilles, or enclosures in garages or carports attached to the building (NFPA 54-1984). This type of installation presents special considerations in determining efficiency.

Generally, the following three sources of installation information must be followed to ensure the safe operation of a heating system: (1) the equipment manufacturer's installation instructions, (2) local installation code requirements, and (3) national installation code requirements. Local code requirements may or may not be available, but the other two are always available. Depending on the type of fuel being used, one of the following national code requirements will apply:

NFPA 54-1984 (also ANSI Z223.1-84) *National Fuel Gas Code*
NFPA 70-84 *National Electric Code*
NFPA 31-83 *Oil Burning Equipment Code*

Comparable Canadian *Standards* bear the numbers CSA Z240.4, Z240.5, Z240.6.1, and Z240.6.2. These regulations provide complete information about construction materials, gas line sizes, flue pipe sizes, wiring sizes, and the like.

Proper design of the air-distribution system is necessary for both comfort and safety. Chapter 33 of the 1985 FUNDAMENTALS Volume and Chapter 17 of the 1987 HVAC Volume provide information on the design of ductwork for forced-air heating systems. Forced-air furnaces may provide design airflow at a static pressure as low as 0.12 in. of water (30 Pa) for a residential unit to above 1.0 in. of water (250 Pa) for a commercial unit. The air-distribution system must handle the required volume flow rate within the pressure limits of the equipment. If the system is a combined heating-cooling installation, the air-distribution system must meet the cooling requirement because more air is required for cooling than for heating. It is also important to include the pressure drop of the cooling coil. The ARI maximum allowable pressure drop for residential cooling coils is 0.3 in. of water (75 Pa).

### AGENCY LISTINGS

The construction and performance of furnaces is controlled by several agencies.

The Gas Appliance Manufacturers Association (GAMA), in cooperation with its industry members, sponsors a certification program relating to gas and oil-fired residential furnaces and boilers. This program uses an independent laboratory to verify the furnace and boiler manufacturer's stated annual fuel utilization efficiency (AFUE) and heating capacity, as determined by testing in accordance with the Department of Energy's *Uniform Method for Measuring the Energy Consumption of Furnaces and Boilers.* Gas and oil furnaces with input ratings less than 225,000 Btu/h (66 kW), and gas and oil boilers with input ratings less than 300,000 Btu/h (88 kW) are currently included in the program.

Also included in the program is the semiannual publication of the GAMA *Consumers Directory of Certified Efficiency Ratings for Residential Heating and Water Heating Equipment,* which identifies certified products and lists the input rating, certified heating capacity, and Annual Fuel Utilization Efficiency (AFUE) for each. Participating manufacturers are entitled to use the GAMA Certification Symbol (seal). These directories are distributed to the reference departments of the public libraries in the United States.

ANSI *Standard* Z21.47-83, *Gas-Fired Central Furnaces,* (sponsored by the American Gas Association) gives minimum construction, safety, and performance requirements for gas furnaces. The AGA maintains laboratories to certify furnaces and operates a factory inspection service. Furnaces tested and found to be in compliance are listed in the AGA Directory and carry the Blue Star Seal of Certification. Underwriters Laboratory and certain other approved laboratories can also test and certify equipment in accordance with ANSI *Standard* Z21.47.

Gas furnaces may be certified for standard, alcove, closet, or outdoor installation. Standard installation requires clearance between the furnace and combustible material of at least 6 in. (15.2 cm). Furnaces certified for alcove or closet installation can be installed with reduced clearance, as listed. Furnaces certified for outdoor installation must operate properly in a 40 mph (64.4 km/h) wind. Construction materials must be able to withstand natural elements without degradation of performance and structure. Horizontal furnaces are normally certified for installation on combustible floors and for attic installation and are so marked, in which case they may be installed with point or line contact between the jacket and combustible constructions. Upflow and downflow furnaces are normally certified for alcove or closet installation. Gas furnaces may be listed to burn natural, mixed, manufactured, LP, or LP-air gases. A furnace must be ordered for the specific gas to be used, since different burners and controls may be required, as well as orifice changes. In Canada, similar requirements have been established by the Canadian Standards Association and the Canadian Gas Association. The testing agency is the Canadian Gas Association and certified products are marked with the CGA seal.

Sometimes oil burners and control packages are sold separately; however, they are normally sold as part of the furnace package. Pressure-type or rotary burners should bear the Underwriters Laboratories label showing compliance with UL 296. In addition, the complete furnace should bear markings indicating compliance with UL 727 and be listed as such by the Laboratory. Vaporizing burner furnaces should be listed under UL 727.

The *Underwriters Laboratory Standards* UL 883, UL 1025 and UL 1096 give requirements for the listing and labeling of electric furnaces. Similarly, UL 559 gives the requirements for heat pumps.

The following table summarizes those important standards that apply to space-heating equipment and are issued by the American Gas Association, Underwriters Laboratories, and the Canadian Gas Association:

| | |
|---|---|
| ANSI/ASHRAE 103-1982 | Methods of Testing for Heating Seasonal Efficiency of Central Furnaces and Boilers |
| ANSI Z83.8-85 | Gas Unit Heaters |
| ANSI Z83.9-82 | Gas-Fired Duct Furnaces |
| ANSI Z21.47-83 | Gas-Fired Central Furnaces |
| ANSI Z21.64-85 | Direct-Vent Central Furnaces |
| ANSI Z21.66-85 | Electrically Operated Automatic Vent-Damper Devices for Use with Gas-Fired Appliances |
| ANSI Z21.67-85 | Mechanically Actuated Automatic Vent-Damper Devices for Use with Gas-Fired Appliances |
| ANSI Z21.68-85 | Thermally Actuated Automatic Vent-Damper Devices for Use with Gas-Fired Appliances |
| ANSI Z83.4-85 | Direct Gas-Fired Makeup Air Heaters |
| ANSI Z83.6-82 | Gas-Fired Infrared Heaters |
| UL 296-80 | Oil Burner |
| UL 307-78 | Heating Appliances for Mobile Homes and Recreational Vehicles |
| UL 559-85 | Heat Pumps |
| UL 727-80 | Oil-Fired Central Furnaces |
| UL 883-80 | Fan-Coil Units and Room Fan-Heater Units |
| UL 1096-81 | Electric Central Air Heating Equipment |
| UL 1025-80 | Electric Air Heaters |
| CGAS 2.3 | Gas-Fired Gravity and Forced-Air Central Furnace with Inputs up to 400 000 Btu/h |
| CGAS 3.2 | Gas-Fired Gravity and Forced-Air Central Furnaces with Inputs over 400 000 Btu/h |
| CGAS 3.7 | Makeup Air Heaters |
| CGAS B2.5 | Vented Wall Furnaces |
| CGAS B2.6 | Unit Heaters |
| CGAS B2.8 | Duct Furnaces |
| CGAS B2.16 | Infrared Radiant Heaters |
| CGAS B2.19 | Sealed Combustion System Wall Furnaces |
| CGAS B140.4 | Oil-Fired Gravity and Forced-Air Central Furnaces |

## REFERENCES


Benton, R. 1983. Computer Predictions and Field Test Verification of Energy Savings With Improved Control. ASHRAE *Transactions*, 89(1).

Bonne, U.; J.E. Janssen; A.E. Johnson; and W.T. Wood. 1976. Residential Heating Equipment HFLAME Evaluation of Target Improvements. *National Bureau of Standards,* Final Report, Contract No. T62709, Center of Building Technology, Honeywell, Inc.

Bonne, U.; J.E. Janssen; and R.H. Torborg, 1977. Efficiency and relative operating cost of central combustion heating systems. IV. Oil-fired residential systems. ASHRAE *Transactions*, 83(1).

Bullock, E.C. 1977. Energy Saving through Thermostat Setback with Residential Heat Pumps (Workshop on Thermostat Setback). *National Bureau of Standards.*

Chi, J. 1977. DEPAF—A Computer Model for Design and Performance Analysis of Furnaces. AICHE-ASME *Heat Transfer Conference*, Salt Lake City, UT, August 15-17.

Deppish, J.R., and D.W. DeWerth. 1986. GATC Studies Common Venting. *Gas Appliance and Space Conditioning Newsletter,* Number 10, September.

Fischer, R.D.; F.E. Jakob; L.J. Flanigan; D.W. Locklin; and R.A. Cudnik. 1984. Dynamic Performance of Residential Warm-Air Heating Systems—Status of ASHRAE Project SP43. ASHRAE *Transactions*, 90(2B): 573-590.

Gable, G.K., and K. Koenig. 1977. Seasonal operating performance of gas heating systems with certain energy saving features. ASHRAE *Transactions*, 83(1).

Herold, K.E.; R.D. Fischer; and L.J. Flanigan. 1987. Measured Cooling Performance of Central Forced-Air Systems and Validation of the SP43 Simulation Model. ASHRAE *Transactions*, 93(1):1443-1457.

Herold, K.E.; L.J. Flanigan; R.D. Fischer; and R.A. Cudnik. 1987. Update on Experimental Validation of the SP43 Simulation Model for Warm Air Heating Systems. ASHRAE *Transactions*, 93(1).

Herold, K.E.; F.E. Jakob; and R.D. Fischer. 1986. The SP43 Simulation Model: Residential Energy Use. *Proceedings* of ASME Conference at Anaheim, pp. 81-87.

Hise, E.C., and A.S. Holman. 1977. Heat balance and efficiency measurements of central, forced-air, residential gas furnaces. ASHRAE *Transactions*, 83(1).

Jakob, F.E.; R.D. Fischer; L.J. Flanigan; D.W. Locklin; K.E. Herold; and R.A. Cudnik. 1986. Validation of the ASHRAE SP43 Dynamic Simulation Model for Residential Forced-Warm-Air Systems. ASHRAE *Transactions*, 92(2B):623-643.

Jakob, F.E.; D.W. Locklin; R.D. Fischer; L.J. Flanigan; and R.A. Cudnik. 1986. SP43 Evaluation of System Options for Residential Forced-Air Heating. ASHRAE *Transactions*, 92(2B): 644-673.

Jakob, F.E.; R.D. Fischer; and L.J. Flanigan. 1987. Experimental Validation of the Duct Submodel for the SP43 Simulation Model. ASHRAE *Transactions*, 93(1):1499-1515.

Koenig, K. 1978. Gas furnace size requirements for residential heating using thermostat night setback. ASHRAE *Transactions*, 84(2).

Locklin, D.W.; K.E. Herold; R.D. Fischer; F.E. Jacob; and R.A. Cudnik. 1987. Supplemental Information From SP43 Evaluation of

System Options for Residential Forced-Air Heating. ASHRAE *Transactions*, 93(2):1934-1958.

Nelson, L.W., and W. MacArthur. 1978. Energy Saving through Thermostat Setback. ASHRAE *Journal*, September.

NFPA. 1984. *National Fuel Gas Code. National Fire Code* 54-1984. National Fire Protection Association. Quincy, MA.

Schade, G.R. 1978. Saving energy by night setback of a residential heat pump system. ASHRAE *Transactions*, 84(1).

Stickford, G.H., *et al.* 1985. Technology Development for Corrosion-Resistant Condensing Heat Exchangers. *Battelle Columbus Laboratories* to *Gas Research Institute*, GRI-85/0282.

CHAPTER 25

# RESIDENTIAL IN-SPACE HEATING EQUIPMENT



IN-SPACE heating equipment differs from central heating in two ways: (1) ducts are not needed to convey heat from the point of generation to the room that is to be heated, and (2) gravity-type basic models (fossil-fueled) do not require electrical connections. They depend on convection for circulation from the heat source to the room.

## GAS IN-SPACE HEATERS (NATURAL AND LP)

Gas in-space heating equipment comes in a wide variety of styles and models.

### Room Heaters

A **vented circulator room heater** (see Figure 1) is a self-contained, freestanding, nonrecessed, gas-burning appliance that furnishes direct warm air to the space in which it is installed, without ducting. It converts the energy in the fuel gas to convected and radiant heat by transferring heat from flue gases to a heat exchanger surface without mixing flue gases and circulating heated air.

A **vented radiant circulator** is equipped with high-temperature glass panels and radiating surfaces to increase radiant heat transfer. Separation of flue gases from circulating air must be maintained. Vented radiant circulators range from 10,000 to 75,000 Btu/h (3 to 22 kW).

Gravity-vented radiant circulators may also be equipped with an optional circulating air fan but perform satisfactorily with or without the fan in operation.

Fan-type vented radiant circulators are equipped with an integral circulating air fan, which is necessary for satisfactory performance.

Vented room heaters must be connected to a vent, chimney, or single-wall metal pipe venting system engineered and constructed to develop a positive flow to the outside atmosphere. Room heaters should not be used in a room that is virtually airtight.

**Unvented radiant or convection heaters** are relatively inexpensive to install. They range in size from 10,000 to 40,000 Btu/h (3 to 12 kW) and can be freestanding or wall-mounted nonrecessed units of either the radiant or closed-front type.

Unvented room heaters require an outside air intake. The size of the fresh air opening required is marked on the heater. To ensure adequate fresh air supply, unvented gas-heating equipment must, according to voluntary standards, include a device that shuts the heater off if the air supply becomes inadequate.

Unvented room heaters may not be installed in hotels, motels, or rooms of institutions, such as hospitals, sanitariums, etc.

**Catalytic room heaters** are fitted with a fibrous material impregnated with a catalytic substance that accelerates the oxidation of a gaseous fuel to produce heat without flames. The design distributes the fuel throughout the fibrous material so that oxidation occurs on the surface area in the presence of a catalyst and room air.

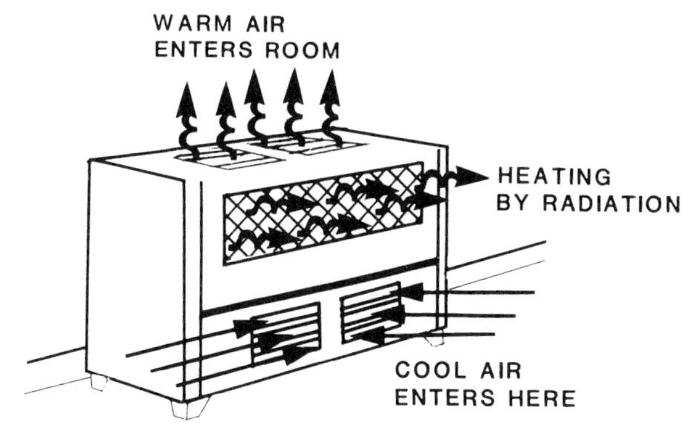


**Fig. 1 Room Heater**

The preparation of this chapter is assigned to TC 6.4, In-Space Convective Heating.

Catalytic heaters transfer heat both by low-temperature radiation and by convection. The surface temperature is below a red heat and is generally below 1200°F (650°C) at the maximum fuel input rate. Since they operate without a flame, catalytic heaters offer the inherent safety feature of flameless combustion, as compared with a conventional flame-type gas fueled burner. This advantage was recognized in the marketing of catalytic camp heaters fueled either with petroleum gas or liquid fuels. Other applications for catalytic heaters have been as agricultural heaters and for industrial applications in combustible atmospheres.

Unvented household catalytic heaters are used in Europe. Most of these are portable and are typically mounted on casters in a casing that includes a cylinder of liquified petroleum gas (LPG) so they may be rolled from one room to another. LPG gas cylinders of greater than two pounds of fuel are not permitted for indoor use in the United States. As a result, catalytic room heaters sold in the United States are generally permanently installed and fixed as wall-mounted units. Local codes and the *National Fuel Gas Code* (ANSI/AGA Z223.1-84) should be reviewed for accepted combustion air requirements and venting practices.

## Wall Furnaces

A wall furnace (see Figure 2) is a self-contained vented appliance with grilles that is designed to be a permanent part of the structure of a building. It furnishes heated air that is circulated by either gravity or a fan. A wall furnace may have boots, which cannot extend 10 in. (25 mm) beyond the horizontal limits of the casing through walls of normal thickness, to provide heat to adjacent rooms. Wall furnaces range from 10,000 to 90,000 Btu/h (3 to 26 kW). Wall furnaces are classified as conventional- or direct-vent.

**Conventional-vent units** require approved B-1 vent pipes and are installed to comply with *The National Fuel Gas Code* (ANSI *Standard* Z223.1-84).

Some wall furnaces are available as counterflow units, which reverse the natural flow of air across the heat exchanger. Air enters at the top of the furnace and discharges at or near the floor. Since warm air is buoyant and will rise, counterflow systems reduce heat stratification in a room. As with any vented unit, a minimum of inlet air for proper combustion must be supplied. A fan is required for proper operation.

**Vented-recessed wall furnaces** are recessed in the wall. The decorative grille work extending into the room allows more usable area in the room being heated. Dual-wall furnaces are two units that fit between the studs of adjacent rooms, thereby using a common vent.

Both vented-recessed and dual-wall furnaces are usually gravity-type units. Cool room air enters at the bottom and is warmed by conduction as it passes across the heat exchanger. As it warms, the air expands and rises, entering the room through the grillwork at the top of the heater. This action sets up a convection current that draws in additional air to be heated at the bottom. As long as the thermostat calls for the burners to be on, this process continues. Accessory fans can assist in the movement of air across the heat exchanger and help minimize air stratification.

**Direct-vent wall furnaces** are constructed so that combustion air comes from outside, and all flue gases discharge into the outside atmosphere. These appliances are complete with grilles, or the equivalent, and are designed to be attached to the structure permanently. Direct-vent wall heaters are normally mounted on walls with outdoor exposure.

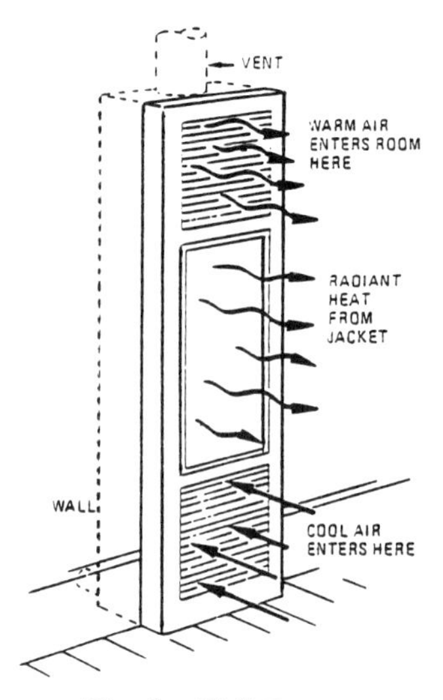


Fig. 2 Wall Furnace

Direct-vent wall furnaces can be used in extremely tight (well insulated) rooms because combustion air is drawn from outside the room. There are no infiltration losses for dilution or combustion air. Most direct-vent heaters are of the gravity-type; fan-type wall furnaces are also available. Accessory fan kits are available for the gravity models. Direct-vent furnaces are available from 6000 to 65,000 Btu/h (1.8 to 19 kW) input.

## Floor Furnaces

Floor furnaces are self-contained units (see Figure 3) suspended from the floor of the heated space. Combustion air is taken from outside, while flue gases are also vented outside. The cold air returns at the periphery of the floor register; warm air to the room comes up through the center portion of the register.

## Combustion Air Requirements

All gas-burning appliances, including gas in-space heaters, require oxygen for complete combustion of the gas being supplied.

Ten cubic feet of air at sea level to 2000 ft (610 m) contains sufficient oxygen to burn one cubic foot of natural gas [approximately 1000 Btu (11.4 $MJ/m^3$)]. Ratings shown on the manufacturer's rating plates are for elevations up to 2000 ft (610 m). For ratings above 2000 ft (610 m), the rate should be reduced at the rate of 4% for each 1000 ft (13%/km) of elevation. (For additional information on this subject, see Appendix B of ASHRAE *Standard* 62-1981, *Ventilation for Acceptable Indoor Air Quality.*)

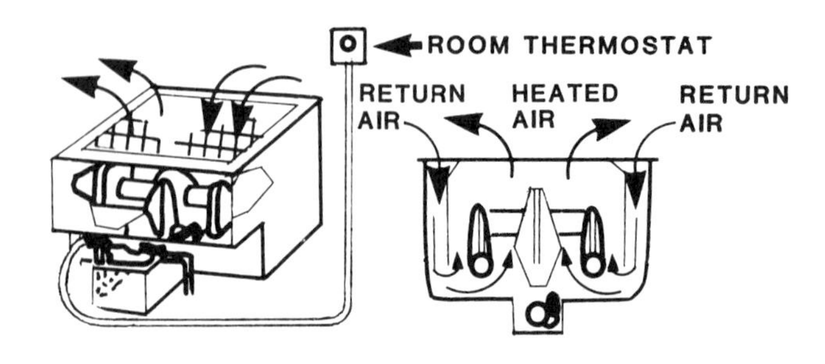


Fig. 3 Floor Furnace

## Types of Controls

Gas in-space heaters are controlled by the following four different types of valves.

**Single-Stage Control Valve.** The full on-off single-stage valve is controlled by a wall thermostat. Models are available that are powered by a 24-V supply or from energy supplied by the heat of the pilot light on the thermocouple (self-generating).

**Two-Stage Control Valve.** The two-stage type (with hydraulic-type thermostat) fires at either a full input (100% of rating) or at some reduced step, which can be as low as 20% of the heating rate. The amount of time at the reduced firing rate depends on the heating load and the relative oversizing of the heater.

**Step-Modulating Control Valve.** The step-modulating type (with hydraulic-type thermostat) steps on to a low fire and then either cycles off and on at the low fire, if the heating load is light, or it gradually increases its heat output to meet any higher heating load that cannot be met with the low firing rate. There is an infinite number of fuel firing rates between low fire and high fire with this control.

**Manual-Control Valve.** The manual-control valve is user-controlled rather than thermostat-controlled. The user adjusts the fuel flow and, thus, the level of fire to suit heating requirements.

Thermostats used for temperature control with gas in-space heaters are of the following two types:

**Wall Thermostats** are available in 24-V and millivolt systems. The 24-V requires an external power source and a 24-V transformer. Wall thermostats respond to temperature changes and turn the automatic valve to either full-on or full-off. The millivolt system requires no external power. The power is generated by multiple thermocouples and may be either 250 or 750 millivolt depending on the distance to the wall thermostat. This system also turns the automatic valve to either *full-on* or *full-off.*

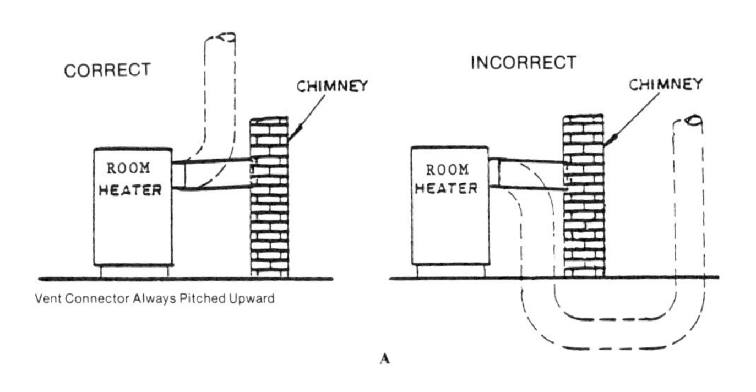


A

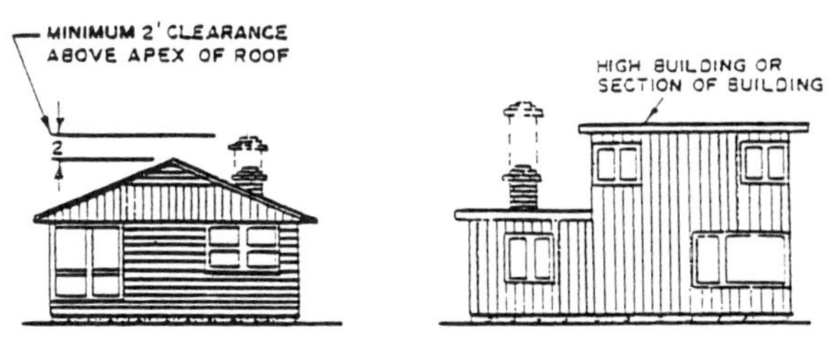


Correct Chimney Designs Illustrated By Dashed Lines. Chimney Should Extend Above Roofline Of Adjacent Buildings. Illustrated Conditions May Promote A Downdraft.

B

**Fig. 4 Vent Connection Methods**

**Built-In Hydraulic Thermostats** are also available in two types: (1) a snap-action built with a liquid-filled capillary tube, which responds to changes in temperature and turns the valve to either *full-on* or *full-off*; and (2) a modulating-type, which is similar to the first type, except the valve comes on and goes off at a preset minimum input. Temperature alters the input anywhere from *full-on* to the minimum input. When the heating requirements are satisfied, the unit goes off.

## Selecting Gas-Heating Equipment

The size of the unit depends on the size of the room, the number and direction of exposures, the amount of insulation in ceilings and walls, and the geographical location. Heat loss requirements can be calculated from Chapter 25 in the 1985 FUNDAMENTALS Volume.

## Vent Connection Methods

The installation of any vented gas-fired appliance requires that the correct procedures be followed for venting combustion products. Figures 4A and 4B show typical vent connection methods. A detailed description of proper venting techniques is found in the *National Fuel Gas Code* (ANSI *Standard* Z223.1-84) and Chapter 26 of this volume.

# OIL IN-SPACE CONVECTIVE HEATERS

Four types of oil in-space convective heaters are available.

## Vaporizing Pot Heaters

These heaters use burners with an oil vaporizing bowl (or other receptacle), which admits liquid fuel and air in controllable quantities; the fuel is vaporized by the heat of combustion and mixed with the air in appropriate proportions. Combustion air may be induced by natural draft or forced into the vaporizing bowl by a fan. Indoor air is generally used for combustion and draft dilution. Window-installed units have the burner section outdoors. Both natural-convection or forced-convection heating units are available. A small blower is sold as an option on some models. The heat exchanger, usually cylindrical, is made of steel (see Figure 5). These heaters are available as room units (both radiant- and circulation-type), floor furnaces, and recessed wall heaters. They may also be installed in a window, depending on the cabinet construction. Exterior venting is always used. A 3 to 5 gal (11.37 to 18.05 L) fuel tank may be attached to the heater. A larger, outside tank can be used.

Vaporizing burners of the pot-type (see Figure 5) are equipped with a single constant-level and metering valve. Fuel flows by gravity to the burner through the adjustable metering valve. Control can be manual, with an *off* pilot and variable settings up to *maximum,* or it can be thermostatically controlled, with the burner operating at a selected firing rate between pilot and *high.*

## Powered Atomizing Heaters

Wall furnaces, floor furnaces, and freestanding room heaters are also available with a powered gun-type burner using No. 1 or No. 2 fuel oil. For more information, see Chapter 22 of this volume.

## Portable Kerosene Heaters

Only kerosene is used to fuel these heaters and they are not vented. Precautions must be taken to provide sufficient ventilation. Kerosene heaters are of two basic types: a radiant type

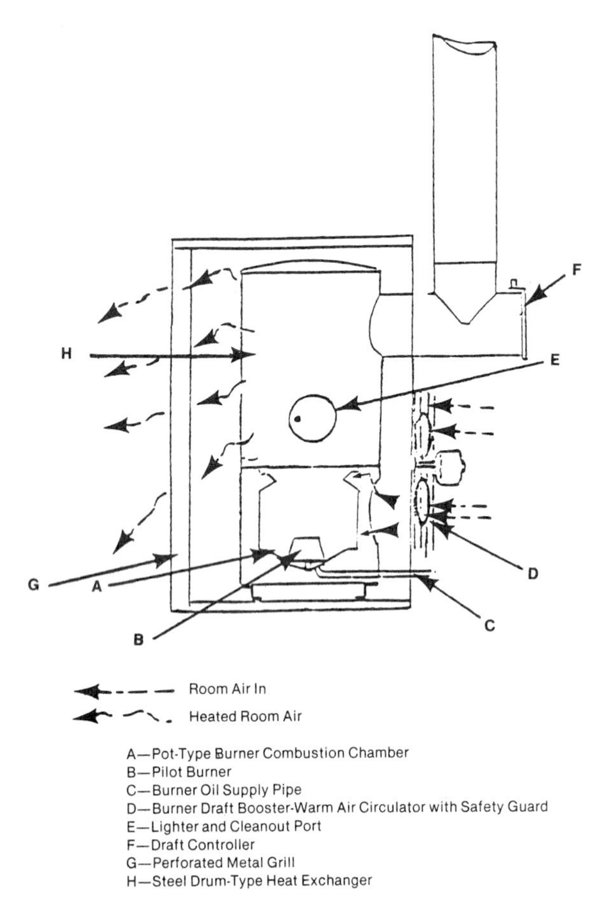


**Fig. 5 Oil-Fueled Heater with Vaporizing Pot-Type Burner**

employing a reflector and producing primarily radiant heat and a convection type of cylindrical shape, which produces convection heat. Fuel vaporizes from the surface of a wick, which is immersed in an integral fuel tank of up to 2 gallon capacity (7.6 L), similar to that of a kerosene lamp. Fuel-burning rates range from about 5000 to 22,500 Btu/h (1.5 to 3.4 kW). Radiant heaters usually have a removable fuel tank to facilitate re-fueling.

A third type of portable kerosene heater has a vaporizing burner and heat circulating fan. These heaters are available with thermostatic control and variable heat output.

Another kerosene heater is the catalytic type, which uses a metal catalyst to oxidize the fuel. It is started by lighting kerosene at the surface; however, after a few moments, the catalyst surface heats to the point that flameless oxidation of the fuel begins.

### Waste Oil Heaters

These simple drum heaters, which may burn dirty oil, transmission fluid, etc., are vented and equipped with a barometric draft regulator. Application is limited to garages or commercial facilities with access to waste oil.

## ELECTRIC IN-SPACE HEATERS

### Wall and Floor Heaters

Heaters for recessed or surface wall-mounting are made with bare wire or metal-sheathed elements. An inner liner or reflector is usually placed between elements and casing to promote circulation and minimize the rear casing temperature. Heat is distributed by both convection and radiation; the proportion of each depends on unit construction.

Ratings are usually 1000 to 5000 W. Voltages are standard values of 120, 208, 240, and 277. Models with air circulation fans are available. Other types with similar components can be floor-recessed. Electric convectors should be placed so that air moves freely across the elements.

### Baseboard Heaters

A metal casing, similar in appearance to a baseboard, contains one or more horizontal heating elements. The elements may be finned, sheathed, cast grid, or have ceramic extended surfaces. The vertical dimension is usually less than 9 in. (230 mm), with the projection from the wall less than 3.5 in. (90 mm). Units are available from 1 to 12 ft (300 to 3700 mm) in length, with ratings from 100 to 400 W/ft (330 to 1310 W/m), and they fit together to make up any desired continuous length or rating. Electric hydronic baseboard heaters containing immersion heating elements and an antifreeze solution are made with ratings of 300 to 2000 W. The placement of any type of electric baseboard heater follows the same principles that apply to baseboard installations (described in Chapter 28 of this volume), since baseboard heating is primarily perimeter heating.

### Radiant Convector Wall Panel

Glass electric heating units, sometimes described as radiant panels, depend on the heating effect produced by passage of current through a thin coating of conductive material fused to one face of a panel of 0.25 in. (6.4 mm) thick special glass. The conductive layer may be sprayed-on aluminum, printed metallic oxide patterned to form a grid several thousandths of an inch thick, or material fused-on to give a uniform coating less than 0.0001 in. (2.5 $\mu$m) over the entire active face of the panel. Normal glass operating temperatures are 300 to 400°F (150 to 200°C).

Electrical leads, connected in various ways, allow easy connection in the field. Insulators within a metal frame support the panel, and a reflector behind the glass provides space for air circulation. Protective guards are usually included to reduce hazards.

Glass units are usually rated between 500 and 6000 W for standard voltages of 120, 208, 240, and 277. Frame sizes vary from about 30 by 24 in. (762 by 610 mm) for 1000 W size to 42 by 6 in. (1070 by 150 mm) for baseboard models and are arranged for recessed or surface-mounting. Thermostats integral with the unit are optional.

Radiant panels using tubular elements welded to or cast into extended aluminum panels have emissivity characteristics similar to glass panels.

### Imbedded Cable Heat

Chapter 7 of the 1987 HVAC Volume covers resistance wire embedded in ceilings, walls, and floors on site or in prefabricated panels.

### Cord-Connected Portable Heaters

Portable electric heaters are often used in areas not accessible to a central heating system. They are also used to maintain an occupied room at a comfortable level independent of the rest of the residence.

Portable electric heaters for connection to 120-V, 15-A outlets are available with outputs of 2050 to 5100 Btu/h (600 to 1500 W), with the most common being 1320 and 1500 W. Many heaters are available with a selector switch for three wattages, *i.e.,* 1100-1250-1500 W. Heavy-duty heaters are usually connected to 240-V, 20-A outlets with outputs up to 13,700 Btu/h (4000

W), while those for connection to 240-V, 30-A outlets have outputs up to 19,100 Btu/h (5600 W). All electric heaters of the same wattage produce the same amount of heat.

There are two basic types of portable electric heaters: radiant and convection.

**Radiant heaters** are used to provide heat for people or to apply heat to objects such as furniture and walls. An element in front of a reflector radiates heat outward in a direct line. Conventional radiant heaters have ribbon or wire elements that produce heat. Quartz radiant heaters have coil wire elements encased in quartz tubes. The temperature of a radiant wire element usually ranges between 1200 to 1800°F (650 to 1000°C).

**Convection heaters** warm the air in rooms or zones. Air flows directly over the hot elements and mixes with room air. Convection heaters are available with and without fans. The temperature of a convection element is usually less than 930°F (500°C).

Portable electric heaters are categorized as (1) low silhouette—lower than 10 in. (250 mm) and longer than 30 in. (760 mm); (2) upright; (3) heavy-duty (more than 150 V); and (4) quartz.

An adjustable, built-in bimetal thermostat usually controls the power to portable electric heaters. Fan-forced heaters usually produce better temperature control because the fan, in addition to cooling the case, forces room air past the thermostat for fast response. One built-in control uses a thermistor to signal a solid-state logic circuit that adjusts wattage and fan speed. Most quartz heaters use an adjustable control that operates the heater a percent of total cycle time from 0 (*off*) to 100% (*full-on*).

## Control of In-Space Electric Heaters

Low-voltage and line-voltage thermostats of on-off operation are the basic types used for control of in-space electric heaters. Low-voltage thermostats, operating at 30 V or less, control relays or contactors that carry the rated voltage and current load of the heaters. Since the control current load is small (usually less than one amp), the small switch can be controlled by a highly responsive sensing element.

Line-voltage thermostats carry the full load of the heaters at rated voltage directly through their switch contacts. Most switches carry a listing by Underwriters Laboratories at 22 A (resistive), 277 V rating. While most electric in-space heating systems are controlled by remote wall-mounted thermostats, many types are available with integral or built-in line-voltage thermostats.

Most low-voltage and line-voltage thermostats use small internal heaters, either fixed or adjustable in heat output that are energized when the thermostat contacts close to provide heat anticipation. The cycling rate of the thermostat is increased by the use of anticipation heaters, resulting in more accurate control of the space temperature.

*Droop* is an apparent shift or lowering of the control point and is associated with line-voltage thermostats. In line-voltage thermostats, switch heating caused by large currents can add materially to the amount of droop. Most line-voltage thermostats in residential use control room heaters of 3 kW (12.5 A at 240 V) or less. With this moderate load and with properly sized anticipation heaters, the droop experienced in those applications is acceptable. Cycling rates and droop characteristics are significant factors describing thermostat performance.

# SOLID-FUEL IN-SPACE HEATERS

The classification solid-fuel in-space heater (see Table 1) refers to most wood-burning and coal-burning devices, except central wood-burning furnaces and boilers. An in-space heater can be either a fireplace or a stove.

## Simple Fireplaces

Simple fireplaces, especially all-masonry and noncirculating metal built-ins, produce little useful heat. They lend atmosphere

**Table 1 Solid-Fuel In-Space Heaters**

| Type[a] | Approximate Efficiency[a] | Features | Advantages | Disadvantages |
|---|---|---|---|---|
| Simple Fireplaces, Masonry or prefabricated | −10% to +10% | Open front. Radiates heat in one direction only. | Visual beauty | Low efficiency. Heats only small areas. |
| High Efficiency Fireplaces | 25% to 45% | Freestanding or built-in with glass doors, grates, duct, and blowers. | Visual beauty. More efficient. Heats larger areas. Long service life. Max. safety. | Medium efficiency |
| Box Stoves | 20% to 40% | Radiates heat in all directions. | Low initial cost. Heats large areas. | Fire hard to control. Short life. Wastes fuel. |
| Airtight Stoves | 40% to 55% | Radiates heat in all directions. Sealed seams, effective draft control. | Good efficiency. Long burn times, high heat output. Longer service life. | Can create creosote problems. |
| High Efficiency Catalytic Wood Heaters | 65% to 75% | Radiates heat in all directions. Sealed seams, effective draft control. | Highest efficiency. Long burn times, high heat output. Long life. | Creosote problems. High purchase price. |

[a]Product categories are general; product efficiencies are approximate.

and a sense of cozyness to a room. Freestanding fireplaces are slightly better heat producers. Simple fireplaces have an average efficiency of about 10%. In extreme cases, more heated air is drawn up the chimney than is produced by the fire.

The addition of glass doors has both a positive and negative effect. The glass doors restrict the free flow of indoor heated air up the chimney, but at the same time, they restrict the radiation of the heat from the fire into the room.

### High Efficiency Fireplaces

High efficiency fireplaces (almost always steel factory-built units) are simple fireplaces with additional energy-saving features that boost heat output to 25 to 45%. All high efficiency fireplaces have doors, usually transparent glass panels, though not all fireplaces that have glass doors are high efficiency.

High efficiency fireplaces also have a duct system built around the firebox. Cool indoor air enters along the floor, circulates up and around the firebox, and reenters the room from an outlet(s). This circulation is often enhanced by a blower(s) or fan(s). In addition, ducts to the outside supply outside combustion air to the fireplace.

### Freestanding Fireplaces

The freestanding fireplace can be energy efficient if it has glass doors, air circulation fan(s), etc. Only a few have every feature. This fireplace is efficient because it radiates heat in all directions. The freestanding fireplace may complicate interior decorating and can be an obstacle depending on its placement.

### Simple Box Stoves

Box stoves come in a wide variety of sizes, shapes, and styles. Some burn wood, coal, or both. Box stoves are built of lightweight materials and are normally the least expensive choice. They do not control the combustion air, so they are only 25 to 40% efficient. Because they tend to leak air and are difficult to damper, they can overfire, warp, and have a short service life. Most box stoves cannot hold a fire overnight.

### Controlled Combustion (Airtight) Heaters

Controlled combustion or airtight heaters are more expensive than box stoves, but they have a much longer life and operate at better efficiencies. Airtight heaters typically operate at combustion efficiencies of 40 to 55%. Improved versions of airtight heaters may obtain combustion efficiencies of up to 65%. In addition, overfiring is eliminated, and long burn times are possible.

But creosote formed from products of incomplete combustion is a safety concern of these heaters. Volatile gases released by combustion must be thoroughly oxygenated and above about 1100°F (590°C) to burn completely. When an airtight heater is operated at or near full draft, so that combustion is complete and chimney temperatures remain above 270°F (130°C), creosote is of little concern. At lower draft settings (during an overnight burn, for example), the volatile gases may cool to the point that they will not burn, even though they are oxygenated fully. The unburned gases then pass up the chimney, where they can condense as corrosive, ignitable, flammable creosote. Dense smoke from a smoldering fire is most likely to form large amounts of creosote.

### Catalytic Wood Heaters

Several companies manufacture catalytic stoves and add-on combusters for existing airtight heaters. A precious metal catalyst makes the flammables in wood smoke burn at lower-than-normal temperatures. The temperature of combustion can be as low as 450°F (230°C). Once the reaction begins, the temperature of the catalytic unit rises dramatically. Smoke is reduced, little creosote is formed, and the efficiency of combustion increases as much as 25% above a comparable airtight heater.

### Circulating Stoves

Circulating stoves transfer most of their heat by convection. Typically, these stoves are constructed by surrounding a cast-iron or steel firebox with a sheet metal cabinet. Vents in the top and bottom of the cabinet allow natural airflow to circulate heat from the stove. A major advantage to circulating stoves is that their exposed surfaces do not become too hot to touch; therefore, they can be placed closer to combustible materials than radiant stoves.

## GENERAL INSTALLATION PRACTICES

The criteria to ensure safe operation are normally covered by local codes and ordinances or by state or federal requirements in the rare instances where local jurisdictions do not apply. Most codes, ordinances, or regulations reference the following building codes or standards for in-space heating:

| **Building Codes** | **Reference** |
|---|---|
| *BOCA/National Building Code,* 10th ed. (1985) | BOCA |
| *CABO One- and Two-Family Dwelling Code* (1986) | CABO |
| *Standard Building Code* (1985) | SBCCI |
| *Uniform Building Code* (1985) | ICBO |
| **Mechanical Codes** | |
| *BOCA/National Mechanical Code,* 6th ed. (1987) | BOCA |
| *Uniform Mechanical Code* (1985) | ICBO/IAPMO |
| *Standard Mechanical Code* (1985) | SBCCI |
| **Electrical Codes** | |
| *National Electric Code* | ANSI/NFPA 70-84 |
| *Canadian Electrical Code* | CSA C22.1-1986 |
| **Chimneys** | |
| *Chimneys, Fireplaces, Vents and Solid Fuel-Burning Appliances* | ANSI/NFPA 211-84 |
| *Chimneys, Factory-Built Residential-Type and Building Heating Appliance* | UL 103-83 |

Chapter 45, "Codes and Standards," has further information, including the names and addresses of the above agencies. Safety and performance criteria are furnished by the manufacturer.

### Equipment Using Utility-Furnished Energy

Those systems that rely on energy furnished by a utility are usually required to comply with local utility service rules and regulations. The utility will usually provide information on the installation and use of the equipment that uses their energy. Bottled gas (LPG) equipment is generally listed and tested under the same standards as natural gas. LPG equipment may be identical to natural gas equipment, but it always has a different orifice and, sometimes, a different burner and controls. The listings and examinations are usually the same for natural, mixed, manufactured, or LPG.

### Equipment Involving Products of Combustion

Equipment producing products of combustion normally must be provided with a closed piping system from the combustion chamber outdoors. Gas-fired equipment located in large-volume spaces and residential in-space unvented heaters are exceptions.

Oil- and gas-fired equipment may be vented through masonry stacks, chimneys, specifically designed venting systems, or, in some cases, venting systems incorporating forced- or induced-draft fans. Chapter 26 covers chimneys, gas vents, and fireplace systems in more detail.

## Solid-Fuel Safety

The evacuation of the gases of combustion are a prime concern in the installation of solid fuel-burning equipment. NFPA 211, *Chimneys, Fireplaces, Vents, and Solid Fuel-Burning Appliances,* contains requirements that should be followed. Because safety requirements for stovepipes (connections between stoves and chimneys) are not always readily available, safety considerations are summarized here.

A stovepipe is usually a black (or blue) steel single-wall pipe of a thickness shown in Table 2. Stainless steel is a corrosion-resistant alternative that need not meet thicknesses listed in Table 2. Stovepipe should be installed with the crimped (male) end of the pipe toward the stove (so that creosote and water drips back to the stove). The pipe should be as short as practical with a minimum of turns and horizontal run. Horizontal runs should be pitched 1/4 in./ft (20 mm/m) up toward the chimney.

The stovepipe should not pass through a ceiling, closet, alcove, or concealed space, so chimney material is required. When passing through a combustible interior or exterior wall, the builder must convert to an approved chimney, run stovepipe through a ventilated thimble (at least 18 in. [450 mm] larger than 6 in. [150 mm] stovepipe, 20 in. [500 mm] larger than 8 in. [200 mm]), or cut away the combustible material to allow 18 in. (450 mm) clearance all around the stovepipe.

Coal should be burned only in fireplaces or stoves designed specifically for coal burning. The chimney used in coal-fired applications must also be designed and approved for coal as well as for wood.

The safe operation of wood-burning stoves begins with an understanding of differences between wood and other fuels. Gas, for example, is held closely to a constant heating value, is furnished at consistent pressure, is burned at a constant rate, and has consistent excess air. Wood, on the other hand, varies in composition, size, shape, density, and moisture content. Burning is complicated in the *controlled combustion* or *airtight heater* because it operates mostly in an *air-starved* environment. Wood fuel is not continuously fed in gaseous or atomized form, which would stimulate a consistent burning rate. Instead, a supply is introduced periodically into the combustion chamber.

When operated manually, the air supply to the stove varies considerably. Preheating the fuel prior to ignition is necessary so that there is enough material burning to start a draft, which will continue to increase (unless reduced). The stronger the draft, the more air is supplied and the faster the fuel burns, though not always with a corresponding increase in usable heat.

Two control methods are widely used. First, control of the draft at the stove or by damper operation between the stove and the stovepipe; and secondly, control of the rate of burning by control of the air admitted to the stove (sometimes adjusting only the primary air opening, sometimes adjusting both primary and secondary openings). Experimentation determines what works best for a given installation. However, there are some principles of operation that are beneficial to any system.

**Creosote** forms on all wood-burning systems. Thin deposits in the stovepipe and chimney do not interfere with the operation, but thick deposits may ignite and cause a dangerous chimney fire.

**Table 2 Stovepipe Wall Thickness[a]**

| Size Dia in. | Gauge | Thickness in. | Size Dia mm | Thickness mm |
|---|---|---|---|---|
| 4 | 26 | 0.018 | 100 | 0.50 |
| 6 | 24 | 0.024 | 150 | 0.60 |
| 8 | 24 | 0.024 | 200 | 0.60 |
| 10 | 22 | 0.030 | 250 | 0.80 |

[a]Do not use thinner pipe. The recommended thickness will not last as long as desired under wood stove operations. At least 18 in. clearance must be left between stovepipe and wall or ceiling.

A clean chimney contributes to safe and effective operation. If sound installation and operation practices are followed, a clean chimney can be maintained without frequent cleaning. Ideally, the rear (rather than the top) discharge opening from the stove is used. It is located so that a capped tee can be used immediately off the stove or off the tapered increasing-diameter adapter (if one is used) to discharge directly upward through the stovepipe into the chimney. The bottom leg of the tee has a cap that can be easily removed to see up through the chimney. Some manufacturers make a stainless steel fitting with a convenient removable cap for the bottom leg.

Chimney cleaning is done best with wire brushes that are made to fit various sizes of round manufactured chimneys and standard tile liners of masonry chimneys. Unlined or large masonry chimneys should not be used for wood-burning equipment unless lined with chimney tile or metal stovepipe the same size as the stove discharge. Oversized chimneys may increase creosote deposits.

Creosote problems can be reduced by (1) opening the air inlet to establish a hot fire quickly and then closing it to the desired position; (2) adding only a partial load (about one-third capacity) when refueling to keep the fire hot; (3) frequently inspecting and cleaning the chimney; and (4) learning how to burn all types of wood, although well-seasoned hardwood burns best.

# AGENCY TESTING

The construction and performance of in-space heaters are contained in the standards of several agencies. The following summarizes the important standards that apply to residential in-space heating and are coordinated or sponsored by ASHRAE, the American National Standards Institute (ANSI), Underwriters Laboratories (UL), American Gas Association (AGA), and the Canadian Gas Association (CGA). Some CGA standards have a CANI prefix.

| | |
|---|---|
| ANSI Z21.11.1-83 | *Gas-Fired Room Heaters, Vented* |
| ANSI Z21.11.2-83 | *Gas-Fired Room Heaters, Unvented* |
| ANSI Z21.44-85 | *Gas-Fired Gravity and Fan-Type Direct-Vent Wall Furnaces* |
| ANSI Z21.48-86 | *Gas-Fired Gravity and Fan-Type Floor Furnaces* |
| ANSI Z21.49-86 | *Gas-Fired Gravity and Fan-Type Vented Wall Furnaces* |
| CANI 2.1-86 | *Gas-Fired Vented Room Heaters* |
| CGA 2.5-86 | *Gravity and Fan-Type Vented Wall Furnaces* |
| CGA 2.19-81 | *Gravity and Fan-Type Sealed Combustion System Wall Furnaces* |
| ANSI/UL 127-85 | *Factory-Built Fireplaces* |
| ANSI/UL 574-80 | *Electric Oil Heaters* |
| UL 647-82 | *Unvented Kerosene-Fired Heaters and Portable Heaters* |
| ANSI/UL 729-76 | *Oil-Fired Floor Furnaces* |
| ANSI/UL 730-78 | *Oil-Fired Wall Furances* |
| ANSI/UL 737-82 | *Fireplace Stoves* |
| UL 896-73 | *Oil-Burning Stoves* |

ANSI/UL 1025-80 *Electric Air Heaters*
ANSI/UL 1042-79 *Electric Baseboard Heating Equipment*
ANSI/UL 1482-83 *Heaters, Room Solid-Fuel Type*
ASHRAE 106-1984 *Rating Wood-Burning Fireplaces and Fireplace Stoves*
ASHRAE 62-1981 *Ventilation for Acceptable Indoor Air Quality*

## BIBLIOGRAPHY

ANSI. 1984. *National Fuel Gas Code.* American National Standards Institute *Standard* ANSI Z223.1-84. (Also NFPA 54-84.)

Gas Appliance Manufacturers Association. Directory of Gas Room Heaters, Floor Furnaces and Wall Furnaces. Arlington, VA.

MacKay, S., L.D. Baker, J.W. Bartok, and J.P. Lassoie. 1985. *Burning Wood and Coal.* Northeast Regional Agricultural Engineering Service, Cornell University, Ithaca, NY.

National Fire Protection Association. 1984. Chimneys, Fireplaces, Vents and Solid Fuel Burning Appliances. NFPA 211-84. Quincy, MA.

U.S. Department of Energy. 1984. Uniform Test Method for Measuring the Energy Consumption of Vented Home Heating Equipment. *Federal Register,* Vol. 49, p. 12,169, March 28.

Wood Heating Education and Research Foundation. 1984. *Solid Fuel Safety Study Manual for Level I Solid Fuel Safety Technicians.* Washington, DC.

CHAPTER 26

# CHIMNEY, GAS VENT, AND FIREPLACE SYSTEMS



A properly designed chimney controls draft and removes flue gases. This chapter describes the design of chimneys that discharge flue gases from appliance-chimney and fireplace-chimney systems.

In this chapter, *appliance* refers to any furnace, boiler, or incinerator (including the burner). The term *chimney* includes specialized vent products such as gas vents, unless the context indicates otherwise. *Draft* is negative gauge pressure, measured relative to atmospheric pressure, so positive draft is negative gauge pressure. *Flue gas* is the mixture of gases discharged from the appliance and conveyed by the chimney or vent system.

Appliances have the following Draft Configurations (Stone 1971):

1. Those that require draft applied at the appliance flue gas outlet to induce air into the appliance
2. Those that operate without draft applied at the appliance flue gas outlet, *e.g.*, a gas appliance with a draft hood in which the combustion process is isolated from chimney draft variations
3. Those that produce positive pressure at the appliance outlet collar so no chimney draft is needed; appliances that produce some positive outlet pressure but also need some chimney draft

In the first two configurations, hot flue gas buoyancy, induced draft chimney fans, or a combination of both produce draft. The third configuration may not require a chimney draft, but it should be considered in the design if a chimney is used. If the chimney system is undersized, draft inducers in the connector or chimney may supply appliance-chimney system draft needs. If the connector or chimney pressure requires control, draft control devices must be used.

The draft needed to overcome chimney flow resistance ($\Delta p$) is as follows:

$$\Delta p = \text{Theoretical Draft} - \text{Available Draft} = D_t - D_a$$

The above three appliance draft configurations use this equation: in the second configuration with zero draft requirement at the appliance outlet, available draft is zero, and theoretical draft of the chimney equals the chimney flow resistance.

*Available Draft*, $D_a$, is the draft needed at the appliance outlet. If increased chimney height and flue gas temperatures provide surplus available draft, draft control is required.

*Theoretical Draft*, $D_t$, is the natural draft produced by the buoyancy of hot gases in the chimney relative to cooler gases in the atmosphere. It depends on chimney height and the mean gas temperature difference, $t_m$, which is a weighted average temperature difference of the flue and atmospheric gases. Therefore, cooling by heat transfer through the chimney wall is a key variable in chimney design. Precise evaluation of theoretical draft is not necessary for most design calculations due to the availability of design charts and capacity tables in the building codes and manufacturers' data sheets.

Chimney temperatures and acceptable combustible material temperatures must be known to determine safe clearances between the chimney and combustible materials. Safe clearances for some chimney systems, such as Type *B* gas vents, are determined by standard tests and/or specified in building codes.

The following sections cover the basis of chimney design for average operating conditions but do not include a rigorous evaluation of the mean flue gas temperature in the chimney.

## CHIMNEY FUNCTIONS— GENERAL CONSIDERATIONS

The proper chimney can be selected by evaluating such factors as draft configuration, size, and operating conditions of the appliance; construction of surroundings; appliance usage classification; residential, low, medium, or high heat (NFPA *Standard* 211); and height of building. The chimney designer should know the applicable codes and standards to ensure acceptable construction.

In addition to chimney draft, the following factors must to be considered for safe and reliable operation: air supply; draft control devices; chimney materials (corrosion and temperature


The preparation of this chapter is assigned to TC 3.7, Fuels and Combustion.

resistance); flue gas temperatures, composition, and dew point; wind eddy zones; and particulate dispersion. Chimney materials must resist oxidation and condensation at both high and low fire levels.

**Startup**. The Design Chart may be used to determine vent or chimney size based on steady-state operating conditions. The chart, however, does not consider modulation, cycling, or time to achieve equilibrium flow conditions from a cold start. While mechanical draft systems have no problem starting gas flow, gravity systems rely on the bouyancy of hot gases as the sole force to displace the cold air present in the chimney. Priming follows Newton's Laws of motion. The time to fill a system with hot gases, displace the cold air, and start flow is reasonably predictable and is usually a minute or less. But unfavorable thermal differentials, building-chimney interaction, mechanical (exhaust fans, etc.), or wind forces that oppose the normal flow of vent gases can overwhelm the buoyancy force. Then, rapid priming cannot be obtained solely from correct system design.

**Air Intakes**. All rooms or spaces containing fuel-burning equipment must have a constant combustion air supply at adequate static pressure to ensure proper combustion. In addition, outside air is required to replace the air entering chimney systems through draft hoods and barometric draft regulators and to ventilate closely confined boiler and furnace rooms.

Because of the variable air requirements, the presence of other air-moving equipment in the building, and the variations in building construction and arrangement, no universally accepted rule specifies outdoor air openings. The minimum combustion air opening size depends on burner input with a practical minimum being the vent connector or chimney area. Any design must consider flow resistance of the combustion air supply, including register-louvre resistance. Air supply openings that meet most building code requirements have a negligible flow resistance.

**Vent Size**. Small residential and commercial natural draft gas appliances have draft hoods and vent diameters of 3 to 12 in. (75 to 305 mm). NFPA *Standard* 54 (also ANSI *Standard* Z223.1), *National Fuel Gas Code*, recommends sizes or input capacities for most acceptable gas appliance venting materials. These sizes also apply to gas appliances with integral automatic vent dampers, as well as to appliances with field-installed automatic vent dampers. Field-installed automatic vent dampers should be certified for use with the specific appliance by a recognized testing agency or evaluated for safe application by qualified installers.

**Draft Control**. Pressure, temperature, and other draft controls have replaced draft hoods in many residential furnaces and boilers to attain higher steady-state and seasonal efficiencies. Appliances that use pulse combustion or forced- or induced-draft fans, as well as those designed for sealed or direct venting, do not have a draft hood but may require special venting or special vent terminals. If fan-assisted burners deliver fuel and air to the combustion chamber and also overcome most of the appliance flow resistance, draft hoods or other control devices may be installed, depending on the design of the appliance. Category III and IV and some I and II appliances, as defined in ANSI *Standard* Z21.47, do not use draft hoods, so the manufacturer's vent system design requirements should be followed. The section on "Vent and Chimney Accessories" has more information on draft hoods, barometric regulators, draft fans, and other draft-control devices.

Frequently, a chimney must produce excess flow or draft. For example, dangerously high flue gas outlet temperatures from an incinerator may be reduced by diluting the air in the chimney with excess draft. The section on "Mass Flow Based on Fuel and Combustion Products" briefly addresses draft-control conditions.

**Pollution Control**. Where control of pollutant emissions is impossible, the chimney should be tall enough to ensure dispersion over a wide area to prevent objectionable ground level concentrations. The chimney can also serve as a passageway to carry flue gas to pollution-control equipment. This passageway must meet the building code requirements of a chimney, even at the exit of pollution-control equipment, because of possible exposure to heat and corrosion. A bypass chimney also should be provided to allow continued exhaust in the event of pollution-control equipment failure, repair, or maintenance.

**Equipment Location**. Chimney materials may permit installing appliances at intermediate or all levels of a high-rise building without imposing weight penalties. Some gas vent systems permit individual apartment-by-apartment heating systems.

**Wind Effects**. Wind and eddy currents affect the discharge of gases from vents and chimneys. They must expel flue gas beyond the cavity or eddy zone surrounding a building to prevent reentry through openings and fresh air intakes. A chimney and its termination can stabilize the effects of wind on appliances and their equipment rooms. In many locations, the equipment room air supply is not at neutral pressure under all wind conditions. Locating the chimney outlet well into the undisturbed wind stream and away from the cavity and wake zones around a building counteracts wind effects on the air-supply pressure. It also prevents reentry through openings and contamination of fresh air intakes.

Chimney outlets below parapet or eave level, nearly flush with the wall or roof surface, or in known regions with stagnant air may be subjected to downdrafts and are undesirable. Caps for downdraft and rain protection must be installed according to their listings and the cap manufacturer's instructions, or the applicable building code.

Wind effects can be minimized by putting the chimney terminal and the combustion air inlet terminal close together in the same pressure zone.

**Safety Factors**. Safety factors allow for uncertainties of vent and chimney operation. For example, flue gas must not spill from a draft hood or barometric regulator, even when the chimney has very low available draft. The Table 2 design condition for gas vents, namely 300°F (150°C) rise in the system at 5.3% $CO_2$, allows gas vents to operate with reasonable safety above or below the suggested temperature and $CO_2$ limits.

Safety factors may also be added to the system friction coefficient to account for a possibile extra fitting, soot accumulation, and air supply resistance. Specific gravity of flue gases can vary depending on the fuel burned. Natural gas flue gas, for example, has a density as much as 5% less than air, while coke flue gas has a density as much as 8% greater. However, these density changes are insignificant relative to other uncertainties, so no compensation is needed.

## THE CHIMNEY DESIGN EQUATION

Chimney design balances the forces that produce flow against those that retard flow (friction). *Theoretical draft* is the force that produces flow in gravity or natural draft chimneys. It is defined as the static pressure resulting from the difference in densities between a stagnant column of hot flue gases and an equal column of ambient air. In the design or balancing process, theoretical draft may not equal friction loss, because the appliance is frequently built to operate with some specific pressure (positive or negative) at the appliance flue-gas exit. This exit pressure, or *available draft*, depends on appliance operating characteristics, the fuel, and the type of draft control.

Flow losses caused by friction may be estimated by several formulas for flow in pipes or ducts, such as the equivalent length method or the loss coefficient or velocity head method. Chapter 33 of the 1985 FUNDAMENTALS Volume covers computation of flow losses. This chapter emphasizes the loss coefficient method, because fittings usually cause the greater portion of system pressure drop in chimney systems, and conservative loss coefficients (which are almost independent of piping size) provide an adequate basis for design. Paul *et al.* (1985) developed a computer program (available from the Gas Research Institute) entitled *Vent-II (Revision 2): A Microcomputer Program for Designing Vent Systems for Gas Appliances.*

For large gravity chimneys, available draft may be calculated (in I-P units) from Eq. (1) or Eq. (2). Both equations use the equivalent length approach, as indicated by the symbol $L_e$ in the flow-loss term (ASHVE 1941). These equations permit considering density difference between chimney gases and ambient air, as well as compensating shape factors. Mean flue-gas temperature must be estimated separately.

For a cylindrical chimney:

$$D_a = 2.96HB\left(\frac{\varrho_o}{T_o} - \frac{\varrho_c}{T_m}\right) - \frac{0.000315w^2T_m fL_e}{1.3 \times 10^7 d_f^5 B\varrho_c} \quad \text{(1 I-P)}$$

and, for a rectangular chimney:

$$D_a = 2.96HB\left(\frac{\varrho_o}{T_o} - \frac{\varrho_c}{T_m}\right) - \frac{0.000097w^2T_m fL_e(x+y)}{1.3 \times 10^7 (xy)^3 B\varrho_c} \quad \text{(2 I-P)}$$

In these expressions, $\varrho_o/T_o$ determines theoretical draft, based on applicable gas and ambient density. The term, $\varrho_c/T_m$, defines draft loss based on the factors for flow in a circular or rectangular duct system. To use these equations, the quantity of flow, $w$, for a variety of fuels and situations and the available draft needs of various types of appliances must be determined.

Equations (1) and (2) may be rearranged into a form that is more readily applied to the problems of chimney design, size, and capacity by considering the following factors:

1. mass flow of combustion products
2. chimney gas temperature and density
3. theoretical and available draft
4. allowable system pressure loss because of flow
5. chimney gas velocity
6. system resistance coefficient
7. the final input-volume relationships

For applications to system design, the chimney gas velocity is eliminated; however, actual velocity can be found readily, if needed.

**Mass Flow of Combustion Products (Step 1).** Mass flow in a chimney or venting system may differ from that in the appliance, depending on the type of draft control or number of appliances operating in a multiple appliance system. Mass flow (rather than volume flow) is preferred because it remains constant in any continuous portion of the system, regardless of changes in temperature or pressure. For the chimney gases resulting from any combustion process, mass flow can be expressed as:

$$w = IM/1000 \quad (3)$$

where

$w$ = mass flow, lb/h (g/s)
$I$ = appliance heat input, Btu/h (kW)
$M$ = mass flow input ratio, lb of chimney products per 1000 Btu of fuel burned (kg/MJ). $M$ depends on the composition of the fuel and excess air (or $CO_2$) percentage in the chimney.

**Mean Chimney Gas Temperature (Step 2).** Chimney gas temperature, which is covered in a later section, depends on the fuel, appliance, draft control, chimney size, and configuration.

Density of gas within the chimney and theoretical draft both depend on gas temperature. While the gases flowing in a chimney system lose heat continuously, from the entrance to the exit, a single mean gas temperature must be used either in the design equation or the chart. Mean chimney gas density is essentially the same as that of air density at the same temperature. Thus, density may be found as:

$$\varrho_m = C_1 B/T_m \quad (4)$$

where

$\varrho_m$ = gas density, lb/ft$^3$ (kg/m$^3$)
$C_1$ = constant, 1.325 (0.00348)
$B$ = local barometric pressure, in. Hg (Pa)
$T_m$ = mean chimney gas temperature at average conditions in the system, °R (K)

The constant in Eq.(4) is a compromise value for typical humidity. The subscript $m$ for density and temperature requires that these properties be calculated at mean gas temperature or vertical midpoint of a system (inlet conditions can be used where temperature drop is not significant).

**Theoretical and Available Draft (Step 3).** The theoretical draft of a gravity chimney or vent is the difference in weight (mass) between a given column of warm (light) chimney gas and an equal column of cold (heavy) ambient air. Chimney gas density or temperature, chimney height, and barometric pressure determine theoretical draft; flow is not a factor. The equation for theoretical draft assumes chimney gas density is the same as that of air at the same temperature and pressure, so:

$$D_t = C_2 BH\left(\frac{1}{T_o} - \frac{1}{T_m}\right) \quad (5)$$

where

$D_t$ = Theoretical Draft, in. of water (Pa)
$C_2$ = constant, 0.2554 (0.03413)
$H$ = height of chimney above datum of draft measurement, ft (m)
$T_o$ = ambient temperature, °R (K)

Theoretical draft thus increases directly with height and with the difference in density between the hot and cold columns.

**System Pressure Loss Resulting from Flow (Step 4).** In any chimney system, flow losses, expressed as pressure drop $\Delta p$ in in. of water (Pa), absorb the energy difference between theoretical and available draft, so:

$$\Delta p = D_t - D_a \quad (6)$$

Available draft, $D_a$, is the static pressure defined by the appliance operating requirements as follows:

1. Negative (below atmospheric) draft appliances: $D_a$, Available Draft is positive in Eq. (6) so $\Delta p < D_t$.
2. Draft hood (neutral) draft appliances: $D_a = 0$, and $\Delta p = D_t$.
3. Positive (above atmospheric) forced draft appliances: $D_a$ is negative in Eq. (6) so $\Delta p = D_t - (-D_a) = D_t + D_a$ and $\Delta p > D_t$.

Regardless of the sign of $D_a$, $\Delta p$ is always positive.

In any duct system, flow losses resulting from velocity and resistance can be determined from the Bernoulli equation as:

$$\Delta p = \frac{k\varrho_m V^2}{5.2(2g)} \quad \text{(7 I-P)}$$

$$\Delta p = k\varrho_m V^2/2 \quad \text{(7 SI)}$$

where

$k$ = dimensionless system resistance coefficient of the piping and fittings
$V$ = system gas velocity at mean conditions, ft/s (m/s)
$g$ = gravitational constant 32.1740 ft/s$^2$

Pressure losses are thus directly proportional to the resistance factor and to the square of the velocity.

**Chimney Gas Velocity (Step 5).** Velocity in a chimney or vent varies inversely with gas density, $\varrho_m$, and directly with hourly mass flow, $w$. The equation for gas velocity at mean gas temperature in the chimney is:

$$V = \frac{w}{C_3 \varrho_m d_i^2} \quad (8)$$

*where*

$V$ = gas velocity, ft/s (m/s)
$d_i$ = inside diameter, in. (mm)
$\varrho_m$ = gas density, lb/ft$^3$ (kg/m$^3$)
$C_3$ = constant, (3600/144)($\pi$/4) = 19.63 ($\pi/4 \times 10^{-3} = 7.85 \times 10^{-4}$)

To express velocity as a function of input and chimney gas composition, $w$, in Eq. (8) is replaced by using Eq. (3).

$$V = \frac{IM}{C_3 \varrho_m d_i^2} \quad (9)$$

Thus, chimney velocity depends on the product of heat input, $I$, and mass flow input, $M$.

**System Resistance Coefficient (Step 6).** The velocity head method for resistance losses assigns a fixed numerical coefficient (independent of velocity) or $k$ factor to every fitting or turn in the flow circuit, as well as to piping.

**Input, Diameter, and Temperature Relationships (Step 7).** To obtain a design equation in which all terms are readily defined, measured, or predetermined, the gas velocity and density terms must be eliminated. Thus, using Eq. (4) to replace $\varrho$ and Eq. (9) to replace $V$ in Eq. (7) gives:

$$\Delta p = \frac{k \varrho_m V^2}{5.2(2g)} = \frac{k}{5.2(2g)} \left(\frac{T_m}{C_1 B}\right) \left(\frac{IM}{C_3 d_i^2}\right)^2 \quad (10\ \text{I-P})$$

$$\Delta p = \frac{k}{2} \left(\frac{T_m}{C_l B}\right) \left(\frac{IM}{C_3 d_i^2}\right)^2 \quad (10\ \text{SI})$$

Rearranging to solve for $I$ and including the value of $2g$ and $C_3$ gives:

$$I = C_4 \frac{(d_i)^2}{M} \left(\frac{\Delta p B}{k T_m}\right)^{0.5} \quad (11)$$

*where*

$C_4$ = constant, 4.13 × 10$^5$ (0.0655)

Solving for input using Eq. (11) is a one-step process, given the diameter and configuration of chimney. More frequently, however, input, available draft, and height are given and the diameter, $d_i$, must be found. Because system resistance is a function of chimney diameter, a trial resistance value must be assumed to calculate a trial diameter. This method allows for a second (and usually accurate) solution for the final required diameter.

**Volume of Flow in Chimney or System (Step 8).** Volume flow may be calculated in a chimney system for which Eq. (11) can be solved by solving Eq. (7) for velocity at mean density (or temperature) conditions:

$$V = C_5 \sqrt{\Delta p / k \varrho_m} \quad (12)$$

*where*

$C_5$ = constant, 18.9 ($\sqrt{2}$)

This equation can be expressed in the same terms as Eq. (11) by using the density value $\varrho$ of Eq. (4) in Eq. (12) and then substituting Eq. (12) for velocity, $V$. Area is expressed in terms of $d_i$.

$$Q = C_6 d_i^2 (\Delta p\, T_m / kB)^{0.5} \quad (13)$$

*where*

$Q$ = volume flow rate, cfm (m$^3$/s)
$C_6$ = constant, 5.2 (18.83)

The volume flow obtained from Eq. (13) is at mean gas temperature, $T_m$, and at local barometric pressure, $B$.

Equation (13) is useful in the design of forced and induced draft systems because draft fans are usually rated in cfm (L/s) flowing at some standard ambient or selected gas temperature. An induced draft fan is necessary for chimneys that are undersize, too low, or those that must be operated with negative static draft in the manifold under all conditions.

Figure 1 is a graphic solution for Eqs. (11) and (13), which is accurate enough for most problems. However, to use either the equations or the chart, the details in the sections to follow should be understood so proper choices can be made for mass flow, pressure loss, and heat-transfer effects. Neither the Design Chart nor the equations contain the same number or order of steps as the derivation; for example, a step disappears when theoretical and available draft are combined into $\Delta p$. Similarly, the examples selected vary in their sequence of solution, depending on which parameters are known and on the need for differing answers such as diameter for a given input, diameter versus height, or the amount of pressure boost from a forced-draft fan.

## MASS FLOW BASED ON FUEL AND COMBUSTION PRODUCTS

In chimney system design, the composition and flow rate of the flue gases must be assumed to determine the mass flow input ratio, $M$. The mass flow equations in Table 1 illustrate the influence of fuel composition; however, additional guidance is needed for system design. The information provided for many heat-producing appliances is limited to whether they have been tested, certified, listed, or approved to comply with applicable standards. From this information and from the type of fuel and draft control, certain inferences can be drawn regarding the flue gases. Table 2 suggests typical values for the vent or chimney systems for gaseous and liquid fuels when specific outlet conditions for the appliance are not known. When combustion conditions are given in terms of excess air, Figure 2 can be used to estimate $CO_2$.

Figure 3 can be used with Figure 1 to estimate mass and volume flow for certain flow volume and forced-draft designs. Flow conditions within the chimney connector, manifold, vent, and chimney vary with configuration and appliance design and are not necessarily the same as boiler or appliance outlet conditions. The Design Chart (Figure 1) is based on the fuel combustion products and temperatures within the chimney system. If a gas appliance with draft hood is used, Table 2 recommends that dilution air through the draft hood reduce $CO_2$ percentage to 5.3%. For equipment using draft regulators, the dilution and temperature reduction is a function of the draft regulator gate opening, which depends on excess draft. If the chimney system produces the exact draft necessary for the appliance, little dilution takes place.

For manifolded gas appliances that have draft hoods, the dilution through draft hoods of inoperative appliances must be considered in precise system design. However, with forced-draft appliances having wind box or inlet air controls, dilution through inoperative appliances may be unimportant, especially if pressure at the outlet of inoperative appliances is neutral (atmospheric level). Mass flow within incinerator chimneys must account for the probable heating value of the waste, its moisture content, and the use of additional fuel to initiate or sustain combustion. Classifications of wastes and corresponding values of $M$ in Table 3 are based on recommendations from the Incinerator Institute of America. Combustion data given for Types 0, 1, and 2 wastes do not include any additional fuel. Where constant burner operation accompanies the combustion of waste, the additional quantity of products should be considered in the chimney design.

The system designer should obtain exact outlet conditions for the maximum rate operation of the specific appliance. This information can save chimney construction costs. For appliances with higher seasonal or steady-state efficiencies, however, special attention should be given to the manufacturer's venting recommendations because the flue products may differ in composition and temperature from conventional values.

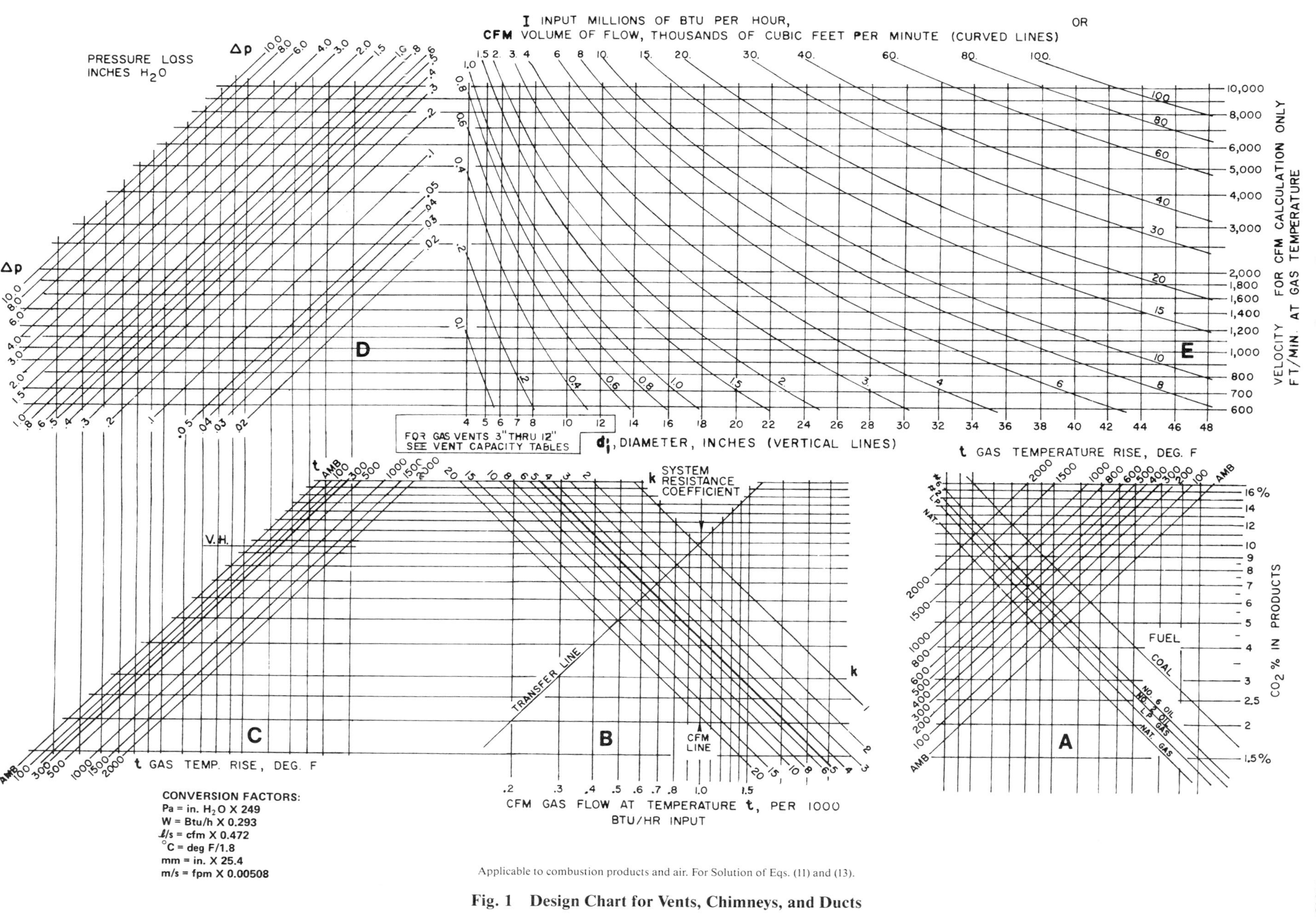


Applicable to combustion products and air. For Solution of Eqs. (11) and (13).

**Fig. 1 Design Chart for Vents, Chimneys, and Ducts**

**Table 1 Mass Flow Equations for Common Fuels**

| Fuel | Mass Flow Input Ratio, M: lb Total Products / 1000 Btu Fuel Input | kg / MJ |
|---|---|---|
| Natural gas | $0.705\left(0.159 + \frac{10.72}{\%CO_2}\right)$ | $0.303\left(0.159 + \frac{10.72}{\%CO_2}\right)$ |
| LP (propane, butane, or mixture) | $0.706\left(0.144 + \frac{12.61}{\%CO_2}\right)$ | $0.304\left(0.144 + \frac{12.61}{\%CO_2}\right)$ |
| No. 2 Oil (light) | $0.72\left(0.12 + \frac{14.4}{\%CO_2}\right)$ | $0.31\left(0.12 + \frac{14.4}{\%CO_2}\right)$ |
| No. 6 Oil (heavy) | $0.72\left(0.12 + \frac{15.8}{\%CO_2}\right)$ | $0.31\left(0.12 + \frac{15.8}{\%CO_2}\right)$ |
| Bituminous coal (soft) | $0.76\left(0.11 + \frac{18.2}{\%CO_2}\right)$ | $0.327\left(0.11 + \frac{18.2}{\%CO_2}\right)$ |
| Type O waste or wood | $0.69\left(0.16 + \frac{19.7}{\%CO_2}\right)$ | $0.30\left(0.16 + \frac{19.7}{\%CO_2}\right)$ |

$\%CO_2$ is determined in products with water condensed (dry basis).
Total products includes combustion products and excess air.

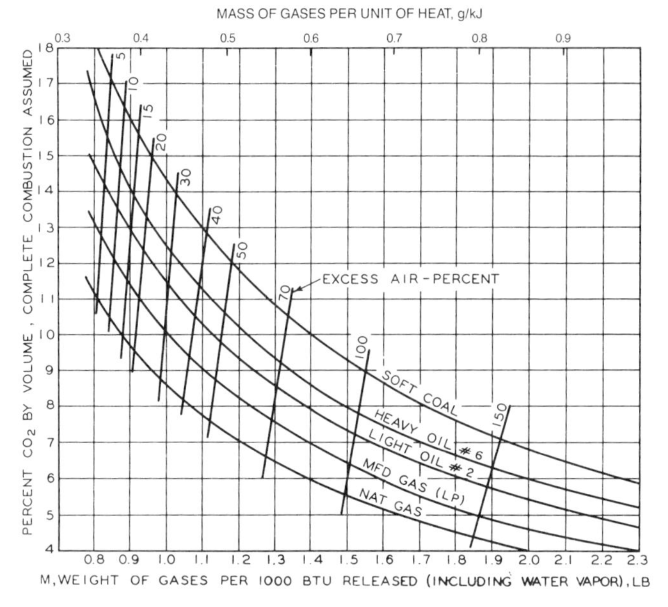


**Fig. 2 Graphical Evaluation of Rate of Flue Gas Flow from Percent $CO_2$ and Fuel Rate**

## CHIMNEY GAS TEMPERATURE AND HEAT TRANSFER

Figure 1 is based on a design ambient temperature of 60°F (15.6°C), and all temperatures given are in terms of rise above this ambient. Thus, the 300°F (166.7°C) line indicates a 360°F (182.2°C) observed flue gas temperature. Using a reasonably high ambient (such as 60°F [15.6°C]) for design ensures improved operation of the chimney when ambient temperatures drop, because of greater temperature differentials and greater draft.

A design requires assuming an initial or inlet chimney gas temperature. In the absence of specific data, Table 4 is conservative and may be used to select a temperature. For appliances capable of operating over a range of temperatures, size should be calculated at both extremes to ensure an adequate chimney.

The drop in flue gas temperature from appliance to exit reduces capacity, particularly in sizes of 12 in. (305 mm) or less. In gravity Type *B* gas vents, which may be as small as 3 in. (76 mm) in diameter, and in other systems used for venting gas appliances, capacity is best determined from the *National Fuel Gas Code* (ANSI *Standard* Z223.1). In this code, the tables compensate for the particular characteristics of the chimney material involved, except for very high single-wall metal pipe. Over 12 through 18-in. (305 through 457-mm) diameters, the effect of heat loss diminishes greatly, because there is greater gas flow relative to system surface area. For 20-in. (508-mm) diameters and above, cooling has little effect on final size or capacity.

A straight vertical vent or chimney directly off the appliance requires little compensation for cooling effects, even with smaller sizes. However, a horizontal connector running from appliance to the base of the vent or chimney has enough heat loss to diminish draft and capacity. Figure 4 is a plot of temperature correction, $C_u$, which is a function of connector size, length, and material, for either conventional single-wall metal, or double-wall metal.

To use Figure 4, estimate connector size and length and read the temperature multiplier. For example, 16 ft of single wall connector, 7 in. in diameter has a multiplier of 0.61. If inlet temperature rise is 300°F above ambient, operating mean temperature rise will be 0.61 (300) = 183°F rise. This factor adequately corrects the temperature at the midpoint of the vertical vent for heights up to 100 ft (30 m).

The temperature multiplier must be applied to grids A, C, and D in the Design Chart, Figure 1, as follows:

1. In grid A, the temperature must be multiplied by 0.61. For an appliance with outlet temperature rise above ambient of 300°F, flow in the vent is based on 300(0.61), or 183°F rise.
2. This same 183°F rise must be used in grid C.

**Table 2 Typical Chimney and Vent Design Conditions[a]**

| Fuel | Appliance | $\%CO_2$ | Temperature Rise °F | Temperature Rise (°C) | Mass Flow Input Ratio, M: lb Total Products[b] / 1000 Btu Fuel Input | kg/MJ | Gas Density[c] lb/ft³ | Gas Density[c] (kg/m³) | Flow Rate/Unit Heat Input[c] at Gas Temperature: cfm/1000 Btu/h | L/(s • kW) |
|---|---|---|---|---|---|---|---|---|---|---|
| Natural gas | Draft hood | 5.3 | 300 | (167) | 1.60 | (0.688) | 0.0483 | (0.7728) | 0.552 | (0.889) |
| LP gas | Draft hood | 6.0 | 300 | (167) | 1.64 | (0.705) | 0.0483 | (0.7728) | 0.566 | (0.912) |
| Gas | No draft hood | 8.0 | 400 | (222) | 1.10 | (0.473) | 0.0431 | (0.6896) | 0.425 | (0.684) |
| No. 2 oil | Residential | 9.0 | 500 | (278) | 1.24 | (0.533) | 0.0389 | (0.6224) | 0.531 | (0.855) |
| Oil | Forced draft over 400,000 Btu/h (117 kW) | 13.5 | 300 | (167) | 0.86 | (0.370) | 0.0483 | (0.7728) | 0.297 | (0.478) |
| Waste, Type O | Incinerator | 9.0 | 1340 | (744) | 1.62 | (0.697) | 0.0213 | (0.3408) | 1.267 | (2.040) |

[a] The values tabulated are for appliances with flue losses of 20% or more. For appliances with lower flue losses (high-efficiency-types), see the appliance installation instructions or ask the manufacturer for operating data.
[b] Total products include combustion products and excess air.
[c] At sea level.

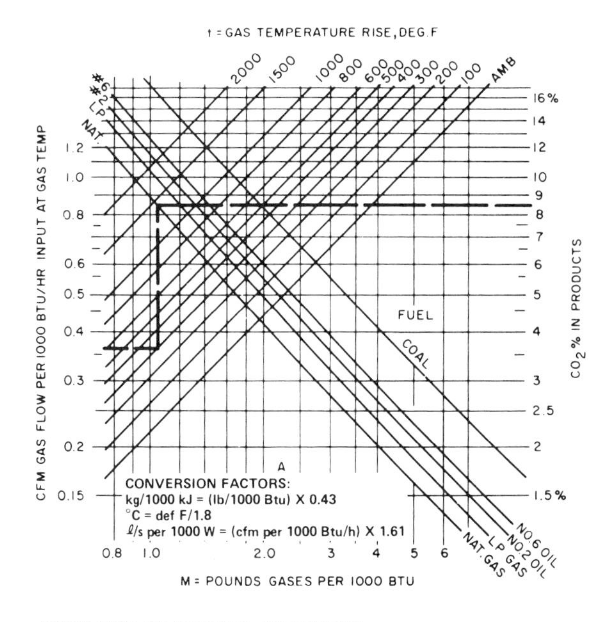


**Fig. 3 Flue Gas Weight and Volume Flow**

3. $\Delta p$ must be determined using a 183 °F rise for theoretical draft to be consistent with the other two temperatures. (It is incorrect to multiply theoretical draft pressure by the temperature multiplier.)

The first trial solution for diameter, using Figure 1 or Eqs. (11) and (13), need not consider the cooling temperature multiplier, even for smaller sizes. A first approximate size can be used for the temperature multiplier for all subsequent trials because capacity is insensitive to small changes in temperature.

The correction procedure assumes the overall heat-transfer coefficient of a vertical chimney is about 0.6 or less—the value for double-wall metal. This procedure does not correct for cooling in very high stacks constructed entirely of single-wall metal, especially those exposed to cold ambient temperatures. For severe exposures or excessive heat loss, a trial calculation assuming a conservative operating temperature shows whether capacity problems will be encountered.

For more precise heat loss calculations, Table 5 suggests overall heat-transfer coefficients for various constructions (Segeler 1965), as installed in typical environments at usual flue gas flow velocities. For masonry, any additional thickness beyond the single course of brick plus tile liner used in residential chimneys decreases the coefficient.

**Table 4 Chimney Gas Temperatures for Various Appliances**

| Appliance Type | Initial Temperature in Chimney °F | (°C) |
|---|---|---|
| Natural gas-fired heating appliance with draft hood (low efficiency) | 360 | (182) |
| L.P. gas-fired heating appliances with draft hood (low efficiency) | 360 | (182) |
| Gas-fired heating appliance, no draft hood (low efficiency) | 460 | (238) |
| Oil-fired heating appliances (low efficiency) | 560 | (293) |
| Conventional incinerators | 1400 | (760) |
| Controlled air incinerators | 1800 to 2400 | (980 to 1320) |
| Pathological incinerators | 1800 to 2800 | (980 to 1540) |
| Turbine exhaust | 900 to 1400 | (480 to 760) |
| Diesel exhaust | 900 to 1400 | (480 to 760) |
| Ceramic kilns | 1800 to 2400 | (980 to 1320) |

Subtract 60 °F (15.6 °C) ambient to obtain temperature rise for use with Figure 1.

Design Eqs. (11) and (13) do not account for effects of heat transfer or cooling on flow, draft, or capacity. Equations (14) and (15) describe flow and heat transfer, respectively, within a draft hood-type of gas venting system at standard conditions of $T_o$ = 520 °R (289 K) and $B$ = 29.92 in. Hg (101.325 kPa).

$$q_m = C_7\sqrt{2g}\, c_p \frac{B}{C_8}\frac{A}{T_m}\left(\frac{H}{k\,T_o}\right)^{0.5}(T_m - T_o)^{1.5} \qquad (14)$$

$$\frac{q}{q_m} = \frac{t_e}{t_m} = \exp\left(\pi \bar{U} d_f \frac{L_m t_m}{q_m}\right) \qquad (15)$$

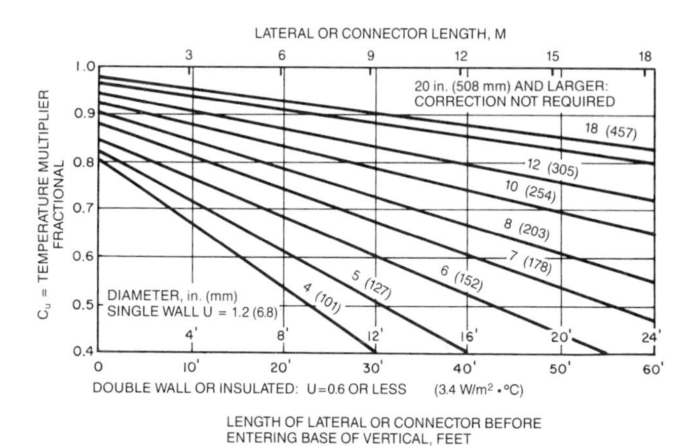


**Fig. 4 Temperature Multiplier, $C_u$, for Compensation of Heat Losses in Connector**

**Table 3 Mass Flow for Incinerator Chimneys**

| Type of Waste | Heat Value of Waste[a] Btu/lb | Heat Value of Waste[a] (MJ/kg) | Auxiliary Fuel[a] per Unit of Waste Btu/lb | Auxiliary Fuel[a] per Unit of Waste (MJ/kg) | Combustion Products cfm/lb waste @ 1400 °F[b] | (L/s • kg) @ 760 °C[b] | lb/h per lb waste[c] | (g/s per kg)[c] | lb products / 1000 Btu waste | kg / MJ |
|---|---|---|---|---|---|---|---|---|---|---|
| Type 0 | 8500 | (19.8) | 0 | 0 | 10.74 | (11.17) | 13.76 | (3.82) | 1.62 | (0.697) |
| Type 1 | 6500 | (15.1) | 0 | 0 | 8.40 | (8.74) | 10.80 | (3.00) | 1.66 | (0.714) |
| Type 2 | 4300 | (10.0) | 0 | 0 | 5.94 | (6.18) | 7.68 | (2.13) | 1.79 | (0.770) |
| Type 3 | 2500 | (5.8) | 1500 | (3.5) | 4.92 | (5.12) | 6.25 | (1.74) | 2.50 | (1.075) |
| Type 4 | 1000 | (2.3) | 3000 | (7.0) | 4.14 | (4.31) | 5.33 | (1.48) | 5.33 | (2.292) |

[a] Auxiliary fuel may be used with any type of waste, dependent on the design of the incinerator.
[b] Specialized units may produce higher or lower outlet gas temperatures, which must be considered in sizing the chimney, using Eqs. (11) or (13) or the Design Chart.
[c] These values must be multiplied by pounds (kg) of waste burned per hour to establish mass flow.

**Table 5 Overall Heat-Transfer Coefficients, U, of Various Chimneys and Vents**

| | U, Btu/h • ft² • °F (W/m² • °C)[a] | | |
|---|---|---|---|
| **Material** | **Observed** | **Design** | **Remarks** |
| Industrial steel stacks | — | 1.3 (7.4) | Under wet wind |
| Clay or iron sewer pipe | 1.3-1.4 | 1.3 (7.4) | Used as single-wall material |
| Asbestos-cement gas vent | 0.72-1.42 | 1.2 (6.8) | Tested per UL *Std.* 441 |
| Black or painted steel stove pipe | — | 1.2 (6.8) | Comparable to weathered galvanized steel |
| Single-wall galvanized steel | 0.31-1.38 | 1.0 (5.7) | Depends on surface condition and exposure |
| Single-wall unpainted pure aluminum | — | 1.0 (5.7) | No. 1100 or other bright surface aluminum alloy |
| Brick chimney, tile lined | 0.5-1.0 | 1.0 (5.7) | Residential construction as used for gas appliances |
| Double-wall gas vent, 1/4-in. (6.35-mm) air space | 0.37-1.04 | 0.4 (2.27) | Galvanized steel outer pipe, pure aluminum inner pipe; tested per UL *Std.* 441 |
| Double-wall gas vent, 1/2-in. (12.70-mm) air space | 0.34-0.7 | 0.4 (2.27) | |
| Insulated prefabricated chimney | 0.34-0.7 | 0.4 (2.27) | Solid insulation meets UL *Std.* 103 when fully insulated |

[a] U values based on inside area of chimney.

*where*

$A$ = area of passage cross section, ft² (m²)
$H$ = height of chimney above grade or inlet, ft (m)
$t_m$ = $(T_m - T_o)$
$L_m$ = Length from inlet to location (in the vertical) of the mean gas temperature, ft (m)
$d_f$ = inside diameter, ft (m)
$t_c$ = $(T - T_0)$ = temp. difference entering the system, °F (°C)
$\bar{U}$ = heat transfer coefficient, Btu/s • ft² • °F (W/m² • °C)
$C_7$ = 520 (289)
$C_8$ = 29.92 (101.325)

Assuming reasonable constancy of $\bar{U}$, the overall heat-transfer coefficient of the venting system material, Eqs (14) and (15) provide a solution for maximum gas vent capacity. They also can be used to develop cooling curves or calculate the length of pipe where internal moisture condenses. Kinkead (1962) details methods of solution and application to both individual and combined gas vents.

## THEORETICAL DRAFT, AVAILABLE DRAFT, AND ALTITUDE CORRECTION

Equation (5) for theoretical draft is the basis for Figure 5, which can be used up to 1000 °F (540 °C) and 7000 ft (2130 m) elevation. Theoretical draft should be estimated and included in system calculations, even for appliances producing considerable positive outlet static pressure, to achieve the economy of minimum chimney size. Equation (5) may be used directly to calculate exact values for theoretical draft at any altitude. For ease of application and consistency with the Design Chart, Table 6 lists approximate theoretical draft for typical gas temperature rises above 60 °F (15.6 °C) ambient.

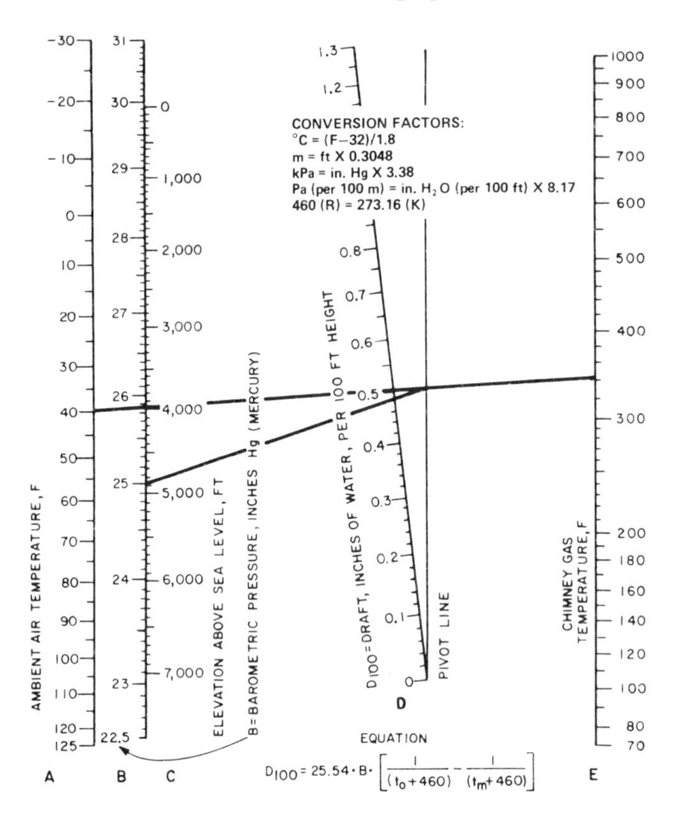


Fig. 5 Theoretical Draft Nomograph

Appliances with fixed fuel beds, such as hand-fired coal stoves and furnaces, require negative available draft. Small oil heaters with pot-type burners, as well as residential air furnaces with pressure atomizing oil burners, need negative available draft, which can usually be set by following the manufacturer's instructions for setting the draft regulator. Available draft requirements for larger packaged boilers or equipment assembled from components may be negative, zero (or neutral), or positive.

Compensation of theoretical draft for altitude or barometric pressure is usually necessary for appliances and chimneys functioning at elevations greater than 2000 ft (610 m). Figure 5, which

**Table 6 Approximate Theoretical Draft of Chimneys**

| Gas Temp. Rise, °F | $D_t$ per 100 ft in. of water | Temp. Rise, °C | $D_t$/m Pa |
|---|---|---|---|
| 100 | 0.2 | 75 | 2 |
| 150 | 0.3 | 100 | 3 |
| 200 | 0.4 | 150 | 4 |
| 300 | 0.5 | 200 | 5 |
| 400 | 0.6 | 300 | 6 |
| 500 | 0.7 | 400 | 7 |
| 600 | 0.8 | 600 | 8 |
| 800 | 0.9 | 900 | 9 |
| 1100 | 1.0 | 1500 | 10 |
| 1600 | 1.1 | | |
| 2400 | 1.2 | | |

*Notes*: Ambient temp. = 60 °F (16 °C) = 520 °R (289 K)
Chimney gas density = air density
Sea level barometric pressure = 29.92 in. Hg (101.325 kPa)
Equation (5) may be used to calculate exact values for $D_t$ at any altitude.

Table 7 Altitude Correction

| Altitude | | Barometric Pressure, B | | |
|---|---|---|---|---|
| ft | (m) | in. Hg | (kPa) | Factor |
| Sea level | | 29.92 | (101.325) | 1.00 |
| 2,000 | (610) | 27.8 | (94.0) | 1.08 |
| 4,000 | (1220) | 25.8 | (87.2) | 1.16 |
| 6,000 | (1830) | 24.0 | (81.1) | 1.25 |
| 8,000 | (2440) | 22.3 | (75.4) | 1.34 |
| 10,000 | (3050) | 20.6 | (69.6) | 1.45 |

Multiply operating input by the factor to obtain design input.

is applicable from sea level to 7000 ft (2130 m), may be used, or pressure versus elevation data may be consulted. Depending on the design, the following approaches to pressure or altitude compensation are necessary for chimney sizing.

1. Design Chart: Use sea level theoretical draft.
2. Equation (11): Use local theoretical draft with exact input or use sea level theoretical draft with Btu input multiplied by ratio of sea level to local barometric pressure (Table 7 factor).
3. Equation (13): Use local theoretical draft and barometric pressure with volume flow at the local density.

Figure 1 may be corrected for altitude or reduced air density by multiplying the operating input by the factor in Table 7. Gas appliances with draft hoods, for example, are usually derated 4% per 1000-ft (13%/km) elevation above sea level when they are operated at 2000-ft (610-m) altitude or above. The altitude correction factor derates the design input so the vent size at altitude for derated gas equipment is effectively the same as at sea level. For other appliances where burner adjustments or internal changes might be used to adjust for reduced density at altitude, the same factors produce an adequately compensated chimney size. For example, an appliance operating at 6000-ft elevation at 10,000,000 Btu/h input, but requiring the same draft as at sea level, should have a chimney selected on the basis of 1.25 times the operating input, or 12,500,000 Btu/h (3.7 MW).

Theoretical draft must be estimated at sea level to calculate $\Delta p$ for use with Figure 1. The altitude correction multiplier for input (Table 7) is the sole method of correcting to other elevations. Reducing theoretical draft imposes an incorrect compensation on the chart.

## SYSTEM FLOW LOSSES

Theoretical draft is always positive (unless chimney gases are colder than ambient air); however, available draft can be negative, zero, or positive. The pressure difference, $\Delta p$, or theoretical minus available draft, overcomes the flow losses. Table 8 lists the pressure components for three draft configurations. The table applies to still-air (no wind) conditions. The effect of wind on capacity or draft may be included by imposing a static pressure (either positive or negative) or by changing the vent terminal resistance loss. However, a properly designed and located vent terminal should cause little change in $\Delta p$ at typical wind velocities.

Table 8 Pressure Equations for $\Delta p$

| Required Appliance Outlet Pressure or Available Draft, $D_a$ | $\Delta p$ Equation | |
|---|---|---|
| | Gravity Only | Gravity Plus Inducer |
| 1. Negative, needs natural draft | $\Delta p = D_t - D_a$ | $\Delta p = D_t - D_a + D_b$ |
| 2. Zero, vent with draft hood or balanced forced draft | $\Delta p = D_t$ | $\Delta p = D_t + D_b$ |
| 3. Positive, forced draft | $\Delta p = D_t + D_a$ | $\Delta p = D_t + D_a + D_b$ |

Notes: Equations use the absolute value of $D_a$.
$D_b$ = static pressure boost of inducer at flue gas temperature and rated flow.

Although small static draft pressures can be measured at the entrance and exit of gas appliance draft hoods, available draft at the appliance is effectively zero. Therefore, all theoretical draft energy produces chimney flow velocity and overcomes chimney flow resistance losses.

## CHIMNEY GAS VELOCITY

Input capacity or diameter of a chimney may usually be found without determining flow velocity. Internal or exit velocity must occasionally be known, however, to ensure effluent dispersal or avoid flow noise. Also, the flow velocity of incinerator chimneys, turbine exhaust systems, and other appliances with high outlet pressures or velocities is needed to estimate piping loss coefficients.

Equations (7), (8), (9), and (12) can be applied to find velocity. Figure 1 may also be used to determine velocity; the right-hand scale of *Group E* reads directly in velocity for any combination of flow and diameter. For example, at 10,000 cfm and 34-in. diameter (6.305-ft$^2$ area), the indicated velocity is about 1600 fpm. The velocity may also be calculated by dividing volume flow rate by chimney area.

A similar calculation may be performed when the energy input is known. For example, a 34-in. chimney serving a 10 million Btu/h natural gas appliance, at 8.5% $CO_2$ and 300°F above ambient in the chimney, produces (from Figure 3) 0.36 cfm per 1000 Btu/h. Chimney gas flow rate is:

$$10{,}000{,}000(0.36/1000) = 3600 \text{ cfm } (1.7 \text{ m}^3/\text{s})$$

Dividing by area to obtain velocity, V = 3600/6.305 = 571 fpm (2.9 m/s). The right-hand scale of *Group E*, Figure 1, may also be multiplied by the cfm per 1000 Btu to find velocity. For the same chimney design conditions, the scale velocity value of 1600 fpm is multiplied by 0.36 to yield a velocity 576 fpm (2.9 m/s) in the chimney.

Chimney gas velocity affects the piping friction factor and also the roughness correction factor. The section on "Resistance Coefficients" has further information, and Example 2 illustrates how these factors are used in the velocity equations.

In a typical chimney design study, gas flow volume and temperature are known, while height, diameter, and velocity are unknown. These unknowns lead to a family of solutions; the final selection depends on the economics of chimney material and equipment operating cost. The typical velocity in vents and chimneys ranges from 300 to 3000 fpm (1.5 to 15 m/s).

Chimney systems can operate over a wide range of velocities, depending on modulation characteristics of the burner equipment or the number of appliances in operation. A chimney design developed for maximum input and maximum velocity should be satisfactory at reduced input because theoretical draft is roughly proportional to flue gas temperature, while flow losses are proportional to the square of the velocity. So, as input is reduced, flow losses decrease more rapidly than system motive pressures.

Effluent dispersal may occasionally require a minimum upward chimney outlet velocity, such as 3000 fpm (15 m/s). A tapered exit cone can best meet this requirement. For example, to increase an outlet velocity from 1600 to 3000 fpm from the 34 in. chimney ($A$ = 6.305 ft$^2$ area), the discharge area of the cone must be 6.305 x 1600/3000 = 3.36 ft$^2$ (0.31 m$^2$), or 24.8 in. (630 mm) in diameter.

An exit cone avoids excessive flow losses because the entire system operates at the lower velocity, and a resistance factor is only

**Table 9 Resistance-Loss Coefficients**

| Component | Suggested Design Value, Dimensionless[a] | Estimated Span and Notes |
|---|---|---|
| Inlet—acceleration | | |
| Gas vent with draft hood | 1.5 | 1.0 to 3.0 |
| Barometric regulator | 0.5 | 0.0 to 0.5 |
| Direct connection | 0.0 | Also dependent on blocking damper position |
| Round Elbow, 90° | 0.75 | 0.5 to 1.5 |
| Round Elbow, 45° | 0.3 | — |
| Tee or 90° connector | 1.25 | 1.0 to 4.0 |
| Y connector | 0.75 | 0.5 to 1.5 |
| Cap, top | | |
| Open straight | 0.0 | — |
| Low resistance (UL) | 0.5 | 0.0 to 1.5 |
| Other | — | 1.5 to 4.5 |
| Spark Screen | 0.5 | — |
| Converging exit cone | $(d_{i1}/d_{i2})^4 - 1$ | System designed using $d_{i1}$ |
| Tapered reducer ($d_{i1}$ to $d_{i2}$) | $1 - (d_{i2}/d_{i1})^4$ | System designed using $d_{i2}$ |
| Increaser | | See Chapter 2, Fluid Flow, 1985 FUNDAMENTALS Volume. |
| Piping $k_L$ | $0.4 \dfrac{L\text{, ft}}{d_i\text{, in.}}$ <br> $0.033 \dfrac{L\text{, mm}}{d_i\text{, mm}}$ | Numerical coefficient from 0.2 to 0.5; see Figure 13, Chapter 2, 1985 FUNDAMENTALS Volume for size, roughness, and velocity effects. |

[a] Initial assumption, when size is unknown:
$k$ = 5.0 for entire system, for first trial.
$k$ = 7.5 for combined gas vents only.
*Note*: For *combined gravity gas vents* serving two or more appliances (draft hoods), multiply total $k$ (of components + piping) by 1.5 to obtain gravity system design coefficient. (This rule does not apply to forced- or induced-draft vents or chimneys.)

added for the cone (Table 9). In this case, the added resistance for a gradual taper nearly equals:

$$k = (d_{i1}/d_{i2})^4 - 1 = (34/24.8)^4 - 1 = 2.53$$

Noise in chimneys may be caused by turbulent flow at high velocity or by combustion-induced oscillations or resonance. Noise is seldom encountered in gas vent systems or in systems producing negative available draft, but it may be a problem with forced-draft appliances. Turbulent flow noise may be avoided by designing for lower velocity, which may entail increasing the chimney size above the minimum recommended by the appliance manufacturer. Chapter 52 of the 1987 HVAC Volume has more information on noise control.

## RESISTANCE COEFFICIENTS

The resistance coefficient, $k$, which appears in Eqs. (11) and (13) and Figure 1 summarizes the friction loss of the entire chimney system, including piping, fittings, and configuration or interconnection factors. Capacity of the chimney varies inversely with the square root of $k$, while diameter varies as the fourth root of $k$. The insensitivity of diameter and input to small variations in $k$ simplifies design. Analyzing such details as pressure regain, increasers and reducers, and gas cooling junction effects is unnecessary if slightly high resistance coefficients are assigned to any draft diverters, elbows, tees, terminations, and, particularly, piping.

The flow resistance of a fitting such as a tee with gases entering the side and making a 90° turn is assumed to be constant at $k$ = 1.25, independent of size, velocity, orientation, inlet or outlet conditions, or whether the tee is located in an individual vent or in a manifold. Conversely, if the gases pass straight through a tee, as in a manifold, assumed resistance is zero, regardless of any area changes or flow entry from the side branch. For any chimney with fittings, the total flow resistance is a constant plus variable piping resistance—the latter being a function of center line length divided by diameter. Table 9 suggests moderately conservative resistance coefficients for common fittings. Elbow resistance may be lowered by long radius turns; however, corrugated 90° elbows may have resistance values at the high end of the scale. The table shows resistance as a function of inlet diameter, $d_{i1}$, and outlet diameter, $d_{i2}$.

Expressed mathematically, system resistance, $k$, may be written:

$$k = k_1 + n_2k_2 + n_3k_3 + k_4 + k_L + \text{etc.}$$

*where*

$k_1$ = inlet-acceleration coefficient
$k_2$ = elbow loss coefficient, $n_2$ = number of elbows
$k_3$ = tee loss coefficient, $n_3$ = number of tees
$k_4$ = cap, top, or exit cone loss coefficient
$k_L$ = piping loss = $FL/d$ (Figure 13, Chapter 2, of the 1985 FUNDAMENTALS Volume)

For combined gas vents using appliances with draft hoods, the summation, $k$, must be multiplied by 1.5 (see Example 4).

The resistance coefficient method adapts well to systems whose fittings cause significant losses. Even for extensive systems, an initial assumption of k = 5.0 gives a tolerably accurate vent or chimney diameter in the first trial solution. Using this diameter with the piping-resistance function in a second trial normally yields the final answer.

The minimum system resistance coefficient in a gas vent with a draft hood is always 1.0 because all gases must accelerate through the draft hood from almost zero velocity to vent velocity.

For a system connected directly to the outlet of a boiler or other appliance where the capacity is stated as full-rated heat input against a positive static pressure at the chimney connection, minimum system resistance is zero, and no value is added for existing velocity head in the system.

When size is unknown, the following $k$ values may be used to run a first trial estimate.

$k$ = 5.0 for the entire system
$k$ = 7.5 for combined gas vents only

*Note*: For combined gas vents serving two or more appliances (draft hoods), multiply total $k$ by 1.5 to obtain the gravity system design coefficient. (This multiplier does not apply to forced or induced draft vents or chimneys.)

For simplified design in I-P units, the piping resistance loss function, $k_L = 0.4L/d_i$ applies for all sizes of vents or chimneys and for all velocities and temperatures. As diameter increases, this function becomes increasingly conservative, which is desirable, because larger chimneys are more likely to be made of rough masonry construction or other materials with higher pressure losses. The 0.4 constant also introduces an increasing factor of safety for flow losses at greater lengths and heights.

Figure 6 is a plot of friction factor, $F$, versus velocity and diameter for commercial iron and steel pipe at a gas temperature of 300°F (150°C) above ambient (Lapple 1949). The figure shows, for example, a 48-in. diameter chimney with 80 ft/s gas velocity may have a constant as low as 0.2. In most cases, $0.3L/d_i$ gives reasonable design results for chimney sizes 18 in. (460 mm) and

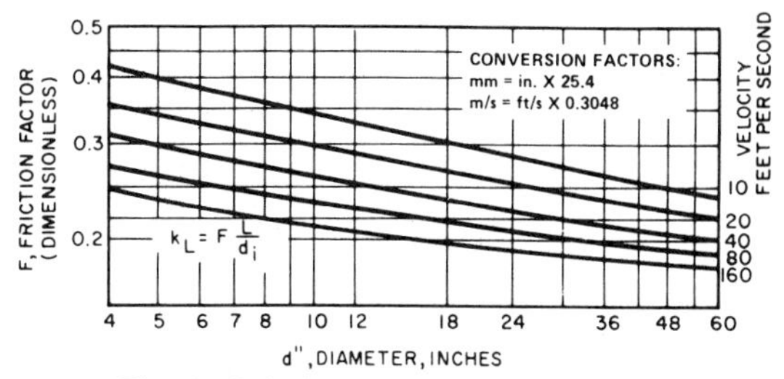


**Fig. 6 Friction Factor for Piping—Commercial Iron and Steel Pipe (Lapple 1949)**

above because systems of this size usually operate at gas velocities greater than 10 ft/s (3 m/s).

At 1000°F (540°C) or over, the factors in Figure 6 should be multiplied by 1.2. Because Figure 6 is for commercial iron and steel pipe, an additional correction for greater or less surface roughness may be imposed from Figure 3, Chapter 33, of the 1985 FUNDAMENTALS Volume. For example, the factor for a very rough 12 in. (305 mm) diameter pipe may be doubled at a velocity as low as 2000 fpm (10 m/s).

For most chimney designs, a friction factor, $F$, of 0.4 (I-P or SI units) gives a conservative solution for diameter or input for all sizes, types, and operating conditions of prefabricated and metal chimneys; alternately, a $F = 0.30$ is reasonable if the diameter is 18 in. (460 mm) or over. Because neither input nor diameter is particularly sensitive to the total friction factor, the overall value of $k$ requires little correction.

Masonry chimneys, including those lined with clay flue tile, may have rough surfaces, tile shape variations that cause misalignment, and joints at frequent intervals with possible mortar protrusions. In addition, the inside cross-sectional area of liner shapes may be less than expected because of local manufacturing variations, as well as differences between claimed and actual size. To account for these characteristics, the estimate for the piping-loss coefficient should be on the high side, regardless of chimney size or velocity.

Computations should be made by assuming smooth surfaces and then adding a final size increase to compensate for shape factor and friction loss. It follows that performance or capacity of metal and prefabricated chimneys is generally superior to site-constructed masonry.

## Configuration and Manifolding Effects

The most common configuration is the individual vent, stack, or chimney, in which one continuous system carries the products from appliance to terminus. Other configurations include the combined vent serving a pair of appliances, the manifold serving several, and branched systems with two or more lateral manifolds connected to a common vertical system. As the number of appliances served by a common vertical vent or chimney increases, the precision of design decreases because of diversity factors (variation in the number of units in operation) and the need to allow for maximum and minimum input operation (Stone 1957). For example, the vertical common vent for interconnected gas appliances must be larger than for a single appliance of the same input to allow for operating diversity and draft-hood dilution effects. Connector rise, headroom, and configuration in the equipment room must be designed carefully to avoid draft-hood spillage and related oxygen depletion problems.

For typical combined vents, the diversity effect must be introduced into Figure 1 by multiplying system resistance by 1.5 (see Table 9 and Example 5). This multiplier compensates for junction effect and part-load operation.

Manifolds for appliances with barometric draft regulators can be designed without allowing for dilution by inoperative appliances. In this case, because draft regulators remain closed until regulation is needed, dilution under part-load is negligible. In addition, flow through any inoperative appliance is negligible because the combustion air inlet dampers are closed and the multiple-pass heat exchanger has a high internal resistance.

Multiple oil-burning appliances, for example, have a lower flow velocity and, hence, lower losses. As a result, they produce reasonable draft at part-load or with only one of several appliances in operation. Therefore, diversity of operation has little effect on chimney design. Some installers set each draft regulator at slightly different settings to avoid oscillations or hunting, possibly caused by burner or flow pulsations.

To determine the resistance coefficient of any portion of a manifold, the calculation begins with the appliance most distant from the vertical portion. All coefficients are then summed from its outlet to the vent terminus. The resistance of a series of tee joints to flow passing horizontally straight through them (not making a turn) is the same as that of an equal length of piping (as if all other appliances were *off*). This assumption holds, whether the manifold is tapered (to accommodate increasing input) or of constant size large enough for the accumulated input.

Coefficients are assigned only to inlet and exit conditions, to fittings causing turns, and to the piping running from the affected appliance to the chimney exit. Piping shape may initially be ignored, whether it is round, square, or rectangular and whether it is for connectors, vertical piping, or both.

Certain high-pressure, high-velocity packaged boilers require special manifold design to avoid turbulent flow noise. In such cases, manufacturers' instructions usually recommend increaser *Y* fittings, as shown in Figure 7. The loss coefficients listed in Table 9 for standard tees and elbows are higher than necessary for long radius elbows or *Y* entries. Occasionally, on equipment with high chimneys augmenting boiler outlet pressure, it may appear feasible to reduce the diameter of the vertical portion to below that recommended by the manufacturer. However, any reduction may cause turbulent noise, even though all normal design parameters have been considered.

Manufacturers' sizing recommendations, shown in Figure 7, apply to the specific appliance and piping arrangement shown. The values are conservative for long radius elbows or *Y* entries. Frequently, the boiler room layout forces the use of additional elbows. In such cases, the size must be increased to avoid excessive flow losses.

The Design Chart (Figure 1) can be used to calculate the size of a vertical portion smaller in area than the manifold, or a chimney connector smaller than the vertical, but with one simplifying assumption. The maximum velocity of the flue gas, which exists in the smaller of the two portions, is assumed to exist throughout the entire system. This assumption leads to a conservative design, as true losses in the larger area is lower than assumed. Further, if the size change is small, either as a contraction or enlargement, the added loss coefficient for this transition fitting (see Table 9) is compensated for by reduced losses in the enlarged part of the system.

These comments on size changes apply more to individual than to combined systems because it is undesirable to reduce the vertical area of the combined type and, more frequently, it is desirable to enlarge it. If an existing vertical chimney is slightly undersized for the connected load, the complete chart method must be applied to determine whether a pressure boost is needed, as size is no longer a variable.

Sectional gas appliances with two or more draft hoods do not pose any special problems if all sections fire simultaneously. In this case, the designer can treat them as a single appliance. The appliance installation instructions either specify the size of manifold for interconnecting all draft hoods or require a combined area equal to the sum of all attached draft hood outlet areas. Once the

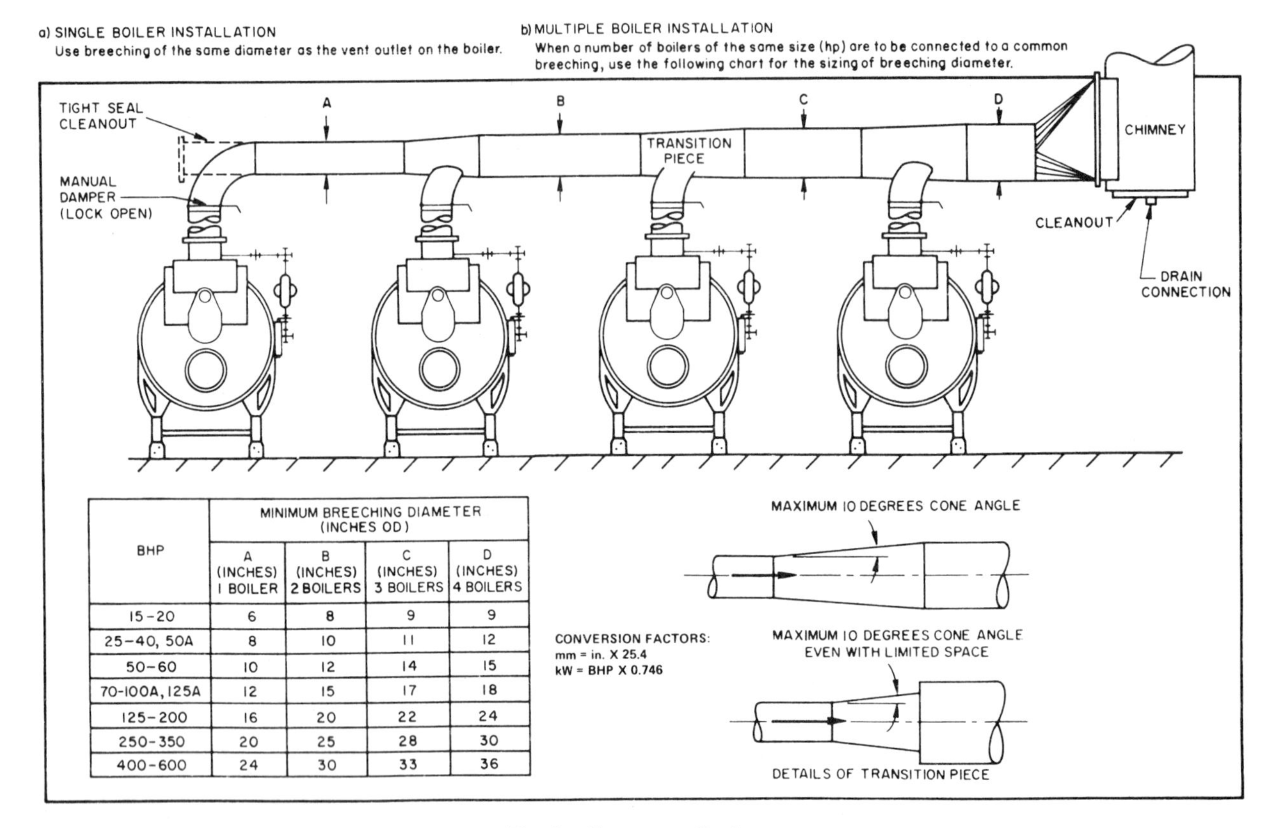


| BHP | MINIMUM BREECHING DIAMETER (INCHES OD) | | | |
|---|---|---|---|---|
| | A (INCHES) 1 BOILER | B (INCHES) 2 BOILERS | C (INCHES) 3 BOILERS | D (INCHES) 4 BOILERS |
| 15-20 | 6 | 8 | 9 | 9 |
| 25-40, 50A | 8 | 10 | 11 | 12 |
| 50-60 | 10 | 12 | 14 | 15 |
| 70-100A, 125A | 12 | 15 | 17 | 18 |
| 125-200 | 16 | 20 | 22 | 24 |
| 250-350 | 20 | 25 | 28 | 30 |
| 400-600 | 24 | 30 | 33 | 36 |

**Fig. 7 Connector Design**

manifold has been designed and constructed, it can be connected to a properly sized chimney connector, vent, or chimney. If the connector and chimney size is computed as less than manifold size (as may be the case with a tall chimney) the operating resistance of the manifold will be lower than the sum of the assigned component coefficients because of reduced velocity.

The general rule for conservative system design in which manifold, chimney connector, vent, or chimney are different sizes, can be stated as: *Always assign full resistance coefficient values to all portions carrying combined flow and determine system capacity from the smallest diameter carrying the combined flow.* In addition, horizontal chimney connectors or vent connectors should pitch upward toward the stack, 1/4 inch per foot minimum (20 mm/m).

The following sample calculation, using Eq. (11)—which differs from the original order of steps used with Figure 1—illustrates the direct solution for input, velocity, and volume. A calculation for input that is derived from the chart differs because of the chart arrangement.

---

**Example 1**. Find the input capacity (Btu/h) of a vertical, double-wall *Type B* gas vent, 24 in. in diameter, 100 ft high at sea level. This vent is used with draft hood natural gas-burning appliances.

*Solution*:

Step 1. Mass Flow from Table 2. $M = 1.60$ lb/1000 Btu for natural gas, if no other data are given.

Step 2. Temperature from Table 4. Temperature rise = 300°F and T = 360 + 460 = 820°R for natural gas.

Step 3. Theoretical draft from Table 6. For 100 ft height at 300°F rise, $D = 0.537$ in. of water.
Available draft for draft hood appliances: zero.

Step 4. Flow Losses from Table 8. $\Delta p = D = 0.537$; or flow losses for a gravity gas vent equal theoretical draft at mean gas temperature.

Step 5. Resistant Coefficients from Table 9. For a vertical vent:

| | | |
|---|---|---|
| Draft Hood | | $k_1 = 1.5$ |
| Vent Cap | | $k_4 = 1.0$ |
| For 100 ft piping, | $k_1 = 0.4\ (100/24)$ | $= 1.67$ |
| | System $k$ | $= 4.17$ |

Step 6. Solution for Input.

Altitude: Sea level, $B = 29.92$; from Table 7, $d_i = 24$ in. These values are substituted into Eq. (11) as follows:

$$I = 4.13 \times 10^5 \frac{(d_i)^2}{M}\left(\frac{\Delta pB}{kT_m}\right)^{0.5}$$

$$= 4.13 \times 10^5 \frac{(24)^2}{1.60}\left(\frac{(0.537)(29.92)}{(4.17)(820)}\right)^{0.5}$$

$I = 10.2 \times 10^6$ Btu/h input capacity.

Step 7. A solution for velocity requires a prior solution for input to apply to Eq. (9). First using Eq. (4):

$$\varrho_m = 1.325\,\frac{29.92}{820} = 0.0483 \text{ lb/ft}^3$$

$$V = \frac{(10.2 \times 10^6)(1.60)}{1000} \times \frac{1}{(19.63)(0.0483)(24)^2}$$

$= 29.9$ ft/s (9.1 m/s)

Step 8. Volume flow can now be found, because velocity is known. The flow area of 24 in. diameter is 3.14 ft$^2$, thus:

$$Q = (60 \text{ s/h})(3.14 \text{ ft}^2)(29.9 \text{ ft/s}) = 5633 \text{ cfm}$$

No heat loss correction is needed to find the new gas temperature because the size is greater than 20 in. (510 mm), and this vent is vertical, with no horizontal connector.

---

For this same problem, Figure 1 requires a different sequence of solution. Mass flow input ratio for a given fuel (with parameter *M*) is not used directly; the chart requires selecting a $CO_2$ percentage in the chimney, either from Table 2 or from operating data on the appliances. Then, the temperatures are entered only as rise above ambient. The solution path is as follows:

1. Enter *Group A* at 5.3% $CO_2$, and construct line horizontally to left intersecting Nat Gas.
2. From Nat Gas intersection, construct vertical line to $t = 300$.
3. From 300 intersection in *A*, go horizontally left to transfer line to *B*.
4. Transfer line vertically to $k = 4.17$.
5. From $k = 4.17$ run horizontally left to $t = 300$ in *C*.
6. From 300 go up vertically to $\Delta p = 0.54$ in *Group D.*
7. From $\Delta p = 0.54$ in *D*, go horizontally right to intersection $d_i = 24$ in. in *Group E.*
8. Read capacity or input at 24 in. intersection in *E* as 10.2 million Btu/h (3 MW) between curved lines.

If input is known and diameter must be found, the procedure is the same as with Eq. (11). A preliminary *k*, usually 5.0 for an individual vent or chimney, must be estimated to find a trial diameter. This diameter is used to find a corrected *k*, and the chart is solved again for diameter.

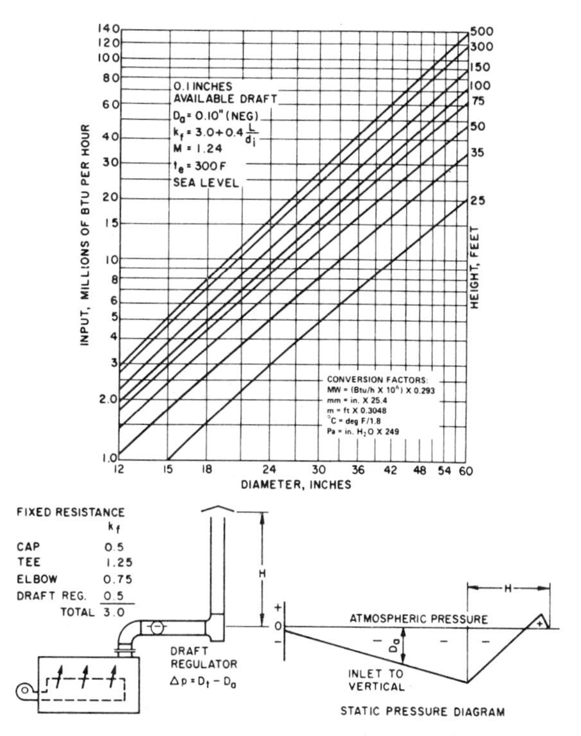


**Fig. 9 Draft Regulated Appliance with 0.10 in. (water guage) (25 Pa) Available Draft Required**

## SIMPLIFIED CHIMNEY CAPACITY CALCULATION EXAMPLES

Figures 8, 9, 10, and 11 show chimney capacity for individually vented appliances computed by the methods presented. These capacity curves may be used to estimate input or diameter for the Design Chart (Figure 1) or Eqs. (11) or (13). These capacity curves apply primarily to individually vented appliances with a lateral chimney connector; systems with two or more appliances or additional fittings require a more detailed analysis. Figures 8 through 11 assume the length of the horizontal connector is no less than 10 ft (3 m), or no longer than 50% of the height, or 50 ft (15 m), whichever is less. Chimney heights of 10 to 20 ft (3 to 6 m) assume a fixed 10-ft (3-m) long connector. Between 20 and 100 ft (6 and 30 m), the connector is 50% of the height. If the chimney height

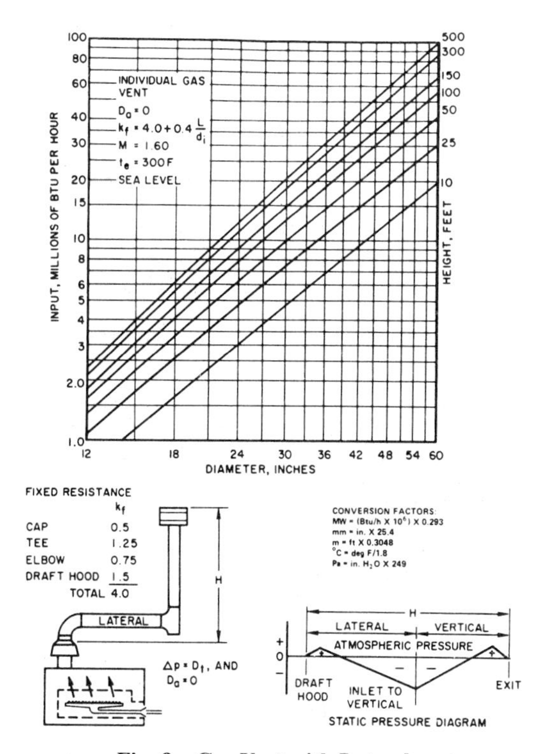


**Fig. 8 Gas Vent with Lateral**

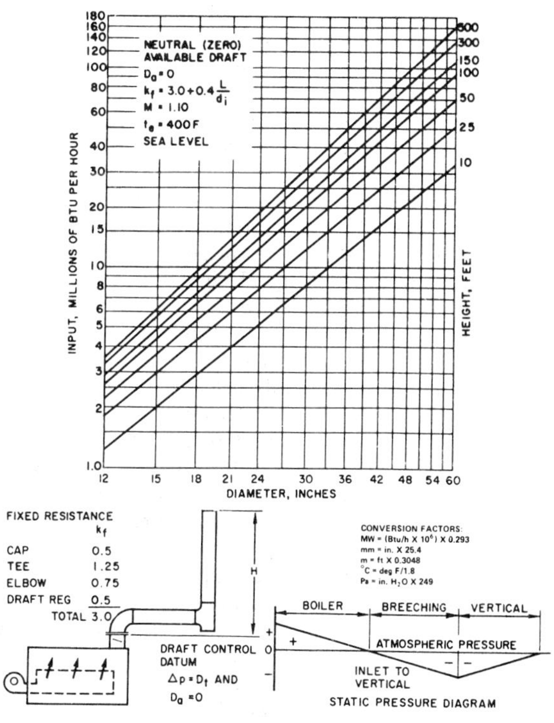


**Fig. 10 Forced-Draft Appliance with Neutral (Zero) Draft Negative Pressure Connector**

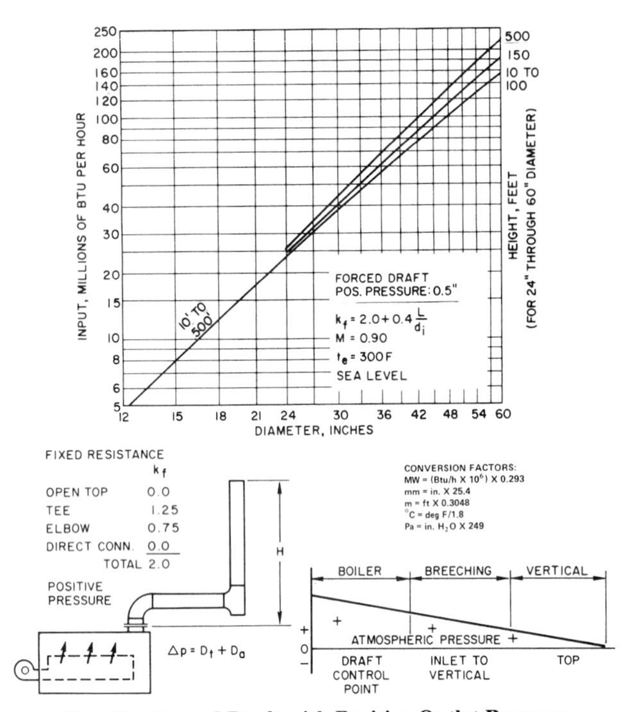


**Fig. 11 Forced Draft with Positive Outlet Pressure**

exceeds 100 ft (30 m), the connector is fixed at 50-ft (15-m) long.

For a chimney of similar configuration but with a shorter connector, the size indicated is slightly larger than necessary. In deriving the data for Figures 8 to 11, additional conservative assumptions were used, including the temperature correction, $C_u$, for double-wall laterals (see Figure 4), and a constant friction factor for all sizes (0.4 $L/d$).

The loss coefficient for a low resistance cap is included in Figures 8 to 10. If no cap is installed, these figures indicate a larger size than needed.

Figure 8 applies to a gas vent with draft hood and a lateral that runs to the vertical section. Maximum static draft is developed at the base of the vertical, but friction reduces the observed value to less than the theoretical draft. Areas of positive pressure may exist at the elbow above the draft hood and at the inlet to the cap. The height of the system is the vertical distance from the draft hood outlet to the vent cap.

Figure 9 applies to a typical boiler system requiring negative overfire, as well as negative static outlet draft. The chimney static pressure is below atmospheric pressure, except for the minor outlet reversal caused by cap resistance. Height of this system is the difference in elevation between the point of draft measurement (or control) and the exit. (Chimney draft should not be based on the height above the boiler room floor).

Figure 10 illustrates the use of a negative static pressure connector serving a forced-draft boiler. This system minimizes flue gas leakage in the equipment room. The draft is balanced or neutral, which is similar to a gas vent, with zero draft at the appliance outlet and pressure loss, $\Delta p$, equal to theoretical draft.

Figure 11 applies to a forced-draft boiler capable of operating against a positive static outlet pressure of up to 0.50 in. (125 Pa). The chimney system has no negative pressure, so outlet pressure may be combined with theoretical draft to get minimum chimney size. For chimney heights or system lengths less than 100 ft (30 m), the effect of adding 0.50 in. (125 Pa) positive pressure to theoretical draft causes all curves to fall into a compressed zone, best shown as a single line. For sizes less than 24 in. (610 mm), height has no effect on capacity, while for sizes of 24 in. (610 mm) or more, heights greater than 100 ft (30 m) add to capacity. The capacity line for heights from 10 to 100 ft (3 to 30 m) applies also to the size of a straight horizontal discharge or exhaust system with no vertical rise and up to 100 ft (30 m) long. An appliance that can produce 0.50 in. (125 Pa) positive forced draft is adequate for venting any simple arrangement with up to 100 ft (30 m) of flow path.

## Design Chart Examples

The following examples illustrate the use of the Design Chart (Figure 1) and the corresponding equations. All examples are solved in I-P units.

**Example 2.** Individual Gas Appliance with Draft Hood at Sea Level (see Figure 12).

Input = 980,000 Btu/hr; natural gas; 80-ft high with 40-ft lateral; Double-Wall $U$ = 0.6. Find the vent diameter.

*Solution*: Assume $k$ = 5.0. The following factors are used successively: $CO_2$ = 5.3% for Nat Gas; Gas temp. rise = 300°F; Transfer line: $k$ = 5.0; $\Delta p$ = 0.537 (80/100) = 0.43.

*Prelimiary Solution*: At $I = 0.98 \times 10^6$, read diameter, $d_i$ = 8.5 in. Use next largest diameter, 9 in., to correct temperature and theoretical draft; compute new system $k$ and $\Delta p$. From Figure 4, $C_U$ = 0.70, $t$ = (0.70)(300) = 210°F. This temperature determines the new value of theoretical draft, found by interpolation between 200 and 250°F in Table 6 = 0.422, $D_t$ = 0.8(0.422) = 0.34 in. of water. The four fittings of the system have a total fixed $k_f$ = 4.0. Add $k_L$, the piping component for 120 ft of 9 in. diameter, to $k_f$, or $k$ = 4.0 + 0.4 (120/9) = 9.3 for the system. System losses, $\Delta p$, equal theoretical draft = 0.34.

*Final Solution:* Returning to the chart, the factors are: $CO_2$ = 5.3%; Nat. Gas; Gas temp. rise = 210°F; Transfer line: $k$ = 9.3; Gas temp rise = 210°F; $\Delta p$ = 0.34. At $I = 0.98 \times 10^6$, read the diameter = 10 in., which is the correct answer. Had system resistance been found using 10 in. rather than 9 in., the final size would be less than 10 in., based on a system $k$ less than 9.3.

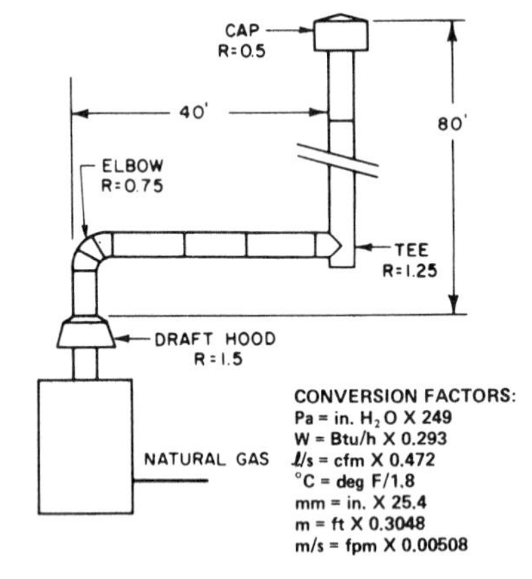


**Fig. 12 Illustration for Example 2**

**Example 3.** Gravity Incinerator Chimney (see Figure 13).

Located at 8000 ft elevation, the appliance burns 600 lb/h of Type O waste with 100% excess air at 1400°F outlet temperature. Ambient temperature is 60°F. Outlet pressure is zero at low fire, +0.10 in. of water at high fire. The chimney will be prefabricated medium heat-type with a 60-ft connector, a roughness factor of 1.2 (see Figure 3, Chapter 33 of the 1985 FUNDAMENTALS Volume). The incinerator outlet is 18-in. in diameter, and it normally uses a 20-ft vertical chimney. Find the diameter of the chimney, the connector, and the height required to overcome flow and fitting losses.

*Solution*:

1. Mass flow from Table 3 is 13.76 lb gases per lb of waste or $w$ = 600(13.76) = 8256 total lb/h.
2. Mean chimney gas temperature is based on 60-ft length of 18-in diameter, double wall chimney (see Figure 4), $C_u$ = 0.83. Temperature

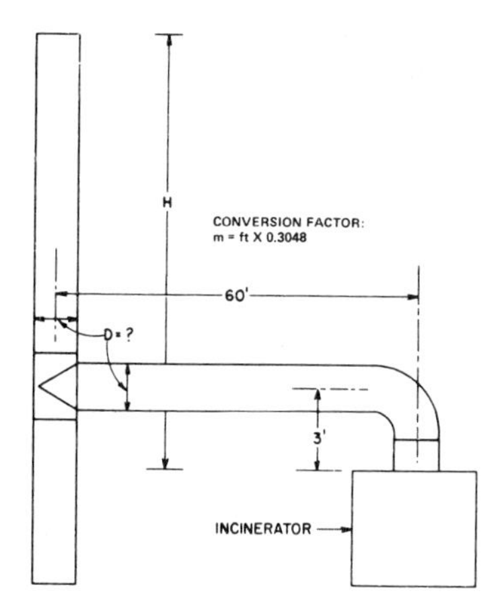


**Fig. 13 Illustration for Example 3**

rise = 1400 − 60 = 1340°F; thus, $t_m$ = 1340(0.83) = 1112 rise above 60°F ambient. $T_m$ = 1112 + 60 + 460 = 1632°R. Use this temperature to find gas density at 8000 ft elevation (from Table 7B = 22.3 in. Hg):

$$\varrho_m = 1.325(22.3/1632) = 0.0181 \text{ lb/ft}^3$$

3. Find the required height by finding theoretical draft per foot from Eq. (5) or Figure 5, because Table 6 applies only to sea level.

$$D_t/H = (0.2554)(22.3)\left(\frac{1}{520} - \frac{1}{1632}\right)$$

$$= 0.0075 \text{ in. of water per ft of length}$$

4. Find $\Delta p$, allowable pressure loss, as if the incinerator chimney is for a positive pressure appliance (see Table 8) having an outlet pressure of +0.1 in. of water. Thus, $\Delta p = D_t + D_a$ where $D_t = 0.0075H$, $D_a$ = 0.10 and $\Delta p = 0.0075H + 0.1$ in. of water.
5. Calculate flow velocity at mean temperature from Eq. (8) to balance flow losses against diameter/height combinations:

$$V = \frac{8256}{(19.63)(0.018)(18)^2} = 72 \text{ ft/s}$$

This velocity exceeds the capability of a gravity chimney of moderate height and may require a draft inducer if an 18-in. chimney must be used. Verify by calculating resistance and flow losses by the following steps.

6. Resistance coefficients for fittings from Table 9 are:

| | |
|---|---|
| 1 Tee | 1.25 |
| 1 Elbow | 0.75 |
| Spark screen | 0.50 |
| Total fitting | 2.50 |

The piping resistance, adjusted for length, diameter, and a roughness factor of 1.2 must be added to the total fitting resistance. From Figure 6, find the friction factor at 18-in. diameter and 72 ft/s as 0.22. Assuming 20 ft of height with a 60 ft lateral, piping friction loss is:

$$\frac{FL}{d_i} = \frac{(0.22)(1.2)(80)}{18} = 1.17$$

or total friction, $k$ = 1.17 + 2.50 = 3.67.

Use Eq. (7) to find $\Delta p$, which will determine if this chimney height and diameter are suitable.

$$\Delta p = \frac{(3.67)(0.0181)(72)^2}{(5.2)(64.4)} = 1.03 \text{ in. of water flow losses}$$

For these operating conditions, available draft plus theoretical yields:

$\Delta p$ = 0.0075(20) + 0.1 = 0.15 + 0.1 = 0.25 in. of water driving force

Flow losses of 1.03 in. exceed the 0.25-in. driving force; thus, the selected diameter, height, or both are incorrect, and this chimney will not work. This can also be shown by comparing draft per foot with flow losses per foot for the 18-in. diameter configuration. Flow losses per foot of 18-in chimney are:

(0.22)(1.2)/18 = 0.0147 in. of water

Draft per foot of height = 0.0075 in. of water

Regardless of how high the chimney is made, losses caused by a 72 ft/s velocity build up faster than draft.

7. A draft inducer could be selected to make up the difference between losses of 1.03-in. and the 0.25-in. driving force. Operating requirements are:

$w/(60\,\varrho)$ = 8256/(60 • 0.0181) = 7600 cfm

$\Delta p$ = 1.03 − 0.25 = 0.78 in. of water at 7600 cfm and 1112°F.

If the inducer selected injects single or multiple air jets into the gas stream (see Figure 23C), it should be placed only at the chimney top or outlet. This location requires no compensation for additional air introduced by an enlargement downstream from the inducer.

Because 18 in. is too small, assume that a 24-in. diameter may work at 20-ft height and recalculate.

1. As before, $w$ = 8256 lb/h of gases.
2. At 24-in. diameter, no temperature correction is needed for the 60-ft connector. Thus, $T_m$ = 1400 + 460 = 1860°R (see Table 5) and density is:

$$\varrho_m = (1.325)(22.3/1860) = 0.0159 \text{ lb/ft}^3$$

3. Theoretical draft per foot of chimney height:

$$D_t = 0.2554(22.3)\left(\frac{1}{520} - \frac{1}{1860}\right) = 0.00789 \text{ in. of water}$$

4. Velocity is:

$$V = \frac{8256}{(19.63)(0.0159)(24)^2} = 45.9 \text{ ft/s}$$

and the friction factor from Figure 6 is 0.225 for the piping used, (roughness factor 1.2) is 0.27.

5. For the entire system with $k$ = 2.50 for fittings and 80 ft of piping, find $k$ = 2.5 + 0.27(80/24) = 3.4. Calculating:

$$\Delta p = \frac{(3.4)(0.0159)(45.9)^2}{(5.2)(64.4)} = 0.340 \text{ in. of water}$$

$D_t + D_a$ = (20)(0.00789) + 0.1 = 0.258 in. of water and driving force is less than losses. The small difference indicates that 20 ft of height is insufficient, but that additional height may solve the problem. The added height must make up for 0.084 additional draft. As a first approximation:

Added Height = Additional Draft/Draft per Foot
= 0.084/0.00789 = 10.6 ft.

Total height = 20 + 10.6 = 30.6 ft. This is less than the actual height needed because resistance changes have not been included. For an exact solution for height, the driving force can be equated to flow losses as a function of $H$. The complete equation is:

$$0.10 + 0.00789H = \frac{(3.175 + 0.01125H)(0.0159)(45.9)^2}{(5.2)(64.4)}$$

Solving, $H$ = 32.15 ft at 24-in. diameter.

Checking by substitution, total driving force = 0.354 in. of water and total losses, based on a system with 92.15 ft of piping, also equals 0.354 in. of water.

The value of H = 32.15 ft is the minimum necessary for proper system operation. Because of the great variation in fuels and firing rate with incinerators, greater height for assurance of adequate draft and combustion control should be used. An acceptable height would be from 40 to 50 ft.

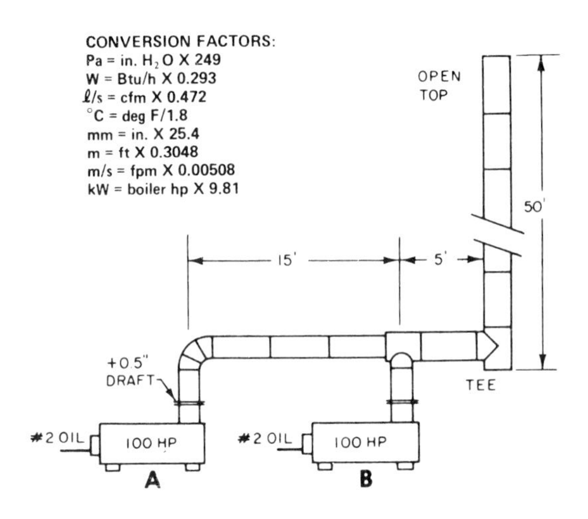


**Fig. 14 Illustration for Example 4**

**Example 4**. Two Forced-Draft Boilers (see Figure 14).

This example shows how multiple-appliance chimneys can be separated into subsystems. Each boiler is rated 100 boiler horsepower on No. 2 oil. Manufacturer states operation at 13.5% $CO_2$ and 300°F rise, against 0.50 in. positive static pressure at outlet. Chimney has 20-ft single-wall manifold and is 50-ft high above sea level. Find size of connector, manifold, and vertical.

*Solution*: First, find the capacity or size of the piping and fittings from boiler A to the tee over boiler B. Then, size the boiler B tee and all subsequent portions to carry the combined flow of A and B. Also, check the subsystem for boiler B; however, because its shorter length compensates for greater fitting resistance, its connector may be the same size as for boiler A.

Find the size for combined A and B flows either by assuming $k = 5.0$ or estimate that the size will be twice that found for boiler A operating by itself. Estimate system resistance for the combined portion by including those fittings in the B connector, as well as those in the combined portion.

Data needed for solution for boiler A using Eq. (11) for No. 2 oil at 13.5% $CO_2$ is:

$$M = 0.72\left(0.12 + \frac{14.4}{13.5}\right) = 0.854$$

For a temperature rise of 300°F and an ambient of 60°F, $T_m = 300 + 60 + 460 = 820$°R.

Theoretical draft from Table 6 = 0.537 in. of water for 100 ft of height; for 50-ft height, $D_t = (0.537)50/100 = 0.268$ in. of water.

Using positive outlet pressure of 0.5 in. of water as available draft, $\Delta p = 0.5 + 0.268 = 0.768$ in. of water.

Assume $k = 5.0$, and for 80% efficiency, input is 100 times boiler horsepower or:

$I = 100(42{,}000) = 4.2 \times 10^6$

Substitute in Eq. (11):

$$4.2 \times 10^6 = 4.13 \times 10^5 \frac{(d_i)^2}{0.854}\left(\frac{(0.768)(29.92)}{(5.0)(820)}\right)^{0.5}$$

Solving, $d_i = 10.77$ in. as a first approximation. Find correct $k$ using next largest diameter, or 12 in.

| | |
|---|---|
| Inlet acceleration | = 0.0 |
| 90° Elbow | = 0.75 |
| Tee | = 1.25 |
| Piping 0.4(70/12) | = 2.33 |
| System $k$ | = 4.33 |

*Note*: Assume tee over Boiler *B* has $k = 0$ in Subsystem *A*.

Corrected temperature rise (see Figure 4 for single wall) = 0.75 (300) = 225 or $T_m = 225 + 60 + 400 = 745$°R. This temperature changes $D_t$ to 0.44 (Table 6) per 100 ft or 0.22 for 50 ft. $\Delta p$ becomes $0.22 + 0.50 = 0.72$ in. of water.

$$4.2 \times 10^6 = 4.13 \times 10^5 \frac{(d_i)^2}{0.854}\left(\frac{(0.72)(29.92)}{(4.33)(745)}\right)^{0.5}$$

$d_i = 10.3$ in., or a 12 in. diameter is adequate.

For size of manifold and vertical, starting with the tee over boiler B, assume 16 in. (see also Figure 11).

System $k = 3.9$ for the 55 ft of piping from B to outlet.

| | |
|---|---|
| Inlet-acceleration | = 0.0 |
| Two tees (boiler B subsystem) | = 2.5 |
| Piping 0.4(55/16) | = 1.4 |
| Total | = 3.9 |

Temperature and $\Delta p$ will be as corrected (a conservative assumption) in the second step for boiler A. Having assumed a size, find input.

$$I = 4.13 \times 10^5 \frac{16^2}{0.854}\left(\frac{(0.72)(29.92)}{(3.9)(745)}\right)^{0.5} = 10.66 \times 10^6 \text{ Btu/h}$$

A 16 in. diameter manifold and vertical is more than adequate. Solving for the diameter at the combined input, $d_i = 14.3$ in., so a 15 or 16 in. chimney must be used.

*Note*: Regardless of calculations, do not use connectors smaller than the appliance outlet size in any combined system. Applying the temperature correction for a single-wall connector has little effect on the result because positive forced draft is the predominate motive force for this system.

**Example 5**. Six Gas Boilers Manifolded at 6000-ft Elevation.

Each boiler is fired at $1.6 \times 10^6$ Btu/h, A.G.A. certified, with draft hoods and an 80-ft long manifold connecting into a 400-ft high vertical, as per Figure 15. Each boiler is controlled individually. Find the size of the constant-diameter manifold, vertical, and connectors with a 2-ft rise. All are double-wall.

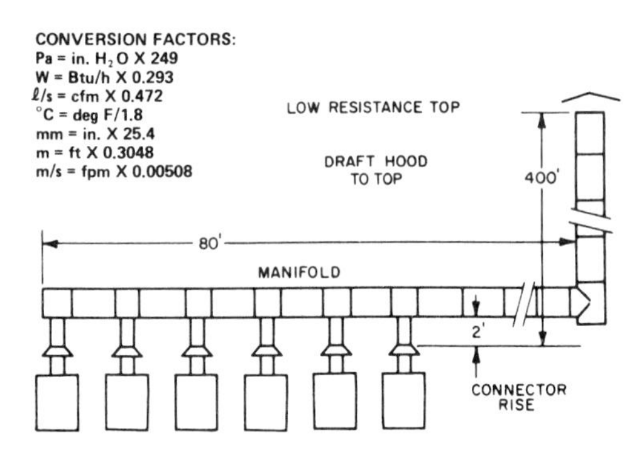


**Fig. 15 Illustration for Example 5**

*Solution*: Simultaneous operation determines both the vertical and manifold sizes. Assume the same appliance operating conditions as in Example 1: $CO_2 = 5.3$%, natural gas, temperature rise = 300°F. Initially assume $k = 5.0$ is multiplied by 1.5 for combined vent (see note at bottom of Table 9); thus, design $k = 7.5$. For gas vent $\Delta p$ at 400-ft height = 4(0.537) = 2.15 in. At 6000-ft elevation, operating input must be multiplied by altitude correction (Table 7) of 1.25. Total design input is $1.6 \times 10^6(6)(1.25) = 12 \times 10^6$ Btu/h. From Table 2, $M = 1.60$ at operating conditions. $T_m = 300 + 60 + 460 = 820$°R, and $B = 29.92$ because the 1.25 input multiplier corrects back to sea level. Using Eq. (11):

$$12 \times 10^6 = 4.13 \times 10^5 \frac{(d_i)^2}{1.60}\left(\frac{(2.15)(29.92)}{(7.5)(820)}\right)^{0.5}$$

$$d_i = 21.3 \text{ in.}$$

Because the diameter is greater than 20 in., no temperature correction is needed.

Recompute $k$ using 22 in. diameter:

| | |
|---|---|
| Draft hood inlet-acceleration | = 1.5 |
| Two tees (connector and base of chimney) | = 2.5 |

| | |
|---|---|
| Low resistance top | = 0.5 |
| Piping 0.4(480/22) | = 8.8 |
| System $k$ | = 13.3 |

Combined gas vent design $k$ = multiple vent factor 1.5(13.3) = 19.9. Substitute again in Eq. (11):

$$12 \times 10^6 = 4.13 \times 10^5 \frac{(d_i)^2}{1.60}\left(\frac{(2.15)(29.92)}{(19.9)(820)}\right)^{0.5}$$

$d_i$ = 27.2 in., so use 28 in.

| | |
|---|---|
| System $k$ (fixed) | = 4.5 |
| Piping 0.4(480/28) | = 6.9 |
| | 11.4 |

Design $k$ = 11.4(1.5) = 17.1

Substitute in Eq. (11) to obtain the third trial:

$$12 \times 10^6 = 4.13 \times 10^5 \frac{(d_i)^2}{1.60}\left(\frac{(2.15)(29.92)}{(17.1)(820)}\right)^{0.5}$$

$d_i$ = 26.2 in.

The third trial is less than the second and again shows the manifold and vertical chimney diameter to be between 26 and 28 in.

For connector size, see the *National Fuel Gas Code* for double-wall connectors of combined vents. The height limit of the table is 60 ft—do not extrapolate and read the capacity of 18 in. diameter as 1,740,000 at 2-ft rise. Use 18-in. connector or draft hood outlet, whichever is larger. No altitude correction is needed for connector size; the draft hood outlet size considers this effect.

*Note*: Equation (11) can also be solved at local elevation for exact operating conditions. At 6000 ft, local barometric pressure is 24 in., and assumed theoretical draft must be corrected in proportion to the reduction in pressure:

$D_t$ = 2.15 (24.0/29.92) = 1.72 in. Operating input of $9.6 \times 10^6$ Btu/h is used to find $d_i$ again taking final $k$ = 17.1

$$9.6 \times 10^6 = 4.13 \times 10^5 \frac{(d_i)^2}{1.60}\left(\frac{(1.72)(24.0)}{(17.1)(820)}\right)^{0.5}$$

$d_i$ = 26.2 (same as before)

This example illustrates the equivalence of the chart method of solution with the solution by Eq. (11). Equation 11 gives the correct solution using either Method 1, with the input only corrected back to sea level condition, or Method 2, using $\Delta p$ at the elevation and local barometric pressure with operating input at altitude. Method 1, correcting input only, is the only choice with Figure 1 because the Design Chart cannot correct to local barometric pressure.

**Example 6.** Pressure Boost for Undersized Chimney (not illustrated).

A natural gas boiler (no draft hood) is connected to an existing 12-in. diameter chimney. Input is $4 \times 10^6$ Btu/h operating at 10% $CO_2$ and 300°F rise above ambient. System resistance loss coefficient, $k$ = 5.0 with 20-ft vertical height. Appliance operates with neutral outlet static draft; thus, $D_a$ = 0.

*a.* How much draft boost is needed at operating temperature?
*b.* What fan rating is required at 60°F ambient temperature?
*c.* Where should the fan be located in the system?

*Solution*: Combustion data: 10% $CO_2$ at 300°F rise indicates 0.31 cfm per 1000 Btu, using Figure 3.

Total flow rate = (0.33/1000)(4)(10) = 1240 cfm at chimney gas temperature. Then, using Eq. (13), the only unknown is $\Delta p$.

Substituting:

$$1240 = 5.2(12)^2 \left(\frac{\Delta p(300 + 60 + 460)}{(5.0)(29.92)}\right)^{0.5}$$

$\Delta p$ = 0.50 in. of water needed at 300°F rise. Pressure boost supplied by fan must equal $\Delta p$ minus theoretical draft (Table 8) when available draft is zero. For 20 ft of height at 300°F rise,

$$D_t = 0.537(20/100) = 0.11 \text{ in. of water}$$

*a.* Draft boost = $\Delta p - D_t$ = 0.5 − 0.11 = 0.39 in. of water at operating temperature.
*b.* The draft fans may be rated with ambient air at 60°F and 1240 cfm. Flow rate is (see Fan Laws) inversely proportional to absolute gas temperature. Thus, for ambient air,

$$D_b = 0.39 \frac{T_m}{T_o} = 0.39 \left(\frac{300 + 60 + 460}{60 + 460}\right) = 0.61 \text{ in. of water}$$

This pressure is needed to produce 0.39 in. at operating temperature. In specifying power ratings for draft fan motors, a safe policy is to select one that operates at the required cfm (L/s) at ambient temperature and pressure (see Example 7).

*c.* A fan can be located anywhere from boiler outlet to chimney outlet. Regardless of location, the amount of boost is the same; however, chimney pressure relative to atmosphere will change. At boiler outlet, the fan pressurizes the entire connector and chimney. Thus, the system should be gas tight to avoid leaks. At the chimney outlet, the system is below atmospheric, so any leaks will flow into the system and seldom cause problems. With an ordinary sheet metal connector attached to a tight vertical chimney, the fan may be placed close to the vertical chimney inlet. Thus, the connector leaks safely inward, while the vertical chimney is under pressure.

**Example 7.** Draft Inducer Selection (see Figure 16).

A third gas boiler must be added to a two-boiler system with an 18-in. diameter, 15-ft horizontal, with 75 ft of total height. Outlet conditions for natural gas draft hood appliances are 5.3% $CO_2$ at 300°F rise. Boilers are controlled individually, each with $1.6 \times 10^6$ Btu/h or $4.8 \times 10^6$ total input. System is now undersized for gravity full-load operation. Find capacity, pressure, size, and power rating of a draft inducer fan installed at the outlet.

*Solution*: Using Eq. (13) requires evaluating two operating conditions: (1) full input at 300°F rise and (2) no input with ambient air. Because the boilers are controlled individually, the system may operate at nearly ambient temperature (100°F rise or less) when only one boiler operates at part load. Use the system resistance $k$ for boiler 3 as the system value for simultaneous operation. It needs no compensating increase as with gravity multiple venting, because a fan induces flow at all temperatures. The resistance summation is:

| | |
|---|---|
| Inlet-acceleration (draft hoods) | = 1.50 |
| Tee above boiler | = 1.25 |
| Tee at base of vertical | = 1.25 |
| 90 feet of 18 in. dia. = 0.4(90/18) | = 2.00 |
| System total | = 6.00 |

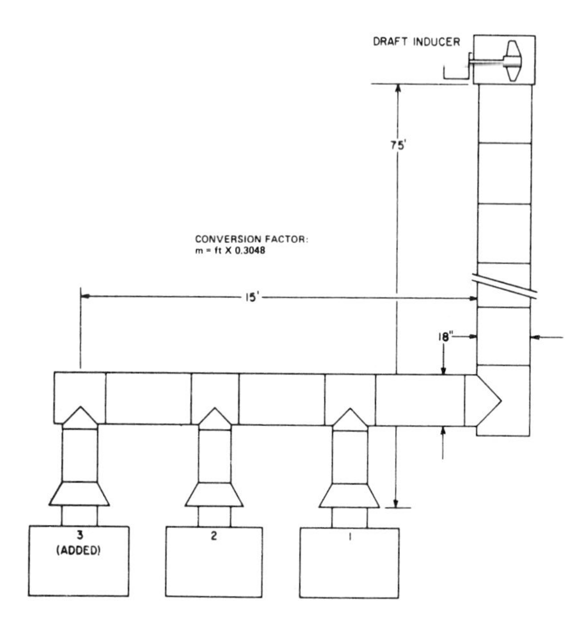


**Fig. 16 Chimney System**

At full load, $T_m = (300 + 60 + 460) = 820\,°R$, $B = 29.92$ in. Hg. Flow in cfm must be found for operating conditions of 1.60 lb per 1000 Btu at density $\varrho_m$ and full input $(4.8 \times 10^6)/60$ Btu/min.

$$\varrho_m = 1.325\,B/T_m = 1.325(29.92/820) = 0.0483\ \text{lb/ft}^3$$

$$Q = \left(\frac{1.60}{1000}\right)\left(\frac{1}{0.0483}\right)\left(\frac{4.8 \times 10^6}{60}\right) = 2650\ \text{cfm}$$

For 300 °F rise, $D_t = 0.537(75/100) = 0.403$ in. of water theoretical draft in the system. Solving Eq. (13) for $\Delta p$:

$$\Delta p = \frac{(6)(29.92)(2650)^2}{(5.2)^2(18)^4(820)} = 0.541\ \text{in. of water}$$

Thus, a fan is needed because $\Delta p$ exceeds $D_t$. Required static pressure boost (Table 8) is:

$$D_b = 0.541 - 0.403 = 0.137\ \text{in. of water at } 300\,°\text{F}$$
(or a density of 0.0483 lb/ft$^3$).

Fans are rated for ambient or standard air (60 to 70 °F) conditions. Pressure is directly proportional to density or inversely proportional to absolute temperature. Moving 2650 cfm at 300 °F against 0.137 in. of water pressure requires the ambient pressure with standard air to be:

$$D_b = 0.137(820/520) = 0.22\ \text{in. of water.}$$

Thus, a fan that delivers 2650 cfm at 0.22 in. of water at 60 °F is required. Figure 17 shows the operating curves of a typical fan that meets this requirement. The exact volume and pressure developed against a system $k$ of 6.0 can be found for this fan by plotting system airflow rate versus $\Delta p$ from Eq. (13) on the fan curve. The solution, at Point C, occurs at 1950 cfm, where both the system $\Delta p$ and fan static pressure equal 0.46 in. of water.

While some fan manufacturers' ratings are given at standard air conditions, the motors selected will be overloaded at temperatures below 300 °F. Figure 17 shows that power required for two conditions with ambient air are:

1. 1950 cfm at 0.46 in. static pressure requires 0.51 horsepower
2. 2650 cfm at 0.25 in. static pressure requires 0.50 horsepower

Thus, the minimum-size motor will be at least 1/2 horsepower, running at 1590 rpm.

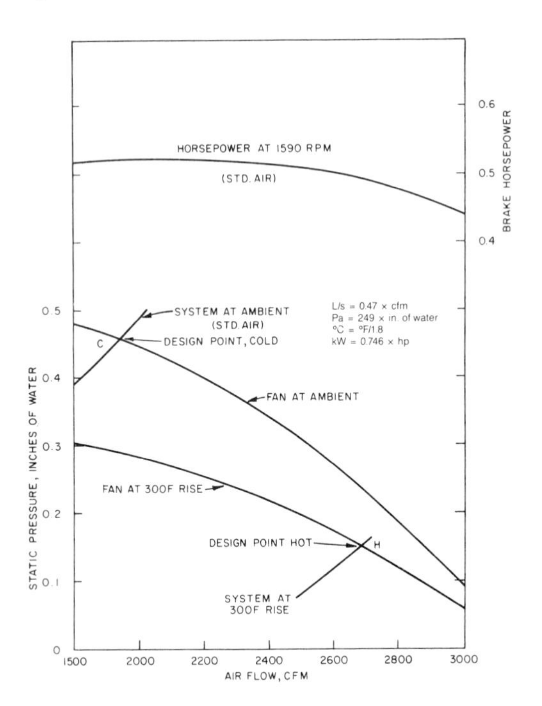


**Fig. 17 Typical Fan Operating Data and System Curves**

Manufacturers' literature must be analyzed carefully to discover if the sizing and selection method is consistent with appliance and chimney operating conditions. Final selection requires both a thorough analysis of fan and system interrelationships and consultation with the fan manufacturer to verify the capacity and power ratings.

## GAS APPLIANCE VENTING

Gas-burning equipment requiring venting of combustion products must be installed and vented in accordance with ANSI *Standard* Z223.1 (NFPA *Standard* 54), the *National Fuel Gas Code.* This standard includes capacity data and definitions for commonly used gas vent systems.

### Vent Connectors

Vent connectors connect gas appliances to the gas vent, chimney, or single-wall metal pipe, except when the appliance flue outlet or draft hood outlet is connected directly to one of these. Materials for vent connectors for conversion burners or other equipment without draft hoods must resist corrosion and heat not less than that of galvanized 0.0276 in. (24 gauge or 0.701 mm) thick sheet steel. Where a draft hood is used, the connector must resist corrosion and heat not less than that of galvanized 0.0187 in. (28 gauge or 0.475 mm) thick sheet steel or *Type B* vent material.

### Masonry Chimneys Used for Gas Appliances

A masonry chimney serving a gas-burning appliance should have a tile liner and should comply with applicable building codes. An additional chimney liner may be needed to avoid slow priming and/or condensation, particularly an exposed masonry chimney with high mass and low flue gas temperature. A low temperature chimney liner may be a single-wall passage of pure aluminum or stainless steel or a double-wall *Type B* vent.

### Type B and L Factory-Built Venting Systems

Factory prefabricated vents are listed by Underwriters' Laboratories for use with various types of fuel-burning equipment. These should be installed according to the manufacturer's instructions and their listing.

*Type B* gas vents are listed for vented gas-burning appliances. They should not be used for incinerators, appliances readily converted to the use of solid or liquid fuel, and combination gas-oil burning appliances. They may be used in multiple gas appliance venting systems.

*Type B-W* gas vents are listed for vented wall furnaces, certified as complying with the pertinent ANSI *Standard.*

*Type L* venting systems are listed by Underwriters' Laboratories in 3 through 6 in. (76 through 152 mm) sizes and may be used for those oil- and gas-burning appliances (primarily residential) that are certified or listed as suitable for this type of venting. Under the terms of listing, a single-wall connector may be used in open accessible areas between the appliance's outlet and *Type L* material in a manner analogous to *Type B. Type L* piping material is recognized in NFPA *Standards* 54 and 211 for certain connector uses between appliances such as domestic incinerators and chimneys.

### Gas Equipment without Draft Hoods

Figure 1 or the equations may be used to calculate chimney size for nonresidential gas appliances with the draft configurations listed as (1) and (3) at the beginning of this chapter. Draft configurations (1) and (3) for residential gas appliances, such as boilers and furnaces, may require special vent systems. The appliance test and certification standards include evaluation of the manufacturer's appliance installation instructions (including the vent system) and of operating and application conditions that affect venting. The appliance instructions must be followed strictly.

### Conversion to Gas

The installation of conversion-burner equipment requires evaluating for proper chimney draft and capacity by the methods in this chapter or by conforming to local regulations. The physical condition and suitability of an existing chimney must be checked before converting it from a solid or liquid fuel to gas. For masonry chimneys, local experience may indicate how well the construction withstands the lower temperature and higher moisture content of natural or liquified petroleum gas combustion products. The "Masonry Chimneys Used for Gas Appliances" section under "The Gas Appliance Vent" section of this chapter has more details.

The chimney should be relined, if required, with corrosion resistant masonry or metal to prevent deterioration. A chimney should be inspected and, if needed, cleaned. The chimney drop-leg (bottom of the chimney) must be at least 4 in. (100 mm) below the bottom of the connection to the chimney. A liner must extend beyond the top of the chimney. The chimney should also have a cleanout at the base.

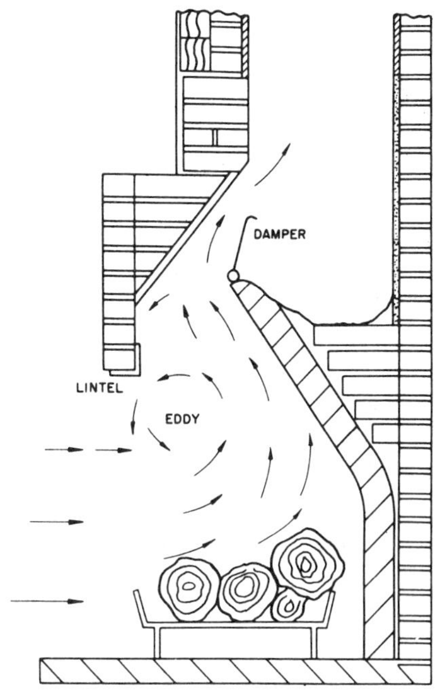


Fig. 18 Eddy Formation

## THE FIREPLACE CHIMNEY

Fireplaces with natural draft chimneys follow the same gravity fluid flow law as gas vents and thermal flow ventilation systems (Stone 1969). All thermal or buoyant energy is converted into flow, and no draft exists over the fire or at the fireplace inlet. Formulas have been developed to study a wide range of fireplace applications, but the material here covers general cases only.

Mass flow of hot flue gases through a vertical pipe up to some limiting value is a function of rate of heat release and the chimney area, height, and system pressure loss coefficient. A fireplace may be considered as a gravity duct inlet fitting, with a characteristic entrance-loss coefficient and an internal heat source. A fireplace functions properly (does not smoke) when adequate intake or face velocity across those critical portions of the frontal opening nullifies external drafts or internal convection effects.

The mean flow velocity into a fireplace frontal opening is nearly constant from 300°F (150°C) gas temperature rise up to any higher temperature. Local velocities vary within the opening, depending on its design, because the air enters horizontally along the hearth, flows into the fire and upward, and clings to the back wall, as shown in Figure 18. A recirculating eddy forms just inside the upper front half of the opening, induced by the high velocity or flow along the back. Restrictions or poor construction in the throat area between lintel and damper also increase the eddy. Because the eddy moves smoke from the zone of maximum velocity, its tendency to escape must be counteracted by some minimum air movement inward over the entire front of the fireplace, particularly under the lintel.

A minimum mean frontal inlet velocity of 0.8 ft/s (0.24 m/s), in conjunction with a chimney gas temperature of at least 300 to 500°F (150 to 260°C) above ambient, should control smoking in a well-constructed conventional masonry fireplace. Figure 19 shows fireplace and chimney dimensions for the specific conditions of circular flues at 0.8 ft/s (0.24 m/s) frontal velocity. The chart solves readily for maximum frontal opening for a given chimney, as well as for chimney size and height with a predetermined opening. Figure 19 assumes no wind or air supply difficulties.

A damper in standard masonry construction with a free area equal to or less than the required flue area is too restrictive. Most damper information does not list damper-free area or effective throat-opening dimensions. Further, interference with lintels or other parts may cause dimensions to vary. For best results, a damper should have a free area at least twice the chimney area.

Indoor-outdoor pressure differences caused by winds, building stack effects, and operation of forced-air heating systems or mechanical ventilation affect the operation of a fireplace. Thus, smoking during start-up can be caused by factors seldom related to the chimney. Frequently, in new homes (especially in high-rise multiple family construction), fireplaces of normal design cannot cope with mechanically induced reverse flow or shortages of

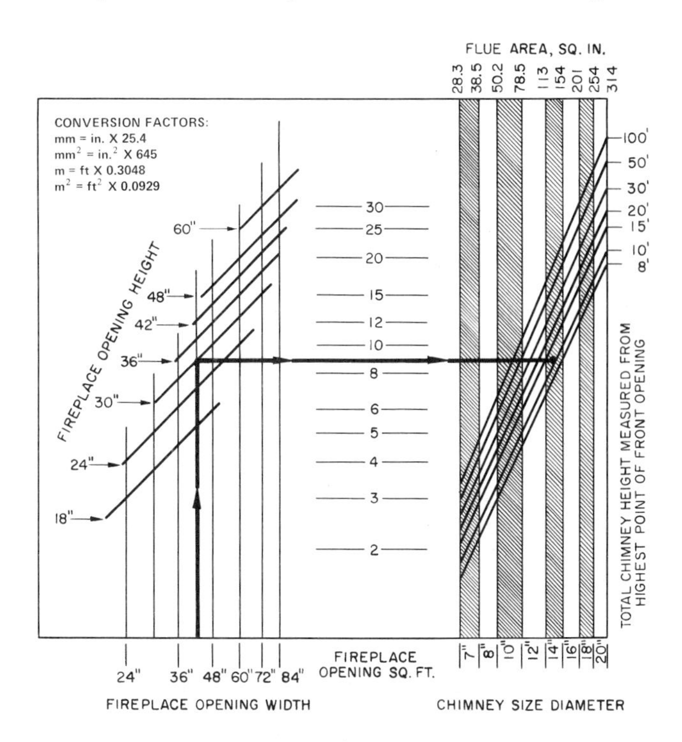


Fig. 19 Fireplace Sizing Chart for Circular Chimneys
Mean Face Velocity = 0.8 ft/s (0.24 m/s)

combustion air. In such circumstances, fireplaces should include induced draft blowers of sufficient capability to overpower other mechanized air-consuming systems. An inducer for this purpose is best located at the chimney outlet and should produce 0.8 to 1.0 ft/s (0.24 to 0.30 m/s) fireplace face velocity of ambient air in any individual flue or a chimney velocity of 10 to 12 ft/s (3 to 3.7 m/s).

Remedies to increase frontal velocity include the following: (1) increase chimney height (using the same flue area) and extend the last tile 6 in. (150 mm) upward, or more; (2) decrease frontal opening by lowering the lintel or raising the hearth (glass doors may help); and (3) increase free area through the damper (ensure that it opens fully without interferences).

## AIR SUPPLY TO FUEL-BURNING EQUIPMENT

A failure to supply outdoor air may result in erratic or even dangerous operating conditions. A correctly designed gas appliance with a draft hood can function with short vents (5 ft [1.5 m] high) using an air supply opening as small in area as the vent outlet collar. Such an orifice, when equal to vent area, has a resistance coefficient in the range of 2 to 3. If the air supply opening is as much as twice the vent area, however, the coefficient drops to 0.5 or less.

The following rules may be used as a guide:

1. Residential heating equipment installed in unconfined spaces in buildings of conventional construction does not ordinarily require ventilation other than normal air infiltration. In any residence or building that has been built or altered to conserve energy or to minimize infiltration, the heating appliance area should be considered as a confined space. The air supply should be installed in accordance with the *National Fuel Gas Code* or the recommendation given below.
2. Residential heating equipment installed in a confined space having unusually tight construction requires two openings to the outside. Free area must be greater than 1 $in^2$ per 4000 Btu/h (550 $mm^2/kW$) input with vertical ducts or 1 $in^2$ per 2000 Btu/h (1100 $mm^2/kW$) with horizontal ducts. Upper openings should be in the ceiling or as high as possible in the side wall; lower openings should be in the floor or wall below the combustion area inlet to the appliance.
3. Complete combustion of gas or oil requires approximately 1 $ft^3$ of air, at standard conditions, for each 100 Btu (270 L/s • MJ) of fuel burned, but excess air for proper burner operation is usually required. When the building space in which fuel-burning equipment is installed does not have adequate air infiltration to ensure proper combustion, two permanent openings to the outdoors or to spaces freely communicating to the outdoors are required. One opening should be located above the highest draft hood relief opening or draft regulator opening and the other below the combustion air inlet to the heating equipment. Such openings should have a minimum free area of 1 $in^2$ for every 5000 Btu/h (440 $mm^2/kW$).
4. The size of these air openings may be modified when special engineering ensures an adequate supply of air for combustion, dilution, and ventilation, or where local ordinances apply to legal boiler and machinery rooms.
5. In calculating free area of air inlets, the blocking effect of louvers, grilles, or screens protecting openings should be considered. Screens should not be smaller than 1/4-in. (6.4-mm) mesh. If the free area through a particular louver or grille is known, it should be used in calculating the size opening required to provide the free area specified. If the free area is not known, it may be assumed that wood louvers have 20 to 25% free area and metal louvers and grilles have 60 to 75% free area.
6. Mechanical ventilation systems serving the fuel-burning equipment room or adjacent spaces should not be permitted to create negative equipment room air pressure. The equipment room may require tight self-closing doors and provisions to supply air to the spaces under negative pressure so the fuel-burning equipment and venting operate properly.
7. Fireplaces may require special consideration. For example, a residential attic fan can be hazardous if it is inadvertently turned on while a fireplace is in use.
8. In buildings where large quantities of combustion and ventilation or process air are exhausted, a sufficient supply of fresh uncontaminated makeup air, warmed, if necessary, to the proper temperature should be provided. Good practice provides about 5 to 10% more makeup air than the amount exhausted.

## VENT AND CHIMNEY MATERIALS

Factors to be considered when selecting chimney materials include (1) the temperature of gases; (2) their composition and propensity to condense (dew point); (3) the presence of sulfur, halogens, and other fuel and air contaminants that lead to corrosion; and (4) the operating cycle of the appliance.

Figure 1 covers materials for vents and chimneys in the 4- to 48-in. (100- to 1220-mm) size range; these include single-wall metal, various multi-wall air and mass insulated types, and precast and site constructed masonry. While each has different characteristics such as frequency of joints, roughness, heat loss, etc., the type of materials used for systems 14 in. (360 mm) and larger is relatively unimportant in determining draft or capacity. This does not preclude selecting a safe product or method of construction that minimizes heat loss and fire hazard in the building.

National codes and standards classify heat-producing appliances as low, medium, and high heat, with appropriate reference to chimney and vent constructions permitted with each. These classifications are primarily based on size, process use, or combustion temperature. In many cases, the appliance classification gives little information about the outlet gas temperature or venting need. The designer should, wherever possible, obtain gas outlet temperature conditions and properties that apply to the specific appliance, rather than going by code classification only.

Where building codes permit engineered chimney systems, selection based on gas outlet temperature can save space, as well as reduce structural and material costs. For example, in some jurisdictions, approved gas-burning appliances with draft hoods operating at inputs over 400,000 Btu/h (117 kW) may be placed in a heat-producing classification that prohibits use of *Type B* gas vents. An increase in input may not cause an increase in outlet temperature or in venting hazards, and most building codes recommend correct matching of appliance and vent.

Single-wall uninsulated steel stacks can be protected from condensation and corrosion internally with refractory fire-brick liners or by guniting calcium aluminate cement over a suitable interior expanded metal mesh or other reinforcement. Another form of protection applies proprietary silica or other prepared refractory coatings to pins or a support mesh on the steel. The material must then be suitably cured for moisture and heat resistance.

Moisture condensation on interior surfaces of connectors, vents, stacks, and chimneys is a more serious cause of deterioration than heat. Chimney wall temperature and flue gas velocity, temperature, and dew point affect condensation. Contaminants such as sulfur, chlorides, and flourides in the fuel and combustion air raise the flue gas dew point. Studies by Yeaw and Schnidman (1943), Pray *et al.* (1942-53), Mueller (1968), and Beaumont *et al.* (1970) indicate the variety of analysis methods, as well as difficulties in predicting the causes and probability of actual condensation.

Combustion products from any fuel containing hydrogen condenses onto cold surfaces or condenses in bulk if the main flow of flue gases is cooled sufficiently. Because flue gas loses heat through walls, condensation, which first occurs on interior wall surfaces cooled to the flue gas dew point, forms successively a dew, then a liquid film, and, with further cooling, causes liquid to flow down into zones where condensation would not normally occur.

Start-up of cold interior chimney surfaces is accompanied by transient dew formation, which evaporates upon heating above the dew point. Little corrosion results from this phenomenon when very low sulfur fuels are used. Proper selection of chimney dimension and materials minimizes condensation and, thus, corrosion.

Experience shows a correlation between sulfur content of the fuel and deterioration of interior chimney surfaces. Figure 20 illustrates one cause, which applies to any fuel gas. The figure shows that the flue gas dew point increases at 40% excess air from 127 °F (53 °C) with zero sulfur to 220 °F (104 °C) with 15 grains of sulfur in fuel gas having a heating value of 1000 Btu/ft$^3$ (37.3 MJ/m$^3$). This corresponds to about 0.45% sulfur by weight (mass) in the fuel gas. A minimum flue gas dew point temperature of 220 °F (104 °C) has been cited for oil containing 3.0% sulfur, indicating considerable uncertainty regarding flue gas dew point temperature as a function of sulfur content of the fuels. This difference may be caused by the difficulty in predicting the physical state of the $H_2O$-$SO_2$-$SO_3$ excess air condition as these flue gases flow through a chimney.

Because the corrosion mechanism is not completely understood, judicious use of resistant materials, suitably insulated or jacketed to reduce heat loss, is preferable to low-cost single-wall construction. Refractory materials and mortars should be acid resistant, while steels should be resistant to sulfuric, hydrochloric, and hydroflouric acids; pitting; and oxidation. Where low flue gas temperatures are expected, together with low ambients, an air space jacket or mineral fiber lagging, suitably protected against water entry, helps maintain surface and flue gas temperatures above the dew point. In addition to reducing corrosion, the use of low-sulfur fuel reduces air pollution (which is required in many localities).

Type 1100 aluminum alloy or any other non-copper-bearing aluminum alloy of 99% purity or better provides satisfactory performance in prefabricated metal gas-vent products. For chimney service, gas temperatures from appliances burning oil or solid fuels may exceed the melting point of aluminum; therefore, steel is required. Stainless steels such as *Type 430* or *Type 304* give good service in residential construction and are recognized in *UL*-listed prefabricated chimneys. Where more corrosive environments are anticipated, such as high sulfur fuel or chlorides from solid fuel,

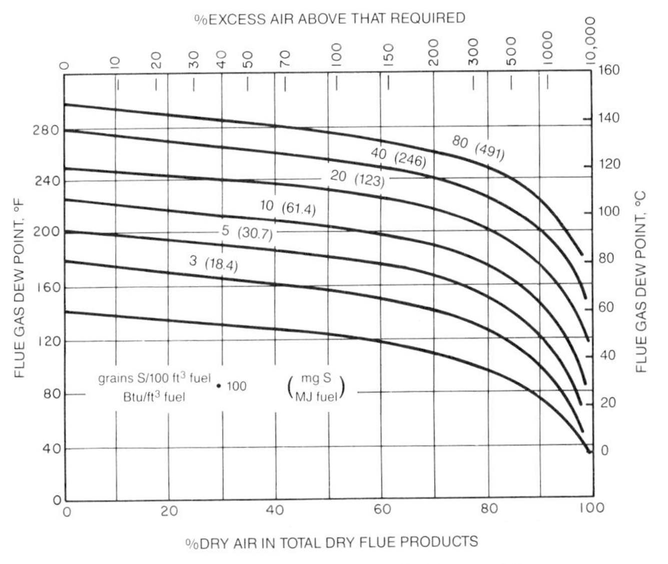


**Fig. 20 Influence of Sulfur Oxides on Flue Gas Dew Point (Stone 1969)**

contaminating air, or refuse, *Type 29-4E* or equivalent stainless steel offers a good match of corrosion resistance and mechanical properties.

As an alternative to stainless steel, porcelain enamel offers good resistance to corrosion, if two coats of acid resistant enamel are used on all surfaces. A single coat, which always has imperfections, will allow base metal corrosion, spalling, and early failure.

Prefabricated chimneys and venting products are available that use light-weight, corrosion-resistant materials, both in metal and masonry. The standardized, prefabricated, double-wall metal *Type B* gas vent has an aluminum inner pipe and a coated steel outer casing, either galvanized or aluminized. Standard air space from 1/4 to 1/2 in. (6.4 to 12.7 mm) is adequate for applicable tests and a wide variety of exposures.

Air-insulated all-metal chimneys are available for low-heat use in residential construction. Thermosiphon air circulation or multiple reflective shielding with three or more walls keep these units cool. Insulated, double-wall, residential chimneys are also available. The annulus between metal inner and outer walls is filled with insulation and retained by coupler end structures for rapid assembly.

Prefabricated, air-insulated, double-wall metal chimneys for multifamily residential and larger buildings, classed as *building-heating appliance chimneys*, are available (Figure 21). Refactory lined prefabricated chimneys (medium heat-type) are also available for this use.

Commercial and industrial incinerators, as well as heating appliances, may be vented by prefabricated metal-jacketed cast refractory chimneys, which are listed in the medium heat category and are suitable for intermittent gas temperatures to 2000 °F (1100 °C). All prefabricated chimneys and vents carrying a listing by a recognized testing laboratory have been evaluated for class of service regarding temperature, strength, clearance to adjacent combustible materials, and suitability of construction in accordance with applicable national standards.

Underwriters' Laboratories standards (as listed in Table 10) describe the construction and temperature testing of various classes of prefabricated vent and chimney materials. Standards for some related parts and appliances are also included in Table 10, because a listed factory-built fireplace, for example, must be used with a specified type of factory-built chimney. The temperature given for the steady-state operation of chimneys is the lowest in the test sequence. Factory-built chimneys under UL *Standard* 103 are also required to demonstrate adequate safety during a one-hour test at 1330 °F (720 °C) rise and to withstand a 10-minute simulated soot burnout at either a 1630 °F (906 °C) or 2030 °F (1128 °C) rise.

These product tests determine minimum clearance to combustible surfaces or enclosures, based on allowable temperature rise on combustibles. They also ensure that the supports, spacers, and parts of the product that contact combustible materials remain at safe temperatures during operation. Product markings and installation instructions of listed materials are required to be consistent with test results, refer to types of appliances that may be used, and explain structural and other limitations.

## VENT AND CHIMNEY ACCESSORIES

The design of a vent or chimney system must consider the existence or need for such accessories as draft diverters, draft regulators, induced-draft fans, blocking dampers, expansion joints, and vent or chimney terminals. Draft regulators include barometric draft regulators and furnace sequence draft controls, which monitor automatic flue dampers during operation. The design, materials, and flow losses of chimney and vent connectors are covered in other sections of the chapter.

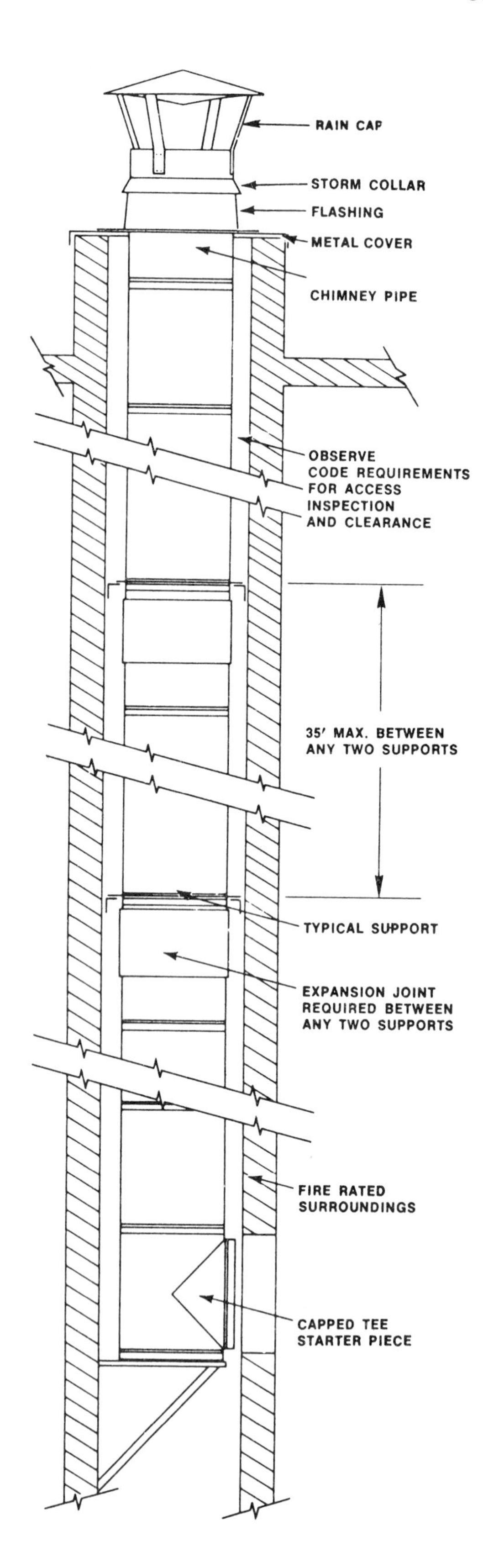


**Fig. 21 Building Heating Equipment or Medium-Heat Chimney**

**Table 10 List of Underwriters Laboratories Test Standards**

| No. | Subject | Steady-State Appliance Flue Gas Temp. Rise, °F | (°C) | Fuel |
|---|---|---|---|---|
| 103 | Chimneys, factory-built, residential type (includes building heating appliances) | 930 | (517) | All |
| 127 | Fireplaces, factory-built | 930 | (517) | Solid or gas |
| 311 | Roof jacks for mobile homes | 930 | (517) | Oil, gas |
| 378 | Draft equipment (such as regulators and inducers) | — | — | All |
| 441 | Gas vents (Type B, BW) | 480 | (267) | Gas only |
| 641 | Low-temperature venting systems (Type L) | 500 | (278) | Oil, gas |
| 959 | Chimneys, factory-built medium heat | 1730 | (961) | All |

## Draft Hoods

The draft hood isolates the appliance from venting disturbances (updrafts, downdrafts, or blocked vent) and allows combustion to start without venting action. Suggested general dimensions of draft hoods are given in ANSI *Standard* Z21.12, which describes certification test methods for draft hoods. In general, inlet and outlet flue pipe sizes of the draft hood should be the same as that of the appliance outlet connection. The vent connection at the draft hood outlet should have a cross-sectional area at least as large as that of the draft hood inlet.

Draft hood selection comes under the following two categories:

1. Draft hood supplied with a design-certified gas appliance—certification of a gas appliance design under pertinent ANSI Standards includes its draft hood (or draft diverter). Consequently, a draft hood should not be altered, nor should it be replaced without consulting the manufacturer and local code authorities.
2. Draft hoods supplied separately for gas appliances— installation of listed draft hoods on existing vent or chimney connectors should be made by experienced installers in accordance with accepted practice standards.

Every design-certified gas appliance requiring a draft hood must be accompanied by a draft hood or provided with a draft diverter as an integral part of the appliance. The draft hood is a vent inlet fitting, as well as a safety device for the appliance, and certain assumptions can be made regarding its interaction with a vent. First, when it is operating without spillage, the heat content of gases (enthalpy relative to dilution air temperature) leaving the draft hood is almost the same as that entering. Second, safe operation is obtained with from 40 to 50% dilution air. It is unnecessary to assume 100% dilution air for gas venting conditions. Third, during certification tests, the draft hood must function without spillage, using a vent with not over 5 ft (1.5 m) of effective height and one or two elbows. Therefore, if vent heights appreciably greater than 5 ft (1.5 m) are used, an individual vent of the same size as the draft hood outlet may be much larger than necessary.

When vent size is reduced, as with tall vents, draft hood resistance is less than design value relative to the vent, and the vent tables in the National Fuel Gas Code give adequate guidance for such size reductions.

Despite its importance as a vent inlet fitting, the draft hood designed for a typical gas appliance primarily represents a compromise of the many design criteria and tests solely applicable to that appliance. This permits considerable variation in resistance loss and thus, catalog data on draft hood resistance loss coefficients does not exist. The span of draft hood loss coefficients, including inlet acceleration, varies from the theoretical minimum of 1.0 for certain low-loss bell or conical shapes to 3 or 4, where the draft hood relief opening is located within a hot-air discharge (as with wall furnaces), and high resistance is needed to limit sensible heat loss into the vent.

Draft hoods must not be used on Draft Configurations (1) or (3) equipment operated with either power burners or forced venting, unless the equipment has fan-assisted burners that overcome some or most of the appliance flow resistance and create a pressure inversion ahead of the draft (or barometric) hood.

Gas appliances with draft hoods must have excess chimney draft capacity to draw in adequate draft hood dilution air. Failure to provide adequate combustion air can cause oxygen depletion and cause flue gases to spill from the draft hood.

## Draft Regulators

Appliances requiring negative static draft at the appliance flue gas outlet generally make use of barometric regulators for combustion stability. A balanced hinged gate in these devices bleeds air into the chimney automatically when pressure decreases. This action simultaneously increases gas flow and reduces temperature. Well-designed barometric regulators provide constant static pressure over a span of impressed draft of about 0.2 in. of water (50 Pa), where impressed draft is that which would exist without regulation. A regulator can maintain a 0.06 in. (15 Pa) draft for impressed drafts from 0.06 in. through 0.26 in. of water (15 through 65 Pa). If the chimney system is very high or otherwise capable of Available Draft in excess of the pressure span capability of a single regulator, additional or oversize regulators may be used. Figure 22 shows proper locations for regulators in a chimney manifold.

Commercial and industrial furnaces and boilers are frequently installed in multiples, as illustrated at the right.

For this type of installation, it is best to use a draft control for each boiler located on the uptake between the smoke outlet and the breaching—location A. When this uptake is too short to permit the installation of a control, locate a separate control for each boiler on the main breaching, as illustrated in location B. If, because of general crowding or other reasons, neither of these locations is possible, use a single large control in the breaching between the boiler nearest the chimney and the chimney, as shown in location C.

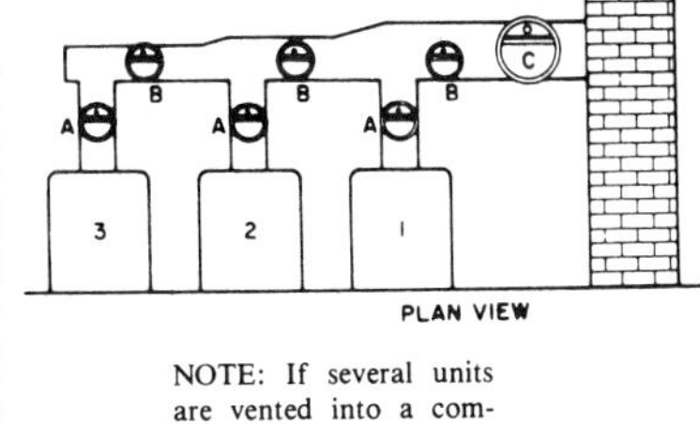


NOTE: If several units are vented into a common breaching, the most draft-critical unit should vent highest in, or be placed closest to, the chimney. Incinerators should be placed farthest from, or vented lowest in, the chimney.

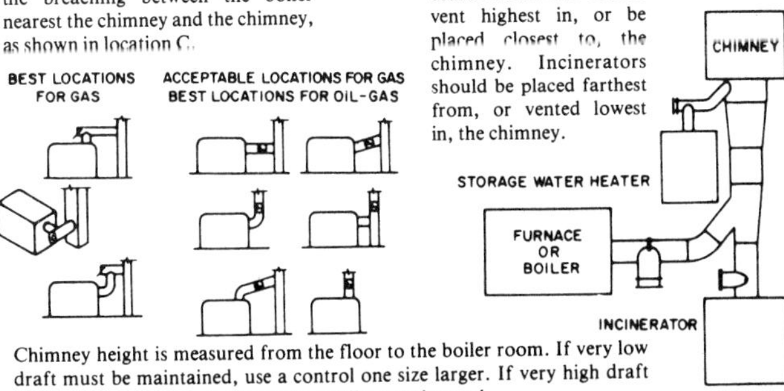


Chimney height is measured from the floor to the boiler room. If very low draft must be maintained, use a control one size larger. If very high draft will be necessary, a control one size smaller may be used.

**Fig. 22 Use of Barometric Draft Regulators**

Barometric regulators are available with double-acting dampers, which also swing out to relieve momentary internal pressures or divert continuing downdrafts. In the case of downdrafts, temperature safety switches, actuated by hot gases escaping at the regulator, sense and limit malfunctions.

## Vent Dampers

Electrically, mechanically, and thermally actuated automatic vent dampers can reduce energy consumption and improve the seasonal efficiency of gas- and oil-burning appliances. Vent dampers reduce the loss of heated air through gas appliance draft hoods and the loss of specific heat from the appliance after the burner has ceased firing. These dampers may be retrofit devices or integral components of some appliances.

Electrically and mechanically actuated dampers must open prior to main burner gas ignition and must not close during operation. These safety interlocks, which electrically interconnect with existing control circuitry, include an additional main control valve, if called for, or special gas pressure-actuated controls.

Vent dampers that are thermally actuated with bimetallic elements are available with spillage-sensing interlocks with burner controls for a draft-hood-type gas appliance. These dampers open in response to gas temperature after burner ignition. Because thermally actuated dampers may exhibit some flow resistance, even at equilibrium operating conditions, instructions regarding allowable heat input and minimum required vent or chimney height should be observed.

Special care must be taken to ensure safety interlocks with appliance controls are installed according to instructions. Spillage-free gas venting after the damper has been installed must be verified with all types.

The energy savings of a vent damper can vary widely. Dampers reduce energy consumption under one or a combination of the following conditions:

1. Heating appliance is oversized.
2. Chimney is too high or oversized.
3. Appliance is located in heated space.
4. Two or more appliances are on the same chimney (but a damper must be installed on each appliance connected to that chimney).
5. Appliance is located in building zone at higher pressure than outdoors. This positive pressure can cause steady flow losses through the chimney.

Energy savings may not justify the cost of installation if one or more of the following conditions exist:

1. All combustion and ventilation air is supplied from outdoors to direct vent appliances or to appliances located in a separate unheated room.
2. Appliance is in an unheated basement that is insulated from the heated space.
3. A short vent in a one-story flat-roof house is unlikely to carry away a significant amount of heated air.

For vents or chimneys serving two or more appliances, dampers should be installed on all attached appliances for maximum effectiveness. If only one damper is installed in this instance, loss of heated air through an open draft hood may negate a large portion of the potential energy savings.

## Heat Exchangers or Flue Gas Heat Extractors

The sensible heat available in the flue products of properly adjusted furnaces burning oil or gas is about 10 to 15% of the rated input. Small accessory heat exchangers that fit in the connector between the appliance outlet and the chimney can recover a portion of this heat for localized use; however, they may cause some adverse effects.

All gas vent and chimney size or capacity tables assume the gas temperature or heat available to create Theoretical Draft is not reduced by a heat-transfer device. In addition, the tables assume flow resistance for connectors, vents, and chimneys, comprising typical values for draft hoods, elbows, tees, caps, and piping with no allowance for added devices placed directly in the stream. Thus, heat exchangers or flue gas extractors should offer no flow resistance or negligible resistance coefficients when they are installed.

A heat exchanger that is reasonably efficient and offers some flow resistance may adversely affect the system by reducing both flow and gas temperature. This may cause moisture condensation in the chimney, draft hood spillage, or both. Increasing heat-transfer efficiency increases the probability of both effects occurring simultaneously. An accessory heat exchanger in a solid fuel system, especially a wood stove or heater, may collect creosote or cause its formation downstream.

The retrofit of heat exchangers in gas appliance venting systems requires careful evaluation of heat recovered versus both installed cost and the potential for chimney safety and operating problems. Every heat exchanger installation should undergo the same spillage tests given a damper installation. In addition, the gas temperature should be checked to ensure it is high enough to avoid condensation between the exchanger outlet and the chimney outlet.

## DRAFT FANS

The selection of draft fans, blowers, or inducers must consider (1) types and combinations of appliances, (2) types of venting material, (3) building and safety codes, (4) control circuits, (5) gas temperature, (6) permissible location, (7) noise, and (8) power cost. Besides specially designed fans and blowers, some conventional fans can be used if the wheel and housing materials are heat and corrosion resistant, and if blower and motor bearings are protected from adverse effects of the gas stream.

Small draft inducers for residential gas appliance and unit heater use are available with direct-drive blower wheels and an integral sail switch to sense flow, as shown in Figure 23A. The control circuit for such applications must prove adequate vent flow both before and while fuel flows to the main burner. Other types of small inducers are either saddle-mounted blower wheels (Figure 23B) or venturi ejectors that induce flow by jet action (Figure 23C). An essential safety requirement for inducers serving draft hood gas appliances does not permit appliance interconnections on the discharge or outlet side of the inducer unless the upstream induced-draft appliance is design-certified as ANSI *Standard* Z21.47, Category I (or the equivalent for other than central furnaces) for common-venting with draft hood appliances. This requirement prevents backflow through an inoperative appliance draft hood.

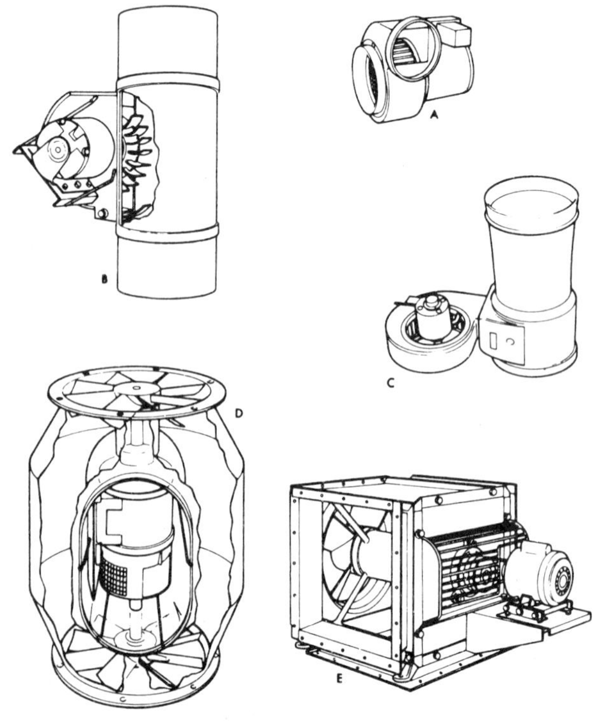


**Fig. 23 Draft Inducers**

With prefabricated sheet metal venting products such as *Type B* gas vents, the draft inducer should be located at or downstream from the point the vent exits the building. This placement keeps the indoor system below atmospheric pressure and prevents gases from escaping through seams and joints. If the inducer can not be placed on the roof, metal joints must be reliably sealed in all pressurized parts of the system.

Pressure capability of residential draft inducers is usually less than 1 in. of water (250 Pa) static pressure at rated flow. Larger inducers of the fan, blower, or ejector types have greater pressure capability and may be used to reduce system size, as well as supplement available draft. Figure 23D shows one specialized axial flow fan capable of higher pressures. This unit is structurally self-supporting and can be mounted in any position in the connector or stack, because the motor is in a well, separated from the gas stream. A right-angle fan, as shown in Figure 23E, is supported by an external bracket and adapts to several inlet and exit combinations. The unit uses the developed draft and an insulated tube to cool the extended shaft and bearings.

Pressure, volume, and power curves should be obtained to match an inducer to the application. For example, in an individual chimney system (without draft hood) in which a directly connected inducer only handles combustion products, power required for continuous operation need only consider volume at operating gas temperature. An inducer serving multiple, separately controlled, draft hood gas appliances must be powered for ambient temperature operation at full flow volume in the system. At any input, the inducer for a draft hood gas appliance must handle about 50% more standard chimney gas volume than a directly connected inducer. These demands follow the *Fan Laws* (Chapter 3) applicable to power venting as follows; at constant volume with a given size inducer or fan:

1. Pressure difference developed is directly proportional to gas density.
2. Pressure difference developed is inversely proportional to absolute gas temperature.
3. Pressure developed diminishes in direct proportion to drop in absolute atmospheric pressure, as with altitude.
4. Required power is directly proportional to gas density.
5. Required power is inversely proportional to absolute gas temperature.

Centrifugal and propeller draft inducers in vents and chimneys are applied and installed the same as any heat-carrying duct system. Venturi ejector draft boosters involve some added consideration. An advantage of the ejector is that motor, bearings, and blower blades are outside the contaminated gas stream, thus eliminating a major source of deterioration. This advantage causes some loss of efficiency and can lead to reduced capacity because undersized systems, having considerable resistance downstream, may be unable to handle the added volume of the injected airstream without loss of performance. Ejectors are best suited to use at the exit or where there is an adequately sized chimney or stack to carry the combined discharge.

If total pressure defines outlet conditions or is used for fan selection, the relative amounts of static and velocity pressures must be factored out; otherwise, the velocity head method of calculation does not apply. To factor total pressure into its two components, either the discharge velocity in an outlet of known area or flow rate must be known. For example, if an appliance or blower produces a total pressure of 0.25 in. of water at 1500 ft/min discharge velocity, the velocity pressure component can be found from Figure 1. Enter at the horizontal line marked "VH" in grid C and move horizontally to operating temperature. Read up from the appropriate temperature rise line to the 1500 ft/min horizontal velocity line in the $\Delta p$ block. Here the velocity pressure reads 0.14 in. (calculates to 0.143 at ambient standard conditions) so static pressure is $0.25 - 0.14 = 0.11$ in. (27 Pa). This static pressure is part of the system driving force, so it combines with theoretical draft to overcome losses in the system.

Draft-inducer fans can be operated either continuously or on demand. In either case, a safety switch that senses flow or pressure is needed to interrupt burner controls, if adequate draft fails. Demand operation links the thermostat with the draft-control motor. Once flow starts, as sensed with a flow or pressure switch, the burner can be started. With a venting system serving several separately controlled appliances, a single draft inducer in the common vent, operating continuously, simplifies draft-proving control circuitry. However, the single fan increases boiler standby loss and heat losses via ambient air drawn through inoperative appliances.

## TERMINATIONS—CAPS AND WIND EFFECTS

The vent or chimney height and method of termination is governed by a variety of considerations, including fire hazard, wind effects, entry of rain, debris, and birds, as well as by operating considerations such as draft and capacity. For example, the long-accepted 3-ft (0.9-m) height required for residential chimneys above a roof is required so small sparks will burn out before they fall on the roof shingles.

Many vent and chimney malfunctions are attributed to interactions of the chimney termination or its cap with winds acting on the roof or with adjoining buildings, trees, or mountains. Because winds fluctuate, no simple method of analysis or reduction to practice exists for this complex situation. Figures 24, 25, and 26 show some of the complexities of wind flow contours around simple structural shapes.

Figure 24 shows three zones with differing degrees of flue gas dispersion around a rectangular building: the cavity or eddy zone, the wake zone, and the undisturbed flow zone (Clark 1967). In addition, a fourth flow zone of intense turbulence is located downwind of the cavity. Chimney gases discharged into the wind at a point close to the roof surface in the cavity zone may be recirculated locally. Higher in the cavity zone, wind eddies can carry

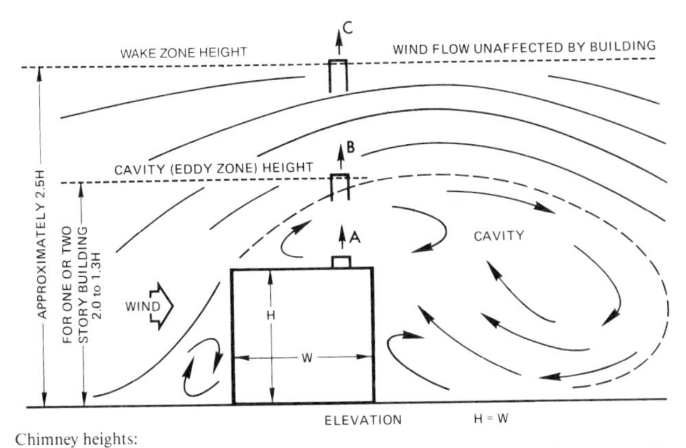


Chimney heights:
A—Discharge into cavity should be avoided because reentry will occur. Dispersion equations do not apply.
B—Discharge above cavity is good. Reentry is avoided, but dispersion may be marginal or poor from standpoint of air pollution. Dispersion equations do not apply.
C—Discharge above wake zone is best; no reentry; maximum dispersion.

**Fig. 24 Wind Eddy and Wake Zones for One- or Two-Story Buildings and Their Effect on Chimney Gas Discharge**

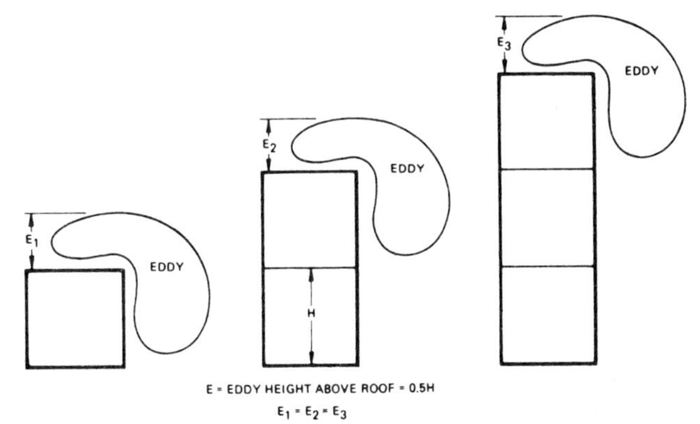


Studies found for a single cube-shaped building (length equals height) that (1) the height of the eddy above grade is 1.5 times the building height and (2) the height of unaffected air is 2.5 times height above grade. The eddy height above the roof equals 0.5H, and it does not change as building height increases in relation to building width.

**Fig. 25 Height of Eddy Currents Around Single High-Rise Buildings**

more dilute flue products to the lee side of the building. Gases discharged into the wind in the wake zone do not recirculate into the immediate vicinity, but may soon descend to ground level. Above the wake zone, dispersal into the undisturbed wind flow carries and dilutes the flue gases over a wider area. The boundaries of these zones vary with building configuration and wind direction and turbulence; they are influenced strongly by surroundings.

The possibility of air pollutants reentering the cavity zone due to plume spread, or air pollution intercepting downwind cavities associated with adjacent structures or downwind buildings should be considered. Thus, a design criteria of elevating the stack discharge above the cavity is not valid for all cases. A meteorologist, experienced with dispersion processes near buildings, should be consulted for complex cases and for cases involving air contaminants.

As chimney height increases through these zones, draft performance improves dispersion, while additional problems of gas cooling, condensation, and structural wind load are created. As building height increases (Figure 25), the eddy forming the cavity zone no longer descends to ground level. For a low, wide building (Figure 26), wind blowing parallel to the long roof dimension can reattach to the surface; thus, the eddy zone becomes flush with the roof surface (Evans 1957). For satisfactory dispersion with low, wide buildings, chimney height must still be determined as if $H = W$ (Figure 24).

Evans (1957) and Chien *et al.* (1951) studied pitched roofs in relation to wind flow and surface pressures. Because the typical residence has a pitched roof and probably uses natural gas or a

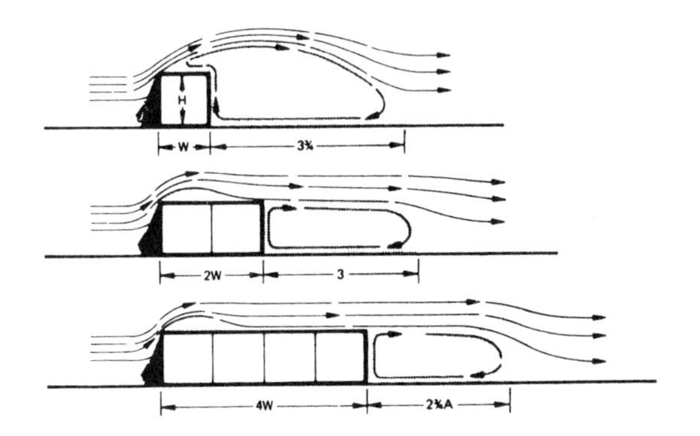


**Fig. 26 Eddy and Wake Zones for Low, Wide Buildings**

low-sulfur fossil fuel, dispersion is not important because flue products are relatively free of pollutants. For example, a precaution found in installation standards of the *National Fuel Gas Code* (NFPA 54) controls minimum distance of the gas vent termination from an air intake (9 in. [230 mm] gravity, 12 in. [305 mm] forced draft), but it does not require penetration above the cavity zone.

Flow of wind over a chimney termination can assist or impede draft. In regions of stagnation, on the windward side of a wall or a steep roof, winds create positive static pressures that impede established flow or cause backdrafts within inoperative vents and chimneys.

Conversely, location near the surface of a low flat roof can aid draft because the entire roof surface is under negative static pressure. Velocity is low, however, due to the cavity formed as wind sweeps up over the building. With greater chimney height, termination above the low-velocity cavity or negative pressure zone subjects the chimney exit to greater wind velocity, thereby increasing draft from two causes: (1) height and (2) wind aspiration over an open top. As the termination is moved from the center of the building to the sides, its exposure to winds and pressure also varies.

Terminations on pitched roofs may be exposed to either negative or positive static pressures, as well as to variation in wind velocity and direction. On the windward side, pitched roofs vary from complete to partial negative pressure as pitch increases from approximately flat to 30° (Chien *et al.* 1951). At a 45° pitch, the windward pitched roof surface is strongly positive, and, beyond this slope, pressures approach those observed on a vertical wall facing the wind. Wind pressure varies with its' horizontal direction on a pitched roof and on the lee or sheltered side, wind velocity is very low and static pressures are usually negative. Wind velocities and pressures vary, not only with pitch, but with position between ridge and eave. Reduction of these observed external wind effects to simple rules of termination for a wide variety of chimney and venting systems requires many compromises.

In the wake zone or any higher location exposed to full wind velocity, an open top can create strong venting updrafts. The updraft effect relative to wind dynamic pressure is related to the Reynolds number. Open tops, however, are sensitive to the wind angle, as well as to rain (Clark 1967), and many proprietary tops have been designed to stabilize wind effects, and improve the performance. Because of the many compromises made in the design of a vent termination, this stability is usually achieved by sacrificing some of the updraft created by the wind. Further, the location of a vent cap in a cavity region frequently removes it from the zone where wind velocity could have a significant effect.

Studies of vent cap design undertaken to optimize performance of residential types indicate that the following performance features are important: (1) still-air resistance, (2) updraft ability with no flow, and (3) discharge resistance when vent gases are carried at low velocity in a typical wind (10 ft/s [3 m/s] vent velocity in a 20 mph [9 m/s] wind). Tests described in UL *Standard* 441 for proprietary gas vent caps consider these three aspects of performance to ensure adequate vent capacity.

Frequently, the air supply to an appliance room is difficult to orient to eliminate wind effects. Therefore, the vent outlet must have a certain updraft capability, which can help balance a possible adverse wind. When wind flows across an inoperative vent termination, a strong updraft develops. Appliance start-up reduces this updraft and, in typical winds, the vent cap may develop greater resistance than it would have in still air. Certain vent caps can be made with very low still-air resistance, yet exhibit excessive wind resistance, which reduces capacity. Finally, because the appliance operates whether or not there is a wind, still-air resistance must be low.

Some proprietary air ventilators have excessive still-air resistance and should be avoided on vent and chimney systems, unless a considerably oversized vent is specified. Vertical-slot ventilators, for example, have still-air resistance coefficients of about 4.5. To achieve low still-air resistance on vents and chimneys, the vertical-slot ventilator must be 50% larger than the diameter of the chimney or vent, unless it has been specifically listed for such use.

Free-standing chimneys high enough to project above the cavity zone require structurally adequate materials or guying and bracing for prefabricated products. The prefabricated metal building/heating equipment chimney places little load on the roof structure, but guying is required at 8 to 12 ft (2.4 to 3.6 m) intervals to resist both overturning and oscillating wind forces. Various other expedients, such as spiral baffles on heavy gauge free-standing chimneys, have been used to reduce oscillation.

The chimney height needed to carry the effluent into the undisturbed flow stream above the wake zone can be reduced by increasing the effluent discharge velocity. A 3000 ft/min (15 m/s) discharge velocity avoids downward eddying along the chimney and expels the effluent free of the wake zone. Velocity this large can be achieved only with forced or induced draft.

The entry of rain is a problem for open, low-velocity, or inoperative systems. Good results have been obtained with drains that divert the water onto a roof or into a collection system leading to a sump. Figure 27 shows several configurations (Clark 1967 and Hama and Downing 1963). The runoff from stack drains contains acids, soot, and metallic corrosion products, which can cause roof staining. Therefore, these methods are not recommended for residential use. An alternate procedure is to allow all water to drain to the base of the chimney, where it is piped from a capped tee to a sump.

Rain caps prevent the vertical discharge of high-velocity gases. However, caps are preferred for residential gas-burning equipment because it is easier to exclude rain than to risk rain water leakage at horizontal joints or to drain it. Also, caps keep out debris and bird nests, which can block the chimney. Satisfactory gas vent cap

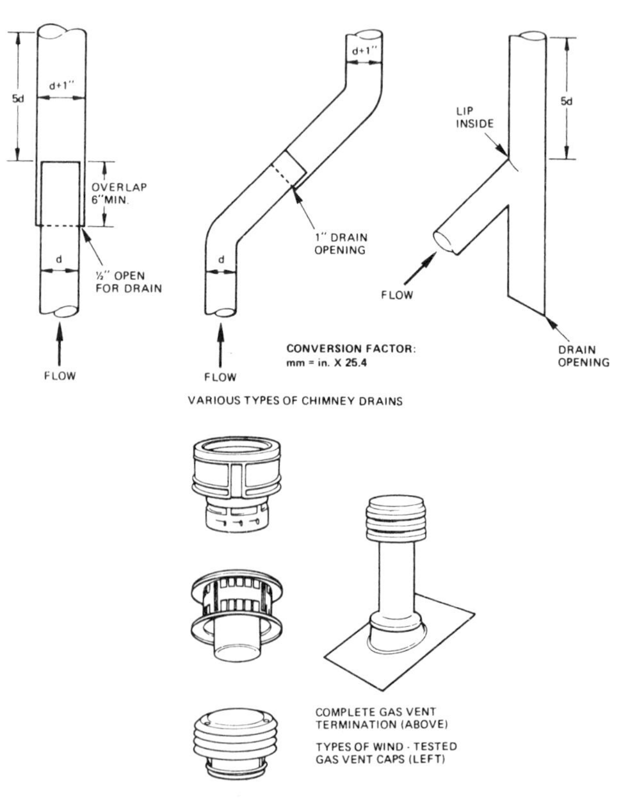


**Fig. 27 Vent and Chimney Rain Protection**

CHAPTER 27

# UNIT VENTILATORS, UNIT HEATERS, AND MAKEUP AIR UNITS



THIS chapter defines unit ventilators, unit heaters, and makeup air units. The types, characteristics, method of rating, and procedures for selection and application of these products are described, as well.

## PART I: UNIT VENTILATORS

The term *unit ventilator* describes a type of unit heater that is applied primarily to classrooms, where controlled ventilation is required. The *heating unit ventilator* heats, ventilates, and cools a space by introducing outdoor air in quantities up to 100% of its rated capacity. The heating medium may be steam, hot water, gas, or electricity. The components of a heating unit ventilator are fans, motor, heating element, dampers, filters, and outlet grilles (or diffusers), encased in an acoustically and thermally insulated cabinet.

An *air-conditioning unit ventilator,* in addition to the normal winter function of heating, ventilating, and cooling with outdoor air, also cools and dehumidifies during the summer season.

The typical unit ventilator is controlled to permit the heating, ventilating, and cooling effect to be varied while the fans are operating continuously. Usually, the discharge air temperature is varied in accordance with the space requirements. The heating unit ventilator brings in outdoor air whenever the outdoor temperature is below the room temperature for *ventilation cooling.*

The preparation of this chapter is assigned to TC 6.1 Hot-Water and Steam-Heating Equipment and Systems.

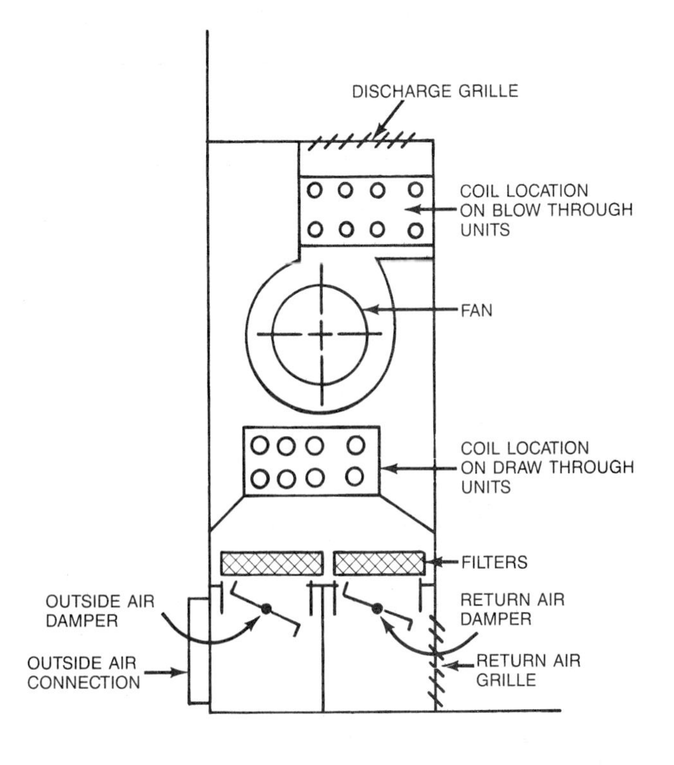


**Fig. 1 Unit Ventilator Arrangement**

**Table 1 Typical Unit Ventilator Capacities**

| cfm (L/s) | Total Heating Capacity, Btu/h (kW),[a] Heating Unit Ventilator | Total Cooling Capacity, Btu/h (kW),[b] Air-Conditioning Unit Ventilators |
|---|---|---|
| 500 (240) | 38,000 (11) | 19,000 (5.5) |
| 750 (350) | 50,000 (15) | 28,000 (8) |
| 1000 (470) | 72,000 (21) | 38,000 (11) |
| 1250 (590) | 85,000 (25) | 47,000 (14) |
| 1500 (710) | 100,000 (29) | 56,000 (16) |

[a]20% outdoor air at 0°F (−17.8°C), 70°F (21.1°C) room air, 180°F (82.2°C) water temperature.

[b]20% outdoor air at 95°F (35°C) db, 78°F (25.6°C) wb, room air 80°F (26.7°C) db, 67°F (19.4°C) wb, 45°F (7.2°C) entering water temperature.

Air-conditioning unit ventilators can provide mechanical cooling when the outdoor air temperature is too high for ventilation cooling.

## DESCRIPTION

Unit ventilators are available for floor mounting, recessed applications, and with various airflow and capacity ratings. They can be arranged for either blow-through or draw-through operation. Figure 1 describes the general arrangement of floor-mounted unit ventilators. Table 1 shows typical nominal unit ventilator capacities.

## HEATING CAPACITY REQUIREMENTS

Since a unit ventilator has a dual function of introducing outdoor air for ventilation and maintaining a specified room condition, the required heating capacity is the sum of the heat required to bring outdoor air up to room temperature and the heat required to offset room losses. The ventilation cooling capacity of a unit ventilator is determined by the air volume delivered by the unit and the temperature difference between the unit discharge and the room temperature.

*Example.* A room has a heat loss of 24,000 Btu/h at a winter outdoor design condition of 0°F and an indoor design of 70°F, with 250 cfm, 20% outdoor air. Minimum air discharge temperature from the unit is 60°F. To obtain the specified number of air changes, a 1250 cfm unit ventilator is required. Determine the ventilation heat requirement, the total heating requirement, and the ventilation cooling capacity of this unit with outdoor air temperature below 60°F.

*Solution:*

$$H_v = BQ_v(t_i - t_o)$$

*where*

$H_v$ = heat required to heat ventilating air, Btu/h (W)
$B$ = 1.085 (1.212)
$Q_v$ = ventilating air, cfm (L/s)
$t_i$ = required room air temperature, °F (°C)
$t_o$ = outdoor air temperature, °F (°C)

$H_v = 1.085(250)(70 - 0) = 19{,}000$ Btu/h

$$H_t = H_v + H_s$$

*where*

$H_t$ = total heat requirement, Btu/h (W)
$H_s$ = heat required to make up heat losses, Btu/h (W)

$H_t = 19{,}000 + 24{,}000 = 43{,}000$ Btu/h

$$H_c = BQ(t_i - t_f)$$

*where*

$H_c$ = ventilation cooling capacity of unit, Btu/h (W)
$B$ = 1.085 (1.212)
$Q$ = volume of air handled by the unit ventilator, cfm (L/s)
$t_i$ = required room temperature, °F (°C)
$t_f$ = unit discharge air temperature, °F (°C)
$H_c$ = 1.085 (1250)(70 − 60) = 13,500 Btu/h

## SELECTION

Items to be considered in the application of unit ventilators are (1) unit air capacity, (2) percent minimum outdoor air, (3) heating and cooling capacity, (4) control, and (5) location of unit.

Primary considerations in selecting the unit air capacity are the ventilation cooling capacity and the number of occupants in the space. Other factors to be considered are legal requirements, volume of the room, density of occupancy, and the usage of the room. The number of air changes required for a specific application also depends on window area, orientation, and the maximum outdoor temperature at which the unit will be able to cool by ventilation.

For classrooms, air changes per hour to prevent overheating at 55°F (13°C) outdoor air varies from six (for rooms with small north windows) to 12 (for rooms with large south windows). Factories and kitchens require 30 to 60 air changes or more. Office areas may need from 10 to 15 air changes per hour.

The minimum amount of outdoor air for ventilation is determined after the total air capacity has been established. It may be governed by codes or calculated by the engineer to meet the ventilating air needs of the particular application (see ASHRAE *Standard* 62-1981, *Ventilation for Acceptable Air Quality*).

In the absence of other criteria, 5 cfm (2.36 L/s) per occupant is used frequently as minimum outdoor air for classroom applications. Heating capacity should always be selected after selecting the unit air capacity for mild-weather cooling.

## CONTROL

The principal difference in the many different controls is the amount of outdoor air delivered. Usually, a room thermostat simultaneously controls a valve, damper, or step controller to regulate the heat supply, and a damper to regulate the outdoor air supply. An airstream thermostat in the unit prevents the discharge of air below the desired minimum temperature. Unit ventilator controls provide the proper sequence for the following stages:

**Warmup Stage.** Full heat is generated with the outdoor damper closed, and 100% room air is recirculated and heated until the room temperature approaches the desired level.

**Heating and Ventilating Stage.** As the room temperature rises into the operating range of the thermostat, ventilation is accomplished through opening of the outdoor air damper. As the room temperature continues to rise, the unit ventilator heat supply is throttled.

**Cooling and Ventilating Stage.** When the room temperature rises above normal, cool air is discharged into the room. The room thermostat does this by throttling the heat supply, finally shutting it off, and then opening the outdoor air damper to prevent the room from overheating. The airstream thermostat frequently takes control during this stage to keep the discharge temperature from falling below a set level.

The three basic cycles of control commonly used are as follows:

**Cycle I.** 100% of outdoor air is admitted at all times, except during the warmup stage.

**Cycle II.** A minimum amount of outdoor air (normally 25 to 50%) is admitted during the heating and ventilating stage. This percentage is gradually increased to 100%, if needed, during the ventilation cooling stage.

**Cycle III.** Except during the warmup stage, a variable amount of outdoor air is admitted, as needed to maintain a fixed temperature of the air entering the heating element. Air admission is controlled by the airstream thermostat, which is set low enough—often at 55°F (13°C)—to provide cooling when needed.

Air-conditioning unit ventilators also have a mechanical cooling stage in which a fixed amount of outdoor air is introduced, with cooling capacity controlled by the room thermostat.

For economy, the building may be maintained at reduced temperature at night and during weekends and vacations. One typical control uses the natural convective capacity of the unit, with the fans off, supplemented by cycling the fan, when required, to maintain desired room temperatures.

## LOCATION

Unit ventilators are normally installed on an outside wall near the center line of the room. They often incorporate provisions to prevent window downdraft, a problem in classrooms with large window areas in cold climates, where air that is cooled by contact with cold glass flows down into the occupied space. Window downdraft can be combated by directing a portion of the unit ventilator discharge air into a delivery duct along the sill of the window, with upwardly directed air.

## MAINTENANCE

Unit ventilators have maintenance requirements similar to those for unit heaters as described in Part II of this chapter.

# PART II: UNIT HEATERS

A *unit heater* is an assembly of a fan and motor, a heating element, and an enclosure whose function is to heat a space. Filters, dampers, directional outlets, duct collars, combustion chambers, and flues may also be included. Some of these are shown in Figures 2 to 10.

## CLASSIFICATION

Unit heaters can be classified according to one or more of the following methods.

1. **Heating Medium.** Five heating media can be considered: *steam, hot water, gas indirect-fired, oil indirect-fired, and electric.*
2. **Type of Fan.** Three types of fans can be considered: *propeller, centrifugal,* and *remote air mover.* Propeller fan units may be *horizontal-blow* or *downblow.* Centrifugal fan units may be of the smaller *cabinet* type or larger *industrial* type. Units with remote air movers are known as *duct unit heaters.*

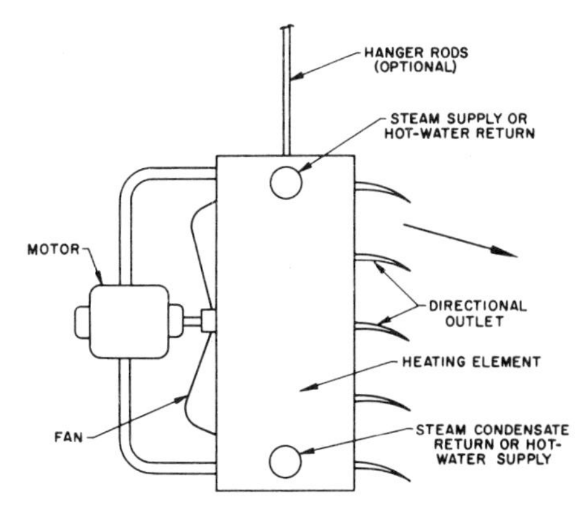


**Fig. 2 Propeller Fan Unit Heater, Horizontal-Blow-Type**

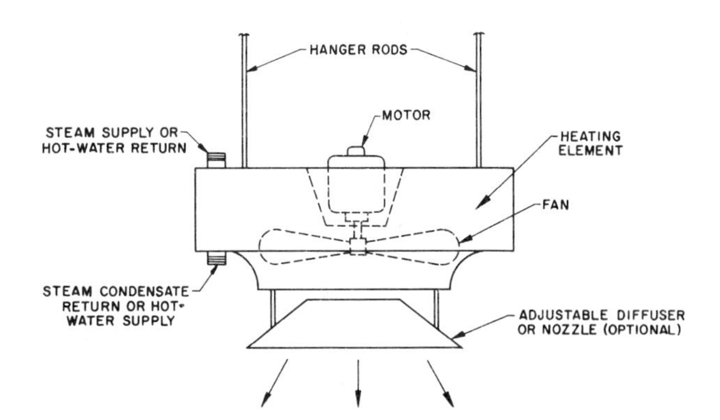


**Fig. 3 Propeller Fan Unit Heater, Downblow-Type**

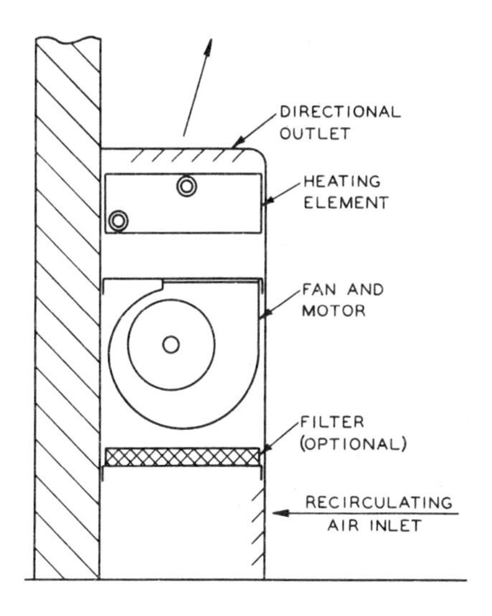


**Fig. 4 Floor-Mounted Centrifugal Fan Unit Heater, Cabinet-Type**

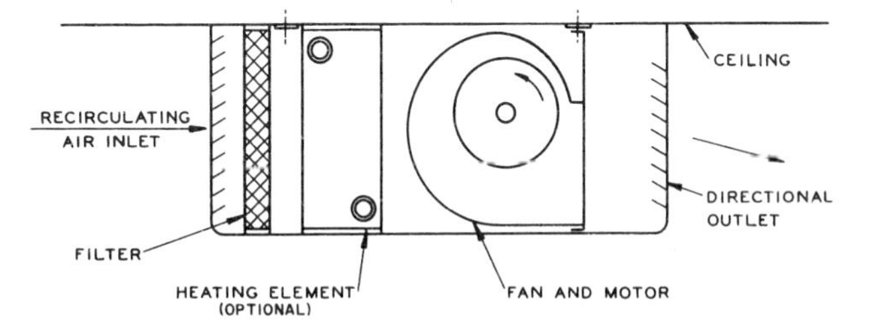


**Fig. 5 Suspended Centrifugal Fan Unit Heater, Cabinet-Type**

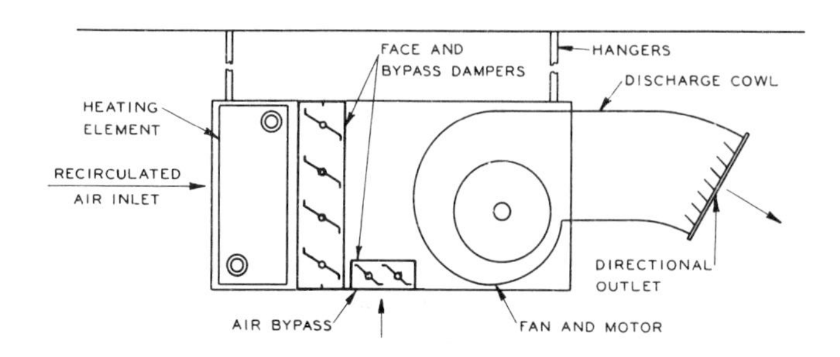


**Fig. 6 Suspended Centrifugal Fan Unit Heater, Industrial-Type, Bypass Controlled**

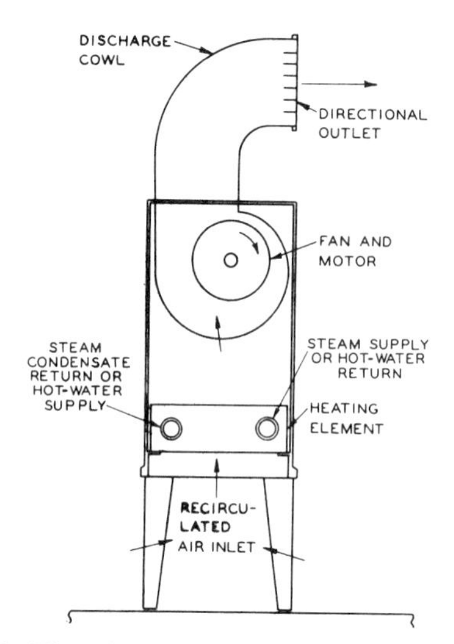


**Fig. 7 Floor-Mounted Centrifugal Fan Unit Heater, Industrial-Type**

3. **Arrangement of Elements.** Two types of units can be considered: the *draw-through*, in which the fan draws air through, and the *blow-through*, in which the fan blows air through the heating element. Indirect-fired unit heaters are always of the blow-through type.

## APPLICATION

Unit heaters have relatively large heating capacities in compact casings, the ability to project heated air in a controlled manner over a considerable distance, and a relatively low installed cost. They are, therefore, usually applied where the required heating capacity, the volume of the heated space, or both, are too large to be handled adequately or economically by other means. They eliminate extensive duckwork.

Unit heaters are used for spot or intermittent heating, such as blanketing outside doors. They are also used where filtration of heated air is required.

Unit heaters are used to heat garages, factories, warehouses, showrooms, stores, and laboratories, as well as corridors, lobbies, vestibules, and similar auxiliary spaces in buildings.

Unit heaters may be applied to industrial processes, such as drying and curing, in which the use of heated air in rapid circulation with uniform distribution is of particular advantage. They may be used for moisture absorption applications, such as the prevention of condensation on ceilings or other cold surfaces of buildings in which process moisture is released. When such conditions are severe, unit ventilators or makeup air units may be required.

## SELECTION

Factors to be considered in the selection of unit heaters include (1) the heating medium to be employed, (2) the type of unit, (3) the location of the unit for proper heat distribution, (4) the permissible sound level, (5) the need for filtration, and (6) the heating capacity.

### Heating Medium

**Steam or hot-water unit heaters** are relatively inexpensive, but require a boiler and piping system. Therefore, they are most frequently used in new installations involving a relatively large number of units and in existing systems that have capacity for the additional load. High-pressure steam or high temperature

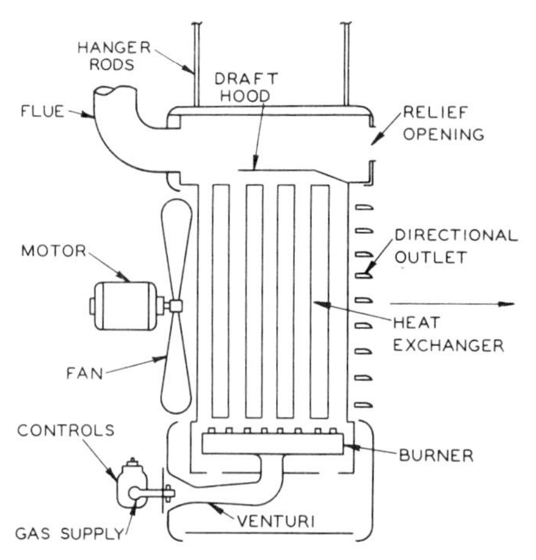


**Fig. 8 Indirect Gas-Fired Propeller Fan Unit Heater**

hot-water units are normally used only in very large installations or when a high-temperature medium is required for process work. Low-pressure steam and conventional hotwater units are usually selected for smaller installations and for those concerned primarily with comfort heating.

**Gas and oil indirect-fired unit heaters** may be preferred when the number of units does not justify the expense or space requirements of a new boiler system or where individual metering of the fuel supply is required. Gas indirect-fired units are usually either the horizontal propeller-type or the industrial centrifugal-type. Oil indirect-fired units are largely the industrial centrifugal-type. Some codes limit the use of indirect-fired unit heaters in certain applications.

**Electric unit heaters** are used when electric power cost is competitive with other fuels or for isolated locations, intermittent use, supplementary heating, or temporary service. Typical applications are ticket booths, watchmen's offices, factory offices, and locker rooms. Electric units are particularly useful in isolated and untended pumping stations or pits, where they may be thermostatically controlled to prevent freezing.

### Type of Unit

**Propeller fan units** are used in free-delivery applications where the heating capacity and distribution requirements can best be

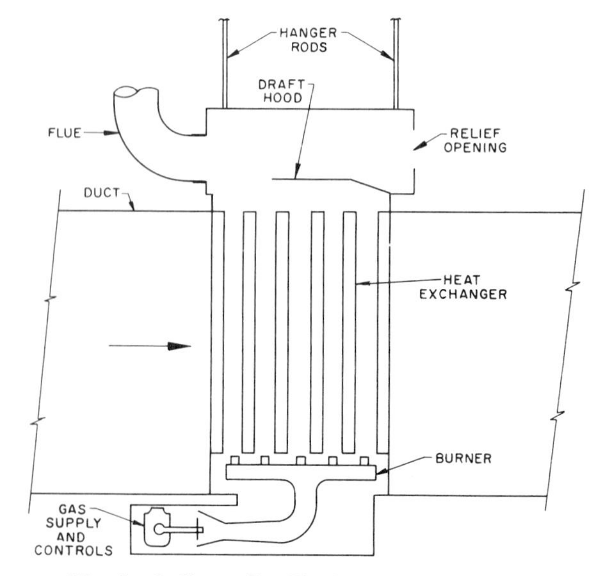


**Fig. 9 Indirect Gas-Fired Duct Unit Heater**

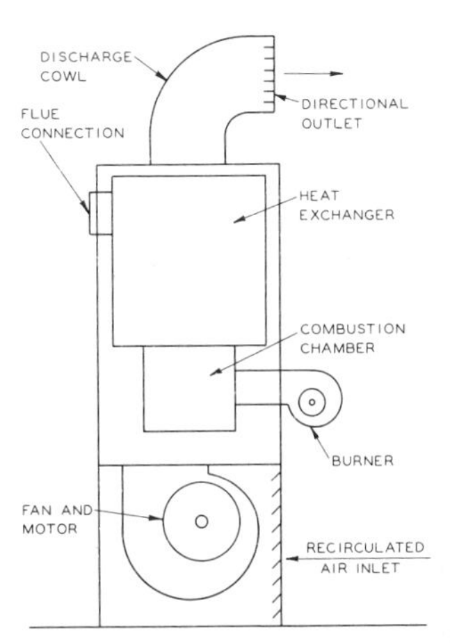


**Fig. 10 Oil or Gas Indirect-Fired, Floor-Mounted, Centrifugal Fan Unit Heater, Industrial-Type**

met by single or multiple units of moderate output and where filtration of the heated air is not required. **Horizontal-blow units** are usually associated with low to moderate ceiling heights. **Downblow units** are used in high-ceiling spaces and where floor and wall space limitations dictate an out-of-the-way location for the heating equipment. Downblow units may be supplied with an adjustable diffuser to vary the air-discharge pattern from a high velocity vertical jet (to achieve the maximum distance of downward throw) to a horizontal discharge of lower velocity (to prevent excessive air motion in the zone of occupancy). Revolving diffusers are also available.

**Duct heaters** are used in systems where there is an air-handling unit remote from the heater. These heaters provide an economical means of adding heating to existing cooling or ventilating systems with ductwork.

**Industrial centrifugal fan units** are applied where heating capacities and space volumes are large or where filtration of the heated air or operation against static resistance are required. Downblow or horizontal-blow units may be used, depending on the requirements.

**Cabinet unit heaters** are used where a more attractive appearance is desired. They are suitable for free delivery or low static pressure duct applications, may be equipped with filters, and can be arranged to discharge either horizontally or vertically up or down.

## Location for Proper Heat Distribution

Units must be selected, located, and arranged to provide complete heat coverage with acceptable air motion and temperature in the occupied zone. Proper application depends on size, number, type, and direction of blow of the units; mounting height; outlet velocity and temperature; air volume; and type of directional outlet used. Manufacturers publish data for specific units, showing heat coverage and mounting heights for various outlet velocities and outlet arrangements.

The mounting height may be governed by space limitations or by the presence of such equipment as display cases or machinery. Noise and air velocity in the occupied zone generally increase with increased outlet velocities and must, therefore, be considered.

The blow depends on the temperature of air leaving the heater, as well as on its velocity and volume. Increased discharge temperature reduces the area of heat coverage and the mounting height. Downblow unit heaters must be applied carefully. The higher the mounting height, the lower the outlet temperature of the air must be to force the heated air down into the occupied zone.

To obtain the desired air distribution and heat diffusion, unit heaters are commonly equipped with directional outlets, adjustable louvers, or fixed or revolving diffusers. For a given unit with a given discharge temperature and outlet velocity, the mounting height and heat coverage can vary widely with the type of outlet.

Obstructions, such as columns, beams, partitions, or machinery, in either the discharge airstream or the approach area to the unit or the presence of strong drafts or other air currents reduce the coverage. Exposures such as large glass areas or outside doors, especially on the windward side of the building, require special attention; units should be arranged so that they blanket exposures with a curtain of heated air that intercepts the cold drafts.

For area heating, horizontal-blow unit heaters in exterior zones should be located to blow either along the exposure or toward it at a slight angle. When possible, multiple units should be arranged so that the discharge airstreams support each other and create a general circulatory motion in the space. Interior zones under exposed roofs or skylights should be blanketed completely. Downblow units should be arranged so that the heated areas from adjacent units overlap slightly to provide complete coverage.

For spot heating of individual spaces in larger unheated areas, single unit heaters may be used, but allowance must be made for the inflow of unheated air from adjacent spaces and the consequent reduction in heat coverage.

Horizontal unit heaters should discharge well above head level. Both horizontal and vertical units should be placed so that the heated airstream is delivered to the occupied zone at acceptable temperatures and velocities. When possible, units should be located so that the discharge airflow is into open spaces, such as aisles, and not directly on the occupants. Generally, outlet air temperature of free-delivery unit heaters used for comfort heating should be from 50 to 60°F (10 to 15.6°C) higher than the design room temperature. For further information on air distribution, see Chapter 32 of the 1985 FUNDAMENTALS Volume.

Manufacturers' catalogs usually give suggestions for arrangements, mounting heights, heat coverages, final temperatures, and directional outlets.

## Sound Level

Sound pressure levels in workplaces are now limited by federal regulations to 90 dbA for personnel occupancy of eight hours per day. Unit heaters may contribute a significant portion of noise level. An analysis of both the diverse sound sources and the locations of personnel stations establishes the limit to which the unit heaters must be held. (See Chapter 52 of the 1987 HVAC Volume.)

## Ratings of Unit Heaters

It is common to rate unit heaters on the amount of heat delivered by the air above an entering air temperature of 60°F (16°C). Rating of steam unit heaters has been based upon dry saturated steam at 2 psig (13.8 kPa) pressure at the heater coil, air at 60°F (16°C) (29.92 in. [101.13 kPa] Hg barometric pressure) entering the heater, and the heater operating free of external resistance to airflow. The capacity of a heater increases as the steam pressure increases and decreases as the entering air temperature increases.

Rating of hot water unit heaters is usually based on water at 200°F, water temperature drop of 20°F, entering air at 60°F (29.92 in. Hg barometric pressure) (93°C, 11°C drop of 16°C entering air at 101.325 kPa), and the heater operating free of external resistance to airflow. Variations in entering water temperature, entering air temperature, and water flow rate affect capacity.

Gas-fired unit heaters are rated in both input and output, in accordance with approval requirements of the American Gas Association. Ratings of oil-fired unit heaters are based on heat delivered at heater outlet. Electric unit heaters are rated on the energy input to the heating element.

**Effect of Resistance on Capacity.** Unit heaters are customarily rated at free-delivery. If outdoor air intakes, air filters, or ducts on the inlet or discharge are used, a reduction in air and heating capacity will result due to this added resistance to airflow, unless fan speed is increased. This reduction in capacity depends on the characteristics of the heater, and on the type, design, and speed of the fans, so that no specific percentage reduction can be assigned at a given added resistance. The heat output to be expected under other than free-delivery conditions should be secured from the manufacturer.

**Effect of Inlet Temperature.** Changes in entering air temperature influence the total heating capacity and the final temperature. Since many unit heaters are located some distance from the occupied zone, possible differences between the temperature of the air entering the unit and that of air being maintained in the heated area should be considered, especially with downblow unit heaters.

Higher velocity units and units with lower vertical discharge air temperature maintain lower floor-to-ceiling temperature gradients than units with higher discharge temperatures. Valve-controlled or bypass-controlled units with continuous fan operation maintain lower temperature gradients than units with intermittent fan operation. Directional control of discharge air can also assist in effecting satisfactory heat distribution.

Heaters for high-pressure steam or high-temperature hot water should be designed to produce approximatley the same leaving air temperature as would be obtained from a lower temperature heating medium.

### Filtration

Propeller-type unit heaters are designed for operating with heater friction loss only. If the internal building contaminants require filtration, centrifugal fan units or cabinet units should be used.

## AUTOMATIC CONTROL

A steam or hot-water unit heater can have *on-off operation* of the unit fan or *continuous fan operation* with modulation of heat output.

For *on-off* operation, a room thermostat starts and stops the fan motor. A limit thermostat, often strapped to the supply or return pipe, prevents fan operation if heat is not being supplied to the unit. An auxiliary switch that energizes the fan only when power is applied to open the motorized supply valve may also prevent undesirable cool air from being discharged.

Continuous fan operation eliminates intermittent blasts of hot air resulting from *on-off* operation, as well as the stratification of temperature from floor to ceiling, which often occurs during the *off* periods. A proportional room thermostat controls a valve modulating the heat supply to the coil or a bypass around the heating element. A limit thermostat or auxiliary switch stops the fan when heat is no longer available.

Indirect-fired and electric units are usually controlled by intermittent operation of the heat source under control of the room thermostat, with a separate fan switch to run the fan when heat is being supplied. (For a general discussion of automatic control, refer to Chapter 51 of the 1987 HVAC Volume.)

Unit heaters may be used in summer to provide air motion. The source of heat is shut off, and the thermostat has a bypass switch that permits independent fan operation.

## PIPING CONNECTIONS

Piping connections for steam unit heaters are similar to those for other types of fan blast heaters. Basic piping principles for steam systems are covered in Chapter 11 of the 1987 HVAC Volume.

Rapid condensation of steam is characteristic of steam unit heaters, especially during heating-up periods. The return piping must be planned to keep the heating coil free of condensate during periods of maximum heat output, and the steam piping must be able to carry a full supply of steam to the unit to take the place of condensed steam. Adequate sizes of piping are especially important when a unit heater fan is cycled *on* and *off* because the condensate rate fluctuates rapidly. Recommended piping connections for unit heaters are shown in Figures 11 to 14.

In steam systems, the branch from the supply main to the heater must pitch toward the main and be connected to its top to prevent condensate in the main from draining through the heater, where it might reduce capacity and cause noise.

The return piping from steam unit heaters should provide a minimum drop of 10 in. (250 mm) below the heater, so the head of water required to overcome resistances of check valves, traps, and strainers will not cause condensate to remain in the heater.

Dirt pockets at the outlet of unit heaters are essential, and strainers with 0.063 in. (1.6 mm) perforations prevent rapid plugging by retaining dirt and scale, which might affect operation of check valves and traps. Strainers should always be installed in the steam supply line if the heater is equipped with the steam-distributing type of coils or is valve-controlled.

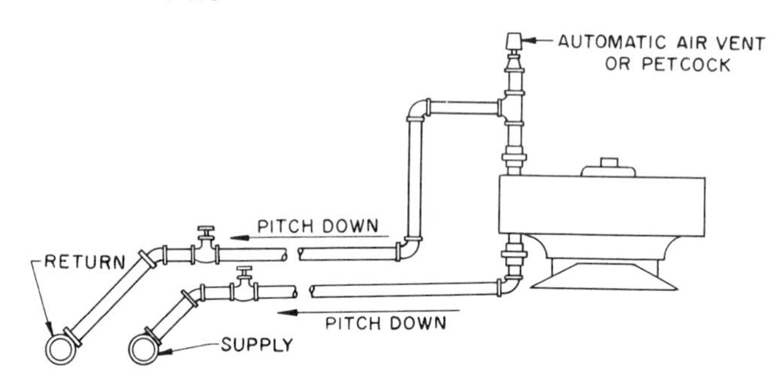


**Fig. 11 Vertical Unit Heater Connections to Lower Hot-Water Mains**

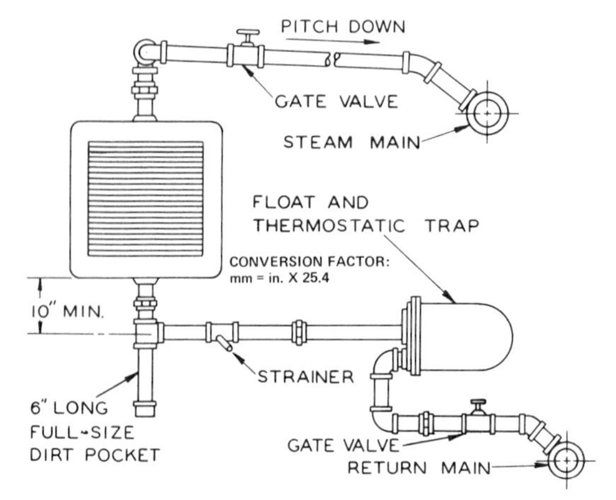


**Fig. 12 Unit Heater Connection for Low-Pressure Steam, Open Gravity or Vacuum Return System**

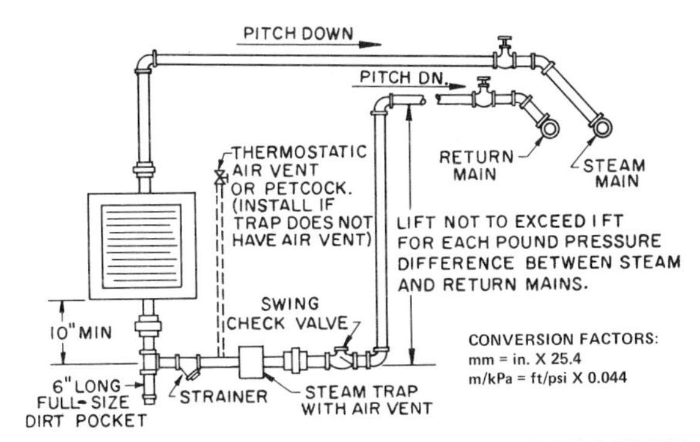


**Fig. 13 Horizontal Unit Heater Connections to Overhead Steam and Return Mains**

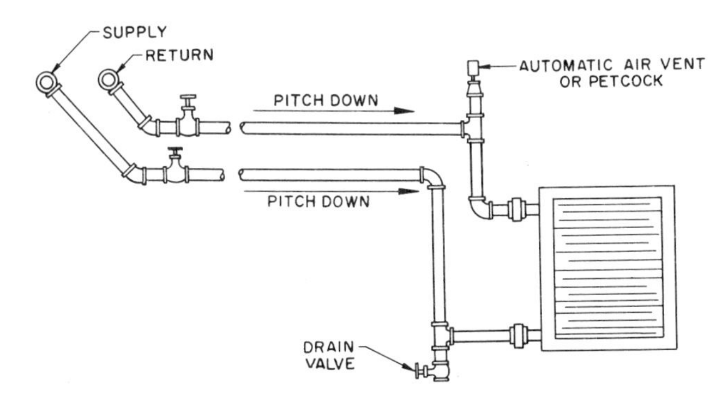


**Fig. 14 Horizontal Unit Heater Connections to Overhead Hot-Water Mains**

An adequate air vent is required for low-pressure closed-gravity systems. The vertical pipe connection to the air vent should be at least 0.75 in. (20 mm) IPS to permit separation of the water from the air passing to the vent. If thermostatic instead of float-and-thermostatic traps are used in vacuum systems, a cooling leg must be provided ahead of the trap.

In high-pressure systems, it is common to vent the air continuously through a petcock, unless the steam trap has a provision for venting air. Most high-pressure return mains terminate in flash tanks that are vented to the atmosphere. Pressure-reducing valves should be installed to permit operation of the heaters at low pressure. Traps used must be suitable for the operating pressure encountered.

When piping is connected to hot water unit heaters, it must be pitched to permit the escape of air to the high point in the piping, where it can be vented to the atmosphere. An air vent at the heater is used to remove air or vent the top of the heater. Provision must be made for complete drainage of the system, including nipple and cap drains on drain cocks, when units are located below mains.

## MAINTENANCE

Regularly scheduled inspections will maintain operating economy and heating capacity. Heating elements should be cleaned by brushing or blowing with high-pressure air or by using a steam spray. A portable sheet metal enclosure may be used to enclose the smaller heaters partially for cleaning in place with air or steam jets. The heating element may need to be removed and washed with a mild alkali solution, followed by a thorough rinsing with water. Propeller units do not have filters and are, therefore, more susceptible to dust buildup on the coils.

Dirty fan blades cause capacity loss and possible imbalance and vibration, with consequent noise and damage to bearings. Fan blades should be inspected and cleaned when necessary. Vibration and noise may also be caused by improper fan position or loose set screws. A fan guard is recommended on downblow unit heaters that have no diffuser or other device to catch the fan blade if it falls from the unit.

A considerable difference exists in the amounts of attention required by motors used with unit heaters. This is particularly true of lubrication, the instructions for which must be followed carefully for trouble-free operation. Excessive lubrication may cause the lubricant to damage the motor. An improper lubricant may cause failure of bearings. Instructions for care of the motor on any unit heater should be obtained from the manufacturer and kept at the unit.

Fan bearings and drives must be given the lubrication and other attention specified by the manufacturer. Couplings should be inspected periodically for wear and alignment. V-belt drives should have all belts replaced with a matched set if one belt shows wear.

Periodic inspections of traps, inspections of check and air valves, and the replacement of worn parts are important. Strainers should be cleaned regularly.

When filters are incorporated in the unit, they must be cleaned or replaced when dirty.

Chapter 53 of the 1987 HVAC Volume and *Bulletin 60, The Installation and Maintenance of Steam and Hot Water Unit Heaters, 1972,* published by the Air Moving and Conditioning Association, Inc., cover corrosion prevention in unit heaters.

# PART III: MAKEUP AIR UNITS

A *makeup air unit* is an assembly of elements that conditions makeup air introduced into a space for ventilation to replace air exhausted from a building. If the temperature or humidity (or both) within the structure are controlled, the environmental control system must have the capacity to condition the replacement air. One way to condition the replacement air is to use a makeup air unit.

Makeup air units may, therefore, be required to heat, ventilate, cool, humidify, dehumidify, and filter. Makeup air units may be required to replace the air at the space conditions or may be used for part or all of the heating, ventilating, or cooling load.

Packaged makeup air units are offered by manufacturers in capacities from 2000 to 100,000 cfm (0.9 to 47.2 $m^3/s$). Equipment ranges from basic units of fans, heating elements, and controls for truss mounting to packaged penthouse units, which include filters, air washers, or cooling coils as complete assemblies. Walk-in service compartments, which are heated, ventilated, and lighted, can also be included.

## SYMPTOMS OF AN AIR-STARVED BUILDING

Buildings under negative pressure caused by inadequate makeup air have the following symptoms:

1. Gravity stacks from unit heaters and processes back-vent.
2. Exhaust systems do not perform at rated volume.
3. The perimeter of the building is cold in the winter due to high infiltration.
4. There are severe indrafts at exterior doors.
5. Exterior doors are hard to open.

6. Heating systems cannot maintain uniform comfort conditions throughout the building: the center core area becomes overheated.

## APPLICATIONS

Traditionally, makeup air has been considered a means of supplying replacement air that is heated to approximate room temperature to compensate for the air being exhausted from the building. Equipment developments have proved that makeup air units can also provide adequate ventilation and control infiltration. This is done by controlled outside air and return air dampers (Figure 15). Prior to these product developments, little consideration was given to air distribution alternatives other than ductwork. Like most heating systems, makeup air heaters were considered to be a system requiring ductwork for proper distribution.

Recent applications show that properly sized heaters can achieve good air distribution through pressurization with little or no ductwork. The building walls and roof form a duct structure that allows the heater to distribute the air at building pressures of 0.05 in. water gauge (12 Pa) or less, depending on the building porosity and design. Since the air requirements within the building are subject to the building activity and the effect caused by wind and temperature variation, makeup air heaters are available with modulating dampers that are controlled by automatic pressure switches. By maintaining a constant pressure, the tempered or untempered outside air delivered by the heater travels toward the natural relief in the perimeter walls.

Infiltration that would normaly enter around door and window openings becomes low-velocity exfiltration. Mechanical exhaust increases the natural ventilation rate by requiring additonal outside air from the heater. Variable outside air heaters can be sized to handle both the winter and summer air requirements. In the summer, heat buildup can be held to a minimum by constantly flushing a high positive outside air change through the building space. In the winter, only minimum outside air is introduced into the building to meet ventilation and replacement air requirements.

Another benefit of makeup air heating comes from the low temperature differential between the room or space ambient temperature and the heater discharge temperature. Based on the heat loss of the building and its ventilation requirement, the burner and fan can be sized so that the average discharge difference would be between 5 and 10 °F (3 and 6 °C). Maximum allowable discharge temperature should be below 100 °F (38 °C). The combination of pressure and low-temperature discharge prevents the normally expected stratification of air caused by standard heating systems that discharge air at 120 °F (50 °C) or more. Even in a tall building, a properly designed makeup air system produces stratification of only 2 to 4 °F (1 to 2 °C).

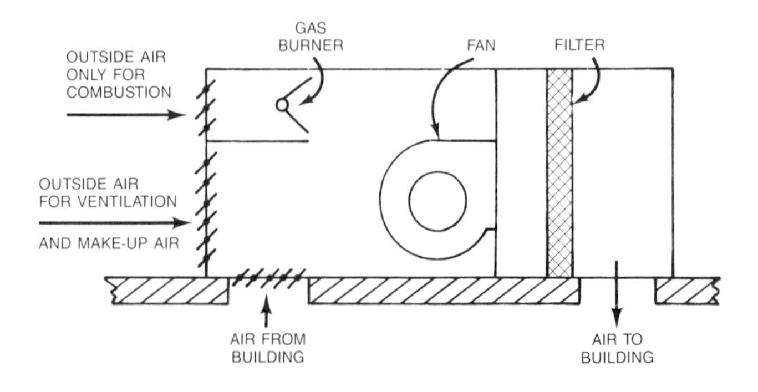


**Fig. 15 Direct Fired, Make-up Air Unit**

Throw or velocity of discharge air is not necessarily a critical factor in a pressure system. The system relies, instead, on pressure to move the air from the point of entrance to the perimeter relief. The advantage of pressure is that it is not affected by obstacles such as strong stacks or machinery. The air flows around the objects on its way to the relief in the perimeter walls. It is more economical to use the makeup air heater in the summer mode to produce a high ventilation rate in order to reduce internal heat buildup than to run ductwork from the heater to the work station. Floor or column fans can create a wind chill cooling effect in the summer when the internal heat load is not allowed to build up. Duct is expensive and is fixed in place. Work stations often move and change. Duct also adds static pressure to the fan load, which requires horsepower motors to move the same air. In the heating season, a high-velocity, low-temperature air discharge can produce an uncomfortable wind chill condition at work stations.

Improved summer cooling can be obtained by adding an air washer to the makeup air unit. Chapter 4 of this volume and Chapter 56 of the 1987 HVAC Volume have further details.

Makeup air for environmental control may be provided by unit ventilators with the air-handling capacity to meet summer requirements.

Makeup air units are also used as door heaters. The units operate continuously, delivering outside air to the building at, or slightly above, room temperature. When the door opens, a door switch signals the unit to discharge the air at a higher temperature. Units used for this purpose must have the capacity to provide final temperatures of about 120 °F (50 °C) at the local design temperature. The supply duct must produce an air curtain across the face of the door opening. The purpose of the air curtain is not to prevent the entry of outside air but to mix hot air with air entering the doorway.

Spot cooling in high heat areas can be done with makeup air. Velocity is an important component of the design, and special attention should be given to the effective temperature of the air supply. In the winter, sensible temperatures are elevated to prevent discomfort.

Process air requirements for closed systems, such as spray booths and ovens, are provided by the same type of equipment as for the general area. In enclosed spray booths, the air-change rate is generally so great that the entire environment of the booth is that of the air supply. Again, it is necessary to consider the effective temperature for personnel comfort.

## SELECTION

Functions to consider in the selection of makeup air units are type of unit, heating and cooling, filters, automatic controls, sound levels, and capacities.

### Types

Makeup air units are available in blow-through or draw-through configurations for either indoor or outdoor installation. The components used with other types of air-handling units are also used with makeup air units.

Fans are centrifugal (single or double width, forward-or backward-inclined, flat-plate or airfoil) or axialflow (propeller, ducted-propeller, vaneaxial, tube axial, or axial-centrifugal).

**Indoor Units.** Indoor units may be truss-mounted, floor-mounted, or wall-mounted.

**Outdoor (Weatherproof) Units.** Rooftop locations for makeup air units, which remove them from truss spaces, eliminate many problems of crowding and conflict with piping, conveyors, or lighting. But even with extensive catwalks, access platforms, and ladders, maintenance of overhead equipment is often difficult and, thus, neglected.

Moving the equipment to the outside also eliminates many of the restrictions on size and enables the manufacturer to design greater flexibility into the equipment (see Figure 16).

**Packaged Makeup Air Equipment.** Makeup air units are available in packages, using any heat source with optional components and configurations, which permit custom design. Equipment manufacturers can provide a packaged, engineered product, including power and control wiring, piping, and temperature controls.

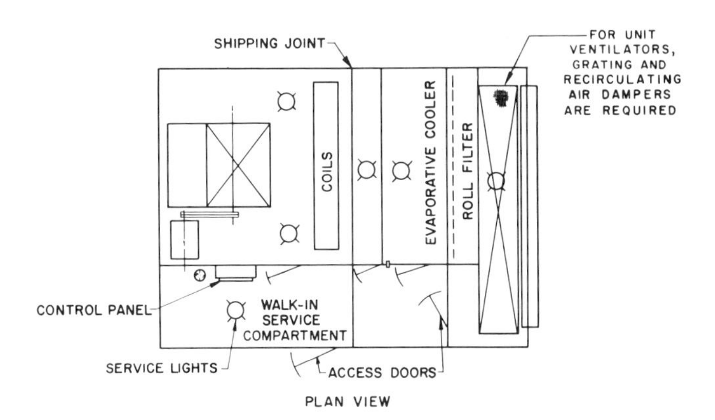


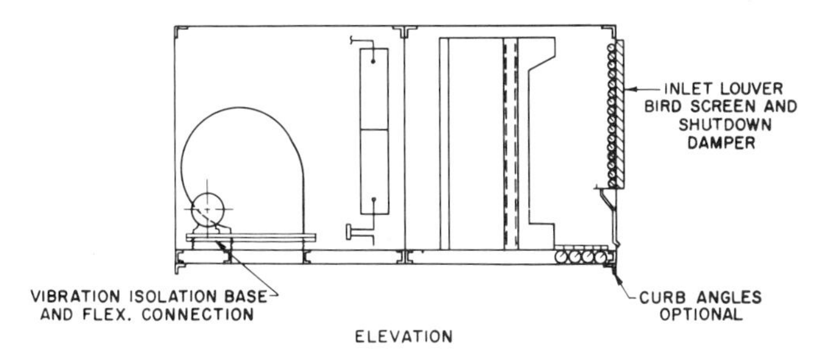


**Fig. 16 Typical Packaged Penthouse Makeup-Air Heater**

## Heating and Cooling Media

Heat sources include gas (direct- or indirect-fired), oil (indirect-fired), steam, hot water, electric, heat-transfer fluids, and exhaust air (air-to-air heat exchangers).

Cooling sources include refrigerants and chilled-water coils, evaporative cooling, sprayed coils, air washers, and air-to-air heat exchangers.

**Gas Direct-Fired Units.** The rapid response and ability to modulate over the complete range (turn down from 25 or 30 to 1) makes the direct-fired burner particularly suited for heating makeup air. The burner releases all of the heat from combustion into the airstream. As shown in Figure 15, outside air should pass through the burner but not return air from the building. This practice eliminates health hazards caused by chemicals released in the building from being converted in the burner to toxic gases. Typical of this is freon, which, when burned, releases phosgene gas.

**Gas and Oil Indirect-Fired Units.** Gas and oil indirect-fired units (with air-to-air heat exchangers) are used for makeup air, but their limited turndown of approximately 3 to 1 and the condensation danger of flue gases in tubes have limited their use as 100% makeup air heaters. However, their use as recirculating heaters, with a percentage of outside air, has made them an important part of the makeup air field (see Part I).

**Steam Units.** To permit cleaning, coil-fin spacing should not exceed 0.13 in. (3 mm). Steam makeup air units should always be specified with filters to protect the coils from dirt and capacity loss.

Steam makeup air heaters require careful design to prevent freezeup of coils. Design considerations include (1) large tubes of the steam-distributing-type, preferably mounted vertically; (2) adequately sized steam traps located for easy servicing; (3) sufficient static head on traps; and (4) an adequate condensate return system.

An effective temperature control is the use of preheat and reheat coils, with full steam pressure on the preheat coil and modulating control on the reheat coil. In areas where design temperatures reach −30 °F (−40 °C), it may be necessary to use two preheat coils, each controlled separately.

Effective freeze protection may be achieved by face and bypass control systems with full steam pressure in the coils at all times. Face and bypass dampers should be designed to prevent stratification of air to the fans and to eliminate spin at the fan inlets. Bypass dampers must be sized to handle the full air capacity of the unit. It is important to limit radiation and air wipage of the coils on light loads. This can be done by installing face dampers on the downstream side of the coils.

**Hot-Water Units.** The design of hot-water heating coils is described in Chapter 9 of this volume. Hot-water coils are not as prone to freezing as steam coils because the cooler water can be removed from the hot-water coil more readily. Nonetheless, freezeups do occur, and precautions must be taken.

When freezing is a possibility, parallel flow circuiting should be specified. Counterflow circuiting, particularly with long coils, is dangerous because the coldest air is in contact with the coldest water.

Proper control is important in freeze protection. Overly hot water and oversized valves and pumps may cause hunting of the control. Freezing can occur when the flow-control valve is shut.

Hot water flow must be maintained through the unit during freezing weather, even if the unit is not in operation or the coil should be drained. If outside air dampers do not close tightly, some leakage of cold air across the coil can be expected.

**Air-to-Air Heat Exchangers.** Because of recent increases in fuel costs, these types of energy recovery makeup air units are important. For a description, see Chapter 34 of this volume.

## Filters

Filters may be automatic or manual roll types, throwaway or cleanable panels, bag type, electrostatic, or a combination of these types (see Chapter 10 of this volume).

## Automatic Controls

Makeup air units require modulating temperature controls to maintain the discharge temperature near room temperature with a large change in intake temperatures. Design conditions may vary from −40 to 35 °F (−40 to 2 °C). Outdoor air imposes an instantaneous load similar to solar heat loads on windows; therefore, capacity must be based on maximum loads.

The basic temperature control system for a makeup air heater is discharge control. Because of the wide load variation, reset is necessary to prevent offset or droop. In direct-fired gas heaters, the discharge control sensor must act quickly because of the rapid response of the burner. Turndown ratios of 25 or 30 to 1 are needed to prevent overheating as outdoor temperatures rise close to the ambient.

While control of discharge temperature is basic, other control systems may include a room thermostat control, where the control point of the discharge controller is changed to satisfy room conditions or a room pressure control to maintain a slight building pressure.

### Sound Level

Noise in the work environment is limited by law. See the section "Sound Level" in Part II.

### Capacities and Ratings

The capacity of a makeup air unit is primarily rated by air volume. Unlike heating systems where loads change with the seasons of the year, exhaust systems usually operate at a constant volume. Standard air should be used in capacity ratings, and all supply and exhaust volumes should be reduced to standard conditions. Fan data, as shown on fan curves, are for bare fans and must be corrected to account for location in an enclosure. Performance testing of the assembly of the components in a complete unit is necessary to have true ratings. The sum of the individual resistances of components as tested under ideal conditions may be less than the system resistance when the components are placed together.

Ratings of a makeup air unit include the following: (1) air volume in scfm (L/s); (2) total static pressure; (3) external static pressure; (4) fan static efficiency at point of rating; (5) sound power level at point of rating; and (6) heating and cooling capacities in Btu/h (W), including inlet and outlet dry- and wet-bulb temperatures.

## CODES AND STANDARDS

Government health departments are becoming more aware of the health problems associated with buildings under negative air pressure. Back-venting of combustion equipment has become a serious health hazard, and several cities have passed ordinances requiring the installation of tempered makeup air equipment.

Design engineers must know codes, particularly those regarding the use of direct gas-firing, which is subject to special regulation in most places. Some codes limit direct gas-firing to areas not containing sleeping quarters; prohibit recirculation of air from the space through the burner; and stipulate adequate exhaust, proper filtration, and specific clearances from combustible materials.

## MAINTENANCE

Maintenance requirements are similar to those of unit heaters (see Part II of this chapter).

## REFERENCES

ACGIH. 1985. *Industrial Ventilation Guide.* American Conference of Government Industrial Hygenists, Cincinatti, OH.

ASHRAE. 1981. Ventilation for Acceptable Air Quality. ASHRAE *Standard* 62-1981.

ICBO. 1985. *Uniform Mechanical Code.* International Conference of Building Officials, Whittier, CA. Also International Association of Plumbing and Mechanical Officials, Los Angeles.

## CHAPTER 28

# RADIATORS, CONVECTORS, BASEBOARD AND FINNED-TUBE UNITS



RADIATORS, convectors, and baseboard and finned-tube units are heat-distributing devices used in steam and low-temperature water-heating systems. They supply heat through a combination of radiation and convection and maintain the desired air temperature in the space. These units should be placed at the points of greatest heat loss of the space to offset or counteract these losses. For example, such units are commonly located under windows, along exposed walls, and at door openings.

The term *radiator* is generally confined to sectional cast-iron column, large-tube, or small-tube units.

The term *convector* refers to a heat-distributing unit that operates with gravity-circulated air. It has a heating element with a large amount of secondary surface and contains two or more tubes headered at both ends. The heating element is surrounded by an enclosure with an air inlet opening below and an air outlet opening above the heating element.

The term *baseboard* (or *baseboard radiation*) refers to heat-distributing units designed for installation along the bottom of walls, in place of the conventional baseboard. They may be made of cast iron, with a substantial portion of the front face directly exposed to the room, or with a finned-tube element in a sheet-metal enclosure. They operate with gravity-circulated room air.

The term *finned tube* (or *fin-tube*) refers to heat-distributing units fabricated from metallic tubing, with metallic fins bonded to the tube. They operate with gravity-circulated room air and may be installed bare, with an expanded metal grille, a cover, or an enclosure with top, front, or inclined outlets.

## HEAT EMISSION

These heat-distributing units emit heat by a combination of radiation to the space and convection to the air within the space. Chapter 3 of the 1985 FUNDAMENTALS Volume covers the heat-transfer processes and the factors that influence them. Those units with a large portion of their heated surface exposed to the space (*i.e.,* radiator and cast-iron baseboard) emit more heat by radiation than do units with completely or partially concealed heating surfaces (*i.e.,* convector, finned tube, and finned-tube-type baseboard). Also, finned-tube elements constructed of steel emit a larger portion of heat by radiation than do finned-tube elements constructed of nonferrous materials.

The output ratings of these heat-distributing units are expressed in Btu/h (W), MBh (1000 Btu/h), or in square feet ($ft^2$) equivalent direct radiation (EDR). For example, 240 Btu/h = 1 $ft^2$ EDR with 1 psig steam.

## RADIATORS

*Small-tube* radiators, with a length of only 1.75 in. (44 mm) per section, occupy less space than the older column and large-tube units and are particularly suited to installation in recesses.

The Institute of Boiler and Radiator Manufacturers (I=B=R, now Hydronics Institute), in cooperation with the National Bureau of Standards, established the *Simplified Practice Recommendation* R174-65, *Cast-Iron Radiators,* for small-tube radiators. This data is no longer available, but Table 1 shows the dimensions and ratings of the units currently manufactured.

*Column, wall-type,* and *large-tube radiators* are no longer manufactured but, as many of these units are still in use, Tables 2, 3, and 4 provide principal dimensions and average ratings. See also *Hydronic Rating Handbook* (Byrley 1978).

## PIPE COILS

Pipe coils have largely been replaced by finned-tube radiation. The heat emission of such pipe coils is shown in Table 5 (for estimating replacement requirements and boiler loads).

## CONVECTORS

Convectors are made in a variety of depths, sizes, lengths, and in enclosure- or cabinet-types. The heating elements are available in fabricated ferrous and nonferrous metals. The air enters the

The preparation of this chapter is assigned to TC 6.1 Hot-Water and Steam-Heating Equipment and Systems.

Table 1 Small-Tube Cast-Iron Radiators

| Number of Tubes per Section | Catalog Rating per Section[a] | | A Height[c] | Section Dimensions B Width | | C Spacing[b] | D Leg Height[c] |
|---|---|---|---|---|---|---|---|
| | | | | Min | Max | | |
| | $ft^2$ | Btu/h (W) | in. (mm) | in. (mm) | in. (mm) | in. (mm) | in. (mm) |
| 3 | 1.6 | 384 (113) | 25 (635) | 3.25 ( 83) | 3.50 ( 89) | 1.75 (44) | 2.50 (64) |
| 4 | 1.6 | 384 (113) | 19 (483) | 4.44 (113) | 4.81 (122) | 1.75 (44) | 2.50 (64) |
| | 1.8 | 432 (127) | 22 (559) | 4.44 (113) | 4.81 (122) | 1.75 (44) | 2.50 (64) |
| | 2.0 | 480 (141) | 25 (635) | 4.44 (113) | 4.81 (122) | 1.75 (44) | 2.50 (64) |
| 5 | 2.1 | 504 (148) | 22 (559) | 5.63 (143) | 6.31 (160) | 1.75 (44) | 2.50 (64) |
| | 2.4 | 576 (169) | 25 (635) | 5.63 (143) | 6.31 (160) | 1.75 (44) | 2.50 (64) |
| 6 | 2.3 | 552 (162) | 19 (483) | 6.81 (173) | 8 (203) | 1.75 (44) | 2.50 (64) |
| | 3.0 | 720 (211) | 25 (635) | 6.81 (173) | 8 (203) | 1.75 (44) | 2.50 (64) |
| | 3.7 | 888 (260) | 32 (813) | 6.81 (173) | 8 (203) | 1.75 (44) | 2.50 (64) |

[a]These ratings are based on steam at 215 °F (101.7 °C) and air at 70 °F (21.1 °C). They apply only to installed radiators exposed in a normal manner, not to radiators installed behind enclosures, grilles, or under shelves. For Btu/h (W) ratings at other temperatures, divide table values by factors found in Table 7.

[b]Length equals number of sections multiplied by 1.75 in. (44 mm).

[c]Overall height and leg height, as produced by some manufacturers, are 1 in. (25 mm) greater than shown in Columns A and D . Radiators may be furnished without legs. Where greater than standard leg heights are required, this dimension shall be 4.5 in. (114 mm).

Table 2 Column-Type Cast-Iron Radiator (Discontinued)

| Height in. (mm) | One Column $ft^2$ | One Column Btu/h (W) | Two Column $ft^2$ | Two Column Btu/h (W) | Three Column $ft^2$ | Three Column Btu/h (W) | Height in. (mm) | Four Column $ft^2$ | Four Column Btu/h (W) | Five Column $ft^2$ | Five Column Btu/h (W) | Six Column $ft^2$ | Six Column Btu/h (W) |
|---|---|---|---|---|---|---|---|---|---|---|---|---|---|
| | Generally Accepted Rating per Section[a] | | | | | | | | | | | | |
| 15 (381) | | | 1.50 | 360 (106) | | | 13 (330) | | | | | 3.00 | 720 (211) |
| 18 (457) | | | | | 2.25 | 540 (158) | 16 (406) | | | | | 3.75 | 900 (264) |
| 20 (508) | 1.50 | 360 (106) | 2.00 | 480 (141) | | | 18 (457) | 3.0 | 720 (211) | 4.67 | 1120 (328) | 4.50 | 1080 (316) |
| 22 (559) | | | 2.25 | 540 (158) | 3.00 | 720 (211) | 20 (508) | | | | | 5.00 | 1200 (352) |
| 23 (584) | 1.67 | 400 (117) | 2.33 | 560 (164) | | | 22 (559) | 4.0 | 960 (281) | | | | |
| 26 (660) | 2.00 | 480 (141) | 2.67 | 640 (188) | 3.75 | 900 (264) | 26 (660) | 5.0 | 1200 (351) | 7.00 | 1680 (492) | | |
| 32 (813) | 2.50 | 600 (176) | 3.33 | 800 (235) | 4.50 | 1080 (316) | 32 (813) | 6.5 | 1560 (457) | | | | |
| 38 (965) | 3.00 | 720 (211) | 4.00 | 960 (281) | 5.00 | 1200 (352) | 38 (965) | 8.0 | 1920 (563) | 10.00 | 2400 (703) | | |
| 45 (1143) | | | 5.00 | 1200 (352) | 6.00 | 1440 (422) | 45 (1143) | 10.0 | 2400 (703) | | | | |

[a]These ratings are based on steam at 215 °F (101.7 °C) and air at 70 °F (21.1 °C). They apply only to installed radiators exposed in a normal manner, not to radiators installed behind enclosures, grills, or under shelves. For ratings at other temperatures, divide table values by factors found in Table 7.

enclosure below the heating element, is heated in passing through the element, and leaves the enclosure through the outlet grille located above the heating element. Factory-assembled units comprised of a heating element and enclosure are widely used. These may be freestanding, wall-hung, or recessed (Figure 1) and may have outlet grilles and arched inlets or inlet grilles, as desired.

When cabinets or enclosures are used without being furnished by the manufacturer, their proportions should be designed so they do not impair the performance of the assembled convector. The cabinet or enclosure for the convector should fit as snugly as possible, so the air passing through cannot bypass the heating element.

Table 3 Cast-Iron Wall Radiators (Discontinued)

| Approximate Dimensions, in. (mm) | | | Heat Output[a] | |
|---|---|---|---|---|
| Height | Length or Width | Thickness | $ft^2$ | Btu/h (W) |
| 13.25 (337) | 16.50 (419) | 3 (76) | 6.50 | 1560 (457) |
| 13.25 (337) | 22 (559) | 3 (76) | 8 | 1920 (563) |
| 22 (559) | 13.25 (337) | 3 (76) | 8 | 1920 (563) |
| 13.25 (337) | 29 (737) | 3 (76) | 11 | 2640 (774) |
| 29 (737) | 13.25 (337) | 3 (76) | 11 | 2640 (774) |

[a]These ratings are based on steam at 215 °F (101.7 °C) and air at 70 °F (21.1 °C). They apply only to installed radiators exposed in a normal manner, not to radiators installed behind enclosures, grills, or under shelves. For ratings at other temperatures, divide table values by factors found in Table 7.

## BASEBOARD UNITS

Baseboard heat-distributing units are divided into two types: radiant-convector and finned-tube.

The *radiant-convector-type* baseboard is made of cast iron or steel. The units have air openings at the top and bottom to permit circulation of room air over the wall side of the unit, which has extended surface to provide increased heat output. A large portion of the heat emitted is transferred by convection.

The *finned-tube-type* baseboard has a finned-tube heating element concealed by a long, low sheet-metal enclosure or cover. A major portion of the heat is transferred to the room by convection. The output varies over a wide range, depending on the physical dimensions and the materials used. A unit with too high an output per unit length should be avoided. Optimum comfort for room occupants is obtained when units are installed along as much of the exposed wall as possible.

The basic advantage of the baseboard unit is that its normal placement is along the cold walls and under areas where the greatest heat loss occurs. Other advantages claimed are (1) it is

**Table 4 Large-Tube Cast-Iron Radiators (Discontinued)**

**Sectional, cast-iron, tubular-type radiators of the large-tube pattern, that is, having tubes approximately 1 3/8 in. (35 mm) in diameter, 2 1/2 in. (64 mm) on centers.**

| Number of Tubes per Section | Catalog Rating per Section[a] ft² | Btu/h (W) | Height in. (mm) | Width in. (mm) | Section Center Spacing[b] in. (mm) | Leg Height[c] to Tapping in. (mm) |
|---|---|---|---|---|---|---|
| | 1.75 | 420 (123) | 20 (508) | | | |
| | 2.00 | 480 (141) | 23 (584) | 4.63 | 2.5 | 4.5 |
| 3 | 2.33 | 560 (164) | 26 (660) | (117) | (64) | (114) |
| | 3.00 | 720 (211) | 32 (813) | | | |
| | 3.50 | 840 (246) | 38 (965) | | | |
| | 2.25 | 540 (158) | 20 (508) | | | |
| | 2.50 | 600 (176) | 23 (584) | 6.25 to | 2.5 | 4.5 |
| 4 | 2.75 | 660 (193) | 26 (660) | 6.81 | (64) | (114) |
| | 3.50 | 840 (246) | 32 (813) | (159 to | | |
| | 4.25 | 1020 (299) | 38 (965) | 173) | | |
| | 2.67 | 640 (188) | 20 (508) | | | |
| | 3.00 | 720 (211) | 23 (584) | 8.0 to | 2.5 | 4.5 |
| 5 | 3.50 | 840 (246) | 26 (660) | 8.56 | (64)[d] | (114) |
| | 4.33 | 1040 (305) | 32 (813) | (203 to | | |
| | 5.00 | 1200 (352) | 38 (965) | 217) | | |
| | 3.00 | 720 (211) | 20 (508) | | | |
| | 3.50 | 840 (246) | 23 (584) | 9.0 to | | |
| 6 | 4.00 | 960 (281) | 26 (660) | 10.38 | 2.5 | 4.5 |
| | 5.00 | 1200 (352) | 32 (813) | (229 to | (64) | (114) |
| | 6.00 | 1440 (422) | 38 (965) | 264) | | |
| | 2.50 | 600 (176) | 14 (356) | 11.38 to | 2.5 | 3 (76) |
| 7 | 3.00 | 720 (211) | 17 (432) | 12.81 | (64) | 3 (76) |
| | 3.67 | 880 (258) | 20 (508) | (289 to | | 3 or 4.5 |
| | | | | 325) | | (76 or 114) |

[a]These ratings are based on steam at 215 °F (101.7 °C) and air at 70 °F (21.1 °C). They apply only to installed radiators exposed in a normal manner, not to radiators installed behind enclosures, grills, or under shelves. For ratings at other temperatures, divide table values by factors found in Table 7.

[b]Maximum assembly is 60 sections. Length equals number of sections multiplied by 2.5 in. (64 mm).

[c]Where greater than standard leg heights are required, this dimension shall be 6 in. (152 mm), except for seven-tube sections, in heights from 13 to 20 in. (330 to 508 mm), inclusive, for which this dimension shall be 4.5 in. (114 mm). Radiators may be furnished without legs.

[d]For five-tube hospital-type radiation, this dimension is 3 in. (76 mm).

inconspicuous, (2) it offers minimal interference with furniture placement, and (3) it distributes the heat near the floor. This last characteristic reduces the floor-to-ceiling temperature gradient to about 2 to 4 °F (1 to 2 °C) and tends to produce uniform temperatures throughout the room. It also makes baseboard heat-distributing units adaptable to homes without basements, where cold floors are common (Kratz and Harris 1945).

Heat-loss calculations for baseboard heating systems are the same as those used for other types of heat-distributing units. The procedure for designing baseboard heating systems is given in I=B=R *Installation Guide No.* 200 (1966). Ratings for baseboard heat-distributing units are expressed in Btu/h per linear foot (W/m).

**Table 5 Heat Emission of Pipe Coils Placed Vertically on a Wall (Pipes Horizontal) Containing Steam at 215 °F (101.7 °C) and Surrounded with Air at 70 °F (21.1 °C)**

**Btu per linear foot (W/m) of coil per hour [not linear feet (metre) of pipe]**

| Size of Pipe | 1 in. (25 mm) | 1¼ in. (32 mm) | 1½ in. (40 mm) |
|---|---|---|---|
| Single row | 132 (127) | 162 (156) | 185 (178) |
| Two | 252 (242) | 312 (300) | 348 (335) |
| Four | 440 (423) | 545 (524) | 616 (592) |
| Six | 567 (545) | 702 (675) | 793 (762) |
| Eight | 651 (626) | 796 (765) | 907 (872) |
| Ten | 732 (704) | 907 (872) | 1020 (981) |
| Twelve | 812 (781) | 1005 (966) | 1135 (1091) |

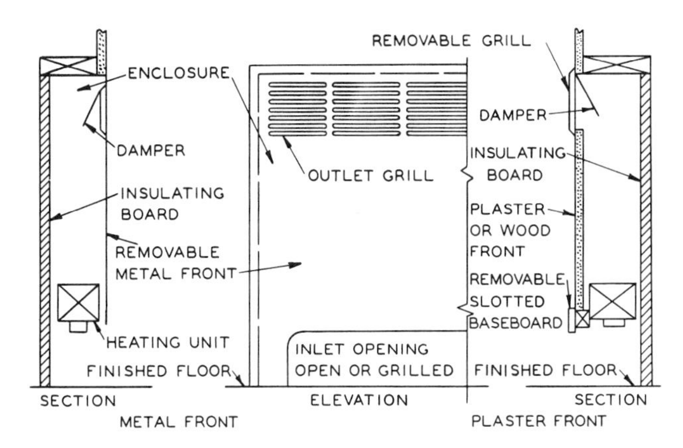


**Fig. 1 Typical Recessed Convector**

## FINNED-TUBE UNITS

A finned-tube, heat-distributing unit is a room air heater composed of a finned-tube element fabricated from a metallic tube to which metallic fins have been bonded; it is generally installed in an enclosure or cover. Finned-tube elements are available in several tube sizes, in either steel or copper—1 to 2 in. IPS (25 to 50 mm) or 3/4 to 1 1/4 in. (20 to 32 mm) nominal copper—with various fin sizes, spacings, and materials. The resistance to the flow of steam or water is the same as that through standard distribution piping of equal size and type.

The finned-tube unit can be used with steam or hot-water systems. It is advantageous for heat distribution along the entire outside wall, thereby preventing downdrafts along the walls in buildings such as schools, churches, hospitals, offices, airports, and factories. It may be the principal source of heat in a building or a supplementary heater to combat downdrafts along the exposed walls in conjunction with a central conditioned air system.

Normal placement of a finned tube is along the walls where the heat loss is greatest. If necessary, the units can be installed in two or three tiers along the wall. Hot-water installations requiring two or three tiers should run a serpentine water flow because a header connection with parallel flow may (1) permit the water to short circuit along the path of least resistance, (2) suffer reduced capacity because of low-water velocity in each tier, or (3) become air-bound in one or more of the tiers.

Finned-tube elements installed in occupied spaces generally have *covers* or *enclosures* in a variety of designs. When human contact is unlikely, they are sometimes installed bare or are provided with an expanded metal grille for minimum protection. A cover is a fabricated shield that has a portion of the front skirt made of solid material. It can be mounted with clearance between the wall and the cover, and without completely enclosing the rear of the finned-tube element. A cover may have a top, front, or inclined outlet. An enclosure is a shield of solid material that completely encloses both the front and rear of the finned-tube element. An enclosure may have an integral back or may be installed tightly against the wall so that the wall forms the back, and it may have a top, front, or inclined outlet.

Many enclosures have been developed to meet building design requirements. The wide variety of finned-tube elements (tube size and material, fin size, spacing, fin material, and multiple tier installation), along with the various heights and designs of enclosures, give great flexibility of selection for finned-tube units that meet the needs of load, space, and appearance.

Heat loss calculations for finned-tube heating systems are the same as those used for other types of radiation. Ratings are expressed in Btu/h per linear foot (W/m) or square feet Equivalent

Direct Radiation (EDR) per linear foot for steam, and Btu/h per linear foot (W/m) for water.

## OTHER HEAT-DISTRIBUTING UNITS

A variety of radiators and convectors in heating systems are used outside of the United States. The most common types are sectional radiators, panel radiators, tubular steel radiators, and specialty radiators.

Sectional radiators are fabricated from welded sheet metal sections (generally 2, 3, or 4-tube wide), resemble free-standing cast-iron radiators, and distribute heat economically.

Panel radiators consist of fabricated flat panels (generally 1, 2, or 3 deep), with or without exposed extended fin surface attached to the rear for increased output. These radiators are most common in Europe.

Tubular steel radiators consist of supply and return headers with interconnecting parallel steel tubes in a wide variety of lengths and heights. Some are used in bathroom towel-heating applications, whereas others are used in shapes specially adapted to coincide with the building structure.

Specialty radiators are fabricated of welded steel or extruded aluminum for installation in ceiling grids or floor-mounting arrangements. An array of shapes different from conventional radiators are available.

## RATINGS OF HEAT-DISTRIBUTING UNITS

### Radiators

Current methods for testing radiators were established by the *Simplified Practice Recommendation,* R174-65, *Cast-Iron Radiators,* which has now been withdrawn (see Table 1).

### Convectors

The generally accepted method of testing and rating ferrous and nonferrous convectors was given in *Commercial Standard* CS 140-47, *Testing and Rating Convectors* (U.S. Dept. of Commerce 1947), but it has been withdrawn by the Department of Commerce. This standard contained details covering construction and instrumentation of the test booth or room, and procedures for determining steam and water ratings.

Steam ratings are expressed in sq. ft. EDR and Btu/h (W). Water ratings are expressed in Btu/h (W) at specified water temperature drops and average water temperatures.

Under the provisions of *Commercial Standard* CS 140-47, the rating of a top outlet convector was established at a value not in excess of the test capacity (which is the heat extracted from the steam or water in the convector under standard test conditions). For convectors with other types of enclosures or cabinets, a percentage that varies up to a maximum of 15% (depending on the height and type of enclosure or cabinet) was added for heating effect (Bradbee 1927 and Willard *et al.* 1929). The additions made for heating effect must be shown in the manufacturer's literature.

The testing and rating procedure set forth by *Commercial Standard* CS 140-47 does not apply to finned-tube or baseboard radiation.

### Baseboard

The generally accepted method of testing and rating baseboards is covered in the I=B=R *Testing and Rating Code for Baseboard Radiation* (Hydronics Institute 1981). This *Code* contains details covering construction and instrumentation of the test booth or room, procedures for determining steam and hot water ratings, and licensing provisions for obtaining approval of these ratings.

Baseboard ratings include an allowance for heating effect of 15%, added to the test capacity. The addition made for heating effect must be shown in the manufacturer's literature.

Steam ratings are expressed in Btu/h per linear foot (W/m) or square feet EDR per linear foot. Water ratings are expressed in Btu/h per linear foot (W/m) at specified water flow rates and average water temperatures.

### Finned-Tube

The generally accepted method of testing and rating finned-tube units is covered in the I=B=R *Testing and Rating Code for Finned-Tube (Commercial) Radiation* (Hydronics Institute 1973). This *Code* contains details covering construction and instrumentation of the test booth or room, procedures for determining steam and water ratings, and licensing provisions for obtaining approval of these ratings. Steam ratings are expressed in Btu/h per linear foot (W/m) or square feet EDR per linear foot. Water ratings are determined from steam ratings and are expressed in Btu/h per linear foot (W/m) at specified water flow rates and average water temperatures. Table 6 gives factors for converting steam ratings to hot water ratings at various average water temperatures. These factors apply only when the water velocity in the element is 3.0 ft/s (0.9 m/s). (See Figure 2.)

The rating of a finned-tube unit in an enclosure that has a top outlet is established at a value not in excess of the test capacity (which is the heat extracted from the steam or water in the unit under standard test conditions). For finned tube with other types of enclosures or covers, a percentage is added for heating effect, which varies up to a maximum of 15%, depending on the height and type of enclosure or cover. The additions made for heating effect must be shown in the manufacturer's literature (Pierce 1963).

### Other Heat-Distributing Units

Unique and imported radiators generally are tested and rated for heat emission in accordance with prevailing standards. These other testing and rating methods are basically the same procedures as the IBR Code, which is the standard in the United States.

### Corrections for Nonstandard Conditions

The heat output of a radiator, convector, baseboard, or finned-tube heat-distributing unit is an exponential function of the tem-

**Table 6 Factors to Convert I=B=R Finned-Tube Steam Ratings to Hot-Water Ratings at Temperatures Indicated**

| Avg. Water Temperature, °F (°C) | Factor | Avg. Water Temperature, °F (°C) | Factor |
|---|---|---|---|
| 100 (37.8) | 0.15 | 185 (85.0) | 0.73 |
| 110 (43.3) | 0.20 | 190 (87.8) | 0.78 |
| 120 (48.9) | 0.26 | 195 (90.6) | 0.82 |
| 130 (54.4) | 0.33 | 200 (93.3) | 0.86 |
| 140 (60) | 0.40 | 205 (96.1) | 0.91 |
| 150 (65.6) | 0.45 | 210 (98.9) | 0.95 |
| 155 (68.3) | 0.49 | 215 (101.7) | 1.00 |
| 160 (71.1) | 0.53 | 220 (104.4) | 1.05 |
| 165 (73.9) | 0.57 | 225 (107.2) | 1.09 |
| 170 (76.7) | 0.61 | 230 (110.0) | 1.14 |
| 175 (79.4) | 0.65 | 235 (112.8) | 1.20 |
| 180 (82.2) | 0.69 | 240 (115.6) | 1.25 |

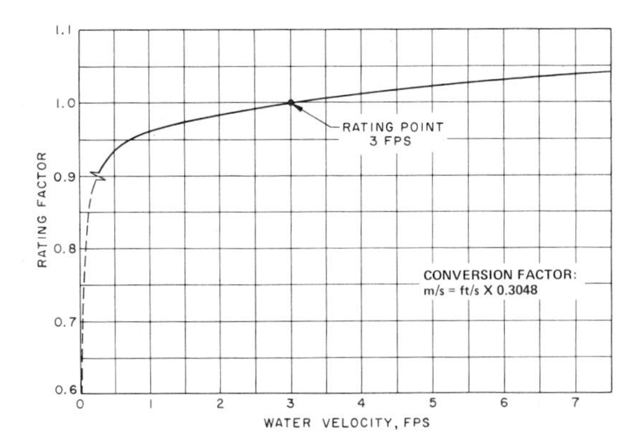


**Fig. 2 Effect of Water Velocity on Finned-Tube Output**

perature difference between the air in the room and the heating medium in the room-heating unit, shown as:

$$H = c\,(t_s - t_a)^n \quad (1)$$

*where*

$H$ = heat output, Btu/h (W)
$c$ = a constant determined by test
$t_s$ = average temperature of heating medium, °F (°C). For hot water, the arithmetic average of the entering and leaving water temperatures is used
$t_a$ = room air temperature, °F (°C). Air temperature 60 in. (1.52 m) above the floor is generally used for radiators, while the entering air temperature is used for all other types of heating units
$n$ = an exponent that equals 1.3 (1.2) for cast-iron radiators, 1.4 (1.31) for baseboard radiation, and 1.5 (1.42) for convectors. For finned-tube units, $n$ varies with air and heating medium temperatures. Correction factors to convert outputs at standard rating conditions to outputs at other conditions are given in Tables 6 and 7.

## APPLICATIONS

Radiators, convectors, and finned-tube and baseboard units are installed in areas of greatest heat loss—under windows, along cold walls, or at doorways. These units may be used in any steam or low-temperature water system, and in some medium-temperature water systems. However, units with different performance characteristics should not be installed in the same zone, *i.e.,* cast-iron radiators should not be mixed with finned-tube-type baseboards. See Chapter 13 of the 1987 HVAC Volume. On one-pipe steam systems, it is recommended that only the larger tube sizes—1.25 in. (32 mm) and larger—be used to allow drainage of condensate counterflow to the steam.

The increasing popularity of medium- and high-temperature water systems has changed the application of finned-tube and baseboard units (see Chapter 15 of the 1987 HVAC Volume). Associated with this is the use of relatively large temperature drops through the system—drops of as much as 60 to 80 °F (33 to 44 °C) in low-temperature systems and as high as 200 °F (110 °C) in high-temperature systems. These higher temperature drops result in lower water velocities in the finned-tube elements, which require an understanding of the resulting effect on heat output.

Figure 2 plots the effect of water velocity on heat output of various sizes of finned-tube elements. There is little significant variation in output over the range from 0.5 to 6.0 ft/s (0.15 to 1.83 m/s)—factors range from 0.93 to 1.03 where 1.0 is at 3.0 ft/s (0.91 m/s). However, at a point between 0.4 and 0.25 ft/s (0.12 and 0.08 m/s), the factor drops rapidly. This is the critical velocity range at which flow changes from turbulent to streamline. Therefore, velocities below these points should be avoided in a hot-water heating system, since results could not be predicted accurately.

The designer should check water velocity throughout the system and select finned-tube or baseboard elements on the basis of velocity, as well as average temperature. Manufacturers of finned-tube and baseboard elements offer a variety of tube sizes—ranging from 0.5 in. (15 mm) copper tubes for small baseboard elements to 2 in. IPS (50 mm) for large finned-tube units—to aid in maintenance of turbulent flow conditions over a wide range of flows.

In areas where zone control rather than individual room control can be applied, all finned-tube units in the zone should be in series. In such a series loop installation, however, temperature drop must be considered in selecting the element for each separate room in the loop.

Primary-secondary pumping or the use of a secondary heat exchanger (see Chapters 13 and 15 of the 1987 HVAC Volume)

**Table 7 Correction Factors for Various Types of Heating Units**

| Steam Pressure (Approximate) | | | | Temp. of Steam or Water °F (°C) | Cast-Iron Radiators | | | | | Convectors | | | | | Finned-Tube | | | | | Baseboard | | | | |
|---|---|---|---|---|---|---|---|---|---|---|---|---|---|---|---|---|---|---|---|---|---|---|---|---|
| Gauge | | Absolute | | | Room Temp, °F (°C) | | | | | Inlet Air Temp, °F (°C) | | | | | Inlet Air Temp, °F (°C) | | | | | Inlet Air Temp, °F (°C) | | | | |
| in. Hg. | (kPa) | psi | (kPa) | | 80 (26.7) | 75 (23.9) | 70 (21.1) | 65 (18.3) | 60 (15.6) | 75 (23.9) | 70 (21.1) | 65 (18.3) | 60 (15.6) | 55 (12.8) | 75 (23.9) | 70 (21.1) | 65 (18.3) | 60 (15.6) | 55 (12.8) | 75 (23.9) | 70 (21.1) | 65 (18.3) | 60 (15.6) | 55 (12.8) |
| | | | | 100 (37.8) | | | | | | | | | | | 0.10 | 0.12 | 0.15 | 0.17 | 0.20 | 0.08 | 0.10 | 0.13 | 0.15 | 0.18 |
| | | | | 110 (43.3) | | | | | | | | | | | 0.15 | 0.17 | 0.20 | 0.23 | 0.26 | 0.13 | 0.15 | 0.18 | 0.21 | 0.25 |
| | | | | 120 (48.9) | | | | | | | | | | | 0.20 | 0.23 | 0.26 | 0.29 | 0.33 | 0.18 | 0.21 | 0.25 | 0.28 | 0.31 |
| | | | | 130 (54.4) | | | | | | | | | | | 0.26 | 0.29 | 0.33 | 0.36 | 0.40 | 0.25 | 0.28 | 0.31 | 0.34 | 0.38 |
| Vacuum | | | | 140 (60) | | | | | | | | | | | 0.33 | 0.36 | 0.40 | 0.42 | 0.45 | 0.31 | 0.34 | 0.38 | 0.42 | 0.45 |
| 22.4 | (−75.8) | 3.7 | (25.5) | 150 (65.6) | 0.39 | 0.42 | 0.46 | 0.50 | 0.54 | 0.35 | 0.39 | 0.43 | 0.46 | 0.50 | 0.40 | 0.42 | 0.45 | 0.49 | 0.53 | 0.38 | 0.42 | 0.45 | 0.49 | 0.53 |
| 20.3 | (−68.9) | 4.7 | (32.4) | 160 (71.1) | 0.46 | 0.50 | 0.54 | 0.58 | 0.62 | 0.43 | 0.47 | 0.51 | 0.54 | 0.58 | 0.45 | 0.49 | 0.53 | 0.57 | 0.61 | 0.45 | 0.49 | 0.53 | 0.57 | 0.61 |
| 17.7 | (−59.9) | 6.0 | (41.4) | 170 (76.7) | 0.54 | 0.58 | 0.62 | 0.66 | 0.69 | 0.51 | 0.54 | 0.58 | 0.63 | 0.67 | 0.53 | 0.57 | 0.61 | 0.65 | 0.69 | 0.53 | 0.57 | 0.61 | 0.65 | 0.69 |
| 14.6 | (−49.6) | 7.5 | (51.7) | 180 (82.2) | 0.62 | 0.66 | 0.69 | 0.74 | 0.78 | 0.58 | 0.63 | 0.67 | 0.71 | 0.76 | 0.61 | 0.65 | 0.69 | 0.73 | 0.78 | 0.61 | 0.65 | 0.69 | 0.72 | 0.78 |
| 10.9 | (−37.2) | 9.3 | (64.1) | 190 (87.8) | 0.69 | 0.74 | 0.78 | 0.83 | 0.87 | 0.67 | 0.71 | 0.76 | 0.81 | 0.85 | 0.69 | 0.73 | 0.78 | 0.81 | 0.86 | 0.69 | 0.73 | 0.78 | 0.82 | 0.86 |
| 6.5 | (−22 ) | 11.5 | (79.3) | 200 (93.3) | 0.78 | 0.83 | 0.87 | 0.91 | 0.95 | 0.76 | 0.81 | 0.85 | 0.90 | 0.95 | 0.77 | 0.81 | 0.86 | 0.90 | 0.95 | 0.81 | 0.86 | 0.92 | 0.95 | 1.00 |
| psi | kPa | | | | | | | | | | | | | | | | | | | | | | | |
| 1 | (6.2) | 15.6 | (107.6) | 215 (101.7) | 0.91 | 0.95 | 1.00 | 1.04 | 1.09 | 0.90 | 0.95 | 1.00 | 1.05 | 1.10 | 0.91 | 0.94 | 1.00 | 1.06 | 1.11 | 0.91 | 0.95 | 1.00 | 1.05 | 1.09 |
| 6 | (41.9) | 21 | (143.3) | 230 (110.0) | 1.04 | 1.09 | 1.14 | 1.18 | 1.23 | 1.05 | 1.10 | 1.15 | 1.20 | 1.26 | 1.03 | 1.08 | 1.14 | 1.19 | 1.24 | 1.04 | 1.09 | 1.14 | 1.19 | 1.25 |
| 15 | (104.3) | 30 | (205.6) | 250 (121.1) | 1.23 | 1.28 | 1.32 | 1.37 | 1.43 | 1.27 | 1.32 | 1.37 | 1.43 | 1.47 | 1.20 | 1.26 | 1.31 | 1.37 | 1.43 | 1.22 | 1.27 | 1.32 | 1.37 | 1.43 |
| 27 | (187.3) | 42 | (288.6) | 270 (132.2) | 1.43 | 1.47 | 1.52 | 1.56 | 1.61 | 1.47 | 1.54 | 1.59 | 1.67 | 1.72 | 1.38 | 1.44 | 1.50 | 1.56 | 1.62 | 1.43 | 1.47 | 1.52 | 1.59 | 1.64 |
| 52 | (360.7) | 67 | (462. ) | 300 (148.9) | 1.72 | 1.75 | 1.82 | 1.89 | 1.92 | 1.85 | 1.89 | 1.96 | 2.04 | 2.08 | 1.67 | 1.73 | 1.79 | 1.86 | 1.92 | 1.75 | 1.82 | 1.89 | 1.92 | 1.96 |

These correction factors provide means of determining output ratings for radiators, convectors, and finned-tube and baseboard units at operating conditions other than standard and, hence, also provide means of selecting these units to satisfy heating loads for a given space at any given set of operating conditions. Standard conditions for a radiator are 215 °F (101.7 °C) heating medium temperature and 70 °F (21.1 °C) room temperature (at the center of the space and at the 5-ft [1.52-m] level). Standard conditions for convectors and finned-tube and baseboard units are 215 °F (101.7 °C) heating medium temperature and 65 °F (18.3 °C) inlet air temperature. Inlet air at 65 °F (18.3 °C) for convectors and finned-tube or baseboard units represent the same room comfort conditions as 70 °F (21.1 °C) room air temperature for a radiator.

To determine the output of a heating unit under conditions other than standard, mutiply the standard heating capacity by the appropriate factor for the actual operating heating medium and room or inlet air temperatures.

can also help finned-tube units applied to systems with high temperature drops. The designer should also consider the effects of high temperature and large drops on the problems of control, balancing, expansion, and installation of finned-tube and baseboard units.

Unique and imported radiators are generally not suitable for steam applications although they have been used extensively in low-temperature water systems with valves and connecting piping left exposed. Various combinations of supply and return locations are possible, which may alter the heat output. Although long lengths may be ordered for linear applications, lengths may not be reduced or increased by field modification. The small cross-sectional areas often inherent in unique radiators requires careful evaluation of flow requirements, water temperature drop, and pressure drops.

### Performance at Low Water Temperatures

Tables 6 and 7 summarize the performance of baseboard and finned-tube units with an average water temperature down to 100 °F (37.8 °C). Solar-heated water or heat pump system cooling water are typical applications in this range.

### Enclosure, Paint, Humidity Effect

An enclosure placed around a direct radiator restricts the airflow and diminishes the proportion of output resulting from radiation. Enclosures of proper design may, however, improve the heat distribution within the room as compared to the heat distribution obtained with an unenclosed radiator (Willard *et al.* 1929, Allcut 1933).

For a radiator or cast-iron baseboard, the finish coat of paint affects the heat output. Oil paints of any color give about the same results as unpainted black or rusty surfaces, but an aluminum or a bronze paint reduces the heat emitted by radiation. The net effect may reduce the total heat output of the radiator by 10% or more (Rupert 1937, Severns 1927, Allen 1920).

Some commercial enclosures and shields for use on direct radiators are equipped with water pans for adding moisture to the air in the room. An average evaporative rate of about 0.235 lb/h • ft$^2$ (1.15 kg/h • m$^2$) of water surface may be obtained from such pans, when a radiator is steamheated and the relative humidity in the room is between 25 and 40% (Kratz).

## REFERENCES


Allcut, E.A. 1933. Heat Output of Concealed Radiators. *School of Engineering Research Bulletin* 140. University of Toronto.

Allen, J.R. 1920. Heat Loss from Direct Radiation. ASHVE *Transactions*, Vol. 26, p. 11.

Brabbee, C. 1927, The Heating Effect of Radiators. ASHVE *Transactions*, Vol. 26, p. 11.

Byrley, T.R. 1978. *Hydronic Rating Handbook*. Color Art Inc., St. Louis, MO.

Department of Commerce. 1947. *Commercial Standard for Testing and Rating Convectors*, CS 140-47. Washington, DC.

Hydronics Institute. 1981. *I=B=R Testing and Rating Code for Baseboard Radiation*, 5th ed. Cleveland, OH.

Hydronics Institute. 1973. *I=B=R Testing and Rating Code for Finned Tube (Commercial) Radiation*, 4th ed. Cleveland, OH.

Hydronics Institute. 1966. *I=B=R Installation Guide for Residential Hydronic Heating Systems*, No. 200, 1st ed. Cleveland, OH.

Kratz, A.P. and W.S. Harris. 1945. A Study of Radiant Baseboard Heating in the I=B=R Research Home. *Engineering Experiment Station Bulletin* 358. University of Illinois.

Kratz, A.P. Humidification for Residences. *Engineering Experiment Station Bulletin* 230, p. 20. University of Illinois.

Pierce, J.D. 1963. Application of Fin Tube Radiation to Modern Hot Water Systems. ASHRAE *Journal*, Feb., p. 72.

Rubert, E.A. 1937. Heat Emission from Radiators. *Engineering Experiment Station Bulletin* 24. Cornell University, Ithaca, NY.

Severns, W.H. 1927. Comparative Tests of Radiator Finishes. ASHVE *Transactions*, Vol. 33, p. 41.

Willard, A.C; A.P. Kratz; M.K. Fahnestock; and S. Konzo. 1929. Investigation of Heating Rooms with Direct Steam Radiators Equipped with Enclosures and Shields. ASHVE *Transactions*, Vol. 35, p. 77 or *Engineering Experiment Station Bulletin* 192, University of Illinois.

Willard, A.C; A.P. Kratz; M.K. Fahnestock; and S. Konzo. Investigation of Various Factors Affecting the Heating of Rooms with Direct Steam Radiators. *Engineering Experiment Station Bulletin* 223.

## CHAPTER 29

# INFRARED HEATERS



---

LOW- and medium-intensity infrared heaters are compact, self-contained, direct-heating devices used in hangars, factories, warehouses, foundries, greenhouses, and gymnasiums, and for areas such as loading docks, racetrack stands, under marquees, outdoor restaurants, and around swimming pools. Infrared heaters are also used for snow control, condensation control, and industrial process heating.

Low-temperature radiant heaters are often used in office buildings and other commercial buildings. These heaters can be used in conjunction with VAV systems.

Infrared heating units may be electric, gas fired, or oil fired. They consist of an infrared source or generator operating in a temperature range of from 350 to 5000°F (180 to 2760°C), with the specific temperature determined by energy source, configuration, size, etc. Reflectors can be used to control distribution of radiation in specific patterns.

## ENERGY CONSERVATION

Infrared heating units are effective for spot heating. However, because they use less energy than conventional space-heating systems, a primary application is for total heating of large areas and entire buildings. Radiant heaters use electromagnetic waves to transfer energy directly to solid objects. Little energy is lost during the transmission because air is a poor absorber of infrared energy. Since an intermediate transfer medium (such as air or water) is not needed, fans or pumps are not required.

As floors and objects are warmed by the infrared energy, they, in turn, reradiate heat to the air by convection. Reradiation to surrounding objects is like convection to ambient air. An energy saving advantage of radiant heat is it can be turned off when it is not needed; when it is turned on again, it is effective in minutes.

The preparation of this chapter is assigned to TC 6.5, Radiant Space Heating and Cooling.

Dry-bulb temperatures are slightly less than the mean radiant temperature. When convective heat is used, the opposite is true. Since human comfort is determined by the average of mean radiant and dry-bulb temperatures, dry-bulb temperature for a given comfort level is lower when heating with radiation (ASHRAE 1981). Heat lost to ventilating air and by transmission is correspondingly lower, as is energy consumption.

Buckley *et al.* (1987) show data validated by application records that compares energy savings of infrared heating systems with other types of comfort heating systems. A 1973 report by the New York State Interdepartmental Fuel and Energy Committee identifies annual fuel savings as high as 50%. Recognizing the reduction in fuel requirement for a given application, it is common to install 80 to 85% of the ASHRAE calculated heat loss in buildings that are heated by infrared radiant systems.

## INFRARED ENERGY GENERATORS

### Gas Infrared

Modern gas-fired infrared heaters burn gas to heat a specific radiating surface. The surface is heated by direct flame contact or with combustion gases. Studies by the Gas Research Board of London (1944), Plyler (1948), and Haslam *et al.* (1925) reveal that only 10 to 20% of the energy produced by open combustion of a gaseous fuel is infrared radiant energy, whereas wavelength span can be controlled by design. The specific radiating surface of a properly designed unit increases radiant release efficiency and directs radiation toward the load. Heaters are available in the following types (see Table 1 for characteristics):

**Indirect infrared radiation units** (Figures 1a, 1b, and 1c) are internally fired and have the radiating surface between the hot gases and load. Combustion takes place within the radiating elements, which operate with surface temperatures up to 1200°F (650°C). The elements may be tubes or panels, or they may have metal or ceramic components. **Indirect infrared radiation units** are usually vented and may require eductors, as is shown in Figure 1b.

Table 1 Characteristics of Typical Gas-Fired Infrared Heaters

| Characteristics | Type 1 | Type 2 | Type 3 |
|---|---|---|---|
| Operating Temperature | To 1200 °F (650 °C) | 1600 to 1800 °F (870 to 980 °C) | 650 to 700 °F (340 to 370 °C) |
| Relative heat Intensity,[a] Btu/h • ft$^2$ (kW/m$^2$) | Low to 7500 (24) | Medium 17,000 to 32,000 (54 to 100) | Low 800 to 3000 (2.5 to 9.5) |
| Response time (heatup) | 180 s | 60 s | 300 s |
| Radiation generating ratio[b] | 0.35 to 0.55 | 0.35 to 0.60 | No data |
| Thermal shock resistance | Excellent | Excellent | Excellent |
| Vibration resistance | Excellent | Excellent | Excellent |
| Color blindness[c] | Excellent | Very good | Excellent |
| Luminosity (visible light) | To dull red | Yellow red | None |
| Mounting height | 9 to 50 ft (3 to 15 m) | 12 to 50 ft (4 to 15 m) | To 10 ft (3 m) |
| Wind or draft resistance | Good | Fair | Very good |
| Venting | Optional | Nonvented | Nonvented |
| Flexibility | Good | Excellent-wide range of heat intensities and mounting possibilities available | Limited-to-low heat intensity applications |

[a]Heat intensity emitted at burner surface.
[b]Ratio of radiant output to Btu (kJ) input.
[c]Color blindness refers to absorptivity by various loads of energy emitted by the different sources.

**Porous matrix infrared radiation units** (Figure 1d) have a refractory material, which may be porous ceramic, drilled port ceramic, stainless steel, or a metallic screen. The units are enclosed except for the major surface facing the load. A combustible gas-air mixture enters the enclosure, flows through the refractory material to the exposed face, and is distributed evenly by the porous character of the refractory. Combustion occurs evenly on the exposed surface. The flame recedes into the matrix, which adds radiant energy to the flame. If the refractory porosity is suitable, an atmospheric burner can be used, resulting in a surface temperature approaching 1650 °F (900 °C). Power burner operation may be required if refractory density is high. However, the resulting surface temperature may also be higher (1800 °F or 980 °C).

**Catalytic oxidation infrared radiant units** (Figure 1e) are similar to the porous matrix units in construction, appearance, and operation, but the refractory material is usually glass wool and the radiating surface is a catalyst that causes oxidation to proceed without visible flames.

## Electric Infrared

Electric infrared heaters use heat produced by current flowing in a high-resistance wire or ribbon. The following types are most commonly used (see Table 2 for characteristics):

**Metal sheath infrared radiation elements** (Figure 2a) are composed of a nickel-chromium heating wire embedded in an electrical insulating refractory, which is encased by a metal tube. These elements have excellent resistance to thermal shock, vibration, and impact, and they can be mounted in any position. At full voltage, the elements attain a sheath surface temperature of 1200 to 1800 °F (650 to 980 °C). Higher temperatures are obtained by such configurations as a hairpin shape. These units generally contain a reflector, which directs radiation to the load. Higher efficiency is obtained if these elements are shielded from wind.

**Reflector lamp infrared radiation units** (Figure 2b) have a coiled tungsten filament, which approximates a point source radiator. The filament is enclosed in a heat-resistant, clear, frosted, or red glass envelope, which is partially silvered inside to form an efficient reflector. Common units may be screwed into a 120-V light socket.

**Quartz-tube infrared radiant units** (Figure 2c) have a coiled nickel-chromium wire lying unsupported within an unevacuated fused quartz tube, which is capped (not sealed) by porcelain or metal terminal blocks. These units are easily damaged by impact and vibration, but stand up well to thermal shock and splashing. They must be mounted in a horizontal position to minimize coil sag, and they are usually mounted in a fixture that contains a reflector. Normal operating temperatures range from 1300 to 1800 °F (700 to 980 °C) for the coil and about 1200 °F (650 °C) for the tube.

**Tubular Quartz Lamp Units** (Figure 2d) consist of a 0.38 in. diameter (9.5 mm) fused quartz tube containing an inert gas and a coiled tungsten filament held in a straight line and away from the tube by tantalum spacers. Filament ends are embedded in sealing material at the ends of the envelope. Lamps must be mounted horizontally, or nearly so, to minimize filament sag and overheating of the sealed ends. At normal design voltages, quartz lamp filaments operate at about 4050 °F (2230 °C), while the envelope operates at about 1100 °F (590 °C).

**Low-temperature radiant heating panels** (Figure 2e) consist of a one-inch (25-mm) thick galvanized steel panel with a graphite or nichrome-wire heating element. Panels come in various dimensions ranging in widths from 10 to 30 in. (250 to 760 mm) and lengths from 12 to 96 in. (300 to 2440 mm). Maximum watt density is 95 W/ft$^2$ (1020 W/m$^2$). Normal operating temperatures of radiating surface are 200 to 300 °F (95 to 150 °C). Units can be laid in a T-bar grid system or a recessed frame, or they can be surface mounted.

## Oil Infrared

Oil-fired infrared radiation-type heaters are similar to gas-fired indirect infrared radiation units (Figures 1a, 1b, and 1c). Oil-fired units are vented.

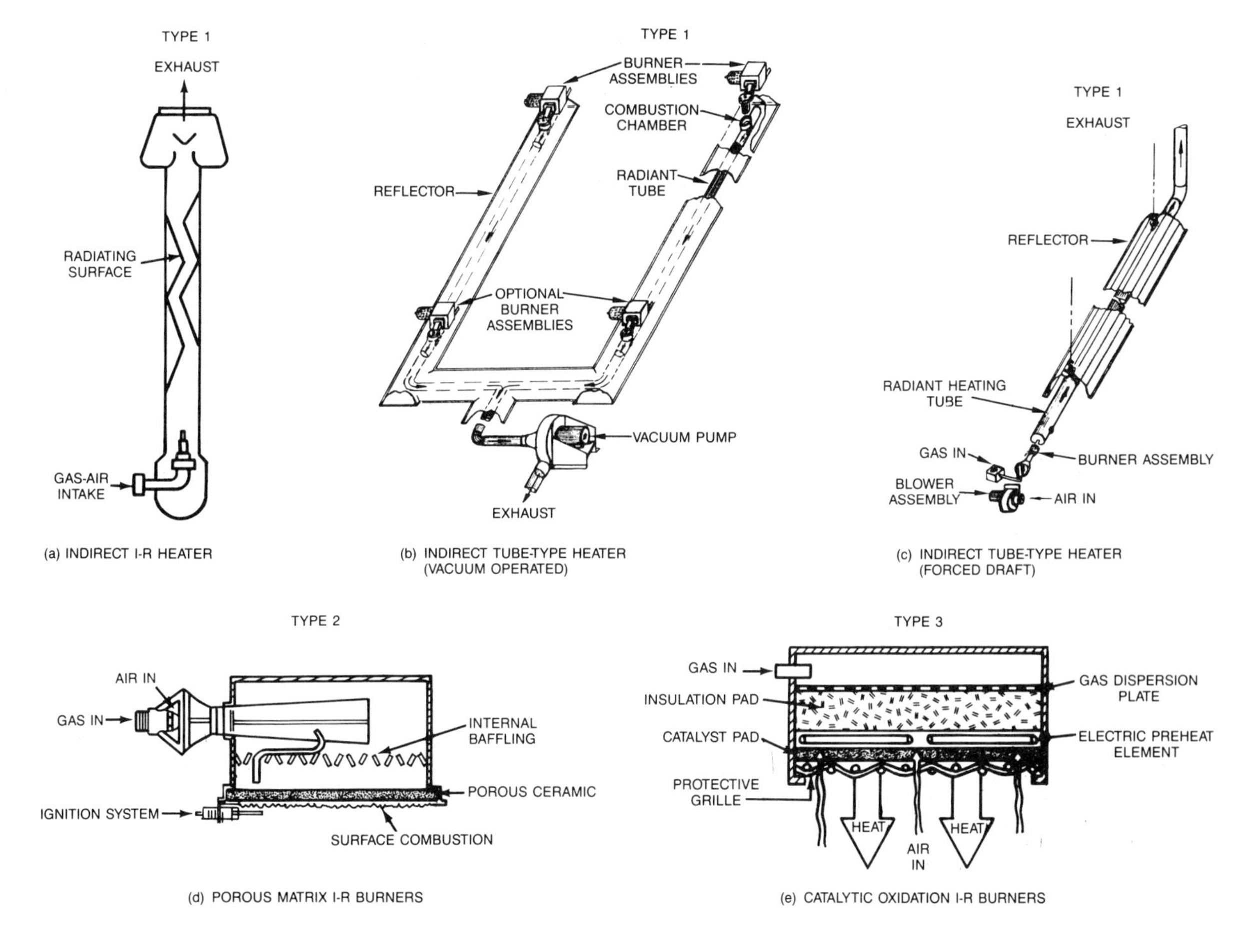


**Fig. 1 Types of Gas-Fired Heaters**

## SYSTEM EFFICIENCY

Because many factors contribute to the specific performance of an infrared system, a single criterion should not be used to evaluate comparable systems. Therefore, use at least two of the following indicators when evaluating system performance:

**Radiation generating ratio** equals infrared energy generated divided by total energy input.

**Fixture efficiency** is an index of a fixture's ability to emit the radiant energy developed by the infrared source; it is usually based on total energy input. The housing, reflector, and other parts of a fixture absorb some infrared energy and convert it to heat, which is lost through convection. A fixture that controls direction and distribution of energy effectively may have a lower fixture efficiency.

**Pattern efficiency** is an index of a fixture's effectiveness in directing the infrared energy into a specific pattern. This effectiveness, plus effective use of the pattern, influences the total effectiveness of the system (Boyd 1963). Typical radiation-generating ratios of gas infrared generators are indicated in Table 1. Limited test data indicate that the amount of radiant energy emitted from gas infrared units, compared to the amount of convective energy, ranges from 35 to 60%. The Stefan-Boltzmann Law can be used to estimate the infrared output capability, if reasonably accurate values of true surface temperature, emitting area, and surface emittance are available (DeWirth 1960). DeWirth (1962) also addresses the spectral distribution of energy curves for several gas sources.

Table 2 lists typical radiation-generating ratios of electric infrared generators. Fixture efficiencies are typically 80 to 95% of the radiation-generating ratios.

Infrared heaters should be operated at rated output. A small reduction in input causes a larger decrease in radiant output because of the fourth power dependence of radiant output on radiator temperature. Because a great variety of infrared units with a variety of reflectors and shields are available, the manufacturer's information should be consulted.

## REFLECTORS

Emitted radiation in most infrared heating devices is directed or concentrated by reflectors. Mounting height and the choice between spot or total heating usually determines what type of reflector will achieve the desired heat-flux pattern at floor level. Four types of reflectors can be used: (1) Parabolic, which produces essentially parallel beams of energy; (2) Elliptical, which directs all energy that is received or generated at the first focal point through a second focal point; (3) Spherical, a special class of elliptical reflectors with coincident foci; and (4) Flat, which redirects the emitted energy without concentrating or collimating the rays.

Energy data furnished by the manufacturer should be consulted to apply a heater properly.

Table 2 Characteristics of Five Electric Infrared Elements

| Characteristic | Type 1: Metal Sheath | Type 2: Reflector Lamp | Type 3: Quartz Tube | Type 4: Quartz Lamp | Type 5: Low-Temperature |
|---|---|---|---|---|---|
| Resistor material | Nickel-chromium alloy | Tungsten wire | Nickel-chromium alloy | Tungsten wire | Graphite or nichrome-wire |
| Relative heat intensity | Med, 60 W/in. (2.4 kW/m), 0.5 in. (13 mm) dia | High, 125-375 W/spot | Medium to high, 75 W/in. 0.5 in. (13 mm) dia | High, 100 W/in. (3.9 kW/m), 3/8 in. (9.5 mm) dia | Low, 95 W/ft$^2$ (1020 W/m$^2$) |
| Resistor temperature | 1750°F (950°C) | 4050°F (2230°C) | 1700°F (930°C) | 4050°F (2230°C) | 200 to 350°F (93 to 177°C) |
| Envelope temperature (in use) | 1550°F (840°C) | 525-575°F (275-300°C) | 1200°F (650°C) | 1100°F (590°C) | 160 to 300°F (70 to 150°C) |
| Radiation generating ratio[a] | 0.58 | 0.86 | 0.81 | 0.86 | 0.7 to 0.8 |
| Response time (heatup) | 180 s | A few seconds | 60 s | A few seconds | 600 s |
| Luminosity (visible light) | Very low (dull red) | High 8 lumens/W | Low (orange) | High 7.5 lumens/W | None |
| Thermal shock resistance | Excellent | Poor to excellent (heat-resistant glass) | Excellent | Excellent | Excellent |
| Vibration resistance | Excellent | Medium | Medium | Medium | Excellent |
| Impact resistance | Excellent | Medium | Poor | Poor | Excellent |
| Resistance to drafts or wind[b] | Poor | Excellent | Medium | Excellent | Poor |
| Mounting position | Any | Any | Horizontal[c] | Horizontal | Any |
| Envelope material | Steel Alloy | Regular or heat-resistant glass | Translucent quartz | Clear, translucent or frost quartz and integral red filter glass | Steel alloy |
| Color blindness | Very good | Fair | Very good | Fair | Very good |
| Flexibility | Good-wide range of watts density, length, and voltage practical | Limited to 125-250 and 375 watts at 120 V | Excellent-wide range of watts density, diameter, length, and voltage practical | Limited. 1 to 3 wattages for each voltage; 1 length for each capacity | Good-wide range of watts density, length, and voltage practical |
| Life expectancy | Over 5000 h | 5000 h | 5000 h | 5000 h | Over 10,000 h |

[a]Ratio or radiant output to watt input (elements only).
[b]May be shielded from wind effects by louvers, deep-drawn fixtures, or both.
[c]May be provided with special internal supports for other than horizontal use.

## CONTROLS

Normally, all controls (except the thermostat) are built into gas-fired infrared heaters, whereas electric infrared fixtures usually do not have built-in controls. Because of the effects of direct radiation, higher mean radiant temperature (MRT), and decreased ambient temperature compared to warm air systems, infrared heating requires careful selection and location of the thermostat sensor. Some installers recommend placing the thermostat or sensor in the radiation pattern, while others do not recommend it. The nature of the system, the type of infrared heating units used, and the nature of the thermostat or sensor dictate the appropriate approach. Furthermore, no single location appears to be equally effective during the periods after a cold start and after substantial operation. The most desirable cycling rate for thermostats controlling infrared heaters has not been defined fully (Walker 1962).

An infrared heating system controlled by low-limit thermostats can be used for freeze protection. A thermostat usually controls an automatic valve on gas-fired infrared units to provide *on-off* control of gas flow to all burners. If a unit has a pilot flame, a sensing element prevents the flow of gas to burners only, or both burners and pilot, when the pilot is extinguished. Electrical ignition may be used with provision for manual or automatic reignition of the pilot if it goes out. Electric spark ignition may also be used.

Gas infrared systems for full building heating may have a zone thermostatic control system in which a thermostat representative of one outside exposure operates heaters along that outside wall. Two or more zone thermostats may be required for extremely long wall exposures. Heaters for an internal zone may be grouped around a thermostat representative of that zone. Manual switches are usually used for spot or area heating, but input controllers may also be used.

Input controllers control electric infrared heating units effectively with metal sheath or quartz tube elements. An input controller is a motor-driven cycling device whose *on* time per cycle can be set. A 30-second cycle is normal. When a circuit's capacity exceeds an input controller's rating, the controller can be used to cycle a pilot circuit of contactors adequate for the load.

Input controllers work well with metal sheath heaters because the sheath mass smooths the pulses into even radiation. The control method decreases the efficiency of infrared generation slightly. Quartz tube elements, which have a warmup time of several seconds, have perceptible, but not normally disturbing, pulses of infrared, with only moderate reduction in generation efficiency when controlled with these devices.

Input controllers should not be used with quartz lamps because the cycling luminosity would be distracting. Instead, output from a quartz lamp unit can be controlled by changing the voltage to the lamp element. This control can be done by modulating transformers or by switching the power supply from *hot-to-hot* to *hot-to-ground* potential.

Power drawn by the tungsten filament of the quartz lamp varies approximately as the 1.5 power of the voltage, while that of the metal sheath or quartz tube elements (using nickel-chromium wire) varies as the square of the voltage. Multiple circuits for electric infrared systems can be manually or automatically switched to provide multiple stages of heat.

Three circuits or control stages are usually adequate. For areas with fairly uniform radiation, one circuit should be controlled with input control or voltage variation control on electric units,

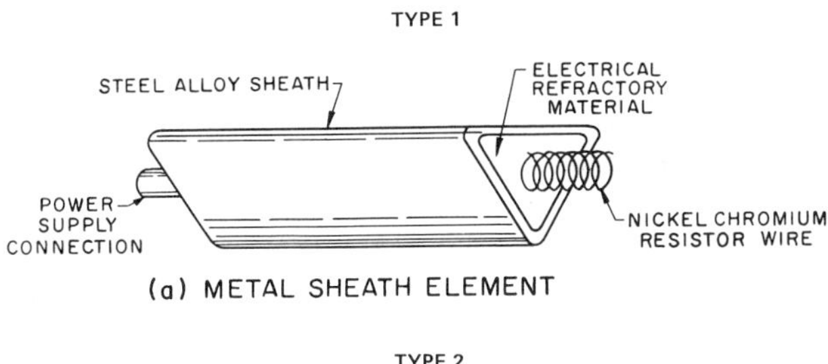


(a) METAL SHEATH ELEMENT

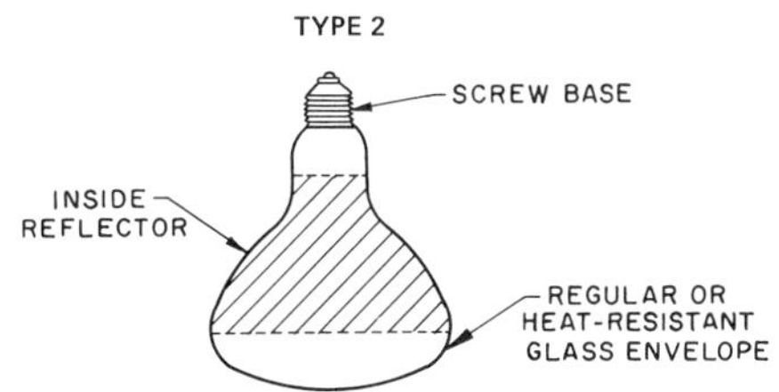


(b) REFLECTOR LAMP ELEMENT

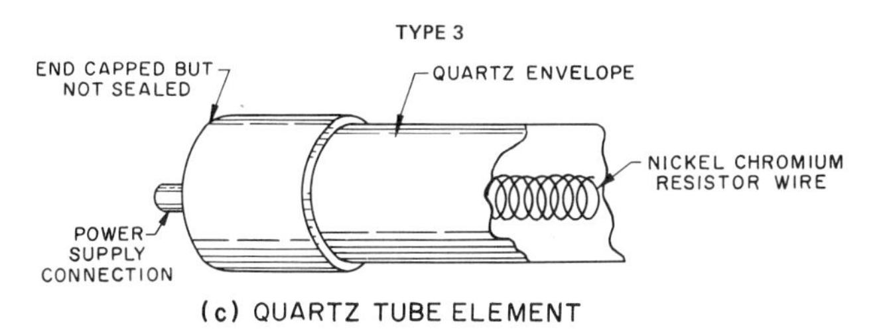


(c) QUARTZ TUBE ELEMENT

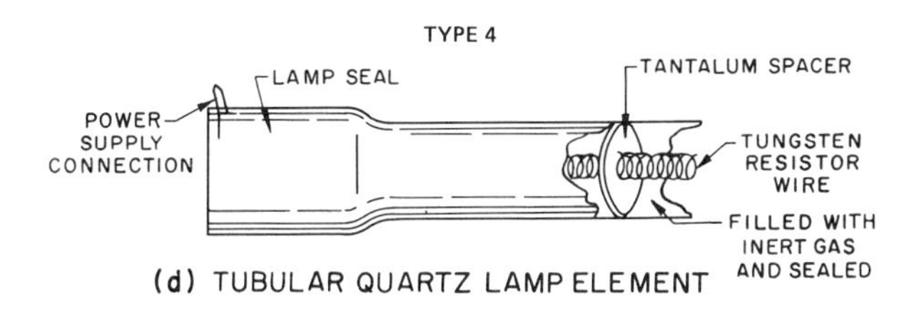


(d) TUBULAR QUARTZ LAMP ELEMENT

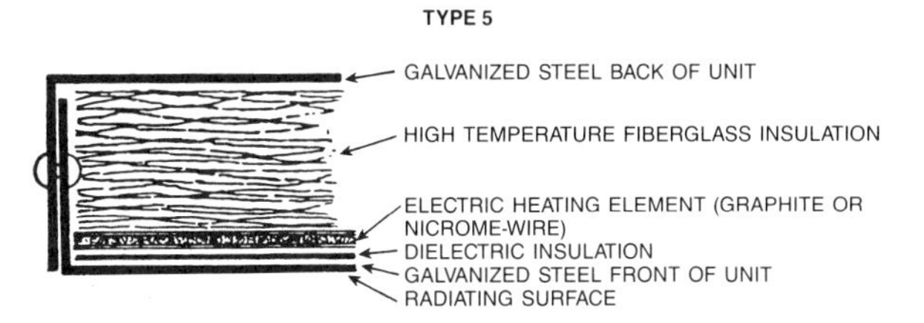


(e) LOW-TEMPERATURE RADIANT HEATING PANEL

**Fig. 2 Common Electric Infrared Heaters**

while the other two are on *full on* or *off* control. This arrangement gives flexible, staged control with maximum efficiency of infrared generation. The variable circuit alone provides zero to one-third capacity. Adding another circuit at *full on* provides one-third to two-thirds capacity, and adding the third circuit provides two-thirds to full capacity.

## PRECAUTIONS

Precautions for the application of infrared heaters include the following:

1. All infrared heaters covered in this chapter have high surface temperatures when they are operating and, therefore, should not be used when the atmosphere contains ignitable dust, gases, or vapors in hazardous concentrations.
2. Manufacturer's recommendations for clearance between a fixture and combustible material should be followed. If combustible material is stored without adequate clearance between it and a fixture, warning notices defining proper clearances should be posted near the fixture.
3. Manufacturer's recommendations for clearance between a fixture and personnel areas should be followed to prevent personnel stress from local overheating.
4. Infrared fixtures should not be used if the atmosphere contains gases, vapors, or dust that decompose to hazardous or toxic materials in the presence of high temperature and air. For example, infrared units should not be used in an area with a degreasing operation that uses trichlorethylene, unless the area has a suitable exhaust system that isolates the contaminate. Trichlorethylene, when heated, forms phosgene (a toxic compound) and hydrogen chloride (a corrosive compound).
5. Humidity must be controlled in areas with unvented gas-fired infrared units because water formed by combustion increases humidity. Sufficient ventilation, direct venting, or insulation on cold surfaces helps control the moisture problems.
6. Adequate makeup air must be provided to replace the air used by combustion-type heaters, regardless of whether units are direct-vented or not.
7. If unvented combustion-type infrared heaters are used, the area must have adequate ventilation to ensure that products of combustion in the air are held to an acceptable level (Prince 1962).
8. Personnel, to be kept comfortable with infrared heating equipment, should be protected from substantial wind or drafts. Suitable wind shields seem to be more effective than increased radiation density (Boyd 1960).

## MAINTENANCE

Electric infrared systems require little care beyond the cleaning of reflectors.

Quartz and glass elements must be handled carefully because they are fragile, and fingerprints must be removed (preferably with alcohol) to prevent etching at operating temperature, which causes early failure.

Gas-fired and oil-fired infrared heaters require periodic cleaning to remove dust, dirt, and soot. Reflecting surfaces must be kept clean to remain efficient. An annual cleaning of heat exchangers, radiating surfaces, burners, and reflectors with compressed air is usually sufficient. Chemical cleaners must not leave a film on aluminum surfaces.

Both main and pilot air ports of gas-fired units should be kept free of lint and dust. The nozzle, draft tube, and nose cone of burners of oil-fired units are designed to operate in a particular combustion chamber, so they must be replaced carefully when they are removed.

## REFERENCES

ASHRAE. 1981. *ASHRAE Standard* 55-1981, Thermal Environmental Conditions for Human Occupancy.

Boyd, R.L. 1960. What Do We Know About Infrared Comfort Heating? *Heating, Piping and Air Conditioning,* November, p. 133.

Boyd, R.L. 1963. Control of Electric Infrared Energy Distribution. *Electrical Engineering*, February, p. 103.

Buckley, N.A. *et al.* 1987. Engineering Principles Support An Adjustment Factor When Sizing Gas-Fired Low-Intensity Infrared Equipment. ASHRAE *Transactions* 93(1B).

Buckley, N.A. and T. Seel. 1986. Gas-Fired Low-Intensity Radiant Heating Provides A Cost-Effective Efficient Space Conditioning Alternate. ASHRAE *Transactions* 92(1B):616-627.

DeWerth, D.W. 1960. Literature Review of Infra-Red Energy Produced with Gas Burners. *Research Bulletin* No. 83, American Gas Association, Cleveland, OH.

DeWirth, D.W. 1962. A Study of Infra-Red Energy Generated by Radiant Gas Burners. *Research Bulletin* No. 92, American Gas Association.

Gas Research Board of London. 1944. The Use of Infra-Red Radiation in Industry. *Information Circular* No. 1.

Haslam, W.G. *et al.* 1925. Radiation From Non-luminous Flames. *Industrial and Engineering Chemistry*, March.

Plyler, E.K. 1948. Infra-red Radiation From Bunsen Flames. *Journal of Research*, National Bureau of Standards, Vol. 40, February, p. 113.

Prince, F.J. 1962. Selection and Application of Overhead Gas-fired Infrared Heating Devices. ASHRAE *Journal*, October, p. 62.

Walker, C.A. 1962. Control of High Intensity Infrared Heating. ASHRAE *Journal*, October, p. 66.

## CHAPTER 30

# CENTRIFUGAL PUMPS



CENTRIFUGAL pumps recirculate hot water in heating systems and chilled water cooling systems to establish a predetermined rate of flow between the boiler/chiller (for thermal storage tanks) and the space conditioning terminal units. The influence of pump performance on installation, system controllability, and seasonal operating costs is covered in Chapter 14 of the 1987 HVAC Volume.

Other pump applications on hydronic systems include (1) condenser water circuits to cooling towers and water source heat pumps, (2) boiler feed, and (3) condensate return. Pumps are required with boiler feed and condensate return only if a steam boiler is included in the system. In such cases, the boiler manufacturer defines the specific pumping requirements. When a cooling tower rejects heat for a chilled water plant, the condenser water pumps are selected on the basis of the flow rate specified by the refrigeration equipment manufacturer and the location of the tower relative to the condenser.

In centrifugal pumps, a driver converts part of the output torque into pressure energy by centrifugal force, which is a function of the impeller vane peripheral velocity. Impeller rotation adds energy to a liquid after it enters the eye of the impeller. The casing collects the liquid as it leaves the impeller and guides it out the discharge nozzle. The pressure energy added by the pump (1) overcomes the friction caused by the flow through heating and air-conditioning equipment, *i.e.,* piping, valves, coils, chillers, or boilers; and (2) raises the water to higher elevations such as to the top of a cooling tower.

## PUMP TYPES

Most centrifugal pumps used in hydronic systems are single stage with a single- or double-entry impeller. Double suction pumps are generally used for high-flow applications, but either form is available with similar performance characteristics and efficiencies. Selection can be based on installed cost and personal preferences.

These pumps have either volute or diffuser types of casings. The volute types include all pumps that collect the water from the impeller and discharge it perpendicularly to the pump shaft. Diffuser-type casings collect the water from the impeller and discharge it parallel to the pump shaft. All pumps described here are the volute type, except the vertical turbine pump, which is a diffuser type.

Pumps can be classified by method of connection to the electric motor and can be close-coupled or flexible-coupled. The close-coupled pump has the impeller mounted directly on a motor shaft extension, while the flexible-coupled pump has an impeller shaft supported by a frame or bracket that is connected to the electric motor through a flexible coupling.

Pumps are also classified by their mechanical features and installation arrangement. *Circulator* is a generic term for pipe-mounted, low-head, low-flow units, which usually have fractional horsepower, and may be either wet rotor or conventional flexible-coupled open-type motor driven. In addition to their application in residential and small commercial buildings, circulators recirculate flow of terminal unit coils to enhance heat-transfer efficiencies and improve the management of large systems.

Integral-horsepower pumps are available as close-coupled or base mounted. The close-coupled pumps are *end suction* for horizontal mounting or *vertical inline* for direct installation in the piping. The base-mounted pumps are *end-suction-frame mounted* or *double suction* horizontally split case units. Double-suction pumps can also be arranged in a vertical position on a support frame with the motor vertically mounted on a bracket above the pump unit.

Pumps are labeled by their mounting position, either horizontal or vertical. Significant types of pumps used in heating and air conditioning or hydronic systems are (1) circulator; (2) close-coupled, end suction; (3) frame-mounted or flexible-coupled, end suction; (4) double suction, horizontal split case, single-stage; (5) horizontal split case, multistage; (6) vertical in-line; and (7) vertical turbine. Table 1 lists these pump types and summarizes their design features. Figure 1 shows the general configuration of these pumps and lists their typical applications. Many variations of these pumps are offered by pump manufacturers for particular applications.

## GENERAL CONSTRUCTION FEATURES

The following are important construction features of centrifugal pumps.

*Materials.* Centrifugal pumps are generally offered in bronze-

The preparation of this chapter is assigned to TC 8.10, Pumps and Hydronic Piping.

**Table 1 Mechanical Features of Centrifugal Pumps**

| Pump Type | Impeller Type | Number of Impellers | Casing | Motor Connection | Motor Mounting Position |
|---|---|---|---|---|---|
| Circulator | Single suction | One | Volute | Flexible-coupled | Horizontal |
| Close-coupled, end suction | Single suction | One or two | Volute | Close-coupled | Horizontal |
| Frame-mounted, end suction | Single suction | One or two | Volute | Flexible-coupled | Horizontal |
| Double suction, split casing | Double suction | One | Volute | Flexible-coupled | Horizontal or vertical |
| Vertical inline | Single suction | One | Volute | Flexible- or close-coupled | Vertical |
| Vertical turbine | Single suction | One to twenty | Diffuser | Flexible-coupled | Vertical |

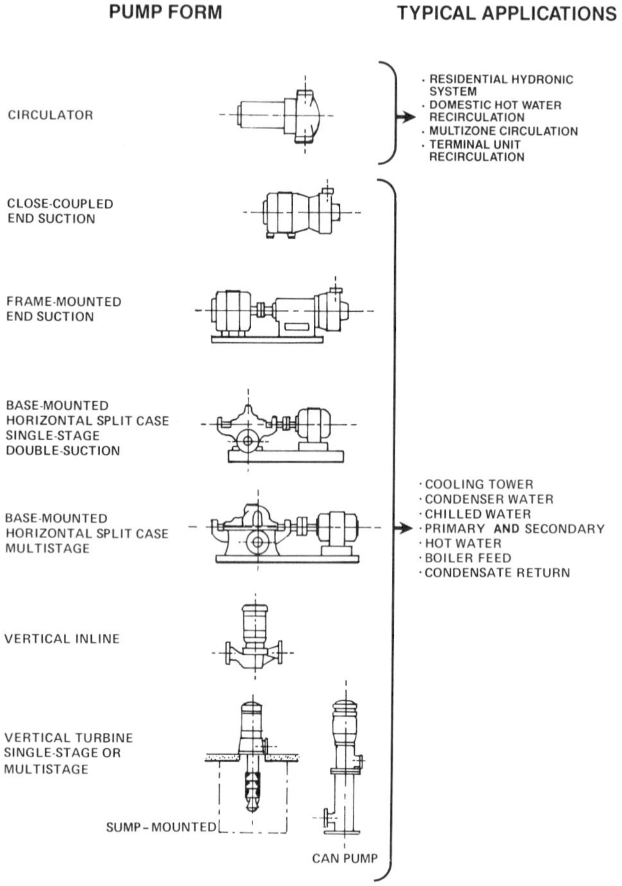

PUMP FORM

TYPICAL APPLICATIONS

CIRCULATOR

- RESIDENTIAL HYDRONIC SYSTEM
- DOMESTIC HOT WATER RECIRCULATION
- MULTIZONE CIRCULATION
- TERMINAL UNIT RECIRCULATION

CLOSE-COUPLED END SUCTION

FRAME-MOUNTED END SUCTION

BASE-MOUNTED HORIZONTAL SPLIT CASE SINGLE-STAGE DOUBLE-SUCTION

BASE-MOUNTED HORIZONTAL SPLIT CASE MULTISTAGE

- COOLING TOWER
- CONDENSER WATER
- CHILLED WATER
- PRIMARY AND SECONDARY
- HOT WATER
- BOILER FEED
- CONDENSATE RETURN

VERTICAL INLINE

VERTICAL TURBINE SINGLE-STAGE OR MULTISTAGE

SUMP-MOUNTED

CAN PUMP

**Fig. 1 Types of Centrifugal Pumps Used in Hydronic Systems**

fitted, all bronze, or iron-fitted construction. In bronze-fitted construction, the impeller, shaft-sleeve (if used), and wear-rings are bronze, and the casing is cast iron. These construction materials refer to the liquid end of the pump (those parts of the pump that contact the liquid being pumped).

The *stuffing box* is that portion of the pump where the rotating shaft enters the pump casing. To seal leaks at this point, a mechanical seal or packing is used in the stuffing box.

*Mechanical seals* are used predominately in hydronic applications. As with pumps, many styles and types of seals are available. There are unbalanced and balanced (for higher pressures) seals. Inside seals operate inside the stuffing box while outside seals have their rotating element outside the box. Pressure and temperature limitations vary depending on the liquid being pumped and the style of seal. The seal material and style are supplied by the manufacturer after he has been informed of the kind of liquid to be pumped and the temperature and pressure limitations.

*Packing* is used also, particularly where abrasive substances included in the water are not detrimental to system operation. Some leakage at the packing gland is needed to lubricate and cool the area between packing material and shaft.

*Shaft-sleeves* protect the motor or pump shaft.

*Wearing-rings* are for the impeller and/or casing. They are replaceable and prevent wear to the impeller or casing.

*Ball bearings* are most frequently used, except in circulators, where motor and pump bearings are the sleeve-type.

The *balance-ring* is placed on the back side of a single-inlet, enclosed impeller to reduce the axial load. Double-inlet impellers are inherently axially balanced.

Nominal *operating speeds* of motors may be selected in the range between 600 and 3600 rpm (10 and 60 r/s). (Pump manufacturers should be consulted for optimum pump speed for each specific pumping requirement, with due consideration for efficiency, cost, noise, and maintenance.)

Figure 2 (supplied by the Hydraulic Institute) shows most of the parts previously described.

## PUMP TERMS, EQUATIONS, AND LAWS

Table 2 lists terms and equations for pumping and Table 3 lists the affinity laws for pumps. These laws describe the relationships among the changes of pump impeller diameter, speed, and specific gravity. Without knowledge of the system head curve, the laws should not be used to predict the pump performance of a particular hydronic system. Figure 3 describes pump performance at 1750 and 1150 rpm (29 and 19 r/s), in accordance with the affinity laws for constant impeller diameter and viscosity.

If the hydronic system has a system head curve as shown in Figure 3, curve A, the pump at 1150 rpm (19 r/s) will operate at point 1, not at point 2, as the affinity laws predict. If the system head curve is the same as curve B in this figure, at 1150 rpm (19 r/s), the pump will run at shutoff head and will not deliver water, thus demonstrating that the affinity laws should be used to develop new pump curves, but not to predict performance unless the system curve is known. (System head curves are covered in "Hydronic System Characteristics.")

## PUMP PERFORMANCE CURVES

Performance of a pump is most commonly shown by graphs, as in Figure 4, which relate the flow (gpm and L/s), the head produced (ft and m), the power required (BHP and kW), the efficiency (%), the shaft speed (rpm and r/s), and the net positive suction head (ft and m, absolute) required for pumps with various impeller diameters. (See the section Pump Suction Characteristics for a description of NPSH [net positive suction head].) For many small pumps, some of this information is omitted. Pump curves present the average results obtained from testing several pumps of the same design under standardized test conditions. Consult manufacturers for pump applications that differ considerably from ordinary practice.

The head-capacity curve for a centrifugal pump describes the head produced by the pump, from maximum flow to the shutoff or no-flow condition. The pump curve is considered flat if the shutoff head is about 1.10 to 1.20 times the head at the best ef-

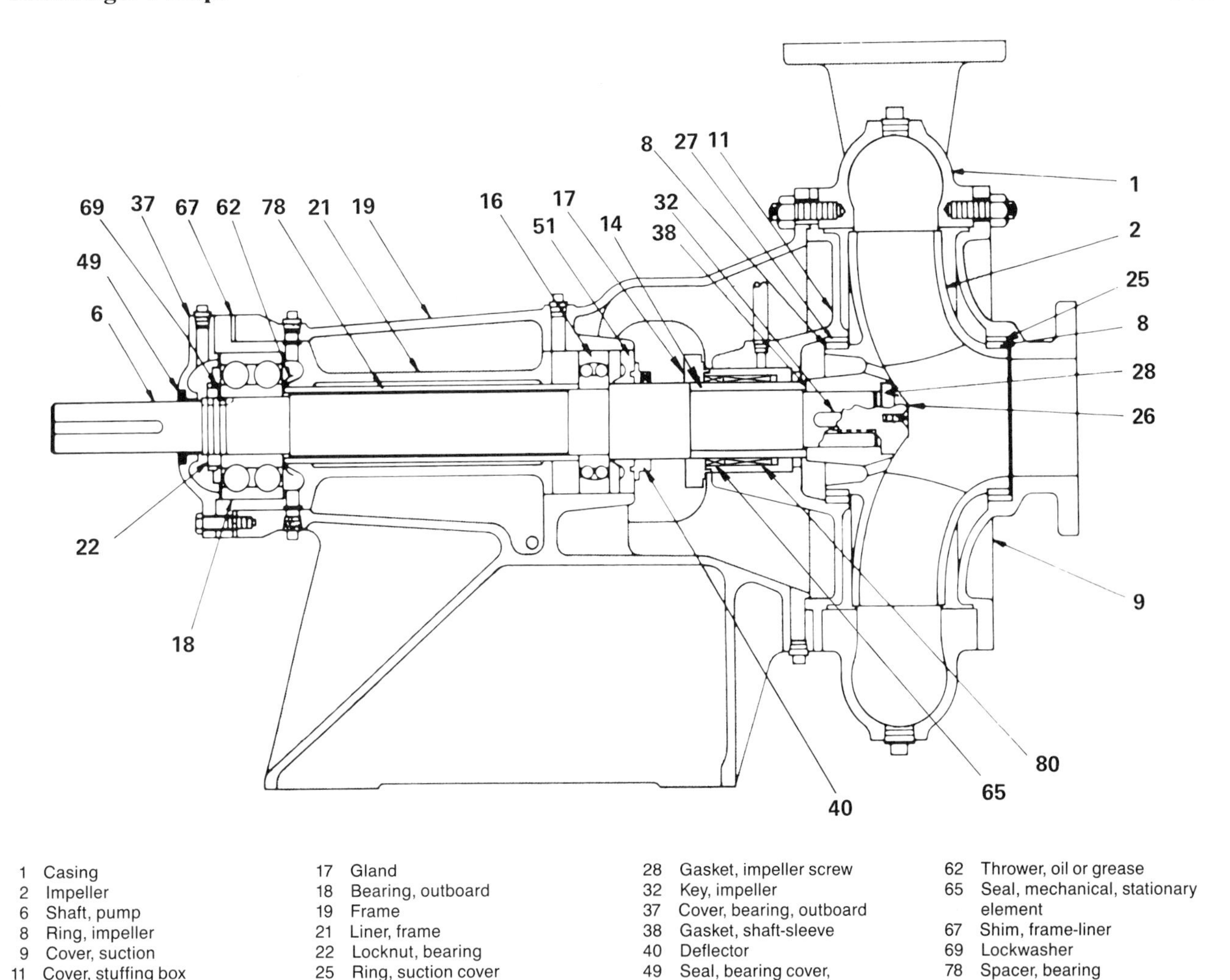


Numbers do not represent standard part numbers in use by any manufacturer.

**Fig. 2 Typical Cross Section of an End Suction Pump**

ficiency point. If the head at shutoff exceeds 1.20 times the head at the best efficiency point, it is called a steepcurved pump.

The need for energy conservation and the use of two-speed and variable-speed pumps require particular attention to the efficiency curves of a pump. As shown in Figure 4, the best efficiency point is 400 gpm and 44 ft (25 L/s and 13.5 m or 130 kPa) of head. The efficiency curves form an *eye,* with the maximum efficiency being the inner circle. The affinity laws can be used to predict the best efficiency point at other pump speeds; this best efficiency point follows a parabolic curve to zero as the pump speed is decreased. The best efficiency curve for the pump of Figure 4 is described in Figure 5, which shows the movement of the best efficient point from high to low speed.

## PUMP SUCTION CHARACTERISTICS (NPSH)

Particular attention must be given to the condition of the liquid as it enters a pump in condenser, condensate, and boiler feed systems. If the absolute pressure on the liquid at the suction nozzle approaches the vapor pressure of the liquid, cavitation occurs, and vapor pockets form in the impeller passages. The collapse of the vapor pockets could progressively damage the impeller.

The amount of pressure in excess of the vapor pressure required to prevent the formation of vapor pockets is the *net positive suction head required* (NPSHR). NPSHR is a characteristic of a given pump, and it varies considerably with pump speed and flow. NPSHR is determined by testing individual pumps; it increases rapidly at high flows, as shown in Figure 4.

Particular attention must be given to NPSHR when a pump is operating with hot liquids. The vapor pressure increases with water temperature and reduces the *net positive suction head available* (NPSHA). While each pump has its own particular NPSHR, each installation has its own particular NPSHA, which is the total useful energy above the vapor pressure of the liquid available to the pump at the suction connection. To calculate the NPSHA, use one of the following:

$$\text{NPSHA in proposed installation} = h_p + h_z - h_{vpa} - h_f$$

$$\text{NPSHA in existing installation} = p_a + p_s + v^2/2g - h_{vpa} - h_f$$

*where*

NPSHA = net positive suction head available at pump suction nozzle

**Table 2 Common Pump Terms, Abbreviations, and Formulas**

| Term | Abbreviation | IP Units | SI Units | I-P Formula | SI Formula |
|---|---|---|---|---|---|
| Velocity | $v$ | ft/s | m/s | | |
| Volume | $V$ | ft$^3$ | m$^3$ | | |
| Flow Rate | $Qv$ | gpm | L/s | | |
| Pressure | $p$ | psi | kPa | | |
| Density | $\varrho$ | lb/ft$^3$ | kg/m$^3$ | | |
| Acceleration of Gravity | $g$ | 32.17 ft/s$^2$ | 9.807 m/s$^2$ | | |
| Specific Gravity | SG | — | — | $\frac{\text{Mass of Liquid}}{\text{Mass of Water at 4° C}}$ | |
| Speed | $n$ | rpm | r/s | | |
| Head | $H$ | ft | m | 2.31 $p$/SG | $p/\varrho g$ |
| Net Positive Suction Head (NPSH) | $H$ | ft | m | | |
| Efficiency | | | | | |
| Pump | $\eta_p$ | | | | |
| Electric Motor | $\eta_m$ | | | | |
| Variable Speed Drive | $\eta_v$ | | | | |
| Equipment (Constant Speed Pumps) | $\eta_e$ | | | $\eta_e = \eta_p \cdot \eta_m/100$ | |
| Equipment (Variable Speed Pumps) | $\eta_e$ | | | $\eta_e = \eta_p \cdot \eta_m \cdot \eta_v/10{,}000$ | |
| Utilization<br>*where:*<br>$Q_D$ = Design flow<br>$Q_A$ = Actual flow<br>$H_D$ = Design head<br>$H_A$ = Actual head | $\eta_u$ | | | $\eta_u = 100\,\frac{Q_D H_D}{Q_A H_A}$ | |
| System Efficiency Index (decimal) | | | | SEI = $\eta_\varepsilon \cdot \eta_u/10{,}000$ | |
| Output Power (pump) | $P_o$ | hp | kW | $Q_v H$SG/3960 | $Q_v\, p/1000$ |
| Shaft Power | $P_s$ | hp | kW | 100 $P_o/\eta_p$ | 100 $P_o/\eta_p$ |
| Input Power | $P_i$ | kW | kW | 74.6 $P_s/\eta_m$ | 100 $P_s/\eta_m$ |

**Table 3—Affinity Laws for Pumps**

| Impeller Diameter | Speed | Specific Gravity (SG) | To Correct for | Multiply by |
|---|---|---|---|---|
| Constant | Variable | Constant | Flow | $\left(\frac{\text{New Speed}}{\text{Old Speed}}\right)$ |
| | | | Head | $\left(\frac{\text{New Speed}}{\text{Old Speed}}\right)^2$ |
| | | | BHP (or kW) | $\left(\frac{\text{New Speed}}{\text{Old Speed}}\right)^3$ |
| Variable | Constant | Constant | Flow | $\left(\frac{\text{New Diameter}}{\text{Old Speed}}\right)$ |
| | | | Head | $\left(\frac{\text{New Diameter}}{\text{Old Speed}}\right)^2$ |
| | | | BHP (or kW) | $\left(\frac{\text{New Diameter}}{\text{Old Speed}}\right)^3$ |
| Constant | Constant | Variable | BHP (or kW) | $\frac{\text{New SG}}{\text{Old SG}}$ |

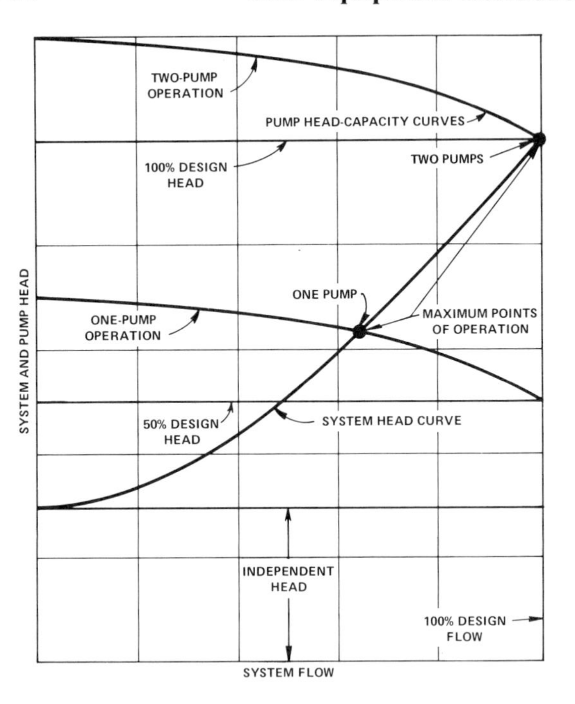


**Fig. 3 Application of Affinity Laws**

$h_p$ = absolute pressure on surface of liquid where pump takes suction, ft (m) of liquid

$h_z$ = static elevation of the liquid above centerline of pump, ft. (m) of liquid; $h_z$ is minus if liquid level is below pump centerline

$h_f$ = friction and entrance head losses in the suction piping, ft. (m) of liquid

$h_{vpa}$ = absolute vapor pressure at pumping temperature, ft (m) of liquid

$p_a$ = atmospheric pressure for the elevation of the installation, ft (m) of liquid

$p_s$ = gauge pressure at suction flange of pump corrected to centerline of pump, ft. (m) of liquid; $p_s$ is minus if below atmospheric pressure

$v^2/2g$ = velocity head at point of measurement of $p_s$, ft (m) of liquid

If NPSHA is less than NPSHR, cavitation, noise, inadequate pumping, and mechanical problems occur. NPSH is normally not a factor with hot-water and chilled-water pumps when sufficient system fill pressure is usually exerted on the pump suction.

## HYDRONIC SYSTEM CHARACTERISTICS

Hydronic systems in HVAC are all of the loop type (open or closed), which circulate water through the system and return it to the pumps. No appreciable amount of water is lost from the system, except in the cooling tower, where evaporative cooling occurs.

Hydronic systems are either full-flow or throttling-flow. Full-flow systems are usually found on residential or small commercial systems where pump motors are small and the energy waste caused by constant flow is not appreciable. Larger systems use flow control valves that control flow in the system in accordance with the heating or cooling load imposed. Hydronic systems can also operate continuously or intermittently. Most hot-water or chilled-water systems operate continuously as long as a heating or cooling load exists on the system. Condensate and boiler feed pumps are often of the intermittent type, starting and stopping as the water level changes in condensate tanks or boilers.

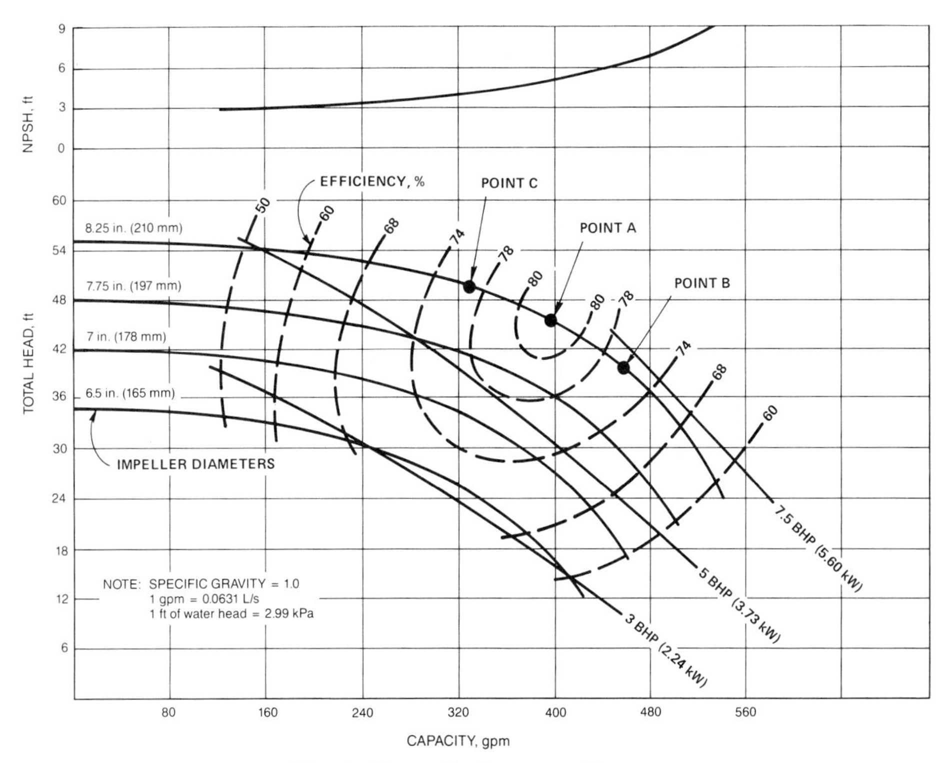


Fig. 4 Pump Performance Curves

## SYSTEM HEAD CURVES

The friction head loss of a piping system depends on the flow rate through it. If one set of friction head/flow rate data is available, a characteristic curve may be developed for the system by using the principle that friction head varies directly with the square of the flow.

After the heat loss or heat gain of a building has been calculated, the appropriate temperature difference is determined, and the design flow rate is established. The form of the piping circuits and their piping sizes are developed next, and the design friction head loss is calculated. When these two values are known, friction heads corresponding to other flow rates are calculated using the formula $H_1/H_2 = (Q_1/Q_2)^2$. Plotting the known and calculated points generates the *system head curve* as shown in Figure 6. This curve is typical for a constant flow recirculated system (friction head loss only), and no static or minimum head is maintained.

The second example of the system curve is a system that has static head as well as friction loss. Such a system might be the piping circuit between a refrigeration plant condenser and its cooling tower, with the elevation difference between the water level in the tower pan and spray header pipe creating the static head. With this pumping system, the head curve datum point is the value of the static condition rather than zero. A friction head does not develop until a flow exists, and flow does not begin until the head developed by the pump exceeds the static head. To calculate the points from which the static head is plotted, the same head/flow statement is used, but it is based on the calculated friction loss value. The static head is not part of

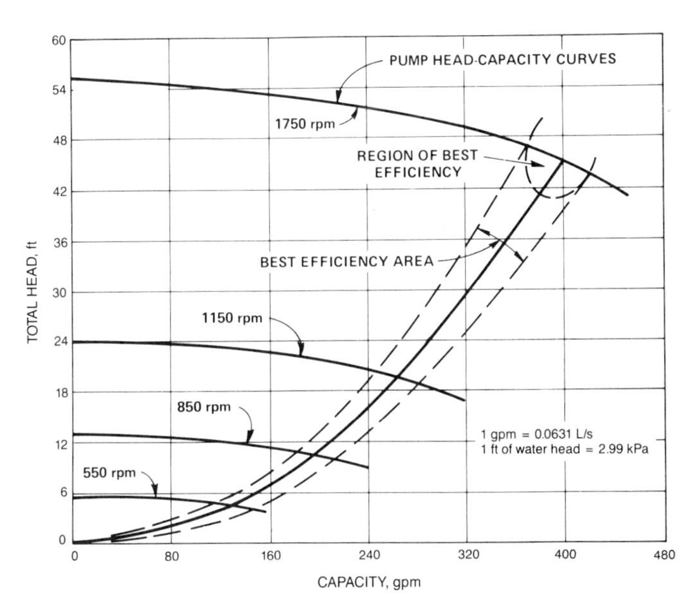


Fig. 5 Pump Best Efficiency Curves

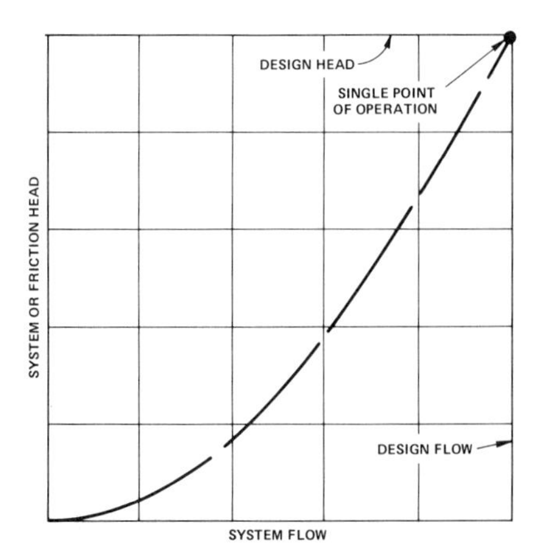


Fig. 6 Constant Flow System Head Curve

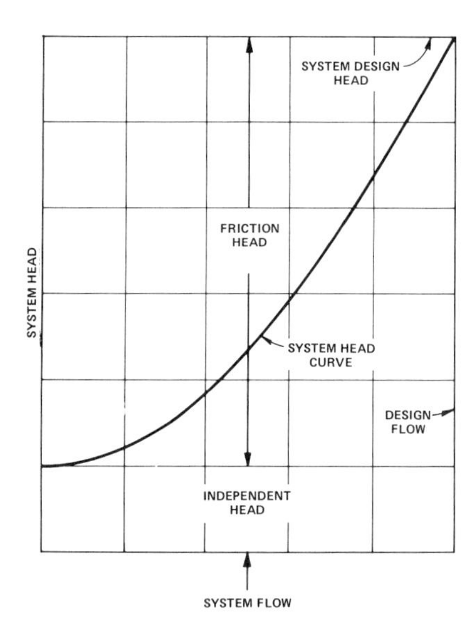


**Fig. 7 Typical System Head Curve**

the system curve computation and is, therefore, termed *independent head* as shown by Figure 7.

In some hydronic systems, flow rates are changed by variable-speed pumping or by the sequencing of constant-speed multiple pumps. Such systems must be accurately balanced, so all circuits and all terminal units receive equivalent, proportional flow. If individual control valves are not used on the terminal units, the static head curve is similar to the one shown in Figure 6. This system is seldom used because control at part-load operation is difficult. Diversity is not provided, so the system is not economical for a large multi-use or multi-building system.

Most variable-volume systems have individual two-way valves on each terminal unit that permit full diversity or random loading from zero to full load. Regardless of the load required, the designer establishes a minimum pressure difference to ensure any terminal and its control valve receive design flow at full output demand. When graphing a system curve for a *non-symmetrically* loaded variable-volume system the $\Delta_p$ (minimum maintained head) is treated like a static head condition and becomes the starting datum for the system curve (Figure 7).

A variable-volume system such as the one described, can ran-

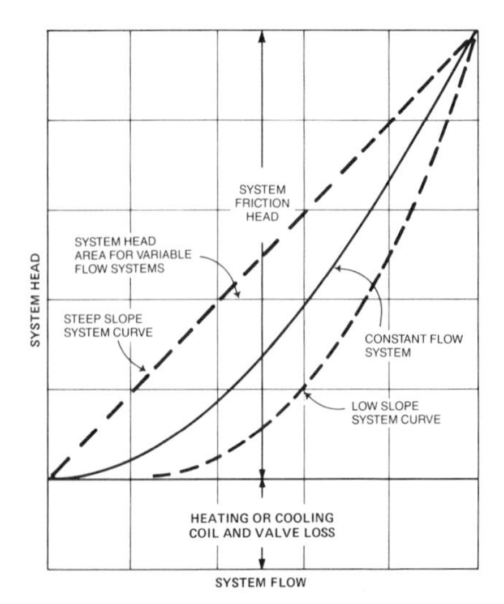


**Fig. 8 System Head for Heating or Cooling System with Control Valves**

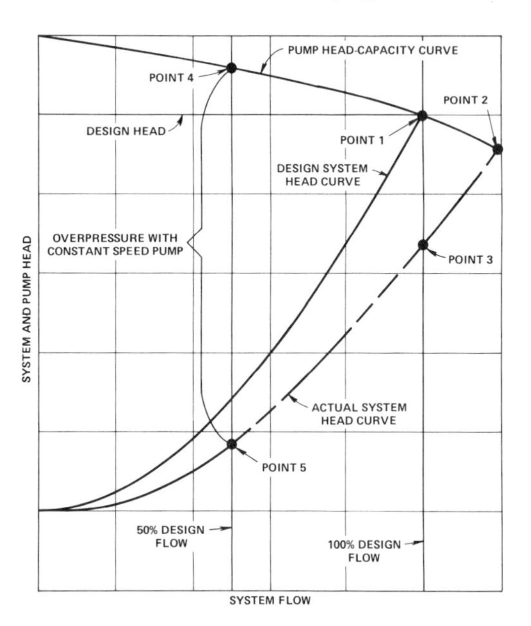


**Fig. 9 Pump Operating Points**

domly load or unload its terminals from zero to full load in any sequence. A system that picks up terminal loads in the order of the distance of the terminals from the main equipment room operates with a *low slope* head curve (Figure 8). A system that starts loading progressively with terminals remote from the equipment room establishes a *steep slope* curve.

The net vertical difference between the low and high curves in Figure 9 is the difference in friction loss developed by the distribution mains for the two extremes of possible load patterns. The low and high curves represent the boundaries for the operation of the system. The area in which the system operates depends on the diverse loading or unloading imposed on the terminal units.

Friction head losses for pipe and fittings are available in other chapters and volumes of the ASHRAE Handbook as well as in the *Pipe Friction Manual* of the Hydraulic Institute.

## SELECTION AND ARRANGEMENT OF PUMPS

The selection of pumps for a particular hydronic system requires a substantial amount of data to ensure an adequate, reliable, and efficient pump is selected. The following is some of the information required:

1. Maximum and minimum flow in system.
2. System head at maximum and minimum flows.
3. Continuous or intermittent operation.
4. System operating pressures and temperatures.
5. Pump environmental conditions, including ambient temperature.
6. Number of pumps desired and percent of standby required for emergency operation.
7. Electrical current characteristics.
8. Electrical service starting limitations.
9. Special electrical controls.
10. Water chemistry that may affect materials selection.

A specific application may have conditions that require more information.

The selection of pumps should consider changes of flow in the system. Figure 4 describes a typical head-capacity curve and efficiency pattern for a centrifugal pump. For a pump running with little change in volume, the desirable selection point is Point

A in the region of best efficiency. If the pump application is throttling where the flow is normally less than design, the pump should be selected to the right at Point B, so the pump performance passes through the region of best efficiency as flow is reduced in the system. Other design criteria such as NPSH must be considered when a pump is selected at Point B.

In some cases, operation at flows greater than design may occur; for such applications, the pump should be selected at Point C, so the performance passes through the region of best efficiency as flow increases in the system. This discussion is meant as a guide to evaluating pump efficiency as flow in a system changes, not as a set of rules. The designer should evaluate the effects that flow or head variation have on pump efficiency.

The selection of pumps for hydronic systems must consider the point of operation of a pump on its head-capacity curve. Figure 9 describes pump-operating points that may result when a pump is selected for a specific hydronic system. Point 1 indicates the selection point for the pump; at this point, the capacity of the pump is equal to the design flow and head of the system. Design contingencies may have included friction head factors that actually do not exist when the system is put into operation. This fact is displayed in Figure 9 by the two system head curves, one for design and the other for the actual system head curve.

In Figure 9, if the system is free flowing without control valves and, with an actual system head curve as shown, the pump will operate at Point 2, not Point 1; the pump produces a higher flow rate than design flow rate. If the system is controlled with two-way valves on all heating or cooling coils, the pump will operate at Point 1 at design flow and will create an overpressure on the coils and control valves equal to the head difference between Points 1 and 3. If the system flow is reduced to 50% of design on such a system, the overpressure will increase to the amount between Points 4 and 5. Pump operation can be summed up as follows:

1. On a system without control valves, the pump always operates at the point of intersection of the pump head-capacity curve and the system head curve.
2. On controlled flow systems, the pump follows its head capacity curve. The difference between pump head and system head is converted into overpressure, which is consumed by the control valves.

Because overpressure can occur in controlled flow systems when coils are equipped with two-way control valves, selection and application of pumps for such systems must keep overpressure to a minimum by the following methods:

1. Multiple pumps operating in parallel.
2. Multiple pumps operating in series.
3. Multispeed pumps.
4. Variable-speed pumps.

The actual method used on a specific hydronic system depends on the economics of that system. The effects of these methods on overpressuring are described within the chapter. The actual results for a particular system can be determined by developing the system head curve and plotting pump head-capacity curves on the same graph with the system head curve.

Overpressure is most commonly eliminated by using multiple pumps in parallel. Figure 10 describes two pumps piped in parallel, while Figure 11 includes a system head curve as well as the head-capacity curves for single-pump and two-pump operation. Figure 11 shows one pump at 50% system flow will reduce the overpressure caused by two-pump operation or one pump designed to handle maximum design flow and head.

Figure 12 illustrates two pumps piped in series with bypasses

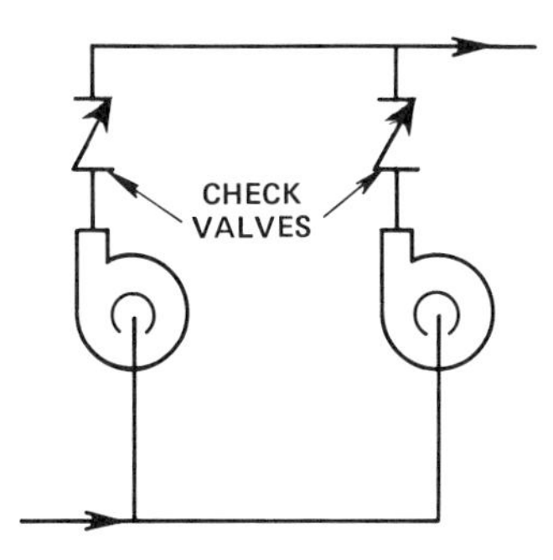


**Fig. 10 Parallel Pumping**

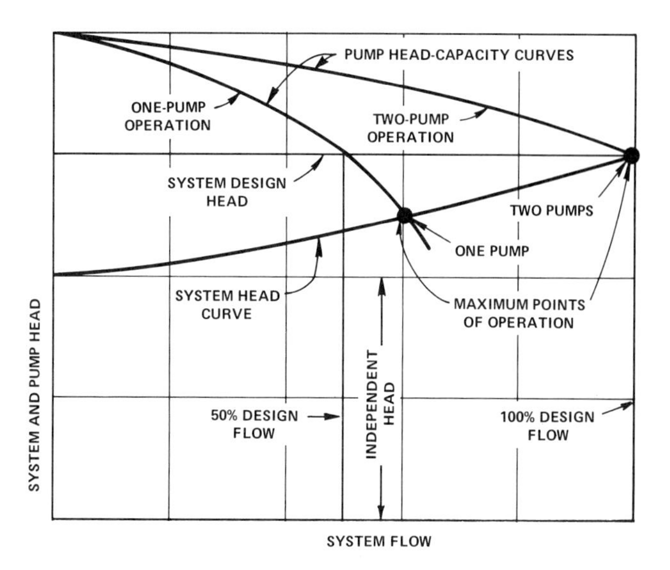


**Fig. 11 Pump and System Curves for Parallel Pumping[a]**

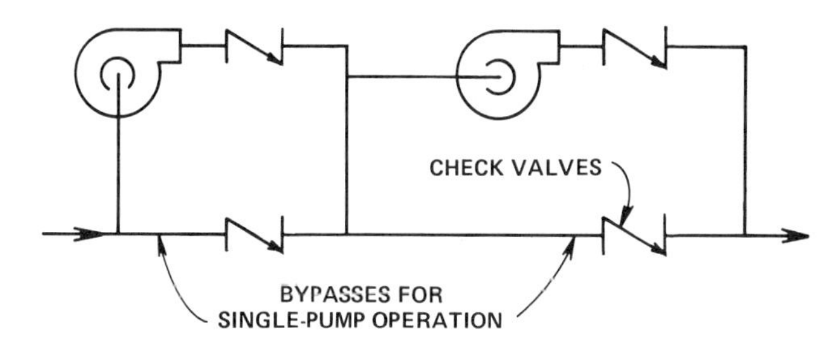


**Fig. 12 Series Pumping**

for single-pump operation. Figure 13 shows series pumping on a hydronic system with a large amount of system friction. For such systems, series pumping can greatly reduce the overpressure on a controlled flow system. Series pumping should not be used on hydronic systems with *flat* system head curves similar to the one shown in Figure 11. For such a system, one pump operation with series connection results in the pump running at shutoff head and producing no flow in the system.

Two-speed motors, which are standard units, are available in 1750/1150 rpm (29 and 19 r/s), 1750/850 rpm (29 and 14 r/s), 1150/850 rpm (19 and 14 r/s), and 3500/1750 rpm (58 and 29 r/s) speeds. These motors can reduce overpressuring at reduced system flow. Figure 5 describes the effect of two-speed motors on pump-head capacity curves, particularly for a 1750/1150 rpm (29 and 19 r/s) motor.

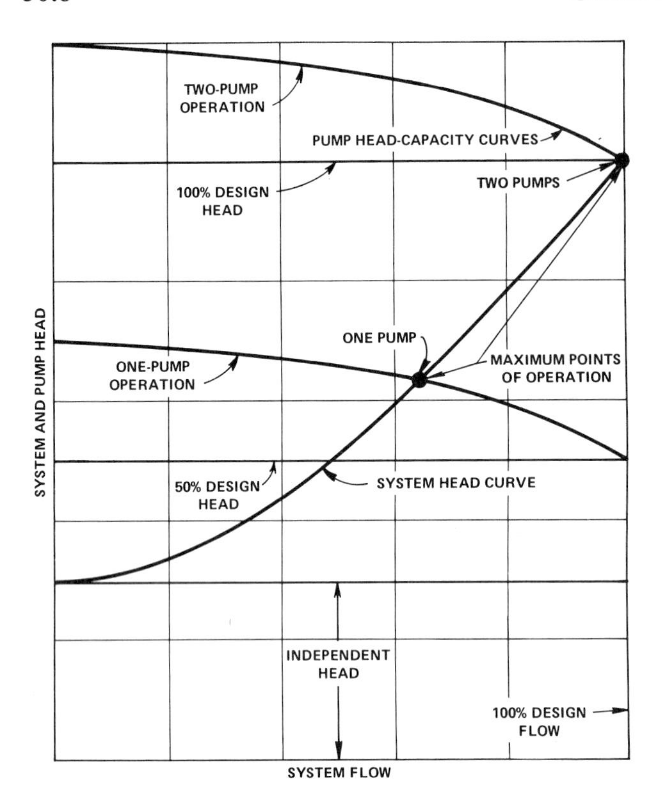


**Fig. 13 Pump and System Curves for Series Pumping[a]**

Variable-speed drives have a similar effect on pump-head capacity curves as two-speed motors. These drives normally have an infinitely variable speed range, so the pump, with proper controls, can follow the system head curve without any overpressure. Figure 5 describes the operation of a variable speed pump at 1750, 1150, 850, and 550 rpm (29, 19, 14, and 9 r/s).

The selection and arrangement of pumps and controls for a specific installation requires additional evaluation. This discussion is a brief outline and must be complemented with additional economic evaluations.

## MOTIVE POWER

Electic motors drive most centrifugal pumps for hydronic systems. Internal combustion engines or turbines power some pumps, especially in central power houses for large installations; they are discussed in detail in Chapter 32, Engine and Turbine Drives. The following description discusses electric motor drives.

Electric motors for centrifugal pumps can be any of the horizontal or vertical electric motors described in Chapter 31, Motors and Motor Protection. As discussed earlier, many of the centrifugal pumps for hydronic systems are close-coupled with the pump impeller mounted on a motor shaft extension; others are flexibly-coupled through a pump mounting bracket or frame to the electric motor.

The sizing of electric motors is critical because of the increased costs of electrical power. In the past, recommended practice specified the use of nonoverload motors; *i.e.*, the motor nameplate rating did not exceed the pump brake horsepower (kW) at any point on the pump head-capacity curve for that particular speed and impeller. A centrifugal pump on a hydronic system with variable flow has a broad range power requirements, resulting in low motor loading at low flow conditions. Figure 14 is a typical efficiency curve for an induction motor; it demonstrates that operation at low loads can result in low motor efficiencies and high operating costs. To lessen the effect of low-efficiency motors operating at low loads, pump motors should operate within the service factor of the motor, providing the service factor is acceptable to the pump manufacturer.

A number of variable-speed drive devices are available for operating centrifugal pumps on hydronic systems. These include fluid coupling, SCR variable frequency, direct current, wound rotor, and eddy-current drives. Each drive has specific design features, which should be evaluated for use with hydronic pumps; in each case, the efficiency range should be investigated from minimum to maximum speed. Figure 15 describes the overall range of efficiencies for variable-speed drives.

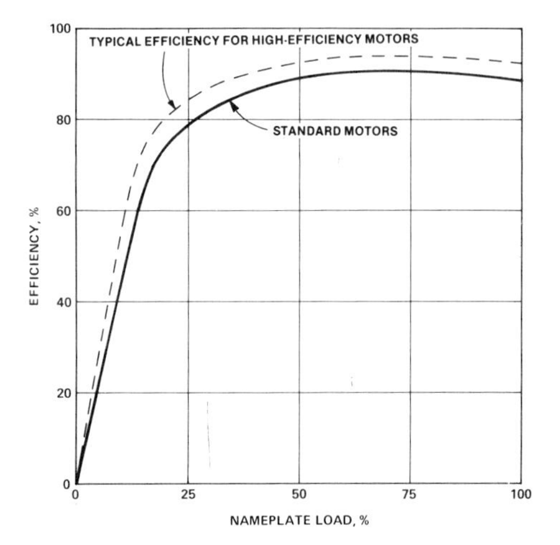


**Fig. 14 Typical Efficiency Curve for Induction Motor**

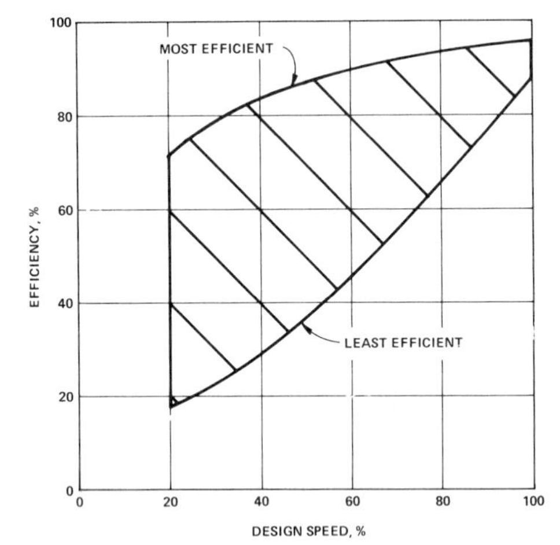


**Fig. 15 Efficiency Range of Variable Speed Drives**

## HYDRONIC APPLICATIONS

Four major types of hydronic systems that use centrifugal pumps are (1) chilled water, (2) hot water, (3) condenser water, and (4) condensate or boiler feedwater systems. The following is a brief description of system head curves affecting pump selection; more detailed information is included in Chapter 13 of the 1987 HVAC Volume.

Chilled-water and hot-water pumps are used in closed recirculating systems. Hot-water pumps, particularly high-temperature hot-water units, require special pump features, such as high-temperature seals and thermal expansion means. The hydraulic application is similar for chilled- and hot-water pumps.

Figure 8 illustrates a typical system head curve for a variable-volume, chilled- or hot-water system. Flow is regulated through the heating or cooling coils by two-way valves. The independent head, as shown, is the pressure drop of the heating and cooling coils and their control valves. The system friction head is much greater than the coil and valve loss, resulting in a *steep* system head curve. Such a system lends itself to multiple pumps, two-speed motors, or variable-speed drives on the pumps to avoid overpressuring the system.

Three-way valves on the heating and cooling coils also eliminate overpressuring, but they waste energy because the pumps must deliver system design flow at all times, even though a small heating or cooling load exists on the system. Pumps and controls should be selected and installed so pump flow and head match, as closely as is economically feasible, the required head and flow of the system from minimum to maximum load.

Condenser water systems with cooling towers are open loop recirculating systems. The independent head is the static rise to the top of the cooling tower, as shown in Figure 16. The suction conditions of condenser water pumps must be considered to ensure the NPSHA from the system is always greater than the NPSHR by the pumps. To avoid the negative effects of suction, turbulence, and friction, the condenser protection strainer should be placed on the discharge side of the pump. Also, condenser water pipes may become rusted, eroded, or coated with material, so pipe friction must be evaluated on a new and old pipe basis.

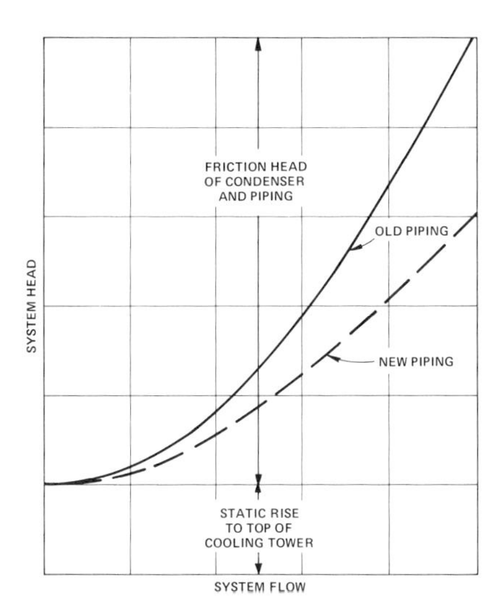


**Fig. 16 System Head Curve for Condenser Water System**

Figure 16 describes system head curves for both new and old pipes, demonstrating that a possible change in the system head curve should be considered as the system ages.

Condensate and boiler feed systems, unlike chilled-, hot-, or condenser water systems, have *flat* system head curves as shown in Figure 17. The boiler pressure is the independent head and, in most cases, is much greater than the system friction head. These systems lend themselves to parallel pumping and seldom require two-speed or variable-speed pumps. Because these units pump hot condensate from open tanks, care should be taken to ensure the NPSHA from the system is always greater than the NPSHR by the pumps.

## ENERGY CONSERVATION IN PUMPING

The continuous operation of many pumps in heating and air

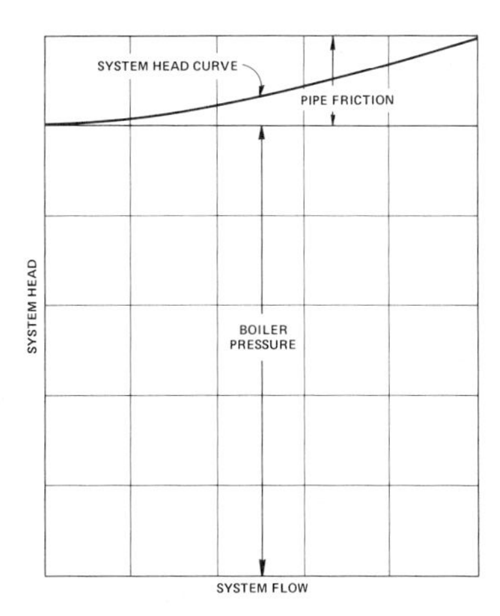


**Fig. 17 System Head Curve for Condensate or Boiler Feed System**

conditioning requires appreciable energy consumption, even with relatively small pumps. Therefore, pump efficiency must be studied when pumps are selected for a particular application.

Economical use of energy depends on the efficiency of pumping equipment and drivers, as well as the use of pumping energy in the water system. To clarify these two phases of energy use, two different efficiencies are used in the pump industry. One is the overall efficiency of the pumping equipment, called the *equipment efficiency* or *wire-to-water efficiency.* For electric motor-driven, constant-speed pumps, the equipment efficiency is the product of the pump efficiency $\eta_p$ and the motor efficiency, $\eta_m$. The equation for equipment efficiency, in percent, is

$$\eta_e = \eta_p \bullet \eta_m /100$$

If the pumps are variable speed, the variable-speed drive efficiency, $\eta_v$, must be included in the equipment efficiency equation, which becomes

$$\eta_e = \eta_p \bullet \eta_m \bullet \eta_v /10{,}000$$

The equipment efficiency shows how much of the energy applied results in useful energy in the water. A more difficult to define efficiency is that of energy use delivered to the water by the pump. To understand the efficiency of utilization, $\eta_u$, for a hydronic system, the perfect pump and control system must be defined. Such a system would deliver the needed amount of water at the right pressure to all parts of the hydronic system at all times. The system head curve would describe the flow-pressure relationship for a hydronic system at all flow rates in the system.

If control valves are used on all heating or cooling coils so that the water flow in the system is the same as the design water flow, and if the pump delivers only the pressure required by the hydronic system at all flow conditions, then the system head curve would represent the perfect hydronic system, and the efficiency of use would be 100%. Unfortunately, calculated system head curves using (1) friction tables for piping and fittings and (2) manufacturers' data for friction losses of control valves and equipment often contain contingency factors. Thus, the calculated friction head is greater than the actual system friction head. The true system head curve, therefore, can be determined only by testing the system after it is installed, which makes exact energy calculations difficult. For practical purposes, the designer of hydronic systems can accept calculated friction losses as those for a perfect hydronic system and use the following equation to determine the efficiency of utilization, in percent:

**Table 4—Pumping System Trouble Analysis Guide**

| Complaint | Possible Cause | Recommended Action |
|---|---|---|
| | Shaft misalignment | • Check and realign. |
| | Worn coupling | • Replace and realign. |
| | Worn pump/motor bearings | • Replace, check manufacturer's lubrication recommendations.<br>• Check and realign shafts. |
| | Improper foundation or installation | • Check foundation bolting or proper grouting.<br>• Check possible shifting because of piping expansion/contraction.<br>• Realign shafts. |
| | Pipe vibration and/or strain caused by pipe expansion/contraction | • Inspect, alter or add hangers and expansion provision to eliminate strain on pump(s). |
| Pump or system noise | Water velocity | • Check actual pump performance against specified and reduce impeller diameter as required.<br>• Check for excessive throttling by balance valves or control valves. |
| | Pump operating close to or beyond end point of performance curve | • Check actual pump performance against specified and reduce impeller diameter as required. |
| | Entrained air or low suction pressure | • Check expansion tank connection to system relative to pump suction.<br>• If pumping from cooling tower sump or reservoir, check line size.<br>• Check actual ability of pump against installation requirements.<br>• Check for vortex entraining air into suction line. |
| | Pump running backward (3-phase) | • Reverse any two-motor leads. |
| | Broken pump coupling | • Replace and realign. |
| | Improper motor speed | • Check motor nameplate wiring and voltage. |
| | Pump (or impeller diameter) too small | • Check pump selection (impeller diameter) against specified system requirements. |
| | Clogged strainer(s) | • Inspect and clean screen. |
| | Clogged impeller | • Inspect and clean. |
| | System not completely filled | • Check setting of PRV fill valve.<br>• Vent terminal units and piping high points. |
| Inadequate or no circulation | Balance valves or isolating valves improperly set | • Check settings and adjust as required. |
| | Air-bound system | • Vent piping and terminal units.<br>• Check location of expansion tank connection line relative to pump suction.<br>• Review provision for air elimination. |
| | Air entrainment | • Check pump suction inlet conditions to determine if air is being entrained from suction tanks or sumps. |
| | Insufficient NPSHR | • Check NPSHR of pump.<br>• Inspect strainers and check pipe sizing and water temperature. |

$$\eta_u = 100\ Q_s H_s / (Q_a H_a)$$

where $Q_s$ and $H_s$ are the flow and head required by the system at a specific heating or cooling load, while $Q_a$ and $H_a$ are the actual flow and head of the pump at that heating or cooling load. This equation demonstrates three-way or bypass valves that waste energy by increasing the flow through the system do reduce

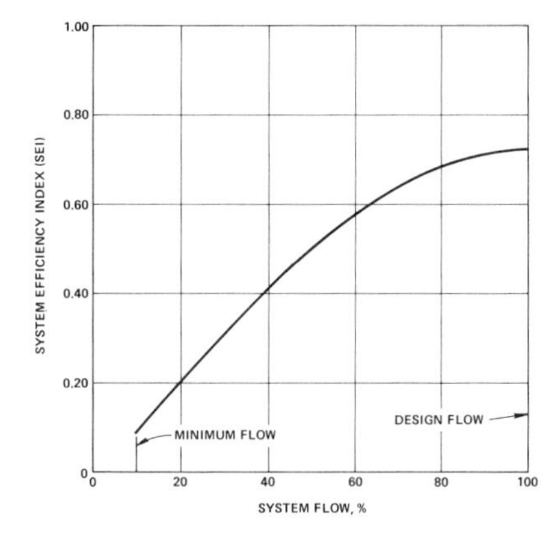


**Fig. 18 Curve for System Efficiency Index (SEI)**

the efficiency of utilization. Pumps reduce this efficiency if they provide more head than is needed by the system. Overflow and overpressure must be avoided to increase the efficiency of utilization.

The hydronic *System Efficiency Index, SEI,* is the useful energy or system load divided by the input power. It is often described as the *wire-to-system* efficiency, which indicates the ratio of the useful energy to the power input to an electric motor. Therefore, the following equation applies for pumping systems.

$$SEI = \eta_\omega \eta_u / 10^4 \text{ with } SEI \text{ expressed as a decimal.}$$

This equation can be a guideline for achieving energy conservation in pumping and control systems. For proper evaluation of a hydronic system, the *SEI* must be calculated from minimum to maximum flow, not just at design condition. Figure 18 describes one *SEI* curve for a hydronic system from minimum to maximum flow. The *SEI* curve may be as smooth as in Figure 18 or it may be jagged, representing changes in pump operation.

For pumping systems in heating and air conditioning, the *SEI* can be as low as 0.10 with maximums of about 0.75 to 0.78. This variation demonstrates the need for energy evaluation of pumping systems to ensure the maximum *SEI* is achieved at all possible loads on the hydronic system, not just at design load.

## INSTALLATION AND OPERATION

The installation of a centrifugal pump should consider ambient, hydraulic, electrical, chemical, metallurgical, and acoustical conditions that exist for a specific installation. The instructions for installation, operation, and maintenance of centrifugal pumps, 14th edition of the *Hydraulic Institute Standards,* provide detailed information that should help the designer of hydronic systems. Table 4 also describes common pumping problems and solutions to them.

## BIBLIOGRAPHY


ASHRAE. 1983. *Water System Design and Retrofit for Energy/Cost Effectiveness.* Professional Development Seminar Text.

Hicks and Edwards. 1970. *Pump Application Engineering.* McGraw-Hill Book Company, Inc., New York.

Hydraulic Institute. 1978. *Hydraulic Institute Standards,* 14th ed. Cleveland, OH.

Karassik, I.J., W.C. Krutzch, J.P. Messina, and W.H. Fraser. 1986. *Pump Handbook,* 2nd ed. McGraw Hill Book Company, Inc., New York.

Stepanoff, A.J. 1957. *Centrifugal and Axial Flow Pumps: Theory, Design and Application,* 2nd ed. John Wiley and Sons, Inc., New York.

# CHAPTER 31

# MOTORS AND MOTOR PROTECTION



THE alternating-current (ac) motor is available in many different types. The direct-current (dc) motor is also used, but to a limited degree. Complete technical information is available on all types of ac and dc motors in NEMA *Standard MG* 1-78, *Motors and Generators.*

## ALTERNATING-CURRENT POWER SUPPLY

Following are important characteristics of an ac power supply: (1) voltage, (2) the number of phases, (3) frequency, (4) voltage regulation, and (5) continuity of power.

Various voltage systems are used, such as 115-V, 230-V, etc. Standard voltage ratings for 60-Hz electric power systems and equipment have been established by the American National Standards Institute (ANSI C84.1-77).

The *nominal voltage* (or *service voltage*) of a circuit or system is the value assigned to the circuit or system to designate its voltage class. It is the voltage at the connection between the electric systems of the supplier and the user. The *utilization voltage* is the voltage at the line terminals of the equipment.

The single-phase and three-phase motor and control voltage ratings shown in Table 1 are adapted to the nominal system voltages indicated. Motors with these ratings are considered suitable for ordinary use on their corresponding systems; for example, a 230-V motor should generally be used on a nominal 240-V system. Operation of 230-V motors on a nominal 208-V system is not recommended because the utilization voltage is commonly below the tolerance on the voltage rating for which the motor is designed. Such operation generally results in excessive overheating and serious reduction in torque.

Motors are usually guaranteed to operate satisfactorily and to deliver their full power at the rated frequency and at a voltage 10% above or below rating, or at the rated voltage and plus or minus 5% frequency variation. Table 2 shows the effect of voltage and frequency variation on induction-motor characteristics.

The phase voltages of three-phase motors should be balanced. If not, a small voltage imbalance can produce a greater current imbalance and a much greater temperature rise, which can result in nuisance overload trips or motor failures. Motors should not be operated where the voltage imbalance is greater than 1% without checking with the manufacturer. Voltage imbalance is defined in the NEMA *Standard* MG 1-78, para. 14.34 as:

$$\% \text{ Voltage Imbalance} = 100\left[\frac{\text{Maximum Voltage Deviation from Average Voltage}}{\text{Average Voltage}}\right]$$

In addition to voltage imbalance, current imbalance can be present in a system where Y-Y transformers without tertiary windings are used, even if the voltage is in balance. As stated previously, this current imbalance is not desirable. If this current imbalance exceeds either 10% or the maximum imbalance recommended by the manufacturer, corrective action should be taken (see NFPA 70-84).

$$\% \text{ Current Imbalance} = 100\left[\frac{\text{Maximum Current Deviation from Average Current}}{\text{Average Current}}\right]$$

Another cause of current imbalance is normal winding impedance imbalance, which adds or subtracts from the current imbalance caused by voltage imbalance.

**Table 1 Motor and Motor-Equipment Voltages**

| Applicable to All Nominal System Voltages Containing This Voltage | All Motor and Motor-Control Equipment Nameplate Voltage Ratings Containing This Voltage | | | |
|---|---|---|---|---|
| | Integral hp (kW) | | Fractional hp (small kW) | |
| | Three-Phase | Single-Phase | Three-Phase | Single-Phase |
| 120 | — | 115 | — | 115 |
| 208 | 200 | — | 200 | — |
| 240 | 230 | 230 | 230 | 230 |
| 277 | — | 265 | — | 265 |
| 480 | 460 | — | 460 | — |
| 600[a] | 575 | | 575 | — |
| 2400 | 2300 | — | — | — |
| 4160 | 4000 | — | — | — |
| 4800 | 4600 | — | — | — |
| 6900 | 6600 | — | — | — |
| 13,800 | 13,200 | — | — | — |

[a]Certain control and protective equipment have a maximum voltage limit of 600 V; the manufacturer or power supplier or both should be consulted to ensure proper application.

The preparation of this chapter is assigned to TC 8.11, Electric Motors—Open and Hermetic.

**Table 2 Effect of Voltage and Frequency Variation on Induction Motor Characteristics[a]**

| Voltage and Frequency Variation | | Starting and Maximum Running Torque | Synchronous Speed | % Slip | Full-Load Speed | Efficiency | | |
|---|---|---|---|---|---|---|---|---|
| | | | | | | Full Load | 0.75 Load | 0.5 Load |
| Voltage Variation | 120% Voltage | Increase 44% | No change | Decrease 30% | Increase 1.5% | Small increase | Decrease 0.5 to 2% | Decrease 7 to 20% |
| | 110% Voltage | Increase 21% | No change | Decrease 17% | Increase 1% | Increase 0.5 to 1% | Practically no change | Decrease 1 to 2% |
| | Function of voltage | $(\text{Voltage})^2$ | Constant | $\frac{1}{(\text{Voltage})^2}$ | (Synchronous speed slip) | — | — | — |
| | 90% Voltage | Decrease 19% | No change | Increase 23% | Decrease 1.5% | Decrease 2% | Practically no change | Increase 1 to 2% |
| Frequency Variation | 105% Frequency | Decrease 10% | Increase 5% | Practically no change | Increase 5% | Slight increase | Slight increase | Slight increase |
| | Function of frequency | $\frac{1}{(\text{Frequency})^2}$ | Frequency | — | (Synchronous speed slip) | — | — | — |
| | 95% Frequency | Increase 11% | Decrease 5% | Practically no change | Decrease 5% | Slight decrease | Slight decrease | Slight decrease |

| Voltage and Frequency Variation | | Power Factor | | | Full-Load Current | Starting Current | Temperature Rise, Full Load | Maximum Overload Capacity | Magnetic Noises, No Load in Particular |
|---|---|---|---|---|---|---|---|---|---|
| | | Full Load | 0.75 Load | 0.5 Load | | | | | |
| Voltage Variation | 120% Voltage | Decrease 5 to 15% | Decrease 10 to 30% | Decrease 15 to 40% | Decrease 11% | Increase 25% | Decrease 9 to 11°F (5 to 6°C) | Increase 44% | Noticeable increase |
| | 110% Voltage | Decrease 3% | Decrease 4% | Decrease 5 to 6% | Decrease 7% | Increase 10 to 12% | Decrease 5.4 to 7.2°F (3 to 4°C) | Increase 21% | Increase slightly |
| | Function of voltage | — | — | — | — | Voltage | — | $(\text{Voltage})^2$ | — |
| | 90% Voltage | Increase 3% | Increase 2 to 3% | Increase 4 to 5% | Increase 11% | Decrease 10 to 12% | Increase 11 to 13°F (6 to 7°C) | Decrease 19% | Decrease slightly |
| Frequency Variation | 105% Frequency | Slight increase | Slight increase | Slight increase | Decrease slightly | Decrease 5 to 6% | Decrease slightly | Decrease slightly | Decrease slightly |
| | Function of frequency | — | — | — | — | $\frac{1}{\text{Frequency}}$ | — | — | — |
| | 95% Frequency | Slight decrease | Slight decrease | Slight decrease | Increase slightly | Increase 5 to 6% | Increase slightly | Increase slightly | Increase slightly |

[a]These variations are general and will differ somewhat for specific ratings.

## CODES AND STANDARDS

*Codes* are requirements for public health or safety; *standards* are uniform methods of rating, sizing, and measuring the performance of electrical equipment to provide a guideline for use and comparison. Codes and standards are sponsored at a national level and are usually voluntary. Those for electrical equipment used in the air-conditioning and refrigerating industries are listed in Chapter 45 of this volume. The *National Electrical Code* (NEC) (NFPA 70-87) and the *Canadian Electrical Code,* Part I (CSA *Standard* C22.1-82) are important in the United States and Canada.

The *National Electrical Code* contains the minimum recommendations considered necessary to ensure safety of electrical installations and equipment. It is also referred to in Sub-Part *S* (Electrical) of the Occupational Safety and Health Acts (OSHA) of 1970 and, therefore, becomes part of the OSHA requirements where they apply. In addition, practically all communities in the United States have adopted it as a minimum electrical code.

Underwriters Laboratories, Inc. (UL) promulgates standards for various types of equipment. Underwriters' standards for electrical equipment cover construction and performance for the safety of such equipment and interpret requirements to ensure compliance with the intent of the NEC. A complete list of available standards may be obtained from UL, which also publishes lists of equipment that comply with their standards. These listed products bear the UL label and are recognized by local authorities.

The *Canadian Electrical Code,* Part I, is a standard of the Canadian Standards Association. It also is a voluntary code with minimum requirements for electrical installations in buildings of every kind.

The *Canadian Electrical Code,* Part II, contains specifications for the construction and performance of electrical equipment, in compliance with Part I. Underwriters Laboratories' standards and standards of the Canadian Standards Association for electrical equipment are similar, so equipment designed to meet the requirements of one code may also meet the requirements of the other. However, there is not complete agreement between the codes, so individual standards must be checked when designing equipment for use in both countries. The Canadian Standards Association examines and tests material and equipment for compliance with the *Canadian Electrical Code.*

Copies of electrical standards may be obtained directly from the sponsoring organizations (see Chapter 45 of this volume).

## MOTOR EFFICIENCY

The many factors affecting motor efficiency include (1) sizing of the motor to the load, (2) type of motor specified, (3) motor design speed, and (4) type of bearing specified.

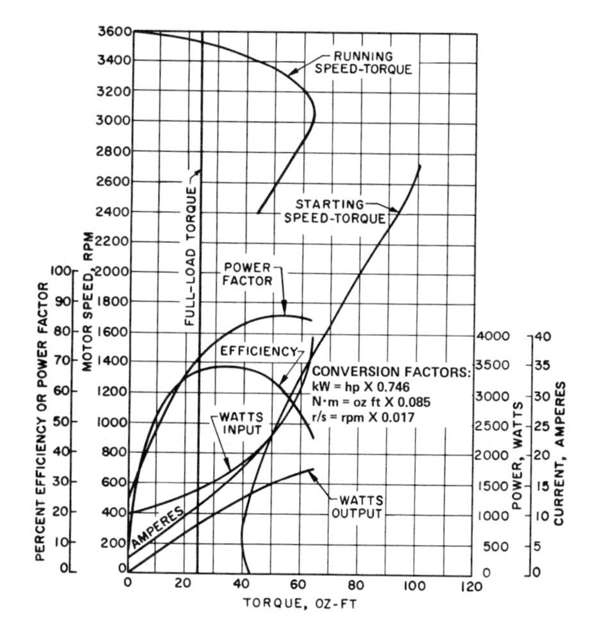


**Fig. 1 Typical Performance Characteristics of Capacitor Start/Induction-Run Two-Pole General Purpose Motor, 1 hp ( 0.75kW)**

Oversizing of motor to load may result in inefficiency. As shown in the performance characteristic curves for single-phase motors in Figures 1, 2, and 3, the efficiency falls off rapidly at loads lighter than the rated full load. Polyphase motors usually reach peak efficiency at loads slightly lighter than full load (see Figure 4). Motor performance curves are available from the motor manufacturer and may help in applying motors optimally for an application. Larger output motors are more efficient at rated load than are smaller motors. Also, higher speed induction motors are more efficient.

The type of motor specified is significant because a permanent split-capacitor motor is more efficient than a shaded-pole fan motor. A capacitor-start/capacitor-run motor is more efficient than either a capacitor-start or a split-phase motor. In polyphase motors, the lower the locked rotor torque specified, the higher the efficiency obtained in the design.

The motor industry now offers high-efficiency design motors. These generally incorporate the more efficient type of motor for an application and more material than a standard efficiency design of the same type. More information on motor efficiency can be found in NEMA *Standards* No. MG 10-83 and No. MG 11-77.

## NON-HERMETIC MOTORS

### Types

The electrical industry classifies motors as *small kilowatt* (*fractional horsepower*) or *integral kilowatt* (*integral horsepower*). [Note: in this context, *kilowatt* refers to power output of the motor.] Small kilowatt motors have ratings of less than 1 hp (0.746 kW) at 1700 to 1800 rpm (28.3 to 30 r/s) for four-pole and 3500 to 3600 rpm (58.3 to 60 r/s) for two-pole machines. These motors are available in a wide variety of types and sizes, depending on their application.

Single-phase motors are available through 5 hp (3.7 kW) and are most common through 0.75 hp (0.56 kW). Motors larger than 0.75 hp (0.56 kW) are usually polyphase.

Table 3 lists motors by types indicating the normal kilowatt range and the type of power supply used. All motors listed are suitable for either direct or belt drive, except shaded-pole motors (limited by low starting torque). Application of permanent split-capacitor motors to belt drives must be done carefully because of lower starting torque.

### Application

When applying an electric motor, the following characteristics are important: (1) mechanical arrangement, including position of the motor and shaft, type of bearing, portability desired, drive connection, mounting, and space limitations; (2) speed range desired; (3) power requirement; (4) torque; (5) inertia; (6) frequency of starting; and (7) ventilation requirements. Motor characteristics that are frequently applied are generally presented in curves (see Figures 1 through 4).

**Torque.** The torque required to operate the driven machine at all times between initial breakaway and final shutdown is important in determining the type of motor. The torque available at zero speed or standstill (the *starting torque*) may be less than 100% or as high as 400% of full-load torque, depending on motor design. The *starting current,* or *locked-rotor current*, is usually 400 to 600% of the current at rated full load.

**Full-load torque** is the torque developed to produce the rated hp (kW) at the rated speed. **Full-load speed** also depends on the design of the motor. For induction motors, a speed of 1725 rpm (28.8 r/s) is typical for four-pole motors and a speed of 3450

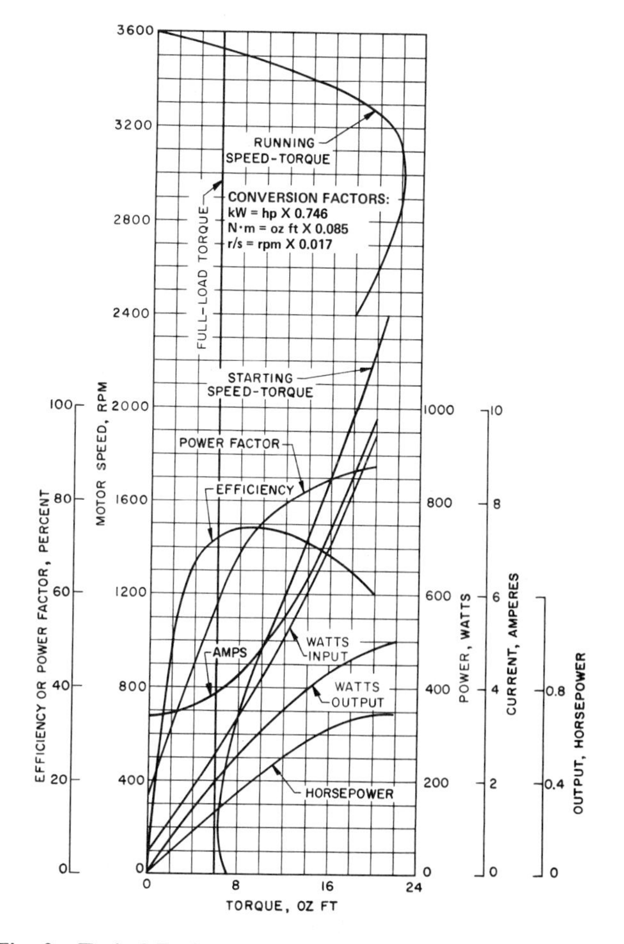


**Fig. 2 Typical Performance Characteristics of a Resistance-Start Split-Phase Two-Pole Hermetic Motor, 0.25 hp (0.19 kW)**

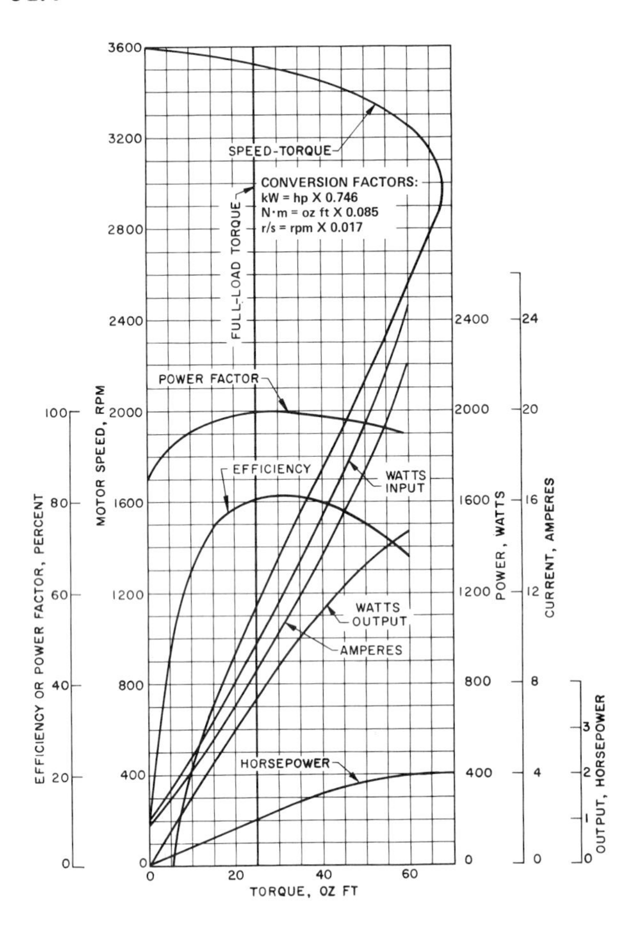


Fig. 3 Typical Performance Characteristics of a Permanent Split-Capacitor Two-Pole Motor, 1 hp (0.75 kW)

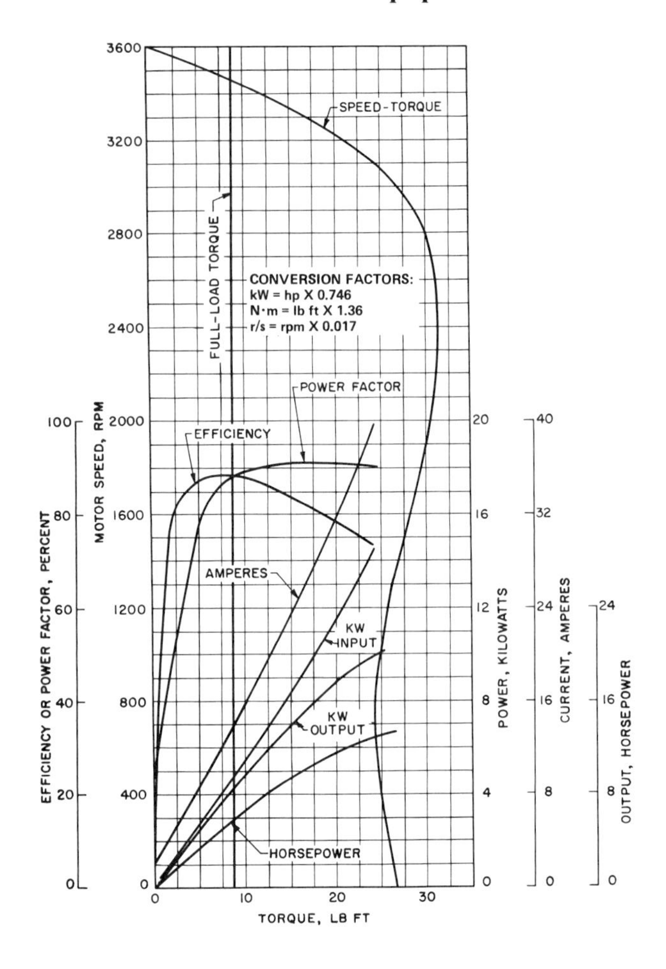


Fig. 4 Typical Performance Characteristics of a Polyphase Two-Pole Motor, 5 hp (3.7 kW)

rpm (57.5 r/s) is typical for two-pole motors (60 Hz).

Motors have a **maximum** or **breakdown torque**, which cannot be exceeded. The relation between breakdown torque and full-load torque varies widely, depending on motor design.

**Power.** The power delivered by a motor is a product of its torque and speed. Since a given motor delivers increasing power up to maximum torque, a basis for power rating is needed. The National Electrical Manufacturers Association (NEMA) bases power rating (hp or kW) on breakdown torque limits for single-phase motors, 10 hp (7.5 kW) and less. All others are rated at their power capacity within voltage and temperature limits as listed by NEMA.

Full-load rating is based on maximum winding temperature. If the nameplate marking includes the maximum ambient temperature for which the motor is designed and the insulation system designation, the maximum temperature rise of the winding may be determined from the appropriate section of NEMA *Standard* MG-1.

**Service factor** is the maximum overload that can be applied to general-purpose motors and certain definite purpose motors without exceeding the temperature limitation of the insulation system. When the voltage and frequency are maintained at the values specified on the nameplate and the ambient temperature does not exceed 104°F (40°C), the motor may be loaded up to the power obtained by multiplying the rated hp (kW) by the service factor shown on the nameplate.

The power rating is normally established on the basis of tests run in still air. However, most direct-drive air-moving applications are checked with air over the motor. If motor nameplate marking does not specify a service factor, see the appropriate section of NEMA *Standard* MG-1. Characteristics of ac motors are given in Table 4.

## HERMETIC MOTORS

A hermetic motor consists of a stator and a rotor, without shaft, end shields, or bearings, for installation in hermetically

**Table 3 Motor Types**

| Type | Range hp (kW) | Type of Power Supply |
|---|---|---|
| *Fractional Sizes:* | | |
| Split-phase | 0.05-0.5 (0.04-0.4) | single-phase |
| Capacitor-start | 0.05-1.5 (0.04-1.1) | single-phase |
| Repulsion-start | 0.13-1.5 (0.1-1.1) | single-phase |
| Permanent split-capacitor | 0.05-1.5 (0.04-1.1) | single-phase |
| Shaded-pole | 0.01-0.25 (0.007-0.19) | single-phase |
| Squirrel cage induction | 0.17-1.5 (0.13-1.1) | polyphase |
| Direct current | 0.5-1.5 (0.4-1.1) | dc |
| *Integral Sizes:* | | |
| Capacitor-start/capacitor-run | 1-5 (0.75-3.7) | single-phase |
| Capacitor-start | 1-5 (0.75-3.7) | single-phase |
| Squirrel cage induction (normal torque) | 1-up (0.75-up) | polyphase |
| Slip-ring | 1-up (0.75-up) | polyphase |
| Direct current | 1-up (0.75-up) | dc |
| Permanent split-capacitor | 1-5 (0.75-3.7) | single-phase |

Table 4 Characteristics of AC and DC Motors (Non-hermetic)

| | Split-Phase | Permanent Split-Capacitor | Capacitor-Start Induction-Run | Capacitor-Start Capacitor-Run | Shaded-Pole | Polyphase, 60-Hz |
|---|---|---|---|---|---|---|
| Connection diagram | RUN START | RUN CAP. START | RUN CAP. START | RUN START | RUN | |
| Speed torque curves | RUN START F.L. 250% 125% | RUN F.L. 250% 25% | RUN START F.L. 250% 350% | RUN START F.L. 275% | RUN F.L. 125% 25% | START & RUN |
| Starting method | Centrifugal Switch | None | Centrifugal Switch | Centrifugal Switch | None | Motor Controller |
| Ratings, hp (kW) | 0.05-0.5 (0.04-0.4) | 0.05-5 (0.04-3.7) | 0.05-5 (0.04-3.7) | 0.05-5 (0.04-3.7) | 0.01-0.25 (0.04-0.19) | 0.5 up (0.4 up) |
| Full-load speeds at 60-Hz (two-pole, four-pole) | 3450-1725 | 3450-1725 | 3450-1725 | 3500-1750 | 3100-1550 1000 | 3500-1750 |
| Torque[a] Locked rotor Breakdown | 125-150% 250-300% | 25% 250-300% | 250-350% 250-300% | 250% 250% | 25% 125% | 150-350% 250-350% |
| Speed classification | Constant | Constant | Constant | Constant | Constant or adjustable | Constant |
| Full-load power factor | 60% | 95% | 65% | 95% | 60% | 80% |
| Efficiency | Medium | High | Medium | High | Low | High-Medium |

[a]Expressed as percent of rated horsepower torque.

sealed refrigeration compressor units. With the motor and compressor sealed in a common chamber, the insulation system must be impervious to the action of the refrigerant and lubricating oil. Hermetic motors are used in both welded and accessible hermetic (semihermetic) compressors.

## Application

**Domestic Refrigeration.** Hermetic motors up to 0.33 hp (0.25 kW) are used. They are split-phase, permanent split-capacitor, or capacitor-start motors for medium or low starting torque compressors and capacitor-start and special split-phase motors for high starting torque compressors.

**Room Air Conditioners.** Motors from 0.33 to 3 hp (0.25 to 2.24 kW) are in use. They are permanent split-capacitor or capacitor-start/capacitor-run types. These designs have high power factor and efficiency and meet the need for low current draw, particularly on 115-V circuits.

**Central Air-Conditioning Systems** (including **Heat Pumps**). Both single-phase (5 hp [3.7 kW] and below) and polyphase (1.5 hp [1.1 kW] and above) motors are used. The single-phase motors are permanent split-capacitor or capacitor-start/capacitor-run types.

**Small Commercial Refrigeration.** Practically all are below 5 hp (3.7 kW), with single-phase being the most common. Capacitor-start/induction-run motors are normally used up to 0.75 hp (0.56 kW) because of starting torque requirements. Capacitor-start/capacitor-run motors are used for larger sizes because they provide high starting torque and high full-load efficiency and power factor.

**Large Commercial Refrigeration.** Most motors are three-phase and larger than 5 hp (3.7 kW).

Horsepower (kilowatt) ratings of motors for hermetic compressors do not necessarily have a direct relationship to the thermodynamic output of a system. Designs are tailored to match the compressor characteristics and specific applications. For additional information pertaining to hermetic motor application, see Chapter 12 of this volume.

## INTEGRAL THERMAL PROTECTION

The *National Electrical Code* and UL standards cover motor protection requirements. Separate, external protection devices include the following:

**Thermal protectors** are protective devices for assembly as an integral part of a motor or hermetic refrigerant motor compressor. They protect the motor against overheating caused by overload, failure to start, or excessive operating current. Thermal protectors are required to protect polyphase motors from overheating because of an open phase in the primary circuit of the supply transformer. Thermal protection is accomplished by either line-break devices or control circuit thermal sensing systems.

The protection of hermetic motor-compressors has some unique aspects compared to non-hermetic motor protection. The refrigerant cools reciprocating compressors, so the thermal protector may be required to prevent overheating from loss of refrigerant charge, low suction pressure and high superheat at the compressor, obstructed suction line, or malfunction of the condensing means.

The *National Electrical Code, Article* 440, limits maximum continuous currents on motor-compressors to 156% of rated load current if an integral thermal protector is used. NEC *Article* 430 limits maximum continuous current on non-hermetic motors to different percentages of full-load current as a function of size. If separate overload relays or fuses are used for protection, *Article* 430 limits maximum continuous current respectively to 140% and 125% of rated load.

Underwriters Laboratories (UL) *Standard* 984 specifies that the compressor enclosure must not exceed 300°F (150°C) under any conditions. The motor winding temperature limits are set by the compressor manufacturer based on individual compressor design considerations. UL *Standard* 547 sets the limits for motor winding temperature for open motors as a function of the class of the motor insulation used.

**Line-break thermal protectors,** integral with a motor or motor-compressor, sensing both current and temperature, are connected electrically in series with the motor; their contacts interrupt the total motor line-current. These protectors are used in fractional and small integral single-phase and polyphase motors up through 15 hp (11 kW).

Protectors installed inside a motor-compressor are hermetically sealed, since exposed arcing cannot be tolerated in the presence of refrigerant. They provide better protection than the external type for loss of charge, obstructed suction line, or low voltage on the stalled rotor. This is due to low currents associated with these fault conditions, hence the need to sense the motor temperature increase by thermal contact. Protectors used inside the compressor housing must withstand pressure requirements established by UL.

Protectors mounted externally on motor-compressor shells, sensing only shell temperature and line current, are typically used on smaller compressors, such as those used in household refrigerators and small room air conditioners. One benefit occurs during high head pressure starting conditions, which can occur if voltage is lost momentarily or if the user inadvertently turns off the compressor with the temperature control and then turns it back on immediately. Usually, these units will not start under these conditions. When this happens, the protector takes the unit off the line and resets automatically when the compressor cools and pressures have equalized to a level that allows the compressor to start.

Protectors installed in non-hermetic motors may be attached to the winding or may be mounted off the windings but within the motor housing. Those protectors placed on the winding are generally installed prior to varnish dip and bake, and their construction must prevent varnish from entering the contact chamber.

Since the protector carries full motor line current, its size is based on adequate contact capability to interrupt the stalled current of the motor on continuous cycling for periods specified in UL *Standards* 984 and 547.

The compressor or motor manufacturer applies and selects appropriate motor protection in cooperation with the protector manufacturer. Any change in protector rating, by other than the specifying manufacturer after the proper application has been made, may result in either overprotection and frequent nuisance tripouts or underprotection and burnout of the motor windings. Connections to protector terminals, including lead wire sizes, should not be changed, and no additional connections should be made to the terminals. Any change in connection changes the terminal conditions and affects protector performance.

Control circuit thermal protection systems approved for use with a motor or motor compressor, either sensing both current and temperature or sensing temperature only, are used with integral hp (integral kW) single-phase and three-phase motors.

The current and temperature system uses bimetallic temperature sensors installed in the motor winding in conjunction with thermal overload relays. The sensors are connected in series with the control circuit of a magnetic contactor that interrupts the motor current. Thermostat sensors of this type, depending on their size and mass, are capable of tracking motor winding temperature for running overloads. On locked rotor, in cases where the rate of change in winding temperature is rapid, the temperature lag is usually too great for such sensors to provide protection when they are used alone. However, when they are used in conjunction with separate thermal overload or magnetic time-delay relays that sense motor current, the combination provides protection; on locked rotor, the thermal or magnetic relays protect for the initial heating cycle, and the combined functioning of relay and thermostat protects for subsequent cycles.

Thermistor sensors may also be used with electronic current-sensing devices. The temperature-only system uses sensors that undergo a change in resistance with temperature. The resistance change provides a switching signal to the electronic circuit, whose output is in series with the control circuit of a magnetic contactor used to interrupt the motor current. The output of the electronic protection circuitry (module) may be an electromechanical relay or a power triac. The sensors may be installed directly on the stator winding end turns or buried inside the windings. Their small size and good thermal transfer allow them to track the temperature of the winding for locked rotor, as well as running overload.

The sensors may be of three types. One type uses a ceramic material with positive temperature coefficient of resistance; the material exhibits a large abrupt change in resistance at a particular design temperature. This change occurs at what is known as the *anomaly point* and is inherent in the sensor. The anomaly point remains constant once the sensor is manufactured; sensors are produced with anomaly points at different temperatures to meet different application requirements. However, a single module calibration can be supplied for all anomaly temperatures of a given sensor type.

Another sensor type uses a metal wire, which has a linear increase in resistance with temperature. The sensor assumes a specific value of resistance corresponding to each desired value of response or operating temperature. It is used with an electronic protection module calibrated to a specific resistance. Modules supplied with different calibrations are used to achieve various values of operating temperature.

A third type is a negative temperature coefficient of resistance sensor, which is integrated with electronic circuitry similar to that used with the metal wire sensor.

More than one sensor may be connected to a single electronic module in parallel or series, depending on design. However, the sensors and modules must be of the same system design and intended for use with the particular number of sensors installed and the wiring method employed. Electronic protection modules must be paired only with sensors specified by the manufacturer, unless specific equivalency is established and identified by the motor or compressor manufacturer.

## MOTOR CONTROL

In general, motor control equipment may (1) disconnect the motor and controller from the power supply, (2) start and stop the motor, (3) protect against short circuits, (4) protect from overheating, (5) protect the operator, (6) control motor speed, and (7) protect motor branch circuit conductors and control apparatus.

### Separate Motor Protection

Most air-conditioning and refrigeration motors or motor compressors, whether open or hermetic, are equipped with integral motor protection by the equipment manufacturer. If this is not the case, separate motor-protection devices, sensing current only, must be used. These consist of thermal or magnetic relays, similar to those used in industrial control, that provide running overload and stalled-rotor protection. Because hermetic motor windings heat rapidly due to the loss of the cooling effect of refrigerant gas flow when the rotor is stalled, *quick-trip* devices must be used.

Thermostats or thermal devices are sometimes used to supplement current sensing devices. Such supplements are necessary (1) when automatic restarting is required after trip or (2) to protect from abnormal running conditions that do not increase motor current. These devices are covered in the section "Thermal Protection Integral to the Motor or Motor-Compressor."

## Protection of Control Apparatus and Branch Circuit Conductors

In addition to protection for the motor itself, Articles 430 and 440 of the *National Electric Code* require the control apparatus and branch circuit conductors to be protected from overcurrent resulting from motor overload or failure to start. This protection can be given by some thermal protective systems that do not permit a continuous current in excess of required limits. In other cases, a current-sensing device, such as an overload relay, a fuse, or a circuit breaker, is used.

**Circuit breakers** are used for disconnecting, as well as circuit protection and are available in ratings for use with small household refrigerators, as well as in large commercial installations. Manual switches for disconnecting and fuses for short-circuit protection are also used. For single-phase motors up to 3 hp (2.2 kW), 230-V, an attachment plug is an acceptable disconnecting device.

**Controllers.** The motor control used is determined by the size and type of motor, the power supply, and the degree of automation. Control may be manual, semiautomatic, or fully automatic.

Central system air conditioners are generally located at a distance from the controlled space environment control, such as room thermostats and other control devices. Therefore, **magnetic controllers** must be used in these installations. Also, all dc and all large ac installations must be equipped with in-rush **current-limiting controllers,** which are discussed later in this chapter. **Synchronous motors** are sometimes used to improve the power factor. **Multispeed motors** provide flexibility for many applications.

**Manual control** for an ac or dc motor is usually located near the motor. When so located, an operator must be present to start and stop or change the speed of the motor by adjusting the control mechanism.

Manual control is the simplest and least expensive control method for small ac motors, both single-phase and polyphase, but it is seldom used with hermetic motors. The manual controller usually consists of a set of main line contacts, which are provided with thermal overload relays for motor protection.

Manual speed controllers can be used for large air-conditioning systems using **slip-ring motors;** they may also provide reduced-current starting. Different speed points are used to vary the amount of cooling provided by the compressor.

**Across-the-line magnetic controllers** are widely used in central air-conditioning systems. They are applicable to motors of all sizes, provided power supply and motor are suitable to this type of control. Across-the-line magnetic starters may be used with automatic control devices for starting and stopping. Where push buttons are used, they may be wired for either low-voltage release or low-voltage protection.

**Full-voltage starting** for motors is preferable because of its lower initial cost and simplicity of control. Except for dc machines, most motors are mechanically and electrically designed for full-voltage starting. The starting in-rush current, however, is limited in many cases by power company regulations made because of voltage fluctuations, which may be caused by heavy current surges. It is, therefore, often necessary to reduce the starting current below that obtained by across-the-line starting in order to meet the limitations of power supply. One of the simplest means of accomplishing this reduction is by the use of resistors in the primary circuit. As the motor accelerates, the resistance is cut out by the use of timing or current relays.

Another method of reducing the starting current for an ac motor uses an *autotransformer* motor controller. Starting voltage is reduced, and, when the motor accelerates, it is disconnected from the transformer and connected across-the-line by means of timing or current relays. Primary resistor starters are generally smaller and less expensive than autotransformer starters for motors of moderate size. However, primary resistor starters require more line current for a given starting torque than do autotransformer starters.

**Star-delta motor controllers** limit current efficiently, but they require motors designed for this type of starting. They are particularly suited for centrifugal, rotary screw, and reciprocating compressor drives starting without load.

**Part-winding motor controllers,** sometimes called **incremental start controllers,** are used to limit line disturbances by connecting only part of the motor winding to the line and connecting the second motor winding to the line after a time interval of one to three seconds. If the motor is not heavily loaded, it accelerates when the first part of the winding is connected to the line; if it is too heavily loaded, the motor may not start until the second winding is connected to the line. In either case, the voltage *dip* will be less than the dip that would result if a standard squirrel cage motor with a cross-the-line starter were used. Part-winding motors may be controlled either manually or magnetically. The magnetic controller consists of two contactors and a timing device for the second contactor.

**Multispeed motor controllers.** Multispeed motors provide flexibility in many types of drives in which variation in capacity is needed. Two types of multispeed motors are used: (1) motors with one reconnectable winding and (2) motors with two separate windings. Motors with separate windings need a contactor for each winding and only one contactor can be closed at any time. Motors with a reconnectable winding are similar to motors with two windings, but the contactors and motor circuits are different.

**Slip-Ring motor controllers.** Slip-ring ac motors provide variable speed. The *wound rotor* of these motors functions in the same manner as in the squirrel cage motor, except that the rotor windings are connected through slip rings and brushes to external circuits with resistance to vary the motor speed. Increasing the resistance in the rotor circuit reduces motor speed, and decreasing the resistance increases motor speed. When the resistance is shorted out, the motor operates with maximum speed, efficiency, and power factor. On some large installations, manual drum controllers are used as speed-setting devices. Complete automatic control can be provided with special control devices for selecting motor speeds.

**Controllers for direct-current motors.** These motors have favorable speed-torque characteristics, and their speed is easily controlled. Controllers for dc motors are more expensive than those for ac motors, except for very small motors. Large dc motors are started with resistance in the armature circuit, which is reduced step by step until the motor reaches its base speed. Higher speeds are provided by weakening the motor field.

## Single-Phase Motor-Starting Methods

Motor-starting switches and relays for *single-phase motors* must provide a means for disconnecting the *starting winding* of split-phase or capacitor-start/induction-run motors or the start capacitor of capacitor-start/capacitor-run motors. Open machines usually have a centrifugal switch mounted on the motor shaft, which disconnects the starting winding at about 70% of full-load speed.

The starting methods by use of relays are as follows:

**Thermally-Operated Relay.** When the motor is started, a contact that is normally closed applies power to the starting winding. A thermal element that controls these contacts is in series with the motor and carries line current. Because of the current flow through this element, it is heated until, after a definite period of time, it is warmed sufficiently to open the contacts and remove power from the starting winding. The running current then heats the element enough to keep the contacts open. The setting of the time for the starting contacts to open is determined by tests on the system components, *i.e.,* the relay, the motor, and the compressor, and is based on a prediction of the time delay required to bring the motor up to speed.

An alternate form of a thermally operated relay is a PTC (*Positive Temperature Coefficient* of resistance) starting device. This uses a ceramic element of low resistance, at room temperature, whose value increases about 1000 times the room temperature resistance when it is heated to a predetermined temperature. It is placed in series with the start winding of split-phase motors and allows current flow when power is applied, permitting the motor to start. After a definite period of time, the self-heating of the PTC resistive element causes it to reach its high-resistance state, reducing current flow in the start winding to milliamp level during running. The residual current maintains the PTC element in the high-resistance state while the motor is running. A PTC starting device may also be connected in parallel with a run capacitor and the combination may be connected in series with the starting winding. It allows the motor to start like a split-phase motor and, when the PTC element reaches the high-resistance state, it allows the motor to operate as a capacitor-run motor.

**Current-Operated Relay.** In this type of connection, a relay coil carries the line current going to the motor. When the motor is started, the inrush current to the running winding passes through the relay coil, causes the normally open contacts to close, and applies power to the starting winding. As the motor comes up to speed, the current decreases until, at a definite calibrated value of current corresponding to a preselected speed, the magnetic force of the coil diminishes to a point that allows the contacts to open to remove power from the starting winding. This relay takes advantage of the *main winding current* versus *speed* characteristics of the motor. The current-speed curve varies with line voltage, so that the starting relay must be selected for the voltage range likely to be encountered in service. Ratings established by the manufacturer should not be changed because this may result in undesirable starting characteristics. They are selected to disconnect the starting winding or start capacitor at approximately 70 to 90% of synchronous speed for four-pole motors.

**Voltage-Operated Relay.** Capacitor-start and capacitor-start/capacitor-run hermetically sealed motors above 0.5 hp (0.373 kW) are usually started with a normally closed contact voltage relay. In this method of starting, the relay coil is connected in parallel with the starting winding. When power is applied to the line, the relay does not operate because it is calibrated to operate at a higher voltage. As the motor comes up to speed, the voltage across the starting winding and relay coil increases in proportion to the motor speed. At a definite voltage corresponding to a preselected speed, the relay operates and opens its contacts, thereby opening the starting-winding circuit or disconnecting the starting capacitor. The relay then keeps these contacts open because there is sufficient voltage induced in the starting winding, when the motor is running, to hold the relay in the open contact position.

## REFERENCES

ANSI. 1982. *Electrical Power Systems and Equipment—Voltage Ratings (60 Hz).* ANSI C84.1-82. American National Standards Institute, New York.

CSA. 1982. *Canadian Electrical Code.* Canadian Standards Association *Standard* C22.1-82, p. 8. Rexdale, Ontario, Canada.

IEEE. 1986. *Procedures for Testing Single-Phase and Polyphase Induction Motors for Use in Hermetic Compressors.* Institute of Electrical and Electronic Engineers *Standard* 839-1986. New York.

NEMA. 1977. *Energy Management Guide for Selection and Use of Single-Phase Motors.* MG 11-77. National Electrical Manufacturers Association, Washington, DC.

NEMA. 1978. *Motors and Generators.* MG 1-78. National Electrical Manufacturers Association, Washington, DC.

NEMA. 1983. *Energy Management Guide for Selection and Use of Polyphase Motors.* National Electrical Manufacturers Association *Standard* MG 10-83. Washington, DC.

NFPA. 1987. *National Electrical Code.* NFPA-70-87. National Fire Protection Association, Quincy, MA.

## CHAPTER 32

# ENGINE AND TURBINE DRIVES



---

THIS chapter addresses drives (other than electric motors) for reciprocating and centrifugal compressors, including gas engines (Part I) and steam and gas turbines (Part II). Engines are also used to drive generators in standby power and continuous-duty power applications, either with or without the useful application of recoverable engine heat (cogeneration). These systems are covered in Chapter 8 of the 1987 HVAC Volume. Electric motors are detailed in Chapter 31 of this volume.

# PART I: ENGINES

## ENGINE APPLICATIONS

### Engine-Driven Reciprocating Compressors

Engine-driven, reciprocating-compressor, water-chiller units are usually field-assembled from commercially available equipment for comfort service, low-temperature refrigeration, and heat pump applications. Both direct-expansion coils and liquid chillers are used. Some models achieve a low operating cost and a high degree of flexibility by combining speed variation with cylinder unloading. These units achieve capacity control by reducing engine speed to about 30 to 50% of rated speed; further capacity modulation may be achieved by unloading the compressor in increments. Engine speed should not be reduced below the minimum that the manufacturer specifies is required for adequate lubrication.

Most engine-driven reciprocating compressors are equipped with a cylinder loading mechanism for idle (unloaded) starting. This arrangement may be required because the starter may not have sufficient torque to crank both the engine and the loaded compressor. With some compressors, not all the cylinders unload (*e.g.*, four out of twelve) and, in this case, a bypass valve has to be installed for a fully unloaded start. The engine first speeds to one-half or two-thirds of full speed. Then, a gradual cylinder load is added and the engine speed increases over a period of 120 to 180 seconds. In some applications, such as an engine-driven heat pump, low-speed starting may cause oil accumulation and sludge. As a result, a high-speed start is required.

These systems operate on approximately 8 to 13 ft$^3$ of 1000 Btu/ft$^3$ natural gas per hp-h (84 to 137 mL/s of pipeline quality, [37.3 kJ/L] natural gas per kW) in sizes down to 25 tons (88 kW). Comparable heat rates for diesel engines run from 7000 to 9000 Btu/hp-h (2.7 to 3.5 W/W). Smaller units are not commercially available. Cooling tower pumps can also be driven by the engine. Figure 1 illustrates the fuel economy effected by varying prime mover speed with load (reciprocating compressor) until the machine is operating at about half capacity. Below this level, the load is reduced at essentially constant engine speed by unloading the compressor cylinders.

Frequent operation of the system at low engine-idling speed may require an auxiliary oil pump for the compressor. To reduce wear and assist in starts, a tank-type lube oil heater or a crankcase heater and a motor-driven auxiliary lubricating oil pump should

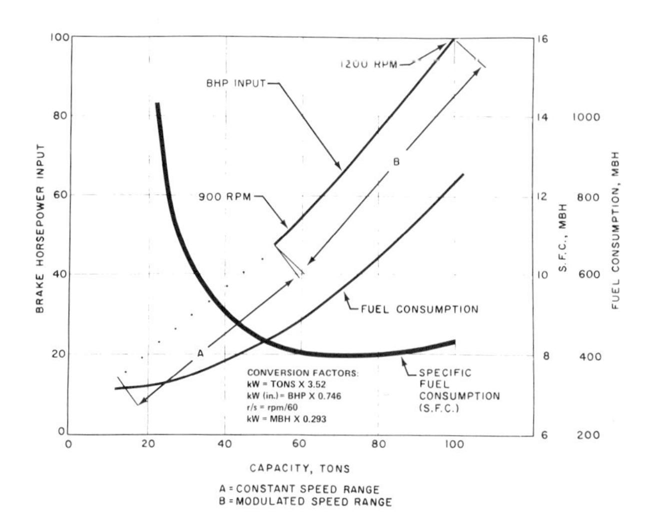


**Fig. 1 Performance Curves for Typical 100-Ton (350-kW), Gas Engine-Driven Reciprocating Chiller**

The preparation of this chapter is assigned to TC 9.5, Cogeneration Systems and TC 8.2, Centrifugal Machines.

be installed to lubricate the engine with warm oil when it is not running. Special suction line arrangements are also required when vertical lifts are used.

## Engine-Driven Centrifugal Compressors

Complete systems are not commercially available and must be field-assembled. These units may be equipped with either manual or automatic start-stop systems and engine-speed controls.

The centrifugal compressor is driven through a speed increaser. Many of these compressors operate at about six times the speed of the engines; compressor speeds of up to 14,000 rpm (233 r/s) have been used. To effect the best compromise between the initial cost of the equipment (engine, couplings, and transmission) and the maintenance cost, engine speeds between 900 and 1200 rpm (15 and 20 r/s) are generally used.

Engine output can be modulated by reducing engine speed. If the operation at 100% of rated speed produces 100% of the rated output, approximately 60% of rated output is available at 75% of rated speed. A vane-type control that regulates flow to the compressor is usually employed to effect any further reductions in output.

## Engine-Driven Heat Pumps

An additional economic gain can result from operating an engine-driven air-conditioning unit as a heat pump to provide heat during the winter months. Using the same equipment for both heating and cooling reduces capital investment. A gas-engine drive for heat pump operation also makes it possible to recover heat from the engine exhaust gas and jacket water.

Figure 2 shows the total energy available from a typical engine-driven heat pump operating at 1 bhp/ton (0.2 shaft kW/kW). Table 1 lists the coefficients of performance for various configurations.

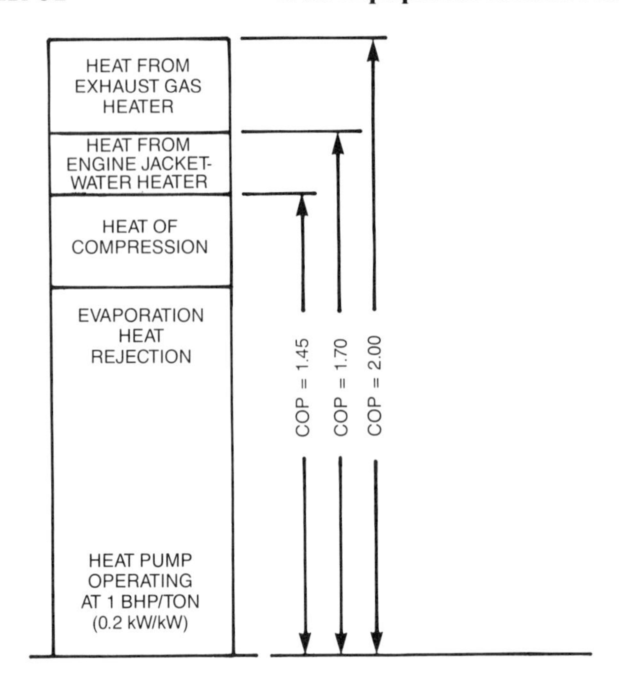


**Fig. 2 Heat Balance for Engine-Driven Heat Pump**

# ENGINE FUELS

## Fuel Selection

Engines may be fueled with gasoline, natural gas, propane, sludge gas, or diesel oil. Multifueled engines, using diesel oil as one of the fuels, are available from several manufacturers in engine sizes larger than 220 hp (165 kW). When gas is used as the fuel in a diesel, a small amount of diesel oil is used as pilot oil or as the ignitor.

Gasoline engines are not generally used because of fuel storage hazards, fuel cost, and maintenance problems caused by combustion product deposits on internal parts such as pistons and valves.

Methane-rich sludge gas obtained from sewage treatment processes at treatment plants can be used as a fuel for both engines and other heating services. The fuel must be dried prior to its injection into the engine. In addition, because of the lower heat content of the fuel gas (approximately 600 to 700 Btu/ft$^3$ [22.4 to 26.1 kJ/L]) the engine must be fitted with a larger carburetor. The large amount of hydrogen sulfide in the fuel requires the use of special materials, such as aluminum in the bearings and bushings and Teflon in the O-rings and gaskets. The final choice of fuels should be based on fuel availability, cost, storage requirements, and fuel rate, because, except for gasoline engines, maintenance costs tend to be comparable.

# FUEL HEATING VALUE

Fuel consumption data may be reported in terms of either *higher heating value* (HHV) or *lower heating value* (LHV). HHV is preferred by the gas utility industry and is the basis for most gaseous fuel considerations. Most natural gases have an LHV/HHV factor of 0.9. For other gaseous fuels however, this factor may range from 0.87 for hydrogen to 1.0 for carbon monoxide. For fuel oils, this factor ranges from 0.96 (heavy oils) to 0.93 (light oils); the HHV is customarily used for oil (including pilot oil in dual-fuel engines).

The engine power ratings of many manufacturers are based on an LHV of approximately 900 Btu/ft$^3$ (33.7 kJ/L). Manufacturers sometimes suggest derating engine output about 2% per 10% decrease in fuel heating below the base specified by the manufacturer.

# ENGINE SIZING

In sizing an engine, the size and nature of the load and economics require analysis.

Proper evaluation of the various heat gains and their time of occurrence avoids overestimating the actual load. Chapter 26 of the 1985 FUNDAMENTALS Volume shows methods for calculating cooling load.

Proper evaluation of operating cost requires knowledge of consumption rates. Specific data can be obtained from the manufacturer; however, a range of consumption rates is given in Table 2. Atmospheric corrections included in Eq. (1) may be used to match prime movers to loads at various ambient conditions.

$$kW_u = kW_{max} \cdot \text{rating } (\textit{derating}) \text{ factor} \qquad (1)$$

*where*

$kW_u$ = usable kW of power
$kW_{max}$ = kW power rating under ideal conditions, based on individual manufacturers' laboratory performance data (usually a dynamometer test), corrected in accordance with manufacturers' specifications; for example:

60°F (15.5°C) and 29.92 in. Hg (101 kPa) or
80°F (26.5°C) and 1000 ft (305 m) elevation (NEMA) or
85°F (29.5°C) and 500 ft (152 m) elevation (SAE) or
90°F (32.0°C) and 1500 ft (457 m) elevation (DEMA)

**Table 1 Coefficient of Performance for Engine-Driven Heat Pump**

| Item | Heat Source: Refrigerant Condenser only (Btu/ton-h) | (W/W) | Refrigerant Condenser and Jacket Water (Btu/ton-h) | (W/W) | Refrigerant Condenser, Jacket Water, Exhaust Gas (Btu/ton-h) | (W/W) |
|---|---|---|---|---|---|---|
| Total heat input to engine | 10,000 | 0.83 | 10,000 | 0.83 | 10,000 | 0.83 |
| Cooler heat rejection to condenser (from building load) | 12,000 | 1.000 | 12,000 | 1.000 | 12,000 | 1.000 |
| Heat of compression | 2,545 | 0.212 | 2,545 | 0.212 | 2,545 | 0.212 |
| Heat from engine jacket water heater | — | — | 2,500 | 0.208 | 2,500 | 0.208 |
| Heat from exhaust gas heater | — | — | — | — | 3,000 | 0.250 |
| Total heat to heating circuit | 14,545 | 1.212 | 17,045 | 1.420 | 20,045 | 1.670 |
| Coefficient of performance | 1.45 | 1.45 | 1.70 | 1.70 | 2.00 | 2.00 |

rating factor = [100 − (% altitude correction + % temperature correction + % heating value correction + % reserve)]/100

*where*

altitude correction = 3% per 1000 ft (305 m) above a specified level for naturally aspirated engines; 2% per 1000 ft (305 m) for turbocharged engines.

temperature correction = 1% per 10°F (5.5°C) rise above a specified base temperature for air intake.

heating value correction = see section on fuel heating value.

reserve = a percentage allowance (safety factor) to permit design output under unforeseen operating conditions that would reduce output, such as a dusty environment, poor maintenance, higher ambient temperature, or lowered cooling efficiency. Recommended values can be found in Table 3.

**Table 2 Fuel Consumption Rates**

| Fuel | Heating Value: Btu/gal | MJ/L | Range of Consumption: Btu/hp-h | W/W |
|---|---|---|---|---|
| Fuel oil | 137,000 to 156,000 | 38.1 to 43.4 | 7000 to 9000 | 0.58 to 0.75 |
| Gasoline | 130,000 | 36.2 | 10,000 to 14,000 | 0.83 to 1.17 |

| Typical consumption for different types of gas engines | Compression Ratio | Gas Consumption: Btu/hp-h | W/W |
|---|---|---|---|
| Turbocharged | 10.5:1 | 8,100 | 0.67 |
| Naturally aspirated | 10.5:1 | 9,200 | 0.76 |
| Naturally aspirated | 7.5:1 | 10,250 | 0.85 |

## DESIGN CONSIDERATIONS

The following should be considered in designing any engine-driven system:

1. Exhaust lines must be vented outdoors.
2. Noise levels must be acceptable by OSHA standards. (See Chapter 52 of the 1987 HVAC Volume.)

**Table 3 Minimum Engine Reserves for Air Conditioning and Refrigeration, Percent**

| Altitude ft | Altitude m | Naturally Aspirated: Air Conditioning | Naturally Aspirated: Refrigeration | Turbocharged Aftercooled: Air Conditioning | Turbocharged Aftercooled: Refrigeration |
|---|---|---|---|---|---|
| Sea level | Sea level | 15 | 20 | 20 | 30 |
| 1000 | 305 | 12 | 17 | 18 | 28 |
| 2000 | 610 | 10 | 14 | 16 | 26 |
| 3000 | 915 | 10 | 11 | 14 | 24 |
| 4000 | 1220 | 10 | 10 | 12 | 22 |
| 5000 | 1525 | 10 | 10 | 10 | 20 |
| 10,000 | 3050 | 10 | 10 | 10 | 10 |

3. Cooling towers should be sized to handle the engine water-jacket heat rejection. Exact figures for engine heat rejection should be obtained from the manufacturer. See Chapter 20 of this volume.
4. Inlet air for gas turbines and engines should be taken from outside.
5. Engine rooms should have adequate ventilation to prevent excessive room temperatures caused by radiated engine heat. If radiated heat is 8 to 10% of the fuel input, an air cooler may be required.

## INSTALLATION

### Noise Control

Engine-driven machines installed indoors, even where the background noise level is high, usually require noise attenuation and isolation from adjoining areas. Air-cooled radiators, noise radiated from surroundings, and exhaust heat recovery boilers may also require silencing. Boilers that operate dry do not require separate silencers. Installations in more sensitive areas may be isolated, receive sound treatment, or both.

Basic attenuation includes (1) turning air intake and exhaust openings away (usually up) from the potential listener; (2) limiting blade-tip speed (if forced-draft air cooling is used) to 12,000 fpm (60 m/s) for industrial applications, 10,000 fpm (50 m/s) for commercial applications, and 8000 fpm (40 m/s) for critical locations; (3) acoustically treating the fan shroud and plenum between blades and coils; (4) isolating (or covering) moving parts, including the unit, from its shelter (where used); (5) properly selecting the gas meter and regulator(s) to prevent singing; and (6) adding sound traps or silencers on ventilation air intake, exhaust, or both.

Further attenuation means are (1) lining the intake and exhaust manifolds with sound-absorbing materials; (2) mounting the unit, particularly a smaller engine, on vibration isolators, thereby reducing foundation vibration; (3) installing a barrier between the prime mover and the listener (often, a concrete block enclosure); (4) enclosing the unit with a cover of absorbing material; and (5) locating the unit in a building constructed of massive materials, paying particular attention to the acoustics of the ventilating system and doors.

### Exhaust Systems

Engine exhaust must be safely conveyed from the engine through piping and any auxiliary equipment to the atmosphere and at an allowable pressure drop and noise level. Allowable back pressures, which vary with engine design, run from 2 to 25 in. of water (0.5 to 6.2 kPa gauge). For low-speed engines, this limit is typically 6 in. of water (1.5 kPa); for high-speed engines, it is typically 12 in. of water (3.0 kPa). Adverse effects of excessive pressure drops include power loss, poor fuel economy, and excessive valve temperatures; all of these result in shortened service life and jacketwater overheating.

General installation recommendations include the following:

1. Installing a high-temperature, flexible connection between engine and exhaust piping—exhaust gas temperature does not normally exceed 1200°F (650°C); however, 1400°F (760°C) may be reached for short periods. An appropriate stainless-steel connector may be used.
2. Adequately supporting the exhaust system weight downstream of the connector—at maximum operating temperature, no weight should be exerted on the engine or the exhaust outlet.
3. Minimizing the distance between silencer and engine.
4. Using a 30 to 45° tail-pipe angle to reduce turbulence.
5. Specifying tail-pipe length (in the absence of other criteria) in odd multiples of 16.8 $T_e^{0.5}/P$, *where* $T_e$ is the temperature of the exhaust gas (Kelvin), and $P$ is exhaust frequency, pulses per second. (In the I-P system, this equation would be odd multiples of 12.5 $T_e^{0.5}/P$, *where* $T_e$ is temperature of the exhaust gas (Rankine), and $P$ is exhaust frequency, pulses per second.) The value of $P$ is calculated as follows:
   $P$ = rpm/120 (r/s/2) and rpm/60 (r/s) for four-stroke and twostroke engines, respectively, *where* rpm (r/s) is equal to the engine speed for V-engines with two exhaust manifolds. A second but less desirable exhaust arrangement is a Y-connection with branches entering the single pipe at about 60°; a T-connection should never be used, as the pulses of one branch would interfere with the pulses from the other.
6. Using an engine-to-silencer pipe length that is 25% of the tailpipe length.
7. Using a separate exhaust for each engine to reduce the possibility of condensation in the off-engine.
8. Using individual silencers to reduce condensation that results from an off-engine.
9. Limiting heat radiation from exhaust piping by means of a ventilated sleeve surrounding the pipe or high-temperature insulation.
10. Using fittings large enough to maintain adequate pressure.
11. Providing for thermal expansion in exhaust piping, about 0.09 in./ft (7.5 mm/m) of length.
12. Specifying muffler pressure drops to be within engine backpressure limits.
13. Not connecting engine exhaust pipe to chimney that serves natural-draft gas appliances.
14. Sloping the exhaust away from the engine to prevent condensate back flow. Drain plugs in silencers and drip legs in long, vertical exhaust runs may also be required. The use of rain caps may prevent the entrance of moisture, but might add back pressure.

The exhaust pipe may be routed between an interior engine installation and a roof-mounted muffler through (1) an existing unused flue or one serving powervented gas appliances only (a flue should not be used when exhaust gases may be returned from the interior); (2) an exterior fireproof wall, with provision for condensate drip to the vertical run; or (3) the roof, provided that a galvanized thimble with flanges having an annular clearance of 4 to 5 in. (100 to 125 mm) is used. A clearance of 1 to 2 in. (25 to 50 mm) between the flue terminal and the rain cap on the pipe permits the venting of the flue. A clearance of 30 in. (750 mm) between the muffler and roof is common. Vent passages and chimneys should be checked for resonance.

When interior mufflers must be used, the practice in recommendation 3 should be followed, except that the inside of the muffler portion of the flue should be insulated. Flue runs exceeding 25 ft (7.6 m) may require power venting.

**Table 4 Minimum Exhaust Pipe Diameter (in. and mm) to Limit Engine Exhaust Back Pressure to 8 in. of Water (1.99 kPa Gauge)**

| Output Power | | Equivalent Length of Exhaust Pipe | | | | | | | |
|---|---|---|---|---|---|---|---|---|---|
| hp | kW | 25 ft | 7.6 m | 50 ft | 15.2 m | 75 ft | 22.9 m | 100 ft | 30.5 m |
| 25 | 19 | 2 | 50 | 2 | 50 | 3 | 80 | 3 | 80 |
| 50 | 37 | 2.5 | 65 | 3 | 80 | 3 | 80 | 3.5 | 90 |
| 75 | 56 | 3 | 80 | 3.5 | 90 | 3.5 | 90 | 4 | 100 |
| 100 | 75 | 3.5 | 90 | 4 | 100 | 4 | 100 | 5 | 125 |
| 200 | 150 | 4 | 100 | 5 | 125 | 5 | 125 | 6 | 150 |
| 400 | 300 | 6 | 150 | 6 | 150 | 8 | 200 | 8 | 200 |
| 600 | 450 | 6 | 150 | 8 | 200 | 8 | 200 | 8 | 200 |
| 800 | 600 | 8 | 200 | 8 | 200 | 10 | 250 | 10 | 250 |

**Flexible Connections.** The following design and installation features are recommended:

*Material*—Convoluted steel (Grade 321 stainless steel is recommended). Stainless steel is favored for interior installations.

*Location*—Principal imposed motion (vibration) should be at right angles to the connector axis.

*Assembly*—The connector (not an expansion joint) should not be stretched or compressed; it should be secured without bends, offsets, or twisting (the use of float flanges is recommended).

*Anchor*—The exhaust pipe should be rigidly secured immediately downstream of connector in line with the downstream pipe.

**Exhaust Piping.** Yoloy, Cor-Ten, or steel pipe of standard weight may be joined by fittings of malleable cast iron. Table 4 shows exhaust pipe size. The exhaust pipe should be at least as large as the engine exhaust connection. Stainless steel double-wall liners may be used.

## Ventilation Air

Sufficient ventilation offsets heat losses from engine-driven equipment, muffler, and exhaust piping, and protects against fuel supply leaks except rupture of the supply line. Equation (2) may be used to calculate ventilation air requirements.

$$V = UA_T(t_1 - t_2)/K(t_2 - t_3) \quad (2)$$

*where*

$V$ = ventilation air required, cfm (L/s)
$U$ = 2 Btu/h • ft² • °F (11.4 W/m² • °C)
$A_T$ = total engine surface, $2[h(l + w) + lw]$, plus insulated exhaust piping area, ft²(m²)
$t_1$ = mean engine surface temperature
$t_2$ = engine room temperature, °F (°C)
$t_3$ = maximum outdoor air temperature, generally between 90 and 105°F (32 and 40°C)
$h$ = height, ft (m)
$l$ = length, ft (m)
$w$ = width, ft (m)
K = 1.08 I-P units (1.21 SI units)

Also note that:

$(t_2 - t_3)$ = 20°F when $t_3$ = 90°F; 10°F when $t_3$ = 120°F (generally) [11°C when $t_3$ = 32°C; 5.5°C when $t_3$ = 49°C]

Table 5 is sometimes used for minimum ventilation air requirements. Ventilation may be provided by a metal hood, preferably one with a fan that induces the draft through a sleeve surrounding the exhaust pipe. A slight positive pressure is maintained in the engine room.

**Table 5 Ventilation Air for Engine Equipment Room**

| Engine Room Air Temp Rise[a] | | Muffler and Exhaust Pipe[b] | | Muffler and Exhaust Pipe[c] | | Engine, Air-Cooled or Radiator-Cooled[d] | |
|---|---|---|---|---|---|---|---|
| °F | °C | cfm/hp-h | L/kW•s | cfm/hp-h | L/kW•s | cfm/hp-h | L/kW•s |
| 10 | 5.5 | 140 | 89 | 280 | 177 | 550 | 348 |
| 20 | 11.0 | 70 | 44 | 140 | 89 | 280 | 177 |
| 30 | 16.5 | 50 | 32 | 90 | 57 | 180 | 114 |

[a]Exhaust minus inlet
[b]Insulated or enclosed in ventilated duct
[c]Not insulated
[d]Heat discharged in engine room

## Combustion Air

The following factors apply to combustion air requirements:

1. Supply an airflow of 2 to 5 cfm/bhp (1.3 to 3.2 L/s per shaft kW), depending on type, design, and size. Two-cycle units consume about 40% more air than four-cycle units.
2. Avoid heated air because power output varies by $(T_r/T)^{0.5}$, *where* $T_r$ is the temperature at which the engine is rated, and $T$ is the engine air intake temperature, both in °R or K.
3. Locate the intake away from air contamination sources.
4. Install properly sized air cleaners that can be readily inspected (indicators are available) and maintained.

**Engine Room Air-Handling Systems** may include exhaust fans, louvers, shutters, bird screens, and air filters. The total static pressure opposing the fan should be 0.35 in. of water (87 Pa) maximum.

**Shutters and Louvers** are either manually or motor operated and control the quantity of intake or exhaust air. The vent area should be increased 25 to 50% to account for them. Thermostatically controlled shutters regulate airflow to maintain the desired temperature range. In cold climates, louvers should be closed when the engine is shut down to help maintain engine ambient temperature at a safe level. A crankcase heater can be installed on backup systems located in unheated spaces.

**Air Cleaners** minimize cylinder wear and piston ring fouling. About 90% of valve, piston ring, and cylinder wall wear is the result of dust. Both dry and wet cleaners are used. Wet cleaners, if oversized or operated below their capacity, are generally inefficient; if they are too small, the resultant oil pullover reduces filter life. Filters may also serve as flame arresters.

## Foundations

Modern multicylinder, medium-speed engines may not require massive concrete foundations, although concrete offers advantages in cost and in maintaining alignment for some driven equipment (see Chapter 52 of the 1987 HVAC Volume).

Fabricated steel bases are satisfactory for direct-coupled, self-contained units, such as electric sets. Steel bases mounted on vibration isolators, steel-spring type or equal, are adequate and need no special foundation other than a floor designed to accommodate the weight.

Concrete bases are also satisfactory for such units, provided the bases are equally well-isolated from the supporting floor or subfloor. Glass-fiber blocks are effective as isolation material for concrete bases. Concrete bases should be thick enough to prevent deflection. Excessively thick bases only serve to increase subfloor or soil loading. Such bases should always be supported by a concrete subfloor, and some acceptable isolation material should be placed between the base and the floor. To avoid the transmission of vibration, an engine base or foundation should never rest directly on natural rock formations.

## Engine Cooling

An engine converts fuel to shaft power and heat. Methods of dissipating this heat include (1) a jacket water system; (2) exhaust gas, which includes latent heat; (3) lubrication and piston cooling oil; (4) turbocharger and air intercooler; and (5) radiation from engine surfaces.

The manner and amount of heat rejection varies with the type, size, make of engine, and the extent of engine loading.

Installation includes the following:

1. A fresh air entrance at least as large as the radiator face and 25 to 50% larger if protective louvers impede airflow.
2. Auxiliary means, *e.g.,* a hydraulic, pneumatic, or electric actuator, to open the louvers blocking the heated air exit, as cooling fan pressure is insufficient for this purpose.
3. Control of jacket water temperature by louvers in lieu of a bypass arrangement.
4. Positioning the engine so the face of the radiator is in a direct line with an air exit leeward to the prevailing wind.
5. An easily removable shroud so exhaust air cannot reenter the radiator.
6. Separating the units in a multiple installation to avoid air recirculation among them.
7. Low temperature protection against snow and ice formation.

Propeller fans cannot be attached to long ducts because they can only achieve low static pressure. Radiator cooling air directed over the engine promotes good circulation around the engine, so it runs cooler than for airflows in the opposite direction.

**Water Cooling.** In most installations, heat pumps, and where noise tolerance is low, heat in the engine coolant is removed by heat exchange with a separate water system.

Recirculated water is cooled in cooling towers. Water is added to make up for evaporation. Closed systems should be treated with a rust inhibitor and/or antifreeze to protect the engine jacket from corrosion.

Since engine coolant travels in a closed loop, it is usually circulated on the shell side of the exchanger. A minimum fouling factor of 0.002 should be assigned to the tube side.

**Operating Temperatures.** Water jackets outlet and inlet temperature ranges of 175 to 190°F (80 to 88°C) and 165 to 175°F (74 to 80°C), respectively, are generally recommended, except when the engines are used with a heat-recovery system. These temperatures are maintained by one or more thermostats that act to bypass water, as required. A 10 to 15°F (5.6 to 8.5°C) temperature rise is usually accompanied by a circulating water rate of about 0.5 to 0.7 gpm per engine hp (42 to 60 mL/s per engine kW).

**Installation** involves (1) Sizing water piping according to the engine manufacturer's recommendations, (2) avoiding restrictions in the water pump inlet line, (3) never connecting piping rigidly to engine, and (4) providing shutoffs to facilitate maintenance.

# MAINTENANCE

Maintenance of engine-driven refrigeration compressors helps ensure continuous and economical system operation. Maintenance for the compressor section is similar to the maintenance of electric-drive units, with the exception of gearbox periodic oil changing and the operation and care of external lubricating pumps, which may be required if the compressor is operated at varying speeds.

Engines require periodic servicing and replacement of parts, depending on usage and the type of engine. Records should be kept of all servicing done; checklists should be used for this purpose.

**Table 6 Recommended Engine Maintenance**

| Procedure | Hours between Procedure | |
|---|---|---|
| | Diesel Engine | Gas Engine |
| 1. Take lubricating oil sample | 1 per month plus 1 at each oil change | 1 per month plus 1 at each oil change |
| 2. Change lubricating oil filters | 350 to 750 | 500 to 1000 |
| 3. Clean air cleaners, fuel | 350 to 750 | 350 to 750 |
| 4. Clean fuel filters | 500 to 750 | N.A. |
| 5. Change lubricating oil | 500 to 1000 | 1000 to 2000 |
| 6. Clean crankcase breather | 350 to 700 | 350 to 750 |
| 7. Adjust valves | 1000 to 2000 | 1000 to 2000 |
| 8. Lubricate tachometer, fuel priming pump, and auxiliary drive bearings | 1000 to 2000 | 1000 to 2000 (fuel pump N.A.) |
| 9. Service ignition system, adjust breaker gap, timing, spark plug gap, and magneto | N.A. | 1000 to 2000 |
| 10. Check transistorized magneto | N.A. | 6000 to 8000 |
| 11. Flush lubrication oil piping system | 3000 to 5000 | 3000 to 5000 |
| 12. Change air cleaner | 2000 to 3000 | 2000 to 3000 |
| 13. Replace turbocharger seals and bearings | 4000 to 8000 | 4000 to 8000 |
| 14. Replace piston rings, cylinder liners (if applicable), connecting rod bearings, and cylinder heads; recondition or replace turbochargers; replace gaskets and seals | 8000 to 12,000 | 8000 to 12,000 |
| 15. Same as item 14, plus recondition or replace crankshaft, replace all bearings | 24,000 to 36,000 | 24,000 to 36,000 |

Table 6 shows ranges of typical maintenance routines for both diesel and gas-fired engines, based on the number of hours run. The actual intervals vary according to cleanliness of the combustion air, cleanliness of the engine room, the manufacture of the engine, the number of engine starts and stops, and lubricating conditions indicated by the oil analysis. With some engines and some operating conditions, the intervals between procedures, as listed in Table 6, may be extended.

A preventive maintenance program should also include inspections for (1) leaks (a visual inspection facilitated by a clean engine); (2) abnormal sounds and odors; (3) unaccountable speed changes; (4) condition of fuel and lubricating oil filters; (5) water and lubricating oil temperatures; (6) individual cylinder compression pressures, which are useful in indicating blow-by; and (7) changes in valve tappet clearance, which indicate the extent of wear in the valve system.

Lubricating oil analysis is a low-cost method of determining the physical condition of the engine and a guide to maintenance procedures. Commercial laboratories providing this service are widely available. The analysis should measure the concentration of various elements found in the lubricating oil, such as bearing metals, silicates, and calcium. It should also measure the field dilution into the oil, suspended and nonsuspended solids, water, and oil viscosity. The laboratory can often assist in interpreting the readings and alert the user to impending problems.

### Lubrication

Manufacturers' recommendations should be followed. Both the crankcase oil and oil filter elements should be changed at least once every six months.

## CODES AND STANDARDS

In addition to applicable local codes, the following National Fire Protection Association Standards should be consulted.

*National Electrical Code*

Article 440, Air Conditioning and Refrigeration Equipment
Article 445, Generators
Article 700-12(b), Emergency System Set
Articles 701 and 702, Stand-by Power

*National Fire Codes*

30 Flammable and Combustible Liquids Code
31 Installation of Oil Burning Equipment
37 Installation and Use of Stationary Combustion Engines and Gas Turbines
54 National Fuel Gas Code
58 Storage and Handling of Liquefied Petroleum Gases
59 Storage and Handling of Liquefied Petroleum Gases at Utility Gas Plants
59A Production, Storage and Handling of Liquefied Natural Gas (LNG)
90A Installation of Air Conditioning and Ventilation Systems
211 Chimneys, Fireplaces, Vents and Solid Fuel Burning Appliances

Gas codes cover service at all pressures. ASHRAE *Standard* 15-1978, *Safety Code for Mechanical Refrigeration*, covers the refrigeration section of the system and should be followed.

## CONTROLS AND ACCESSORIES

### Line-Type Gas-Pressure Regulator

Turbocharged (and aftercooled) engines, as well as many naturally aspirated units, are equipped with line regulators designed to control the gas pressure to the engine regulator, as shown in Table 7. The same regulators (both line and engine) used on naturally aspirated gas engines may be used on turbocharged equipment.

Line-type gas-pressure regulators are commonly called *service regulators* (and *field regulators*). They are usually located just upstream of the engine regulator to ensure the required pressure range exists at the inlet to the latter control. A remote location is sometimes specified; authorities having jurisdiction should be consulted. Although this intermediate regulation does not constitute a safety device, it does permit initial regulation (by the gas utility at meter inlet) at a higher outlet pressure, thus

**Table 7 Line Regulator Pressures**

| Line Regulator | Turbocharged Engine (Gauge) | | Naturally Aspirated Engines (Gauge) | |
|---|---|---|---|---|
| | psi | kPa | psi | kPa |
| Inlet[a] | 14 to 20 | 96 to 138 | 2 to 50 | 14 to 345 |
| Outlet[b] | 12 to 15[c] | 83 to 103[c] | 7-10 in. of water | 1.7 to 2.5 |

[a]Overall ranges, not the variation for individual installations.
[b]Also inlet to engine regulator.
[c]Turbocharger boost plus 7 to 10 in. of water (1.7 to 2.5 kPa gauge).

allowing an extra cushion of gas between the line regulator and the meter for both full gas flow at engine start-ups and delivery to any future branches from the same supply line. The engine manufacturer will specify the size, type, orifice size, and other regulator characteristics based on the anticipated gas-pressure range.

### Engine-Type Gas-Pressure Regulator

This engine-mounted pressure regulator, also called a *carburetor regulator* (and sometimes a *secondary regulator* or a *B regulator*), controls the fuel pressure to the carburetor. Regulator construction may vary with the fuel passed. The unit is similar to a zero governor.

### Air-Fuel Control

The quantity flow of air-fuel mixtures in definite ratios must be controlled under all of the load and speed conditions required of engines.

**Air-Fuel Ratios.** High-rated, naturally aspirated, spark-ignited engines require closely controlled air-fuel ratios. Excessively lean mixtures cause excessive oil consumption and engine overheating. Engines using pilot oil ignition can run at rates above 0.18 ppm • hp (1.8 g/s • kW) without misfiring. Air rates may vary with changes in compression ratio, valve timing, and ambient conditions.

**Carburetors.** In these venturi-type devices, the airflow mixture is controlled by a governor-actuated butterfly valve. This air-fuel control has no moving parts other than the butterfly valve.

The motivating force in naturally aspirated engines is the vacuum created by the intake strokes of the pistons. Turbocharged engines, on the other hand, supply the additional energy of pressurized air and pressurized fuel.

### Ignition

An electrical system or pilot oil ignition may be used. Electrical systems are either low-tension (make-and-break) or high-tension (jump spark). Systems with breakerless ignition distribution are also in use.

### Governors

A governor senses speed (and sometimes load), either directly or indirectly, and acts by means of linkages to control the flow of gas and air through engine carburetors or other fuel-metering devices to maintain a desired speed. Speed control gained from electronic, hydraulic, or pneumatic governors extends engine life by minimizing forces on engine parts, permits automatic throttle response without operator attention, and prevents destructive overspeeding.

Use of a separate overspeed device (*e.g.,* a maximum speed-type of governor, sometimes called an *overspeed stop*) prevents runaway in the event of a failure that renders the governor inoperative.

Both constant and variable engine speed controls are available. For constant speed, the governor is set at a fixed position. These positions can be reset manually.

### Safety Controls and Considerations

Both a low-lubrication pressure switch and a high jacket water temperature cutout are standard for most gas engines. Other safety controls used include (1) an engine speed governor; (2) ignition current failure shutdown (battery-type ignition only); and (3) the safety devices associated with a driven machine. These devices shut down the engine to protect it against mechanical damage. They do not necessarily shut off the gas fuel supply unless they are specifically set to do so. Shutting off fuel to the problem engine may be desirable. For example, a control may stop one engine and simultaneously energize the cranking circuit of its standby.

## GAS LEAKAGE PREVENTION

The first method of avoiding gas leakage that results from engine regulator failure is a solenoid shutdown valve with a positive cutoff, installed either upstream or downstream of the engine regulator. The second method is a sealed combustion system that carries any leakage gas directly to the outdoors (*i.e.,* all combustion air is ducted to the engine directly from the outdoors).

## EXHAUST HEAT RECOVERY

Exhaust heat may be used to make steam or used directly for drying or other processes. The steam provides space heating, hot water, and absorption refrigeration, which may supply air conditioning and process refrigeration. Heat-recovery systems generally involve equipment specifically tailored for the job, although conventional firetube boilers are sometimes used.

Exhaust-heat recovery from reciprocating engines is usually accomplished by ebullient cooling of the jacket water and the use of a muffler-type of exhaust-heat recovery unit. Table 8 gives the temperature levels normally required for various heat-recovery applications.

**Ebullient Systems.** These pressurized reciprocating engine cooling-heat recovery systems are also called *high temperature,* because they operate with a few degrees of temperature differential in the range of 212 to 250 °F (100 to 120 °C), sometimes as high as 270 °F (132 °C).

Engine builders have approved various conventionally cooled models for ebullient cooling application. Azeotropic antifreezes should be used to ensure constant coolant composition in boiling. However, the engine power outputs are generally derated by the manufacturer when the jackets are cooled ebulliently.

Ebullient cooling usually replaces the radiator, the belt-driven fan, and the gear-driven water pump.

**Heat from Exhaust Gases.** In some engines, exhaust heat rejection exceeds jacket water rejection. Generally, gas engine exhaust temperatures run from 700 to 1200 °F (370 to 650 °C), as shown in Table 9. Cooling exhaust gases to about 300 °F (150 °C) prevents condensation in the exhaust line. Thus, about 50 to 75% of the sensible heat in the exhaust may be considered recoverable.

The economics of exhaust-heat boiler design often limit the temperature differential between exhaust gas and generational steam to a minimum of 100 °F (56 °C). Therefore, in low-pressure

**Table 8 Temperature Levels Normally Required for Various Heating Applications**

| Application | Temperature °F | Temperature °C |
|---|---|---|
| Absorption refrigeration machines | 190 to 245 | 88 to 118 |
| Space heating | 120 to 250 | 50 to 121 |
| Water heating (domestic) | 120 to 200 | 50 to 93 |
| Process heating | 150 to 250 | 66 to 121 |
| Evaporation (water) | 190 to 250 | 88 to 121 |
| Residual fuel heating | 212 to 330 | 100 to 166 |
| Auxiliary power producers, with steam turbines or binary expanders | 190 to 350 | 88 to 177 |

**Table 9 Approximate Full-Load Exhaust Mass Flows and Temperatures for Various Types of Engines**

| | Mass Flow | | Temperature | |
|---|---|---|---|---|
| Types of Engines | lb/bhp-h | g/kW • s | ° F | °C |
| Two-Cycle | | | | |
| Blower charged gas | 16 | 2.7 | 700 | 370 |
| Turbocharged gas | 14 | 2.4 | 800 | 430 |
| Blower charged diesel | 18 | 3.0 | 600 | 320 |
| Turbocharged diesel | 16 | 2.7 | 650 | 340 |
| Four-Cycle | | | | |
| Naturally aspirated gas | 9 | 1.5 | 1200 | 650 |
| Turbocharged gas | 10 | 1.7 | 1200 | 650 |
| Naturally aspirated diesel | 12 | 2.0 | 750 | 400 |
| Turbocharged diesel | 13 | 2.2 | 850 | 450 |
| Gas Turbine, nonregenerative | 60 | 10.1 | 1050 | 570 |

steam boilers, gas temperature can be reduced to 300 to 350°F (150 to 175°C); the corresponding final exhaust temperature range in high-pressure steam boilers is 400 to 500°F (200 to 260°C). Table 8 gives application data.

Exhaust heat from the engine may be used directly in a waste heat absorption chiller. Units are available in sizes ranging from 100 to 1500 tons (350 to 5000 kW).

## HEAT RECOVERY ABSORPTION CHILLERS

Two-stage absorption chillers have been developed to take advantage of the dual temperatures available from engine exhaust and jacket water. The use of the engine exhaust heat to provide cooling was described in the previous section on "Exhaust Heat Recovery." A class of absorption machines has been designed specifically for heat recovery in cogeneration applications. These units make direct use of both the jacket water and the exhaust gas. Ebullient cooling of the engine is not required (Figure 3). These chillers range in size from approximately 50 to 200 tons (175 to 700 kW).

## DUAL-SERVICE APPLICATIONS

An engine that drives a refrigeration compressor can be switched over, automatically, to drive an electrical generator in the event of a power failure (Figures 4 and 5). This plan assumes loss of the compressor service can be tolerated in an emergency.

If the engine of a dual-service system equals 150 or 200% of the compressor load, the power available from the generator can be delivered to the utility grid during normal operation. While induction generators may be used for this application, a synchronous generator is required for emergency operation if there is a grid outage.

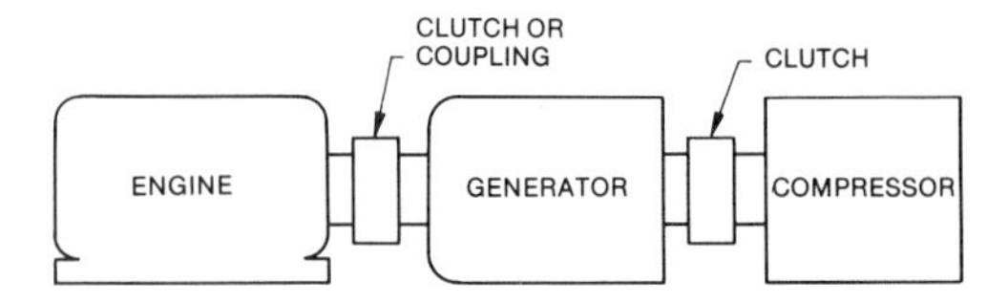


[a]Generator rotor requires little energy while generator is under no load. In essence, the rotor is a flywheel during this period.

**Fig. 4 Dual-Service Application with Generator Rotor Attached to Engine Crankshaft[a]**

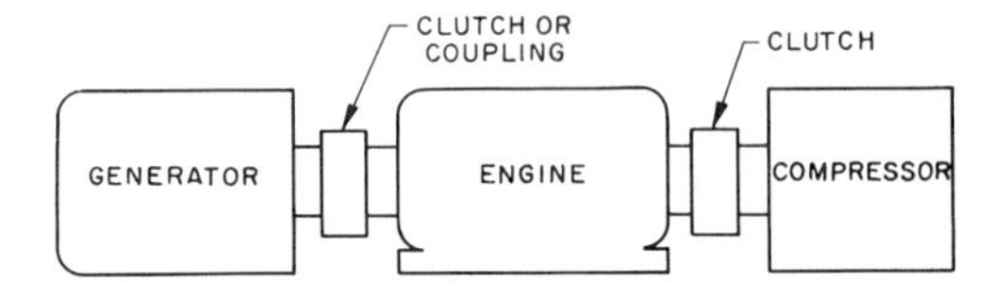


**Fig. 5 Dual-Service Application Using Double-Ended Engine**

Dual-service arrangements have the following advantages:

1. Lower capital investment and reduced space and maintenance are required when compared to two engines in single service, even after allowing for the additional controls needed with dual-service installations.
2. Dual-service engines are more reliable than emergency (reserved) single-service engines because of continuous or regular use.
3. Engines that are in service and running can be switched over to emergency power generation with a minimum loss of continuity.

## EXPANSION RECIPROCATING ENGINES

These engines are used mainly for cryogenic applications to about −320°F (−196°C), *e.g.,* oxygen for steel mills, low-temperature chemical processes, and the space program.

Relatively high-pressure air or gas expanded in an engine drives a piston and is cooled in the process. At the shaft, about 42

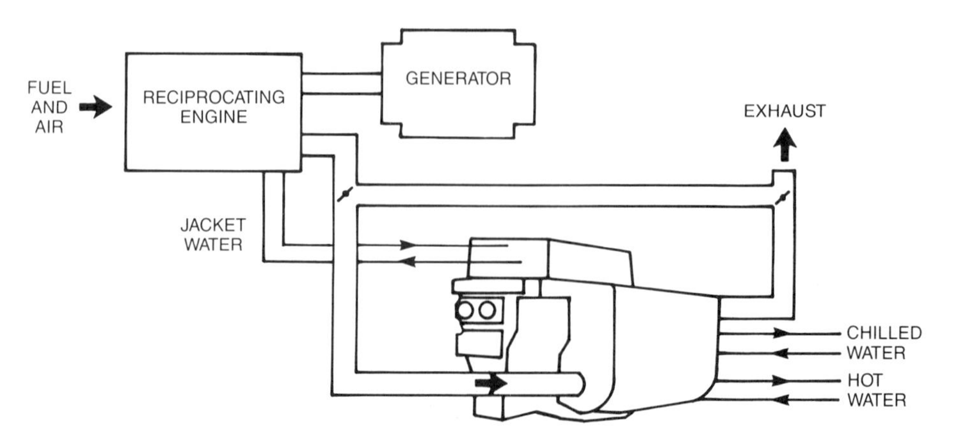


**Fig. 3 Exhaust Gas Chiller-Heater with Jacket Heat Reclaim**

Btu/hp (59.4 kJ/kW) are removed. Available units, developing as much as 600 hp (450 kW), handle flows ranging from 100 to 10,000 scfm (47 to 4700 L/s). Throughput at a given pressure is controlled by varying the cutoff point, the engine speed, or both. The conversion efficiency of heat energy to shaft work ranges from 65 to 85%. A 5 to 1 pressure ratio and an inlet pressure of 3000 psig (21 MPa) are recommended. Outlet temperatures of −450 °F (−268 °C) have been handled satisfactorily.

# PART II: STEAM AND GAS TURBINES

## STEAM TURBINES

### Application

Steam turbines are principally used in the air-conditioning and refrigeration field to drive centrifugal compressors. Such compressors are usually part of a water- or brine-chilling system using one of the halogenated hydrocarbon refrigerants. In addition, many industrial processes employ turbine-driven centrifugal compressors with a variety of other refrigerants such as ammonia, propane, and butane.

Associated applications of steam turbines include driving chilled-water and condenser-water circulating pumps and serving as prime movers for electrical generators in total-energy systems.

The most fundamental factor in deciding to use steam turbines may be cost; comparative initial and operating cost analyses point to the use of steam turbines as being the least expensive prime movers. Particularly in industrial applications, the steam turbine may be advantageous, serving either as a work-producing steam-pressure reducer or as a *scavenger,* employing otherwise wasted low-pressure steam. Many steam turbines are used in urban areas where commercial buildings are served with steam from a central public utility or municipal source. Finally, the institutional field, where large central plants serve a multitude of buildings with heating and cooling, uses steam turbine-driven equipment. Chapter 8 of the 1987 HVAC Volume covers the application of steam turbines in total energy and cogeneration systems.

### Types

Mechanical-drive steam turbines are generally condensing and noncondensing, denoting the condition of the exhaust steam. Figure 6 shows basic types of turbines. NEMA *Publication* SM23 (1985) defines these and further subdivisions of their basic families.

**Noncondensing Turbine.** A steam turbine designed to operate with an exhaust steam pressure equal to, or greater than, atmospheric pressure.

**Condensing Turbine.** A steam turbine designed to operate with an exhaust steam pressure below atmospheric pressure.

**Automatic-Extraction Turbine.** A steam turbine that has both an opening(s) in the turbine casing for the extraction of steam and means for directly regulating the steam flow to the turbine stages below the extraction opening.

**Nonautomatic-Extraction Turbine.** A steam turbine that has an opening(s) in the turbine casing for the extraction of steam but does not have means for controlling the pressure of the extracted steam.

**Induction (Mixed-Pressure) Turbine.** A steam turbine with separate inlets for steam at two pressures, has an automatic device for controlling the pressure of the secondary steam inducted into the turbine, and means for directly regulating the flow of steam to the turbine stages below the induction opening.

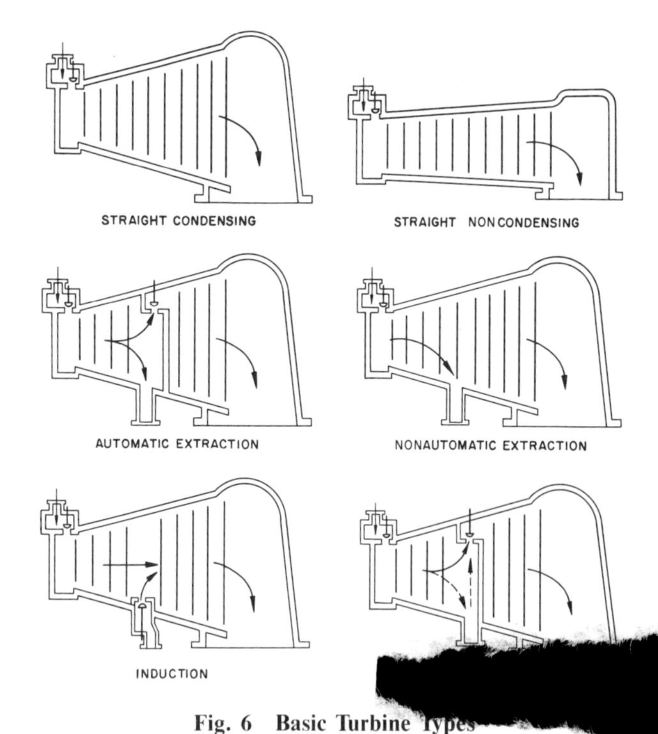


Fig. 6 Basic Turbine Types

**Induction-Extraction Turbine.** A steam turbine with the capability of either exhausting or admitting a supplemental flow of steam through an intermediate port in the casing, thereby maintaining a process heat balance.

Turbines of the extraction and induction-extraction type may have several casing openings, each passing steam at a different pressure.

Most steam turbines driving centrifugal compressors for air conditioning are of the multistage condensing type (Figure 7). Such a turbine gives good steam economy at reasonable initial cost. Usually, steam is available at a gauge pressure of 50 psig (350 kPa) or higher, and there is no demand for exhaust steam. However, turbines may work equally well where an abundance of low-pressure steam is available. The wide range of application of this turbine is shown by at least one industrial firm that drives a sizable capacity of water-chilling centrifugals with an initial steam-gauge pressure of less than 4 psig (28 kPa), thus

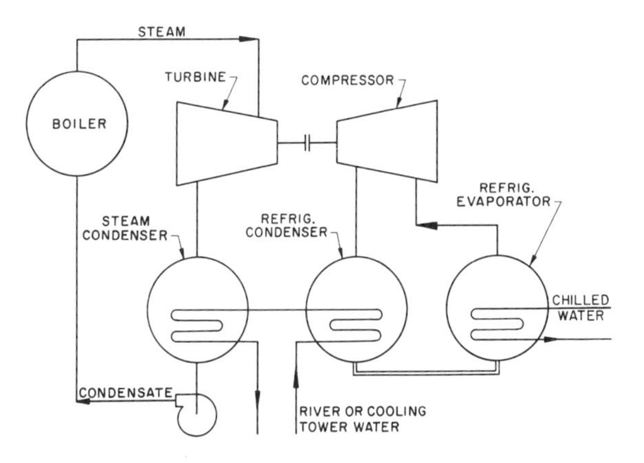


Fig. 7 Condensing Turbine-Driven Centrifugal Refrigeration Machine

balancing summer cooling against winter heating with steam from generator-turbine exhausts.

Aside from rather wide industrial use, the noncondensing (or *back-pressure*) turbine is most often used in water-chilling plants for driving a centrifugal compressor, which shares the cooling load with one or more absorption units (Figure 8). The exhaust steam from the turbine, commonly at about 15 psig (100 kPa), serves as the heat source for the absorption unit's generator (concentrator). This dual use of the heat energy in the steam generally results in a lower energy input per unit of refrigeration effect output than is attainable, operating alone, by either the turbinedriven centrifugal or the absorption unit. An important aspect in the design of such combined systems is the need to balance the turbine exhaust steam flow with the absorption input steam requirements over the full range of load.

Extraction and mixed-pressure turbines find their principal use in industry or in large central plants. Extracted steam is often used for boiler feedwater heating or where steam with lower heat content is needed.

Steam energy in a turbine produces shaft power by the following methods: the necessary rotative force may be imposed on the turbine through the velocity, pressure ene[illegible] ...rgy is [illegible]

INDUCTION-EXTRACTION

[illegible] little or no steam pressure [illegible] ...their blades. Such combinations of nozzles and velocity-powered wheels are characteristic of an *impulse* turbine. A *reaction* turbine uses alternate rows of fixed and moving blades, generally of an airfoil shape, with steam velocity increasing in the fixed portion and dropping in the movable portion and steam pressure dropping through both the fixed and movable blades.

The power capabilities of a reaction turbine are at a maximum when the moving blades are traveling at about the velocity of the steam passing through them; in the impulse turbine, maximum power is produced with a blade velocity of about 50% of steam velocity. Steam velocity is related directly to pressure drop. To achieve the desired relationship between steam velocity and blade velocity without resorting to large wheel diameters or high rotative speeds, most turbines include a series of impulse or reaction stages, or both, thus dividing the total steam pressure drop into manageable increments. A typical commercial turbine may have two initial rows of rotating impulse blading with an intervening stationary row (called a *Curtis stage*), followed by several alternating rows of fixed and movable blading of either the impulse or reaction type. Most multistage turbines use some degree of reaction.

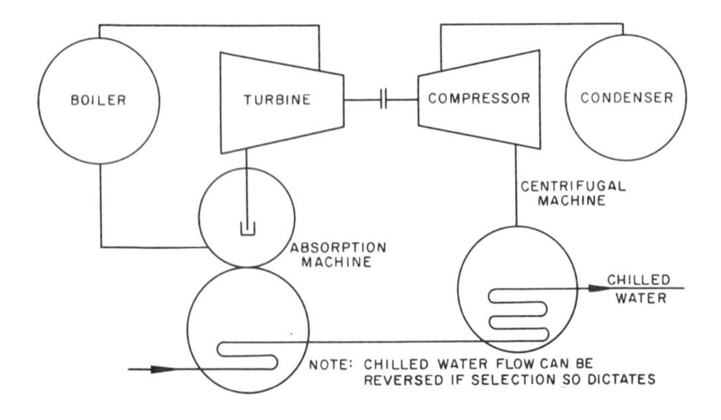


**Fig. 8 Combination Centrifugal-Absorption System with Back-Pressure Turbine**

## Performance

The energy input to a steam turbine is denoted by steam rate (or *water rate*). Actual steam or design rate is the theoretical steam rate of the turbine, divided by its efficiency. Typical efficiencies for mechanical steam turbines range from 55 to 80%. As defined by NEMA, theoretical steam rate is the quantity of steam per unit of power required by an ideal Rankine cycle heat engine. Therefore:

$$w = c/(h_1 - h_2)$$

*where*

$w$ = theoretical steam rate, lb/hp • h (kg/kWh)
$h_1$ = enthalpy of steam at its initial temperature and pressure, Btu/lb (kJ/kg)
$h_2$ = enthalpy of steam at exhaust steam pressure and initial entropy, Btu/lb (kJ/kg)
$c$ = 2545 in I-P units (3600 in SI units)

Turbine performance tests should be conducted in accordance with appropriate ASME Performance Test Code 6 (1985), 6S (1970), and 6A (1982).

The steam rate of a turbine is reduced with higher turbine speeds, greater number of stages, larger turbine size, and a higher difference in heat content between entering and leaving steam conditions. Often, one or more of these factors can be improved with only a nominal increase in initial capital cost. Centrifugal water-chilling compressor applications range from approximately 100 to 10,000 hp (75 to 7500 kW) and 3000 to 10,000 rpm (50 to 170 r/s), with the higher speeds generally associated with lower power outputs, and lower speeds with higher power outputs.

Higher compressor speeds cannot always be handled by direct-connected steam turbines, without severe cost penalty. Turbines with their wheels shrunk or keyed to the shaft (or both) are suited to the lower and intermediate compressor-speed requirements. Most turbine designs, however, change markedly above a high-speed threshold of about 7500 to 8500 rpm (125 to 140 r/s). The resultant stresses and critical-speed characteristics demand solid-rotor construction, with the wheels acting as an integral part of the shaft material. To avoid this cost penalty, an intermediate gear speed-changer is often used. Such gears may be either a separate component, connected by flexible couplings to the compressor and turbine, or built integrally with the steam turbine. In the latter case, the manufacturer rates power output at the gearshaft, thus taking into account the 2 to 3% power lost by the gear. (Some typical characteristics of turbines driving centrifugal water chillers are shown in Figures 9 to 12.)

While initial steam pressures commonly fall in the 100 to 250 psig (700 to 1700 kPa) range, wide variations are possible. Back pressures associated with noncondensing turbines generally range from 50 psig (350 kPa) to atmospheric, depending on the use for the exhaust steam. Raising the initial steam temperature by superheating improves steam rates.

NEMA standards govern allowable deviations from design steam pressures and temperatures. Because of possible unpredictable variations in steam conditions and load requirements, turbines are selected for a power capability of 105 to 110% of design shaft output and a speed capability of 105% of design rpm (r/s).

## Governors

One important aspect of turbine application and specification is the type and quality of speed-governing apparatus. The wide variety of available governing systems permits selecting a governor ideally matched to the characteristics of the driven machine and the load profiles.

The principal and most common function of a steam-turbine governing system is to maintain constant turbine speed despite

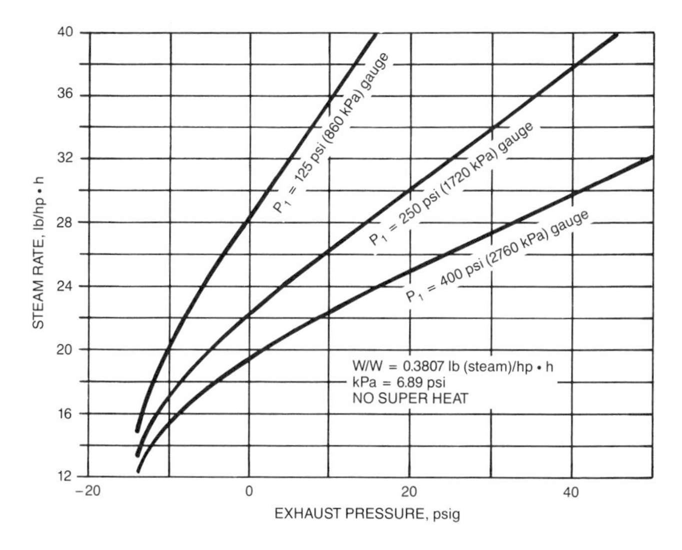


Fig. 9 Typical Effect of Exhaust Pressure on Noncondensing Turbine

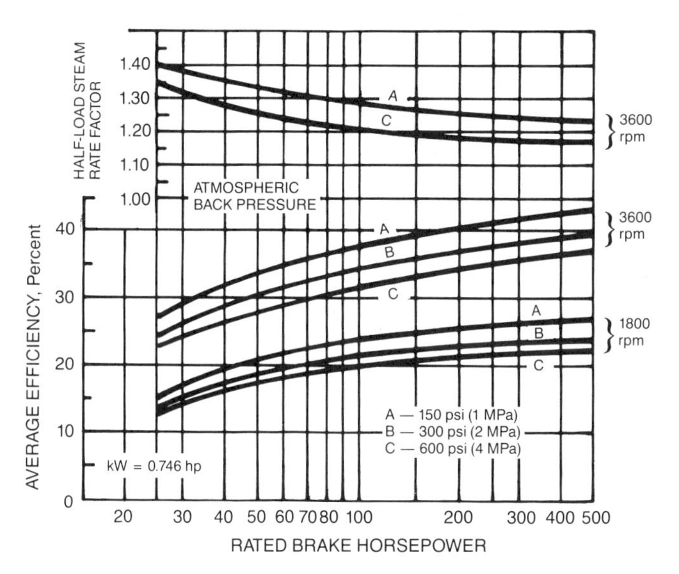


Fig. 10 Efficiency of Typical Single-Stage Noncondensing Turbine

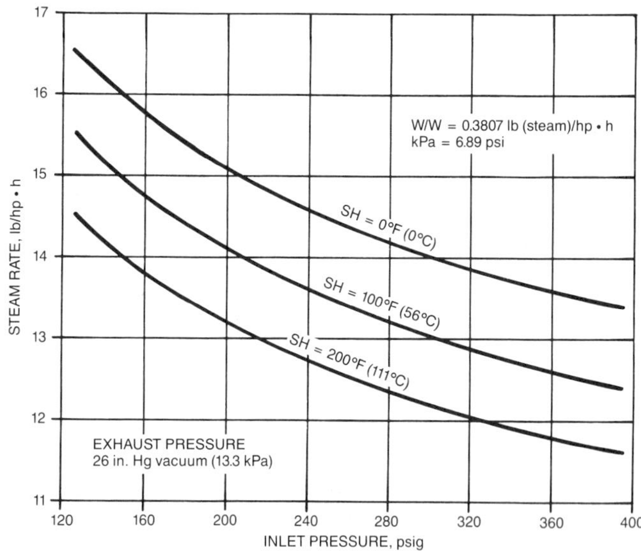


Fig. 11 Effect of Inlet Pressure and Superheat on Condensing Turbine

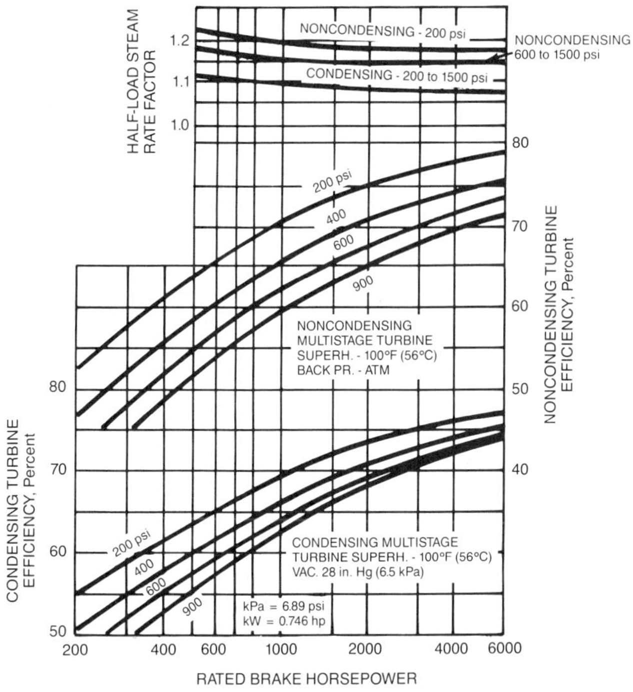


Fig. 12 Efficiency of Typical Multistage Turbines

load fluctuations or minor variations in supply steam pressure. This arrangement assumes that close control of the output of the driven component, such as a centrifugal compressor in a water-chilling system, is primary to plant operation, and that the compressor can adjust its capacity to varying loads.

Often it is desirable to vary the turbine speed in response to an external signal. In centrifugal waterchilling systems, for example, reduced speed generally reduces steam consumption at partial load. This control is usually an electric or pneumatic device responsive to the temperature of the fluid leaving the water-chilling heat exchanger (evaporator). Because compressor capacity varies with speed, the compressor's internal capacity-controlling devices are usually kept inoperative when speed is controlled. Process applications frequently require placing an external signal on the turbine-governing system, overriding its inherent constant-speed capability. Such external signals may be necessary to maintain a fixed compressor discharge pressure, regardless of load or condenser-water temperature variations. Plants relying on a closely maintained heat balance may control turbine speed to maintain an optimum pressure level of steam entering, being extracted from, or exhausting from the turbine. One example is the combination turbine absorption plant, where control of pressure of the steam exhausting from the turbine (and feeding the absorption unit) is an integral part of the plant-control system.

The steam-turbine governing system consists of (1) a speed governor (mechanical, hydraulic, electrical); (2) a speed-control mechanism (relays, servomotors, pressure or power amplifying devices, levers, and linkages); (3) governor-controlled valve(s); (4) a speed changer; and (5) external control devices, as required.

The speed governor responds directly to turbine speed and initiates action of the other parts of the governing system. The simplest speed governor is the direct-acting flyball type, which depends on changes in centrifugal force for proper action. Capable of adjusting speeds through an approximate 20% range, it is widely used on single-stage mechanical-drive steam turbines with speeds up to 5000 rpm (85 r/s) and steam pressures up to 600 psig (4.1 MPa).

The speed governor used most frequently on centrifugal water-chilling system turbines is the oil-pump type. In its direct-acting form, oil pressure, produced by a pump either directly mounted on the turbine shaft or in some form responsive to turbine speed, actuates the inlet steam valve.

The oil-relay hydraulic governor, as shown in Figure 13, has greater sensitivity and effective force. Here, the speed-induced oil-pressure changes are amplified in a servomotor or pilot-valve relay to produce the motive effort required to reposition the steam inlet valve or valves.

The least expensive turbine has a single governor-controlled steam admission valve, perhaps augmented by one or more small auxiliary valves (usually manually operated), which close off nozzles supplying the turbine steam chest for better part-load efficiency. Figure 14 shows the effect of auxiliary valves on part-load turbine performance.

For more precise speed-governing and maximum efficiency without manual valve adjustment, multiple automatic nozzle control is used (Figure 15). Its principal application is in larger turbines where a single governor-controlled steam admission valve would be too large to permit sensitive control. The greater power required to actuate the multiple-valve mechanism dictates the use of hydraulic servomotors.

Speed-changers adjust the setting of the governing system while the turbine is in operation. Usually, they comprise either a means of changing spring tension or of regulating oil flow by a needle valve. The upper limit of a speed-changer's capability should not exceed the rated turbine speed. Such speed-changers, while usually mounted on the turbine, may sometimes be remotely located at a central control point.

As stated previously, external control devices are often used when some function other than turbine speed is controlled. In such cases, an electric, hydraulic, or pneumatic signal overrides the turbine speed governor's action, and the latter assumes a speed-limiting function. The external signal controls steam admission either by direct inlet-valve positioning or by adjustment of the speed-governor setting. The valve-positioning method either exerts mechanical force on the valve-positioning mechanism or, if power has to be amplified, regulates the pilot valve in a hydraulic servomotor system. The speed-governor adjusting method is preferable where more precise control is required. Although the external signal continually resets the governor, as required, the speed governor always provides ideal turbine speed control. Thus, it maintains the particular set speed, regardless of load or steam-pressure variations.

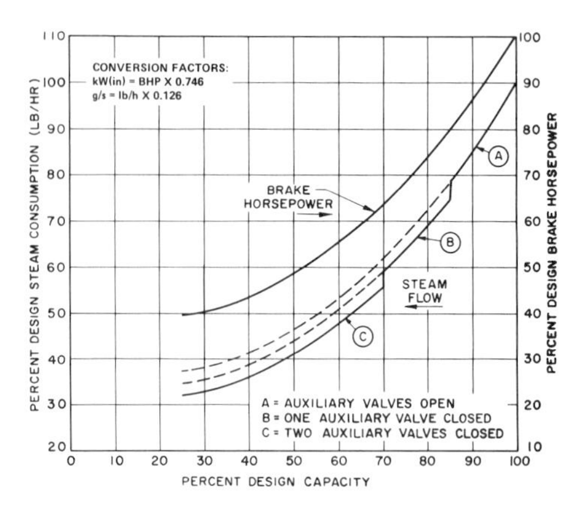


**Fig. 14 Typical Part-Load Turbine Performance Showing Effect of Auxiliary Valves**

NEMA classifies steam-turbine governors as shown in Table 10.

**Range of Speed-Changer Adjustment,** expressed as a percentage of rated speed, is the range through which the turbine speed may be adjusted downward from rated speed by the speed-changer, with the turbine operating under the control of the speed governor and passing a steam flow equal to the flow at rated power output and speed.

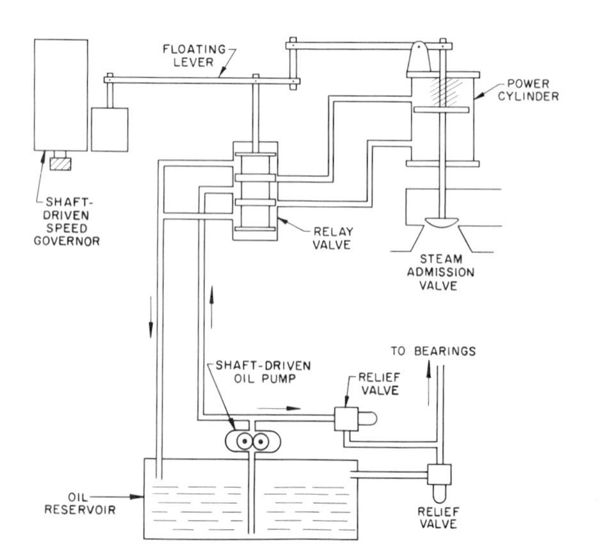


**Fig. 13 Oil Relay Governor**

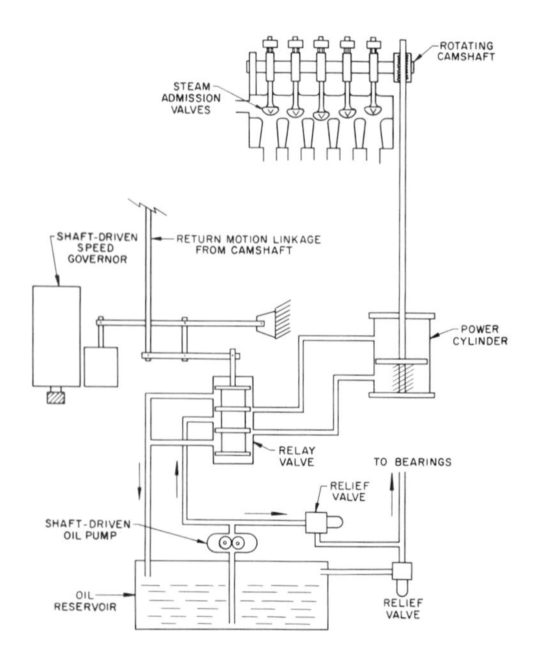


**Fig. 15 Multivalve Oil Relay Governor**

**Table 10 NEMA Classification of Speed Governors**

| Class of Governor | Adjustable Speed Range, % | Maximum Steady-State Speed Regulation, % | Maximum Speed Variation, % Plus or Minus | Max'm Speed Rise, % | Trip Speed, % (Above Rated Speed) |
|---|---|---|---|---|---|
| A | 10 to 65 | 10 | 0.75 | 13 | 15 |
| B | 10 to 80 | 6 | 0.50 | 7 | 10 |
| C | 10 to 80 | 4 | 0.25 | 7 | 10 |
| D | 10 to 90 | 0.50 | 0.25 | 7 | 10 |

The range of the speed-changer adjustment, expressed as a percentage of rated speed, is derived from the following equation:

$$\text{Range (Percent)} = \frac{\left(\text{Rated speed}\right) - \left(\text{Minimum speed setting}\right)}{\text{Rated speed}} 100$$

**Steady-State Speed Regulation,** expressed as a percentage of rated speed, is the change in sustained speed when the power output of the turbine is gradually changed from rated power output to zero power output under the following conditions:

1. The steam conditions (initial pressure, initial temperature, and exhaust pressure) are set at rated values and held constant.
2. The speed-changer is adjusted to give rated speed with rated power output.
3. Any external control device is rendered inoperative and blocked in the *open* position to allow the free flow of steam to the governor-controlled valve(s).

The steady-state speed regulation, expressed as a percentage of rated speed, is derived from the following equation:

$$\text{Regulation (Percent)} = \frac{\left(\text{Speed at zero power output}\right) - \left(\text{Speed at rated power output}\right)}{\text{Speed at rated power output}} 100$$

Steady-state speed regulation of automatic-extraction or mixed-pressure-type turbines is derived with zero extraction or induction flow and with the pressure-regulating system(s) inoperative and blocked in the position corresponding to rated extraction or induction pressure(s) at rated power output.

**Speed variation,** expressed as a percentage of rated speed, is the total magnitude of speed change or fluctuations from the speed setting. It is defined as the difference in speed variation between the governing system in operation and the governing system blocked to be inoperative, with all other conditions constant. This characteristic includes dead band and sustained oscillations. The speed variation, expressed as a percentage of rated speed, is derived from the following equation:

$$\text{Speed Variation (Percent)} = \frac{\left(\text{Change in speed above set speed}\right) + \left(\text{Change in speed below set speed}\right)}{\text{Rated speed}} 100$$

*Dead Band* is the characteristic of the speed-governing system (referred to as *wander*). It is the insensitivity of the speed-governing system and the total speed change during which the governing valve(s) does not change position to compensate for the speed change.

*Stability* is a measure of the ability of the speed-governing system to position the governor-controlled valve(s) so sustained oscillations of speed are not produced during a sustained load demand or following a change to a new load demand.

*Speed Oscillations* are the characteristics of the speed governing system (referred to as *hunt*). The ability of a governing system to keep sustained oscillations to a minimum is measured by its stability.

**Maximum Speed Rise,** expressed as a percentage of rated speed, is the maximum momentary increase in speed obtained when the turbine is developing rated power output at rated speed and the load is suddenly and completely reduced to zero. The maximum speed rise, expressed as a percentage of rated speed, is derived from the following equation:

$$\text{Speed Rise (Percent)} = \frac{\left(\text{Maximum speed at zero power output}\right) - \left(\text{Rated speed}\right)}{\text{Rated speed}} 100$$

**Trip Speed** is the speed at which the overspeed protective device operates.

## Other Controls

In addition to speed-governing controls, certain safety devices are required on steam turbines. These include an overspeed mechanism, which acts through a quick-tripping valve, independent of the main governor valve, to shut off the steam supply to the turbine, and a pressure-relieving valve in the turbine casing. Overspeed trip devices may act directly through linkages to close the steam valve or hydraulically by relieving oil pressure, allowing the valve to close. Also, the turbine must shut down should other safety devices, such as oil pressure failure or any of the refrigeration protective controls, so dictate. These devices usually act through an electrical interconnection to close the turbine trip valve mechanically or hydraulically.

To shorten the coast-down time of a tripped condensing-type turbine, a vacuum breaker in the turbine exhaust opens to admit air on receiving the trip signal.

## Lubrication

Small turbines may often have only simple oil rings to handle bearing lubrication, but most turbines in refrigeration service have a complete pressure-lubrication system. Basic components include a shaft-driven oil pump, an oil filter, an oil cooler, a means of regulating oil pressure, a reservoir, and interconnecting piping. Turbines having a hydraulic governor may use oil from the lubrication circuit or, with certain types of governors, have a self-contained governor oil system. To ensure an adequate supply of oil to bearings during acceleration and deceleration periods, many turbines include an auxiliary motor-driven or turbine-driven oil pump. Oil pressure-sensing devices act in two ways: (1) to stop the auxiliary pump once the shaft-driven pump has attained proper flow and pressure; or (2) to start the auxiliary pump if the shaft-driven pump fails or loses pressure when decelerating.

In some industrial applications, the lubrication systems of the turbine and the driven compressor are integrated. Proper oil pressure, temperature and compatability of the lubricant qualities must be maintained.

## Construction

Turbine manufacturers' standards prescribe casing materials for various limits of steam pressure and temperature. The use of built-up or solid rotors is a function of turbine speed or inlet steam temperature. Water must drain from pockets within the turbine casing to prevent damage caused by condensate accumulation. Carbon rings or closely fitted labyrinths prevent leakage of steam between pressure stages of the turbine, outward steam leakage, and inward air leakage at the turbine glands. Erosive and corrosive effect of moisture entering with the supply steam must be considered. Heat loss is prevented (often at the manufacturer's plant) by thermal insulation and protective metal jacketing on the hotter portions of the turbine casing.

## GAS TURBINES

The gas turbine has achieved an increasingly important position as a prime mover for centrifugal refrigeration compressor drives in applications ranging from 800 to 20,000 hp (0.5 to 15 MW).

It has the following advantages over other internal combustion engine drivers:

1. Small size
2. High power-to-weight ratio
3. Ability to burn a variety of fuels
4. Ability to meet stringent pollution standards
5. High reliability
6. Available in self-contained packages
7. Instant power; no warmup required
8. No cooling water required
9. Vibration-free operation
10. Easy maintenance
11. Low installation cost
12. Clean, dry exhaust
13. Lubrication oil not contaminated by combustion oil

### Application

Figure 16 shows a typical gas turbine refrigeration cycle. A gas turbine must be brought up to speed by an auxiliary starter. With a single-shaft turbine, the air compressor turbine, speed reducer gear, and refrigeration compressor must all be started and accelerated by this starter. The refrigeration compressor must be unloaded to ease this starting requirement. Sometimes, this may be done by making sure the capacity control vanes close tightly. At other times, it may be necessary to depressurize the refrigeration system to get started.

With a split-shaft design, only the air compressor and the gas-producer turbine must be started and accelerated. The rest of the unit starts rotating whenever enough energy is supplied to the blades of the power turbine. At this time, the gas-producer turbine is up to speed, and the fuel supply is ignited.

Electric starters are usually available as standard equipment. Reciprocating engines, steam turbines, and hydraulic or pneumatic motors may also be used.

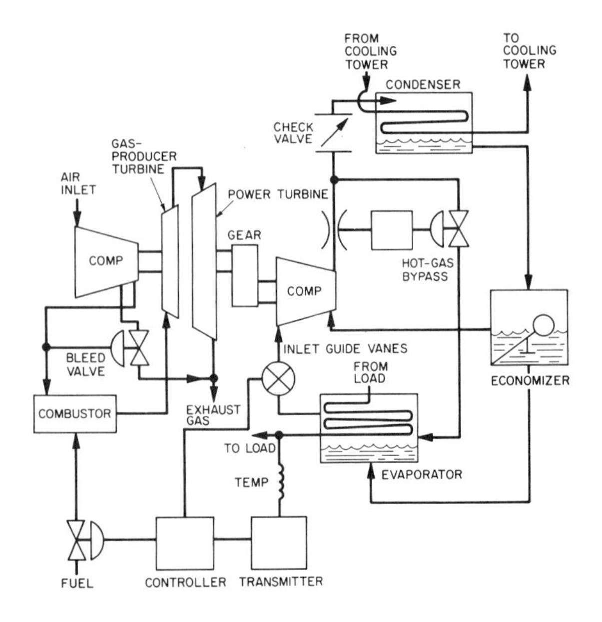


Fig. 16 Typical Gas Turbine Refrigeration Cycle

### Theory

The basic gas-turbine cycle (Figure 17) is the *Brayton Cycle* (open cycle), which consists of adiabatic compression, constant pressure heating, and adiabatic expansion. Figure 17 shows that the thermal efficiency of a gas turbine falls below the ideal value because of inefficiencies in the compressor and turbine, and because of duct losses. Increases in entropy occur during the compression and expansion processes, and a reduction in area enclosed by points 1, 2, 3, and 4 exists. This loss of area is a direct measure of the loss in efficiency of the cycle.

Nearly all turbine manufacturers present gas turbine engine performance in power terms and specific fuel consumption in terms of Btu/hp-h (W/W). A comparison of fuel consumption in specific terms is the quickest way to compare gas turbine overall thermal efficiencies.

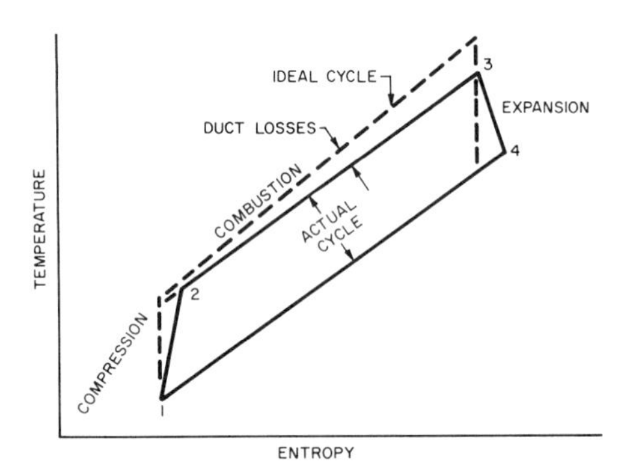


[a]Ideal cycle = dashed lines; solid lines = actual cycle.

Fig. 17 Temperature-Entropy Diagram for Brayton Cycle[a]

### Components

Figure 18 presents the major components of the gas-turbine unit: the air compressor, the combustor, and the turbine. Atmospheric air is compressed to about 175 psi (1200 kPa) by the air compressor. Fuel is then injected into the airstream and ignited in the combustor, with the leaving gases attaining temperatures between 1400 and 1800°F (760 and 980°C). These high-pressure hot gases are then expanded through a turbine, which provides not only the power required by the air compressor, but also power to drive the refrigeration machine.

Gas turbines are available in two major classifications—*single-shaft* (Figure 18) and *split-shaft* (Figures 19 and 20). The single-shaft turbine has the air compressor, the gas-producer turbine, and the power turbine on the same shaft. If the turbine is split,

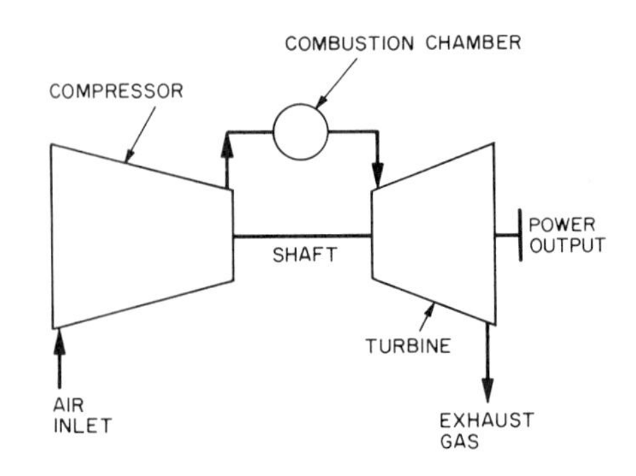


THERMAL EFFICIENCY RANGE 18-36%

Fig. 18 Single-Shaft Turbine

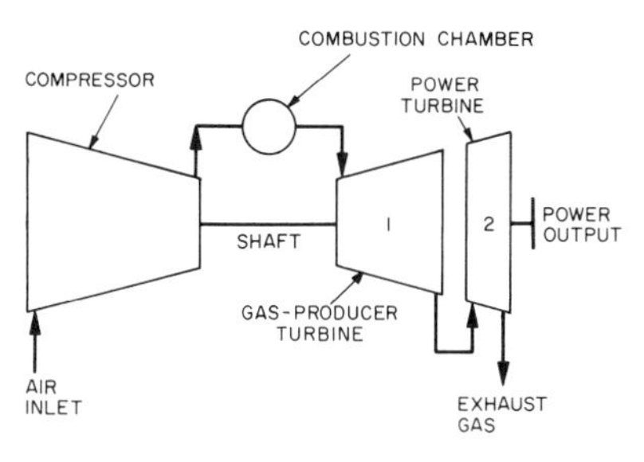


**Fig. 19 Split-Shaft Turbine**

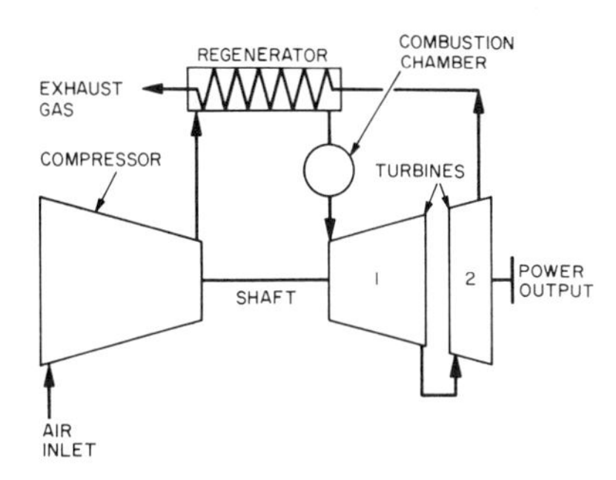


**Fig. 20 Split-Shaft Turbine with Regenerator**

with the section required for the air compression on one shaft and the section producing output power on a separate shaft, it is called a split-shaft or dual-shaft design. For a split-shaft turbine, the portion that includes the compressor, combustion chamber, and first turbine section is the *hot gas generator* or *gas producer.* The second turbine section is the *power turbine.*

The turbine used depends on job requirements. Single-shaft engines are usually selected when constant-speed drive is required, as in generator drives, and when starting torque requirements are low. A single-shaft engine can be used to drive centrifugal compressors, but the starting system and the compressor match point must be considered.

Split-shaft engines allow for variable speed at full load. An additional advantage of the split-shaft engine is that it can easily be started with a high torque load connected to the power output shaft.

## Performance

Most centrifugal refrigeration compressors are driven by constant-speed drivers, such as electric motors, and refrigeration capacity is controlled by variable inlet guide vanes on the compressor. Governors can maintain steam turbines at a constant speed to control the steam supply. Also, a fuel-supply regulator can maintain single-shaft gas turbines at a constant speed.

With the split-shaft design, the output shaft of the turbine unit runs at the speed required by the refrigeration compressor. The temperature of the chilled water or brine leaving the cooler of the refrigeration machine controls the fuel.

The rating of a gas turbine is affected by both inlet pressure to the air compressor and exhaust pressure from the turbine. In most applications, filters and silencers have to be installed in the air inlet. Silencers, waste heat boilers, or both are used on the exhaust. The pressure drop of these accessories and piping losses must be considered when determining the power output of the unit. The power output is also affected by altitude and atmospheric temperature. Gas turbine ratings are usually given at ISO conditions (59 °F [15 °C]) and sea level pressure) at the inlet flange of the air compressor and the exhaust flange of the turbine. Corrections for other conditions must be obtained from the manufacturer, as they will vary with each model. Some approximations follow:

1. Inlet temperature. Each 18 °F (10 °C) rise in inlet temperature decreases the power output by 9%, and vice versa.
2. An increase of 1000 ft in altitude decreases the power output by approximately 3.5%. (Each increase of 100 m in altitude decreases the power output by approximately 1.15%.)
3. Inlet pressure loss in filter, silencer, and ducting decreases power output by approximately 0.5% for each in. (2% per kPa) of water pressure loss.
4. Discharge pressure loss in exhaust heat boiler, silencer, and ducting decreases power output approximately 0.3% for each in. (1.2% per kPa) of water pressure loss.

Figure 21 shows a typical performance curve for a 10,000 hp (8 MW) turbine engine. For a given air inlet temperature, 86 °F (30 °C), for example, the engine develops its maximum power at about 82% of maximum speed.

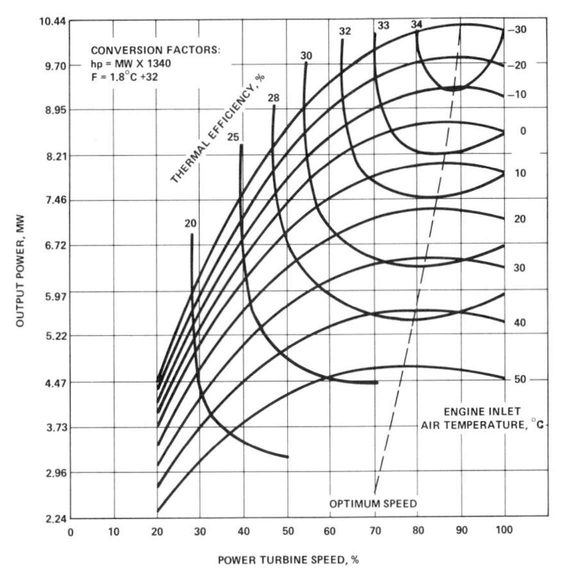


**Fig. 21 Typical Turbine Engine Performance Characteristics**

## EXHAUST HEAT USE

The overall thermal efficiency of the prime mover is 18 to 36%, with exhaust gases from the turbine ranging from 806 to 986 °F (430 to 530 °C). If the exhaust heat can be used, overall thermal efficiencies can be increased.

Figure 17 shows a regenerator that uses the heat of the exhaust gases to heat the air from the compressor prior to combustion. Overall thermal efficiency can be increased to between 28 and 38% by using a regenerator.

If process heat is required, the exhaust can satisfy a portion of that heat, and the combined system is a *cogeneration* system. The exhaust can be used (1) directly as a source of hot air, (2) in a large boiler or furnace as a source of preheated combustion air (as the exhaust contains about 16% oxygen), or (3) to heat a process or working fluid such as the steam system shown

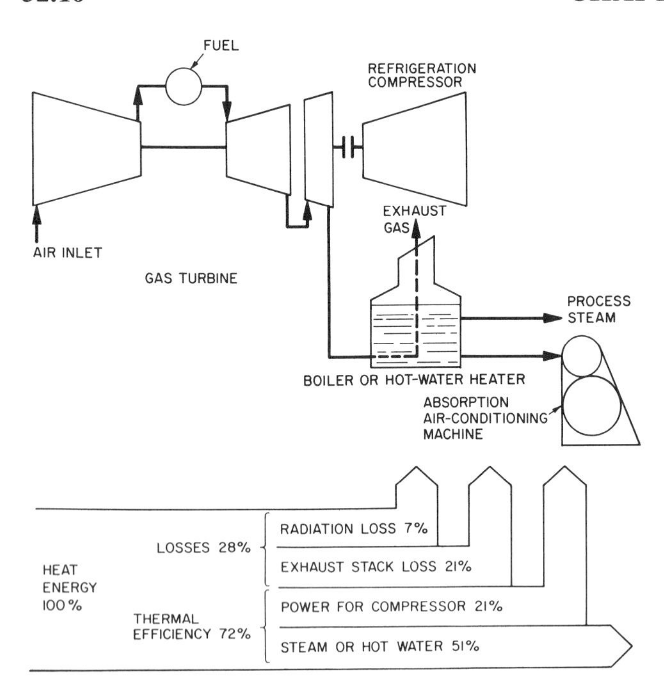


**Fig. 22 Gas Turbine Refrigeration System Using Exhaust Heat**

in Figure 22. Thermal efficiencies of these systems vary from 50 to 90% and over. The exhaust of a gas turbine has about 4000 to 8000 Btu/hp (5.7 to 11.3 MJ/kW) of available heat.

Additionally, because of the high oxygen content, the exhaust stream can support the combustion of an additional 30,000 Btu/hp (42.4 MJ/kW) of fuel. This additional heat can then be used in general manufacturing processes.

## Operation

Gas turbines operate with a wide range of fuels. For refrigeration service, a natural gas system is usually provided with an option of a standby No. 1 or No. 2 grade fuel oil system.

When selecting gas turbines, the output shaft must rotate in the direction required of the refrigeration compressor (in many cases, the manufacturer of split-shaft engines can provide the power turbines section with either direction of rotation).

At low loads, both the gas-turbine unit and the centrifugal refrigeration machine are affected by surge, which is a characteristic of all centrifugal- and axial-type compressors. At a certain pressure ratio or head, a minimum flow through the compressor is necessary to maintain stable operation. In the unstable area, a momentary backward flow of gas occurs through the compressor. Stable operation can be maintained, however, by the use of a hot-gas bypass valve.

The turbine manufacturer normally includes automatic surge protection, either as a bleed valve that bypasses a portion of the air directly from the axial compressor into the exhaust duct or by changing the position of the axial compressor stator vanes. Both methods are employed in some cases.

The assembly should be prevented from rotating backward, which may occur if the unit is suddenly stopped by one of the safety controls. The difference in pressure between the refrigeration condenser and cooler can make the compressor suddenly become a turbine and cause it to rotate in the opposite direction. This rotation can force hot turbine gases back through the air compressor, causing considerable damage. Reverse flow through the refrigeration compressor may be prevented in a variety of ways, depending on the system's components. When there is no refrigerant receiver, quick-closing inlet guide vanes are usually satisfactory, because there is very little high-pressure refrigerant to cause reverse rotation. A refrigeration system with a receiver has a substantial amount of energy available to cause reverse rotation. This can be reduced by opening the hot-gas bypass valve on shutdown and by employing a discharge check valve on the compressor.

Safety controls usually supplied with a gas turbine include the following:

1. Overspeed
2. Compressor surge
3. Overtemperature during operation under load
4. Low oil pressure
5. Failure to light off during start cycle
6. Underspeed during operation under load

**Noise.** On air-conditioning installations with gas turbine drivers, the control of noise cannot be overlooked. Gas turbine manufacturers have developed sound-attenuated enclosures that cover the turbine and gear package. Turbine drivers, when properly installed with a sound-attenuated enclosure, an inlet silencer, and an exhaust silencer, will meet the strictest noise standards. However, the turbine manufacturer should be consulted for detailed noise level data and recommendations on the least expensive method of attenuation for a particular installation.

**Pollution.** Gas turbine power plants have relatively low emission levels compared to other internal combustion engines; however, for each application, the gas turbine manufacturer should be consulted to ensure that local and state codes are met.

## Maintenance

Industrial gas turbines are designed to operate 12,000 to 30,000 hours between overhauls, with normal maintenance. Normal maintenance includes checking filter and oil level, inspection for leaks, etc., all of which can be done by the operator with ordinary mechanics' tools. However, the engine requires factory-trained service personnel to inspect engine components such as combustors and nozzles. These inspections, depending on the manufacturer's recommendations, are required as frequently as every 4000 hours of operation.

Most gas turbines are maintained by condition-monitoring and inspection rather than by specific overhaul intervals. Gas turbines specifically designed for industrial applications may have an indefinite life for the major housings, shafts, and low-temperature components. Hot section repair intervals for combustor and turbine components can vary from 10,000 to 100,000 hours.

The total cost of maintaining a gas turbine includes (1) cost of operator time, (2) normal parts replacement, (3) lubricating oil, (4) filter changes, (5) overhauls, and (6) factory service time (to conduct engine inspections). The cost of all of these items can be estimated by the manufacturer and must be taken into account to determine the total operating cost.

## REFERENCES

ASME. 1985. Guidance for Evaluation of Measurement Uncertainty in Performance Tests of Steam Turbines. ASME PTC 6-85 *Report*, American Society of Mechanical Engineers, New York.

ASME. 1982. Appendix A to Test Code for Steam Turbines. PTC 6A-82 *Addendum*.

ASME. 1970. Simplified Procedures for Routine Performance Tests of Steam Turbines. PTC 6S-70 *Report*.

NEMA. 1985. Steam Turbines for Mechanical Drive Service. NEMA *Standard* SM 23. National Electrical Manufacturers Association, Washington, DC.

Transamerica Delaval, Inc. 1983. *Transamerica Delaval Engineering Handbook*, 4th ed., McGraw-Hill, New York.

# CHAPTER 33

# PIPES, TUBES, AND FITTINGS



THIS chapter covers the selection, application, and installation of pipe, tubes, fittings, and valves commonly used for heating, air-conditioning, and refrigerating systems. Pipe hangers and pipe expansion are also addressed. When selecting and applying these components, applicable local codes, state or provincial codes, and voluntary industry standards (some of which have been adopted by code jurisdictions) must be followed.

The following U.S. organizations issue codes and standards for piping systems and components:

ASME — American Society of Mechanical Engineers
ASTM — American Society for Testing Materials
NFPA — National Fire Protection Association
BOCA — Building Officals & Code Administrators, International, Inc.
MSS — Manufacturers Standardization Society of the Valve & Fitting Industry, Inc.
ANSI — American National Standards Institute
AWWA— American Water Works Association

Parallel Federal Specifications also have been developed by government agencies and are used for many public works projects. Chapter IV of ASME *Standard* B31.9, *Building Services Piping*, lists applicable U.S. codes and standards for HVAC piping. In addition, it gives the requirements for safe design and construction of piping systems for building heating and air conditioning.

## PIPE

### Steel Pipe

Steel pipe is manufactured by several processes. Seamless pipe, made by piercing or extruding, has no longitudinal seam. Other manufacturing methods roll skelp into a cylinder and weld a longitudinal seam. A continuous-weld (CW) furnace-butt-welding process forces and joins the edges together at high temperature. An electric current welds the seam of electric resistance welded (ERW) pipe. ASTM *Standards* A106 and A53 specify steel pipe. Both specify an *A* and *B* grade. The *A* grade has a lower tensile strength and is not widely used.

The ASME pressure piping codes require that a longitudinal joint efficiency factor, *E*, be applied to each type of seam when calculating the allowable stress, as listed in Table 1. ASME *Standard* B36.10 specifies the dimensional standard for steel pipe. Through 12 in. (300 mm) diameter, nominal pipe sizes (NPS) are used, which do not match the internal or external diameters. For 14 in. (355 mm) and larger pipe, the size corresponds to the outside diameter.

Steel pipe is manufactured with wall thicknesses identified by schedule and weight. Although schedule numbers and weight designations are related, they are not constant for all pipe sizes. Standard weight (STD) and Schedule 40 pipe have the same wall thickness through 10 in. (250 mm) NPS. For 12 in. (300 mm) and larger standard weight pipe, wall thicknesses increase with each size. A similar equality exists between Extra Strong (XS) and Schedule 80 pipe up through 8 in. (200 mm); afterward, XS pipe has a 0.500 in. (12.7 mm) wall, while Schedule 80 increases in wall thickness. Table 2 lists properties of representative steel pipe.

Joints in steel pipe are made by welding or using threaded, flanged, grooved, or welded outlet fittings. Unreinforced branch connections weaken a main pipeline, and added reinforcement is necessary, unless the excess wall thickness of both mains and branches are sufficient to sustain the pressure.

ASME *Standard* B31.1, *Power Piping*, gives formulas for determining when reinforcement is required. Such calculations are seldom needed in HVAC applications because (1) standard weight pipe through 20 in. (510 mm) NPS at 300 psig (2.1 MPa) requires no reinforcement, but full-size branch connections are not recommended; and (2) fittings such as tees and reinforced outlet fittings provide inherent reinforcements for sizes above 20 in. (510 mm) and pressures above 200 psig (1.4 MPa).

The preparation of this chapter is assigned to TC 8.10, Pumps and Hydronic Piping.

Table 1 Allowable Stresses for Pipe and Tube[a]

| Specification | Grade | Type | Manufacturing Process | Available Sizes, in. | Minimum Tensile Strength ksi | Basic Allowable Stress, $S$ ksi | Joint Efficiency Factor, $E$ | Allowable Stress, $SE^b$, ksi | Allowable Stress Range, $S_A{}^c$, ksi |
|---|---|---|---|---|---|---|---|---|---|
| ASTM A53 Steel | — | F | Cont. Weld | 1/2 to 4 | 45.0 | 11.25 | 0.6 | 6.8 | 16.9 |
| ASTM A53 Steel | B | S | Seamless | 1/2 to 26 | 60.0 | 15.0 | 1.0 | 15.0 | 22.5 |
| ASTM A53 Steel | B | E | ERW | 2 to 20 | 60.0 | 15.0 | 0.85 | 12.8 | 22.5 |
| ASTM A106 Steel | B | S | Seamless | 1/2 to 26 | 60.0 | 15.9 | 1.0 | 15.0 | 22.5 |
| ASTM B-88 Copper | — | — | Hard Drawn | 1/4 to 12 | 36.0 | 9.0 | 1.0 | 9.0 | 13.5 |

[a] Listed stresses are for temperatures to 650 °F (343 °C) for steel pipe and 250 °F (121 °C) for copper tubing.
[b] To be used for Internal Pressure Stress Calculations in Eqs. (1) and (2).
[c] To be used only for piping flexibility calculations; Eqs. (3) and (4).
Conversion Factors: mm = 25.4 × in. MPa = 6.89 × ksi ksi = 1000 lb

## Copper Tube

Because of their inherent resistance to corrosion and ease of installation, copper and copper alloys are often used in heating, air-conditioning, refrigeration, and water supply installations. There are two principal classes of copper tube. ASTM *Standard* B88 includes Type *K*, *L*, *M*, and *DWV* for water and drain service. ASTM *Standard* B280 specifies ACR (Air-Conditioning and Refrigeration) tube for refrigeration service.

Type *K*, *L*, *M*, and *DWV* designate descending wall thicknesses for copper tube. All types have the same outside diameter for corresponding sizes. Table 3 lists properties of ASTM B88 copper tube. In the plumbing industry, tube of nominal size approximates the inside diameter. The heating and refrigeration trades specify copper tube by the outside diameter (OD). ACR tubing has a different set of wall thicknesses. Types *K*, *L*, and *M* tube may be hard drawn or annealed (soft) temper.

Copper tubing is joined with soldered or brazed wrought or cast copper capillary socket-end fittings. Table 4 lists the pressure-temperature ratings of soldered and brazed joints. Small copper tube is also joined by flare or compression fittings.

Hard-drawn tubing has a higher allowable stress value than annealed, but if hard tubing is joined by soldering or brazing, the annealed allowable stress value should be used.

Brass pipe and copper pipe are also made in steel pipe thicknesses for threading. High cost has eliminated these materials from the market.

The heating and air-conditioning industry generally uses Type *L* and *M* tubing, which have a higher internal working pressure rating than the solder-joints used at fittings. Type *K* may be used with brazed joints for higher pressure-temperature requirements or for direct burial. Type *M* should be used with care where exposed to potential external damage.

Copper and brass should not be used in ammonia refrigerating systems. The "Special Requirements" section in this chapter covers other limitations on refrigerant piping.

## Ductile Iron and Cast Iron

Cast-iron soil pipe comes in two thicknesses—extra-heavy and service weight. It is not used under pressure because the pipe is not suitable and the joints are not restrained. Cast-iron pipe and fittings typically have bell and spigot ends for lead and oakum joints or elastomer push-on joints. Cast-iron pipe and fittings are also furnished with *no-hub* ends for joining with *no-hub* clamps. Local plumbing codes specify permitted materials and joints.

Ductile iron has now replaced cast iron for pressure pipe. Ductile iron is stronger, less brittle, and similar to cast iron in corrosion resistance. It is commonly used for buried pressure water mains or in other locations where internal or external corrosion is a problem. Joints are made with flanged fittings, mechanical joint (MJ) fittings, or elastomer gaskets for bell and spigot ends. Bell and spigot and MJ joints are not self restrained. Restrained MJ systems are available. Ductile iron pipe is made in seven thickness classes for different service conditions. AWWA C150/A21.50, *Thickness Design of Ductile Iron Pipe*, covers the proper selection of pipe classes.

## Non-Metallic

Plastic pipe falls into two main categories: (1) thermoplastic and (2) thermosetting resins. Thermoplastic piping is not recommended for air or compressed gas service. Thermoplastics are the most commonly used in HVAC and plumbing practice. These plastics are light, low priced, and corrosion resistant and offer many advantages for systems that operate at close-to-ambient temperatures. The pressure ratings fall off rapidly as the fluid temperature rises above 100 °F (40 °C). Allowable hydrostatic design stresses are listed in ASME *Standard* B31.9. The manufacturer's corrosion-resistance charts should be used to select the proper material for each application. Polyvinyl chloride (PVC), made to ASTM *Standard* D1785; chlorinated polyvinyl chloride (CPVC), made to ASTM *Standard* F441; and polybutylene (PB), made to ASTM *Standards* D3309, D3000, and D2666 are most often used. PV and CPVC are available in sizes up to 12 in. (300 mm) and PB in sizes up to 6 in. (150 mm). CPVC and PB (hot water resin) retain their strength better at elevated temperatures. PB is joined by socket heat fusion, mechanical, and insert/crimp joining methods. PVC and CPVC are joined with threaded, solvent cemented, or flanged fittings. For especially corrosive waste or pressure piping systems, Polypropylene (PP) or Polyvinylidene Flouride (PVDF-Kynar) thermoplastics can be used. PP and PVDF use fittings with threaded, flanged, or heat fusion joints. Threaded joints require the use of Schedule 80 pipe.

Reinforced thermosetting resin (RTR) piping systems are made by several manufacturing methods using fiberglass reinforcing in epoxy or vinyl ester resins. Also known as fiberglass reinforced plastic (FRP), this pipe provides high strength and temperature resistance and is made in sizes up to 48 in. (1220 mm). The piping is made with bell and spigot ends for use with *0*-rings and tapered bell and spigot ends that are joined with adhesive. Flanged and mechanical joint connections are also used. Many thermosetting pipe systems are available, and the manufacturer's recommendations should be followed for installation method and joining techniques.

# FITTINGS

Table 5 lists standards giving dimensions and pressure ratings for fittings, flanges, and flanged fittings. This data is also available from manufacturers' catalogs.

# JOINING METHODS

Threading is the most commonly used method for joining small

**Table 2 Steel Pipe Data**

| Nom. Size and Pipe O.D., *D*, in. | Schedule Number or Weight[a] | Wall Thick-ness, in. *t* | Inside Diameter in. *d* | Surface Area | | Cross-Sectional | | Weight of | | Working Pressure[b] ASTM A53 B to 400 °F | | |
|---|---|---|---|---|---|---|---|---|---|---|---|---|
| | | | | Outside $ft^2/ft$ | Inside $ft^2/ft$ | Metal Area $in^2$ | Flow Area $in^2$ | Pipe lb/ft | Water lb/ft | Mfr. Pro-cess | Joint Type | psig |
| 1/4 | 40 ST | 0.088 | 0.364 | 0.141 | 0.095 | 0.125 | 0.104 | 0.424 | 0.045 | CW | Thrd | 188 |
| D = 0.540 | 80 XS | 0.119 | 0.302 | 0.141 | 0.079 | 0.157 | 0.072 | 0.535 | 0.031 | CW | Thrd | 871 |
| 3/8 | 40 ST | 0.091 | 0.493 | 0.177 | 0.129 | 0.167 | 0.191 | 0.567 | 0.083 | CW | Thrd | 203 |
| D = 0.675 | 80 XS | 0.126 | 0.423 | 0.177 | 0.111 | 0.217 | 0.141 | 0.738 | 0.061 | CW | Thrd | 820 |
| 1/2 | 40 ST | 0.109 | 0.622 | 0.220 | 0.163 | 0.250 | 0.304 | 0.850 | 0.131 | CW | Thrd | 214 |
| D = 0.840 | 80 XS | 0.147 | 0.546 | 0.220 | 0.143 | 0.320 | 0.234 | 1.087 | 0.101 | CW | Thrd | 753 |
| 3/4 | 40 ST | 0.113 | 0.824 | 0.275 | 0.216 | 0.333 | 0.533 | 1.13 | 0.231 | CW | Thrd | 217 |
| D = 1.050 | 80 XS | 0.154 | 0.742 | 0.275 | 0.194 | 0.433 | 0.432 | 1.47 | 0.187 | CW | Thrd | 681 |
| 1 | 40 ST | 0.133 | 1.049 | 0.344 | 0.275 | 0.494 | 0.864 | 1.68 | 0.374 | CW | Thrd | 226 |
| D = 1.315 | 80 XS | 0.179 | 0.957 | 0.344 | 0.251 | 0.639 | 0.719 | 2.17 | 0.311 | CW | Thrd | 642 |
| 1 1/4 | 40 ST | 0.140 | 1.380 | 0.435 | 0.361 | 0.669 | 1.50 | 2.27 | 0.647 | CW | Thrd | 229 |
| D = 1.660 | 80 XS | 0.191 | 1.278 | 0.435 | 0.335 | 0.881 | 1.28 | 2.99 | 0.555 | CW | Thrd | 594 |
| 1 1/2 | 40 ST | 0.145 | 1.610 | 0.497 | 0.421 | 0.799 | 2.04 | 2.72 | 0.881 | CW | Thrd | 231 |
| D = 1.900 | 80 XS | 0.200 | 1.500 | 0.497 | 0.393 | 1.068 | 1.77 | 3.63 | 0.765 | CW | Thrd | 576 |
| 2 | 40 ST | 0.154 | 2.067 | 0.622 | 0.541 | 1.07 | 3.36 | 3.65 | 1.45 | CW | Thrd | 230 |
| D = 2.375 | 80 XS | 0.218 | 1.939 | 0.622 | 0.508 | 1.48 | 2.95 | 5.02 | 1.28 | CW | Thrd | 551 |
| 2 1/2 | 40 ST | 0.203 | 2.469 | 0.753 | 0.646 | 1.70 | 4.79 | 5.79 | 2.07 | CW | Weld | 533 |
| D = 2.875 | 80 XS | 0.276 | 2.323 | 0.753 | 0.608 | 2.25 | 4.24 | 7.66 | 1.83 | CW | Weld | 835 |
| 3 | 40 ST | 0.216 | 3.068 | 0.916 | 0.803 | 2.23 | 7.39 | 7.57 | 3.20 | CW | Weld | 482 |
| D = 3.500 | 80 XS | 0.300 | 2.900 | 0.916 | 0.759 | 3.02 | 6.60 | 10.25 | 2.86 | CW | Weld | 767 |
| 4 | 40 ST | 0.237 | 4.026 | 1.178 | 1.054 | 3.17 | 12.73 | 10.78 | 5.51 | CW | Weld | 430 |
| D = 4.500 | 80 XS | 0.337 | 3.826 | 1.178 | 1.002 | 4.41 | 11.50 | 14.97 | 4.98 | CW | Weld | 695 |
| 6 | 40 ST | 0.280 | 6.065 | 1.734 | 1.588 | 5.58 | 28.89 | 18.96 | 12.50 | ERW | Weld | 696 |
| D = 6.625 | 80 XS | 0.432 | 5.761 | 1.734 | 1.508 | 8.40 | 26.07 | 28.55 | 11.28 | ERW | Weld | 1209 |
| 8 | 30 | 0.277 | 8.071 | 2.258 | 2.113 | 7.26 | 51.16 | 24.68 | 22.14 | ERW | Weld | 526 |
| D = 8.625 | 40 ST | 0.322 | 7.981 | 2.258 | 2.089 | 8.40 | 50.03 | 28.53 | 21.65 | ERW | Weld | 643 |
| | 80 XS | 0.500 | 7.625 | 2.258 | 1.996 | 12.76 | 45.66 | 43.35 | 19.76 | ERW | Weld | 1106 |
| | 30 | 0.307 | 10.136 | 2.814 | 2.654 | 10.07 | 80.69 | 34.21 | 34.92 | ERW | Weld | 485 |
| 10 | 40 ST | 0.365 | 10.020 | 2.814 | 2.623 | 11.91 | 78.85 | 40.45 | 34.12 | ERW | Weld | 606 |
| D = 10.75 | XS | 0.500 | 9.750 | 2.814 | 2.552 | 16.10 | 74.66 | 54.69 | 32.31 | ERW | Weld | 887 |
| | 80 | 0.593 | 9.564 | 2.814 | 2.504 | 18.92 | 71.84 | 64.28 | 31.09 | ERW | Weld | 1081 |
| | 30 | 0.330 | 12.090 | 3.338 | 3.165 | 12.88 | 114.8 | 43.74 | 49.68 | ERW | Weld | 449 |
| 12 | ST | 0.375 | 12.000 | 3.338 | 3.141 | 14.58 | 113.1 | 49.52 | 48.94 | ERW | Weld | 528 |
| D = 12.75 | 40 | 0.406 | 11.938 | 3.338 | 3.125 | 15.74 | 111.9 | 53.48 | 48.44 | ERW | Weld | 583 |
| | XS | 0.500 | 11.750 | 3.338 | 3.076 | 19.24 | 108.4 | 65.37 | 46.92 | ERW | Weld | 748 |
| | 80 | 0.687 | 11.376 | 3.338 | 2.978 | 26.03 | 101.6 | 88.44 | 43.98 | ERW | Weld | 1076 |
| | 30 ST | 0.375 | 13.250 | 3.665 | 3.469 | 16.05 | 137.9 | 54.53 | 59.67 | ERW | Weld | 481 |
| 14 | 40 | 0.437 | 13.126 | 3.665 | 3.436 | 18.62 | 135.3 | 63.25 | 58.56 | ERW | Weld | 580 |
| D = 14.00 | XS | 0.500 | 13.000 | 3.665 | 3.403 | 21.21 | 132.7 | 72.04 | 57.44 | ERW | Weld | 681 |
| | 80 | 0.750 | 12.500 | 3.665 | 3.272 | 31.22 | 122.7 | 106.05 | 53.11 | ERW | Weld | 1081 |
| 16 | 30 ST | 0.375 | 15.250 | 4.189 | 3.992 | 18.41 | 182.6 | 62.53 | 79.04 | ERW | Weld | 421 |
| D = 16.00 | 40 XS | 0.500 | 15.000 | 4.189 | 3.927 | 24.35 | 176.7 | 82.71 | 76.47 | ERW | Weld | 596 |
| | ST | 0.375 | 17.250 | 4.712 | 4.516 | 20.76 | 233.7 | 70.54 | 101.13 | ERW | Weld | 374 |
| 18 | 30 | 0.437 | 17.126 | 4.712 | 4.483 | 24.11 | 230.3 | 81.91 | 99.68 | ERW | Weld | 451 |
| D = 18.00 | XS | 0.500 | 17.000 | 4.712 | 4.450 | 27.49 | 227.0 | 93.38 | 98.22 | ERW | Weld | 530 |
| | 40 | 0.562 | 16.876 | 4.712 | 4.418 | 30.79 | 223.7 | 104.59 | 96.80 | ERW | Weld | 607 |
| 20 | ST | 0.375 | 19.250 | 5.236 | 5.039 | 23.12 | 291.0 | 78.54 | 125.94 | ERW | Weld | 337 |
| D = 20.00 | 30 XS | 0.500 | 19.000 | 5.236 | 4.974 | 30.63 | 283.5 | 104.05 | 122.69 | ERW | Weld | 477 |
| | 40 | 0.593 | 18.814 | 5.236 | 4.925 | 36.15 | 278.0 | 122.82 | 120.30 | ERW | Weld | 581 |

[a] Numbers are schedule numbers per ASTM B36.10; ST = Standard Weight; XS = Extra Strong.

[b] Working pressures have been calculated per ASME/ANSI B31.9 using furnace butt weld (continuous weld, CW) pipe through 4 in. and electric resistance weld (ERW) thereafter. The allowance, A, has been taken as:

(a) 12.5% of *t* for mill tolerance on pipe wall thickness, plus

(b) An arbitrary corrosion allowance of 0.025 in. for pipe sizes through NPS 2 and 0.065 in. from NPS 2½ through 20, plus

(c) A thread cutting allowance for sizes through NPS 2.

Because the pipe wall thickness of threaded standard weight pipe is so small after deducting the allowance, A, the mechanical strength of the pipe is impaired. It is good practice to limit standard weight threaded pipe pressures to 90 psig for steam and 125 psig for water.

Conversion Factors:
mm = 25.4 × in.
$m^2/m$ = 0.3048 × $ft^2/ft$
$mm^2$ = 645 × $in.^2$
kg/m = 1.49 × lb/ft
kPa gauge = 6.89 × psig

**Table 3 Copper Tube Data**

| Nominal Diameter | Type | Wall Thickness, $t$, in. | Diameter Outside, $D$, in. | Diameter Inside, $d$, in. | Surface Area Outside $ft^2/ft$ | Surface Area Inside $ft^2/ft$ | Cross-Sectional Metal Area $in^2$ | Cross-Sectional Flow Area $in^2$ | Weight of Tube lb/ft | Weight of Water lb/ft | Working Pressure[a,b,c] ASTM B88 to 250 °F Annealed psig | Working Pressure[a,b,c] ASTM B88 to 250 °F Drawn psig |
|---|---|---|---|---|---|---|---|---|---|---|---|---|
| 1/4 | K | 0.035 | 0.375 | 0.305 | 0.098 | 0.080 | 0.037 | 0.073 | 0.145 | 0.032 | 851 | 1596 |
| | L | 0.030 | 0.375 | 0.315 | 0.098 | 0.082 | 0.033 | 0.078 | 0.126 | 0.034 | 730 | 1368 |
| 3/8 | K | 0.049 | 0.500 | 0.402 | 0.131 | 0.105 | 0.069 | 0.127 | 0.269 | 0.055 | 894 | 1676 |
| | L | 0.035 | 0.500 | 0.430 | 0.131 | 0.113 | 0.051 | 0.145 | 0.198 | 0.063 | 638 | 1197 |
| | M | 0.025 | 0.500 | 0.450 | 0.131 | 0.008 | 0.037 | 0.159 | 0.145 | 0.069 | 456 | 855 |
| 1/2 | K | 0.049 | 0.625 | 0.527 | 0.164 | 0.138 | 0.089 | 0.218 | 0.344 | 0.094 | 715 | 1341 |
| | L | 0.040 | 0.625 | 0.545 | 0.164 | 0.143 | 0.074 | 0.233 | 0.285 | 0.101 | 584 | 1094 |
| | M | 0.028 | 0.625 | 0.569 | 0.164 | 0.149 | 0.053 | 0.254 | 0.203 | 0.110 | 409 | 766 |
| 5/8 | K | 0.049 | 0.750 | 0.652 | 0.196 | 0.171 | 0.108 | 0.334 | 0.418 | 0.144 | 596 | 1117 |
| | L | 0.042 | 0.750 | 0.666 | 0.196 | 0.174 | 0.093 | 0.348 | 0.362 | 0.151 | 511 | 958 |
| 3/4 | K | 0.065 | 0.875 | 0.745 | 0.229 | 0.195 | 0.165 | 0.436 | 0.641 | 0.189 | 677 | 1270 |
| | L | 0.045 | 0.875 | 0.785 | 0.229 | 0.206 | 0.117 | 0.484 | 0.455 | 0.209 | 469 | 879 |
| | M | 0.032 | 0.875 | 0.811 | 0.229 | 0.212 | 0.085 | 0.517 | 0.328 | 0.224 | 334 | 625 |
| 1 | K | 0.065 | 1.125 | 0.995 | 0.295 | 0.260 | 0.216 | 0.778 | 0.839 | 0.336 | 527 | 988 |
| | L | 0.050 | 1.125 | 1.025 | 0.295 | 0.268 | 0.169 | 0.825 | 0.654 | 0.357 | 405 | 760 |
| | M | 0.035 | 1.125 | 1.055 | 0.295 | 0.276 | 0.120 | 0.874 | 0.464 | 0.378 | 284 | 532 |
| 1 1/4 | K | 0.065 | 1.375 | 1.245 | 0.360 | 0.326 | 0.268 | 1.217 | 1.037 | 0.527 | 431 | 808 |
| | L | 0.055 | 1.375 | 1.265 | 0.360 | 0.331 | 0.228 | 1.257 | 0.884 | 0.544 | 365 | 684 |
| | M | 0.042 | 1.375 | 1.291 | 0.360 | 0.338 | 0.176 | 1.309 | 0.682 | 0.566 | 279 | 522 |
| | DWV | 0.040 | 1.375 | 1.295 | 0.360 | 0.339 | 0.168 | 1.317 | 0.650 | 0.570 | 265 | 497 |
| 1 1/2 | K | 0.072 | 1.625 | 1.481 | 0.425 | 0.388 | 0.351 | 1.723 | 1.361 | 0.745 | 404 | 758 |
| | L | 0.060 | 1.625 | 1.505 | 0.425 | 0.394 | 0.295 | 1.779 | 1.143 | 0.770 | 337 | 631 |
| | M | 0.049 | 1.625 | 1.527 | 0.425 | 0.400 | 0.243 | 1.831 | 0.940 | 0.792 | 275 | 516 |
| | DWV | 0.042 | 1.625 | 1.541 | 0.425 | 0.403 | 0.209 | 1.865 | 0.809 | 0.807 | 236 | 442 |
| 2 | K | 0.083 | 2.125 | 1.959 | 0.556 | 0.513 | 0.532 | 3.014 | 2.063 | 1.304 | 356 | 668 |
| | L | 0.070 | 2.125 | 1.985 | 0.556 | 0.520 | 0.452 | 3.095 | 1.751 | 1.339 | 300 | 573 |
| | M | 0.058 | 2.125 | 2.009 | 0.556 | 0.526 | 0.377 | 3.170 | 1.459 | 1.372 | 249 | 467 |
| | DWV | 0.042 | 2.125 | 2.041 | 0.556 | 0.534 | 0.275 | 3.272 | 1.065 | 1.416 | 180 | 338 |
| 2 1/2 | K | 0.095 | 2.625 | 2.435 | 0.687 | 0.637 | 0.755 | 4.657 | 2.926 | 2.015 | 330 | 619 |
| | L | 0.080 | 2.625 | 2.465 | 0.687 | 0.645 | 0.640 | 4.772 | 2.479 | 2.065 | 278 | 521 |
| | M | 0.065 | 2.625 | 2.495 | 0.687 | 0.653 | 0.523 | 4.889 | 2.026 | 2.116 | 226 | 423 |
| 3 | K | 0.109 | 3.125 | 2.907 | 0.818 | 0.761 | 1.033 | 6.637 | 4.002 | 2.872 | 318 | 596 |
| | L | 0.090 | 3.125 | 2.945 | 0.818 | 0.771 | 0.858 | 6.812 | 3.325 | 2.947 | 263 | 492 |
| | M | 0.072 | 3.125 | 2.981 | 0.818 | 0.780 | 0.691 | 6.979 | 2.676 | 3.020 | 210 | 394 |
| | DWV | 0.045 | 3.125 | 3.035 | 0.818 | 0.795 | 0.435 | 7.234 | 1.687 | 3.130 | 131 | 246 |
| 3 1/2 | K | 0.120 | 3.625 | 3.385 | 0.949 | 0.886 | 1.321 | 8.999 | 5.120 | 3.894 | 302 | 566 |
| | L | 0.100 | 3.625 | 3.425 | 0.949 | 0.897 | 1.107 | 9.213 | 4.291 | 3.987 | 252 | 472 |
| | M | 0.083 | 3.625 | 3.459 | 0.949 | 0.906 | 0.924 | 9.397 | 3.579 | 4.066 | 209 | 392 |
| 4 | K | 0.134 | 4.125 | 3.857 | 1.080 | 1.010 | 1.680 | 11.684 | 6.510 | 5.056 | 296 | 555 |
| | L | 0.110 | 4.125 | 3.905 | 1.080 | 1.022 | 1.387 | 11.977 | 5.377 | 5.182 | 243 | 456 |
| | M | 0.095 | 4.125 | 3.935 | 1.080 | 1.030 | 1.203 | 12.161 | 4.661 | 5.262 | 210 | 394 |
| | DWV | 0.058 | 4.125 | 4.009 | 1.080 | 1.050 | 0.741 | 12.623 | 2.872 | 5.462 | 128 | 240 |
| 5 | K | 0.160 | 5.125 | 4.805 | 1.342 | 1.258 | 2.496 | 18.133 | 9.671 | 7.846 | 285 | 534 |
| | L | 0.125 | 5.125 | 4.875 | 1.342 | 1.276 | 1.963 | 18.665 | 7.609 | 8.077 | 222 | 417 |
| | M | 0.109 | 5.125 | 4.907 | 1.342 | 1.285 | 1.718 | 18.911 | 6.656 | 8.183 | 194 | 364 |
| | DWV | 0.072 | 5.125 | 4.981 | 1.342 | 1.304 | 1.143 | 19.486 | 4.429 | 8.432 | 128 | 240 |
| 6 | K | 0.192 | 6.125 | 5.741 | 1.603 | 1.503 | 3.579 | 25.886 | 13.867 | 11.201 | 286 | 536 |
| | L | 0.140 | 6.125 | 5.845 | 1.603 | 1.530 | 2.632 | 26.832 | 10.200 | 11.610 | 208 | 391 |
| | M | 0.122 | 6.125 | 5.881 | 1.603 | 1.540 | 2.301 | 27.164 | 8.916 | 11.754 | 182 | 341 |
| | DWV | 0.083 | 6.125 | 5.959 | 1.603 | 1.560 | 1.575 | 27.889 | 6.105 | 12.068 | 124 | 232 |
| 8 | K | 0.271 | 8.125 | 7.583 | 2.127 | 1.985 | 6.687 | 45.162 | 25.911 | 19.542 | 304 | 570 |
| | L | 0.200 | 8.125 | 7.725 | 2.127 | 2.022 | 4.979 | 46.869 | 19.295 | 20.280 | 224 | 421 |
| | M | 0.170 | 8.125 | 7.785 | 2.127 | 2.038 | 4.249 | 47.600 | 16.463 | 20.597 | 191 | 358 |
| | DWV | 0.109 | 8.125 | 7.907 | 2.127 | 2.070 | 2.745 | 49.104 | 10.637 | 21.247 | 122 | 229 |
| 10 | K | 0.338 | 10.125 | 9.449 | 2.651 | 2.474 | 10.392 | 70.123 | 40.271 | 30.342 | 304 | 571 |
| | L | 0.250 | 10.125 | 9.625 | 2.651 | 2.520 | 7.756 | 72.760 | 30.054 | 31.483 | 225 | 422 |
| | M | 0.212 | 10.125 | 9.701 | 2.651 | 2.540 | 6.602 | 73.913 | 25.584 | 31.982 | 191 | 358 |
| 12 | K | 0.405 | 12.125 | 11.315 | 3.174 | 2.962 | 14.912 | 100.554 | 57.784 | 43.510 | 305 | 571 |
| | L | 0.280 | 12.125 | 11.565 | 3.174 | 3.028 | 10.419 | 105.046 | 40.375 | 45.454 | 211 | 395 |
| | M | 0.254 | 12.125 | 11.617 | 3.174 | 3.041 | 9.473 | 105.993 | 36.706 | 45.863 | 191 | 358 |

Notes for Table 3:

[a] When using soldered or brazed fittings, the joint determines the limiting pressure.
[b] Working pressures calculated using ASME B31.9 allowable stresses. A 5% mill tolerance has been used on the wall thickness. Higher tube ratings can be calculated using the allowable stress for lower temperatures.
[c] If soldered or brazed fittings are used on hard drawn tubing, use the annealed ratings. Full-tube allowable pressures can be used with suitably rated flare or compression-type fittings.

Conversion Factors: mm = 25.4 × in.
$m^2/m = 0.3048 \times ft^2/ft$
$mm^2 = 645 \times in.^2$
kg/m = 1.49 × lb/ft
kPa gauge = 6.89 × psig

diameter steel or brass pipe, as shown in ASME *Standard* B1.20.1. Pipe with a wall thickness less than standard weight should not be threaded. ASME *Standard* B31.5, *Refrigeration Piping*, limits the threading for various refrigerants and pipe sizes.

## Soldering and Brazing

Copper tube is usually joined by soldering or brazing socket end fittings. Brazing materials melt at temperatures over 1000 °F (540 °C) and produce a stronger joint than solder. Table 4 lists soldered and brazed joint strengths. Lead-based solders should not be used for potable water systems.

## Flared and Compression Joints

Flared and compression fittings can be used to join copper, steel, stainless steel, and aluminum tubing. Properly rated fittings can keep the joints as strong as the tube.

## Flanges

Flanges can be used for large pipe and all piping materials. They are commonly used to connect to equipment, valves, and wherever it may be necesssary to open the joint to permit service or replacement of components. For steel pipe, flanges are available in pressure ratings to 2500 psig (17 MPa). For welded pipe, weld neck, slip-on, or socket weld connections are available. Thread-on flanges are available for threaded pipe.

Flanges are generally flat faced or raised face. Flat-faced flanges with full-faced gaskets are generally used with cast iron and materials that cannot take high bending loads. Raised-face flanges with ring gaskets are preferred with steel pipe because they facilitate increasing the sealing pressure on the gasket to help prevent leaks. Other facings, such as *O*-ring and ring joint, are available for special applications.

All flat-faced, raised-face, and lap-joint flanges require a gasket between the mating flange surfaces. Gaskets are made from rubber, synthetic elastomers, cork, fiber, plastic, teflon, metal, and a combination of these materials. The gasket must be compatible with the flowing media and the temperatures at which the system is operating.

## Welding

Welding steel pipe joints over 2 in. (50 mm) in diameter offer the following advantages:

1. Welded joints do not age, dry out, or deteriorate as do gasketed joints.
2. Welded joints can accommodate greater vibration and water hammer and higher temperatures and pressures than other joints.
3. For critical service, pipe joints can be tested by any of several non-destructive examination (NDE) methods, such as by radiography or ultrasound.
4. Welded joints provide maximum long-term reliability.

The applicable section of ASME *Standard* B31, *Code for Pressure Piping*, and the ASME *Boiler and Pressure Vessel Code* specify the proper welding methods. The *Code for Pressure Piping* series requires that all welders and welding procedure specifications (WPS) be qualified. Separate WPS are needed for different welding methods and materials. The qualifying tests and the variables requiring separate procedure specifications are set forth in the ASME *Boiler and Pressure Vessel Code*—Section IX. The manufacturer, fabricator, or contractor is responsible for the welding procedure and welders. Section B31.9, *Building Services Piping*, of the *Code for Pressure Piping* requires visual examination of welds and outlines limits of acceptability.

The following welding processes are often used in the HVAC industry:

SMAW—Shielded Metal Arc Welding (stick welding). The molten weld metal is shielded by the vaporization of the electrode coating.

GMAW—Gas Metal Arc Welding, also called MIG. The elec-

**Table 4 Internal Working Pressure for Copper Tube Joints**
(Based on ANSI/ASME *Standard* B31.9, *Building Services Piping*)

| Alloy Used for Joints | Service Temperature, °F | Water and Non-Corrosive Liquids and Gases[a] Internal Working Pressure, psi — Nominal Tube Size, Type K, L, M, in. 1/4 to 1 | 1 1/4 to 2 | 2 1/2 to 4 | 5 to 8[a] | 10 to 12[a] | Sat. Steam and Condensate 1/4 to 8 |
|---|---|---|---|---|---|---|---|
| 50-50 | 100 | 200 | 175 | 150 | 130 | 100 | — |
| Tin-Lead[b] | 150 | 150 | 125 | 100 | 90 | 70 | — |
| Solder | 200 | 100 | 90 | 75 | 70 | 50 | — |
| ASTM B32 Gr 50A | 250 | 85 | 75 | 50 | 45 | 40 | 15 |
| 95-5 | 100 | 500 | 400 | 300 | 270 | 150 | — |
| Tin-Antimony | 150 | 400 | 350 | 275 | 250 | 150 | — |
| Solder | 200 | 300 | 250 | 200 | 180 | 140 | — |
| ASTM B32 Gr 50TA | 250 | 200 | 175 | 150 | 135 | 110 | 15 |
| Brazing Alloys | 100-200 | c | c | c | c | c | — |
| Melting at or | 250 | 300 | 210 | 170 | 150 | 150 | — |
| above 1000 °F | 350 | 270 | 190 | 150 | 150 | 150 | 120 |

[a] Solder joints are not to be used for:
(a) Flammable or toxic gasses or liquids.
(b) Gas, vapor, or compressed air in tubing over 4 in., unless maximum pressure is limited to 20 psi.

[b] Lead solders should not be used in potable water systems.
[c] Rated pressure for temperatures up to 200 °F is that of the tube being joined.
Conversion Factors: °C = (°F − 32)/1.8 mm = 25.4 × in. kPa = 6.89 × psi

**Table 5 Applicable Standards for Fittings and Valves**

| | ASME Std. |
|---|---|
| **Steel[a]** | |
| Flanges and Flanged Fittings (Classes 150 and 300) | B16.5 |
| Factory-made Wrought Steel Butt-Welding Fittings | B16.9 |
| Forged Steel Fittings, Socket Welded and Threaded | B16.11 |
| Wrought Steel Butt-Welding Short Radius Elbows | B16.28 |
| **Cast Iron, Malleable Iron, Ductile Iron[b]** | |
| Cast-Iron Pipe Flanges and Flanged Fittings | B16.1 |
| Malleable Iron Threaded Fittings, Classes 150 and 300 | B16.3 |
| Cast-Iron Threaded Fittings, Classes 125 and 250 | B16.4 |
| Cast-Iron Screwed Drainage Fittings | B16.12 |
| Ductile Iron Pipe Flanges and Flanged Fittings, Classes 150 and 300 | B16.42 |
| **Copper and Bronze[c]** | |
| Cast Bronze Threaded Fittings, Classes 125 and 250 | B16.15 |
| Cast Copper Alloy Solder Joint Pressure Fittings | B16.18 |
| Wrought Copper and Copper Alloy Solder Joint Pressure Fittings | B16.22 |
| Cast Copper Alloy Solder Joint Drainage Fittings, *DWV* | B16.23 |
| Bronze Pipe Flanges and Flanged Fittings, Classes 150 and 300 | B16.24 |
| Cast Copper Alloy Fittings for Flared Copper Tubes | B16.26 |
| Wrought Copper Solder Joint Drainage Fittings | B16.29 |
| | **ASTM Std.** |
| **Non-Metallic[d]** | |
| Threaded PVC Plastic Pipe Fittings, Schedule 80 | D2464 |
| Threaded PVC Plastic Pipe Fittings, Schedule 40 | D2466 |
| Socket-Type PVC Plastic Pipe Fittings, Schedule 80 | D2467 |
| Reinforced Epoxy Resin Piping Gas Pressure Fittings | D2517 |
| Threaded CPVC Plastic Pipe Fittings, Schedule 80 | F437 |
| Socket-Type CPVC Plastic Pipe Fittings, Schedule 40 | F438 |
| Socket-Type CPVC Plastic Pipe Fittings, Schedule 80 | F439 |
| Socket Heat Fusion Polybutylene Fittings | D3309 |
| Insert Fittings for Polybutylene Tubing | F845/D3309 |
| PVC Solvent Cements | D2564 |
| CPVC Solvent Cements | F493 |

[a] Wrought steel butt-welding fittings are made to match steel pipe wall thicknesses and are rated at the same working pressure as seamless pipe. Flanges and flanged fittings are rated by working steam pressure classes. Forged steel fittings are rated from 2000 to 6000 psi (14 to 41 MPa) in classes and are used for high temperture and pressure service for small pipe sizes.
[b] The class numbers refer to the maximum working saturated steam guage pressure (in psi). For liquids at lower temperatures, higher pressures are allowed. Groove-end fittings of these materials are made by various manufacturers who publish their own ratings.
[c] The classes refer to maximum working steam pressure (in psi). At ambient temperatures, higher liquid pressures are allowed. Solder joint fittings are limited by the strength of the soldered or brazed joint (Table 4).
[d] Ratings of plastic fittings match the pipe of corresponding schedule number.

trode is a continuously fed wire, which is shielded by argon or $CO_2$ gas from the welding gun nozzle.

GTAW—Gas Tungsten Arc Welding, also called TIG or Heliarc. This process uses a non-consumable tungsten electrode surrounded by a shielding gas. The weld material may be provided from a separate non-coated rod.

### Reinforced Outlet Fittings

Reinforced outlet fittings are used to make branch and take-off connections and are designed to permit welding directly to pipe without supplemental reinforcing. Fittings are available with threaded, socket, or butt-weld outlets.

### Mechanical Joints

Grooved joint systems require that a shallow groove be cut into the pipe end. These joints can be used with steel, cast iron, ductile iron, and plastic pipes. A segmented clamp engages the grooves, and the seal is provided by a special gasket designed so internal pressure tightens the seal. Some clamps are designed with clearance between tongue and groove to accommodate misalignment and thermal movements, while others are designed to limit movement and provide a nearly rigid system. Manufacturers' data gives temperature and pressure limitations.

Another form of mechanical joint consists of a sleeve slightly larger than the outside diameter of the pipe. The pipe ends are inserted into the sleeve, and gaskets are packed into the annular space between the pipe and coupling and held in place by retainer rings. This type of joint can accept some axial misalignment, but it must be anchored or otherwise restrained to prevent axial pull-out or lateral movement. Manufacturers provide pressure-temperature data.

Ductile iron pipe is furnished with a spigot end adapted for a gasket and retainer ring. This joint is also not restrained.

### Unions

Unions allow disassembly of a joint. Unions are three-part fittings with a mating machined seat on the two parts that thread onto the pipe ends. A threaded locking ring holds the two ends tightly together. A union also allows threaded pipe to be turned at the last joint connecting two pieces of equipment. Companion flanges (a pair) for small pipe serve the same purpose.

Although most PB plastic piping systems use socket fusion and insert/crimp-type fittings, larger diameter pipe can be joined by butt fusion. For transition to metallic piping, thread joints or fittings such as flanges, unions, and mechanical compression devices are used to connect plastic piping to metal components. Where a plastic threaded part is needed, a molded male thread adaptor should be used. Due to the difference in the coefficient of thermal expansion of metals and the various plastic materials, the fitting manufacturer should be consulted to determine the temperature range for which it is acceptable. For systems having a greater temperature range, transition fittings that incorporate an elastomeric seal should be used.

Flanges of plastic or other materials designed for use with plastic piping are available. Consult the pipe manufacturer for details of design, rating, and use. For softer plastics, flange backup plates should be used.

## VALVES

The types of valves used in heating, plumbing, and air-conditioning systems are listed below. Manufacturers' catalogs list the pressure-temperature ratings.

**Gate Valves.** This valve is used to open or close service and not for throttling. The open valve has a low resistance to flow. These valves are made in many configurations of bronze, cast iron, ductile iron, and steel.

**Globe Valves.** These valves are used for throttling service and have a high resistance to flow. Globe valves are made of the same materials as gate valves.

**Check Valves.** Check valves prevent reversal of flow. The moving fluid holds the valve open, and reversal closes the valve. These valves can also be spring loaded to close before an actual reversal of flow takes place. Check valves are made of the same materials as gate valves. Flow resistance of most configurations is low.

**Plug Valves.** Plug valves operate from *fully open* to *fully closed* with a quarter turn. Certain types may be used for throttling control. Larger sizes are gear operated. These valves are available with full size or restricted ports, with the pressure resistance being higher through restricted ports. Plug valves have cast iron, ductile iron, or steel bodies with a variety of trim.

**Butterfly Valves.** These valves operate from *open* to *closed* with a 90 degree turn. They can be used for throttling and are made in sizes 2 in. (50 mm) and over. This valve is available in many configurations; it is made with a body of cast iron, ductile iron, or steel and with disks of various corrosion-resistant materials. These valves have low resistance to flow when fully open.

**Ball Valves.** Ball valves operate from *full open* to *full closed* with a quarter turn, have low flow resistance, and may be used for throttling control. The ball material is corrosion resistant, and the other trim varies with the manufacturers' standards.

**Pressure-Reducing Valves.** These special valves control downstream pressure to a constant level. PRVs must be selected carefully to hold the pressure level while passing fluctuating quantities of fluid.

**Pressure-Relief Valves.** Relief valves open when the system pressure reaches a preset level to prevent over-pressuring the system components. These are special valves made by many manufacturers for specific applications.

## SPECIAL SYSTEMS

Certain piping systems are governed by separate codes or standards, which are summarized below. Generally, any failure of the piping in these systems is dangerous to the public, so local areas have adopted laws enforcing the codes.

**Boiler Piping.** ASME *Standard* B31.1 and the ASME *Boiler and Pressure Vessel Code* (Section I) specify the piping inside the code required stop valves on boilers that operate above 15 psig (100 kPa) with steam or 160 psig or 250°F (1.1 MPa or 121°C) with water. These codes require fabricators and contractors to be certified for such work. The field or shop work must also be inspected while it is in progress by inspectors commissioned by the National Board of Boiler and Pressure Vessel Inspectors.

**Refrigeration Piping.** ASME *Standard* B31.5, *Refrigerant Piping*, and ASHRAE *Standard* 15-78, *Standard Safety Code for Mechanical Refrigeration* cover the requirements for refrigerant piping.

**Plumbing Systems.** Local codes cover piping for plumbing systems.

**Sprinkler Systems.** NFPA *Standard* 13, *Installation of Sprinkler Systems*, covers this field.

**Fuel Gas.** ANSI Z223.1, *National Fuel Gas Code*, prescribes fuel gas piping in buildings.

## SELECTION OF MATERIALS

Each HVAC system and, under some conditions, portions of a system require a study of the conditions of operation to determine suitable materials. For example, because the static pressure of water in a high-rise building is higher in the lower levels than in the upper levels, different materials may be required along vertical zones.

The following factors should be considered when selecting material for a piping system:

1. Code requirements
2. Working fluid in the pipe
3. Pressure and temperature of the fluid
4. External environment of the pipe
5. Cost of the installed system

Table 6 lists materials currently used for heating and air-conditioning piping systems. The pressure and temperature rating of each component selected must be considered; the lowest rating establishes the operating limits of the system.

## PIPE WALL THICKNESS

The primary factors determining pipe wall thickness are the hoop stresses due to internal pressure and the longitudinal stresses due to pressure, weight, and other sustained loads. Detailed stress calculations are seldom required for HVAC applications because standard pipe has ample thickness to sustain the pressure and longitudinal stresses due to weight (assuming hangers are spaced in accordance with Table 9).

## STRESS CALCULATIONS

Although stress calculations are seldom required, the factors involved must be understood. The main areas of concern are (1) internal pressure stresses, (2) longitudinal stresses due to pressure and weight, and (3) stresses due to expansion and contraction.

The ASME B-31 Standards establish a basic allowable stress, $S$, equal to 1/4 minimum tensile strength of the material. This value is adjusted, as indicated below, because of the nature of certain stresses and manufacturing processes.

Hoop stress caused by internal pressure is the major stress on pipes. As certain forming methods form a seam that may be weaker than the base material, the standard specifies a joint efficiency factor, $E$, which, multiplied by the basic allowable stress, establishes a maximum allowable stress value in tension, $SE$. (Table A-1 in ASME B-31.9 lists values for $SE$ for commonly used pipe materials.) The joint efficiency factor can be significant; for example, seamless pipe has a joint efficiency factor of 1, so it can be used to the full allowable stress (one quarter of tensile strength). In contrast, butt-welded pipe has a joint efficiency factor of 0.60, so its maximum allowable stress value must be derated ($SE = 0.6S$).

$$t_m = \frac{PD}{2SE} + A \tag{1}$$

$$P = \frac{2SE\,(t_m - A)}{D} \tag{2}$$

*where*

$t_m$ = minimum required wall thickness, in. (mm)
$SE$ = Maximum Allowable Stress, psi (kPa)
$D$ = Outside Pipe Diameter, in. (mm)
$A$ = Allowance for manufacturing tolerance, threading, grooving and corrosion, in. (mm)

Equations (1) and (2) determine minimum wall thickness for a given pressure and maximum pressure allowed for a given wall thickness. Both equations incorporate an Allowance Factor, $A$, to compensate for manufacturing tolerances, material removed in threading or grooving, and corrosion. For the seamless, butt welded and electric resistance welded (ERW) pipe most commonly used in HVAC work, the standards apply a manufacturing tolerance of 12.5%. Working pressures for steel pipe, as indicated in Table 1, have been calculated using a manufacturing tolerance of 12.5%; standard allowance for depth of thread, where applicable; and a corrosion allowance of 0.065 in. (1.65 mm) for pipes 2-1/2 in. (65 mm) and larger and 0.025 in. (0.64 mm) for pipes 2 in. (50 mm) and smaller. Where corrosion is known to be greater or lesser, pressure rating can be recalculated in accordance with Eq. (2). Higher pressure ratings than shown can be obtained by using ERW or seamless pipe in lieu of CW pipe 4 in. (100 mm) and under; and seamless pipe in lieu of ERW pipe 5 in. (125 mm) and over due to higher joint efficiency factor, as well as by using heavier wall pipe.

Table 6 Application of Pipe, Fittings, and Valves for Heating and Air-Conditioning

| Application | Pipe Material | Weight | Joint Type | Fitting Class | Fitting Material | System Temperature | System Maximum Pressure at Temperature[a] |
|---|---|---|---|---|---|---|---|
| **Recirculating Water** | | | | | | °F | psig |
| 2 in. (50 mm) and smaller | Steel (CW) | Standard | Thread | 125 | Cast Iron | 250 | 125 |
| | Copper, Hard | Type *L* | 95-5 Solder | — | Wrought Copper | 250 | 150 |
| | PVC | Sch 80 | Solvent | Sch 80 | PVC | 75 | 350 |
| | CPVC | Sch 80 | Solvent | Sch 80 | CPVC | 150 | 150 |
| | PB | SDR-11 | Heat Fusion | — | PB | 160 | 115 |
| | | | Insert Crimp | — | Metal | 160 | 115 |
| 2.5 to 12 in. (65 to 300 mm) | A53 B ERW Steel | Std | Weld | Std | Wrought Steel | 250 | 400 |
| | | | Flange | 150 | Wrought Steel | 250 | 250 |
| | | | Flange | 125 | Cast Iron | 250 | 175 |
| | | | Flange | 250 | Cast Iron | 250 | 400 |
| | | | Groove | — | MI or Ductile Iron | 230 | 300 |
| | PB | SDR-11 | Heat Fusion | | PB | 160 | 115 |
| **Steam and Condensate** | | | | | | | |
| 2 in. (50 mm) and smaller | Steel (CW) | Std[b] | Thread | 125 | Cast Iron | | 90 |
| | | | Thread | 150 | Malleable Iron | | 90 |
| | A53 B ERW Steel | Std[b] | Thread | 125 | Cast Iron | | 100 |
| | | | Thread | 150 | Malleable Iron | | 125 |
| | A53 B ERW Steel | XS | Thread | 250 | Cast Iron | | 200 |
| | | | Thread | 300 | Malleable Iron | | 250 |
| 2.5 to 12 in. (65 to 300 mm) | Steel | Std | Weld | Std | Wrought Steel | | 250 |
| | | | Flange | 150 | Wrought Steel | | 200 |
| | | | Flange | 125 | Cast Iron | | 100 |
| | A53 B ERW Steel | XS | Weld | XS | Wrought Steel | | 700 |
| | | | Flange | 300 | Wrought Steel | | 500 |
| | | | Flange | 250 | Cast Iron | | 200 |
| **Refrigerant** | | | | | | | |
| | Copper, Hard | Type *L* or *K* | Braze | — | Wrought Copper | | |
| | A53 B SML Steel | Std | Weld | | Wrought Steel | | |
| **Underground Water** | | | | | | | |
| Through 12 in. (300 mm) | Copper, Hard | Type *K* | 95-5 Solder | — | Wrought Copper | 75 | 350 |
| Through 6 in. (150 mm) | Ductile Iron | Class 50 | MJ | MJ | Cast Iron | 75 | 250 |
| | PB | 5 DR 9 & 11 | Heat Fusion | | PB | 75 | 250 / 160 |
| | | 5 DR 7 & 11.5 | Insert Crimp | | Metal | 75 | 250 / 160 |
| **Potable Water, Inside Building** | | | | | | | |
| | Copper, Hard | Type *L* | 95-5 Solder | — | Wrought Copper | 75 | 350 |
| | Steel Galvanized | Std | Thread | 125 | Galv. Cast Iron | 75 | 125 |
| | | | | 150 | Galv. Mall. Iron | 75 | 125 |
| | PB | SDR-11 | Heat Fusion | | PB | 75 | 200 |
| | | | Insert Crimp | | Metal | 75 | 200 |

[a] Maximum allowable working pressures have been de-rated in this table. Higher system pressures can be used for lower temperatures and smaller pipe sizes. Pipe, fittings, joints, and valves must all be considered.

[b] Extra strong pipe is recommended for all threaded condensate piping to allow for corrosion.

Conversion Factors: °C = (°F − 32)/1.8  kPA = 6.89 × psi

Longitudinal stresses due to pressure, weight, and other sustained forces are additive, and the sum of all such stresses must not exceed the basic allowable stress, *S*, at the highest temperature at which the system will operate. Longitudinal stress due to pressure equals approximately half the hoop stress caused by internal pressure, which means at least half the basic allowable stress, *S*, is available for weight and other sustained forces. This factor is taken into account in Table 9.

Stresses due to expansion and contraction are cyclical in nature and because creep allows some stress relaxation, the ASME B-31 standards permit designing to an allowable stress range, $S_A$, as shown in Eq. (3). Table 1 lists allowable stresses for most commonly used piping materials.

$$S_A = 1.25\ S_c + 0.25\ S_h \quad (3)$$

*where*

$S_A$ = allowable stress range, psi (kPa)
$S_c$ = allowable cold stress, psi (kPa)
$S_h$ = allowable hot stress, psi (kPa)

## PIPE EXPANSION AND FLEXIBILITY

Changes in temperature causes dimensional changes in all materials. Table 7 shows the coefficients of expansion for piping materials most commonly used in HVAC systems. For systems operating at high temperatures, such as steam and hot water, the rate of expansion is high and significant movements can occur in short runs of piping. Even though rates of expansion may be low for systems operating in the range of 40 to 100 °F (5 to 40 °C), such as chilled and condenser water, they can cause large movements in long runs of piping, as commonly occur in distribution systems and high-rise buildings. Therefore, in addition to design requirements for pressure, weight, and other loadings, piping systems must accommodate thermal and other movements to prevent the following:

1. Failure of pipe and supports from overstress and fatigue
2. Leakage of joints
3. Detrimental forces and stresses in connected equipment

The anchor forces and bowing of pipe anchored at both ends is generally too high, so general practice is to *never anchor a straight run of steel pipe at both ends.* Piping systems must be allowed to expand or contract due to thermal changes. Ample flexibility can be attained by designing pipe bends and loops or supplemental devices, such as expansion joints, into the system.

End reactions transmitted to rotating equipment, such as pumps or turbines, may deform the equipment case and cause bearing misalignment, which may ultimately cause the component to fail. Consequently, manufacturers' recommendations on the allowable forces and movements that may be placed on their equipment must be followed.

## PIPE BENDS AND LOOPS

Detailed stress analysis requires involved mathematical analysis and is generally performed by computer programs. However, such involved analysis is not required for most HVAC systems because the piping arrangements and temperature ranges at which they operate lend themselves to simple analysis.

### *L* Bends

The guided cantilever beam method of evaluating *L* bends can be used to design *L* bends, *Z* bends, pipe loops, branch take-off connections, and some more complicated piping configurations.

Equation (4) may be used to calculate the length of leg BC needed to accommodate thermal expansion or contraction of leg AB for a guided cantilever beam (Figure 1).

$$L = \sqrt{\frac{\Delta D E}{C_1 S_A}} \quad (4)$$

*where*

$L$ = length of leg required to accommodate thermal growth of long leg BC, ft (mm)
$\Delta$ = thermal expansion or contraction of leg AB, in. (mm)
$D$ = actual pipe outside diameter, in. (mm)
$E$ = modulus of elasticity, psi (kPa)

**Table 7 Thermal Expansion of Metal Pipe**

| Saturated Steam Pressure, psig | Temperature °F | Linear Thermal Expansion, in./100 ft | | |
|---|---|---|---|---|
| | | Carbon Steel | Type 304 Stainless Steel | Copper |
| | −30 | −0.19 | −0.30 | −0.32 |
| | −20 | −0.12 | −0.20 | −0.21 |
| | −10 | −0.06 | −0.10 | −0.11 |
| | 0 | 0 | 0 | 0 |
| | 10 | 0.08 | 0.11 | 0.12 |
| | 20 | 0.15 | 0.22 | 0.24 |
| −14.6 (Vacuum) | 32 | 0.24 | 0.36 | 0.37 |
| −14.6 | 40 | 0.30 | 0.45 | 0.45 |
| −14.5 | 50 | 0.38 | 0.56 | 0.57 |
| −14.4 | 60 | 0.46 | 0.67 | 0.68 |
| −14.3 | 70 | 0.53 | 0.78 | 0.79 |
| −14.2 | 80 | 0.61 | 0.90 | 0.90 |
| −14.0 | 90 | 0.68 | 1.01 | 1.02 |
| −13.7 | 100 | 0.76 | 1.12 | 1.13 |
| −13.0 | 120 | 0.91 | 1.35 | 1.37 |
| −11.8 | 140 | 1.06 | 1.57 | 1.59 |
| −10.0 | 160 | 1.22 | 1.79 | 1.80 |
| −7.2 | 180 | 1.37 | 2.02 | 2.05 |
| −3.2 (Vacuum) | 200 | 1.52 | 2.24 | 2.30 |
| 0 | 212 | 1.62 | 2.38 | 2.43 |
| 2.5 | 220 | 1.69 | 2.48 | 2.52 |
| 10.3 | 240 | 1.85 | 2.71 | 2.76 |
| 20.7 | 260 | 2.02 | 2.94 | 2.99 |
| 34.6 | 280 | 2.18 | 3.17 | 3.22 |
| 52.3 | 300 | 2.35 | 3.40 | 3.46 |
| 75.0 | 320 | 2.53 | 3.64 | 3.70 |
| 103.3 | 340 | 2.70 | 3.88 | 3.94 |
| 138.3 | 360 | 2.88 | 4.11 | 4.18 |
| 181.1 | 380 | 3.05 | 4.35 | 4.42 |
| 232.6 | 400 | 3.23 | 4.59 | 4.87 |
| 294.1 | 420 | 3.41 | 4.83 | 4.91 |
| 366.9 | 440 | 3.60 | 5.07 | 5.15 |
| 452.2 | 460 | 3.78 | 5.32 | 5.41 |
| 551.4 | 480 | 3.97 | 5.56 | 5.65 |
| 666.1 | 500 | 4.15 | 5.80 | 5.91 |
| 797.7 | 520 | 4.35 | 6.05 | 6.15 |
| 947.8 | 540 | 4.54 | 6.29 | 6.41 |
| 1118 | 560 | 4.74 | 6.54 | 6.64 |
| 1311 | 580 | 4.93 | 6.78 | 6.92 |
| 1528 | 600 | 5.13 | 7.03 | 7.18 |
| 1772 | 620 | 5.34 | 7.28 | 7.43 |
| 2045 | 640 | 5.54 | 7.53 | 7.69 |
| 2351 | 660 | 5.75 | 7.79 | 7.95 |
| 2693 | 680 | 5.95 | 8.04 | 8.20 |
| 3079 | 700 | 6.16 | 8.29 | 8.47 |
| | 720 | 6.37 | 8.55 | 8.71 |
| | 740 | 6.59 | 8.81 | 9.00 |
| | 760 | 6.80 | 9.07 | 9.26 |
| | 780 | 7.02 | 9.33 | 9.53 |
| | 800 | 7.23 | 9.59 | 9.79 |
| | 820 | 7.45 | 9.85 | 10.07 |
| | 840 | 7.67 | 10.12 | 10.31 |
| | 860 | 7.90 | 10.38 | 10.61 |
| | 880 | 8.12 | 10.65 | 10.97 |
| | 900 | 8.34 | 10.91 | 11.16 |
| | 920 | 8.56 | 11.18 | 11.42 |
| | 940 | 8.77 | 11.45 | 11.71 |
| | 960 | 8.99 | 11.73 | 11.98 |
| | 980 | 9.20 | 11.00 | 12.27 |
| | 1000 | 9.42 | 12.27 | 12.54 |

Conversion Factors: kPa = 6.89 × psig + 101.3
°C = (°F − 32)/1.8
mm/m = (in./100 ft)/1.2

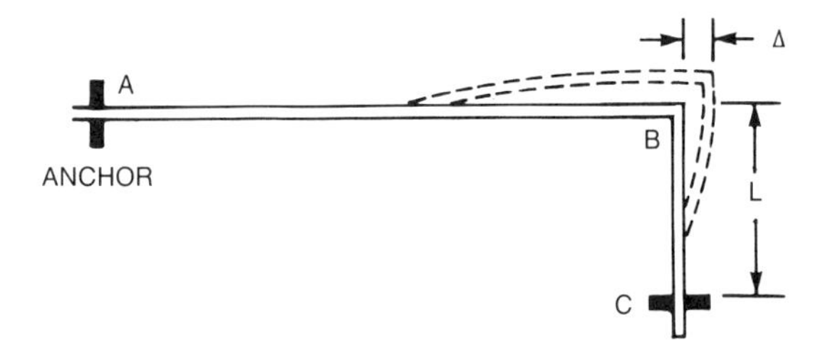


**Fig. 1 Guided Cantilever Beam**

$S_A$ = allowable stress range, psi (kPa)
$C_1$ = constant, 48 (0.33)

For the commonly used A53 Grade *B* seamless or ERW pipe, an allowable stress, $S_A$, of 22,500 psi (155 MPa) can be used without overstressing the pipe. However, this can result in very high end reactions and anchor forces, especially with large diameter pipe. Designing to a stress range, $S_A$, of 15,000 psi results in Eq. (5), which provides reasonably low end reactions without requiring too much extra pipe. In addition, Eq. (5) may be used with A53 butt welded pipe and B88 drawn copper tubing.

For A53 continuous (butt) welded, seamless, and ERW pipe, and B88 drawn copper tubing:

$$L = C_2\sqrt{\Delta D} \quad (5)$$

*where*

$C_2$ = constant, 6.225 (74.7)

The guided cantilever method of designing *L* bends assumes no restraints; therefore, care must be taken in supporting the pipe. For horizontal *L* bends, it is usually necessary to place a support near point *B*, and any supports between points *A* and *C* must provide minimal resistance to piping movement; this is done by using slide plates or hanger rods of ample length with hanger components selected to allow for swing of no greater than 4 degrees.

For *L* bends containing both vertical and horizontal legs, any supports on horizontal leg must be spring hangers designed to support the full weight of pipe at normal operating temperature with a maximum load variation of 25%.

The force developed in an *L* bend that must be sustained by anchors or connected equipment is determined by the following equation:

$$F = E_c I\Delta / C_3 L^3 \quad (6)$$

*where*

$F$ = force, lbs (kN)
$E_c$ = modulus of elasticity, psi (kPa)
$I$ = moment of inertia, in.$^4$ (mm$^4$)
$L$ = length of offset leg, ft (mm)
$\Delta$ = deflection of offset leg, in. (mm)
$C_3$ = constant, 144 (12/1000)

In lieu of using Eq. (6), for *L* bends designed in accordance with Eq. (5) for 1 in. (25 mm) or more of offset, a conservative estimating value for force is 500 lb. per diameter inch (88 N/mm); *eg*., a 3 in. pipe would develop 1500 lbs. force.

## *Z* Bends

*Z* bends, as shown in Figure 2, are very effective in accommodating pipe movements. A simple and conservative method of sizing *Z* bends is to design the offset leg to be 65% of the values used for an *L* bend in Eq. (2), which results in Eq. (7) as follows:

$$L = C_4\sqrt{\Delta D} \quad (7)$$

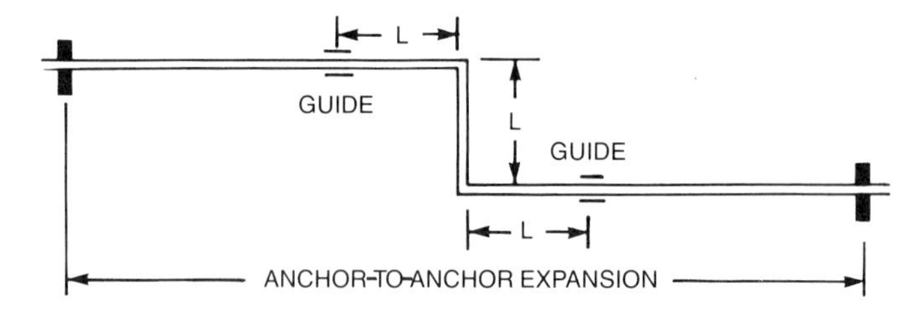


The distance from the guides, if used, to the offset should equal or exceed the length of the offset.

Offset piping must be supported with hangers, slide plates, and spring hangers similar to those for *L* bends.

**Fig. 2 *Z* Bend in Pipe**

*where*

$L$ = length of offset leg, ft (mm)
$\Delta$ = anchor-to-anchor expansion, in. (mm)
$D$ = pipe outside diameter, in. (mm)
$C_4$ = constant, 4.0 (48.6)

The force developed in a *Z* bend can be calculated with acceptable accuracy by Eq. (8).

$$F = C_5\Delta(D/L)^2 \quad (8)$$

*where*

$F$ = force, lbs (N)
$D$ = pipe outside diameter, in. (mm)
$L$ = length of offset leg, ft (mm)
$\Delta$ = anchor-to-anchor expansion, in. (mm)
$C_5$ = constant, 4000 (101 000)

## *U* Bends and Pipe Loops

Pipe loops or *U* bends are commonly used in long runs of piping. A simple method of designing pipe loops is to calculate the anchor-to-anchor expansion and, using Eq. (5), determine length, *L*, necessary to accommodate this movement and then determine pipe loop dimensions: $W = L/5$ and $H = 2W$.

*Note*: Guides must be spaced no closer than twice the height of the loop, and piping between guides must be supported, as described for *L* bends, when the length of pipe between guides exceeds maximum allowable spacing for size pipe.

Table 8 provides pipe loop dimensions for pipe sizes 1 through 24 in. (25 through 610 mm) and anchor-to-anchor expansion (contraction) of 2 through 12 in. ( 50 through 300 mm).

No simple method has been developed to calculate pipe loop forces; however, they are generally low. A conservative estimate is 200 lb per diameter inch (35 N/mm); *eg*., a 2 in. pipe will develop 400 lb force and a 12 in. pipe will develop 2400 lb force.

## Cold Springing of Pipe

Cold springing or cold positioning of pipe consists of offsetting or springing the pipe in a direction opposite the expected movement. Cold springing is not recommended for most HVAC piping systems. Further, *cold springing does not permit designing a pipe bend or loop for twice the calculated movement*. For example, if a particular *L* bend can accommodate 3 in. (75 mm) of movement from a neutral position, cold springing does not permit the *L* bend to accommodate 6 in. (150 mm) of movement.

## Analyzing Existing Piping Configurations

Piping systems are best analyzed by computer stress analysis programs because these provide all pertinent data including stress,

**Table 8 Pipe Loop Design for A-53 Grade B Carbon Steel Pipe Through 400 °F**

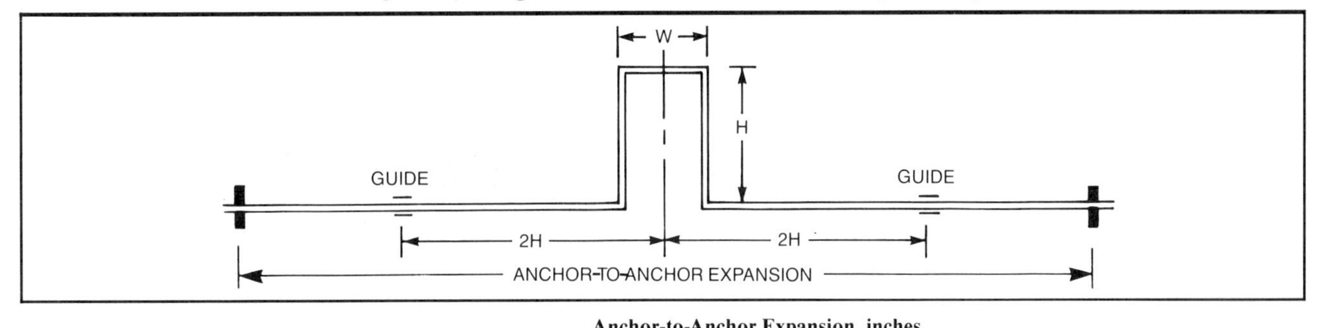


| Pipe Size in. | Anchor-to-Anchor Expansion, inches: 2 | | 4 | | 6 | | 8 | | 10 | | 12 | |
|---|---|---|---|---|---|---|---|---|---|---|---|---|
| | W | H | W | H | W | H | W | H | W | H | W | H |
| 1 | 2 | 4 | 3 | 6 | 3.5 | 7 | 4 | 8 | 4.5 | 9 | 5 | 10 |
| 2 | 3 | 6 | 4 | 8 | 5 | 10 | 5.5 | 11 | 6 | 12 | 7 | 14 |
| 3 | 3.5 | 7 | 5 | 10 | 6 | 12 | 6.5 | 13 | 7.5 | 15 | 8 | 16 |
| 4 | 4 | 8 | 5.5 | 11 | 6.5 | 13 | 7.5 | 15 | 8.5 | 17 | 9 | 18 |
| 6 | 5 | 10 | 6.5 | 13 | 8 | 16 | 9 | 18 | 10 | 20 | 11 | 22 |
| 8 | 5.5 | 11 | 7.5 | 15 | 9 | 18 | 10.5 | 21 | 12 | 24 | 13 | 26 |
| 10 | 6 | 12 | 8.5 | 17 | 10 | 20 | 11.5 | 23 | 13 | 26 | 14 | 28 |
| 12 | 6.5 | 13 | 9 | 18 | 11 | 22 | 12.5 | 25 | 14 | 28 | 15.5 | 31 |
| 14 | 7 | 14 | 9.5 | 19 | 11.5 | 23 | 13 | 26 | 15 | 30 | 16 | 32 |
| 16 | 7.5 | 15 | 10 | 20 | 12.5 | 25 | 14 | 28 | 16 | 32 | 17.5 | 35 |
| 18 | 8 | 16 | 11 | 22 | 13 | 26 | 15 | 30 | 17 | 34 | 18.5 | 37 |
| 20 | 8.5 | 17 | 11.5 | 23 | 14 | 28 | 16 | 32 | 18 | 36 | 19.5 | 39 |
| 24 | 9 | 18 | 12.5 | 25 | 14.5 | 29 | 17.5 | 35 | 19.5 | 39 | 21 | 42 |

W and H dimensions in feet. 2H + W = L L Determined from Eq. (4).
H = 2W
5W = L

Approximate force to deflect loop = 200 lb/diam. in. (35 N/mm).
Example: 8 in. pipe creates a 1600 lb. force.
Conversion Factors: mm = 25.4 × in. mm = 304.8 × ft

movements, and loads. Services can perform such analysis if programs are not available in-house. However, many situations do not require such detailed analysis. A simple, yet satisfactory method for single and multi-plane systems is to divide the system with real or hypothetical anchors into a number of single-plane units, as shown in Figure 3, which can be evaluated as *L* and *Z* bends.

## EXPANSION JOINTS AND EXPANSION COMPENSATING DEVICES

Although the inherent flexibility of the piping system should be used to the maximum extent possible, expansion joints must be used where movements are too large to accommodate with pipe bends or loops or where insufficient room exists to construct a loop of adequate size. Typical situations are tunnel piping and risers in high-rise buildings, especially for steam and hot water pipes where large thermal movements are involved.

Packed and packless expansion joints and expansion compensating devices are used to accommodate movement, either axially or laterally.

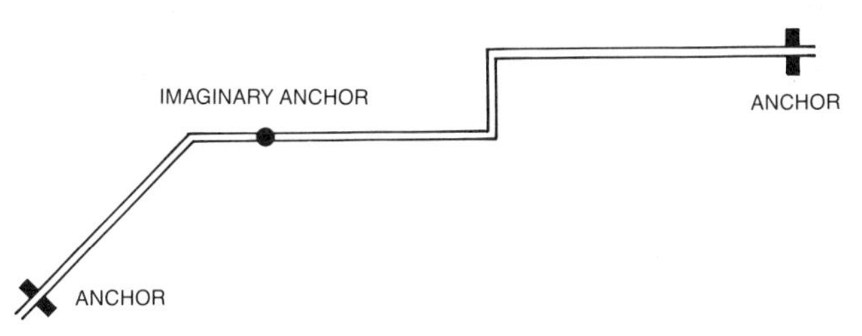


**Fig. 3 Multiplane Pipe System**

In the *axial method* of accommodating movement, the expansion joint is installed between anchors in a straight line segment and accommodates axial motion only. This method has high anchor loads, primarily due to pressure thrust. It requires careful guiding, but expansion joints can be spaced conveniently to limit movement of branch connections. The axial method finds widest application for long runs without natural offsets, such as tunnel and underground piping and risers in tall buildings.

The lateral or offset method requires the device to be installed in a leg perpendicular to the expected movement and accommodates lateral movement only. This method generally has low anchor forces and minimal guide requirements. It finds widest application in lines with natural offsets, especially where there are few or no branch connections.

**Packed expansion joints** depend on slipping or sliding surfaces to accommodate the movement and require some type of seals or packing to seal the surfaces. Most such devices require some maintenance but are not subject to catastrophic failure. Further, with most packed expansion joint devices, any leaks that develop can be repacked under full line pressure without shutting down the system.

**Packless expansion joints** depend upon the flexing or distortion of the sealing element to accommodate movement. They generally do not require any maintenance, but maintenance or repair is not usually possible. If a leak occurs, the system must be shut off and drained, and the entire device must be replaced. Further, catastrophic failure of the sealing element can occur and, although likelihood of such failure is remote, it must be considered in certain design situations.

Packed expansion joints are preferred where long-term system reliability is of prime importance (using types that can be repacked under full-line pressure) and where major leaks can be life threatening or extremely costly. Typical applications are risers, tunnels, underground pipe, and distribution piping systems. Packless expansion joints are generally used where even small leaks cannot be tolerated, such as for gas and toxic chemicals, where temperature limitations preclude use of packed expansion joints, and for very large diameter pipe where packed expansion joints cannot be constructed or cost would be excessive.

In all cases, expansion joints should be installed, anchored, and guided in accordance with expansion joint manufacturers' recommendations.

## Packed Expansion Joints

There are two types of packed expansion joints—packed slip expansion joints and flexible ball joints.

**Packed slip expansion joints** are telescoping devices designed to accommodate axial movement only. Some sort of packing seals the sliding surfaces. The original packed slip expansion joint used multiple layers of braided compression packing, similar to the stuffing box commonly used with valves and pumps; this arrangement requires shutting and draining the system for maintenance and repair. Advances in design and packing technology have eliminated these problems, and most current packed slip joints use self-lubricating semi-plastic packing, which can be injected under full-line pressure without shutting off system. (Many manufacturers use asbestos-based packings, unless requested otherwise. Asbestos-free packings, such as flake graphite, are available and, although more expensive, should be specified in lieu of products containing asbestos.)

Standard packed slip expansion joints are constructed in sizes 1.5 to 36 in. (40 to 915 mm) of carbon steel with weld or flange ends for pressures to 300 psig (2.1 MPa) and temperatures to 800 °F (425 °C). Larger sizes, higher temperature, and higher pressure designs are available. Standard single joints are generally designed for 4, 8, or 12 in. (100, 200, or 300 mm) axial traverse; double joints with an intermediate anchor base can accommodate twice these movements. Special designs for greater movements are available.

**Flexible ball joints** are used in pairs to accommodate lateral or offset movement and must be installed in a leg perpendicular to the expected movement. The original flexible ball joint design incorporated only inner and outer containment seals that could not be serviced or replaced without removing the ball joint from the system. The packing technology of the packed slip expansion joint, as explained above, has been incorporated into the flexible ball joint design; now, packed flexible ball joints have self-lubricating semi-plastic packing that can be injected under full-line pressure without shutting off the system.

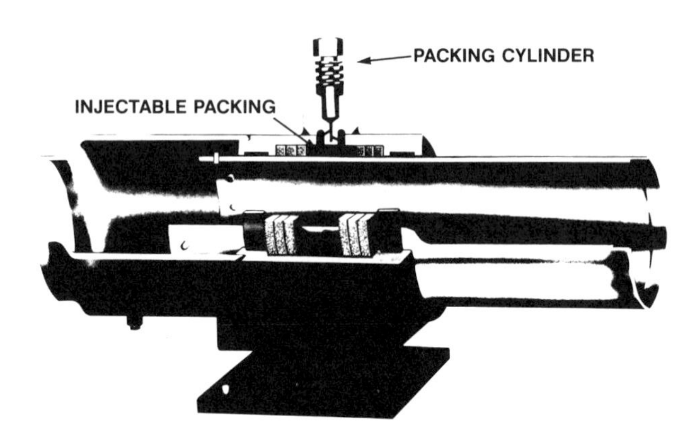


**Fig. 4 Packed Slip Expansion Joint**

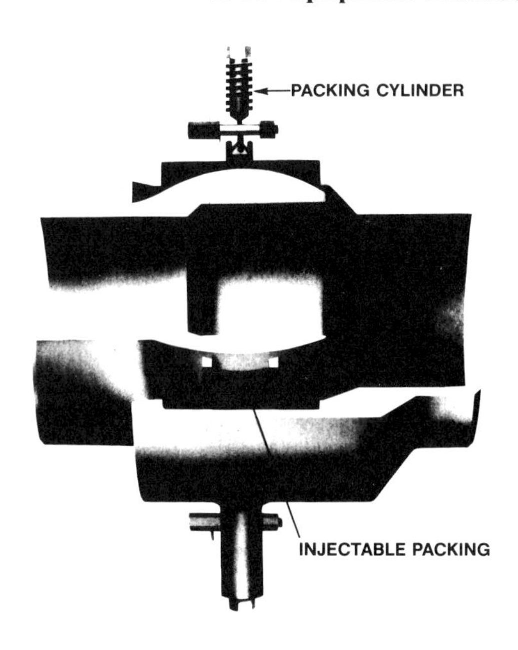


**Fig. 5 Flexible Ball Joint**

Standard flexible ball joints are available in sizes 1¼ through 30 in. (32 through 760 mm) with threaded [1¼ to 2 in. (32 to 50 mm)], weld, and flange ends for pressures to 300 psig (2.1 MPa), and temperatures to 750 °F (400 °C). Larger sizes, higher temperature, and higher pressure ranges are available.

## Packless Expansion Joints

Metal bellows expansion joints, rubber expansion joints, and flexible hose or pipe connectors are some of the packless expansion joints available.

**Metal bellows expansion joints** have a thin-wall convoluted section that accommodates movement by bending or flexing. The bellows material is generally Type 304, 316, or 321 stainless steel; but monel, inconel, and other materials are commonly used to satisfy service conditions. Small diameter expansion joints in sizes 3/4 through 3 in. (20 through 80 mm) are generally called *expansion compensators* and are available in all bronze or steel construction. Metal bellows expansion joints can generally be designed for the pressures and temperatures commonly encountered in HVAC systems and can also be furnished in rectangular configurations for ducts and chimney connectors.

Overpressurization, improper guiding, and other forces can distort the bellows element. For low-pressure applications, such distortion can be controlled by the geometry of the convolution or thickness of the bellows material. For higher pressures, internally pressurized joints require reinforcing. Externally pressurized designs are not subject to such distortion and are not generally furnished without supplemental bellows reinforcing.

Single and double bellows expansion joints primarily accommodate axial movement only, similar to packed slip expansion joints. Although bellows expansion joints can accommodate some lateral movement, the *universal tied bellows expansion joint* accommodates large lateral movement. This device operates much like a pair of flexible ball joints, except bellows elements are used instead of flexible ball elements. The tie rods on this joint contain the pressure thrust, so anchor loads are much lower than with axial-type expansion joints.

**Rubber expansion joints** are similar to single metal bellows expansion joints, except they incorporate a non-metallic elastomeric bellows sealing element and generally have more stringent temperature and pressure limitations. Although rubber expansion joints can be used to accommodate expansion and contraction of the piping system, they are primarily used as flexible connectors at equipment to isolate sound and vibration and eliminate stress at equipment nozzles.

**Flexible hose** can be constructed of elastomeric material or corrugated metal with an outer braid for reinforcing and end restraint. Flexible hose is primarily used as a flexible connector at equipment to isolate sound and vibration and eliminate stress at equipment nozzles; however, flexible metal hose is well suited for use as an *offset-type expansion joint*, especially for copper tubing and branch connections off risers.

## PIPE-SUPPORTING ELEMENTS

Pipe-supporting elements consist of hangers, which support from above; supports, which bear load from below; and restraints, such as anchors and guides, which limit or direct movement, as well as support loads. Pipe-supporting elements withstand all static and dynamic conditions including the following:

1. Weight of pipe, valves, fittings, insulation, and fluid contents, including test fluids if heavier-than-normal flow media
2. Occasional loads such as ice, wind, and seismic forces
3. Forces imposed by thermal expansion and contraction of pipe bends and loops
4. Frictional, spring, and pressure thrust forces imposed by expansion joints in the system
5. Frictional forces of guides and supports
6. Other loads as might be imposed such as water hammer, vibration, and reactive forces of relief valves
7. Test loads and forces

In addition, pipe-supporting elements must be evaluated in terms of stress at point of connection to the pipe and the building structure. Stress at the point of connection to the pipe is especially important for base elbow and trunnion supports, as the limiting and controlling parameter is usually not the strength of the structural member, but the localized stress and the point of attachment to the pipe. Loads on anchors, cast in place inserts, and other attachments to concrete should not be more than 1/5 the ultimate strength of attachment, as determined by manufacturers' tests. In addition, all loads on the structure should be given to and coordinated with the structural engineer.

The *Code for Pressure Piping* (ASME B-31) establishes criteria for the design of pipe-supporting elements and the Manufacturers Standardization Society of the Valve and Fitting Industry (MSS) has established standards for the design, fabrication, selection, and installation of pipe hangers and supports based on these codes.

MSS *Standard* SP-69 and the catalogs of many manufacturers illustrate the various hangers and components and provide information on the types to use with various pipe systems. Table 9 shows suggested pipe support spacing and Table 10 provides maximum safe loads for threaded steel rods.

The loads on most pipe-supporting elements are moderate and can be selected safely in accordance with above information and manufacturers' catalog data; however, some loads and forces can be very high, especially in multi-story buildings and for large diameter pipe, especially where expansion joints are used at high operating pressures. Consequently, a qualified engineer should design or review the design of all anchors and pipe-supporting elements, especially for the following:

1. Steam systems operating above 15 psig (100 kPa)
2. Hydronic systems operating above 160 psig or 250°F (1.1 MPa or 121°C)
3. Risers over 10 stories or 100 ft (30 m)
4. Systems with expansion joints, especially for pipe diameters 3 in. (80 mm) and greater
5. Pipe sizes over 12 in. (300 mm) diameter
6. Anchor loads greater than 10,000 lbs. (10 kips) (4.5 Mg)
7. Moments on pipe or structure in excess of 1000 ft lbs (1.4 kN•m)

**Table 9 Suggested Hanger Spacing and Rod Size for Straight Horizontal Runs (MSS Standard SP-69)**

| | Feet | | | |
|---|---|---|---|---|
| | Std. Wt. Steel Pipe[a,b] | | | |
| Inches | Water | Steam | Copper Tube Water | Rod Size Inches |
| 1/2 | 7 | 8 | 5 | 1/4 |
| 3/4 | 7 | 9 | 5 | 1/4 |
| 1 | 7 | 9 | 6 | 1/4 |
| 1 1/2 | 9 | 12 | 8 | 3/8 |
| 2 | 10 | 13 | 8 | 3/8 |
| 2 1/2 | 11 | 14 | 9 | 3/8 |
| 3 | 12 | 15 | 10 | 3/8 |
| 4 | 14 | 17 | 12 | 1/2 |
| 6 | 17 | 21 | 14 | 1/2 |
| 8 | 19 | 24 | 16 | 5/8 |
| 10 | 20 | 26 | 18 | 3/4 |
| 12 | 23 | 30 | 19 | 7/8 |
| 14 | 25 | 32 | | 1 |
| 16 | 27 | 35 | | 1 |
| 18 | 28 | 37 | | 1 1/4 |
| 20 | 30 | 39 | | 1 1/4 |

[a] Adapted from MSS *Standard* SP-69.
[b] Spacing does not apply where span calculations are made or where concentrated loads are placed between supports such as flanges, valves, specialties, etc.
Conversion Factors: mm = 25.4 × in. = 304.8 × ft

**Table 10 Capacities of ASTM A36 Steel Threaded Rods**

| Rod Diameter in. | Root Area of Coarse Thread in² | Maximum Load[a] lb |
|---|---|---|
| 1/4 | 0.027 | 240 |
| 3/8 | 0.068 | 610 |
| 1/2 | 0.126 | 1130 |
| 5/8 | 0.202 | 1810 |
| 3/4 | 0.302 | 2710 |
| 7/8 | 0.419 | 3770 |
| 1 | 0.552 | 4960 |
| 1 1/4 | 0.889 | 8000 |

[a] Based on an allowable stress of 12,000 psi reduced by 25% using the root area in accordance with ASME B31.1 and MSS SP-58.
Conversion Factors: mm = 25.4 × in.
$mm^2 = 645 \times in^2$
kg = 0.454 × lb

## BIBLIOGRAPHY

ANSI. 1984. *National Fuel Gas Code*. ANSI *Standard* Z223.184. American National Standards Institute, New York, NY.
ASHRAE. 1978. *Safety Code for Mechanical Refrigeration. ASHRAE Standard* 15-1978.
ASME. 1986. *Boiler and Pressure Vessel Codes.* American Society of Mechanical Engineers, New York, NY.
ASME. 1983. *Power Piping*. ASME *Standard* B31.1-83.
ASME. 1987. *Refrigeration Piping.* ASME *Standard* B31.5-87.

ASME. 1982. *Building Services Piping.* ASME *Standard* B31.9-82.

ASME. 1985. *Welded and Seamless Wrought Steel Pipe.* ASME *Standard* B36.10M-85.

ASTM. 1987. *Standard Specification for Pipe, Steel, Black and Hot-Dipped, Zinc-Coated Welded and Seamless (Rev A).* ASTM *Standard* A53-87. American Society for Testing and Materials, Philadelphia, PA.

ASTM. 1987. *Standard Specification for Seamless Carbon Steel Pipe for High Temperature Service.* ASTM *Standard* A106-87.

ASTM. 1986. *Standard Specification for Seamless Copper Water Tube.* ASTM *Standard* B88-86.

ASTM. 1986. *Standard Specification for Seamless Copper Tube for Air Conditioning and Refrigeration Field Service.* ASTM *Standard* B280-86.

ASTM. 1986. *Standard Specification for Poly(Vinyl Chloride) (PVC) Plastic Pipe, Schedules 40, 80, and 120.* ASTM *Standard* D1785-86.

ASTM. 1983. *Specification for Polybutylene (PB) Plastic Pipe (SIDR-PR) Based on Controlled Inside Diameter.* ASTM *Standard* D2662-83.

ASTM. 1981. *Specification for Polybutylene (PB) Plastic Pipe (SIDR-PR) Based on Outside Diameter.* ASTM *Standard* D300073 (Rev. 1981).

ASTM. 1985. *Specification for Polybutylene (PB) Plastic Hot and Cold-Water Distribution Systems (Rev. B).* ASTM *Standard* D3309-85.

ASTM. 1986. *Standard Specification for Chlorinated Poly (Vinyl Chloride) (CPVC) Plastic Pipe, Schedules 40 and 80 (Rev. A).* ASTM *Standard* F441-86.

ASTM. 1986. *Standard Specification for Plastic Inserts Fittings for Polybutylene (PB) Tubing..* ASTM *Standard* F845-86.

AWWA. 1986. *Thickness Design of Ductile-Iron Pipe.* AWWA *Standard* C150-81 (Rev. 1986). American Water Works Association, Denver, CO.

MSS. *Pipe Hangers and Supports—Materials, Design and Manufacturer.* MSS *Standard* SP-58. Manufacturers Standardization Society of the Valve and Fitting Industry, Vienna, VA.

MSS. *Pipe Hangers and Supports—Selection and Application.* MSS *Standard* SP-59.

NFPA. 1987. *Standard for the Installation of Sprinkler Systems.* NFPA *Standard* 13-87. National Fire Protection Association, Quincy, MA.

CHAPTER 34

# AIR-TO-AIR ENERGY-RECOVERY EQUIPMENT



THIS chapter addresses the objectives and applications of air-to-air energy-recovery equipment and describes the various types of equipment available. The purpose of all air-to-air energy-recovery equipment is to reduce the amount of energy consumed by a building or process in a cost-effective manner by transferring energy between the supply and exhaust airstreams. This equipment may also reduce the size and capital cost of the supporting utility equipment (*e.g.*, boilers, chillers, and burners). Properly applied air-to-air energy-recovery equipment should add little or no first cost, while providing long term benefits through reduced energy consumption.

## TYPES OF EQUIPMENT

The air-to-air energy-recovery applications include (1) process-to-process, (2) process-to-comfort, and (3) comfort-to-comfort.

### Process-to-Process

In process-to-process applications, heat is captured from the process exhaust airstream and transferred to the process supply airstream. Economics favor higher temperature processes, because more heat is available for recovery. Equipment is available to handle process exhaust temperatures as high as 1600 °F (870 °C).

Typical applications include driers, kilns, and ovens. Figure 1 shows a typical oven application for an air-to-air energy-recovery device of 70% sensible heat effectiveness, operating under typical winter design conditions.

When applying air-to-air energy-recovery from process exhaust, the following should be evaluated:

1. *Effects of Corrosives.* Process exhaust frequently contains substances requiring corrosion-resistant construction materials for the energy-recovery device.
2. *Effects of Condensables.* The process exhaust may contain vapor that could condense upon cooling in the energy-recovery device. In some cases, the condensed compound can be recovered.
3. *Effects of Contaminants.* If the process exhaust contains particulate contaminants or condensables, the energy-recovery device should be accessible for cleaning. Air filtration and a recovery device with an open structure may be necessary to reduce the frequency of cleaning.
4. *Effects on Other Equipment.* Removing heat from the process exhaust may reduce the cost of pollution control by allowing less expensive bags in bag houses or by improving the efficiency of electronic precipitators. As a result, energy recovery and pollution control can often be coupled with beneficial effects.

Process-to-process recovery devices generally recover only sensible heat and do not transfer latent heat (humidity); in most cases, moisture transfer is detrimental to the process. Process-to-process

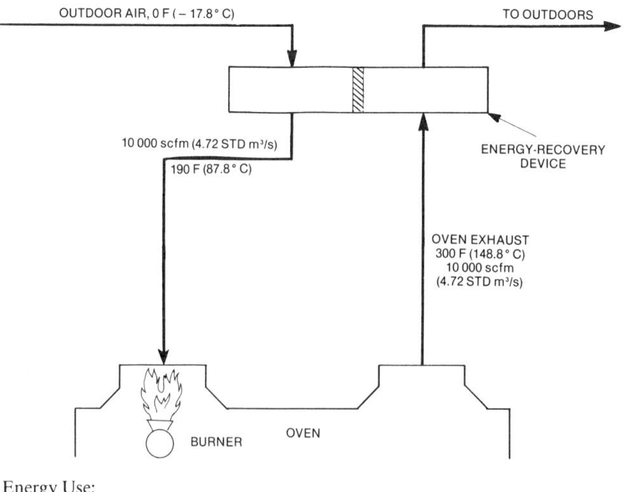


Energy Use:
With Recovery: 2,050,000 Btu/h (601 kW)
Without Recovery: 4,320,000 Btu/h (1266 kW)

**Fig. 1 Process-to-Process Sensible Heat Device**

The preparation of this chapter is assigned to TC 5.5, Air-To-Air Energy Recovery.

applications usually involve maximum recovery; in cases involving condensables, less recovery may be desired to prevent condensation and possible corrosion.

## Process-to-Comfort

In process-to-comfort applications, waste heat captured from a process exhaust is used to heat building makeup air during winter. Typical applications include foundries, strip coating plants, can plants, plating operations, pulp and paper plants, and other processing areas with heated process exhaust and large makeup air volume requirements. Figure 2 shows a typical process-to-comfort application.

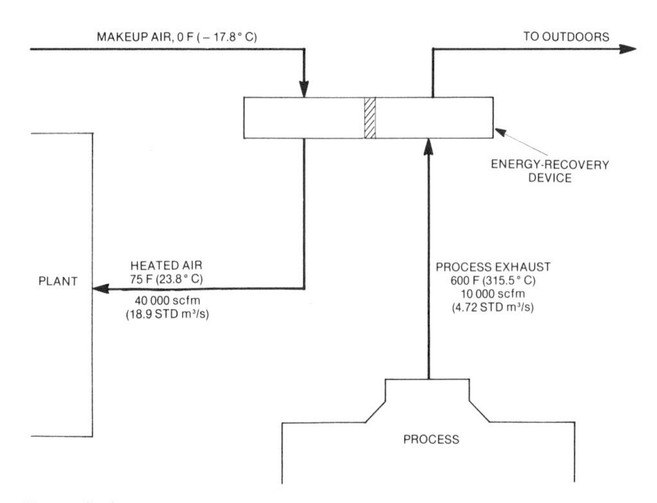


Energy Savings:
With Recovery: 3,240,000 Btu/h (950 kW).

**Fig. 2 Process-to-Comfort Sensible Heat Device**

Although full recovery is desired in process-to-process applications, recovery for process-to-comfort applications must be modulated during warm weather to prevent overheating the makeup air. During summer, no recovery is required. Because energy is saved only in the winter and recovery is modulated during moderate weather, process-to-comfort applications save less energy over a year than process-to-process applications.

When applying process-to-comfort energy-recovery, the following effects should be evaluated:

1. *Effects of Corrosives.* Process exhaust frequently contains compounds that require compatible construction materials for the energy-recovery device.
2. *Effects of Condensables.* The process exhaust may contain compounds with concentration high enough to cause them to condense upon cooling in the energy-recovery device.
3. *Effects of Contaminants.* If the process exhaust contains particulate contaminants or condensables, the energy-recovery device should be accessible for cleaning. Air filtration and a recovery device with an open structure should be selected to reduce the frequency of cleaning.

Process-to-comfort recovery devices generally recover sensible heat only and do not transfer moisture between the airstreams.

## Comfort-to-Comfort

In comfort-to-comfort applications, the equipment transfers energy from the exhaust air to the supply air of the building. Also, it lowers the enthalpy of the building supply air during warm weather and raises it during cold weather.

Air-to-air energy-recovery devices available for comfort-to-comfort applications fall into two categories: sensible heat devices and total heat devices.

Sensible heat devices only transfer sensible heat (dry-bulb temperature) between supply and exhaust airstreams, except where the exhaust airstream is cooled to below its dew point. Then, condensation occurs and effects system performance, depending on the device used. Total heat devices transfer both sensible heat and latent heat (humidity) between supply and exhaust airstreams. Unlike process-to-process and process-to-comfort applications, latent transfer is preferred in comfort-to-comfort applications.

Figure 3 shows a typical comfort-to-comfort application of a device with 70% effectiveness of sensible heat only; Figure 4 shows a device of 70% effectiveness of both sensible heat and latent heat operating under the same conditions. When these examples operate under typical summer design conditions, the total heat device recovers nearly three times as much energy as the sensible heat device. Under the typical winter design conditions, the total heat device recovers over 25% more energy than the sensible heat device. Figure 5 shows these example systems for both summer and winter conditions on a psychrometric chart, which permits comparing the energy transfers directly.

When applying comfort-to-comfort energy recovery, the following should be evaluated:

1. *Effect of Particulate Contaminants.* If the building exhaust site contains a large amount of particulate matter, such as dust, lint, and animal hair, the building exhaust air should be filtered to prevent plugging of the energy-recovery device. The building supply air should be filtered, as required, to protect the air-conditioning equipment.
2. *Effect of Gaseous or Vaporous Contaminants.* If the building exhaust air contains gaseous or vaporous contaminants such

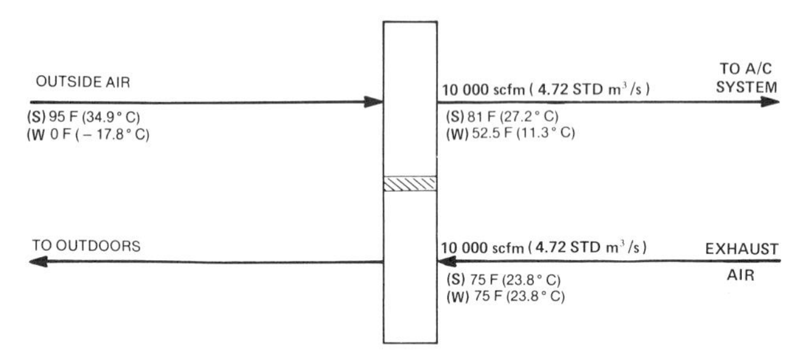


Energy Saved:
Summer: 151,000 Btu/h (44 kW)
Winter: 567,000 Btu/h (166 kW).

**Fig. 3 Comfort-to-Comfort Sensible Heat Device**

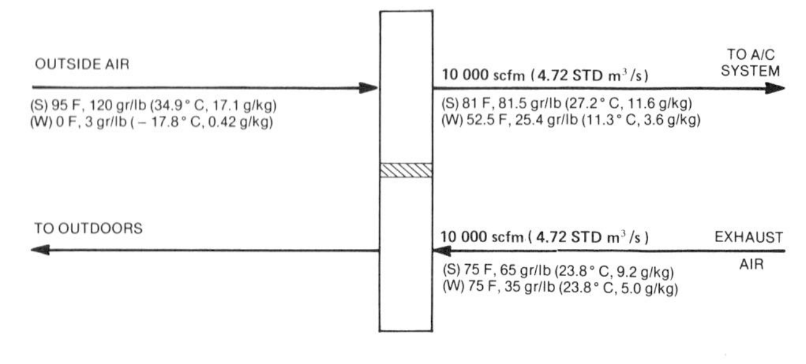


Energy Saved:
Summer: 428,000 Btu/h (125 kW)
Winter: 729,000 Btu/h (214 kW).

**Fig. 4 Comfort-to-Comfort Total Heat Device**

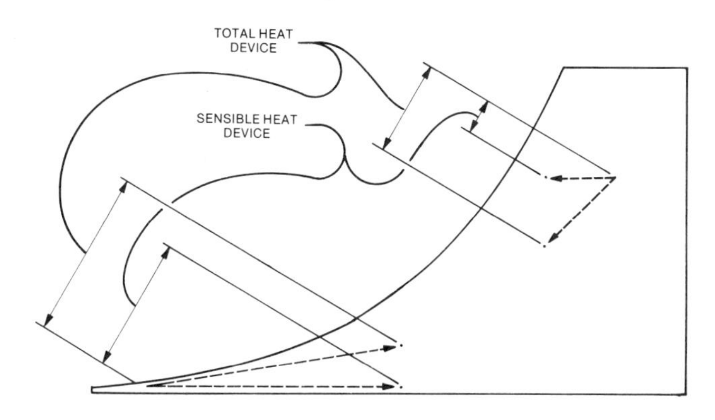


**Fig. 5 Comfort-to-Comfort Sensible Heat versus Total Heat Recovery**

as hydrocarbons, sulfur compounds, and water-soluble chemicals, their effect on the energy-recovery device should be investigated.

3. *Effect of Indoor Humidity Level.* Winter performance of the energy-recovery device should be investigated if the humidity in the building is controlled during the winter months or if the building is located in a cold climate. Special controls, preheating of the supply airstream, or auxiliary heating of the heat-transfer media may be required to prevent frosting of the energy-recovery device or to prevent operating problems during winter. Total heat devices are less likely to frost than sensible heat devices.
4. *Comparison of Recovery Devices.* All air-to-air energy-recovery devices, even sensible-heat-only devices, should be investigated on the basis of their individual features, recovery effectiveness, and cost.

## CORROSION AND FOULING

ASHRAE-sponsored research examined the effects of corrosion and fouling on various metals used in heat-recovery devices. The study examined the exhausts from pulp and paper processes, industrial metal painting, fast-food kitchens, and areas of heavy smoking.

In many industrial applications, the end-user knows which materials are most corrosion-resistant for the process. When the end-user is inexperienced, an examination of on-site ductwork and/or available literature may help in selecting materials. If the capital cost of a heat-exchanger installation is large or if the life of selected materials must be estimated, a field installation of material samples may be justified. ASHRAE (1982) describes experimental procedures for such investigations. Any study should recognize that other contaminants not directly related to the process may be present in the exhaust stream (*e.g.*, chemicals in the water wash for spray-painting, or welding fumes from an adjacent area).

Moderate corrosion generally occurs over time. However, severe corrosion can form pinholes and cause mechanical failure of the heat exchanger. Exhaust leaking into the supply stream can also become a problem, particularly if the exhaust constituents are toxic or odorous. Different heat exchangers may be more or less sensitive to cross leakage.

Moderate corrosion roughens the surface of a metal and increases its heat-transfer coefficient initially. However, continual corrosion of the surface may not continue to increase surface roughness, and the heat-transfer coefficient will approach a maximum. Severe fouling, in contrast, eventually reduces the overall heat-transfer coefficient.

The thermal effectiveness established under laboratory procedures and conditions, as described in ASHRAE *Standard* 84-78, is a reliable measure of the thermal performance of a heat exchanger installed under adverse conditions.

The increase in resistance to airflow through the exchanger as fouling increases is more critical to performance. This increased resistance reduces the fluid flow through the exchanger or increases the fan power. Air-side pressure loss can be measured accurately in the field, and the measurements can indicate excessive fouling. Experience can then help in setting cleaning schedules. Similarly, frosting of heat-transfer surfaces can be detected by an increase in the operating pressure drop across the exchanger. The pressure loss must be detected before the unit is blocked totally. The cleaning of heat exchanger surfaces is important when selecting heat-recovery equipment.

## ECONOMIC EVALUATION

Chapter 49 of the 1987 HVAC Volume gives information on owning and operating costs of a system. The following costs and benefits should be considered for any air-to-air recovery application.

### Initial Cost

1. Of energy recovery devices
2. Of installation, including additional ductwork, piping, and electrical services
3. Of larger fans and/or motors that may be needed to overcome the increased static pressure loss caused by the energy-recovery device
4. Of additional air filtration equipment and installation, if required
5. Of capacity controls, if required
6. Of auxiliary heaters for frost control, if required

### Initial Savings

1. Due to the reduced size of boilers or heating equipment, along with associated coils, piping, pumps, etc., because of the reduced heating load
2. Due to the reduced size of the chiller or cooling plant, and their associated coils, piping, pumps, etc., because of the reduced cooling load
3. Due to the reduced the size of electrical supply equipment

### Future Saving Value

1. Of the energy-recovery system
2. Of the associated equipment affected by the installation minus the salvage value of the associated equipment without the energy-recovery device

### Periodic Costs

1. Maintenance of the energy-recovery device
2. Operating cost of the energy-recovery device, along with any drives, motors, controls, and defrost heaters
3. Additional fan power due to increased static pressure drop
4. Maintenance of the additional filtration, if any

### Periodic Savings

1. Due to reduced energy consumption for the heating season
2. Due to reduced energy consumption for the cooling season

## CONTROLS

Energy-exchanger controls regulate the amount of energy transferred to supply air at specified conditions. When the desired supply air sensible and/or latent heat is the same as, or higher than, that of the exhaust air, maximum recovery capability is necessary. However, most ventilation systems are selected to maintain specific indoor conditions at extreme outside design conditions, which occur only a few hours a year. Thus, recovery modulation may be required during both summer and winter to prevent supply air overheating or overhumidification.

The designer should refer to (1) the following sections that describe the different types of air-to-air energy-recovery devices and (2) manufacturer's literature regarding the control methods for specific equipment.

## PERFORMANCE RATING OF EQUIPMENT

Performance of an air-to-air heat exchanger is usually expressed as its effectiveness in transferring (1) sensible heat (dry-bulb temperature), (2) latent heat (humidity ratio), and (3) total heat (enthalpy). ASHRAE *Standard* 84-78 covers the rating and testing of Air-to-Air Heat-Recovery Equipment. In the standard, effectiveness, $\varepsilon$, of a heat exchanger is defined as follows:

$$\varepsilon = \frac{\text{Actual transfer for the given device}}{\text{Maximum possible transfer between airstreams}}$$

Referring to Figure 6,

$$\varepsilon = \frac{W_s(X_1 - X_2)}{W_{min}(X_1 - X_3)} = \frac{W_e(X_4 - X_3)}{W_{min}(X_1 - X_3)} \quad (1)$$

*where*

$\varepsilon$ = sensible heat, latent heat, or total heat effectiveness
$X$ = dry-bulb temperature, humidity ratio, or enthalpy at the location indicated in Figure 6

For the latent and total heat effectiveness:

$W_s$ = mass flow rate of supply
$W_e$ = mass flow rate of exhaust

For the sensible heat effectiveness:

$W_s$ = (specific heat) (mass flow rate) for the supply
$W_e$ = (specific heat) (mass flow rate) for the exhaust
$W_{min}$ = smaller of $W_s$ and $W_e$

The leaving supply air condition is then

$$X_2 = X_1 - \varepsilon \frac{W_{min}}{W_s}(X_1 - X_3) \quad (2)$$

and the leaving exhaust air condition is

$$X_4 = X_3 + \varepsilon \frac{W_{min}}{W_e}(X_1 - X_3) \quad (3)$$

Effectiveness is important because it includes the mass flow ratio.

The assumptions for Equation (1) include the following:

1. Negligible heat transfer between heat exchanger case and surroundings
2. No internal sources of energy from fans or defrost devices
3. Negligible changes in kinetic and potential energy

The first two assumptions are generally not satisfied for residential air-to-air heat-recovery ventilators. When volume flows are small and residence times are large, the heat transfer through the case may be significant, unless the case is insulated heavily. Moreover, both internal fans and defrost devices are commonly used.

The effectiveness of a particular air-to-air energy-recovery device is a function of such variables as the supply and exhaust mass flow rates and the energy-transfer characteristics of the device. Because of the many combinations of these variables, performance data must be established for each device. Manufacturers normally have detailed performance data on their equipment. Different devices are detailed in the following sections.

## ROTARY AIR-TO-AIR ENERGY EXCHANGER

A rotary air-to-air energy exchanger, or *heat wheel*, has a revolving cylinder filled with an air permeable medium with a large internal surface area. Adjacent supply and exhaust airstreams each flow through half the exchanger in a counterflow pattern (Figure 6). Heat-transfer media may be selected to recover only sensible heat or total heat (sensible heat plus latent heat).

Sensible heat is transferred as the medium picks up and stores heat from the hot airstream and gives it up to the cold one. Latent heat is transferred as the medium (1) condenses moisture from the airstream that has the higher humidity ratio (by means of absorption for liquid desiccants and adsorption for solid desiccants), with a simultaneous release of heat; and (2) releases the moisture through evaporation (and heat pickup) into the airstream that has the lower humidity ratio. Thus, the moist air is dried while the drier air is humidified. In total heat transfer, both sensible and latent heat transfer occur simultaneously.

Figure 7 illustrates a typical sensible heat-recovery process between supply and exhaust airstreams. Cold air is heated from 1 to

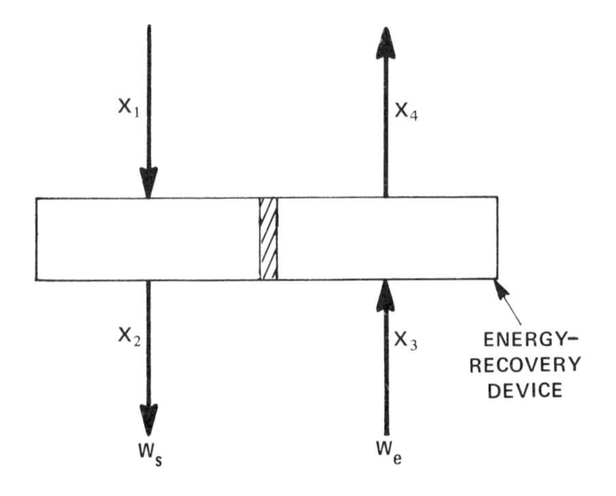


**Fig. 6 Effectiveness Ratings (Countercurrent Flow)**

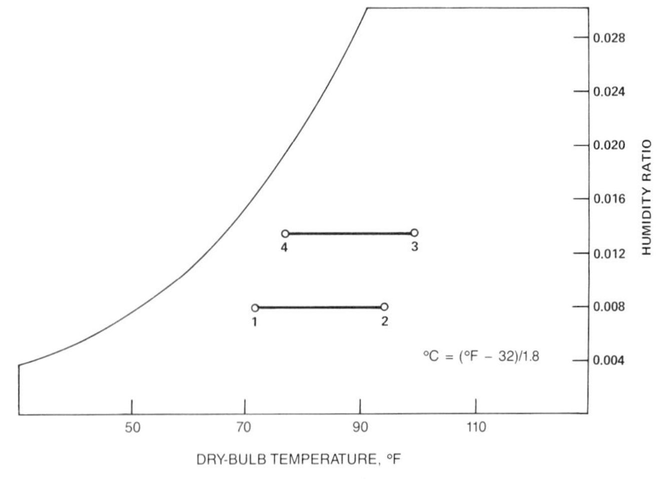


**Fig. 7 Sensible Heat Recovery**

2, while hot air is cooled from 3 to 4. In this case, the cold-air temperature is above the dew point of the hot air, and no condensation takes place in the media. Figure 8 illustrates a sensible heat-recovery process in which condensation occurs in the hot airstream, along with evaporation in the cold one. In this case, latent heat transfer enhances overall wheel effectiveness.

Figure 9 illustrates a total heat-recovery process for when mass flow rates and the latent and total heat effectiveness are equal. For this case, outlet states 2 and 4 lie on the straight line through cold-air inlet state 1 and hot-air inlet state 3.

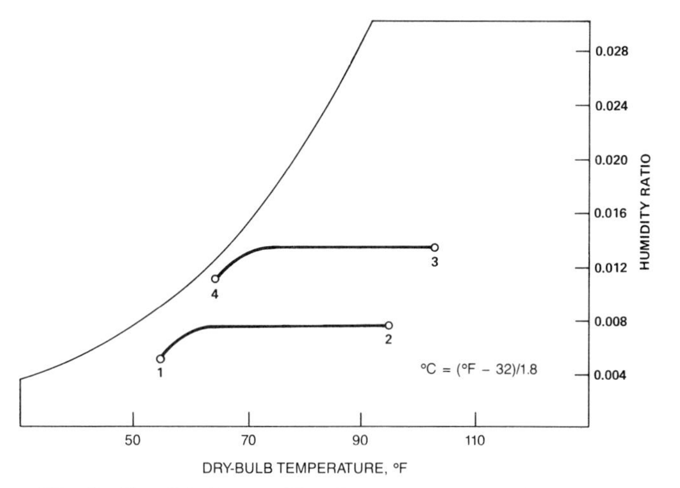


**Fig. 8 Sensible Rotary Heat Exchanger Recovering Latent Heat**

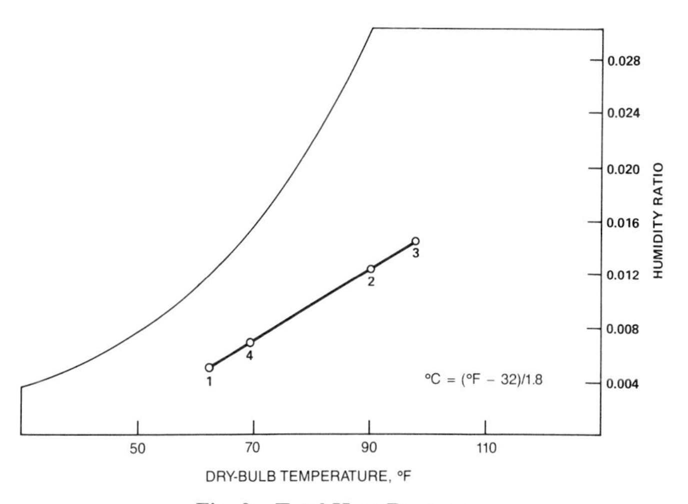


**Fig. 9 Total Heat Recovery**

## Construction

The air contaminants, dew point, exhaust air temperature, and supply air properties influence the choice of structural materials for the casing, rotor structure, and medium of a rotary energy exchanger. Aluminum and steel are the usual structural, casing, and rotor materials for normal comfort ventilating systems. Exchanger media are fabricated from metal, mineral, or man-made materials and are classified as providing either random flow or directionally oriented flow through their structures.

Random flow media, consisting of knitted corrugated mesh, is made by knitting 8 to 12 mil (0.2 to 0.3 mm) diameter wire into an open woven cloth, which is layered to the desired configuration. Aluminum mesh, commonly used for comfort ventilation systems, is packed in pie shaped wheel segments to a density of 4 lb/ft$^3$ (64 kg/m$^3$). Stainless steel and monel mesh, used for high temperature and corrosive atmosphere applications, are packed to a density of 11 lb/ft$^3$ (176 kg/m$^3$). This media should only be used with clean, filtered airstreams, because it plugs easily.

Directionally oriented media consist of small 0.0625 in. (1.6 mm) triangular air passages parallel with the direction of airflow. The triangular shape gives the largest exposed surface for air contact per unit of face area, is strong, and is easily produced by interleaving layers of flat and corrugated material. Aluminum foil, inorganic sheet, treated organic sheet, and synthetic materials are used for low and medium temperatures. Stainless steel and ceramics are used for high temperatures and corrosive atmospheres.

Media surface areas exposed to airflow vary from 100 to 1000ft$^2$/ft$^3$ (330 to 3300 m$^2$/m$^3$), depending on the medium type and physical configuration. Media may also be classified according to their ability to recover only sensible heat or total heat. Media for sensible heat recovery are fabricated from aluminum, copper, stainless steel, and monel. Media for total heat recovery are fabricated from any of a number of materials (aluminum, mineral, or man-made) and treated to give them specific moisture-recovery capability. Aluminum is given a surface of alumina or other desiccant material. Mineral and man-made materials are impregnated with a desiccant—typically lithium chloride.

## Performance

The performance of a rotary energy exchanger is defined by the exchanger's effectiveness and the pressure drop through its media. Face velocities for most energy-recovery applications range from 500 to 800 fpm (2.5 to 4.0 m/s). Low face velocities give lower pressure drop, higher effectiveness, and lower operating costs, but require larger units with higher capital cost and more installation space. High face velocities give the reverse.

Typical pressure drops for various media at 500 fpm (2.5 m/s) vary from 0.4 to 0.7 in. of water (100 to 175 Pa). The manufacturer's catalog includes actual data for each case. Average effectiveness values for sensible and total heat exchangers range from 70 to 80% for equal supply and exhaust air mass flow rates and for usual exchanger face velocities.

Rotary energy exchangers are available in single units to 68,000 cfm (32 m$^3$/s) capacity; they are not usually larger than 14 ft (4270 mm) in diameter because of difficulty in shipping, erecting, and fitting into buildings. Multiple units provide greater single-system capacities. Units are available for temperatures of −70 to 1500 °F (−60 to 800 °C). When installed horizontally (vertical airflow), units greater than 8 ft (2440 mm) in diameter require special structural considerations because of their size and weight.

## Cross Contamination

Cross contamination, or *mixing*, of air between supply and exhaust airstreams occurs in all rotating energy exchangers through two mechanisms—*carry over* and *leakage*. Carry over occurs as air in each airstream is entrained within the volume of the rotating medium and is carried into the other airstream. Leakage occurs because of the different static pressures in the two airstreams, which drive air from a higher to a lower static pressure region. Contamination by leakage can be reduced by placing the blowers so they promote leakage of outside air to the exhaust airstream. Carry over occurs each time a portion of the matrix passes the seals dividing the supply and exhaust airstreams. Carry over from exhaust to supply may be undesirable, so a purge section can be installed on the heat exchanger to prevent cross-contamination. In some applications, a small recirculating volume is not a concern, so a purge section is unnecessary. For example, many systems are designed to recirculate a portion of exhaust air as part of the operating cycle.

Critical applications, however, such as hospital operating rooms, laboratories, and clean rooms, require stringent control of carry over and a purge section, followed by high efficiency filtration. Research (ASHRAE 1974) has shown that carry over can be reduced to below 0.1% of the exhaust airflow with a purge section.

The theoretical carry-over of a wheel without a purge is directly proportional to the speed of wheel and the void volume of the media (75 to 95% void, depending on type and configuration). For example, a 10 ft (3.0 m) diameter, 8 in.- (200-mm) deep wheel, with a 90% void volume operating at 12 rpm, has a carry-over volume of:

$$(10)^2(0.7854)(8/123)(0.9)(12) = 568 \text{ cfm}$$

If the wheel is handling 20,000 cfm (9.44 $m^3/s$) balanced flow, the percent carry over is:

$$(568/20{,}000)(100) = 2.84\%$$

The exhaust fan, which is usually located at the exit of the exchanger, should be sized to include leakage, purge, and carry-over airflows.

## Filtration

Because the exchanger medium mainly consists of voids, most fine, dry, inert airborne contaminants pass through and present no problems. Even so, filters should be placed in both supply and exhaust airstreams to reduce the amount of contaminants entering the exchanger and the frequency of cleaning. Exhaust filters are especially important when contaminants are sticky, greasy, fibrous, or present in large quantities. Supply filters eliminate insects, leaves, and other large particles or foreign materials. Wet snow can also cause severe plugging problems. Steps that ensure a continuous flow of supply air should be included in the design.

## Condensation and Freeze-up

In a sensible heat exchanger, moisture condenses on the medium when the hot airstream is cooled below its dew point. In moderate cases, the water condensed in one airstream evaporates into the other and enhances exchanger effectiveness. In severe cases, some of the condensed water reevaporates, while the rest accumulates on the medium until the excess drips off the wheel into a drain pan and exits through a drain connection. If the wheel is installed horizontally, the condensate that falls from the wheel into the duct below must be removed.

Frosting (sublimation of water vapor) and icing (freezing of subcooled condensate) on sensible exchanger media can occur at below freezing supply air temperatures when the medium temperature drops below the frost point of the exhaust air. The rate of frost accumulation depends on the temperature of the supply air, humidity ratio of the exhaust air, exchanger effectiveness, and duration of frosting conditions. Generally, frost forms on the discharge face of the exhaust airstream and then increases in both thickness and depth of media penetration consistent with the duration and intensity of the frosting conditions. In extreme cases, the airflow can be blocked, or the medium may be structurally damaged.

Frosting and icing usually occur at lower temperatures on total heat wheels than on sensible heat wheels. The manufacturer's literature should include information on frosting.

An increase in the pressure drop across the wheel indicates the onset of frosting or icing. It can be prevented either by preheating the cold supply air or by reducing wheel energy transfer. A heating coil, located in the supply airstream ahead of the wheel, can preheat the air to prevent frosting and icing. Controlling wheel speed or bypassing a portion of the cold supply air around the wheel (see the following section on "Controls") reduces wheel energy transfer. This method requires additional reheat coil capacity to compensate for the reduced amount of heat recovered from the exhaust air. Normally, frosting does not occur when the cold supply air is above 21 °F (−6 °C). In some cases, heat is added to the exhaust airstream entering the wheel. This procedure eliminates the preheat coil and its attendant problems if the coil fails.

## Controls

Most wheels operate under varying recovery conditions, so their controls must be flexible. For example, when the desired supply air temperature (or humidity or dew point) is the same as or higher than the exhaust air temperature (or humidity or dew point), maximum recovery capability is needed. However, most ventilation systems are engineered to produce specific conditions at extreme outside design conditions (which occur only a few days a year), and they must be modulated during most of the time.

Two controls methods are commonly used to regulate wheel energy recovery. In the first method, *supply air bypass control*, the amount of supply air allowed to pass through the wheel establishes the supply air temperature. An air bypass damper, controlled by a wheel supply air discharged temperature sensor, regulates the proportion of supply air permitted to bypass the exchanger.

The second method regulates the energy recovery rate by varying wheel rotational speed (Figure 10). The most frequently used

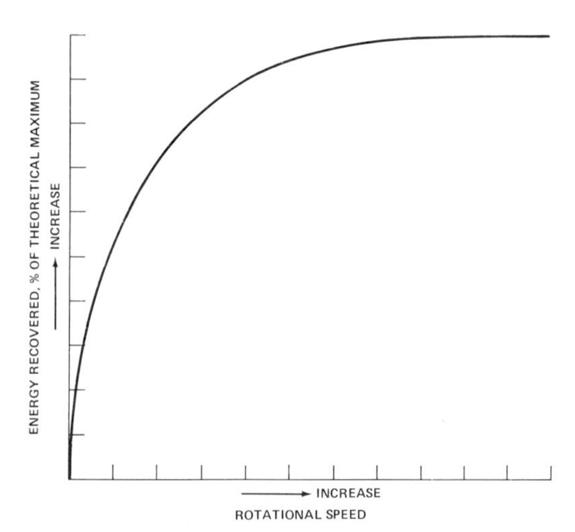


**Fig. 10 Typical Wheel Energy-Recovery Capacity versus Rotational Speed**

variable speed drives are (1) a silicon controlled rectifier (scr) with variable speed dc motor, (2) a constant speed ac motor with hysteresis coupling, and (3) an ac frequency inverter with an ac induction motor. A modulating sensor at the desired location controls all these devices.

A deadband control, which stops or limits the exchanger, may be necessary when no recovery is desired (*e.g.*, when outside air temperature is higher than the required supply air temperature but below the exhaust air temperature). When the outside air temperature is above the exhaust air temperature, the equipment operates at full capacity to cool the incoming air.

## Maintenance

Energy exchanger wheels require little maintenance. The following maintenance procedures ensure best performance:

1. Clean the medium when lint, dust, or other foreign materials build up. Cleaning methods suitable for one type of medium are not necessarily suitable for other types. Media treated with a liquid desiccant for total heat recovery may not be wetted; they are cleaned by vacuuming the wheel face or by blowing dry, compressed air through the passages. Metallic and some nonmetallic media may be vacuumed, blown out with compressed air, or cleaned with steam, hot water (with or without detergent), or a suitable solvent. The manufacturer's cleaning instructions should be followed.
2. Maintain drive motor and train in accordance with the manufacturer's recommendations. Give particular attention to speed control motors that have commutators and brushes; these require more frequent inspection and maintenance than do induction motors. Brushes should be replaced, and the commutator should be turned and undercut periodically.
3. Inspect wheels regularly for proper belt or chain tension.
4. Refer to the manufacturer's recommendations for spare and replacement parts.

## COIL ENERGY-RECOVERY LOOP

A typical coil energy-recovery loop system (Figure 11), places extended surface, finned-tube water coils in the supply and exhaust airstreams of a building or process. The coils are connected in a closed loop via counterflow piping, and an intermediate heat-transfer fluid of water (typically) or a freeze-preventive solution is pumped through the coils.

This system transfers energy from the warmer to the cooler airstream. In a typical comfort-to-comfort application, the system is seasonally reversible—the supply air is preheated when the outdoor air is cooler than the exhaust air and precooled when the outdoor air is warmer. This system operates generally for sensible heat recovery.

As with other air-to-air energy-recovery equipment, moisture must not freeze in the exhaust coil air passage. A dual purpose, three-way temperature-control valve prevents the exhaust coil from freezing. The valve is controlled to maintain the entering solution temperature to the exhaust coil to not less than 30 °F (−1 °C). This condition is maintained by bypassing some of the warmer solution from the supply air coil. The valve can also ensure that a prescribed air temperature from the supply air coil is not exceeded for those applications in which the energy recovered must be limited.

### System Characteristics

This highly flexible system is well-suited for renovation and industrial applications. It accommodates remote location of supply and exhaust ducts. It also allows the simultaneous recovery of energy from multiple supply and exhausts. A closed expansion tank must be included to allow for fluid expansion and contraction when the system is operated under different conditions. A closed expansion tank also minimizes oxidation when ethylene glycol is used.

Standard finned-tube water coils may be used. The manufacturer's curves give pressure drops for a specific design. Coil face velocities typically range from 300 to 600 fpm (1.5 to 3 m/s). Lower face velocities require larger coils, which may have a higher first cost but lower (fan) operating costs.

### Effectiveness

The coil energy-recovery loop is primarily a heating device, because it cannot transfer moisture from one airstream to another. For the most cost-effective operation, with equal airflow rates and no condensation, typical effectiveness values range from 60 to

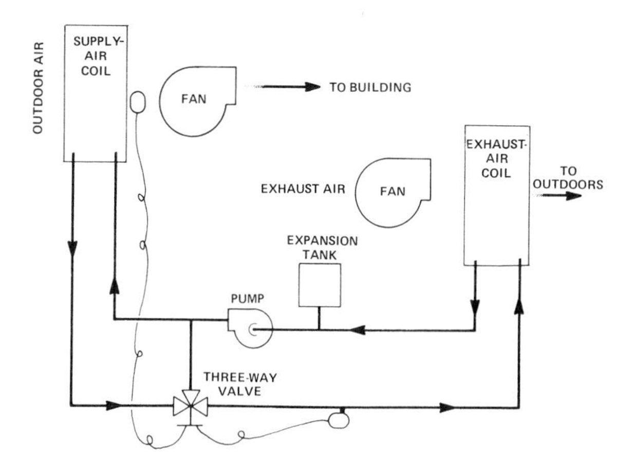


**Fig. 11 Arrangement of Coil Energy-Recovery Loop**

65%. Highest effectiveness does not necessarily give the greatest net cost savings. Cost savings can be more accurately determined by considering the cost of heating and cooling, the operating costs of the system, the operating times of the system, the geographic location, and the schedule of system capacities (including the effect of controls) against outside air temperatures. Manufacturers use computer programs to provide this design analysis.

The following example illustrates the capacity of a typical system:

**Example.** A waste heat recovery system heats 10,000 cfm of air from a 0 °F design outdoor temperature to an exhaust temperature of 75 °F dry bulb/60 °F wet bulb. The air flows through identical eight-row coils at a 400 fpm face velocity. A 30% ethylene glycol solution flows though the coils at 26 gpm.

Figure 12 shows the effect of the outside air temperature on capacity, including the effects of the three-way temperature-control valve. The capacity is constant when the valve controls the entering fluid temperature of the exhaust coil to prevent frosting—below 18.5 °F for this example—because the exhaust coil is the source of heat and has a constant airflow rate, entering air temperature, liquid flow rate, entering fluid temperature (as set by valve), and fixed coil parameters.

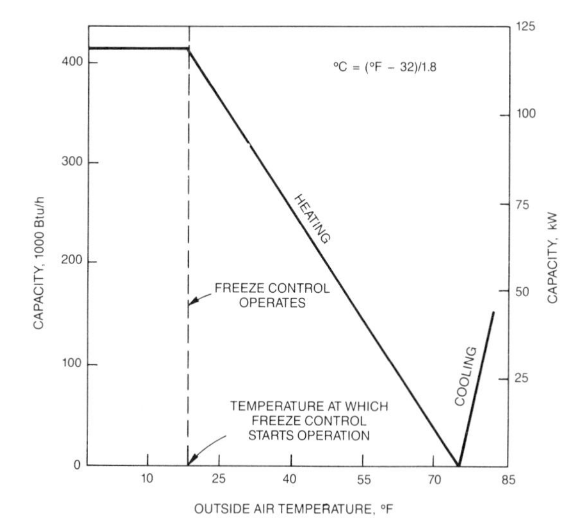


**Fig. 12 Energy-Recovery Capacity versus Outside Air Temperature for Typical Loop**

At the 0 °F design temperature, with the control valve operating, 414,000 Btu/h is recovered. Equation (1) may be used to calculate the sensible heat effectiveness as 51% ($\varepsilon$ = 414,000/810,000) and the leaving air dry-bulb temperature of the supply coil as 38.3 °F when the outside air is 0 °F. When the three-way control valve operates at 18.5 °F outside air temperature, 414,000 Btu/h is recovered. At this point, the sensible heat effectiveness is 67.2%. Above 75 °F outside air temperature, the system begins to cool the supply air.

Typically, the sensible heat effectiveness of a coil energy-recovery loop is independent of the outside air temperature. However, when the capacity is controlled (as in the proceeding example) the sensible heat effectiveness decreases.

### Construction Materials

Coil energy-recovery loops incorporate coils constructed to suit the environment and operating conditions to which they are exposed. For typical comfort-to-comfort applications, standard coil construction usually suffices. Process-to-process and process-to-comfort applications require considering such factors as the effect of high temperature, condensables, corrosives, and contaminants on the coil(s).

At temperatures, above 400 °F (204 °C), special construction may be required to ensure a permanent fin-to-tube bond. The effects of condensables and other adverse factors may require special coil construction and/or coatings.

### Cross Contamination

Complete separation of the airstreams eliminates cross contamination between the supply and exhaust air.

### Maintenance

Coil energy-recovery loops require little maintenance. The only moving parts are the circulation pump and the three-way control valve. However, the following items must be considered to ensure optimum operation: air filtration, cleaning of the coil surface, and periodic maintenance of the pump and valve. Energy-recovery coils require the same filtration that a dehumidification coil would need in the same environment.

Coils can be cleaned with steam, compressed air, hot soapy water, or suitable solvents. If the exhaust air dictates frequent cleaning, automatic wash systems may be installed. The thermal-transfer fluid may require maintenance. Fluid manufacturers or their representatives should be contacted for specific recommendations.

### Thermal-Transfer Fluids

The thermal-transfer fluid used in a closed-loop application depends largely on the application and temperatures of the two airstreams. Table 1 lists common thermal-transfer fluids and their characteristics.

An inhibited ethylene glycol solution in water is commonly used when freeze protection is required. Such solutions should not be used at temperatures greater than 275 °F (135 °C), where it has a tendency to break down to an acidic sludge. If a freeze-preventive solution is needed and exhaust air temperatures are greater than 275 °F (135 °C), a nonaqueous synthetic heat-transfer fluid may be used. Heat-transfer fluid manufacturers or their representatives can recommend specific fluids.

## TWIN-TOWER ENTHALPY RECOVERY LOOP

### Method of Operation

This air-to-liquid, liquid-to-air enthalpy recovery system consists of one or more contractor towers to handle the building or process supply air and one or more contractor towers to handle the building or process exhaust air. A sorbent liquid circulates continuously between the supply and exhaust air contractor towers; alternately, it contacts both supply and exhaust airstreams directly (Figure 13). This liquid also transports water vapor, as well as heat, between the two airstreams. In a typical comfort-conditioning application, building supply air is cooled and dehumidified during summer operation and heated and humidified during winter. The sorbent solution is usually a halogen salt solution, such as lithium chloride and water. Pumps circulate the solution between supply and exhaust towers.

Both vertical and horizontal airflow contractor towers are manufactured. In the vertical configuration, the supply or exhaust air passes vertically through the contact surface counterflow to the sorbent liquid to achieve high contact efficiency. In the horizontal configuration, air passes horizontally through the contact surface crossflow to the sorbent liquid, yielding a slightly lower contact efficiency. Vertical and horizontal contractor towers can be combined in a common system. Contactor towers of both configurations are supplied with airflow capacities of up to 100,000 scfm (47.2 $m^3/s$).

The contractor surface is usually made of nonmetallic materials, while the tower casing is made of steel with protective coatings. Air leaving the contactor tower passes through demister pads, which remove any droplets entrained in the sorbent solution.

### Design Considerations

**Operating Temperature Limits.** Twin-tower enthalpy recovery systems operate primarily for in the comfort-conditioning range. They are not suitable for high-temperature applications, such as industrial ovens. During summer, these systems operate with building supply air temperatures as high as 115 °F (46 °C). Winter supply air temperatures as low as −40 °F (−40 °C) can be tolerated without freezing or frosting problems, because the sorbent solution is an effective antifreeze at all useful concentrations.

**Enthalpy Recovery Effectiveness.** Figure 14 shows enthalpy recovery effectiveness as a function of contractor tower air face velocity for a twin-tower enthalpy recovery loop.

**Airflow and Pressure Drop.** Figure 15 shows a typical air-side pressure drop versus air face velocity curve for a twin-tower system. Contactor towers generally operate between 300 and 425 fpm (1.5 and 2.2 m/s) air face velocity. Air-side pressure drops generally range from 0.7 to 1.2 in. of water (170 to 300 Pa).

**Static Pressure Effects.** Because contactor supply and exhaust air towers are independent units connected only by piping, supply and exhaust air fans can be located wherever desired; the contactor towers generally operate with any air inlet static pressure from −6 to 6 in. of water (−1.5 to 1.5 kPa). The exhaust contactor tower may be operated at a higher internal static pressure than the supply contactor tower without danger of exhaust-to-supply cross contamination.

**Cross Contamination.** Particulate cross contamination does not occur in twin-tower systems because particulates, once wetted, remain contained within the sorbent solution until they are filtered. Supply and exhaust contactors are independent units connected only by the solution piping. Limited gaseous cross contamination may occur, depending on the solubility of the gas in the sorbent solution. Tests of gaseous cross contamination using sulfur hexafluoride tracer gas techniques have shown typical gaseous cross contamination rates of 0.025% for the twin-tower system.

Sorbent solutions—especially the halide brines—are bactericidal at all useful concentrations. Lithium chloride, as used in twin-tower systems, has a virucidal effect against some viruses. Microorganism testing of the twin-tower system, using the six-plate Anderson Sampling Technique, has shown that micro-

Table 1 Heat-Transfer Fluids

| Registered Tradename of Fluid | Composition | Producer | Usable Temperature Range | Flash Point, °F | Fire Point, °F | Specific Heat Btu/lb, °F 200°F | Specific Heat 400°F | Viscosity Lb/ft h 200°F | Viscosity 400°F | Thermal Conductivity Btu/h ft² °F 200°F | Thermal Conductivity 400°F | Density Lb/ft³ 200°F | Density 400°F |
|---|---|---|---|---|---|---|---|---|---|---|---|---|---|
| Dowtherm A | diphenyl-diphenyl oxide eutectic | Dow Chemical Co. | 60 to 750°F | 255°F c.o.c. | 275°F c.o.c. | 0.43 | 0.50 | 2.57 | 0.90 | 0.077 | 0.068 | 62.5 | 56.4 |
| Dowtherm G | di and tri aryl ethers | Dow Chemical Co. | 20 to 700°F | 305°F c.o.c. | 315°F c.o.c. | 0.42 | 0.48 | 7.02 | 1.65 | 0.074 | 0.072 | 65.5 | 60.0 |
| Dowtherm J | alkylated aromatic | Dow Chemical Co. | 100 to 575°F | 145°F c.o.c. | 155°F c.o.c. | 0.49 | 0.60 | 0.90 | 0.34 | 0.073 | 0.068 | 50.4 | 44.1 |
| Dowtherm LF | aromatic blend | Dow Chemical Co. | −25 to 600°F | 260°F c.o.c. | 280°F c.o.c. | 0.43 | 0.51 | 2.66 | 0.92 | 0.077 | 0.067 | 61.2 | 55.9 |
| Dowtherm SR-1 | inhibited ethylene glycol | Dow Chemical Co. | −40 to 300°F | 250°F (100%) p.m.c.c. | 250°F (100%) c.o.c. | 0.88 | — | 2.18 | — | 0.243 | — | 62.1 | — |
| Dowfrost | inhibited propylene glycol | Dow Chemical Co. | −40 to 300°F | 214°F t.c.c. | 220°F c.o.c. | 0.92 | — | 2.18 | — | 0.220 | — | 64.1 | — |
| Mobiltherm 600 | mineral oil | Mobil Oil Co. | 10 to 600°F | 415°F c.o.c. | N/A | 0.46 | 0.56 | 16.2 | 2.78 | 0.066 | 0.062 | 52.0 | 47.9 |
| Mobiltherm 603 | mineral oil | Mobil Oil Co. | 20 to 600°F | 405°F c.o.c. | N/A | 0.54 | 0.65 | 9.92 | 2.27 | 0.074 | 0.069 | 44.0 | 40.2 |
| Mobiletherm Light | mineral oil | Mobil Oil Co. | −15 to 400°F | 250°F c.o.c. | N/A | 0.44 | 0.53 | 3.82 | 1.38 | 0.065 | 0.061 | 55.6 | 52.0 |
| Therminol 44 | modified ester base | Monsanto | −50 to 425°F | 405°F c.o.c. | 438°F c.o.c. | 0.51 | 0.57 | 2.34 | 0.66 | 0.076 | 0.065 | 54.1 | 48.7 |
| Therminol 55 | synthetic hydrocarbon | Monsanto | 0 to 600°F | 355°F c.o.c. | 410°F c.o.c. | 0.52 | 0.62 | 10.3 | 2.18 | 0.075 | 0.069 | 52.4 | 47.8 |
| Therminol 60 | polyaromatic compound | Monsanto | −60 to 600°F | 310°F c.o.c. | 320°F c.o.c. | 0.44 | 0.54 | 4.0 | 1.3 | 0.073 | 0.068 | 59.1 | 54.6 |
| Therminol 66 | modified terphenyl | Monsanto | 15 to 650°F | 355°F c.o.c. | 382°F c.o.c. | 0.44 | 0.53 | 10.0 | 2.1 | 0.067 | 0.061 | 59.8 | 55.0 |
| Therminol 88 | mixed terphenyl | Monsanto | 300 to 750°F | 375°F c.o.c. | 460°F c.o.c. | Solid | 0.50 | Solid | 2.0 | Solid | 0.071 | Solid | 59.9 |
| Therminol VP-1 | diphenyl oxide eutectic | Monsanto | 60 to 750°F | 255°F c.o.c. | 275°F c.o.c. | 0.42 | 0.49 | 2.66 | 0.87 | 0.075 | 0.065 | 62.7 | 56.7 |
| Thermia A | mineral oil | Shell Oil Co. | −50 to 350°F | 300°F c.o.c. | N/A | 0.51 | — | N/A | — | 0.072 | — | 52.4 | — |
| Thermia C | mineral oil | Shell Oil Co. | 15 to 600°F | 455°F c.o.c. | N/A | 0.51 | 0.60 | N/A | N/A | 0.073 | 0.069 | 51.8 | 47.5 |
| Syltherm 444 | silicone base | Dow-Corning Corp. | −50 to 400°F | 450°F c.o.c. | 500°F | 0.39 | 0.42 | 15.5 | 5.6 | 0.08 | 0.074 | 55.5 | 49.9 |
| Syltherm 800 | silicone base | Dow-Corning Corp. | −40 to 750°F | 310°F c.o.c. | 380°F | 0.42 | 0.46 | 7.8 | 2.6 | 0.076 | 0.072 | 54.6 | 48.4 |

organism transfer does not occur between the supply and exhaust airstreams. Tests of actual contactor towers have shown that they effectively remove as much as 94% of the atmospheric bacteria in the supply or exhaust air.

**Effect of Building or Process Contaminants.** If the building or process exhaust contains large amounts of lint, animal hair, or other solids, filters should be placed upstream of the exhaust air contactor tower.

If the building or process exhaust contains large amounts of gaseous contaminants, such as chemical fumes and hydrocarbons, the possibility of cross-contamination should be investigated, as well as the possible effects of contaminants on the sorbent solution.

**Winter Operation.** When applying twin-tower systems on controlled-humidity buildings or areas located in colder climates, possible saturation effects should be considered. These saturation effects, which can cause condensation, frosting, and icing in other types of equipment, may overdilute the sorbent solution in the twin-tower system. Heating the sorbent liquid supplied to the supply air contactor tower, as shown in Figure 16, can prevent dilution. Heating raises the discharge temperature and humidity of the supply air contactor tower, thus balancing the humidity of the system and preventing overdilution.

A thermostat, sensing the temperature of the air leaving the supply air contactor tower, controls the solution heater. Thus, the twin-tower system can deliver air at a constant temperature all winter, regardless of the outside air temperature. Automatic addition of makeup water (Figure 16) to maintain a fixed concentration of the sorbent solution enables the twin-tower system to deliver supply air at fixed humidity during cold weather. Thus, the

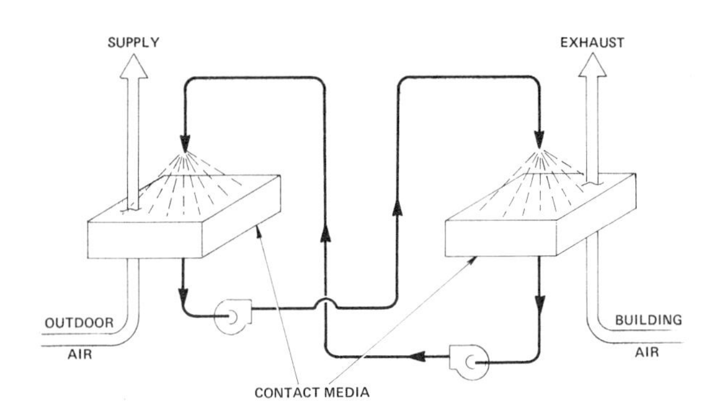


**Fig. 13 Twin-Tower Enthalpy Recovery Loop**

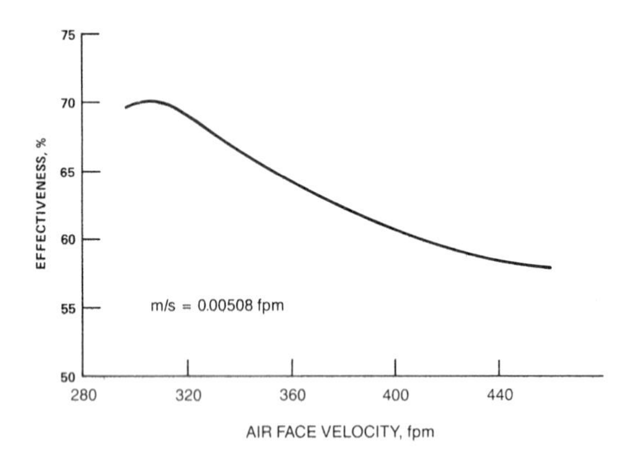


**Fig. 14 Total Energy-Recovery Effectiveness**

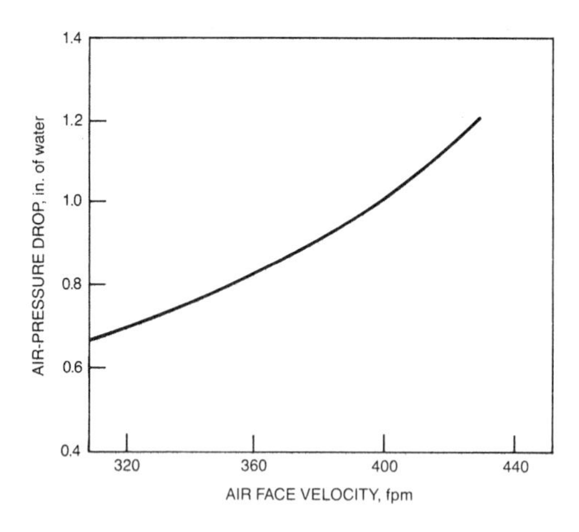


**Fig. 15 Typical Air-Side Pressure Drop**

system provides a constant supply air temperature and humidity without preheat or reheat coils or humidifiers. When auxiliary heat is added, the definition of winter effectiveness becomes:

$$\varepsilon = \frac{W_e(X_3 - X_4)}{W_{min}(X_3 - X_1)} \quad (4)$$

**Multiple-Towers.** Any number of supply air towers can be combined with a single exhaust air tower, or any number of exhaust air towers can be combined with a single supply air tower, as shown in Figure 17. If sufficient elevation difference exists between supply and exhaust towers, gravity returns the sorbent solution from the upper tower(s) (Figure 17).

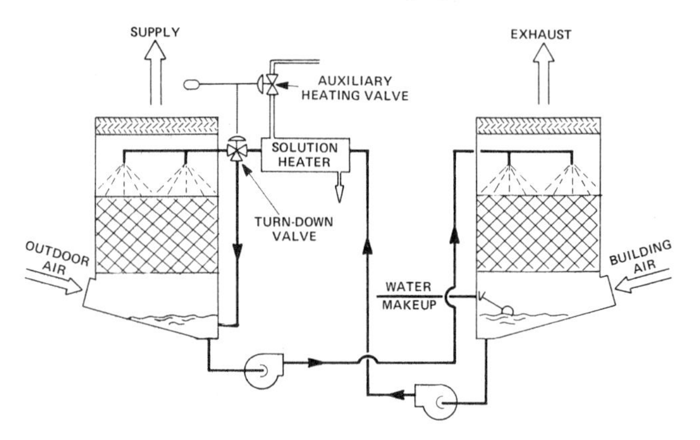


**Fig. 16 Twin-Tower Enthalpy Recovery Loop (Winter Operation and Control)**

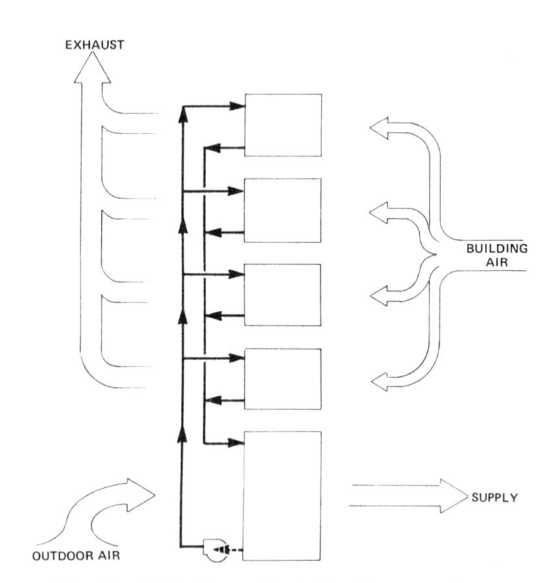


**Fig. 17 Twin-Tower Enthalpy Recovery Loop (Multiple-Unit System)**

## Maintenance

Twin-tower enthalpy recovery systems operate with only routine maintenance. Complete instructions on procedures, as well as spare-parts lists, are included in operating manuals relevant to each installation. Periodically, as with comparable air washer systems, the circulation pumps, spray nozzles, liquid transfer controls, and mist eliminators will need checking and adjusting (or maintenance).

Inhibited halide brine solutions are the normal energy-transfer media in these systems. The manufacturer provides a free bimonthly solution sampling service for the life of the system to monitor and report any changes in concentration, inhibitor level, pH, etc., that are necessary to ensure continued maximum performance. Total system inspections can also be provided.

## HEAT PIPE HEAT EXCHANGER

A heat pipe heat exchanger has no moving parts; it is a passive energy-recovery device. Although it appears similar to a standard steam or chilled-water coil, it differs in two major aspects. First, as shown in Figure 18, each individual tube is a heat pipe or individual conductor of energy compared to standard coils with return bends or headers. Second, the heat pipe heat exchanger is

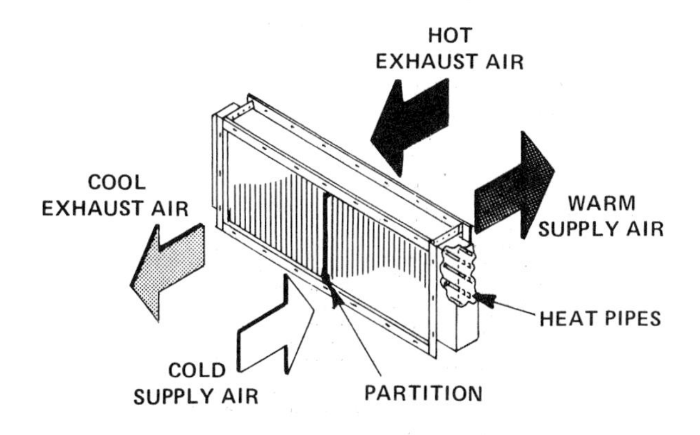


**Fig. 18 Heat Pipe Heat Exchanger**

divided into two airflow paths. Hot air passes through one side of the exchanger and cold air through the other side in the opposite direction, *i.e.*, in a counterflow arrangement. The heat pipes transfer sensible energy from the hot air to the other side of the exchanger, where it warms the cold air. Although the heat pipes span the width of the unit, a sealed partition separates the two airstreams to prevent cross-contamination.

A heat pipe is a tube fabricated with a capillary wick structure, charged with a suitable two phase heat-transfer fluid and sealed permanently (Figure 19). The fluid is normally a Class I refrigerant in HVAC applications, but other fluorocarbons, water, and other compounds offer suitable heat transfer for particular temperature applications. Thermal energy applied to either end of the heat pipe causes the fluid at the end to vaporize. The medium, now in vapor form, travels to the other end of the pipe, where the thermal energy is removed. The vapor condenses into liquid again, thus giving up the latent heat of condensation. The condensed liquid then flows back to the evaporator section (*i.e.*, the hot end) to complete the cycle. In essence, the heat pipe operates on a condensation/evaporation cycle, which is continuous as long as temperature difference drives the process. The heat pipe transfers energy with a small temperature drop; consequently, the process is often considered isothermal. There is, however, a small temperature drop through the tube wall, wick, and fluid medium. Heat pipes have a finite heat-transfer capacity that is affected by such factors as wick design, tube diameter, working fluid, and tube orientation relative to horizontal.

## Face Velocity and Pressure Drop

Design face velocities for heat pipe heat exchangers range from 400 to 800 fpm (2.0 to 4.1 m/s); the most common velocity is 450 to 550 fpm (2.3 to 2.8 m/s). Design face velocities are established on allowable pressure drop, rather than on recovery performance.

Pressure drops at 60% effectiveness range from 0.4 to 0.7 in. of water (100 to 175 Pa) at 400 fpm (2.0 m/s) and up to 1.5 to 2.0 in. of water (375 to 500 Pa) at 800 fpm (4.1 m/s). Recovery performance (effectiveness) decreases with increasing velocity, but the decrease in effectiveness is not as rapid as the increase in pressure drop.

Available fin designs include continuous corrugated plate fin, continuous plain plate fin, and spiral fins. These various fin designs and tube spacings cause the variation in pressure drop, noted above, at a given face velocity.

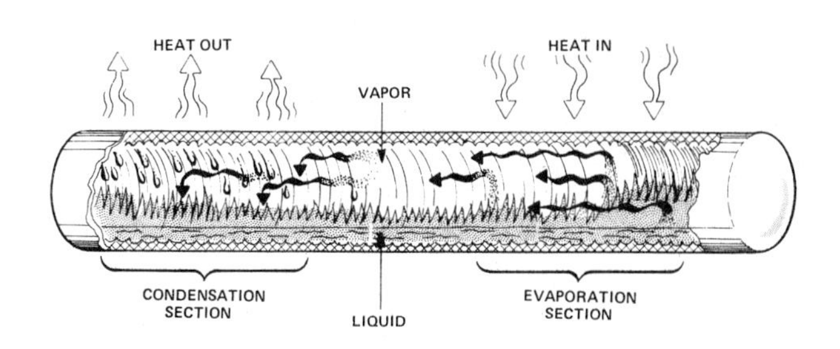


**Fig. 19 Heat Pipe**

## Airflow Arrangements

Heat pipe heat exchangers should operate with counterflow airstreams for maximum effectiveness. However, they can be operated with parallel airstreams at reduced effectiveness. For example, a heat exchanger operating with equal mass flows at 60% sensible effectiveness in a counterflow arrangement operates at 48% sensible effectiveness when placed in a parallel-flow arrangement.

## Effectiveness

The sensible effectiveness of a heat pipe heat exchanger is a function of several variables: (1) the heat-transfer surface area of the combination of rows and fins, (2) the ratio of the heat capacities of the two airstreams, and (3) the rate at which the two airstreams pass through the heat exchanger.

The capacity of a heat pipe heat exchanger to transfer energy is a function of design and orientation. Air-side coefficients of fin design, surface area, and velocity dictate the effectiveness or recovery factor. A secondary surface area (fins), offers equal heat transfer rates in a shallow depth in the direction of airflow compared to an exchanger with only primary surface for heat exchange. The tube diameter and orientation or tilt then determines the capacity of the heat pipe to transfer the energy exchanged by the finned surface.

Figure 20 presents typical effectiveness curves for various face velocities with a counterflow airflow arrangement. As the total number of rows of tubes increase, the effectiveness increases at a lower rate. For example, at a given velocity, with a fin spacing of 14 fins/in. (1.8 mm spacing), it takes six rows of tubes to produce 60% effectiveness; increasing the number of rows to 12 increases the effectiveness to 75%. The effectiveness of a heat pipe heat ex-

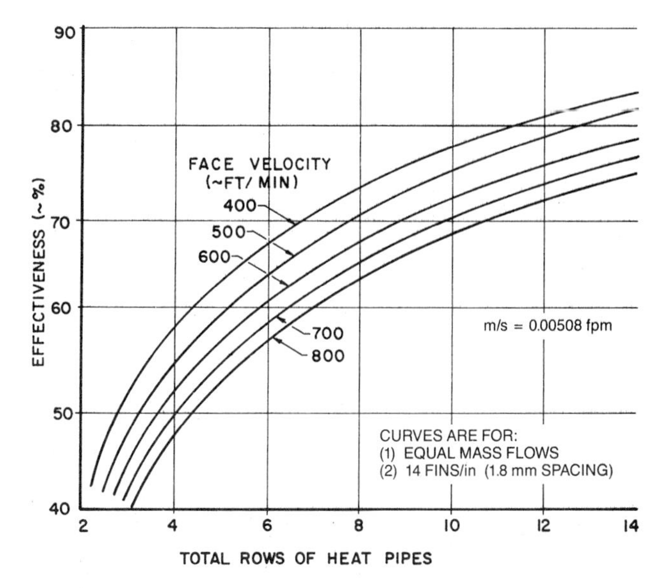


**Fig. 20 Typical Curve for Various Face Velocities**

changer depends on the total number of rows and not on the number of rows in a single unit. As all tubes are independent, two units in series yield the same effectiveness as a single unit with the same total number of rows of tubes. Series units are often used in industrial applications to facilitate handling and cleaning maintenance.

Fin design and spacing should be based on the dirtiness of the two airstreams and the resulting cleaning maintenance required. For HVAC systems, 14 fins/in. (1.8 mm) is suitable fin spacing. Industrial applications use wider fin spacing of 5, 8, or 11 fins/in. (5.1, 3.2, or 2.3 fins/mm). Plate fin heat pipes can be constructed with different fin spacings on the exhaust and supply ends, allowing a wider fin spacing on the dirty exhaust side, if necessary, than on the cleaner supply air side. This feature provides flexibility where pressure drop constraints exist and also maintains performance as dirt accumulates on the exhaust-side surface.

## Construction Materials

Heat pipe heat exchangers for exhaust temperatures below 425 °F (218 °C) are most often constructed with aluminum tubes and fins. Copper tubes and fins are available for HVAC use. The materials selected are usually determined by the contaminants present and their respective resistance to corrosion. Tubes and fins are usually constructed of the same material to avoid problems with the differing thermal expansion of different materials. Protective coatings designed for finned-tube heat exchangers have permitted inexpensive aluminum to replace exotic metals in corrosive atmospheres.

Copper tubes and fins are used in the same temperature range as aluminum units, although to a lesser extent. Because they are more expensive than aluminum units of comparable effectiveness, copper units are generally used where aluminum is unsuitable due to corrosion or cleaning problems.

Heat pipe heat exchangers for use above 425 °F (218 °C) are constructed with steel tubes and fins. Often, the fins are aluminized to prevent rusting. Composite systems for special applications may be created by assembling units having different materials of construction and/or different working fluids.

## Cross Contamination

Heat pipes have no cross contamination for pressure differential between airstreams of up to 10 in. of water (2.5 kPa). A sealed partition separates the two airstreams and prevents leakage from one to the other. For additional insurance against cross contamination, the unit may be constructed with two separating partitions having a small space between them. Any leakage would seep into the space between the two ducts, rather than from one duct to the other.

## Fan Requirements

Heat pipes do not require that the fans be located in any special manner, because they are insensitive to pressure differential between airstreams within certain limits.

## Condensation and Freeze-up

Under some conditions of supply temperature and exhaust relative humidity, condensation of moisture from comfort exhaust air can occur. Condensation increases the effectiveness of the heat pipe heat exchanger as it does with other sensible heat exchangers, but it also increases the pressure drop. (Manufacturer's recommendations should be followed.) When outside air temperature is very low, this condensation may freeze. While freeze-up does not damage the heat exchanger, it reduces and eventually stops the flow of exhaust air, thus stopping energy transfer, unless a control system is used to prevent blockage.

Industrial process exhaust vapors often condense when they are cooled. The amount of condensation depends on the initial concentration of the vapors and the amount of cooling that the exhaust undergoes. This condensate seldom freezes because the exhaust temperature is too warm. Condensation in either HVAC or industrial applications is transferred as sensible energy to the supply airstream. The improved heat recovery performance is gained without the transfer of contaminant-laden moisture.

## Evaporative Air Cooling

Increased cooling performance may be achieved during summer by spraying water into the exhaust air ahead of the heat pipe heat exchanger to lower its dry-bulb temperature to near the wet-bulb temperature. The spray creates a greater temperature difference, which increases effectiveness. This indirect evaporative cooling phenomenon, as applied to heat pipes and fixed-plate exchangers is addressed in the section on "Fixed-Plate Exchangers."

## Controls

Gravity can assist in returning the condensate within a heat pipe to the evaporator section by placing the heat pipe on a slope with the hot end below horizontal. Conversely, placing the heat pipe on a slope with the hot end above horizontal retards condensate flow. Consequently, changing the slope (tilt) of a heat pipe controls the effectiveness or amount of heat that it transfers.

Slope is controlled by pivoting the exchanger about the center of its base and attaching a temperature-controlled actuator to one end of the exchanger (Figure 21). Pleated flexible connectors attached to the ductwork allow freedom for the small tilting movement (7° maximum).

Tilt control may be desired to regulate the recovery performance of a heat pipe heat exchanger for the following reasons:

1. Maximize energy transfer for both the heating and cooling seasons.
2. Control the supply air temperature at some desired level. This regulation is often required for large buildings to avoid overheating the air supplied to the interior zone.
3. Limit frost formation on the weather face of the exhaust side at low outside air temperatures. By reducing the recovery of the unit, the exhaust air leaves the unit at a warmer temperature and stays above frost-forming conditions.

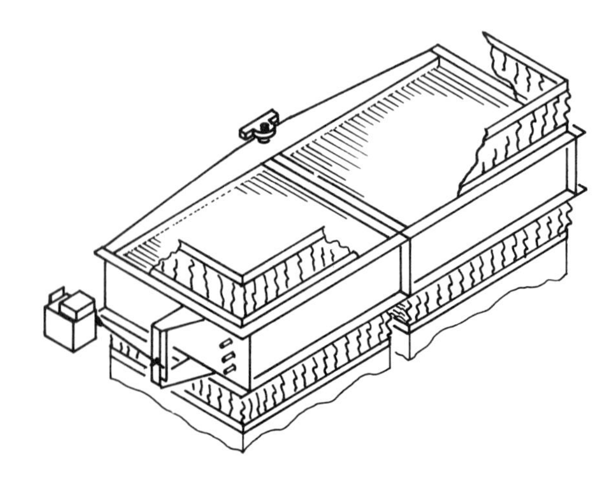

**Fig. 21 Heat Pipe Heat Exchanger with Tilt Control**

While tilt control is the only way to accomplish all three of the above functions, other methods are available for individual functions. Supply air temperature can be regulated with face and bypass dampers. Similarly, frost formation can be prevented by these devices. Preheating the supply air upstream of the heat-recovery unit provides still another approach to frost prevention.

### Maintenance

How the heat pipe exchanger is cleaned depends on the nature of the material to be removed. Grease buildup from kitchen exhaust, for example, is often removed with an automatic water-wash system. Other dirt requires manually spraying the units, soaking the units in a cleaning tank, or using soot blowers. The cleaning method should be decided during design so the appropriate heat pipe exchanger can be selected.

Cleaning frequency depends on the quality of the exhaust airstream. HVAC systems require little cleaning; industrial systems usually require more.

## FIXED-PLATE EXCHANGERS

Fixed-surface plate exchangers can be classified into two categories: (1) the pure-plate heat exchanger, consisting of only primary heat-transfer surface; and (2) the plate-fin heat exchanger, made up of alternate layers of separate plates and interconnecting fins. The pure-plate exchanger is of a counterflow design, whereas the basic plate-fin exchanger is a crossflow design, which is sometimes arranged to approach a counterflow unit. Counterflow gives the greatest temperature difference for maximum heat transfer, but crossflow can sometimes give more convenient air connections. The typical plate exchanger transfers sensible heat only, except when the temperature of one airstream is low enough to cause condensation in the opposing airstream.

The plate heat exchanger is a static device, which, with proper care in design and manufacture, has no leakage between airstreams.

### Design Considerations

Plate exchangers are of many proprietary designs; weights, sizes, and flow patterns (Figure 22) depend on the manufacturer. Many are built so they can be added and stacked as modules to handle almost any airflow and pressure-drop requirement. Sizes range from 7 to 18 $ft^3$ per 1000 cfm (1.4 to 3.5 $m^3$ per $m^3/s$), with the pressure drop ranging between 1.5 to 0.5 in. of water (375 and 125 Pa), respectively. Plate spacing ranges from 2 to 10 plates/in. (12.7 to 2.5 mm). The most popular construction material is aluminum due to its corrosion resistance and ease of fabrication. Because the fabrication materials are relatively thin, their conductivity has little influence on the effectiveness of a pure-plate exchanger; so these exchangers are fabricated from various steel alloys for applications up to 1600 °F (870 °C).

One of the advantages of the plate exchanger is that it is a static device, which can be built so little or no leakage occurs between airstreams. Some units are built with a continuous sheet of metal formed to produce two divided air passages, some with sheets formed and welded, and some with tubes rolled into crown sheets similar to the construction of fire-tube boilers.

Most manufacturers offer plate exchangers in modular design. Modules range in capacity from 1000 to 10,000 cfm (0.5 to 4.7 $m^3/s$) and can be arranged in the field into installations up to 100,000 cfm (47.2 $m^3/s$). Some units offer easy access to the heat-transfer surface for cleaning after installation. Automatic wash systems for remote cleaning are also available.

Modular units are assembled and sealed with gasketing, molded seals, welded seams, etc., depending on the application and the airstream temperatures. Several stand-alone modular designs re-

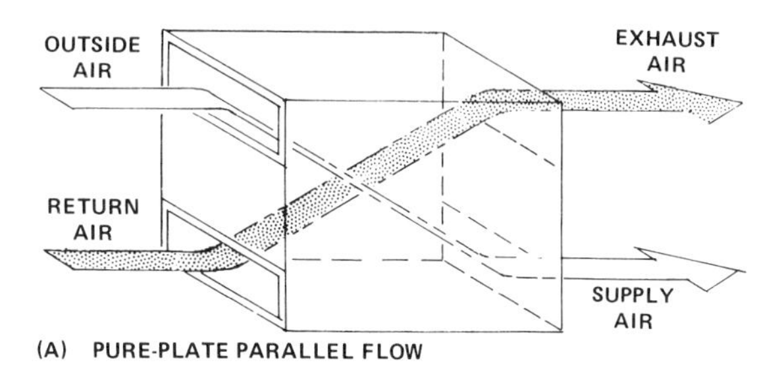


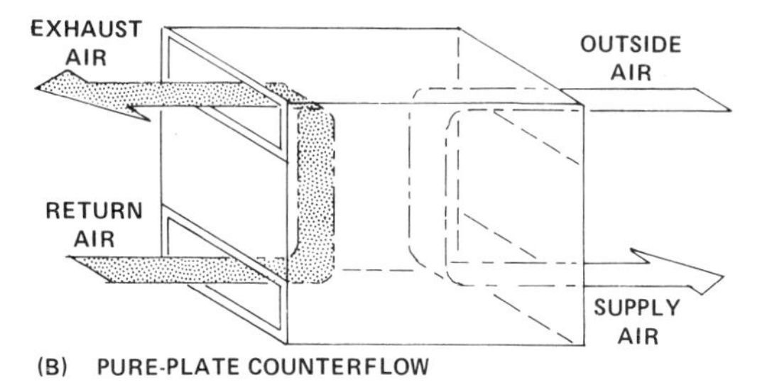


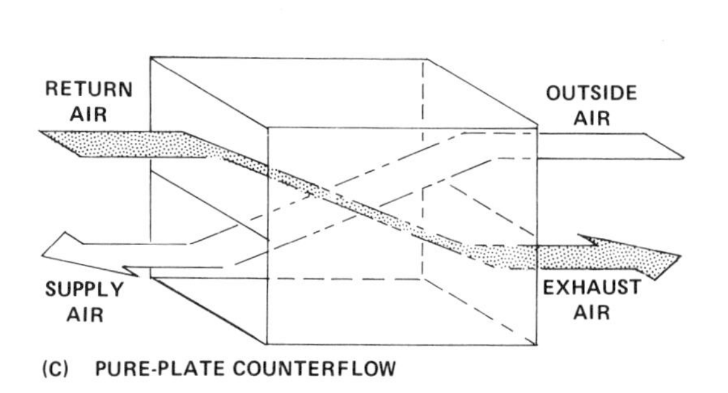


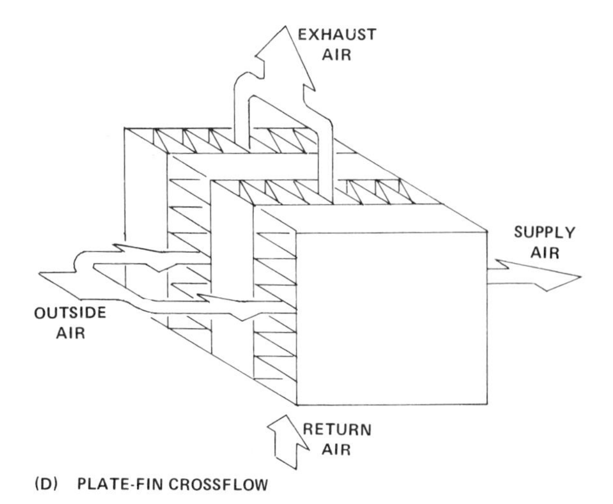


**Fig. 22 Basic Plate Exchanger Types**

quire only multiple connections between the air ducts and the heat exchanger modules.

## Performance

The plate heat exchanger achieves a high recovery sensible heat effectiveness beeause it only has a primary surface area and does not depend on fin effectiveness to enhance its heat-transfer capabilities. Figure 23 shows a typical effectiveness curve. As with all air-to-air heat exchangers, the efficiency with which energy can be removed from the waste gas stream depends on the intake and exhaust flow ratios (Figure 24).

## Pressure Drop

Pressure drop through a plate heat exchanger with a given design is a function of gas temperature and mass flow rate. Figure 25 shows a typical relationship in a dimensionless flow parameter, *R*. As indicated by the upward sloping curves, flow within the typical plate heat exchanger is in the turbulent region.

## Condensing Within an Exhaust Airstream

Many heat-recovery applications—such as drying ovens, curing ovens, swimming pools, and cooking areas—have high moisture content exhaust gases from which heat must be extracted. Plate exchangers, as well as other sensible heat exchangers, are ideally suited to such applications because the air to be heated can benefit by the transfer of the latent heat of condensation. For every pound (kilogram) of moisture condensed on the exhaust side, about 1050 Btu (2450 kJ) is transferred to the incoming air. As a result, the same incoming airflow temperature rise may be maintained with as little as one-third the exhaust gas flow when the system is condensing as when it is noncondensing—an important feature when waste process heat is used to heat makeup air.

Most plate exchangers are equipped with condensate drains, which remove both the condensate and waste water when a water-wash system is used. Heat recovered from a high humidity exhaust is better returned to a building or process by a sensible exchanger than by an enthalpy exchanger, if moisture is unwanted in the supply airstream. Figures 26 and 27 show typical condensing situa-

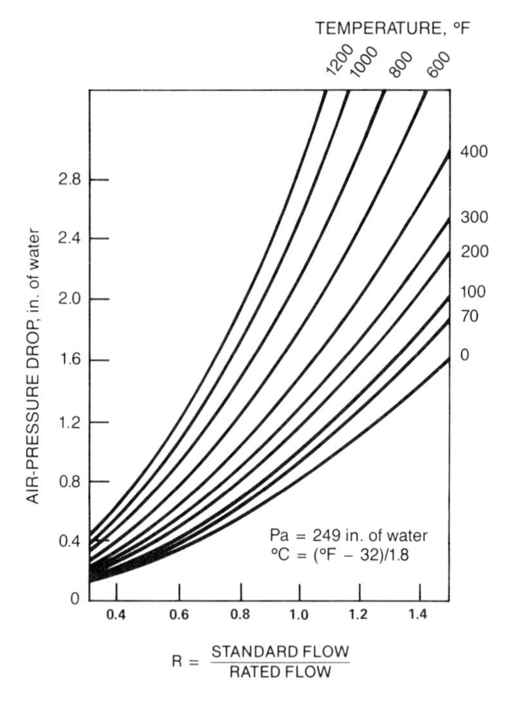


**Fig. 25 Pressure Drop versus Flow at Various Temperatures for a Typical Plate Exchanger**

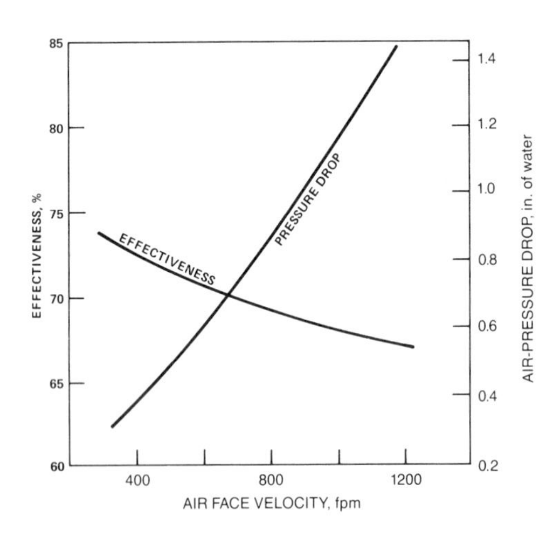


**Fig. 23 Typical Effectiveness and Air Pressure Drop for Fixed-Plate Exchangers**

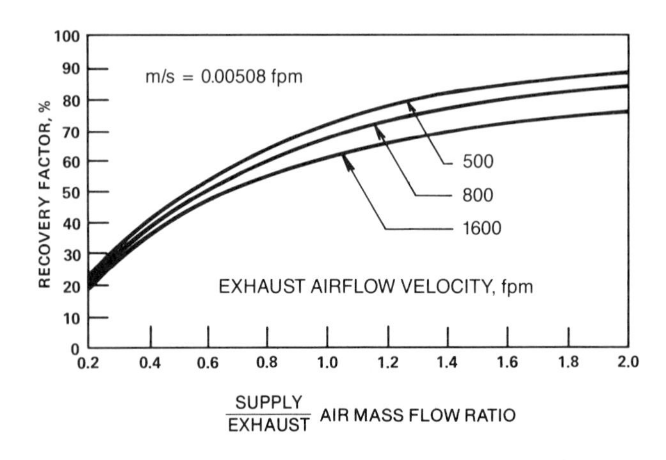


**Fig. 24 Typical Unbalanced Airflow Recovery Factors**

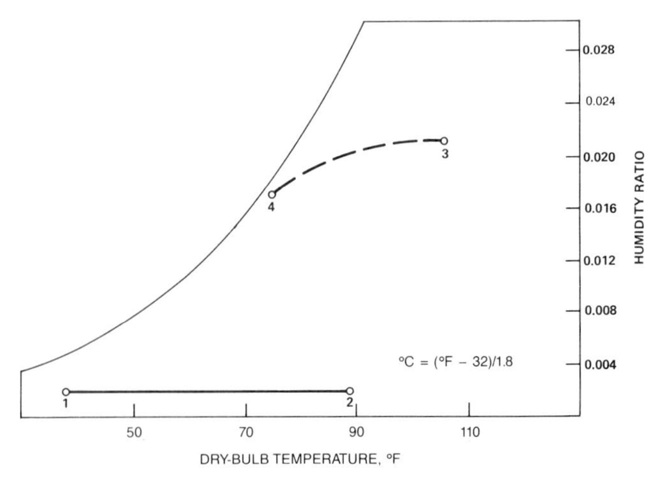


**Fig. 26 Condensing within a Sensible Plate Heat Exchanger for Balanced Flow**

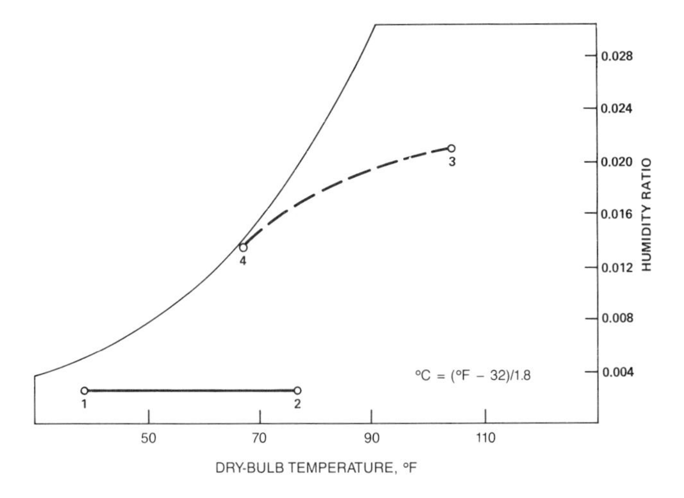


**Fig. 27 Condensing within a Sensible Plate Heat Exchanger for Unbalanced Flow**

tions within a sensible heat exchanger. Figure 26 illustrates the thermodynamic process for balanced flow, and Figure 27 illustrates the process when the supply air rate is twice that of the exhaust.

Figures 26 and 27 show the higher the moisture content of the exhaust, the less likely freezing will occur in the exhaust air passageways. Table 2 illustrates the effect of moisture content on the frost threshold. Frosting can be controlled by either preheating the incoming supply air or bypassing a portion of the incoming air. Bypassing reduces the ratio, K, in Table 2 and is generally better than preheating.

## Evaporative Air Cooling

Air passing over water absorbs the water until it becomes saturated or until it reaches 100% relative humidity. Air gives up about 1050 Btu in sensible heat per pound of water evaporated (2450 kJ/kg), which lowers the air temperature.

Evaporative air cooling equipment may be placed in two general classes: *direct* and *indirect*. In direct evaporative coolers, air passes through a wetted media. As the water evaporates it absorbs sensible energy from the air, while simultaneously carrying the same amount of energy in latent form to the airstream. Thus, the energy level of the airstream remains, constant, while the moisture content increases.

A plate-type or heat pipe air-to-air heat exchanger can evaporatively cool air indirectly. A scavenger airstream (either outside air or building return air) is directed though Side A (Figure 28) of the plate heat exchanger. Evaporation lowers the plate temperature to the wet-bulb temperature of the scavenger air. Heat then passes from the supply air (which enters through Side B) to the cooled plate, and its dry-bulb temperature is reduced with no addition of moisture.

Figure 29 shows that the indirect evaporative cooling process lowers the dry-bulb temperature and reduces the enthalpy of the supply airstream. The solid line between Points 3 and 4 is supply air being cooled without moisture addition. The dashed line between Points 1 and 2 is scavenger air. System wet-bulb depression efficiencies can be about 80 to 90%. Wet-bulb depression efficiency is defined as the dry-bulb depression of the supply air temperature leaving the heat exchanger divided by the difference between the dry-bulb and wet-bulb temperature of the entering air. For example, a system with 95 °F dry bulb and 58 °F wet bulb with 90% efficiency would have a leaving dry bulb of 61.7 °F.

$$\text{LDB} = 95 - 0.90(95 - 58) = 61.7\,°\text{F}$$

(A system with 35 °C dry bulb and 15 °C wet bulb with 90% efficiency would have a leaving dry bulb of $35 - 0.90(35 - 15) = 17\,°\text{C}$.)

In areas with low or moderate wet-bulb temperatures, indirect evaporative cooling is being used for HVAC applications because its operation, compared to mechanical cooling schemes, is more cost effective as a result of energy savings.

Typically, an indirect packaged cooler reduces the sensible load on the evaporated side of a conventional refrigeration system. The energy used by the indirect evaporative cooler (pumps and fans) is significantly less than that required by the refrigeration system. Energy efficiency ratios (EERs) of IEC systems can range from 30 to 70, depending on available wet-bulb depression. Computer simulation using hourly weather data tapes, and verified by field experience, has shown the operating costs of conventional cooling systems can be reduced 20 to 40% by the application of indirect

**Table 2 Frost Threshold Temperature, $T_1$, for Various Exhaust Air Conditions**

| Exhaust Air Conditions | | | Ratio of Supply Airflow to Exhaust Airflow, $K$ | | | | | | | |
|---|---|---|---|---|---|---|---|---|---|---|
| $T_3$ | | RH, | 0.5 | | 0.7 | | 1.0 | | 2.0 | |
| °F | °C | % | °F | °C | °F | °C | °F | °C | °F | °C |
| 60 | 16 | 30 | 2 | −16 | 15 | −9 | 23 | −5 | 32 | 0 |
| 60 | 16 | 40 | 2 | −16 | 15 | −9 | 23 | −5 | 32 | 0 |
| 60 | 16 | 50 | −4 | −20 | 9 | −13 | 18 | −7 | 32 | 0 |
| 60 | 16 | 60 | −9 | −22 | 4 | −15 | 13 | −10 | 32 | −5 |
| 70 | 21 | 30 | −13 | −25 | 5 | −15 | 17 | −8 | 28 | −2 |
| 70 | 21 | 40 | −21 | −29 | −3 | −19 | 10 | −12 | 21 | −6 |
| 70 | 21 | 50 | −27 | −33 | −9 | −22 | 3 | −16 | 15 | −9 |
| 70 | 21 | 60 | −32 | −35 | −13 | −25 | −1 | −18 | 10 | −12 |
| 75 | 24 | 30 | −25 | −31 | −4 | −20 | 10 | −12 | 23 | −5 |
| 75 | 24 | 40 | −33 | −36 | −12 | −24 | 2 | −17 | 15 | −9 |
| 75 | 24 | 50 | −40 | −40 | −20 | −29 | −6 | −21 | 7 | −14 |
| 75 | 24 | 60 | −47 | −44 | −26 | −32 | −12 | −24 | 1 | −17 |
| 80 | 27 | 30 | −35 | −37 | −11 | −24 | 4 | −15 | 19 | −8 |
| 80 | 27 | 40 | −44 | −42 | −20 | −29 | −5 | −20 | 10 | −12 |
| 80 | 27 | 50 | −53 | −47 | −30 | −34 | −14 | −25 | 1 | −17 |
| 80 | 27 | 60 | −62 | −52 | −39 | −39 | −23 | −30 | −8 | −22 |
| 90 | 32 | 30 | −58 | −50 | −30 | −34 | −11 | −24 | 5 | −15 |
| 90 | 32 | 40 | | | | | −24 | −31 | −8 | −22 |
| 90 | 32 | 50 | | | | | | | −20 | −29 |

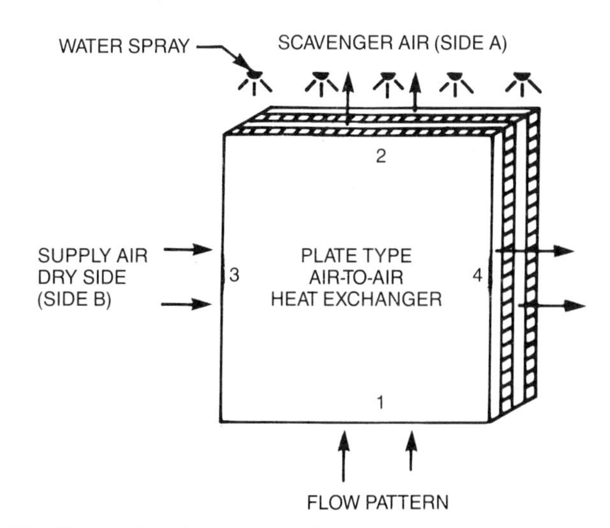


**Fig. 28 Indirect Evaporative Cooling with a Plate Exchanger**

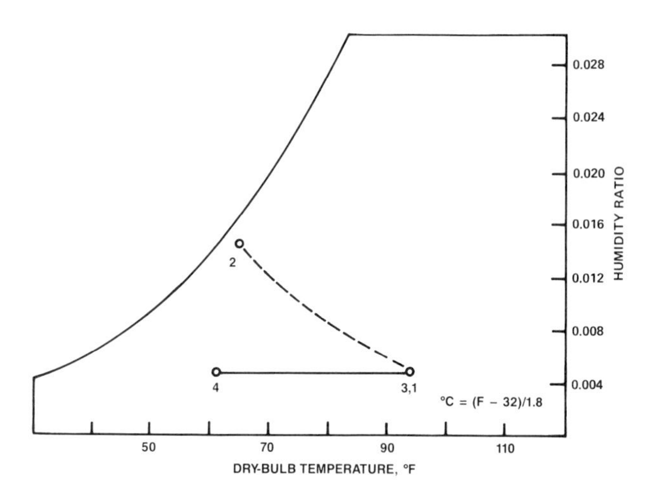


**Fig. 29 Psychrometrics of Evaporative Cooling**

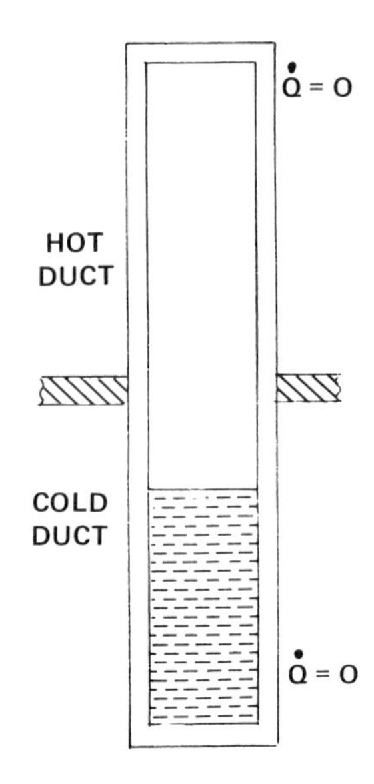


**Fig. 31 Unidirectional Sealed-Tube Thermosiphon with Tube Not Conducting**

evaporative cooling. When a second indirect evaporative cooler is used, the overall system operating costs can be reduced by as much as 60 to 75%.

Besides reduced operating costs, overall mechanical refrigeration system requirements are reduced, so smaller mechanical refrigeration systems can be selected. In some cases, the mechanical system may be eliminated. Chapter 4 of this volume and Chapter 56 of the 1987 HVAC Volume has further information on evaporative cooling.

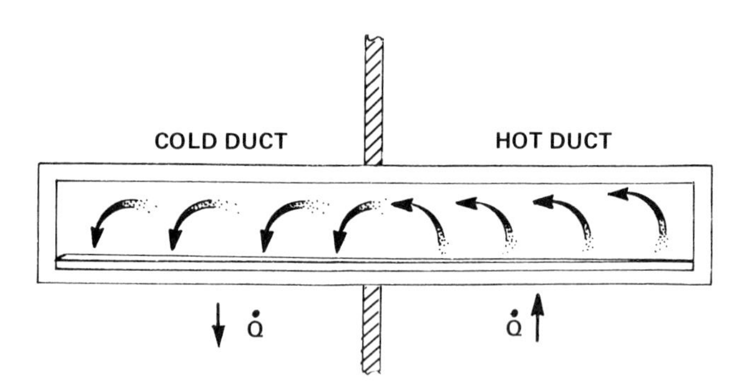


**Fig. 32 Bidirectional Sealed-Tube Thermosiphon in Which Tube Conducts Equally Well in Either Direction**

## THERMOSIPHON HEAT EXCHANGERS

Thermosiphon heat exchangers use the natural gravity circulation of a boiling and condensing intermediate fluid to transfer energy between exhaust and supply airstreams. They may be classified as *sealed-tube thermosiphons* and *coil-loop thermosiphons.*

The sealed-tube thermosiphon (shown in Figures 30, 31, and 32), although similar in form and operation to a heat pipe, has two distinct differences: (1) heat pipes depend primarily on capillary forces to force the intermediate liquid to flow from the cold to the hot end of the tubes, whereas thermosiphon tubes rely only on gravity; and (2) heat pipes vaporize fluid from the wick surface, whereas the thermosiphon tube depends, at least initially, on nucleate boiling. The coil-loop thermosiphon (Figures 33 and 34) is similar to the coil energy-recovery loop outlined earlier. The most obvious difference is the absence of a circulating pump in the thermosiphon loop and the need for evaporator and condenser coils, rather than single-phase liquid coils.

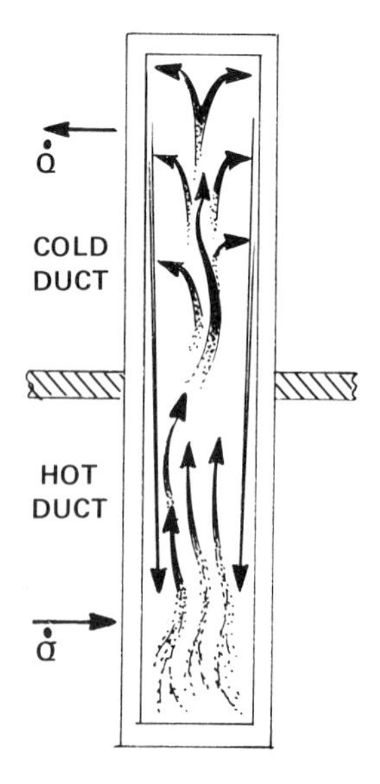


**Fig. 30 Unidirectional Sealed-Tube Thermosiphon with Tube Conducting**

### Principle of Operation

A thermosiphon is a sealed system containing a two-phase working fluid. Because part of the system contains vapor and part contains liquid, the pressure in a thermosiphon is governed by the liquid temperature at the liquid-vapor interface. If the surroundings cause a temperature difference between two regions in a thermosiphon where liquid vapor interfaces are present, the resulting vapor-pressure difference makes vapor flow from the warm to the cool region. The flow is sustained by condensation in the cool region and evaporation in the warm region. When the cool and warm regions of the thermosiphon are suitably oriented, the condensate returns to the evaporator region by gravity and completes the cycle.

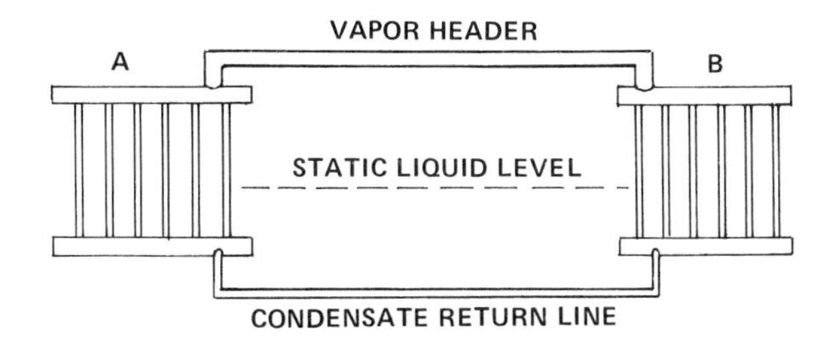


**Fig. 33 Bidirectional Thermosiphon Coil Loop Which Can Transfer Energy in Either Direction**

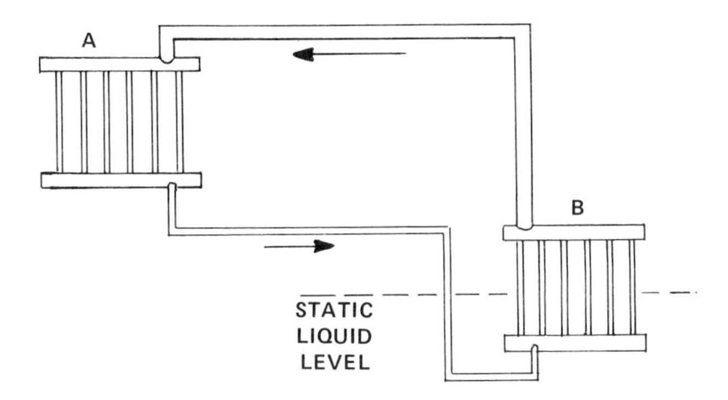


**Fig. 34 Unidirectional Thermosiphon Coil Loop Which Can Transfer Energy Only from B to A**

## Characteristics of Two-Phase Thermosiphons

The geometric orientation of the thermosiphon, the temperature difference between the two ducts and the tube diameter(s) and length(s), govern the operating characteristics of these units. If the orientation is such that liquid is present within the tube(s) in both ducts at all times (Figures 32 and 33), the unit can transfer heat in either direction (bidirectionally). If the orientation is such that the liquid can drain by gravity from the heat exchanger tube(s) in one of the ducts (Figures 30, 31, and 34), that portion cannot act as an evaporator. In this case, heat can be transferred in one direction only (unidirectionally) toward the liquid deficient region. This unidirectional behavior can benefit some applications. For example, where the evaporator operates as a solar collector, it absorbs energy when the sun is shining but shuts off whenever the collector temperature drops below the sink temperature, without the need of auxiliary controls.

Another characteristic arises because a large temperature difference may be required to initiate nucleate boiling within the evaporator tube(s). Once initiated, however, evaporation continues down to a much smaller temperature difference. Due to this phenomenon, two-phase thermosiphons are not appropriate in applications where the hot-to-cold duct temperature difference does not exceed some minimum threshold value. Correlations exist to predict the temperature difference required to sustain boiling, but further research is required to determine values of the temperature difference required to initiate nucleate boiling on various surfaces with various fluids. In the absence of specific data for the desired surface fluid combination, a minimum temperature difference of 27 to 36°F (15 to 20°C) is suggested.

## Sealed-Tube Thermosiphons

Sealed-tube thermosiphons are used only when the supply and exhaust air ducts are adjacent to one another. Their operating characteristics, applications, and limitations are similar to those described in the section on "Heat Pipe Heat Exchangers." For a bidirectional system in which the tubes are mounted horizontally, however, thermosiphon performance is more sensitive to tube misalignment than heat pipe performance. Capillary forces ensure that liquid returns to the evaporator end of the heat pipe, while the thermosiphon relies only on gravity. If all the liquid in a tube resides at the cold end of the tube, no heat transfer can take place. This problem is not one for unidirectional heat-flow installations in which the evaporator (hot) end of the tube is always at a lower elevation than the condenser end. However, these units require a threshold temperature difference to initiate operation.

## Coil-Loop Thermosiphons

Coil-loop thermosiphons may be used when the supply and exhaust air ducts are not adjacent to one another. Although similar in appearance and application to the coil energy-recovery loop, the coil-loop thermosiphon has several unique features. (The section on "Coil Energy-Recovery Loop" has further information.)

As illustrated in Figures 33 and 34, a single-coil loop thermosiphon consists of two coils interconnected by a vapor and a condensate line. The loop is charged with working fluid at its saturation state, so part of it is filled with liquid and part with vapor. The pressure within a sealed loop depends on the working fluid used and the fluid temperature at the liquid-vapor interface. The maximum loop pressure corresponds to the saturation pressure at the maximum temperature the unit experiences. Similarly, the minimum pressure corresponds to the saturation pressure at the minimum temperature possible. As each working fluid has an optimum operating range, the proper working fluid should be selected for the intended application.

The coil in the warmer air duct must have some liquid, working fluid present for vaporization to take place. If liquid is present in both coils, the system can transfer heat in either direction. If it is present only in one coil, energy can be transferred only from the wet to the dry coil. Operation stops if the supply and exhaust temperature reverse in this unidirectional system because no liquid is available to vaporize. Condensation and evaporation are affected in different ways by the diameter, length, and orientation of the coil tubes, as well as by the amount of liquid charge present in each coil. As a result, systems may be designed without external controls to provide a different effectiveness for each energy-flow direction.

Unidirectional coil-loop thermosiphons are more efficient than bidirectional units operating under the same conditions because the coils and loop charge may be selected for optimum performance for only one function, rather than for two functions (evaporation and condensation). In either case, to recover a higher

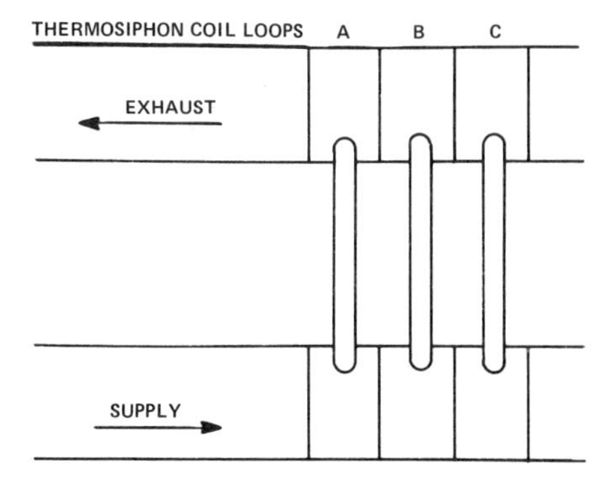


**Fig. 35 Plan View of Multiple Thermosiphon Coil Loops in a Counterflow Configuration**

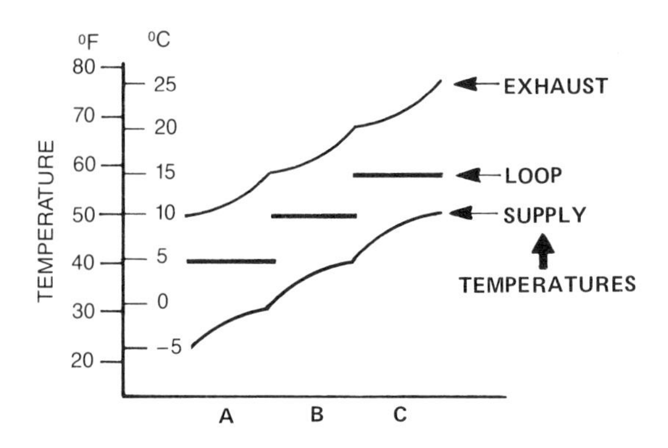


**Fig. 36 Performance of Multiple Thermosiphon Coil Loops in Series**

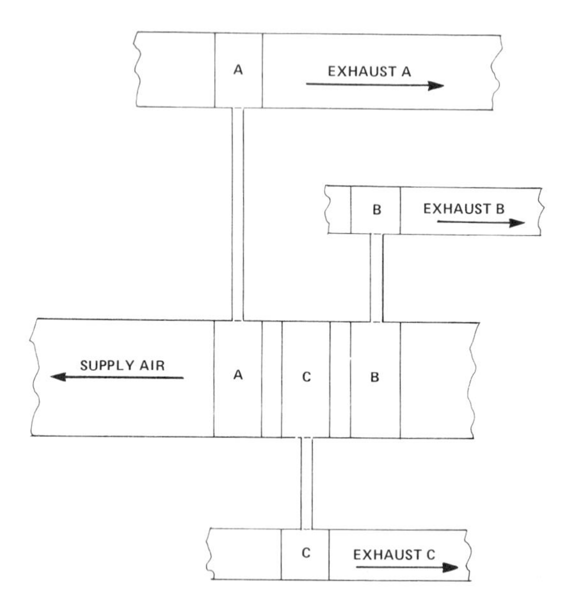


**Fig. 37 Multiple Exhaust Single-Supply Installation**

proportion of the available energy, several coil-loop thermosiphons may be mounted like wafers along the supply and exhaust ducts (Figure 35) to create a counterflow heat exchanger with an overall effectiveness greater than that of a single loop. Figure 36 shows temperature distribution for a system with equal exhaust and supply airflow rates and assumes each individual coil has an effectiveness of 50%. Consequently, the effectiveness of each separate thermosiphon loop is about 25%.

The effectiveness of any two loops is 40%; the effectiveness of any three loops is 50%. If the effectiveness of each individual coil was 80%, the effectiveness of three loops mounted in series, as shown, would be 67%. The maximum possible effectiveness for these systems is limited by the minimum temperature difference required to sustain nucleate boiling. For example, if an overall temperature difference of 40°F is available between the hot and cold ducts and a 8°F temperature difference is required to sustain boiling, the maximum possible effectiveness is $(40 - 8)/40 = 0.80$ or 80%.

The most economical combination of coil sizes and number of loops depends on the specified design criteria, the lifetime of the system, and the economic criteria.

As with heat pipes, each loop is independent, so different working fluids can be used in each loop if it is advantageous. In addition, thermosiphon coil loops may be used with multiple supply or exhaust ducts. Figure 37 schematically illustrates such a case, in which a common supply duct is being heated by three different exhaust ducts.

## BIBLIOGRAPHY


ASHRAE. 1974. Symposium on Air-to-Air Heat Recovery. ASHRAE *Transactions* 80(2): 307-332.

ASHRAE. 1975. ASHRAE Symposium on Rotary Air-to-Air Heat Exchangers for Energy Recovery. ASHRAE *Transactions* 81(2): 389-420.

ASHRAE. 1978. Method of Testing Air-to-Air Heat Exchangers. ASHRAE *Standard* 84-78.

ASHRAE. 1982. Symposium on Energy Recovery from Air Pollution Control. ASHRAE *Transactions* 88(1): 1197-1225.

ASHRAE. 1982. ASHRAE Symposium on Condensible Gases in the Effluent Air for Moderate Temperatures. ASHRAE *Transactions* 89(2B): 263-374.

ASHRAE. 1977-1986. Research Reports for RP 140:

Ali, A.F.M. and T.W. McDonald. 1977. Thermosiphon Loop Performance Characteristics: Part 2, Simulation Program. ASHRAE *Transactions* 83(2): 260-278.

Mather, G.D. and T.W. McDonald. 1986. Simulation Program for a Two-Phase Thermosiphon-Loop Heat Exchanger. ASHRAE *Transactions* 92(2a): 473-485.

McDonald, T.W. and A.F.M. Ali. 1977. Thermosiphon Loop Performance Characteristics: Part 3, Simulated Performance. ASHRAE *Transactions* 83(2): 279-287.

McDonald, T.W.; K.S. Hwang; and R. DiCiccio. 1977. Thermosiphon Loop Performance Characteristics: Part 1, Experimental Study. ASHRAE *Transactions* 83(2): 250-259.

Stauder, F.A. and T.W. McDonald. 1986. Experimental Study of a Two-Phase Thermosiphon-Loop Heat Exchanger. ASHRAE *Transactions* 92(2a): 486-487.

ASHRAE. 1978-1986. Research Reports for RP-188:

Mather, G.D. and T.W. McDonald. 1986. Simulation Program for a Two-Phase Thermosiphon-Loop Heat Exchanger. ASHRAE *Transactions* 92(2a): 473-485.

McDonald, T.W.; A.F.M. Ali; and S. Sampath. 1978. The Unidirectional Coil Loop Thermosiphon Heat Exchanger. ASHRAE *Transactions* 84(2): 27-37.

McDonald, T.W.; M. Kosnik; and G. Bertoni. 1985. Performance of a Two-Phase Thermosiphon Air-to-Air Heat Exchanger. ASHRAE *Transactions* 91(2a): 209-215.

McDonald, T.W. and S. Sampath. 1980. The Bidirectional Coil Loop Thermosiphon Heat Exchanger. ASHRAE *Transactions* 86(2): 37-47.

McDonald, T.W. and S. Raza. 1984. Effect of Unequal Evaporator Heating and Charge Distribution on the Performance of a Two-Phase Thermosiphon Loop Heat Exchanger. ASHRAE *Transactions* 90(2a): 431-440.

Stauder, F.A. and T.W. McDonald. 1986. Experimental Study of a Two-Phase Thermosiphon-Loop Heat Exchanger. ASHRAE *Transactions* 92(2a): 486-487.

Bosch, J.J.; G.J. Gudac; R.H. Howell; M.A. Mueller; and H.J. Sauer. 1981. Effectiveness and Pressure Drop Characteristics of Various Types of Air-to-Air Energy Recovery Systems. ASHRAE *Transactions* 87(1): 199-210.

Bowlen, K.L. 1974. Energy Recovery from Exhaust Air. ASHRAE *Journal*, April, pp. 49-56.

Howell, R.H. and H.J. Sauer, Jr. 1981. Promise and Potential of Air-to-Air Energy Recovery Systems. ASHRAE *Transactions* 87(1): 167-182.

Howell, R.H.; H.J. Sauer; and J.R. Wray. 1981. Frosting and Leakage Testing of Air-to-Air Energy Recovery Systems. ASHRAE *Transactions* 87(1): 211-221.

Moyer, R.C. 1978. Energy Recovery Performance in the Research Laboratory, ASHRAE *Journal*, May.

Ruch, M.A. 1976. Heat Pipe Exchangers as Energy Recovery Devices. ASHRAE *Transactions* 82(1): 1008-1014

Scofield, M. and J.R. Taylor. 1986. A Heat Pipe Economy Cycle. ASHRAE *Journal*, October.

# CHAPTER 35

# RETAIL FOOD STORE REFRIGERATION EQUIPMENT



THE modern retail store is a high-volume food sales outlet with maximum inventory turnover. Almost half of retail food sales is made up of perishable or semi-perishable foods requiring refrigeration. These foods include fresh meats, dairy products, perishable produce, frozen foods, ice cream and frozen desserts, and various special items such as bakery products. These foods are displayed in highly specialized and flexible storage, handling, and display apparatus.

These products need to be kept under safe temperatures during storage and processing, as well as when they are on display. The back room of a food store is both a processing plant and a warehouse distribution point. It includes specialized refrigerated rooms, which must be coordinated during construction planning because of the interaction between the store's environment and the refrigeration equipment (Figure 1). Chapter 18 of the 1987 HVAC Volume and Chapter 32 of the 1986 REFRIGERATION Volume cover, in more detail, the importance of coordination.

## DISPLAY REFRIGERATORS

Each category of perishable food has its own physical characteristics, handling logistics, and display requirements that dictate specialized display shapes and flexibility required for merchandising. Also, the same food product requires different display treatment in different locations, depending on such things as neighborhood preferences, neighborhood income level, size of store, volume of sales, and the local availability of food items by type.

The preparation of this chapter is assigned to TC 10.7, Commercial Food Display and Storage Equipment.

Table 1 summarizes a study of ambient conditions in retail food stores. Individual store ambient readings showed that only 5% of all readings (including those when the air conditioning was not working or not turned on) exceeded 75 °F (24 °C) dry bulb or 0.0113 lb (11.3 g) of moisture per lb (kg) of dry air. Based on these data, the industry has chosen 75 °F (24 °C) dry bulb and 64 °F (18 °C) wet bulb (55% rh) as *summer design conditions.*

Manufacturers sometimes publish ratings for open refrigerators at lower ambient conditions than this standard because the milder conditions significantly reduce the heat load on the refrigerators. In addition, lower ambient conditions permit fewer defrosts and reductions in anti-sweat heaters, which make it possible to save substantial amounts of energy.

Attention should be given to the opposite condition, whereby store environments higher than the industry standard will dramatically raise the refrigeration requirements and consequently the energy demand.

### Meat Display

Most meat is sold in prepackaged form. It is usually cut and packaged on the store premises. Control of temperature, time, and sanitation from the truck to the checkout counter is important. Surface temperature on the meat in excess of 40 °F (4.4 °C) shortens the salable life of meat products significantly and increases the rate of discoloration.

Sanitation is also important. Even if all else is kept equal, good sanitation can as much as double the salable life of meat in a display refrigerator. In this chapter, sanitation includes the control of the time during which meat is exposed to temperatures above 40 °F (4.4 °C).

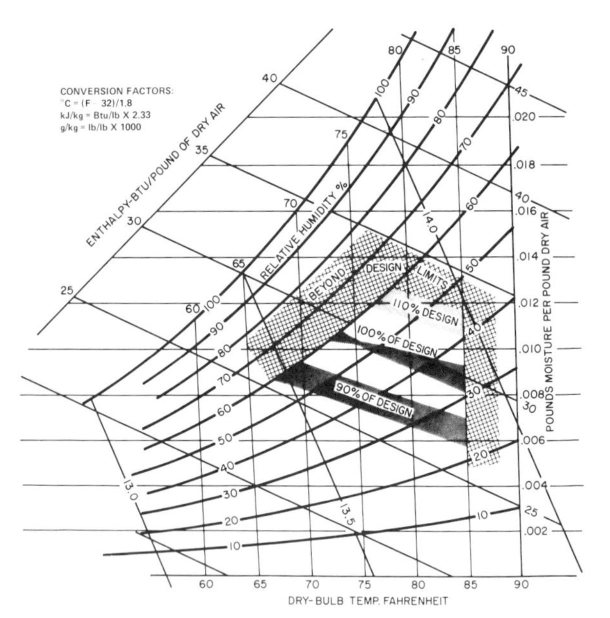


**Fig. 1 Open Display Refrigerator Load Factors Superimposed on Psychrometric Chart[a]**

[a]Percentages show approximate change in heat load as a result of changes in store ambients. Store designs at the lower conditions can result in energy savings. (Chart is not intended for calculations.)

The design of open meat display refrigerators, either well-type or vertical, is limited by the freezing point of meat. Ideally, the refrigerators are set to operate as cold as possible without freezing the meat. The temperatures are maintained with a minimum of fluctuations (with the exception of defrost) to ensure the coldest possible stable internal and meat surface temperatures.

Along with molds and natural chemical changes, bacteria also discolors meat. With good control of sanitation and refrigeration, experiments in stores have produced shelf life of one week and more. Bacterial population is greatest on the exposed surface of displayed meat because it becomes warmer than the interior. Although the cold airflow refrigerates each package, the surface temperature cumulatively increases (along with bacterial growth) by the following:

1. Infrared rays from lights
2. Infrared rays from the ceiling surface
3. High stacking of meat products
4. Voids in display
5. Store drafts that disturb refrigerator air

Where these factors are handled improperly, the surface temperature of the meat will often be above 50°F (10°C). It takes great care in every building and equipment detail, as well as refrigerator loading, to maintain meat surface temperature below 40°F (4.4°C). However, the required diligence is rewarded by excellent shelf life.

The surface temperature will rise significantly during defrost. However, tests have compared matched samples of meat—one goes through normal defrost and the other is removed from the refrigerator during its defrosting cycles. While defrosting characteristics of refrigerators vary, such tests have shown that the effects of properly handled defrosts on shelf life are negligible. Tests for a given installation can easily be run to prove the effects of defrosting on shelf life for that specific set of conditions.

**Table 1 Average Store Conditions**

| Season | Dry Bulb °F (°C) | Wet Bulb °F (°C) | Lb (g) Moisture per Lb (kg) Dry Air | rh, % |
|---|---|---|---|---|
| Winter | 69 (20.6) | 54 (12.2) | 0.0054 (5.4) | 36 |
| Spring | 70 (21.1) | 58 (14.4) | 0.0079 (7.9) | 50 |
| Summer | 71 (21.7) | 61 (16.1) | 0.0091 (9.1) | 56 |
| Fall | 70 (21.1) | 58 (14.4) | 0.0079 (7.9) | 50 |

Store Conditions Survey conducted by Commercial Refrigerator Manufacturers' Association from December 1965 through March 1967. Approximately 2000 store readings in all parts of the country, in all types of stores, during all months of the year reflected the above ambient store conditions.

**Self-Service Meat Refrigerators.** The meat department planner can select from a wide variety of available meat display possibilities.

1. Single-shelf refrigerators with or without rear or front access storage doors (Figure 2).
2. Two, three, or more shelves, with or without rear access (Figures 3 and 4).
3. Any of the above with or without glass fronts.
4. Processed meat versions of the above, often designed for somewhat higher temperatures but including special merchandising shelves or accessories.

All of the above refrigerators are available with a variety of lighting, superstructures, and other accessories tailored to the special merchandising needs.

**Closed Meat or Deli Refrigerators.** Generally, closed refrigerators are definable in one of the following categories:

1. Fresh red meat, with or without storage compartment (Figure 5).
2. Delicatessen and smoked or processed meats, with or without storage.
3. Fresh fish and poultry, usually without storage but designed to display the fresh fish and/or poultry on a bed of cracked ice.

These refrigerators are offered in a variety of configurations. They are available with gravity or forced convection coils, and their glass fronts may be nearly vertical or angled as much as 20 deg. Gravity coils are usually preferred for the more critical products. However, the described types can be used for red meats, delicatessen items, or fish products in a variety of design combinations.

These refrigerators typically use sliding rear-access doors, which are sometimes removed during busy periods. There is a

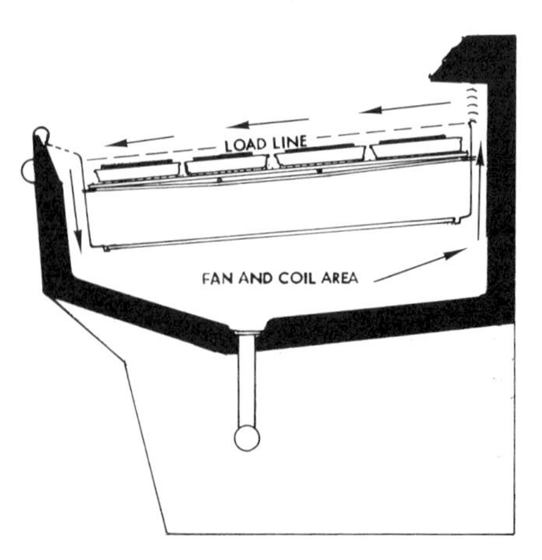


**Fig. 2 Single-Shelf Meat Display Refrigerator**

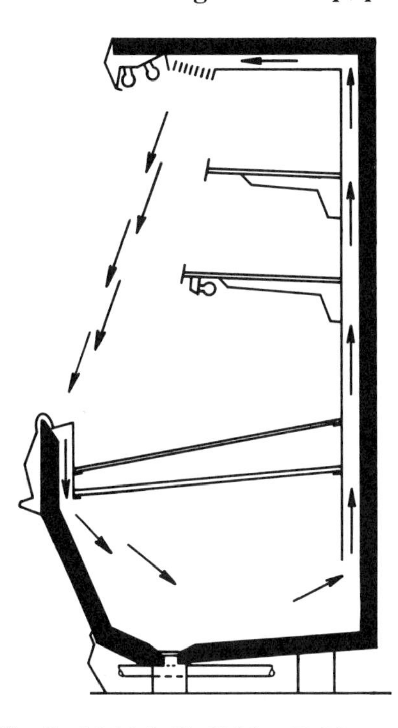

Fig. 3 Multiple-Shelf Meat Refrigerator

conflict between the need for highlighting the product in a closed refrigerator and the need for maintaining a high moisture level in the atmosphere within the refrigerator. Even the lighting within the store is involved. Bright store lighting requires even brighter refrigerator lighting to highlight the product in the refrigerator. Lights in the refrigerator produce heat. Reheating refrigerated air produces dehumidification.

The best solution for this problem is to subdue ambient lighting around the refrigerators so that minimum lighting within the refrigerator is not dominated. Meat and deli refrigerators usually produce the longest shelf life when the air temperatures

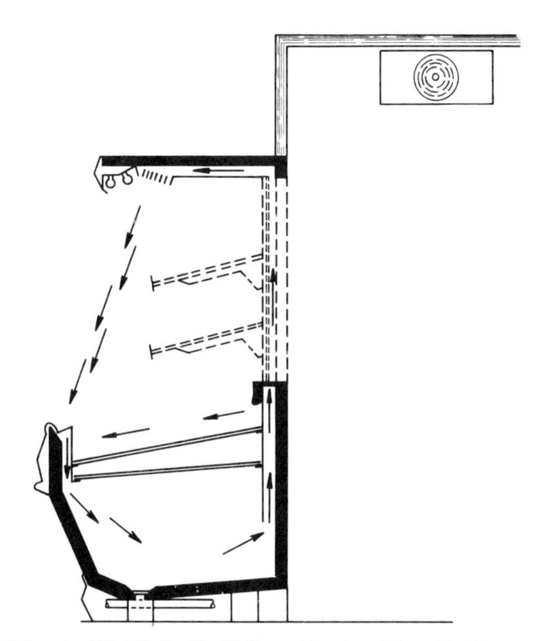

Fig. 4 Multiple-Shelf Rear-Service Meat Refrigerator

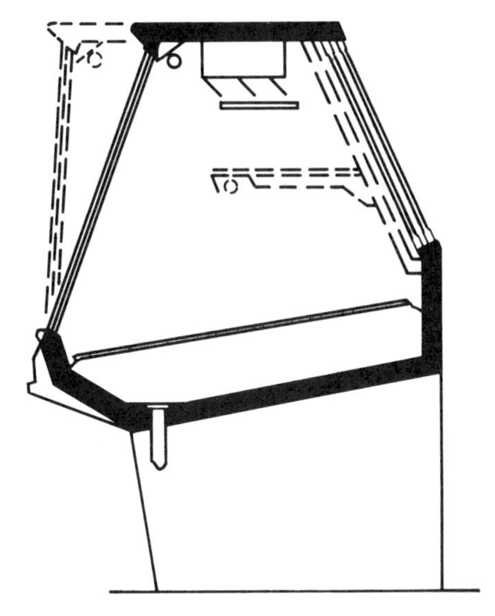

Fig. 5 Closed Meat Display Refrigerator

in the refrigerator are between 34 and 40°F (1 and 4°C). At the same time, the velocity of the refrigerated air over the product should be minimized. It is desirable to maintain high relative humidity with frequent off cycling (from the controls) so that minimum dehydration will occur.

## Dairy Display

Dairy products include such products with great sales volume as fresh milk, butter, eggs, and margarine. They also include the myriad of small items of fresh (and sometimes processed) cheeses, special above-freezing pastries, and other perishables. The available equipment includes the following:

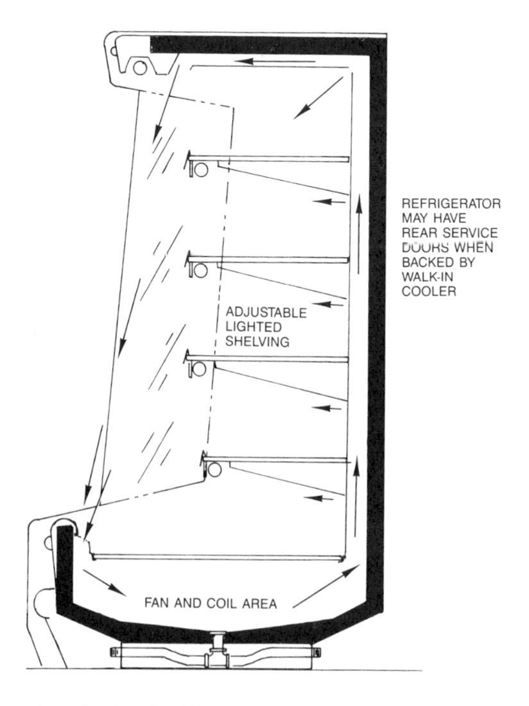


Fig. 6 Vertical-Type Dairy Refrigerator

1. Full-height, full-adjustable shelved displayers without doors in back for use against a wall or with doors in back for rear service or for service from the rear through a dairy cooler (Figure 6). The effect of rear service openings on the surrounding refrigeration must be considered (Figure 7).
2. Two-, three-, four-, or five-deck front service, center aisle, or wall-type refrigerators (Figure 8).
3. A variety of other special displayers, including single-deck and island-type displayers, some of which are self-contained and reasonably portable for seasonal, perishable specialities.
4. The refrigerator, similar to Item 1, but able to receive either conventional shelves and a base shelf and front or pre-made displays on pallets or carts. This version comes with either front-load capability only or rear-load capability only (Figures 8 and 9).

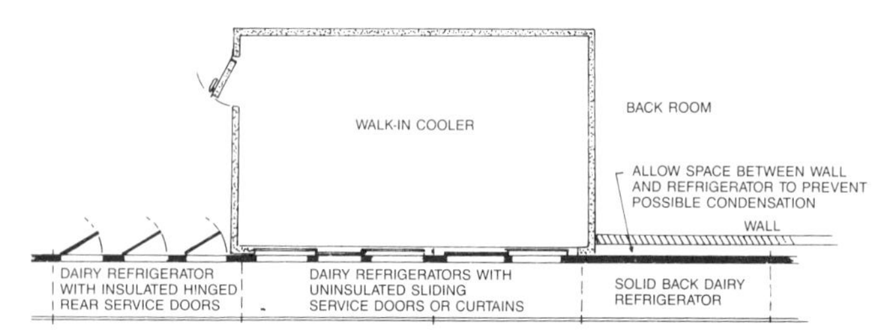


**Fig. 7 Possible Dairy Display Arrangements**

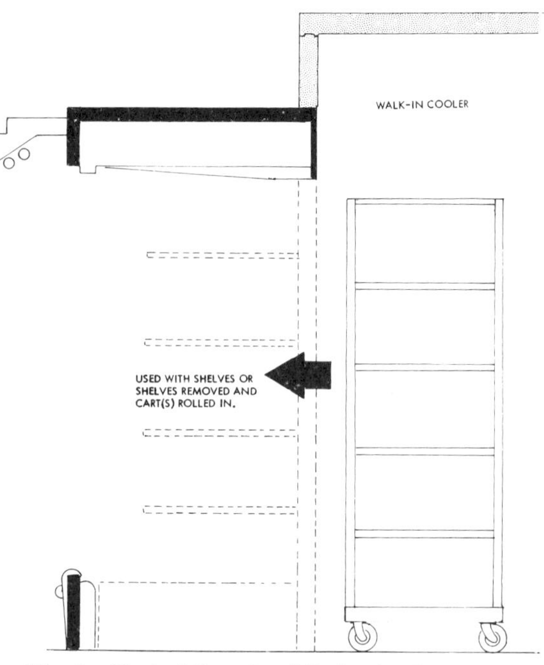


**Fig. 9 Vertical Rear-Load Dairy (or Produce) Refrigerator with Roll-In Capability**

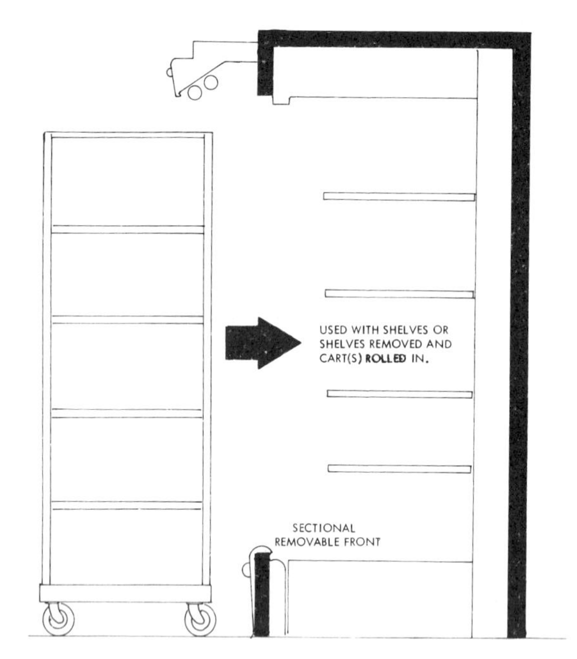


**Fig. 8 Vertical Front-Load Dairy (or Produce) Refrigerator with Roll-In Capability**

## Fruit and Produce

Wrapped and unwrapped produce is often intermixed in the same display refrigerator. Ideally, unwrapped produce should have low velocity refrigerated air forced up through the loose product. Water is usually also sprayed on the leafy vegetables to retain their crispness and freshness. However, packaging prevents this air from circuiting through wrapped produce and requires higher velocity air. To display both packaged and unpackaged produce in the same display refrigerator, the available equipment is usually a compromise between these two desired features and is suitable for both types of product. The equipment available includes the following:

1. Wide or narrow single-deck displayers with or without mirrored superstructures.
2. Two- or three-deck displayers, similar to the ones in Figure 10, usually for multiple hookup with the above display refrigerators.

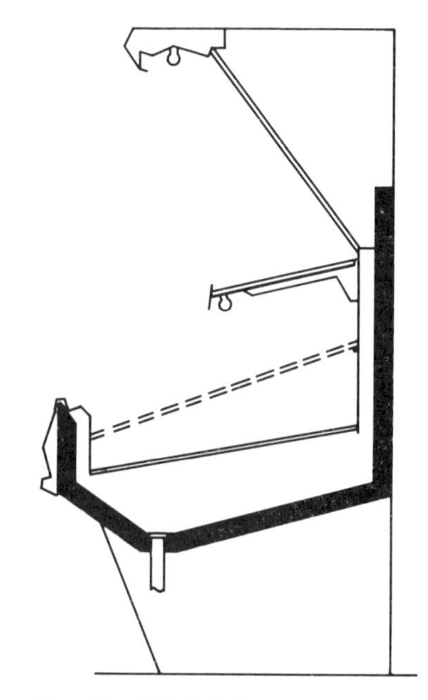

**Fig. 10 Multiple-Shelf Fruit and Produce Refrigerator**

3. Because of the nature of produce merchandising, there is a variety of non-refrigerated displayers of the same family design, which are usually designed for connection in continuous lineup with the refrigerators.
4. The refrigerator, like Item 2, but able to receive either conventional shelves and a base shelf and front or pre-made displays on pallets and carts. This version comes with either front-load capability only or rear-load capability only (Figures 8 and 9).

This produce equipment is generally available with a variety of merchandising and other accessories, including bag compartments, sprayers for wetting the produce, night covers, scale racks, sliding mirrors, and other display shelving and apparatus.

## Frozen Food and Ice Cream

To display frozen foods most effectively (depending on varied need), many types of display refrigerators have been designed and are available. These include the following:

1. Single-deck well-type refrigerators for one-side shopping. Many types of merchandising superstructures for related non-refrigerated foods are available. Configurations of these refrigerators are designed for matching lineup with fresh meat refrigerators, and there are similar refrigerators for matching lineup of ice cream refrigerators with their frozen food counterparts. These refrigerators are offered with or without glass fronts (Figure 11).
2. Single-deck island for shop-around. These are available in widths ranging from the above single-deck refrigerators to refrigerators of double width, with various sizes in between. Some across-the-end increments are available to complete the shop-around configuration. They are available with or without various merchandising superstructures for selling related non-refrigerated food items (Figure 12).
3. Freezer shelving in two to six levels with many refrigeration system configurations (Figure 13).
4. Glass door, front reach-in refrigerators, usually of a continuous lineup design. This style allows for maximum inventory volume and variety in minimum floor space. These advantages must be weighed against the barrier produced by the doors and the greater difficulty in restocking. The front-to-back dimension of these cabinets is usually about 24 in. (600 mm); greater attention must be given to the back product to provide the desired rotation. While some believe that these refrigerators cost less and consume less energy in their operation than open multi-deck low-temperature refrigerators, specific comparisons should be made by models to determine advantages in original and operating costs (Figure 14).
5. Spot merchandising refrigerators, usually self-contained and sometimes arranged for quick change from the nonfreezing to freezing temperature to allow for promotional items of either type (*i.e.*, fresh asparagus or ice cream).
6. Versions of most of the above items are available for ice cream and usually have modified coils, flues, defrost heaters, and other changes necessary for the approximately 10 °F (6 °C) colder temperature required. As display temperature lowers to below 0 °F (−18 °C) (product temperature), the problem of frost and ice accumulation in flues and product zone increases drastically. Proper rotation and grooming minimize frost accumulation.

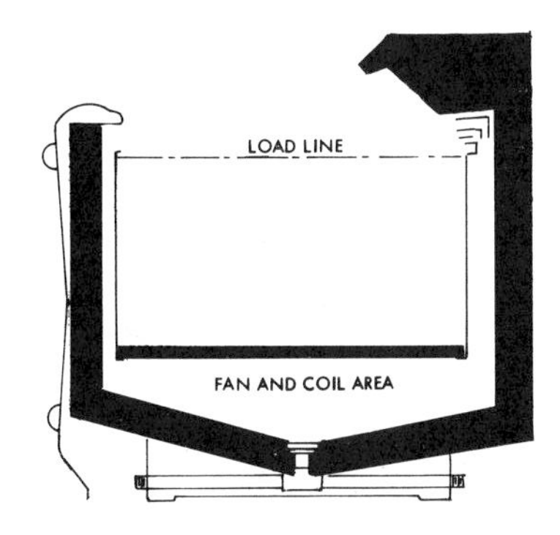


**Fig. 11 Single-Deck Well-Type Frozen Food Refrigerator**

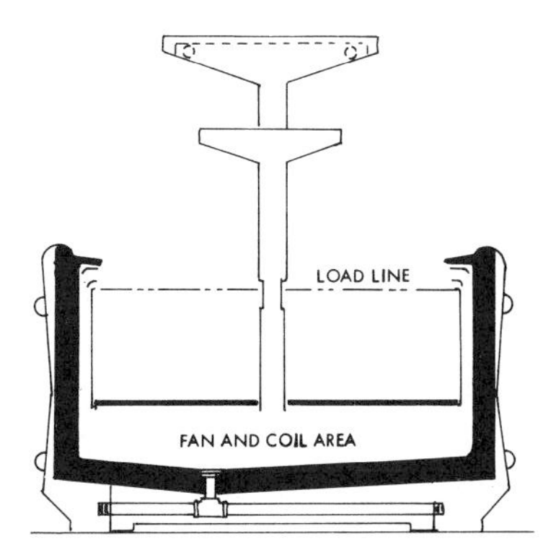


**Fig. 12 Single-Deck Island-Type Frozen Food Refrigerator**

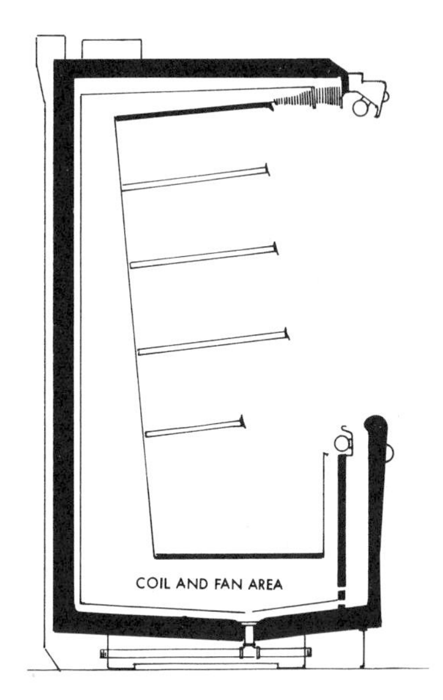


**Fig. 13 Multiple-Deck Frozen Food Refrigerator**

## Refrigerator Construction

Commercial refrigerators for market installations are usually of the endless construction-type, which allows a continuous display as refrigerators are joined. Separate end sections are provided for the first and last units in a continuous display. Methods of joining self-service refrigerators vary, but they are usually bolted, cam locked, or wedged together.

All refrigerators are constructed with surface zones of transition between the refrigerated area and the room atmosphere.

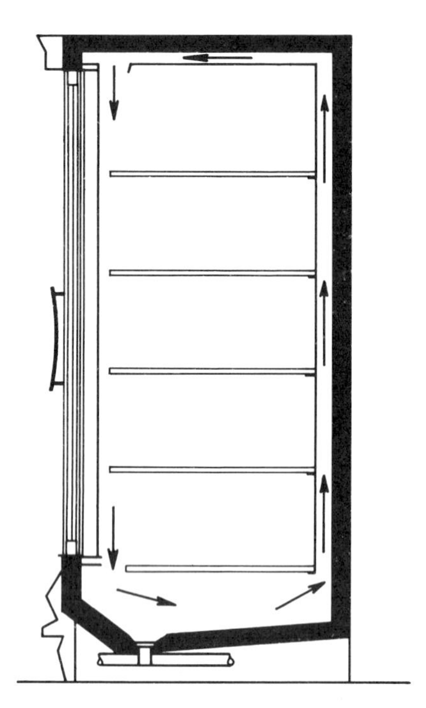

**Fig. 14 Glass Door, Frozen Food Reach-In**

Thermal breaks of various designs separate the zones to minimize the amount of surface on the refrigerator below the dew point. Surfaces in front of discharge air nozzles, sometimes on the nose of the shelving and sometimes at the refrigerator's front rails, may be below the dew point. In glass door reach-in freezers or medium-temperature refrigerators, the frame jambs and the glass are below the dew point. In these locations, resistance heat is used effectively to raise the surface temperature above the dew point.

With the current emphasis on energy efficiency, the designer has developed means other than resistance heat to raise the temperature of surfaces above the dew point. However, when no other technique is known, resistance heating becomes a necessity. That heat may sometimes be controlled by cycling and/or proportional controllers to reduce the annual energy consumption.

The designers of the stores can do a great deal to promote energy efficiency. Not only does controlling atmosphere within a store reduce the refrigeration requirements, it also holds down the heating of the equipment described here. This heat not only consumes energy; it also places added demand on the refrigeration load.

Evaporators and air-distribution systems for display refrigerators are highly specialized and are usually fitted precisely into the particular display refrigerator. As a result, they are inherent in the fixture and are not standard independent evaporators. The design of the air-circuit system, the evaporator, and the means of defrosting are the result of extensive testing to produce the particular display results desired.

### Defrost Methods

The defrosting of refrigerators is primarily an application matter, even though the defrosting mechanism is built into the refrigerator. Refer to the section "Methods of Defrost" in Chapter 32 of the 1986 REFRIGERATION Volume.

Defrosting is accomplished by electric heaters, cycling off the compressor, selective ingestion of store air, or latent heat-reverse cycle (hot gas) defrosting. In the defrost operation, not only must the evaporator be defrosted, particularly in the low-temperature equipment, but also frost in the flues and around the fan blades in various areas of the air-distribution system must be melted and completely drained. The design details are the result of laboratory testing; calculations can give only an approximate indication of the heat required or how it must be distributed.

Defrosting is usually controlled by a variety of clocks. A recent device varies the defrost frequency either according to need or based on the store environment. Some energy reduction and quality improvement are thus possible. Regardless of the controls, the manufacturer's recommendations should be followed.

### Cleaning and Sanitizing Equipment

Since the evaporator coil is the most difficult item to clean, the operator should consider the judicious use of high-pressure, low-liquid-volume sanitizing equipment. This type of equipment enables personnel to spray cleaning and sanitizing solutions into the duct, grill, coil, and waste outlet areas with a minimum of disassembly and a maximum of effectiveness.

## STORAGE COOLERS

Each category of food product displayed is usually backed up by storage in the back room. This storage equipment usually consists of refrigerated rooms made up of sectional walls and ceilings equipped with the necessary storage racks or hanging rails (or both) for a particular food product.

Walk-in coolers, which serve a dual purpose of storage and display, are equipped with either sliding or hinged glass doors on the front. These door sections are often a prefabricated assembly set into a door opening within the front of the cooler. Allowance must be made in computing the refrigeration load for the extra service load.

## RETAIL FOOD STORE CONDENSING UNITS

In the last decade, important technology advances in condensing unit systems and components have reduced energy costs. The following choices are available for display refrigerators and walk-in coolers.

### Air-Cooled Condensing Unit

A single-unit compressor with air-cooled condenser systems can be mounted in racks up to three high to save space (Figure 15). These units may have condensers sized so that the temperature differential (TD) is in the 10 to 25°F (6 to 14°C) range. Optionally available next-larger-size condensers are often used to achieve lower TDs and higher Energy Efficiency Ratios (EER) in some supermarkets, convenience stores, and other simple applications. Single compressors with heated crankcases and heated insulated receivers and other suitable outdoor controls are assembled into weatherproof racks for outdoor installations. Sizes range from 0.5 to 30 horsepower (0.4 to 23 kW).

### Water-Cooled Condensing Unit

Water-cooled units range in size from 0.5 to 30 horsepower (0.4 to 23 kW) and are best for hot, dry climates (Figure 16). The city water-cooled condensing unit is no longer economical

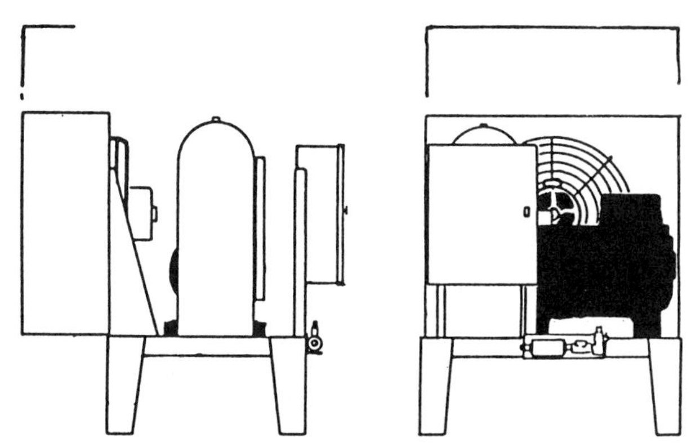

Fig. 15 Single-Compressor Air-Cooled Condensing Unit
(Reprinted by Permission of Tyler Refrigeration)

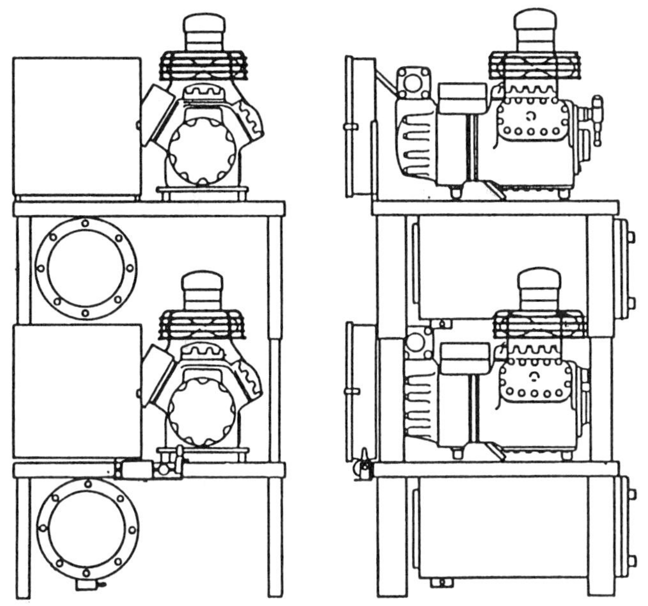

Fig. 16 Water-Cooled Condensing Units
(Reprinted by Permission of Tyler Refrigeration)

due to the high cost of water and sewer fees. Cooling towers, in which one tower cools the water for all compressors, have been used instead. Closed-water systems are also used in areas that can use an evaporative water cooler.

## Remote Condenser-Compressor Unit

Remote units operate efficiently with minimum condenser maintenance. Evaporative condensers are also used in some areas (Figure 17). Sizes range from 0.5 to 30 HP (0.4 to 23 kW).

Recommended sizing for remote air-cooled condensers is 10°F (6°C) TD for low-temperature application and 15°F (8°C) TD for medium-temperature application. Remote water-cooled condensers are often used in areas with abundant water.

## Single-Compressor Control

Of the three classifications described above, single-compressor units make up half of the supermarket compressor equipment presently used. A recently developed solid-state pressure control for single units has helped control excess capacity when the ambient temperature drops. The control senses the pressure and adjusts the cutout point to eliminate short cycling, which ruins many compressors in low load conditions. This control also saves energy by maintaining a higher suction pressure than would otherwise be possible and by reducing overall running time.

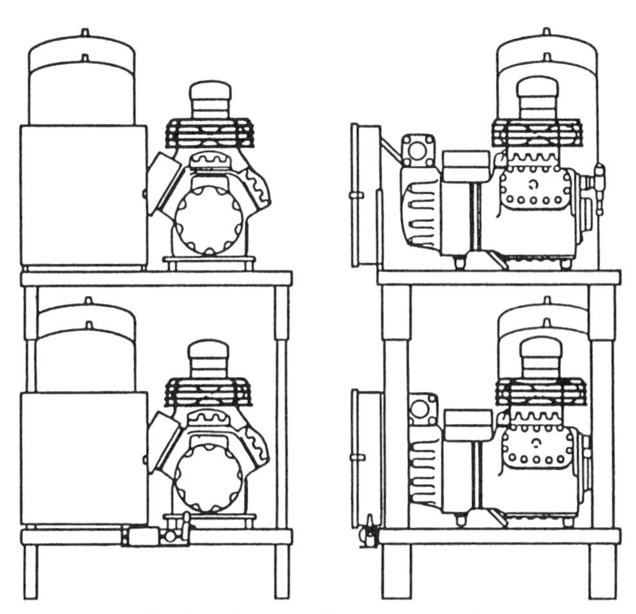

Fig. 17 Remote Compressor Units
(Reprinted by Permission of Tyler Refrigeration)

## Parallel Compressors

Up to six compressors may be operated in parallel on a single rack with a large receiver and multi-station manifolds for liquid, suction, and hot gas (Figure 18). These systems have a remote condenser and almost always include heat recovery from a secondary condenser coil in the air handler. A separate compressor for ice cream refrigerators on a low-temperature system or for meat refrigerators on a medium-temperature system can be on the rack and piped so that only the heat removed from the lower-temperature refrigerators is supplied at the less efficient rate. Paralleled compressors may be equal or unequal in size.

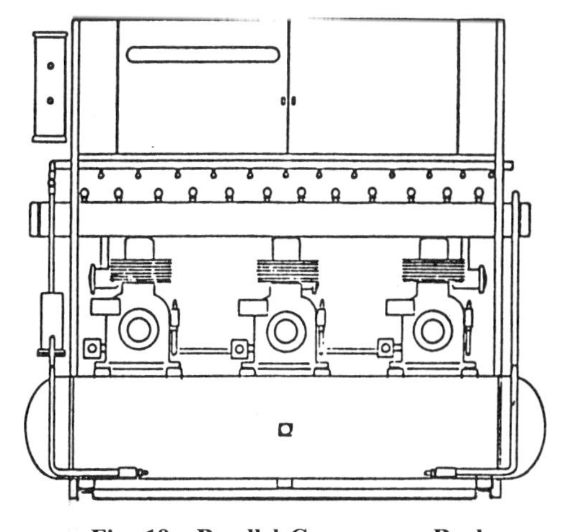

Fig. 18 Parallel Compressor Rack
(Reprinted by Permission of Tyler Refrigeration)

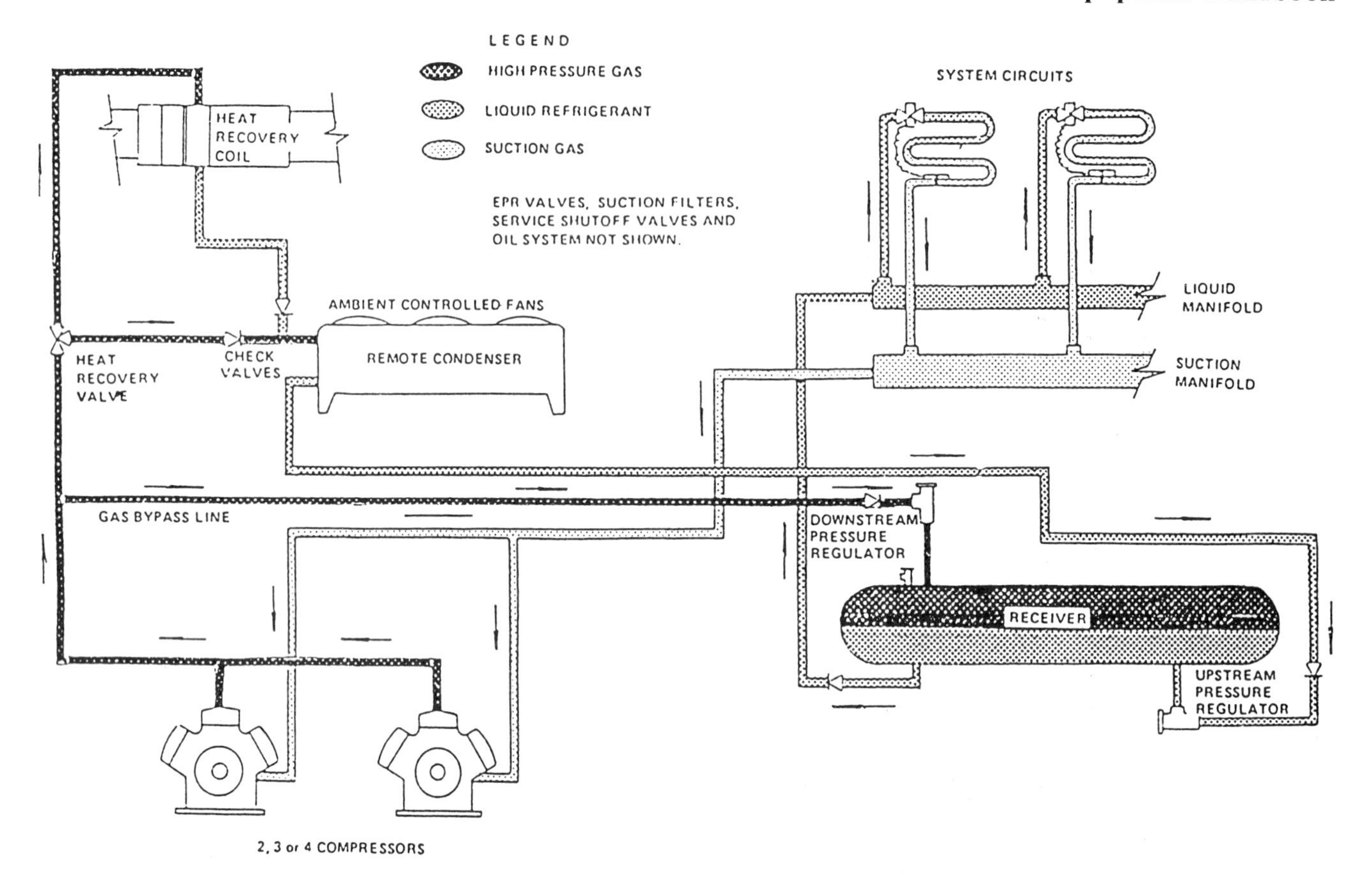


**Fig. 19 Basic Parallel System with Remote Air-Cooled Condenser and Heat Recovery**
**(Reprinted by Permission of Tyler Refrigeration)**

## Remote Air-Cooled Condenser and Heat-Recovery Coil

Remote air-cooled condensers are popular for use with parallel compressors. The condenser coil TD is in the range of 10 to 25 °F (6 to 14 °C). For energy conservation, generous sizing of the condenser with a lower TD is recommended. Figure 19 diagrams a basic parallel system with an air-cooled condenser and heat-recovery coil.

### Evaporative Condenser

Evaporative condensers have a coil in which refrigerant is condensed and has a means to supply air and water over its external surfaces (Figure 20). Heat is transferred from the condensing refrigerant inside the coil to the external wetted surface and then into the moving airstream principally by evaporation. In areas where the wet-bulb temperature is about 30 °F (17 °C) below the dry-bulb temperature, the condensing temperature can run from 10 to 30 °F (11 to 17 °C) above the wet-bulb temperature. This lower condensing temperature saves energy. One evaporative condenser can be installed for the entire store.

## MATCHING REFRIGERATOR REQUIREMENTS WITH COMPRESSOR CAPACITY

The designer matches requirements of refrigerator lineups to the capacity of a single-compressor unit or divides them into manageable circuits for parallel compressor assemblies. The parallel assembly adapts easily to hot gas defrost. Gas directly from the compressor discharge, or, in some instances, from the top of the receiver at a lower temperature, flows through a manifold. Electric defrost, air defrost, and off-cycle defrost can also be used on both parallel and single-unit systems. Liquid and/or suction line solenoid valves control the circuits for defrosting. Often, a suction stop is combined with the temperature controlling Evaporator Pressure Regulator (EPR). The en-

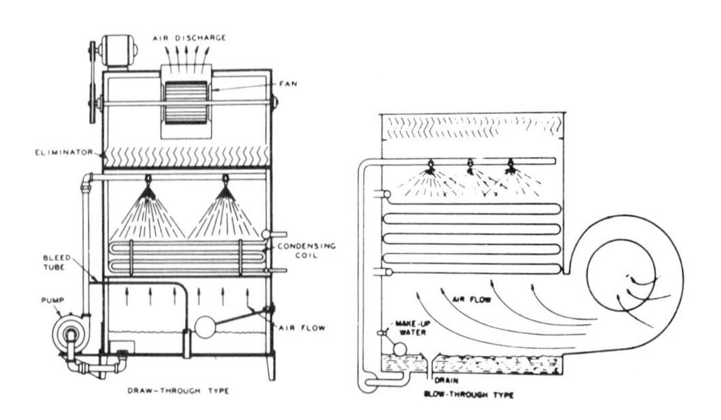


**Fig. 20 Functional View of Evaporative Condenser**

tire system with its individual circuits is controlled by a multi-circuit time clock. Temperature control on the branch circuits is also achieved by refrigerator thermostats actuating liquid line solenoid valves. Ball-type shutoff valves isolate each circuit for service convenience. Refrigerator requirements are now often given as a refrigeration load per unit length, with a lower value sometimes allowed for parallel systems. The rationale for this lower value is that peak loads are less with programmed defrost. Refrigerator temperature recovery after a defrost period is less of a strain than it would be to a single-compressor system.

The published refrigerator load requirements allow for extra capacity for temperature pull-down after defrost, per ASHRAE *Standard* 72-1983. The industry considers a standard store condition to be 75 °F and 55% rh, which is maintained with air conditioning. Much of this air-conditioning load is carried by the open refrigerators, and credit for the heat removed by them should be considered in sizing the air-conditioning system.

## Typical Parallel Compressor Operation

A typical supermarket includes one or more medium-temperature parallel systems for dairy and produce refrigerators and the medium-temperature walk-in coolers. The system may have a separate compressor for the meat or deli refrigerators, or all units may have a single compressor. Energy Efficiency Ratios (EERs) typically run from 8 to 9 Btu/h per W (2.3 to 2.6 W/W) for the main load. Low-temperature refrigerators and coolers are grouped on one or more parallel systems with ice cream refrigerators on a satellite or on a single compressor. EERs range from 4 to 5 Btu/h per W (1.17 to 1.47 W/W) for the frozen foods and ice cream units as low as 3.5 to 4.0 Btu/h per W (1.03 to 1.17 W/W). Cutting and preparation rooms are most economically placed on a single unit, since the refrigeration EER is at nearly 10 Btu/h per W (2.9 W/W). Air-conditioning compressors are also separate, since their EER can range up to 11 Btu/h per W (3.2 W/W) (Figure 21).

## Capacity Control of Parallel Systems

This system must be designed to maintain proper refrigerator temperatures under peak summer load. During the remainder of the year, store conditions can be easily maintained at a more ideal condition, and refrigeration load will be lower. In the past, refrigeration systems were operated at 90 °F condensing conditions or above to maintain enough high-side pressure to feed the expansion valves properly. Recent observations have determined that when outdoor ambient conditions allow, the condensing temperature can be allowed to follow the ambient down to about 70 °F (21 °C). When proper liquid-line piping practices are observed, the valves feed the evaporators properly under these low condensing temperatures. Therefore, at partial load, the system has excessive capacity to perform adequately.

Multiple compressors may be controlled or staged based on a drop in system suction pressure. If the compressors are equal in size, a mechanical device can turn off one compressor at a time until only one is running. The suction pressure will be perhaps 5 psi or more (34 kPa) below optimum. Newer, solid-state control devices can cycle units *on* and *off* while the system remains at one economical pressure setting, and the run time for each compressor motor can be equalized. Satellite compressors can also be controlled accurately with one control that also monitors other functions such as oil loss, alarm functions, etc.

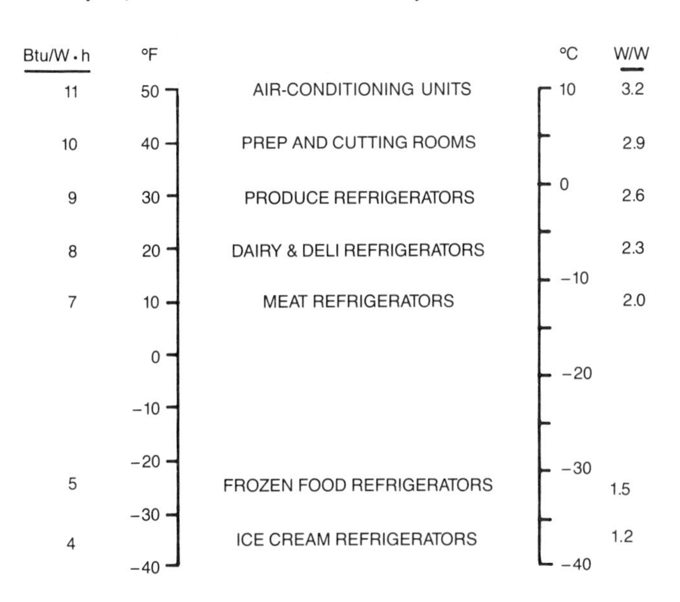


Fig. 21 Typical Compressor Efficiency

## Staging Unequally Sized Compressors

Unequally sized compressors can also be staged to obtain a greater range of control. Figure 22 shows seven stages of capacity from a 5, 7, and 10 HP (3.8, 5.2, and 7.5 kW) compressor parallel arrangement. Improved unloaders on multi-bank compressors also promise to provide greater flexibility in the future.

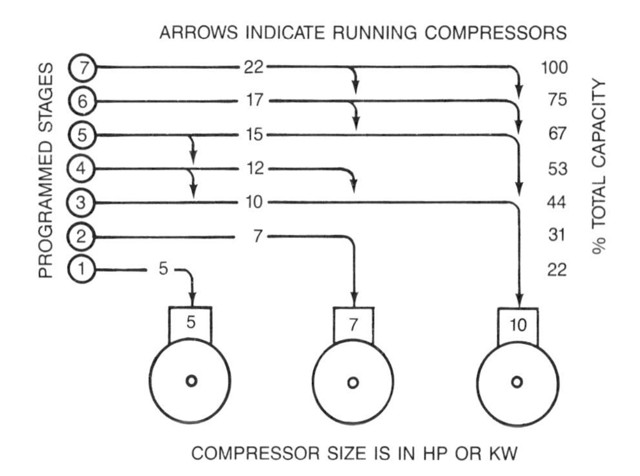


Fig. 22 Stages with Mixed Compressors

## Subcooling Liquid Refrigerant

Allowing refrigerant to *naturally subcool* in cool weather as it returns from the remote condenser also saves energy. This can be done by several means. The most common is to flood the condenser and allow the liquid refrigerant to cool close to the ambient temperature. Mechanical subcooling may be economical in the southern United States. A subcooling compressor can be combined on a parallel system. Another method is to have a separate parallel system with branches to subcoolers on the other parallel and single system in the store. The mechanical subcooling would be set to operate when the ambient temperature raises the refrigerant temperature above the desired subcooled temperature. The advantage to this method is that the mechanical subcooling compressor can run at twice the efficiency of the main system, thus saving energy.

### Heat Recovery

Heat recovery is an important feature of virtually every compressor system, parallel or single. A heat-recovery coil is simply a second condenser coil placed in the air handler. If the store needs heat, this coil is energized and run in series with the regular condenser (Figure 19). The heat-recovery coil can be sized for a 30 to 50 °F TD (17 to 28 °C). Lower head pressures in parallel systems permit little heat recovery. However, when heat is required in the store, simple controls create a higher head pressure for heat recovery. When compared with the cost of auxiliary gas or electric heat, the slightly higher energy consumption is more than compensated for by the value of the heat gained.

Water can also be heated by a de-superheater installed on one large single unit.

### Factory-Assembled Equipment

Factory assembly of the necessary compressor systems with either direct air-cooled or any style of remote condenser is common practice. Both single and parallel systems can be housed, pre-piped, and pre-wired at the factory. The complete unit is then delivered to the job site for placement on the roof or beside the store.

## ENVIRONMENTAL EQUIPMENT AND CONTROL

Major components of common store environmental equipment include (1) central air handler with fresh makeup air-mixing box, (2) air-cooling coils, (3) heat-recovery coils, (4) supplemental heat equipment, (5) connecting ducts, and (6) termination units such as air diffusers and return grilles.

Environmental control is the heart of energy management. Control panels, designed for the unique heating, cooling, and humidity control requirements for food stores, provide several stages of heating (up to 8) and cooling (up to 3), plus a dehumidification stage. When high humidity exists in the store, cooling is activated to remove moisture. The controller receives input from temperature and dew point sensors that are located in the sales area.

Some controllers include night setback for cool climates and night setup for warm climates. This feature saves energy by turning the air handler off and allowing the store to cool or warm up several degrees, normally 6 to 10 °F (3 to 6 °C) beyond the store temperature set point.

Following are rules for good air distribution in food stores:

**Air Circulation.** Operate 100% of the time the store is open at a volume of 0.7 to 1 cfm/ft$^2$ (3.5 to 5 L/[s • m$^2$]) of sales area.

**Fresh Air.** Introduce whenever the air handler is operating. It should equal the required air change per hour or an allowance for all exhaust fans, whichever is greater.

**Discharge Air.** Discharge most or all of the air in areas where heat loss or gain occurs. This load normally is at the front of the store and around glass areas and doors.

**Return Air.** Locate return air registers as low as possible. With low registers, return air temperature may be 50 to 55 °F (10 to 13 °C). This practice reduces heating and cooling requirements and temperature stratification compared to high returns.

## BIBLIOGRAPHY

Biggers, J.A., Jr. 1960. Condensation in and on open display equipment as related to air-conditioned vs non-air-conditioned stores. ASHRAE *Journal,* September, p. 64.

Commercial Refrigerator Manufacturers' Association. 1986. Retail Food Store Refrigerators, Health and Sanitation. CRS-SI-86.

Naumann, H.D., W.C. Stringer, and P.F. Gould. 1965. Guidelines for Handling Pre-packaged Meat in Retail Stores. University of Missouri, *Manual* 64.

Rainwater, J.H. 1959. Five defrost methods for commercial refrigerators. ASHRAE *Journal,* March, p. 61.

Tyler, T.L. 1960. Air conditioning in food stores and its effect on the operation of open refrigerated display equipment. ASHRAE *Journal,* September, p. 63.

## CHAPTER 36

# FOOD SERVICE AND GENERAL COMMERCIAL REFRIGERATION



FOOD service requires refrigerators that meet a variety of needs. This chapter covers refrigerators available for (1) restaurants, (2) fast-food stores, (3) cafeterias, (4) commissaries, (5) hospitals, (6) schools, and (7) other specialized applications.

Most of the refrigerated products used in food service applications are self-contained, and the corresponding refrigeration systems are conventional. Some systems, however, do employ ice for fish, salad pans, or specialized preservation and/or display. (Further information on some of these products will be found in Chapters 35, 37, 39, and 40.)

Generally, electrical and sanitary requirements of refrigerators are covered by criteria, standards, and inspections of Underwriters Laboratories, the National Sanitation Foundation, and the U.S. Public Health Service.

Frame construction is usually of metal. Occasionally, wood is used in minor amounts, but it is rarely visible in the final assembly. Plastics are increasingly used for decoration, finish, trim, breaker strips, and gaskets.

Nickel-chromium stainless steel is the most common sheet metal used for these refrigerators. Other sheet metals used are aluminum and carbon steel, either hot or cold rolled. Stainless steel usually has a grained or polished finish when exposed in final assembly. Aluminum generally is unfinished, although it may occasionally be lacquered or coated with some other synthetic finish. Carbon steel may be black, plated, porcelain enameled, aluminized, or coated with either a synthetic enamel or one of the vinyl finishes.

Insulation is usually one of the following: (1) foam polyurethane, poured in place; (2) slab polyurethane; (3) slab polystyrene; and (4) glass fiber batts.

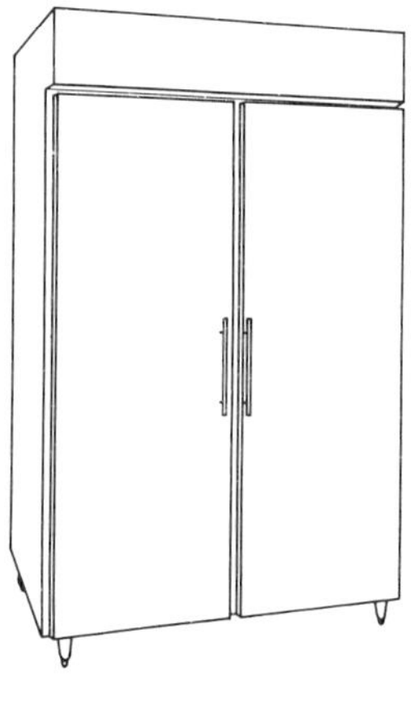

## REACH-IN AND SPECIALTY CABINETS

The **reach-in refrigerator** is an upright, box-shaped cabinet with straight vertical front(s) and hinged or sliding doors (Figure 1). It is usually about 2.5 to 3 ft (750 to 900 mm) deep, about 6 ft (1800 mm) high, and ranges in width from about 3 to 10 ft (900 to 3000 mm). Capacity ranges are from about 20 to 90 ft$^3$ (500 to 3000 L). These capacities and dimensions are standard from most manufacturers.

The typical reach-in cabinet (Figure 1) is available in many styles and combinations. Other shapes, sizes, and capacities are available on a custom basis from some manufacturers (Figure 2).

There are many varied adaptations of refrigerated spaces for storing perishable food items. Reach-ins, by definition, are

1. LOW AND MEDIUM TEMPERATURE
2. 1, 2, OR 3 DOOR
3. STAINLESS STEEL, ALUMINUM, OR ORGANIC FINISHES ON MILD STEEL.
4. COMBINATIONS OF FINISHES ON EXTERIOR AND INTERIOR. (SEE #3)
5. MANY HEIGHTS TO FIT SPECIFIC APPLICATIONS, *i.e.* UNDERCOUNTER

NOTE:
A. TOP MOUNT COMPRESSOR SHOWN. OTHER STYLES MAY HAVE COMPRESSOR LOCATED IN LOWER SECTION.
B. LEGS SHOWN ON REFRIGERATOR. MOST CODES ALSO PERMIT SEALING REFRIGERATOR TO FLOOR.

**Fig. 1 Reach-In Food Storage Cabinet Features**

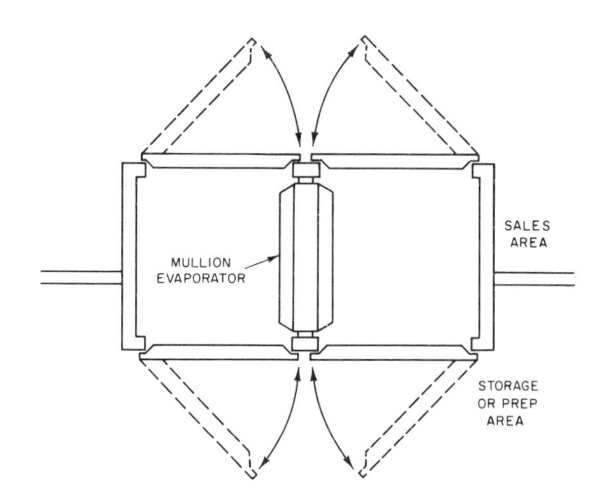


**Fig. 2 Pass-Through Styles Facilitate Some Handling Situations**

The preparation of this chapter is assigned to TC 10.7, Commercial Food-Display and Storage Equipment.

normal- or low-temperature refrigerators small enough to be moved into a building. This definition also includes refrigerators and freezers built for special purposes, such as mobile cabinets or refrigerators on wheels, display refrigerators for such products as pies and cakes, and display refrigerators for bakery goods. The latter two cabinets usually have glass doors. Candy refrigerators also have specialized size, shape, and temperature.

Refrigerated vending machines satisfy the general definition of reach-ins, but they also receive coins and dispense products. Beverage-dispensing units dispense a measured portion into a cup, rather than in a bottle or can. These and other machines, such as soft ice cream freezers, are covered in Chapter 39.

## TYPES OF CONSTRUCTION

Reach-in refrigerators are available in two basic types of construction. The older style is a wood frame substructure clad with a metal interior and exterior. The newer style is a welded assembly of exterior panels with insulation and liner inserts.

In descending order of cost, the materials used on exteriors and interiors are stainless steel, porcelain enamel on steel, aluminum, and synthetic enamel on steel. The requirements are for a material that (1) matches or blends with that used on nearby equipment; (2) is easy to keep clean; (3) is not discolored or etched by commonly used cleaning materials; (4) is strong enough to resist denting, scratching, and abrasion; and (5) provides the necessary frame strength. The material chosen by an individual purchaser depends a great deal on layout and budget.

### Temperature Ranges

Reach-in refrigerators are available for medium- or low-temperature ranges. The medium-temperature range has a maximum of 45 °F (7 °C) and a minimum of 32 °F (0 °C), with the most desirable average temperature being close to 38 °F (3.5 °C). The low-temperature range need not be below −10 °F (−23.5 °C) and should not exceed 10 °F (−12 °C). The desirable average is 0 °F (−18 °C) for frozen foods and −5 °F (−20.5 °C) for ice cream. Both temperature ranges are available in cabinets of many sizes, and some cabinets combine both ranges.

### Refrigeration Systems

Remote refrigeration systems are often used when cabinets are installed in a hot or otherwise unfavorable location, where the noise or heat of the condensing units would be objectionable, or under other special circumstances.

Self-contained systems, in which the condensing unit and controls are built into the refrigerator structure, are of two general types. The first has the condensing unit beneath the cabinet; in some designs it takes up the entire lower part of the refrigerator, while in others it occupies only a corner at one lower end. The second type has the condensing unit on top. There is no advantage to locating a self-contained condensing unit beneath the refrigerator; although the air near the floor is generally cooler and thus beneficial to the condensing unit, it is usually dirtier. Putting the condensing unit on top of the cabinet allows full use of cabinet space, and, although the air passing over the condenser may be warm, it is clean and more abundant. In addition, top mounting of both condensing unit and coil offers the same physical and constructional advantages. Frequently, bare-tube condensers are used to minimize dust clogging.

### Styles

Reach-ins have doors on the front. Refrigerators that have doors on both front and rear are called *pass-through* or *reach-through refrigerators.* Doors are either full height (one per section) or half height (two per section). Doors may be glazed or solid, hinged or sliding.

### Interiors

Shelves are standard interior accessories and are usually furnished three or four per full height section.

### Modifications and Adaptations

**Food service applications** often require extra shelves or tray slides, pan slides, or other interior accessories to increase food-holding capacity or make operation more efficient. With increasing use of foods prepared off-premises, specialization of on-premise storage cabinets is growing. This is developing an increasing pressure for designs that consider new food shapes, as well as in-and-out handling and storage.

**Beverage service applications** use standard cabinets if reach-ins are used, except when glass doors and special interior racks are used for chilled wine display.

**Meal factories,** such as airline or central feeding commissaries, require rugged, heavy-duty equipment, often fitted for bulk in-and-out handling.

**Retail bakeries** also have special requirements. The dough retarder refrigerator and the bakery freezer permit the baker to spread the work load over the entire week and to offer a greater variety of products. The recommended temperature for a dough retarder is 36 to 40 °F (2.0 to 4.5 °C). The relative humidity should be in excess of 80% (never lower) to prevent crusting or other undesirable effects. In the freezer, the temperature should be held at 0 °F (−18 °C). All cabinets should be equipped with racks to hold the 18 by 26 in.(450 by 660 mm) bun pans, which are standard throughout the baking industry.

**Retail stores** use reach-ins for many different nonfood applications. Drug stores often have refrigerators with special drawers for storage of biologicals. (See later section on "Nonfood installations.")

Retail florists use reach-in refrigerators for displaying and storing flowers. Although a few floral refrigerator designs are considered conventional in the trade, the majority are custom built. The display refrigerator located in the sales area at the front of the shop may include a picture window display front and have one or more display-type access doors, either swinging or sliding. A variety of open refrigerators may also be used.

For the general range of flowers in a refrigerator, most retail florists have found that best results are obtained at temperatures from 40 to 45 °F (4.5 to 7 °C). The refrigeration coil and condensing unit should be selected to maintain a high relative humidity. Some florists favor a gravity-type cooling coil because the circulating air velocity is low. Others, however, choose forced-air cooling coils, which develop a positive but gentle airflow through the refrigerator. The forced-air coil has an advantage when the in-and-out service is especially heavy because it provides quick temperature recovery during these peak conditions.

**Nonfood installations** use a wide range of reach-ins, some standard, except for accessory or temperature modifications, and some completely special. Examples are (1) biological and pharmaceutical cabinets; (2) blood bank refrigerators; (3) low and ultralow temperature cabinets for bone, tissue, and red-cell storage; and (4) special-shape refrigerators to hold column chromatography and other test apparatus.

Blood bank refrigerators for whole blood storage are usually standard models, ranging in size from under 20 to 45 ft$^3$ (500 to 1300 L) but with the following modifications:

1. Temperature is controlled at 37 to 41 °F (3 to 5 °C).
2. Special shelves and/or racks are sometimes used.

3. A temperature recorder, usually hand-wound, with a 24-hour or seven-day chart is furnished.
4. An audible and/or visual alarm system is supplied to warn of unsafe blood temperature variation.
5. Sometimes an additional alarm system is provided to warn of power failure.

Most biological serums and vaccines require refrigeration for proper preservation and to retain highest potency. In hospitals and laboratories, the refrigerator temperatures should be 34 to 38°F (1 to 3.5°C). In other locations, temperatures may be slightly higher. The refrigerator should provide low humidity and, of course, should not freeze.

Products in these refrigerators are kept in specially designed stainless steel drawers sized for convenient storage, labeled for quick and safe identification, and perforated for good air circulation.

Biological refrigerators, those for laboratory use, and mortuary refrigerators involve the same technology as that for food preservation. Storage in mortuary refrigerators is usually short-term, normally 12 to 24 hours at 34 to 38°F (1 to 3.5°C). Refrigeration is obtained by a standard air- or water-cooled condensing unit with a forced-air cooling coil.

Mortuary refrigerators are built in various sizes and arrangements, the most common being two- and four-cadaver self-contained models. In the former cabinet are two individual storage compartments, one above the other, with the condensing unit compartment located above and indented into the upper front of the cabinet; ventilation grills are on the front and top of this section. The four-cadaver cabinet is equivalent to two two-cadaver cabinets set together, the storage compartments being two cabinets wide by two cabinets high with the compressor compartment above. Six- and eight-cadaver cabinets are built along the same lines. The two-cadaver refrigerator is approximately 38 in. (950 mm) wide by 94 in. (2400 mm) deep by 77 in. (1900 mm) high and is usually built and shipped set up.

Each compartment contains a mortuary rack consisting of a carriage supporting a stainless steel tray. The carriage is telescoping, equipped with roller bearings so that it slides out through the door opening and is self-supporting even when extended. The tray is removable. Some specifications call for a dial thermometer to be mounted on the exterior front of the cabinet to show the inside temperature reading.

## ROLL-IN CABINETS

These cabinets are very similar in style and appearance to reach-ins. *Roll-ins* (Figure 3) are usually part of a food-handling or other specific-purpose system (Figure 4). Pans, trays, or other specially sized and/or shaped receptacles are used to serve a specific system need, such as the following:

1. Schools, hospitals, cafeteria, and other food-handling facilities
2. Meal manufacturing
3. Bakery processing
4. Pharmaceutical products
5. Body parts preservation (*i.e.,* blood)

The roll-in differs from the reach-in in the following ways:

1. The inside floor is at about the same level as the surrounding room floor, so wheeled racks of product can be rolled directly from the surrounding room into the cabinet interior.
2. Cabinet doors are full height, with drag gaskets on their bottoms.
3. Cabinet interiors have no shelves or other similar accessories.

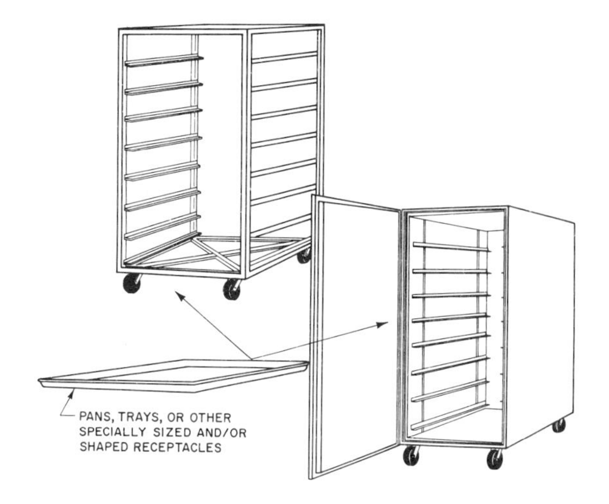


**Fig. 3 Roll-In Cabinet**

The racks that roll in and out of these cabinets are generally fitted with slides to handle 18 by 26 in. (450 by 660 mm) pans, although some newer systems call for either 12 by 20 in. (300 by 500 mm) or 12 by 18 in. (300 by 450 mm) steam table pans.

Manufacturers and contractors offer various methods of insulating the floor area. This problem must not be ignored, particularly if the roll-in is to hold frozen food.

## FOOD FREEZERS

Section II of the 1986 REFRIGERATION Volume covers the commercial freezing of food products. However, some hospitals, schools, commissaries, and other mass-feeding operations use on-premises freezing to level work loads and operate kitchens on normal schedules. Industrial freezing equipment is usually too large for these applications, so operators are using either regular frozen food storage cabinets for limited amounts of freez-

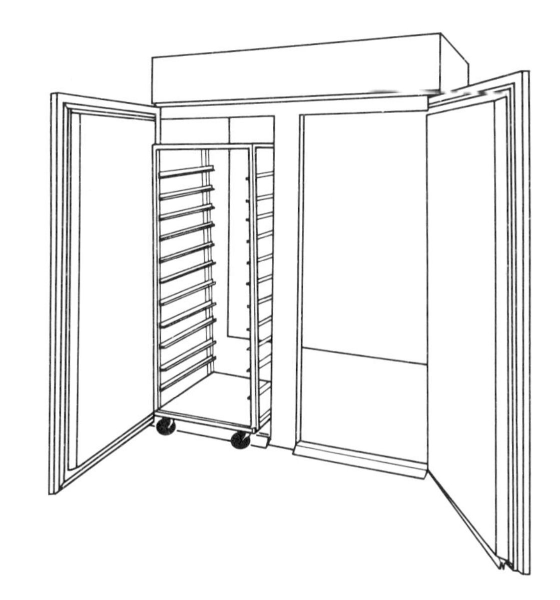

**Fig. 4 Roll-Ins—Usually Part of a Food-Handling or Other Special-Purpose System**

ing or special reach-ins that are designed and refrigerated to operate as batch-type blast freezers.

## WALK-IN COOLERS/FREEZERS

This type of commercial refrigerator is a factory-made, prefabricated, sectional version of the built-in, large-capacity cooling room. It nearly matches the reach-in type in meeting a wide variety of applications.

Its function is to store foods and other perishable products in larger quantities and for longer periods than the reach-in refrigerator. Good refrigeration practice dictates that dissimilar foods be stored in separate rooms because they require different temperatures and moisture conditions and because odors from some foods are absorbed by others. Large food operations usually require three rooms: one for fruits and vegetables, one for meats and poultry, and one for dairy products. A fourth room, at 0°F (−18°C), may be added for frozen foods.

The sectional cooler offers flexibility over the built-in type: it is easily erected, easily moved, and, by adding standard sections, can be readily altered to meet changing requirements, uses, or layouts. Also, the sectional walk-in cooler can be erected outside a building, providing more refrigerated storage with no expensive building costs except for footings and a low-cost roof supported by the cooler. Further advantage results from the high degree of skill applied in the design and construction of the factory-built product, which is subject to close inspection of materials and vapor-sealing techniques.

### Self-Contained Sectional Walk-In Coolers

The versatility of sectional walk-in coolers was greatly increased by the introduction of self-contained models. There are various methods of application (Figure 5).

Foam plastic materials, both self-contained and remote sectional coolers, have further improved. Polyurethane foamed-in-place between two skins of metal suitably box-formed makes a lightweight, waterproof panel. Additionally, the foam is a more efficient insulator, allowing slimmer panels for equal insulation value compared to most other insulations.

### Walk-in Floors

Sectional walk-in coolers, termed *floorless* by the supplier, are furnished with floor splines to fasten to the existing floor to form a base for the wall sections. Models with an insulated floor are also available.

A normal temperature (above freezing) cooler can be erected on an uninsulated concrete floor (on the ground) with only about 5% greater energy requirements than a cooler on an insulated floor. About 1.3 Btu/h • ft$^2$ (4.1 W/m$^2$) enters from the ground. Generally, floor losses are considered negligible.

Level entry is becoming more important as the use of hand and electric trucks increases. The advantage of recessed floors and convenience of level entry afforded by a floorless cooler can also be obtained by recessing a sectional insulated floor.

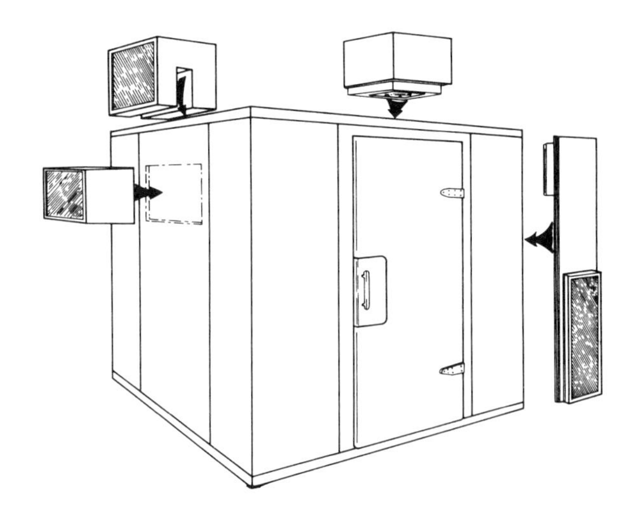

**Fig. 5 Equipment Added to Make a Walk-In Cooler Self-Contained**

### Design Characteristics

The factory-made walk-in cooler consists of standard top, bottom, wall, door, and corner sections, which are shipped to the user and erected on the job. The frames are filled with insulation and are covered on the inside and outside with metal. The edges of these frames are usually of tongue-and-groove construction and either fitted with a gasket material or provided with suitable caulking material to ensure a tight vapor seal when assembled. These sections are assembled on the site with either lag screws or hooks operated from inside of the cooler.

Exterior and interior surfaces are made of one or more of the following:

1. Stainless steel
2. Enamel, baked on steel
3. Vinyl, fused on steel
4. Aluminum
5. Zinc coated steel

At one time, coolers were used primarily to hold sides or quarters of beef, lamb carcasses, crates of vegetables, and other bulky items. Food operations now rarely, if ever, use such items. If they do, the items are broken down, trimmed, or otherwise processed before going into refrigerated storage. The modern cooler is not a storage room for large items, but a temporary place for quantities of small, partially or totally processed products. The food cooler, therefore, is likely to be equipped with sturdy, adjustable shelving about 18 in. (450 mm) deep and arranged in tiers, three or four high, around the inside walls. Alternatively, the cooler is often provided with rolling racks that are actually shelving on wheels. These racks are rolled directly into (and out of) the cooler.

## CHAPTER 37

# HOUSEHOLD REFRIGERATORS AND FREEZERS



THIS chapter covers the design and construction of full-sized household refrigerators and freezers—the most common of which are illustrated in Figure 1. Small portable and secondary refrigerators are not addressed here specifically. Some of these small refrigerators use absorption systems, special forms of compressors, and, in some cases, thermoelectric refrigeration. A major application has developed in the recreational vehicle market.

The section of this chapter on "Refrigeration Systems" only covers the vapor-compression cycle, which is almost universally used for full-sized household refrigerators and freezers. In these applications, where several hundred Btu/h (W) are pumped through perhaps 100°F (38°C) differential from freezer to room temperature, other *electrically powered* systems compare unfavorably to the manufacturing and operating costs of vapor-compression systems. Typical operating efficiencies of the three most practical refrigeration systems are as follows for a 0°F (−18°C) freezer and 90°F (32°C) ambient:

Thermoelectric—approximately 0.3 Btu/watt-hour (0.09 W/W)
Absorption—approximately 1.5 Btu/watt-hour (0.44 W/W)
Vapor-compression—approximately 3.8 Btu/watt-hour (1.1 W/W)

An absorption system may operate from gas at a lower cost per Btu(kJ), but the initial cost, size, and weight of this system has made it unattractive for major appliances where electric power is available. Because of its simplicity, thermoelectric refrigeration could replace other systems if an economical thermoelectric material were developed.

The preparation of this chapter is assigned to TC 7.1, Residential Refrigerators, Food Freezers, and Drinking Water Coolers.

During 1985 in the United States, two-door combination refrigerator/freezers accounted for nearly 94% of the refrigerator market, with the *side-by-side* type growth apparently leveling

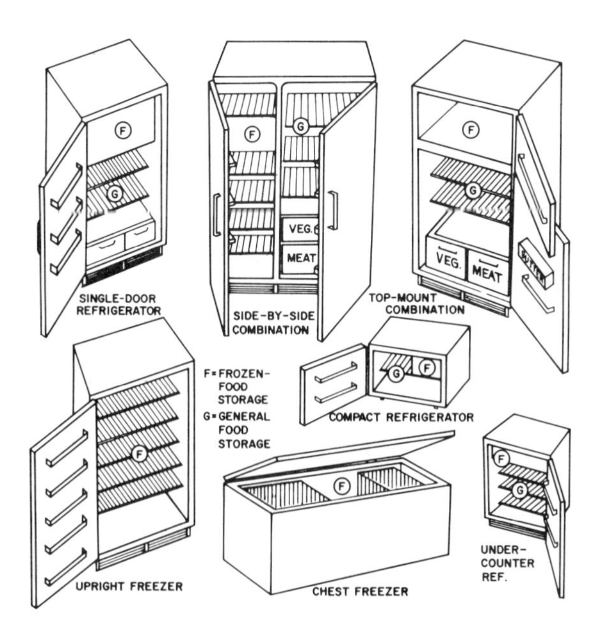


**Fig. 1 Configurations of Contemporary Household Refrigerators and Freezers**

off to 16% of this quantity and the *bottom-mounted freezer* type gaining a share in the market place. The *no-frost* system accounted for more than 84% of those sold.

## PRIMARY FUNCTIONS

Food storage space at reduced temperature is the primary function of a refrigerator or freezer, with *ice making* an essential secondary function. For the preservation of fresh food, a general storage temperature between 32 and 39°F (0 and 4°C) is desirable. Higher or lower temperatures or a humid atmosphere are more suitable for storing certain foods. A discussion of special-purpose storage compartments to provide these conditions is in the "Cabinet" section of this chapter. Food freezers and combination refrigerator-freezers that are used for long-term storage are designed to hold temperatures near 0°F (−18°C) and always below 8°F (−13°C). In single-door refrigerators, the frozen-food space is usually warmer and is not intended for long-term storage. Optimum conditions for food preservation are addressed in more detail in Chapters 9 through 27 of the 1986 REFRIGERATION Volume.

## PERFORMANCE CHARACTERISTICS

A refrigerator or freezer must maintain desired temperatures and have reserve capacity to cool to these temperatures when started on a hot summer day. Most models cool down within a reasonable time in a 110°F (43°C) ambient at rated voltage.

Overall system efficiency has become more important, as rising energy costs have driven cost of operation upward. The challenge for the designer to control noise and vibration has been made more complex with the need for fans for forced-air circulation and new compressors with higher efficiencies and capacities. The need for increased storage volumes and better insulation efficiency has resulted in almost universal use of foam insulation, which is less acoustically absorbent than glass fiber. Vibrations from running or stopping the compressor must be isolated to prevent mechanical transmission to the cabinet or to the floor and walls where it may cause additional vibration and noise.

## SAFETY REQUIREMENTS

American manufacturers comply with Underwriters' Laboratories *Standard* UL-250, *Standards for Safety, Household Refrigerators and Freezers,* which protects the user from electrical shock, fire dangers, and other hazards under normal and some abnormal conditions. Specific areas of product safety addressed include motors, hazardous moving parts, grounding and bonding, stability (cabinet tipping), door-opening force, door-hinge strength, shelf strength, component restraint (shelves and pans), glass strength, cabinet and unit leakage current, leakage current from surfaces wetted by normal cleaning, high voltage breakdown, ground continuity, testing and inspection of polymeric parts, and uninsulated live electrical parts accessible with an articulated probe.

## DURABILITY AND SERVICE

Studies show that the average refrigerator will belong to the original owner for 10 to 15 years and will continue to give satisfactory service for a considerably longer period. There is a wide variation in life span, however; some refrigerators with hermetic refrigeration systems run for over 30 years. Their reliability, particularly of the older and simple refrigerators, has led people to expect a 15- to 20-year life of the hermetic unit, and, in many respects, the appliance must be designed to protect itself over this period. Motor overload protectors are normally incorporated, and an attempt is made to design *fail-safe* circuits so that the hermetic motor of the compressor will not be damaged by a failure of a minor external component, unusual voltage extremes, or voltage interruptions.

Customer-operated devices must withstand frequent use. For example, a refrigerator door may be opened and shut over 300,000 times in its lifetime. To protect the customer against the cost of premature failure, most manufacturers will replace faulty parts within one year and repair or replace a faulty hermetic system within five years at no charge for materials. Beyond these warranties, the terms vary among manufacturers; they are sometimes combined with an extended service contract (a form of insurance paid for by the customer).

In the design of refrigerators and freezers, provisions must be made for economical and effective servicing if damage or malfunction occurs in the field.

## CABINETS

A good cabinet design achieves the optimum blend of the following:

1. maximum food-storage volume for the floor area occupied by the cabinet
2. the best in utility, performance, convenience, and reliability
3. minimum heat gain
4. minimum cost to the consumer

### Use of Space

The fundamental factors in cabinet design are usable food-storage capacity and its external dimensions. Food-storage volume has been considerably increased without a corresponding increase in the external dimensions of the cabinet. This has been accomplished mainly by using thinner but more effective insulation and by reducing the space occupied by the compressor and condensing unit. The method of computing storage volume and shelf area is described in ANSI/AHAM *Standard* HRF-1-1986.

### Frozen-Food Storage

The increased use of frozen foods requires refrigerators with much larger frozen-food storage compartments. In single-door models, the frozen-food storage volume is provided by a freezing compartment across the top of the general food storage compartment. An insulated baffle beneath the evaporator allows it to operate at the low temperatures required for short-term frozen-food storage, while maintaining temperatures above freezing in the general food storage compartment. Sufficient air passes around the baffle to cool the general storage compartment.

In larger refrigerators, the frozen-food space often represents a large part of the total volume, and it usually has a separate exterior door or drawer and a lower temperature capability; in this case, the model is classified as a combination refrigerator-freezer. The frozen-food compartment in these combinations is most often positioned across the top; but on the largest models, there has been a trend toward side-by-side arrangements. A few bottom-mounted models have reappeared in the market. Two-door cabinets are sometimes built with two separate inner liners, housed in a single outer shell. Others use only a single liner divided into two sections by an insulated partition integral with the liner or installed as a separate piece.

Food freezers are offered in two forms: upright (vertical) and chest. Locks are often provided on the lids or doors as protection against accidental door opening or access by children. A power supply indicator light or a thermometer with an external dial may be provided to warn of high storage temperatures.

### Special-Purpose Compartments

Special-purpose compartments provide a more suitable environment for storage of specific foods. For example, a warmer compartment for maintaining butter at a suitable spreading temperature is often found in the door. Some refrigerators have a meat storage compartment that can maintain storage temperatures just above freezing and may include an independent temperature adjustment feature. High-humidity compartments for storage of leafy vegetables and fresh fruit are found in practically all refrigerators. These generally tight-fitting drawers, located within the food compartment, protect vulnerable foods from the drying effects of circulating dry air in the general storage compartment. The desired conditions are maintained in the special storage compartments and drawers by (1) enclosing them to prevent air exchange with the general storage area and (2) surrounding them with cold air to maintain the desired temperatures.

### Ice and Water Service

Through a variety of manual and automatic means, refrigerators provide ice. Ice trays are placed into the freezing compartment in a stream of air that is substantially below 32°F (0°C) or placed in contact with a directly refrigerated evaporator surface for manual operation.

Automatic ice-making equipment in household refrigerators has increased. Almost all automatic defrost refrigerators include factory-installed automatic ice makers or can accept a field installable ice maker.

The ice-maker mechanism is located in the freezer section of the refrigerator and requires an attachment to a water line. The ice-freezing rate is primarily a function of the system design. Most ice makers are in no-frost refrigerators, and the water is frozen by passing refrigerated air over the ice mold. Because the ice maker has to share the available refrigeration capacity with the freezer and food compartments, the ice production rate is usually limited by design to 4 to 6 lb (2 to 3 kg) per 24 hours. An ice production rate of about 4 lb (2 kg) per 24 hours, coupled with an ice storage container capacity of 7 to 10 lb (3 to 5 kg) is adequate for most users.

When designing an ice maker, the various methods of automatically accomplishing the basic functions must be considered to determine if they meet the design objectives. These five basic functions are as follows:

1. **Initiating** the ejection of the ice as soon as the water is frozen is necessary to obtain a satisfactory production rate. Ejection before complete freezing causes wet cubes to freeze together in the storage container or may cause the ice mold to overfill. One method is to initiate ejection in response to the temperature of a selected location in the mold when complete freezing is indicated. Another successful method is to initiate ejection based on the time required to freeze the water under normal freezer temperatures. In either method, the temperature or time required may vary in different applications, depending on the cooling air temperature, as well as the rate and direction of the airflow.
2. **Ejecting** the ice from the mold must be a reliable operation. In several designs, this is accomplished by freeing the ice from the mold with an electric heater and pushing it from the tray into an ice storage container. In other designs, water is frozen in a plastic tray by passing refrigerated air over the top so that the water freezes from the top down. The natural expansion that takes place during this freezing process causes the ice to partially *freeze free* from the tray. By twisting and rotating the tray, the ice can be completely freed and ejected into a container.
3. **Driving** the ice maker in most designs is done by a gear motor, which operates the ice ejection mechanism and may also be used to time the freezing cycle and the water-filling cycle, and to operate the stopping means.
4. **Filling** the ice mold with a constant volume of water, regardless of the variation in line water pressure, is necessary to ensure uniform-sized ice cubes and prevent overfilling. This is done by timing a solenoid flow-control valve or by using a solenoid-operated, fixed-volume slug valve.
5. **Stopping** is another necessary function when the ice storage container is full until some ice is used. This is accomplished by using a *feeler* type ice level control or a weight control.

Ice service has become more convenient in some models by dispensing ice through the freezer door. In one case, the ice is ejected into a storage container, which is accessible from the outside of the freezer door as a tilt-out compartment. In another design, ice is delivered through a trap door in the freezer door by an auger mechanism operating in the ice storage container. The auger motor is energized when a push-button switch is contacted by the action of placing a glass under the trap door. An additional selector switch is available to actuate an ice crusher as the cubes pass through the door on some designs. Chilled water and/or juice dispensing are provided on still other designs.

### Thermal Considerations

The total heat load imposed on the refrigerating system comes from heat sources that are external and internal to the cabinet. The relative values of the basic or predictable portions of the heat load (which are independent of usage) are shown in Figure 2. A large portion of the peak heat load may result from door openings, food loading, and ice making, which are variable and unpredictable quantities dependent on customer use. As the beginning point for the thermal design of the cabinet, the significant portions of the heat load are normally calculated and then confirmed by test.

The major predictable heat load is the heat passing through the cabinet walls. Table 1 shows the insulating value of fibrous and foam insulations commonly used to insulate the cabinet; for more detailed information, see Chapter 20 of the 1985 FUNDAMENTALS Volume. Note that foamed insulation permits a significant reduction in wall thickness for the same heat leakage.

*External sweating* can be avoided by keeping exterior surfaces warmer than the dew point of the environment. Condensation occurs most likely around the hardware, on door mullions, along the edge of door openings, and on any cold refrigerant tubing that may be exposed outside the cabinet. In a 90°F (32°C) room, no external surface temperature on the cabinet should be more than 5 or 6°F (3.3°C) below the room temperature. If it is necessary to raise the exterior surface temperature to avoid sweating, this can be done by either routing a loop of the condenser tubing under the front flange of the cabinet outer shell or by locating low-wattage wires or ribbon heaters behind the critical surfaces. Some refrigerators have power-saving electrical switches that allow the user to de-energize these electrical heaters when the environmental conditions do not require their use.

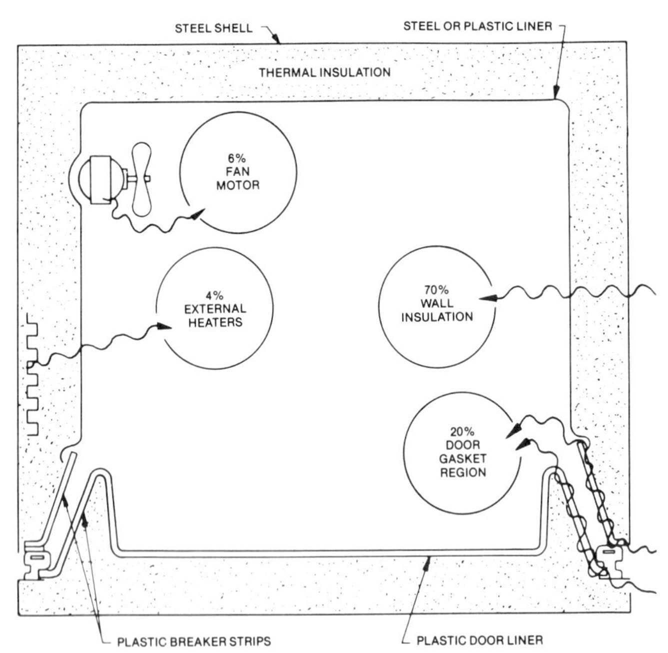


**Fig. 2 Cabinet Cross Section, Showing Typical Contributions to Total Heat Load**

Temporary condensation on internal surfaces may occur with frequent door openings, so the interior of the general storage compartment must be designed to avoid objectionable accumulation or drippage.

Figure 2 shows the design features of the throat section where the door meets the face of the cabinet. On products with metal liners, metal-to-metal contact between inner and outer panels is avoided through the use of thermal breaker strips. Since the air gap between the breaker strip and the door panel provides a low-resistance heat path to the door gasket, the clearance should be kept as small as possible and the breaker strip as wide as practical. If the inner liner is made of plastic rather than steel, there is no need for separate plastic breaker strips because these would be formed as an integral part of the liner.

*Structural supports,* used to support and position the food compartment liner from the outer shell of the cabinet, are usually constructed of a combination of steel and plastics to provide adequate strength with maximum thermal insulation.

*Internal heat loads,* which must be overcome by the system's refrigerating capacity, are generated by fan motors used for air circulation and by heaters used to prevent undesirable internal cabinet sweating or frost buildup or to modify storage temperatures, where required.

## Structure and Materials

The external shell of the cabinet is usually a single fabricated steel structure, which supports the inner food compartment liner, the door, and the refrigeration system. The space between the inner and outer walls of the cabinet may be filled with fibrous insulation, foamed slabs, or foamed-in-place insulation. In general, the door and breaker strip construction is similar to that shown in Figure 2, although breaker strips and food liners formed of a single plastic sheet are also common. The doors cover the whole front of the cabinet, and plastic sheets become the inner surface for the doors, so no separate door breaker strips are required. The door liners are usually formed to provide an array of small door shelves and racks. Cracks and crevices are avoided, and edges are rounded and smooth to facilitate cleaning. Interior lighting uses incandescent lamps controlled by mechanically operated switches, actuated by the opening of the refrigerator door(s) or chest freezer lid. Table 2 summarizes the most common materials and manufacturing methods used in the construction of refrigerator and freezer cabinets.

The cabinet design must provide for the special requirements of its refrigerating system. For example, it may be desirable to refrigerate the freezer sections by attaching evaporator tubing directly to the food compartment liner. Also, it may be desirable, particularly with food freezers, to attach the condenser tubing directly to the shell of the cabinet to prevent external sweating. Both designs influence the cabinet heat leakage and the amount of insulation required.

The method of installing the refrigerating system into the cabinet is also important. Frequently, the system is installed in two or more component pieces and then assembled and processed in the cabinet. Unitary installation of a completed system directly into the cabinet allows the system to be tested and charged beforehand. The cabinet design must be compatible with the method of installation chosen. In addition, systems using forced air frequently require ductwork in the cabinet or insulation spaces.

Mechanically, the overall structure of the cabinet must be strong enough to withstand shipping, in which case it will be strong enough to withstand daily usage. Porcelain-enameled inner food liners must be designed to prevent porcelain chipping or crazing at the points of support. Plastic food liners must withstand the thermal stresses they are exposed to during shipment and usage, and they must be unaffected by the common contaminants they might encounter in a kitchen environment. Shelves must be designed not to deflect excessively under the heaviest anticipated load. Refrigerator doors and associated hardware must withstand a total of about 300,000 door openings.

Expanded-in-place foam insulation has had an important influence on cabinet design and assembly procedures. Not only is the wall thickness reduced, but the rigidity and bonding ac-

**Table 1 Effect of Thermal Insulation on Cabinet Wall Thickness**

| | | Wall Thickness, inches (mm) | | | |
|---|---|---|---|---|---|
| | | For Threshold of External Sweating in 90°F (32°C) at 75% r.h. | | Common Practice | |
| Insulation | Thermal Conductivity, Btu • in./ft² • h • °F (W • mm/m² • K) | 0°F (−18°C) | 38°F (3°C) | 0°F (−18°C) | 38°F (3°C) |
| Mineral or Glass Fiber, Air Filled | 0.22 to 0.28 (32 to 40) | 2.0 to 2.7 (50 to 70) | 1.3 to 1.75 (33 to 45) | 3.0 to 3.5 (75 to 90) | 2.3 to 2.75 (60 to 70) |
| Foamed-in-place Urethane Foam, Heavy Gas Filled | 0.13 (19) | 1.25 (32) | 0.85 (22) | 1.9 (48) | 1.6 (40) |
| Foamed Slab Urethane Foam, Heavy Gas Filled | 0.16 (23) | 1.5 (38) | 1.0 (25) | 2.0 (50) | 1.5 (38) |

Table 2 Cabinet Materials and Manufacturing Methods

| Component | Material | Common Thickness, in. (mm) | Manufacturing Method | Finish |
|---|---|---|---|---|
| Outer cabinet assembly | | | Welded assembly | |
| Wrapper and top | Low carbon cold-rolled steel | 0.030-0.036 (0.76-0.91) | Roll form and bend | Organic finish |
| Back | Low carbon cold-rolled steel | 0.024-0.033 (0.61-0.84) | Draw and stamp | (polyester, |
| Bottom | Low carbon cold-rolled steel | 0.016-0.025 (0.41-0.64) | Draw and stamp | alkyd, acrylic, etc.) |
| Inner cabinet liner | Enameling iron | 0.024-0.036 (0.61-0.91) | Bend and weld | Vitreous enamel |
| | or | | | |
| | Low carbon cold-rolled steel | 0.024-0.036 (0.61-0.91) | Bend and weld | Organic finish |
| | or | | | |
| | Aluminum | 0.024-0.036 (0.61-0.91) | Bend and weld | Anodized or organic |
| | or | | | |
| | Plastic | 0.050-0.200 (1.3-5.1) | Injection molded or vacuum formed | None |
| Inner door liner | Plastic | 0.075-0.095 (1.9-2.4) | Vacuum formed or injection molded | None |
| Breaker strips | Plastic | 0.075-0.095 (1.9-2.4) | Extruded or injection molded | None |

tion of the foam usually eliminates the need for structural supports. The foam is normally expanded directly into the insulation space, adhering to the food compartment liner and the outer shell. Unfortunately, this does not permit disassembly of the cabinet for service or repairs. Alternative constructions that overcome this objection use prefoamed slabs or expand the foam against an inner mold, which is later withdrawn and replaced with the food compartment liner. In either case, the liner is not bonded to the foam, and it can be easily removed if required. However, the foam no longer provides the structural tie between the liner and the outer shell.

The steel outer shell of refrigerators and freezers is usually finished with a synthetic baked enamel, applied over a chemically treated surface, which assures good paint adhesion and prevents the spread of rust if scratches should extend through the finish. Steel food compartment liners are commonly finished in acid-resistant porcelain enamel, which usually consists of one ground coat and one or more finish coats, each separately fired. Organic finishes are available for use on steel and aluminum liners, and prefinished stock has been used where edges are lock seamed.

**Use of Plastics.** As much as 15 or 20 lb (7 or 9 kg) of plastic is incorporated in a typical refrigerator or freezer, and the usage is increasing, largely because of the following:

1. Wide range of physical properties
2. Good bearing qualities
3. Electrical insulating ability
4. Moisture and chemical resistance
5. Low thermal conductivity
6. Ease of cleaning
7. Pleasing appearance with or without an applied finish
8. Potential of multifunctional design in a single part
9. Transparency, opacity, and colorability
10. Ease of forming and molding
11. Lower cost

A few examples illustrate the versatility of plastics. High impact polystyrene and ABS plastics are used for inner door liners and food compartment liners (see Table 2). In these applications, no applied finish is necessary. These and similar thermoplastics such as polypropylene and polyethylene are also selected for evaporator doors, baffles, breaker strips, drawers, pans, and many small items. The phenolics are used for decorative door panels, terminal boards and terminal covers, and as a binder for the glass fiber insulation. The good bearing qualities of nylon and acetal are used advantageously in such applications as hinges, latches, and rollers for sliding shelves. Gaskets, both for the refrigerator and the evaporator doors, are generally made of vinyl or rubber.

Many items (such as ice cubes and butter) readily absorb odors and tastes from materials to which they are exposed. Manufacturers, accordingly, take particular care to avoid using any plastics or other materials in the interior of the cabinet that will impart an odor or taste.

## Moisture Sealing

To retain the original insulating qualities of the cabinet, the insulation must be kept dry. Moisture may get into the insulation through leakage of water from the food compartment liner, the defrost water disposal system, or, most commonly, through vapor leaks in the outer shell.

Outer shell construction is generally seam or spot welded and carefully sealed against vapor transmission by using mastics and hot melt asphaltic or wax compounds at all joints and seams. In addition, door gaskets, breaker strips, and other parts should provide maximum barriers for vapor flow from the room air to the insulation. When refrigerant evaporator tubing is attached directly to the food compartment liner, as is generally done in chest freezers, moisture will not migrate from the insulation space, and special efforts must be made to vapor-seal this space.

Although urethane foam insulation tends to inhibit moisture migration, it does have a tendency to trap water when migrating vapor reaches a temperature below its dew point. The foam then becomes permanently wet, and its insulation value is decreased. For this reason, a vaportight exterior cabinet is equally important with foam insulation.

## Door Latching

Latching of doors is accomplished by mechanical or magnetic latches that act to compress relatively soft compression gaskets made of extruded rubber or vinyl compounds. Gaskets with magnetic materials embedded in the gasket are generally used.

Chest freezers are sometimes designed so that the weight of the lid acts to compress the gasket, although most of the weight is counterbalanced by springs in the hinge mechanism.

In 1956, the Refrigerator Safety Act, Public Law 84-930, was enacted prohibiting shipment in interstate commerce of any household refrigerator that was not equipped with a device or system permitting the door to be opened from the inside. Although freezers are not included under P.L. 930, Underwriters' Laboratories *Standard* UL-250 has adopted the requirements of P.L. 930 to prevent entrapment in freezers.

In addition, since freezers are sometimes located in areas that have public access, they are often equipped with key locks to prevent pilferage. UL requires these key locks to be the non-self-engaging type, and the key must be self-ejected from the slot when not in place. This type of locking mechanism prevents entrapment caused by the accidental locking of the door.

### Cabinet Testing

Specific tests necessary to establish the adequacy of the cabinet as a separate entity include (1) structural tests, such as repeated twisting of the cabinet and door; (2) door slamming test; (3) tests for vapor-sealing of the cabinet insulation space; (4) odor and taste transfer tests; (5) physical and chemical tests of plastic materials; and (6) heat leakage tests. Cabinet testing is also discussed later in this chapter.

## REFRIGERATING SYSTEMS

The vapor compression refrigerating systems used with modern refrigerators vary considerably in capacity and complexity, depending on the cabinet application. They are hermetically sealed and normally require no replenishment of refrigerant or oil during the useful life of the appliance. The components of the system must provide optimum overall performance and reliability at minimum cost. In addition, all safety requirements of the UL *Standard* 250 must be met. Fluorinated hydrocarbon refrigerants such as R-12 are nonflammable, nontoxic, odorless, and nonirritating, and are universally used in household refrigeration appliances. However, there is concern about the eventual environmental impact of fluorinated hydrocarbons on the atmosphere.

The design of refrigerating systems for refrigerators and freezers has improved because of new refrigerants, wider use of aluminum, smaller and more efficient motors and compressors, universal use of capillary tubes, and simplified electrical components. These refinements have kept the vapor compression system in the best competitive position for household application.

### The Refrigerating Circuit

Figure 3 shows the refrigerant circuit for a vapor compression refrigerating system. The refrigeration cycle is as follows: (1) electrical energy supplied to the motor drives a positive displacement compressor, which draws cold, low-pressure refrigerant vapor from the evaporator and compresses it; (2) the resulting high-pressure, high-temperature discharge gas then passes through the condenser, where it is condensed to a liquid, when the heat rejected to the ambient air; (3) the liquid refrigerant passes through a metering capillary tube to the evaporator at a reduced pressure; and (4) the low-pressure, low-temperature liquid in the evaporator absorbs heat from its surroundings, evaporating to a gas, which is again withdrawn by the compressor.

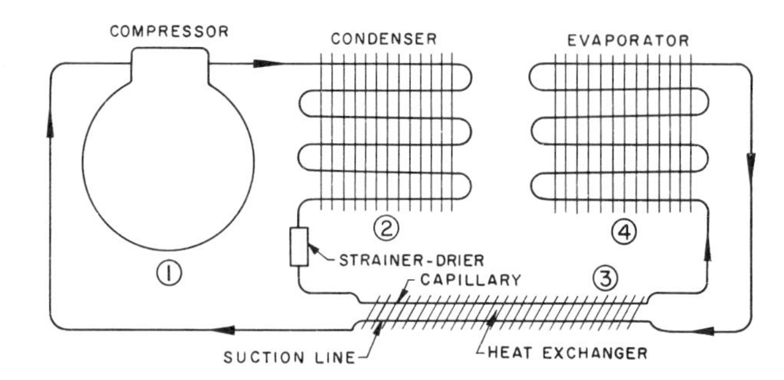


**Fig. 3 The Refrigeration Circuit**

Note that energy enters the system through the evaporator (heat load) and through the compressor (electrical input). Thermal energy is rejected to the ambient by the condenser and the compressor shell. A portion of the capillary tube is usually soldered to the suction line for heat exchange. By using the effect of the cool suction gas, system capacity is further increased.

A *strainer-drier* is usually placed ahead of the capillary tube to remove foreign material and moisture. Refrigerant charges of 1 lb (0.45 kg) or less of R-12 are common. A thermostat (or cold-control) cycles the compressor to provide the desired temperatures within the refrigerator. During the off cycle, the capillary tube permits the pressures to equalize throughout the system.

Materials used in refrigeration circuits are selected for (1) their mechanical properties, (2) their compatibility with the refrigerant and oil on the inside, and (3) their resistance to oxidation and galvanic corrosion on the outside. Evaporators are usually made of aluminum tubing, either with integral extruded fins or with extended surfaces mechanically attached to the tubing. Condensers are usually made of steel tubing with an extended surface of steel sheet or wire. Steel tubing is used on the high-pressure side of the system, which is normally dry, and copper is used for suction tubing, where condensation can occur. Because of its ductility, corrosion resistance, and ease of brazing, copper is used for capillary tubes and often for small connecting tubing. Wherever aluminum tubing comes in contact with copper or iron, it must be protected against moisture to avoid electrolytic corrosion.

### Defrosting

A small portion of the industry is still using, in some models, simple manual defrost in which the cooling effect is generated by gravity circulation of air over a refrigerated surface (evaporator) located at the top of the food compartment. The refrigerated surface forms some of the walls of a frozen food space, which usually extends across the width of the food compartment. Defrosting is typically accomplished by manually turning off the temperature-control switch.

**Cycle Defrosting (Partial Automatic Defrost).** Combination refrigerator-freezers sometimes use two separate evaporators for the fresh food and freezer compartments. The fresh-food compartment evaporator defrosts during each off cycle of the compressor, with the energy for defrosting provided mainly by the heat leakage into the fresh-food compartment. The cold control senses the temperature of the fresh-food compartment evaporator and cycles the compressor *on* when the evaporator surface is above 32°F (0°C). The freezer evaporator requires infrequent manual defrosting.

**No-Frost Systems (Automatic Defrost).** Most combination refrigerator-freezers and some upright food freezers are often refrigerated by air that is fan-blown over a single evaporator, concealed from view. Because the evaporator is colder than the freezer compartment, it collects practically all of the frost, and there is little or no permanent frost accumulation on the frozen food or on exposed portions of the freezer compartment. The evaporator is defrosted automatically by electric heat or hot refrigerant gas, and the defrosting period is of short duration to limit food temperature rise. The resulting water is disposed of automatically by draining to the exterior, where it is evaporated in a pan located in the warm condenser airstream or on warm condenser coils. Defrosting is usually initiated by a timer at intervals of up to 24 hours. If the timer operates only when the compressor runs, the accumulated time tends to reflect the probable frost load.

Developments in electronics have allowed introduction of micropressor-based control systems to some household refrigerators. An adaptive defrost function is usually included in the software. Various parameters are monitored so that the period between defrosts varies according to actual conditions of use. Adaptive defrost tends to reduce energy consumption and improve food preservation.

**Forced Heat for Defrosting.** All no-frost systems add heat to the evaporator to accelerate melting during the short defrosting cycle. The most common method uses a 300 to 1000 W electric heater. The typical defrost cycle is initiated by a timer, which stops the compressor and energizes the heater.

When the evaporator has melted all of the frost, a defrost termination thermostat opens the heater circuit. In most cases, the compressor is not restarted until the evaporator has drained for a few minutes and the system pressures have subsided, thus reducing the applied load for restarting the compressor. Commonly used defrost heaters include *metal-sheathed heating elements* in thermal contact with evaporator fins and *radiative heating elements* positioned to radiate heat to the evaporator.

## The Evaporator

The *manual defrost* evaporator usually is a box with three or four sides refrigerated. Refrigerant may be carried in tubing brazed to the walls of the box, or the walls may be constructed from double sheets of metal, which are brazed or metallurgically bonded together with integral passages for the refrigerant. These constructions usually use aluminum, and special attention is required to avoid (1) contamination of the surface with other metals that would promote galvanic corrosion and (2) configurations that may be readily punctured in use.

The *cycle defrost* evaporator for the fresh-food compartment is designed for natural defrost operation and is characterized by its low thermal capacity. It may be a lightweight vertical plate, usually made from bonded sheet with integral refrigerant passages, or a serpentine coil with or without fins. In either case, it should be located near the top of the compartment and arranged for good water drainage during the defrost cycle. In some designs, this cooling surface has been located in an air duct remote from the fresh-food space, with air circulated continuously by a small fan.

The *no-frost forced-convection* evaporator is usually a forced-air fin-and-tube arrangement designed to minimize the effect of frost accumulation, which tends to be relatively rapid in a single evaporator system. The coil is usually arranged for airflow parallel to the long dimension of the fins.

The fins may be more widely spaced at the air inlet to provide for preferential frost collection and to minimize the air restriction effects of the frost. All surfaces must be heated adequately during the defrost cycle to ensure complete defrosting, and provision must be made for draining and evaporating the defrost water outside the food-storage spaces.

Evaporators for *chest freezers* usually consist of tubing that is in good thermal contact with the exterior of the food compartment liner. The tubing is usually concentrated near the top of the liner with wider spacing near the bottom to take advantage of gravity circulation of the air inside. *Upright food freezers* usually have refrigerated shelves and a refrigerated surface at the top of the food compartment. These are commonly connected in series with an accumulator at the exit end. No-frost freezers usually incorporate a fin-and-tube evaporator and an air circulating fan as used in the no-frost combination refrigerator-freezers.

## The Condenser

The condenser is the main heat-rejecting component in the refrigerating system. It may be cooled by natural draft on free-standing refrigerators and freezers or fan-cooled on larger models and on models designed for built-in applications.

The **natural draft condenser** is located on the back wall of the cabinet and is cooled by natural air convection under the cabinet and up the back. The most common form of natural draft condenser consists of a flat serpentine of steel tubing with steel cross wires welded on 0.25-in. (6-mm) centers on one or both sides perpendicular to the tubing. Tube-on-sheet construction may also be used.

The **hot wall condenser**, another natural draft arrangement, is used principally with food freezers. It consists of condenser tubing attached to the inside surface of the cabinet shell. The shell thus acts as an extended surface for heat dissipation. With this construction, external sweating is seldom a problem.

The **forced-draft condenser** may be of fin-and-tube folded banks of tube-and-wire or tube-and-sheet construction. Various forms of condenser construction are used to minimize clogging caused by household dust and lint. The compact, fan-cooled condensers are usually designed for low airflow rates because of noise limitations. Air ducting is often arranged to use the front of the machine compartment for the entrance and exit of air. This makes the cooling air system largely independent of the location of the refrigerator and permits built-in applications.

A portion of the condenser may be located under the defrost water evaporating pan to promote water evaporation. The condenser may also incorporate a section for *compressor cooling* from which the partially condensed refrigerant is routed to an oil-cooling loop in the compressor, where the liquid refrigerant, still at high pressure, absorbs heat and is reevaporated. The vapor is then routed through the balance of the condenser, to be condensed in the normal manner. In some designs, as noted previously, a portion of the condenser tubing is routed internally in contact with the outer case in place of anti-sweat heaters.

Condenser performance may be evaluated directly on calorimeter test equipment similar to that used for compressors. However, the final design of the condenser must be determined by performance tests on the refrigerator under a variety of operating conditions.

Generally, the most important design requirements for a condenser include (1) sufficient heat dissipation at peak-load conditions, (2) storage volume that adequately prevents excessive pressures during pulldown or in the event of a restricted or plugged capillary tube, (3) good refrigerant drainage to minimize the off cycle losses and the time required for equalization of system pressures, (4) external surface easily cleaned or designed to avoid dust and lint accumulation, and (5) adequate safety factor against bursting.

## The Capillary Tube

The refrigerant metering device in general usage is the capillary tube, a small-bore tube connecting the outlet of the condenser to the inlet of the evaporator. The *regulating effect* of this simple control device is based on the principle that a given weight of liquid passes through a capillary more readily than the same weight of gas at the same pressure. Hence, if uncondensed refrigerant vapor enters the capillary, the mass flow will be reduced, giving the refrigerant more cooling time in the condenser. On the other hand, if liquid refrigerant tends to back up in the condenser, the condensing temperature and pressure rises, resulting in an increased mass flow of refrigerant. Under extreme conditions, the capillary either passes considerable uncondensed gas or backs liquid refrigerant well up into the condenser. Under normal operating conditions, a capillary tube gives good performance and efficiency. Figure 4 shows the typical effect of capillary refrigerant flow rate on system performance.

A capillary tube has the advantage of extreme simplicity and no moving parts. It also lends itself well to being soldered to the suction line for heat-exchange purposes. This position prevents sweating of the otherwise cold suction line, and, at the same time, increases the refrigerating capacity. Another advantage is that the pressure equalizes throughout the system during the off cycle and reduces the starting torque required of the compressor motor. The capillary is the narrowest passage in the refrigerant system and the place where low temperature first occurs. For that reason, a combination strainer-drier is usually located directly ahead of the capillary to prevent it from being plugged by ice or any foreign material circulating through the system.

**Selection.** The optimum metering action can be obtained by variations in either the diameter or the length of the tube. Such factors as the physical location of the system components and the heat exchanger length (36 in. [0.9 m] or more is desirable) may be determining factors in the selection of the length and bore of the capillary tube for any given application.

Capillary selection is covered in detail in Chapter 19. Once a preliminary selection is made, an experimental unit can be equipped with three or more different capillaries that can be activated independently. System performance can then be evaluated by using each capillary with slightly different flow characteristics in turn.

Final selection of the capillary requires an optimization of performance under both no load and pulldown conditions, with maximum and minimum ambient and load conditions. The optimum refrigerant charge can also be determined during this process.

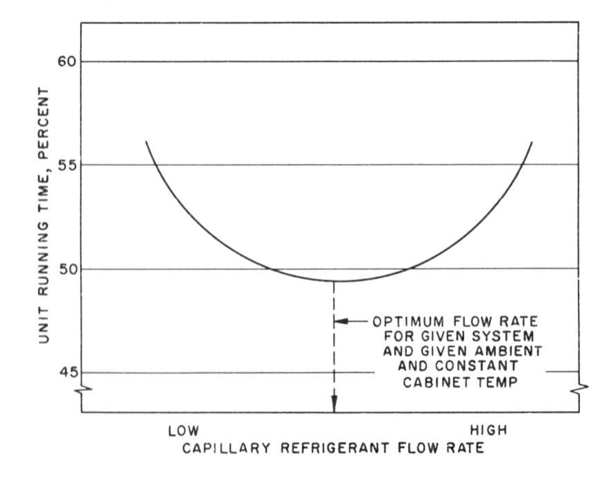


**Fig. 4 Typical Effect of Capillary Tube Selection on Unit Running Time**

## The Compressor

While a more detailed description of compressors can be found in Chapter 12, a brief discussion of the small compressors used in household refrigerators and freezers is included here.

These products use positive displacement compressors in which the entire motor-compressor is hermetically sealed in a welded steel shell. Capacities range from about 300 Btu/h (90 W) to about 2000 Btu/h (600 W) when measured at the usual rating conditions of −10°F (−23°C) evaporator, 130°F (54°C) condenser, 90°F (32°C) ambient, with the suction gas superheated to 90°F (32°C) and the liquid subcooled to 90°F (32°C).

Design emphasis is placed on ease of manufacturing, reliability, low cost, quiet operation, and efficiency. Figure 5 illustrates

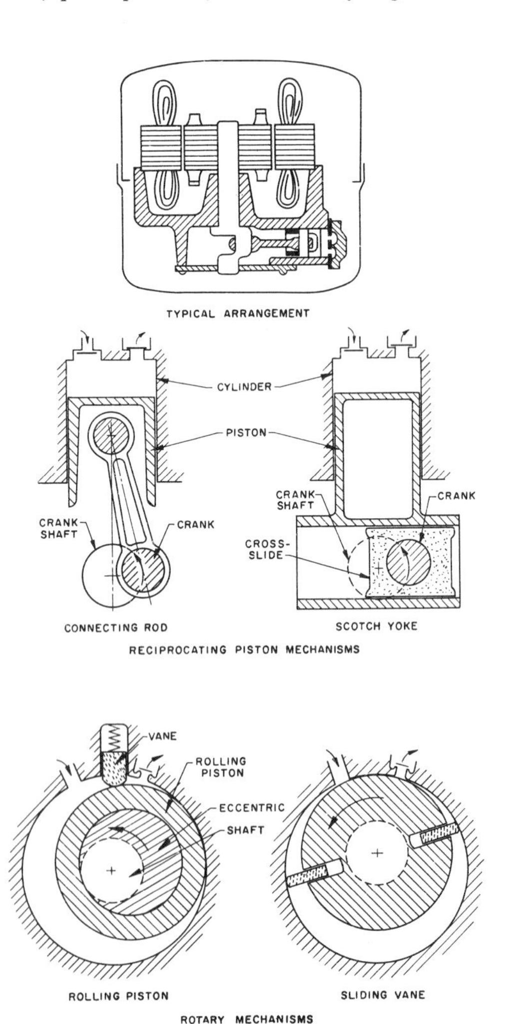


**Fig. 5 Refrigerator Compressor Types**

the two types of rotary compressors and two reciprocating piston compressor mechanisms, which are used in virtually all conventional refrigerators and freezers, with no one type being much less costly than the others. While rotary compressors are somewhat more compact than reciprocating compressors, a greater number of close tolerances are involved in their manufacture.

These compressors are directly driven by squirrel cage induction motors, of the two-pole, 3450 rpm type, although some four-pole, 1750 rpm motors are also used. Field windings are insulated with special wire enamels and plastic slot and wedge insulation; all are chosen for their compatibility with the refrigerant and oil. During continuous runs at rated voltage, motor winding temperatures may be as high as 250 °F (120 °C) for 110 °F (43 °C) ambient temperature. In addition to maximum operating efficiency at normal running conditions, the motor must provide sufficient torque for starting and temporary peak loads because of start-up and pulldown of a warm refrigerator, and for the load conditions associated with the defrosting action. These conditions should be met at the anticipated extremes of line voltage.

Starting torque is provided by a split-phase winding circuit, which may include a starting capacitor in the larger motors. When the motor comes up to speed, an external electromagnetic relay or positive temperature coefficient (PTC) device disconnects the start winding. A run capacitor sometimes is employed for greater motor efficiency. Motor overload protection is provided by an automatic resetting switch, which is sensitive to a combination of motor current and compressor case temperature, or internal winding temperature.

The compressor is cooled by rejecting heat to the surroundings. This is easily accomplished with a fan-cooled system. However, an oil-cooling loop carrying partially condensed refrigerant may be necessary when the compressor is used with a natural-draft condenser and in some forced-draft systems above 1000 Btu/h (300 W).

## Temperature-Control System

The temperature-control thermostat is normally an electromechanical switch. It is actuated by a temperature-sensitive power element that has a condensable gas charge, which operates a bellow or diaphragm. At operating temperature, this charge is in a two-phase state, and the temperature at the gas-liquid interface determines the pressure on the bellows. To maintain temperature control at the bulb end of the power element, the bulb must be the coldest point at all times.

The thermostat must have an electrical switch rating for the inductive load of the compressor and other electrical components carried through the switch. The thermostat is usually equipped with a shaft and knob for adjustment of the operating temperature.

In the simple gravity-cooled system, the sensing bulb of the thermostat is normally clamped to the evaporator. The location of the bulb and the degree of thermal contact is selected to produce both a suitable cycling frequency for the compressor and the desired refrigerator temperature. Small refrigerators sold in Europe sometimes are equipped with a manually operated push button to prevent the control from coming on until defrost temperatures are reached; afterward, normal cycling is resumed.

In the combination refrigerator-freezer with the split air system, the location of the thermostat sensing bulb depends on whether an automatic damper control is used to regulate the airflow to the fresh-food compartment. When such an auxiliary control is used, the sensing bulb is usually located to sense the temperature of the air leaving the evaporator. In manual damper controlled systems, the sensing bulb is usually placed in the cold airstream to the fresh-food compartment. The sensing bulb location is frequently related to the damper effect on the airstream. Depending on the design of this relationship, the damper may become the freezer temperature adjustment, or it may serve the fresh-food compartment, with the thermostat being the adjustment for the other compartment. The temperature-sensing bulb should be located to provide a large enough temperature differential to drive the switch mechanism, while (1) avoiding excessive cycle length; (2) avoiding short cycling time, which can cause compressor starting problems; and (3) avoiding annoyance to the user from frequent noise level changes.

In some refrigerators, microprocessor-based control systems have replaced the electromechanical thermostat switch and, in some cases, both compartment controls, with thermistor-sensing devices that relay electronic signals to the microprocessor. Electronic control systems provide a higher degree of independent temperature adjustments for the two main compartments.

## System Design and Balance

A principal design consideration is the selection of components that will operate together to give the optimum system performance when the total cost is considered. Normally, a range of combinations of values for these components will meet the performance requirements, and the lowest cost is only obtained through a careful analysis or a series of tests—usually both. For instance, for a given cabinet configuration, food-storage volume, and temperatures, the following can be traded off against one another: (1) insulation thickness and overall shell dimensions, (2) insulation material, and (3) system capacity. Each of these variables affects the total cost, and most of them can be varied only in discrete steps.

The **experimental procedure** involves a series of tests. Calorimeter tests may be made on the compressor and condenser, separately or together, and on the compressor and condenser operating with the capillary tube and heat exchanger. Final selection of the components requires performance testing of the system when installed in the cabinet. These tests also determine the refrigerant charge, airflows for the forced draft condenser and evaporator, temperature control means and calibration, necessary motor protection, etc. The "Evaluation" section of this chapter covers the final evaluation tests to be made on the complete refrigerator. The interaction between components is further addressed in Chapter 18. This method assumes a knowledge (equations or graphs) of the performance characteristics of the various components, including the heat leakage of the cabinet and the heat load imposed by the customer. The analysis may be performed manually point by point, as shown below. If enough component information exists, it can be programmed into a computer simulation capable of responding to various design conditions or statistical situations. Although the available information may not always be adequate for an accurate analysis, this procedure is often useful, though it must be followed by confirming tests. *Example 1* illustrates the analytical design procedure.

---

**Example 1** (I-P units): Assume that a 16 $ft^3$, two-door, no-frost refrigerator, with 4 $ft^3$ of freezer located at the top and 12 $ft^3$ of general refrigerated space located at the bottom, is to be refrigerated by a single fin-and-tube evaporator located against the rear wall of the freezer. Assume that the small fan produces an airflow of 60 cfm through the evaporator, and a system of ducts delivers proportioned amounts of refrigerated air to the two compartments. A compressor and a fin-and-tube condenser with a fan delivering 90 cfm is located in the machine compartment at the bottom of the cabinet. Refrigerant 12 is used. The design objective for the performance of the system is to achieve average air temperatures of −3°F in the freezer and 30°F in the general refrigerated space with continuous operation of the system in a 110°F ambient room. For simplification, calculations are based on dry air.

**Solution:**

**Heat Load.** With a nominal insulated wall thickness of 1.75 in. and an assumed k-factor for the R-11 blown foam insulation of 0.135 Btu • in./ft$^2$ • h • °F, the following calculations of the heat load may be typical (U.S. Dept. of Commerce 1975).

| Source | Freezer Compartment, Btu/h | General Storage Compartment, Btu/h | Total, Both Compartments Btu/h | Total, % |
|---|---|---|---|---|
| Through walls and doors | 150 | 190 | 340 | 62 |
| Allowance for gasket and front construction | 70 | 80 | 150 | 28 |
| Allowance for mullion and surface heaters | 10 | 6 | 16 | 3 |
| Input to fan motor | 40 | — | 40 | 7 |
| Total | 270 | 276 | 546 | 100 |

**Airflow and Air Temperature.** Assume the space allowed in the freezer permits installing a fin-and-tube coil 2 in. deep, 22 in. wide, and 10 in. high. Air enters the lower edge of the coil and flows parallel to the height of the fins, so the effective face area, $A$, is 44 in. or 0.306 ft. The airflow of 60 cfm produces a face velocity of 60/0.306 = 196 fpm. To minimize frost clogging, select a fin spacing of 4 per inch. The chart in Figure 6 shows the basic coil capacity, at the calculated face velocity, as 81 Btu/ft$^2$ per °F temperature difference between the entering air and saturated refrigerant temperature. Figure 6 also shows the correction factor, $c_h$, for a 10 in. fin height to be 1.375 and the correction factor, $c_s$, for 4 fins/in. to be 1.0. The temperature change, $\Delta t$, in the air flowing through the coil may be calculated by the relationship:

$$\Delta t = q/V c_v$$

*where*

$q$ = heat load, Btu/h
$V$ = air volume, ft$^3$/h
$c_v$ = specific heat at constant volume, 0.0181 Btu/ft$^3$ • °F
$\Delta t$ = 546/(3600)(0.0181) = 8.4°F

The temperature of the air entering the coil is assumed to equal the average freezer temperature plus 0.5 $\Delta t$. Thus, the temperature of the air entering the coil will be (−3 + 8.4/2) = 1.2°F. The following relationship gives the temperature difference, $TD$, between the evaporator coil entering air and the saturated refrigerant.

(Heat Flow) = (Heat Flux per unit area per unit temp. diff. ) (Area)(Temp Diff, $TD$)

$$TD = \frac{q_d}{q_c A C_h C_s}$$

*where*

$q_d$ = total design capacity, Btu/h
$q_c$ = basic coil capacity, Btu/h
$A$ = face area, ft$^2$
$C_h$ = correction factor for fin height
$C_s$ = correction factor for fin spacing

$$TD = \frac{546}{(81)(0.305)(1.375)(1.0)} = 16°F$$

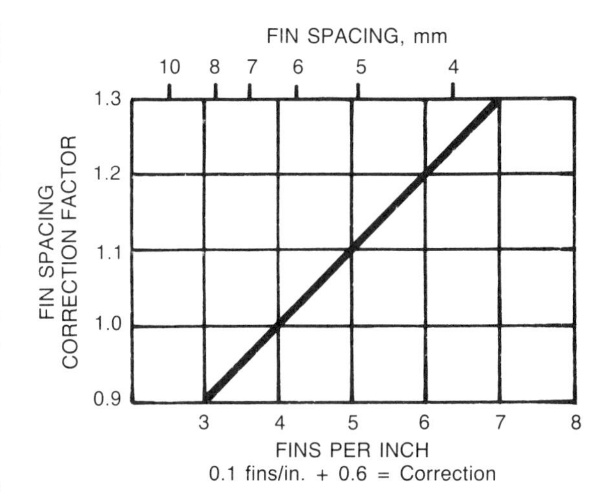


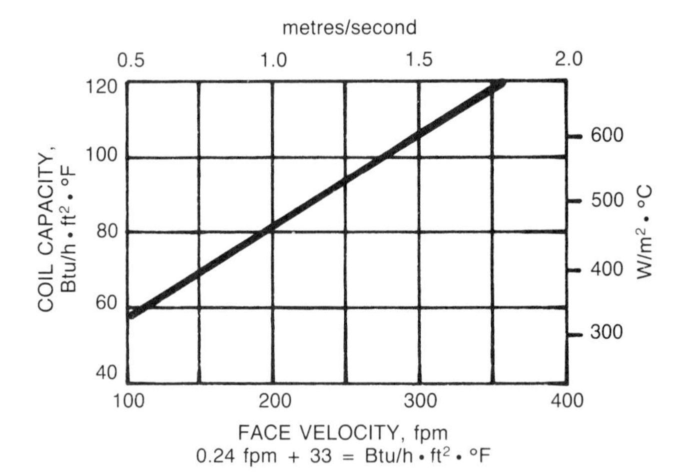


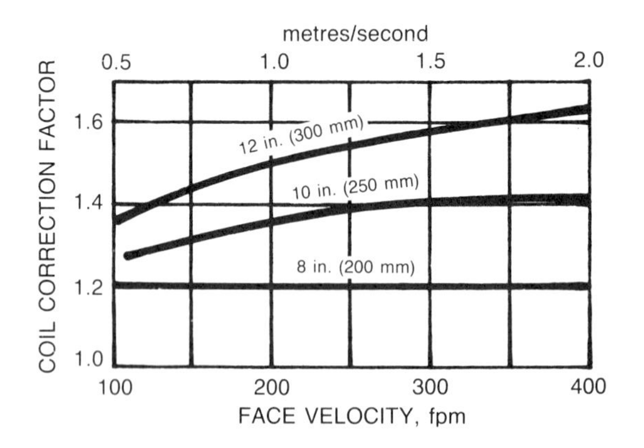


**Fig. 6 Finned Coil Capacity Curves**

**Refrigerant Temperatures and Pressures.** It follows that the saturated refrigerant temperature will be 16°F below the inlet air temperature of 1.2°F, or −14.8°F. The general refrigerated space is cooled by conduction through the insulated partition between the compartments (and through the metal walls if a single food compartment liner is used), as well as by a small percentage of the air circulated through the evaporator coil. A typical range would be 5 to 10% of the total airflow. Since this portion of the air changes in enthalpy to a much greater extent than that circulating within the freezer, the refrigerating capacity of the evaporator coil will be somewhat larger than that indicated by these simplified calculations.

**System and Compressor Capacity.** Using R-12, a pressure drop in the evaporator and suction line is assumed to be the equivalent of 3°F change in saturation temperature, which results in a compressor suction pressure equivalent to (−14.8°F − 3°F) = −17.8°F. From Figure 7, a compressor calorimeter curve based on no liquid subcooling and an assumed

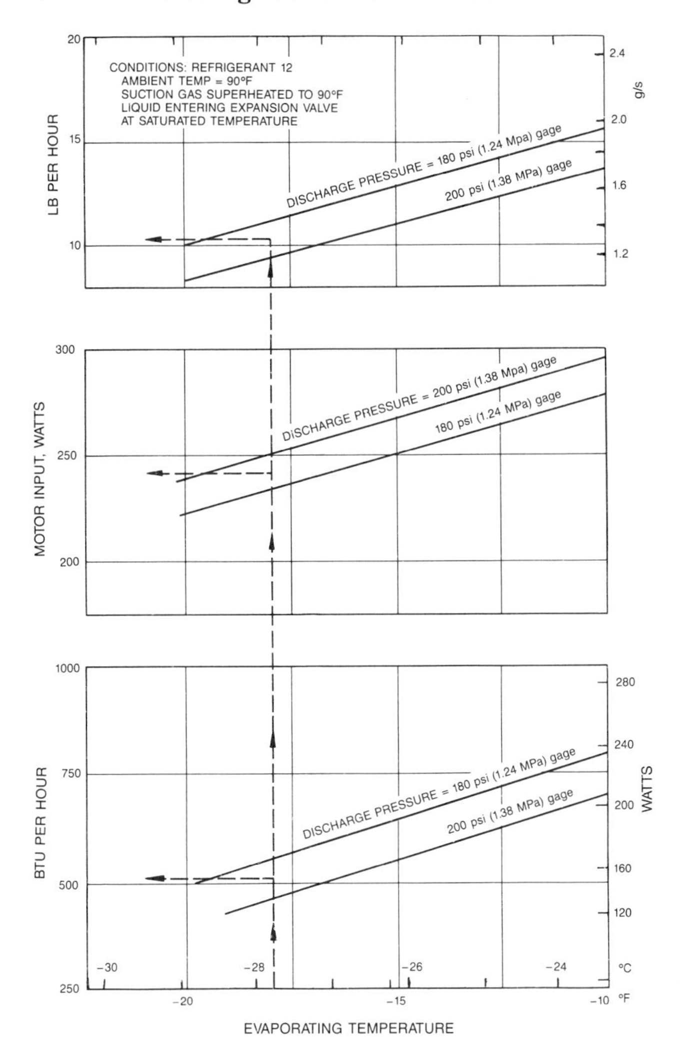


**Fig. 7 Compressor Calorimeter Curves**

discharge pressure of 190 psig, a compressor capacity of 550 Btu/h with 250 motor watts may be estimated. This will be adequate for the design load of 546 Btu/h. A more accurate estimate of refrigerating capacity may be made by using the pounds per hour of refrigerant pumped from Figure 7 and calculating the change in the enthalpy of the refrigerant between the saturated liquid entering the evaporator and the saturated gas leaving the evaporator, with an estimated allowance for the heat recovered by superheating the suction gas approximately to room temperature in the capillary heat exchanger.

**Condenser Capacity.** Although there will be considerable variations for different designs, it may be assumed that the compressor casing and tubing dissipates to the room 80% of the heat equivalent of the electrical input to the motor and that the condenser must provide the balance of the total heat to be rejected. The motor power is the equivalent of 854 Btu/h. Thus, 171 Btu/h added to the cooling load of 550 Btu/h makes the heat to be rejected by the condenser equal to 721 Btu/h. For the assumed 190 psig discharge pressure and estimating a 2 psi pressure drop in the condenser tubing, the condensing temperature will be found from a table or curve of R-12 vapor pressures to be about 132°F. The temperature difference between the refrigerant at 132°F and the 110°F ambient air entering the condenser will be 22°F. Assume that a fin-and-tube condenser is to be used with fins 2 in. by 6 in. and a finned length of 12 in. The air will enter the face of the condenser, which is 6 in. by 12 in., or 0.5 $ft^2$. With the condenser fan delivery of 90 cfm, this provides a face velocity of 180 fpm. To minimize dust clogging, the fins are spaced at 3 per inch. An approximation of the cooling capacity of this condenser may be obtained from Figure 6, where the basic coil capacity for 180 fpm is found to be 76 Btu/$ft^2$ • h • °F. The correction factor for fin height will be unity, and for fin spacing will be 0.9. Thus, the net coil capacity = (76)(0.5)(1)(0.9)(22) = 752 Btu/h. This may be assumed to be adequate to meet the expected condenser load of 721 Btu/h.

The knowledge of the components is limited, so simplifying assumptions must be used in such an analysis. Hence, it represents only the beginning point for confirming tests and, sometimes, a more careful study of specific variables.

## Processing and Assembly Procedures

All parts and assemblies that are to contain refrigerant are processed to avoid or remove unwanted substances from the final sealed system and to charge the system with refrigerant and oil. Each component should be thoroughly cleaned and then stored in a clean, dry condition prior to assembly. The presence of free water in stored parts produces harmful compounds such as rust or aluminum hydroxide, which are not removed by the normal final assembly process. Procedures for dehydration, charging, and testing may be found in Chapter 21.

Assembly procedures are somewhat different, depending on whether the sealed refrigerant system is completed as a unit before being assembled to the cabinet, or whether components of the system are first brought together on the cabinet assembly line. Using the unitary installation procedure, the system may be tested for its ability to refrigerate and then stored or delivered to the cabinet assembly line.

# EVALUATION

Preceding sections of this chapter concerned the design and testing of the cabinet and the refrigerating system as components of the refrigerator or freezer. Once the unit is assembled, laboratory tests, supplemented by field testing, are necessary to determine actual performance. The following aspects are considered in this section:

1. test facilities required
2. established test procedures, published by standard, technical, and industry organizations
3. special performance testing
4. materials testing
5. life testing of components
6. field testing

## Environmental Test Rooms

Controlled temperature and humidity test rooms are essential for performance testing of refrigerators and freezers. ANSI/AHAM *Standard* HRF-1-1986 describes the environmental conditions to be maintained. The rooms should be capable of providing ambient temperatures ranging from 70 to 110°F (21 to 43°C) accurate to within 1°F (0.5°C) of the desired value. The temperature gradient and the air circulation within the room should also be maintained closely. To provide more flexibility in testing, it may be desirable to have an additional test room that can also cover the range between 0 and 70°F (−18 and 21°C).

At least one test room should have the capability of maintaining a desired relative humidity within ±2%, up to 85%.

### Instrumentation

All instruments should be calibrated at regular intervals. The instruments for adequate performance testing of a refrigerator or freezer are described in ANSI/AHAM *Standard* HRF-1-1986. Instrumentation should have accuracy and response capabilities of sufficient quality to measure the dynamics of the systems tested.

Thermocouple recorders, indicating and recording wattmeters, ammeters, voltmeters, and potentiometers are used in testing refrigerators and freezers. Refrigerator test laboratories have developed automated means of data acquisition (with computerized data reduction output) and the automated programming of tests.

### Standard Performance Test Procedures

ANSI/AHAM *Standard* HRF-1-1986 describes tests for determining the performance of refrigerators and freezers. It specifies the standard ambient conditions, power supply, and means for selecting samples and measuring temperatures. Test procedures include the following:

1. **No-Load Pulldown Test.** This tests the ability of the refrigerator or freezer in a 110°F (43°C) ambient temperature to pull down from a stabilized warm condition to design temperatures within an acceptable period.
2. **Simulated Load Test (*Refrigerators*) or Storage Load Test (*Freezers*).** This test determines the electrical energy (kWh) consumption rate per 24-hour period, the percent operating time of the compressor motor, and temperatures at various locations within the cabinet at 70, 90, and 110°F (21, 32, and 43°C) ambient temperatures for a range of temperature control settings. The cabinet doors remain closed during the test. The freezer compartment is loaded with filled frozen packages. Each test point may take eight hours or more to ensure steady-state condition and accuracy of data. The data taken are usually plotted as shown in Figure 8 for a combination refrigerator-freezer with only a fresh-food temperature control. If there is a separate control for freezer temperature, these graphs can carry additional curves for high and low freezer control settings. Freezers are tested similarly but in a 90°F (32°C) ambient. Under actual operating conditions in the home (with frequent door openings and ice making) the performance may not be as favorable as that shown by this test. However, the test indicates general performance, which can serve as a basis for comparison.
3. **Ice-Making Test.** This test performed in a 90°F (32°C) ambient determines the rate of making ice with the ice trays or other ice-making equipment furnished with the refrigerator.
4. **External Surface Condensation Test.** This test determines the extent of moisture condensation on the external surfaces of the cabinet in a 90°F (32°C), high-humidity ambient when the refrigerator or freezer is operated at normal cabinet temperatures. Although ANSI/ AHAM *Standard* HRF-1-1986 calls for this test to be made at a relative humidity of 75 ±2%, it is customary to determine the sweating characteristics through a wide range of relative humidity up to 85%. This test also determines the need for, and the effectiveness of, anti-condensation heaters in the cabinet shell and door mullions.

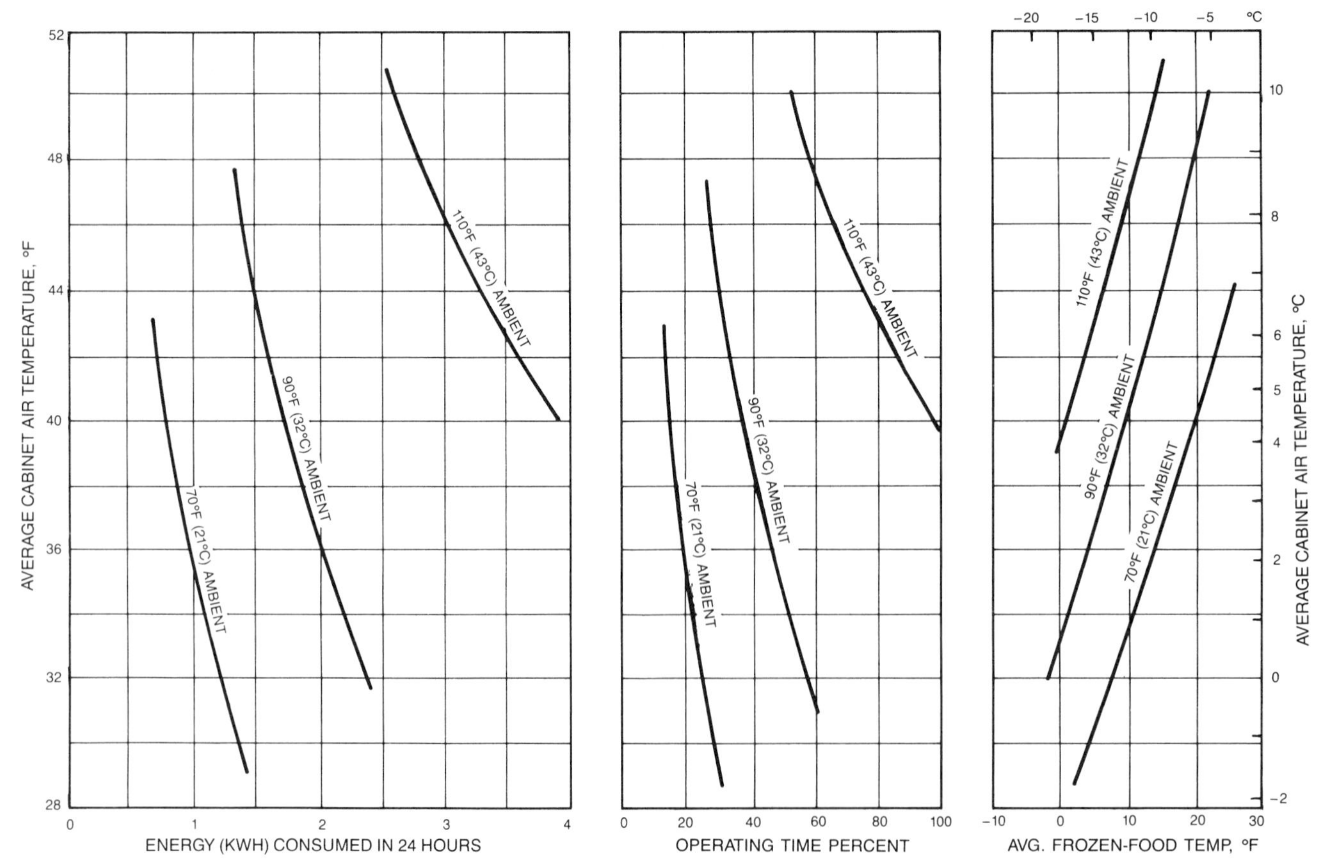


**Fig. 8 Sample Plot of Test Results of Simulated Load Tests**

5. **Internal Moisture Accumulation Test.** This dual-purpose test is also run under high-temperature, high-humidity conditions. First, it determines the effectiveness of the moisture-sealing of the cabinet in preventing moisture from getting into the insulation space and degrading the performance and life of the refrigerator. Secondly, it determines the rate of frost buildup on refrigerated surfaces, the expected frequency of defrosting, and the effectiveness of any automatic defrosting features, including means for defrost water disposal.

   This test is made in ambient conditions of 90°F (32°C) and 75% rh with the cabinet temperature control set for normal temperatures. The test period extends over a 21-day period with a rigid schedule of door openings over the first 16 hours of each day. For a refrigerator, the test calls for 96 door openings per day for the general refrigerated compartment; 24 per day for the freezer compartment and food freezers.
6. **Current Leakage Test.** This test determines the electrical current leakage through the entire electrical insulating system under severe operating conditions. To eliminate the possibility of a shock hazard, a current leakage of 0.75 mA at rated line voltage, in accordance with ANSI/UL *Standard* 250-1983, is considered the maximum allowable industry limit.
7. **Handling and Storage Test.** In common with most other major appliances, it is during shipping and storage that a refrigerator is exposed to its most severe impact forces, to vibration, and to the extremes of temperature. When packaged, it should withstand, without damage, a drop of several inches onto a concrete floor, the impact experienced in a freight car coupling at 10 mph (4.5 m/s), and jiggling equivalent to a trip of several thousand miles by rail or truck. The widespread use of plastic parts makes it important to select materials that also withstand the high and low temperature extremes that may be experienced.

   This test determines the ability of the cabinet, when packaged for shipment, to withstand handling and storage conditions in extreme temperatures. It involves raising the crated cabinet 6 in. (150 mm) off the floor and suddenly releasing it on one corner. This is done for each of the four corners. This procedure is carried out at stabilized temperature conditions—first in a 140°F (60°C) ambient temperature, and then in a 0°F (−18°C) ambient. At the conclusion of the test, the cabinet is uncrated, operated, and all accessible parts are examined for damage.

## Special Performance Testing

To ensure customer acceptance, several additional performance tests are customarily performed, as follows:

1. **Usage Test.** This is similar to the *Internal Moisture Condensate Test,* except that additional performance data will be taken during the test period, including (a) electrical energy consumption, kW • h(kJ), per 24-hour period; (b) percent running time of the compressor motor; and (c) cabinet temperatures. These data give an indication of the reserve capacity of the refrigerating system and the temperature recovery characteristics of the cabinet.
2. **Low Ambient Temperature Operation.** It is customary to make a *Simulated Load Test* and an *Ice-Making Test* at ambient temperatures of 55°F (12.8°C) or below. This test determines performance under unusually low temperature conditions.
3. **Food Preservation Tests.** This test determines the food-keeping characteristics of the general refrigerated compartment and is useful for evaluating the utility of special compartments such as vegetable crispers, meat keepers, high-humidity compartments, and butter keepers. This test is made by loading the various compartments with food, as recommended by the manufacturer, and periodically observing the condition of the food.
4. **Noise Tests.** The complexity and increased size of refrigerators has made it difficult to keep the sound level within acceptable limits. Thus, sound testing is important to ensure customer acceptance.

   A meaningful evaluation of the sound characteristics may require a specially constructed room with a background sound level of 30 dB or less. The wall treatment may be either reverberant, semi-reverberant, or anechoic, with the reverberant room construction usually favored in making an instrument analysis. A listening panel is most commonly used for the final evaluation, with most manufacturers striving to correlate instrument readings with panel judgment.
5. **High- and Low-Voltage Tests.** The ability of the compressor to start and to pull down the system after an *ambient soak* is tested with applied voltages that are at least 10% above and below the rated voltage. (The starting torque is reduced at low voltages; the motor tends to overheat at high voltage.)
6. **Special Functions Tests.** Refrigerators and freezers with special features and functions may require additional testing. In the absence of formal procedures for this purpose, test procedures are usually improvised, as required.
7. **Method of Determining Energy Consumption.** This is a special 90°F (32°C) ambient, no-door-opening test using the test procedure for Electrical Refrigerators and Electric Refrigerators-Freezers and Freezers published by the Department of Energy (DOE) in the *Federal Register,* Vol. 47, No. 154, August 10, 1982. This test procedure specifies a statistical sampling plan, which must be followed to establish the estimated annual cost of energy for labeling under the FTCs Energyguide Program. The DOE test procedure presently references AMAH *Standard* HRF-1-1979 for methods of testing.

## Materials Testing

The materials used in a refrigerator or freezer should meet certain test specifications. All materials in contact with foods must meet U.S. Food and Drug Administration requirements. Metals, paints, and surface finishes may be tested according to procedures specified by ASTM and others. Plastics may be tested according to procedures formulated by the SPI Appliance Committee. In addition, the following tests on materials, as applied in the final product, are assuming importance in the refrigeration industry (Federal Specification AA-R-00211 H[GL]):

1. **Odor and Taste Contamination.** This test determines the intensity of odors and tastes imparted by the cabinet air to uncovered, unsalted butter stored in the cabinet at operating temperatures.
2. **Stain Resistance.** The degree of staining is determined when the cabinet exterior surfaces and the surface of plastic interior parts are coated with a typical staining food (for example, Prepared Cream Salad Mustard).
3. **Environmental Cracking Resistance Test.** This tests the cracking resistance of the plastic inner door liners and breaker strips at operating temperatures when coated with a 50/50 mixture of oleic acid and cottonseed oil. The cabinet door shelves are loaded with weights, and the doors are slammed on a prescribed schedule extending over an eight-day test period. The parts are then examined for cracks and crazing.

4. **Breaker Strip Impact Test.** This test determines the impact resistance of the breaker strips at operating temperature when coated with a 50/50 mixture of oleic acid and cottonseed oil. The breaker strip is impacted by a 2-lb (0.9-kg) dart dropping from a prescribed height. The part is then examined for cracks and crazing.

## Component Life Testing

Various components of a refrigerator and freezer cabinet are subject to continual use by the consumer throughout the life of the product, and must be adequately tested to ensure their durability for at least a 10-year life. Some of these items are (1) hinges, (2) latch mechanism, (3) door gasket, (4) light and fan switches, and (5) door shelves. These components may be checked by an automatic mechanism, which opens and closes the door in a prescribed manner. A total of 300,000 cycles is generally accepted as the standard for design purposes. Door shelves should be loaded as they would be for normal home usage. Several other important characteristics may be checked during the same test: (1) retention of door seal, (2) rigidity of door assembly, (3) rigidity of cabinet shell, and (4) durability of inner door panels.

Life tests on the electrical and mechanical components of the refrigerating system may be made, as required.

## Field Testing

Additional valuable information may be obtained from a program of field testing in which test models are placed in selected homes for observation. Since high temperature and humidity are the most severe conditions encountered, the Gulf Coast is a popular field test area. Laboratory testing has limitations in the complete evaluation of a refrigerator design, and field testing can provide the final assurance of customer satisfaction.

Field testing is only as good as the degree of policing and the completeness and accuracy of reporting. However, if done properly, the information is important, not only in product evaluation, but in providing criteria for more realistic and timely laboratory test procedures and acceptance standards.

## REFERENCES

AHAM. 1986. Household Refrigerators, Combination Refrigerator Freezers, and Household Freezers. ANSI/AHAM *Standard* HRF-1-1986. Chicago: Association of Home Appliance Manufacturers.

*Federal Specification.* 1977. Refrigerators, Mechanical, Household (Electrical, Self-Contained), AA-R-00211H(GL).

Office of Business Research and Analysis. 1975. Economic Significance of Fluorocarbons, *U.S. Department of Commerce.*

Refrigeration Safety Act. *Public Law* 84-930.

UL. 1983. Household Refrigerators and Freezers. UL *Standard* 250-83. Northbrook, IL: Underwriters Laboratories.

CHAPTER 38

# DRINKING WATER COOLERS AND CENTRAL SYSTEMS



## UNITARY COOLERS

A MECHANICALLY refrigerated drinking water cooler consists of a factory-made assembly in one structure. This cooler, by means of a complete mechanical refrigeration system, has the primary function of cooling potable water and dispensing it by integral and/or remote means.

*Water coolers* differ from *water chillers*. Water coolers dispense potable water. Water chillers are used in air-conditioning systems for residential, commercial, and industrial applications, as well as for cooling water for industrial processes.

The capacity of a water cooler is expressed in gallons (litres) per hour and is the quantity of water cooled in one hour from a specified inlet temperature to a specified dispensing temperature (see the following section on "Ratings"). Normal standard capacities of water coolers range from 1 to 30 gph (3.8 to 114 L/h).

### Types

Figure 1 shows the three basic types of water coolers.

A **bottle water cooler** uses a bottle or reservoir for storing the supply of water to be cooled and a faucet or similar means for filling glasses, cups, or other containers. It also includes a waste-water receptacle.

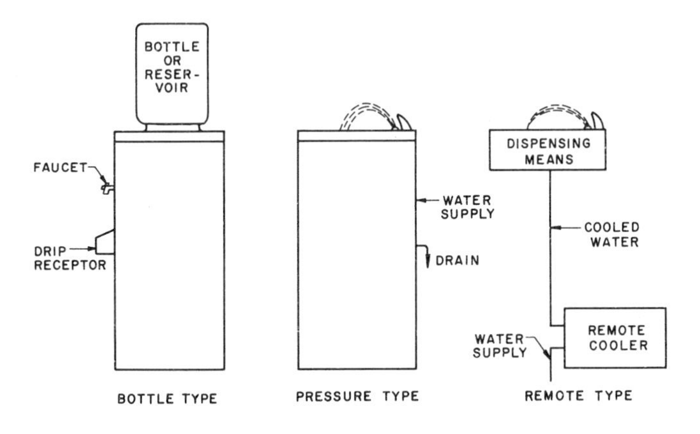


**Fig. 1 Basic Drinking Water Coolers**

A **pressure-type water cooler** is supplied with potable water under pressure and includes a waste-water receptacle or means for disposing of water to a plumbing drainage system (see Figure 2). These coolers use a faucet or similar means for filling glasses or cups, or a valve to control the flow of water as a projected stream from a bubbler so that water may be consumed without using glasses or cups.

A **remote-type cooler** is a factory assembly in one structure that uses a complete mechanical refrigeration system and functions primarily to cool potable water for delivery to a separately installed dispensing means.

Coolers, in addition to the basic descriptions, are also described by (1) specialized conditions of use, (2) additional functions they perform, or (3) type of installation.

1. **Specialized Uses**
   a. An explosion-proof water cooler is constructed for safe operation in hazardous locations, as classified in Article 500 of the *National Electrical Code*.
   b. A cafeteria cooler is one that is supplied with water under pressure from a piped system and is intended primarily for use in cafeterias and restaurants for dispensing water rapidly and conveniently into glasses or pitchers. It includes a means for disposing waste water to a plumbing drain system.
2. **Additional Functions**
   a. A water cooler may also have a *refrigerated compartment* with or without provisions for making ice.

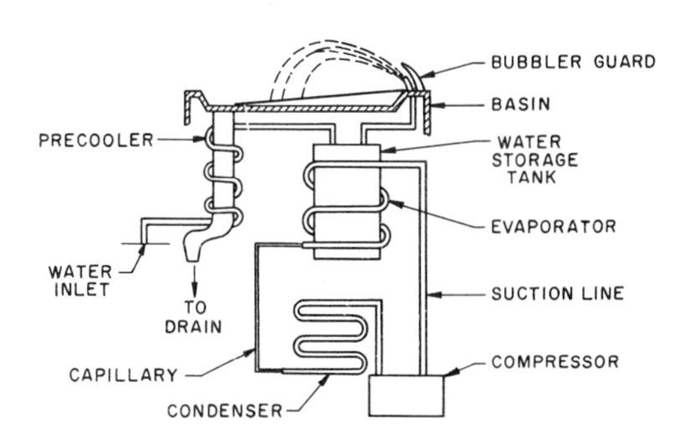


**Fig. 2 Pressure Water Cooler**

The preparation of this chapter is assigned to TC 7.1, Residential Refrigerators, Food Freezers and Drinking Water Coolers.

b. A water cooler may also include a means for heating and dispensing *potable water* for making instant hot beverages and soups.

3. **Type of Installation** (see Figure 3)
   a. Freestanding
   b. Flush-to-wall
   c. Wall-hung
   d. Semirecessed
   e. Fully recessed

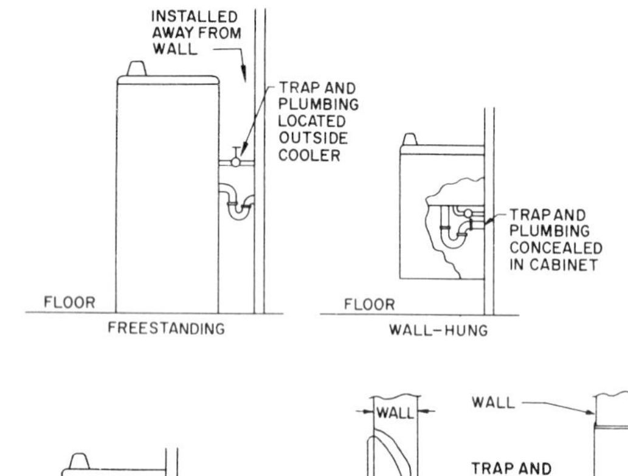


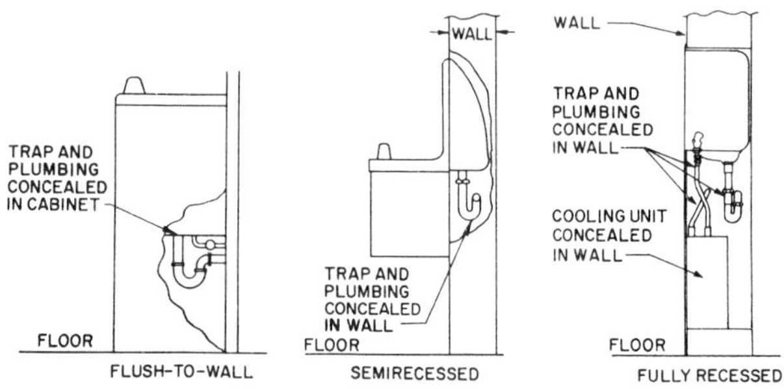


**Fig. 3 Types of Installation for Drinking Water Coolers**

## Refrigeration Systems

Hermetically sealed motor *compressors* are commonly used for ac applications, both 50 and 60 Hz. Belt-driven compressors are generally used only for dc and 25-Hz supply. Compressors are similar to those used in household refrigerators and range from 0.05 to 0.75 hp (0.04 to 0.56 kW).

Forced-air-cooled *condensers* are most commonly used. In coolers rated less than 5 gph (19 L/h), natural convection, air-cooled (static) condensers are sometimes included. Water-cooled condensers are used on models intended for high ambient temperatures or where lint and dust in the air make air-cooled types impractical.

Capillary tubes are used almost exclusively for *refrigerant flow control* in hermetically sealed systems. In belt-driven systems and some hermetically sealed systems, expansion valves are used.

Most water coolers manufactured today have the evaporator formed by refrigerant tubing bonded to the outside of a water circuit. However, some are made with an immersed evaporator.

The water circuit is usually a tank or a coil of large tubing. Materials used in the water circuit are usually nonferrous or stainless steel. Since the coolers dispense water for human consumption, sanitary requirements are essential. (See ARI *Standard* 1010-84.)

Pressure coolers are often equipped with *precoolers* to exchange heat from the supply water to the waste water. When drinking from a bubbler stream, the user wastes about 60% of the cold water, which runs down the drain. A precooler puts the incoming water in heat exchange relationship with the waste water. Sometimes the cold waste water subcools liquid refrigerant in an arrangement called a *subcooler*. Coolers intended only to dispense water into cups are not equipped with precoolers, since there is no appreciable quantity of waste water.

Water coolers providing a refrigerated storage space are commonly referred to as *compartment coolers*. The compartment cooler refrigeration systems vary from a single-series system to two independent systems.

Compartment coolers use a single-series system to feed the water cooling evaporator first and then the compartment evaporator. When the compressor operates, both water and compartment cooling take place. The thermostat is usually located where it is most affected by the compartment temperature. Therefore, the water cooling is affected greatly by compartment loading and usage.

Some compartment coolers with a single compressor have two temperature controls. These controls operate the compressor in conjunction with the solenoid valve(s), which direct refrigerant flow through the desired evaporator, as required. Also, some compartment coolers use two independent refrigeration systems.

## Stream Regulators

Since the principal function of a pressure water cooler is to provide a drinkable stream of cold water from a bubbler, it is usually provided with a valve to maintain a constant stream height, independent of supply pressure. A flow rate of 0.5 gpm (0.03 L/s) from the bubbler is generally accepted as giving an optimum stream for drinking.

## Ratings

Water coolers are rated on the basis of continuous flow capacity under specified water temperature and ambient conditions. ARI *Standard* 1010-84, *Drinking Fountains and Self-Contained Mechanically Refrigerated Drinking Water Coolers,* gives the generally accepted rating conditions and references test methods as prescribed in ASHRAE *Standard* 18-79, *Methods of Testing for Rating Drinking Water Coolers with Self-Contained Mechanical Refrigeration Systems.* Table 1 gives the standard rating conditions.

**Table 1 Standard Rating Conditions**
(Adapted from ARI *Standard* 1010-84)

| | Temperature, °F (°C) | | | | |
|---|---|---|---|---|---|
| Type of Cooler | Ambient | Inlet Water | Cooled Water | Heated Potable Water* | Spill, Percent |
| Bottle types | 90 (32) | 90 (32) | 50 (10) | 165 (74) | None |
| Pressure types | | | | | |
| Using Precooler or Non-Precooler Drain (Bubbler Service) | 90 (32) | 80 (27) | 50 (10) | 165 (74) | 60 |
| Not using Precooler or Other Heat-Transferring Device | 90 (32) | 80 (27) | 50 (10) | 165 (74) | None |
| Compartment Coolers | Standard Rating Conditions for water cooling noted above apply. During the Standard Rating Test, there shall be no melting of ice, nor shall the average temperature exceed 46°F (7.8°C) in the refrigerated compartment. | | | | |

*This temperature shall be referred to as the Standard Rating Temperature (Heating).

NOTE:
1. For water-cooled condenser water coolers, the established flow of water through the condenser shall not exceed 2.5 times the Base Rate Capacity, and the outlet condenser water temperature shall not exceed 130°F (54.4°C).
2. Voltage for above tests shall be at namplate voltages in accordance with ARI *Standard* 110 for *Air-Conditioning and Refrigerating Equipment Nameplate Voltages.*

**Table 2 Water Cooler Requirements**
(Adapted from ARI *Standard* 1020-84)

| | Persons Served Per Gallon Per Hour (mL/s) |
|---|---|
| Offices, Schools, Hospitals, Retail Stores, Buildings, Office Building Lobbies, Theater Lobbies, Airline Terminals | 25 |
| Light Manufacturing | 15 |
| Heavy Manufacturing | 12 |
| Hot, Heavy Manufacturing | 10 |

## Applications

ARI *Standard* 1020-84, *Application and Installation of Drinking Fountains and Drinking Water Coolers,* includes the following guidelines for placement and selection of a water cooler:

1. **General Locations.** Drinking fountains or water coolers should be available within 200 ft (61 m) of any location where persons are regularly engaged in work.
2. **Prohibited Locations.** Drinking fountains or water coolers should not be installed in toilet rooms or any location where the equipment is exposed to contamination from toxic or otherwise hazardous materials.
3. **Ventilation.** Adequate ventilation should be provided for water coolers equipped with air-cooled condensers in accordance with the manufacturer's installation instructions.
4. **Minimum Requirements.** Table 2 presents minimum requirements for the application of water coolers based on recognized industry practice.

**Example.** A manufacturing facility employs 625 people. There are 95 office personnel: 60 have access to one water cooler and 35 have access to another. There are 51 people in a hot, heavy-manufacturing environment with access to one water cooler. The remaining 479 people are involved in light manufacturing operations and are equally divided among 5 water coolers. Required water capacities are calculated as follows:

$$\text{Capacity per Water Cooler} = \frac{\text{Persons served}}{\text{Value from Table 2} \cdot \text{No. of water coolers}}$$

Capacity for Office Water Coolers:

$$\frac{60}{25 \cdot 1} = 2.4 \text{ gph}$$

$$\frac{35}{25 \cdot 1} = 1.4 \text{ gph}$$

Capacity for Hot, Heavy-Manufacturing Coolers:

$$\frac{51}{10 \cdot 1} = 5.1 \text{ gph}$$

Capacity for Light Manufacturing Coolers:

$$\frac{479}{15 \cdot 5} = 6.4 \text{ gph}$$

## Standards and Codes

In addition to ARI *Standard* 1010-84 and ASHRAE *Standard* 18-79, Underwriters Laboratories *Standard* 399 covers safety requirements. Federal Specification 00-D-00566 had been prescribed by government purchasers. Four commercial item descriptions are A-A-1151, A-A-1152, A-A-1153, and A-A-1154, which supersede 00-D-0056.

Many local plumbing codes apply directly or indirectly to water coolers. These codes are directed primarily toward eliminating any possibility of cross-connection between the potable water system and the waste-water (or refrigerant) system. Most coolers are made with double-wall construction to eliminate the possibility of conflict with any code.

# CENTRAL SYSTEMS

A central chilled drinking water system may be considered in a multi-story office building where drinking fountains are stacked one floor level above the other. The system should be designed to provide 50°F (10°C) water to the drinking fountains. To allow for heat gain in the distribution system, the chiller should be sized to provide 45°F (7.2°C) outlet water. Water system working pressures should be limited to 80 psi (550 kPa) gauge.

The components in a central chilled drinking water system can be broken down into the following categories: (1) chiller, (2) distribution piping, and (3) drinking fountains.

## Chillers

The chiller may be a field built-up system or a factory-assembled unit; it is common practice to use factory-assembled units. In either event, the chiller consists of the following components:

**Compressor.** A semi-hermetic or hermetic direct-driven compressor, using R-12 or R-22, is used.

**Condenser.** A condenser that is either water-cooled or air-cooled is used. Large air-cooled condensers are often remotely located.

**Evaporator.** A direct-expansion-type evaporator is used. It may be of the shell-and-tube type construction with a separate storage tank or an immersion-type coil in a storage tank. If a separate tank is used, a circulating pump is needed to circulate the water between the evaporator and storage tank. (Check local codes for allowable construction.) Most areas require two thicknesses of metal between potable water and refrigerant circuits.

**Storage.** A storage capacity of approximately one-half the rated chiller capacity is a good starting point for sizing the tank. Unique and unusual situations require the storage tank size to be modified. Care must be taken in selecting tank materials to reduce the possibility of galvanic action. (See the section "Distribution Piping System" for systems designed without a storage tank.)

**Pumps.** System water-circulating pumps are normally bronze-fitted, close-coupled, single-stage pumps with mechanical seals. Pump sizing is determined from the size of the recirculating system. The pump flow rate should be such that there is no more than a 5°F (3°C) rise through the loop.

**Controls.** Typical system controls include (1) high- and low-pressure cutouts, (2) freeze control, (3) water temperature control, and (4) a flow switch to ensure that there is water flowing before the compressor is permitted to operate.

**Options.** The following options should also be considered: (1) duplex system pumps with timers, (2) split refrigeration systems with timers, (3) tank anodes, and (4) water filtration combined with activated carbon treatment. These are some of the more common options.

## Distribution Piping System

The distribution piping system delivers chilled water to the drinking fountains. The piping can be galvanized steel, copper, or brass designed for a working pressure of 80 psi (550 kPa) gauge. Recommendations on the use of piping materials that minimize corrosion and galvanic action are presented in Chapter 54 of the 1987 HVAC Volume.

Table 3 Refrigeration Load[a]

| Water Inlet Temp. °F | Btu per gal cooled to 45°F | Water Inlet Temp. °C | J/mL cooled to 7°C |
|---|---|---|---|
| 65 | 167 | 18 | 46 |
| 70 | 208 | 21 | 58 |
| 75 | 250 | 24 | 71 |
| 80 | 291 | 27 | 83 |
| 85 | 333 | 30 | 96 |
| 90 | 374 | 33 | 108 |

[a]Multiply values by flow rate in gph (mL/s) to calculate the refrigeration load in Btu/h (W).
Refr. Load = (Specific heat)(Density)(Temp. Diff.)

The makeup cold-water lines are made of the same material as the distribution piping. When the water supply has objectionable characteristics, such as high iron or calcium content, or contains odoriferous gases in solution, filtration and/or activated carbon treatment should be applied in the makeup water.

Some manufacturers choose not to use storage tanks in their chiller design. These manufacturers suggest using large diameter manifolds in the distribution system. The chiller manufacturer should be contacted for assistance in manifold sizing.

Insulation is necessary on all of the distribution piping and the storage tanks. The insulation should be glass fiber insulation, such as that normally used on chilled-water piping, with a conductivity of 0.22 Btu • in./h • ft$^2$ • °F (0.032 W/[m • K]) at a 50°F (10°C) mean temperature, and with a vapor barrier jacket, or equal. All valves and piping, including the branch to the fixture, should be insulated. The waste piping from the drinking fountain, including the trap, should be insulated. This insulation is the same as that recommended for use on cold-water lines.

## Drinking Fountains

Any standard drinking fountain can be used on a central drinking water system. They are made of stainless steel, bronze, vitreous china, or marble. It is important, however, that the automatic volume or stream regulator provided with the fountain provide a constant stream height from the bubbler, with inlet pressures up to 80 psi (550 kPa) gauge.

## System Design

The load is based on building population, the population's activity level, and the environment in which the activity takes place. [Table 2 lists the appropriate factor(s).] Table 3 converts the load in gph (L/h) to refrigeration load in Btu/h (W). The heat gain from the distribution piping system is based on a circulating water temperature at 45°F (7.2°C). Table 4 lists the heat gains for various ambient temperatures. The length of all lines must be included when calculating the heat gain in the distribution piping. Table 5 tabulates the heat input from various sizes of circulating pump motors.

Table 4 Circulating System Line Loss

| Nominal Pipe Size, In. (mm) | Btu/h per ft per °F (W • °C/m) | Btu/h per 100 ft (Watt per 100 m) (45°F [7.2°C] circulating water) Room Temperature, °F (°C) 70 (21) | 80 (27) | 90 (32) |
|---|---|---|---|---|
| 1/2 (15) | 0.110 (0.190) | 280 (269) | 390 (374) | 500 ( 480) |
| 3/4 (20) | 0.119 (0.206) | 300 (288) | 420 (403) | 540 ( 518) |
| 1 (25) | 0.139 (0.240) | 350 (336) | 490 (470) | 630 ( 605) |
| 1 1/4 (32) | 0.155 (0.268) | 390 (374) | 550 (528) | 700 ( 672) |
| 1 1/2 (40) | 0.174 (0.301) | 440 (422) | 610 (586) | 790 ( 758) |
| 2 (50) | 0.200 (0.346) | 500 (480) | 700 (672) | 900 ( 864) |
| 2 1/2 (65) | 0.228 (0.394) | 570 (547) | 800 (768) | 1030 ( 989) |
| 3 (80) | 0.269 (0.465) | 680 (653) | 940 (902) | 1210 (1162) |

Table 5 Circulating Pump Heat Input

| Motor Hp (kW) | 1/4 (0.19) | 1/3 (0.25) | 1/2 (0.37) | 3/4 (0.56) | 1 (0.75) |
|---|---|---|---|---|---|
| Btu/h (W) | 636 (186) | 850 (249) | 1272 (373) | 1908 (559) | 2545 (746) |

The total cooling load consists of the heat removed from the makeup water, the heat gains from the piping, the heat gains from the storage tank, and the heat input from the pumps. Oversizing of the chiller may be included if there is a potential for either future building expansion or a higher population. Otherwise, the cold water storage tank should compensate for any abnormal water draw requirements.

The *circulating pump* is sized to circulate a minimum of 3 gpm (0.2 L/s) per branch or the gpm (L/s) necessary to limit the temperature rise of the circulatory water to 5°F (2.8°C), or whichever is greater. Table 6 lists the circulating pump capacity to limit the temperature rise of the circulated water to 5°F (2.8°C). If a separate pump circulates water between the evaporator and the storage tank, the energy input to this pump must be included in the heat gain.

The *storage tank* capacity should be at least 50% of the hourly usage. The hourly usage may be selected from ARI *Standard* 1020-84.

General criteria for *sizing* distribution piping for a central chilled drinking water system are as follows:

1. Limit the maximum velocity of the water in the circulating piping to 3 fps (0.9 m/s) to avoid giving the water a milky appearance.
2. Avoid excessive friction head losses. Energy required to circulate water enters the water as heat and requires additional capacity in the water chiller. Accepted practice limits the maximum friction loss to 10 ft of head per 100 ft of pipe (1 kPa/m).
3. Dead-end piping, such as from main riser to fountain, should be kept as short as possible and should not exceed 15 ft (7.6 m) in length. Maximum diameter of such dead-end piping should not exceed 3/8 in. (10 mm) IPS, except on very short runs.
4. Size piping on total number of gallons circulated. This includes gallons consumed plus gallons (litres) necessary for heat leakage.

General criteria for *design layout* of piping for a central chilled drinking water system are as follows:

Table 6 Circulating Pump Capacity[a]

Gph per 100 ft (L/h per 100 m) of pipe including all branch lines necessary to circulate to limit temperature rise to 5°F (2.8°C) (water at 45°F [7.2°C])

| Nominal Pipe Size, In. (mm) | Room Temperature, °F (°C) 70 (21) | 80 (27) | 90 (32) |
|---|---|---|---|
| 1/2 (15) | 8.0 (99) | 11.1 (138) | 14.3 (177) |
| 3/4 (20) | 8.4 (104) | 11.8 (146) | 15.2 (188) |
| 1 (25) | 9.1 (113) | 12.8 (159) | 16.5 (205) |
| 1 1/4 (32) | 10.4 (129) | 14.6 (181) | 18.7 (232) |
| 1 1/2 (40) | 11.2 (139) | 15.7 (195) | 20.2 (250) |

[a]Add 20% for safety factor. For pump head figure longest branch only. Install pump on the return line to discharge into the cooling unit. Makeup connection should be between the pump and the cooling unit.

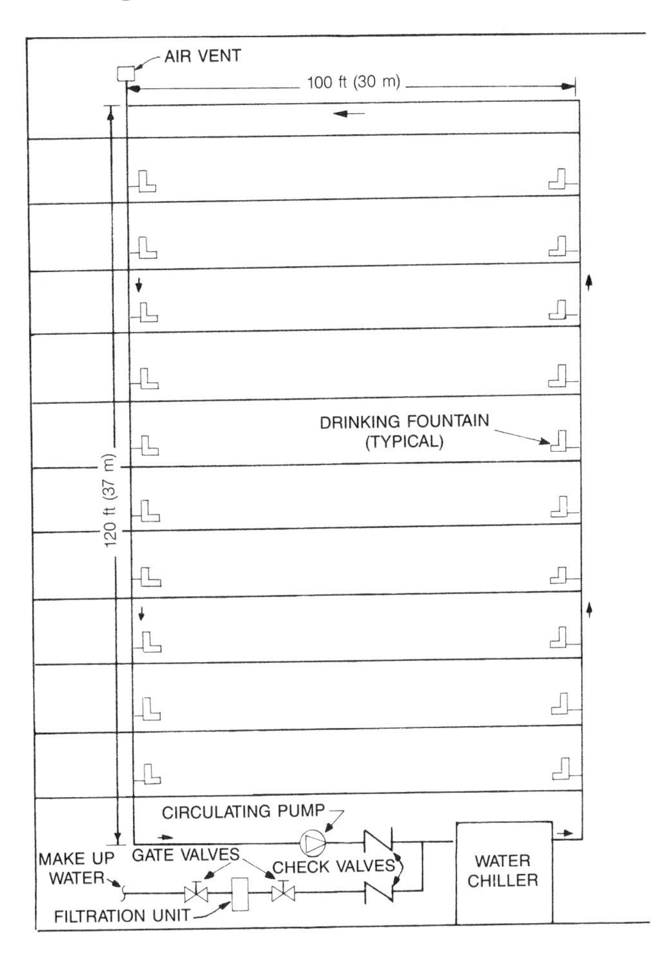


**Fig. 4 Typical Central System**

1. Keep pipe runs as straight as possible with a minimum of offsets.
2. Use long sweep fittings wherever possible to reduce friction loss.
3. In general, limit maximum pressure developed in any portion of the system to 125 psi (860 kPa). If the height of a building should cause pressures in excess of 125 psi (860 kPa), divide the building into two or more systems.
4. If more than one branch line is used, install balancing cocks on each branch.
5. Provide pressure-relief valve and air vents at high points in chilled-water loop.

**Example:** Design a central chilled drinking water system for the 10-story building in Figure 4. The net floor area is 14,600 ft$^2$ per floor, and occupancy is assumed at 100 ft$^2$ per person. The entering (makeup) water is 70°F maximum, and the ambient temperature is 90°F maximum. Applicable codes are the *Uniform Plumbing Code, Uniform Building Code, National Standard Plumbing Code,* and all local plumbing codes.

**Solution:**

1. Number of Drinking Fountains Required:
   Occupancy = 14,600/100 = 146 people per floor.
   The *National Standard Plumbing Code* requires one fountain for every 100 people, or 146/100 = 1.46 fountains per floor. Therefore, use 2 fountains per floor, for a total of 20 fountains.
2. Estimated Usage Load:
   Makeup water = 146 people per floor • 10 floors/25 persons per gph (from Table 2) = 58.4 gph.
3. Refrigeration Load to Cool Makeup Water:
   Refrigeration Load = 58.4 gph • 208 Btu/h per gallon (from Table 3) = 12,150 Btu/h.
4. Refrigeration Load Resulting from Piping Heat Gain:
   Assume a 3/4 in. diameter chilled water circuit. Then, the heat gain from the piping system of Figure 4 is (from Table 4):

| | | |
|---|---|---|
| Risers: | 120 ft • 540 Btu/100 ft • 2 risers | = 1296 |
| Distri. Mains: | 100 ft • 540 Btu/100 ft • 2 mains | = 1080 |
| | Total Piping Heat Gain | = 2376 Btu/h |

The chilled water must be circulated at a minimum of 3 gpm.

5. Refrigeration Load Resulting from Circulating Pump Input:
   In some chiller systems, one pump circulates water through the evaporator heat exchanger and the building plumbing system. In this case, there is no need to account for the heat gain from the pump, since it has already been included in the chiller's capacity. If there are two separate pumps, the heat input from the circulating pump needs to be considered. Assume this system uses two pumps.

   Pump Selection:
   15.2 gph/100 ft (Table 6, 3/4 in. pipe at 90°F ambient) • 440 ft = 67 gph or slightly over 1 gpm. Since this is less than the established 3 gpm minimum circulating rate, the 3 gpm figure will be used in determining the plumbing system pressure drop.

   Pressure Drop Calculations:
   Pipe length = 440 ft
   Increase pipe length 50% to allow for fittings. If an unusually large number of fittings are used, consider each for its actual contribution to pressure drop. Therefore:
   Equivalent pipe length = 440 • 1.5 = 660 ft
   Water flow = 3 gpm
   Pipe size = 0.75 in.
   Pressure drop per 100 ft of pipe (Table 7) = 4.1 ft of head
   Total pressure = 4.1 • 660/100 = 27.1 ft of head
   Pump manufacturer's literature shows that a 1/3 hp (0.25 kW) pump motor is required. From Table 5, the heat input of the pump motor is 850 Btu/h.
6. Refrigeration Load Resulting from Storage Tank Heat Gain:
   Normally, a tank sized for 50% of the anticipated hourly demand is specified. In this example, 60 gph is the hourly demand; therefore, a 30-gallon storage tank is used. This is approximately the capacity of a 16 in. diameter by 48-in. long tank. Assume 1.5 in. insulation,

**Table 7 Friction of Water in Pipes**

Loss of head in feet from friction per 100 ft of smooth straight pipe (Pa/m)

| Gpm | 1/2 in. (15 mm) Pipe | | 3/4 in. (20 mm) Pipe | | 1 in. (25 mm) Pipe | | 1 1/4 in. (32 mm) Pipe | | 1 1/2 in. (40 mm) Pipe | |
|---|---|---|---|---|---|---|---|---|---|---|
| (L/h) | Vel.[a] | Head[b] | Vel.[a] | Head[b] | Vel.[a] | Head[b] | Vel.[a] | Head[b] | Vel.[a] | Head[b] |
| 1 ( 227) | 1.05 (0.32) | 2.1 (206) | — | — | — | — | — | — | — | — |
| 2 ( 454) | 2.10 (0.64) | 7.4 (726) | 1.20 (0.37) | 1.9 (186) | — | — | — | — | — | — |
| 3 ( 681) | 3.16 (0.96) | 15.8 (1550) | 1.80 (0.55) | 4.1 (402) | 1.12 (0.34) | 1.26 (124) | — | — | — | — |
| 4 ( 912) | — | — | 2.41 (0.73) | 7.0 (687) | 1.49 (0.45) | 2.14 (210) | 0.86 (0.26) | 0.57 (56) | — | — |
| 5 (1135) | — | — | 3.01 (0.92) | 10.5 (1030) | 1.86 (0.57) | 3.25 (320) | 1.07 (0.33) | 0.84 (82) | 0.79 (0.24) | 0.40 (39) |
| 10 (2270) | — | — | — | — | 3.72 (1.13) | 11.7 (1150) | 2.14 (0.65) | 3.05 (300) | 1.57 (0.48) | 1.43 (140) |
| 15 (3405) | — | — | — | — | — | — | 3.20 (0.98) | 6.50 (637) | 2.36 (0.72) | 3.0 (294) |
| 20 (4540) | — | — | — | — | — | — | — | — | 3.15 (0.96) | 5.2 (510) |

[a]Velocity, ft/s (m/s). [b]Head = friction head per 100 ft of pipe (Pa/m).

45°F water, with the tank being in a 90°F room. Assume the insulation to have an overall heat transfer coefficient of 0.13 Btu/h • ft$^2$ • °F. The surface area of the tank is about 20 ft$^2$. Therefore, the heat gain is:

$$20 \bullet 0.13\ (90 - 45) = 117 \text{ Btu/h}$$

7. Load Summary:

| Item | Heat Gain, Btu/h |
|---|---|
| Makeup water | = 12,150 |
| Piping | = 2376 |
| Circulating Pump Heat | = 850 |
| Storage tank | = 117 |
| Required Chiller Capacity | = 15,493 |

## Codes and Regulations

Most mechanical installations are regulated by local codes. Most local codes are based on guide codes prepared by state or national code-writing organizations. Usually, one of these guide codes has been selected by a particular municipality or other governmental body and has been modified to suit local conditions. For this reason, it is important to refer to the actual code used in the locality. Other codes that require careful review include refrigeration codes. Many follow ASHRAE *Standard* 15-78, *Safety Code for Mechanical Refrigeration,* but they may have exceptions. Electrical regulations, as they apply to control and power wiring, and ASME requirements for tanks and piping must also be followed.

CHAPTER 39

# BOTTLED BEVERAGE COOLERS AND REFRIGERATED VENDING MACHINES



## BOTTLED BEVERAGE COOLERS

THE types of machines discussed here are (1) horizontal cold-wall, (2) horizontal forced air, (3) vertical forced air, and (4) air curtain. All references to bottles apply to cans, as well.

### Horizontal Cold-Wall Coolers

Cold-wall coolers may be set up for vending or non-vending operation. The evaporator, in the form of plates or tubing, is often located on the outside surface of the bottle compartment liner. For effective heat transfer, the design should provide good contact between the evaporator and the liner.

Coolers made this way are vulnerable to moisture migration into the insulation space from both the bottle compartment and the exterior of the cabinet. Frost results if moisture-laden air reaches the evaporator surface. During defrost periods, this frost melts and soaks the insulating material, thus reducing the insulation effect. The exposed surfaces of the inner and outer shell are also susceptible to corrosion. An airtight seal, both inside and outside, is desirable but difficult to achieve.

Whereas servicing or replacing the evaporator is difficult, cleaning the cooling compartment is relatively easy. In some cold-wall coolers, the evaporator is located inside the bottle compartment. The moisture migration problem is reduced greatly, and access to the evaporator is improved, but cleaning the cooling compartment is more difficult. The usable space in the cooler is reduced by the presence of the evaporator and the necessary protective racks. The evaporator usually occupies the entire inside perimeter of the liner.

In any cold-wall construction, complete defrost must be effected on each cycle of the cooler; otherwise, a frost buildup can occur, which impedes heat transfer and reduces bottle capacity. The bottles are usually placed upright in the cooling compartment for better temperature control. The bottles are cooled primarily by conduction (direct contact) and secondarily by convection. Thus, the bottles in contact with the cold-wall surface cool more rapidly than those that must rely, to a greater extent, on convection.

Cold-wall coolers are limited mainly to such applications as small offices, service stations, and markets. An advantage is the small heat gain that occurs when the lid is opened to obtain the product, since the cooled air does not spill from the cooling compartment. The absence of a circulating fan in the cooling compartment further reduces heat gain and power consumption.

The preparation of this chapter is assigned to TRG 7.2, Beverage Coolers.

### Horizontal Forced-Air Coolers

The storage capacity of this cooler is relatively high compared to the cold-wall type, since the product can be stacked horizontally while obtaining fairly uniform cooling. Capacities range from 5 to 60 cases of twenty-four 12 oz (355 mL) bottles, with the most common capacities being from 10 to 30 cases.

The evaporator coil and fan assembly generally are located as shown in Figure 1. The circulating fan normally operates continuously, but, in some cases, the fan circuit is opened when the lid of the cooler is opened. The cold air is blown over the top layer of the bottles, then is pulled down through the lower layers to the bottom of the cooler, where a rack provides a space for return air. When loaded fully, the densely packed bottles offer considerable resistance to airflow. While this arrangement provides a fairly even airflow throughout the load, it is difficult to cool the bottles to the desired temperature in the extreme corners of larger coolers.

Operating costs are low with this type of cooler because of the low cabinet heat gain. The temperature differential between bottle and refrigerant is held to a minimum, thereby ensuring a relatively high refrigeration suction-pressure and consequent improvement in efficiency.

Dependable stock rotation is not ensured, since a bottle can be removed from any section of the cooler. Easy access to the refrigeration unit for replacing and servicing is a benefit, while the compact design simplifies replacement of the complete unit.

### Vertical Forced-Air Vendors

These vendors are manufactured in capacities of 40 to 450 or more bottles or cans (see Figure 2). The vertical stack-serpentine or slant-shelf arrangement is used in most cases. In the vertical stack-serpentine arrangement, the bottles are placed in columns above the vending mechanism; in the inclined shelf arrangement the bottles roll toward the lower end of the shelf to an access door.

Stable product temperatures require forced-air circulation, usually with a finned-tube evaporator. Air should be directed from the evaporator to the bottles, which are in position to be vended. This approach keeps the bottles colder and, when the vendor is reloaded, allows for quicker cooling.

Bottles can be refrigerated a few hours before they are vended, so the refrigeration unit capacity need not be sufficient to cool a full load at one time. Generally, the product load moves through the refrigerated zone as the bottles progress toward the vending position. Operation of the fan during the off cycle of the refrigeration system aids both in keeping the bottles at approximately the same temperature and in defrosting the evaporator.

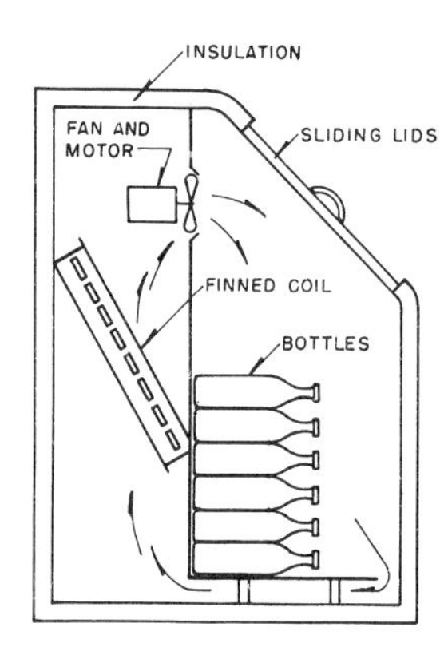


**Fig. 1 Horizontal Forced-Air Bottle Cooler**

## Vertical Forced-Air Coolers

These coolers are known as *visa coolers, glass door merchandisers,* or *cold carton coolers*. The equipment is full height, is available with adjustable shelves, and has glass doors that either slide or swing open. The refrigeration compressor is located at either the top or bottom of the cabinet, depending on the manufacturer's design.

Bottles placed on the shelves either singly or in cartons are chilled by forced-air circulation. This air is directed to achieve uniform cooling throughout the cabinet so that the bottles are cooled more quickly when the cabinet is reloaded. Airflow control prevents the fogging of glass doors.

## Air Curtain Coolers

Open-faced air curtain display coolers for retailing bottled and canned product are becoming more prevalent. The design and application of this type of cooler is discussed in Chapter 35.

## Refrigeration Capacity

Sales rate is an important consideration in the refrigeration capacity of the vendor. If sales are steady, a high degree of zone cooling is desirable. However, heavy peak demands require more refrigeration capacity and cooling of a greater percentage of the bottles to ensure that the demand can be met.

A knowledge of the anticipated sales per hour and day is needed to determine the machine capacity, as well as its refrigeration capacity. The vendor must refrigerate bottles as rapidly as they are sold, except that peak sales may be taken care of by a reserve of cold bottles in the vendor.

The anticipated sales rate and the anticipated reloading schedule determines the capacity of the machine. Reloading once a day is considered normal, but reloading twice a day is common. When a considerable distance must be traveled to reload, it is often economical to provide a vendor with enough product to require reloading only once or twice a week.

One method of ensuring adequate cooling with minimum compressor size is to use the precool area in the refrigeration compartment, which is in relatively poor heat-transfer relationship to both the evaporator and the bottles in the vending mechanism. Products placed in this area cool gradually and, if the heat-transfer characteristics are balanced, will be cold by the time those in the vending section have been used. The precooled bottles are then loaded into the vending columns and are replaced with more warm bottles. One advantage of this arrangement is that warm product can be loaded into the vendor without greatly increasing the temperature of the product in the vending position. The main disadvantage is the lack of positive stock rotation, since extra work is involved in moving the bottles from the precooling area. Thus, the attendant often loads directly into the vending mechanism.

The trend has been to incorporate as much available vendor space as possible into the vending mechanism, which is usually designed for first-in, first-out delivery, thus ensuring positive stock rotation. When this is done, a reserve of cold bottles is usually left in each column by means of a sold-out device. This arrangement ensures that cold product can be vended immediately on reloading and also provides time for the additional product to cool. The total capacity of a machine should be slightly more than the expected peak sales between reloadings, plus capacity to last until the reloaded product is cooled.

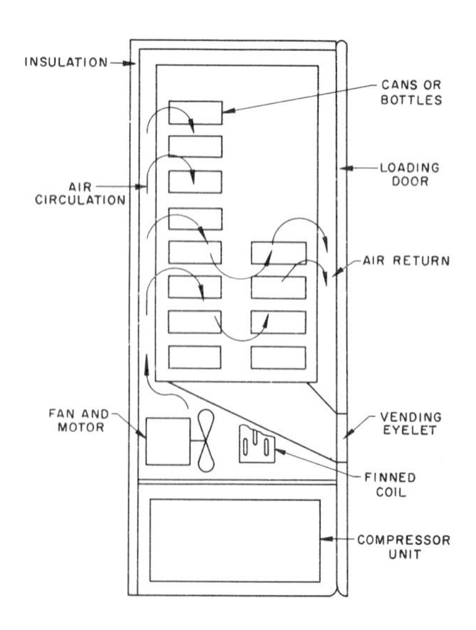


**Fig. 2 Typical Vertical Forced-Air Bottler or Double Depth Can Vendor**

## Temperatures and Controls

The expected range of ambient conditions is an important factor in design. Outdoor locations can cause exterior cabinet surfaces to reach 160°F (71°C) in direct sunlight or require operation at ambient conditions below 40°F (4°C). The product must often be cooled from above 100°F (38°C) to the desired temperature of 34°F (1°C) or lower.

Control requirements include the capability of holding bottle temperatures within 2 or 3°F (1 or 2°C) of the desired point and ensuring adequate defrosting of the evaporator, all under widely varying ambient conditions. Defrosting is accomplished by several methods: (1) the hot gas system; (2) a timing device operating heaters or ensuring an adequate cutoff period every 24 hours; or (3) the use of a constant cut-in control with a fixed cut-in temperature of approximately 35°F (2°C). This latter control must have a wide differential, as the cutoff temperature is usually 12 to 25°F (−11 to −4°C).

To ensure operation under low ambient conditions, controls with a cross ambient feature (such as liquid-charged bellows) are sometimes used, or the entire control is located inside the cooling compartment (see Chapter 19).

Because higher ambient temperatures cause longer running times and bottle temperatures are not usually sensed directly, higher ambients often result in colder bottles. Careful control-bulb location and refrigeration unit design reduces this characteristic to acceptable levels. If a capillary tube expansion system is used, an additional control is sometimes provided to stop the condenser fan at low ambients to provide better functioning of the capillary tube.

## Size and Shape of Product Containers

**Bottles.** Many different sizes and shapes of bottles are used for beverages, with each brand usually having a distinctive size and shape. Bottle dimensions are critical in some vending mechanisms. Information regarding the bottle length and major diameter; weights of the bottle, liquid, and cap; the specific heat of the filled bottle; and the freezing point of the beverage may be obtained from the bottling plant. The wide variety of surface irregularities has not caused detectable cooling rate differences in bottle coolers tested.

**Cans.** Cans used for beverages for vending machines are usually 8 to 12 oz (237 to 355 mL) in capacity. Since these cans are compact, they can be stacked tightly. The capacity of a given cabinet can be increased nearly 50% over that obtained with a bottle load if the internal cabinet space can be used. This greater product load has some effect on the refrigeration load. Although the specific heat of steel is less than glass (0.11 *versus* 0.19), the conductivity of steel is higher. Therefore, the initial heat rejection by the refrigeration system has to be much greater for the can vendor than for a comparable bottle vendor. This difference, while technically significant, has not been noted by any applicable agencies, so tests used for can vendors are the same as those used for bottle vendors. A convertible bottle/can vendor's refrigeration system must satisfy the higher requirement of the different load.

## Freezing Point of Beverages

The freezing point of sugar-based beverages under carbonation pressure in the bottle varies considerably. While some beverages with low carbonation and low sugar form heavy ice crystals near 31°F (−1°C), others remain fluid in the capped bottle at temperatures as low as 21°F (−6°C). Also, sample bottles of the same beverage vary over a wide range, with some bottles showing large amounts of ice crystals at 27°F (−3°C) and others showing no ice crystals when sustained at 21°F (−6°C), even after being shaken and jarred.

When the cap (or crown) is removed from a bottled beverage, the freezing point changes radically. Release of the carbon dioxide pressure allows the beverage to assume a consistent freezing point, which, for many beverages, is approximately 29°F (−2°C). If the liquid is colder than this when the cap is removed, ice crystals form rapidly as the temperature immediately jumps to the normal freezing point. This may be desirable in some warmer climates of the United States, where customers want ice crystals or flake ice to form in their bottle when the cap is removed. To accomplish this, the bottle must be vended at a temperature within a narrow range of from 1 to 4°F (0.5 to 2°C) below the beverage's freezing point at atmospheric pressure. If the temperature of the bottle is lowered further, the ice forming causes the liquid and ice to spill out of the bottle.

The specific heat of most carbonated beverages is close to that of water. Dietetic beverages made from sugar-free syrups should be considered as having a specific heat equal to water; they freeze at 32°F (0°C). For determining cooling rates, test bottles may be filled with water to the usual beverage level without undue errors in performance values. To prevent freezing in the test bottles, it is standard to add enough ethylene glycol antifreeze to the water to lower the freezing point to approximately 20°F (−7°C).

## Air Circulation

Circulating air, used almost universally for cooling bottles, may be varied in quantities and velocities for particular applications. While a pound of air carries only approximately 0.2 Btu for each °F change in temperature as the air moves through the cabinet (0.84 kJ/kg per°C), the quantity of air is determined by the velocity needed around the bottles. Vending principles determine bottle positions in machines, so that the air circulation spaces around and between the bottles are predetermined. High velocities over the bottles can be obtained by concentrating the chosen quantity of air over fewer bottles. Air quantities of 1.5 cfm (0.7 L/s) per bottle to be cooled at about 150 fpm (0.76 m/s) velocity have been found satisfactory for nominal cooling rates. The use of 2.5 cfm (0.1 L/s) per bottle at a velocity of approximately 250 fpm (1.3 m/s) has resulted in an increased cooling rate of nearly 40%.

Static pressure differences across the fan blade seldom exceed 0.1 in. of water (25 Pa) in the typical vendor. Accurate static pressure readings are difficult to obtain in most vendors because of the turbulence in the confined spaces used for air passageways. Adequate space for the desired fan size allows use of conventional fan motors of 1400 to 1500 rpm (23.3 to 25 r/s), while a small fan space may require 2800 to 3000 rpm (46.7 to 50 r/s) motors. Since the power required increases as the cube of the fan speed, unwanted heat is the penalty for not allowing room for the desired fan and motor. Shaded-pole motors delivering 0.0125 hp or less (4 to 9 W output) have been satisfactory.

Condensate from the evaporator usually drains from the cooled compartment into a container or pan. A trap in the drain line aids in preventing air and moisture from entering the cabinet through the drain. Provision must be made for the occasional breakage of a bottle; therefore, the pan or container is usually removable for cleaning. Evaporative surfaces may be provided to dispose of the normal condensate water. Sufficient evaporative capacity should be provided to dispose of condensate collected during operation at 90°F (32°C) dry-bulb temperature and 90% relative humidity.

## Equipment Evaluation

The following considerations aid in evaluating beverage coolers:

1. The cooling rate should be balanced with the anticipated vending rates and load schedules.
2. The time required for cold bottles to become available after reloading and whether or not warm bottles should be vended immediately after reloading.
3. Temperature of product as vended should be uniform.
4. The product temperature should not vary excessively with changes in ambient temperature.
5. Refrigeration equipment should be readily accessible.
6. The ratio of evaporator and condenser volumes should be satisfactory for operation in the entire range of ambient temperatures anticipated in a particular location.
7. Cabinet and exterior tubing should not sweat excessively during periods of high relative humidity.
8. Cost of operation should be factored in.
9. The unit should be capable of operating satisfactorily at reduced voltages, which may be encountered.
10. Evaporator defrosting and condensate disposal should be satisfactory when the unit is operated during periods of high relative humidity.
11. Electrical components should meet the requirements of Underwriters' Laboratories.

### Refrigeration Calculations

The total load of a vendor is the sum of the product load and the heat gain through the cabinet. Conduction values calculated for the walls may be increased 50 to 100% to allow for losses across breaker strips, gaskets, etc. Heat removed from bottles of beverage may be calculated by considering the glass and the beverage separately. The specific heat of glass is 0.19 Btu/lb • °F (0.80 kJ/kg • °C), while the specific heat of the beverage may be taken as unity.

The product load for cooling bottled beverages is:

$$Q_b = (t_1 - t_2)(W_g C_g + W_p C_p)N/T$$

*where*

$Q_b$ = average load for cooling bottles and beverage, Btu/h (W)
$t_1$ = initial temperature of bottles, °F (°C)
$t_2$ = final temperature of bottles, °F (°C)
$W_g$ = weight of each bottle, lb (kg)
$W_p$ = weight of product in each bottle, lb (kg)
$C_g$ = specific heat of glass, Btu/lb • °F (kJ/kg • °C)
$C_p$ = specific heat of product, Btu/lb • °F (kJ/kg • °C)
$N$ = number of bottles
$T$ = time, h (s)

Heat leakage through the cabinet walls can be determined by using the appropriate conductivity and thickness of the insulation, cabinet wall area, and temperature difference. Additional losses through door seals and from other leakage must be approximated. For a vendor with a 100-bottle capacity, the total heat leakage would be approximately 300 Btu/h (90 W).

The cooling load may be reduced considerably in cabinets with zone cooling. Air may be directed so that two or three zones are cooled; then, the average bottle cooling load is reduced by an amount proportional to the amount of cooling done in each zone.

Water-cooling coils with drinking fountains located on the side of the vendor are popular in some areas. This additional refrigeration load is not usually considered in the design and can affect bottle cooling rate, depending on the water usage. The water-cooling possibilities of such an arrangement are usually limited.

## BULK VENDORS AND DISPENSERS

### Postmix Equipment

The term *postmix* applies to bulk carbonated-drink machines, connected directly to city water or some other source of potable water. These machines contain bulk syrup and a means for carbonating water. The actual mixing of the drink is achieved in the cup, since both water and syrup are measured and dispensed through independent hydraulic circuits. The refrigeration of the water in these machines is accomplished by various methods.

One method passes the potable water through a coil in a water bath in which an ice bank is built around refrigerant coils. An agitator in the water bath provides rapid heat exchange. Ice bank size is maintained by commercial controls designed for this purpose. This system provides reserve for peak vending periods until the ice bank is depleted. Another method uses a cast aluminum direct-expansion heat exchanger to cool the potable water as it passes through the coil.

The occasional drink vended from the machine should be as cold as drinks served on a continuous basis. To get a cold occasional drink, it is necessary that both the water and syrup be refrigerated before delivery into the cup. The syrup reservoirs need not be refrigerated, but storage life of the syrup may be extended if it is kept cool.

Installations requiring the refrigeration unit to be remotely located from the dispensing valves must have cooling between the two to avoid warmup with its resultant loss of drink quality. Warmup is overcome by placing the carbonated water and syrup lines in an insulated line through which cold water is circulated. The recirculating system ensures that even the occasional drink is delivered within established temperature standards.

Standard carbonators, which introduce a controlled amount of $CO_2$ gas into the potable water supply, accomplish the carbonation. The quantity of water that may be carbonated at one time can vary from one to several drinks. This incoming potable water should be filtered to remove the chlorine and impurities for proper carbonation and suitable taste.

A variable orifice-type regulating valve controls the flow of both water and syrup, although in some machines, a positive displacement pump supplies the syrup to the cup. However, the cost and complexity of such a system is greater than that for the metering-type regulating valve. When the regulating-valve system is used, the syrup storage tanks must be pressurized to provide motive force for the dispensing of the syrup.

In a pressurized $CO_2$ system, the syrups are exposed to and absorb gas. This pressure must be regulated to prevent excessive foaming when the drink is dispensed. Dietetic syrups usually absorb more gas than sugar-based syrups; when both syrups are dispensed from the same machine, each requires its own pressure regulation.

The *Bag-in-Box* is a more recent method of handling bulk syrup. The syrup is typically filled into a 5-gallon (19-liter) plastic bag, which is then placed in a cardboard box for shipping and handling. Since these bags cannot be pressurized, the syrup must be pumped out of the bags to the beverage-dispensing system. Gas- or electric-driven diaphragm pumps are typically used for this purpose. Carbonation of the syrup is not a problem with this system.

### Premix Equipment

The term *premix* means that the beverage is mixed and packaged in bulk quantities in sanitary, sealed containers before delivery. Premixed beverages are distributed in 2.5, 5, and 10 gal (9.5, 19, and 38 L) stainless steel tanks, properly mixed and carbonated for ready consumption after cooling. An approved polyethylene hose is used to connect the tanks to the machine. It usually has stainless steel braid reinforcement and sanitary shutoff valves, which close when the connection is broken. The cooling coil carrying the beverage is made of an approved stainless steel, and all joints are welded. Tanks are usually connected in series with a regulated $CO_2$ pressure outlet connected to the tanks to force the liquid through the circuit.

The number of tanks that can be connected in this manner is limited by the foaming characteristics of the beverage. The more tanks connected, the greater the pressure required to force the beverage out of the dispensing valve at a flow rate suitable for the particular metering system used. As pressure is increased, the beverage in all tanks absorbs more $CO_2$. A typical beverage containing more than 3.7 volumes of $CO_2$ will foam excessively after passing through the uniform flow control and dispensing valve into the cup. The machine must also be capable of dispensing the first drink cold even after long periods of standing in the sun.

A typical dispensing cycle can be explained as follows: Gas entering the gas port of the product tank pressurizes the product in the tank. When the dispensing valve is opened, this pressure forces the product up through the dip tube, through the coupling tubing, and into the next product tank. This process continues until the product is forced out of the last tank, through a cooling mechanism, and then through the tubing to the valve. The dispensing valve can be opened either automatically or manually to deliver the product into a cup.

Current practice is to cool the beverage after it leaves the tank by a flash cooler, which consists of a sanitary coil immersed in a eutectic solution or water, located adjacent to refrigerant coils. An ice bank is usually provided for additional capacity for periods of peak dispensing. The required rate of heat removal from the walls of the beverage coil during peak flow periods is so great that methods of conducting this heat to the ice or eutectic may be employed. Various types of extended surfaces are used to overcome the resistance of the water film adjacent to the beverage coil.

The evaporator coil is placed in relation to the beverage coil such that the water or solution being frozen is in the most suitable position. If the evaporator is located too far from the beverage coil, lag in rebuilding the latent-heat bank occurs; if the beverage and evaporator tubes are too closely in contact, the beverage can freeze. A dependable control must be used to keep the evaporator temperature above the beverage freezing point. The amount of ice needed for cooling a given amount of beverage can be calculated from the dispensing rate required.

Placing the evaporator in a heat-exchange relation with the beverage coil has the advantage of adding the refrigeration system capacity to that of the latent storage; this aids in heat transfer. To prevent freezing the beverage, however, the critical temperature control must be considered carefully.

Carbonation and freezing points vary with the amount of carbonation and different flavors. Freezing points ordinarily range from 31 to 27 °F (−1 to −3 °C) for sugar-based products. Nonsugar or diet products freeze at 32 °F (0 °C). When beverages of two or more flavors are dispensed in one machine, the temperature must be adjusted for the one with the highest freezing temperature.

### Beverage-Dispensing Equipment

A *beverage dispenser* is usually defined either as a self-contained, hand operated, integral refrigerated unit or as a remote dispensing installation, whose refrigeration unit and dispensing system are located separately. This equipment is usually assembled to dispense one of the following products through either an electrical or mechanical valve:

1. *Premix,* which is a mixed, finished drink stored in bulk quantities in sanitary sealed containers.
2. *Postmix,* which is a finished drink when drawn from the dispensing valve, mixed from potable water and syrup cooled within the dispenser, and carbonated (where applicable) by a separate assembly, which is either cooled within the dispenser or operated at room ambient from an external location.
3. *Water or other soft drinks,* juice, fruit drinks, etc., either carbonated or noncarbonated.

The integral dispenser has a self-contained refrigeration system consisting of a condensing unit coupled to a refrigerant coil placed in a water bath. An ice bank is built around the refrigerant coils to handle peak loads. An agitator in the bath provides rapid heat exchange. The syrup lines on postmix equipment are usually routed through the chilled water bath.

The refrigerant coil and product circuits are sometimes located within a dry-type flash cooler, which uses metals or plastic as a direct heat-transfer medium between them. This equipment sometimes uses ice as a primary refrigeration source. The ice may be manufactured either by auxiliary equipment located adjacent to the dispenser or at a remote location. The ice that cools the cold-plate evaporator must be repacked and replenished, as required by the dispensing rate of the beverage equipment.

At locations requiring more than one dispensing station, satellites containing only dispensing valves are coupled to a remote flash cooler installation. This remote cooler is usually of the water bath-type, using water circulation for heat transfer. The cooling rate of the flash cooler and the size of the ice bank are determined by the number of dispensing stations and the total quantity of cooled product dispensed per unit of time. The dispensing valves and product lines connecting the station to the remote cooler also require cooling, which can be accomplished by circulating the water used in the flash cooling unit through tubes adjacent to or near the tubes and valves containing the product to be vended. These cooled areas or lines must be insulated to ensure maximum cooling of the product circuits and control the formation of external condensation on the chilled lines.

A refinement in cooling remote stations on postmix installations is the use of an external pump to recirculate carbonated water constantly through the remote tower. This water is subsequently bled off into each drink during the dispensing operation. This system provides water for *casual drinks* (occasional drinks drawn at extended intervals) that are cooled to the same temperature as drinks drawn at a rapid rate.

### Placement of Equipment

The placement of premix and postmix equipment is predicated on the ability of the equipment to satisfy a particular location's *peak product draw requirements,* which may be defined as the number of drinks of a prescribed size that can be vended at a desired rate to a maximum desired temperature before the reserve cooling capacity is depleted. Both the time required to rebuild the ice bank and the draw rate the machine can sustain after the reserve cooling has been depleted must be considered. This sustained rate of vending would be the peak rate on equipment that does not include an ice bank or some other type of reserve cooling.

The vending capacity of postmix beverage vendors is limited by the machine's cup capacity. Vendors can dispense up to 1800 cups of 7, 9, and 10 fluid oz (207, 234, and 296 mL). With such flexibility, this type of vending equipment has been readily accepted for use in field locations where the use of bottles or cans is impractical and bottle loss or damage has been excessive. These machines also have a high product storage capacity. However, they must be serviced frequently to ensure proper sanitation.

## OTHER REFRIGERATED VENDING MACHINES

### Ice Makers

Ice makers are commonly included as an integral part of cup drink vendors to provide from 1 to 2 oz (30 to 60 g) of flaked ice (see Chapter 40). They are compact systems using a 0.025 or 0.33 hp (20 to 250 W) compressor and have a capacity of 100 to 200 lb (45 to 90 kg) of ice per day. The complete system includes a water reservoir with means to maintain proper water level in the ice maker, the ice machine, an insulated storage-dispensing hopper, and the ice chute to the cup.

The ice making is usually controlled by an ice level switch assembly at the top of the storage-dispensing hopper, which shuts off the refrigeration unit when the hopper is full. The dispensing operation is controlled by the vendor's dispensing cycle. An agitator in the storage-dispensing hopper keeps the ice in a suitable condition for vending. Various methods are used to govern the amount of ice dispensed in each cup. One method uses a variable timed opening of the hopper's dispensing port. Another method uses ice-measuring cups, which may be changed for different ice throws.

Melted water from the storage-dispensing hopper is refrozen in some machines, while others use various heating and air-movement devices to evaporate it. Because the vendor is unattended, some systems have additional controls that shut off the ice maker if the water supply is cut off. Shutoff controls are also used to sense an inoperative ice-scraping auger or gear motor failure.

### Refrigerated Food Vendors

Packaged ice, milk, and other perishable food products are now vended in various sized containers from automatic merchandisers. Machine capacities range from about 70 to 300 cartons. The wide variations in the configuration of stored cartons and vending mechanisms require careful consideration in the selection of the type of refrigeration system to be used. Some configurations can lend themselves to forced-air circulation; others may require the use of plate-type evaporators. The forced-air systems are usually more effective in controlling temperature uniformity of the stored product. Perishable food should be stored at temperatures between 34 and 40°F (1 and 5°C). These vendors must be manufactured to conform to the requirements of the U.S. Public Health Service Sanitation Code.

In frozen food vendors, the product must be stored at temperatures of −10 to 10°F (−23 to −12°C) and in ice vendors at approximately 20°F (−7°C). When forced-air circulation systems are used in these vendors, provisions for adequate defrosting of the evaporator must be provided that will not affect the quality of the product.

On vendors using plate-type evaporators, space must be provided for frost buildup. Since this system does not readily lend itself to automatic defrost, provision must be made for sealing moisture out of the refrigerated compartment. The prevention of moisture migration into machines used to store food products below 32°F (0°C) is important during loading, when it may be necessary to use baffles or inner doors to prevent moisture migration into the cold evaporator.

## FROZEN BEVERAGE EQUIPMENT

*Frozen beverage* is the popular name for beverages that are dispensed in a semifrozen state and can be either consumed through a straw or with a spoon. Equipment is presently available to serve both carbonated and noncarbonated frozen beverages. Carbonated frozen beverage machines are usually of the sealed, pressurized type and use carbon dioxide pressure to force the frozen beverage from the freezing cylinder. However, some carbonated frozen-beverage machines and most noncarbonated beverage machines use an unpressurized storage vat, along with gravity and augering, to dispense the frozen beverage into a cup.

Carbonated frozen-beverage machines are normally of the postmix design; they incorporate equipment to carbonate the water, blend it with the syrup, and supply the mixture to the freezing cylinder automatically. Most noncarbonated frozen-beverage machines are designed to be filled by pouring noncarbonated premixed product manually into the holding reservoir.

Most frozen-beverage machines refrigerate the walls of the freezing chamber, in which the beverage is frozen and stored prior to dispensing. As the beverage is frozen, continuously rotating blades scrape it from the walls as fast as it accumulates. When sufficient frozen beverage has accumulated, the compressor cycles *off* by means of either a torque control that senses the viscosity of the frozen beverage or a temperature control that senses its temperature. Torque controls are the most popular because beverages freeze at different temperatures, depending on their sugar content. Torque controls function independent of temperature. Normal temperatures of the frozen product vary between 26 and 28°F (−3 and −2°C).

All frozen-beverage machines dispense a drink with a degree of *overrun* (expansion). Overrun of noncarbonated frozen drinks occurs primarily because of expansion from the freezing of ice; overrun of carbonated frozen drinks occurs from both freezing and breakout of $CO_2$ gas. Dispensing valves are normally designed to promote the amount of overrun, which usually runs between 50 to 100%. Syrups used in frozen-beverage equipment are usually prepared to enhance the quality of the drinks dispensed and to improve their freezing characteristics. Adding stabilizers and foaming agents to the syrup promotes the desired overrun by both keeping the ice crystals from growing too large and keeping the $CO_2$ bubbles small.

## APPLICABLE STANDARDS

The following standards have been developed to ensure the safe, sanitary, and proper operation of beverage vending equipment.

ASHRAE *Standard 32-1982, Methods of Testing and Rating Beverage Coolers,* describes a uniform procedure for testing refrigeration systems, determining performance data, and rating bottle capacity for self-contained, mechanically refrigerated bottled beverage coolers and vendors.

Underwriters Laboratories *Standard* 541-79, *Refrigerated Vending Machines.*

ASHRAE *Standard* 29-78, *Methods of Testing Automatic Icemakers.*

ASHRAE *Standard* 91-76, *Method of Testing for Rating Pre-Mix and Post-Mix Soft Drink Vending and Dispensing Equipment.*

U.S. Public Health Service *Publication* 546, *Sanitation Ordinance and Code for Vending of Foods and Beverages.* The U.S. Department of Health and Human Services adopted the *Sanitation Ordinance and Code,* which specifically applies to the vending of foods and beverages. While compliance is voluntary except at government installations, many municipal, county, and state health agencies require food-vending operators to meet this code. The code requires screens to vermin-proof a machine, a manual-reset thermostat to stop all vending if the food temperature exceeds 45°F (7°C), and other sanitary provisions.

NSF *Standard* 25-80, *Vending Machines for Food & Beverages.* National Sanitation Foundation, Ann Arbor, MI.

ASME *Standard* F2.1-82, *Food, Drug, and Beverage Equipment.* American Society of Mechanical Engineers, New York.

The National Automatic Merchandising Association, Chicago, has established standard of sanitation, safety, design, and construction of various vendors.

The Canadian Standards Association has developed standards that must be met by machines sold in Canada.

Many larger cities have their own health codes, which must be individually considered.

# CHAPTER 40

# AUTOMATIC ICE MAKERS



THIS chapter addresses commercial-size automatic ice makers—their construction, operation, and application. Specifically omitted are ice makers used in domestic refrigerators (covered in Chapter 37, "Household Refrigerators and Freezers") and large plant units requiring trained personnel (covered in Chapter 34 of the 1986 REFRIGERATION Volume). This chapter is concerned with machines that make small, fairly uniform pieces of ice, such as ice flakes or ice cubes, up to approximately 2 oz (60 g) in mass. These units usually include a bin or ice storage facility and are arranged and controlled to maintain a certain stated amount of product in storage. Unit capacities range from production rates of 15 lb (6.8 kg)/24 h to several tons (hundred kilograms) a day.

Fundamentally, ice means spot cooling. The presence of water as a result of meltage is often used in applications where products, especially vegetables, tend to lose weight (through dehydra tion) and become less attractive in appearance. Today, the largest markets for ice makers are bars, restaurants, cafeterias, soft drink parlors, motels, hotels, hospitals, fish and vegetable markets, concession stands, fast-food services, and prepackaged ice for home use.

Although ice can be made from sea water (with use usually confined to fresh fish preservation) or almost any other water, ice in the automatic ice-maker industry is customarily made from potable water and is kept relatively uncontaminated. Further classifications are *clear ice* and *cloudy ice;* there are different degrees of cloudiness, depending on the ice-maker design.

Machines using a batch process make clear ice particles or chunks; machines using a continuous process either produce cloudy flakes or they compact the flakes into more or less opaque pieces of various geometric shapes. An appreciation of the characteristics and source material (water) of product ice is important not only to the design engineer but also to anyone wishing to apply, specify, or purchase ice-making equipment for specific applications.

Fundamentally, ice is crystalline in structure and follows the physical laws of crystalline material. However, water is seldom pure $H_2O$ and, depending on its source, is generally made up of a complex mixture of layers of tiny water crystals interspersed with layers of an amorphous matrix of other chemicals. This composition accounts not only for the clarity (or lack of it), but also for a number of other qualities of ice, such as hardness or, particularly, the shearing quality.

Factors influencing the actual character of the ice formed are not only the water of which it is made but also the freezing rate and degree of washing of the interface between the already frozen ice and the water to be frozen.

The process of removing heat from a body of water to turn it into ice is normally a concentrating process that tends to freeze pure $H_2O$ and leave the remaining liquid water with a higher percentage of extraneous chemicals than it had at the beginning of the process. This concentrating occurs at the interface between the newly formed ice and the water surrounding it. If the freezing rate is too fast, the rejected chemicals are frozen into a matrix surrounding the pure $H_2O$ crystals, and the remaining water is consumed or converted into ice at the original concentration. If, however, freezing is slow enough or the ice interface is washed by a flow of water, these rejected chemicals can be removed and the resulting ice is significantly purer than the water from which it is made.

Continuous ice makers make ice that contains most of the chemicals in the original water, although some rejection of extraneous chemicals may result if sufficient bleeding of the makeup water mixture is continuously maintained. Units operating on a batch cycle, however, tend to produce ice that is purer than its source water.

Absolutely pure water is a rare substance where ice makers are concerned. Since water easily dissolves many substances, many impurities are always present in natural water, the most common of which are compounds of sodium, calcium, magnesium, and iron. In addition, the water may carry such suspended impurities as fine clay and sand and fragments of vegetation, as well as entrapped air and microscopic organisms of all sorts, including bacteria.

Calcium and magnesium salts make water hard and tend to come out of solution and deposit in an ice maker, eventually making the ice maker inoperative if periodic cleaning and lime removal are not performed conscientiously. Typhoid, cholera, and dysentery are primarily spread by infected water supplies, so only water that is bacteriologically and chemically safe must be used for product ice that is to be used in beverages or is to come in direct contact with food.

Practically all domestic water systems have different mixtures of the same common chemicals. Whenever the total chemical content exceeds 400 ppm (mg/kg), or if especially objectionable gases or compounds are present, auxiliary water treatment is indicated.

## DEFINITIONS

The following terms are used in the ice-making industry:

**Pounds (Kilograms) of Corrected Products,** or **144 Btu (334 kJ) Ice** describes an imaginary product that absorbs 144 Btu/lb (334 kJ/kg) while melting at 32°F (0°C).

To convert the product of any particular machine to 144 Btu (334 kJ) ice, a sample of the actual product as it leaves the evaporator is tested in a calorimeter. The reading in Btu/lb (kJ/kg) of product is then divided by 144 (334), which gives a multiplier to be used with that machine to establish an equitable comparison of lb (kg) of ice/24 h among all types of machines.

**Cube Ice** normally refers to a fairly uniform product that is hard, solid, usually clear, and generally weighs less than 2 oz (60 g) per piece—as distinguished from flake, crushed, or fragmented ice.

**Flake Ice** or **Flaked Ice** is made in a thin sheet anywhere from approximately 0.06 to 0.24 in. (1.6 to 6.4 mm) thick. The sheet may be flat or curved, but the thin ice is generally broken into random-sized flakes when harvested. The term also refers to machines that produce ice in this manner and, additionally, compress or extrude a product in larger chunks that either resemble pebbles or are in irregular but fairly uniform masses.

**Crushed Ice** is made in hard masses that are later crushed into a smaller size. This ice is characterized by the amount of fine or slush ice mixed in with the more uniform larger chunks.

**Blowdown** or **Bleedoff** is the rejection of a certain amount of recirculated or ingredient water to control the amount of chemicals that are present because of the concentrating effect of water frozen into ice.

**Harvesting** or **Harvest Cycle** (sometimes called the **defrost cycle**) is the removal or separation of the manufactured ice from the evaporator.

**Auger** originally referred to the scraper or helical wedging device that rotated inside a cylindrical evaporator, removed the ice from the interior wall, and pushed the separated ice up and out of the evaporator section. Later, the term was applied to the ice remover (usually helical), even when the ice was made on the outside of a small, vertical, cylindrical evaporator.

**Density,** as used in this chapter, refers to the weight per unit volume ratio, *e.g.,* lb/ft$^3$ (kg/m$^3$), of a sample of the product from a given machine and not to the density of an individual particle of product. The definition is useful in determining the amount of ice that can be stored in a bin having a known usable storage volume.

## ICE-MAKER CONSTRUCTION

### Types of Product and Evaporators

Flaked ice is uniformly thin, randomly shaped in its perimeter, and sometimes clear but more often cloudy or actually white. Evaporators currently used to produce flaked ice are of three main forms: cylinders, flat plates, and flexible belt.

The most common evaporator is a cylinder, usually brass, with refrigerated tubing wrapped on the outside and bonded to it. The evaporator may also be two concentric cylinders with water on either the outside of the outside cylinder or the inside of the inside cylinder, and with the annular space between cylinders occupied by the refrigerant. The simplest form, however, is a cylinder with the refrigerant on the inside and the water to be frozen on the outside. These evaporators require augers, helixes, or equivalent components to remove the ice.

Both cylinder types can be made with an extended auger section in which the ice flakes removed are compressed and the resultant product extruded and ejected. This compressed product differs considerably in both density and appearance from the uncompressed flaked ice and can be binned and dispensed far more readily.

The cylinder evaporator is subject to many variations. Both clear and opaque ice can be produced, depending on the means of distributing the water. The cylinder can also be stationary or rotated, as well as placed in a vertical or horizontal position. The harvesting means is also open to many variations, depending on whether ice is formed on the inside or the outside of the cylinder.

The cylinder evaporator has been produced to meet capacities of from 100 lb to 20 tons/day (45 to over 18,000 or more kg/day), using the whole range of refrigerants from the halogenated fluorocarbons and methyl chloride to ammonia. Larger flexible-cylinder machines have also been refrigerated with circulated cold brine. In general, if the type of ice produced is satisfactory for the intended application, this evaporator form can be the least expensive to produce and the most efficient to operate thermodynamically. These machines are almost exclusively operated on a continuous cycle rather than in a batch process, although batch-process machines are available when clear ice is sought. Cracked ice can be produced on a flat evaporator, using a batch process and either an ice crusher or some other ice-breaking feature.

The *belt-ice* method is used mostly in industrial-size machines. Belt ice is produced from a metal belt turning over two pulleys—the belt sliding upward across a slightly-curved inclined refrigerated surface. Flowing water strikes the belt near the upper end of its length and flows downward, forming ice as it flows. The ice thus formed continues past the point of impact of the water and leaves the belt in a ribbon or sheet as the belt passes over the upper pulley.

Cube ice is produced from a great variety of evaporators. Most of these machines fall into five main types: flat plates (using either the whole surface or selected spots); multiple cells or molds; tube machines; rod machines; and channel machines.

**Flat-plate evaporators** may be vertical or in an inclined position; it may have water flowing on either the top or the underside of the evaporator. The main characteristics are that the ice is formed in a rectangular slab during the freezing cycle, and it is usually quite clear. During the harvest cycle, the evaporator plate is heated, and the ice drops or slides off. This slab is then cut into square or rectangular pieces, most often with a grid of either electrically heated wires or small tubes carrying a heated fluid. The slab thickness can usually be varied at will, but the cube sizes are determined by the previously selected grid pattern.

There are several other versions of the flat-plate evaporator in which grids of metal or plastic are pressed against the flat evaporator and a third movable element closes the side opposite the evaporator plate, thereby forming individual cells. Recirculated water is supplied to each cell. In these versions, the individual cubes are ejected during the harvest cycle, and no cutting is required.

**Cell-** or **mold-type evaporators** make the individual cubes without any additional cutting. The cups or cells may be any shape from round to square to polysided. They are usually inverted, and water is sprayed up into the cavities to produce clear ice. During the harvest cycle, the water is shut off, and the cavities are heated in various ways to free the ice cubes, which fall onto a grill that directs the ice to a chute or bin below.

**Tube machines** make ice inside the tubes, with the refrigerant outside the tubes. The ice is made in a long cylindrical shape, usually with a hole in the center, and is broken into short lengths as it emerges from the evaporator. The ice is kept clear by making sure that an excess of water is continuously passing through the tubes.

In one version, a series of tubes is held in place by tube sheets within a vertical shell. The shell and tubes act as a flooded evaporator during the freezing process and as a condenser during the harvesting cycle. Water is pumped to the top of the shell and flows by gravity down the inside walls of the freezing tubes.

Another version is a single tube within a tube. The ice is formed on the inside of the inner tube, and the refrigerant is in the annular space between the two tubes. Water is pumped through the inner tube until the resistance to flow becomes greatly increased, at which time hot gas is introduced into the annular space. When ice is free, it is forced out by water pump pressure and cut or broken into lengths of two or three diameters.

A third cube machine of the tube type forms separate ice bodies in square, vertical, stainless steel tubes, which are banded at uniformly spaced intervals with copper heat conductors, which, in turn, are refrigerated. Water flows down the inside of the tubes and turns to ice in the region of the bands, thereby producing clear ice cubes with an hourglass-shaped hole through the center. Hot-gas harvesting is used, and the cubes fall by gravity into a storage bin below.

**Rod units** usually employ a series of short refrigerated rods (or, in some cases, short tubes) sealed at the bottom, which drop into a tank of flowing or agitated water. Clear, thimble-like pieces of ice form, and at a predetermined time, the water tank is removed (or the fingers are raised). When the hot gas thaws the ice loose, the ice falls free and is guided into an opening that leads to an ice bin. Both the length of the ice piece and its outer diameter can be varied.

**Channel-type evaporators** are made by forming thin, stainless steel sheets into a series of channels. These sheets are then mounted upright so that the channels are vertical. On one side, water flows down the sides and bottom. On the opposite side and at right angles to the channels is a refrigerant-carrying serpentine tube, which is bonded to the stainless steel at equally spaced intervals, thus forming a series of cold spots over which the water flows. When hot gas replaces the cold refrigerant, individual pieces of ice are freed and fall into a bin below.

Almost as many types of evaporators and ice machines have been built as there are makes of automobiles. The machines that have survived have the fewest number of moving parts, are economical to manufacture, and have controls as close to fail-safe as possible. As long as the ice is compatible with its ultimate use, the acceptance of any specific design is based as much on reliability as on efficiency or even first cost.

## The Complete Ice-Maker Package

Automatic ice-making equipment can be purchased in a wide variety of models, with some companies offering as many as three or four different ice shapes. Some units are made up of separate modules capable of being assembled into complete units to satisfy a wide range of applications. These modules often include an automatic ice-making section for either cubes or flakes; the section shuts off when it is set on a container or bin of some kind. Crusher sections are offered that can be set to produce crushed ice from ice cubes and direct uncrushed ice into one bin section and crushed into another. Various bin sizes with different access ports are often available. Additionally, sanitary dispensing units are offered as an alternate to a simple ice bin. Other add-on accessories include water stations for filling drinking water glasses with ice and water and soda fountain dispensing units.

In addition to the modular approach, units are specifically designed for many special applications. These special units may be designed to conserve floor space or fit under counters. Hospital units must protect the product from contamination, so the units must be able to withstand thorough cleaning and sanitizing by cleaning personnel.

Coin-operated machines that make, bag, and vend ice (all automatically) have not proved to be economical because the cost and complication of the machines are too great for the yearly volume of ice sold. However, vending machines that simply vend bags of ice are gaining acceptance. In this case, the ice is made and bagged elsewhere. This operation is especially economical in connection with merchandising requiring refrigerated holding rooms for storing backup supplies.

## Performance and General Operating Characteristics

Typically, 100 lb (45 kg) of opaque ice flakes produced by a 0.5-hp (373-W) unit requires approximately 3 kWh (10.8 MJ); 100 lb (45 kg) of clear ice cubes by a 0.5-hp (373-W) unit requires approximately twice as much. These values are for 90 °F (32 °C) ambient, 70 °F (21 °C) water, and air-cooled units. Water consumption of air-cooled units is approximately a lb (kg) of water/lb (kg) of ice on flaked machines and up to 50% more for clear ice makers. Heat rejection rates for all 0.5-hp (373-W) units are approximately 8000 Btu/h (2.3 kW) for an air-cooled unit.

## System Design

All automatic ice-making systems can be broken down into the following subsystems:

1. Refrigeration circuit
2. Water (for ice making) circuit
3. Ice removal and/or harvest system
4. Unit electrical controls
5. Ice storage and dispensing systems, where included
6. General safety and sanitation codes as they apply to each of the above systems
7. Special application and specifications that may bear on individual system designs

**Refrigeration Circuit.** Any refrigeration system is a balance of the compressor capacity, the condenser capacity, and the evaporator capacity. In the previous sections, the types of evaporators in current production were reviewed. The thinner the ice produced, the more thermally efficient the production method; as a corollary, the more prime refrigerant surface per in.$^3$ (mm$^3$) of ice, the more efficient the heat transfer. Also, a continuous process without a harvesting and a freezing cycle is thermally more efficient than a batch process. The determining factors in any evaporator design are (1) production of the kind of ice required for the specific end use; (2) manufacturing economics; and (3) the ease of cleaning and extent of service required to keep the evaporator running for at least eight or ten years.

The best heat-transfer metals are the most desirable for an evaporator. The surface on which ice is made, however, must be corrosion resistant, nontoxic, nonporous, and readily cleanable; most important of all, the refrigerant passages must be clean and clear of all oxides and corrosive fluxes. The cross-sectional areas must be uniform so gas velocities do not fall below those necessary to prevent oil separation or accumulation.

The high side (compressor, condenser, and receiver) is generally made up by the ice maker manufacturer from separately selected components suitable for the specific design, rather than from stock high-side units furnished by compressor manufacturers.

Both air- and water-cooled condensers have their particular application advantages. The air-cooled units are the simplest; however, proper ventilation of the installation area may not be practical, or the rejected heat may be too much of a load for the air-conditioning system if the unit is installed in an air-conditioned space. Thus, water-cooled units may be the simplest way of conducting all the rejected heat from the installation area. An alternative is a remotely located air-cooled condenser, which rejects heat away from the air-conditioned space and yet retains the economy of the air-cooled system.

Commercial high sides are usually low-temperature units for flaked-ice machines and medium-temperature units for cube-ice machines (batch process systems where hot gas is used for defrost). The compressor motors are generally of high-torque design.

Refrigerant 12 is most commonly selected, although both Refrigerants 22 and 502 are also used, especially in the larger capacities.

The liquid control means may be (1) an automatic expansion valve (especially good if the evaporator load is fairly steady); (2) a fixed superheat thermal expansion valve (the easiest to apply); or (3) a capillary restrictor tube (the least expensive to provide, but the most difficult to apply).

If a thermal expansion valve is used, a liquid receiver generally is required, but if an automatic expansion valve or a capillary tube is used for liquid control, a suction line accumulator is generally used.

Unless the manufacturer has sophisticated process equipment and a reliable quality-control system is in operation, all systems should have strainer-driers in the high side. Ice-making systems are essentially low-temperature systems and are, therefore, very sensitive to excess moisture in the refrigerant circuit, particularly in systems using Refrigerant 12.

For most cubers and other batch-process units, a hot-gas solenoid valve bypasses hot compressor discharge gas around the condenser and moves it directly into the evaporator during the harvest cycle. This is normally an electrically operated closed valve, which must be large enough to (1) provide as short a harvest cycle as practical without creating such a thermal shock to the ice that the ice tends to break up and (2) prevent causing excessive compressor overload.

High- and low-pressure controls may or may not be used in the refrigerant circuit. Most water-cooled condensing units use automatic-reset high-pressure controls. Depending on the extent of system safety, the low-pressure cutout is most often a manual reset control; this arrangement ensures that the unit will be investigated to determine the cause when an excessively low pressure develops.

**Ingredient Water Circuit.** The potable water used to make ice is normally a part of the city water supply. In the majority of flaked-ice units, a constant volume of water is maintained in the system by a float valve. This mechanically operated valve is usually mounted in a separate float chamber.

A flexible line runs from this float chamber to the evaporator section. A constant and fixed water level is maintained in the evaporator because of the vertical relationship between it and the water level in the float chamber.

Where clear ice is made, a water pump usually circulates water, taken from a water tank or sump, through the evaporator, from which the water returns to the sump. A float valve normally maintains a constant water level in the sump. The float valve in this case may be either mounted directly in the sump or have a separate float chamber.

The float valve and its enclosure must meet sanitary code requirements so that no back siphonage can occur to contaminate the primary water supply. This is usually accomplished by ensuring that the valve-outlet orifice is at least 1 in. (25 mm) above the float chamber or any possible flood level within the machine proper. Also, if such a flood develops (for instance, because of a failure of the float valve to shut off), the exposed electrical equipment must not be shorted out by splash or free-running water.

Unit protection from water failure with air-cooled condensers is normally limited to a low-pressure cutout in the refrigerant circuit. Not all units use water pumps. In some machines, water may be kept moving by a stirring device. Compressed air has also been used to agitate water adjacent to the ice.

The trend in ice makers is to have a minimum quantity of water in the water system at any one time. If the unit is of the continuous-process type and is making clear or fairly clear ice, a constant bleedoff must be used to control the mineral concentration. With this type of operation, no relationship exists between quantity of bleedoff and total system water quantity. In batch processes, where clear ice is made, either a constant bleedoff or a batch dumping of a certain amount of water at the end of each cycle is used. As stated earlier, keeping the water quantity to a minimum maintains a good flushing action at the end of each cycle. This dumping can be obtained by a siphon effect, initiated by stopping the water pump and allowing all the system water to flow back into the water sump, thus increasing the water level to the point where the siphon starts. Alternatively, it may be simply a matter of energizing a solenoid valve in the water sump drain line, allowing any given amount of water to escape during the harvest cycle.

In all cases, drainage must be provided at the lowest point in the water system so the unit can be drained completely after any routine cleaning operation, before shipping, and when it is prepared for winter storage. All materials that come into contact with the ingredient water or the ice must be nontoxic, noncorrosive, smooth, impervious, non odor imparting, cleanable, and durable. Minimum standards for sanitary design are specified in the National Sanitation Foundation *Standard* 12; listing by NSF is almost mandatory.

**Ice Removal** and/or **Harvesting Systems.** Usually an auger removes the ice from flaked-ice continuous-process machines.

The design of most of these machines is based on breaking the ice away from the evaporator at the interface between the ice and the metal freezing surface. This bond is not consistent in shear strength, however, and this, as well as the irregular pattern of the ice leaving the freezing surface, causes a highly irregular torque requirement on the auger drive shaft. As a result, side thrusts on the auger vary considerably. As designed, cutting or breaking of the bond is balanced equally between cutters approximately 180 degrees apart. When ice breaks loose irregularly, this balancing effect is lost. Side thrusts become of major proportion; thus, if close clearances are to be maintained between cutter and evaporator, the evaporator shell and auger must be quite sturdy.

The strength of the primary bond or fracture plane between evaporator and ice is influenced by many variables. Strength can vary with the temperature of the ice at the bonding plane, the water composition, the conditions of the evaporator surface, and the angle and shape of the cutting blade. These conditions can change from day to day or season to season. Thus, the ice-harvesting design becomes crucial, and this factor alone has been the downfall of many promising evaporator designs.

Most ice-cube machines are batch-process units that use hot gas to melt the bond between the ice and the evaporator. If the ice is made in cups, the adhesion or surface tension can create a vacuum and prevent the ice from falling. Either vacuum-breaking holes or some means of exerting pressure on the ice may be required to release the ice. When ice forms on ice rods, it must reach a critical mass to overcome the adhesive forces at the interface between evaporator and ice so it will drop off. Ice is abrasive, and this fact must be considered on all ice-handling components.

**Unit Electrical Controls.** The control systems for continuous-process machines are relatively simple, requiring only that the unit be shut down manually or automatically when an ice bin has been filled. For the protection of the machine, controls must shut the unit down whenever the water supply fails, an auger overloads its drive motor, excessively high discharge pressures are experienced, or abnormally low suction pressures are encountered.

A batch ice maker, which goes through a freezing and harvesting cycle, requires a more complicated control system than one that freezes continuously. A batch cycle must terminate the freezing at the proper time to start the thawing cycle and to cause the freezing operation to resume when harvesting is complete. Timers, pressure-operated switches, ambient-compensated thermostats, water-overflow actuators, two-element thermostats, ice thickness feelers, and combinations of these have been used. Selection depends on what is most suitable for the particular evaporator design. Additional control is needed to shut off the unit when the storage bin is full of ice; a temperature-sensing element, which can be contacted by the ice, is often used. Mechanical feelers and photoelectric devices have also been applied successfully.

All controls must be effective over a range of ambient air temperatures from 40 to 110°F (4 to 43°C) and supply water temperatures from 40 to 100°F (4 to 38°C).

Numerous safety devices help prevent damage to the apparatus or injury to personnel. These include (1) high-pressure cutout, (2) low-pressure cutout, (3) motor overload protector, (4) overfreeze protection, (5) fusable plugs in receiver shells, (6) safety switches interlocking with access panels, and (7) other devices that a particular design may require.

**Ice Storage** and **Ice-Dispensing Systems.** Most automatic ice makers provide an ice storage bin designed to hold approximately a 10- to 12-hour ice production. A certain demand does exist for units of a medium-to-large capacity of both flaked ice and ice cubes with no integral bins. These units are ideal for applications where high demands occur less frequently than every 12 hours. For example, country clubs have high weekend demands. Also, supermarkets often use ice beds to display vegetables, meats, or fish; these beds are remade only once or twice a week. Often, a standard bin is made by the ice maker manufacturer and holds the usual 12-hour production of ice. Bins made by independent ice bin manufacturers are better for larger quantities of ice for special applications.

Ice bins are usually built to NSF construction standards, with from 1.5 to 3.0 in. (40 to 75 mm) of insulation on the sides and top and from 2 to 4 in. (50 to 100 mm) of insulation on the bottom. These bins are seldom refrigerated because of the problem of the ice refreezing together, so drainage must be provided. Drains should never be less than 0.5 in. (12 mm) and preferably should be 0.75 in. (20 mm) in internal diameter. A strainer should be included at the drain inlet. Access to the ice can be achieved at the top, bottom, or side of the storage bin.

Ice is quite impervious to outside attack. However, as air near the ice cools, moisture and any contaminants present in the air are condensed onto the ice surface. Except for unsanitary areas, however, the greatest source of contamination lies in scoops, shovels, or other instruments introduced into the ice and then withdrawn and left in other unsanitary areas.

Many sanitary dispensers are available, but the first cost has held sales down, except in those applications where sanitation is crucial. Such an ice dispenser (1) must protect the ice from outside contamination while it is in storage; (2) must be readily cleanable so that frequent cleaning can be done by regular housekeeping help; and (3) should cost less than the ice machine. These objectives are met by many current designs, and sanitary ice dispensers are an increasing part of the standard product line for major manufacturers. A need for storing and dispensing uncontaminated ice continues to be of major interest to hospitals, motels, restaurants, and operations where ice may come in direct contact with food or beverages.

**Safety and Sanitary Standards.** The following standards apply to ice-making equipment.

National Sanitation Foundation *Standard* 12, *Automatic Ice Making Equipment.*

Underwriters Laboratories, Inc. (UL) *Standard* 563-84, *Ice Makers.* This standard covers electrical safety.

Canadian Standards Association (CSA) *Standard* 313 covers electrical safety of ice-maker products for the Canadian market.

Air-Conditioning and Refrigeration Institute (ARI) *Standard* 810-79, *Automatic Commercial Ice-Makers*, details the conditions and methods of rating. These ratings are based on ASHRAE *Standard* 29-78, *Method of Testing Automatic Ice Makers.*

Other standards, such as the ASME *Unfired Pressure Vessel Code,* may apply, as well as individual state and city plumbing and sanitary codes.

**Special Concerns.** Noise and vibration, while of little concern in a busy kitchen, should be considered for equipment installed in a dining room or quiet hospital area. A noise level equal to that of a window air conditioner is acceptable for most applications, although some specifications may quote specific NC sound power levels.

Condensation or sweating on the bin or ice-dispensing part of the equipment may be a concern. Generally, equipment is satisfactory for most applications if no moisture drips from the unit when it operates for four hours in an ambient of 90°F (32°C) dry bulb, 78°F (25.6°C) wet bulb.

Installation, operation, and owner maintenance determines the eventual life of the unit, no matter how well it is designed and built. Therefore, the designer should make the unit as simple and easy to clean as possible. In addition, operating and maintenance instructions that are clear and positive will help the owner extend the life of the machine.

**Special Applications and Specifications.** Several special applications require additional nonstandard design features. Shipboard application usually requires that units continue to produce ice when they are subjected to a 15-degree pitch and a 30-degree roll; for naval vessels, however, 15-degree permanent list and high shock are additional requirements. Also, a no-radio interference requirement is often added.

Explosion-proof units are sometimes required, but this is a special feature that normally is not a production item. Usually, odd electrical current characteristics can be handled by transformers, since direct current is now rarely encountered. For the international market, many manufacturers provide specific models that meet specific local voltage and frequency requirements. Occasionally, measures are required to prevent fungus and high humidity.

## APPLICATION

A wide selection of ice-maker designs, capacities, and ice shapes is available. The primary factors in selecting the best equipment for any specific application include the following considerations:

1. What type (or types) of ice best satisfies the application?
2. What storage or dispensing methods will be required?
3. Is the cost of ice of primary importance?
4. Is the cost of electricity and/or water of primary importance?
5. What is the quality of the available water supply?
6. Are there noise-level restrictions that must be met?
7. Is reliability and long life of primary importance?
8. Are service and maintenance facilities readily available?
9. What are the space limitations (if any) for the installation?
10. Are there any general environmental conditions or restrictions that must be met?

### Choice of Ice Types

Differentiating among the types of ice, two general physical characteristics are apparent: (1) ice can be clear, white, or have varying degrees of opacity; and (2) the individual ice pieces can come in a myriad of sizes and shapes. Generally, the shape description is cubes, flakes, crushed, or aggregate. This last designation is not generally used, but it is meant to cover the more or less uniform ice that results from thin ice flakes that have been compressed and extruded and then broken again as the extrusion is forced against a stop of some kind. (Trade designations are *pebble ice, granular ice, nugget ice,* or *cracked ice.*) This ice differs from crushed ice in that it has relatively few fines and is neither entirely clear nor entirely white, but has various degrees of cloudiness. Also, ice that is stored for a few hours may vary greatly from ice just coming out of the ice-maker discharge chute. The stored ice, unless cubed, is said to cure. In addition to having been drained of most of its excess water, it has changed its structure so that both its density and its thermal capacity (Btu/lb or kJ/kg) have changed. This change can be important when figuring bin capacities.

Sometimes, maximum ice size is important. For instance, it must be small enough to readily enter the opening of a patient's carafe for hospital use. It must not have so many fines that it clogs straws or glass drinking tubes.

Shape can also be important; for instance, ice flakes are suited for packing with flowers because they do not bruise the petals. This shape is also ideal for fish, because it tends to conform to the surface on which it lies. On the other hand, an egg shape has long-lasting qualities because it presents a minimum of outside surface to the liquid to be cooled. Also, various forms of ice react differently in bins. Some do not pack well, and some do not lend themselves to deep bins. Ice cubes generally flow more easily.

Probably the most important consideration for the selection of one ice form over another is user preference. In many applications, one type of ice is clearly superior to others, but, frequently, any form will do the job. When ice is to be delivered to specification, however, all ice-making capacities must be quoted from ASHRAE *Standard* 29-78, which converts all product poundage to 144 Btu/lb ice per 24 hours at stated ambient and water conditions. Also, ice bin capacities generally are understood to mean lb (kg) of cured ice and not 144 Btu/lb (334 kJ/kg) ice, unless the latter is specifically mentioned. Normally, these are not significant points, but they become important if strict specifications are involved or actual cooling ability is guaranteed.

In the beverage industry, the ice must not impart any flavor to the beverage. If the local water contains gases or solids that impart a distinctive odor or taste, a flake-ice machine may not be satisfactory, whereas a cube ice maker would eliminate most of the objectionable flavor. On the other hand, proper water pretreatment may make flake ice acceptable.

### Ice Storage

Most automatic ice makers not only produce ice, they also store it and provide some means of dispensing it. Storage is expected to be an insulated container with 2 to 3 in. (50 to 75 mm) of insulation designed for periodic cleaning or sanitizing. (Ideally, it will meet NSF standards). Additionally, it must be designed so that ice can be discharged from the bin with a minimal hazard that the withdrawal will affect the sanitary conditions of the remaining ice. (This ideal prevents the use of scoops or shovels that touch the ice remaining in storage.)

This design also precludes any openings into the ice-containing area in which packages or bottles can be left to cool and possibly contaminate the ice in storage. Whether or not the equipment is maintained in a sanitary condition is entirely up to the owner. The equipment supplier, however, must ensure that the equipment is capable of being maintained in a sanitary condition when straightforward cleaning instructions supplied by the manufacturer are followed.

Ice-storage capacities are commonly stated in terms of how much ice, or product, can be accommodated when the bin is full. Since ice has a natural angle of repose, the machine generally turns off before reaching the full bin capacity. This factor should be considered when selecting equipment. In any application, ice bin and machine production capacities must balance to fit the ice-demand cycle. Equipment must also be selected for maximum demand, and machine capacities at lowest production rates must be compatible with the installation environment.

### Ice Costs

Ice is a relatively inexpensive commodity. If ice is used in small quantities, the cost of making it is secondary to the space the ice maker occupies, the convenience of keeping the ice as close as possible to the point of use, or the original first cost. This relative importance must be kept in mind when choosing the ice form or size and when considering reliability. Thus, two small machines are safer than one large one, and the machine that makes ice for the least amount of money is the logical choice.

## Equipment Life

Unit life depends on installation, environment, and regular cleaning and maintenance. Unless these factors are considered at the time of installation, ice-making equipment must be classified as relatively short-lived. Where ice is added to food or beverages, the regular cleaning and inspection required for sanitary purposes should promote longer life. An awareness of the necessity for regular cleaning, not only of the ice-storage facilities but also of the ice-making parts of the ice maker, must be impressed on the user. Neglected equipment in areas of poor makeup water may have a life as short as three years, whereas if water treatment is used and the equipment is regularly cleaned and maintained, it could extend to a ten-year life.

## Installation

Proper installation is the foremost requirement of economical, maintenance-free operation. Manufacturer's recommendations should be followed closely.

**Table 1 Ice-Maker Application Data[a]**

| Classification | Application | Preferred Form of Ice | Use Cycle[b] | Consumption[c] |
|---|---|---|---|---|
| | | **Food and Drink Service** | | |
| Fast-food drive-in | Water, soft drinks | Crushed/flake | 7 days; 150% 2 consec. days | 0.25 to 0.5 lb (0.11 to 0.23 kg)/customer |
| Night clubs | Food-drinks | Cube | 7 days; 200% 2 consec. days | 2 to 4 lb (0.9 to 1.8 kg)/seat |
| Bar cocktail lounges | Drinks | | | |
| Carry home | Retail stores | | | |
| | Recreation areas | Cube/crushed | 7 days; 200% 3 consec. days | Varies by season, available in 5 and 10 lb (2.3 and 4.6 kg) size bags |
| | Marinas, fishing piers | | | |
| Caterers | Banquets | All | Varies | 1 lb (0.46 kg) per meal |
| | Truck (mobile) | Flaked/crushed | 7 days; 140% 5 consec. days | 200 to 400 lb (90 to 180 kg)/truck |
| | In-plant feeding | All | 7 days; 140% 5 consec. days | 0.5 to 1.0 lb (0.23 to 0.46 kg)/meal |
| Auditoriums, stadiums | | Crushed/flake | Varies | 0.5 lb (0.23 kg)/customer |
| Churches | Banquets | Cube | 7 days; 200% 2 consec. days | 0.5 lb (0.23 kg)/meal |
| Theaters | Snack bar | Crushed/flake | 7 days; 200% 2 consec. days | 0.5 lb (0.23 kg)/snack bar patron |
| Hotel-motel, resorts | Banquet, meeting room | Cube | 7 days; 200% 2 consec. days | 1.0 lb (0.46 kg)/meal |
| | Dining room | Cube/flake | 7 days; 200% 2 consec. days | 1.0 lb (0.46 kg)/meal |
| | Coffee shop | Cube/flake | Daily, 100% | 1.0 lb (0.46 kg)/meal |
| | Guest room service | Cube | Daily, 100% | 5 lb (2.3 kg)/room |
| | Kitchen | Flake/crushed | Daily, 100% | 0.5 lb (0.23 kg)/meal |
| | Buffets-display | Flake/crushed | Varies Daily, 100% | 10 lb/ft$^2$ (49 kg/m$^2$) of display area |
| Restaurants, cafeterias | Dining room | All | 7 days; 150% 2 consec. days | 0.5 lb (0.23 kg)/meal |
| | Serving line display | Flake/crushed | 7 days; 100% | 10 lb/ft$^2$ (49 kg/m$^2$) of display |
| | Kitchen | Flake/crushed | 7 days; 100% | 0.5 lb (0.23 kg)/meal |
| Military bases | Mess halls | Flake/crushed | 7 days; 100% | 0.5 lb (0.23 kg)/meal |
| | Post exchange clubs | Cube/crushed | Daily, 100% | 1.5 lb (0.68 kg)/customer |
| Prisons | Dining halls | Flake/crushed | Daily, 100% | 0.33 lb (0.15 kg)/meal |
| Airlines, airline caterers | In-flight feeding | Cube | 7 days; 150% 2 consec. days | 1.0 lb (0.46 kg)/meal |
| Clubs, country and private | Banquet, meeting room | Cube | 7 days; 200% 2 consec. days | 0.5 lb (0.23 kg)/meal |
| | Dining room | Cube | 7 days; 200% 2 consec. days | 0.5 lb (0.23 kg)/meal |
| | Bar, cocktail lounge | Cube | 7 days; 200% 2 consec. days | 0.5 to 1.0 lb (0.23 to 0.46 kg)/customer |
| | Kitchen | Crushed/flake | 1 day; 100% | 0.5 lb (0.23 kg)/meal |
| | Golf course | All | 1 day; 100% | 1 to 4 lb/gallon (0.1 to 0.5 kg/liter) of water cooler capacity |
| | | **Food Preservation** | | |
| Seafood markets | Merchandising (display) | Crushed/flake | 7 days; 140% 5 consec. days | 5 lb/ft$^2$ (24.5 kg/m$^2$) of display |
| Florists | Shipping | Flakes | Varies | 2 lb (0.9 kg)/box of flowers |
| Groceries | Display | Crushed/flake | Daily | 5 lb/ft$^2$ (25 kg/m$^2$) of display |
| | | **Medical** | | |
| Hospitals | Dietary | Crushed/flake | Daily, 100% | 0.5 lb (0.23 kg)/meal |
| | Nursing service | Crushed/flake | Daily, 100% | 5.0 lb (2.3 kg)/bed |
| Nursing homes | Dietary | Crushed/flake | Daily, 100% | 0.33 lb (0.15 kg)/meal |
| | Nursing service | Crushed/flake | Daily, 100% | 3 lb (1.4 kg)/bed |
| Athletic fields | Ice packs and drinking water | Crushed/flake | Daily, 100% | 3 to 5 lb (1.4 to 2.3 kg)/person per day |
| Construction | Drinking water coolers | Crushed/cubes | Daily, 100% | 3 to 5 lb (1.4 to 2.3 kg)/person per shift |

[a]Additional application information other than food, beverages, or display counters may be found under general refrigeration requirements by industry or product.
[b]These values assist in balancing bin capacity and ice-making capability to obtain the most economical combinations. The length of the normal use cycle is shown first. The next figure is an approximation of how much ice might be consumed during a peak period in the use cycle; it is expressed as a percentage of the average daily production. The last figure indicates how long the period of high consumption might last.
[c]Ice consumption is generally cyclic. Proper storage capacity helps ensure an adequate supply of ice during high-consumption periods. The values shown are average consumption for the indicated applications.

Often, the necessary drain and its pitch is overlooked when a location is chosen. The drain must take care of blowdown, ice meltage, and (in the case of water-cooled condensers) waste water. Bin drains should be insulated to prevent moisture from condensing on their cold exteriors; this condensation is often mistaken for a leaking bin.

Bad water conditions can create innumerable problems for ice-making equipment. When melted, the ice may leave a scum on water in a glass, or it may have an objectionable taste. Bad water also may have deleterious effects on the equipment. When the local water conditions are known to cause problems, water treatment must be considered at the outset.

Water treatment firms in the vicinity of the installation are the best source of water treatment recommendations. In general, demineralization offers the best overall water conditioning preparation for ice-making equipment. Softening by ion exchange (sodium cycle zeolite) eliminates most scaling problems, but when dissolved solids content of makeup water exceeds 400 ppm, cloudy, mushy ice is apt to result. Treatment of makeup water with polyphosphates reduces only the scaling tendencies, but this lengthens equipment life.

Ambient conditions have an important influence on performance and maintenance requirements. By avoiding locations where temperatures are extreme—either high or low—better performance may be ensured. Also, ease of accessibility for cleaning and maintenance helps prolong life. Most ice makers create some noise, so it is important to locate them where noise is not objectionable.

Table 1 lists ice-maker application data and is a digest of the experiences of many individuals and firms recommending and installing ice-making equipment in diverse areas of the country under various local conditions. The figures, therefore, are averages, and on any particular application they must be modified up or down according to local conditions.

### Environmental Considerations

Many areas experience severe shortages of electric power and supply water, so it is important for ice-making systems to operate with maximum efficiency. Some measures of operating efficiency are as follows:

1. Lb (kg) of ice per gallon ($m^3$) of supply water
2. Lb (kg) of ice per kWh (MJ) of power consumed by the ice-making system
3. Lb (kg) of ice per gallon ($m^3$) of condensing water
4. Lb (kg) of ice per $ft^2$ ($m^2$) of floor space

## BIBLIOGRAPHY

ASHRAE. 1978. *Method of Testing Ice Makers.* ASHRAE *Standard* 29-78.

CSA. 1964. *Standards for Safety, Ice Makers.* CSA *Standard* 133, Canadian Standards Association.

Eddy, D.E. 1965. Manufacture, Storage, Handling and Uses of Fragmentary Ice. ASHRAE *Journal,* September 1965. p. 66.

FAO. 1968. Ice in Fisheries. *FAO Fisheries Report.* Food and Agriculture Organization of the United Nations, Rome.

Hardenberg, R.E., A.E. Watada, and C.Y. Wang. 1986. Commercial Storage of Fruits, Vegetables and Florist and Nursery Stocks. United States Department of Agriculture, *Agriculture Handbook* Number 66, Agricultural Reserch Service, Washington, DC.

NSF. 1977. *Automatic Ice Making Equipment.* NSF *Standard* 12-77. National Sanitation Foundation, Ann Arbor, MI.

Proctor, W.T. 1977. Ice Vending—A Growing Market. *American Automatic Merchandisers,* May 1977, p. 36.

UL. 1984. *Ice Makers.* UL *Standard* 563-84. Underwriters Laboratories Inc., Northbrook, IL.

CHAPTER 41

# ROOM AIR CONDITIONERS AND DEHUMIDIFIERS



## PART I: ROOM AIR CONDITIONERS

A room air conditioner is an encased assembly designed as a unit primarily for mounting in a window or through a wall. These units are designed for the delivery of cool or warm conditioned air to the room, either without ducts or with very short ducts up to a maximum of about 48 in. (1200 mm). Each unit includes a prime source of refrigeration, dehumidification, and means for circulating and cleaning air, and may also include means for ventilating, and/or exhausting and heating. The ANSI/AHAM standard on *Room Air Conditioners* (*Standard* RAC 1-82) has further details.

The basic function of a room air conditioner is to provide comfort by cooling, dehumidifying, filtering or cleaning, and circulating the room air. It may also provide ventilation by introducing outdoor air into the room and/or exhausting room air to the outside. Also, comfort may be provided by controlling the room temperature through selection of the desired thermostat setting. The conditioner may provide heating by heat pump operation, electric resistance elements, or by a combination of both.

Figure 1 shows a typical room air conditioner. Warm room air passes over the cooling coil and, in the process, gives up its sensible and latent heat. The conditioned air is then recirculated in the room by a fan or blower.

The heat from the warm room air vaporizes the cold (low-pressure) liquid refrigerant flowing through the evaporator. The vapor then carries the heat to the compressor, which compresses the vapor and increases its temperature to a value higher than the temperature of the outdoor air. In the condenser, the hot (high-pressure) refrigerant vapor liquefies and gives up the heat from the room air to the outdoor air. The high-pressure liquid refrigerant then passes through a restrictor, which reduces its pressure and temperature. The cold (low-pressure) liquid refrigerant then reenters the evaporator to repeat this refrigeration cycle.

## SIZES AND CLASSIFICATIONS

The cooling capacities of commercially available room air conditioners range from 4000 to 36,000 Btu/h (1.2 to 10.5 kW).

Room air conditioners are equipped with line cords, which may be plugged into standard or special electric circuits. Most units are designed to operate at 115, 208, or 230 V, single-phase, 60-Hz power. Some units are rated at 265 V or 277 V, and the chassis or chassis assembly must provide for permanent electrical connection, in accordance with the *National Electrical*

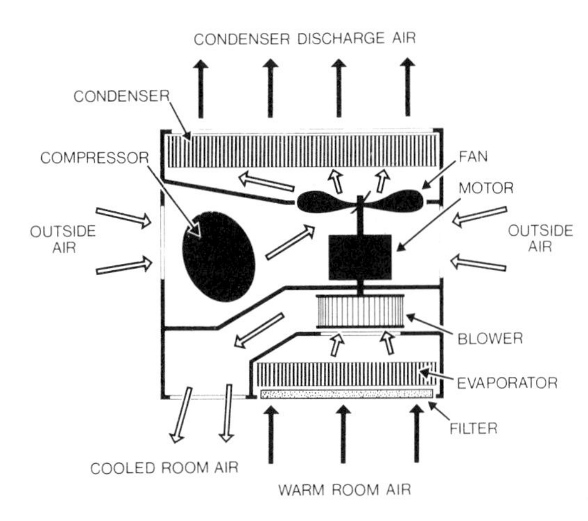


**Fig. 1 Schematic View of Typical Room Air Conditioner**

The preparation of this chapter is assigned to TC 7.5, Room Air Conditioners and Dehumidifiers.

*Code* (NEC). These units are typically used in multiple room installations (see Chapters 42 and 43 of this volume). The maximum rating of 115-V units is generally limited to 12 A, since this is the maximum allowable current for a single-outlet, 15-A circuit permitted by the NEC.

A popular 115-V model is one rated at 7.5 A. The largest capacity 115-V units are in the 12,000 to 14,000 Btu/h (3.5 to 4.1 kW) range. Capacities for 230-, 208-, or 230/208-V units range from 8000 to 36,000 Btu/h (2.3 to 10.5 kW). Capacities for 265- or 277-V units range from 6000 to 17,000 Btu/h (1.8 to 5.0 kW). (For a more detailed discussion of wiring limitations, see the section on "Installation and Service.")

Models designed for countries outside the United States are generally designed for 50- or 60-Hz systems, with typical design voltage ranges of 100 to 120 and 200- to 240-V, single-phase. Many manufacturers apply 60-Hz units on a 50-Hz power supply, provided the supply voltage of the 50-Hz power supply is 87% of the nameplate voltage rating at 60 Hz. For example, 200 V at 50 Hz would be required for a unit rated 230 V at 60 Hz. The capacity of a unit rated at 60 Hz will be reduced by approximately the ratio 50/60 (0.833) when operated at 50 Hz.

Heat pump models are also available and are usually designed for 208- or 230-V applications. These units are generally designed for reversed refrigerant-cycle operation as the normal means of supply heat, but they may incorporate electrical resistance heat to either supplement the heat pump capacity or provide the total heating capacity when outdoor temperatures drop below a predetermined value. Another type of heating model incorporates electrical heating elements in regular cooling units so that heating is provided entirely by electrical resistance heat.

## SYSTEMS DESIGN

The design of a room air conditioner is usually based on one or more of the following criteria, any one of which automatically limits the freedom of the designer in overall system design:

1. Lowest initial cost
2. Lowest operating cost (highest efficiency)
3. Minimum energy efficiency ratio (MEER), as legislated by the federal and/or state governments
4. Low sound level
5. Physical chassis size
6. An unusual chassis shape (minimal depth, height, etc.)
7. An amperage limitation (7.5 A, 12 A, etc.)
8. Weight

*Minimum energy efficiency ratio* (MEER) is defined as the unit capacity in Btu/h divided by the power input to the unit in watts at the standard rating condition:

$$\frac{\text{Btu/h (capacity)}}{\text{Watts (input)}}$$

The optimum design results from a carefully selected group of components consisting of an evaporator, a condenser, a compressor, one or more fan motors, air impellers for evaporator and condenser airflow, and a restrictor device.

The following combinations illustrate the effect on the various components as a result of an initial design parameter:

**Low Initial Cost.** High airflow with minimum heat exchanger surface keeps the initial cost low. These units have a low cost compressor, which is selected by analyzing various compressor and coil combinations and choosing the one that achieves optimum performance consistent with passing all tests required by UL, AHAM, etc. For example, a high capacity compressor might be selected to meet the capacity requirement with a minimum heat-transfer surface, but frost tests under maximum load may not be acceptable. These tests set the upper and lower limits on what is acceptable when low initial cost is the prime consideration.

**Low Operating Cost.** Low air volumes with large heat exchanger surfaces keep operating costs low. A compressor with a low compression ratio operates at low head and high suction pressures, which results in a high Energy Efficiency Ratio.

### Compressors

Room air conditioner compressors range in capacity from approximately 4000 to 48,000 Btu/h (1.2 to 14 kW). Design data are available from compressor manufacturers with rating conditions, as follows, for standard and high-efficiency compressors:

| | Standard | | High Efficiency | |
|---|---|---|---|---|
| | °F | °C | °F | °C |
| Evaporating temperature | 45 | (7.2) | 49 | (9.4) |
| Evaporator suction temperature | 95 | (35.0) | 51 | (10.6) |
| Compressor suction temperature | 95 | (35.0) | 65 | (18.3) |
| Condensing temperature | 130 | (54.4) | 120 | (48.9) |
| Liquid temperature | 115 | (46.1) | 100 | (37.8) |
| Ambient temperature | 95 | (35.0) | 95 | (35.0) |

Compressor manufacturers offer complete performance curves at various evaporating and condensing temperatures to aid in selection for a given design specification.

### Evaporator and Condenser Coils

These coils are generally of either the tube-and-plate-fin variety or the tube-and-spine-fin variety. Information on the performance of such coils is available from suppliers, and original equipment manufacturers usually develop data for their own coils. Design parameters to be considered when selecting coils are Btu/h • ft$^2$ (W/m$^2$); dry-bulb and moisture content of the entering air; air-side friction loss; internal refrigerant pressure drop; coil surface temperature; air volume; and air velocity.

### Restrictor Application and Sizing

Essentially three types of restrictor devices are available to the designer: (1) the *thermostatic expansion valve,* which maintains a constant amount of superheat from a point near the outlet of the evaporator to a point on the suction line; (2) the *automatic expansion valve,* which maintains a constant suction pressure; and (3) the *restrictor tube* (*capillary*). The capillary is the most popular device for room air-conditioner applications because of its low cost and high reliability, even though the control of refrigerant over a wide range of ambient temperatures is not optimum. A recommended procedure for optimizing charge balance, condenser subcooling, and restrictor sizing is as follows:

1. Use an adjustable restrictor, so a series of tests may be run with a flooded evaporator coil and various refrigerant charges to determine the optimum point of system operation.
2. Reset the adjustable restrictor to the optimum setting, remove from the unit, and measure flow pressure with a flow comparator similar to that described in ASHRAE *Standard* 28-78 (*Method of Testing Flow Capacity of Refrigerant Capillary Tubes*).
3. Install a restrictor tube having the same flow rate as the adjustable restrictor. Usually, restrictor tubes are selected on the basis of cost, with shorter tubes generally being less expensive.

Chapter 19 in this volume has further information about restrictors.

### Fan-Motor and Air-Impeller Selections

The two types of motors generally used on room air conditioners are (1) the low efficiency, shaded-pole type; and (2) the more efficient permanent split-capacitor type, which requires the use of a run capacitor. Air impellers are usually of two types: (1) the forward-curved blower wheel and (2) the axial- or radial-flow fan blade. In general, *blower wheels* are used to move small to moderate amounts of air in a high resistance system, and *fan blades* move moderate to high air volumes in low resistance applications.

The combination of the fan motor and the air impeller is such an important part of the overall design that the designer should work closely with the manufacturers of both components. Performance curves are available for motors, blower wheels, and fans; unfortunately, however, the data are of ideal systems not usually found in practice because of physical size, motor speed, and component placement limitations.

## PERFORMANCE DATA

Industry standards are published for the performance of room air conditioners summarize all existing standards for rating, safety, and recommended performance levels. An industry certification program, under the sponsorship of AHAM, covers the majority of room air conditioners and certifies the cooling and heating capacities and electrical input (in amperes) of each for adherence to nameplate rating.

The following tests are specified by ANSI/AHAM *Standard* RAC-1-1982:

1. Cooling capacity test
2. Heating capacity test
3. Maximum operating conditions test (heating and cooling)
4. Enclosure sweat test
5. Freezeup test
6. Recirculated air quantity test
7. Moisture removal test
8. Ventilating air quantity and exhaust air quantity test
9. Electrical input test (heating and cooling)
10. Power factor test
11. Condensate disposal test
12. Application heating capacity test
13. Outside coil de-icing test

### Efficiency

Efficiency for room air conditioners may be shown in either of two forms:

1. Coefficient of performance (COP—generally for heating)

$$\frac{\text{capacity in Btu/h}}{\text{input in watts} \cdot 3.413} \text{ or } \left(\frac{\text{capacity in watts}}{\text{input in watts}}\right)$$

2. Energy efficiency ratio (EER—generally for cooling)

$$\frac{\text{capacity in Btu/h}}{\text{input in watts}} \text{ or } \left(\frac{\text{capacity in watts}}{\text{input in watts}}\right)$$

### Sensible Heat Ratio

The ratio of sensible heat to total heat removal is a performance characteristic that is useful in evaluating units for specific conditions. A low ratio provides more dehumidification, and certain areas, such as New Orleans and Phoenix, might best be served with units having lower and higher ratios, respectively.

### Energy Conservation and Efficiency

The rising cost of energy, the need for energy conservation, and mandated efficiency standards are key factors in the growing availability of high-efficiency units in all capacity sizes. Also, two federal energy programs have increased the demand for higher efficiency room air conditioners.

First, the *Energy Policy and Conservation Act* passed in December, 1975, *Public Law* 94-163, requires the Federal Trade Commission (FTC) to prescribe an energy usage label for many major appliances, including room air conditioners. The program provides consumers with operating cost data at the point of sale.

Second, the *National Energy Conservation Policy Act* (NECPA), passed in November, 1978, *Public Law* 95-619, directs the Department of Energy to establish minimum efficiency standards for major appliances, including room air conditioners. Subsequently, the *National Appliance Energy Conservation Act* of 1987 (NAECA) provided a single set of minimum efficiency standards for major appliances, including room air conditioners. The room air conditioner portion provided for minimum efficiencies for 12 classes, based on physical conformation. The minimums range from 8 to 9 EER (Energy Efficiency Ratio) and apply for all units built after January 1, 1990. These requirements will be reviewed prior to January 1995. All state and local minimum efficiency standards are automatically superceded.

Whether estimating potential energy savings associated with appliance standards or estimating consumer operating costs, the annual hours of operation of a room air conditioner are of vital importance. These figures have been compiled from various studies commissioned by DOE and AHAM for every major city and region in the United States. The national average is estimated to be 750 hours per year.

The basic formula for computing cost of operation is as follows:

$$C = R \cdot H \cdot W/(1000)$$

*where*

$C$ = cost of operation, dollars
$R$ = average cost, dollars per kWh (MJ)
$H$ = hours of operation
$W$ = watt input

### High Efficiency Design

The EER can be affected by three design parameters. The first is *electrical efficiency.* Fan motors range from 25 to 65% in efficiency; compressor motors range from 60 to 85% in efficiency. The second parameter, *refrigerant cycle efficiency,* is increased by increasing the heat-transfer surface to minimize the temperature differential between the refrigerant and the air. This allows the use of a smaller displacement compressor with a high-efficiency motor. The third parameter is *air circuit efficiency*, which can be increased by minimizing the airflow pressure drop across the heat-transfer surface to reduce the load on the fan motor.

Increasing the EER affects the unit characteristics. Table 1 shows how room air-conditioner capacity ranges are affected by increased EER, from 6.5 Btu/h/W (1.9 W/W) to Table 1 values. The results of this tabulation are approximate values.

**Table 1 Effect of Increased EER over Base Value of 6.5**

| Capacity Range | | EER | | Weight Increment, | Volume Increment, | Price Increment, |
|---|---|---|---|---|---|---|
| Btu/h | (W) | Btu/h | W/W | Approx., % | Approx., % | Approx., % |
| 4000 to 10,000 | (1170 to 2930) | 10.0 | 2.93 | 39 | 48 | 50 |
| 12,000 to 20,000 | (3520 to 5860) | 10.5 | 3.08 | 52 | 60 | 37 |
| 24,000 to 27,000 | (7030 to 7910) | 11.0 | 3.22 | 58 | 80 | 40 |

The increases range from 30 to 80% for weight increases and 30 to 100% for volume increases; small capacity units would lose their portability, and units larger than 20,000 Btu/h (6 kW) would no longer be adaptable to window mounting.

Higher EERs are not the complete answer to reducing energy costs. More efficient use of power for an application can be accomplished by proper sizing of the unit, keeping infiltration and leakage losses to a minimum, increasing building insulation, reducing unnecessary internal loading, providing effective system maintenance, and balancing load by use of a thermostat and thermostat setback.

## SPECIAL DESIGN FEATURES

Room air conditioners extend beyond the building when mounted flush with the inside wall; some models are designed to minimize this extension. Low capacity models are usually smaller and less obtrusive than higher capacity models. Units are often installed through the wall, where they do not interfere with windows. Exterior cabinet grilles are often designed to harmonize with the architecture of various buildings.

Most units have adjustable louvers or deflectors to distribute the air into the room with satisfactory throw and without drafts. The louver design should minimize recirculation of discharge air into the air inlet. Some units employ motorized deflectors for changing the air direction continuously. Discharge air velocities range from 300 to 1200 fpm (1.5 to 6 m/s), with low velocities preferred in rooms where people are at rest.

Most room air conditioners are designed for bringing in outdoor air, exhausting room air, or both. Controls usually permit these features to function independently.

Temperature is controlled by an adjustable built-in thermostat. The thermostat and unit controls may operate in one of the following modes:

1. The unit is set to the cool position and the thermostat setting is adjusted, as needed. The circulation blower runs without interruption while the thermostat cycles the compressor on and off.
2. Some air conditioners may use a two- or three-stage thermostat, which reduces the blower speed as the room temperature approaches the set temperature, cycles the compressor off on further temperature drop, and, finally, cycles the blower off on still further room temperature drop. As the room temperature rises, the sequence reverses. Any combination of these sequences are available.
3. Some air conditioners use, in addition to the control sequence above, an optional *automatic fan mode* when both the blower and the compressor are cycled simultaneously by the thermostat; this mode of operation requires close attention to proper thermostat sensitivity. One of the advantages with this arrangement is improved humidity control caused by not reevaporating moisture from the evaporator coil into the room on the *off* cycle. Another advantage is lower operating cost because the blower motor does not operate during the *off* cycle. The effective EER may be increased an average of 10% by using the feature (ORNL-NSF-EP-85).

Disadvantages to cycling fans with the compressor may be (1) changing noise level from fan cycling and (2) deterioration in room temperature control.

Room air conditioners are simple to operate. Usually, one control operates the unit electrically, while a second controls the temperature. Additional knobs or levers operate louvers, deflectors, the ventilation system, exhaust dampers, and other special features. The controls are usually arranged on the front of the unit or concealed behind a readily accessible door; however, they may also be arranged on the top or sides of the unit.

Filters on room air conditioners remove airborne dirt to provide clean air to the room and keep dirt off the cooling surfaces. Filters are made of expanded metal (with or without a viscous oil coating), glass fiber, or synthetic materials, and may be either throwaway or reusable. A dirty filter reduces cooling and air circulation and frequently allows frost to accumulate on the cooling coil; therefore, filter location should allow for easy checking, cleaning, and replacement.

Some units have louvers or grilles at the rear to enhance appearance and protect the condenser fins. Sometimes, these louvers separate the airstreams to and from the condenser and reduce recirculation. Side louvers on the outside portion of the unit, when provided, are an essential part of the condenser air system. Care should be taken not to obstruct the air passages through these louvers.

The sound level of a room air conditioner (see the following section on "Noise Standards") is an important factor, particularly when the unit is installed in a bedroom. A certain amount of sound can be expected because of the movement of air through the unit and the operation of the compressor. However, the sound level of a well designed room air conditioner is relatively low, and the emitted sound is relatively free from high pitched and metallic noise. Usually, fan motors with two or more speeds are used to provide a slower fan speed for quieter operation. To avoid rattles and vibration in the building structure, units must be installed correctly. (See the section on "Installation and Service.")

Room air conditioners require a durable finish, especially for those parts exposed to the weather. Some manufacturers use a special grade of plastic for weather-exposed parts; if metal is used, good practice calls for baked finishes over phosphatized or zinc-coated steels, and/or the use of corrosion-resistant materials such as aluminum or stainless steel.

## NOISE STANDARDS

Sound reduction (both indoor and outdoor) is becoming increasingly important. Whereas indoor sound can be displeasing to the owner, excessive outdoor sound (condenser fan and compressor) can be irritating to the owner's neighbors. Many local and state outdoor noise ordinances are being considered, and some have been passed into law.

AHAM is developing an application standard that will allow the prediction of both outdoor sound pressure levels at distances that may be specified in a property line ordinance and the indoor sound pressure level that would exist in a typical room containing sound-absorbing furnishings.

## SAFETY CODES AND STANDARDS

### In the United States

The *National Electrical Code*, NFPA *Standard* 70; Underwriters Laboratories *Standard* UL 484-82, *Room Air Conditioners;* and ASHRAE *Standard* 15-78, *Safety Code for Mechanical Refrigeration* pertain to room air conditioners. Local regulations may differ with these standards to some degree, but the basic requirements are generally accepted throughout the United States.

*The National Electrical Code,* sponsored by The National Fire Protection Association, covers the broad area of electrical conductors and equipment installed within or on public and private buildings and other premises. Its purpose is the practical safeguarding of persons, buildings, and building contents from hazards that arise from the use of electricity for light, heat, power, radio, signaling, and other purposes. The *Code* contains basic minimum provisions considered necessary for safety. Proposals for modification of the *Code* can be made by any interested person or organization. Of primary interest to the room air-conditioner designer is Article 440, *Air Conditioning and Refrigerating Equipment.*

*The Safety Code for Mechanical Refrigeration,* sponsored by ASHRAE, covers field-erected equipment and is intended to ensure the safe design, construction, installation, operation, and inspection of every refrigerating system that vaporizes and normally liquefies a fluid in its refrigeration cycle. It provides reasonable safeguards to life, health, and property; corrects practices inconsistent with safety; and prescribes safety standards that will influence future progress and developments in refrigerating systems.

Both codes recognize equipment and materials submitted for test by an independent organization under uniform conditions; such testing agencies also have follow-up inspection services of current production runs. Underwriters Laboratories prepares standards for safety, applies such requirements to products voluntarily submitted by manufacturers, and, by means of follow-up inspections, determines that production of the equipment continues to comply with requirements. The UL *Standard for Room Air Conditioners* is divided into two sections: Construction and Performance; these set minimum requirements necessary to protect against casualty, fire, and electrical shock hazards.

### In Canada

The Canadian Standards Association has developed its *Standard for Room Air Conditioners* (C22.2 No. 117-70), which forms part of the *Canadian Electrical Code.*

### International (Other than Canada)

Two useful documents that assist the designer are (1) the International Electrotechnical Commission *Publication No.* 378, *Safety Requirements for the Electrical Equipment Room Air Conditioners* (IEC 328-ANSI); and (2) the International Organization for Standards *Recommendation* R859, *Testing and Rating Room Air Conditioners* (ISO R859). These publications are available through the American National Standards Institute (ANSI).

### Product Standards

The CSA *Standard* C22.2 No. 117-70 and UL *Standard* 484-82 are similar in content. Using the UL standard as an example, safety from a casualty viewpoint involves such items as the unit enclosure (including materials), the unit's ability to protect against contact with moving and uninsulated live parts, and the means for unit installation or attachment. Attention is also given to the refrigeration system's ability to withstand operating pressures, pressure relief of the system in the event of fire, and toxicity of the refrigerant. Electrical considerations include supply connections, grounding, internal wiring and wiring methods, electrical spacings, motors and motor protection, uninsulated live parts, motor controllers and switching devices, air-heating components, and electrical insulating materials.

The performance section of the standard includes a rain test for determining the unit's ability to stand a beating rain without creating a shock hazard because of current leakage or insulation breakdown. Other tests include (1) leakage current limitations based on the ANSI *Standard* C101.1 for leakage current for appliances; (2) measurement of input currents for the purpose of establishing nameplate ratings and for sizing the supply circuit for the unit; (3) temperature tests to determine whether or not components exceed their recognized temperature limits and/or electrical ratings (ANSI/AHAM *Standard* RAC-1-1982); and (4) pressure tests to ensure that excessive pressures do not develop in the refrigeration system.

Abnormal conditions are also considered, such as (1) failure of the condenser fan motor, which may result in excessive pressures being developed in the system; and (2) air heater burnout, to determine whether combustibles within or adjacent to the unit may be ignited. A static load test is also conducted on window-type room air conditioners to determine whether or not the mounting hardware can support the unit adequately. As part of normal production control, tests are conducted for leakage of the refrigeration system, dielectric withstand, and grounding continuity.

Plastic materials are receiving increased consideration in the design and fabrication of room air conditioners because of their ease in forming inherent resistance to corrosion and their decorative qualities. When considering the use of plastic, the engineer should consider the tensile, flexural, and impact strength of the material; the flammability characteristics; and—from the standpoint of degradation—the resistance to water absorption, exposure to ultraviolet light, ability to operate at elevated temperatures, and thermal aging characteristics. From a product safety standpoint, some of these factors are of lesser importance, since failure of the part will not result in a hazardous condition. For other parts, such as the bulkhead, base pan, and unit enclosure, which either support components or provide structural integrity, all of the preceding factors must be considered, and a complete analysis of the material must be made to determine that it is suitable for the application.

## INSTALLATION AND SERVICE

Installation procedures vary because units can be mounted in various ways. It is important to select the particular mounting for each installation that best satisfies the user and complies with existing building codes. Common mounting methods include the following:

**Inside Flush Mounting.** Interior face of conditioner is approximately flush with inside wall

**Balance Mounting.** Unit is approximately half inside and half outside the window

**Outside Flush Mounting.** Outer face of unit is flush or slightly beyond outside wall

**Special Mounting.** Includes casement windows, horizontal sliding windows, office windows with swinging units (or swinging windows) to permit window washing, and transoms over doorways

**Through-the-Wall Mounts** or **Sleeves.** Used for installing window-type chassis, complete units, or consoles in walls of apartment buildings, hotels, motels, and residences

Room air conditioners have become more compact in size because of consumer preference for minimum loss of window light and minimum projection both inside and outside the structure. Several types of expandable mounts are now available for fast, dependable installation in single- and double-hung windows, as well as windows of the horizontal sliding type. Installation kits provide all parts needed for structural mounting, such as gaskets, panels, and seals for weathertight assembly.

Adequate wiring and proper fuses must be provided for the service outlet. The necessary information is usually given on instruction sheets or stamped on the air conditioner near the service cord or on the serial plate. It is important to follow the manufacturer's recommendation for size and type of fuse. All units are equipped with grounding-type plug caps on the service cord as received from the manufacturer. Receptacles with a grounding contact correctly designed to fit the air-conditioner service cord plug cap should be used when units are installed.

Units rated 265 or 277 V must provide for permanent electrical connection with armored cable or conduit to the chassis or chassis assembly. It is customary to provide an adequate cord and plug cap within the chassis assembly to facilitate installation and service.

One type of room air conditioner is the *integral chassis design,* with the outer cabinet fastened permanently to the chassis. Most of the electrical components can be serviced by partially dismantling the control area without removing the unit from the installation.

Another type of room air conditioner is the *slide-out chassis design,* which allows the outer cabinet to remain in place while the chassis is removed for service.

## FUTURE DEVELOPMENTS

Energy conservation and higher efficiency requirements will have increasing influence in future room air-conditioner designs. Higher capacities and higher efficiency for 115-V units are particularly important because of the cost of installing additional 230-V wiring in existing homes.

Material improvements will continue to be important for reducing weight and cost while maintaining the basic functional requirements of room air conditioners.

## ELECTRONICS

Microprocessors with the capability of monitoring and controlling numerous functions at essentially the same time have been incorporated into exceptionally capable control systems for room air conditioners. These microelectronic controls offer improved appearance with digital displays and touch panels that allow simple fingertip programming of desired temperature, on-off timing, modulated fan speeds, bypass capabilities, and more sophisticated sensing for humidity, temperature, and airflow control.

Future capabilities include diagnostics, the monitoring of outside ambient air conditions for more precise indoor programming and, possibly, audio control, as already demonstrated in other products. More effective and practical energy saving is also possible with the inherent precision timing and temperature control.

# PART II: DEHUMIDIFIERS

This section deals with domestic dehumidifying units that are selfcontained, electrically operated, and mechanically refrigerated. Dehumidifiers remove moisture from the air that is circulating through the unit. The resulting reduction of relative humidity helps prevent rust, rot, mold, and mildew on surfaces within the room or other enclosed space where the dehumidifier is used.

A dehumidifier consists of a motor-compressor unit, a refrigerant condenser, an air-circulating fan, a refrigerated surface (refrigerant evaporator), a means for collecting and/or disposing of the condensed moisture, and a cabinet to house these components. A typical dehumidifier unit is shown in Figure 2.

The fan draws the moist room air through the cold coil and cools it below its dew point, removing moisture that either drains into the water receptacle or passes through a drain into the sewer. The cooled air then passes through the condenser, where it is reheated. Then, with the addition of other unit-radiated heat,

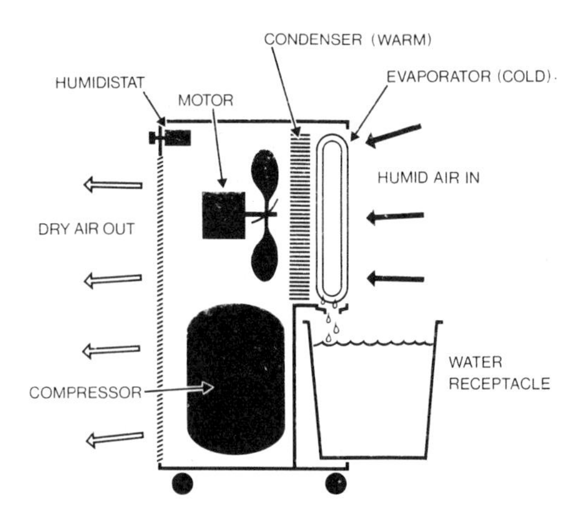


**Fig. 2 Typical Dehumidifier Unit**

the air is discharged into the room at a higher temperature but at a lower relative humidity. Continuous circulation of room air through the dehumidifier gradually reduces the relative humidity of the room.

## DESIGN AND CONSTRUCTION

Dehumidifiers use hermetic-type motor-compressors of various ratings, depending on the designed output of the overall unit. The refrigerant condenser is usually a conventional finned-tube coil. Refrigerant flow metering is usually provided by a capillary tube, although some high-capacity dehumidifiers may use an expansion valve.

A propeller-type fan, direct-driven by a shaded-pole motor, moves the air through the dehumidifier. The airflow rate through typical dehumidifiers ranges from 125 to 250 cfm (60 to 120 L/s), depending on the unit's moisture output capacity. In general, the output capacity of a dehumidifier increases with an increase in airflow. Extremely high rates of airflow, however, may cause objectionable sound. Domestic dehumidifiers ordinarily maintain satisfactory humidity levels in an enclosed space when the airflow rate and unit placement permit the entire air volume of the space to be moved through the dehumidifier once an hour. As an example, an airflow rate of 200 cfm (94 L/s) would provide one air change per hour in a room having a volume of 12,000 $ft^3$ (340 $m^3$).

The refrigerated surface (evaporator) is usually a bare-tube coil structure, although finned-tube coils can be used if they are spaced to permit rapid runoff of water droplets. Vertically disposed bare-tube coils tend to collect smaller drops of water, promote quicker runoff, and cause less water reevaporation than do finned-tube cooling coils or horizontally arranged bare-tube coils. Continuous bare-tube coils, wound in the form of a flat circular spiral (sometimes consisting of two coil layers) and mounted with the flat dimension of the coil in a vertical plane, are a good design compromise because they have most of the advantages of the vertical-tube coil.

Evaporators are protected against corrosion by such processes as painting, waxing, and anodizing (on aluminum). Waxing reduces the wetting effect that promotes condensate formation; tests on waxed evaporator surfaces versus nonwaxed surfaces show a negligible loss in capacity. Thin paint films do not have an appreciable effect on capacity.

Removable water receptacles, provided with most dehumidifiers, hold from 16 to 24 pints (7.6 to 11.3 litres) and are usually made of plastic to withstand corrosion. Ease of removal and handling without spilling are important. Most dehumidifiers also provide a means of attaching a flexible hose to either the water receptacle or a fitting specially provided for that purpose. A flexible hose permits direct gravity drainage to a sewer.

Dehumidifier cabinet enclosures vary with respect to aesthetic design. The more expensive models usually have higher output capacity, are more attractively styled, and are provided with various auxiliary features. The addition of a humidity-sensing control (to cycle the unit automatically) maintains a preselected relative humidity. These humidistats are normally adjustable within a range of 30 to 80% relative humidity. The humidistat may also provide a detent setting that causes continuous running of the unit. Most humidistats provide an on-off line switch; some models also include an additional sensing and switching device, which automatically turns the unit off when the water receptacle is full and requires emptying. This second device may or may not include an indicating light.

Dehumidifiers are designed to provide optimum performance at standard rating conditions of 80°F (26.7°C) dry-bulb room ambient and 60% relative humidity. When the room ambient condition causes the thermal loading to drop to less than that occasioned by a condition of 65°F (18.3°C) dry bulb and 60% relative humidity, then the refrigerant pressure and corresponding evaporator surface temperature usually decrease to the point where frost forms on the cooling coil. This effect is especially noticeable on units having a capillary tube for refrigerant flow metering.

Some dehumidifier models are equipped with special defrost controls, which cycle the compressor *off* under frosting conditions. This type of control is generally a bimetal thermostat that is strategically attached to the evaporator tubing, allowing the dehumidifying process to continue at a reduced rate when frosting conditions exist. The humidistat can sometimes be adjusted to a higher relative humidity setting, which reduces the number and duration of running cycles and permits reasonably satisfactory operation at low-load conditions. In many instances, especially in the late fall and early spring, supplemental heat must be provided from other sources to maintain conditions within the space above those at which frosting can occur.

Dehumidifiers are usually equipped with rollers or casters, so they move easily from place to place. Typical sizes and weights of dehumidifiers vary, as follows:

Width: 10 to 20 in. (250 to 500 mm).
Height: 12 to 24 in. (300 to 600 mm).
Depth: 10 to 20 in. (250 to 500 mm).
Weight: 35 to 75 lb (16 to 34 kg).

## CAPACITY AND PERFORMANCE RATING

Dehumidifiers are available with moisture removal capacities ranging from 11 to 50 pints (5 to 24 litres) per 24-hour day. The input to domestic dehumidifiers varies from 200 to 700 W, depending on the output capacity rating. Dehumidifiers are operable from ordinary household electrical outlets (115-V, single-phase, 60 Hz).

ANSI/AHAM *Standard* DH-1-1986, *Dehumidifiers,* establishes a uniform procedure for determining the rated capacity of dehumidifiers under certain specified test conditions and establishes other recommended performance characteristics, as well. An industry certification program, under the sponsorship of AHAM, covers the great majority of dehumidifiers and certifies the dehumidification capacity (Directory of Certified Dehumidifiers). The following tests are specified by the dehumidifier standard.

**Dehumidifier Capacity.** Units must be tested in a room maintained at the following test conditions:

Dry-bulb temperature, 80°F (26.7°C).
Wet-bulb temperature, 69.6°F (21°C).
Relative humidity, 60%.

The capacity is to be stated in pints/24 h(L/24 h).

**Maximum Operation Conditions.** The test is conducted in conditions of 90°F (32.2°C) dry bulb, 74.8°F (24°C) wet bulb, and 50% relative humidity. The unit runs continuously at this condition at 90 and 110% of rated voltage for two hours, after which the power is cut off for 120 seconds. The unit should restart within 300 seconds and run continuously for one hour.

**Low-Temperature Test.** The unit is operated in conditions of 65°F (18.3°C) dry bulb, 56.6°F (14°C) wet bulb, and 60% relative humidity for a period of eight consecutive hours. If the unit is not equipped with a defrost control, there should not be ice or frost remaining on the portion of the evaporator coil that is exposed to the entering airstream. If the unit is equipped with a defrost control, it should operate for at least 50% of the test period with no solid ice or frost remaining on the portion of the evaporator coil that is exposed to the entering airstream at the end of each cycle.

## CODES

Dehumidifiers are designed to meet the safety requirements of UL *Standard* 474-81, *Dehumidifiers;* Canadian Standards Association, *Canadian Electrical Code,* Part II, *Specification C22.2, No. 32, Construction and Test of Electrically Operated Refrigerating Machines;* and ASHRAE *Standard* 15-1978, *Safety Code for Mechanical Refrigeration.*

It is customary for UL-listed and CSA-approved equipment to have a label or data plate that indicates approval. UL also publishes the *Electric Appliance and Utilization Equipment List,* which covers this type of appliance.

## ENERGY CONSERVATION AND EFFICIENCY

The *Energy Policy and Conservation Act* of 1976 (EPCA), amended in 1978 by the *National Energy Conservation Policy Act* (NECPA), also applies to dehumidifiers but on a lower priority basis. A test procedure has been issued by DOE, including an energy factor determination in pints/kWh (L/kh) to evaluate efficiency. Dehumidifiers have been exempted from the FTC Labeling Requirements (F.R.,November 19, 1979) and the DOE Minimum Standards Program.

## REFERENCES

AHAM. 1982. Room Air Conditioners. ANSI/AHAM *Standard* RAC-1-1982, Association of Home Appliance Manufacturers, Chicago.

AHAM. 1987. *Directory of Certified Dehumidifiers.* Association of Home Appliance Manufacturers, Chicago, IL.

ASHRAE. 1978. Safety Code for Mechanical Refrigeration. ASHRAE *Standard* 15-78.

ASHRAE. 1978. Method of Testing Flow Capacity of Refrigerant Capillary Tubes. ASHRAE *Standard* 28-78.

CSA. 1982. Room Air Conditioners. CSA *Standard* C22.2, No. 117, Canadian Standards Association, Rexdale, Ontario, Canada.

IEC. *Safety Requirements for the Electrical Equipment Room Air Conditioners.* IEC378, International Electrotechnical Commission. Available from American National Standards Institute, New York.

ISO. Testing and Rating Room Air Conditioners. International Standards Organization. *ISO* R859, Available from American National Standards Institute, New York.

NFPA. 1987. *National Electrical Code*. NFPA No. 70-87, National Fire Protection Association, Quincy, MA.

Oak Ridge National Laboratory. *Report ORNL-NSF-EP-85*.

UL. 1982. Room Air Conditioners. UL *Standard* 484-82, Underwriters Laboratories, Inc., Northbrook, IL.

UL. 1981. Dehumidifers. UL *Standard* 474-81, Underwriters Laboratories, Inc., Northbrook, IL.

## CHAPTER 42

# UNITARY AIR CONDITIONERS AND UNITARY HEAT PUMPS



# PART I: UNITARY AIR CONDITIONERS

THE Air Conditioning and Refrigeration Institute (ARI) defines a unitary air conditioner as one or more factory-made assemblies that normally include an evaporator or cooling coil and a compressor and condenser combination. It may include a heating function, as well.

ARI further defines air source unitary heat pumps as consisting of one or more factory-made assemblies, which normally include an indoor conditioning coil, compressor(s), an outdoor coil, including means to provide a heating function, and perhaps an optional cooling function.

## UNITARY EQUIPMENT CONCEPT

Unitary equipment consists of factory-matched refrigerant cycle components that are applied in the field to fulfill the requirements of the user. Often, a heating function compatible with the cooling system and a control system that requires a minimum of field wiring are incorporated by the manufacturer. Cooling capacities for unitary equipment range from 0.75 to 115 tons (2.5 to 400 kW).

A variety of products is available to meet the objectives of nearly any system. Condensers (air, water, or evaporatively cooled) are matched with various ventilation cycles. Many different heating sections (gas- or oil-fired, electric, or condenser reheat), air filters, and heat pumps, which are a specialized form of unitary product, are available. Such matched equipment, selected with compatible accessory items, requires little field design and field installation work. Most factory-produced unitary products are available in sizes up to 40 tons (140 kW). Larger equipment and units fitted with specialized options are usually custom built.

The preparation of this chapter is assigned to TC 7.6, Unitary Air Conditioners and Heat Pumps.

In the development of unitary equipment, the following design objectives are considered: (1) user requirements, (2) application requirements, (3) installation, and (4) service.

### User Requirements

The user primarily needs either comfort or a controlled environment for products or manufacturing processes. Usually cooling, dehumidification, filtration, and air circulation will meet those needs, although heating, humidification, and ventilation may also be required in some applications.

### Application Requirements

Unitary equipment is available in many supporting configurations, such as the following:

**Single Zone, Constant Volume** consists of one controlled space with one thermostat that varies unit running time to match the sensible load.

**Multizone, Constant Volume** has several controlled spaces, usually two to ten, served by one unit that supplies air of different temperatures to different zones as demanded (Figure 6).

**Single Zone, Variable Volume** consists of several controlled spaces, perhaps as many as 40 or 50, served by one unit. Supply air from the unit is at a constant temperature, with air volume to each space varied to satisfy space demands (Figure 5).

Such factors as size, shape, and use of the building, availability and cost of fuels, building aesthetics (equipment located outdoors), and space available for equipment are considered to determine the type of unitary equipment best suited to a given application. In general, roof-mounted single-packaged unitary equipment is limited to five or six stories because duct space and available blower power become burdensome in higher buildings. Variable-air-volume units should not be used to condition a space that contains widely divergent load demands, since the unit is capable of supplying only a constant supply air temperature. Indoor, single-zone equipment is generally less expensive to maintain and service than multizone units located outdoors.

The building load and airflow requirements determine equipment capacity and the availability and cost of energy, and equipment costs determine the equipment energy source. Control system requirements must be established, and any unusual operating conditions must be considered early in the planning stage. In some cases, custom-designed equipment may be necessary. Of course, suitable, standard equipment is more economical.

Manufacturers' literature has detailed information about geometry, performance, electrical characteristics, application, and operating limits. The designer then focuses on selecting suitable equipment with the capacity for the application.

### Installation

Most manufacturers' installation instructions describe step-by-step procedures that allow orderly planning of labor, materials, and use of tools. A wiring diagram and installation and service manual(s) are often included to help the installer. Unitary equipment is designed to keep installation costs low. The installer must ensure that equipment is installed properly so that it functions effectively and in accordance with the criteria contained in the manufacturers' literature. Adequate planning for the installation of large, roof-mounted equipment is important, since special rigging equipment is frequently required.

A refrigerant system must be clean, dry, and leak free. An advantage of unitary equipment is that proper installation minimizes the risk of field contamination of the refrigerant system.

In the installation of split systems, which require field installing of piping from the evaporator to the condenser, particular care must be taken to remove contaminants and moisture from the system. (Chapters 6 and 7 of the 1986 REFRIGERATION Volume have more detail.)

Larger unitary equipment must be located to avoid noise and vibration problems. Single-piece equipment with over 20 tons (70 kW) of capacity should be mounted on concrete pads when vibration is critical. Large-capacity equipment should be roof mounted only after the structural adequacy of the roof has been evaluated. Roof-mounted units with return fans that use ceiling space for the return plenum should always have a minimum 8 ft. (2.4 m) long, lined return plenum, if they are located over occupied space. Duct silencers should be used where low sound levels are desired. Weight (mass) and sound data are available from many manufacturers.

Unitary equipment should be listed or certified by nationally recognized testing laboratories to ensure safe operation and compliance with government and utility regulations. The equipment should also be installed to comply with the rating and application requirements of the agency standards to ensure that it performs according to industry criteria.

Larger and more specialized equipment often does not carry agency labeling. However, power and control wiring practices should comply with the *National Electrical Code* to facilitate acceptance by local inspection authorities. Local inspectors should be consulted before installation.

### Service

A clear and accurate wiring diagram and a well written service manual are essential to the installer and service personnel. Easy and safe service access must be provided in the equipment for periodic maintenance of filters and belts, cleaning, and lubrication. In addition, access for replacement of major components must also be provided and preserved.

The availability of replacement parts aids to proper service. Equitable warranty policies, covering one year of operation after installation, are offered by most manufacturers. Extended compressor warranties are often optional.

Service personnel must be qualified to repair or replace equipment components. They must also understand the importance of controlling moisture and other contaminants within the refrigerant system; they should know how to clean the hermetic system if it has been opened for service (see Chapter 7 of the 1986 REFRIGERATION Volume).

## TYPES OF UNITARY EQUIPMENT

Table 1 shows the types of unitary air-conditioning equipment available, and Table 2 shows the types of unitary heat pumps available. The following variations apply to some types and sizes of unitary equipment:

**Arrangement**. Major unit components for various types of unitary air conditioners are arranged as shown in Table 1 and unitary heat pumps as shown in Table 2.

**Heat Rejection**. Unitary air-conditioner condensers may be air cooled, evaporatively cooled, or water cooled; the letters A, E, or W follow the ARI-type system designation.

**Heat Source/Sink**. Unitary heat-pump outdoor coils are air-sourced or water-sourced as designated by A or W following the ARI designation. The same coils that act as a heat sink in the cooling mode act as the heat source in the heating mode.

**Unit Exterior**. The unit exterior should be decorative for in-space application, functional for equipment room and ducts, and weatherproofed for outdoors.

**Placement**. Unitary equipment can be mounted on floors, walls, ceilings, or roofs.

**Indoor Air**. Equipment with fans may have airflow arranged for vertical upflow or downflow, horizontal flow, 90 or 180 degree turns, or multizone. Indoor coils without fans are intended for forced-air furnaces or blower packages. Variable-volume blowers may be incorporated with any system.

**Location**. Unitary equipment intended for indoor use may be placed in exposed locations with plenums or furred-in ducts, or concealed in closets, attics, crawl spaces, basements, garages, utility rooms, or equipment rooms. Wall-mounted equipment may be attached to or built into a wall or transom. Outdoor equipment may be mounted on roofs or placed on the ground.

**Heat**. Unitary systems may incorporate gas-fired, oil-fired, electric, or hot-water or steam-coil heating sections. In unitary heat pumps, these heating sections provide supplemental heating capability.

**Ventilation Air**. Outdoor air dampers may be built into the equipment to provide outdoor air for cooling or ventilation.

Unitary equipment is usually designed with fan capability for ductwork, although some units may be designed to discharge directly into the conditioned space.

Figures 1 through 6 show single-package equipment. Figure 7 shows a typical installation of a split-system, air-cooled condensing unit with indoor coil, the most widely used unitary cooling system. Referring to the numbers in Figure 7, hot moist household air, 1, is drawn through a filter, 2, by means of a blower, 3, located in a furnace, 4. The air then passes through a cooling coil, 5, which absorbs the heat. This absorbed heat is passed on by refrigerant tubes, 6, to an outdoor condensing unit, 7, where the heat is rejected to the outside air. At the same time the heat is being removed from the house, the cooling coil removes excessive moisture and drains it away through pipe, 8. Clean, cool, dehumidified air, 9, is then circulated into every room of the home through the ductwork. Figures 8 and 9 also show split system condensing units with coils and blower coil units.

**Table 1 Classification of Unitary Air Conditioners**

| Types of Unitary Air Conditioners | | | |
|---|---|---|---|
| **System Designation** | **ARI-Type** | **Heat Rejection** | **Arrangement** |
| Single Package | SP-A<br>SP-E<br>SP-W | Air<br>Evap Cond<br>Water | Fan / Comp<br>Evap / Cond |
| Refrigeration Chassis | RCH-A<br>RCH-E<br>RCH-W | Air<br>Evap Cond<br>Water | Comp<br>Evap / Cond |
| Year-Round Single Package | SPY-A<br>SPY-E<br>SPY-W | Air<br>Evap Cond<br>Water | Fan<br>Heat / Comp<br>Evap / Cond |
| Remote Condenser | RC-A<br>RC-E<br>RC-W | Air<br>Evap Cond<br>Water | Fan, Evap, Comp — Cond |
| Year-Round Remote Condenser | RCY-A<br>RCY-E<br>RCY-W | Air<br>Evap Cond<br>Water | Fan, Evap, Heat, Comp — Cond |
| Condensing Unit Coil Alone | RCU-A-C<br>RCU-E-C<br>RCU-W-C | Air<br>Evap Cond<br>Water | Evap — Cond, Comp |
| Condensing Unit Coil and Blower | RCU-A-CB<br>RCU-E-CB<br>RCU-W-CB | Air<br>Evap Cond<br>Water | Fan, Evap — Cond, Comp |
| Year-Round Condensing Unit Coil and Blower | RCUY-A-CB<br>RCUY-E-CB<br>RCUY-W-CB | Air<br>Evap Cond<br>Water | Fan, Evap, Heat — Cond, Comp |

A suffix of "-O" following any of the above classifications indicates equipment not intended for use with field-installed duct systems.

Many types of special unitary equipment include the single-packaged air conditioner for use with variable-air-volume systems, as shown in Figure 5. These units are often equipped with a factory-installed system for controlling air volume in response to supply duct pressure.

Another example of a specialized unit is the multizone unit shown in Figure 6. The manufacturer usually provides all controls, including zone dampers. The air path in these units is designed so that supply air may flow through a hot deck containing a means of heating or through a cold deck, which usually contains a direct-expansion evaporator coil.

To make multizone units more efficient, a control is commonly provided that locks out cooling by refrigeration when the heating unit is in operation and vice versa. Such a means of control does not allow one unit to be applied to zones that are subject to widely different loads.

Another variation to improve efficiency is the 3-deck multizone. This unit has a hot deck, a cold deck, and a mixed-air deck. This approach prevents a satisfied zone from requiring heating and cooling energy to maintain its condition.

## REFRIGERATION SYSTEM DESIGN

The packaged product concept of unitary equipment permits optimum system design. Factory assembly of components and control devices help ensure proper functioning and optimum reliability. Laboratory testing and continual monitoring of field performance result in reliable designs, customer satisfaction, and the attainment of cost and efficiency goals.

Chapters 6, 12, and 15 describe compressor and coil designs. Chapter 18 covers balancing components. Chapter 21 of this volume and Chapters 6 and 7 of the 1986 REFRIGERATION Volume contain information on refrigerant system purity and cleanliness. Chapter 8 of the 1986 REFRIGERATION Volume has information on lubrication. Proper coil circuiting is a criterion for adequate oil return to the compressor. Proper selection of refrigerant piping diameters is also important for proper oil return (see Chapter 3 of the 1987 HVAC Volume). Crankcase heaters may be used to prevent refrigerant migration to the compressor crankcase during shutdown. Oil pressure switches and

Table 2 Classification of Unitary Heat Pumps

| Types of Unitary Heat Pumps | | | |
|---|---|---|---|
| System Designation | ARI-Type[a] Heating and Cooling | ARI-Type[a] Heating Only | Arrangement |
| Single Package | HSP-A<br>HSP-W | HOSP-A<br>HOSP-W | Fan / Comp<br>Indoor Coil / Outdoor Coil |
| Remote Outdoor Coil | HRC-A-CB | HORC-A-CB | Fan<br>Indoor Coil — Outdoor Coil<br>Comp |
| Remote Outdoor Coil with No Indoor Fan | HRC-A-C | HORC-A-C | Indoor Coil — Outdoor Coil<br>Comp |
| Split System | HRCU-A-CB<br>HRCU-W-CB | HORCU-A-CB<br>HORCU-W-CB | Fan — Comp<br>Indoor Coil — Outdoor Coil |
| Split System No Indoor Fan | HRCU-A-C | HORCU-A-C | Comp<br>Indoor Coil — Outdoor Coil |

[a]A suffix of "-O" following any of the above classifications indicates equipment not intended for use with field-installed duct systems.

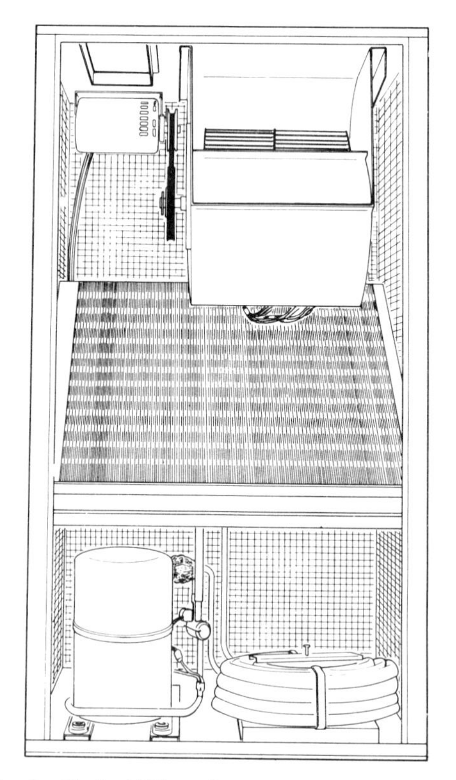

Fig. 1 Typical Water-Cooled Single-Package Air Conditioner

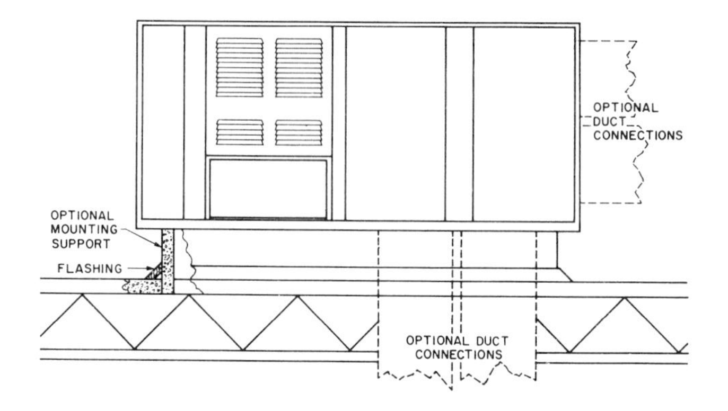


Fig. 2 Rooftop Installation of An Air-Cooled Single-Package Unit

pumpdown and pumpout controls are used when additional assurance of reliability is economical and required.

High-pressure limiting devices, internal-pressure bypasses, and limited compressor motor torque are used to prevent excessive mechanical and electrical stresses. Low-pressure cutout controllers may be used to protect against loss of charge, coil freezeup, or loss of evaporator airflow.

Refrigerant flow is most commonly controlled by either a fixed metering device, such as an orifice or capillary tube or by thermal expansion valves. Capillaries are simple, reliable, and economical, and they can be sized for peak performance at rating conditions. The evaporator may be overfed at high condensing temperatures and underfed at low condensing temperatures because of changing pressure differential across the capillary. When such conditions exist, a less-than-optimum cooling capacity usually results. However, the degree of loss varies with the design of the condenser, volume of the system, and the total refrigerant charge. The amount of unit charge is critical and a capillary-controlled evaporator must be matched to the specific condensing unit.

Properly sized thermal expansion valves provide constant superheat and good control over a range of operating condi-

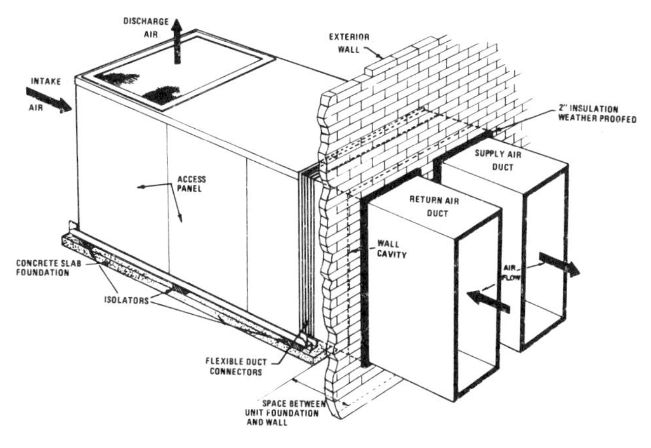


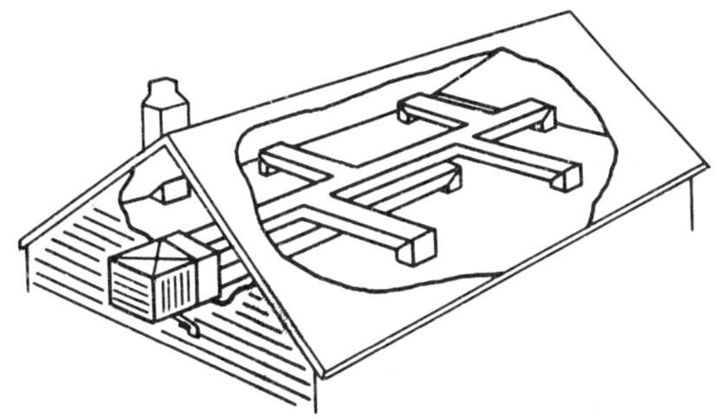

Fig. 3 Through-the-Wall Installation of an Air-Cooled Single-Package Unit

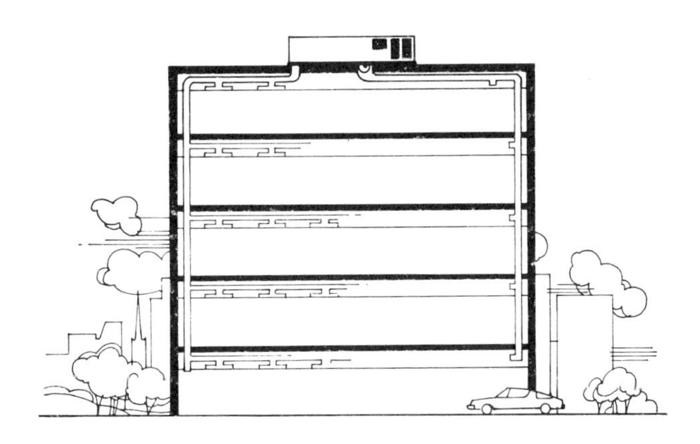

Fig. 4 Multistory Rooftop Installation of Single-Package Unit

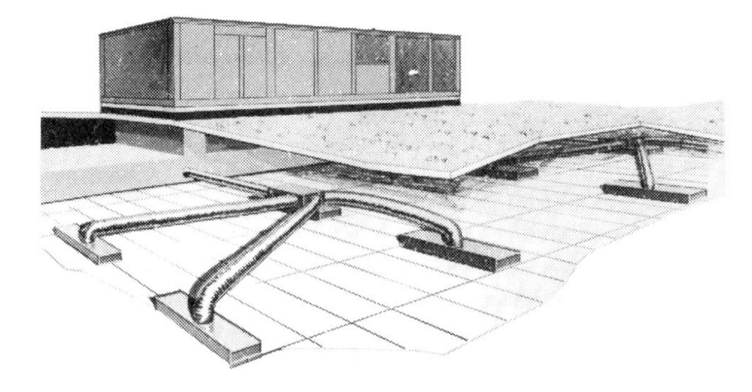

Fig. 5 Single-Package Air Equipment with Variable Air Volume

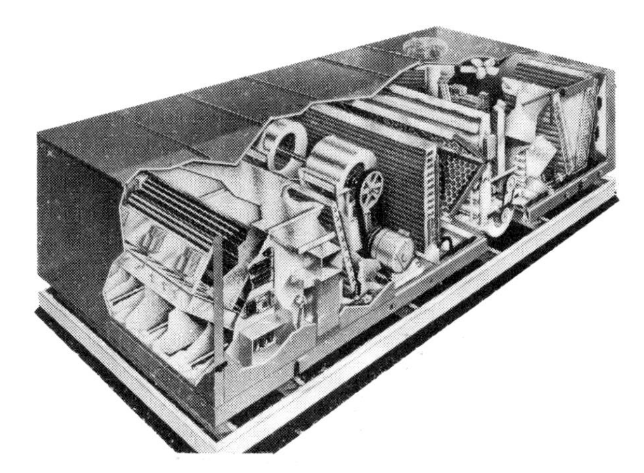

Fig. 6 Typical Rooftop Multizone Air-Cooled Single-Package Air-Conditioner

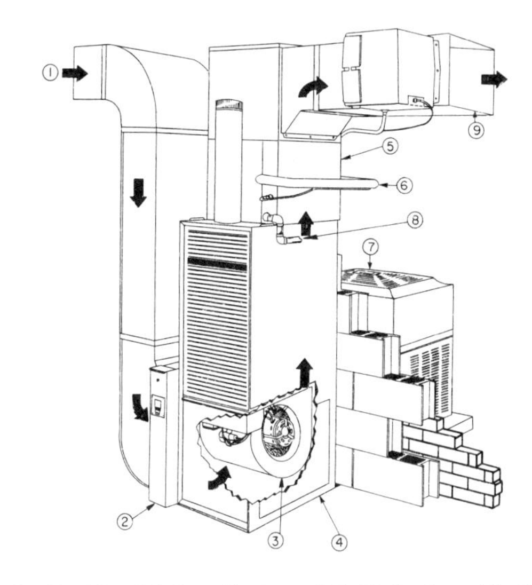

Fig. 7 Residential Installation of Split-System Air-Cooled Condensing Unit with Indoor Coil and Upflow Furnace

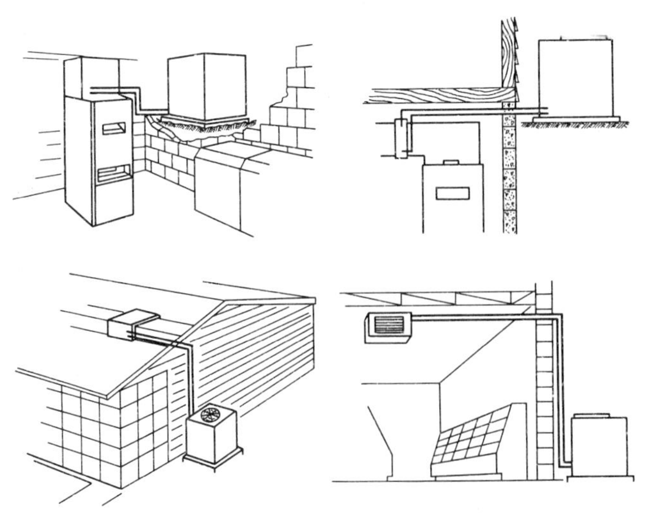

Fig. 8 Outdoor Installations of Split-System Air-Cooled Condensing Units with Coil and Upflow Furnace, or with Indoor Blower Coils

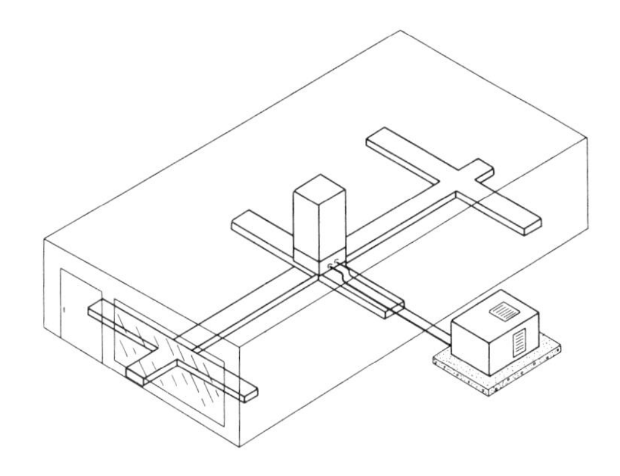

**Fig. 9 Outdoor Installation for Split-System Air-Cooled Condensing Unit with Indoor Coil and Downflow Furnace**

tions. Superheat is adjusted for minimum stable operation, usually 7 to 14 °F (4 to 8 °C). Compressor loading can be limited with vapor-charged expansion valves. Low discharge pressure (low ambient) operation causes a diminished pressure across valves and capillaries so that full flow is not maintained. Decreased capacity, low coil temperatures, and freezeup can result unless low ambient control is provided. (Chapter 20 has additional details about capillaries and valves.)

Properly designed unitary equipment allows a minimum amount of liquid refrigerant to return through the suction line to the compressor during abnormal operation. Normally, the heat absorbed in the evaporator vaporizes all the refrigerant and adds a few degrees of superheat. However, any conditions that increase refrigerant flow beyond the heat-transfer capabilities of the evaporator can cause liquid carryover into the compressor return line. Such an increase may be caused by a poorly positioned thermal element of an expansion valve or by an increase in condensing pressure of a capillary system, which may be caused by fouled condenser surfaces, excessive refrigerant charge, reduced flow of condenser air or water, or the higher temperature of the condenser cooling media. Heat transfer at the evaporator may be reduced by dirty surfaces, low-temperature differentials across the evaporator, or reduced airflow caused by a blockage in the system.

Transient flow conditions are a special concern. During *off* periods, refrigerant migrates and condenses in the coldest part of the system. In an air conditioner, this area is typically the evaporator, if it is within a cooled space. When the compressor starts, the liquid tends to return to the compressor in slugs. The severity of slugging is affected by temperature differences, *off* time, component positions, and traps formed in suction lines. Various methods such as suction-line accumulators, specially designed compressors, the refrigerant pumpdown cycle, or limited refrigerant charge are used to cope with equipment problems associated with excessive liquid return. Chapter 3 of the 1987 HVAC Volume has further information on refrigerant piping.

Strainers minimize the risk of restricting capillary tubes and expansion valves by foreign material, such as small quantities of solder, flux, and varnish. Overheated and oxidized oil may dissolve in warm refrigerant and deposit at lower temperatures in capillaries, expansion valves, and evaporators.

Buildings with high internal heat loads require cooling even at low outdoor temperatures. The capacity of air-cooled condensers can be controlled by changing airflow or flooding tubes with refrigerant. Airflow can be changed by using dampers, adjusting fan speed, or by stopping some of the fan motors in a multifan system.

Suction pressures drop momentarily during start-up. Systems with a low-pressure, cutout controller may require a time-delay relay to bypass the low-pressure controller momentarily to prevent nuisance tripping.

In cool weather, air conditioners only operate for short periods. If the weather is also damp, high levels and wide variance in humidity may also occur. Properly designed capacity-controlled units operate for longer periods, which may improve humidity control and comfort. In any case, cooling equipment should not be oversized.

Units with two or more separate refrigerant systems permit operation of individual units, which reduces capacity while matching the changing load conditions better.

Larger, single-compressor systems may offer capacity reduction through the use of cylinder-unloading compressors or multispeed compressors or by the addition of hot-gas bypass controls. At full-load operation, efficiency is unimpaired. However, reduced capacity operation may increase or decrease system efficiency, depending on the type of capacity reduction method used.

Multispeed compressors can improve efficiency at part-load operation. Cylinder unloading can decrease system performance, with the extent of the decrease depending on the particular method used. Hot-gas bypass does not reduce capacity efficiently, although it generally provides a wider range of capacity reduction. Systems with capacity-reduction compressors usually have capacity-controlled evaporators, otherwise the resultant evaporator coil temperatures may be too high to provide dehumidification.

Capacity-controlled evaporators are usually split, with one of the expansion valves controlled by a solenoid valve. Evaporator capacity is reduced by closing the solenoid valve. The compressor capacity-reduction controls or the hot-gas bypass system then provides maximum dehumidification, while the evaporator coil temperature is maintained above freezing to avoid coil frosting. Chapter 3 of the 1986 REFRIGERATION Volume has more details on hot-gas bypass systems.

## AIR-HANDLING SYSTEMS

High airflow, low static pressure performance, simplicity, economics, and compact arrangement are characteristics that make propeller fans particularly suitable for nonducted air-cooled condensers. Small diameter fans are direct-driven by four-, six, or eight-pole motors. Low starting torque requirements allow the use of single-phase shaded pole and permanent split-capacitor (PSC) fan motors and simplify speed control for low outdoor temperature operation. Many larger units use multiple fans and three-phase motors. Larger diameter fans are belt-driven at a lower rpm to maintain low tip speeds and quiet operation.

Centrifugal fans meet the higher static pressure requirements of ducted, air-cooled condensers, forced-air furnaces, and evaporators. Indoor airflow must be adjusted to suit duct systems and plenums. Some small blowers are direct-driven with multi-speed motors. Large blowers are always belt driven and may have variable-pitch motor pulleys for airflow adjustment. Vibration isolation reduces the amount of noise transmitted by bearings, motors, and fans into cabinets. (See Chapter 3 for details of fan design and Chapters 1 and 2 for information on air distribution systems.)

Disposable fiberglass filters are popular because they are available in standard sizes at low cost. Cleanable filters offer economic advantages when cabinet dimensions are not compatible with common sizes. Electronic or other high efficiency air cleaners are used when a high degree of cleaning is desired. Larger equipment may frequently be provided with automatic roll filters or high-efficiency bag filters. (See Chapter 10 for additional details about filters.)

Provision for introducing outdoor air is made in many unitary units; rooftop units are particularly adaptable for receiving outdoor air. Some units have automatically controlled dampers to permit cooling by outdoor air, which increases system efficiency.

## ELECTRICAL DESIGN

Electrical controls for unitary equipment are selected and tested to perform their individual and interrelated functions properly and safely within the entire range of operating conditions. Internal linebreak thermal protectors provide overload protection for most single-phase motors, smaller sizes of three-phase motors, and hermetic compressor motors. These rapidly responding temperature sensors, imbedded in motor windings, can provide precise locked rotor and running overload protection regardless of ambient temperatures.

Branch-circuit, short-circuit, and ground-fault protection is commonly provided by fused disconnect switches. Time-delay fuses allow selection of fuse ratings closer to running currents and thus provide backup motor overload protection, as well as short-circuit and ground-fault protection. Circuit breakers may be used in lieu of fuses where their use conforms to appropriate code requirements.

Some larger compressor motors have dual windings and contactors for step starting. A brief delay between contactor energization reduces the intensity of inrush current disturbance.

The use of 24-V (NEC Class 2) control circuitry is common for room thermostats and interconnecting wiring between split systems. It offers advantages in temperature control, safety, and ease of installation.

Motor speed control is used to vary evaporator airflow of direct-drive fans, air-cooled condenser airflow for low outdoor temperature operation, and to vary compressor speed to match load demand. Multitap motors and autotransformers provide one or more speed steps. Solid state speed control circuits can provide a widely adjustable speed range. However, motor bearings, windings, overload protection, and the motor suspension system must be suitable for operation over the full speed range.

In addition to speed control, solid-state circuits provide reliable temperature control, motor protection, and expansion valve refrigerant control. Complete temperature control systems are frequently included with the unit. Solid-state electronic control devices permit unitary equipment to be fitted with sophisticated temperature control systems, previously only installed in the field. Control system features such as automatic night setback, economizer control sequence, and zone demand control of multizone equipment contribute to improved comfort and energy savings.

## MECHANICAL DESIGN

Cabinet height dimensions are important for rooftop and ceiling-suspended units. Large unit design must consider the size limitations of truck bodies, freight cars, doorways, elevators, and the various rigging practices. In addition, structural strength of both the unit and the crate must be adequate for handling, warehouse stacking, shipping, and rigging.

The following criteria are also important: (1) cabinet insulation must prevent sweating in high-humidity ambient conditions, (2) insulated surfaces exposed to moving air should withstand air erosion, (3) air leakage around panels and at cabinet joints must be prevented, and (4) the cabinet insulation must be adequate to reduce energy transfer losses from the circulating air stream.

Also, cooling coil air velocities must be low enough to prevent blowoff of condensate. The drain pan must be sized to contain the condensate and must also be protected from high-velocity air. Service access must be provided for installation and repair. Versatility of application, such as multifan discharge direction and either-hand piping is another consideration. Weatherproofing requires careful attention and testing.

## ACCESSORIES

Using standard, cataloged accessories, the system designer can apply the unitary product concept to solve special application problems. Typical examples (Figures 2, 3, 7, and 8) are plenum coil housings, return air-filter grilles, and diffuser-return grilles for single-outlet units. Air duct kits offered for rooftop units (Figure 2) permit concentric or side-by-side ducting, as well as horizontal or vertical connections. Mounting frames are available to facilitate unit support and roof flashing. Other accessories include high-static-pressure fan drives, controls for low outdoor temperature operation, and duct damper kits for control of outdoor air intakes and exhausts.

## HEATING

For unitary air-conditioning systems it is important that cooling coils be installed downstream of furnaces so that condensation will not form inside the combustion and flue passages. Upstream cooling coil placement is permissible when the furnace has been approved for this type of application and designed to prevent corrosion. Burners, pilot flames, and controls must be protected from the condensate.

Chapter 9 describes hot water and steam coils used in unitary equipment, as well as the prevention of coil freezing from ventilation air in cold weather. Chapters 22, 24 and 27 discuss forced-air and oil- and gas-fired furnaces commonly used with, or included as part of, year-round equipment.

## EQUIPMENT AND SYSTEM STANDARDS

### Performance Rating—DOE Requirements

The United States Department of Energy has mandated a rating procedure for residential unitary equipment. Residential equipment is defined as single phase products with cooling capacities less than 65,000 Btu/h (19 kW). The testing and rating procedure is documented in Appendix M to subpart 430 of Section 10 of the *Code of Federal Regulations—Uniform Test Method for Measuring the Energy Consumption of Central Air Conditioners*.

The testing procedure provides a seasonal measure of operating efficiency. Seasonal Energy Efficiency Ratio (SEER) is an estimate of the ratio of the total seasonal cooling requirements measured in Btu and the total seasonal kilowatt hours of energy. This efficiency value is developed in the laboratory by conducting up to four separate tests at various indoor and outdoor conditions, including a measure of performance under cyclic operation. The SEER rating is expressed in Btu per watt hour.

Heating mode efficiencies of heat pumps are also expressed as the ratio of the total heating requirement, measured in Btu, and the total seasonal kilowatt hours used. This measure of efficiency is expressed as a Heating Seasonal Performance Factor (HSPF). In the laboratory, this efficiency measurement is determined from the results of up to four different testing conditions (five for two-speed or two-compressor systems), including a measure of cyclic performance. The magnitude of the HSPF measurement depends not only on the equipment performance, but also on the climatic conditions and the heating load relative to the equipment capacity.

For rating purposes, the Department of Energy has divided the United States into six climatic regions and has defined a range of maximum and minimum design loads. This division has the effect of producing 12 different HSPF ratings for a given piece of equipment. For comparison purposes, DOE has established Region 4 (moderate northern climate) as being a typical climatic region.

The SEER and HSPF varies appreciably with equipment design and size and from manufacturer to manufacturer. Present cooling SEER ratings range from 6 to 14. Heating HSPF at standard rating conditions for heat pumps range from 5 to 9.

### Performance Rating—ARI Certification Programs

The Air Conditioning and Refrigeration Institute (ARI) conducts three certification programs relating to unitary equipment, which are covered in the following standards:

ARI *Standard* 210, *Unitary Air-Conditioning Equipment*

ARI *Standard* 210/240, *Unitary Air-Conditioning and Air-Source Heat Pump Equipment*

ARI *Standard* 270, *Sound Rating of Outdoor Unitary Equipment*

These standards include the safety codes and performance requirements that are necessary for good equipment design. They also include the methods of testing established by ANSI/ASHRAE *Standard* 37-1978, *Methods of Testing for Rating Unitary Air-conditioning and Heat Pump Equipment.*

As part of its certification program, ARI publishes the Directory of Certified Unitary Products. Issued twice a year, the Directory identifies certified products enrolled in one or more programs and lists the certified capacity, energy efficiency, and sound rating for each.

Certification involves the annual audit testing of approximately 30% of the basic unitary equipment models of each participating manufacturer. These programs are limited to equipment up to 135,000 Btu/h (40 kW) cooling capacity.

ARI *Standard* 210/240 for unitary equipment establishes definitions and classifications, testing and rating methods, and performance requirements. Ratings are determined at ARI standard rating conditions, rated nameplate voltage, and prescribed discharge duct static pressures with rated evaporator airflow not exceeding 37.5 cfm per 1000 Btu/h (0.06 L/s W). The standard requires units to dispose of condensate properly and prohibits them from sweating under cool, humid conditions. Also, the ability to operate satisfactorily and restart at high ambient temperatures with low voltage is tested.

For certification under ARI *Standard* 270, outdoor equipment is tested in accordance with ANSI *Standard* S1.21-72. Test results obtained on a one-third octave band basis are converted to a single number for application evaluation. Application principles are covered in ARI *Standard* 275.

### Performance Rating—Equipment Over 135,000 Btu/h (40 kW)

Unitary air-conditioners exceeding 135,000 Btu/h (40 kW) can be tested in accordance with ARI *Standard* 360, *Commercial Industrial Air-Conditioning Equipment.*

Unitary heat pumps with capacities of 135,000 Btu/h (40 kW) or larger are covered by ARI *Standard* 340, *Commercial and Industrial Heat Pump Equipment.*

Approval agencies list unitary air conditioners complying with a standard like UL 465-82, *Central Cooling Air Conditioners.* An evaluation of the product determines that its design complies with the construction requirements specified in the standard and that the equipment can be installed in accordance with the applicable requirements of NFPA 70, *The National Electric Code*; ASHRAE 15-1978, *Safety Code for Mechanical Refrigeration*; NFPA 90A, *Installation of Air Conditioning and Ventilating Systems*; and NFPA 90B, *Installation of Warm Air Heating and Air Conditioning Systems.*

Tests determine that the equipment and all components will operate within their recognized ratings, including electrical, temperature, and pressure, when the equipment is energized at rated voltage and operated at specified environmental conditions. Stipulated abnormal conditions are also imposed, wherein the product must perform in a safe manner. The evaluation covers all operational features (such as electric space heating) that may be used in the product.

Products complying with the applicable requirements may bear the agency listing mark. An approval agency program includes the auditing of continued production at the manufacturer's factory.

Similarly, a standard like UL 559-85, *Heat Pumps*, is used for agency approvals of heat pumps.

# PART II: UNITARY HEAT PUMPS

Capacities of unitary heat pumps range from about 1.5 to 30 tons (5 to 105 kW), although there is no specific limitation. This equipment is used in residential, commercial, and industrial installations. The multi-unit installation, with a number of individual units, is particularly advantageous because it permits zoning, which provides the opportunity for heating or cooling in each zone on demand.

Application factors unique to unitary heat pumps include the following:

1. The unitary heat pump normally fulfills a dual function—heating and cooling; therefore, only a single piece of equipment is required for year-round comfort. Some manufacturers offer heating-only heat pumps for special applications.
2. A single source of energy can supply both heating and cooling requirements.
3. Heat output can be as much as two to four times that of the purchased energy input.
4. Vents of chimneys may be eliminated, thus reducing building costs.
5. A moderate supply air temperature in the heating cycle provides even heat, and close adherence to air-distribution principles ensures proper airflow and distribution while providing comfort.

In an air-source heat pump an air-to-refrigerant heat exchanger (outdoor coil in Figure 10) rejects heat to outdoor air when cooling indoor air and extracts heat from outdoor air when heating indoor air. Figure 10 shows a flow diagram of this cycle.

Most residential applications consist of an indoor fan and coil unit, either vertical or horizontal, and an outdoor fan-coil unit. The compressor is usually located in the outdoor unit. Electric heaters are commonly included within the indoor unit to provide heat during defrost cycles and during periods of high heating demand that cannot be satisfied by the heat pump alone.

Air-source heat pumps can be added to new or existing gas- or oil-fired furnaces. This unit, typically called a dual-fuel or hybrid heat pump, normally operates as a conventional heat pump. During extremely cold weather the refrigeration system is turned off and the oil or gas provides the auxiliary heat usually provided by electric heaters. These add-on heat pumps share the same air distribution system with the gas- or oil-fired warm air furnace and may be arranged in either parallel or series airflows.

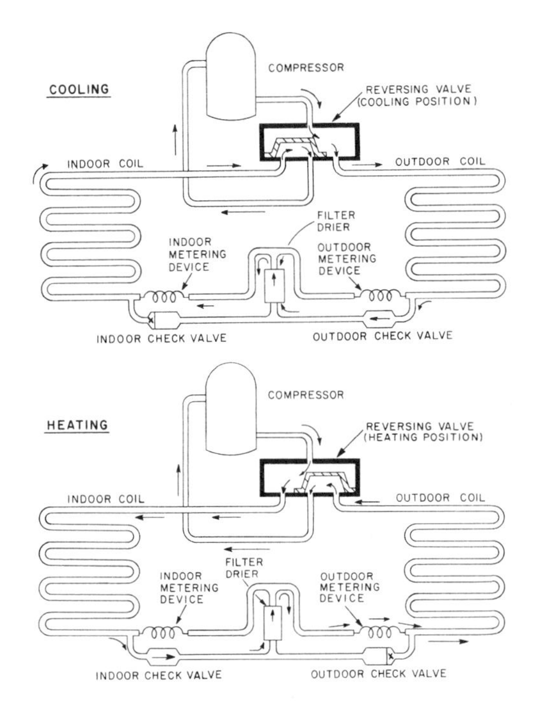


**Fig. 10 Typical Schematic of an Air-to-Air Heat Pump System**

However, the gas or oil furnace should never be upstream of the indoor coil when both systems are operated together. This arrangement raises the condensing temperature of the refrigeration system, which could cause a compressor failure.

## HEAT PUMP SELECTION

Figure 11 shows the performance characteristics of a typical air-source heat pump. Also shown are the heating and cooling loads for a typical building.

The heat pump heating capacity decreases as the ambient temperature decreases. This characteristic is opposite to the trend of the building load requirement. The outdoor temperature at which the heat pump capacity equals the building load is called the *balance point*. When the outdoor temperature is below the balance point, supplemental heat must be added to make up the difference, as shown by the shaded area. The COP shown in Figure 11 below the balance point is for the refrigeration system only and does not include the electric resistance heat. The electric heat has a COP of unity; thus, if it were included, the COP below the balance point would be lower than that shown.

In selecting the proper size heat pump, the cooling load for the building is calculated using standard practice. This usually results in a heating capacity that has a balance point near 32 °F (0 °C). The balance point may be lowered to use less resistance heat by improving the thermal qualities of the structure or by choosing a heat pump larger than the cooling load requires. In practice oversized cooling capacity causes excessive cycling, which results in an uncomfortable temperature and humidity level.

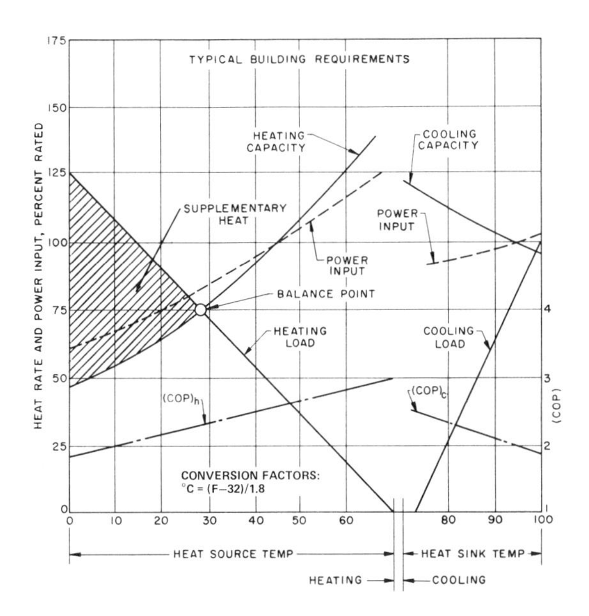


**Figure 11 Operating Characteristics of Single-Stage Unmodulated Heat Pump**

## HEAT PUMP REFRIGERANT SYSTEMS AND COMPONENTS

The reversing valve is the critical additional component required to make a heat pump air-conditioning system. Chapter 19 describes this valve and other refrigerant components in more detail.

Heat pump yearly operating hours are often up to five times those of a cooling unit. In addition, heating extends over a greater range of system operating conditions at higher stress conditions, so the design must be thoroughly analyzed to ensure maximum reliability. Improved components and protective devices contribute to better reliability, but the equipment designer must select components that are approved for the specific application. In all operating conditions, compressors should be protected against loss of lubrication, liquid floodback, and high discharge temperatures.

For a reliable and efficient heat pump system, the following factors must be considered: (1) outdoor coil circuitry, (2) defrost and water drainage, (3) refrigerant flow controls, (4) refrigerant charge management, and (5) compressor selection.

### Outdoor Coil Circuitry

The outdoor coil operates as an evaporator when the heat pump is used for heating. The refrigerant in the coil is less dense than when the coil operates as a condenser. For this reason, higher flow velocities prevail, unless the number of circuits is increased to reduce the pressure drop to acceptable levels. The circuitry may be a compromise between optimum performance as an evaporator and condenser, but for the best overall annual performance, the circuitry should favor the evaporator performance.

### Defrost and Water Drainage

During colder outdoor temperatures, usually below 40 to 45 °F (4 to 7 °C), the outdoor coil operates below freezing temperatures and frost builds up on the surface. The usual method of defrosting the coil is to reverse the refrigerant flow and operate the outdoor coil as a warm condenser, which melts the frost or ice. This normally takes four to eight minutes. The outdoor fan is usually turned off during the defrost period. The operation of the system in the defrost transient is a complex phenomenon in which the suction and discharge pressures are changing as the refrigerant is temporarily rerouted (Domingorena and Ball 1980).

For improved defrosting and water removal, some of the hot discharge gas should be directed near the bottom portions of the coil, and the water should drain freely from the cabinet. In cold climates, the cabinet should be raised above the ground to provide good drainage during defrost and to minimize snow and ice buildup around the cabinet. During prolonged periods of severe weather, it may be necessary to clear ice and snow from the unit.

Several methods are used to determine the need to defrost. One of the more common, simple, and reliable control methods is to initiate defrost at predetermined time intervals (usually 90 minutes). Other systems detect a need for defrosting by measuring changes in air pressure drop across the outdoor coil or changes in temperature difference between the outdoor coil and the outdoor air. Microprocessors are applied to control this function, along with numerous other functions (Mueller and Bonne 1980).

### Refrigerant Flow Controls

Separate refrigerant flow controls are usually used for the indoor and outdoor coils. Since the refrigerant flow reverses its direction between the heating and cooling mode of operation, a check valve bypasses in the appropriate direction around each expansion device. Either capillaries or thermostatic expansion valves may be used; however, capillaries require that greater care be taken to prevent excessive flooding of refrigerant into the compressor.

### Refrigerant Charge Management

Refrigerant charge management requires extra care to control compressor flooding and the storage of refrigerant in the system during both heating and cooling. The mass flow of refrigerant during cooling is greater than during heating. Consequently, the amount of refrigeration stored may be greater in the heating mode than in the cooling mode, depending on the relative internal volumes of the indoor and outdoor coils. Furthermore, the internal volumes of the indoor and outdoor coils are usually different. Usually, the internal volume of the indoor coils ranges from 110% to 70% of the outdoor coil volume, and their relative volumes can be adjusted so that the coils not only transfer heat but also manage the charge.

When using capillaries, the refrigerant may be stored in an accumulator in the suction line or in charge-modulating vessels that can remove the refrigerant charge from circulation when compressor floodback is imminent. Thermostatic expansion valves reduce the flooding problem, but storage may be required in the condenser.

To maintain performance reliability, the amount of refrigerant in the system must be checked and adjusted in accordance with the manufacturer's recommendations, particulary when charging a heat pump.

### Compressor Selection

The compressor should be selected on the basis of the compressor manufacturer's recommendations. Compressors in a heat pump operate over a wide range of suction and discharge pressures; thus, their design parameters, such as refrigerant discharge temperatures, pressure ratios, clearance volume, and motor-overload protection, require special concern. In all operating conditions, compressors should be protected against loss of lubrication, liquid floodback, and high discharge temperatures.

## HEAT PUMP SYSTEM CONTROL AND INSTALLATION

The installation should always follow the manufacturer's instructions. As mentioned in the discussion under defrost, the outdoor unit should be raised from the ground in severe climate areas to permit free drainage of defrost water and to minimize the potential blocking by snow. Since the supply air from a heat pump is at a lower temperature, 90 to 100°F (32 to 38°C), than most heating systems, the duct and supply register should control velocity and throw to minimize cool drafts.

Low voltage heating/cooling thermostats control heat pump operation. Both automatic models that switch automatically from heating to cooling operation and manual selection models are available. Usually, there are two stages of heating control. The first stage controls heat pump operation, and the second stage controls supplementary heat. When the outdoor temperature drops below the balance point, supplementary heat is added to provide the required heating. The amount of supplementary heat is often controlled by an outdoor thermostat that allows additional stages of heat to be turned on only when required by the colder outdoor temperature.

Microprocessor technology has led to the use of night setback modes and intelligent recovery schemes for morning warmup on heat-pump systems.

### Desuperheaters

Desuperheaters are being applied to unitary air conditioners and heat pumps. These devices recover heat from the compressor discharge gas and use it to heat domestic hot water. The desuperheater usually consists of a pump, heat exchanger, and controls, and it can produce about 15 gallons of heated water per hour per ton of air conditioning (4.5 mL/s per kW). Because desuperheaters improve cooling performance and reduce the degrading effect of cycling during heating, they are best applied where cooling requirements are high and in climates where a significant number of heating hours occur above the building's balance point (Counts 1985).

## BIBLIOGRAPHY

Alabama Power Company. 1983. Guidelines for Application and Installation of Desuperheaters (Discharge Line Water Heaters) Used with Heat Pump Systems. Birmingham, AL.

ARI. 1981. Unitary Air-Conditioning Equipment. ARI *Standard* 210-81. Air Conditioning and Refrigeration Institute, Arlington, VA.

ARI. 1984a. Unitary Air-Conditioning and Air-Source Heat Pump Equipment. ARI *Standard* 210/240-84.

ARI. 1984b. Sound Rating of Outdoor Unitary Equipment. ARI Standard 270-84.

ARI. 1984c. Application of Sound Rated Outdoor Unitary Equipment. ARI *Standard* 275-84.

ARI. 1985a. Commercial and Industrial Unitary Heat Pump Equipment. ARI *Standard* 340-85.

ARI. 1985b. Commercial and Industrial Unitary Air-Conditioning Equipment. ARI *Standard* 360-85.

ARI. 1987. Directory of Certified Unitary Products. Issued by ARI.

Counts, D. 1985. Performance of Heat Pump/Desuperheater Water Heating Systems. ASHRAE *Transactions* 91 (25): 1473-1487.

Domingorena, A.A. and S.J. Ball. 1980. Perfomance Evaluation of a Selected Three-Ton Air-to-Air Heat Pump in the Heating Mode. Oak Ridge National Laboratory, ORNL/CON-34.

Mueller, D. and U. Bonne. 1980. Heat Pump Controls: Microelectronic Technology. ASHRAE *Journal*, September.

# CHAPTER 43

# APPLIED PACKAGED EQUIPMENT



AN applied packaged unit is a factory-designed, self-contained air-conditioning unit that can be integrated into an applied air-conditioning system. Products included in this definition are packaged terminal air conditioners (including the subclass, packaged terminal heat pumps) and water-source heat pumps, which may be used in water-loop, ground-water, surface-water, or earth-coupled systems.

## PACKAGED TERMINAL AIR CONDITIONERS

The Air Conditioning and Refrigeration Institute (ARI) defines a packaged terminal air conditioner (PTAC) as a wall sleeve and a separate unencased combination of heating and cooling assemblies specified by the builder and intended for mounting through the wall. It includes a prime source of refrigeration, separable outdoor louvers, forced ventilation, and heating by hot water, steam, or electricity as chosen by the builder. PTAC units with direct-fired gas heaters are also available from some manufacturers. A packaged terminal heat pump (PTHP) is a heat pump version of a PTAC, and it is included in the heating-by-electricity portion of the ARI definition.

Packaged terminal air conditioners are designed primarily for commercial installations to provide the total heating and cooling functions for a room or zone and are specifically for through-the-wall installation. The units are mostly used in relatively small zones on the perimeter of buildings such as hotel and motel guest rooms, apartments, hospitals, nursing homes, and office buildings. In larger buildings, they may be used with nearly any system selected for environmental control of the building core.

Packaged terminal air conditioners (PTAC) and packaged terminal heat pumps (PTHP) are similar in design and construction. The most apparent difference between a PTAC and a PTHP is the addition of a refrigerant reversing valve in the PTHP. Additional components that control the heating functions of the heat pump could include an outdoor thermostat to signal the need for changes in heating operating modes, and, in the more complex designs, frost sensors, defrost termination devices, and base pan heaters.

The preparation of this chapter is assigned to TC 7.6, Unitary Air Conditioners and Heat Pumps.

### Sizes and Classifications

Packaged terminal air conditioners are available in a wide range of capacities from 5000 to 18,000 Btu/h (1.5 to 5.3 kW) cooling and 2500 to 35,000 Btu/h (0.7 to 10.3 kW) heating.

The units are available as sectional type or integrated type. Both types include the following:

1. Heating elements available in hot water, steam, electric, or gas heat
2. Integral or remote temperature and operating controls
3. Wall sleeve or box
4. Removable (or separable) outdoor louver
5. Room cabinet
6. Means for controlled forced ventilation
7. Means for filtering all air delivered to the room
8. Ductwork

The assembly is intended for free conditioned-air distribution, but a particular application may require minimal ductwork with a total external static resistance up to 0.1 in. of water (25 Pa).

A sectional-type unit (Figure 1) has a provision for the addition of a cooling option; the integrated type unit (Figure 2) has a provision for the addition of a heating option.

### General System Design Considerations

Packaged terminal air conditioners and packaged terminal heat pumps represent a unique challenge to the HVAC system designer. Since the product becomes part of the building's facade, the architect must consider the product during the conception of the building. The installation of wall sleeves is more of an ironworker's, mason's, or carpenter's craft. All electric PTAC

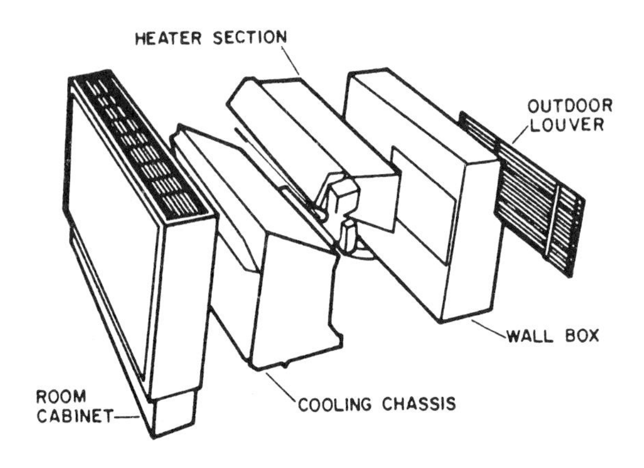


**Fig. 1 Packaged Terminal Air Conditioner—Sectional Type**

units dominate the PTAC market. Recent market statistics indicate that 50% are PTHP, 45% are PTAC with electric heat, and 5% involve the other forms of heating.

All the energy of the all-electric versions is dispersed through the building through electrical wiring, so the electric designer and the electrical contractor play a major role. The final installation is reduced to sliding in the chassis and plugging the unit into an adjacent receptacle. For these all-electric units, the traditional HVAC contractor's installation skills involving ducting, piping, and refrigeration systems are bypassed, so the conventional HVAC contractor may not be used.

When comparing the selections of a gas-fired PTAC to a PTAC with electric heat or a PTHC, both operating and installation costs should be evaluated. Generally, a gas-fired PTAC is more expensive to install but less expensive to operate in the heating mode. A life cycle cost comparison is recommended.

One main advantage of the PTAC/PTHP concept is that it provides excellent zoning capability. Units can be shut down or operated in a holding condition during periods of unoccupancy. Present equipment efficiency-rating criteria only recognize the benefits of single units, so an efficiency comparison to other approaches may suffer.

The designer must also consider that the total capacity is the sum of the peak loads of each zone rather than the peak load of the building. Therefore, the total cooling capacity of the zonal system must exceed the total capacity of a central system.

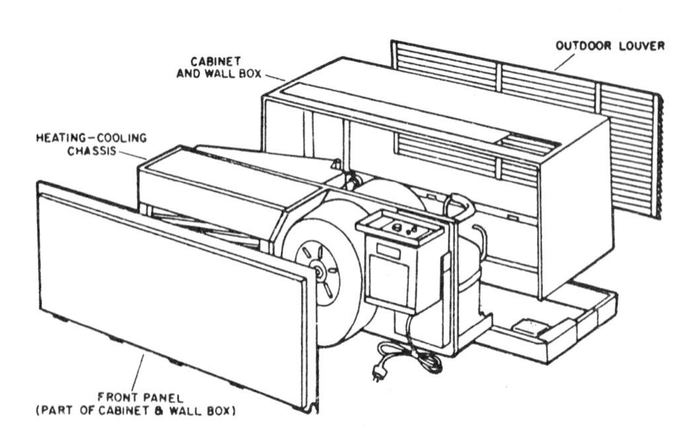


**Fig. 2 Packaged Terminal Air Conditioner—Integrated Type**

Because PTAC units are located within the conditioned space, both appearance and sound level of the equipment are important considerations. Sound attenuation in ducting is not available with the free-discharge PTAC units.

The designer must also consider the added infiltration and thermal leakage load resulting from the perimeter wall penetrations. These losses are accounted for during the on-cycle in the equipment cooling ratings and PTHP heating ratings, but during the off-cycle or with other forms of heating, they could be significant.

The widely dispersed PTAC units also present challenges relative to effective condensate disposal and the handling of extreme situations when designed condensate disposal systems are inadequate.

Most packaged terminal equipment is designed to fit into a wall aperture of approximately 40 in. wide and 16 in. high (1000 by 400 mm). While unitary products can increase in size with increasing cooling capacity, all PTAC/PTHP units, regardless of cooling capacity, are usually constrained to one cabinet size. The exterior of the equipment must be essentially flush with the exterior wall to meet most building codes. In addition, cabinet structural requirements and the slide-in chassis reduces the area that outdoor air must enter and exit to less than 3.5 $ft^2$ (0.33 $m^2$).

Provision must be made to minimize outdoor air recirculation, and attention must be given to architectural appearance needs. This may increase parasitic air-side pressure drops. With a capacity range that usually spans about a 3 to 1 ratio, this causes difficulty in maintaining the efficiency of units at the higher levels of cooling capacity.

## PTAC/PTHP Equipment Design

**Compressor.** PTAC units are designed with single-speed compressors. Both reciprocating and rotary types are used. They operate on a single-phase power supply and are available in 208, 230, and 265 volt versions. The compressors usually are protected with electromechanical protective devices, with some of the more advanced models employing electronic protective systems.

**Fan Motor(s).** Some PTAC units employ a single double-shafted direct-drive fan motor that provides the motive force for both the indoor and outdoor air-moving devices. This motor usually has two speeds that affect equipment sound level, throw of the conditioned air, cooling capacity, efficiency, and the sensible-to-total capacity ratio.

Full-featured models have two fan motors: one for the indoor air mover and the other for the outdoor air mover. Two motors provide greater flexibility in locating components, since the indoor and outdoor fans are no longer constrained to the same rotating axis. They also allow for the possibility of different fan speeds for the indoor and outdoor systems. In this case, the outdoor fan motor is usually single speed, and the indoor fan motor has two or more speeds. Also, the designer has a broader selection of air-moving devices and can provide the user with a wider range of sound level and conditioned air throw options. Efficiency can be maintained at lower indoor fan speeds. When heating (other than a heat pump in the heating mode), the outdoor fan motor can be switched off automatically to reduce electrical energy consumption and decrease infiltration and heat transmission losses through the PTAC unit.

**Indoor Air Mover.** The airflow quantity, air-side pressure rise, available fan motor speeds, and sound level requirements of the indoor air system of a PTAC indicate that a centrifugal blower wheel provides reasonable indoor air performance. In some cases, proprietary mixed-flow blowers are used. Dual-fan motor units permit the use of two centrifugal blower wheels or a cross-flow blower to provide a more even discharge of the conditioned air.

**Indoor Air Circuit.** PTACs have an air filter of fiberglass, metal, or plastic foam, which removes large particulate matter from the circulating airstream. In addition to improving the quality of the indoor air, this filter also reduces fouling of the indoor heat exchanger. The PTAC also provides mechanical means of introducing outdoor air into the indoor airstream. This air, which may or may not be filtered, controls infiltration and pressurization of the conditioned space and, during portions of the year, can serve as an outdoor air economizer.

**Outdoor Air Mover.** Outdoor air movers may be either centrifugal blower wheels, mixed-flow blowers, or axial flow fans.

**Heat Exchangers.** PTAC units may use conventional plate-fin heat exchangers, which have either copper or aluminum smooth-bore tubes. The fins are usually aluminum, which may, in some cases, be coated to retard corrosion. Since PTACs are generally restricted in physical size, performance improvements based on increasing heat-exchanger size is limited. Therefore, some manufacturers, to improve performance or reduce costs, employ heat exchangers with performance enhancements on the air side (lanced fins, spine fin, etc.) and/or the refrigerant side (internal finning or rifling).

**Refrigerant Expansion Device.** Most PTACs use a simple capillary as an expansion device. Off- rating-point performance is improved if thermal expansion valves are used.

**Condensate Disposal.** Condensate forms on the indoor coil when cooling. Some PTACs require that a drain system be installed to convey the condensate to a disposal point. Other units spray the condensate on the outdoor coil where it is evaporated and dispersed to the outdoor ambient air. This evaporative cooling of the condenser enhances performance, but the potentially negative effects of fouling and corrosion of the outdoor heat exchanger must be considered. This problem is especially severe in a coastal installation where salt spray could mix with the condensate and, after repeated evaporation cycles, build a corrosive, salt-water solution in the condensate sump.

A PTHP also produces a condensate in the heating mode. If the outdoor coil operates below freezing, the condensate forms frost, which is melted during defrost. This water must be disposed of in some manner. If drains are used and the heat pump operates in below freezing weather, the drain lines must be protected from freezing. Outside drains cannot be used in this case, unless they have drain heaters. Some PTHPs spray the condensate formed during heating onto the indoor coil, which humidifies the indoor air. Inadequate condensate disposal can lead to overflow at the unit and potential staining of the building facade.

**Controls.** PTAC units have a built-in manual mode selector (cool, heat, fan only, and off) and a manual fan-speed selector. A thermostat adjustment is provided with set points usually identified in subjective terms such as *high*, *normal*, and *low*. Some units incorporate electronic controls, which provide room temperature limiting, evaporator freeze-up protection, compressor lockout in the case of actual or impending compressor malfunction, and service diagnostic aids. Advanced master controls at a central location are also used. This enables an operator at the master control to override the control settings registered by the occupant. These master controls may limit operation when certain room temperature limits are exceeded, adjust thermostat set points during unoccupied periods, and turn off certain units to limit peak electrical demand.

**Wall Sleeves.** A wall sleeve is a required part of a PTAC unit. It becomes an integral part of the building structure and must be designed with sufficient strength to maintain its dimensional integrity after installation. It must withstand the potential corrosive effects of other building materials, such as mortar, and must endure the long-term exposure to the outdoor elements.

**Outdoor Grille.** The outdoor grille or louvers must be compatible with the architecture of the structure. Most manufacturers provide options in this area. A properly designed grille prevents birds, vermin, and outdoor debris from entering, impedes the entry of rain and snow and, at the same time, provides adequate free area for the outdoor airstream to enter and exit with a minimum of recirculation.

**Slide-in Components.** The interface between the wall sleeve and slide-in components allows them to be easily inserted and later removed for service and/or replacement. In the event of serious malfunction, the offending slide-in component can be quickly replaced with a spare, and the repair can be made off the premises. An adequate seal at the interface is essential to seal out wind, rain, snow, and insects without jeopardizing the slide-in/slide-out feature.

**Indoor Appearance.** Since the PTAC units are located within the conditioned space, the indoor appearance must blend in with the indoor decor of the room. Manufacturers provide a variety of indoor treatments, which include variations in shape, style, and materials. In addition to metal convector-style fronts, many provide wood or simulated wood front alternatives.

## Heat Pump Operation

Basic PTHP units operate in the heat pump mode down to an outdoor temperature just above the point that frosting of the outdoor heat exchanger would occur. When that outdoor temperature is reached, the heat pump mode is locked out, and other forms of heating are required. Some PTHPs use control schemes that extend the operation of the heat pump to lower temperatures. One approach permits heat pump operation down to outdoor temperatures just above the freezing point. If the outdoor coil frosts, it is defrosted by shutting down the compressor and allowing the outdoor fan to continue circulating outdoor air over the coil. Another approach permits heat pump operation to even lower outdoor temperatures by using a reverse cycle defrost sequence. In those cases, the heat pump mode is usually locked out for outdoor temperatures below 10°F (−12°C).

## Performance and Safety Testing

Packaged terminal air conditioners may be rated in accordance with ANSI/ARI *Standard* 310-85. Packaged terminal heat pumps are rated in accordance with ARI *Standard* 380-85, *Packaged Terminal Heat Pumps*.

The ARI periodically issues a *Directory of Certified Applied Air-Conditioning Products*, which lists the cooling capacity, EER, heating capacity, and airflow rate in cfm for each participating manufacturer's PTAC models. The listings of PTHP models also include the heating COP.

Cooling and heating capacities, as listed in the ARI Directory, must be established in accordance with ANSI/ASHRAE *Standard* 37-1978, *Methods of Testing for Rating Unitary Air-Conditioning and Heat Pump Equipment* or with ASHRAE *Standard* 16-1983, *Method of Testing for Rating Room Air Conditioners and Packaged Terminal Air Conditioners*. All standard heating ratings should be established in accordance with ANSI

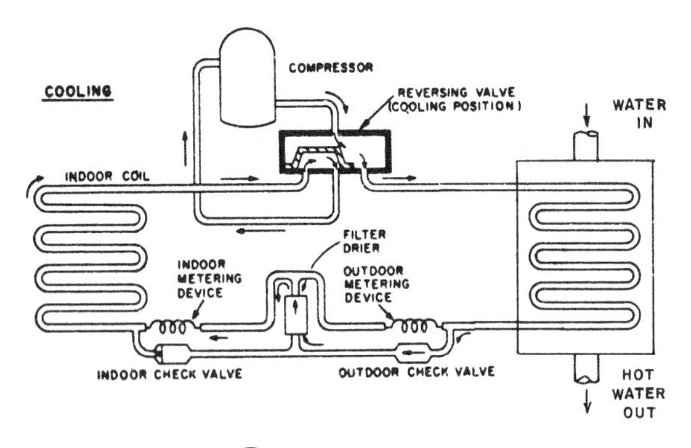


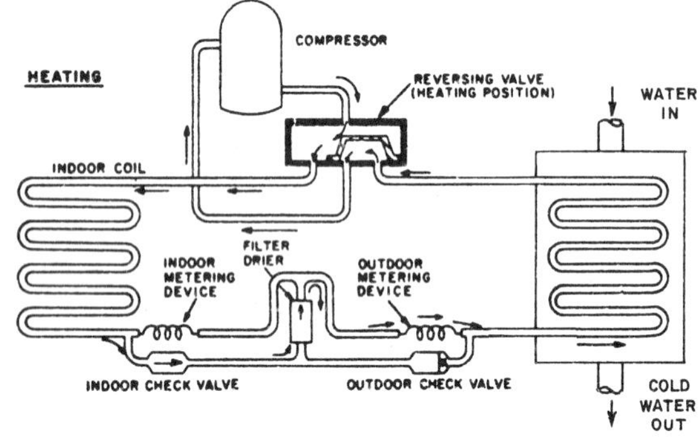


**Fig. 3 Typical Schematic of a Water-Source Heat-Pump System**

/ASHRAE *Standard* 58-74, *Method of Testing Room Air-Conditioner Heating Capacity.*

Additionally, PTAC units should be constructed in accordance with ANSI/ASHRAE *Standard* 15-1978, *Safety Code for Mechanical Refrigeration* and should comply with the safety requirements of ANSI/UL *Standard* 484-82, *Room Air Conditioners.*

## WATER-SOURCE HEAT PUMPS

Water-Source Heat Pumps (WSHP) are single-package reverse-cycle heat pumps that use water as the heat source when in the heating mode and as a heat sink when in the cooling mode. The water supply may be a recirculating closed loop, a well, a lake, or a stream. Water is usually supplied at 2 to 3 gpm per ton of cooling capacity (36 to 54 mL/s per kW). The main components of a WSHP refrigeration system are a compressor, a refrigerant-to-water heat exchanger, a refrigerant-to-air heat exchanger, refrigerant expansion devices, and a refrigerant-reversing valve. Figure 3 shows a typical schematic of a WSHP system.

Designs of water-source heat pumps range from horizontal units located primarily above the ceiling or on the roof to vertical units usually located in basements or equipment rooms, and console units located in the conditioned space. Figures 4 and 5 illustrate typical designs.

### Types of Water-Source Heat Pump Systems

Water source heat pumps are used in a variety of systems. These include the following:

1. Water Loop Heat Pump Systems
2. Ground Water Heat Pump Systems
3. Surface Water Heat Pump Systems
4. Earth Coupled Heat Pump Systems

Water-Loop Heat Pumps (WLHP) use a circulating water loop as the heat source and the heat sink. When the loop water temperatures exceed a certain level due to heat added as the result of heat pump cooling, a cooling tower dissipates heat from the water loop into the atmosphere. When the loop water temperatures drop below a prescribed level due to heat being removed as a result of heat-pump heating, heat is added to the circulating loop water, usually with a boiler. In multiple-unit installations, under conditions when some heat pumps operate in the cooling mode while others operate in the heating mode, the loop water temperatures remain between the prescribed limits.

Ground-Water Heat Pumps (GWHP) pump ground water from a nearby well and pass it through the heat pump's water-to-refrigerant heat exchanger where it is warmed or cooled, depending on the operating mode. It is then discharged to a drain, a stream, a lake, or it is returned to the ground through a reinjection well.

Many state and local jurisdictions have enacted ordinances relating to the use and discharge of ground water. Since aquifers, the water table, and ground water availability vary from region to region, these regulations cover a wide spectrum.

Surface Water Heat Pumps (SWHP) pump water from a nearby lake, stream, or canal. After passing through the heat pump

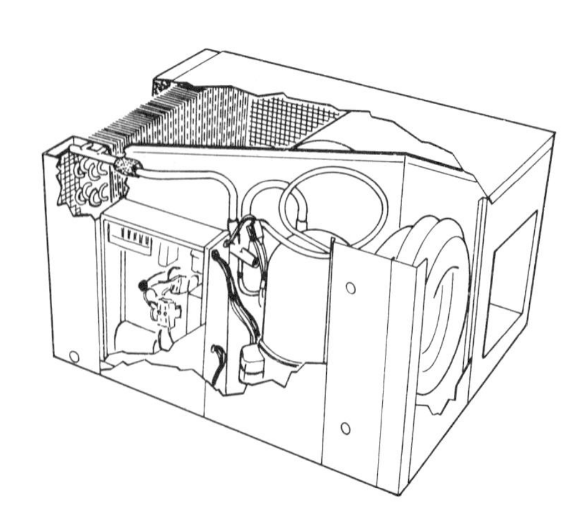

**Fig. 4 Typical Horizontal Water-Source Heat Pump**

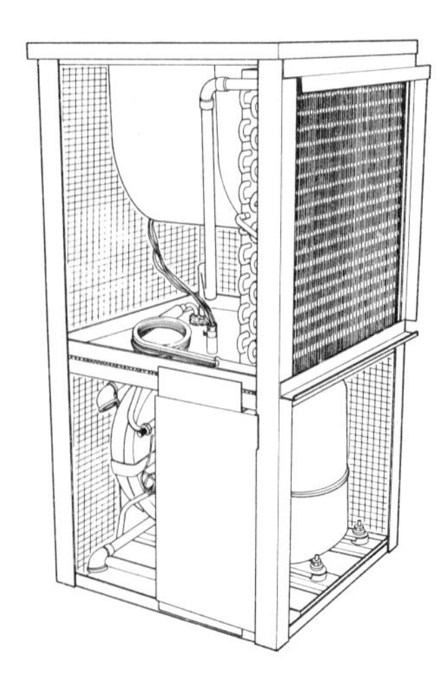

**Fig. 5 Typical Vertical Water-Source Heat Pump**

heat exchanger, it is returned to the source or a drain at a temperature level several degrees warmer or cooler, depending on the operating mode of the heat pump.

Earth Coupled Heat Pumps (ECHP), sometimes called a ground-coupled heat pump system, use a closed water loop that passes through a heat exchanger buried in the earth. Usually, piping is installed in either a shallow horizontal or deep vertical array to form the heat exchanger. The massive thermal capacity of the earth provides a temperature stabilizing effect on the circulating loop water. Installing this type of system requires detailed knowledge of the climate; the site; the soil thermal characteristics; and the performance, design, and installation of water-to-earth heat exchangers. The ASHRAE publication, *Design/Data Manual for Closed-Loop Ground-Coupled Heat Pump Systems* has detailed information on design and installation of ECHP systems. Additional information on these and other water-source heat pump systems is presented in Chapter 9, "Applied Heat Pump Systems," in the 1987 HVAC Volume.

## Entering Water Temperatures

These various applications provide a wide range of entering water temperatures to water source heat pumps. Not only do the entering water temperatures vary by type of application, but they also vary by climate and time of year. Due to the wide range of entering water temperatures encountered, it is not feasible to design a universal packaged product that can handle the full range of possibilities effectively. Within systems that experience a wider range of entering water temperatures (such as surface-water and earth-coupled), specific WSHP models optimized for the region are sometimes required to achieve optimum performance.

## Performance Certification Programs

ARI maintains a certification program for water source heat pumps that is aimed primarily at heat pumps used in WLHP systems. Units are tested in accordance with ARI *Standard* 320, which provides standard cooling ratings at 85°F (29.4°C) entering water temperature and standard heating ratings at 70°F (21.1°C) entering water temperature. Maximum operating conditions are checked at the upper temperature level at 90°F (32.2°C) entering water temperature and at the lower temperature level at 65°F (8.3°C) leaving water temperature. This latter temperature corresponds to approximately 55°F (13°C) entering water when cooling. The range of test conditions covers the extremes of entering water temperatures typically encountered in WLHP systems.

ARI also has a companion certification program for ground-water-source heat pumps. In this program, units are tested in accordance with ARI *Standard* 325, which provides standard cooling ratings at both 50 and 70°F (10.0 and 21.1°C) entering water temperature and standard heating ratings at these same two entering water temperatures. Maximum operating conditions are checked at the upper temperature level at 75°F (23.9°C) entering water temperature and at the lower temperature level at 45°F (7.2°C) entering water temperature. These entering water temperatures bracket the range of ground water temperatures found across the United States.

The ARI *Directory of Certified Applied Air-Conditioning Products* lists the cooling capacity, cooling EER, heating capacity, and heating COP of units rated in accordance with ARI *Standard* 320. For ground-water-source heat pumps rated in accordance with ARI *Standard* 325, the directory also lists these same parameters for the two standard entering water temperatures, along with water flow rates and heat exchanger water-pressure-drop data.

There are no ARI certification programs for water-source heat pumps designed specifically for surface water or earth-coupled applications.

## Water Source Heat Pump Equipment Design

From a descriptive standpoint, the components used in water-source heat pumps optimized for the various systems are essentially the same. Optimizing the performance of WSHP units to match differing levels of entering water temperatures is accomplished by the relative sizing of the indoor and the refrigerant-to-water heat exchangers and matching the expansion devices to the refrigerant flow rates.

**Compressors.** WSHP units usually have single-speed compressors. Higher capacity equipment may use multiple compressors, which provide capacity modulation. The compressors may be of the reciprocating or rotary type. Single-phase units are available at voltages of 115, 208, 230, and 265. All larger equipment is for three-phase power supplies with voltages of 230, 460, or 550. The compressors are usually protected with eletromechanical protective devices.

**Indoor Air System.** Console WSHP models are designed for free delivery of the conditioned air and other WSHP units having ducting capability. Smaller WSHPs have multi-speed, direct-drive centrifugal blower wheel fan systems. Large capacity equipment has belt-drive systems. All units have built-in air filters of fiberglass, metal, or plastic foam.

**Indoor Heat Exchanger.** The indoor heat exchanger of WSHP units is a conventional plate fin coil of copper tubes and aluminum fins. The tubing in the coil is circuited so that it can function effectively as an evaporator with the refrigerant flow in one direction and as a condenser when the refrigerant flow is reversed.

**Refrigerant-to-Water Heat Exchanger.** The heat exchanger, which couples the heat pump to source/sink water, is either of the tube-in-tube or tube-in-shell type. This heat exchanger must also function in either the condensing or evaporating mode, so special attention is given to refrigerant-side circuitry. The construction of the heat exchanger is usually of copper and steel where the source/sink water is exposed only to the copper portions. Cupronickel options to replace the copper are usually available for use with brackish or corrosive water.

**Refrigerant Expansion Devices.** Most WSHPs rated in accordance with ARI *Standard* 320, a single rating point standard for each operating mode, use simple capillaries as expansion devices. Units with capillaries do not perform well when rated at the dual rating points of ARI *Standard* 325. Off-rating-point performance is improved if thermal expansion valves are used, since they provide peak performance over a broader range of inlet water temperatures.

**Refrigerant-Reversing Valve.** The refrigerant-reversing valves in water-source heat pumps are identical to those used in unitary air-source heat pumps.

**Condensate Disposal.** Condensate forms on the indoor coil when cooling is collected and conveyed to a drain system.

**Controls.** Console WSHP units have built in operating mode selector and thermostatic controls. Ducted units use low-voltage remote heat/cool thermostats.

**Special Features.** A number of special features are available on certain models of WSHP of some manufacturers. These features include the following:

*Capacity Modulation.* Either multiple compressors or hot gas bypass may be used.

*Variable Air Volume (VAV).* Reduces fan energy usage and requires some form of capacity modulation.

*Automatic Water Valve.* Closes off water flow through unit when compressor is off and permits variable water volume in the loop, which reduces pumping energy usage.

*Outdoor-Air Economizer.* Cools directly with outdoor air to reduce or eliminate the need for mechanical refrigeration during mild or cold weather when outdoor humidity levels and outdoor air quality are appropriate.

*Water-Side Economizer.* Cools with loop water to reduce or eliminate the need for mechanical refrigeration during cold weather and requires a hydronic coil in the indoor air circuit, which is valved into the circulating loop when loop temperatures are relatively low and there is a call for cooling.

*Electric Heaters.* Used in WLHP systems that do not have a boiler as a source for loop heating.

## BIBLIOGRAPHY

ARI. 1985a. Packaged Terminal Air Conditioners. ARI *Standard* 310-85. Air-Conditioning and Refrigeration Institute, Arlington, VA.
ARI. 1985b. Ground Water-Source Heat Pumps. ARI *Standard* 325-85.
ARI. 1985c. Packaged Terminal Heat Pumps. ARI *Standard* 380-85.
ARI. 1986. Water-Source Heat Pumps. ARI *Standard* 320-86.
ARI. 1987a. Directory of Certified Applied Air-Conditioning Products.
ARI. 1987b. Directories of Certified Unitary Products.

# CHAPTER 44

# SOLAR ENERGY EQUIPMENT



COMMERCIAL and industrial solar energy systems are generally classified according to the heat-transfer fluid used in the collector loop—namely air or liquid. While both share the basic fundamentals associated with conversion of solar radiant energy, the equipment used in each is entirely different. Air systems are primarily limited to forced-air space heating and industrial and agricultural drying processes. Liquid systems are suitable for a broader range of applications, such as hydronic space heating, service water heating, industrial process water heating, energizing absorption air- conditioning, pool heating, and as a heat source for series-coupled heat pumps. Because of this wide range in capability, liquid systems are more commonly found in commercial industrial applications.

## AIR-HEATING SYSTEMS

Air-heating systems circulate air through ducting to and from an air-heating collector (Figure 1). Air systems are effective for space heating applications because a heat exchanger is not required and the collector inlet temperature is low throughout the day (approximately room temperature). Air systems do not need protection from freezing, overheat, or corrosion. Furthermore, air costs nothing and does not cause disposal problems. However, air ducts and air-handling equipment require more space than pipes and pumps, ductwork is hard to seal, and leaks are difficult to detect. Fans consume more power than the pumps of a liquid system, but if the unit is installed in a facility that uses air distribution, only a slight power cost is chargeable against the solar space-heating system.

Most air space-heating systems also preheat domestic hot water through an air-to-liquid heat exchanger. In this case, tightly fitting dampers are required to prevent reverse thermosiphoning at night, which could freeze water in the heat exchanger coil. When this system only heats water in the summer, the parasitic power consumption must be charged against the solar energy system, because no space heating is involved and there are no comparable energy costs are associated with conventional water heating. In some situations, solar water-heating air systems could be more expensive than conventional water heaters, particularly if electrical energy costs are high. To reduce the parasitic power consumption, some systems use the low speed of a two-speed fan.

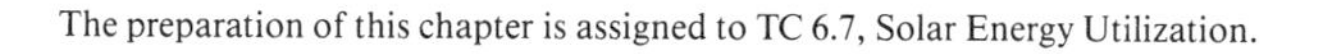

The preparation of this chapter is assigned to TC 6.7, Solar Energy Utilization.

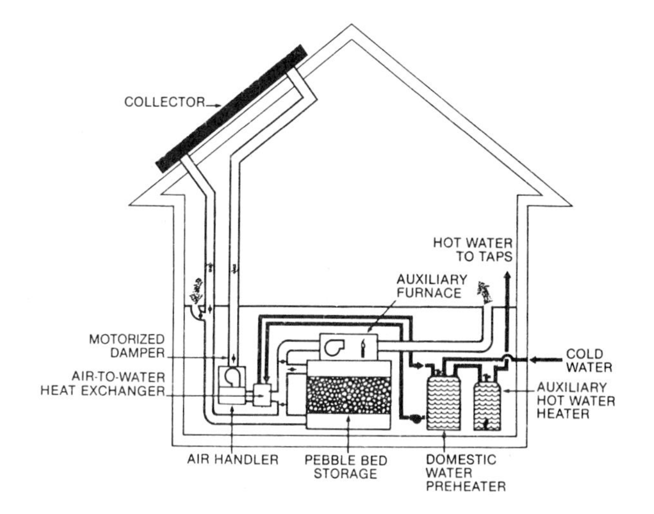


**Fig. 1 Air-Heating Space and Domestic Water Heater System**

## LIQUID HEATING SYSTEMS

Freezing is the principal cause of liquid system failure. For this reason, freeze tolerance dominates the design and selection of equipment. A solar collector radiates heat to the cold sky and will freeze at air temperatures well above 32°F (0°C). Where freezing conditions are rare, small solar heating systems may be equipped with low-cost, low reliability protection devices that depend on electrical and/or mechanical components such as electronic controllers and automatic valves. With increasing consumer/government demands for extended warranties, solar designers and installers must consider reliable designs. Because of the large investment associated with most commercial and industrial installation, reliable freeze protection is essential, even in the warmest climates.

### Open and Closed Systems

In the *open* liquid solar energy system, the collector loop is open to the city water supply, and fresh water circulates through the collector. In the *closed* system, the collector loop is isolated from the high-pressure city water supply by a heat exchanger. In areas of poor water quality, isolation protects the collectors from fouling due to minerals in the water. Closed-loop systems also offer greater freeze protection, so they are used almost exclusively in commercial and industrial applications.

Closed-loop systems use two methods of freeze protection: (1) non-freezing fluids and (2) drainback.

**Nonfreezing Fluid Freeze Protection.** The most popular solar energy system for commercial application is the closed-loop system. It contains a nonfreezing heat-transfer fluid to transmit heat from the solar collectors to storage (Figure 2). The most common heat transfer fluids are water/ethylene glycol and water/propolyene glycol, although other heat-transfer fluids such as silicone oils, hydrocarbon oils, or refrigerants can be used. Because the collector loop is closed and sealed, the only contribution to pumping head is friction loss; therefore, the location of the solar collectors relative to the heat exchanger and storage tank is not critical. Traditional hydronic sizing methods can be used for selecting pumps, expansion tanks, heat exchangers, and air-removal devices, as long as the thermal properties of the heat-transfer liquid are considered.

When the control system senses an increase in solar panel temperature, the pump circulates the heat-transfer liquid and energy is collected. The same controller also activates a pump on the domestic water side that circulates water through the heat exchanger where it is heated by the heat-transfer fluid. This mode continues until the temperature differentials between the collector and the tank are too slight for meaningful energy to be collected. At this point, the control system shuts the pumps off. During low-temperature conditions, the non-freezing fluid protects the solar collectors and related piping from bursting. Because the heat transfer fluid can affect system performance, reliability, and maintenance requirements, the fluid selected should be carefully considered.

Because the collector loop of the nonfreezing system remains filled with fluid, it gives flexibility for routing pipes and locating components. However, a double-separation heat exchanger is generally required (by local building codes) to prevent contamination of the domestic water in the event of a leak. The double-wall heat exchanger also protects the collectors from freeze damage if water leaks into the collector loop. However, the double-wall heat exchanger reduces efficiency by forcing the collector to operate at higher temperatures. The heat exchanger can be placed inside the tank, or an external heat exchanger can be used, as shown in Figure 2. The collector loop is closed and, therefore, requires an expansion tank and pressure-relief valve. Air purge is also necessary to expell air during filling and to remove air that has been absorbed into the heat-transfer fluid.

Over-temperature protection is necessary to prevent the collector fluid from corroding or decomposing. For maximum reliability, the glycol should be replaced every few years. In other cases, systems have failed because the collector fluid in the loop thermosyphoned and froze the water in the heat exchanger. Such a situation is particularly disastrous and must be avoided by design if the water side is exposed to the city water system because the collector loop eventually fills with water and all freeze protection is lost.

## Drainback Freeze Protection

A *drainback* solar water-heating system uses ordinary water as the heat transport medium between the collectors and thermal energy storage. Reverse draining (or backsyphoning) the water into a drainback tank protects the system from freezing whenever the controls turn off the circulator pump (Figure 3) or a power outage occurs.

The drainback tank can be a sump with a volume slightly greater than the collector loop or it can be the thermal energy storage tank. The collector loop can be vented to the atmosphere or not. Many designers prefer the non-vented approach because makeup water is not required and the corrosive effects of the air that would otherwise be ingested into the collector loop are eliminated. However, the pressure is still low enough to avoid the use of ASME code tanks.

Service hot water may be warmed through a single separation (single-wall) heat exchanger, because the system contains only potable fluids. The drainback system is virtually failsafe because it automatically reverts to a safe condition whenever the circulator pump stops. Furthermore, a 20 to 30% glycol solution can be added to drainback loops for added freeze protection in case of controller or sensor failure. Because the glycol is not exposed to stagnation temperatures, it does not decompose.

A drainback system requires space for the pitching of collectors and pipes necessary for proper drainage. Also, a nearby heated area must have a room for the pumps and the drainback tank. Plumbing exposed to freezing conditions drains to the drainback tank, making the drainback design unsuitable for sites where the collector cannot be elevated above the storage tank.

Both dynamic and static head losses comprise the pumping power required of a drainback system. The dynamic head loss is due to the friction of fluid motion in the pipes and is a function of fluid viscosity, pipe diameter, and fluid velocity. In drainback systems, an additional static head loss is associated with the distance the water must drop. It is this differential head in the downcoming pipe that creates the potential between the collector supply and return leg and causes the fluid in the collector loop to syphon back through the collector pump when it is turned off. In many systems, this drop is relatively short—the distance between the top of the drainback tank and the surface of the fluid in the tank. Upon startup, an open drop will occur in the downcomer until steady flow is established, thus requiring a greater pumping head for a few minutes until the

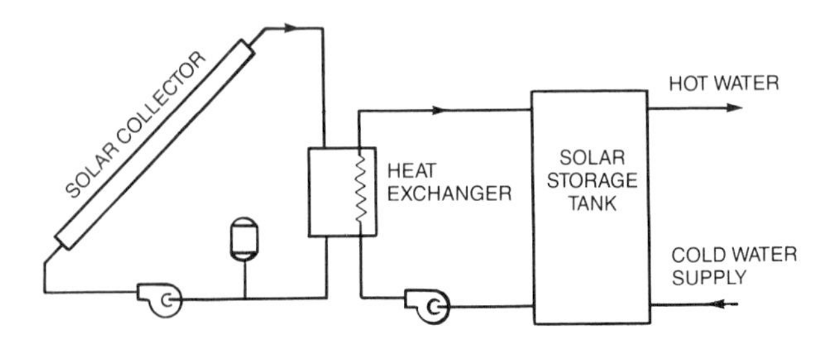


**Fig. 2 Simplified Schematic of Non-Freezing System**

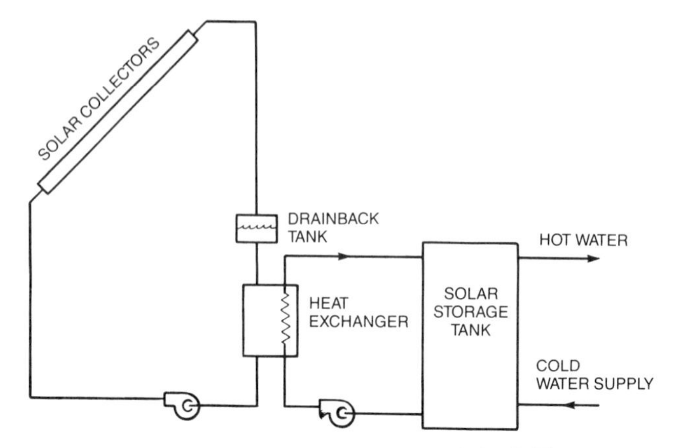


**Fig. 3 Simplified Schematic of Drainback Freeze Protection System**

collector loop is charged. To ensure that the collector loop will drain, some systems have an oversized downcomer, which results in a relatively long drop. These *open drop systems* consume a greater amount of pump energy because of the continued presence of this high hydrostatic head loss.

Drainback performs better than other systems in areas that are marginal for solar applications. Drainback has the advantage because time and energy are not lost in reheating a fluid mass left in the collector and associated piping (as in the case of antifreeze systems). Also, water has a higher heat-transfer capacity and is less viscous than other heat-transfer fluids, resulting in smaller parasitic energy use and higher overall system efficiency. Closed-return designs also consume less parasitic energy for pumping because water is the heat-transfer fluid. Drainback systems can be worked on safely under stagnation conditions, but they should be restarted at night to avoid unnecessary thermal stress on the collector.

## SOLAR ENERGY COLLECTORS

Solar energy system design requires careful attention to detail. Solar radiation is a low intensity form of energy, and the equipment to collect and use it is expensive. Imperfections in design and installation can lead to poor cost effectiveness or to complete system failure. Chapter 47 of the 1987 HVAC Volume covers solar energy use. An ASHRAE manual on solar energy system design is also available (Newton 1982).

Solar energy and HVAC systems often use the same components and equipment. Because other chapters of the handbook cover this equipment, this chapter only covers the following elements that are exclusive to solar energy applications:

- Collectors and collector arrays
- Thermal energy storage
- Heat exchangers
- Controls

### Collectors

Solar Collectors can be divided into liquid heating and air heating. The most common type for commercial, residential, and low-temperature industrial applications (<200°F or 95°C) is the flat-plate collector. Figure 4 shows cross sections of flat-plate air and liquid collectors. A flat-plate collector contains an absorber plate covered with a black surface coating and one or more transparent covers. The covers are transparent to incoming solar radiation and relatively opaque to outgoing (long wave) radiation, but their principal purpose is to reduce heat losses by convection. The collector box is insulated to prevent conduction heat loss from the back and edge of the absorber plate. This type of collector can supply hot water or air at temperatures up to 200°F (95°C), although relative efficiency diminishes rapidly at temperatures above 160°F (70°C). The advantages of flat-plate collectors are simple construction, low relative cost, no moving parts, relative ease of repair, and durability. They also absorb diffuse radiation, which is a distinct advantage in cloudy climates.

Collectors also take the form of concentric glass cylinders, with the space between cylinders evacuated (Figure 5). This vacuum envelope reduces convection and conduction losses, so the cylinders can operate at higher temperatures than flat-plate collectors. As with flat-plate collectors, they collect both direct and diffuse radiation. However, their efficiency is higher at low-incidence angles than at the normal position of the sun. This effect tends to give an evacuated tube collector an advantage in day-long performance over the flat-plate collector. Because of its high temperature capability, the evacuated tube collector is favored for energizing absorption air-conditioning equipment.

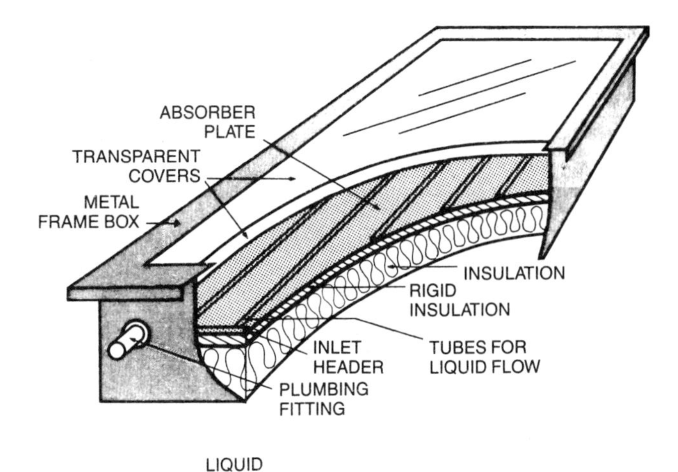


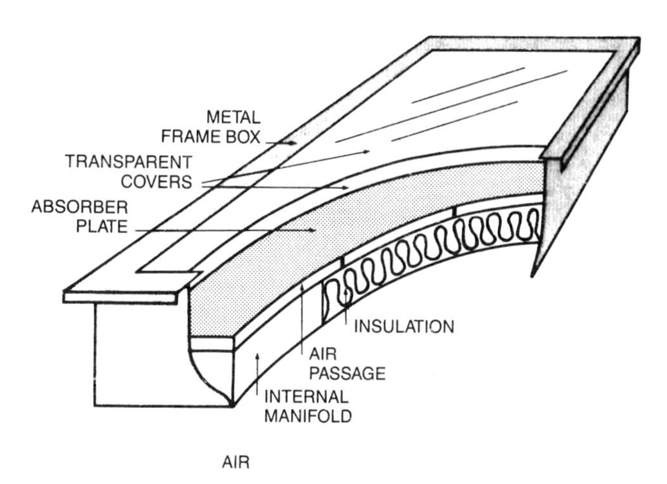


**Fig. 4 Solar Flat-Plate Collectors**

The flat-plate and evacuated-tube collectors are usually mounted in a fixed position. Concentrating collectors are available that must be arranged to track the movement of the sun. These are mainly used for high-temperature industrial applications above 240°F (116°C) and are not discussed in this handbook.

Air-heating collectors, similar to their liquid-heating counterparts, are contained in a box, covered with one or more glazings, and insulated on the sides and back. The primary differences are in the design of the absorber plate and flow passages. Because the working fluid (air) has poor heat-transfer characteristics, it flows over the entire absorber plate, and sometimes on both the front and back of the plate in order to provide greater heat-transfer surface. In spite of the larger surface area, air collectors generally have poorer overall heat transfer than

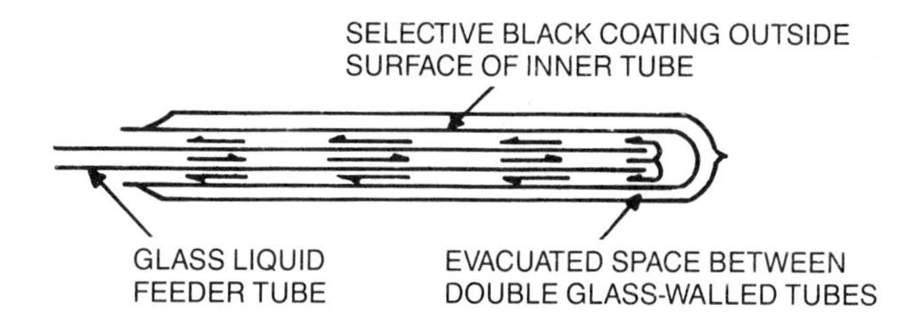


**Fig. 5 Evacuated Tube Collector**

liquid collectors. However, they are usually operated at a lower temperature for space heating applications because they require no intervening heat exchangers.

As with liquid collectors, there is a trend toward the use of spectrally selective surface treatments in combination with a single glazing of low-iron glass. Also, air collectors are being designed for flow rates in the range of 4 scfm/ft$^2$ (20 m/s), whereas early models had suggested flows of 2 scfm/ft$^2$ (10 m/s). While these modifications increase air collector efficiency and lower manufacturing costs, they also increase fan power consumption and lower output air temperature. In some applications, natural-convection air collectors are cost effective. These collectors are self regulating, can be designed not to reverse at night, and use no fan power.

### Collector Performance

Under steady conditions, the useful heat delivered by a solar collector is equal to the energy absorbed in the heat-transfer fluid minus the heat losses from the surface directly and indirectly to the surroundings. This principle can be stated in the relationship

$$Q_\mu = A_c \, [I_t \tau\alpha - U_L \, (\bar{t}_p - t_a)] \qquad (1)$$

*where*

$Q_u$ = useful energy delivered by collector, Btu/h (W)
$A_c$ = total collector area, ft$^2$ (m$^2$)
$I_t$ = total (direct plus diffuse) solar energy incident on the upper surface of the sloping collector structure, Btu/h • ft$^2$ (W/m$^2$)
$\tau$ = fraction of the incoming solar radiation that reaches the absorbing surface, transmissivity (dimensionless)
$\alpha$ = fraction of the solar energy reaching the surface that is absorbed, absorptivity (dimensionless)
$U_L$ = overall heat loss coefficient, Btu/h • ft$^2$ • °F (W/m$^2$ • °C)
$\bar{t}_p$ = average temperature of the absorbing surface of the absorber plate, °F (°C)
$t_a$ = atmospheric temperature, °F (°C)

With the exception of average plate temperature, these terms can be readily determined. For convenience, Eq. (1) can be modified by substituting inlet fluid temperature for the average plate temperature, if a suitable correction factor is applied. The resulting equation is

$$Q_\mu = F_R A_C \, [I_t \tau\alpha - U_L \, (t_i - t_a)] \qquad (2)$$

*where*

$t_i$ = temperature of the fluid entering the collector, °F (°C)
$F_R$ = correction factor or *heat-removal efficiency factor* having a value less than 1.0

The heat removal factor, $F_R$, can be considered the ratio of the heat actually delivered to that delivered if the collector plate were at uniform temperature equal to that of the entering fluid. A $F_R$ of 1.0 is theoretically possible if (1) the fluid is circulated at such a high rate that its temperature rises a negligible amount and (2) the heat-transfer coefficient and fin efficiency is so high that the temperature difference between the absorber surface and the fluid is negligible.

In Eq. (2), the temperature of the inlet fluid depends on the characteristics of the complete solar heating system and the heat demand of the building. $F_R$, however, is affected only by the solar collector characteristics, the fluid type, and its flow rate through the collector.

Solar air heaters remove substantially less heat than liquid collectors. However, their lower collector inlet temperature makes their system efficiency comparable to liquid systems for space heating applications.

Eq. (2) may be rewritten in terms of the efficiency of total solar radiation collection by dividing both sides of the equation by $I_t A_c$. The result is

$$\eta = F_R \tau\alpha - F_R U_L \frac{(t_i - t_a)}{I_t} \qquad (3)$$

Eq. (3) plots as a straight line on a graph of efficiency versus the heat loss parameter $(t_i - t_a)/I_t$. The intercept (intersection of the line with the vertical efficiency axis) equals $F_R \tau\alpha$ and the slope of the line. That is, any efficiency difference divided by the corresponding horizontal scale difference equals $-F_R U_L$. If experimental data on collector heat delivery at various temperatures and solar conditions are plotted on a graph, with efficiency as the vertical axis and $(t_i - t_a)/I_t$ as the horizontal axis, the best straight line through the data points correlates collector performance with solar and temperature conditions. The intersection of the line with the vertical axis is the ambient temperature, where collector efficiency is at its maximum. At the intersection of the line with the horizontal axis, collection efficiency is zero. This condition corresponds to such a low radiation level, or to such a high temperature of the fluid into the collector, that heat losses equal solar absorption, and the collector delivers no useful heat. This condition is normally called *stagnation* and usually occurs when no coolant flows to a collector.

Equation (3) includes all important design and operational factors affecting steady-state performance, except collector flow rate and solar incidence angle. Flow rate indirectly affects performance through the average plate temperature. If the heat-removal rate is reduced, the average plate temperature increases, and more heat is lost. If the flow is increased, collector plate temperature and heat loss decrease.

These relationships assume that the sun is perpendicular to the plane of the collector, which rarely occurs. For glass cover plates, specular reflection of radiation occurs, thereby reducing the $t_a$ product. The *incident angle* modifier $K_{\tau\alpha}$, defined as the ratio of $t_a$ at some incidence angle, $\theta$, to $\tau\alpha$ at normal radiation $(\tau\alpha)_n$, is described by the following expression for specular reflection:

$$K_{\tau\alpha} = \frac{(\tau\alpha)}{(\tau\alpha)_n} = 1 + b_o \left[\frac{1}{\cos\theta} - 1\right] \qquad (4)$$

For a single glass cover, $b_o$ is approximately $-0.10$. Many flat plate collectors, particularly evacuated tubes, have some limited focusing capability. The incident angle modifiers for these collectors are not modeled well by the simple expression for specular reflection, which is a linear function of $(1/\cos\theta) - 1$.

Equation (3) is not convenient for air collectors when it is desirable to present data based on collector outlet temperature rather than inlet temperature, which is common for liquid systems. The relationship between the heat-removal factors for these two cases follows:

$$F_R \, (\tau\alpha) = \frac{(F_R \tau\alpha)'}{1 + (F_R U_L)'/\dot{m}c_p} \qquad (5a)$$

$$F_R U_L = \frac{(F_R U_L)'}{1 + (F_R U_L)'/\dot{m}c_p} \qquad (5b)$$

where $\dot{m}c_p$ is mass times specific heat of air, $F_R U_L$ and $F_R(\tau\alpha)$ applies to $t_i - t_a$ in Eq. (3), and $(F_R U_L)'$ and $(F_R \tau\alpha)'$ applies to $t_{out} - t_a$ in Eq. (3).

### Testing Methods

ASHRAE *Standard* 93-1986 gives information on testing solar energy collectors using single-phase fluids and no significant internal storage. The data can be used to predict performance in any location and under any conditions where load, weather, and insolation are known.

The standard presents efficiency in a modified form of Eq. (3). It specifies that the efficiency be reported in terms of *gross collector area, $A_g$*, rather than *aperature collector area, $A_c$*. The reported efficiency is lower than the efficiency based on net area, but the total energy collected does not change by this simplification. Therefore, *gross collector area must be used* when analyzing performance of collectors based on experiments that determine $F_R\tau\alpha$ and $F_RU_L$, according to ASHRAE *Standard* 93-1986.

Also, the standard suggests testing be done at 0.03 gpm/ft$^2$ (20 mL/s m$^2$ of gross collector area for liquid systems and that the test fluid be water. While it is acceptable to use lower flow rates or a heat-transfer fluid other than water, the designer must adjust the $F_R$ for a different heat-removal rate based on mass and specific heat, $mc_p$. The following *approximate* approach may used to estimate *small* changes in $mc_p$.

$$\frac{(F_RU_L)_2}{(F_RU_L)_1} = \frac{1 - exp\,[-\,A_c(F_RU_L)'/(\dot{m}c_p)_2]}{1 - exp\,[-\,A_c(F_RU_L)'/(\dot{m}c_p)_1]} \tag{6}$$

Air collectors are tested at a flow rate of 2 scfm/ft$^2$ (10 mL/s), and the same relationship applies for adjusting to other flow rates.

Annual compilations of collector test data that meet the criteria of ASHRAE *Standard* 93-1986 may be obtained from the following organizations.

Unitary/Solar Directory
Air Conditioning and Refrigeration Institute
815 North Fort Meyer Drive
Arlington, VA 22209

Solar Rating and Certification Corporation
1001 Connecticut Avenue, N.W.
Suite 800
Washington, D.C. 20036

Another source of this information is the collector manufacturer. However, a manufacturer sometimes publishes efficiency data at a much higher flow rate than the recommended design value, so collector data should be obtained from an independent laboratory qualified to conduct the testing prescribed by ASHRAE *Standard* 93-1986.

## Operational Results

Logee and Kendall (1984) analyzed data from actual sites with regard to collector and collector control performance. The systems were classified with respect to load (service hot water, building space heating, air conditioning), collector design (glazing, absorber surface, evacuated tube, concentrating), and heat-transfer fluid (air, water, oil or glycerol). Collector efficiency curves generated from the field data represented one month's operation during each of the four seasons. Most of the collectors (18) in the 33 systems studied performed below ASHRAE test results. Eleven systems had similar collector efficiencies as the test panel results, and four were slightly more efficient.

## Collector Construction

**Absorber Plates**. The key component of a flatplate collector is the absorber plate. It contains the heat-transfer fluid and serves as a heat exchanger by converting radiant energy into thermal energy. It must maintain structural integrity at temperatures ranging from below freezing to well above 300°F (150°C). Chapter 47 in the 1987 HVAC Volume illustrates typical liquid collectors and shows the wide variety of absorber plate designs in use.

Materials for absorber plates and tubes are usually highly conductive metals such as copper, aluminum, and steel, although low-temperature collectors for swimming pools are usually made from extruded elastomeric material such as EPDM and PVC. Flow passages and fins are usually copper, but aluminum fins are sometimes inductively welded to copper tubing. Occasionally, fins are mechanically attached, but the potential for corrosion exists with this design. A few manufacturers produce all-aluminum collectors, but they must be checked carefully to see if they have corrosion protection in the collector loop.

Figure 6 shows a plan view of typical absorber plates. The serpentine design is rarely used because it is difficult to drain and imposes a high pressure drop. Most manufacturers use absorber plates similar to the example shown in Figures 6d and 6e.

In liquid collectors, manifold selection is important because the design can restrict the array piping configuration. The manifold must be drainable and free floating, with generous allowance for thermal expansion. Some manufacturers provide a choice of manifold connections to give designers flexibility in designing arrays. One such product can be obtained with side, back, or end connectors or a combination of these.

In air collectors, most manufacturers increase the heat transfer area via fins, matrices, or corrugated surfaces. Many of these designs increase air turbulence, which improves the collector efficiency (at the expense of increased fan power).

Figure 7 shows cross sections of typical air collectors. Fins on the back of the absorber (Figure 7a) increase the convection heat transfer surface. Air flowing across a corrugated absorber plate (Figure 7b) creates turbulence along the plate, which increases the convective heat-transfer coefficient. A box frame (Figure 7c) creates airflow passages between the vanes. The vanes conduct from the absorbing surface plate to the back plate. Heat is transferred to the air by all of the surfaces of each boxed airflow channel. A matrix absorber plate (Figure 7d) is formed by stacking several sheets of metal mesh such as expanded metal plastering lath. Placing the mesh diagonally in the collector

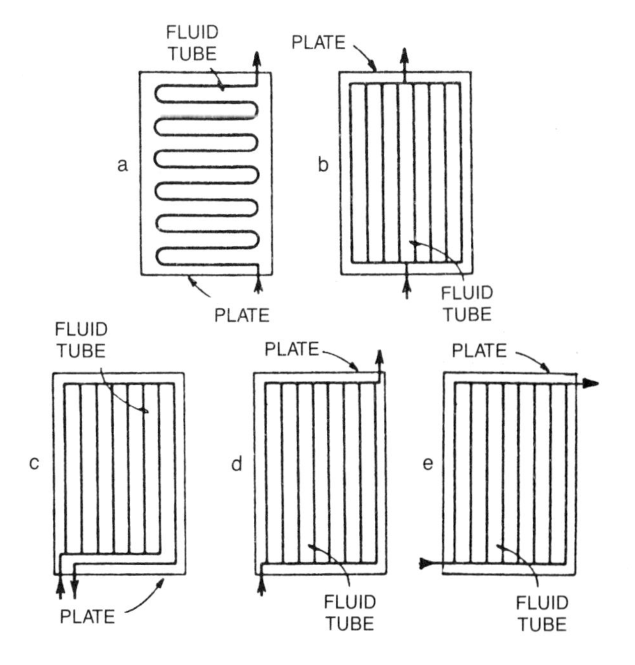


**Fig. 6 Plan View of Liquid Collector Absorber Plates**

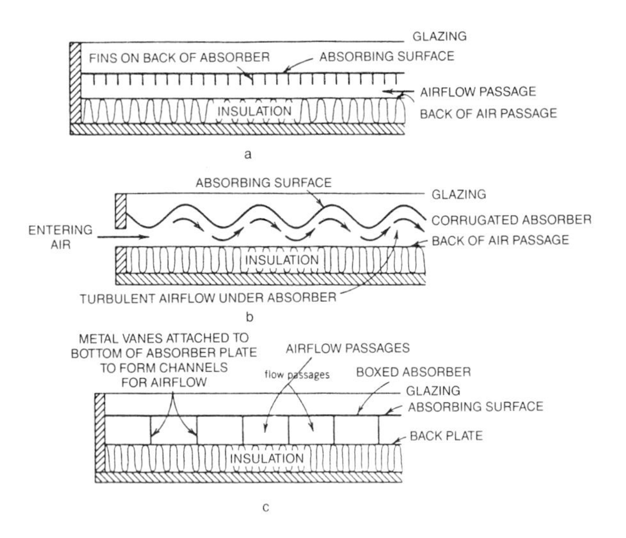


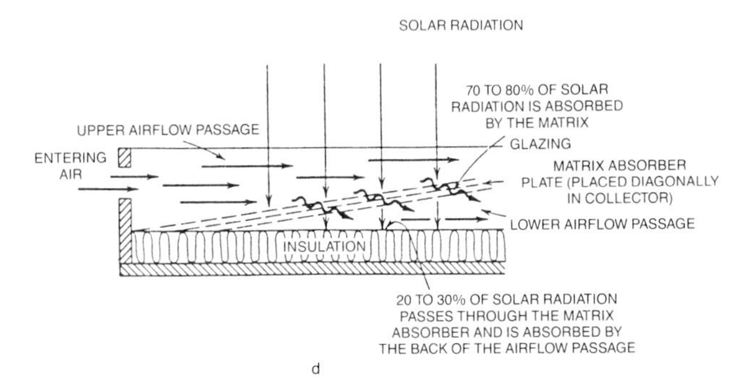


**Fig. 7 Cross-Section of Air Collectors**

forces the air through the matrix so it does not contact the glazing after being heated.

A well-designed collector manifold with a series/parallel connection minimizes leaks and reduces the operating costs of the fans. Manufactured collectors are made in modular sizes, typically 3 by 7 ft (910 by 2130 mm), and these are connected to form an array. Because most commercial-scale systems involve upwards of 1000 ft$^2$ (90 m$^2$) of collector, there can be numerous ducting connections, depending on design.

Figure 8 shows the simplest collector manifolding. Each collector has its own inlet and outlet, and each requires two branch

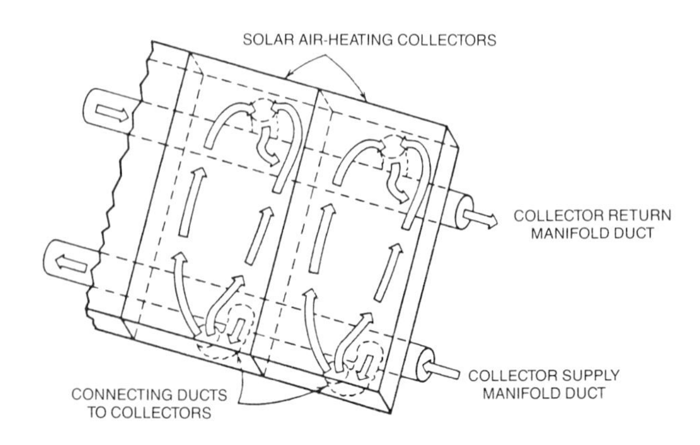


**Fig. 8 Air Collectors with External Manifolds**

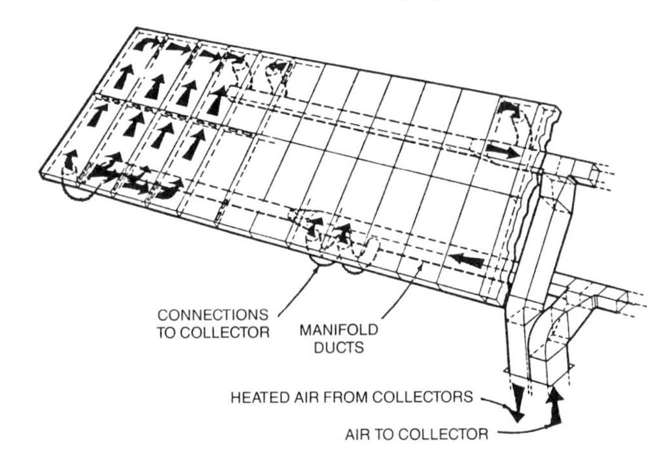


**Fig. 9 Air Collectors with Internal Manifolds (Solaron)**

connections per collector (and possibly a balancing damper). Some models are specifically designed for either series or parallel connections, and some collectors have built-in manifold connections to simplify their connection as an array. Figure 9 shows an example of an internally manifolded collector in a combination series-parallel arrangement. The number of ducts connecting the collectors to the trunk ducts are reduced to minimize leakage and reduce ducting and installation costs.

Absorber plates may be coated with spectrally selective and spectrally nonselective materials. Selective coatings are more efficient, but they also cost more than nonselective, or flat black, coatings. The Argonne National Laboratory has published a detailed discussion of the various coating materials used in solar applications (ANL 1979).

**Housing**. The collector housing is the container that provides structural integrity for the collector assembly. The housing must be structurally sound, weather-tight, fire resistant, and capable of being connected mechanically to a substructure to form an array. Collector housing materials include the following:

- Galvanized or painted steel
- Aluminum folded sheet stock or extruded wall materials
- Various plastics, either molded or extruded
- Composite wood products
- Standard elements of the building

Extruded anodized alumimum offers great durability and ease of fabrication. Grooves are sometimes included in the channels to accommodate proprietary mounting fixtures. The high temperatures of a solar collector deteriorates wood housings; consequently, they are often forbidden by fire codes.

**Glazing**. Solar collectors for domestic hot water are usually single glazed to reduce absorber plate convective and radiative losses. Some collectors have double glazing to further reduce these losses; however, these should be restricted to applications at the higher value of $(t_i - t_a)/I$, *e.g.* for space heating or activating absorption refrigeration. Glazing materials are either plastic, plastic film, or glass. Glass can absorb the long-wavelength thermal radiation emitted by the absorber coating, but it is not affected by ultraviolet radiation. Because of their impact tolerance, only tempered, low-iron glass covers should be considered. These covers have a solar transmission rating of 86%. With acid etching, this transmission rating can be increased to 90%.

If the probability of vandalism is high, polycarbonate, which has high impact resistance, should be considered. Unfortunately, its transmittance is not as high as low-iron glass and it is susceptable to long term ultraviolet (UV) degradation.

**Insulation**. Collector enclosures must be well insulated to minimize heat losses. The insulation must withstand temperatures up to 400 °F (200 °C) and, most importantly, must not produce volatiles within this range. Many insulation materials designed for construction applications are not suitable for solar collectors because the binders outgas volatiles at normal collector operating temperatures.

Solar collector insulation is typically made of mineral fiber, ceramic fiber, glass and plastic foam, and fiberglass. Fiberglass is the least expensive insulation and is widely used in solar collectors. For high-temperature applications, rigid fiberglass board with a minimum of binder is recommended. Also, a layer of polyisocyanurate foam in collectors is often used because of its superior $R$ factor. Because it can outgas at high temperatures, the foam must not be allowed to contact the collector plate.

Figure 10 illustrates the preferred method of combining fiberglass and foam insulations to combine high efficiency with durability. Note that the absorber plate should be free-floating to avoid thermal stresses. Regardless of attempts to make collectors watertight, moisture is always present in the interior. This moisture can physically degrade mineral wool and reduce the $R$ value of fiberglass, so drainage and venting are crucial.

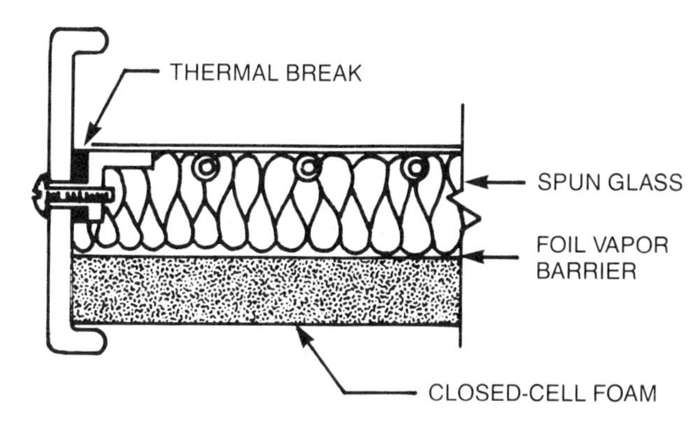


**Fig. 10 Cross-Section of Suggested Insulation to Reduce the Heat Loss from the Back Surface of the Absorber**

## Collector Test Results and Initial Screening Methods

Final selection of a collector should be made only after energy analyses of the complete system, including realistic weather conditions and loads, have been conducted for one year. Also, a preliminary screening of collectors with various performance parameters should be conducted to identify those that best match the load. The best way to accomplish this is to identify the expected range of $(t_i - t_a)/I$ for the load and climate on a plot of the efficiency as a function of heat loss parameter, as indicated in Figure 11.

Ambient temperature during the swimming season may vary from 18 °F (10 °C) below pool temperature to 18 °F (10 °C) above. The corresponding parameter values range from 0.15 on cool overcast days (low $I$) to as low as −0.15 on hot overcast days. For swimming pool heating, the unglazed collector offers the highest performance and is the least expensive collector available.

The heat loss parameter for service water heating can range from 0.05 to 0.35, depending on the climate at the site and desired hot-water delivery temperature. Space heating requires an even greater collector inlet temperature than hot water and the primary load coincides with lower ambient temperature. In many areas of the United States, space heating operation coincides with low radiation values, which further increases the heat-loss parameter. However, many space-heating systems are accompanied by water heating at a lower value of $(t_i - t_a)/I_t$.

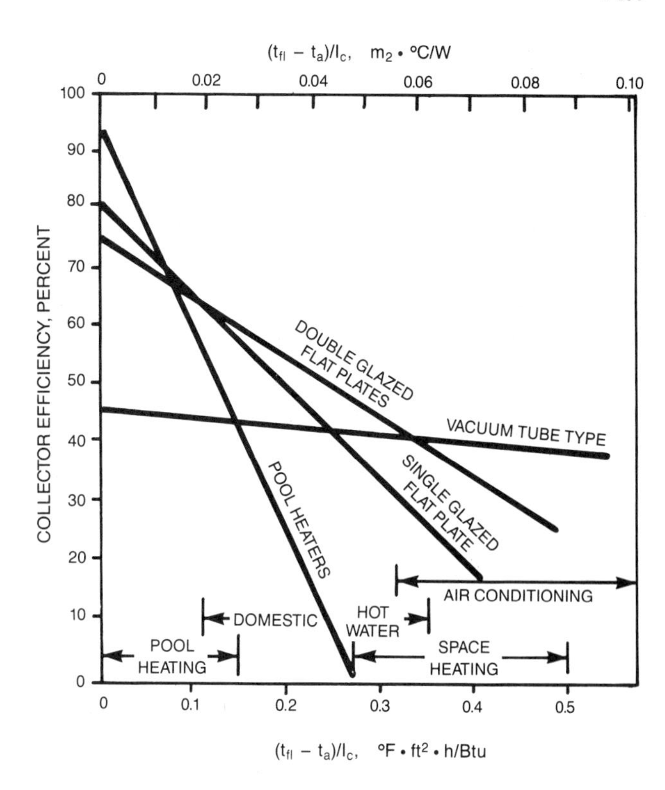


**Fig. 11 Collector Efficiencies of Various Liquid Collectors**

Air conditioning with solar activated absorption equipment is only economical when the cooling season is long and solar radiation levels are high. These devices require at least 180 °F (80 °C) water. Thus, on an 80 °F (27 °C) day with radiation at 300 Btu/h • ft² (950 W/m²), the heat loss parameter is 0.3. Higher operating temperatures are desirable to prevent excessive derating of the air conditioner. In this application, only the most efficient (low $F_R U_L$) collector is suitable.

Collector efficiency curves may be used as an initial screening device. However, efficiency curves only illustrate the instantaneous performance of a collector. They do not include incidence angle effects (which vary throughout the year), heat exchanger effects, probabilities of occurance of $t_i$, $t_a$, and $I_t$, system heat loss, or control strategies. The final selection requires determining the long-term energy output of a collector combined with cost effectiveness studies. Estimating the annual performance of a particular collector and system requires the aid of appropriate analysis tools such as FCHART (Beckman *et al.* 1977) or TRNSYS (SEL 1983).

## Generic Test Results

While details of construction may vary, it is possible to group collectors into generic classifications. Using the performance characteristics of each classification, the designer can select the category best suited to a particular application.

Huggins and Block (1983) identified eight such classifications for liquid collectors from a sample of 270 collectors tested by various laboratories. Following is a list of the classifications:

**Number of Cover Plates**. The majority of collectors used in the South and far West of the United States have one cover plate. Double glazed, *i.e.*, two-cover plates, are more common in colder climates. The trend in cold climates has been toward single glazing combined with a selective surface. However, with the advent of reflectivity coatings on plastic films, triple and quadruple glazings may become common.

**Cover Plate Material**. The following is a list of common cover plate materials:

*Glass* is the most widely used glazing material. It has a high transmittance and long-term durability.

*Fiber reinforced plastic* (FRP) is the second most widely used material.

*Thin film plastics* include such trade names as Tedlar, Mylar, Teflon, and Lexan.

**Absorber Plate Coating**. The most common coatings are listed below:

*Selective surface coatings* such as black chrome, black nickel, and copper oxide have high absorptance and low emittance properties. The infrared emissivity of these surfaces is below 0.2.

*Moderately selective surface coatings* are special paints that have moderately selective surface properties. The emissivity of the surfaces ranges from 0.2 to 0.7.

*Flat black paints* are non-selective and have high heat resistance. The emissivity of these surfaces ranges from 0.7 to 0.98.

**Absorber Materials**. Copper, aluminum, and stainless steel may be used in combinations of tubes and fins or integral tubes in plates.

**Absorber Configuration**. The absorber may be configured with parallel pipes, series or serpentine pipes, a parallel and series combination, or plate flow.

**Enclosure**. The frame holding the collector components may be either metallic or non-metallic.

**Insulation Materials**. Fiberglass and foam insulation, or a combination of both, may be used to keep heat from escaping from the back and sides of the collector.

Table 1 lists the eight generic solar collectors in which five or more collectors were tested. The first three columns specify the generic type; the fourth column shows the number of collectors in each category, and the final columns lists the mean, the standard deviation, and the maximum and minimum values of the intercept and slope. Logee and Kendall (1984) showed collectors in actual applications can perform differently than they do on test stands. This difference emphasizes the need for good system designs and installation practices.

## MODULE DESIGN

### Piping Configuration

Most commercial and industrial systems require a large number of collectors. Connecting the collectors with one set of manifolds makes it difficult to ensure drainability, balanced flow, and low-pressure drop. An array usually includes many individual groups of collectors, called modules, to provide the necessary flow characteristics. Modules can be grouped into (1) parallel flow or (2) combined parallel/series flow. Parallel flow is the most frequently used because it is inherently balanced, has low pressure drop, and is drainable. Figure 12 illustrates various collector header designs and the reverse return method of forming a parallel flow module (the flow is parallel but the collectors are connected in series). When arrays must be greater than one panel high, a combination of series/parallel flow may be used. Figure 13 illustrates one method of connecting collectors in a two-high module.

Generally, flat-plate collectors are made to connect to the main piping in one of two methods shown in Figure 12. The *external manifold* collector has a small-diameter connection meant to carry only the flow for one collector. It must be connected individually to the manifold piping, which is not part of the collector panel, as depicted in Figure 12a and Figure 13.

The *internal manifold* collector incorporates the manifold piping integral with each collector (Figures 12b and 12c). Headers at either end of the collector distribute flow to the risers. Several collectors with large headers can be placed side by side to form a continuous supply and return manifold. With 1 in. (25 mm) headers, four to six 40-ft$^2$ (3.7-m$^2$) collectors can be placed side by side. Collectors with 1-in. headers can be mounted in a module without producing unbalanced flow. Most collectors have four plumbing connections, some of which may be capped if the collector is located on the end of the array. Internally manifolded collectors have the following advantages:

- Piping costs are lower because fewer fittings and less insulation is required
- Heat loss is less because less piping is exposed
- Installation is more attractive

**Table 1 Collector Intercept and Slope by Generic Type (Huggins and Block 1983)**

| Glazing & Cover Material | Absorber Material & Type | Absorber Coating | Number Of Collectors | Intercept, $F_R\tau\alpha$ | | Slope, $(F_RU_L)$, Btu/h • ft$^2$ • °F (W/m$^2$ • °C) | |
|---|---|---|---|---|---|---|---|
| | | | | Mean | Standard Deviation | Mean | Standard Deviation |
| Single Glass | Copper Tubes and Fins | Flat Black Paint | 47 | 67.2 | 5.0 | -115 (-653) | 14 (79) |
| Single Glass | Copper Tubes and Fins | Moderately Selective | 9 | 73.0 | 3.6 | -112 (-636) | 11 (62) |
| Single Glass | Copper Tubes and Fins | Selective Surface | 58 | 71.7 | 3.3 | -83 (-471) | 11 (62) |
| Single Glass | Copper Tubes and Aluminum Fins | Flat Black Paint | 22 | 69.1 | 6.0 | -116 (-659) | 12 (68) |
| Single Glass | Copper Sheet Integral Tubes | Selective Surface | 6 | 70.5 | 5.1 | -89 (-505) | 17 (97) |
| Single FRP | Copper Tubes and Fins | Flat Black Paint | 26 | 61.9 | 5.5 | -117 (-664) | 15 (85) |
| Single FRP | Copper Tubes and Aluminum Fins | Flat Black Paint | 11 | 57.1 | 6.2 | -114 (-647) | 10 (57) |
| Double Glass | Copper Tubes and Fins | Flat Black Paint | 9 | 59.7 | 6.7 | -84 (-477) | 9 (51) |

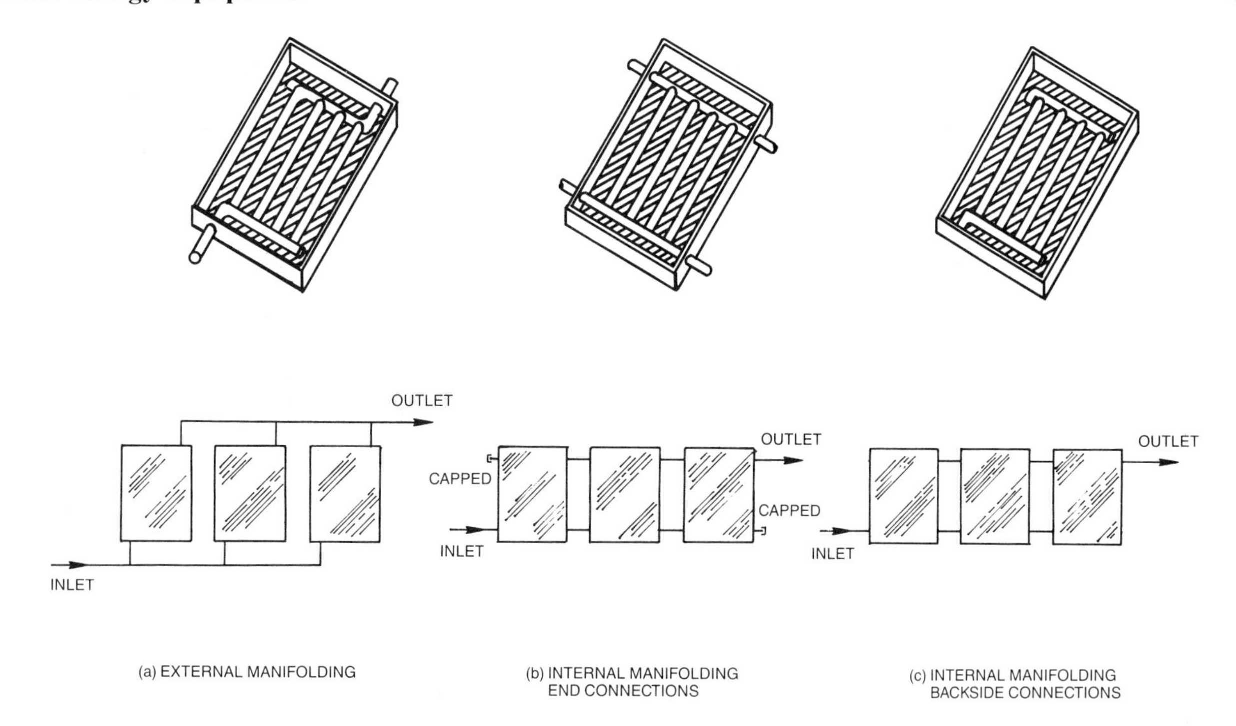


**Fig. 12 Collector Manifolding Arrangements for Parallel Flow Module**

Some of the disadvantages of the internally manifolded collectors are as follows:

- The entire module must be pitched for drainback systems, thus complicating mounting
- Flow may be imbalanced if too many collectors are connected in parallel
- Removing the collector for servicing may be difficult
- Stringent thermal expansion requirements must be met if too many collectors are combined in a module

## Velocity Limitations

Fluid velocity limits the number of internally manifolded collectors that can be contained in a module. For 1-in. (25-mm) headers, up to eight 20-ft$^2$ (1.9-m$^2$) collectors can usually be connected for satisfactory performance. If too many are connected in parallel, the middle collectors will not receive enough flow, and performance will decrease. Also, connecting too many collectors increases pressure drop. Figure 14 illustrates the effect of collector number on performance and pressure drop for one particular design. Newton (1983) describes a general method to determine the number of internally manifolded collectors that can be connected.

Flow restrictors can be used to accommodate a large number of collectors in a row. The flow distribution in the twelve collectors of Figure 15 would not be satisfactory without the flow restrictors shown at the interconnections. The flow restrictors are barriers with a hole drilled the size of the diameter indicated. Some manufacturers calculate the required hole diameters and provide the predrilled restrictors.

Chapter 34 of the 1985 FUNDAMENTALS Volume gives information on sizing piping for self-balancing flow in external

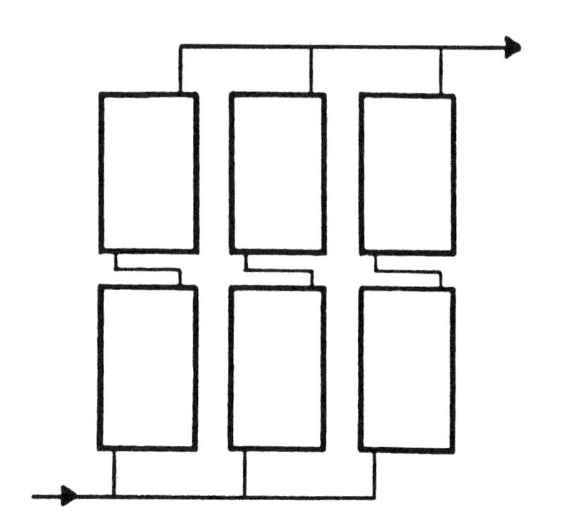

**Fig. 13 Example of Collector Manifolding Arrangements for Combined Series/Parallel Flow Modules**

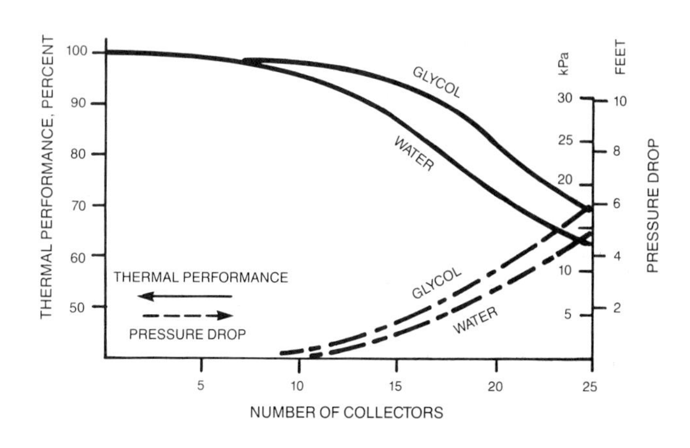


**Fig. 14 Pressure Drop/Thermal Performance of Collectors with Internal Manifolds**

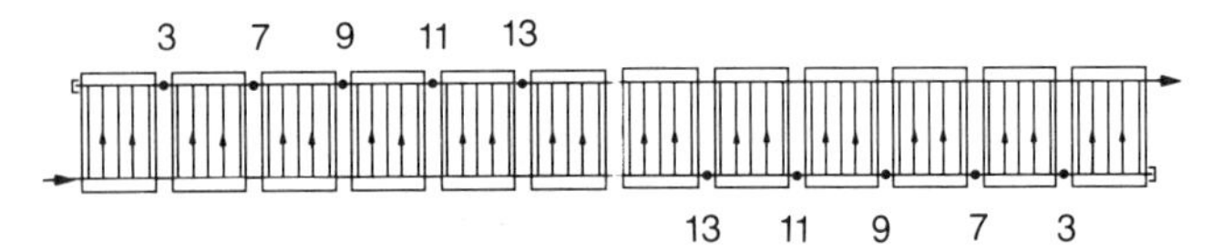


**Fig. 15 Flow Pattern in a Long Collector Row with Restrictions**

manifolded collectors. Knowles (1980) provides the following expression for the minimum acceptable header diameter:

$$D = c\,(Q/p)^{0.45}\,N^{0.64} \qquad (7)$$

*where*

$D$ = header diameter, in. (mm)
$c$ = Constant multiplier, 0.24 (19.5)
$N$ = number of collectors in the module
$Q$ = recommended flow rate for the collector, gpm (L/s)
$p$ = pressure drop across the collector at the recommended flow rate, psi (kPa)

Because piping is available in a limited number of diameters, selection of the next larger size ensures balanced flow. Usually, the size of supply and return are graduated to maintain the same pressure drop while minimizing piping cost. Complicated configurations may require a hydraulic static regain calculation.

## Thermal Expansion

Thermal expansion will affect the module shown in Figure 16. Thermal expansion (or contraction) of a module of collectors in parallel may be estimated by the following equation.

$$\Delta = cn\,(t_c - t_i) \qquad (8)$$

*where*

$\Delta$ = expansion or contraction of the collector array, in. (mm)
$c$ = expansion coefficient, 0.000335 (0.0153)
$n$ = number of collectors in an array
$t_c$ = collector temperature, °F, (°C)
$t_i$ = installation temperature of the collector array, °F (°C)

Because absorbers are rigidly connected, the absorber must have sufficient clearance from the side frame to allow the expansion indicated in Eq. (8).

# ARRAY DESIGN

## Piping Configuration

To maintain balanced flow, an array or field of collectors should be built from identical modules configured as described in previous sections. Whenever possible, modules must be connected in reverse return fashion (Figure 16). The reverse return ensures the array is self balanced. With proper care, an array can drain, which is an essential requirement for drainback freeze protection.

Piping to and from the collectors must be sloped properly in a draindown system. Typically, piping and collectors must slope to drain at 1/4 in. per lineal foot (20 mm/m). Elevations throughout the array should be noted on the drawings, especially the highest and lowest point of the piping.

The external manifold collector has different mounting and plumbing considerations than the internal manifold collector (Figure 17). A module of external manifolded collectors can be mounted horizontally, as shown in Figure 17. The lower header

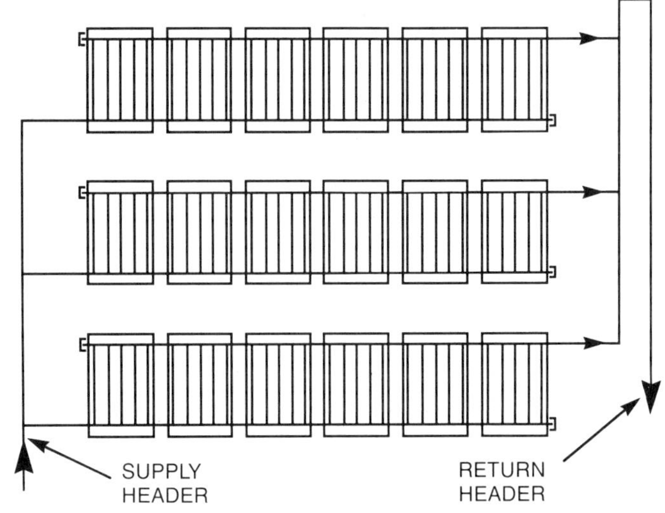


**Fig. 16 Reverse Return Array Piping**

*must* be pitched as shown. The pitch of the upper header can be horizontal or pitched toward the collectors, so it can drain back through the collectors.

Arrays with internal manifolds pose a greater challenge designing and installing the collector mounting system. For these collectors to drain, the entire bank must be tilted, as shown in Figure 17.

Reverse return always implies an extra pipe run. Sometimes, it is more convenient to use a direct return configuration, as shown in Figure 18. In this case, balancing valves are needed to ensure uniform flow through the modules. The balancing valves *must* be connected at the module outlet to provide the flow resistance necessary to ensure filling of all modules on pump start-up.

It is often impossible to configure parallel arrays because of the presence of rooftop equipment, roof penetrations, or other building-imposed constraints. Though the following list is not complete, the following requirements should be considered when developing the array configuration:

- Strive for a self-balancing configuration.
- For drainback systems, design the modules, subarrays, and arrays to be individually and collectively drainable.
- Always locate collectors or modules with high flow resistance at the outlet to improve flow balance and ensure filling of drain down system.
- Minimize flow and heat transfer losses.

In general, it is easier to configure complex array designs for nonfreezing fluid systems. Newton (1983) provides some typical examples. However, with careful attention to the criteria mentioned above, it is also possible to design successful large drainback arrays.

Air distribution is the most critical feature of an air system because pressure drop has a critical effect on fan power. Each collector and the other system components must have proper air distribution for effective operation. Balancing dampers, automatic dampers, backdraft dampers, and fire dampers are usually needed. Air leaks (both into and out of the ducting and from component to component) must be kept to a minimum.

For example, some air collector systems contain a water coil for preheating water. Despite the inclusion of automatic dampers and backdraft dampers, leakage within the system can freeze the coils. One possible solution is to position the coil near the

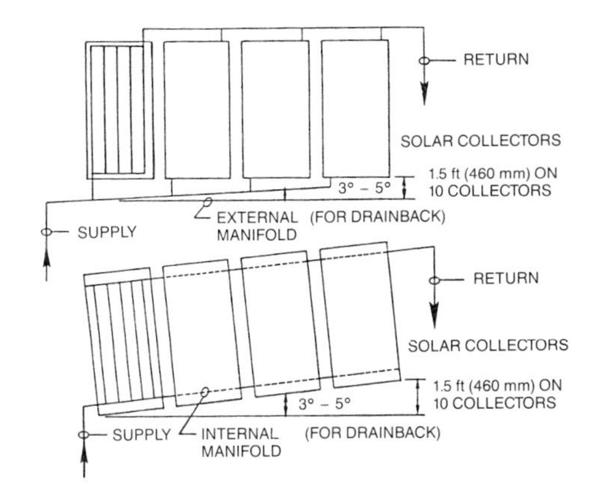


**Fig. 17 Mounting for Drainback Collector Modules**

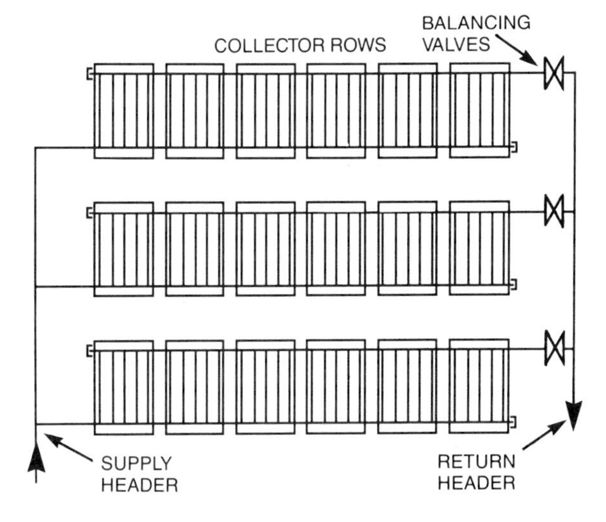


**Fig. 18 Direct Return Array Piping**

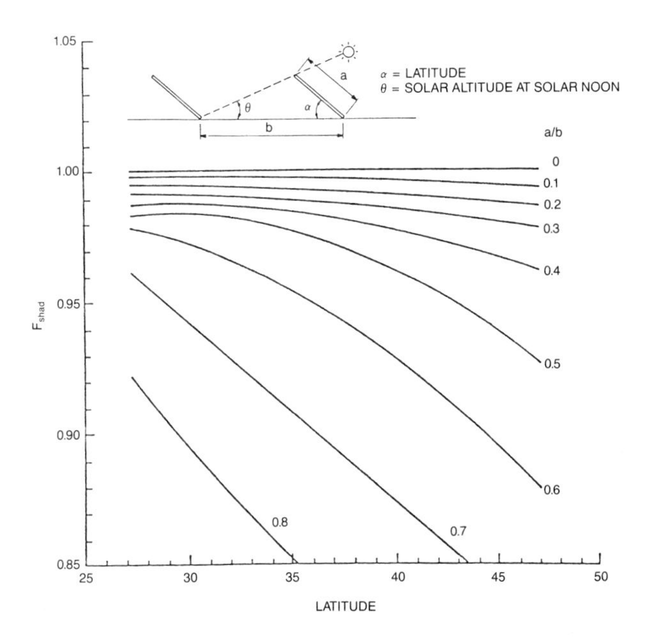


**Fig. 19 Annual Row-to-Row Shading Loss Modifiers for Flat Plate Collectors Tilted at Latitude (Kutscher 1982)**

warm end of the storage bin. Another is to circulate an antifreeze solution in the coil. Whatever the solution, the designer must remember that even the lowest leakage dampers will leak if installed or adjusted improperly.

As with liquid systems, the main supply and return ducts should be connected in a reverse return fashion, with balancing dampers on each supply branch to the collector modules. If reverse return is not feasible, the main ducts should include balancing dampers at strategic locations. Here too, fewer branch ducts reduce balancing needs and costs.

Unlike liquid systems, air collectors can be built on site. While material and cost savings can be substantial with site-built collectors, extreme care must be taken to ensure long life, low leakage, and proper air distribution. Quality control in the field can be a problem, so well-trained designers and installers are critical to the success of these systems.

The impact of the of array and air-distribution system designs must be considered in the overall performance of the system. Beckman *et al.* (1977) give standard procedures for estimating the impact of series connection and duct thermal losses. The impact on fan operation and fan power is more difficult to determine. For unique system designs and more detailed performance estimates, including fan power, an hourly simulation like TRNSYS can be used.

## Shading

When large collector arrays are mounted on flat roofs or level ground, multiple rows of collectors are usually installed in a sawtooth fashion. These multiple rows should be spaced so they do not shade each other at low sun angles. However, it is usually not desirable to avoid mutual shading altogether. It is sometimes possible to add additional rows to a roof or other constrained area; this increases the solar fraction but sacrifices efficiency. Kutsher (1982) presents a method of estimating the energy delivered annually when there is some row-to-row shading within the array. Figure 19 provides a factor, $F_{shad}$, which corrects the annual performance (on an unshaded basis) of the field of shaded collectors. Since the first row is unshaded, the following equation is used to compute the average shading factor for the entire field.

$$F_{shad,\ field} = \frac{1 + (n - 1) F_{shad}}{n} \qquad (9)$$

*where*

$n$ = Number of rows in the field

## Thermal Energy Storage

Design and selection of the thermal storage equipment is one of the most neglected elements of solar energy systems. In fact, the energy storage system has an enormous impact on overall system cost, performance, and reliability. Furthermore, the storage system design is highly interactive with other system elements such as the collector loop and the thermal distribution system. As such, it should be considered within the context of the total system.

Energy can be stored in liquids, solids, or phase change materials (PCM). Water is the most frequently used liquid storage medium for liquid systems, although the collector loop may contain water, oils, or aqueous/glycol as a collection fluid. For service-water heating applications and most building space heating, water is normally contained in some type of tank. Air systems typically store water in rocks or pebbles, but sometimes the structural mass of the building is used. Chapter 46 of the 1987 HVAC Volume and *Thermal Storage* (ASHRAE 1985) cover this topic in more detail.

## Air Storage

The most common storage media for air collectors are rocks or a regenerator matrix made from concrete masonry units (CMUs). Other possible media include phase-change materials (PCMs), water, and the inherent building mass. Gravel is widely used as a storage medium because it is plentiful and relatively inexpensive.

In places where large interior temperature swings are tolerable, the inherent mass of the building may be sufficient for thermal storage. Storage may also be eliminated where the array output seldom exceeds the concurrent demand. These types of applications are usually the most cost-effective applications of air collectors, and heated air from the collectors can be distributed directly to the space.

The three main requirements for a gravel storage system (or any storage system for air collectors) are good insulation, low air leakage, and low pressure drop. Many different designs can fulfill these needs. The container is usually constructed from masonry or frame, or a combination of the two. Airflow can be vertical or horizontal, whichever is most convenient.

Because of the restricted airflow paths in a pebble bed, there is little natural air movement, and temperature can stratify without the effect of gravity. A vertical flow bed that has solar heated air enter at the bottom and exit from the top can work as effectively as a horizontal flow bed. However, it is important to heat the bed with airflow in one direction and to retrieve the heat with flow in the opposite direction. In this manner, pebble beds perform as effective counterflow heat exchangers. Conversely, properly designed and applied *wash through* (one-way flow) beds can be effective, as can underflow beds that charge through airflow and discharge by flow radiation. These are less expensive because they do not require elaborate ductwork or complicated controls.

Rocks for pebble beds range from 0.75 in. nominal to over 4 in. (20 to 100 mm) in size, depending on airflow rates, bed geometry, and desired pressure drops. The volume of rock needed depends on the fraction of the collector output that must be stored. For residential systems, storage volume is typically in the area of 0.5 to 1 $ft^3/ft^2$ (0.15 to 0.3 $m^3/m^2$) of collector area. For commercial systems, these values can be used as guidelines, but a more detailed analysis should be performed. Pebble bed size can be quite large for large arrays, and location and weight can create problems.

Though other storage options exist for air systems, applied knowledge of these techniques is limited. Hollow-core concrete decking appears attractive because of the potential to reduce storage and distribution costs. PCMs are also functionally attractive because of their high volumetric heat storage capabilities, which typically require a tenth of the volume of a pebble bed.

Water can also be used as a storage medium for air collectors through the use of a conventional heating coil to transfer heat from the air to the water in the storage tank. Advantages of water storage include compatability with hydronic heating systems and relatively compact storage (roughly one-third the volume requirement of pebble beds).

## Liquid Storage

For units large enough for commercial liquid systems, the following factors should be considered.

- Pressurized versus unpressurized storage
- External heat exchanger versus internal heat exchanger
- Single versus multiple tanks
- Steel tank versus non-metallic tanks
- Type of service, e.g. Service Hot Water (SHW), Building Space Heating (BSH), or a combination of the two
- Location, space, accessibility constraints imposed by architectural limitations
- Interconnect constraints imposed by the existing mechanical systems
- Limitations imposed by equipment availability

In the following sections, examples of the more common configurations will be presented.

**Pressurized storage**. Defined here as storage that is open to the city water supply, pressurized storage is preferred for small service-water heating systems because it is convenient and provides an economical way of meeting ASME Pressure Vessel Code requirements with off-the-shelf equipment. Typical storage size is about 1.0 to 2 $gal/ft^2$ (40 to 80 $L/m^2$) of collector area. The largest size off-the-shelf water heater is 120 gal (450 L); however, no more than three of these should be connected in parallel. Hence, the largest size system that can can be considered with off-the-shelf water heater tanks is about 360 gal (1360 L). For larger solar hot water and combined systems, the following concerns are important when selecting storage:

- The higher cost per unit volume of ASME rated tanks in sizes greater than 120 gal (450 L)
- Handling difficulties due to weight
- Accessibility to locations suitable for storage
- Interfacing with existing SHW and BSH systems
- Corrosion protection of steel tanks

The choice of pressurized storage for intermediate size systems is based on the availability of suitable, low cost tanks near the site. Identification of a suitable supplier of low-cost tanks can extend the advantages of pressurized storage to larger SHW installations.

Storage pressurized at city water supply pressure is not practical for Building Space Heating (BSH), except for small applications such as residences, apartments, and small commercial buildings.

With pressurized storage, the heat exchanger is always located on the collector side of the tank. Either the internal or the external heat exchanger configuration can be used. Figure 20 illustrates the three principal types of internal heat exchanger concepts, an immersed coil, a wrap-around jacket, and a tube bundle. Small tanks (less than 120 gal [450 L]) are available with either of the first two heat exchangers already installed. For larger tanks, a large assortment of tube bundle heat exchangers are available that can be incorporated into the tank design by the manufacturer.

Sometimes, more than one tank is needed to meet design requirements. Additional tank(s) result in the following benefits:

- added storage volume
- increased heat exchanger surface
- reduced pressure drop in the collection loop

Figure 21 illustrates the multiple tank configuration for pressurized storage. The exchangers are connected in reverse return fashion to minimize flow imbalance. A third tank may be added. Additional tanks have the following disadvantages compared to a single tank of the same volume:

- Installation costs are higher.
- Greater room is required.
- Heat losses are higher so performance is reduced.

An external heat exchanger provides greater flexibility, because the tank and the heat exchanger can be selected independently of each other (Figure 22). Flexibility is not achieved without cost, however, because an additional pump and its parasitic energy requirements are required.

When selecting an external heat exchanger for a system protected by a nonfreezing liquid, the following factors related to

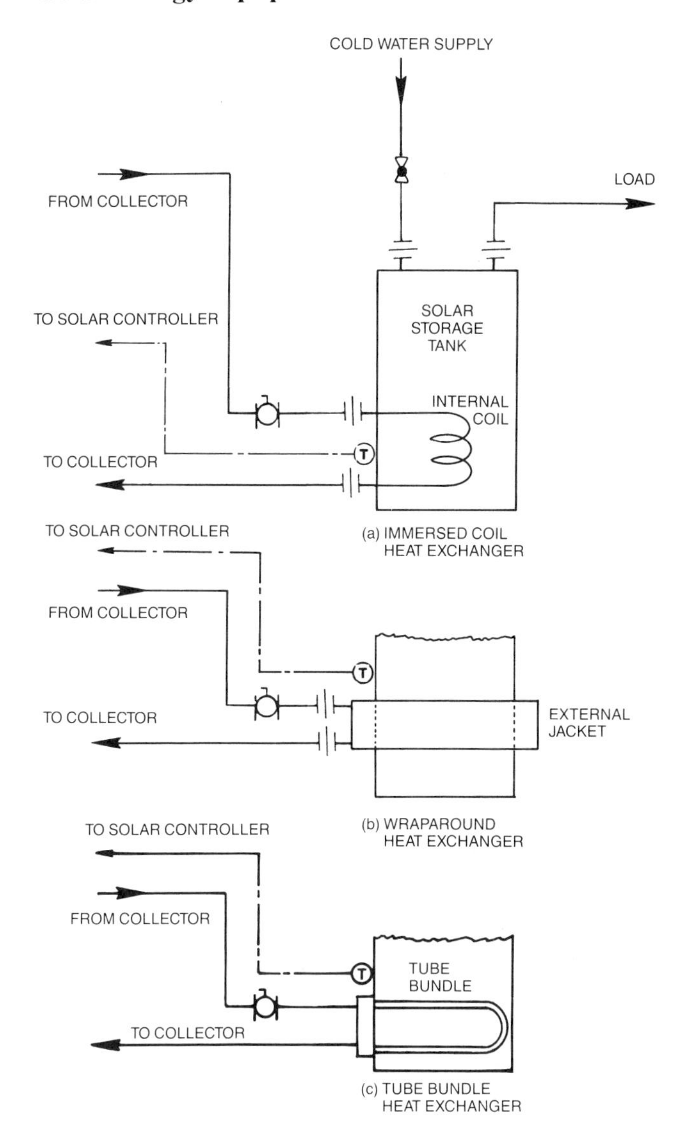


**Fig. 20 Pressurized Storage with Internal Heat Exchanger**

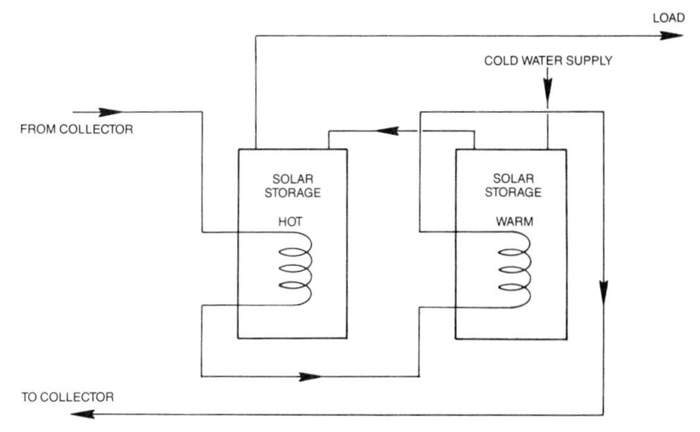


**Fig. 21 Multiple Storage Tank Arrangement with Internal Heat Exchangers**

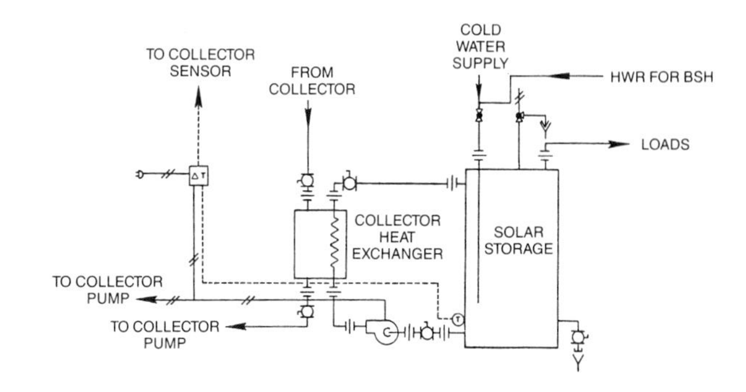


**Fig. 22 External Heat Exchanger for Pressurized Storage**

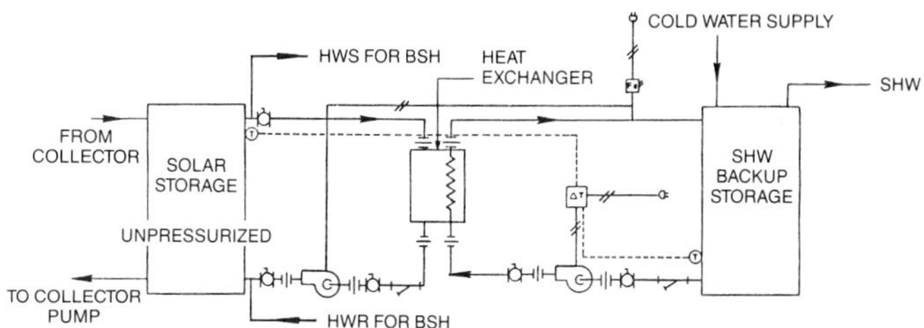


**Fig. 23 Unpressurized Storage System with an External Heat Exchanger**

start-up after at least one night in extrememly cold conditions should be considered:

- Freeze-up of the water side of the heat exchanger
- Performance loss due to extraction of heat from storage

For small systems, an internal heat exchanger/tank arrangement prevents the water side of the heat exchanger from freezing. However, the energy required to maintain the water above freezing must be extracted from storage, thereby decreasing overall performance. With the external heat exchanger/tank combination, a bypass can be arranged to bypass cold fluid around the heat exchanger until it has been heated to an acceptable level, such as 80°F (27°C). When the heat transfer fluid has warmed to this level, it can enter the heat exchanger without causing freezing or extraction of heat from storage. If necessary, this arrangement can also be used with internal heat exchangers to improve performance.

**Unpressurized Storage.** For systems sized greater than about 1000 ft$^2$ (28 m$^2$) (1500 gallon [5.7 m$^3$] storage volume minimum), unpressurized storage is usually more cost effective. As used in this chapter, the term *unpressurized* means tanks at or below the pressure expected in an unvented drainback loop.

Unpressurized storage for water and space heating implies a heat exchanger on the load side of the tank to isolate the high pressure (potable water) loop from the low-pressure collector loop. Figure 23 illustrates unpressurized storage with an external heat exchanger. In this configuration, heat is extracted from the top of the solar storage tank, and the cooled water is returned to the bottom of the tank. On the load side of the heat exchanger, the water to be heated flows from the bottom of the backup storage tank, and heated water returns to the top. The heat exchanger may have a double wall to protect a potable water supply. A differential temperature controller controls the two pumps on either side of the heat exchanger. When small pumps are used, both may be controlled by the same controller without overloading it.

The external heat exchanger shown in Figure 23 provides good system flexibility and freedom in component selection. Occasionally, system cost and parasitic power consumption may be reduced by an internal heat exchanger. At times, heat exchangers fabricated in the field use coiled soft copper tube. For larger systems, where custom fabrication is more feasible, tanks can be supplied with a specified heat exchanger installed at the top.

## Storage Tank Construction

For most liquid solar energy systems, steel is the preferred material for thermal energy storage construction. Steel tanks are (1) relatively easy to fabricate to ASME Pressure Vessel Code requirements, (2) readily available, and (3) easily attached by pipes and fittings.

Steel tanks used for pressures of 30 psi (210 kPa) and above must be ASME rated. Because water main pressure is usually above this level, open systems must use ASME code tanks. These pressure-rated storage tanks are more expensive than nonpressurized types. Significant cost reduction can usually be realized if a nonpressurized tank can be used.

Steel tanks are subject to corrosion, however. Because corrosion rates accelerate at increased temperature, the designer must be particularly aware of corrosion-protection methods for solar energy applications. A steel tank must be protected against (1) electrochemical corrosion, (2) oxidation (rusting), and (3) galvanic corrosion (ANL 1979).

**Electrochemical Corrosion.** The pH of the liquid and the electric potential of the metal primarily govern electromechanical corrosion. A sacrificial anode fabricated from a metal more reactive than steel can protect against this type. Magnesium is recommended for solar applications. Since protection ends when the anode has dissolved, the anode must be inspected annually.

**Oxidation.** Oxygen can enter the tank through the air dissolved in the water entering the tank or through an air vent. In pressurized storage, oxygen is continually replenished by the in-coming water. Besides causing rust, the oxygen catalyzes other types of corrosion. Unpressurized storage systems are not as susceptible to corrosion caused by oxygen because they can be designed as unvented systems and, corrosion is limited to the small amount of oxygen contained in the initial fill water.

Coatings applied to the inside of the tank protect it from oxidation. The following is a list of the most common coatings used:

*Phenolic Epoxy* should be applied in four coats.

*Baked on Epoxy* is preferred over the painted-on epoxy.

*Glass Lining* offers more protection than the epoxies and can be used under severe water conditions.

*Hydraulic Stone* provides the best protection against corrosion and increases the heat-retention capabilities of the tank. The weight may cause handling problems in some installations.

Regardless of the lining used, it should be flexible enough to withstand extreme thermal cycling, or it should have the same coefficient of expansion as the steel tank. All linings used for potable water tanks should be FDA approved for the maximum temperature expected in the tank.

**Galvanic Corrosion.** Dissimilar materials in an electrolyte (water) are in electrical contact with each other and corrode by galvanic action. Copper fittings screwed into a steel tank corrode the steel, for example. Galvanic corrosion can be minimized by using dielectric bushings to connect pipes to tanks.

Fiberglass reinforced plastic (FRP) tanks offer the advantages of light weight, high corrosion resistance, and low cost. Premium quality resins permit operating temperatures as high as 210°F (99°C), well above the temperatures imposed by flat-plate solar collectors. Before accepting delivery of an FRP tank, it should be inspected for damage that may have occurred during shipment. Also, the gel coat on the inside must be intact, and no glass fibers should be exposed.

Concrete vessels, lined with a waterproofing membrane, may also be used to contain a liquid thermal-storage medium. Concrete storage tanks are inexpensive, and they can be shaped to fit almost any retrofit application. Also, they possess excellent resistance to the loading that occurs when they are placed in a below-grade location. Concrete tanks must have *smooth* corners and edges.

Concrete storage vessels have some disadvantages. Seepage often occurs unless a proper waterproofing surface is applied. Waterproofing paints are generally unsatisfactory because the concrete often cracks after settling. Other problems may occur due to poor workmanship, poor location, or poor design. The tanks should typically stand alone and not be integrated with a building or other structure. Also, careful attention should be given to the expansion joints and seams, since they are particularly difficult to seal. Sealing at pipe taps can also be a problem. Penetrations should be above the liquid level if at all possible.

Concrete storage tanks should be used only in vented systems. Finally, concrete tanks are heavy, and may may be more difficult to support in a proper location. The weight may also make insulation of the tank bottom more expensive and difficult.

## Storage Tank Insulation

Heat loss from storage tanks and appurtenances is one of the major causes of poor system performance. The average $R$ value of storage tanks in solar applications is about half the insulation design value because of poorly insulated supports. Different standards recommend various design criteria for tank insulation. The SMACNA (Sheet Metal and Air Conditioning Contractors National Association) standard of a 2% loss in 12 hours is generally accepted because it maintains a more stringent requirement than other standards. The following equation can be used to calculate the insulation $R$ factor for this requirement:

$$\frac{1}{R} = \frac{fQ}{At} \frac{1}{(t_{avg} - t_a)} \qquad (10)$$

*where*

$t_{avg}$ = average temperature of the tank
$t_a$ = air temperature surrounding the tank

The insulation factor, $fQ/At$ is found from Table 2 for various tank shapes (ANL 1979).

Most solar water-heating systems use large steel pressure vessels, which are usually shipped uninsulated. Materials suitable for field insulation include fiberglass, rigid foam, and flexible foam blankets.

Fiberglass is easy to transport and make fire retardent, but it requires significant labor to apply and seal. Another widely used insulation system consists of rigid sheets of polyisocyanurate foam cut and taped around the tank. Material that is 3 to 4 inches thick (75 to 100 mm) can provide R-20 to R-30 (3.5 to 5.3 $m^2 \cdot °C/W$) insulation value.

Rigid foam insulation is sprayed directly onto the tank from a foaming truck or in a shop. It bonds well to the tank surface (no air space between the tank and the insulation), and insulates better than an equivalent thickness of fiberglass. When most foams are exposed to flames, they ignite and/or produce toxic gases. When located in or adjacent to a living space, they often must be protected by a fire barrier and/or a sprinkler system. (See Uniform Building Code, Section 1717).

Some tank manufacturers and suppliers offer custom tank jackets or flexible foam with zipper-like connections that pro-

**Table 2 Insulation Factor ($fQ/At$) for Cylindrical Water Tanks**

| Horizontal Tank Insulation Factor, Btu/h • ft² (W/m²) | | | | |
|---|---|---|---|---|
| SIZE, gal (m³) | D, D | D, 2D | D, 4D | D, 6D |
| 250 (0.95) | 3.63 (11.45) | 3.46 (10.91) | 3.05 (9.62) | 2.77 (8.74) |
| 500 (1.89) | 4.57 (14.42) | 4.36 (13.75) | 4.84 (15.27) | 3.49 (11.01) |
| 750 (2.84) | 5.24 (16.53) | 4.99 (15.74) | 4.40 (13.88) | 3.99 (12.59) |
| 1000 (3.8) | 5.76 (18.17) | 5.49 (17.32) | 4.84 (15.27) | 4.39 (13.85) |
| 1500 (5.7) | 6.60 (20.82) | 6.28 (19.81) | 5.54 (17.48) | 5.03 (15.87) |
| 2000 (7.6) | 7.26 (22.90) | 6.92 (21.83) | 6.10 (19.24) | 5.53 (17.44) |
| 3000 (11.4) | 8.31 (26.21) | 7.92 (24.98) | 6.98 (22.02) | 6.33 (19.97) |
| 4000 (15.1) | 9.15 (28.86) | 8.71 (27.48) | 7.68 (24.23) | 6.97 (21.99) |
| 5000 (18.9) | 9.86 (31.10) | 9.39 (29.62) | 8.28 (26.12) | 7.51 (23.69) |
| **Vertical Tank Insulation Factor, Btu/h • ft² (W/m²)** | | | | |
| SIZE, gal (m³) | D to 3D | D/2, D | D/3, D | D/4, D |
| 80 (0.30) | 2.10 (6.62) | 1.88 (5.93) | 2.15 (6.78) | 1.97 (6.21) |
| 120 (0.45) | 2.39 (7.54) | 2.15 (6.78) | 2.46 (7.76) | 2.26 (7.13) |
| 250 (0.95) | 3.07 (9.68) | 2.74 (8.64) | 2.46 (7.76) | 2.88 (9.09) |
| 500 (1.89) | 3.87 (12.21) | 3.46 (10.91) | 3.96 (12.49) | 3.63 (11.45) |
| 750 (2.84) | 4.43 (13.97) | 3.96 (12.49) | 4.53 (14.29) | 4.16 (13.12) |
| 1000 (3.8) | 4.87 (15.36) | 4.36 (13.75) | 4.99 (15.74) | 4.57 (14.42) |
| 1500 (5.7) | 5.58 (17.60) | 4.99 (15.74) | 5.71 (18.01) | 5.24 (16.53) |
| 2000 (7.6) | 6.13 (19.34) | 5.49 (17.32) | 6.28 (19.81) | 5.76 (18.17) |
| 3000 (11.4) | 7.03 (22.18) | 6.28 (19.81) | 7.19 (22.68) | 6.60 (20.82) |
| 4000 (15.1) | 7.73 (24.39) | 6.92 (21.83) | 7.92 (24.98) | 7.26 (22.90) |
| 5000 (18.9) | 8.33 (26.28) | 7.45 (23.50) | 8.53 (26.90) | 7.82 (24.67) |

vide quick installation and neat appearance. Some of the same fire considerations apply to these foam blankets.

The supports of a tank are a major source of heat loss. To provide suitable load-bearing capability, the supports must be in direct contact with the tank wall or attached to it. Thermal breaks must be provided between the supports and the tank. (Figure 24). If insulating the tank from the support is impractical, then the external surface of the supports must be insulated, and the supports must be placed on insulative material capable of supporting the compressive load. Wood, foamed glass, and closed cell foam can be used, depending on the compressive load.

## Stratification and Short Circuiting

Because hot water rises and cold water sinks in a vessel, the pipe to the collector inlet should always be connected to the bottom of the tank. The collector operates at its best efficiency because it is always at the lowest possible temperature. The return from the collector should always run near the top of the tank, the location of the hottest fluid. Similarly, the load should be extracted from the top of the tank, and cold water makeup should be introduced at the bottom. Because of the increased static head, vertical storage tanks enhance stratification better than horizontal tanks.

If a system has a rapid tank turnover rate or if the buffer tank of the system is closely matched to the load, thermal stratification does not offer any advantages. However, in most solar

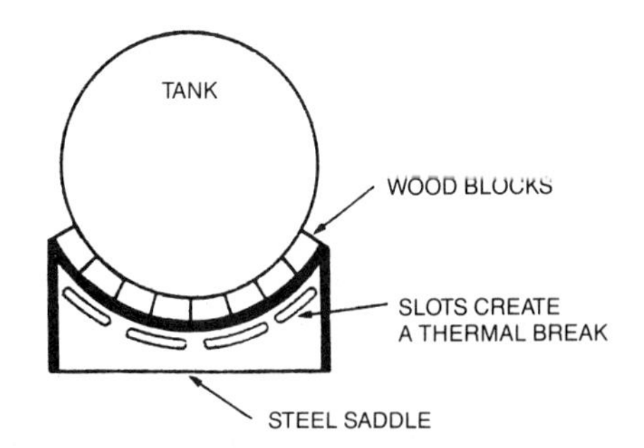


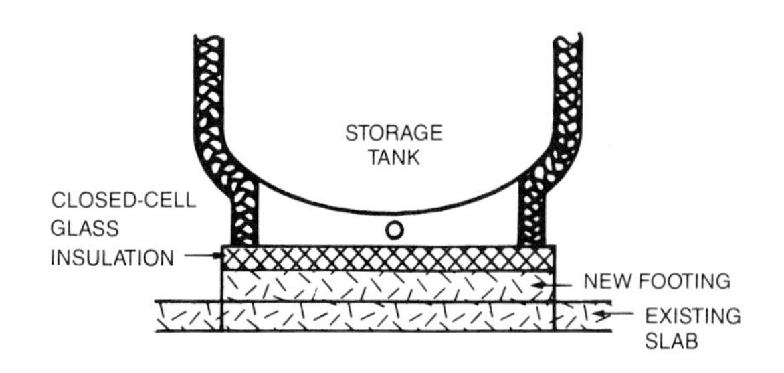


**Fig. 24 Typical Tank Support Detail**

energy systems, thermal stratification increases performance because solar energy-collecting systems operate over relatively small temperature differences. Thus, any enhancement of temperature differentials improves heat-transfer efficiency.

Thermal stratification can be enhanced by the following actions:

- Using a tall vertical tank
- Situating the inlet and outlet piping of a horizontal tank to minimize vertical fluid mixing
- Sizing the inlets and outlets such that exhaust flow velocity is less than 2 fps (0.6 m/s)
- Using flow diffusers (Figure 25)
- Plumbing multiple tanks in series

Multiple tanks generally yield the greatest temperature difference between cold water inlet and hot water outlet. With the piping/tank configuration shown in Figure 21, the difference can approach 30°F (17°C). The additional costs of having two tanks must be considered. Extra footings, insulation, and sensors are required in addition to the cost of the second tank. However, multiple tanks may save labor at the site due to their small diameter and shorter length. The smaller tanks require less demolition to install in retrofit (access through doorways, hallways, windows, etc.).

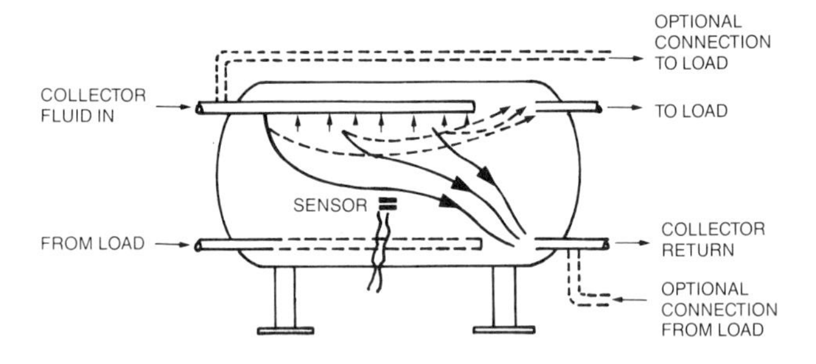


**Fig. 25 Tank Plumbing Arrangements to Minimize Short Circuiting and Mixing (Krieder 1982)**

## Sizing

The following is a list of specific site factors related to system performance or cost:

**Solar Fraction.** Most solar energy systems are designed to produce about 50% of the energy required to satisfy the load. If a system is sized to produce less than this amount, there is greater coincidence between load and available solar energy. Therefore, smaller, lower cost, storage can be used with low-solar-fraction systems without impairing system performance.

**Load Matching.** The load profile of a commercial application may be such that less storage can be used without incurring performance penalties. For example, a company cafeteria or a restaurant serving large luncheon crowds has its greatest hot water usage during, and just after, midday. Because the load coincides with the availability of solar radiation, less storage is required for carryover. For loads that peak during the morning, collectors can be rotated toward the Southeast, sometimes improving output for a given size array.

**Storage Cost.** If a solar storage tank is insulated adequately, performance generally improves with increases in storage volume, but the savings must justify the investment. The cost of storage can be kept at a minimum by using multiple, low-cost water heaters and unpressurized storage tanks made from fiberglass, concrete, etc.

The performance and cost relationships between solar availability collector design, storage design, and load are highly interactive. Furthermore, the designer should maximize the benefits from the investment for at least one year, rather than for one or two months. For this reason, solar energy system performance models, combined with economic analyses, should be used to optimize storage size for a specific site. Models that can be used to determine system performance include FCHART, SOLCOST, and TRNSYS. FCHART (Beckman *et al.* 1977) is readily available in either the work sheet form for hand calculations or as a reasonably priced computer program; however, it contains a *built-in* daily hot-water-load profile specifically for residential applications. This limitation is not very serious; studies have shown the daily hot-water profile has little effect on the system performance.

For daily profiles with extreme deviations from the residential profile, such as the company cafeteria example used previously, FCHART may not provide satisfactory results. SOLCOST can accommodate user selected daily profiles, but it is not widely available. Neither program can accommodate special days, e.g. holidays and weekends, when no service hot-water loads are present. TRNSYS (SEL 1983), which has been modified for operation on personal computers, is readily available and well-supported. It is tedious to set up and run for the first time, but after it is set up, various design parameters, such as collector area, storage size, etc., can be easily evaluated for a given system configuration.

The previous sections have centered on service hot water (SHW) and conventional hydronic space heating loads. Many hydronic solar-energy systems are used with a series-coupled heat pump or an absorption air-conditioning system. In the case of a heat pump, lower storage temperatures are desirable, thereby creating higher collector efficiencies. In this case, the collector size may be greater than recommended for conventional SHW or hydronic Building Space Heating (BSH) systems. In contrast, absorption air conditioners require much higher temperatures than BSH or hydronic BSH to activate the generator (<175°F or 80°C). Therefore, the storage collector ratio can be much less than recommended for BSH. For absorption air conditioning, thermal energy storage may act as a buffer between the collector and generator, which prevents cycling due to frequent changes in insolation level.

# HEAT EXCHANGERS

## Requirements

The heat exchanger transfers heat from one fluid to another. In closed solar-energy systems, it also isolates circuits operating at different pressure levels and separates fluids that must not be mixed. Heat exchangers are used in solar applications to separate the heat-transfer fluid in the collector loop from the domestic water supply in the storage tank (pressurized storage) or the domestic water supply from the storage (unpressurized storage).

The selection of a heat exchanger involves the following considerations:

**Performance.** Heat exchangers always degrade the performance of the solar collector; therefore, the selection of an adequate size is important. When in doubt, an over-sized heat exchanger should be selected.

**Guaranteed Separation of Fluids.** Many code authorities require a vented, double-wall heat exchanger to ensure fluid isolation. System protection requirements may also dictate the need for guaranteed fluid separation.

**Thermal Expansion.** The temperature in a heat exchanger may vary from below freezing to the boiling temperature of water. The design must withstand these thermal cycles without failing.

**Materials.** Galvanic corrosion is always a concern in liquid solar energy systems. Consequently, the piping, collectors, and other hydronic component materials must be compatible.

**Space Constraints.** Often, limited space is available for mounting and servicing the heat exchanger. Physical size and configuration must be considered when selection is made.

**Serviceability.** The water side of a heat exchanger is exposed to the scaling effects of dissolved minerals, so the design must provide access for cleaning and scale removal.

**Pressure Loss.** Energy consumed in pumping fluids reduces system performance. The pressure drop through the heat exchanger should be limited to 1 to 2 psi (7 to 14 kPa) to minimize energy consumption.

**Pressure Capability.** Because the heat exchanger is exposed to cold-water supply pressure, it should be rated for pressures above 75 psig (620 kPa).

## Internal Heat Exchanger

The two basic heat exchangers for solar applications are placed either inside or outside to the storage or drainback tank. The internal heat exchanger can be a coil inside the tank or a jacket wrapped around the pressure vessel (Figure 20). Several manufacturers supply tanks with either type of internal heat exchanger. However, the maximum size of pressurized tanks with internal heat exchangers is usually limited to about 120 gallons (450 L). Heat exchangers may be installed inside non-pressurized tanks that open from the top with relative ease. Figures 26 and 27 illustrate methods of achieving double wall protection with either type of internal heat exchanger.

For installations requiring larger tanks, or heat exchangers, a tube bundle is required. However, it is not always possible to find a heat exchanger of the desired area that will fit within the constraint imposed by the tank diameter. Consequently, a horizontal tank with the tube bundle inserted from the tank end can be used. A second option is to place a shroud around the tube bundle and pump fluid around it (Figure 28). Such an approach combines the performance of an external heat exchanger with the compactness of an internal heat exchanger. Unfortunately, tank mixing and loss of stratification result from this approach.

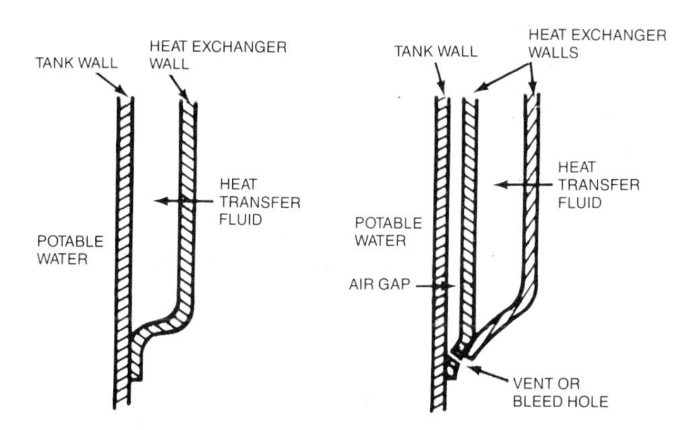


**Fig. 26 Wraparound-Shell Heat Exchangers in Cross-Section**

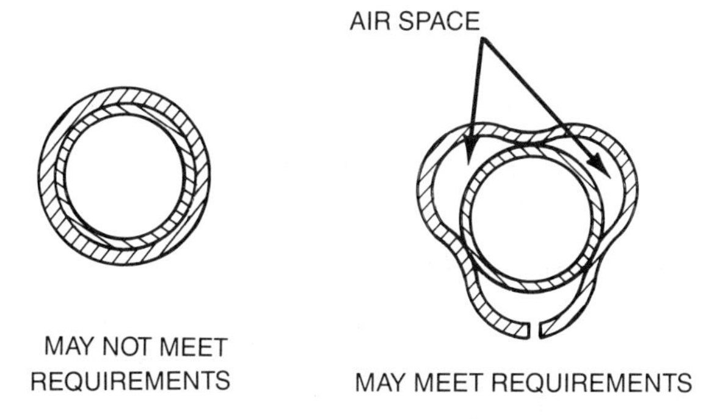


**Fig. 27 Double-Walled Tubing**

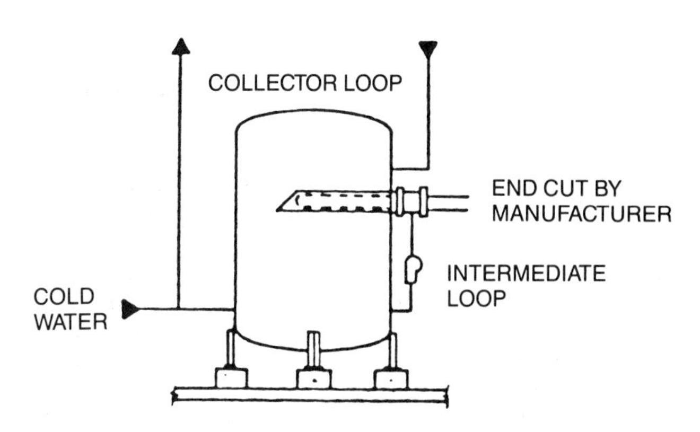


**Fig. 28 Tube Bundle Heat Exchanger with an Intermediate Loop**

## External Heat Exchanger

The external heat exchanger offers a greater degree of design flexibility than the internal heat exchanger because it is detached from the tank. For this reason, it is preferred for most commercial applications. Shell-in-tube, tube-in-tube, and plate-and-frame heat exchangers are used in solar applications. Shell-in-tube heat exchangers are found in many solar designs because they are economical, easy to obtain, and are constructed with suitable material. One limitation of the shell-in-tube is it normally does not have double-wall protection, which is often required for potable water-heating applications. A number of manufacturers produce tube-in-tube heat exchangers that offer high performance and the double wall safety required by many code authorities, but they are usually limited in size. The plate-and-frame heat exchanger is more cost effective for large potable-water applications where positive separation of heat-transfer fluids is required. These heat exchangers are compact and offer excellent heat-transfer performance.

**Plate and Frame.** These heat exchangers are suitable for pressures up to 300 psig (2.2 MPa) and temperatures to 400°F (200°C). They are economically attractive when a quality and heavy-duty construction material is required or if a double-wall shell-and-tube heat exchanger is needed for leak protection. Typical applications include food industries or domestic water heating when any possibility of product contamination must be eliminated. This exchanger also gives added protection to the collector loop.

Contamination is not possible when an intermediate loop is used, as shown in Figure 29 and the integrity of the plates is maintained. A colored fluid in the intermediate loop gives a visual means of detecting a leak through changes in color. The sealing mechanism of the plate-and-frame heat exchanger prevents cross-contamination of heat-transfer fluids. The plate-and-frame heat exchanger cost is comparable to, or cheaper than, equivalent shell-and-tube heat exchangers constructed of stainless steel.

**Shell-and-Tube.** Shell-and-tube heat exchangers accommodate large heat exchanger areas in a compact volume. The number

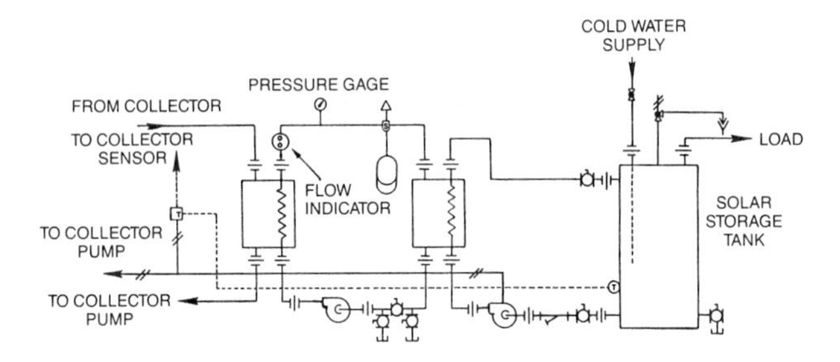


**Fig. 29 Double Wall Protection Using Two Heat Exchangers in Series**

of shell-side passes and tube-side passes (i.e., the number of times the fluid changes directions from one end of the heat exchanger to the other) is a major variable in shell-and-tube heat exchanger selection. Because the exchanger must compensate for thermal expansion, flow in and out of the tube side are generally at the same end of the exchanger. Therefore, the number of tube-side passes is even. By appropriate baffling, two, four, or more tube passes may be created. However, as the number of passes increases, the path length grows, resulting in greater pressure drop of the tube-side fluid. Unfortunately, shell-and-tube heat exchangers are hard to find in the double-wall configuration.

**Tube-and-Tube**. Fluids in the tube-and-tube heat exchanger run counterflow, which gives closer approach temperatures. The exchanger is also compact and only limited by system size. Several may be piped in parallel for higher flow, or in series to provide approach temperatures as close as 15 °F (8.3 °C). Many manufacturers offer the tube-in-tube with double-wall protection.

The counterflow configuration operates at high efficiency. For two heat exchangers in series, the effectiveness may reach 0.80. For single heat exchangers or multiple heat exchangers in parallel, the effectiveness may reach 0.67. The cost of tube-and-tube heat exchangers is low, making them cost-effective for smaller residential systems.

## Heat Exchanger Performance

Because of the high value of the heat being transferred, heat exchangers used in solar energy systems must be over-designed when compared with usual standards. This is especially true with collectors of medium to low performance.

Solar collectors perform less efficiently at high fluid inlet temperatures. Heat exchangers require a temperature difference between the two streams to transfer heat. The smaller the heat transfer surface area, the greater the temperature difference required to transfer the same amount of heat. Consequently, the collector-inlet temperature must be higher for a given tank temperature. As the solar collector is forced to operate at the progressively higher temperatures associated with smaller heat exchangers, its efficiency is reduced.

In addition to size and surface area, heat exchanger configuration is important for achieving maximum performance. Heat exchanger performance is characterized by its *effectiveness* (a type of efficiency), which is defined as:

$$E = \frac{q}{(\dot{m}c_p)_{min}\,(t_{hi} - t_{ci})} \tag{11}$$

where

$q$ = heat transfer rate, Btu/h (W)
$(\dot{m}c_p)_{min}$ = minimum capacitance rate in the heat exchanger (mass flow rate times the fluid specific heat.)
$t_{hi}$ = hot (collector loop) stream inlet temperature
$t_{ci}$ = cold (storage) stream inlet temperature

For heat exchangers located in the collector loop, the minimum flow usually occurs on the collector side rather than the tank side.

The effectiveness is the ratio between the heat actually transferred and the maximum heat that could be transferred for given flow and fluid inlet temperature conditions. The effectiveness is relatively insensitive to temperature, but it is a strong function of heat exchanger design.

A designer must decide what heat-exchanger effectiveness is required for the specific application. A method that incorporates heat exchanger performance into a collector efficiency equation (Eq. 3) uses storage tank temperature, $t_s$, as the collector inlet temperature with an adjusted heat-removal factor, $F_R$. Equation (12) relates $F_R$ and heat exchanger effectiveness:

$$\frac{F_R'}{F_R} = \frac{1}{1 + (F_R U_L/\dot{m}c_p)\,(A\dot{m}c_p/Ec_{min} - 1)} \tag{12}$$

where $c_{min}$ is the smaller of the two fluid capacities ($\dot{m}c_p$) in the heat exchanger.

The heat-exchanger effectiveness must be converted into heat-transfer surface area. *Solar Design Workbook* by SERI provides details of shell-and-tube heat exchangers and heat-transfer fluids.

# CONTROLS

## Applications

The heart of an active solar energy system is the automatic temperature control. Numerous studies and reports of operational systems show faulty controls are usually the cause of poor solar-system performance. Reliable solar-system controllers are available, and with proper understanding of each system function, proper control systems can be designed. In general, control systems should be simple; additional controls are not a good solution to a problem that can be solved by better mechanical design. The following key considerations pertain to control system design:

- Collector sensor location/selection
- Storage sensor location
- Over-temperature sensor location
- Proper on-off controller characteristics
- Selection of reliable solid-state devices, sensors, controllers, etc.
- Control panel location in heated space
- Connection of controller according to manufacturer's instructions
- Design of control system for all possible system operating modes, including heat collection, heat rejection, power outage, freeze protection, auxiliary heating, etc.
- Selection of alarm indicators for critical applications including pump failure, low temperatures, high temperatures, loss of pressure, controller failure, night-time operation.

The five following categories need to be considered when designing automatic controls for solar energy systems:

1. Collection to storage
2. Storage to load
3. Auxiliary energy to load
4. Alarms
5. Miscellaneous (e.g., for heat rejection, freeze protection, draining and over-temperature protection)

## Differential Temperature Controller

Most controls used in solar energy systems are similar to those for HVAC systems. The major exception is the Differential Temperature Controller (DTC), which is the basis of solar energy system control. The DTC is a comparing controller with at least two temperature sensors that controls one or several devices.

Typically, the sensors are located at the solar collectors and storage tank (Figure 30). On unpressurized systems, other DTC's may control the extraction of heat from the storage tank.

The DTC compares the temperature difference and when the temperature of the panel exceeds that of the storage by the predetermined amount (generally by 8 to 20°F [4.5 to 11°C]), the DTC switches to turn on the actuating devices. When the temperature of the panel drops to 3 to 10°F (1.5 to 5.5°C) above the storage temperature, the DTC, either directly or indirectly, stops the pump. Indirect control through a control relay may operate one or more pumps and possibly perform other control functions, such as the actuation of control valves.

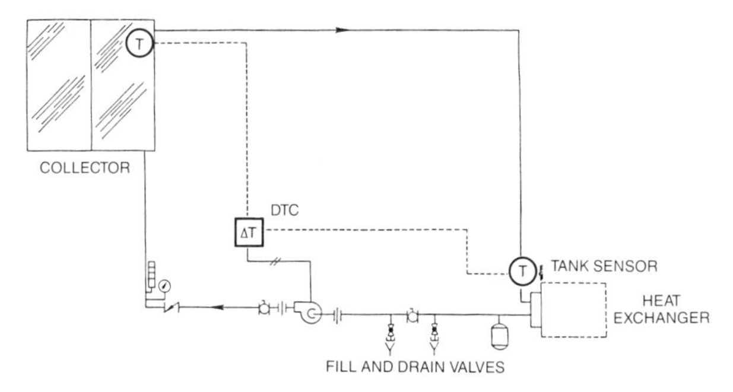


**Fig. 30 Basic Non-Freezing Collector Loop for Building Service Hot Water Heating— Non Glycol Heat Transfer Fluid**

The manufacturer's predetermined set point of the DTC may be adjustable or fixed at a specific temperature differential. If the controller has a fixed differential, the controller selection should correspond to the requirements of the system. An adjustable differential set point makes the controller more flexible and allows it to be adjusted to the specific system. The optimum differential *off* temperature should be the minimum possible; the minimum depends on whether the system circuitry has a heat exchanger between the collectors and storage.

If the system requires a heat exchanger, the energy transferred between two fluids raises the differential temperature set point. The minimum, or *off* temperature differential, is the point at which pumping the energy costs as much as the value of the energy being pumped. For systems with heat exchangers, the *off* set point is generally between 5 and 10°F (3 and 6°C). If the system does not have a heat exchanger, a range of 3 to 6°F (1.5 to 3.5°C) is acceptable for the *off* set point. The heat lost in the piping and the power required to operate the pump should also be considered.

The optimum differential *on* set point is difficult to calculate because of the changing variables and conditions. Typically, the *on* set point is 10 to 15°F (5.5 to 8.5°C) above the *off* set point. The optimum *on* set point trades between optimum energy collection and avoiding short cycling of the pump. ASHRAE's *Active Solar Thermal Design Manual* (Newton 1983) describes techniques for minimizing short cycling.

## Over-Temperature Protection

Overheating may occur during periods of high insolation and low load; thus, all portions of the solar system require protection against overheating. Liquid expansion or excessive pressure may burst piping or storage tanks, and steam or other gases within a system may restrict the liquid flow, making the system inoperable. Glycols break down and become corrosive if subjected to temperatures greater than 240°F (116°C). The system can be protected from overheating by (1) stopping circulation in the collection loop until the storage temperature decreases, (2) discharging the overheated water from the system and replacing it with cold make-up water, or (3) using a heat exchanger coil as a means of heat rejection to ambient air.

The following questions should be answered to determine if over-temperature protection is necessary.

1. Is the load ever expected to be off, such that the solar input will be much higher than the load? The designer must determine possibilities based on the owner's needs and a computer analysis of system performance.
2. Do individual components, pumps, valves, circulating fluids, piping, tanks, and liners need protection? The designer must examine all components and base the over-temperature protection set point on the component that has the lowest maximum operating temperature specification. This may be a valve or pump with a 180 or 300°F (80 or 150°C) maximum operating temperature. Sometimes, this factor may be overcome by selecting components with higher operating temperature capabilities.
3. Is the formation of steam or discharging boiling water at the tap possible? If the system has no mixing valve that mixes cold water with the solar-heated water before it enters the tap, the water must be maintained below boiling temperature. Otherwise, the water will flash to steam as it enters the tap and, most likely, scald the user. Some city codes require a mixing valve be placed in the system for safety.

Differential Temperature Controllers are available that sense over-temperatures. Depending on the controller used, the sensor may be mounted at the bottom or top of the storage tank. If it is mounted at the bottom of the tank, the collector-to-storage differential temperature sensor can be used to sense over-temperatures. Input to a DTC mounted at the top of the tank is independent of the bottom-mounted sensor, and the sensor monitors the true high temperature.

The normal action taken when the DTC senses an over-temperature is to turn off the pump to stop heat collection. After the panels in a drainback system are drained, they will attain stagnation temperatures. While drainback is not desirable, the panels used for these systems should be designed and tested to withstand these conditions. In addition, draindown panels should withstand the thermal shock of start-up when relatively cool water enters the panels while they are at stagnation temperatures. The temperature difference can range from 75 to 300°F (42 to 167°C). Such a difference could warp panels made with two or more materials of different thermal expansion coefficients. If the solar panels cannot withstand the thermal shock, an interlock should be incorporated into the control logic to prevent this situation. One method uses a high-temperature sensor mounted on the collector absorber that prevents the pump from operating until the collector temperature drops below the sensor set point.

If circulation stops in a closed-loop antifreeze system that has a heat exchanger, high stagnation temperatures will occur. These temperatures could breakdown the glycol heat-transfer fluid. To prevent damage or injury due to excessive pressure, a pressure relief valve must be installed in the loop, and a means of rejecting heat from the collector loop must be provided. The next section describes a common way to relieve pressure. When water-based absorber fluids are used, pressure builds up from boiling. However, pressure increases due to the thermal expansion of any fluid.

The pressure-relief valve should be set to relieve at or below the maximum operating pressure of any component in the closed-loop system. Typical settings are around 50 psig (350 kPa), corresponding to a temperature of approximately 300°F (150°C). However, these settings should be checked. When the pressure-relief valve does open, it discharges expensive antifreeze solu-

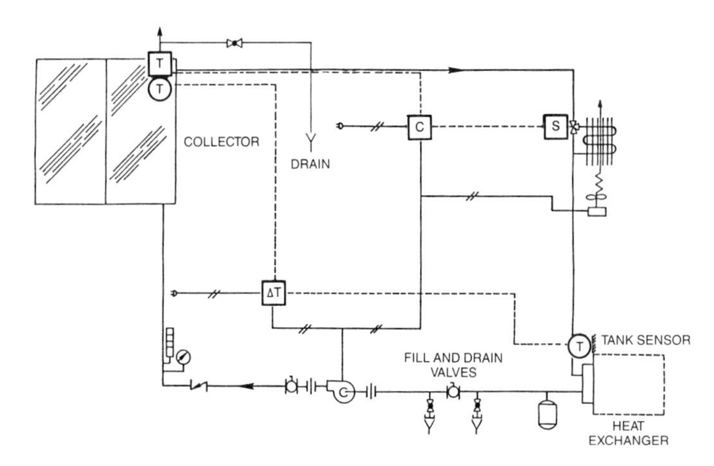


**Fig. 31 Heat Rejection from Non-Freezing System Using a Liquid-to-Air Heat Exchanger**

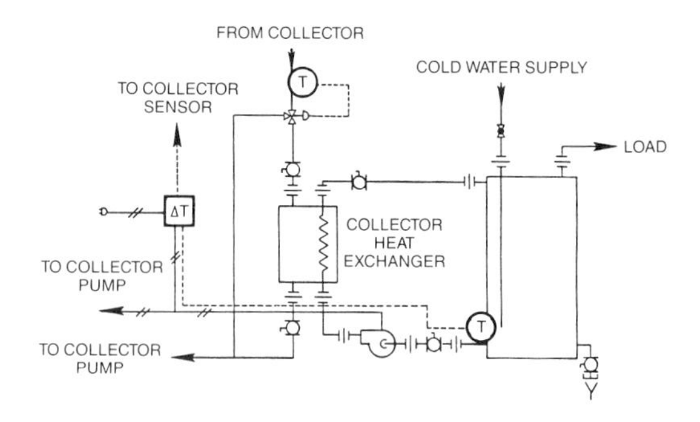


**Fig. 32 Non-Freezing System with H-X Bypass**

tion. Glycol antifreeze solutions damage many types of roof membranes. The discharge can be piped to large containers to save the antifreeze, but this design can create dangerous conditions because of the high pressures and temperatures involved.

If a collector loop containing glycol stagnates, chemical decomposition raises the fusion point of the liquid, and freezing becomes possible. An alternate method continues fluid circulation but diverts the flow from storage to a heat exchanger that dumps heat to the ambient air or other source (Figure 31). This wastes energy, but it protects the system. A sensor on the solar collector absorber plate that turns on the heat-rejection equipment can provide control. The temperature sensor set point is usually 200 to 250 °F (95 to 120 °C) and depends on the system components. When the sensor reaches the high temperature set point, it turns on pumps, fans, alarms, or whatever is necessary to reject the heat and warn of the over-temperature. The dump continues to operate until the over-temperature control in the collector loop DTC senses an acceptable drop in tank temperature and is reset to its normal state.

## Hot Water Dump

If water temperatures above 200 °F (93 °C) are allowed, the standard temperature-pressure (210 °F, 125 psig [99 °C, 960 kPa]) safety relief valve may operate occasionally. If these temperatures are reached, the valve opens, and some of the hot water vents out. However, these valves are designed for safety purposes, and after a few openings, they leak hot water. Thus, they should not be relied upon as the only control device. An aquastat that controls a solenoid, pneumatic, or electrically actuated valve should be used instead.

## Heat Exchanger Freeze Protection

The following factors should be considered when selecting an external heat exchanger for a system protected by a non-freezing fluid that is started after an overnight, or longer, exposure to extremely cold conditions.

- Freeze-up of the water side of the heat exchanger
- Performance loss due to extraction of heat from storage.

An internal heat exchanger/tank has been placed on the water side of the heat exchanger of small systems to prevent freezing. However, the energy required to maintain the water above freezing must be extracted from storage, which decreases overall performance. With the external heat exchanger/tank combination, a bypass can be installed, as illustrated in Figure 32. The controller positions the valve to bypass the heat exchanger until the fluid in the collector loop attains a reasonable level, for example 80 °F (27 °C). When the heat transfer fluid has warmed to this level, it can enter the heat exchanger without freezing or extracting heat from storage. The arrangement illustrated in Figure 32 can also be used with an internal heat exchanger, if necessary, to improve performance.

## REFERENCES

ANL. 1979. *Reliability and Materials Design Guidelines for Solar Domestic Hot Water Systems.* Argonne National Laboratories, ANL/SDP-9.

ANL. 1979. *Design and Installation Manual for Thermal Energy Storage.* Argonne National Laboratory, ANL-79-15.

Beckman, W. A.; S. Klein; and J.A. Duffie. 1977. *Solar Heating Design by the F-Chart Method.* Wiley-Interscience, New York.

Huggins, J.C. and D.L. Block. 1983. Thermal Performance of Flat Plate Solar Collectors by Generic Classification. Proceedings of the ASME Solar Energy Division Fifth Annual Conference, Orlando, Florida.

ICBO. 1985. *Uniform Building Code.* International Conference of Building Officials, Whittier, CA.

Knowles, A.S. 1980. A Simple Balancing Technique for Liquid Cooled Flat Plate Solar Collector Arrays. International Solar Energy Society, Phoenix, AZ.

Krieder, J. 1982. *The Solar Heating Design Process—Active and Passive Systems.* McGraw-Hill, New York.

Kutscher, C.F. 1982. Design Approaches for Industrial Process Heat Systems. Solar Energy Research Institute, SERI/TR-253-1356.

Logee, T.L., and P.W. Kendall. 1984. Component Report Performance of Solar Collector Arrays and Collector Controllers in the National Solar Data Network. SOLAR/0015-84/32 Vitro Corporation.

Newton, A. 1983. *Active Solar Thermal Design Manual.* ASHRAE.

SEL. 1983. TRNSYS a Transient Simulation Program. Engineering Experiment Station Report 38-12, Solar Energy Laboratory, University of Wisconsin-Madison.

CHAPTER 45

# CODES AND STANDARDS

THE Codes and Standards listed in Table 1 represent practices, methods, or standards published by the organizations indicated. They are valuable guides for the practicing engineer in determining test methods, ratings, performance requirements, and limits applying to the equipment used in heating, refrigerating, ventilating, and air conditioning. *Copies can usually be obtained from the organization listed in the Reference column.* These listings represent the most recent information available at the time of publication.

**Table 1 Codes and Standards Published by Various Societies and Associations**

| Subject | Title | Publisher | Reference |
|---|---|---|---|
| *Acoustics* | Standard Acoustical Terminology (reaffirmed 1976) | ASA | ANSI S1.1-1960 |
| *Air Conditioners* | | | |
| Room | Method of Testing for Rating Room Air Conditioners and Packaged Terminal Air Conditioners | ASHRAE | ANSI/ASHRAE 16-1983 |
| | Methods of Testing for Rating Room Fan Coil Air Conditioners | ASHRAE | ASHRAE 79-1984 |
| | Room Air Conditioners (1986) | UL | UL 484 |
| | Room Air Conditioners | AHAM | ANSI/AHAM RAC 1-1982 |
| | Method of Testing Room Air Conditioner Heating Capacity | ASHRAE | ANSI/ASHRAE 58-1986 |
| | Performance Standard for Room Air Conditioners | CSA | C368.1-M1980 |
| | Commercial and Residential Central Air Conditioners | CSA | C22.2 No. 119-M1985 |
| Packaged Terminal | Packaged Terminal Air Conditioners | ARI | ARI 310-85 |
| | Packaged Terminal Heat Pumps | ARI | ARI 380-85 |
| Transport | Air Conditioning of Aircraft Cargo | SAE | SAE AIR 806A |
| | Nomenclature, Aircraft Air-Conditioning Equipment (1978) | SAE | SAE ARP 147C |
| Unitary | Air Conditioners, Central Cooling (1984) | UL | UL 465 |
| | Load Calculation for Commercial Summer and Winter Air Conditioning (Using Unitary Equipment), 2nd ed. (1983) | ACCA | ACCA Manual N |
| | Methods of Testing for Rating Heat Operated Unitary Air Conditioning Equipment for Cooling | ASHRAE | ASHRAE 40-1986 |
| | Methods of Testing for Rating Unitary Air-Conditioning and Heat Pump Equipment | ASHRAE | ANSI/ASHRAE 37-1978 |
| | Methods of Testing for Seasonal Efficiency of Unitary Air-Conditioners and Heat Pumps | ASHRAE | ANSI/ASHRAE 116-1983 |
| | Sound Rating of Outdoor Unitary Equipment | ARI | ARI 270-84 |
| | Application of Sound Rated Outdoor Unitary Equipment | ARI | ARI 275-84 |
| | Unitary Air-Conditioning Equipment | ARI | ARI 210-81 |
| | Unitary Air-Conditioning and Air-Source Heat Pump Equipment | ARI | ARI 210/240-84 |
| | Commercial and Industrial Unitary Air-Conditioning Equipment | ARI | ARI 360-86 |
| *Air Conditioning* | Automotive Air Conditioning Hose (1971) | SAE | SAE J 51-1985 |
| | Environmental System Technology (1984) | NEBB | NEBB |
| | Equipment Selection and System Design Procedures for Commercial Summer and Winter Air Conditioning, First ed. (1977) | ACCA | Manual Q |

Table 1 Codes and Standards Published by Various Societies and Associations

| Subject | Title | Publisher | Reference |
|---|---|---|---|
| | Gas-Fired Absorption Summer Air Conditioning Appliances (with 1982 addenda) | AGA | ANSI Z21.40.1-1981 |
| | Load Calculation for Residential Winter and Summer Air Conditioning, 7th ed. (1986) | ACCA | ACCA Manual J |
| | Equipment Selection and System Design Procedures, 2nd ed. (1984) | ACCA | ACCA Manual D |
| | Installation Standards for Heating and Air-Conditioning Systems, 6th ed. (1986) | SMACNA | SMACNA |
| | HVAC Systems and Applications, First ed. (1986) | SMACNA | SMACNA |
| Transport | Air Conditioning Equipment, General Requirements for Subsonic Airplanes (1961) | SAE | SAE ARP 85D |
| | General Requirements for Helicopter Air Conditioning (1970) | SAE | SAE ARP 292B |
| | Testing of Commercial Airplane Environmental Control Systems (1973) | SAE | SAE ARP 217B |
| *Air Curtains* | Test Methods for Air Curtain Units | AMCA | AMCA 220-82 |
| | Air Outlets and Inlets | ARI | ARI 650-80 |
| | Air Volume Terminals | ARI/ADC | ARI/ADC 880-83 |
| | Selection of Distribution Systems First ed. (1963) | ACCA | ACCA Manual G |
| | Room Air Distribution Considerations First ed. (1963) | ACCA | ACCA Manual E |
| | Methods of Testing for Rating the Air Flow Performance of Outlets and Inlets | ASHRAE | ASHRAE 70-72 |
| | High Temperature Pneumatic Duct Systems for Aircraft (1981) | SAE | SAE ARP 699D |
| | Metric Units and Conversion Factors | AMCA | AMCA 99-0100-76 |
| | Installation Code for Residential Mechanical Exhaust Systems | CSA | C260.1-1975 |
| | Laboratory Certification Manual | ADC | ADC 1062:LCM-83 |
| | Residential Air Exhaust Equipment (1e) | CSA | C260.2-1976 |
| | Test Code for Grilles, Registers and Diffusers | ADC | ADC 1062:GRD-84 |
| *Air Ducts and Fittings* | Flexible Air Duct Test Code | ADC | ADC FD-72 R1-1979 |
| | Installation of Air Conditioning and Ventilating Systems (1981) | NFPA | ANSI/NFPA 90A-1985 |
| | Installation of Warm Air Heating and Air-Conditioning Systems (1980) | NFPA | ANSI/NFPA 90B-1984 |
| | HVAC Duct Construction Standards—Metal and Flexible, First ed. (1985) | SMACNA | SMACNA |
| | HVAC Systems—Duct Design, 2nd ed. (1981) | SMACNA | SMACNA |
| | Round Industrial Duct Construction (1977) | SMACNA | SMACNA |
| | Rectangular Industrial Duct Construction (1980) | SMACNA | SMACNA |
| | Ducted Electric Heat Guide for Air Handling Systems (1971) | SMACNA | SMACNA |
| | Marine Rigid and Flexible Air Ducting (1986) | UL | UL 1136 |
| | Thermoplastic Duct (PVC) Construction Manual (1974) | SMACNA | SMACNA |
| | Marine Rigid and Flexible Air Ducting (1986) | UL | UL 1136 |
| | Pipes, Ducts, and Fittings for Residential-Type Air-Conditioning Systems | CSA | B228.1-1968 |
| | Factory-Made Air Ducts and Connectors (1986) | UL | UL 181 |
| *Air Filters* | Test Performance of Air Filter Units (1987) | UL | UL 900 |
| | Methods of Testing Air-Cleaning Devices Used in General Ventiltion for Removing Particulate Matter | ASHRAE | ASHRAE 52-76 |
| | Commercial and Industrial Air Filter Equipment | ARI | ARI 850-84 |
| | Air Filter Equipment | ARI | ARI 680-86 |
| | Test Performance of High Efficiency, Particulate, Air Filter Units (1985) | UL | UL 586 |
| | Methods of Test for Air Filters Used in Air Conditioning and General Ventilation | BSI | BS 6540 Part 1 |
| | Method for Sodium Flame Test for Air Filters | BSI | BS 3928 |
| | Electrostatic Air Cleaners (1981) | UL | UL 867 |
| *Air-Handling Units* | Central Station Air-Handling Units | ARI | ARI 430-86 |
| | Application of Central Station Air-Handling Unit | ARI | ARI 435-81 |
| *Boilers* | Recommended Design Guidelines for Stoker Firing of Bituminous Coals (1983) | ABMA | ABMA |
| | Boiler Water Limits and Steam Purity Recommendations for Watertube Boilers (3rd ed., 1982) | ABMA | ABMA |
| | Boiler Water Requirements and Associated Steam Purity—Commercial Boilers (1981) | ABMA | ABMA |
| | Boiler and Pressure Vessel Code (eleven sections) (1986) | ASME | ASME |
| | Boiler, Pressure Vessel, and Pressure Piping Code | CSA | B51-M1986 |
| | Lexicon Boiler and Auxiliary Equipment, 5th ed. (1987) | ABMA | ABMA |
| Cast-Iron | Testing and Rating Heating Boilers (1982) (amended 1984) | HYD I | IBR/SBI |
| | Ratings for Cast-Iron and Steel Boilers (1986) | HYD I | IBR/SBI |

**Table 1 Codes and Standards Published by Various Societies and Associations**

| Subject | Title | Publisher | Reference |
|---|---|---|---|
| Gas or Oil | Explosion Prevention of Fuel Oil and Natural Gas-Fired Single-Burner Boiler-Furnaces (1984) | NFPA | ANSI/NFPA 85A-1987 |
| | Explosion Prevention of Natural Gas-Fired Multiple-Burner Boiler-Furnaces (1978) | NFPA | ANSI/NFPA 85B-1984 |
| | Gas-Fired Low-Pressure Steam and Hot Water Boilers (with 1983 addenda) | AGA | ANSI Z21.13-1982 |
| | Gas Utilization Equipment in Large Boilers (with 1972 and 1976 addenda; R-1983) | AGA | ANSI Z83.3-1971 |
| | Oil Fired Boiler Assemblies (1975) | UL | UL 726 |
| | Prevention of Furnace Explosions in Fuel Oil-Fired Multiple-Burner Boiler-Furnaces (1984) | NFPA | ANSI/NFPA 85D-1984 |
| | Control and Safety Devices for Automatically Fired Boilers (CSD.1a is an addenda to 1982 ed.) | ASME | ANSI/ASME CSD.1-1982 CSD.1a-1984 |
| | Oil-Fired Steam and Hot-Water Boilers for Residential Use (3a) | CSA | B140.7.1-1976 |
| | Oil-Fired Steam and Hot-Water Boilers for Commercial and Industrial Use (1a) | CSA | B140.7.2-1967 |
| | Electric, Heating, Water Supply, and Power Boilers (1980) | UL | UL 834 |
| Watertube | Recommended Standard Instrument Connections Manual (1982) | SAMA ABMA | ABMA |
| *Building Codes* | National Building Code, 10th ed. (1987) | BOCA | BOCA |
| | CABO One- and Two-Family Dwelling Code (1986) (with 1987 revisions) | CABO | CABO |
| | Standard Building Code (1988) | SBCCI | SBCCI |
| | Uniform Building Code (1988) | ICBO | ICBO |
| | Uniform Building Code Standards (1985) | ICBO | ICBO |
| | BOCA/National Existing Structures Code, 2nd ed. (1987) | BOCA | BOCA |
| | Model Energy Code (1986) (with 1987 revisions) | CABO | CABO |
| Mechanical | BOCA/National Mechanical Code, 6th ed. (1987) | BOCA | BOCA |
| | Safety Code for Elevators and Escalators (plus two yearly supplements) | ASME | ANSI/ASME A17.1-1984 |
| | Uniform Mechanical Code (1985) (with Uniform Mechanical Code Standards) | ICBO/ IAPMO | ICBO/IAPMO |
| | Standard Mechanical Code (1988) | SBCCI | SBCCI |
| *Burners* | Installation of Domestic Gas Conversion Burners | AGA | ANSI Z21.8-1984 |
| | Domestic Gas Conversion Burners | AGA | ANSI Z21.17-1984 |
| | Oil Burners (1980) | UL | UL 296 |
| | Installation Code for Oil Burning Equipment | CSA | B139-1976 |
| | Installation Code for Oil Burning Equipment, Supplement No. 1 to B139-76 | CSA | B139S1-1982 |
| | General Requirements for Oil Burning Equipment | CAN/ CSA | B140.0-M1987 |
| | Vaporizing-Type Oil Burners | CSA | B140.1-1966 (R1980) |
| | Oil Burners, Atomizing Type | CSA | B140.2.1-1973 |
| | Pressure Atomizing Oil Burner Nozzles | CSA | B140.2.2-1971 (R1980) |
| | Replacement Burners and Replacement Combustion Heads for Residential Oil Burners | CSA | B140.2.3-M1981 |
| *Capillary Tubes* | Method of Testing Flow Capacity of Refrigerant Capillary Tubes | ASHRAE | ASHRAE 28-78 |
| *Chillers* | Methods of Testing Liquid Chilling Packages | ASHRAE | ASHRAE 30-78 |
| | Absorption Water-Chilling Packages | ARI | ARI 560-82 |
| | Centrifugal Water-Chilling Packages | ARI | ARI 550-86 |
| | Reciprocating Water-Chilling Packages | ARI | ANSI/ARI 590-86 |
| *Chimneys* | Chimneys, Fireplaces, and Vents (1977) | NFPA | ANSI/NFPA 211-1984 |
| | Chimneys, Factory-Built, Residential Type and Building Heating Appliance (1983) | UL | UL 103 |
| | Chimneys, Factory-Built, Medium Heat Appliance (1986) | UL | UL 959 |
| *Coils* | Forced-Circulation Air-Cooling and Air-Heating Coils | ARI | ANSI/ARI 410-81 |
| | Methods of Testing Forced Circulation Air Cooling and Air Heating Coils | ASHRAE | ASHRAE 33-78 |
| *Comfort Conditions* | Thermal Environmental Conditions for Human Occupancy | ASHRAE | ANSI/ASHRAE 55-1981 |
| *Compressors* | Compressors and Exhausters (reaffirmed 1979) | ASME | ANSI/ASME PTC 10-74 |

**Table 1 Codes and Standards Published by Various Societies and Associations**

| Subject | Title | Publisher | Reference |
|---|---|---|---|
| | Compressed Air and Gas Handbook, 5th ed. (1988) | CAGI | CAGI |
| | Safety Standard for Compressors for Process Industries | ASME | ASME/ANSI B19.3-1986 |
| | Safety Standard for Air Compressor Systems | ASME | ANSI/ASME B19.1-1985 |
| | Displacement Compressors, Vacuum Pumps and Blowers | ASME | ANSI/ASME PTC 9-1985 |
| | Compressors and Exhausters | ASME | ANSI/ASME PTC 10-1985 |
| Refrigeration | Methods of Testing for Rating Positive Displacement Refrigerant Compressors | ASHRAE | ASHRAE 23-78 |
| | Ammonia Compressor Units | ARI | ANSI/ARI 510-83 |
| | Hermetic Refrigerant Motor-Compressors (1984) | UL | UL 984 |
| | Positive Displacement Refrigerant Compressors and Condensing Units | ARI | ARI 520-85 |
| *Computers* | Protection of Electronic Computer/Data Processing Equipment | NFPA | ANSI/NFPA 75-1987 |
| *Condensers* | Water-Cooled Refrigerant Condensers, Remote Type | ARI | ANSI/ARI 450-79 |
| | Methods of Testing for Rating Remote Mechanical Draft Air-Cooled Refrigerant Condensers | ASHRAE | ASHRAE 20-70 |
| | Method of Testing for Rating Water-Cooled Refrigerant Condensers | ASHRAE | ASHRAE 22-78 |
| | Methods of Testing Remote Mechanical Draft Evaporative Refrigerant Condensers | ASHRAE | ASHRAE 64-74 |
| | Remote Mechanical Draft Air-Cooled Refrigerant Condensers | ARI | ANSI/ARI 460-80 |
| | Standards for Steam Surface Condensers, 8th ed. (1984) | HEI | HEI |
| *Condensing Units* | Methods of Testing for Rating Positive Displacement Condensing Units | ASHRAE | ASHRAE 14-80 |
| | Refrigeration and Air-Conditioning Condensing and Compressor Units (1987) | UL | UL 303 |
| | Commercial and Industrial Unitary Air-Conditioning Condensing Units | ARI | ARI 365-85 |
| *Contactors* | Definite Purpose Magnetic Contactors | ARI | ARI 780-86 |
| | Definite Purpose Contactors for Limited Duty | ARI | ARI 790-86 |
| *Controls* | Limit Controls (1974) | UL | UL 353 |
| | Energy Management Control Systems | ASHRAE | ASHRAE 114-1986 |
| | Primary Safety Controls for Gas- and Oil-Fired Appliances (1985) | UL | UL 372 |
| | Temperature-Indicating and Regulating Equipment (1981) | UL | UL 873 |
| | Industrial Control Equipment (1984) | UL | UL 508 |
| | Temperature-Indicating and Regulating Equipment | CSA | C22.2 No. 24-1987 |
| Residential | Automatic Gas Ignition Systems and Components (with 1987 addenda) | AGA | ANSI Z21.20-1985 |
| | Temperature Limit Controls for Electric Baseboard Heaters | NEMA | NEMA DC 10-83 |
| | Load Control for Use on Central Electric Heating Systems | NEMA | NEMA DC 22-77 (R 1982) |
| | Electric Quick-Connect Terminals (1985) | UL | UL 310 |
| | Quick Connect Terminals | NEMA | ANSI/NEMA DC 2-82 |
| | Line-Voltage Integrally Mounted Thermostats for Electric Heaters | NEMA | NEMA DC 13-79 (R 1985) |
| | Low-Voltage Room Thermostats | NEMA | NEMA DC3-1984 |
| | Hot Water Immersion Controls | NEMA | NEMA DC 12-85 |
| | Warm Air Limit and Fan Controls | NEMA | NEMA DC 4-86 |
| | Gas Appliance Thermostats (with 1985 addenda) | AGA | ANSI Z21.23-1980 |
| | Gas Appliance Pressure Regulators (with 1982 and 1984 addenda) | AGA | ANSI Z21.18-1981 |
| *Coolers* | | | |
| Air | Methods of Testing Forced Convection and Natural Convection Air Coolers for Refrigeration | ASHRAE | ASHRAE 25-77 |
| | Unit Coolers for Refrigeration | ARI | ARI 420-84 |
| Bottled Beverage | Methods of Testing and Rating Bottled and Canned Beverage Vendors and Coolers | ASHRAE | ASHRAE 32-1986 |
| | Refrigerated Vending Machines (1979) | UL | UL 541 |
| Drinking Water | Methods of Testing for Rating Drinking Water Coolers with Self-Contained Mechanical Refrigeration Systems | ASHRAE | ASHRAE 18-79 |
| | Drinking Water Coolers (1987) | UL | UL 399 |
| | Drinking Fountains and Self-Contained, Mechanically Refrigerated Drinking Water Coolers | ARI ANSI | ARI 1010-84 |
| | Application and Installation of Drinking Water Coolers | ARI | ARI 1020-84 |
| | Drinking Water Coolers and Beverage Dispensers | CSA | C22.2 No. 91-1971 (R 1981) |

**Table 1 Codes and Standards Published by Various Societies and Associations**

| Subject | Title | Publisher | Reference |
|---|---|---|---|
| Liquid | Methods of Testing for Rating Liquid Coolers | ASHRAE | ASHRAE 24-78 |
| | Refrigerant-Cooled Liquid Coolers, Remote Type | ARI | ANSI/ARI 480-80 |
| *Cooling Towers* | Water Cooling Towers (1983) | NFPA | ANSI/NFPA 214-1983 |
| | Atmospheric Water Cooling Equipment | ASME | ANSI/ASME PTC 23-1986 |
| | Acceptance Test Code for Water Cooling Towers: Mechanical Draft, Natural Draft Fan Assisted Types, and Evaluation of Results. Addendum-1, For Thermal Testing of Wet/Dry Cooling Towers (1986) | CTI | CTI ATC-105 |
| | Acceptance Test Code for Spray Cooling Systems (1985) | CTI | ATC-133 |
| | Certification Standard for Commercial Water Cooling Towers (1986) | CTI | STD-201 |
| | Code for Measurement of Sound from Water Cooling Towers | CTI | ATC-128 (1981) |
| | Fiberglass-Reinforced Plastic Panels for Application on Industrial Water Cooling Towers | CTI | STD-131 (1983) |
| *Dehumidifiers* | Dehumidifiers | AHAM | ANSI/AHAM DH 1-1980 |
| | Dehumidifiers (3a) | CSA | C22.2 No. 92-1971 |
| | Dehumidifiers (1987) | UL | UL 474 |
| *Desiccants* | Method of Testing Desiccants for Refrigerant Drying | ASHRAE | ASHRAE 35-1983 |
| *Driers* | Liquid-Line Driers | ARI | ARI 710-86 |
| | Method of Testing Liquid-Line Refrigerant Driers | ASHRAE | ASHRAE 63-79 |
| *Electrical* | National Electric Code (1984) | ANSI/ NFPA | ANSI/NFPA 70-1987 |
| | Canadian Electrical Code | CSA | C22.1-1986 |
| | Compatibility of Electrical Connectors and Wiring (1975) | SAE | SAE AIR 1329 |
| | Manufacturers' Identification of Electrical Connector Contacts, Terminals and Splices (1982) | SAE | SAE AIR 1351 A |
| | Voltage Ratings for Electrical Power Systems and Equipment | ANSI | ANSI C84.1-1982 |
| *Energy* | Air Conditioning and Refrigerating Equipment Nameplate Voltages | ARI | ARI 110-80 |
| | Energy Conservation in New Building Design | ASHRAE | ANSI/ASHRAE/IES 90A-1980 |
| | Energy Conservation in Existing Buildings—High-Rise Residential | ASHRAE | ANSI/ASHRAE/IES 100.2-1981 |
| | Energy Conservation in Existing Buildings—Institutional | ASHRAE | ANSI/ASHRAE/IES 100.5-1981 |
| | Energy Conservation in Existing Buildings—Public Assembly | ASHRAE | ANSI/ASHRAE/IES 100.6-1981 |
| | Energy Recovery Equipment and Systems, Air-to-Air (1978) | SMACNA | SMACNA |
| | Energy Conservation Guidelines (1984) | SMACNA | SMACNA |
| | Model Energy Code (MEC) (1986) (with 1987 revisions) | CABO | BOCA/ICBO/NCSBCS/ SBCCI |
| | Energy Management Equipment (1984) | UL | UL 916 |
| | Retrofit of Building Energy Systems and Processes (1982) | SMACNA | SMACNA |
| *Exhaust Systems* | Method of Testing Performance of Laboratory Fume Hoods | ASHRAE | ASHRAE 110-1985 |
| | Installation of Blower and Exhaust Systems for Dust, Stock, Vapor Removal or Conveying (1983) | NFPA | ANSI/NFPA 91-1983 |
| | Fundamentals Governing the Design and Operation of Local Exhaust Systems | AIHA | ANSI Z9.2-1979 |
| | Practices for Ventilation and Operation of Open-Surface Tanks | AIHA | ANSI Z9.1-1977 |
| | Safety Code for Design, Construction, and Ventilation of Spray Finishing Operations (reaffirmed 1971) | ANSI | ANSI Z9.3-1985 |
| | Ventilation and Safe Practices of Abrasives Blasting Operations | ANSI | ANSI Z9.4-1985 |
| | Compressors and Exhausters | ASME | ANSI/ASME PTC 10-1985 |
| | Mechanical Flue-Gas Exhausters | CAN | 3-B255-M81 |
| *Expansion Valves* | Method of Testing for Capacity Rating of Thermostatic Refrigerant Expansion Valves | ASHRAE | ANSI/ASHRAE 17-1986 |
| | Thermostatic Refrigerant Expansion Valves | ARI | ARI 750-81 |
| *Fan Coil Units* | Room Fan-Coil Air Conditioners | ARI | ARI 440-84 |
| | Fan Coil Units and Room Fan Heater Units (1986) | UL | UL 883 |
| | Methods of Testing for Rating Room Fan-Coil Air Conditioners | ASHRAE | ASHRAE 79-1984 |

**Table 1 Codes and Standards Published by Various Societies and Associations**

| Subject | Title | Publisher | Reference |
|---|---|---|---|
| *Fans* | Standards Handbook | AMCA | AMCA 99-83 |
| | Fans | ASME | ANSI/ASME PTC 11-1984 |
| | Electric Fans (1977) | UL | UL 507 |
| | Laboratory Methods of Testing Fans for Rating | ASHRAE | ANSI/ASHRAE 51-1985 ANSI/AMCA 210-85 |
| | Methods of Testing Dynamic Characteristics of Propeller Fans—Aerodynamically Excited Fan Vibrations and Critical Speeds | ASHRAE | ANSI/ASHRAE 87.1-1983 |
| | Laboratory Methods of Testing Fans for Rating | AMCA | ANSI/AMCA 210-85 |
| | Drive Arrangements for Centrifugal Fans | AMCA | AMCA 99-2404-78 |
| | Designation for Rotation and Discharge of Centrifugal Fans | AMCA | AMCA 99-2406-83 |
| | Motor Positions for Belt or Chain Drive Centrifugal Fans | AMCA | AMCA 99-2407-66 |
| | Drive Arrangements for Tubular Centrifugal Fans | AMCA | AMCA 99-2410-82 |
| | Fans and Blowers | ARI | ARI 670-85 |
| | Inlet Box Positions for Centrifugal Fans | AMCA | AMCA 99-2405-83 |
| | Fans and Ventilators | CSA | C22.2 No. 113-M1984 |
| Ceiling | AC Electric Fans and Regulators | AMCA | ANSI-IEC Pub. 385 |
| *Filters* | Flow-Capacity Rating and Application of Suction-Line Filters and Filter Driers | ARI | ANSI/ARI 730-86 |
| | Grease Filters for Exhaust Ducts (1979) | UL | UL 1046 |
| | Grease Extractors for Exhaust Ducts (1981) | UL | UL 710 |
| *Fire Dampers* | Fire Dampers and Ceiling Dampers (1979) | UL | UL 555 |
| *Fire Protection* | Basic/National Fire Prevention Code, 6th ed. (1984, with 1986 amendments) | BOCA | BOCA |
| | National Fire Codes (11 Volumes, issued annually) | NFPA | NFPA |
| | Fire Prevention Code | NFPA | ANSI/NFPA 1-1987 |
| | Fire Protection Handbook, 16th ed. | NFPA | NFPA |
| | Flammable and Combustible Liquids Code | NFPA | ANSI/NFPA 30-1987 |
| | Heat Responsive Links for Fire Protection Service (1987) | UL | UL 33-1987 |
| | Test Method for Surface Burning Characteristics of Building Materials | ASTM/ NFPA | ASTM E 84-87 |
| | Fire Doors and Windows | NFPA | ANSI/NFPA 80-1986 |
| | Life Safety | NFPA | ANSI/NFPA 101-1985 |
| | Standard Fire Prevention Code (1985) with 1986 revisions | SBCCI | SBCCI |
| | Standard Method of Fire Tests of Door Assemblies | NFPA | ANSI/NFPA 252-1984 |
| | Uniform Fire Code (1985) | ICBO/ WFCA | ICBO/WFCA |
| | Uniform Fire Code Standards (1985) | ICBO/ WFCA | ICBO/WFCA |
| *Fireplace Stoves* | Fireplace Stoves (1982) | UL | UL 737 |
| | Factory-Built Fireplaces (1981) | UL | UL 127 |
| *Flow Capacity* | Method of Testing Flow Capacity of Suction Line Filters and Filter Driers | ASHRAE | ASHRAE 78-1985 |
| *Freezers* | | | |
| Household | Household Refrigerators and Freezers (1983) | UL | UL 250 |
| | Household Refrigerators, Combination Refrigerator-Freezers, and Household Freezers | AHAM | ANSI/AHAM HRF 1-1979 |
| | Capacity Measurement and Energy Consumption Test Methods for Household Refrigerators and Combination Refrigerator-Freezers | CSA | C300 M1984 |
| | Energy Consumption, Freezing Capability, and Capacity Measurement Test Methods for Household Freezers | CSA | C359 M1979 |
| Commercial | Ice Cream Makers (1986) | UL | UL 621 |
| | Soda Fountain and Luncheonette Equipment | NSF | NSF-1 |
| | Dispensing Freezers | NSF | NSF-6 |
| | Commercial Refrigerators and Freezers (1985) | UL | UL 471 |
| | Food Service Refrigerators and Storage Freezers | NSF | NSF-7 |
| *Furnaces* | Gas-Fired Gravity and Fan Type Direct Vent Wall Furnaces (with 1985 and 1987 addenda) | AGA | ANSI Z21.44-1985 |
| | Gas-Fired Central Furnaces (except Direct Vent Central Furnaces) (with 1985 and 1986 addenda) | AGA | ANSI Z21.47-1983 |
| | Direct Vent Central Furnaces (with 1986 and 1988 addenda) | AGA | ANSI Z21.64-1985 |
| | Gas-Fired Gravity and Fan Type Floor Furnaces (with 1987 addenda) | AGA | ANSI Z21.48-1986 |

**Table 1 Codes and Standards Published by Various Societies and Associations**

| Subject | Title | Publisher | Reference |
|---|---|---|---|
| | Solid-Fuel and Combination-Fuel Central and Supplementary Furances (1981) | UL | UL 391 |
| | Gas-Fired Gravity and Fan Type Vented Wall Furnaces (with 1987 addenda) | AGA | ANSI Z21.49-1986 |
| | Methods of Testing for Rating Non-Residential Warm Air Heaters | ASHRAE | ASHRAE 45-1986 |
| | Methods of Testing for Heating Seasonal Efficiency of Central Furnaces and Boilers | ASHRAE | ANSI/ASHRAE 103-1982 |
| | Installation of Oil Burning Equipment | NFPA | NFPA 31-1983 |
| | Oil-Fired Central Furnaces (1986) | UL | UL 727 |
| | Gas-Fired Duct Furnaces | AGA | ANSI Z83.9-1986 |
| | Oil-Fired Floor Furnaces (1976) | UL | UL 296.4 |
| | Oil-Fired Wall Furnaces (1987) | UL | UL 730 |
| | Standard Gas Code (1985) with 1986 revisions | SBCCI | SBCCI |
| | Oil Burning Stoves and Water Heaters (2a) | CSA | B140.3-1962 (R1980) |
| | Oil-Fired Warm Air Furnaces (8a) | CSA | B140.4-1974 |
| | Installation Code for Solid-Fuel-Burning Appliances and Equipment | CAN/ CSA | 3-B365-M87 |
| | Solid Fuel-Fired Appliances for Residential Use | CAN/ CSA | 3-B366.1-M87 |
| | Residential Gas Detectors (1983) | UL | UL 1484 |
| | Electric Central Warm-Air Furnaces | CSA | C22.2 No. 23-1980 |
| *Heat Exchangers* | Standard Methods of Test for Rating the Performance of Heat Recovery Ventilators | CSA | C439-M1985 |
| | Standards of Tubular Exchanger Manufacturers Association, 6th ed. (with 1982 addenda) | TEMA | TEMA |
| | Sample Problem Book Supplement (1980) | TEMA | TEMA |
| | Liquid Suction Heat Exchangers | ARI | ANSI/ARI 490-79 |
| | Method of Testing Air-to-Air Heat Exchangers | ASHRAE | ASHRAE 84-78 |
| *Heat Pumps* | Heat Pumps (1985) | UL | UL 559 |
| | Air-Source Unitary Heat Pump Equipment | ARI | ARI 240-81 |
| | Water-Source Heat Pumps | ARI | ARI 320-86 |
| | Ground Water-Source Heat Pumps | ARI | ARI 325-85 |
| | Commercial and Industrial Heat Pump Equipment | ARI | ARI 340-86 |
| | Central Forced-Air Unitary Heat Pumps with or without Electric Resistance Heat | CSA | C22.2 No. 186.1 M1980 |
| | Add-on Heat Pumps | CSA | C22.2 No. 186.2-M1980 |
| | Performance Standard for Unitary Heat Pumps (1a) | CSA | C273.3-M1977 |
| | Installation Requirements for Air-to-Air Heat Pumps (2a) | CSA | C273.5-1980 |
| *Heat Recovery* | Energy Recovery Equipment and Systems, Air-to-Air (1978) | SMACNA | SMACNA |
| | Gas Turbine Heat Recovery Steam Generators | ASME | ANSI/ASME PTC 4.4-1987 |
| *Heaters* | Infrared Application Manual | NEMA | NEMA HE 3-1983 |
| | Air Heaters | ASME | ANSI/ASME PTC 4.3-1985 |
| | Desuperheater/Water Heaters | ARI | ANSI/ARI 470-80 |
| | Electric Heaters for Use in Hazardous Locations, Class I, Groups A, B, C, and D, and Class II, Groups E, F, and G (1985) | UL | UL 823 |
| | Standards for Closed Feed Water Heaters, 4th ed. (1984) | HEI | HEI |
| | Oil-Fired Air Heaters and Direct-Fired Heaters (1975) | UL | UL 733 |
| | Oil-Fired Room Heaters (1973) | UL | UL 896 |
| | Solid Fuel-Type Room Heaters (1983) | UL | UL 1482 |
| | Gas-Fired Room Heaters, Vol. I, Vented Room Heaters (with 1985 and 1987 addenda) | AGA | ANSI Z21.11.1-1983 |
| | Gas-Fired Room Heaters, Vol. II, Unvented Room Heaters (with 1984 addenda) | AGA | ANSI Z21.11.2-1983 |
| | Gas-Fired Infrared Heaters (with 1984 and 1985 addenda) | AGA | ANSI Z83.6-1982 |
| | Gas-Fired Construction Heaters (with 1981 and 1984 addenda) | AGA | ANSI Z83.7-1974 |
| | Direct Gas-Fired Make-Up Air Heaters (with 1986 addenda) | AGA | ANSI Z83.4-1985 |
| | Gas-Fired Unvented Commercial and Industrial Heaters (with 1984 addenda) | AGA | ANSI Z83.16-1982 |
| | Gas-Fired Pool and Spa Heaters (with 1987 addenda) | AGA | ANSI Z21.56-1986 |
| | Motor Vehicle Heater Test Procedure (1982) | SAE | SAE J638 June 1982 |
| | Electric Heating Appliances (1987) | UL | UL 499 |
| | Gas-Fired Heating Equipment, Commercial-Industrial (1973) | UL | UL 795 |
| | Fuel-Fired Heaters—Air Heating—for Construction and Industrial Machinery (1980) | SAE | SAE J1024 April 1980 |
| | Electric Air Heaters (1980) | UL | UL 1025 |

Table 1 Codes and Standards Published by Various Societies and Associations

| Subject | Title | Publisher | Reference |
|---|---|---|---|
| | Space Heaters for Use with Solid Fuels | CSA | B366.2 M1984 |
| | Unvented Kerosene-Fired Room Heaters and Portable Heaters (1982) | UL | UL 647 |
| *Heating* | Aircraft Electrical Heating Systems (1965) (reaffirmed 1983) | SAE | SAE AIR 860 |
| | Environmental System Technology (1984) | NEBB | NEBB |
| | Installation and Operation of Pulverized Fuel Systems | NFPA | ANSI/NFPA 85F-1982 |
| | Manual for Calculating Heat Loss and Heat Gain for Electric Comfort Conditioning | NEMA | NEMA HE 1-1980 (R1986) |
| | Heat Loss Calculation Guide (1984) | HYD I | IBR H-21 |
| | Installation Guide for Residential Hydronic Heating Systems, 6th ed. (1986) | HYD I | IBR 200 |
| | Advanced Installation Guide for Hydronic Heating Systems, 2nd ed. | HYD I | IBR 250 |
| | Installation Standards for Heating and Air-Conditioning Systems (1984) | SMACNA | SMACNA |
| | HVAC Systems—Applications, First ed. (1986) | SMACNA | SMACNA |
| | Electric Baseboard Heating Equipment (1986) | UL | UL 1042 |
| | Electric Central Air Heating Equipment (1986) | UL | UL 1096 |
| | Portable Industrial Oil-Fired Heaters | CSA | B140.8-1967 (R1980) |
| | Portable Kerosene-Fired Heaters | CAN | 3-B140.9.3 M86 |
| | Oil-Fired Service Water Heaters and Swimming Pool Heaters (7a) | CSA | B140.12-1976 |
| | Automatic Flue-Pipe Dampers for Use with Oil-Fired Appliances | CSA | B140.14-M1979 |
| | Residential Electric Heating | CSA | C273.1-M1980 |
| | Performance Standard for Electrical Baseboard Heaters | CSA | C273.2-1971 |
| | Performance Requirements for Electric Heating Line-Voltage Wall Thermostats | CSA | C273.4-M1978 |
| | Electric Air Heaters | CSA | C22.2 No. 46-1981 |
| | Heater Elements | CSA | C22.2 No. 72-M1984 |
| *Humidifiers* | Humidifiers (1987) | UL | UL 998 |
| | Central System Humidifiers | ARI | ARI 610-82 |
| | Self-Contained Humidifiers | ARI | ARI 620-80 |
| | Selection, Installation, and Servicing of Humidifiers | ARI | ARI 630-82 |
| | Appliance Humidifiers | AHAM | ANSI/AHAM HU 1-1987 |
| *Ice Makers* | Ice Makers (1985) | UL | UL 563 |
| | Methods of Testing Automatic Ice Makers | ASHRAE | ASHRAE 29-78 |
| | Automatic Commercial Ice Makers | ARI | ANSI/ARI 810-79 |
| | Split System Automatic Commercial Ice Makers | ARI | ANSI/ARI 815-79 |
| | Ice Storage Bins | ARI | ANSI/ARI 820-79 |
| | Automatic Ice-Making Equipment | NSF | NSF-12 |
| *Incinerators* | Residential Incinerators (1973) | UL | UL 791 |
| | Incinerators, Waste and Linen Handling Systems | NFPA | ANSI/NFPA 82-1983 |
| | Incinerator Performance | CSA | Z103-1976 |
| *Induction Units* | Room Air-Induction Units | ARI | ARI 445-81 |
| *Industrial Duct* | Round Industrial Duct Construction (1977) | SMACNA | SMACNA |
| | Rectangular Industrial Duct Construction (1980) | SMACNA | SMACNA |
| *Infrared Sensing Devices* | Application of Infrared Sensing Devices to the Assessment of Building Heat Loss Characteristics | ASHRAE | ANSI/ASHRAE 101-1981 |
| *Insulation* | Commercial and Industrial Insulation Standards | MICA | MICA 1983 |
| | Test Method for Steady-State Heat Flux Measurements and Thermal Performance of Building Assemblies by Means of a Guarded Hot Box | ASTM | ASTM C236-87 |
| | Test Method for Steady-State Thermal Transmission Properties by Means of the Guarded Hot Plate Apparatus | ASTM | ASTM C177-85 |
| | Test Method for Steady-State Heat Transfer Properties of Horizontal Pipe Insulations | ASTM | ASTM C335-84 |
| | Test Method for Steady-State Heat Flux Measurements and Thermal Transmission Properties by Means of the Heat Flow Meter Apparatus | ASTM | ASTM C518-85 |
| | Thermal and Acoustical Insutlation (Mineral Fiber, Duct Lining Material) | ASTM | ASTM C1071-86 |
| | Thermal Insulation, Mineral Fibre, for Buildings | CSA | A101 M-1983 |
| *Louvers* | Test Method for Louvers, Dampers, and Shutters | AMCA | AMCA 500-83 |

**Table 1 Codes and Standards Published by Various Societies and Associations**

| Subject | Title | Publisher | Reference |
|---|---|---|---|
| *Lubricants* | Test Method for Carbon-Type Composition of Insulating Oils of Petroleum Origin | ASTM | ASTM D2140-86 |
| | Method for Conversion of Kinematic Viscosity to Saybolt Universal Viscosity or to Saybolt Furol Viscosity | ASTM | ASTM D2161-87 |
| | Method for Calculating Viscosity Index from Kinematic Viscosity at 40 and 100°C | ASTM | ASTM D2270-86 |
| | Method for Estimation of Molecular Weight of Petroleum Oils from Viscosity Measurements | ASTM | ASTM D2502-82 |
| | Test Method for Molecular Weight of Hydrocarbons by Thermoelectric Measurement of Vapor Pressure | ASTM | ASTM D2503-82 |
| | Test Method for Mean Molecular Weight of Mineral Insulating Oils by the Cryoscopic Method | ASTM | ASTM D2224-78 (1983) |
| | Test Methods for Pour Point of Petroleum Oils | ASTM | ASTM D97-87 |
| | Recommended Practice for Viscosity System for Industrial Fluid Lubricants | ASTM | ASTM D2422-86 |
| | Test Method for Dielectric Breakdown Voltage of Insulating Liquids Using Disk Electrodes | ASTM | ASTM D877-87 |
| | Test Method for Dielectric Breakdown Voltage of Insulating Oils of Petroleum Origin Using VDE Electrodes | ASTM | ASTM D1816-84a |
| | Method for Separation of Representative Aromatics and Nonaromatics Fractions of High-Boiling Oils by Elution Chromatography | ASTM | ASTM D2549-85 |
| | Method of Testing for Floc Point of Refrigeration Grade Oils | ASHRAE | ANSI/ASHRAE 86-1983 |
| *Measurements* | Standard Measurements Guide: Section on Temperature Measurements | ASHRAE | ASHRAE 41.1-86 |
| | Standard Measurement Guide: Measurement of Proportion of Oil in Liquid Refrigerants | ASHRAE | ANSI/ASHRAE 41.4-1984 |
| | Standard Measurement Guide: Engineering Analysis of Experimental Data | ASHRAE | ASHRAE 41.5-75 |
| | Standard Method for Measurement of Moist Air Properties | ASHRAE | ANSI/ASHRAE 41.6-1982 |
| | Standard Method for Measurement of Flow of Gas | ASHRAE | ASHRAE 41.7-1984 |
| | Standard Methods of Measurement of Flow of Fluid-Liquids | ASHRAE | ASHRAE 41.8-78 |
| | Standard Methods of Measuring and Expressing Building Energy Performance | ASHRAE | ASHRAE 105-1984 |
| | Guide for Dynamic Calibration of Pressure Transducers | ASME | ANSI B88.1-1978 |
| | Procedure for Bench Calibration of Tank Level Gaging Tapes and Sounding Rules | ASME | ANSI B88.2-1981 |
| | Guide for Dynamic Calibration of Pressure Transducers | ASME | ANSI MC88-1-1978 |
| | Glossary of Terms used in the Measurement of Fluid Flow in Pipes | ASME | ASME/ANSI MFC-1M-1986 |
| | Measurement Uncertainty for Fluid Flow in Closed Conduits | ASME | ANSI/ASME MFC-2M-1983 |
| | Measurement of Fluid Flow in Pipes Using Orifice, Nozzle, and Venturi | ASME | ANSI/ASME MFC-3M-1985 |
| | Measurement of Gas Flow by Turbine Meters | ASME | ASME/ANSI MFC-4M-1986 |
| | Measurement of Liquid Flow in Closed Conduits Using Transit-Time Ultrasonic Flowmeters | ASME | ANSI/ASME MFC-5M-1985 |
| | Measurement of Fluid Flow in Pipes Using Vortex Flow Meters | ASME | ASME/ANSI MFC-6M-1987 |
| | Measurement of Gas Flow by Means of Critical Flow Venturi Nozzles | ASME | ASME/ANSI MFC-7M-1987 |
| | Measurement Uncertainty | ASME | ANSI/ASME PTC 19.1-1985 |
| | Pressure Measurement | ASME | ANSI PTC 19.2-1964 |
| | Temperature Measurement | ANSI | ANSI/ASME PTC 19.3-1985 |
| | Measurement of Rotary Speed | ASME | ANSI PTC 19.13-1961 |
| | Measurement of Industrial Sound | ASME | ANSI/ASME PTC 36-1985 |
| *Mobile Homes and Recreational Vehicles* | Plumbing System Components for Manufactured Homes and Recreational Vehicles | NSF | NSF-24 |
| | Recreational Vehicle Parks | NFPA | NFPA 501C-1987 |
| | Shear Resistance Tests for Ceiling Boards for Mobile Homes (1980) | UL | UL 1296 |
| | Roof Trusses for Mobile Homes (1979) | UL | UL 1298 |
| | Gas Burning Heating Appliances for Mobile Homes and Recreational Vehicles (1965) | UL | UL 307B |
| | Gas Supply Connectors for Mobile Homes | IAPMO | IAPMO TSC 9-1985 |

Table 1 Codes and Standards Published by Various Societies and Associations

| Subject | Title | Publisher | Reference |
|---|---|---|---|
| | Liquid-Fuel-Burning Heating Appliances for Mobile Homes and Recreational Vehicles (1978) | UL | UL 307A |
| | Recreational Vehicle Cooking Gas Appliances (with 1982 and 1984 addenda) | AGA | ANSI Z21.57-1982 |
| | Gas-Fired Cooking Appliances for Recreational Vehicles (1976) | UL | UL 1075 |
| | Oil-Fired Warm-Air Heating Appliances for Mobile Housing and Recreational Vehicles (2a) | CSA | B140.10-1974 (R1981) |
| | Definitions and General Requirements for Mobile Homes | CSA | Z240.0.1-86 |
| | Definitions and General Safety Requirements for Recreational Vehicles | CSA | Z240.0.2-M86 |
| | Vehicular Requirements for Mobile Homes | CSA | Z240.1.1-M86 |
| | Vehicular Requirements for Recreational Vehicles | CSA | Z240.1.2-M86 |
| | Structural Requirements for Mobile Homes | CSA | Z240.2.1-M86 |
| | Plumbing Requirements for Mobile Homes | CSA | Z240.3.1-M86 |
| | Plumbing Requirements for Recreational Vehicles | CSA | Z240.3.2-M86 |
| | Installation Requirements for Gas Burning Appliances in Mobile Homes | CSA | Z240.4.1-M86 |
| | Installation Requirements for Propane Appliances and Equipment in Recreational Vehicles | CSA | Z240.4.2-M86 |
| | Oil Installation Requirements for Mobile Homes | CSA | Z240.5.1-M86 |
| | Electrical Requirements for Recreational Vehicles | CSA | Z240.6.2-M86 |
| | Mobile Home Parks | CSA | Z240.7.1-1972 |
| | Recreational Vehicle Parks | CSA | Z240.7.2-1972 |
| | Light Duty Windows | CSA | Z240.8.1-1979 |
| | Requirements for Load Calculations and Duct Design for Heating and Cooling of Mobile Homes | CSA | Z240.9.1-1979 |
| | Roof Jacks for Mobile Homes and Recreational Vehicles (1986) | UL | UL 311 |
| *Motors and Generators* | Motors and Generators | NEMA | NEMA MG 1-1987 |
| | Impedance Protected Motors (1982) | UL | UL 519 |
| | Electric Motors (1984) | UL | UL 1004 |
| | Energy Efficiency Test Methods for Three-Phase Induction Motors/ Efficiency Quoting Method and Permissible Efficiency Tolerance | CSA | C390-1985 |
| | Steam Generating Units | ASME | ANSI/ASME PTC4.1-1985 |
| | Electric Motors and Generators for Use in Hazardous (Classified) Locations (1978) | UL | UL 674 |
| *Outlets and Inlets* | Method of Testing for Rating the Air Flow Performance of Outlets and Inlets | ASHRAE | ASHRAE 70-72 |
| *Pipe, Tubing, and Fittings* | Power Piping | ASME | ANSI/ASME B31.1-1986 |
| | Plastics Piping Components and Related Materials | NSF | NSF-14 |
| | Scheme for the Identification of Piping Systems | ASME | ANSI/ASME A13.1-1985 |
| | Fuel Gas Piping | ASME | ANSI B31.2-1968 |
| | National Fuel Gas Code (with 1987 addenda) | NFPA | ANSI/NFPA 54-1984 |
| | | AGA | ANSI Z223.1-1984 |
| | Refrigeration Piping | ASME | ANSI B31.5-1983 |
| | Refrigeration Tube Fittings (1977) | SAE | ANSI/SAE J513F |
| | Specification for Seamless Copper Pipe, Standard Sizes | ASTM | ASTM B42-87 |
| | Specification for Acrylonitrile-Butadiene-Styrene (ABS) Plastic Pipe, Schedules 40 and 80 | ASTM | ASTM D1527-77 (1982) |
| | Specification for Poly (Vinyl Chloride) (PVC) Plastic Pipe, Schedules 40, 80, and 120 | ASTM | ASTM D1785-86 |
| | Specification for Polyethylene (PE) Plastic Pipe, Schedule 40 | ASTM | ASTM D2104-85 |
| | Standards of the Expansion Joint Manufacturers Association, Inc., 5th ed. (1980 with 1985 addenda) | EJMA | EJMA |
| | Rubber Gasketed Fittings for Fire Protection Service (1978) | UL | UL 213 |
| | Tube Fittings for Flammable and Combustible Fluids, Refrigeration Service and Marine Use (1978) | UL | UL 109 |
| *Plumbing* | BOCA/National Plumbing Code, 7th ed. (1987) | BOCA | BOCA |
| | Standard Plumbing Code (1988) | SBCCI | SBCCI |
| | Uniform Plumbing Code (1985) | IAPMO | IAPMO |
| *Pumps* | Circulation System Components for Swimming Pools, Spas, or Hot Tubs | NSF | NSF-50 |
| | Pumps for Oil-Burning Appliances (1986) | UL | UL 343 |
| | Motor-Operated Water Pumps (1980) | UL | UL 778 |

Table 1 Codes and Standards Published by Various Societies and Associations

| Subject | Title | Publisher | Reference |
|---|---|---|---|
| | Electric Swimming Pool Pumps, Filters and Chlorinators (1986) | UL | UL 1081 |
| | Hydraulic Institute Standards, 14th ed. (1983) | HI | HI |
| | Hydraulic Institute Engineering Data Book, First ed. (1979) | HI | HI |
| | Performance Standard for Liquid Ring Vacuum Pumps, 1st ed. (1987) | HEI | HEI |
| | Centrifugal Pumps | ASME | ANSI PTC 8.2-1965 |
| | Displacement Compressors, Vacuum Pumps and Blowers | ASME | ANSI/ASME PTC 9-1985 |
| *Radiation* | Testing and Rating Code for Baseboard Radiation, 6th ed. (1981) | HYD I | IBR |
| | Testing and Rating Code for Finned-Tube Commercial Radiation (1966) | HYD I | IBR |
| | Ratings for Baseboard and Fin-tube Radiation | HYD I | IBR |
| *Receivers* | Refrigerant Liquid Receivers | ARI | ANSI/ARI 495-85 |
| *Refrigerant Containing Components* | Refrigerant-Containing Components and Acessories, Non-electrical (1986) | UL | UL 207 |
| *Refrigerants* | Number Designation of Refrigerants | ASHRAE | ANSI/ASHRAE 34-78 |
| | Refrigeration Oil Description | ASHRAE | ANSI/ASHRAE 99-1981 |
| | Methods of Testing Discharge Line Refrigerant-Oil Separators | ASHRAE | ASHRAE 69-71 |
| | Sealed Glass Tube Method to Test the Chemical Stability of Material for Use Within Refrigerant Systems | ASHRAE | ANSI/ASHRAE 97-1983 |
| *Refrigeration* | Safety Code for Mechanical Refrigeration | ASHRAE | ANSI/ASHRAE 15-1978 |
| | Capacity Measurement of Field Erected Compression Type Refrigeration and Air-Conditioning Systems | ASHRAE | ASHRAE 83-1985 |
| | Refrigerated Medical Equipment (1978) | UL | UL 416 |
| | Commercial Refrigerated Equipment | CSA | C22.2 No. 120-1974 (R 1981) |
| | Equipment, Design and Installation of Ammonia Mechanical Refrigeration Systems | ANSI | ANSI/IIAR 2-1984 |
| Steam Jet | Standards for Steam Jet Ejectors, 3rd ed. (1967, 3rd printing, 1980) | HEI | HEI |
| | Ejectors | ASME | ASME PTC 24-76 |
| Transport | Safety Practices for Mechanical Vapor Compression Refrigeration Equipment or Systems Used to Cool Passenger Compartment of Motor Vehicles (1981) | SAE | SAE J639 Oct. 1981 |
| | Mechanical Refrigeration Installations on Shipboard | ASHRAE | ANSI/ASHRAE 26-1985 |
| | Mechanical Transport Refrigeration Units | ARI | ARI 1110-83 |
| | General Requirements for Application of Vapor Cycle Refrigeration Systems for Aircraft (1973) (reaffirmed 1983) | SAE | SAE ARP 731A |
| *Refrigerators* | Methods of Testing Open Refrigerators for Food Stores | ASHRAE | ANSI/ASHRAE 72-1983 |
| | Methods of Testing Self-Service Closed Refrigerators for Food Stores | ASHRAE | ASHRAE 117-1986 |
| Commercial | Commercial Refrigerators and Freezers (1985) | UL | UL 471 |
| | Food Service Refrigerators and Storage Freezers | NSF | NSF 7 |
| | Food Carts | NSF | NSF 59 |
| | Refrigerating Units (1976) | UL | UL 427 |
| | Refrigeration Unit Coolers (1984) | UL | UL 412 |
| | Soda Fountain and Luncheonette Equipment | NSF | NSF 1 |
| | Food Service Equipment | NSF | NSF-2 |
| Household | Refrigerators Using Gas Fuel (with 1984 addenda) | AGA | ANSI Z21.19-1983 |
| | Household Refrigerators, Combination Refrigerator Freezers and Household Freezers | AHAM | AHAM HRF 1-1987 |
| | Household Refrigerators and Freezers (1984) | UL | UL 250 |
| *Roof Ventilators* | Power Ventilators (1984) | UL | UL 705 |
| *Solar Equipment* | Method of Testing to Determine the Thermal Performance of Solar Collectors | ASHRAE | ANSI/ASHRAE 93-1986 |
| | Methods of Testing to Determine the Thermal Performance of Domestic Water Heating Systems | ASHRAE | ASHRAE 95-1981 |
| | Methods of Testing to Determine the Thermal Performance of Unglazed Flat-Plate Liquid-Type Solar Collectors | ASHRAE | ANSI/ASHRAE 96-1980 |
| | Method of Measuring Solar-Optical Properties of Materials | ASHRAE | ASHRAE 74-73 |
| *Solenoid Valves* | Solenoid Valves for Liquid Flow Use with Volatile Refrigerants and Water | ARI | ARI 760-80 |
| *Sound Measurement* | Measurement of Sound from Boiler Units, Bottom-Supported Shop or Field Erected, 3rd ed. | ABMA | ABMA-1973 |
| | Sound Rating of Outdoor Unitary Equipment | ARI | ARI 270-84 |

**Table 1 Codes and Standards Published by Various Societies and Associations**

| Subject | Title | Publisher | Reference |
|---|---|---|---|
| | Sound Rating of Non-Ducted Indoor Air-Conditioning Equipment | ARI | ARI 350-86 |
| | Sound Rating of Large Outdoor Refrigerating and Air-Conditioning Equipment | ARI | ARI 370-86 |
| | Procedural Standards for Measuring Sound and Vibration | NEBB | NEBB-1977 |
| | Sound and Vibration in Environmental Systems | NEBB | NEBB-1977 |
| | Application of Sound Rated Outdoor Unitary Equipment | ARI | ARI 275-84 |
| | Method of Measuring Machinery Sound within Equipment Rooms | ARI | ARI 575-79 |
| | Method of Testing In-Duct Sound Power Measurement Procedure for Fans | ASHRAE | ASHRAE 68-1986 |
| | Specification for Sound Level Meters (reaffirmed 1986) | ASA | ANSI S1.4-1983<br>ANSI S1.4A-1985 |
| | Method for the Calibration of Microphones (reaffirmed 1986) | ASA | ANSI S1.10-1966 (R1986) |
| | Reverberant Room Method for Sound Testing of Fans | AMCA | AMCA 300-85 |
| | Sound Rating of Room Air Conditioners | AHAM | AHAM RAC-2SR |
| | Guidelines for the Use of Sound Power Standards and for the Preparation of Noise Test Codes (reaffirmed 1985) | ASA | ASA 10<br>ANSI S1.30-1979 (R1985) |
| | Measurement of Industrial Sound | ASME | ASME/ANSI PTC 36-1985 |
| *Space Heaters* | Electric Air Heaters (1980) | UL | UL 1025 |
| *Symbols* | Graphic Electrical Symbols for Air-Conditioning and Refrigeration Equipment | ARI | ARI 130-82 |
| | Graphic Symbols for Electrical and Electronic Diagrams | ASME | ANSI Y32.2-1975 |
| | Graphic Symbols for Plumbing Fixtures for Diagrams used in Architecture Building Construction | ASME | ANSI Y32.4-1977 |
| | Symbols for Mechanical and Acoustical Elements as used in Schematic Diagrams | ASME | ANSI/ASME Y32.18-1985 |
| | Graphic Symbols for Pipe Fittings, Valves and Piping | ASME | ANSI Z32.2.3-1953 |
| | Graphic Symbols for Heating, Ventilating, and Air Conditioning | ASME | ANSI/ASME Z32.2.4-1984 |
| *Testing and Balancing* | Procedural Standards for Testing, Adjusting, Balancing of Environmental Systems, 4th ed. (1983) | NEBB | NEBB-1983 |
| | HVAC Systems—Testing, Adjusting and Balancing (1983) | SMACNA | SMACNA |
| *Terminals, Wiring* | Equipment Wiring Terminals for Use with Aluminum and/or Copper Conductors | UL | UL 486E |
| *Thermal Storage* | Methods of Testing Active Latent Heat Storage Devices Based on Thermal Performance | ASHRAE | ANSI/ASHRAE 94.1-1985 |
| | Methods of Testing Thermal Storage Devices with Electric Input and Thermal Output Based on Thermal Performance | ASHRAE | ANSI/ASHRAE 94.2-1981 |
| | Metering and Testing Active Sensible Thermal Energy Storage Devices Based on Thermal Performance | ASHRAE | ANSI/ASHRAE 94.3-1986 |
| *Turbines* | Steam Turbines for Mechanical Drive Service | NEMA | NEMA SM 23-1985 |
| *Unit Heaters* | Oil-Fired Unit Heaters (1975) | UL | UL 731 |
| | Gas Unit Heaters (with 1986 addenda) | AGA | ANSI Z83.8-1985 |
| *Valves* | Methods of Testing Nonelectric, Nonpneumatic Thermostatic Radiator Valves | ASHRAE | ASHRAE 102-1983 |
| | Automatic Gas Valves for Gas Appliances (with 1977 and 1981 addenda) | AGA | ANSI Z21.21-1974 |
| | Manually Operated Gas Valves (with 1981 and 1984 addenda) | AGA | ANSI Z21.15-1979 |
| | Relief Valves and Automatic Gas Shutoff Devices for Hot Water Supply Systems | AGA | ANSI Z21.22-1986 |
| | Refrigerant Access Valves and Hose Connectors | ARI | ARI 720-81 |
| | Refrigerant Pressure Regulating Valves | ARI | ARI 770-84 |
| | Solenoid Valves for Use with Volatile Refrigerants | ARI | ARI 760-80 |
| | Face-to-Face and End-to-End Dimensions of Ferrous Valves | ASME | ASME B16.10-86 |
| | Manually Operated Metallic Gas Valves for Use in Gas Piping Systems up to 125 psig (Sizes 1/2 through 2) | ASME | ASME B16.33-81 |
| | Valves—Flanged and Buttwelding End | ASME | ASME B16.34-81 |
| | Hydrostatic Testing of Control Valves | ASME | ASME B16.37-80 |
| | Large Metallic Valves for Gas Distribution (Manually Operated, NPS-2 1/2 to 12, 125 psig Maximum) | ASME | ASME B16.38-85 |
| | Manually Operated Thermoplastic Gas Shutoffs and Valves in Gas Distribution Systems | ASME | ASME B16.40-85 |
| | Safety and Relief Valves | ASME | ASME PTC 25.3-76 |

**Table 1 Codes and Standards Published by Various Societies and Associations**

| Subject | Title | Publisher | Reference |
|---|---|---|---|
| | Electrically Operated Valves | UL | UL 429 |
| | Valves for Anhydrous Ammonia and LP-Gas (Other than Safety Relief) | UL | UL 125 |
| | Safety Relief Valves for Anhydrous Ammonia and LP-Gas | UL | UL 132 |
| | Pressure Regulating Valves for LP-Gas | UL | UL 144 |
| *Vending Machines* | Refrigerated Vending Machines (1979) | UL | UL 541 |
| | Methods of Testing Pre-Mix and Post-Mix Soft Drink Vending and Dispensing Equipment | ASHRAE | ASHRAE 91-76 |
| | Sanitation Ordinance and Code for Vending of Foods and Beverages (1965) | USDA | USDA 546 |
| | Vending Machines for Food and Beverages | NSF | NSF-25 |
| *Vent Dampers* | Electrically Operated Automatic Vent Damper Devices for Use with Gas-Fired Appliances | AGA | ANSI Z21.66-1985 |
| | Mechanically Actuated Automatic Vent Damper Devices for Use with Gas-Fired Appliances | AGA | ANSI Z21.67-1985 |
| | Thermally Actuated Automatic Vent Damper Devices for Use with Gas-Fired Appliances | AGA | ANSI Z21.68-1985 |
| | Vent or Chimney Connector Dampers for Oil-Fired Appliances (1983) | UL | UL 17 |
| *Venting* | Explosion Prevention Systems | NFPA | ANSI/NFPA 69-86 |
| | Chimneys, Fireplaces, Vents and Solid Fuel Burning Appliances | NFPA | ANSI/NFPA 211-84 |
| | Type L Low-Temperature Venting Systems (1986) | UL | UL 641 |
| | Draft Hoods (with 1983 addenda) | AGA | ANSI Z21.12-1981 |
| | Draft Equipment (1973) | UL | UL 378 |
| | Gas Vents (1986) | UL | UL 441 |
| | National Fuel Gas Code (with 1987 addenda) | AGA | ANSI Z223.1-1984 |
| | Guide for Steel Stack Design and Construction (1983) | SMACNA | SMACNA |
| *Ventilation* | Vapor Removal from Cooking Equipment | NFPA | ANSI/NFPA 96-84 |
| | Parking Structures; Repair Garages | NFPA | ANSI/NFPA 88A; 88B-85 |
| | Ventilation for Acceptable Indoor Air Quality | ASHRAE | ASHRAE 62-1981 |
| | Industrial Ventilation | ACGIH | ACHIH |
| *Water Heaters* | Gas Water Heaters, Vol. I, Storage Water Heaters with Inputs Ratings of 75,000 Btu per Hour or Less (with 1985 and 1986 addenda) | AGA | ANSI Z21.10.1-1984 |
| | Gas Water Heaters, Vol. III, Storage, with Inputs Ratings Above 75,000 Btu per Hour, Circulating and Instantaneous Water Heaters (with 1985 and 1986 addenda) | AGA | ANSI Z21.10.3-1984 |
| | Household Electric Storage Tank Water Heaters (1983) | UL | UL 174 |
| | Oil-Fired Storage Tank Water Heaters (1975) | UL | UL 732 |
| | Electric Booster and Commercial Storage Tank Water Heaters (1982) | UL | UL 1453 |
| | Hot Water Generating and Heat Recovery Equipment | NSF | NSF-5 |
| | Construction and Test of Electric Storage Tank Water Heaters | CSA | C22.2 No. 110 M-1981 |
| | Oil Burning Stoves and Water Heaters | CSA | B140.3-1962 (R1980) |
| | Oil-Fired Service Water Heaters and Swimming Pool Heaters | CSA | B140.12-1976 |
| | CSA Standards on Performance of Electric Storage Tank Water Heaters | CSA | C191-series-M1983 |
| | Methods of Testing to Determine the Thermal Performance of Solar Domestic Water Heating Systems | ASHRAE | ANSI/ASHRAE 95-1981 |
| *Woodburning Appliances* | Installation Code for Solid Fuel Burning Appliances and Equipment | CSA | CAN/CSA3-B365-M87 |
| | Method of Testing for Performance Rating of Woodburning Appliances | ASHRAE | ANSI/ASHRAE 106-1984 |
| | Solid-Fuel-Fired Central Heating Appliances | CSA | CAN/CSA3-B366.1-M87 |
| | Space Heaters for Use with Solid Fuels | CSA | B366.2-M1984 |
| | Solid Fuel Type Room Heaters (1983) | UL | UL 1482 |

## ABBREVIATION AND ADDRESSES

**The Codes and Standards Listed in Table 1 Can Be Obtained from the Organizations Listed in the *Publisher* Column.**

| | |
|---|---|
| ABMA | American Boiler Manufacturers Association, Ste. 160, 950 N. Glebe Rd., Arlington, VA 22203 |
| ACCA | Air Conditioning Contractors of America, 1228 17th St., NW, Washington, DC 20036 (formerly the National Environmental Systems Contractors Association) |
| ACGIH | American Conference of Governmental Industrial Hygienists, 6500 Glenway Ave., Bldg. D-7, Cincinnati, OH 45211 |
| ADC | Air Diffusion Council, 230 N. Michigan Ave., Suite 1200, Chicago, IL 60601 |
| AFS | American Foundrymen's Society, Golf and Wolf Rds., Des Plaines, IL 60016 |

**Table 1 Codes and Standards Published by Various Societies and Associations**

| Subject | Title | Publisher | Reference |
|---|---|---|---|
| AGA | American Gas Association, 1515 Wilson Blvd., Arlington, VA 22209 | | |
| AHAM | Association of Home Appliance Manufacturers, 20 N. Wacker Dr., Chicago, IL 60606 | | |
| AIA | American Insurance Association, 85 John St., New York, NY 10038 | | |
| AIHA | American Industrial Hygiene Association, 475 Wolf Ledges Pkwy., Akron, OH 44311-1087 | | |
| AMCA | Air Movement and Control Association, 30 W. University Dr., Arlington Heights, IL 60004 | | |
| ANSI | American National Standards Institute, 1430 Broadway, New York, NY 10018 | | |
| ARI | Air-Conditioning and Refrigeration Institute, 1501 Wilson Blvd., 6th Fl., Arlington, VA 22209 | | |
| ASA | Acoustical Society of America, 335 E. 45 St., New York, NY 10017 | | |
| ASHRAE | American Society of Heating, Refrigerating and Air-Conditioning Engineers, Inc. 1791 Tullie Circle, N.E., Atlanta, GA 30329 | | |
| ASME | The American Society of Mechanical Engineers, 345 E. 47 St., New York, NY 10017 | | |
| | For ordering publications: ASME Marketing Dept., Box 2350, Fairfield, NJ 07007-2350 | | |
| ASTM | American Society for Testing and Materials, 1916 Race St., Philadelphia, PA 19103 | | |
| BOCA | Building Officials and Code Administrators International, Inc., 4501 W. Flossmoor Rd., Country Club Hills, IL 60477-5795 | | |
| BSI | British Standards Institution, 2 Park St., London, W1A 2BS, England | | |
| CABO | Council of American Building Officials, 5203 Leesburg Pike, Ste. 708, Falls Church, VA 22041 | | |
| CAGI | Compressed Air and Gas Institute, Ste. 1230, Keith Bldg., 1621 Euclid Ave., Cleveland, OH 44115 | | |
| CSA | Canadian Standards Association, 178 Rexdale Blvd., Rexdale, Ont. M9W 1R3, Canada | | |
| CTI | Cooling Tower Institute, P.O. Box 73383, Houston, TX 77273 | | |
| EJMA | Expansion Joint Manufacturers Association, Inc., 25 N. Broadway, Tarrytown, NY 10591 | | |
| HEI | Heat Exchange Institute, Ste. 1230, Keith Bldg., 1621 Euclid Ave., Cleveland, OH 44115 | | |
| HI | Hydraulic Institute, 712 Lakewood Ctr. N., 14600 Detroit Ave., Cleveland, OH 44107 | | |
| HYD I | Hydronics Institute, 35 Russo Pl., Berkeley Heights, NJ 07922 | | |
| IAPMO | International Association of Plumbing and Mechanical Officials, 5032 Alhambra Ave., Los Angeles, CA 90032-3490 | | |
| IBR | Institute of Boiler and Radiator Manufacturers, superseded by Hydronics Institute | | |
| ICBO | International Conference of Building Officials, 5360 S. Workman Mill Rd., Whittier, CA 90601 | | |
| IIAR | International Institute of Ammonia Refrigeration, 111 East Wacker Drive, Chicago, IL 60601 | | |
| MCAA | Mechanical Contractors Association of America, 5530 Wisconsin Ave., Chevy Chase, MD 20815 | | |
| NCSBCS | National Conference of States on Building Codes and Standards, 481 Carlisle Dr., Herndon, VA 22070 | | |
| NEBB | National Environmental Balancing Bureau, 8224 Old Courthouse Rd., Vienna, VA 22180 | | |
| NEMA | National Electrical Manufacturers Association, 2101 L St., N.W., Ste. 300, Washington, DC 20037 | | |
| NFPA | National Fire Protection Association, Batterymarch Park, Quincy, MA 02269 | | |
| NSF | National Sanitation Foundation, Box 1468, Ann Arbor, MI 48106 | | |
| SAE | Society of Automotive Engineers, 400 Commonwealth Dr., Warrendale, PA 15096 | | |
| SBCCI | Southern Building Code Congress International, Inc., 900 Montclair Rd., Birmingham, AL 35213 | | |
| SMACNA | Sheet Metal and Air Conditioning Contractors' National Association, 8224 Old Courthouse Rd., Vienna, VA 22180 | | |
| TEMA | Tubular Exchanger Manufacturers Association, Inc., 25 N. Broadway, Tarrytown, NY 10591 | | |
| UL | Underwriters Laboratories Inc., 333 Pfingsten Rd., Northbrook, IL 60062 | | |
| WFCA | Western Fire Chiefs Association, Inc., 5360 S. Workman Mill Rd., Whittier, CA 90601 | | |

# ERRATA

## 1988 ASHRAE HANDBOOK

The errata section notes technical errors found in the current handbook series. Occasional typographical errors and nonstandard symbol labels will be corrected in future volumes.

The authors and editor encourage you to notify them if you find other technical errors. Please send corrections to: Handbook Editor, ASHRAE, 1791 Tullie Circle NE, Atlanta, GA 30329.

### 1985 FUNDAMENTALS (I-P)

**p. 3.22, Reference 57.** The date of publication is April 1956, not 1946.

**p. 4.9, Table 3.** Reference 66 (Martinelli and Nelson 1948), not Reference 51, refers to frictional two-phase pressure drop listed at the bottom of Table 3.

**p. 5.2, Eqs. (9) and (10).** The dimensional units and constants for these equations are incorrectly listed. TC 1.3 is correcting them for the 1989 FUNDAMENTALS Volume.

**p. 6.6, Eq. (4).** Constants $C_8$ through $C_{11}$ should read:

$C_8 = -10440.397$
$C_9 = -11.2946496$
$C_{10} = -0.027022355$
$C_{11} = -0.12890360\ E-04$

**p. 8.2, Eq. (4).** Replace the variable, $\dot{V}_{D_2}$, with $\dot{V}_{O_2}$.

**p. 8.3, Eq. (15).** Constant should be 0.0173, not 0.173.

**p. 14.4, Eq. (1).** $U_H$ is the wind speed in mph, not fpm.

**p. 22.4, Figure 2.** Captions for Figures 2B and 2C are transposed.

**p. 23.3, Eq. (9).** Add the value $R_s$ to the denominator, so the equation reads:

$$q_s = \frac{t_{is} - t_{os}}{[r_s \log_e(r_1/r_i)]/\lambda_1 + [r_s \log_e(r_s/r_1)]/\lambda_2 + R_s} \quad (9)$$

**p. 25.6, Table 3.** Two columns of heat loss values are shown under each R-value listing. The second column of these column pairs is the cumulative value of the first column.

**p. 25.6, sixth line from the bottom of the RH column.** The phrase in parentheses should read: (above 7200 degree days).

**p. 28.45, Reference 93 should read:**

Ayres, J.M. 1977. Predicting Building Energy Requirements. *Heating, Piping, Air Conditioning.* February.

**p. 33.45, Fitting 6-25.** Add main duct coefficients as follows:

| | Main | | | | | | | | | | |
|---|---|---|---|---|---|---|---|---|---|---|---|
| $Q_s/Q_c$ | 0 | 0.1 | 0.2 | 0.3 | 0.4 | 0.5 | 0.6 | 0.7 | 0.8 | 0.9 | 1.0 |
| $C_{c,s}$ | 0.22 | 0.21 | 0.20 | 0.20 | 0.20 | 0.20 | 0.20 | 0.22 | 0.25 | 0.35 | 0.53 |

**p. 34.3, Figure 1.** The 3 on the Head Loss coordinate is misplaced. It should indicate the horizontal line below the 4.

**p. 34.9, LH column.** The second sentence after the section titled, "Tables for Pipe Sizing for Low Pressure," should read, in part:

The values in Table 13 (taken from the basic chart)...

**p. 37.3.** The conversion factors listed for energy apply to the thermochemical Btu. Conversions for the Btu (International Table) are as follows:

| Btu | ft • lb | Calorie | Joule (W • s) |
|---|---|---|---|
| 1 | 778.17 | 251.996 | 1055.056 |
| 0.001285 | 1 | — | 1.3558 |
| — | — | 1 | 4.1868* |

### 1985 FUNDAMENTALS (SI)

**p. 3.22, Reference 57.** The date of publication is April 1956, not 1946.

**p. 4.9, Table 3.** Reference 66 (Martinelli and Nelson 1948), not Reference 51, refers to frictional two-phase pressure drop listed at the bottom of Table 3.

**p. 5.2, Eqs. (9) and (10).** The dimensional units and constants for these equations are incorrectly listed. TC 1.3 is correcting them for the 1989 FUNDAMENTALS Volume.

**p. 6.4, Eqs. (3) and (4).** Constants $C_6$ and $C_{10}$ should read:

$C_6 = -0.9484024\ E-12$
$C_{10} = -0.048640239$

**p. 8.2, Eq. (4).** Replace the variable, $\dot{V}_{D_2}$, with $\dot{V}_{O_2}$.

**p. 8.3, Eq. (15).** Constant should be 0.0173, not 0.173.

**p. 9.8, Figure 15.** The scale for TOTAL HEAT PRODUCTION should range from 2 to 8 W/kg, instead of from 0 to 6 W/kg of live weight.

**p. 17.70, Figure 34.** Values for $A_3$ and $B_3$ in the equations are as follows:

$A_3$ = 1.97668 E − 05
$B_3$ = −7.95090 E − 04

**p. 22.7, Eq. (13) should read:**

$$Q = C_4 A \sqrt{h\,(T_i - T_o)/T_i} \qquad (13)$$

*where*

$Q$ = airflow, L/s
$A$ = free area of inlets or outlets (assumed equal), $m^2$
$h$ = height from lower opening to NPL, m
$T_i$ = average temperature of indoor air at $h$, K
$T_o$ = temperature of outdoor air, K
$C_4$ = 2880, a conversion factor that includes an opening effectiveness value of 65%. Reduce $C_4$ to 2215, or an opening effectiveness value of 50%, if openings are unfavorable.

**p. 23.3, Eq. (9).** Add the value $R_s$ to the denominator, so the equation reads:

$$q_s = \frac{t_{is} - t_{os}}{[r_s \log_e(r_1/r_i)]/\lambda_1 + [r_s \log_e(r_s/r_1)]/\lambda_2 + R_s} \qquad (9)$$

**p. 25.6, Table 3.** Two columns of heat loss values are shown under each R-value listing. The second column of these column pairs is the cumulative value of the first column.

**p. 26.9 and p. 26.11, footnote 4 in Tables 5 and 7.** In both footnotes, change the first part of the first sentence to read:

For each increase in R-value of 1.23 $m^2 \cdot °C/W$ ...

**p. 28.45, Reference 93 should read:**

Ayres, J.M. 1977. Predicting Building Energy Requirements. *Heating, Piping, Air Conditioning*. February.

**p. 33.45, Fitting 6-25.** Add main duct coefficients as follows:

| | Main | | | | | | | | | | |
|---|---|---|---|---|---|---|---|---|---|---|---|
| $Q_s/Q_c$ | 0 | 0.1 | 0.2 | 0.3 | 0.4 | 0.5 | 0.6 | 0.7 | 0.8 | 0.9 | 1.0 |
| $C_{c,s}$ | 0.22 | 0.21 | 0.20 | 0.20 | 0.20 | 0.20 | 0.20 | 0.22 | 0.25 | 0.35 | 0.53 |

**p. 34.9, LH column.** The second sentence after the section titled, "Tables for Pipe Sizing for Low Pressure," should read, in part:

The values in Table 13 (taken from the basic chart)...

## 1986 REFRIGERATION (I-P)

**p. 3.7, Table 6, Copper Pipe section.** Transpose the columns for 1 °F and 0.5 °F under 0 °F Saturated Suction Temperature.

**p. 7.10, delete Reference 26.**

**p. 13.1, footnote in LH column should read:**

The preparation of this chapter is assigned to TC 11.1, Meat, Fish, and Poultry Products.

**p. 38.20, RH column.** Replace part of the column, starting from Eq. (13), with the following text:

$$W_{gc} = \frac{\gamma + 1}{\gamma - 1}\,\alpha \left(\frac{R}{8\pi MT}\right)^{0.5} P(T_2 - T_1) \qquad (13)$$

*where*

$$\alpha = (\alpha_1 \alpha_2)/[\alpha_2 + \alpha_1(1 - \alpha_2)(A_1/A_2)] \qquad (14)$$

$W_{gc}$ = net energy transfer per unit time per unit area of inner surface, Btu/h • $ft^2$
$\gamma$ = $c_p/c_v$, the specific heat ratio of the gas, assumed constant
$R$ = molar gas constant
$P$ = pressure, mm Hg
$M$ = molecular weight of the gas
$T$ = absolute temperature at the point where $P$ is measured, °R
$\alpha$ = overall accommodation coefficient
$A$ = area, and subscripts 1 and 2 refer to the inner and outer surfaces, respectively, $ft^2$

This expression reduces to

$$W_{gc} = C\alpha P(T_2 - T_1) \qquad (15)$$

*where*

$$C = \frac{\gamma + 1}{\gamma - 1}\left(\frac{R}{8\pi MT}\right)^{0.5} \qquad (16)$$

Values of this constant for cryogenic applications have been calculated by Corruccini[35] and are shown in Table 4. The temperature at the pressure gauge is assumed to be 80 °F.

**Table 4 Constants for the Gas Conduction Equation[35]**

| Gas | $T_2$ and $T_1$, °R | Constant |
|---|---|---|
| $N_2$ | ⩽ 720 | 28.0 |
| $O_2$ | ⩽ 540 | 26.0 |
| $H_2$ | 540 and 139 or 540 and 162 | 93.0 |
| $H_2$ | 139 and 36 | 70.1 |
| He | any | 49.3 |
| Air | ⩽ 720 | 27.5 |

**p. 38.21, Table 5.** At 139 °F, the accommodation coefficient for He is 0.4.

## 1986 REFRIGERATION (SI)

**p. 7.10, delete Reference 26.**

**p. 13.1, footnote in LH column should read:**

The preparation of this chapter is assigned to TC 11.1, Meat, Fish, and Poultry Products.

**p. 38.20, RH column.** Replace part of the column, starting from Eq. (13), with the following text:

$$W_{gc} = \frac{\gamma + 1}{\gamma - 1}\,\alpha \left(\frac{R}{8\pi MT}\right)^{0.5} P(T_2 - T_1) \qquad (13)$$

*where*

$$\alpha = (\alpha_1 \alpha_2)/[\alpha_2 + \alpha_1(1 - \alpha_2)(A_1/A_2)] \quad (14)$$

$W_{gc}$ = net energy transfer per unit time per unit area of inner surface, $W/m^2$
$\gamma$ = $c_p/c_v$, the specific heat ratio of the gas, assumed constant
$R$ = molar gas constant
$P$ = pressure, Pa
$M$ = molecular weight of the gas
$T$ = absolute temperature at the point where $P$ is measured, K
$\alpha$ = overall accommodation coefficient
$A$ = area, and subscripts 1 and 2 refer to the inner and outer surfaces, respectively, $m^2$

This expression reduces to

$$W_{gc} = C\alpha P(T_2 - T_1) \quad (15)$$

*where*

$$C = \frac{\gamma + 1}{\gamma - 1}\left(\frac{R}{8\pi MT}\right)^{0.5} \quad (16)$$

Values of this constant for cryogenic applications have been calculated by Corruccini[35] and are shown in Table 4. The temperature at the pressure gauge is assumed to be 300 K.

**Table 4 Constants for the Gas Conduction Equation[35]**

| Gas | $T_2$ and $T_1$, K | Constant |
|---|---|---|
| $N_2$ | ⩽ 400 | 1.1925 |
| $O_2$ | ⩽ 300 | 1.1074 |
| $H_2$ | 300 and 77 or 300 and 90 | 3.9605 |
| $H_2$ | 77 and 20 | 2.9854 |
| He | any | 2.1003 |
| Air | ⩽ 400 | 1.1706 |

**p. 38.21, Table 5.** At 77 K, the accommodation coefficient for He is 0.4.

## 1987 HVAC

**p. 3.3, LH column.** Delete section titled, "Air-Water Radiant Panel Systems."

**p. 7.2, RH column.** Delete sentence below Item 15, which begins, Disadvantages are...

**p. 12.5, RH column.** Replace equations for $P_{12}$ and $P_{21}$ with the following:

$$P_{12} = \frac{12C_2}{2\pi K_s}\ln\sqrt{\frac{a^2 + (d_1 + d_2)^2}{a^2 + (d_1 - d_2)^2}}$$

$$P_{21} = \frac{12C_1}{2\pi K_s}\ln\sqrt{\frac{a^2 + (d_1 + d_2)^2}{a^2 + (d_1 - d_2)^2}}$$

also units for $K_s$ are Btu • in/h • ft² • °F (W/m • °C)

**p. 17.3, RH column.** Replace sentence starting on the eighth line from the bottom of the column with the following sentence:

ACCA (Air-Conditioning Contractors of America) Manual D and G and Chapter 1 of the 1988 EQUIPMENT Volume cover duct design and installation in detail.

**p. 30.6, LH column.** The first sentence of the third paragraph should read:

Table 4 presents ranges of exhaust velocities in ducts.

**p. 37.9, RH column.** The equation for infiltration heat loss, $Q_i$, should read:

$$Q_i = C_1 VN(t_i - t_o)$$

*where*

$C_1$ = 0.018 in I-P units (0.34 in SI units)

**p. 45.13, Figure 16.** Change the trade names of "Temptite" and "Transite" to "Preinsulated Pipe."

**p. 46.8, LH column, para. 2, line 8.** 392 °F should be 39.2 °F.

**p. 46.9, RH column.** On line 3 under the section titled, "Ice Storage," change 4180 J/kg to 335 kJ/kg.

**p. 46.9, RH column.** Change the first part of the next to last sentence of the paragraph titled "Water/Ice Storage on Refrigerant Coils" to read:

These ice storage units are manufactured in sizes from 100 to 1200 ton-hours (1.3 to 15.2 GJ) capacity and...

**p. 46.11, replace Table 5 with the following tables.**

**Table 5 (I-P) Properties of Selected Phase Change Materials[a]**

| Material | Melting Point °F | Specific Gravity | Specific Heat Btu/lb • °F | Thermal Conductivity Btu • in. h • ft² • °F |
|---|---|---|---|---|
| $MgCl_2 \cdot 6\,H_2O$ | 243 | 1.57 | 0.41 | — |
| | — | 1.44 | 0.67 | — |
| $(MgNO_3)_2 \cdot 6\,H_2O$ | 192 | 1.64 | 0.44 | 4.37 |
| | — | 1.56 | 0.60 | 3.40 |
| Paraffin Wax[b] | 122 | 0.79 | 0.69 | 0.97 |
| $Na_2S_2O_3 \cdot 5\,H_2O$ | 118 | 1.73 | 0.35 | 1.66 |
| | — | 1.67 | 0.57 | — |
| $Na_2SO_4 \cdot 10\,H_2O$ | 89 | 1.46 | 0.46 | 3.54 |
| | — | 1.33 | 0.78 | — |
| $CaCl_2 \cdot 6\,H_20$ | 82 | 1.71 | 0.35 | 7.6 |
| | — | 1.50 | 0.52 | 3.74 |
| $H_2O$ | 32 | 0.92 | 0.5 | 2.43 |
| | — | 1.0 | 1.0 | 4.2 |

**Table 5 (SI) Properties of Selected Phase Change Materials**[a]

| Material | Melting Point °C | Density kg/m³ | Specific Heat kJ/kg • °C | Thermal Conductivity W/(m • K) |
|---|---|---|---|---|
| $MgCl_2 \cdot 6\,H_2O$ | 117 | 1570 | 1.71 | — |
| | — | 1440 | 2.80 | — |
| $(MgNO_3)_2 \cdot 6\,H_2O$ | 89 | 1640 | 1.84 | 0.63 |
| | — | 1560 | 2.51 | 0.49 |
| Paraffin Wax[b] | 50 | 790 | 2.88 | 0.14 |
| $Na_2S_2O_3 \cdot 5\,H_2O$ | 48 | 1730 | 1.46 | 0.24 |
| | — | 1670 | 2.38 | — |
| $Na_2SO_4 \cdot 10\,H_2O$ | 32 | 1460 | 1.92 | 0.51 |
| | — | 1330 | 3.26 | — |
| $CaCl_2 \cdot 6\,H_20$ | 28 | 1710 | 1.46 | 1.1 |
| | — | 1500 | 2.17 | 0.54 |
| $H_2O$ | 0 | 920 | 2.09 | 0.35 |
| | — | 1000 | 4.18 | 0.6 |

[a] For each material, the first line lists the properties of the solid, the second line the properties of the liquid.
[b] A representative sample out of a family of many paraffin waxes.

**p. 46.15, LH column.** Revise sentence that starts on the seventh line from the bottom of the column to read, in part:

Most rock has a specific gravity of about 2.6 or a density of 162 lb/ft³ (2600 kg/m³)...

**p. 46.15, RH column.** Change value in last line to 75 Btu/ft³ • °F. The SI value on the following page (5000 kJ/m³ • K) is correct.

**p. 46.20, change Eq. (8 SI) to read:**

$$t_s = -6.67 - 0.087x$$

**p. 54.18, Eq. (11).** Units for $q_2$ are Btu/h (W), not Btu/h • ft² (W/m²).

# COMPOSITE INDEX

## ASHRAE HANDBOOK SERIES

**This index covers the technical data sections of** *the four current* HANDBOOK volumes published by ASHRAE. Listings from each volume are identified by a code letter preceding the page number:

E = 1988 Equipment
H = 1987 HVAC
R = 1986 Refrigeration
F = 1985 Fundamentals

For example, the first listing, **"Abbreviations,"** is followed by secondary listings, "symbols," "graphical, F36.5-11" and so forth. This indicates that abbreviations for graphical symbols can be found on pages 5 through 11 of Chapter 36 in the 1985 FUNDAMENTALS Volume. Note that the code numbers include the chapter number followed by a decimal point and the page number(s) within the chapter. This index is updated in each Handbook volume.

E

G

## J

K



L



M

U

# Terminology of Heating, Ventilation, Air Conditioning, and Refrigeration

## The reference source that should be as close to your work as your Handbooks.

This useful dictionary lists technical terms appropriate to the industry, promotes a precise and consistent vocabulary for HVAC&R professionals, and helps specialists in related fields who occasionally need an authoritative source for term and definitions.

Only terms considered relevant and in common use in the area of HVAC&R were selected. However, the content reflects national and international terms incorporated from ASHRAE Handbooks and standards as well as terms from: International Institute of Refrigeration, American Institute of Architects, American Society for Testing and Materials, Institute of Electrical and Electronic Engineers and the International Standards Organization.

The cross-referenced format lists terms in three ways: by noun, by first word, and as secondary word.

**Published:** 1986 **Pages:** 176
**List:** $30.00 **Member:** $20.00
**Code:** 90290

If you do not wish to cut page, photocopy form below and mail to ASHRAE

***Yes,*** *send me* **Terminology of Heating, Ventilation, Air Conditioning and Refrigeration.**

☐ List: $30.00 ☐ Member: $20.00

Publication Code: 90290

NAME

STREET (DO NOT USE A POST OFFICE BOX)

CITY

STATE/PROVINCE ZIP

COUNTRY PHONE ( )

Mail to: ASHRAE Publication Sales
1791 Tullie Circle, NE
Atlanta, GA 30329

Phone credit card orders to: ASHRAE Publication Sales **404-636-8400.**

Member No.

**PAYMENT OPTIONS:**

☐ Check enclosed (U.S. funds)

Charge: ☐ VISA ☐ AMEX ☐ MasterCard

Card No.

EXPIRATION DATE

SIGNATURE (REQUIRED)

Purchase order does not constitute prepayment.
Money back guarantee if not satisfied.
All prices subject to change without notice.